The chemistry of
**organic arsenic, antimony and
bismuth compounds**

THE CHEMISTRY OF FUNCTIONAL GROUPS

A series of advanced treatises under the general editorship of
Professors Saul Patai and Zvi Rappoport

The chemistry of alkenes (2 volumes)
The chemistry of the carbonyl group (2 volumes)
The chemistry of the ether linkage
The chemistry of the amino group
The chemistry of the nitro and nitroso groups (2 parts)
The chemistry of carboxylic acids and esters
The chemistry of the carbon–nitrogen double bond
The chemistry of amides
The chemistry of the cyano group
The chemistry of the hydroxyl group (2 parts)
The chemistry of the azido group
The chemistry of acyl halides
The chemistry of the carbon–halogen bond (2 parts)
The chemistry of the quinonoid compounds (2 volumes, 4 parts)
The chemistry of the thiol group (2 parts)
The chemistry of the hydrazo, azo and azoxy groups (2 parts)
The chemistry of amidines and imidates (2 volumes)
The chemistry of cyanates and their thio derivatives (2 parts)
The chemistry of diazonium and diazo groups (2 parts)
The chemistry of the carbon–carbon triple bond (2 parts)
The chemistry of ketenes, allenes and related compounds (2 parts)
The chemistry of the sulphonium group (2 parts)
Supplement A: The chemistry of double-bonded functional groups (2 volumes, 4 parts)
Supplement B: The chemistry of acid derivatives (2 volumes, 4 parts)
Supplement C: The chemistry of triple-bonded functional groups (2 parts)
Supplement D: The chemistry of halides, pseudo-halides and azides (2 parts)
Supplement E: The chemistry of ethers, crown ethers, hydroxyl groups
and their sulphur analogues (2 volumes, 3 parts)
Supplement F: The chemistry of amino, nitroso and nitro compounds
and their derivatives (2 parts)
The chemistry of the metal–carbon bond (5 volumes)
The chemistry of peroxides
The chemistry of organic selenium and tellurium compounds (2 volumes)
The chemistry of the cyclopropyl group (2 parts)
The chemistry of sulphones and sulphoxides
The chemistry of organic silicon compounds (2 parts)
The chemistry of enones (2 parts)
The chemistry of sulphinic acids, esters and their derivatives
The chemistry of sulphenic acids and their derivatives
The chemistry of enols
The chemistry of organophosphorus compounds (3 volumes)
The chemistry of sulphonic acids, esters and their derivatives
The chemistry of alkanes and cycloalkanes
Supplement S: The chemistry of sulphur-containing functional groups
The chemistry of organic arsenic, antimony and bismuth compounds

UPDATES

The chemistry of α-haloketones, α-haloaldehydes and α-haloimines
Nitrones, nitronates and nitroxides
Crown ethers and analogs
Cyclopropane-derived reactive intermediates
Synthesis of carboxylic acids, esters and their derivatives
The silicon–heteroatom bond
Syntheses of lactones and lactams

Patal's 1992 guide to the chemistry of functional groups—*Saul Patai*

C—As, C—Sb, C—Bi

The chemistry of
organic arsenic, antimony and bismuth compounds

Edited by

SAUL PATAI

The Hebrew University, Jerusalem

1994

JOHN WILEY & SONS

CHICHESTER—NEW YORK—BRISBANE—TORONTO—SINGAPORE

An Interscience® Publication

Seplae
Chem

Other Wiley Editorial Offices

John Wiley & Sons, Inc., 605 Third Avenue,
New York, NY 10158-0012, USA

Jacaranda Wiley Ltd, 33 Park Road, Milton,
Queensland 4064, Australia

John Wiley & Sons (Canada) Ltd, 22 Worcester Rd,
Rexdale, Ontario M9W 1L1, Canada

John Wiley & Sons (SEA) Pte Ltd, 37 Jalan Pemimpin #05-04,
Block B, Union Industrial Building, Singapore 2057

Library of Congress Cataloging-in-Publication Data

The Chemistry of organic arsenic, antimony, and bismuth compounds /
edited by Saul Patai.
      p.   cm.—(The Chemistry of Functional groups)
  'An Interscience publication.'
  Includes bibliographical references and index.
  ISBN 0–471–93044–X
  1. Organoarsenic compounds. 2. Organoantimony compounds.
3. Organobismuth compounds. I. Patai, Saul. II. Series.
QD412.A7C53   1994   93-3145
547'.0571—dc20   CIP

British Library Cataloguing in Publication Data

A catalogue record for this book is available from the British Library

ISBN   0 471 93044 X

Typeset in Times 9/10 pt by Thomson Press (India) Ltd, New Delhi
Printed and bound in Great Britain by Biddles Ltd, Guildford, Surrey

Contributing authors

Kin-ya Akiba — Department of Chemistry, University of Hiroshima, 1-3-1 Kagamiyama, Higashi-Hiroshima 724, Japan

T. Berclaz — Department de Chimie Physique, Université de Genève, 30 quai Ernest-Ansermet, 1211 Genève 4, Switzerland

Hans Joachim Breunig — Universität Bremen, Fachbereich 2 (Chemie), Postfach 330440, D-28334 Bremen, Germany

Marvin Charton — Chemistry Department, The Clinton Hill Campus, Pratt Institute, DeKalb Avenue and Hall Street, Brooklyn, NY 11205, USA

T. R. Crompton — Hill Cottage, Pen Marian, near Llangoed, Beaumaris, Anglesey, Gwynedd, LL58 8SU, Wales, UK

Kilian Dill — Molecular Devices Corporation, 4700 Bohannon Drive, Menlo Park, California 94025, USA

M. Geoffroy — Départment de Chimie Physique, Université de Genève, 30 quai Ernest-Ansermet, 1211 Genève 4, Switzerland

Ian Gosney — Department of Chemistry, The University of Edinburgh, Edinburgh, EH9 3JJ, Scotland, UK

Marianna Kanska — Department of Chemistry, The University of Warsaw, Warsaw, Poland

Thomas M. Klapötke — Institut für Anorganische und Analytische Chemie der Technischen Universität Berlin, Strasse des 17 Juni 135, D-10623 Berlin, Germany

Katharina C. H. Lange — Institut für Anorganische und Analytische Chemie der Technischen Universität Berlin, Strasse des 17 Juni 135, D-10623 Berlin, Germany

Joel F. Liebman — Department of Chemistry and Biochemistry, University of Maryland, Baltimore County Campus, 5401 Wilkens Avenue, Baltimore, Maryland 21228-5398, USA

Douglas Lloyd — Department of Chemistry, The Purdie Building, University of St Andrews, St Andrews, Fife, KY16 9ST, Scotland, UK

A. G. Mackie — Department of Chemistry, University of Manchester Institute of Science and Technology, P.O. Box 88, Manchester, M60 1QD, UK

Shigeru Maeda — Department of Applied Chemistry and Chemical Engineering, Faculty of Engineering, Kagoshima University, 1-21-40 Korimoto, Kagoshima 890, Japan

C. A. McAuliffe — Department of Chemistry, University of Manchester Institute of Science and Technology, P.O. Box 88, Manchester, M60 1QD, UK

Evelyn L. McGown   The Chemistry Branch, Blood Research Division, Letter-
                            man Army Institute of Research, Presidio of San Fran-
                            cisco, CA 94129-6800, USA
Shigeru Nagase   Department of Chemistry, Faculty of Education, Yoko-
                            hama National University, Yokohama 240, Japan
A. Nagy   Department of General and Inorganic Chemistry, Eötvös
                            University Budapest, P.O. Box 32, H-1518 Budapest 112,
                            Hungary
Yurii S. Nekrasov   Institute of Organo-Element Compounds, Russian
                            Academy of Sciences, Moscow, Russia 117334
Merete Folmer Nielsen   Department of General and Organic Chemistry, The H.C.
                            Ørsted Institute, Universitetsparken 5, DK-2100 Copen-
                            hagen Ø, Denmark
Isaac Nir   Department of Pharmacology, The Hebrew University–
                            Hadassah Medical School, Jerusalem 91010, Israel
José Artur Martinho Simões Departmento de Engenharia Quimica, Instituto Superior
                            Técnico, 1096 Lisboa Codex, Portugal
Suzanne W. Slayden   Department of Chemistry, George Mason University,
                            4400 University Drive, Fairfax, Virginia 22030-4444,
                            USA
D. B. Sowerby   Department of Chemistry, The University of Nottingham,
                            University Park, Nottingham, NG7 2RD, UK
L. Szepes   Department of General and Inorganic Chemistry, Eötvös
                            University Budapest, P.O. Box 32, H-1518 Budapest 112,
                            Hungary
S. B. Wild   Research School of Chemistry, Australian National
                            University, Canberra, ACT 0200, Australia
Uri Wormser   Department of Pharmacology, The Hebrew University–
                            Hadassah Medical School, Jerusalem 91010, Israel
Yohsuke Yamamoto   Department of Chemistry, University of Hiroshima, 1-3-1
                            Kagamiyama, Higashi-Hiroshima 724, Japan
Dmitri V. Zagorevskii   Institute of Organo-Element Compounds, Russian Aca-
                            demy of Sciences, Moscow, Russia 117334
L. Zanathy   Department of General and Inorganic Chemistry, Eötvös
                            University Budapest, P.O. Box 32, H-1518 Budapest 112,
                            Hungary
Mieczysław Zieliński   Isotope Laboratory, Faculty of Chemistry, Jagiellonian
                            University, 30-060 Kraków, ul. Ingardena 3, Poland

Foreword

The series The Chemistry of Functional Groups contains a set of five volumes describing organometallic compounds. However these volumes do not deal at all (or deal only very briefly) with organometallic compounds of arsenic, antimony and bismuth, or with those of germanium, tin and lead. Hence we decided to embody in the series volumes dealing with organic derivatives of the above six metals. The present volume treats organometallic compounds containing As, Sb and Bi, and a forthcoming volume is now under active preparation and will hopefully be published in 1995.

The authors' literature search in this volume extended in most cases up to the end of 1991 or to the beginning of 1992.

Unfortunately, among the chapters planned for this volume, those on NMR, on Vibrational Spectroscopy and on Rearrangements did not materialize. I hope I will be able to include these in one of the forthcoming supplementary volumes of the series.

I would like to use this opportunity to congratulate my good friend Prof. Zvi Rappoport on the occasion of his joining me as co-editor of The Chemistry of the Functional Groups series.

I will be indebted to readers who will bring to my attention mistakes or omissions in this or any other volume of the series.

Jerusalem
January 1994

SAUL PATAI

The Chemistry of Functional Groups
Preface to the series

The series 'The Chemistry of Functional Groups' was originally planned to cover in each volume all aspects of the chemistry of one of the important functional groups in organic chemistry. The emphasis is laid on the preparation, properties and reactions of the functional group treated and on the effects which it exerts both in the immediate vicinity of the group in question and in the whole molecule.

A voluntary restriction on the treatment of the various functional groups in these volumes is that material included in easily and generally available secondary or tertiary sources, such as Chemical Reviews, Quarterly Reviews, Organic Reactions, various 'Advances' and 'Progress' series and in textbooks (i.e. in books which are usually found in the chemical libraries of most universities and research institutes), should not, as a rule, be repeated in detail, unless it is necessary for the balanced treatment of the topic. Therefore each of the authors is asked not to give an encyclopaedic coverage of his subject, but to concentrate on the most important recent developments and mainly on material that has not been adequately covered by reviews or other secondary sources by the time of writing of the chapter, and to address himself to a reader who is assumed to be at a fairly advanced postgraduate level.

It is realized that no plan can be devised for a volume that would give a complete coverage of the field with no overlap between chapters, while at the same time preserving the readability of the text. The Editor set himself the goal of attaining reasonable coverage with moderate overlap, with a minimum of cross-references between the chapters. In this manner, sufficient freedom is given to the authors to produce readable quasi-monographic chapters.

The general plan of each volume includes the following main sections:

(a) An introductory chapter deals with the general and theoretical aspects of the group.

(b) Chapters discuss the characterization and characteristics of the functional groups. i.e. qualitative and quantitative methods of determination including chemical and physical methods, MS, UV, IR, NMR, ESR and PES—as well as activating and directive effects exerted by the group, and its basicity, acidity and complex-forming ability.

(c) One or more chapters deal with the formation of the functional group in question, either from other groups already present in the molecule or by introducing the new group directly or indirectly. This is usually followed by a description of the synthetic uses of the group, including its reactions, transformations and rearrangements.

(d) Additional chapters deal with special topics such as electrochemistry, photochemistry, radiation chemistry, thermochemistry, syntheses and uses of isotopically labelled compounds, as well as with biochemistry, pharmacology and toxicology. Whenever applicable, unique chapters relevant only to single functional groups are also included (e.g. 'Polyethers', 'Tetraaminoethylenes' or 'Siloxanes').

This plan entails that the breadth, depth and thought-provoking nature of each chapter will differ with the views and inclinations of the authors and the presentation will necessarily be somewhat uneven. Moreover, a serious problem is caused by authors who deliver their manuscript late or not at all. In order to overcome this problem at least to some extent, some volumes may be published without giving consideration to the originally planned logical order of the chapters.

Since the beginning of the Series in 1964, two main developments occurred. The first of these is the publication of supplementary volumes which contain material relating to several kindred functional groups (Supplements A, B, C, D, E and F). The second ramification is the publication of a series of 'Updates', which contain in each volume selected and related chapters, reprinted in the original form in which they were published, together with an extensive updating of the subjects, if possible, by the authors of the original chapters. A complete list of all above mentioned volumes published to date will be found on the page opposite the inner title page of this book.

Advice or criticism regarding the plan and execution of this series will be welcomed by the Editor.

The publication of this series would never have been started, let alone continued, without the support of many persons in Israel and overseas, including colleagues, friends and family. The efficient and patient co-operation of staff-members of the publisher also rendered me invaluable aid. My sincere thanks are due to all of them, especially to Professor Zvi Rappoport who, for many years shares the work and responsibility of the editing of this Series.

The Hebrew University   SAUL PATAI
Jerusalem, Israel

Contents

Contents

List of abbreviations used

Ac	acetyl (MeCO)
acac	acetylacetone
Ad	adamantyl
AIBN	azoisobutyronitrile
Alk	alkyl
All	allyl
An	anisyl
Ar	aryl
Bz	benzoyl (C_6H_5CO)
Bu	butyl (also t-Bu or Bu^t)
CD	circular dichroism
CI	chemical ionization
CIDNP	chemically induced dynamic nuclear polarization
CNDO	complete neglect of differential overlap
Cp	η^5-cyclopentadienyl
Cp*	η^5-pentamethylcyclopentadienyl
DABCO	1,4-diazabicyclo[2.2.2]octane
DBN	1,5-diazabicyclo[4.3.0]non-5-ene
DBU	1,8-diazabicyclo[5.4.0]undec-7-ene
DIBAH	diisobutylaluminium hydride
DME	1,2-dimethoxyethane
DMF	N,N-dimethylformamide
DMSO	dimethyl sulphoxide
ee	enantiomeric excess
EI	electron impact
ESCA	electron spectroscopy for chemical analysis
ESR	electron spin resonance
Et	ethyl
eV	electron volt
Fc	ferrocenyl
FD	field desorption
FI	field ionization

xiii

FT	Fourier transform
Fu	furyl(OC_4H_3)
GLC	gas liquid chromatography
Hex	hexyl(C_6H_{13})
c-Hex	cyclohexyl(C_6H_{11})
HMPA	hexamethylphosphortriamide
HOMO	highest occupied molecular orbital
HPLC	high performance liquid chromatography
i-	iso
Ip	ionization potential
IR	infrared
ICR	ion cyclotron resonance
LAH	lithium aluminium hydride
LCAO	linear combination of atomic orbitals
LDA	lithium diisopropylamide
LUMO	lowest unoccupied molecular orbital
M	metal
M	parent molecule
MCPBA	*m*-choloroperbenzoic acid
Me	methyl
MNDO	modified neglect of diatomic overlap
MS	mass spectrum
n	normal
Naph	naphthyl
NBS	*N*-bromosuccinimide
NCS	*N*-chlorosuccinimide
NMR	nuclear magnetic resonance
Pc	phthalocyanine
Pen	pentyl(C_5H_{11})
Pip	piperidyl($C_5H_{10}N$)
Ph	phenyl
ppm	parts per million
Pr	propyl (also *i*-Pr or Pr^i)
PTC	phase transfer catalysis or phase transfer conditions
Pyr	pyridyl (C_5H_4N)
R	any radical
RT	room temperature
s-	secondary
SET	single electron transfer
SOMO	singly occupied molecular orbital
t-	tertiary
TCNE	tetracyanoethylene

TFA	trifluoroacetic acid
THF	tetrahydrofuran
Thi	thienyl(SC_4H_3)
TLC	thin layer chromatography
TMEDA	tetramethylethylene diamine
TMS	trimethylsilyl or tetramethylsilane
Tol	tolyl(MeC_6H_4)
Tos or Ts	tosyl(p-toluenesulphonyl)
Trityl	triphenylmethyl(Ph_3C)
Xyl	xylyl($Me_2C_6H_3$)

In addition, entries in the 'List of Radical Names' in *IUPAC Nomenclature of Organic Chemistry*, 1979 Edition. Pergamon Press, Oxford, 1979, p. 305–322, will also be used in their unabbreviated forms, both in the text and in formulae instead of explicitly drawn structures.

CHAPTER **1**

General and theoretical aspects

SHIGERU NAGASE

Department of Chemistry, Faculty of Education, Yokohama National University, Yokohama 240, Japan

I. INTRODUCTION

The subject of *ab initio* quantum chemistry has long promised to become a major tool for the study of a variety of chemical problems such as structure, stability and reaction mechanism[1]. As seen in the recent, extensive applications to the compounds containing second- and third-row elements such as N and P, *ab initio* calculations have played an important (and in some cases even a crucial) role in prediction and understanding of diverse problems and provided valuable information for establishing useful and general concepts[1]. This is due to the great current progress in new theoretical methods and efficient computer programs as well as powerful computers.

However, the extensive applications to the compounds containing the heavier As, Sb and Bi elements are rather at an early stage despite the potential usefulness. Even in the experimental field[2] the organic chemistry (especially for Sb[3] and Bi[4]) lags behind the developments for the chemistry of nitrogen and phosphorus compounds. For the further progress in the heavier group 15 chemistry, it is of obvious interest to provide systematic insight into the periodic trends of the chemical properties[5] upon going from N and P to As, Sb and Bi. It is well known that relativistic effects become important in heavier element chemistry while they are small in lighter element chemistry. Thus we will discuss mostly

The chemistry of organic arsenic, antimony and bismuth compounds
Edited by S. Patai © 1994 John Wiley & Sons Ltd

recent *ab initio* calculations at high levels which are carried out at the same level of theory for As, Sb and Bi compounds, putting emphasis on periodic trends and relativistic effects. It will be shown that the periodic trends are rather regular in some cases while they become significantly irregular in other cases owing to the increasing relativistic effects.

II. THEORETICAL METHODOLOGY

A. *Ab Initio* Calculations

Ab initio means 'from the beginning', i.e. accurately solving the Schrödinger equation without any empirical input. In practice, however, the accuracy and reliability of *ab initio* calculations are determined by both how large a basis set is used and what kind of theoretical method is employed. The first step in carrying out *ab initio* calculations with any theoretical method is to choose a basis set[1,6]. In general, the calculations become more reliable with the larger basis sets, though the required computation times increase. The double-zeta basis sets augmented by polarization functions (and diffuse functions in the case of anions) are currently standard, but much larger basis sets may be required for complex problems. As for the calculational methods[1,7], the single determinant Hartree–Fock (HF) self-consistent field (SCF) molecular orbital calculation is generally a starting point, though the valence bond (VB) method has been also applied to smaller systems. The HF-SCF calculations usually provide good results for the properties of equilibrium structures. What is missing in the HF calculations is the effect of electron correlation. In the HF method each electron moves around in the average field of all other electrons. Therefore, correlation between the motions of electrons (especially with opposite spins) tends to be underestimated.

The effect of electron correlation is incorporated with perturbation theory or by using the variational principle. The most widely used perturbation method is the Møller–Plesset (MP) scheme developed by Pople and coworkers[1] from second order (MP2) to fourth order (MP4) in the perturbation expansion. On the other hand, typical variational methods are configuration interaction (CI) or multireference CI (MRCI) and multiconfigurational SCF (MCSCF) or complete active-space SCF (CASSCF) methods[7]. In these methods, the wave function is described as a linear combination of many determinants (configurations). In the CI or MRCI methods, the molecular orbitals obtained from HF calculations are usually used to construct the configurations and only the CI coefficients are optimized variationally. In MCSCF or CASSCF, both molecular orbitals and CI coefficients are variationally determined. In order to incorporate efficiently higher-order electron correlation, molecular orbitals from small-scale MCSCF or CASSCF calculations are often used for large-scale CI or MRCI calculations; these may be denoted by CASSCF/CI. It is important to recognize that the use of a small basis set may limit the fraction of the total correlation energy even with a larger number of determinants.

Probably even more important to computational quantum chemistry is the development in the analytical evaluation of first, second and higher derivatives of the potential energy with respect to nuclear coordinates[8]. These analytical derivative methods are indispensable to the location and characterization of the stationary points (minima or transition states) on the potential energy surface[9] and have greatly advanced the scope of applicability of *ab initio* calculations. *Ab initio* calculations are in a position to predict many new types of the heavier group 15 compounds and provide valuable information for the interpretation of complex experimental data.

B. Relativistic Effects

Relativistic effects on chemical properties have been reviewed by many authors[10–19]. For the present purpose, these may be briefly summarized on the basis of the review articles in the following way. Relativistic effects can be defined as anything arising from the finite speed of light, as compared with the infinite speed of light. According to the special relativity theory by Einstein, the mass (m) of an electron increases as its speed (v) approaches the speed (c) of light; $m = m_0[1 - (v/c)^2]^{-1/2}$ where m_0 is the rest mass. Inner 1s electrons with no angular momentum can approach an atomic nucleus most closely and experience the nuclear positive charge. Therefore, the speed of 1s electrons approaches that of light upon going to the heavier atoms with large nuclear positive charges. Since v is roughly equal to $(Z/137)c$, for example, the speed of the 1s electrons of the Bi atom ($Z = 83$) attains 61% of the speed of light, thereby m becoming $1.26 m_0$. Consequently, the Bohr radius decreases by ca 20% since it is in inverse proportion to m. The result is that 1s orbitals are more concentrated near the nucleus and more stabilized than would be expected nonrelativistically.

As a result of mass–velocity correction, inner 1s orbitals shrink. In general, the 1s orbitals may make little important contribution to chemical bonding. However, the contraction of 1s orbitals, in turn, causes the contraction and energetic stabilization of the outer valence s orbitals because the latter orbitals must be orthogonal to the former. For p electrons the mass–velocity effect is similar to that for s electrons. However, its effect is much smaller since the angular momentum keeps p electrons away from the nucleus. The same is also true for d and f electrons with larger angular momentum. However, it is instructive to note that the shielding of the positive nuclear charge due to the contraction of the s and p shells leads to the orbital expansion and energetic destabilization of d and f electrons.

The strong coupling of the spin of the electron with the orbital angular momentum is another important relativistic effect. The angular (l) and spin (s) quantum numbers become no longer good upon going to the heavier atoms. Instead, a good quantum number (j) is their sum ($j = l + s$), there being now no distinction between 'orbital' and 'spin' momentums. The spin–orbit coupling splits the six p shell spin orbitals into two $p_{1/2}$ and four $p_{3/2}$ spinors labeled by the quantum numbers $j = 1/2$ and 3/2. This spin–orbit effect given by the Dirac equation is purely a quantum-relativistic effect. The net result of the mass–velocity correction and the spin–orbit coupling is that the two effects tend to cancel for the $p_{3/2}$ shell but reinforce for $p_{1/2}$. As expected from the same angular dependence (i.e. the same j values) of the $s_{1/2}$ and $p_{1/2}$ orbitals, the $p_{1/2}$ orbitals are spherically symmetrical (this may be surprising to some organic chemists) and lower in energy than the doughnut-shaped $p_{3/2}$ orbitals.

Thus, the main relativistic effects are: (1) the radical contraction and energetic stabilization of the s and $p_{1/2}$ orbitals which in turn induce the radial expansion and energetic destabilization of the outer d and f orbitals, and (2) the well-known spin–orbit splitting. These effects will be pronounced upon going from As to Sb to Bi. Associated with effect (1), it is interesting to note that the Bi atom has a tendency to form compounds in which Bi is trivalent with the $6s^2 6p^3$ valence configuration. For this tendency of the $6s^2$ electron pair to remain formally unoxidized in bismuth compounds (i.e. core-like nature of the 6s electrons), the term 'inert pair effect'[20] or 'nonhybridization effect' has been often used for a reasonable explanation. In this context, the relatively inert $4s^2$ pair of the As atom (compared with the $5s^2$ pair of Sb) may be ascribed to the stabilization due to the 'd-block contraction', rather than effect (1)[17]. On the other hand, effect (2) plays an important role in the electronic and spectroscopic properties of atoms and molecules especially in the open-shell states. It not only splits the electronic states but also mixes the states which would not mix in the absence of spin–orbit interaction. As an example, it was calculated[21] that even the ground state ($^1\Sigma_g{}^+$) of Bi_2 is 25% contaminated by $^3\Pi_g$. In the Pauli Hamiltonian approximation there is one more relativistic effect called the Dawin term. This will tend to counteract partially the mass–velocity effect.

Compounds made of the heavier atoms contain a large number of electrons, making all-electron calculations difficult in large systems of experimental interest. The relativistic

S. Nagase

mass–velocity effect has the largest influence on inner-core electrons, for which the relativistic (Dirac) wave equation must be solved. Instead, relativistic effective core potential (ECP) methods have been recently developed[13-15]. In these methods inner-core electrons are replaced by the relativistic ECPs and only chemically important valence electrons are explicitly calculated. It is an advantage of the ECP methods that large valence basis sets can be used because of the elimination of core electrons. In addition, large-scale relativistic CI calculations can be carried out with spin–orbit interaction[18].

III. PERIODIC TRENDS

A. Structures and Bond Energies

Table 1 summarizes the trends in the structures and bond dissociation energies of hydrides (MH_n, $n = 1$, 2 and 3 for M = As, Sb and Bi) calculated with the CASSCF/CI method[22] together with the available experimental data. The spectroscopic properties and potential energy curves for monohydrides (MH) have been fully summarized by Balasubramanian[23]. Even for simple hydrides, there are only few experimental data for comparison. As Table 1 shows, however, good agreement is seen between the calculated and limited experimental values. This will permit the periodic trends to be discussed on the basis of the calculated values.

As Table 1 shows, the M—H bond lengths (r) of MH_n increase as the M atom becomes heavier. Upon going from MH to MH_2 to MH_3, the M—H bond lengths decrease, though the decreasing trend is somewhat irregular in the case of M = Bi. MH_2 has a bent C_{2v} structure while MH_3 has a pyramidal C_{3v} structure. The H—M—H angles (θ) of MH_2 and MH_3 all decrease for any M atom in going down the periodic table. However, the decrease is rather small. Therefore, for example, the bond angles of 92.2° (AsH_3), 91.5° (SbH_3) and 90.3° (BiH_3) are quite similar and close to that of 93.6° for PH_3 but substantially different from the value of 107.3° for the lightest member (NH_3) in this

TABLE 1. Structures, bond energies (kcal mol^{-1}) and adiabatic ionization energies (eV) of MH, MH_2 and MH_3 (M = As, Sb, Bi). Lengths (r) and angles (θ) are in Å and degrees, respectively

MH_n	r (M—H)		θ (H—M—H)		Bond energies		Ionization energies	
	calc.^{a}	expt.	calc.^{a}	expt.	calc.^{a}	expt.	calc.^{a}	expt.
As—H	1.528	1.523^b			62.4	64.6^c	9.50	9.64^c
Sb—H	1.731	1.72^d			53.9		8.66	
Bi—H	1.869	1.809^e			41.0		8.56	
AsH—H	1.521	1.518^f	90.7	90.4^f	69.1	66.5^c	9.25	9.44^c
SbH—H	1.726		89.8		55.8		8.38	
BiH—H	1.876		88.9		45.4		8.03	
AsH_2—H	1.517	1.511^f	92.2	92.1^f	74.6	74.9^c	9.5	9.82^c
SbH_2—H	1.719	1.704^g	91.5	91.6^g	63.3		8.90	
BiH_2—H	1.865		90.3		51.8		9.00	

^{a} From Reference 22.
^{b} From J. R. Anacona, P. B. Davis and S. A. Johnson, Mol. Phys., **56**, 989 (1985).
^{c} From J. Berkowitz, J. Chem. Phys., **89**, 7065 (1988).
^{d} From P. Bollmark and B. Lindgren, Chem. Phys. Lett., **1**, 480 (1967).
^{e} From A. M. R. P. Bopegedera, C. R. Brazier and P. F. Bernath, Chem. Phys. Lett., **162**, 301 (1989).
^{f} From R. N. Dixon, G. Duxbury and H. M. Lamberton, Proc. R. Soc. London, Ser. A, **305**, 275 (1968).
^{g} From S. Konaka and M. Kimura, Bull. Chem. Soc. Jpn., **43**, 1693 (1970).

group[24]. This is due to the difference in the nature of chemical bonding. The bond angle of 107.3° for NH_3 suggests considerable sp^3 hybrid character in the N—H bonds. On the other hand, the Mulliken populations on the central atoms in AsH_3, SbH_3 and BiH_3 are $s^{1.72}p^{3.19}$, $s^{1.79}p^{3.02}$ and $s^{1.93}p^{3.01}$, respectively, suggesting little hybridization. Thus, the M—H bonds in the heavier hydrides are made of mostly np (M) and $1s$ (H) atomic orbitals. The 6s population on the Bi atom in BiH_3 is 0.14 larger than the 5s population on SbH_3, while the 5s and 4s populations on SbH_3 and AsH_3 differ by only 0.07. This is mainly due to the relativistic mass–velocity stabilization of the $6s^2$ shell of Bi, also known as the inert-pair effect. As a result, the lightest NH_3 and the heaviest BiH_3 exhibit considerable deviation from the trends given by other members in this group.

As Table 1 shows, the trends of the dissociation energies of the M—H bonds are somewhat irregular in magnitude. However, it can be seen that the M—H bond energies increase for any M upon going from MH to MH_2 to MH_3, as expected from the trend of the shortening of the bond lengths, suggesting a preference of trivalent species, but decrease as M becomes heavier. Thus, the heavier atoms form weaker bonds. This trend is also true for the bond energies for clusters such as M_2, M_3 and M_4. The calculated dissociation energies of M_2 to 2M are $71.6^{[25]}$ and $74.7^{[26]}$ kcal mol^{-1} for As_2, $50.0^{[26]}$ kcal mol^{-1} for Sb_2 and $31.8^{[26]}$ and $43.4^{[27]}$ kcal mol^{-1} for Bi_2, though these values tend to be considerably underestimated compared with the experimentally evaluated values[28] of 91.3 (As_2), 71.3 (Sb_2) and 47.0 (Bi_2) kcal mol^{-1}. Similar trends are also calculated for the bond energies of M_3[29a] and M_4[29b] at the CASSCF/CI level.

Fluorine forms stronger bonds with M than hydrogen. For example, the bond energy of 96.9 kcal mol^{-1} calculated for AsF[30] is 35 kcal mol^{-1} larger than the value of 62.4 kcal mol^{-1} for AsH. This trend decreases with M = Bi. However, the calculated (60.6 kcal mol^{-1})[31] and experimental (61.1 kcal mol^{-1})[32] bond energies of BiF are still ca 20 kcal mol^{-1} larger than the value of 41.0 kcal mol^{-1} for BiH, while the average bond energy in BiF_3 is calculated to be 39 kcal mol^{-1} larger than that in BiH_3[33]. As Table 2 shows[34], fluorine substitution has also an effect on the structures. Upon going from MH_3 to MF_3, the bond angles increase. This trend toward planarization is further enhanced as the F atoms are successively replaced by the heavier halogens such as Cl, Br and I. As a result, the bond angles in AsI_3 and SbI_3 become as large as 101 and 100°, respectively.

The trend toward planarization in MH_3 is also enhanced upon ionization. As shown in

TABLE 2. The calculated and experimental bond lengths (r), bond angles (θ) and dipole moments (D) of MX_3 (X = F, Cl, Br, I)

	r (Å)		θ (deg.)		Dipole moments	
	calc.[a]	expt.[b]	calc.[a]	expt.[b]	calc.[a]	expt.[c]
AsF_3	1.699	1.708	95.5	95.9	3.14	2.82
$AsCl_3$	2.186	2.162	98.9	98.6	2.02	2.1
$AsBr_3$	2.355	2.329	100.0	99.7	1.52	1.7
AsI_3	2.579	2.557	100.0	100.2	0.80	0.96
SbF_3	1.885	1.879	94.4	95.0	4.79	
$SbCl_3$	2.359	2.333	97.3	97.2	3.47	3.9
$SbBr_3$	2.520	2.490	98.4	98.2	2.88	2.8
SbI_3	2.737	2.719	99.6	99.1	2.07	1.58

[a] From Reference 34.
[b] From S. Konaka and M. Kimura, *Bull. Chem. Soc. Jpn.*, **43**, 1693 (1970); **46**, 404 (1973); **46**, 413 (1973).
[c] From P. Kisliuk, *J. Chem. Phys.*, **22**, 86 (1954) and references cited therein.

Table 1, the adiabatic ionization energies are 9.5 (AsH_3), 8.9 (SbH_3) and 9.0 (BiH_3) eV, and do not decrease monotonically: the somewhat larger ionization energy of BiH_3 (compared with SbH_3) may be due to the relativistic stabilization of the 6s orbital of Bi. However, the bond angles of MH_3 are all increased by ca 19° upon ionization[22] to 112.1° in AsH_3^+, 110.5° in SbH_3^+ and 109.5° in BiH_3^+. The bond angle of AsH_3^+ is closer to the angle of 120° for the planar form than is that of AsH_3, while the bond angle of BiH_3^+ corresponds to the tetrahedral angle. The energies required to obtain a planar structure are 9.2, 5.5 and 8.1 kcal mol^{-1} for AsH_3^+, SbH_3^+ and BiH_3^+, respectively showing again no regular decrease probably due to the large relativistic effect on BiH_3^+.

As an example of bicyclic compounds, three isomers of bicyclo[1.1.0]tetraarsane, a bicyclo[1.1.0]butane analogue, have been calculated at the HF level[35]. However, no systematic study is available.

B. Singlet–Triplet Energy Differences

In the neighboring group 14, the heavier divalent hydrides (GeH_2, SnH_2 and PbH_2)[36] as well as the lighter ones (CH_2 and SiH_2) have been the topics of many investigations for a number of years as important reaction intermediates. Especially interesting are the singlet–triplet energy differences. The AsH_2^+, SbH_2^+ and BiH_2^+ ions are the isoelectronic species. Table 3 summarizes the structures and properties of these ions calculated at the CASSCF/CI level with and without the spin–orbit interaction[37]. Thus we are now in a position to discuss the spin–orbit effect as well as the mass–velocity effect on the periodic trend.

TABLE 3. Bond lengths (r), bond angles (θ), relative energies (kcal mol^{-1}) and dipole moments (D) of MH_2^+ (M = As, Sb, Bi) calculated without and with spin–orbit interaction[a,b]

State[c]	r (Å)	θ (deg.)	Relative energies	Dipole moments[d]
AsH_2^+				
$^1A_1$ (A_1)	1.520 (1.520)	91.4 (91.4)	0.13 (0.00)	−0.161 (−0.161)
$^3B_1$ (A_1)	1.500 (1.500)	121.8 (121.8)	22.13 (21.80)	−0.195 (−0.195)
$^3B_1$ (A_2)	1.500 (1.500)	121.8 (121.8)	22.13 (22.06)	−0.195 (−0.195)
$^3B_1$ (B_2)	1.500 (1.500)	121.8 (121.8)	22.13 (22.06)	−0.195 (−0.195)
$^1B_1$ (B_1)	1.526 (1.526)	122.2 (122.2)	48.13 (48.06)	−0.274 (−0.274)
SbH_2^+				
$^1A_1$ (A_1)	1.716 (1.716)	90.7 (90.7)	0.70 (0.00)	0.044 (0.036)
$^3B_1$ (A_2)	1.695 (1.695)	119.8 (119.8)	25.40 (25.00)	−0.056 (−0.054)
$^3B_1$ (B_2)	1.695 (1.695)	119.8 (119.8)	25.40 (25.10)	−0.056 (−0.055)
$^3B_1$ (A_1)	1.695 (1.703)	119.8 (103.8)	25.40 (31.20)	−0.056 (0.006)
$^1B_1$ (B_1)	1.731 (1.731)	120.8 (120.8)	47.90 (47.50)	−0.167 (−0.167)
BiH_2^+				
$^1A_1$ (A_1)	1.850 (1.852)	90.0 (90.2)	7.90 (0.00)	0.229 (0.090)
$^3B_1$ (A_2)	1.840 (1.845)	119.2 (119.3)	39.20 (35.10)	−0.282 (−0.259)
$^3B_1$ (B_2)	1.840 (1.844)	119.2 (119.2)	39.20 (35.10)	−0.282 (−0.266)
$^3B_1$ (A_1)	1.840 (1.846)	119.2 (108.2)	39.20 (47.20)	−0.282 (−0.102)
$^1B_1$ (B_1)	1.909 (1.912)	119.9 (120.1)	55.50 (50.80)	−0.198 (−0.183)

[a] From Reference 37.
[b] In parentheses are the values calculated with spin–orbit interaction.
[c] The states in parentheses are spin–orbit states while the states without parentheses are those calculated in the absence of the spin–orbit interaction.
[d] Positive values imply M^+H^- polarity of bonds.

The AsH_2^+, SbH_2^+ and BiH_2^+ ions have the $^1A_1$ ground states of C_{2v} symmetry, as calculated for the corresponding group 14 analogues[38]. The bond angles (θ) decrease in the order AsH_2^+ (91.4°) > SbH_2^+ (90.7°) > BiH_2^+ (90.0°), approaching 90° as the central atom becomes heavier. The gross populations on the ns atomic orbitals of the central heavy atoms are 1.89 and 1.79 for AsH_2^+ and SbH_2^+, respectively, while it is almost 2.0 for BiH_2^+. These are consistent with the phenomenon of the inert-pair effect. In other words, the heavier atoms have a stronger tendency to preserve the ns^2 shell unhybridized even in compounds. On the other hand, the effect of the spin–orbit interaction on the structures and properties of the $^1A_1$ ground states is small because of the closed-shell character, as is apparent from Table 3. However, it may be noteworthy that the $^1A_1$ ground state of BiH_2^+ is stabilized by 7.9 kcal mol^{-1} and its dipole moment is altered from 0.229 to 0.090 D by the spin–orbit interaction. This is due to contamination with other electronic states which do not mix in the absence of the spin–orbit interaction.

Upon going to the first excited $^3B_1$ state, the bond angles open up from 90 to 120° for all the ions in the absence of the spin–orbit interaction. The singlet–triplet energy difference between $^1A_1$ and $^3B_1$ increases in the order AsH_2^+ (22.0 kcal mol^{-1}) < SbH_2^+ (24.7 kcal mol^{-1}) < BiH_2^+ (31.3 kcal mol^{-1}). The increase is only 3 kcal mol^{-1} upon going from AsH_2^+ to SbH_2^+ but 7 kcal mol^{-1} from SbH_2^+ to BiH_2^+. This larger increase in BiH_2^+ is related to the relativistic stabilization of the 6s orbital due to the mass–velocity contraction since the $^1A_1$ state qualitatively arises from the $6s^26p^2$ shell of the Bi$^+$ atom and the $^3B_1$ state from the $6s6p^3$ shell. The enhanced stability of the $6s^26p^2$ shell of Bi$^+$ leads to stabilization of $^1A_1$ relative to $^3B_1$. Similar trends are also calculated upon going from GeH_2 (23.2 kcal mol^{-1}) to SnH_2 (24.5 kcal mol^{-1}) to PbH_2 (33.2 kcal mol^{-1}) at the CASSCF/CI level[38].

In contrast to the $^1A_1$ ground state, the $^3B_1$ state undergoes a significant spin–orbit effect. The spin–orbit interaction splits the $^3B_1$ state into the A_1, A_2 and B_2 components in the C_{2v}^2 double group. As Table 3 shows, the spin–orbit splitting is only 0.3 kcal mol^{-1} in AsH_2^+ but increases significantly to 6.2 kcal mol^{-1} in SbH_2^+. In BiH_2^+ the spin–orbit interaction lowers the $^3B_1$ (A_2) and $^3B_1$ (B_2) states by 4.1 kcal mol^{-1} and raises the $^3B_1$ (A_1) state by 8.0 kcal mol^{-1} with respect to the pure spin $^3B_1$ state, the resultant spin–orbit splitting becoming as large as 12 kcal mol^{-1}. Thus the spin–orbit effect is quite significant

TABLE 4. The contributions of various states to spin–orbit states[a]

Ion	Spin orbit state:	Weight in percent
AsH_2^+	$^1A_1$ (A_1):	$^1A_1$ (93%)
	$^3B_1$ (A_1):	$^3B_1$ (94%) + $^1A_1$ (1.3%)
	$^3B_1$ (A_2):	$^3B_1$ (95%)
	$^3B_1$ (B_2):	$^3B_1$ (95%)
	$^1B_1$ (B_1):	$^1B_1$ (94%)
SbH_2^+	$^1A_1$ (A_1):	$^1A_1$ (93%) + $^3B_1$ (0.3%)
	$^3B_1$ (A_1):	$^3B_1$ (51%) + $^1A_1$ (43%)
	$^3B_1$ (A_2):	$^3B_1$ (95%) + $^3B_2$ (0.05%)
	$^3B_1$ (B_2):	$^3B_1$ (95%) + $^3A_2$ (0.04%)
	$^1B_1$ (B_1):	$^1B_1$ (93%) + $^3A_2$ (0.05%)
BiH_2^+	$^1A_1$ (A_1):	$^1A_1$ (88%) + $^3B_1$ (3.4%) + $^3A_2$ (2%)
	$^3B_1$ (A_1):	$^3B_1$ (59%) + $^1A_1$ (32%)
	$^3B_1$ (A_2):	$^3B_1$ (92%) + $^3B_2$ (0.6%) + $^1A_2$ (0.4%)
	$^3B_1$ (B_2):	$^3B_1$ (92%) + $^3B_2$ (0.6%)
	$^1B_1$ (B_1):	$^1B_1$ (90%) + $^3A_1$ (0.9%) + $^3B_2$ (0.7%)

[a] From Reference 37.

in BiH_2^+. In the presence of the spin–orbit interaction, the energy difference between $^1A_1$ (A_1) and $^3B_1$ (A_1) increases in the order AsH_2^+ ($21.8\,kcal\,mol^{-1}$) < SbH_2^+ ($31.2\,kcal\,mol^{-1}$) < BiH_2^+ ($47.2\,kcal\,mol^{-1}$). The increase of $16\,kcal\,mol^{-1}$ from SbH_2^+ to BiH_2^+ is twice as large as that from AsH_2^+ to SbH_2^+.

As Table 4 shows, the spin–orbit interaction mixes several states. This mixing is quite large for the $^3B_1$ (A_1) states of SbH_2^+ and BiH_2^+. The $^3B_1$ (A_1) state is 51% $^3B_1$ and 43% A_1 in SbH_2^+ while it is 59% $^3B_1$ and 32% A_1 in BiH_2^+. The mixing of the A_1 component has a substantial effect on the bond angles of the $^3B_1$ (A_1) states of SbH_2^+ and BiH_2^+: the angles decrease by 16 and 11°, respectively, and move closer to the bond angles of the $^1A_1$ ground state. Similar trends are also true for the bond lengths. These are larger in SbH_2^+ than in BiH_2^+, as expected from the stronger mixing in the former due to the avoided crossing of the $^3B_1$ (A_1) and $^1A_1$ (A_1) states. It appears that spin–orbit interaction influences structures as well as energies.

As also shown in Table 3, the energy difference between the ground $^1A_1$ and first singlet excited $^1B_1$ state is $48.0\,kcal\,mol^{-1}$ in AsH_2^+, decreases to $47.2\,kcal\,mol^{-1}$ in SbH_2^+, but increases to $47.6\,kcal\,mol^{-1}$ in BiH_2^+. This trend is a little affected by the spin–orbit interaction because of the closed-shell character of these states.

C. Inversion Barriers

As is well-known for NH_3, the heavier hydrides MH_3 (1) also invert via a trigonal D_{3h} planar transition structure (2).

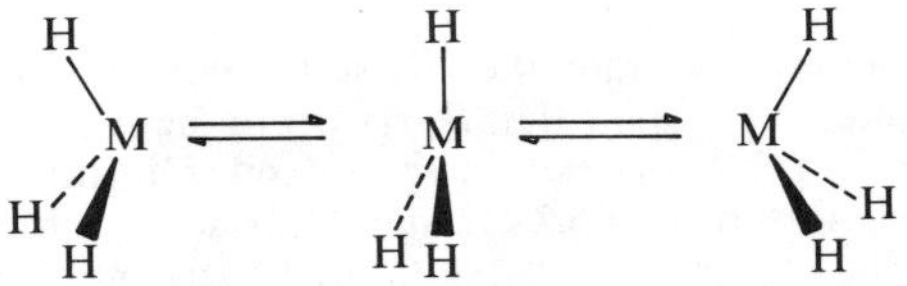

C_{3v} (1) D_{3h} (2) C_{3v} (1)

M = N, P, As, Sb, Bi

SCHEME 1

Table 5 summarizes the calculated barriers to the classical vertex inversion process at several levels of theory[22,39–41]. The barrier height for NH_3 is only ca $6\,kcal\,mol^{-1}$. However, it increases sharply to ca $35\,kcal\,mol^{-1}$ upon going to PH_3. The increase in the barrier becomes relatively small upon going to AsH_3 (barriers = 39–$42\,kcal\,mol^{-1}$) and SbH_3 (barriers = 43–$47\,kcal\,mol^{-1}$), but is significantly enhanced with BiH_3 (barriers = 61–$65\,kcal\,mol^{-1}$). As is apparent from these barriers, the pyramidal C_{3v} structures of the heavier hydrides are rigid in contrast to NH_3 which exhibits rapid umbrella inversion motion at room temperature. This is related to the inert-pair effect: the heavier atoms can preserve the ns^2np^3 valence configuration in the pyramidal C_{3v} structures with bond angles of ca 90°. The unusually high barrier for BiH_3 is due to the relativistically frozen 6s electrons, as is well demonstrated by the fact that its barrier is decreased to 45–$47\,kcal\,mol^{-1}$ by 18–$19\,kcal\,mol^{-1}$ in the absence of the relativistic mass–velocity correction[40].

It should be noted that considerable hybridization is required to attain the planar D_{3h} transition structures. Resultant s–p hybridization is indeed effective for bond making, as is reflected in the calculated results which show that the M—H bonds become shorter in the

TABLE 5. Inversion barriers (kcal mol^{-1}) calculated for MH_3 and MF_3 (M = N, P, As, Sb, Bi)[a]

	Reference 39		Reference 22	Reference 40		Reference 41		
	HF	MP2	CASSCF/SOCI	HF	MP2	HF	MP2	Expt.
Hydrides (MH_3)								
NH_3				5.3	6.0			5.8[b]
PH_3	36.3	35.0		35.9	35.1	37.6	34.7	31.5[c]
AsH_3	42.4	41.3	42.4	40.9	39.2	41.9	39.7	
SbH_3	44.8	42.8	44.7	46.0	43.9	47.8	44.9	
BiH_3			61.8	64.9	63.1	63.4	60.5	
Fluorides (MF_3)								
PF_3	68.4	53.8				67.7	52.4	
AsF_3	57.7	46.3				57.8	45.7	
SbF_3	46.5	38.7				45.4	37.6	
BiF_3						38.7	33.5	

[a] Vertex inversion for MH_3 and edge inversion for MF_3.
[b] From J. D. Swalen and J. A. Ibers, *J. Chem. Phys.*, **36**, 1914 (1962).
[c] From R. E. Weston, *J. Am. Chem. Soc.*, **76**, 2645 (1954).

D_{3h} structures than in C_{3v} structures. However, the hybridization results in a great energy loss. The energy cost for promotion is too large to be compensated just by the shortening of the bonds, leading to high barriers compared with the NH_3 case. Similar discussion has been also conducted as regards somewhat related compounds by Epiotis using MOVB theory in an attempt to generalize the so-called isoelectronic principle[42]. Schwerdtfeger and coworkers explained the increasing trend in barrier heights from NH_3 to BiH_3 in terms of a second-order Jahn–Teller distortion of the trigonal planar D_{3h} structures[40].

For a long time planar D_{3h} structures similar to **2** were believed to be the only possible transition states for the inversion of pyramidal C_{3v} structures. This belief was so deeply rooted that inversion barriers were calculated simply as the energy difference between the D_{3h} and C_{3v} structures without the characterization of the transition state by the vibrational frequency analysis. However, Dixon and Arduengo have recently pointed out that the planar D_{3h} structures of fluorides (MF_3, M = P, As, Sb) have three imaginary frequencies, one out-of-plane (a_2'') and two degenerate in-plane bending (e') modes, and are not the transition states for the inversion[39,43]. It was suggested that the fluorides invert through a planar T-shapped C_{2v} transition state (**3**). This is in sharp contrast to the fact that the lighter analogues such as NF_3 and NCl_3 invert via a D_{3h} structure[44].

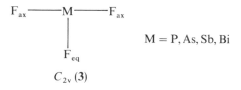

$$F_{ax}\text{——}M\text{——}F_{ax}$$

M = P, As, Sb, Bi

$$F_{eq}$$

C_{2v} (**3**)

SCHEME 2

As Figure 1 shows, the inversion process is visualized as angle deformation motions and called the edge inversion. The T-shaped C_{2v} transition structure (**3**) arises from the C_{3v} pyramidal structure of fluorides by simultaneous opening of the F—M—F and F—M—lone pair angles. Therefore, there are three identical reaction channels[45]. The

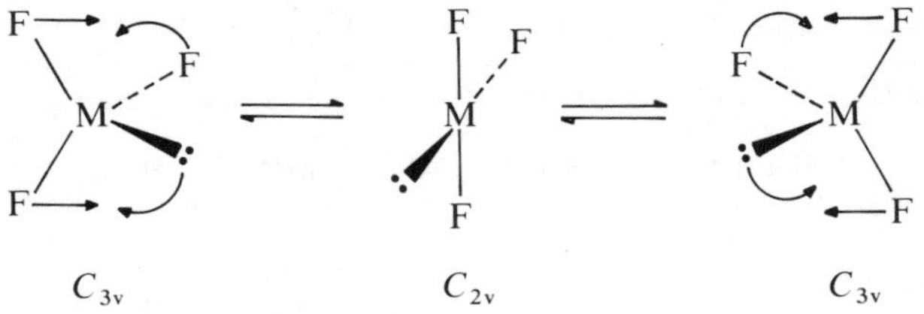

S. Nagase

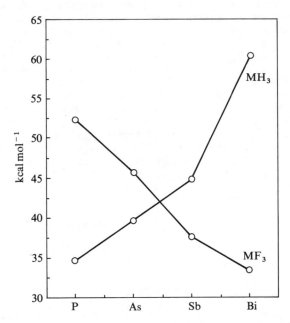

FIGURE 1. The edge inversion process in fluorides (MF_3, M = P, As, Sb, Bi)

preference of the edge inversion is explained by the fact that electrostatic repulsion between the lone pair of the central M atom and the F atoms is considerably less in the T-shaped C_{2v} form than in the trigonal D_{3h} form[39,43]. In fact, the C_{2v} transition structures of PF_3, AsF_3 and SbF_3 are calculated to lie 32, 20 and 19 kcal mol^{-1} lower in energy than the D_{3h} structures at the MP2 level[39]. In contrast, the C_{2v} structures of the hydrides correspond to minima but are 125 (PH_3), 101 (AsH_3) and 69 (SbH_3) kcal mol^{-1} more unstable at the MP2 level than the D_{3h} transition structures[39].

As shown in Table 5, the barriers for PF_3 are *ca* 18 kcal mol^{-1} larger than that for PH_3. This may suggest that electronegative substituents increase inversion barriers. However, it should be noted that the barriers for the edge inversion decrease sharply as the central M atom becomes heavier. As Figure 2 shows, fluorides follow the reverse trend compared with hydrides. As a result, the heavier fluorides such as SbF_3 and BiF_3 invert more easily

FIGURE 2. Comparison of the inversion barriers (kcal mol^{-1}) of MH_3 and MF_3 (M = P, As, Sb, Bi). The values calculated at the MP2 levels are taken from Reference 41

than the corresponding hydrides. Displacement of the F atoms by the heavier halogens leads to a decrease in the barrier height, as was calculated at the HF level[44a] for AsF_3 (55.0 kcal mol^{-1}), $AsCl_3$ (55.5 kcal mol^{-1}) and $AsBr_3$ (52.6 kcal mol^{-1}), though there is a small increase at $AsCl_3$. However, this trend is small compared with the fact that the vertex inversion barrier of ca 78 kcal mol^{-1} for NF_3 is decreased by 55[44b]–65[44a] kcal mol^{-1} in going to NCl_3. It is pointed out that the T-shaped transition state can be stabilized by σ-acceptors in the axial positions and π-donors in the equatorial position[39]. In this context, it is interesting to note that compounds (4) having a T-shaped bond configuration at the M atom have been synthesized and the stability increases as M becomes heavier[46–49].

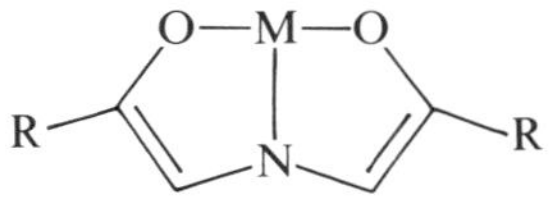

$$M = P, As, Sb$$

(4)

SCHEME 3

D. Multiply Bonded Compounds

For many years the synthesis and characterization of stable compounds featuring double bonding between the heavier group 15 elements have been one of the major challenges of organic chemistry[50]. The breakthrough occurred in 1981 when the first stable P—P doubly bonded compound, a diphosphene (RP=PR) derivative, was synthesized and isolated by Yoshifuji and collaborators[51]. Since then, many kinds of diphosphenes have been reported[50]. In addition, diarsenes (RAs=AsR) have also been successfully synthesized and isolated[52–55]. However, attempts to prepare the still heavier analogues, distibenes (RSb=SbR) and dibismuthenes (RBi=BiR), have all been unsuccessful up to now, except for the metal-coordinated distibene complexes which lose double-bond character to a considerable extent because of side-on η^2 coordination[56–59].

It was also in 1981 that the first stable heavier ethene ($R_2C=CR_2$) analogue in group 14, disilene ($R_2Si=SiR_2$), was synthesized and isolated[60]. Since then, both the experimental and theoretical aspects of the field have grown greatly in an interactive manner[61]. In addition, experimental evidence has accumulated which shows that the Ge atom can also form isolable ethene analogues, digermenes ($R_2Ge=GeR_2$)[62]. However, it has been shown that the heavier Sn atom is reluctant to form distinct double bonds in distannenes ($R_2Sn=SnR_2$)[63,64] while it has been calculated that the heaviest Pb atom does not form a diplumbene ($R_2Pb=PbR_2$) structure at all[65]. This situation may resemble closely the failures to prepare Sb=Sb and Bi=Bi double bonds. In view of this, it is interesting to investigate whether Sb and Bi are also incapable of forming double bonds, as are Sn and Pb, from a theoretical point of view.

Figure 3 shows the calculated structures of the parent compounds (HM=MH, M = P, As, Sb, Bi) at the HF level[66]. For all these compounds, the planar $trans$ structures of C_{2h} symmetry are located as energy minima on the potential energy surfaces. The calculated P–P and As–As distances are in reasonable agreement with the experimental values of 2.001–2.034[51–53,67–70] and 2.224–2.245 Å[52–55] of the derivatives, respectively. In addition, the planar cis structures of C_{2v} symmetry were also located as minima, but less stable than the $trans$ forms by 2–3 kcal mol^{-1} because of steric repulsion between the H atoms.

It is noteworthy that the Sb—Sb and Bi—Bi bond lengths in HSb=SbH. and HBi=BiH are 9.0% (0.257 Å) and 8.9% (0.264 Å) shorter than the single bond lengths in H_2Sb—SbH_2 and H_2Bi—BiH_2, respectively. These bond shortenings are comparable to

S. Nagase

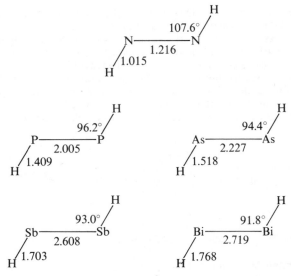

FIGURE 3. Optimized lengths (Å) and angles of the *trans* C_{2h} forms taken from Reference 66

the P—P and As—As bond shortenings of 9.2% (0.203 Å) in HP=PH from H_2P—PH_2 and 9.7% (0.239 Å) in HAs=AsH from H_2As—AsH_2. This is in sharp contrast to the Sn and Pb cases in group 14; distannene and diplumbene structures suffer from strong pyramidalization at the Sn and Pb atoms and the Sn—Sn and Pb—Pb bond lengths become only slightly shorter and much longer, respectively, than the corresponding single bond lengths.

The Sb—Sb and Bi—Bi bond stretching frequencies of 254 and 180 cm^{-1} in HSb=SbH and HBi=BiH are 40.3 and 39.5% higher than those in H_2Sb—SbH_2 and H_2Bi—BiH_2, respectively[66]. These frequency-shifts compare favorably with those of 40.8% in HP=PH (701 cm^{-1}) from H_2P—PH_2 and 44.1% in HAs=AsH (392 cm^{-1}) from H_2As—AsH_2. In addition, the Sb—Sb (1.98) and Bi—Bi (2.03) bond orders in HSb=SbH and HBi=BiH are close to the P—P (2.01) and As—As (1.93) in HP=PH and HAs=AsH[66]. All these suggest that Sb and Bi atoms are capable of forming double bonds, as are P and As atoms. In fact, it is calculated at the two-configuration SCF level that the internal rotations around the Sb—Sb and Bi—Bi bonds in HSb=SbH and HBi=BiH are hindered by significant barriers of 16 and 14 kcal mol^{-1}, respectively, though these barriers are smaller than those in HP=PH and HAs=AsH[66], 31 and 23 kcal mol^{-1}, respectively.

As Figure 3 shows, the H—M—M bond angles in HM=MH (M = P, As, Sb and Bi) are much smaller than the corresponding angle of 107.6° in HN=NH and approach 90° as the M atom becomes heavier. This is because the heavier atom has a lower tendency to form hybrid orbitals and maintains the valence ns^2np^3 configuration even in compounds. However, it is instructive to note that the tendency is not especially unfavourable in forming a double bond in HM=MH, as is apparent from Figure 4. This contrasts with the fact that the group 14 atoms with ns^2np^2 must hybridize to form double bonds in ethene analogues.

The isomerizations of HSb=SbH and HBi=BiH to H_2SbSb and H_2BiBi via 1,2-H shifts are highly endothermic and hindered by sizable barriers of 40 and 36 kcal mol^{-1},

trans   *cis*

FIGURE 4. A simplified description of bond formation by heavier group 15 atoms with the $ns^2 np^3$ valence electron configuration. Doubly occupied ns atomic orbitals are omitted for simplicity

respectively[66], at the MP4 level. HSb=SbH and HBi=BiH are 35 and 37 kcal mol^{-1} more stable than the 1,2-H shifted isomers. This is ascribed to the fact that the Sb and Bi atoms must hybridize in the 1,2-H shifted isomers. Thus, it is not surprising that the energy differences disfavoring the 1,2-H shifted isomers are larger than those of 27 and 34 kcal mol^{-1} for HP=PH and HAs=AsH, respectively. Consequently, H_2SbSb isomerizes back to HSb=SbH with a small barrier while H_2BiBi collapses to HBi=BiH almost with no barrier[66]. This again contrasts with the group 14 case, 1,2-shifted isomers being increasingly favored upon going to distannene and diplumbene[65].

Thus, it seems no wonder that the Sb and Bi atoms can form doubly bonded compounds, as do the lighter P and As atoms. As is apparent from the comparison of the frontier orbital energy levels in Figure 5, however, steric protection by bulky substituents will be essential for the stability enabling isolation. It is also suggested that the Sb and Bi analogues of benzene and cyclobutadiene may be interesting synthetic targets[66]. In this context, it is instructive to note that the stabilization of As_6 (hexaarsabenzene) is successful

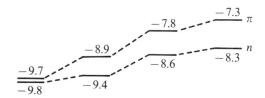

HP=PH   HAs=AsH   HSb=SbH   HBi=BiH

FIGURE 5. Frontier orbital energy levels (eV) for the *trans* C_{2h} forms taken from Reference 66

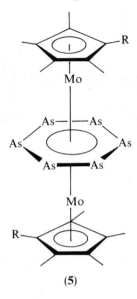

(5)

SCHEME 4

in a triple-decker sandwich complex (**5**, $R = CH_3$, C_2H_5)[71], as in the P_6 (hexaphospha-benzene) case[72]. The X-ray structure shows that the planar As_6 ring is a regular hexagon with the As—As bond lengths between those of single and double bonds. A distinct upfield shift of the ^1H NMR signals of the cyclopentadienyl CH_3 groups in the complex is regarded as evidence of a ring-current effect of the aromatic As_6 ring.

Baldridge and Gordon have investigated the aromaticity (delocalization stabilization) of the monosubstituted benzene analogues (**6**) such as arsabenzene (M = As) and stiba-benzene (M = Sb) including pyridine (M = N) and phosphabenzene (M = P) at the HF level[73]. The stabilization energies due to cyclic delocalization were calculated using the superhomodesmic reactions for the series of compounds. The resultant values are 25.8 (M = N), 23.3 (M = P), 21.6 (M = As) and 18.6 (M = Sb) kcal mol^{-1}. The value for M = N is close to the value of 26.0 kcal mol^{-1} calculated at the same level for benzene. Thus, pyridine is almost as aromatic as benzene. Upon going to the heavier systems, there is a decrease in delocalization stabilization. However, even arsabenzene and stibabenzene are still 93 and 80% as aromatic as phosphabenzene. In contrast, it is shown that in a series of monosubstituted cyclopentadiene compounds (**7**), no net stabilization is obtained with

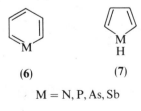

(6)   (7)

M = N, P, As, Sb

SCHEME 5

$M = As$ and Sb due to the induced high strain, except the very small stabilization of 5.2 kcal mol^{-1} for pyrrole ($M = N$)[73]. Gordon and coworkers have also investigated the heavier analogues (**8–10**) of borazine ($B_3N_3H_6$)[74]. It is calculated at the HF level that the planar D_{3h} structures have three imaginary frequencies and are maxima with respect to deformation to chair and boat forms, indicating that their aromaticity is insufficient to maintain planarity.

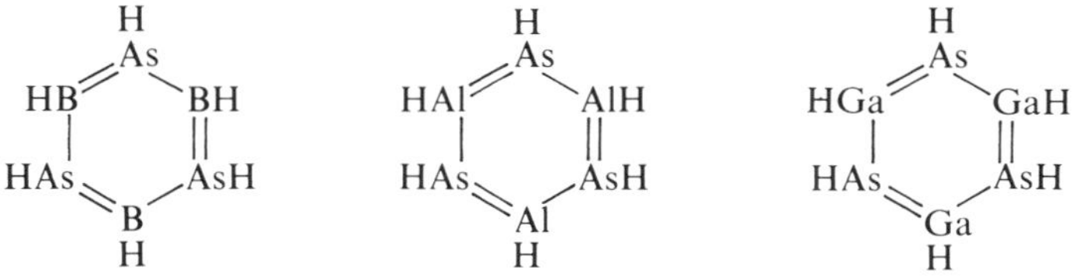

(**8**)   (**9**)   (**10**)

SCHEME 6

Based on the empirical Hückel method, Mishra presented the modified values of the topological resonance energies for monosubstituted cyclobutadiene compounds (**11**)[75]. According to the topological resonance energies, **11** is less antiaromatic than cyclo-butadiene. The antiaromatic destabilization tends to increase slightly upon going from $M = N$ to $M = As$, but decreases strongly as the M atom becomes more heavier. This suggests that the heavier Sb and Bi compounds have less antiaromatic character and are more stable than the lighter analogues. This may be related to the fact that the heavier M atoms form double bonds with bond angles of *ca* 90° at M (see Figure 4), providing a less strained cyclobutadiene framework. On the other hand, Ilic and coworkers pointed out that incorporation of N, P, As, Sb and Bi into the benzene ring does not significantly affect the aromatic character of the six-membered ring[76].

$M = N, P, As, Sb, Bi$

(**11**)

SCHEME 7

Schmidt and collaborators have calculated the effect of substituents ($R = H, CH_3$, F and C_6H_5) on the properties of the triple $M\text{—}C$ bonds in $R\text{—}C\equiv M$ ($M = N, P, As, Sb$) with the coupled electron-pair approximation (CEPA) method[77]. As is apparent from Table 6, the substituents have no drastic effect on the triple bond lengths. However, it should be noted that there is a slight steady increase in the bond lengths, together with more pronounced decrease in the force constants, in the order H, CH_3, F, C_6H_5 for any choice of M. This suggests that the parent compounds ($R = H$) have the strongest triple bond for any M atom. The parent species $HC\equiv As$ and $HC\equiv Sb$ were also calculated at the HF level by Lohr and Scheiner[78] and Cowley and coworkers[79] together with $H_2C\equiv AsH$ and $H_2C\equiv SbH$. The bond-shortening from single bonds, proton affinities and HOMO–LUMO levels are discussed.

The properties of the formal double bonds in the hypervalent compounds H_3AsO and

TABLE 6. The C≡M bond lengths (r, Å), force constants (k, au) and dipole moments (D) of R—C≡M (R = H, CH₃, F, C₆H₅)[a]

M =	r (C≡M)				k (C≡M)				Dipole moments[b]			
	N	P	As	Sb	N	P	As	Sb	N	P	As	Sb
H—C≡M	1.152	1.543	1.649	1.857	1.237	0.592	0.485	0.348	2.87	0.71	0.18	−0.49
CH₃—C≡M	1.154	1.549	1.656	1.866	1.223	0.577	0.468	0.340	3.65	1.67	1.15	0.52
F—C≡M	1.155	1.550	1.660	1.872	1.183	0.553	0.442	0.318	2.13	0.16	−0.42	−1.08
C₆H₅—C≡M	1.157	1.553	1.661	1.872	1.193	0.564	0.455	0.325	4.40	2.18	1.54	0.71

[a] From Reference 77.
[b] Positive values correspond to positive charge on the R side of RCM.

H_3AsS have been investigated by Schneider and collaborators[80]. The dissociation energies of the As—O and As—S bonds are calculated to be 73.2 and 46.3 kcal mol^{-1}, respectively. These values are considerably smaller than those of 102.7 and 54.1 kcal mol^{-1} calculated for the P—O and P—S bonds in H_3PO and H_3PS[81]. In addition, H_3AsO is 33 kcal mol^{-1} less stable than 1,2-H shifted isomer H_2AsOH in which the $4s^2 4p^3$ configuration is probably preserved to a considerable extent. The energy difference of 33 kcal mol^{-1} is much larger than the value of 7 kcal mol^{-1} between H_3PO and H_2POH calculated at comparable levels[82]. These suggest that H_3AsO and H_3AsS are less stable than H_3PO and H_3PS, though H_3AsO has been recently detected in an argon matrix[83]. Upon fluorine substitution, the As—O and As—S stretching force constants increase by 22 and 26%, respectively, reflecting the enhanced role of π back-bonding due to the introduction of three electronegative F atoms. Accordingly, the dissociation energy of the As—O bond increases to 81.2 kcal mol^{-1} in F_3AsO; this compound has been identified and characterized by the vibrational spectrum[84]. However, fluorine substitution has rather an unfavorable effect on the strength of the As—S bond (45.7 kcal mol^{-1} in F_3AsS vs 46.3 kcal mol^{-1} in H_3AsS), the existence of H_3AsS and F_3AsS being still unknown.

In a recent study of the $GaAsH_2$ isomers by Bock and coworkers[85], a doubly bonded structure (12) was located on the potential energy surface. However, it was calculated to be 12 kcal mol^{-1} less stable at the MP4 level than the 1,2-H shifted isomer (13) having a single bond between Ga and As. Simply from the concept of hybridization, the stability difference may be ascribed to the $4s^2$ inert-pair effect. The As atom can preserve the 4s orbital unhybridized in both HAs=GaH and H_2As—Ga. However, the Ga atom must hybridize significantly to form a double bond in HAs=GaH, preferring a monovalent structure.

(12)   (13)

SCHEME 8

E. Hypervalent Compounds

Hypervalent compounds have long attracted both experimental and theoretical attention since they break the octet rule[86] well established in chemistry. To describe the hypervalent bonding, two different models have been proposed. In the first model the nature of hypervalent orbitals was rationalized by incorporating d-orbital participation in hybridization[87]. In another model developed since 1951 the formation of three-center, four-electron bonds without any significant d-orbital contribution has been emphasized to describe the hypervalent bonding[88]. A historical review of the chemical bonding in hypervalent compounds has been recently given by Reed and Schleyer[89]. The hypervalent compounds of the second-row elements have been extensively investigated from both experimental and theoretical points of view. However, there is only little systematic study on the hypervalent compounds containing the heavier As, Sb, Bi atoms.

Moc and Morokuma[41] have recently undertaken the first comparative calculations of the series of the tetravalent MH_4^- and MF_4^- (M = P, As, Sb and Bi) species in an attempt to provide theoretical insight into the periodic trends in the structures and stability: in these species, only AsH_4^- was the subject of previous theoretical work[90], except for the lighter PH_4^-[91,92] and PF_4^-[93]. The lowest energy forms were calculated to be the pseudo-trigonal-bipyramidal structures (14) of C_{2v} symmetry for both MH_4^- and MF_4^- and all

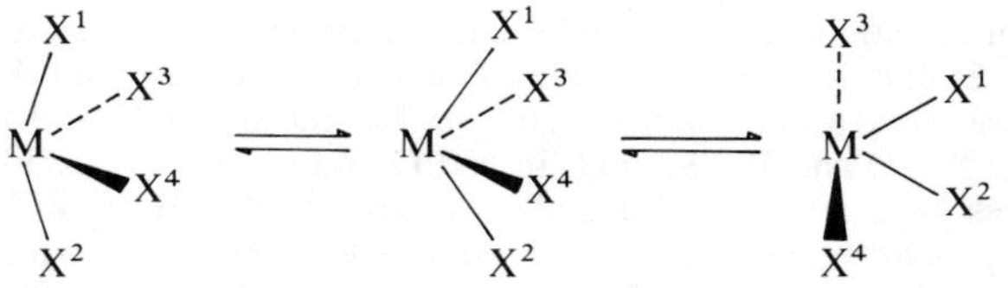

$$C_{2v}\ (\mathbf{14}) \qquad\qquad C_{4v}\ (\mathbf{15}) \qquad\qquad C_{2v}\ (\mathbf{14})$$

SCHEME 9

correspond to minima when electron correlation was taken into consideration at the MP2 level; the C_{2v} structures of $PH_4{}^-$ and $AsH_4{}^-$ do not correspond to energy minima on the potential energy surface without electron correlation[41]. The C_{4v} structures (**15**) of $MH_4{}^-$ and $MF_4{}^-$ characterized by one single b_2 imaginary frequency correspond to the transition states for the exchange of two axial ligands (X^1 and X^2) and two equatorial ligands (X^3 and X^4) in C_{2v} structures. This motion resembles the Berry-like pseudorotation[94] with the equatorially localized lone pair being the pivot 'ligand'.

It is calculated at the MP4 level that the barriers of the hydrides for C_{2v}–C_{4v}–C_{2v} increase in going from $PH_4{}^-$ (4.5 kcal mol^{-1}) to $AsH_4{}^-$ (6.1 kcal mol^{-1}), decrease in going to $SbH_4{}^-$ (2.1 kcal mol^{-1}), and then increase again with $BiH_4{}^-$ (7.2 kcal mol^{-1})[41]. In the fluorides, the barriers decrease monotonically in going from $PF_4{}^-$ (10.6 kcal mol^{-1}) to $AsF_4{}^-$ (7.5 kcal mol^{-1}) to $SbF_4{}^-$ (4.0 kcal mol^{-1}), but increase with $BiF_4{}^-$ (5.4 kcal mol^{-1}) at the MP2 level[41]. The irregularity in the hydride series is explained in terms of relatively strong bonds in $SbH_4{}^-$ (C_{4v}) and relatively weak equatorial bonds in $SbH_4{}^-$ (C_{2v}) as well as relatively small deformation energy for $SbH_4{}^-$ (C_{2v}). On the other hand, it is pointed out that a number of factors contribute to the irregularity in the fluoride series[41]. As Table 7 shows, the thermodynamic stabilities of $MH_4{}^-$ and $MF_4{}^-$ relative to loss of H^- or F^- increase in the order $P < As < Sb = Bi$. In other words, the affinities of MH_3 and MF_3 for H^- and F^- increase in this order. As is apparent from Table 7, electronegative fluorine atoms have a strong effect on the stabilization of the hypervalent structures.

Pentavalent compounds have long been of experimental and theoretical interest in phosphorus chemistry. The parent PH_5, although not yet observed, has been one of the most extensively studied species from a theoretical point of view[90,95-105]. In contrast, no

TABLE 7. The reaction energies (kcal mol^{-1}) for $MX_3 + X^- \rightarrow MX_4{}^{-a,b}$

	X^c	
	H	F
$PX_3 + X^- \rightarrow PX_4{}^-$	−5.8	−48.6
$AsX_3 + X^- \rightarrow AsX_4{}^-$	−11.4	−58.8
$SbX_3 + X^- \rightarrow SbX_4{}^-$	−24.0	−69.7
$BiX_3 + X^- \rightarrow BiX_4{}^-$	−23.6	−70.4

[a] From Refernce 41.
[b] Negative values mean that the reaction is exothermic.
[c] MP4 values for X = H and MP2 values for X = F.

attention was devoted to the heavier analogues except one *ab initio* study of the D_{3h} and C_{4v} structures of AsH_5[90]. Although no vibrational analysis was done in the calculations of AsH_5[90], Moc and Morokuma[106] have recently confirmed that for all the pentavalent species (MH_5, M = P, As, Sb, Bi) the trigonal–bipyramidal D_{3h} structures (16) are minima on the potential energy surfaces while the square-planar C_{4v} structures (17) correspond to the transition states for Berry pseudorotation[94] of the D_{3h} structures. The barries to Berry pseudorotation are calculated to be 1.9, 2.1, 2.3 and 1.9 kcal mol^{-1} for PH_5, AsH_5, SbH_5 and BiH_5 at the MP4 level, respectively[106]. These small barrier heights within *ca* 2 kcal mol^{-1} suggest that all the pentavalent species are nonrigid regardless of the central atoms. However, it should be noted that the changes are again irregular at SbH_5.

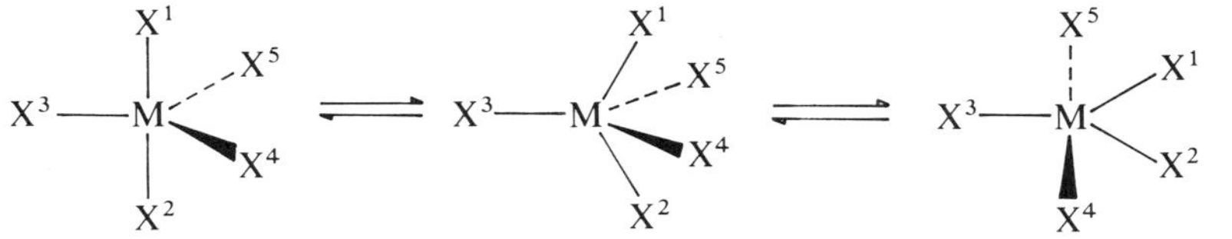

$$D_{3h} \ (16) \qquad\qquad C_{4v} \ (17) \qquad\qquad D_{3h} \ (16)$$

SCHEME 10

It has been pointed out that PH_5 is thermodynamically unstable relative to $PH_3 + H_2$[90,95-97,100-102,104]. As Table 8 shows, this trend is further enhanced upon going to the heavier AsH_5, SbH_5 and BiH_5. This contrasts with the increasing thermodynamic stability in the tetravalent cases but shows similar irregularity. That is, the stability with respect to H_2 loss decreases in going from PH_5 to AsH_5, then increases with SbH_5 and finally decreases significantly with BiH_5, as shown in Figure 6. As an origin of the extra stability at SbH_5, the smaller deformation and more stabilizing back charge transfer ($SbH_3 \rightarrow H_2$) energies are pointed out[106].

However, the zigzag behavior as in Figure 6 due to the tendency of As and Bi to be trivalent may be explained from another point of view. The first difficulty in forming AsH_5 may be ascribed to the so-called 'd-block contraction' caused by an increase in the effective nuclear charge for the 4s electrons due to filling the first d shell ($3d^{10}$)[107]. On the other hand, the strong tendency of BiH_5 to prefer the trivalent BiH_3 is due to the relativistic mass–velocity effect. According to the recent calculations at the MP2 and quadratic CI levels by Schwerdtfeger and collaborators[33], the decomposition of BiH_5 into $BiH_3 + H_2$ becomes 33–35 kcal mol^{-1} less exothermic in the absence of the relativistic effect. This relativistic destabilization for the pentavalent state is further enhanced in AsF_5: the

TABLE 8. The reaction energies (kcal mol^{-1}) for $MH_3 + H_2 \rightarrow MH_5$[a,b]

	HF	MP4
$PH_3 + H_2 \rightarrow PH_5$	44.8	45.3
$AsH_3 + H_2 \rightarrow AsH_5$	55.2	54.6
$SbH_3 + H_2 \rightarrow SbH_5$	50.0	50.2
$BiH_3 + H_2 \rightarrow BiH_5$	75.4	73.0

[a] From Reference 106.
[b] Positive values mean that the reaction is endothermic.

S. Nagase

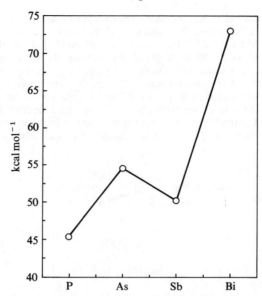

FIGURE 6. The energies (kcal mol^{-1}) of MH$_5$ relative
to MH$_3$ + H$_2$. The values at the MP4 levels are taken
from Reference 106

stability of AsF$_5$ relative to AsF$_3$ + F$_2$ decreases by *ca* 61 kcal mol^{-1} at the MP2 level[33]. Since, however, the stabilization due to the fluorine substitution overcomes the relativistic destabilization, a net result is that AsF$_5$ is *ca* 44 kcal mol^{-1} more stable than AsF$_3$ + F$_2$, unlike the AsH$_5$ case[33].

As in the case of PH$_5$, there is so far no experimental evidence of the existence of AsH$_5$, SbH$_5$ and BiH$_5$. However, the organic derivatives have been synthesized and isolated[108]. The structure of Sb(CH$_3$)$_5$ has been very recently subjected to theoretical as well as experimental studies[109]. Among these, most noteworthy is that pentaphenylbismuth Bi(C$_6$H$_5$)$_5$ shows an intense violet color, as first described by Wittig's group[110], though the lighter P(C$_6$H$_5$)$_5$, As(C$_6$H$_5$)$_5$ and Sb(C$_6$H$_5$)$_5$ are all colorless. The color is an unusual phenomenon in main group chemistry. According to the experimental characterization, P(C$_6$H$_5$)$_5$ and As(C$_6$H$_5$)$_5$ have a trigonal–bipyramidal structure[111] while the structure of Sb(C$_6$H$_5$)$_5$ resembles closely a square pyramid[112]. A recent X-ray structure analysis reveals that Bi(C$_6$H$_5$)$_5$ is best described as a square pyramid[113] and is thus very similar to the colorless Sb(C$_6$H$_5$)$_5$ in structure. Optical investigations show that excitation occurs in the basal plane of the square pyramid[113].

The origin of the violet color of Bi(C$_6$H$_5$)$_5$ has been recently investigated on several model compounds using relativisitic and nonrelativistic extended Hückel and Xα calculations[113b,114]. It is shown that the easy electronic excitation is made possible in the square-pyramidal structure by a lower-lying LUMO; in this structure HOMO is predominantly located in the basal plane while LUMO, having considerable s character of the central atom, is most strongly stabilized in the heaviest bismuth case as a result of the relativisitic 6s stabilization. More important is the symmetry aspect of the HOMO–LUMO transition in the square-pyramidal structure since the transition becomes allowed only relativistically (it is nonrelativistically forbidden). Thus, the violet color of Bi(C$_6$H$_5$)$_5$ is a result of relativistic effects. In other words, 'nonrelativistic' Bi(C$_6$H$_5$)$_5$ would be colorless just from a theoretical point of view. It is interesting to see relativistic effects with the naked eye.

IV. ACKNOWLEDGMENTS

I am thankful to Miss Kaoru Kobayashi for her assistance in preparing this manuscript. I am also thankful to Dr. Moc and Prof. Morokuma for sending preprints prior to publication.

V. REFERENCES

1. W. J. Hehre, L. Radom, P. v. R. Schleyer and J. A. Pople, *Ab initio Molecular Orbital Theory*, Wiley, New York, 1986.
2. J. L. Wardell, in *Comprehensive Organic Chemistry* (Eds. G. Wilkinson, F. G. A. Stone and W. A. Abell), Vol. 2, Chap. 13, Pergamon Press, Oxford, 1982.
3. For the latest annual survey, see: L. D. Freedman and G. O. Doak, *J. Organomet. Chem.*, **442**, 1 (1992).
4. For the latest annual survey, see: G. O. Doak and L. D. Freedman, *J. Organomet. Chem.*, **442**, 61 (1992).
5. For a review of similarity and difference of chemical bonding between lighter and heavier main group elements, see: W. Kutzelnigg, *Angew. Chem., Int. Ed. Engl.*, **23**, 272, (1984).
6. For a review of basis set selection, see: E. R. Davidson and D. Feller, *Chem. Rev.*, **86**, 681 (1986).
7. For books on theoretical methods, see:
   (a) H. F. Schaefer III (Ed.), *Modern Theoretical Chemistry*, Vols. 3 and 4, Plenum Press, New York, 1977.
   (b) K. P. Lawley (Ed.), *Ab Initio Methods in Quantum Chemistry—Parts I and II*, in *Adv. Chem. Phys.*, 1987.
   For a recent guide to *ab initio* quantum chemistry, see:
   (c) J. Simons, *J. Phys. Chem.*, **95**, 1017 (1991).
8. P. Pulay, *Adv. Chem. Phys.*, **69**, 241 (1987).
9. H. B. Schlegel, *Adv. Chem. Phys.*, **67**, 249 (1987).
10. K. S. Pitzer, *Acc. Chem. Res.*, **12**, 271 (1979).
11. P. Pyykkö and J. P. Desclaux, *Acc. Chem. Res.*, **12**, 276 (1979).
12. G. L. Malli (Ed.), *Relativistic Effects in Atoms, Molecules and Solids*, Plenum Press, New York, 1982.
13. M. Krauss and W. J. Stevens, *Ann. Rev. Phys. Chem.*, **35**, 357 (1984).
14. P. A. Christiansen, W. C. Ermler and K. S. Pitzer, *Ann. Rev. Phys. Chem.*, **36**, 407 (1985).
15. K. Balasubramanian and K. S. Pitzer, *Adv. Chem. Phys.*, **67**, 287 (1987).
16. P. J. Dagdigian and M. L. Campbell, *Chem. Rev.*, **87**, 1 (1987).
17. P. Pyykkö, *Chem. Rev.*, **88**, 563 (1988).
18. K. Balasubramanian, *J. Phys. Chem.*, **93**, 6585 (1989).
19. L. J. Norrby, *J. Chem. Educ.*, **68**, 110 (1991).
20. (a) N. Y. Sidgwick, *Some Physical Properties of the Covalent Link in Chemistry*, Cornell University Press, Ithaca, New York, 1933.
   (b) For discussions of the inert-pair effect, see, for example: J. G. Huheey, *Inorganic Chemistry. Principles of Structure and Reactivity*, 3rd ed., Harper & Row, New York, 1983.
21. P. A. Christiansen, *Chem. Phys. Lett.*, **109**, 145 (1984).
22. D. Dai and K. Balasubramanian, *J. Chem. Phys.*, **93**, 1837 (1990).
23. K. Balasubramanian, *Chem. Rev.*, **89**, 1801 (1989).
24. R. N. Dixon, G. Duxbury and H. M. Lamberton, *Proc. R. Soc. London, Ser. A*, **305**, 275 (1968).
25. G. E. Scuseria, *J. Chem. Phys.*, **92**, 6722 (1990).
26. Lai-Sheng Wang, Y. T. Lee, D. A. Shirley, K. Balasubramanian and P. Feng, *J. Chem. Phys.*, **93**, 6310 (1990).
27. K. Balasubramanian and D.-W. Liao, *J. Chem. Phys.*, **95**, 3064 (1991).
28. K. P. Huber and G. Herzberg, *Molecular Spectra and Molecular Structure. IV. Constants of Diatomic Molecules*, Van Nostrand Reinhold, New York, 1979.

29. (a) K. Balasubramanian, K. Sumathi and D. Dai, *J. Chem. Phys.*, **95**, 3494 (1991).
    (b) H. Zhang and K. Balasubramanian, *J. Chem. Phys.*, **97**, 3437 (1992).
    (c) K. Balasubramanian, *Chem. Rev.*, **90**, 93 (1990).
30. P. A. G. O'Hare, A. Batana and A. C. Wahl, *J. Chem. Phys.*, **59**, 6495 (1973).
31. K. Balasubramanian, *Chem. Phys. Lett.*, **127**, 324 (1986).
32. (a) A. G. Gaydon, *Dissociation Energy of Diatomic Molecules*, 3rd ed., Chapman and Hall, London, 1968.
    (b) W. E. Jones and T. D. McLean, *J. Mol. Spectrosc.*, **83**, 317 (1980); **90**, 481 (1981).
33. (a) P. Schwerdtfeger, G. A. Heath, M. Dolg and M. A. Bennet, *J. Am. Chem. Soc.*, **114**, 7518 (1992).
    (b) M. Dolg, W. Küchle, H. Stoll, H. Preuss and P. Schwerdtfeger, *Mol. Phys.*, **74**, 1265 (1991).
34. Y. Sakai and E. Miyoshi, *J. Chem. Phys.*, **89**, 4452 (1988).
35. J. Rubio, *J. Mol. Struct. (Theochem)*, **150**, 283 (1987).
36. W. P. Neumann, *Chem. Rev.*, **91**, 311 (1991).
37. K. Balasubramanian, *J. Chem. Phys.*, **91**, 2443 (1989).
38. K. Balasubramanian, *J. Chem. Phys.*, **89**, 5731 (1988).
39. D. A. Dixon and A. J. Arduengo III, *J. Am. Chem. Soc.*, **109**, 338 (1987).
40. P. Schwerdtfeger, L. J. Laakkonen and P. Pyykkö, *J. Chem. Phys.*, **96**, 6807 (1992).
41. J. Moc and K. Morokuma, *Inorg. Chem.*, in press.
42. N. D. Epiotis, *J. Mol. Struct. (Theochem)*, **153**, 1 (1987).
43. (a) D. A. Dixon, A. J. Arduengo III and T. Fukunaga, *J. Am. Chem. Soc.*, **108**, 2461 (1986).
    (b) A. J. Arduengo III, D. A. Dixon and D. C. Roe, *J. Am. Chem. Soc.*, **108**, 6821 (1986).
    (c) D. A. Dixon and A. J. Arduengo III, *J. Chem. Soc., Chem. Commun.*, 498 (1987).
44. (a) A. Clotet, J. Rubio and F. Illas, *J. Mol. Struct. (Theochem)*, **164**, 351 (1988).
    (b) D. S. Marynick, *J. Chem. Phys.*, **73**, 3939 (1980).
45. R. M. Minyaev, *J. Mol. Struct. (Theochem)*, **262**, 79 (1992).
46. S. A. Culley and A. J. Arduengo III, *J. Am. Chem. Soc.*, **106**, 1164 (1984).
47. S. A. Culley and A. J. Arduengo III, *J. Am. Chem. Soc.*, **107**, 1089 (1985).
48. C. A. Stewart, R. L. Harlow and A. J. Arduengo III, *J. Am. Chem. Soc.*, **107**, 5543 (1985).
49. A. J. Arduengo III, C. A. Stewart, F. Davidson, D. A. Dixon, J. Y. Becker, S. A. Culley and M. B. Mizen, *J. Am. Chem. Soc.*, **109**, 627 (1987).
50. For reviews, see:
    (a) A. H. Cowley, *Polyhedron*, **3**, 389 (1984).
    (b) A. H. Cowley, *Acc. Chem. Res.*, **17**, 386 (1984).
    (c) A. H. Cowley and N. C. Norman, *Prog. Inorg. Chem.*, **34**, 1 (1986).
    (d) L. Weber, *Chem. Rev.*, **92**, 1839 (1992).
51. M. Yoshifuji, I. Shima, N. Inamoto, K. Hirotsu and T. Higuchi, *J. Am. Chem. Soc.*, **103**, 4587 (1981).
52. A. H. Cowley, J. G. Lasch, N. C. Norman and M. Pakulski, *J. Am. Chem. Soc.*, **105**, 5506 (1983).
53. A. H. Cowley, J. E. Kilduff, J. G. Lasch, S. K. Mehrotra, N. C. Norman, M. Pakulski, B. R. Whittlesey, J. L. Atwood and W. E. Hunter, *Inorg. Chem.*, **23**, 2582 (1984).
54. A. H. Cowley, N. C. Norman and M. Pakulski, *J. Chem. Soc., Dalton Trans.*, 383 (1985).
55. C. Couret, J. Escudie, Y. Madaule, H. Ranaivonjatovo and J.-G. Wolf, *Tetrahedron Lett.*, **24**, 2769 (1983).
56. G. Huttner, U. Weber, B. Sigwarth and O. Scheidsteger, *Angew. Chem., Int. Ed. Engl.*, **21**, 215 (1982).
57. A. H. Cowley, N. C. Norman and M. Pakulski, *J. Am. Chem. Soc.*, **106**, 6844 (1984).
58. A. H. Cowley, N. C. Norman, M. Pakulski, D. L. Bricker and D. H. Russell, *J. Am. Chem. Soc.*, **107**, 8211 (1985).
59. U. Weber, G. Huttner, O. Scheidsteger and L. Zsolnai, *J. Organomet. Chem.*, **289**, 357 (1985).
60. R. West, M. J. Fink and J. Michl, *Science (Washington)*, **214**, 1343 (1981).
61. (a) G. Raabe and J. Michl, *Chem. Rev.*, **85**, 419 (1985).
    (b) R. West, *Angew. Chem., Int. Ed. Engl.*, **26**, 1201 (1987).
    (c) G. Raabe and J. Michl, in *The Chemistry of Organic Silicon Compounds* (Eds. S. Patai and Z. Rappoport), Chap. 17, Wiley, New York, 1989.
    (d) Y. Apeloig, in *The Chemistry of Organic Silicon Compounds* (Eds. S. Patai and Z. Rappoport), Chap. 2, Wiley, New York, 1989.

(e) T. Tsumuraya, S. A. Batcheller and S. Masamune, *Angew. Chem., Int. Ed. Engl.*, **30**, 902 (1991).

(f) S. Nagase, *Polyhedron*, **10**, 1299 (1991); *Pure Appl. Chem.*, **65**, 675 (1993).

62. J. Barrau, J. Escudie and J. Satge, *Chem. Rev.*, **90**, 283 (1990).

63. (a) D. E. Goldberg, P. B. Hitchcock, M. F. Lappert, K. M. Thomas, A. J. Thorne, T. Fjeldberg, A. Haaland and B. E. R. Schilling, *J. Chem. Soc., Dalton Trans.*, 2387 (1986).

(b) D. E. Goldberg, D. H. Harris, M. F. Lappert and K. M. Thomas, *J. Chem. Soc., Chem. Commun.*, **261** (1976).

(c) For the first spectral characterization in solution, see: S. Masamune and L. R. Sita, *J. Am. Chem. Soc.*, **107**, 6390 (1985).

64. (a) For a weak dative bond description, see: Reference 63 and K. W. Zilm, G. A. Lawless, R. M. Merrill, J. M. Millar and G. G. Webb, *J. Am. Chem. Soc.*, **109**, 7236 (1987).

(b) For a σ conjugative interaction, see: M. J. S. Dewar, G. L. Grady, D. R. Kuhn and K. M. Merz, *J. Am. Chem. Soc.*, **106**, 6773 (1984).

(c) For a resonating-unshared-pair description, see: L. Pauling, *Proc. Natl. Acad. Sci. USA*, **80**, 3871 (1983).

65. G. Trinquier, *J. Am. Chem. Soc.*, **112**, 2130 (1990).

66. S. Nagase, S. Suzuki and T. Kurakake, *J. Chem. Soc., Chem. Commun.*, 1724 (1990).

67. A. H. Cowley, J. E. Kilduff, N. C. Norman, M. Pakulski, J. L. Atwood and W. E. Hunter, *J. Am. Chem. Soc.*, **105**, 4845 (1983).

68. J. Escudie, C. Couret, H. Ranaivonjatovo and J. Satge, *Phosphorus and Sulfur*, **17**, 221 (1983).

69. J. Jaud, C. Couret and J. Escudie, *J. Organomet. Chem.*, **C25**, 249 (1983).

70. E. Niecke, R. Ruger, M. Lysek, S. Pohl and W. Schoeller, *Angew. Chem., Int. Ed. Engl.*, **22**, 486 (1983).

71. O. J. Scherer, H. Sitzmann and G. Wolmershäuser, *Angew. Chem., Int. Ed. Engl.*, **28**, 212 (1989).

72. O. J. Scherer, H. Sitzmann and G. Wolmershäuser, *Angew. Chem., Int. Ed. Engl.*, **24**, 351 (1985). For the theoretical studies, see: S. Nagase and K. Ito, *Chem. Phys. Lett.*, **126**, 43 (1986); M. T. Nguyen and A. F. Hegarty, *J. Chem. Soc., Chem. Commun.*, 383 (1986).

73. K. K. Baldridge and M. S. Gordon, *J. Am. Chem. Soc.*, **110**, 4204 (1988).

74. N. Matsunaga, T. R. Cundari, M. W. Schmidt and M. S. Gordon, *Theoret. Chim. Acta*, **83**, 57 (1992).

75. R. K. Mishra and B. K. Mishra, *Chem. Phys. Lett.*, **151**, 44 (1988).

76. P. Ilic, B. Sinkovic and N. Trinajstic, *Isr. J. Chem.*, **20**, 258 (1980).

77. H. M. Schmidt, H. Stoll, H. Preuss, G. Becker and O. Mundt, *J. Mol. Struct.* (*Theochem*), **262**, 171 (1992).

78. L. L. Lohr and A. C. Scheiner, *J. Mol. Struct.* (*Theochem*), **109**, 195 (1984).

79. K. D. Dobbs, J. E. Boggs and A. H. Cowley, *Chem. Phys. Lett.*, **141**, 372 (1987).

80. W. Schneider, W. Thiel and A. Komornicki, *J. Phys. Chem.*, **94**, 2810 (1990).

81. W. Schneider, W. Thiel and A. Komornicki, *J. Phys. Chem.*, **92**, 5611 (1988).

82. W. B. Person, J. S. Kwiatkowski and R. J. Barlett, *J. Mol. Struct.* (*Theochem*), **157**, 237 (1987).

83. L. Andrews, R. Withnall and B. W. Moores, *J. Phys. Chem.*, **93**, 1279 (1989).

84. A. J. Downs, G. P. Gaskill and S. B. Saville, *Inorg. Chem.*, **21**, 3385 (1982).

85. C. W. Bock, K. D. Dobbs, G. J. Mains and M. Trachtman, *J. Phys. Chem.*, **95**, 7668 (1991).

86. (a) G. N. Lewis, *J. Am. Chem. Soc.*, **38**, 762 (1916).

(b) J. Langmuir, *J. Am. Chem. Soc.*, **41**, 868 (1919).

87. (a) L. Pauling, *J. Am. Chem. Soc.*, **53**, 1367 (1931).

(b) L. Pauling, *The Nature of the Chemical Bond*, Cornell University Press, Ithaca, New York, 1960.

88. (a) G. C. Pimentel, *J. Chem. Phys.*, **19**, 446 (1951).

(b) R. J. Hach and R. E. Rundle, *J. Am. Chem. Soc.*, **73**, 4321 (1951).

(c) J. I. Musher, *Angew. Chem., Int. Ed. Engl.*, **8**, 54 (1969).

89. A. E. Reed and P. v. R. Schleyer, *J. Am. Chem. Soc.*, **112**, 1434 (1990).

90. G. Trinquier, J.-P. Daudey, G. Caruana and Y. Madaule, *J. Am. Chem. Soc.*, **106**, 4794 (1984).

91. M. T. Nguyen, *J. Mol. Struct.* (*Theochem*), **49**, 23 (1988).

92. J. V. Ortiz, *J. Phys. Chem.*, **94**, 4762 (1990).

93. M. O'keeffe, *J. Am. Chem. Soc.*, **108**, 4341 (1986).

94. R. S. Berry, *J. Chem. Phys.*, **32**, 933 (1960).

95. A. Rauk, L. C. Allen and K. Mislow, *J. Am. Chem. Soc.*, **94**, 3035 (1972).
96. F. Keil and W. Kutzelnigg, *J. Am. Chem. Soc.*, **97**, 3623 (1975).
97. W. Kutzelnigg, H. Wallmeier and J. Wasilewski, *Theoret. Chim. Acta*, **51**, 261 (1979).
98. S.-K. Shih, S. D. Peyerimhoff and R. J. Buenker, *J. Chem. Soc., Faraday Trans.*, **75**, 379 (1979).
99. J. M. Howell, *J. Am. Chem. Soc.*, **99**, 7447 (1977).
100. W. Kutzelnigg and J. Wasilewski, *J. Am. Chem. Soc.*, **104**, 953 (1982).
101. A. E. Reed and P. v. R. Schleyer, *Chem. Phys. Lett.*, **133**, 553 (1987).
102. C. S. Ewig and J. R. Van Wazer, *J. Am. Chem. Soc.*, **111**, 1552 (1989).
103. P. Wang, D. K. Agrafiotis, A. Streitwieser and P. v. R. Schleyer, *J. Chem. Soc., Chem. Commun.*, 201 (1990).
104. P. Wang, Y. Zhang, R. Glaser, A. E. Reed, P. v. R. Schleyer and A. Streitwieser, *J. Am. Chem. Soc.*, **413**, 55 (1991).
105. H. Wasada and K. Hirao, *J. Am. Chem. Soc.*, **114**, 16 (1992); **114**, 4444 (1992).
106. J. Moc and K. Morokuma, *Inorg. Chem.*, in press.
107. (a) B. Lakatos, *Naturwissenschaften*, **15**, 355 (1954).
    (b) B. Lakatos, *Acta Chim. Hung.*, **8**, 207 (1955).
    (c) N. N. Greenwood and K. Wade, *J. Chem. Soc.*, 1527 (1956).
    (d) A. S. Shchukarev, *Zh. Obshch, Khim.*, **24**, 581 (1954).
    (e) P. S. Bagus, Y. S. Lee and K. S. Pitzer, *Chem. Phys. Lett.*, **33**, 408 (1975).
    (f) P. Pyykkö, *J. Chem. Res.*, 380 (1979).
108. L. D. Freedman and G. O. Doak, *Chem. Rev.*, **82**, 15 (1982).
109. C. Pulham, A. Haaland, A. Hammel, K. Rypdal, H. P. Verne and H. V. Volden, *Angew. Chem., Int. Ed. Engl.*, **31**, 1464 (1992).
110. (a) G. Wittig and M. Reiber, *Justus Liebigs Ann. Chem.*, **562**, 187 (1949).
    (b) G. Wittig and K. Clauß, *Justus Liebigs Ann. Chem.*, **577**, 26 (1952); **578**, 136 (1952).
111. (a) P. J. Wheatley and G. Wittig, *J. Chem. Soc.*, 251 (1962).
    (b) P. J. Wheatley, *J. Chem. Soc.*, 2206 (1964).
112. (a) P. J. Wheatley, *J. Chem. Soc.*, 3718 (1964).
    (b) A. L. Beauchamp, M. J. Bennett and F. A. Cotton, *J. Am. Chem. Soc.*, **90**, 6675 (1968).
113. (a) A. Schmuck, J. Buschmann, J. Fuchs and K. Seppelt, *Angew. Chem., Int. Ed. Engl.*, **26**, 1180 (1987).
    (b) A. Schmuck, P. Pyykkö and K. Seppelt, *Angew. Chem., Int. Ed. Engl.*, **29**, 213 (1990).
114. B. D. El-Issa, P. Pyykkö and H. M. Zanati, *Inorg. Chem.*, **30**, 2781 (1991).

CHAPTER 2

Structural chemistry of organic compounds containing arsenic, antimony and bismuth

D. B. SOWERBY

Department of Chemistry, University of Nottingham, Nottingham NG7 2RD, UK

The chemistry of organic arsenic, antimony and bismuth compounds
Edited by S. Patai © 1994 John Wiley & Sons Ltd

I. INTRODUCTION

This chapter is concerned with the solid state structures of arsenic, antimony and bismuth compounds, which in general contain at least one bond between carbon and the Group 15 element. In selecting material for discussion, I have been greatly aided by the availability of the Cambridge Crystallographic Data Base. The period under consideration covers effectively the years between 1981 and 1992.

Like the lighter members of the group, the chemistry of these elements is dominated by species in which the central atom is in either the +3 or the +5 oxidation state, and it is possible to obtain a realistic assessment of the general stereochemistry of compounds by applying the Valence Shell Electron Pair Repulsion method (VSEPR). Thus, neutral compounds in the +3 oxidation state are pyramidal, with the fourth tetrahedral position being occupied by the lone pair of electrons of the Group 15 element. This arrangement approximates to sp^3 hybridization but the bond angles are usually substantially smaller than the ideal 109.5°. This is a consequence of (a) increased repulsion between the lone pair of electrons and the three bond pairs and (b) a decrease in repulsion between the three bond pairs of electrons as a consequence of both the decreased electronegativity of the heavier members of the group and the increased size of these atoms.

The presence of a lone pair of electrons in a compound in the +3 oxidation state means that such organo-arsenic, antimony and bismuth compounds are Lewis bases, but because the compounds have available empty d orbitals, they are also potential Lewis acids and this behaviour is well established for derivatives carrying one or two organic groups. As an example, a compound such as RMX_2, where R is an alkyl or aryl group, M is the Group 15 element and X is a halogen, will accept up to two halide ions to produce anions with the formulae $[RMX_3]^-$ and $[RMX_4]^{2-}$, while monohalides such as R_2MX will accept one further halogen giving the $[R_2MX_2]^-$ anion. Again using the VSEPR approach of electron counting, the shapes of such species can be predicted. A monomeric $[RMX_3]^-$ ion would be pseudo-trigonal bipyramidal with a lone pair of electrons occupying an equatorial site, while an $[RMX_4]^{2-}$ ion if monomeric would be pseudo-octahedral with the lone pair occupying the sixth octahedral position *trans* to the organic group. For the diorgano-derivative $[R_2MX_2]^-$, electron counting would predict again a pseudo-trigonal bipyramidal arrangement with the lone pair occupying an equatorial site.

This simple approach in practice may require modification. In the first instance, the group 15 atom may still not be coordinatively saturated and, in particular with $[RMX_3]^-$ species, further electron density can be accepted into the Group 15 valence shell from the lone pairs on a halogen atom. If this occurred, the simplest product would be a dimer with two bridging halogen groups and such species are well known.

A second problem occurs as a result of the possible stereochemical non-activity of the lone pair of electrons associated with the Group 15 element, and this is a problem common to the heavier elements of the neighbouring main groups in the Periodic Table. The VSEPR approach to molecular structure of the compounds of these heavier elements is tantamount to saying that d orbitals become part of any hybridization scheme and all the valence electrons are stereochemically active.

On the other hand, there are good reasons for believing that this is not always the case and with, for example, antimony and bismuth, the heavier members of the group, there is evidence for the presence of an 'inert pair' of s electrons. Because the angles between the substituents of a neutral antimony(III) or bismuth(III) compound are close to 90°, it is possible to consider that, rather than using sp^3 hydrid orbitals, the substituents are attached via pure p orbitals, with the lone pair of electrons remaining localized in the appropriate s orbital. The concept of an inert pair of electrons has some validity and it allows rationalization of much of the chemistry of these elements in the $+3$ oxidation state. It is, however, difficult to provide direct experimental evidence for the effect, but it is difficult otherwise to rationalize the almost ideal octahedral structure of $[SbCl_6]^{3-}$.

On the basis of an inert pair of s electrons, formation of anionic species from antimony(III) and bismuth(III) halides and their related mono- and di-organo derivatives follows if a halide ion is considered to approach the central atom *trans* to one of the bonds already present and the $X^-/M—X$ system is treated via a three centre–four electron approach. This reproduces not only the observed structures of the anions but also the observed range of metal–halogen distances. An alternative, qualitative, molecular orbital approach is that a second halogen is bonded via the antibonding orbital of an $M—X$ bond already present.

The most common stereochemistry of compounds with five substituents in the $+5$ oxidation state is the trigonal bipyramid, with the more electronegative substituents occupying axial positions in the absence of gross steric effects. Square pyramidal geometry is considered to be an energetically less favourable alternative and, with the exception of pentaphenyl-antimony and -bismuth, is not observed unless there are constraints within the substituents at the central atom. Recently, evidence has accumulated that compounds can be synthesized with geometries distorted various distances along the Berry pseudo-rotation coordinate between the trigonal bipyramidal and square pyramidal extremes.

Tetrahedral structures are expected, and sometimes observed, when four organic groups are attached to the central atom and the large R_4M^+ cations are well known as stabilizing counterions for a range of unusual anions. Problems arise with some R_4MX species with potentially anionic X groups, as covalence persists and the compounds are five-coordinate trigonal bipyramids. Tetrahedral structures might also be expected for molecules with the stoichiometry R_3MO. This is indeed observed for M = As, as π-overlap between arsenic 4d orbitals and oxygen 2p orbitals in sufficiently strong to support arsenic–oxygen double bonding in the same way that effective 3d–2p overlap produces strong phosphorus–oxygen double bonds. Because of the disparity in size between antimony 5d orbitals and oxygen 2p orbitals, π bonding is very weak and thermodynamic stability in antimony(V)–oxygen systems is gained by forming a second σ bond, which leads in the few systems investigated to cyclic dimers.

The more detailed discussion which follows is divided into separate sections on arsenic, antimony and bismuth compounds, and these are further sub-divided first on the basis of the oxidation state and then in terms of the specific elements attached.

II. ARSENIC COMPOUNDS

A. The $+3$ Oxidation State

1. Compounds with three As—C bonds

There are four independent molecules in the unit cell of triphenylarsenic, connected in pairs by elements of supersymmetry[1]; coordination about arsenic is pyramidal with mean values of the As—C separation and C—As—C angle of, respectively, 1.957 Å and 100.1°. There is little change in geometry when the phenyl group carries a *para* substituent, but the *p*-methoxy and *p*-tolyl derivatives have three-fold symmetry and the phenyl groups are in

TABLE 1. Bond lengths and bond angles for R_3As

R =	Ph	$p\text{-ClC}_6\text{H}_4$	$p\text{-MeOC}_6\text{H}_4$	$p\text{-MeC}_6\text{H}_4$	Mesityl	neo-Pent
mean C—As (Å)	1.957	1.958	1.963	1.954	1.967	2.00
mean C—As—C (deg)	100.0	99.5	98.3	99.3	107.6	94.6

a regular propeller arrangement[2]. Torsion angles vary for the *p*-chloro analogue but, as shown in Table 1, substituent change has virtually no effect on the molecular conformation. With tri(neo-pentyl)arsine, the mean Sb—C distance is larger, while the angle at arsenic is substantially lower[3]. Attaching three bulky mesityl groups to arsenic leads, as perhaps expected, to an increase in the As—C distance and a substantial widening of the bond angles[4]. In tris(pentafluorophenyl)arsine, the distance to arsenic is normal (1.959 Å) and although the mean bond angle is 100.5°, the individual values are 96.6, 101.1 and 101.7°[5].

The chiral *R*-ethylmethylpropylarsine can be obtained in 60% optical purity by reductive methylation of ethylpropylarsinic acid and trapped by the palladium complex **1**[6]. Arsenic coordination is completely regioselective giving **2**, in which the arsine occupies a position *trans* to the dimethylamino group.

(1)   (2)

The first arsacyclopropane (arsirane) **3** was synthesized in 1985 from phenylarsenic dichloride and $LiC(SiMe_3)_2Cl$, via the bis(methylene)arsorane, $PhAs[=C(SiMe_3)_2]_2$[7]. The As—C separation is 2.023 Å and the angle of arsenic in the three-membered ring is 46.0°.

Three examples containing arsenic incorporated into four-membered rings are the arsa-cyclo-butene **4**[8], the 1,2-azaarsetene cation **5**[9] and the 1,2-diarsetene **6**[10]. In the first compound, the AsC_3 ring has a very slight envelope conformation with C—As—C angles of 69.9, 102.3 and 101.7° and As—C distances within the ring of 1.989 and 1.949 Å, the latter to the unsaturated ring atom. The cationic arsetene, obtained from $Ph_3PC(CN)[As(NPr_2^i)_2]$ and sodium tetraphenylborate, is also planar with an interesting intramolecular arsenic–carbon interaction shown by the dotted line in **5**. The unit cell of tetraphenyl-1,2-diarsetane **6**, obtained from diphenylacetylene and hexaphenylcyclo-

(3)   (4)

PPh$_3$

Pr$_2^i$N—As⸰⸰⸰—NR$_2^i$

N—As(NPr$_2^i$)$_2$

(5)

Ph

As

Ph—As—Ph

Ph

(6)

hexa-arsine, contains two highly strained independent molecules. The As—As distance (2.471 Å) is long and there is no cyclic delocalization from the C—C separation (1.361 Å). Arsenic–carbon distances (1.962 Å) are normal and the arsenic lone pairs are directed *trans* to each other.

The diphenylarsenic substituents at the ylidic carbon in Ph$_3$P=C(AsPh$_2$)$_2$, obtained from diphenylarsenic chloride and Ph$_3$P=CH$_2$, adopt a *cis–trans* conformation with the arsenic lone pairs lying in the plane of the heavy atoms[11]. Distances from arsenic to the ylidic carbon are 1.936 and 1.938 Å, with an As—C—As angle of 116.4°. The As—C—P angles are unequal at 108.4 and 128.0° and the ylidic carbon lies some 0.28 Å out of the plane of the heavy atoms.

An interesting triphenylarsine derivative, in which the phenyl groups are linked together by *i*-propenyl substituents, has been synthesized during a search for a molecule with an ideal turnstile structure[12]. The structure is shown in Figure 1 and the constraints imposed lead to a more highly pyramidal arrangement about arsenic (C—As—C, 92.7, 92.9, 93.5°) with somewhat shorter As—C bonds (mean 1.944 Å). The C—As—C angle in the ferroceneophane 7 is, however, more highly strained (87.9°) and, in keeping with this, the distances from arsenic to the carbons of the ferrocene rings are increased to 1.986 Å[13]. Distances from arsenic to the carbonyl carbon atoms are definitely elongated in PhAs[C(O)Bu^t]$_2$ but the angle at arsenic between these carbon atoms is unusually low at 91°[14]. The angles involving the phenyl carbon atom have a mean value of 97.3°.

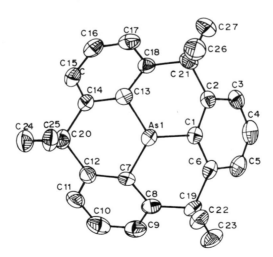

FIGURE 1. The structure of a triphenylarsine derivative triply linked by *i*-propenyl groups. Reproduced by permission of VCH Weinheim from Reference 12

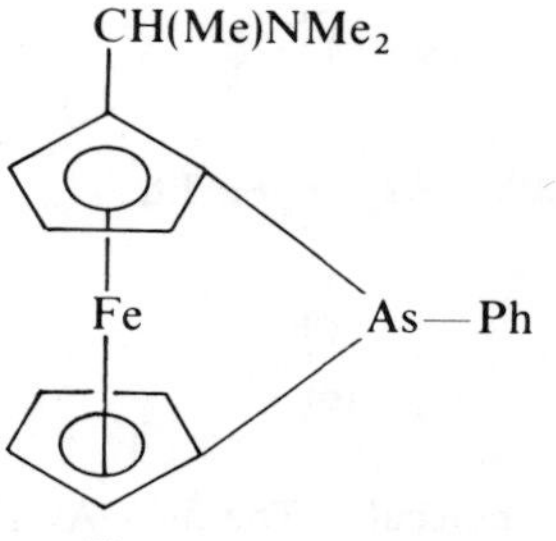

CH(Me)NMe$_2$

(7)

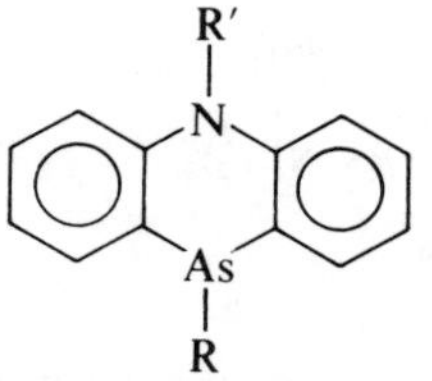

(8)

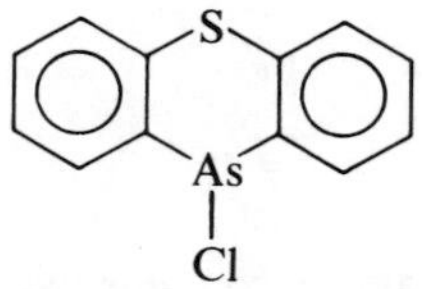

(9)

The hydrate of a natural arseno-betaine, $Me_3AsCH_2COO^-$, crystallizes as a centrosymmetric, hydrogen-bonded dimer with As—C(methyl) distances of 1.92, 1.91 and 1.86 and to the methylene group of 1.89 Å[15].

Phenarsazine structures, such as **8** [R' = H, R = 3-An[16] (**a**), R = 3-Tol[17] (**b**), R = Me[18] (**c**); R' = Ph, R = Ph[19] (**d**)] are of interest in terms of: (a) the conformation of the aromatic ring in relation to the plane bisecting the central ring containing the two heteroatoms and (b) the fold angle of the tricyclic ring. The latter, which varies between 164.6° for **8(a)** and 151.7° for **8(d)**, is related to aromaticity in the central ring, which adopts a flattened boat conformation. In **8(d)** the two phenyl substituents occupy axial positions and are almost perpendicular to each other. The related 10-chlorophenothiarsenin **9** has a similar folded, butterfly structure (fold angle 152.7°) with the chlorine substituent again occupying a quasi-axial position[20]. As—C distances are equal at 1.923 Å and the C—As—C angle at the pyramidal arsenic is 99.8°.

Triorganoarsines are Lewis bases and a number of addition compounds with both main group elements and transition metals have been investigated. These compounds are discussed in the order in which the acceptor element appears in the Periodic Table.

In the series of boron trihalide adducts with trimethylarsine, the As—B distance decreases from 2.07 Å for the chloride to 2.03 Å for the iodide in line with the known Lewis acidities of the boron halides[21]. Arsenic–carbon separations (1.90–1.93 Å) increase slightly from chloride to iodide but all distances are shorter than in the free ligand, in keeping with the change to tetrahedral geometry about arsenic. The compounds all have staggered conformations with effective C_{3v} symmetry. Triethylarsine and arsenic trichloride form a dimeric 1:1 adduct **10** in pentane with a planar As_2Cl_6 unit and As—Cl distances to terminal chlorines of 2.266 and 2.304 Å and 2.714 and 2.814 Å to bridging chlorine atoms[22]. Triethylarsine molecules are bonded to each of these arsenic atoms via As—As bonds (2.469 Å), comparable in length to those in As_4 and $(PhAs)_6$.

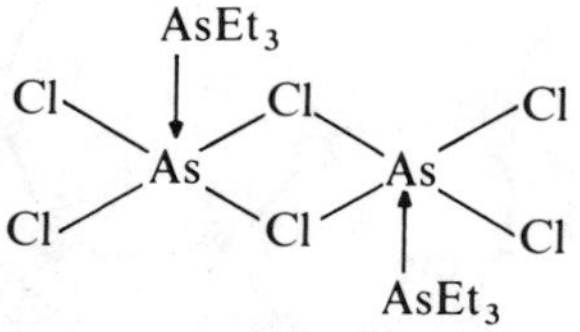

AsEt$_3$

(10)

Rather curiously, in contrast to the behaviour of chlorine and bromine, iodine does not oxidize triphenylarsine to give the related arsenic(V) diiodide but a molecular adduct is formed instead and the iodine–iodine bond remains intact[23,24]. The compound can be obtained in both monoclinic ($P2_1/c$) and orthorhombic ($P2_12_12_1$) modifications and as a 2/3 toluene solvate; molecular parameters in all the forms are very similar. The structure is described as containing a slightly bent As—I—I shaft (angle 174.8°, As—C 2.64 Å) with three phenyl fins projecting from the arsenic atom. As usual, lone-pair donation from arsenic leads to structural changes, for example, in keeping with an increase in s character, there is a decrease in As—C distances (1.921 Å) and an increase in the C—As—C angles. The compound shows a typical iodine charge transfer spectrum and, in agreement with donation of arsenic electron density into an empty iodine σ_u orbital, the iodine–iodine separation increases from 2.660 Å in the free element to 3.005 Å.

There are many examples in which arsine ligands are coordinated *per se* to transition metals and others where the ligand is fragmented on reaction with a low oxidation state transition metal species, which then coordinates the fragment. Some of the former are now considered, while coordination of arsenic fragments is treated below (Section II.A.4). In the dimeric rhodium complex 11, which is a cyclohexane hydroformylation catalyst, coordination about each metal is close to planar, with the tri(*t*-butyl)arsine groups in *cis* positions (As—Rh 2.48 Å)[25]. Both Ph_3As and PhAs groups are incorporated in H_2Ru_3 $(CO)_7(AsPh_3)_2(AsPh)$, the product of reaction between the arsine and the anionic cluster $[HRu(CO)_{11}]^-$ [26]. The two triphenylarsine groups are coordinated to two of the metal atoms in a triangular cluster (As—Ru 2.465, 2.488 Å) which the cluster is almost symmetrically capped by the arsinidine ligand (As—Ru 2.412, 2.418, 2.400 Å).

(11)

Ruthenium carbonyl forms a trigonal bipyramidal triphenylarsine complex, $Ru(CO)_4(AsPh_3)$, where the arsine occupies an axial position (As—Ru 2.461 Å)[27,28], but in $OsBr_4(AsPh_3)_2$, the geometry is octahedral with *trans* arsine ligands (As—Os 2.569 Å)[29]. As previously noted, the C—As—C angles in the ligand increase (99.4, 103.3, 103.3°) on coordination from those in the free molecule. The mean As—Ru distance in the two independent molecules of the dimethylphenylarsine complex, $Ru_3(CO)_9(AsMe_2Ph)_3$, is 2.445 Å[30].

As shown in Figure 2, two of the acetylene groups of the tri(phenylethinyl)arsine bridge between pairs of cobalt atoms in the bis(tetrahedrane) product obtained from a reaction of the arsine with dicobalt octacarbonyl[31]. There is the expected lengthening of the C—C distance in the coordinated groups (1.31 and 1.34 Å) over that in the uncomplexed group (1.28 Å). The As—C distances to coordinated groups (1.92 and 1.95 Å) are also longer than that to the uncoordinated group (1.77 Å) and the angle at arsenic between the coordinated carbon atoms is 103.9° compared with values of 100.5 and 99.7° for the other two angles.

Triphenylarsine and -antimony groups can be incorporated into axial positions in the tetrakis-μ-acetato-dirhodium core at distances of 2.576 and 2.732 Å, respectively[32], and in the centrosymmetric complex, *trans*-IrBr$_4$ (AsEt$_3$)$_2$, the As—Ir distance is 2.489 Å, while As—C distances lie between 1.93 and 1.96 Å and C—As—C angles between 101.4 and 103.9°[33].

The trimesitylarsine complex with mercury(II) nitrate is a centrosymmetric dimer with bridging nitrate groups 12; the asymmetric unit consists of two independent half molecules, differing in nitrate orientation and the configuration of the trimesitylarsine

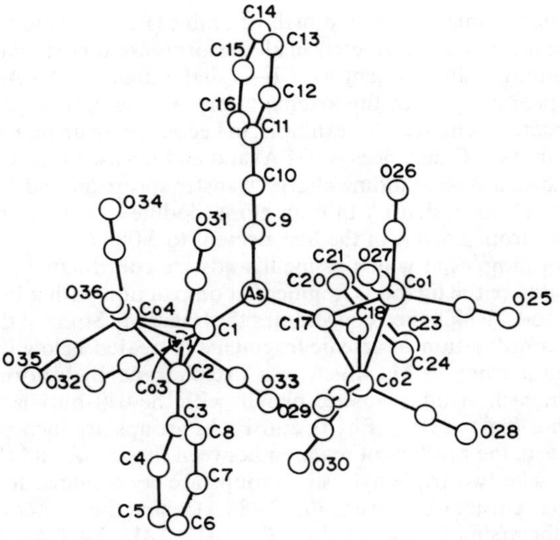

FIGURE 2. The structure of the arsenic-bridged bis(tetra-hedrane) product. Reproduced by permission of Verlag der Zeitschrift der Naturforschung from Reference 31

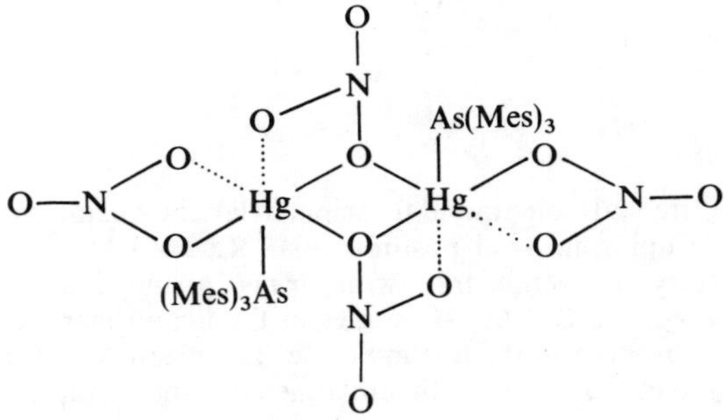

(12)

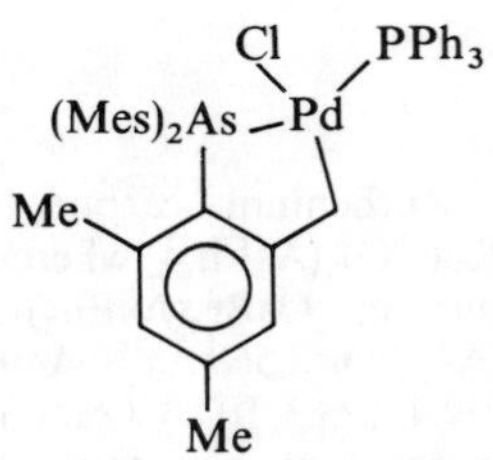

(13)

groups[34]. As—Hg separations are 2.476 and 2.482 Å. Arsenic atoms are in distorted tetrahedral coordination and angles between the carbon atoms are increased to 112.7°, probably the largest value so far known. In addition to coordination via arsenic, one of the methyl groups of trimesitylarsine reacts with $PdCl_2(MeCN)_2$ and triphenylphosphine to give the cyclo-palladated complex 13, which has close to square planar geometry about palladium (As—Pd 2.437 Å)[35].

Arsine ligands coordinate strongly to Ag(I) and complexes with the stoichiometry 1:1, 1:2 and 1:3 have been isolated from silver nitrate and triphenylarsine[36]. Each complex contains direct As—Ag links at distances which increase with increased numbers of co-ordinated arsenic atoms, e.g. 2.471 Å for the 1:1, 2.535 and 2.521 Å for the 1:2 and 2.608, 2.617 and 2.678 Å for the 1:3 complex. The remaining positions in the silver coordination sphere are filled with oxygen atoms of the nitrate groups. Tetrahedral $[Ag(AsPh_3)_4]^+$ cations are also known with $[SnPh_2(NO_3)_3]^-$ and $[SnPh_2(NO_3)_2Cl]^-$ anions and As—Ag separations between 2.65 and 2.70 Å and there is also a discrete, tetrahedral monomeric silver chloride complex, $[Ag(AsPh_3)_3Cl]$, in which the As—Ag bond lengths are 2.622 Å[37]. Finally, in the perchlorate complex, $[Ag\{As(C_5H_9)_3\}_2][ClO_4]$, there are As—Ag contacts of 2.480 and 2.482 Å together with short silver contacts to oxygen atoms of the anion; the silver–arsenic system is non-linear (As—Ag—As 151.2°)[38].

An example of fragmentation of an arsine ligand and its subsequent coordination is provided by two products obtained from tri(p-tolyl)arsine and the acetonitrile substituted Os_3 cluster compounds, $Os_3(CO)_{12-n}(NCMe)_n$ for $n = 1$ and 2[39]. As shown in **14** and **15**, the products both contain both a μ^3-p-tolylarsinidine group and a coordinated benzyne moiety and in **14** there is also a complete ligand molecule. Os—As distances range between 2.40 and 2.54 Å.

(14)   (15)

2. Compounds with multiple bonds to arsenic

Until about fifteen years ago, there was little evidence for the formation of stable main group compounds containing a π bond between p orbitals, other than those which resulted from overlap between pairs of 2p orbitals (the 'double bond' rule). Multiply bonded species were therefore restricted to compounds of first row elements and those where the π interaction was a result of overlap between d and p orbitals. Compounds such as $RAs=CR_2$ and $RAs=AsR$ were unknown and by the 'double bond' rule would not exist. This should be contrasted with compounds such as $R_3As=O$ and the like, which involve d–p π bonding, and can clearly be obtained. This situation has been radically altered by the realization that, although π-bonded species involving 3p–3p, 4p–4p, 3p–2p, 4p–2p, etc. overlap are thermodynamically unstable with respect to polymers or oligomers with σ rather than π bonds, kinetic stability can be achieved by incorporating large substituents, which give steric protection to the 'unstable' π bond. In this fashion, it has now been possible to isolate a range of diphosphenes (—P=P—), diarsenes (—As=As—), disilenes (=Si=Si=), phospha-alkenes (—P=C=), phospha-alkynes (P≡C—), arso-alkenes (—As=C=) and the like. Among the bulky groups that have been used to stabilize diarsenes and related compounds are 2,4,6-tri(t-butyl)phenyl and the bis and tris(trimethylsilyl)methyl groups.

The diarsene compounds structurally investigated so far are $(2,4,6\text{-}Bu_3^tC_6H_2)As=As$ $CH(SiMe_3)_2$[40,41], $(MeSi)_3CAs=AsC(SiMe_3)_3$[42] and $(Me_3Si)_2HCAs=AsCH(SiMe_3)_2$[43] together with the mixed arsenic–phosphorus species $(2,4,6\text{-}Bu_3^tC_6H_2)P=AsCH$ $(SiMe_3)_2$[41,44]. In all cases the substituents occupy *trans* positions about the double bond and the central skeleton is planar with As—As—C angles falling between 93.6 and 106.3°. The As—As distances are between 2.224 and 2.245 Å compared with the 2.43–2.46 Å separation expected for a single bond and represents a shortening from the single bond distance of approximately 10%; there is a similar reduction of the As—P distance in the phospha-arsene (2.124 Å).

Like the related diphosphenes, diarsenes are potential donors towards low oxidation state transition metals via either the arsenic lone pairs of electrons or through the π bond itself. In two chromium pentacarbonyl complexes investigated, one of the arsenic atoms donates a lone pair of electrons and the double bond remains essentially intact, increasing marginally from 2.224 Å in the uncomplexed ligand to 2.246 Å after coordination[45,46].

Phenyl groups do not provide the necessary steric protection and PhAs=AsPh is unknown in the free state, but complexes containing this molecule can be obtained by indirect routes. Among the examples are compounds 16, 17[43] and 18[47] in which the diarsene uses both the arsenic lone pairs and the π bond to attach metal fragments, and 19[48].

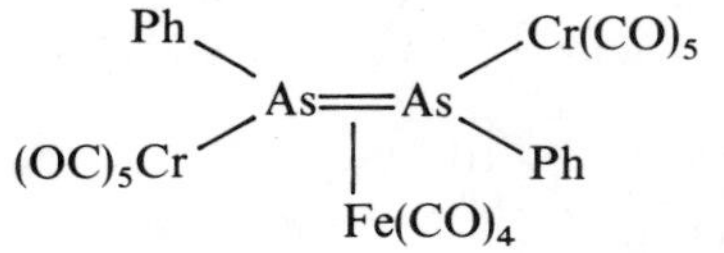

(16)

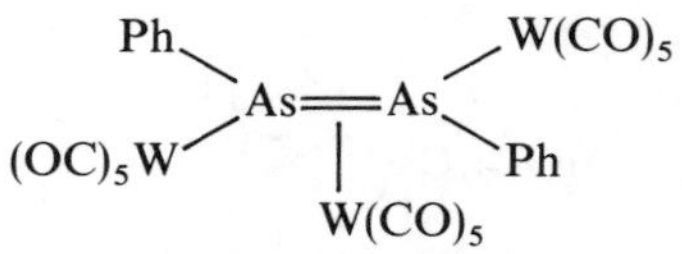

(17)

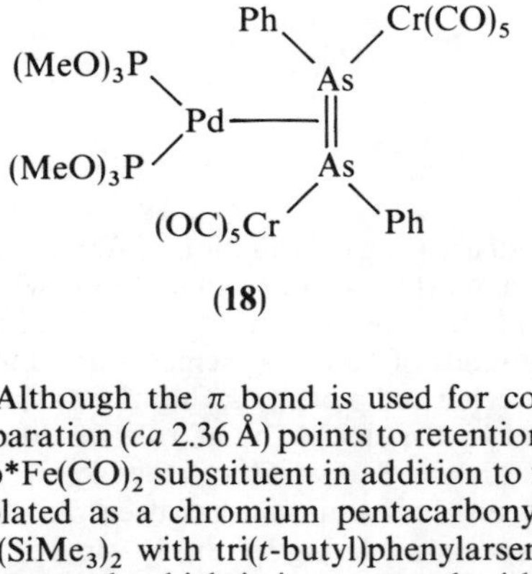

(18)

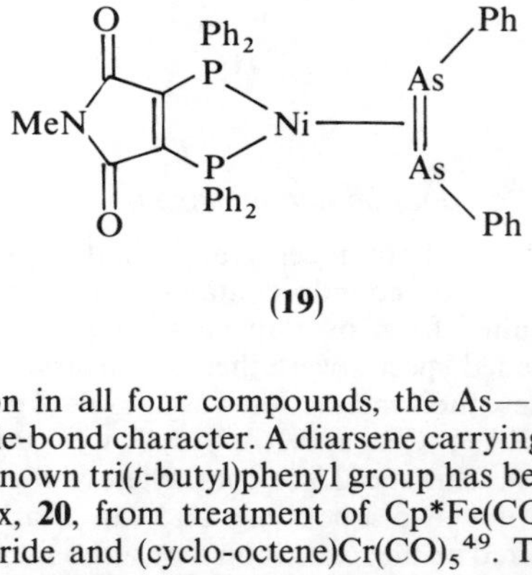

(19)

Although the π bond is used for coordination in all four compounds, the As—As separation (ca 2.36 Å) points to retention of double-bond character. A diarsene carrying a Cp*Fe(CO)₂ substituent in addition to the well known tri(t-butyl)phenyl group has been isolated as a chromium pentacarbonyl complex, 20, from treatment of Cp*Fe(CO)₂As(SiMe₃)₂ with tri(t-butyl)phenylarsenic dichloride and (cyclo-octene)Cr(CO)₅[49]. The compound, which is isostructural with the corresponding diphosphene and phospha-arsene, has an As—As separation of 2.259 Å and As—Cr and As—Fe distances of 2.492 and 2.387 Å, respectively. On the other hand, the As—As distance in [CpFe(CO)₂]-(PhAs—AsPh)[(CO)₂FeCp] is 2.456 Å and the compound is probably best described as a tetrasubstituted diarsine, rather than a diarsene complex[50].

Although arsa-alkene chemistry is much less well developed than that of the phosphorus analogues, there are a number of examples. In the benzazarsole 21, arsenic is bonded to the neighbouring carbon atoms at significantly different distances (1.822 and 1.920 Å), and even the longer bond is shorter than 1.98 Å, the sum of the arsenic and carbon covalent radii[51]. The As=C distance is estimated at 1.78 Å implying that, as shown in the formula, there is substantial double-bond character. Further, the C—N distance (1.38 Å) is short and since the molecule is planar it is considered to be aromatic. There are similar short As—C bonds, 1.833 Å in 22[52], and 1.821 Å in the iron derivative, CpFe(CO)₂—As=C(OSiMe₃)Buᵗ, obtained following a trimethylsilyl shift as shown in equation 1[53,54].

(20)   (21)   (22)

$$CpFe(CO)_2—As(SiMe_3)_2 + Bu^tCOCl \longrightarrow CpFe(CO)_2—As(SiMe_3)(COBu^t) \quad (1)$$

$$CpFe(CO)_2—As=C(OSiMe_3)Bu^t$$

$$2\,MeAs=C(OSiMe_3)Bu^t \longrightarrow \qquad\qquad (2)$$

Not surprisingly, arsenic–carbon double bonds are highly reactive and, as shown in equation 2, UV irradiation promotes dimerization of the substituted propylidene arsine[55]. The four-membered ring is slightly folded with the methyl groups at arsenic and the trimethylsiloxy groups above the ring and the t-butyl groups below. The endocyclic As—C distance is lengthened (2.05 Å) in comparison with the normal As–methyl distances (1.96 Å). The ring angles are 85.9 and 92.5°, respectively, at the arsenic and carbon atoms.

An interesting tetracyclic compound, for which full structural details are available, is obtained by treating a 1,3-aza-arsinine with a phospha-alkyne as shown in equation 3[56].

$$+ 3\,Bu^tC\equiv P \longrightarrow \qquad\qquad + \; PhCN \qquad (3)$$

As found for diarsenes above, it is possible to stabilize species not stable in the free state by coordination to transition metal fragments. This has been achieved with $PhAs=CH_2$ in the rhodium complex $Cp^*Rh(PhAs=CH_2)$, obtained only as a microcrystalline powder[57]. Attempts to recrystallize the material, however, gave the triarsenic complex 23, which has a non-planar ring with normal As—As distances (2.450, 2.433 Å) and an As—As—As angle of 83.0°, similar to that in $(CF_3As)_4$.

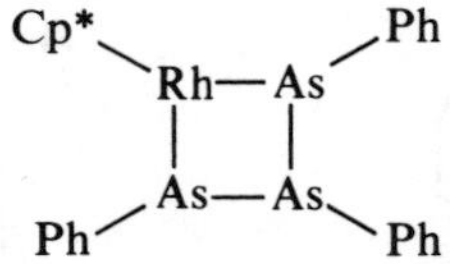

(23)

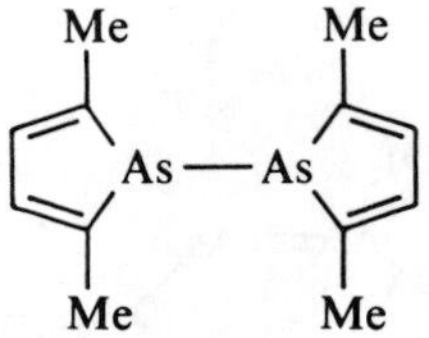

(24)

3. Compounds with As—As single bonds

A number of simple diarsines have been structurally investigated and, although individual molecules are similar to those of the corresponding distibines, the diarsines do not show the extended chain structures and consequent thermochromism often found for the latter. The geometry at the two arsenic atoms is pyramidal, though the angles at arsenic can be asymmetric. The substituents occupy *anti* (*gauche*) conformations and in some cases the molecules have imposed C_2 symmetry. Parameters for four of the compounds are summarized in Table 2.

There are various other diarsines with more complex structures, including the tetramethylbiarsolyl **24**[62] with an As—As bond length of 2.438 Å and the diarsa-hexaphosphine **25**[63], which has a bicyclobutane structure and an As—As separation of 2.441 Å. The molecule has C_i symmetry and two *trans* orientated and slightly folded AsP_3 rings. There is also a new bicyclic $R_2As_2P_2$ species **26** ($R = Bu^t_3C_6H_2$), analogous to the bicyclic tetraphosphines, which has a folded butterfly structure and an *exo–exo* arrangement of aryl groups[64]. The As—As distance (2.383 Å) is substantially shorter than usually observed and probably indicates a degree of multiple bonding. The antimony analogue is not isostructural but has an *exo–endo* substituent arrangement and the Sb—Sb separation is again short (2.723 Å) with the suggestion of a degree of multiple bonding.

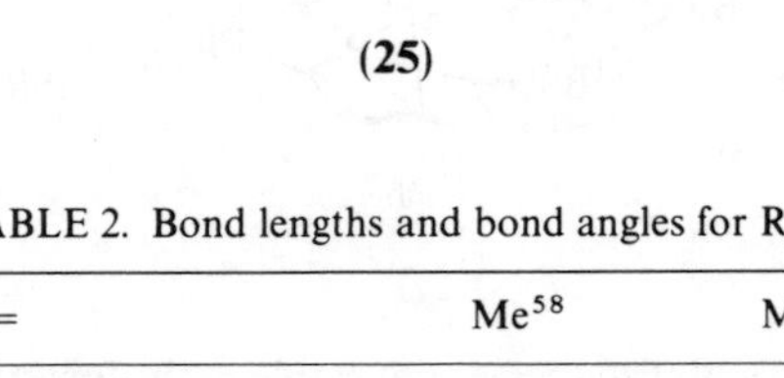

(25)

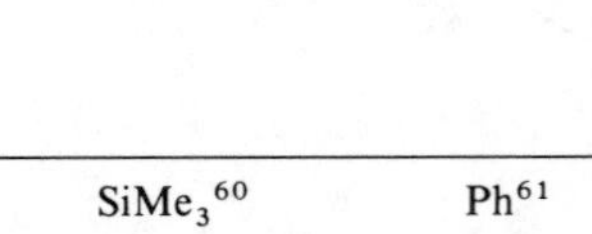

(26)

TABLE 2. Bond lengths and bond angles for R_2AsAsR_2

R =	Me[58]	Mesityl[59]	SiMe$_3$[60]	Ph[61]
As—As (Å)	2.429	2.472	2.458	2.458
C—As (Å)[a]	1.965	1.964/1.995	2.363	
C—As—As (deg)	96.2	94.1/109.6	93.9/113.6	
C—As—C (deg)	96.7		100.9	

[a] For the SiMe$_3$ compound, read Si for C.

An arsenic–arsenic bond (2.427 Å) is generated when dimethylarsenic iodide reacts with Ga_2I_4 to give an ionic product, formulated as $[Me_2As\text{—}AsMe_2I][GaI_4]$[65]. The cation contains both As(III) and As(V); the former is pyramidal (C—As—C 99° with normal bond distances (1.982, 2.028 Å), while at arsenic(V), the geometry is distorted tetrahedral (C—As—C 105–114°) and the As—C distances are substantially shorter (1.916, 1.833 Å).

Tetraorgano-diarsines are electron pair donors and in principle can behave as either unidentate or bidentate ligands. In $CpV(CO)_3[Ph_2AsAsPh_2]$, for example, only one arsenic atom is involved in bonding (As—V 2.536 Å) and the arsenic–arsenic separation (2.472 Å) remains at essentially that of the free ligand[66]. Tetraphenyldiarsine is, however, bidentate in the manganese(I) complex **27**, but again there is little change in the As—As separation (2.489 Å)[67].

Continued investigation of $MeC(CH_2AsI_2)_3$ has led to isolation of two nor-hetero-adamantanes, **28** and **29** (R = COOEt), from reactions with, respectively, diethylmalonate and sodium hydroselenide[68]. As—As distances are 2.469 and 2.483 Å respectively, while in the related cyclic triarsine **30** the separation is reduced to 2.42 Å[69].

In the cyclotetra-arsine, $(Bu^tAs)_4$, mean values for the major parameters are: As—As 2.44, As—C 2.02 Å, As—As—As 86 and As—As—C 101°[70], and a new member of the cyclopentaarsine group, $(Me_3SiCH_2As)_5$, has been obtained by chlorine abstraction using magnesium from $Me_3SiCH_2AsCl_2$[71]. As found for $(MeAs)_5$, there is a small variation in the As—As bond lengths (2.424–2.446 Å) and the same slightly twisted envelope confor-mation of the ring.

A redetermination of the hexaphenylcyclohexa-arsine structure confirms the presence of a highly puckered chair conformation with 91.0° ring angles and one of the three independent As—As distances slightly longer (2.464 Å) than the other two (2.456 and 2.457 Å)[72]. The hexa-arsine coordinates as a 1,4-bidentate group in the $Mo(CO)_4$ complex, obtained simply by treating the arsine with molybdenum hexacarbonyl. The mean As—As distance in the complex is comparable with that in the free ligand, although individual distances vary between 2.426 and 2.471 Å[73]; the variation appears to be unrelated to coordination position. Coordination also brings about a change in ring conformation to the previously unknown twisted boat conformation and the ring angles at arsenic (mean 98.6°) increase substantially from those in the free ligand.

The cyclohexa-arsine is, however, labile and reorganizes on reaction with an
(arene)Mo(CO)$_3$ complex, and depending on conditions, products containing a coordinated cyclo-(PhAs)$_9$ ring (see Figure 3) or an eight-membered flyover chain (see
Figure 4) can be isolated[74]. Reaction of the latter with dimethylacetylene-dicarboxylate
leads to insertion of a C_2 unit into one of the uncoordinated As—As bonds, increasing the

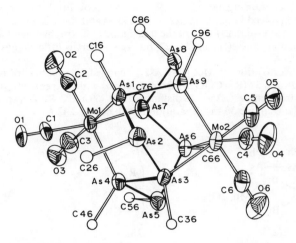

FIGURE 3. The structure of [cyclo-(PhAs)$_9$Mo$_2$(CO)$_6$]
with the phenyl rings shown as ispo atoms only[74].
Reprinted with permission from Rheingold *et al.*, *J. Am.
Chem. Soc.*, **109**, Copyright (1987) American Chemical
Society

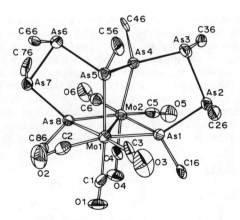

FIGURE 4. The structure of [catena-(PhAs)$_8$-
Mo$_2$(CO)$_6$] with the phenyl rings shown as *ispo*
atoms only[74]. Reprinted with permission from
Rheingold *et al.*, *J. Am. Chem. Soc.*, **109**, 141. Copyright (1987) American Chemical Society

length of the flyover chain by two atoms[75]. Increased fragmentation of the cyclohexa-arsine occurs on reaction with Cp_2TiCl_2, giving **31** containing a coordinated triarsine unit in a $TiAs_3$ ring[76]. The As—Ti distances are 2.655 and 2.668 Å and As—As separations 2.431 and 2.442 Å. There is a similar coordinated triarsine group in $Cp*Rh(CO)(PhAs)_3$, where the four-membered ring is non-planar with *trans*-arranged phenyl groups[77]. Cyclopentamethylarsine is similarly also a potential ligand and provides a coordinated cyclodeca-arsine group in the molybdenum carbonyl complex $(MeAs)_{10}Mo_2(CO)_6$[74].

The cycloarsines $(PhAs)_6$ and $(MeAs)_5$ also behave as sources of the isolobal arsenic analogue of the allyl ligand, i.e. RAsAsAsR, on reaction with $[Cp*M CO)_2]_2$ (for M = Mo or W) to give products with the stoichiometry $Cp*M(CO)_2 (RAsAsAsR)$[78]. Each arsenic atom is bonded to the transition metal and the As—As distances are equal (2.361 Å), representing a bond order of 1.5 as expected for an allylic analogue.

Finally, three other complexes **32–34** with coordinated polyarsine ligands have been isolated. Compounds **32**[79] and **33**[80] are both cobalt(II) compounds containing the tripod ligand $MeC(CH_2PPh_2)_3$ (L), obtained, respectively, from a reaction with phenylarsine and treatment of the corresponding thiodiarsiranediyl complex with diphenyldiazomethane. The third complex **34** is produced when $Na_2Cr_2(CO)_{10}$ and di(t-butyl)chloroarsine react and contains the four-electron, chelating bidentate ligand, $Bu^tAsCl—(AsBu^t)_2—AsBu^tCl$[81]. The $CrAs_4$ ring is virtually planar and the arsenic–arsenic distances are those for single bonds.

(32)   (33)   (34)

4. Other organoarsines as ligands

Organoarsines have been extensively employed as ligands with a range of metals, but full details of such behaviour is beyond the scope of this review. It is worth noting, however, that there is a wide range of compound type and some examples now follow.

The chiral bidentate amino-arsine **35** has been used to form chiral rhodium cationic species[82] and the mixed arsine-phosphines $Ph_2As(CH_2)_{2 \, or \, 3}PPh_2$ have found extensive

(35)

use as bridges between two metal atoms[83–87]. The diarsine analogue $Ph_2AsCH_2AsPh_2$ (L) is similarly versatile and bridges between manganese atoms (As—Mn 2.41 Å) in $Mn(CO)_4LMn(CO)_4$[88], ruthenium atoms in $Ru_3(CO)_8 \cdot L$[89] (As—Ru 2.428–2.551 Å) and rhodium atoms in a 3,5-dimethylpyrazole complex[90]. The chelate ring in products with the methyl substituted diarsine $Me_2AsCH_2CH_2CH_2AsMe_2$ (L) shows a degree of steric flexibility and is in the chair form in $[trans-RhCl_2L_2]ClO_4$ while one ring is chair and the other a distorted skew form in the related *cis* form of the hexafluorophosphate salt[91]. In both cases the As—Rh—As system is linear with As—Rh separations of *ca* 2.42 Å.

The unsaturated fluorocarbon-bridged chelating ligand, $Me_2AsC(CF_3)$=$C(CF_3)$-$AsMe_2$ (L), reacts to give seven-coordinate tungsten compounds in both $W(CO)Br_2$-$[P(OMe)_3]_2L^{92}$ and $W(CO)_2Br_2PPh_3L^{93}$, but on irradiating the ligand in the presence of $Re_2(CO)_{10}$, one dimethylarsine group is cleaved and a CO group is inserted between rhenium and the olefinic carbon atom to give the five-membered ring compound **36**[94]. Although bis(6-methylquinolin-8-yl)phenylarsine **37** is potentially tridentate, only the

(36)   **(37)**

arsenic and one of the nitrogen atoms coordinate in the square-planar palladium(II) complexes obtained from either the dichloride or diiodide (As—Pd 2.32 Å)[95]. Again, only the arsenic atom of the potentially chelating mixed arsenic(III)–phosphorus(V) ligand $Ph_2AsCH_2CH_2P(S)Ph_2$ coordinates in the platinum complex $Pt[S_2P(OEt)_2]_2L$ to give a square-planar arrangement about the metal atom (As—Pt 2.361 Å)[96].

ortho-Phenylenebis(dimethylarsine) (diars) and its phenyl analogues continue to be exploited as chelates. An unusual iron–thioketene complex has been isolated (As—Fe 2.323, 2.379 Å)[97] as has an octahedral rhodium complex, $[Rh(diars)_2Cl(CO_2)]$, containing a molecule of carbon dioxide coordinated via the carbon atom[98]. The more conventional cobalt complex, $[Co(diars)_3][BF_4]_3 \cdot 2H_2O$, is octahedral with As—Co distances ranging between 2.365 and 2.395 Å[99], and the platinum complex, $[Pt(diars)_2I_2]$ $[BF_4]_2$, has *trans*-octahedral geometry (As—Pt 2.45 Å)[100].

A ruthenium complex, $Ru(CO)_2Cl_2L$, has been reported with *ortho*-phenylenebis(methylphenylarsine) (L)[101] and there is a square-planar platinum complex (As—Pt 2.396 Å) with the fully phenylated diarsine in which the metal atom also carries two *p*-(trifluoromethyl)phenyl groups[102].

Only the oxygen atom of the As(V) group is coordinated in UBr_4L_2 (U—O 2.185, As—O 1.644 Å), where L is the partially oxidized ligand, $Ph_2As(O)CH_2CH_2AsPh_2$[103], while in addition to a molecule of triphenylphosphine, all four donor atoms of tris(2-diphenylarsinoethyl)amine, $N(CH_2CH_2AsPh_2)_3$, are coordinated to nickel in $[Ni(PPh_3)L]ClO_4$ giving the complex distorted trigonal bipyramidal geometry[104].

In reactions of bis(diphenylphosphinomethyl)phenylarsine, $PhAs(CH_2PPh_2)_2$,[105-113] with heavier transition metals, arsenic is often uncoordinated, but the ligand shows its full tridentate character in the iridium–gold complex **38**[113]. All three arsenic atoms of the tridentate ligand **39** (L) are also coordinated to cobalt (As—Co 2.280–2.316 Å) in $Co(CO)_3[CF_3C\equiv CCF_3]CoL^{114}$ but in $Co(CO)_3[CF_3C\equiv CCF_3]Co(CO)L$ only the central arsenic and one of the terminal $AsMe_2$ groups are coordinated[115].

Organoarsines, as mentioned previously, often fragment in reactions with low-oxidation-state transition metals and the fragments can be stabilized by coordination to the metal. Many examples are known but the following are representative. Complete cleavage of the As—As bonds in $(PhAs)_6$ and $(MeAs)_5$ can occur to give arsinidine

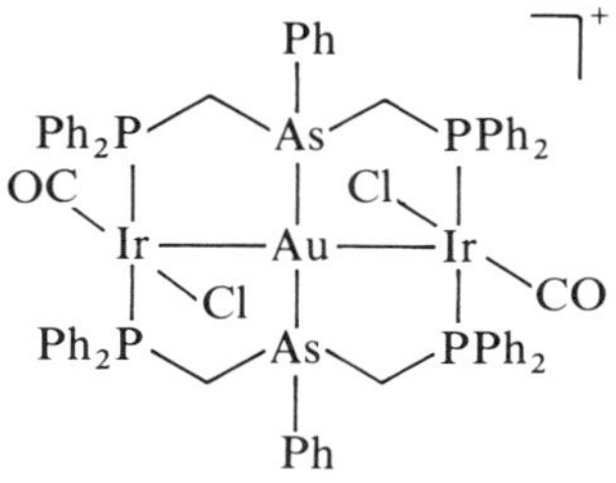

(38)

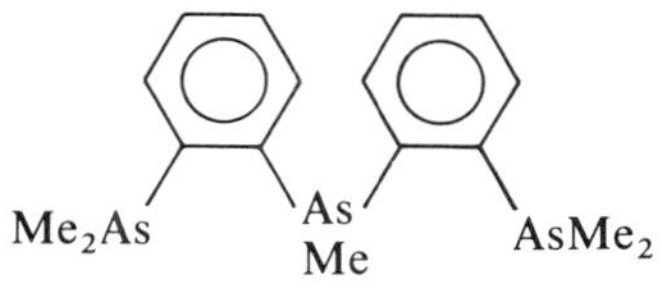

(39)

complexes and among products structurally investigated is $[CpRu(CO)]_2(\mu\text{-}CO)(\mu\text{-}AsR)$, where the two metal atoms are bridged by both a carbonyl group and either a PhAs or MeAs group[116]. Singly bridging groups are present in $Bu^iAs[Cr(CO)]_5]_2$ (As—Cr 2.38, As—C 2.00 Å)[117] and the complex anion $\{Me_2As[Fe(CO)_4]_2\}^-$ (As—Fe 2.395, As—C 1.922 Å)[118].

Reactions of the iron cluster anion **40** with phenylarsenic dichloride[119] and dimethylarsenic chloride[120], respectively, lead to **41** and **42**. In the first case, one edge of the Fe_3 triangle opens and the PhAs group bonds to each of the three iron atoms (As—Fe 2.275, 2.276, 2.359, As—C 1.92 Å), while in the second, the Me_2As group inserts into one of the Fe—Fe bonds (As—Fe 2.382, 2.373, As—C 1.983, 1.968 Å).

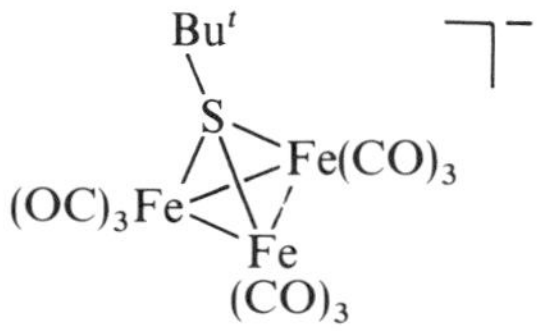

(40)

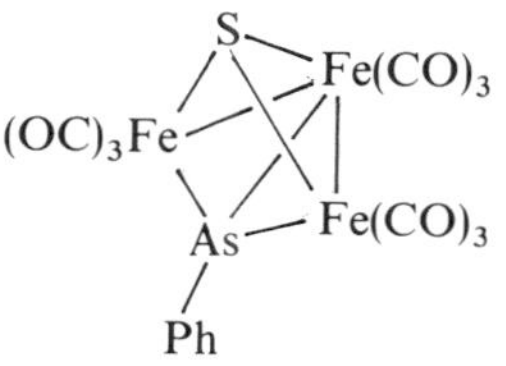

(41)

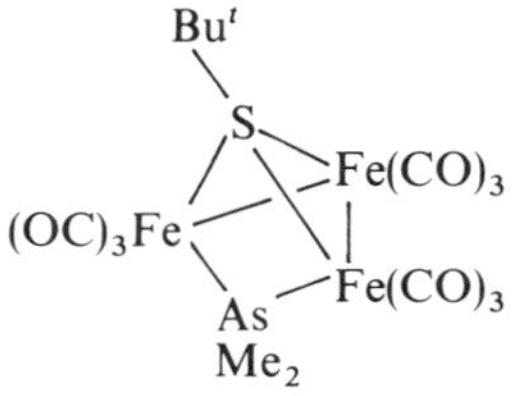

(42)

Bridging by two R_2As groups will give four-membered As_2M_2 rings and examples are known where M = Mo^{121}, Rh^{122} or Re^{123}. The molybdenum compound, $[CpMo(CO)_2$-$(Me_2As)]_2$ (As—Mo 2.595–2.622 Å), is obtained via monomeric $CpMo(CO)_2(:AsMe_2)$ when carbon monoxide is lost from $[CpMo(CO)_3(Me_2As)]$, while a second, $[Rh(CO)_2$-$(Bu^t_2As)_2]_2$, results from a conventional substitution reaction between $[Rh(CO)_2Cl_2]_2$ and Bu^t_2AsLi. The rhenium compound, $[Re(CO)_4(Ph_2As)]_2$ (As—Re 2.60 Å), is obtained by treating $Re_2(CO)_{10}$ with $(PhAs)_6$, providing yet another example of the versatility of this compound as a precursor for unusual arsenic ligands.

Bu^t_2AsH is a precursor for the bulky Bu^t_2As bridging groups, four of which are incorporated in the product from a reaction with the polynuclear iridium carbonyl, $Ir_4(CO)_{12}$[124]. The compound **43** contains a parallelogram of iridium atoms and four slightly asymmetric bridging arsine fragments (As—Ir 2.373, 2.409 Å); the Ir_4As_4 unit is virtually planar.

2,3-Dimethylbutadiene undergoes a 1,4-addition reaction with one of the As—H bonds in the complex $[Cr(CO)_5 \cdot AsPhH_2]$ to give $[Cr(CO)_5 \cdot AsPhHCH_2CMe{=}CMe_2]$[125], while on reaction with $[CpMo(CO)_3]_2$, dimethylarsine loses a hydrogen atom and this together with the remaining Me_2As group are incorporated as bridging groups in the product

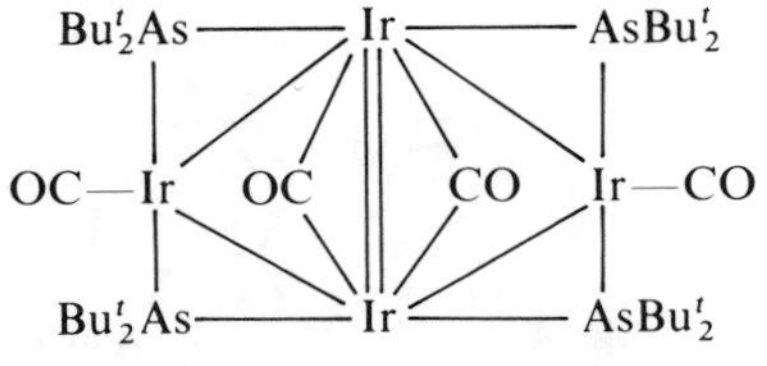

(43)

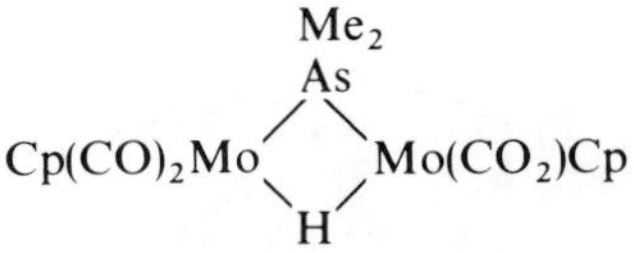

(44)

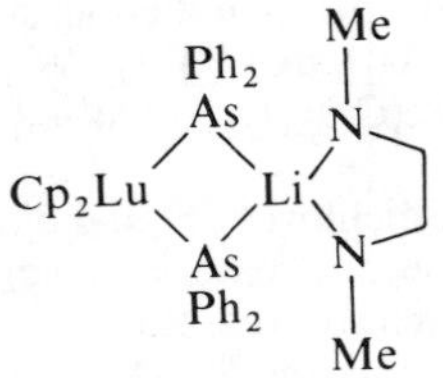

(45)

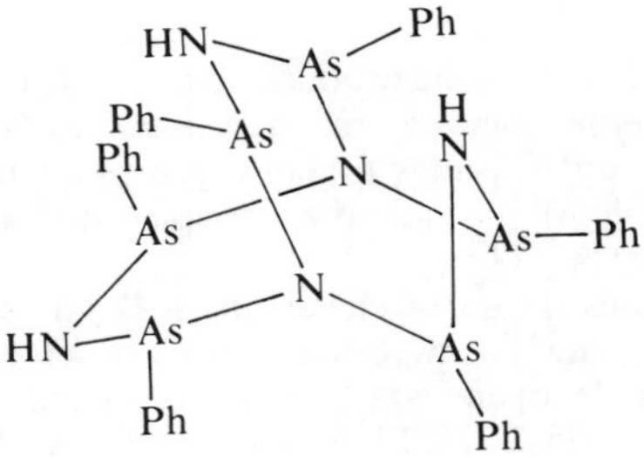

(46)

44^{126}. The molecule has a planar Mo_2AsH core with As—Mo distances of 2.508 Å. Diphenylarsine provides a source of bridging Ph_2As groups in **45**, which has a non-planar four-membered ring with As—Lu distances of 2.870 and 2.896 Å^{127}.

5. Compounds with bonds to nitrogen

Although there are no arsenic–carbon bonds in the imide As(NHAr) (= NAr), obtained by treating arsenic trichloride with the lithium salt of the sterically demanding amine $2,4,6\text{-}Bu_3^tC_6H_2NH_2$ (ArNH$_2$), the product is of interest^{128}. The coordination number of arsenic is two and the compound represents another example of a product whose stability depends on steric protection of an unstable 3p—2p π bond. Both of the arsenic–nitrogen separations are short, 1.714 and 1.745 Å, and the N—As—N angle is 98.8°. The shortness of these As—N bonds can be assessed in comparison with those in tri(pyrrolyl)arsine, which vary from 1.84–1.87 Å with N—As—N angles of 94.7, 96.3 and 98.9°^{129}.

Ammonolysis of phenylarsenic dichloride, originally thought to produce monocyclic $(PhAsNH)_4$, has now been shown by X-ray diffraction to give a tricyclic product $(PhAs)_6N_2(NH)_3$ **46**^{130}. The structure can best be described as resulting from the superposition of a trigonal prism of six arsenic atoms and a trigonal bipyramid of nitrogen atoms. As—C distances are normal (1.91 to 1.97 Å) but the As—NH and As—N distances range between 1.72 and 1.92 Å and 1.83 and 1.92 Å respectively. Hybridization at nitrogen is effectively sp^2 with As—N—As and As—NH—As angles of 117 and 119°, respectively.

Two cyclodiarsazanes (diazadiarsetidines) have been examined by X-ray crystallography. In compound **47**, the ring is almost planar with chlorine groups in *cis* positions and As—N distances ranging between 1.799 and 1.826 Å^{131}, while in the related aryl derivative **48**, although the ring is again planar, the substituents at arsenic are now *trans* to each other^{132}. Arsenic–nitrogen distances are somewhat longer than in the case above (1.847–1.874 Å) with endocyclic angles of 78.1° at arsenic and 101.8° at nitrogen. The

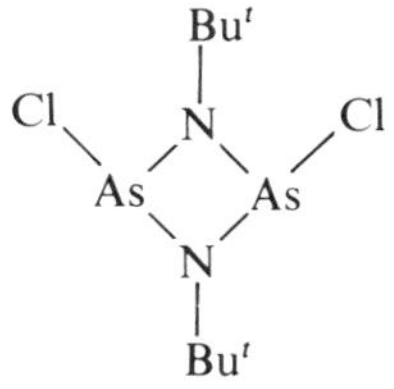

(47)

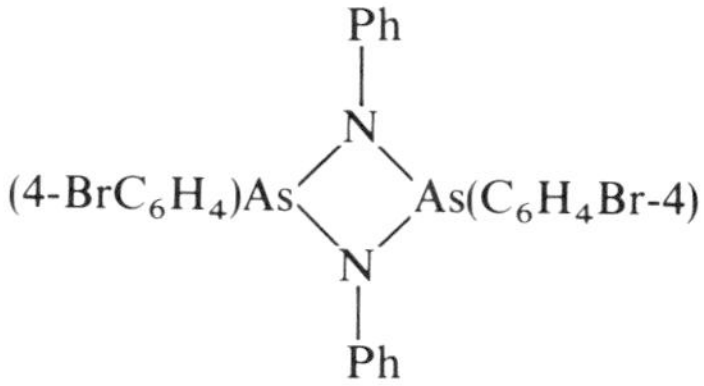

(48)

unusual thionylimide, $Ph_2AsN:S:O$, which coordinates via arsenic to a $Cr(CO)_5$ fragment, shows a normal-length As—N bond (1.89 Å)[133].

Two substituted arsenic sulphurdiimides have been prepared from K_2SN_2 and, respectively, Bu^tAsCl_2 and Ph_2AsCl, with the former leading to **49**, in which the eight-membered ring is in the boat conformation and butyl groups in equatorial positions[134]. The mean As—N separation here is 1.859 Å, which in the second product, $(Ph_2As)_2SN_2$, this distance is increased to 1.888 Å. In the complex formed between two $CpMo(CO)_2$ fragments and the t-butyl analogue, $(Bu^t_2As)_2SN_2$, the configuration is *cis–trans* as found in the isomorphous phosphorus analogue, and from NMR measurements, this persists in solution[135]. The arsenic atoms in **49** bridge between two of the osmium atoms in the cluster compound, $Os_3(CO)_{10} \cdot$ **49** (As—Os 2.403 Å), but there is little change in the ligand on coordination[136]. Reaction of **49** with $Fe_3(CO)_{12}$, on the other hand, leads to loss of one of the NSN units and the two arsenic atoms bridge between iron atoms (As—Fe 2.324 Å) to produce the complex $[Fe(CO)_3]_2(Bu^tAsNSNAsBu^t)$[137]. Finally in this section, a single molybdenum atom can be coordinated by the two arsenic atoms of $PhAs(NSN)_2AsPh$ in **50** (As—Mo 2.55 Å, As—N 1.86 Å)[138].

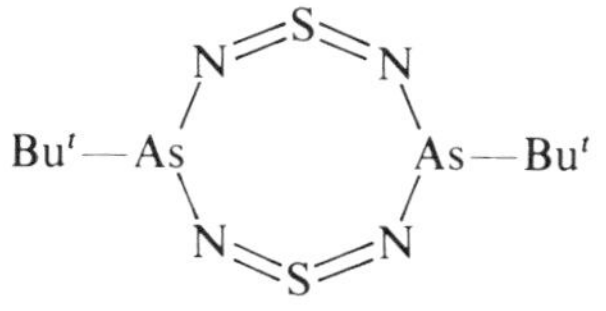

(49)

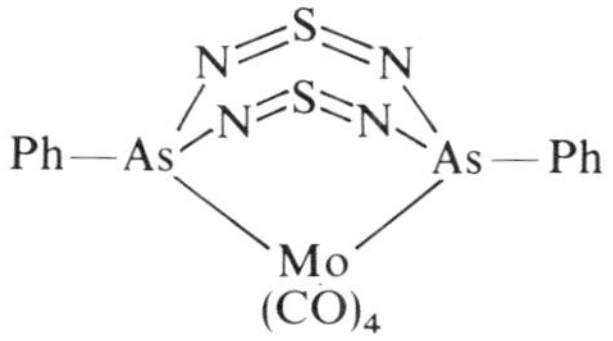

(50)

6. Compounds with bonds to oxygen

Compounds are now known in which there are As_4O_4 and As_6O_6 rings in addition to the better known As_2O_2 systems. Eight-membered systems were obtained in the case of $[(mesityl)AsO]_4$ by a simple hydrolysis of (mesityl)arsenic dichloride with potassium hydroxide in 1,2-dimethoxyethane[139], or for $(MeAsO)_4$ by a metal carbonyl template reaction[140]. The ring has a crown conformation in the mesityl compound with the four oxygen atoms in approximately the same plane, a mean As—O distance of 1.70 Å and values of 98.7 and 117.3° for the ring angles at arsenic and oxygen, respectively. The ring conformation in the methyl compound, which results from oxidation of $(MeAs)_5$, is chair-boat with longer As—O distances (1.778–1.819 Å) and larger oxygen ring angles (118.8–124.0°). A new twelve-membered $(MeAsO)_6$ ring compound, stabilized by coordination to two $Mo(CO)_3$ groups, is obtained when $(MeAs)_5$ is treated with $Mo(CO)_6$ and has a structure consisting of a plane of six oxygen atoms with, above and below, two planes each containing three arsenic atoms[141]. The arsenic planes are positioned for *fac*-coordination to the molybdenum carbonyl groups.

(51)   **(52)**

(53)

More conventional As_2O_2 rings occur in the dimers, $[(CF_3AsO(OH)]_2$ **51** and $[As(CF_3)O(OH)Cl]_2$ **52**, but X-ray studies show an unusual cage structure for $As_4(CF_3)_6O_6(OH)_2$ **53**, obtained by oxidative hydrolysis of a mixture of trifluoromethyl-arsenic iodides[142]. A further unusual tricyclic structure is found in $(CH_2)_3As_4O_4(CH_2)_3$, obtained by hydrolysis of the bis(di-iodide) $I_2As(CH_2)_3AsI_2$[143]. The structure in Figure 5 shows an eight-membered As_4O_4 ring with an *endo–endo* conformation similar to that in N_4S_4 with pairs of arsenic atoms linked by $(CH_2)_3$ chains. The As—O distances, 1.773–1.811 Å, are slightly shorter than expected for single bonds, the geometry about arsenic is distorted tetrahedral and the As—O—As angles are opened to between 123 and 129°.

The seven-membered ring in the 2-phenyldioxaarsepine **54** has a chair conformation with a mean As—O separation of 1.79 Å[144]. There are significant differences in the structures of the related 10,10'-bis(phenoxarsine) oxide and selenide **55**, X = O or Se, particularly in the deviations from planarity of the phenoxarsine rings[145]. Two structurally similar, independent molecules are found in the asymmetric unit for the epoxy-dihydroarsanthrene **56**; both show the butterfly conformation with As—O distances of 1.81 Å and angles at arsenic varying between 88.8 and 93.2°[146].

A six-membered As_3O_3 ring is present in $C(CH_2AsO)_3[CH_2As(OH)_2]$ **57**, the product from hydrolysis of $C(CH_2AsI_2)_4$ with sodium hydroxide in hot THF[147].

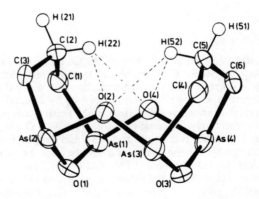

FIGURE 5. The structure of $(CH_2)_3As_4O_4(CH_2)_3$, Reproduced by permission of the Royal Society of Chemistry from Reference 143

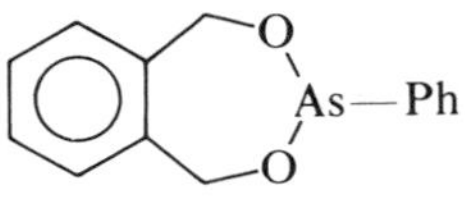

(54)

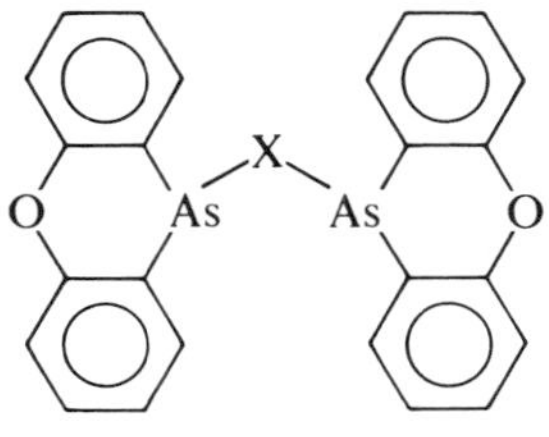

(55)

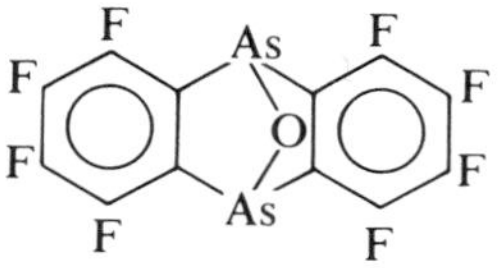

(56)

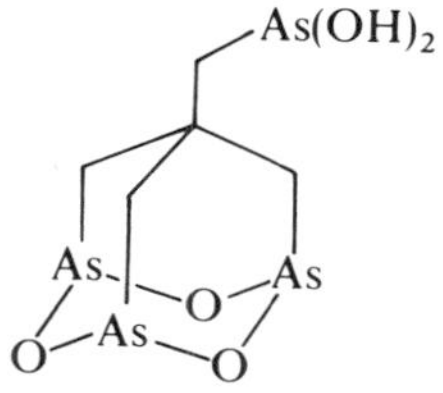

(57)

X-ray studies point to coordination of $R_2As(III)OH$ groups ($R = Ph/Ph$ or Me/Ph) rather than the alternative $R_2As(V)=O$ in the products from reactions between either $MePhAsCl$[148] or Ph_2AsCl[149] and cobaltbis(dimethylglyoximate).

7. Compounds with bonds to sulphur

As in the case of the cyclic arsenic oxides discussed above, it has been possible to obtain ordered cyclic $(MeAsS)_n$ oligomers, where $n = 3$ or 4, by treating $(MeAs)_5$ with elemental sulphur in the presence of molybdenum hexacarbonyl[150]. There is an isolobal relationship between the trimer and cyclohexane and it is thus not surprising that the six-membered ring has a chair conformation with equatorial methyl groups. As—S distances (2.241–2.270 Å) are typical of single bonds and the endocyclic angles are 100.6, 101.1 and 101.9° at arsenic and 92.8, 94.3 and 95.2° at sulphur. Both $(MeAsS)_4$[150] and $(Bu^tAsS)_4$[151] contain rings in the crown conformation with alkyl groups again in equatorial positions, but ring angles at arsenic and sulphur are increased to mean values of, respectively, 102.5 (101.5) and 98.2 (95.8)° for the methyl (t-butyl) compounds. A compound containing a nine-membered boat-chair As_6S_3 ring coordinated to an $Mo(CO)_3$ moiety, but showing substantial As/S disorder, has also been isolated[150].

The reaction between p-tolylarsenic dichloride and BAL, ($HSCH_2CH(SH)CH_2OH$), gives the chelate product **58**, with normal length As—S single bonds (2.225, 2.276 Å) and

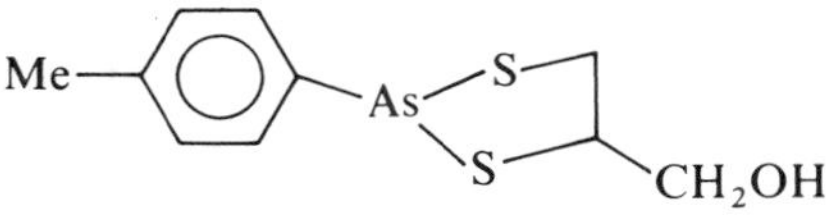

(58)

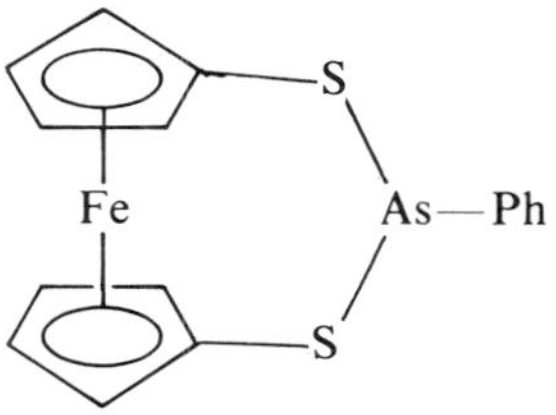

(59)

pyramidal geometry about the arsenic (S—As—S 92.7°)[152]. There are equal As—S bonds (2.25 Å) in the dithia-ferrocenophane **59** and angles between 97.3 and 99.3° at arsenic; one curious feature is a significant difference in C—C bond lengths in the cyclopentadiene rings, with one short (1.407 Å) and one long (1.439 Å) bond in each ring[153].

Dithiophosphate groups in PhAs[S$_2$P(OPr^i)$_2$]$_2$ chelate in an asymmetric fashion with As—S distances of 2.310/2.317 and 3.135/3.187 Å to give a square-pyramidal arrangement about arsenic[154], and when bis(dimethylarsino)sulphide, (Me$_2$As)$_2$S, coordinates to dimeric trimethylplatinum bromide it does so via the arsenic atoms and bridges between the two Pt atoms[155]. The arsenic–sulphur distances are normal with an As—S—As angle of 104.2°.

8. Compounds with bonds to other main group elements

Four-membered ring compounds, analogous to those found in the arsenic–nitrogen system, are formed extensively with other main group elements. This is even the case with lithium in compounds such as **60**, obtained following As—C bond cleavage in the reaction between magnesium bromide and LiAsBu$^t_2$ in THF solution[156]. The ring is planar with equal As—Li distances (2.58 Å) and ring angles of 79.4 and 100.6°, respectively, at the arsenic and lithium atoms. Geometry at the arsenics in the ring is tetrahedral while in the side chain arsenic is pyramidal. Similar structures are found in both the dimeric dimethoxyethane solvate, [Li(DME)As(SiMe$_3$)$_2$]$_2$[157], and in [Li(OEt$_2$)$_2$AsPh$_2$]$_2$[158]. There is a non-planar AsB$_2$N ring in **61**, the product from reaction of PhAsLi$_2$ and Pr$^i_2$NBClNBu^{t}BClNPr$^i_2$; important parameters are: As—B 2.063, 2.069 Å, B—As—B 63.9, As—B—N 95.5 and B—N—B 97.8°[159].

(60)

(61)

Rather surprising, however, are the monomeric structures of PhAsB(mesityl)$_2$· Li(THF)$_3$ and PhAsB(mesityl)$_2$·Li(TMEDA)$_2$, ascribed to stabilization by the sterically demanding mesityl groups[160]. In agreement with multiple As—B bonding, the separations are short (1.93 Å) and these compounds represent the first examples of bonds between sp^2 arsenic centres and planar boron atoms.

The aluminium compound, (Et$_2$AlAsBu$^t_2$)$_2$, obtained from triethylaluminium and dibutylarsine by loss of ethane contains a four-membered ring (As—Al 2.57 Å)[161] and, as shown in Table 3, four-membered, often planar and centrosymmetric rings dominate the related arsenic–gallium species **62**. The geometry at each of the ring atoms is distorted tetrahedral.

Monomeric arsinogallanes are also known, again providing a further example of the stabilizing effect of sterically demanding groups, which prevent oligomerization or polymerization. The tris(arsino)gallane, Ga[As(mesityl)]$_3$, is basically trigonal planar, although the gallium atom lies some 0.15 Å out of the plane, and the As—Ga separations (2.470–2.508 Å) are shorter than those in the four-coordinate compounds mentioned above[168]. Similarly, the monomeric t-butyl derivative, Bu$^t_2$GaAsBu$^t_2$ (Ga—As 2.466 Å),

TABLE 3. Bond lengths and bond angles for **62**

	As—Ga (Å)	Ga—As—Ga (deg)	As—Ga—As (deg)	Reference
R = Bu^t R′ = R″ = Me	2.541 2.558	95.7	84.3	162
R = Me$_3$SiCH$_2$ R′ = R″ = Ph	2.518 2.530	94.92	85.08	163
R = Me$_3$SiCH$_2$ R′ = (Me$_3$SiCH$_2$)$_2$As R″ = Br	2.513 2.521	95.63	84.37	164
R = Me$_3$CCH$_2$ R′ = R″ = Me	2.529 2.533	94.3	85.7	165
R = Me$_3$SiCH$_2$ R′ = R″ = (Me$_3$SiCH$_2$)$_2$As	2.540	95.30	84.81	166
R = Me$_3$Si R′ = R″ = Me$_3$SiCH$_2$	2.572 2.561	93.91	85.85	167

(62)   **(63)**

has trigonal geometry about gallium and pyramidal geometry (C—As—C 110.2°) about arsenic[169]. An anionic species [Ga(AsPh$_2$)$_4$]$^-$, with approximately tetrahedral geometry, is also known[170].

There are two examples of compounds containing six-membered As$_3$Ga$_3$ rings. One of these, [(Me$_3$SiCH$_2$)$_2$AsGaBr$_2$]$_3$ obtained from gallium tribromide and (Me$_3$SiCH$_2$)$_2$-AsSiMe$_3$, contains two independent molecules each lying on a two-fold axis, with As—Ga separations (2.432–2.464 Å) shorter than any previously observed for the atoms in fourfold coordination[171]. The second compound is [Me$_2$GaAsPr$_2^i$]$_3$, where the arsenic–gallium bond distance (mean 2.517 Å) is closer to that in the dimers discussed above[172].

Compounds containing As$_2$In$_2$ and As$_3$In$_3$ rings are also known. The dimer, [(Me$_3$SiCH$_2$)$_2$InAs(SiMe$_3$)$_2$]$_2$, resulting from a reaction between equimolar quantities of (Me$_3$SiCH$_2$)$_2$InCl and (Me$_3$Si)$_3$As, has a planar ring system with a mean As—In distance of 2.728 Å, and angles at indium and arsenic within the ring of, respectively, 85.5 and 95.6°[173]. If the reagents are used in a 2:1 ratio, the product is the heterocycle **63**. The unit cell of the trimer, (Me$_2$InAsMe$_2$)$_3$, contains two independent molecules, one of which is approximately planar and the other puckered, suggesting substantial flexibility in rings containing these heavier atoms[174]. Mean values for the arsenic–indium distances in the planar and puckered molecules are, respectively, 2.68 and 2.67 Å.

Planar four-membered rings are again found in the dimeric species **64**[175] and **65**[176]. In the former, the geometry at tin(II) is pyramidal with As—Sn distances of 2.77 Å and ring angles of 102.22 at arsenic and 77.78° at tin. Refinement of the arsenic selenide structure **65**

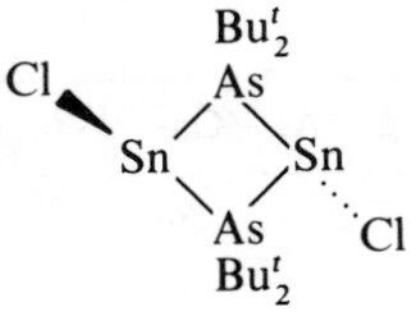

(64)

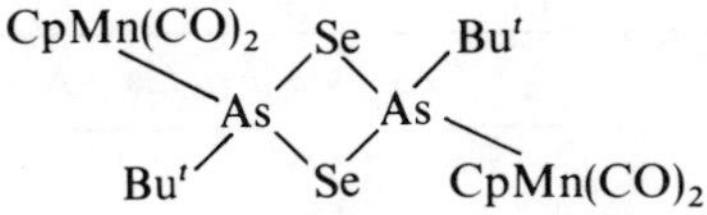

(65)

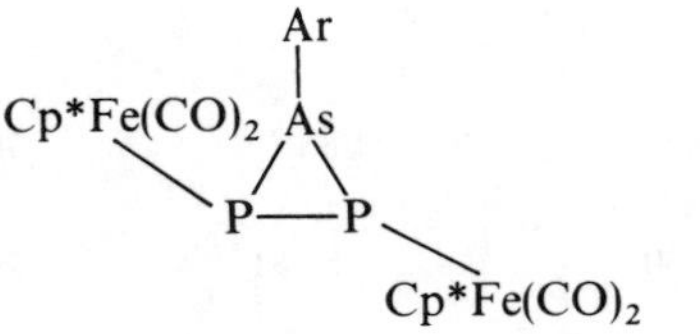

(66)

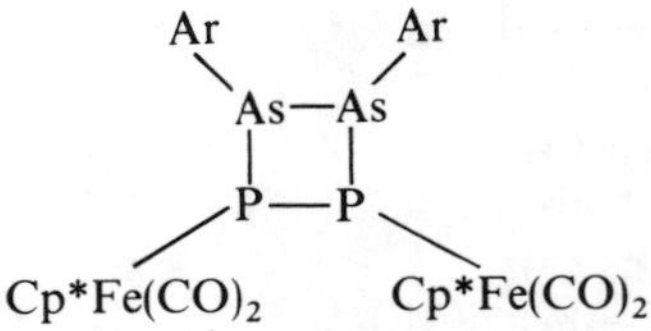

(67)

was only modest, but a planar As_2Se_2 ring is present and the other major structural features were clearly defined.

Two arsenic–phosphorus compounds, **66** and **67**, can be obtained by treating $ArAsCl_2$ ($Ar = 2,4,6-Bu^t_3C_6H_2$) with $Cp*Fe(CO)_2P(SiMe_3)_2$ and feature, respectively, an AsP_2 triangle and a bent diphosphadiarsetane ring in which the substituents are in the all-*trans* arrangement[177]. The As—P distances in **66** (2.316 and 2.350 Å) are slightly different, probably due to steric effects.

9. Miscellaneous compounds

The arsenic–bromine bond distance in *o*-bromophenylarsenic dibromide is 2.347 Å with pyramidal geometry about arsenic (C—As—Br 95.1, 99.4; Br—As—Br 97.8°)[178].

Reaction of butyl-lithium and diphenylarsine in the presence of a crown ether gives an ionic product containing well-separated $(AsPh_2)^-$ anions and $[Li(12\text{-crown-}4)_2]^+$ cations. The As—C distance in the anion (1.972 Å) is normal but the C—As—C angle (108.6°) is widened[179]. The related sodium compound, $NaAsPh_2$(dioxane), has a polymeric structure with chains of alternating dioxane solvated sodium ions and $AsPh_2$ groups arranged parallel to the *a* axis[180]. The chain is almost linear at arsenic (Na—As—Na 173.6°) but bent at sodium (As—Na—As 121.5°); the geometry at arsenic is pseudo-trigonal bi-pyramidal. The arsolyl anion **68** has been identified and is comparable with the pyrrolyl and phospholyl analogues; the As—C distances are 1.90 Å with a C—As—C angle of 87.0°[181].

The pentamethylcyclopentadiene rings in the arsenic metallocene cation, $Cp^*_2As^+$, isolated as the tetrafluoroborate salt, are neither coplanar (angle 36.5°) nor ideally

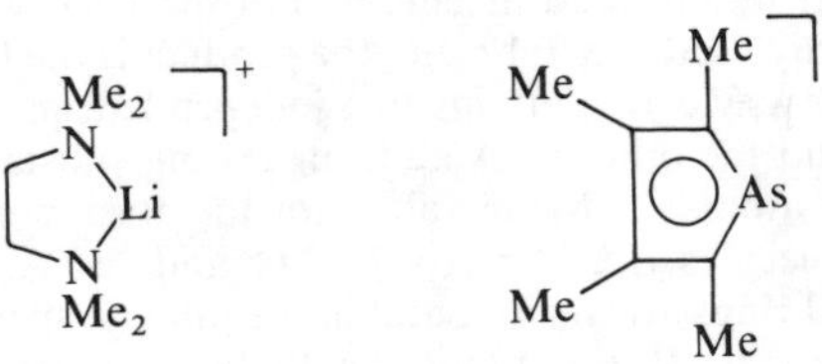

(68)

pentahapto[182]. As—C distances range between 2.188 and 2.742 Å and distortion is thus in the sense of di- or tri-hapto coordination.

Finally here, it is useful to note that arsenic tribromide forms a 2:1 complex with hexaethylbenzene[183], with a structure similar to those of the Menshutkin complexes, obtained from antimony halides and arenes (see Section III.A.6). Complex formation between $AsCl_3$ and both 15-crown-5[184] and [2.2.2]paracyclophane[185] has also been investigated.

B. The +5 Oxidation State

1. Compounds with three or four As—C bonds

Both tris(pentafluorophenyl)arsenic difluoride[186] and tri(neo-pentyl)arsenic dibromide[3] follow the usual pattern for five-coordinate arsenic compounds with unidentate ligands and have trigonal bipyramidal structures with axial halogen atoms. For the former As—F and As—C distances are, respectively, 1.782 and 1.915 Å; in the latter, with sterically demanding alkyl groups, the As—C separation is increased to 1.98 Å and there is asymmetry in the bromine positions (As—Br 2.530 and 2.597 Å).

Large cations, such as tetraphenylarsonium, are useful in stabilizing large and often unusual anions and structure determinations of the hydrogendichloride (Cl...Cl 3.09 Å)[187] and trichloride salts[188] have been carried out. In the latter there are four independent arsenic atoms with As—C distances in the range 1.910–1.921 Å and angles at arsenic between 106.1 and 110.7°. The cation **69** in hydrated 1,1-diphenylarsenanium bromide is also tetrahedral but the *endo* C—As—C angle is reduced to 103.5°[189].

(69)

Diastereoisomers of an asymmetric arsonium cation carrying benzyl, methyl, 4-methylphenyl and naphthalen-1-yl groups have been separated and the structure of the *R*-form determined for the bromide and hexafluorophosphate salts[190]. In each salt there are equal amounts of two conformational isomers, differing with respect to the relative dispositions of tolyl and naphthalenyl substituents, and probably arising because of strong crowding about arsenic. This is confirmed by the spread of As—C distances (1.889–1.949 Å) and C—As—C angles (106.9–112.5°). There is again tetrahedral geometry about arsenic in acetylarsenocholine bromide, $[Me_3AsCH_2CH_2OCMe=O] Br$[191].

2. Compounds with double bonds to carbon

The compounds in this group are triphenylarsonium ylides ($Ph_3As=CRR'$), which are stabilized by the presence of electron-withdrawing groups. Structures of the following compounds have been determined: (a) R = H, R' = COPh[192], (b) R = H, R' = $COCF_3$[193], (c) R = COMe, R' = COOEt[194], (d) R = COMe, R' = CONHPh, (e) R = COMe, R' = NO_2, (f) R = C(O)OMe, R' = $C(O)C_2F_5$[195], (g) R = R' = SO_2Ph[196], (h) R = R' = $C(O)CH_2CMe_2CH_2C(O)$ and (i) R = C(O)Ph, R' = $C(O)C_2F_5$[197]. The As—C bond lengths generally fall in the range 1.86–1.88 Å and are clearly representative of some degree of double bonding (cf As—C 1.98, As=C 1.78 Å). It is clear, however, that in contrast to the phosphorus analogues, the M—C bond here has substantial single-bond character and compounds are perhaps best considered as being dipolar. This is probably a consequence of the relatively poor overlap of arsenic 4d orbitals with carbon 2p orbitals in comparison with phosphorus 3d orbitals overlap. Angles at the ylidic carbon indicate planarity and there are often intramolecular interactions between arsenic and an oxygen atom of the attached R/R' groups.

There are few structures for derivatives of these ylides, but benzoylmethylenetriphenylarsane [compound (a) above] reacts with nickel bis(cyclo-octadiene) in the presence of triphenylphosphine transferring a phenyl group to nickel giving **70**, an ethylene oligomerization catalyst[198]. A methylene ylide, $CH_2:AsPh_2CH_2AsPh_2$, has been stabilized in the chromium carbonyl complex **71**, in which the As—Cr bond. length is 2.450 Å[199,200].

(70)   (71)

3. Compounds with bonds to nitrogen

Arsinimines ($Ph_3As=NR$) are the nitrogen analogues of the ylides considered in the previous section and there is again the potential for multiple bonding between arsenic and nitrogen. In $Ph_3As=NCN$, the As—N distance is 1.739 Å, intermediate in length between As—N single- and double-bond distances (1.91 and 1.71 Å, respectively), again implying a dipolar contribution and again in contrast to the double-bond character of the phosphorus–nitrogen analogue[201]. The As—N—C angle is 118.6° and the two C—N distances are 1.306 and 1.152 Å, for the nitrogens adjacent and remote from the arsenic atom. The related imide **72** has a similar As—N distance (1.760 Å) with an angle of 117.2° at the imide nitrogen[202]. Compound **72** slowly loses N_4S_4 and sulphur in acetone solution

(72)   (73)

(74)

to give an unusual bis(imide) **73** (As—N 1.756, 1.779 Å, As—N—S 120.8, 118.9°) containing a chain of eleven alternating sulphur and nitrogen atoms[203]. A second bis(imide), (Ph$_3$As=N=AsPh$_3$)Cl, provides an example of a new bulky cation and can be obtained by treating triphenylarsenic dichloride with N(SiMe$_3$)$_3$[204]. The As—N separation (1.749 Å) implies multiple bonding and the angle at nitrogen is 123.9°.

Compound **74** provides an example of a four-membered ring compound containing As(V) with trigonal bipyramidal geometry at the five-coordinate arsenic atoms[205]. The nitrogen atoms at a given arsenic occupy both axial (As—N 1.933 Å and equatorial (As—N 1.768 Å) positions. The eight-membered arsenic–nitrogen ring in [(CF$_3$)$_2$AsN]$_4$ has approximately S_4 symmetry and there is a slight alternation in the As—N distances around the ring (1.717, 1.732 Å)[206]. Ring angles at arsenic fall between 124.8 and 125.6° and those at nitrogen between 122.9 and 123.9°.

4. Compounds with bonds to oxygen

In the simple arsonic acid, (o-benzoyl)phenylarsonic acid RAsO(OH)$_2$, the arsenic geometry is distorted tetrahedral with C—As—O angles of 100.7 and 110.3° to hydroxy groups and 116.5° to the doubly bonded oxygen[207]. In keeping with the formulation, there is one short As—O distance (1.652 Å) and two longer ones (1.708 Å); the As—C distance (1.928 Å) is normal and the structure is completed by a short non-bonded contact between arsenic and the benzoyl oxygen atom and hydrogen bonding between both hydroxy groups and symmetry-related terminal oxygen atoms (O...O 2.619, 2.607 Å). As—O distances in the hydrochloride salt of 3-amino-4-hydroxyphenylarsonic acid are similar (As=O 1.652, As—O 1.712, 1.727 Å), but the structure is more complex and features a shortening of the following bonds: As—C (1.879 Å), two of the C—C bonds (1.375, 1.381 Å) and the C—O(H) bond (1.338 Å)[208]. These observations point to some contribution from the zwitterionic form **75**, and the structure is completed by a complex set of hydrogen bonds.

(**75**)   (**76**)

The short As—O separations noted above are also found in triarylarsine oxides and the value in Ph$_3$AsO (1.65 Å) has been ascribed to the presence of one σ and two π bonds[192]. As—O distances in both the triclinic and hexagonal modifications of tri(p-chlorophenyl)arsine oxide (1.64 Å)[209] and in (MeO$_2$CC$_6$H$_4$)Et$_2$AsO (1.651 Å)[210] are similar.

Arsine oxides are good Lewis bases and most work exploring their donor properties has concentrated on the triphenyl derivative; a number of structures have been determined. The monohydrate, Ph$_3$AsO·H$_2$O, has been re-examined, confirming the centrosymmetric, hydrogen bonded, dimeric structure where the As—O separation (1.657 Å) is little changed from that in the free oxide[210]. Other hydrogen-bonded systems examined include, for example, those with p-nitrophenol (As—O 1.668 Å)[211] and both benzene[212] and p-toluene[213] sulphonamides. Centrosymmetric dimers **76** are present in the benzenesulphonamide adduct, but with the toluene analogue three sulphonamide groups bridge between two triphenylarsine oxide molecules. In all these compounds, As—O distances are little changed from that in the free base and this is characteristic of interactions with weak acids.

With strong acids, the pattern is different and a variety of complexes has been isolated with structures based on a hydrogen-bonded $[(Ph_3AsO)_2H]^+$ cation with BF_4^-[214], I_3^-[215] and $AlCl_4^-$[216] counterions. The hydrogen atom was not located in any of these examples but the presence of a linear hydrogen bond (O $\cdots$ O 2.40–2.46 Å) is inferred from the general geometry. An increase in the As—O separation to 1.67–1.69 Å is generally observed.

Species are also known where the proton, instead of being hydrogen bonded to the oxygen atom of an arsine oxide, has been completely transferred to oxygen to produce distinct $[R_3AsOH]^+$ cations. This raises a further point as it is also possible to describe such compounds in terms of a five-coordinate $R_3As(OH)X$ structure. If proton transfer does take place, the arsenic–oxygen bond distance will increase as there is now a single As—OH bond. For $Ph_3As(OH)Cl$, obtained by hydrolysis of a triphenylarsenic imide, X-ray studies do not confirm five-fold coordination about arsenic but the structure contains tetrahedral $[Ph_3AsOH]^+$ cations and chloride anions. Important parameters are: O—As—C 103.2–110.2°, C—As—C 108.7–112.3°, and As—O 1.720 and 1.725 Å[217]. The hydrogen sulphate salt has a similar structure and the As—O distance (1.727 Å) again implies the presence of a single As—O bond[218]; in both cases cations and anions are linked by hydrogen bonds.

Cyclopentamethylene-t-butylhydroxyarsane bromide 77 has a related ionic structure (As—O 1.729 Å) with the six-membered ring in a chair conformation and hydroxy and butyl groups in axial and equatorial positions, respectively[219]. It is curious that hydrolysis of cyclopentamethylenearsenic chloride yields the corresponding dihydroxide, $[(CH_2)_5As(OH)_2]Cl$, with As—O distances of 1.708 and 1.735 Å, rather than the, perhaps expected, arsinic acid, $(CH_2)_5As(O)(OH)$[220]. From the As—O separation (1.733 Å), it appears likely that an As—OH group is also present in the arsenic sugar sulphate 78, obtained from clam kidneys[221]; a related species has also been obtained from algae[222].

(77)   (78)

The arsine oxide 2-$Ph_2As(O)C_6H_4COOH$ provides interesting further information on these species[223], as there is evidence that in chloroform solution it exists unexpectedly as a monocyclic acyloxyhydroxyarsorane 79, while in the solid state it has the novel resonance hybrid structure 80. Here arsenic has five-coordinate trigonal bipyramidal geometry with equatorial carbon atoms (C—As—C 116.4, 118.8 and 119.4 Å) and axial oxygens.

As ligands, triorganoarsine oxides generally behave as conventional unidentate groups, with a small lengthening of the As—O bond on coordination. This is not as great as is observed on coordination of the related trioganophosphine oxides. Examples are, however, known where an arsine oxide bridges between pairs of acceptor elements. Among the addition compounds investigated structurally are the following, in order of the position of the acceptor in the Periodic Table:

$Ph_3AsO \cdot BF_3$, As—O 1.690 Å. As—O—B 125.7°[224];

$(Me_3AsO)_3 \cdot (InX_3)_2$ for X = Cl[225] or Br[226], pairs of indium atoms symmetrically

(79)   (80)

bridged by three oxygen atoms of the ligands, In—O 2.241, 2.234 Å (2.26, 2.29 Å),
As—O 1.710 (1.71 Å), In—O—In 94.5 (94.1°) for X = Cl (Br);

{[SnPh$_2$(Ph$_3$AsO)(NO$_3$)]$_2$(C$_2$O$_4$)}, distorted pentagonal bipyramidal geometry about
tin, As—O 1.674 Å[227];

[SnPh$_2$(ClC$_6$H$_4$)(Ph$_3$AsO)$_2$][Ph$_4$B], trigonal bipyramidal geometry about tin,
As—O 1.670 Å, As—O—Sn 145.9°[228];

[Ni(Ph$_3$AsO)$_4$Br]Br.H$_2$O.1.5(toluene), square-pyramidal geometry about nickel with
oxygen atoms in basal positions (As—O 1.66–1.68 Å), Ni—O 2.00 Å[229];

[Ni(H$_2$O)$_6$]Br$_2$·4(Ph$_3$AsO)·H$_2$O, cation hydrogen bonded to the four arsine oxide
molecules[229];

[Ni(NO$_3$)$_2$(Ph$_3$AsO)$_2$], nickel in distorted octahedral coordination to two *cis* arsine
oxide molecules (As—O 1.655, 1.677 Å) and two bidentate nitrate groups[230];

[Ni(MePh$_2$AsO)$_4$(NO$_3$)][NO$_3$], square-pyramidal coordination about nickel with
oxide oxygen atoms in basal position, As—O 1.661 Å[231];

[M$_2$(acac)$_4$(Ph$_3$AsO)] for M = Co or Ni, triphenylarsine oxide bridges between the two
metal atoms, Co—O 2.16, 2.26 As—O 1.66 Å[232];

[Cu(Ph$_3$AsO)$_4$][CuCl$_2$], square-planar coordination about copper in the cation,
As—O 1.675, 1.691 Å, As—O—Cu 125.5, 128.8°[233];

[Ce(Ph$_3$AsO)$_3$Cl$_3$].MeCN, octahedral geometry in *fac*-isomeric form, As—O
1.650–1.659 Å[234];

[Eu(Ph$_3$AsO)$_3$(NO$_3$)$_3$].4H$_2$O, nine-fold irregular geometry about europium from
three arsine oxide ligands and three bidentate nitrate groups, As—O 1.53, 1.64 and 1.68 Å,
the very short distance probably reflects the sensitivity of the ligand to charged groups in a
trans position[235];

[U(Ph$_3$AsO)$_2$Br$_4$], octahedral geometry about uranium, As—O 1.69, 1.71 Å[236].

Bis(diphenylarsoryl)ethane, Ph$_2$As(O)(CH$_2$)$_2$As(O)Ph$_2$, forms addition compounds in
which each oxygen donates to a molecule of either triphenyltin chloride[237] or nitrate[238],
raising the coordination number of tin to five (trigonal bipyramidal geometry with
equatorial phenyl groups), As—O 1.666 Å, As—O—Sn 134.5 (chloride), 140.2° (nitrate).
The product with dibutyltin dichloride is polymeric with the ligand bridging between
adjacent tin centres, As—O 1.657 Å, As—O—Sn 146.92°[239].

The expected geometry for five-coordinate Group 15 compounds is trigonal bipyramidal, but with phosphorus this geometry can be distorted toward the energetically

less favourable square-pyramidal alternative by incorporating, in particular, five-membered chelating groups into the structure[240]. The trigonal bipyramidal and square pyramidal are actually two extremes, which can be linked by the Berry pseudorotation coordinate, and by examining the angles in observed structures it is possible to assess the extent to which a structure is distorted along this coordinate from the trigonal bipyramidal extreme. A number of arsenic compounds have been examined recently, mainly by Holmes and coworkers, to determine if it is possible to obtain the same range of structures for arsenic as for the lighter analogue, phosphorus. This certainly appears to be the case and, in the following list, the compounds are in order of increasing percentage distortion along the Berry coordinate from trigonal bipyramidal geometry: **81** (11.1%), **82** (20.5%)[241], **83** (21.4%)[242], **84** (21.5%)[243], **85** (39.6%)[242], **86** (45.4%)[241], **87** (95.2%)[243],

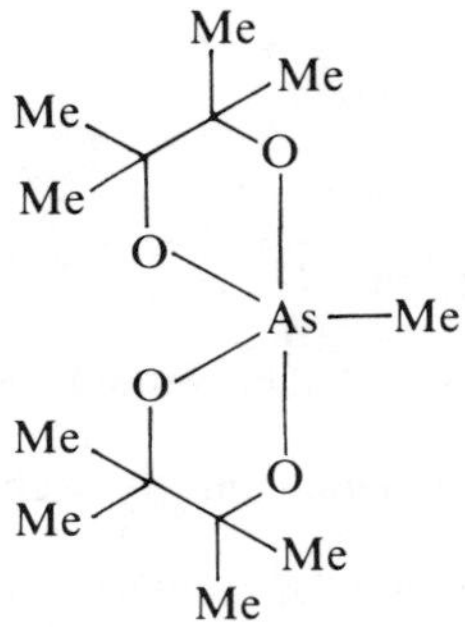

(81)

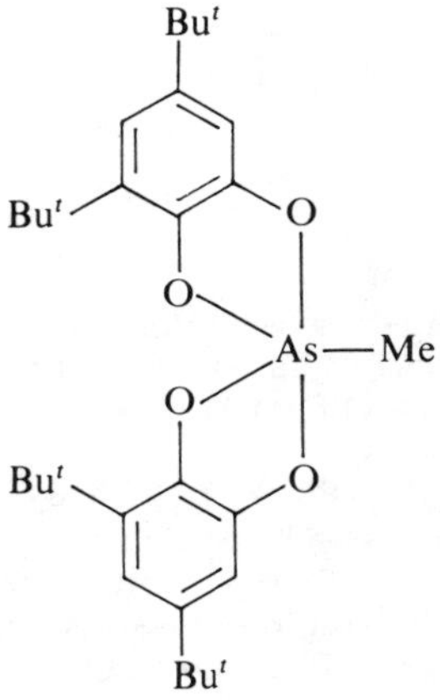

(82)

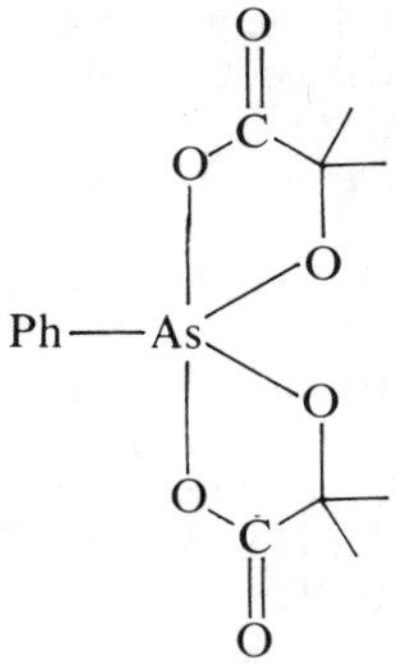

(83)

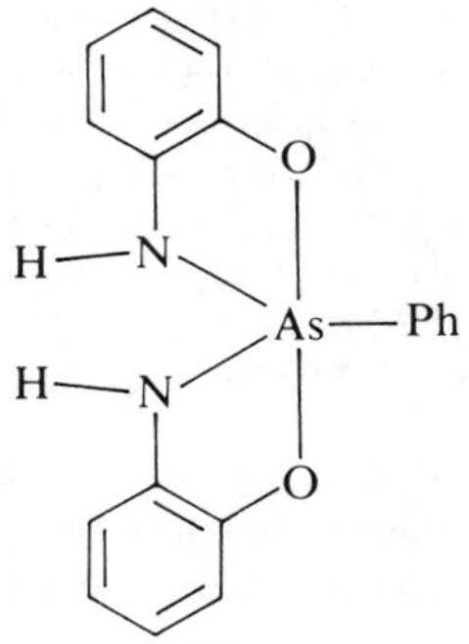

(84)

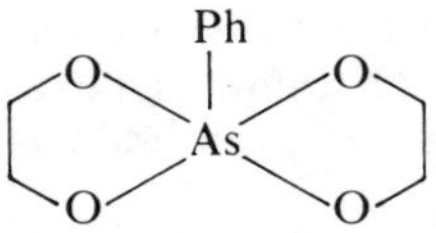

(85)

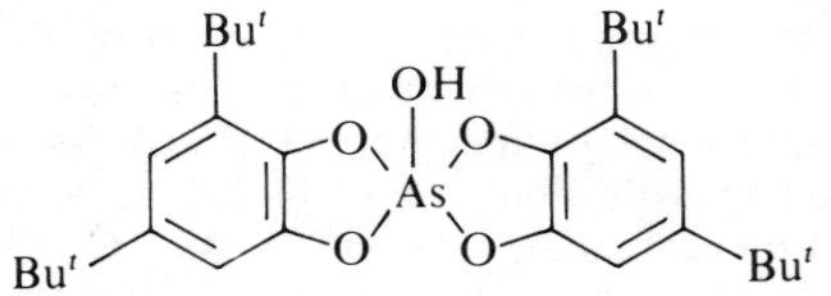

(86)

Ph

(87)

Ph

(88)

Ph

Me   Me

(89)

88 ($\sim 100\%$)[242] and **89** ($\sim 100\%$)[244]. Compounds **87–89** are clearly almost square pyramidal and, for example in compound **88**, the two independent As—O distances are 1.800 and 1.802 Å with *trans* O—As—O basal angles of 150°. This should be compared with As—O distances in **81** of 1.761 and 1.822 Å, which represent bonds to, respectively, the equatorial and axial sites of a basic trigonal bipyramid.

5. Compounds with bonds to sulphur

A number of triorganoarsine sulphide structures have been determined and, in each case, the molecule has slightly distorted tetrahedral geometry with As—S separations of approximately 2.08 Å. This is substantially shorter than the single-bond distance (*ca* 2.25 Å) and clearly represents a degree of π bonding. With triphenylarsine sulphide, the unit cell contains two independent molecules, which lack C_{3v} symmetry from the different orientations of the phenyl groups; the mean As—S and As—C distances are 2.090 and 1.949 Å (although one As—C length is very short, 1.854 Å) and the mean C—As—C and S—As—C angles are 105.2 and 113.5°, respectively[245]. In tri(p-chlorophenyl)arsine sulphide, the As—S distance is 2.074 Å[209], but there is a small increase, representing a weakening of the bond, on substitution of phenyl by alkyl groups in, for example, Ph$_2$EtAsS (As—S 2.081 Å)[246], (p-MeC$_6$H$_4$)Et$_2$AsS (2.090 Å) and PhPr$_2$ AsS (2.079 Å)[247]. The As—S distance (2.069 Å) is not changed from the expected double-bond distance in the solid 1:1 addition compound obtained from two liquids, MePh$_2$AsS and PO(NMe$_2$)$_3$; the structure determination points to no strong attractive forces and compound formation seems to be a result of the fortuitous mutual accommodation of steric and space filling requirements of the two components[248].

Arsine sulphides, like the oxides, are potential ligands but have been far less extensively employed in this role. Structures are, however, available for two indium compounds, InX$_3$(Me$_3$AsS)$_2$ for X = Cl or Br, which have distorted trigonal bipyramidal structures, with the sulphur ligands in axial positions (S—In—S *ca* 162°), and As—S distances in the range 2.130–2.140 Å[249]. The cobalt complex cation, [Co(Me$_3$AsS)$_4$][ClO$_4$], occurs in two conformationally isomeric forms, which differ in the S—Co—S angles and the S—Co—S—As torsion angles[250]. Arsine sulphide fragments can be stabilized by low-oxidation-state transition metals and one example of this is the incorporation of a Bu$_2^t$AsS unit as an η^2 group in Cp(CO)$_2$W(Bu$_2^t$AsS)[251]. The As—S separation, 2.15 Å, falls between the single- and double-bond distances.

Organodithioarsenic(V) ligands, S$_2$AsR$_2^-$, are comparable with the better known dithiophosphorus species but are less extensively investigated. In the tin compound,

$Me_2Sn(S_2AsMe_2)_2$, the ligands are described as unidentate (Sn—S 2.471 Å) with As—S separations (2.17 and 2.089 Å) close to the expected single- and double-bond distances[252], but the ligand forms weak bridges (S ⋯ Sn 3.515 Å) in the solid state to give polymeric chains. Highly unsymmetrical bridging is again observed with the related diphenyl-dithioarsinate in $Ph_2Sb(S_2AsPh_2)$ **90** where As—S distances are 2.050 and 2.214 Å and Sb—S distances are 2.486 and 3.369 Å[253]. In this case, the solid consists of dimeric units based on an eight-membered $Sb_2S_4As_2$ ring with weak transannular interactions, rather than infinite chains. Dimethyldithioarsinate breaks the tetrameric $(Me_3PtCl)_4$ structure to give a dinuclear product $[(Me_3Pt)_2(Me_2AsS_2)_2]$ **91** in which each platinum is coordinated by three methyl groups and three sulphurs, each in *fac*-positions of octahedral coordination about the metal[254]. One sulphur atom of each ligand bonds directly to platinum (Pt—S 2.549 Å) while the second sulphur bridges between the two metal atoms at 2.558 and 2.523 Å; both arsenic sulphur distances are short, indicating some degree of double-bond character.

(90)   (91)

III. ANTIMONY COMPOUNDS

A. The +3 Oxidation State

1. Compounds with two or three Sb—C bonds

These compounds, like the arsenic analogues, have pyramidal structures. In triphenylantimony the asymmetric unit contains two independent molecules with Sb—C distances ranging between 2.143 and 2.169 Å and C—Sb—C angles between 95.1 and 98.0°[255]. The two molecules differ basically in ring conformation. Sb—C distances and C—Sb—C angles in the *p*-tolyl analogue (mean 2.141 Å and 97.3°) are comparable[256], but, because of steric effects in tris(2,6-dimethylphenyl)antimony, there is a large increase in angle to 104.7° and the Sb—C distance also increases to 2.190 Å[257]. In triphenylphos-phonium bis(diphenylstibino)methylide, $Ph_3P=C(SbPh_2)_2$, there are rather shorter Sb—C distances to the ylidic carbon (2.110, 2.136 Å) than to the phenyl groups (2.165, 2.176 Å) but the major feature of this structure is the asymmetry of the angles at the ylidic carbon[258]. The P—C—Sb angles are 115.5 and 130.5°, resulting from the opposite orientation of the diphenylstibino groups relative to the P=C bond.

The first secondary stibine, di(mesityl)stibine, has been prepared as a surprisingly stable solid; for steric reasons, the C—Sb—C angle is also large (101.7°) and the Sb—C distance is 2.168 Å[259]. Successive treatment of this compound with BuLi and copper(I) chloride in the presence of trimethylphosphine leads to the four-membered ring compound **92** in which Sb—Cu separations are 2.669 Å, close to the sum of the covalent radii.

Tri(cyclopentadienyl)antimony contains mono-hapto groups where the Sb—C distances vary from 2.238 to 2.264 Å and the C—Sb—C angles from 94.5 to 98.1°[260]. The orientation of the Cp rings is given by approximately tetrahedral angles between the best

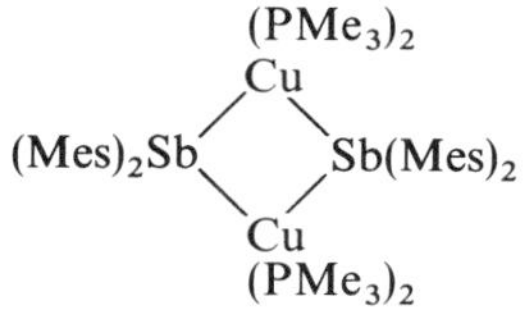

(92)

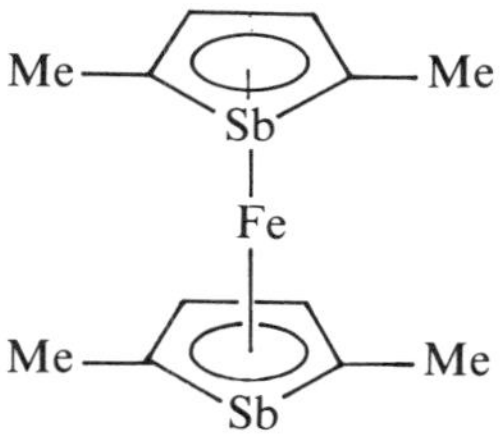

(93)

planes through the five-membered rings and the Sb—C bond directions; as expected for mono-hapto groups, there are three 'long' and two 'short' C—C distances. A monosubstituted cyclopentadiene compound, $CpSbCl_2$, is also known but, as there are three short (2.271, 2.579 and 2.595 Å) and two longer Sb—C distances (ca 2.98 Å), the ring is best considered as being attached in an η^3 fashion[261]. Crystals of the distibaferrocene **93** are similar to the arsenic analogue with a completely eclipsed structure, and although iron is bonded to all five ring atoms, it is closer to the beta carbon atoms (2.05–2.08 Å) than the alpha atoms (2.08–2.14 Å)[262]. The Fe—Sb distance (2.56, 2.57 Å) is naturally longer.

Antimony geometry in the new heterocycle **94** is still highly pyramidal (C—Sb—C 79.9 in the ring, 95.5 and 96.5°) but the stibaindole ring is closely planar and orthogonal to the plane of the phenyl group[263].

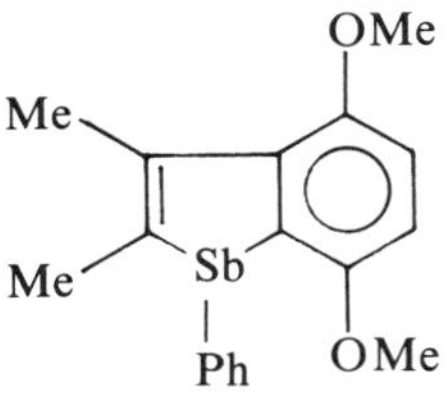

(94)

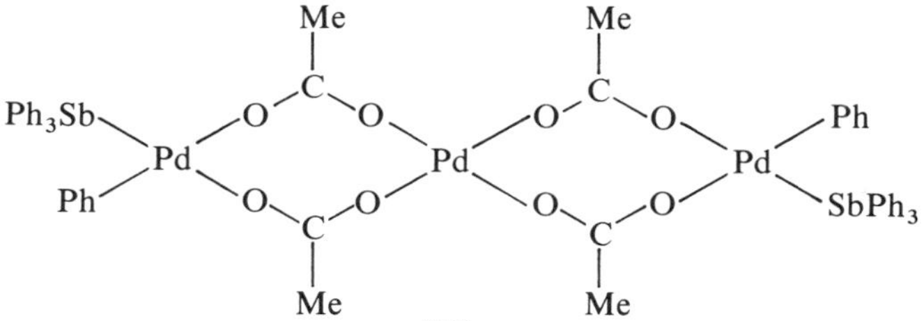

(95)

Trisubstituted stibines have been used as ligands and among the triphenylantimony compounds structurally investigated are the following:

$Re_2H_6(Ph_3Sb)_5$, contains three hydrogen atoms bridging between the two Re atoms, coordinated respectively to two hydrogens and two stibines and one hydrogen and three stibines[264]. Sb—Re distances are longer (mean 2.59 Å) at the second atom than at the first (mean 2.56 Å) from steric crowding.

$[Re_2(CO)_7(Ph_2P)(Ph_2Sb)(Ph_3Sb)]$, the Ph_2Sb group is bridging with longer Sb—Re distances (2.74 Å) than that to the terminal ligand (2.67 Å)[265].

$ReCl_2(NO)_2(Ph_3Sb)_2$, octahedral geometry with trans stibine ligands (Sb—Re 2.71 Å)[266].

$Os(CO)_4(Ph_3Sb)$, trigonal bipyramidal geometry with the stibine in an equatorial position (Sb—Os 2.612 Å)[27].

$OsBr_3(Ph_3Sb)_3$, octahedral geometry with a mer-arrangement of ligands (Sb—Os 2.640–2.654 Å)[267].

$RhCl_2(Ph_3Sb)_2Ph(MeCN)$, octahedral geometry with trans stibine groups (Sb—Rh 2.588 Å)[268].

$Rh_2(Ph_3Sb)_2(PhCONH)_4$, four benzamidato anions bridging between the two metal atoms with axial stibine groups (Sb—Rh 2.681, Rh—Rh 2.463 Å, Sb—Rh—Rh 176.4°[269].

$Pd_2(OAc)_4(Ph_3Sb)_2$, two acetate groups bridge between the metal atoms and four-fold coordination about each palladium is completed by one terminal acetate group and one stibine ligand (Sb—Pd 2.508 Å)[270]. This compound is interesting, as on heating to 60 °C it gives Pd(0) and biphenyl in 96% yield. At 45 °C one aryl group migrates from antimony to palladium to give **95**.

$CuCl(Ph_3Sb)_3 \cdot CHCl_3$, approximately tetrahedral geometry about copper with three relatively long Sb—Cu bonds (mean 2.554 Å) and one short Cu—Cl contact (2.235 Å)[271].

$[Au(Ph_3Sb)_4][Au(C_6F_5)_2]$, the solid contains three independent cations with slight deviations from tetrahedral geometry (Sb—Au 2.585–2.669 Å)[272].

$[Au(Ph_3Sb)_4][Au\{C_6H_2(NO_2)_3\}_2]$, tetrahedral cation as above (mean Sb—Au 2.651 Å, Sb—Au—Sb 107.8–111.0°[273].

The complex, $[Cu\{(C_6H_4\text{-}4\text{-}F)_3Sb\}_4][BF_4]$, contains copper in a C_3 site with Sb—Cu distances of 2.556 and 2.547 Å[274]. A molecule of tri(t-butyl)antimony occupies an axial position (Sb—Fe 2.547 Å) in the trigonal bipyramidal arrangement around iron in $Fe(CO)_4(SbBu^t_3)$[275]. This contrasts with the situation in the osmium–triphenylantimony complex mentioned above[27], where the antimony ligand occupies an equatorial site and shows that sterically demanding and good σ donating ligands favour axial substitution. Complex formation occurs with the antimony analogue of the chelating arsenic ligand, 1,2-bis(dimethylarsino)benzene, and in the *trans* isomer, $[Co\{C_6H_4(SbMe_2)_2\}_2Cl_2]$ $[CoCl_4]$, Sb—Co distances are 2.478 and 2.505 Å[276].

2. Compounds with multiple bonds to antimony

Although there has been great success in stabilizing compounds containing double bonds between atoms of the lighter Group 15 elements, such bonds between pairs of antimony atoms are known only in compounds stabilized by coordination to a transition metal. The first such compound **96** was isolated in 1982[277] from a reaction between phenylantimony dichloride and $Na_2W_2(CO)_{10}$ in THF solution; the Ph_2Sb_2 group is η^2 bonded with an Sb—Sb separation (2.706 Å) which is short in comparison with the single-bond distance (ca 2.84 Å) and represents an Sb—Sb bond order of ca 1.5.

A compound with the same stoichiometry, $\{(Bu^tSb{=}SbBu^t)[Cr(CO)_5]_3\}$, has been obtained from Bu^tSbCl_2 and the analogous chromium decacarbonyl anion in which the Sb—Sb distance is also short (2.720 Å)[278], and there is a third example in the iron carbonyl derivative **97**, where the separation is 2.774 Å[279,280].

Reactions of this type can also give closed stibinidine complexes such as **98**[278] and **99**[279,280], where Sb—W separations are ca 2.84 Å. Stibinidine groups have also been incorporated into a number of high nuclearity metal clusters, among which are

(96)

(97)

(98)

(99)

$[Ni_{10}(PhSb)_2(CO)_{18}]^{2-}$ [281] and $[Os_3(CO)_{10}H(Ph_2Sb)]$ [282]. Rather interestingly, a hydrolysed product containing two $Ni(CO)_2$ fragments doubly bridged by two $Ph_2SbOSbPh_2$ groups was also isolated in the nickel carbonyl reaction; the main feature of the structure is a centrosymmetric eight-membered $(NiSbOSb)_2$ ring with a chair-type conformation.

3. Compounds with antimony–antimony bonds

Most of the compounds in this section are distibines, but molecules containing rings of antimony atoms are now also known. Distibines R_4Sb_2, in contrast to the phosphorus and arsenic analogues, often show thermochromic properties, changing from yellow liquids to red solids on cooling. This phenomenon arises from the formation of chains of molecules, linked together by weak intermolecular Sb...Sb secondary bonds at ca 3.65 Å (cf 4.4 Å for the Sb...Sb van der Waals separation). Thermochromism is observed for R = Me[283,284], SiMe_3[285,286], GeMe_3[287], SnMe_3[288] and in the distibolyl compound **100**[289], but it is significant that with larger substituents such as phenyl[290] or mesityl[291], molecular packing arrangements do not allow the extended Sb...Sb interaction and thermochromism is not observed. In all the compounds, the substituents adopt a staggered *trans* arrangement about the Sb—Sb axis. Important molecular parameters for these compounds are collected in Table 4.

There is a plane of symmetry through compound **101** and the Sb_3 ring forms an almost

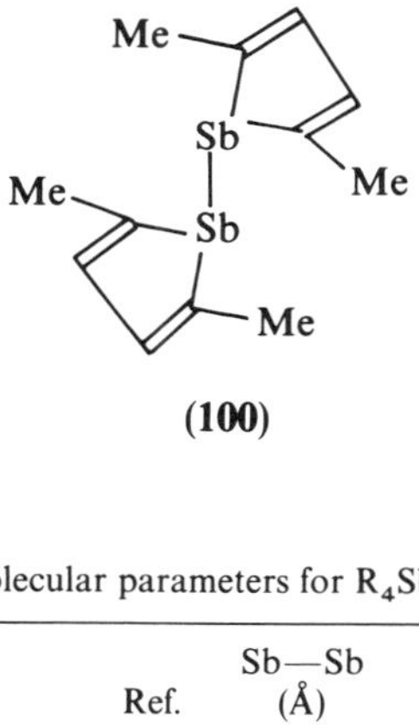

(100)

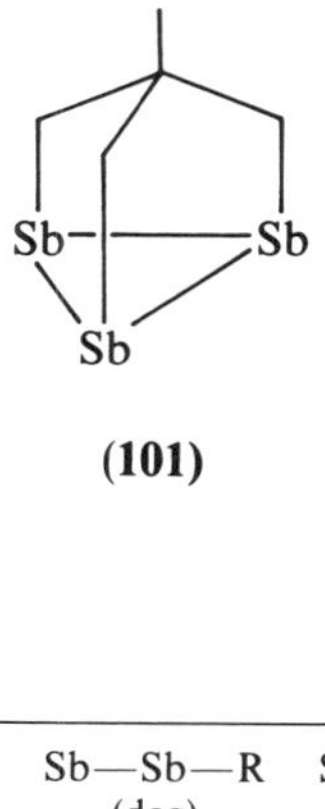

(101)

TABLE 4. Molecular parameters for R_4Sb_2

R	Ref.	Sb—Sb (Å)	Sb—R (Å)	Sb···Sb (Å)	Sb—Sb—R (deg)		Sb—Sb...Sb (deg)
Me(− 160 °C)	283	2.862	2.15	3.645	93.6		180
Me(− 139 °C)	284	2.838	2.134	3.678	92.5,	94.6	179.2
			2.190		94.3,	96.0	
SiMe_3(20 °C)	285	2.867	2.594	3.99	94.4,	98.7	
SiMe_3(− 120 °C)	286	2.863	2.598	3.892	93.4,	98.3	168.4
GeMe_3(− 110 °C)	287	2.851	2.630	3.860	92.7,	97.6	170.5
SnMe_3(20 °C)	286	2.882	2.797	3.879	91.3,	95.6	172.6
SnMe_3(− 120 °C)	286	2.866	2.795	3.811	90.9,	95.4	173.0
SnMe_3	288	2.876	2.80	3.89	91.1,	96.5	173.5
a	289	2.835	2.14	3.625	91.4,	92.2	
Ph	290	2.844	2.175		93.8,	96.7	
Mesityl	291	2.848	2.181–2.201		90.3, 108.4		
					92.6, 109.0		

^{a} Compound **100**.

equilateral triangle (Sb—Sb 2.796, 2.817 Å; Sb—Sb—Sb 59.7, 60.5°); the structure also contains three five-membered Sb_2C_3 rings with mean values for the Sb—C separation and Sb—Sb—C angle of, respectively, 2.17 Å and 86.2°[292]. Two compounds containing an Sb_4 ring[70,293] and one compound containing a six-membered antimony ring system[294,295] have also been studied. The antimony substituents in the puckered four-membered ring compounds, Sb_4R_4, are the sterically demanding t-butyl and mesityl groups and in both cases they occupy all *trans* (pseudo-equatorial) positions to minimize steric interactions. The Sb—Sb distances are almost equal (mean 2.82 Å for the t-butyl and 2.85 Å for the mesityl derivative), with Sb—Sb—Sb angles falling between 77 and 88°. There are further weak contacts in the mesityl structure from both a molecule of benzene of crystallization, which is η^6 attached to one of the antimony atoms at 3.81 Å, and a symmetry-related antimony atom (Sb . . . Sb 3.88 Å), leading to chain formation. Reaction of $PhSb(SiMe_3)_2$ with oxygen in dioxan gives a product containing a centrosymmetric Sb_6 ring in the chair conformation with phenyl groups in equatorial positions[294,295]. The mean Sb—Sb distance is 2.84 Å with ring angles of between 86.8 and 93.6°; again there are short Sb . . . Sb intermolecular contacts (4.20 Å) leading to ring stacking and, in this case, the solid contains a molecule of dioxan of crystallization.

Although these compounds are potential ligands, this aspect has been little investigated but tetraphenyldistibine behaves as a bridging group in $Re_2Br_2(CO)_6(Ph_4Sb_2)$ **102**[296]. Rather surprisingly, the Sb—Sb separation (2.826 Å) is shorter than that in the uncomplexed distibine.

$$(OC)_3Re \underset{\displaystyle Br}{\overset{\displaystyle Br}{<>}} Re(CO)_3$$

$$Ph_2Sb\text{———}SbPh_2$$

(102)

4. Compounds with bonds to sulphur

Antimony–sulphur compounds are quite widespread and stable, probably a consequence of soft acid–soft base interaction, and the solid state structures are characterized by the formation of a number of weaker Sb . . . S secondary bonds. In phenylantimony bis(monothioacetate), the ligands are bonded primarily via sulphur (Sb—S 2.451, 2.471 Å), but, as Sb(III) is coordinatively unsaturated, there is substantial secondary interaction with the oxygen atom of the thioacetate (mean Sb . . . O 2.81 Å) giving a bis chelate structure with antimony in distorted square-pyramidal coordination[297]. The apical position is occupied by the phenyl group and the two sulphur atoms are *cis* to each other (S—Sb—S 84.8°) in the basal plane. Weak dimers are present in the solid state as a result of Sb . . . S bonding to the sulphur forming the longer Sb—S bond.

Bis chelate structures, with weak dimerisation through secondary Sb . . . S bonding in the solid state, are also found for methylantimony bis(ethylxanthate)[298] and the corresponding diethyldithiocarbamate[299]. The structure of the former in Figure 6 shows that the antimony environment is distorted pentagonal pyramidal with an axial methyl group and a basal 'plane' of five sulphur atoms. The ligands are asymmetrically bonded, each with short (2.581, 2.617 Å) and longer (2.834, 2.904 Å) antimony–sulphur bonds and, as in the thioacetate structure, the sulphur atoms attached by the shorter bonds occupy *cis* positions.

On the other hand, phosphinate groups in both $Ph_2Sb(OSPPh_2)$ and $Ph_2Sb(S_2PPh_2)$ are bridging, giving antimony atoms in pseudo-trigonal bypyramidal coordination with

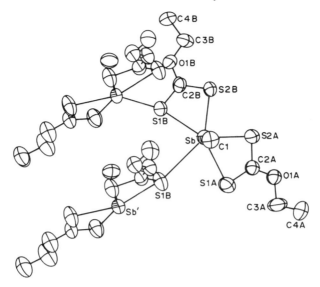

FIGURE 6. The dimeric structure of $MeSb(S_2COEt)_2$. Reproduced by permission of Barth Verlagsgesellschaft mbH from Reference 298

Ph—Sb·······O

(103)

Ph—Sb·······S

(104)

the axial positions occupied by sulphur or oxygen atoms[300]. Both compounds thus have infinite chain structures.

The antimony coordination number in phenyl oxydithia- and trithia-stibocane **103** and **104** is raised to four by a transannular interaction with either the oxygen atom (Sb···O 2.942 Å) or the distant sulphur (Sb···S 3.31 Å)[301]. The eight-membered ring conformations are different in the two cases, with a chair–chair form for **103** and a boat–chair conformation for **104**. Antimony–sulphur separations are *ca* 2.44 Å, but there are further intermolecular Sb...S interactions at distances well within the sum of the van der Waals radii, ranging between 3.56 and 4.04 Å. This raises the coordination number and leads to double chains for the oxa compound and tetrameric units for the thia-analogue. The *p*-nitrophenyl compounds have related structures[302]. The two Sb—S distances in the cyclic stibolane $PhSb(SCH_2CH_2S)$ are 2.43 and 2.46 Å, with an S—Sb—S angle of 88° and a ring conformation midway between the envelope and half-chair forms; intermolecular Sb...S contacts at 3.34 Å here link the molecules into helical chains[303].

In the 8-quinolinethiolate complex **105**, both sulphur (2.444 Å) and nitrogen (2.767 Å) atoms are attached to antimony, the latter as expected at a distance substantially greater

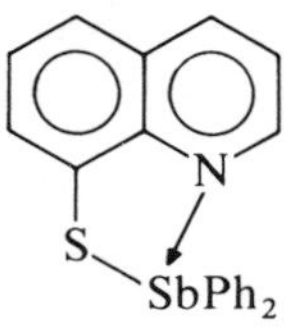

(105)

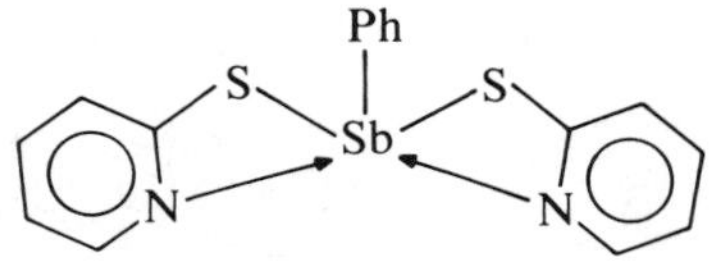

(106)

than the sum of the covalent radii, but an unexpected feature is a close Sb...Sb contact at 3.884 Å *trans* to the sulphur position (S—Sb...Sb 168.3°) giving distorted square-pyramidal geometry about antimony[304]. The related 2-pyridinethiolate group also behaves as a chelating ligand in both the phenylantimony(III) compound 106[305] and its mesityl analogue[306] with the aromatic group occupying the apical position of, again, the distorted square pyramid about antimony. The sulphur atoms, which occupy *cis* basal positions, are effectively equidistant (mean 2.50 Å) from antimony in both compounds. A short intermolecular Sb...Sb contact in the mesityl compound raises the coordination number to six, but none is observed for the phenyl derivative.

A new class of stibathiolanes has been synthesized by transmetallation between a zirconocene and an antimony(III) dihalide as shown in equation 4[307]. Structures for two

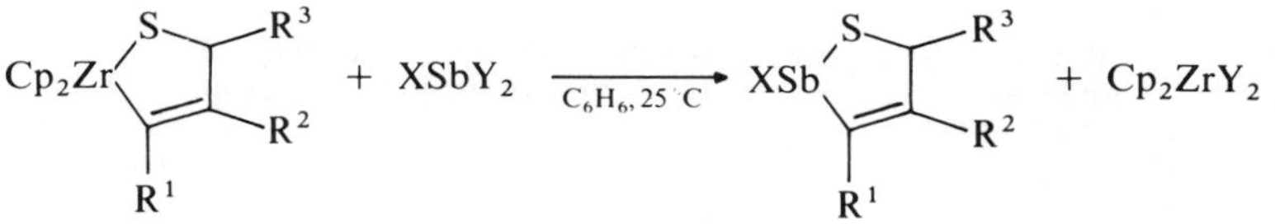

$$X = Ph, Cl$$
$$Y = Cl, Br$$

such products, 107 and 108, point to solid state dimerisation from intermolecular Sb...S bridging (*ca* 3.1 Å) of the type, which as mentioned above is widely observed in antimony(III)–sulphur systems. The antimony geometry is described as 'see-saw', with the antimony atom out of the metallocycle plane.

Diphenyl(phenylthiolato)antimony, Ph_2SbSPh, forms an octahedral molybdenum tricarbonyl complex. $Mo(CO)_3(Ph_2SbSPh)_3$, with *fac*-geometry and the ligands attached via antimony atoms (mean Sb—Mo 2.74 Å, Sb—Mo—Sb 91.6–93.3°) rather than sulphur[308]. Antimony–sulphur distances (mean 2.43 Å) are normal.

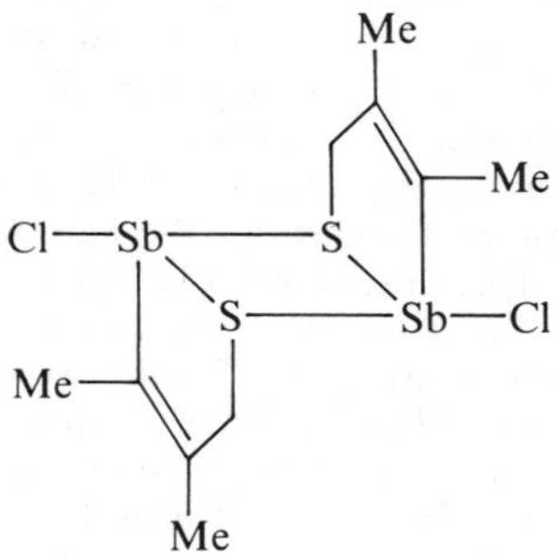

(107)

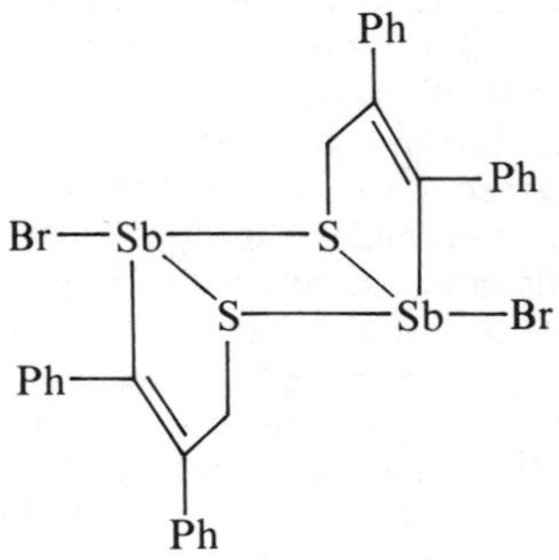

(108)

5. Compounds with bonds to halogens

Unusually, a monomeric dichlorostibine **109** is obtained when antimony trichloride reacts with the highly hindered ligand $LiC(SiMe_3)_2(C_5H_4N-2)$[309]. Geometry about antimony is very distorted pseudo-trigonal bipyramidal with one of the axial positions occupied by chlorine (Sb—Cl 2.469 Å) and the other by nitrogen (Sb—N 2.371 Å). The long Sb—N separation is a consequence of it being a donor bond and strain from incorporation into a four-membered ring.

(**109**)   (**110**)

Organoantimony(III) halides are Lewis acids and yield salts on addition of halide ions. Phenylantimony dichloride, for example, can accept either one or two chloride ions to give ions with the stoichiometries $(PhSbCl_3)^-$ and $(PhSbCl_4)^{2-}$, respectively. Structural studies confirm that the latter is indeed monomeric[311], but there is strong chlorine bridging leading to dimerisation with the former[310,311], to give *trans* square-pyramidal geometry **110** about each antimony. As expected there is a marked difference in bonding distances to the terminal (2.423, 2.540 Å) and bridging (2.657, 3.121 Å) chlorine atoms, but all distances are longer than the statistically determined Sb—Cl single-bond distance (2.32 Å). Again there is square-pyramidal geometry about antimony in the monomeric $(PhSbCl_4)^{2-}$ salt with Sb—Cl distances ranging between 2.537 and 2.770 Å; in this case the anion is stabilized by Cl . . . N—H hydrogen bonding with the pyridinium counterion.

Irrespective of the ratio of chloride ion added, diphenylantimony chloride will accept only one chloride ion to give the $(Ph_2SbCl_2)^-$ anion[310,312], with a distorted pseudo-trigonal bipramidal arrangement of ligands about antimony. Phenyl groups occupy the equatorial positions, with the C—Sb—C angle (97.5°) much reduced from the ideal 120° angle, and chlorine atoms occupy axial sites (Sb—Cl 2.62 Å). The structures of these ions raise a number of problems concerning the stereochemical activity of the lone pair of electrons and the relationship between the lengths of bonds *trans* to each other in these species. A simple approach to these problems was outlined in the Introduction.

Different products appear to be obtained by addition of iodide ions to phenylantimony iodides, although a direct analogue of the diphenylantimony dichloride anion, $(Ph_2SbI_2)^-$, has been isolated with an equivalent pseudo-trigonal bipyramidal structure (Sb—I 2.925, 3.109 Å)[313]. Two other products, i.e. $(Et_4N)_2[Ph_2Sb_2I_6]$, with the same stoichiometry as the chlorine analogue, and $(pyH)_3[Ph_2Sb_2I_7]$, are known but their structures **111** and **112**, respectively, represent new types. In the former, the Sb_2I_6 unit is non-planar (the dihedral angle between the SbI_4 planes is 106.4°) in contrast to a planar Sb_2Cl_6 unit in the chloride. Distances to the two bridging iodides fall between 3.185 and 3.240 Å, with the terminal distances substantially shorter at (2.826–2.890 Å). The second compound is centrosymmetric with a very weak bridge bond (Sb—I 3.404 Å) and terminal Sb—I distances falling between 2.814 and 3.096 Å; the shortest Sb—I distance is that *trans* to the long bridge bond. This paper also presents as analysis of the relationship between the distances of *trans* related antimony(III)-halogen bonds, on the basis of three centre–four electron bonding. One monophenylantimony iodide anion, $[(PhSbI_2)_4I]^-$ **113**, has been prepared by treating $PhSbI_2Cr(CO)_5$ with potassium 18-crown-6 in THF

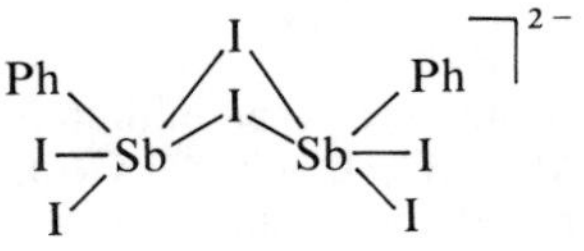

(111)

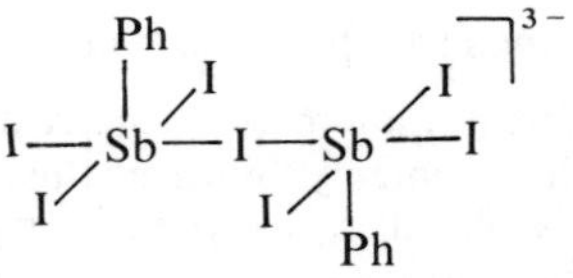

(112)

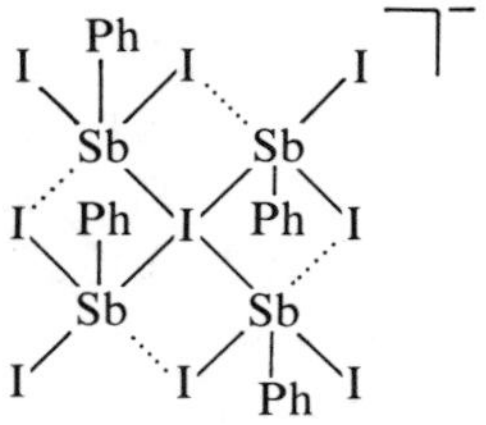

(113)

solution[314]. The structure is based on a central iodide ion lying on a centre of inversion, coordinated to four $PhSbI_2$ units, giving an almost planar Sb_4I_9 unit with phenyl groups disposed on opposite sides of the plane.

6. Miscellaneous compounds

Two anionic organic antimonides, $[Ph_2Sb]^-$ and $[Ph_2Sb-Sb-SbPh_2]^-$, are known, both containing well-separated ions and both as THF solvated $[Li(12\text{-crown-4})_2]^+$ salts[158]. Sb—C distances in the former are 1.972 Å with a bond angle of 108.6°, while, for the latter, the Sb—C distances are increased to 2.166 and 2.192 Å and there is a suggestion of some multiple Sb—Sb bonding from the short separation (2.761 Å) between the atoms. The Sb—Sb—Sb angle is 88.8°. Dimethoxyethane-coordinated lithium atoms alternate with $Sb(SiMe_3)_2$ groups giving long chains of tetrahedrally coordinated lithium and antimony atoms in $Li[Sb(SiMe_3)_2]$. DME, where Sb—Si and Sb—Li distances are 2.532 and 2.933 Å, respectively, and the angles in the chain at antimony and lithium are, respectively, 94.2 and 130.7°[315].

A six-membered ring (Sb—Ga 2.661 Å) with an irregular boat conformation is present in $[Bu_2^tSbGaCl_2]_3$[174] but the ring is the more usual four-membered system in $[Bu_2^tSbInCl(SbBu_2^t)]_2$[316]. The structure shows some disorder but the mean Sb—In distance is 2.844 Å and the ring angles at antimony and indium are 94.9 and 85.1°, respectively.

Molecules of dimethylantimony azide are linked into infinite zig-zag chains, via the alpha nitrogen atoms and, if the antimony lone pair is stereochemically active, coordination about antimony is pseudo-trigonal bipyramidal with phenyl groups in equatorial positions and azide groups (Sb—N 2.32, 2.43 Å) in axial positions[317]. The azide group is linear (178.5°) but the N—Sb—N angle (169.6°) deviates substantially from the 180° ideal; the Sb—N—N angle is 117.4°.

In diphenylantimony thiocyanate, the thiocyanate group similarly behaves in an unexpected way since, although the ligand bridges between pairs of antimony atoms as expected with coordinatively unsaturated antimony(III), it is primarily bonded via nitrogen for two of the three independent thiocyanate groups in the asymmetric unit but via sulphur for the third[318]. As for the azide, the antimony geometry can be considered as pseudo-trigonal bipyramidal with axial thiocyanate groups, but in place of the expected regular arrangement of thiocyanate bridged antimony atoms, coordination at each of the three independent antimony atoms is different. Sb(1), for example, is coordinated to two nitrogen atoms at 2.304 and 2.364 Å, S(2) to one nitrogen atom (2.273 Å) and one sulphur (2.831 Å), while at Sb(3) coordination is to two sulphur atoms (2.700, 2.842 Å). Bridging thiocyanate groups have large Sb—N—C angles (150.7, 164.7 and 174.9°) but much smaller Sb—S—C angles (85.7, 98.6 and 91.4°) and this leads to the development of a curious 'triangular spiral' structure in the solid state, rather than the usual zig-zag chain.

The distibazane **114** has C_i symmetry with a planar four-membered ring and effectively equal Sb—N distances (2.043, 2.051 Å); ring angles at antimony and nitrogen are 78.3 and 101.9° respectively[319].

(114)   (115)

Phthalic acid will displace ethoxy groups from MeSb(OEt)$_2$ to give the diphthalate **115**, which has a centrosymmetric, bridged, dimeric structure in the solid[320]. Each antimony is in distorted octahedral coordination by four oxygen atoms from two asymmetrically bonded carboxylate groups (Sb—O 2.113, 2.652 and 2.088, 2.985 Å), one oxygen from a molecule of ethanol of solvation (Sb—O 2.647 Å) and the carbon atom of the methyl group. But there is also a short (3.132 Å) intermolecular contact to the oxygen of the carboxylate bonded to antimony at 2.652 Å.

The formations of arsenic trihalide complexes with aromatic hydrocarbons was mentioned above, but this type of behaviour is more pronounced with antimony trihalides and leads to Menshutkin complexes, where the ratio of antimony halide to the arene is either 1:1 or 2:1. During the last ten years, these compounds have been extensively investigated by Mootz and Schmidbaur and their coworkers. The structure of the 2:1 complex with benzene is shown in Figure 7 and is best described as an inverse 'sandwich'. Here, the two antimony trichloride molecules are arranged on opposite sides of the aromatic ring at distances of 3.22 and 3.30 Å[321]. As is often found in structures of this type, two of the Sb—Cl bonds are effectively parallel to the plane of the benzene ring while the third is perpendicular and, as in the free trichloride, there are a number of intermolecular Sb⋯Cl interactions which raise the coordination number to give distorted pentagonal bipyramidal coordination about each antimony. Similar 2:1 complexes are found with the trichloride and tribromide and hexamethylbenzene[322], and between the trichloride and both 1,2,4,5-tetramethylbenzene[323] and pentamethylbenzene. In all these compounds, there are tetrameric Sb$_4$X$_{12}$ units and deviations from strict η^6 bonding by the arene.

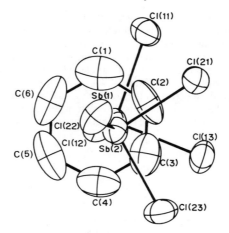

FIGURE 7. The structure of $2SbCl_3 \cdot PhH$. Reproduced by permission of Barth Verlagsgesellschaft mbH from Reference 321

Mesitylene[324] and hexaethylbenzene[325] each give 1:1 complexes with half-sandwich structures with antimony trichloride and a similar 1:1 complex is known for the tribromide with mesitylene. The 2:1 adduct of antimony tribromide with biphenyl is centrosymmetric with a molecule of the antimony halide coordinated to each phenyl group and on opposite sides[326]; a similar picture is found with $2SbCl_3 \cdot (2,2'\text{-dithienyl})$[327], $2SbBr_3 \cdot (9,10\text{-dihyd-roanthracene})$[328] and $2SbCl_3 \cdot (\text{pyrene})$[329]. In the complex of [2.2.2]paracyclophane with two molecules of $SbCl_3$, two of the benzene rings are almost symmetrically coordinated to antimony[185].

Both antimony trichloride[330] and trifluoride[331] form 1:1 complexes with 15-crown-5 in which antimony is sandwiched between almost parallel planes containing the five oxygen atoms (Sb . . . O 2.787–2.997 Å) and, in the former, three chlorines (Sb—Cl 2.405–2.433 Å). Although the antimony–oxygen separations are long compared with the sum of the covalent radii, interaction is sufficiently strong to attach the crown molecule and force the ring into the observed conformation. In the 1:1 complex of 18-crown-6 with antimony trichloride, antimony lies closer to one set of three oxygen atoms (2.989–3.035 Å) than to the other (3.290–3.401 Å)[332].

B. The +5 Oxidation State

1. Compounds with five, four or three Sb—C bonds

Basically trigonal bipyramidal geometry has been confirmed for two pentaorganostibines, $Sb(C:CPh)_5$[333] and biphenyl-2,2'-diyltriphenylantimony **116**[334]. There is little distortion in the former with an axial angle of 179.0 and equatorial angles of 118.2, 119.1 and 122.6°; bonds to the axial carbons are longer (2.15 Å) than those to the equatorial atoms (2.06 Å). The second compound was investigated to determine if the incorporation of a chelating group at antimony would stabilize square-pyramidal geometry, an alternative to the usual trigonal bipyramid. As noted above, this is possible with a number of organo-Group 15 compounds, which contain chelating oxo groups. Pentaphenylanti-

(116)

mony itself has this alternative geometry, and the biphenylyl compound is a constrained version of Ph$_5$Sb. The molecule is, however, trigonal bipyramidal with the biphenylyl group spanning axial and equatorial positions and one phenyl group axial and the other two equatorial. Within experimental error, the two axial distances are equal (2.206 Å) as are the three distances to equatorial atoms (2.146 Å).

Compounds containing four antimony–carbon bonds, R$_4$SbX, are sometimes ionic with a tetrahedral cation, but slightly distorted trigonal bipyramidal geometry with the halogen atom in one of the axial positions is observed with Ph$_4$SbCl[335], (p-tolyl)$_4$ SbCl[336] and Ph$_4$SbBr[337,338]. The axial and equatorial Sb–C distances, of course, differ by some 0.04 Å and the bond to halogen is substantially longer than normal (Sb—Cl *ca* 2.73, Sb—Br 2.96 Å). In addition, the antimony atom is displaced from the centre of the equatorial plane away from the halogen (C$_{eq}$—Sb—C$_{ax}$ *ca* 95.8°) and these observations point to partial ionic character in the Sb–halogen bond and a transition towards tetrahedral character in the R$_4$Sb unit.

Tetraphenylantimony benzenesulphonate, containing a unidentate anion, is very similar with a long Sb—O bond (2.506 Å) and antimony distorted 0.31 Å from the plane of the three equatorial phenyl groups towards the axial phenyl[339]. The compound is actually monohydrated and weak hydrogen bonding leads to centrosymmetric dimers in the solid state. The related tetraphenylantimony acetate is also basically trigonal bipyramidal with the acetate group in one of the axial positions (Sb—O 2.235 Å), but acetate acts as a weak chelate, raising the antimony coordination number via intramolecular Sb $\cdots$ O (2.585 Å) secondary bonding[340]. This has the effect of increasing one of the equatorial angles to 152.6° at the expense of the other two (*ca* 102.5°) and constraining the Sb—O—C angle to 108.1°. The compound will add a molecule of acetic acid, which hydrogen bonds (O $\cdots$ O 2.581 Å) to the C=O of the acetate group. The secondary bonding is thus broken, allowing the equatorial angles to return to close to the ideal 120° and the Sb—O—C angle to open to the more normal 121.8°.

On the other hand, the structure of [Ph$_4$Sb] [N(SO$_2$Me)$_2$] is basically ionic with a tetrahedral cation (mean Sb—C 2.08 Å, C—Sb—C 102.2–116.5°)[341] and tetrahedral geometry is observed about both antimony and rhenium in [Me$_4$Sb] [ReO$_4$] (Sb—C 2.08–2.14 Å, C—Sb—C angles 107.4–112.0°)[342]. The dicarboxylate salts, [Me$_4$Sb] [HX$_2$] for X = benzoate, *o*-phthalate and 4-ethoxysalicylate, are also ionic but anion–cation interactions between antimony and two benzoate oxygens at 3.04 and 3.40 Å in the hydrogendibenzoate salt distort the antimony 'tetrahedron' giving C—Sb—C angles ranging between 91.9 and 124.8°[343].

The planar oxinate group is bidentate in the tetramethylstibonium derivative **117** giving a six-coordinate antimony atom in distorted octahedral coordination, Sb—C distances fall between 2.101 and 2.153 Å while the Sb—O and Sb—N separations are 2.187 and 2.463 Å, respectively[344]. The naphthazarine group in **118** is a bis(chelate) and there is again distorted octahedral geometry about each antimony, although the Sb—O interaction is long (2.255 Å) pointing to bonding intermediate between ionic and covalent[345].

(117)   **(118)**

The methylene-bridged compounds, $(Me_2X_2Sb)_2CH_2$ for $X = Cl^{346}$ or Br^{347}, show trigonal bipyramidal geometry about antimony with halogen atoms in axial positions (Sb—Cl 2.446–2.480, Sb—Br 2.609–2.655 Å). The symmetry of the chloride is close to C_2 and the angle at the bridging methylene is 118.2°. The bromide unit cell contains two independent molecules with Sb—C—Sb angles of, respectively, 112.5 and 120.9°. The stiba-cyclohexane **119** contains a chair-shaped SbC_5 ring, with axial chlorines in a trigonal bipyramidal arrangement about antimony[348]; the *endo* C—Sb—C angle is, not unnaturally, constrained to 106.0° in comparison with *exo* angles at 127°.

(119)

Trigonal bipyramidal triorganoantimony dihalides have very low Lewis acidity but complexes with the following stoichiometries have been isolated: $Me_3SbCl_2\cdot SbCl_3$[349]. $Me_3SbBr_2\cdot SbBr_3$ and $Ph_3SbCl_2\cdot SbCl_3$[350]. In each case, however, complex formation depends on the interaction between axial halogens of the organoantimony(V) compound and the antimony(III) halides. In the methyl compounds, one of the axial halogens forms contacts to two trihalide molecules (at 3.308 and 3.203 Å for the chloride and 3.259 and 3.347 Å bromide) while the second axial halogen is attached to a third Sb(III) centre at 3.154 Å for the chloride and 3.302 Å for the bromide. Consequences of this interaction are significant differences in the Sb—halogen axial distances and the formation of interlinked chains of $Me_3SbX_2\cdot SbX_3$ units parallel to the *a* axis. The phenyl system is similar, but here each axial chlorine is coordinated to one $SbCl_3$ molecule giving infinite chains of $Ph_3SbCl_2\cdot SbCl_3$ units with no-cross-linking. The trigonal bipyramidal arrangement about Sb(V) is little changed on coordination but the Sb(III) geometry becomes square pyramidal with three short (mean 2.337 Å) and two longer (mean 3.262 Å) Sb—Cl contacts.

Antimony(V) chloride also forms a 1:1 complex with Ph_3SbCl_2, but here the increased Lewis acidity of antimony pentachloride leads to strong interaction with one of the axial chlorines and the product, $[Ph_3SbCl][SbCl_6]$, is basically ionic. There is a residual Sb$\cdots$Cl cation–anion interaction at 3.231 Å and the cation geometry is intermediate between the trigonal bipyramidal and tetrahedral extremes.

Diphenylantimony tribromide and acetonitrile form a weak 1:1 complex and, in keeping with the observation that the solvent molecule is readily lost even at room temperature, there is a long Sb$\cdots$N contact (2.53 Å)[351]. The antimony geometry is distorted octahedral with phenyl groups occupying *trans* positions; the Sb—Br bond *trans* to acetonitrile is substantially shorter (2.519 Å) than those *trans* to each other (2.605 Å).

The structure of the more robust diphenylantimony trichloride monohydrate has been redetermined, showing again six-fold coordination about antimony. The Sb—O distance is still rather long (2.311 Å), but the compound is stabilized by a system of O—H···Cl hydrogen bonds, involving both of the water hydrogen atoms.

A bicyclic product **120** is obtained when methylantimony dichloride is treated with the bis(Grignard) reagent $Me_2C[CH_2MgBr]_2$ followed by oxidation with sulphuryl chloride[352]. Each antimony is trigonal bipyramidal with the chlorine (Sb—Cl 2.467, 2.506 Å) and oxygen atoms (Sb—O 2.109, 2.090 Å) in axial sites. The Cl—Sb—O and Sb—O—Sb angles are 177.2 and 118.7°, respectively, and both six-membered rings have boat conformations; the structure is completed by a molecule of hydrogen chloride, which is hydrogen bonded to the bridging oxygen atom.

(120)

2. Compounds with bonds to oxygen

Rather surprisingly, oxidation of trimesitylantimony gives a discrete, non-hydrogen bonded dihydroxide, $(Mes)_3Sb(OH)_2$, rather than a dimeric oxide $(R_3SbO)_2$ usually found when the antimony substituents are sterically less demanding (see below). The geometry is only slightly distorted trigonal bipyramidal, with normal distances (Sb—O 2.027 Å) to the axial hydroxo groups[353]. On reaction with phenylsulphonic acid, the dihydroxide gives an adduct of the oxide, $(Mes)_3SbO·HSO_3SPh$, rather than a sulphonate, where the antimony coordination is distorted tetrahedral[354]. The mean C—Sb—C angle is 114.7°, giving an indication of the distortion caused by the presence of bulky substituents, and the sulphonic acid group is attached to oxygen by a short hydrogen bond (O···O 2.56 Å). The Sb—O separation (1.894 Å) is the shortest recorded distance and contains substantial double-bond character. The related trimesityl antimony hydroxo dichloroacetate[355] and 1-adamantyl carboxylate[356], on the other hand, are both trigonal bipyramidal species with apical oxygen atoms and, while the Sb—O(H) contacts (ca 2.00 Å) are normal, that to the dichloroacetate (Sb—O 2.280 Å) shows some ionic character. The situation is more extreme with $(2,6-Me_2C_6H_3)_3Sb(OH)I$ where the expected trigonal bipyramidal structure is not realised and the compound consists of discrete hydrogen-bonded $[(2,6-Me_2C_6H_3)_3 Sb(OH)]^+ ···I^-$ units[357]. The hydroxystibonium cation has distorted tetrahedral geometry (C—Sb—C ca 115°, O—Sb—C 97.4–106.9°) with Sb—O and O···I distances of 1.907 and 3.315 Å, respectively.

Dicarboxylates with the stoichiometry $R_3Sb(O_2CR')_2$ are similar to the monocarboxylates discussed above, and are based on trigonal bipyramidal geometry with apical oxygen atoms. The geometry is, however, distorted by secondary bonding between antimony and the formally double-bonded oxygen of the carboxylate groups. Among the compounds investigated are: $Ph_3Sb(O_2CPh)_2$ (Sb—O 2.13, 2.14, Sb···O 2.70, 2.81 Å)[358], the bis(2-thenoates) $Ph_3Sb(O_2C-2-C_4H_3S)_2$ (mean Sb—O 2.120, Sb···O 2.744, 2.949 Å[359], $Me_3Sb(O_2C-2-C_4H_3S)_2$ (Sb—O 2.124, 2.136, Sb···O 3.066, 3.093 Å)[360] and the disulphonate $Ph_3Sb(OSO_2Ph)_2$ (Sb—O 2.106, 2.128, Sb $\cdots$ O 3.274, 3.445 Å)[361]. In all cases, the secondary interactions lead to distortions in the equatorial angles, one of which increases at the expense of the other two.

Structures of trimethyl- and triphenylantimony bis(2-pyridinecarboxylates) are complicated, as now there is the possibility of donation to antimony from the pyridine nitrogen lone pair also[362]. As shown in Figure 8 for the phenyl compound, one of the carboxylate groups is asymmetrically chelating to give an Sb $\cdots$ O interaction at 2.721 Å while the second secondary bond is formed with a pyridine nitrogen atom at 2.602 Å. In contrast is the situation with the trimethyl analogue, where each of the two independent molecules behaves in the more conventional fashion and the two secondary bonds are to oxygen atoms of the carboxylate groups.

Monophenylantimony tetraacetate has been successfully prepared and shown to be monomeric in the solid state, with one almost symmetrically chelating (Sb—O 2.152, 2.189 Å) and three unidentate acetate groups[363]. The antimony geometry is very distorted octahedral with one of the chelate oxygen atoms *trans* to the phenyl group. Although three formally unidentate acetate groups are present (Sb—O 1.987, 2.001 and 2.029 Å), each forms a further short intramolecular contact from non-bonded oxygens (Sb $\cdots$ O 2.904, 3.332 and 2.687 Å) raising the effective antimony coordination number to nine.

There has been some progress in stabilizing square-pyramidal geometry about antimony at the expense of the more conventional trigonal bipyramid by the incorporation of chelating groups, as discussed above for related arsenic compounds. Only two examples, pentaphenylantimony itself and triphenylantimony catecholate[364], were previously known to have this unusual geometry. Of the compounds studied recently, there is effectively square-pyramidal coordination in both the phenyl(biphenylyl)tetrachlorocatecholate **121**[365] and the 4-nitrocatecholate **122**[366]. Rather surprisingly, a conventional trigonal bipyramidal structure with two phenyls and one oxygen atom in equatorial positions is found for the tetrachlorocatecholate $Ph_3Sb(O_2C_6Cl_4)$[365]. The maleonitriledithiolate, $Ph_3Sb[S_2C_2(CN)_2]$[366], and the bis(pinacolate), $(o\text{-tolyl})Sb(O_2C_2Me_4)_2$, are also trigonal bipyramidal, and an attempt to form a naphthalenediol derivative from Ph_3SbCl_2 was unsuccessful and gave instead a six-coordinate anionic compound **123**[365].

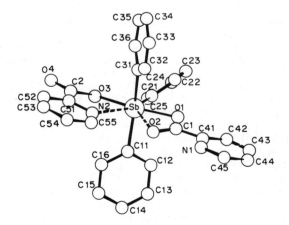

FIGURE 8. The structure of triphenylantimonybis(pyridinecarboxylate). Reproduced by permission of Barth Verlagsgesellschaft mbH from Reference 362

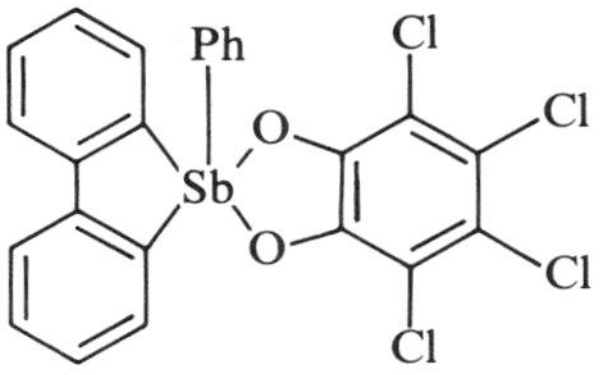

(121)

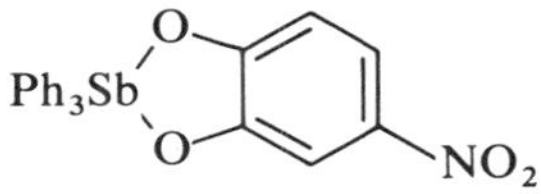

(122)

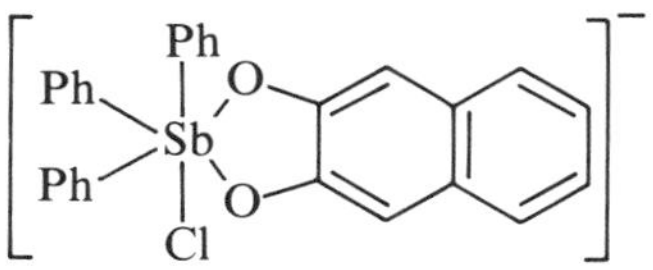

(123)

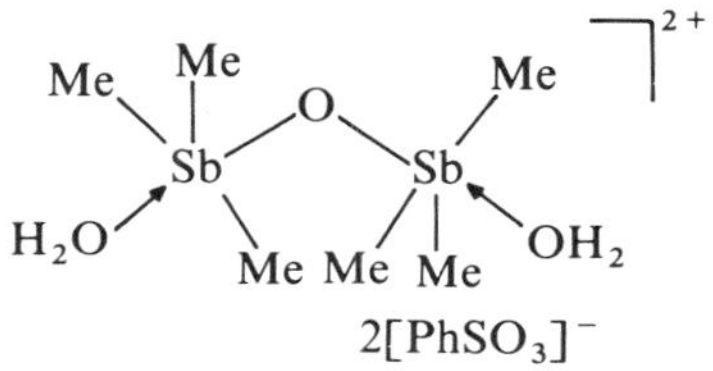

(124)

There are a number of antimony(V) compounds which contain either one or two bridging oxygen atoms and, for compounds in the former category, one of the interesting structural points is the angle at the bridging atom. In the chloride $(Ph_3SbCl)_2O$, which has trigonal bipyramidal geometry with axial oxygen and chlorine atoms, the angle is $139.0°$ [367] but in the corresponding bromide, where there are two independent molecules in the asymmetric unit, it increases to 170.2 and $176.6°$ [368]. Distances to the bridging atom are also different in the two cases, averaging 1.983 Å for the chloride and 1.944 Å for the bromide; analysis of data for a range of these compounds shows that there is no apparent correlation between Sb—O bond lengths and Sb—O—Sb angle.

It is recognized, though, that the nature of the group *trans* to oxygen is likely to have a major influence on the bridging system. The Sb—O—Sb angles in $(Ph_3SbX)_2O$, for $X = PhSO_3$ [369] and CF_3SO_3, are 139.8 and $136.5°$, respectively, and distances to the sulphonate oxygen atoms are long $(2.247, 2.280$ and $2.344, 2.370$ Å$)$ showing these ligands are unidentate with appreciable ionic character. On the other hand, the Sb—O—Sb system is required by symmetry to be linear when $X = 2$-hydroxyethanesulphonate; the Sb—O bridge bonds are short $(1.936$ Å$)$, and perhaps imply some delocalization of oxygen lone-pair density into antimony d orbitals. Bonds to the sulphonate group are again long $(2.276$ Å$)$ in agreement with some degree of ionic character [370].

An unusual ionic structure **124** is reported for a compound with the stoichiometry $(Me_3SbO_3SPh)_2O \cdot 2H_2O$, obtained by treating trimethylantimony dihydroxide with phenylsulphonic acid [371]. The cation consists of two trigonal bipyramidal antimony atoms linked by a common oxygen (Sb—O—Sb $153.2°$) with methyl groups in equatorial positions and the bridging oxygen and water molecules in the axial sites. The latter is only weakly coordinated (Sb—O 2.444 Å) but the system is stabilized by hydrogen bonding with oxygen atoms of the anions.

The first soluble antimony(V) oxide, Ph_2SbBrO, was isolated in 1985 and shown to be dimeric with a central four-membered Sb_2O_2 ring **125** [372]. Antimony is in distorted trigonal bipyramidal coordination with axial positions occupied by bromine atoms (arranged mutually *trans* to each other) and oxygens. Endocyclic angles are 78.6 and $101.5°$, respectively, at the antimony and oxygen atoms and Sb—O distances fall into two sets, the longer ones (mean 2.04 Å) are to the axial atoms and the shorter (mean 1.93 Å) to the equatorial oxygens. Oxidation of triphenylantimony gives a similar dimeric product, although in this case the asymmetric unit contains two independent molecules each lying on a centre of inversion[373]. Again there are two sets of Sb—O distances (mean values 1.928 and 2.075 Å) and mean values for the ring angles at oxygen and antimony are 102.5 and 77.5°, respectively.

The structure of what was previously thought to be diphenylstibinic anhydride and obtained here as a byproduct of the triphenylantimony oxidation reaction, has also been determined. This contains the four-membered Sb_2O_2 ring, which is now bridged as shown in **126** by two Ph_2SbO_2 groups. A similar, quadruply bridged, structure **127** is also found in the centrosymmetric, anionic molybdate, obtained in a curious reaction between tetrabutylammonium molybdate and Ph_3SbI_2[374]. Parameters in the Sb_2O_2 ring are comparable with those mentioned above and the antimony, molybdenum and oxygen atoms of the molybdate group involved in bridging are coplanar. A second centrosymmetric dimer **128**, in this case a hydroxylated anion, is obtained when methylantimony(III) dimethoxide is oxidized with hydrogen peroxide in aqueous alkali[375].

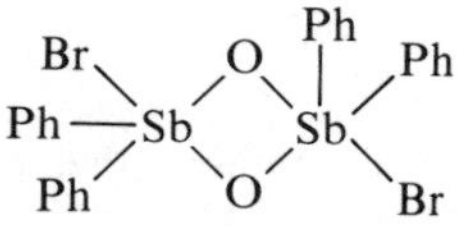

(125)

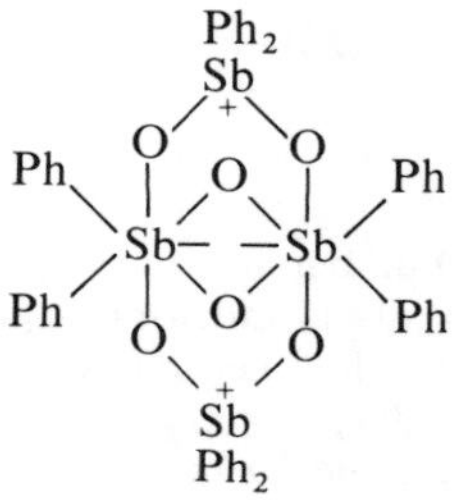

(126)

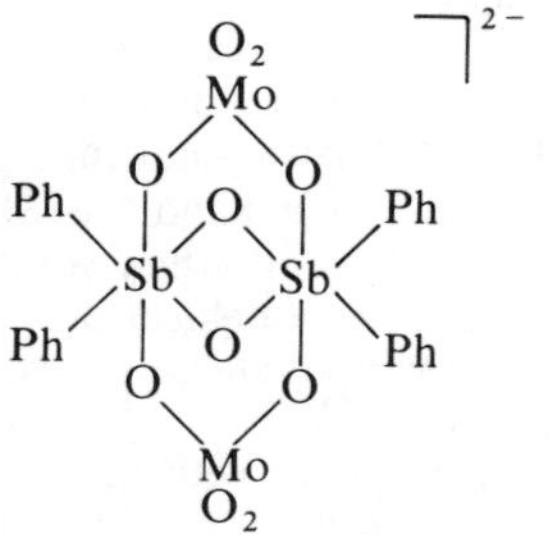

(127)

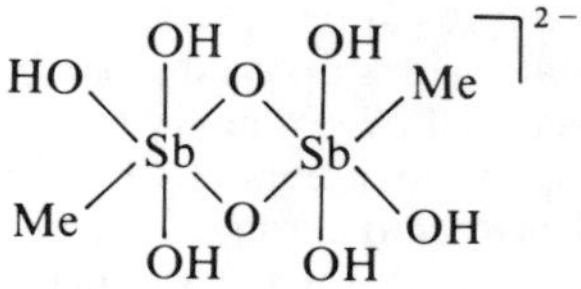

(128)

A linear tristiboxane **129**, in which each antimony is in virtually undistorted trigonal bipyramidal coordination with equatorial phenyl groups, is obtained by treating triphenylantimony oxide with 2, 4-dinitrophenylsulphonic acid[376]. The Sb—O—Sb angles are 140.8° with bridging Sb—O distances of 1.921 and 2.035 Å and, as with other sulphonates, the Sb—O distances (2.509 Å) are long, pointing to a degree of ionic character.

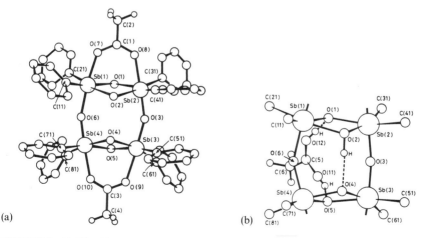

X = OSO$_2$C$_6$H$_3$(NO$_2$)$_2$-2,4

(129)

(130)

Doubly oxygen-bridged diantimony systems where the ring oxygen atoms are either protonated or methylated are also known. An example of the latter is found in **130**, which is the bromine oxidation product of MeSb(OMe)$_2$; Sb—O distances in the four-membered ring are longer (Sb—O 2.135 Å) than in the simple oxygen-bridged compounds, as one of the bonds here is formally a donor linkage, and longer than those to the terminal methoxy groups (1.962 Å). The ring angles (107.8 at oxygen and 72.2° at antimony) are comparable with those above[377].

Protonated four-membered rings are present in two complex products, obtained respectively as the final hydrolysis product of diphenylantimony triacetate[378] and from a reaction between dimeric diphenylantimony bromide oxide and silver oxalate[379]. Partial hydrolysis of diphenylantimony triacetate also gives the single oxygen-bridged [Ph$_2$Sb(OAc)$_2$]$_2$O, in which the acetate groups are unsymmetrically chelating, raising the antimony coordination number to seven. Figure 9a shows that the final hydrolysis product contains four antimony atoms, with pairs of antimony atoms linked alternately by single and double oxygen bridges. From the detail in Figure 9b, one of the oxygens in each of the doubly bridged system is, in fact, protonated and there is a complex hydrogen-bonding system, involving a molecule of acetic acid.

The oxalate product is ionic, [(Ph$_2$Sb)$_6$O$_6$(OH)$_2$(C$_2$O$_4$)] [Ph$_2$Sb(C$_2$O$_4$)$_2$]$_2$, and Figure 10 shows the structure of the complex dication; the anion is the octahedral *cis*-[Ph$_2$Sb(C$_2$O$_4$)$_2$]$^{2-}$ species. The cation is unusual as it contains a planar 12-membered centrosymmetric Sb$_6$O$_6$ ring, cross-linked by two further oxygen atoms to give two

FIGURE 9. (a) The Sb$_4$ cage structure in [Ph$_8$Sb$_4$O$_6$(HOAc)$_3$], (b) Hydrogen bonding across the Sb$_4$O$_6$ cage. Reproduced by permission of The Royal Society of Chemistry from Reference 378

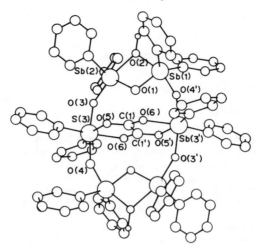

FIGURE 10. The structure of the planar oxalate cation, $[(Ph_2Sb)_6O_6(OH)_2(ox)]^{2+}$. Reproduced by permission of The Royal Society of Chemistry from Reference 379

four-membered $Sb_2(OH)O$ rings. Finally, a quadridentate oxalate group bridges across the centre of the ring. Planarity of the ring is possible as the ring oxygen atoms occupy either *trans* positions at the two octahedrally coordinated antimony atoms or axial positions at the other four antimony atoms, which are in trigonal bipyramidal coordination.

3. Miscellaneous compounds

A number of triphenylantimony ylides have been examined and, contrary to the situation with the arsenic analogues where there are substantial contributions from charged rather than double-bonded forms, the Sb—C bond lengths are significantly closer to those for a double bond. In the bis(phenylsulphonyl)methylide, $Ph_3Sb=C(SO_2Ph)_2$, the Sb—C distance is 2.024 Å (cf the Pauling values of 2.18 for a single and 1.98 Å for a double bond), but there is a short Sb...O contact (2.844 Å) to an oxygen of one of the sulphonyl groups, suggesting that canonical forms such as **131** also contribute[380]. In agreement with this is the shortening of the carbon–sulphur distance and distortion of the antimony geometry from tetrahedral towards trigonal bipyramidal. The picture in a

$$Ph_3\overset{+}{Sb} - \overset{SO_2Ph}{\underset{O^-}{\overset{|}{C}}} \overset{O}{\underset{Ph}{\overset{\nearrow}{S}}}$$

(131)

$$Ph_3Sb=C \begin{array}{c} O \\ \overset{\nwarrow}{C}-C \\ H_2 \\ \overset{\nearrow}{C}-C \\ O \end{array} \begin{array}{c} H_2 \\ CMe_2 \\ H_2 \end{array}$$

(132)

second antimony ylide **132** is very similar, with a short Sb—C bond (2.049 Å) and Sb $\cdots$ O secondary bonds (2.835 and 3.347 Å) to carbonyl oxygen atoms[196].

Antimony(V)–nitrogen compounds are not well known, but structures are available for two such species. In MeSbCl$_2$[N(SiMe$_3$)$_2$]$_2$, obtained by oxidizing MeSb [N(SiMe$_3$)$_2$]$_2$ with sulphuryl chloride, the chlorine atoms occupy axial positions (2.469 Å) and there are short distances (1.992 Å) to nitrogens of the bulky silyl ligands in two of the equatorial positions in the trigonal bipyramidal arrangement about antimony[381]. The second compound is a benzamidine complex **133**, which has distorted octahedral geometry and substantially longer Sb—N distances (2.125, 2.167 Å)[382].

(133)

Although compounds are not known in which antimony forms a stable double bond to oxygen, there is evidence that (p–d)π bonding is effective with sulphur in place of oxygen. This is confirmed by the structure of triphenylantimony sulphide, which forms isolated tetrahedral molecules where the Sb—S distance is reduced to 2.244 Å (other Sb—S distances fall in the range 2.40–2.88 Å)[383]. Angles at antimony between the phenyl groups average 106.4° but those between sulphur and phenyl are larger (mean 112.4°).

Phenylpentahalogenoantimonates, such as Cs[PhSbCl$_5$][384], K[PhSbBr$_5$][384], NH$_4$[PhSbCl$_5$][385], NH$_4$[PhSbBr$_5$][385] and Me$_2$NH$_2$[PhSbCl$_5$][385], are octahedral, with the antimony–halogen bond *trans* to phenyl longer than those *trans* to other halogen atoms. As a specific example, in Cs[PhSbCl$_5$] the *trans* bond length is 2.437 Å, compared with 2.407 and 2.414 Å for the other bonds.

CpFe(CO)(PMe$_3$)SbMe$_2$ is oxidized on treatment with R$_2$PCl to give CpFe(CO)-(PMe$_3$)SbMe$_2$Cl$_2$, which contains an Sb(V) moiety directly attached to iron[386]. The solid contains two, almost identical independent trigonal bipyramidal molecules with the chiral iron group (Sb—Fe 2.487 Å) and two methyl groups (Sb—C 2.15 Å) in equatorial positions; distances to the axial chlorines (2.60 Å) are somewhat longer than normal.

IV. BISMUTH COMPOUNDS

A. The +3 Oxidation State

The first trialkyl bismuth compound examined structurally is Bi[CH(SiMe$_3$)$_2$]$_3$, which shows the expected pyramidal geometry[387]. Both the Bi—C distances (mean 2.328 Å) and the C—Bi—C angles (mean 102.9°) are larger than those determined previously for simple aryl analogues, presumably due to steric demands from the bulky ligands, which also lead to severe ligand distortions. Similar long distances (mean 2.367 Å) and wider angles (mean 106.0°) occur in the hindered tris(trifluoromethyl)phenyl derivative, [(CF$_3$)$_3$C$_6$H$_2$]$_3$Bi[388], but the parameters decrease in the diarylmonochloride, [(CF$_3$)$_3$C$_6$H$_2$]$_2$BiCl (Bi—C mean 2.34 Å, C—Bi—C 106.9°), and revert to more usual values (mean Bi—C 2.25 Å, mean C—Bi—C 94.7°) in (p-toly)$_3$Bi[389].

Triphenylbismuth behaves as a conventional ligand (Bi—Fe 2.570 Å) in the complex cation $[CpFe(CO)_2(BiPh_3)]^+$, where the distance represents basically a simple σ Bi—Fe interaction[390].

The expected diorganobismuth chloride was not obtained from the reaction of bismuth trichloride and the highly hindered ligand $Li[C(SiMe_3)_2(C_5H_4N-2)]$, but instead the product was **134**[309]. Bismuth is in pseudo-trigonal bipyramidal coordination with two carbon atoms and the bismuth lone pair in equatorial positions; the new Bi—C bond is to an allyl ligand, with the THF solvent as the implied reagent.

(134)   **(135)**

Tetraphenyldibismuth, $Ph_2BiBiPh_2$, which is orange at room temperature and yellow at liquid nitrogen temperature, has the same staggered *trans* configuration as the antimony analogue, with pyramidal geometry at the heavy atom (C—Bi—C 98.3, Bi—Bi—C 91.6, 90.9°) and a Bi—Bi separation of 2.990 Å[391,392]. The bond angles suggest that the bismuth lone pair has essentially s character and hindrance from the phenyl groups minimizes any intermolecular Bi···Bi interaction and the compound is not dramatically thermochromic. Reaction with $Co_2(CO)_8$ breaks the Bi—Bi bond giving a product isolated as the triphenylphosphine adduct $[Co(BiPh_2)(CO)_3(PPh_3)]$; cobalt is in trigonal bipyramidal coordination with bismuth (Bi—Co 2.692 Å) and phosphorus atoms in the axial sites[393].

A second dibismuth compound, $Bi_2(SiMe_3)_4$, in which the Bi—Bi separation is 3.035 Å has short intermolecular Bi···Bi interactions leading to chains of molecules which are almost linear (Bi···Bi 3.804 Å, Bi—Bi···Bi 169°)[394]. Infinite chains of alternating $Bi(SiMe_3)_2$ and $Li(MeOCH_2CH_2OMe)$ units, each with approximately tetrahedral geometry, are present in $LiBi(SiMe_3)_2$. DME; angles in the chain at bismuth and lithium are 148 and 132°, respectively.

Carboxylate bridges between pairs of bismuth atoms in diphenylbismuth N-benzoyl-glycinate, $Ph_2Bi(O_2CCH_2NHC(O)Ph)$, give infinite chains with the oxygen atoms (Bi—O 2.396, 2.484 Å) in axial positions and the two phenyl groups (C—Bi—C 95.1°) in the equatorial positions of a pseudo-trigonal bipyramidal arrangement about the central bismuth[395]. The coordination number is, however, effectively raised further by weak interactions to an oxygen of a symmetry-related benzoyl group (3.267 Å) and to a carboxylate oxygen (3.297 Å).

Bismuth–sulphur distances in the oxadithiabismocane **135** are 2.560 and 2.602 Å, but the coordination number of the central atom is raised by both a transannular Bi···S contact at 2.97 Å, and two longer intermolecular Bi···S contacts at 3.440 and 3.509 Å[396]. The eight-membered ring has the chair-chair conformation. Both sulphur atoms of the methylxanthate ligands in $PhBi(S_2COMe)_2$ are attached to bismuth to give a square-pyramidal arrangement, but chelation is asymmetric and there are two short (2.649, 2.670 Å) and two longer Bi—S distances (2.961, 3.079 Å)[397]. The C—Bi—S angles vary between 85.5 and 95.5° and the S_2COC ligand skeletons are close to planarity. There is a similar basic unit in the related dithiocarbamate, $PhBi(S_2CNEt_2)_2$, with asymmetric coordination by each of the sulphur atoms of the ligands (Bi—S 2.778, 2.766 Å and 2.878, 2.895 Å[398]. But here, a weak intermolecular Bi···S contact at 3.421 Å leads to dimers in the solid state giving bismuth pentagonal pyramidal coordination.

The Bi—Se distance in $Ph_2BiSePh$ is 2.704 Å, with trigonal pyramidal coordination (C—Bi—C 90.8, C—Bi—Se 87.4, 97.4°) about bismuth[399].

Bismuth trichloride forms a 1:1 Menshutkin-like complex with mesitylene[324,400] and a 2:1 complex with hexamethylbenzene[322,400], in which the arene is η^6 bonded to bismuth at distances to the ring centre of ca 3.1 Å. As with the related antimony complexes, there are also a number of intermolecular chlorine bridges, raising the bismuth coordination number and leading to either sheets or tetrameric bismuth chloride networks. These compounds differ from the antimony halide analogues where the arene is usually acentrically bonded.

When aluminium trichloride is present, the $BiCl_3$–hexamethylbenzene reaction gives an orange compound with the stoichiometry $BiCl_3 \cdot AlCl_3 \cdot C_6Me_6$[401]. Its structure is based on a centrosymmetric dimer (see Figure 11), with two $AlCl_4$ groups arranged between two η^6 complexed $BiCl_2$ units. The arene group is tilted with respect to bismuth and the distance to the arene centre (2.72 Å) is markedly shorter than in the neutral compounds. If the arene is considered as occupying one coordination site, bismuth is in very distorted octahedral coordination; the compound is probably best considered as an arene-stabilized $[BiCl_2][AlCl_4]$.

There is also a 2:1 complex of bismuth trichloride with pyrene[402] and 1:1 complexes with 18-crown-6[403] and 12-crown-4[402]. [2.2.2]paracyclophane forms a complex with three molecules of the trichloride, involving all three aromatic groups[185].

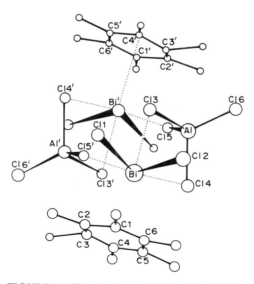

FIGURE 11. The structure of $BiCl_3 \cdot AlCl_3 \cdot C_6Me_6$. Reproduced by permission of VCH Verlags GmbH from Reference 401

B. The +5 Oxidation State

Pentaphenylbismuth is an unusual compound and its purple colour clearly differentiates it from the antimony analogue. It has, however, the same unexpected square-pyramidal structure and the regularity of the structure is shown by angles of 150.3 and 163.5° between opposite basal phenyl groups[405]. The four basal bonds are, as expected for this arrangement, significantly longer (2.32 Å) than the axial bond (2.21 Å). There are no intermolecular interactions and the nature of the chromophore is a matter of some speculation, although a charge transfer transition from the basal phenyls to bismuth is a strong possibility. A range of substituted bismuth aryls has been synthesized including: $Bi(C_6H_4Me-4)_3(C_6F_5)_2$[406], $Bi(C_6H_4F-4)_3(C_6F_5)_2$, $BiPh_3(C_6H_3F_2-2,6)_2$, $BiPh_3(C_6H_4F-2)_2$[407], $Bi(C_6H_4Me-4)_3(C_6H_4F-2)_2$, $Bi(C_6H_4Me-4)_5$[408], $Bi(C_6H_4Me-4)_3(C_6H_3F_2-2,6)_2$ and the biphenylyl derivative, $BiPh_3(C_{12}H_8)$.

Many of the compounds are coloured and dichroic and square-pyramidal geometry dominates. It is interesting to observe, however, that the structure of the *p*-tolyl compound $Bi(C_6H_4Me-4)_5$, which cocrystallizes with a molecule of THF-solvated lithium chloride, is intermediate between square-pyramidal and trigonal bipyramidal[408]. There is also an interesting example of the dramatic effect of small changes within the molecular make-up of these compounds as the colourless *p*-tolyl compound, $Bi(C_6H_4Me-4)_3(C_6H_4F-2)_2$, is trigonal bipyramidal (2-fluorophenyl groups in axial positions, Bi—C 2.358, 2.423 Å and 4-tolyl groups in equatorial sites Bi—C 2.134–2.217 Å) while the related triphenyl derivative $BiPh_3(C_6H_4F-2)_2$ is violet and square pyramidal[407].

Tris(pentafluorphenyl)bismuth difluoride Bi $(C_6F_5)_3F_2$, which cocrystallizes with two molecules of tris(pentafluorophenyl)bismuth (see Figure 12), has an almost regular trigonal bipyramidal structure (equatorial Bi—C 2.28 Å, axial Bi—F 2.088 Å), with short intermolecular contacts (Bi···F 2.759 Å) between the axial fluorine atoms and the bismuth atoms of Bi $(C_6F_5)_3$[408].

In triphenylbismuth di(2-furoate), each of the carboxylate groups is attached to bismuth by one short Bi—O bond (mean 2.296 Å) and one longer bond (mean 2.806 Å), with the weakly bonded atoms *cis* to each other[409]. It is possible to consider that these four oxygen atoms and one of the phenyl groups (Bi—C 2.215 Å) form the equatorial plane of a distorted pentagonal bipyramid about bismuth with the two phenyl groups connected to

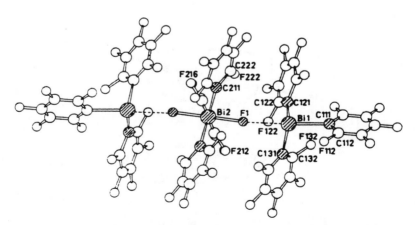

FIGURE 12. The structure of $(C_6F_5)_3BiF_2\cdot2(C_6F_5)_3Bi$. Reproduced by permission of VCH Verlags GmbH from Reference 408

bismuth by shorter bonds (2.206, 2.193 Å) occupying the apical sites. An alternative description would be in terms of a basic trigonal bipyramidal structure with axial oxygens, distorted by secondary Bi ... O bonds.

More conventional octahedral coordination, distorted due to chelation, is observed in the two quinoline derivatives **136**, R = H[410] or Me[411], where *trans* positions are occupied by oxygen (*ca* 2.18 Å) and chlorine (*ca* 2.66 Å). The bond to nitrogen is *ca* 2.75 Å.

(136)

A compound analysing as $Ph_3Bi(O_2CCF_3)$ has been identified by X-ray diffraction as an ionic species $[Ph_4Bi][Ph_2Bi(O_2CCF_3)_2]$, containing both Bi(III) and Bi(V)[410]. The cation is tetrahedral with Bi—C distances of 2.13–2.23 Å, but there is substantial distortion since two of the angles are opened to 114 and 120° as a result of residual Bi ... O interactions with oxygen atoms from the anion. The anion has the expected pseudo-trigonal bipyramidal structure with phenyl groups in equatorial (Bi—C 2.24, 2.26 Å) and oxygen atoms (2.38, 2.41 Å) in axial positions.

The structure of $Ph_4Bi[O_3SC_6H_4\text{-}4\text{-}Me]$ is best described as intermediate between five-coordinate trigonal bipyramidal, with an axial sulphonic acid group, and an ionic tetraphenyl bismuth salt[410]. The Bi—O distance is 2.77 Å and the C—Bi—C angles fall into groups ranging between 98 and 103° (between axial and equatorial sites) and 113 and 123° (between phenyls in equatorial sites).

V. REFERENCES

1. A. N. Sobolev, V. K. Belsky, N. Yu. Chernikova and F. Yu. Akhmadulina, *J. Organomet. Chem.*, **244**, 129 (1983).
2. A. N. Sobolev and V. K. Belsky, *J. Organomet. Chem.*, **214**, 41(1981).
3. J. C. Pazik and C. George, *Organometallics*, **8**, 482 (1989).
4. A. N. Sobolev, I. P. Romm, N. Yu. Chernikova, V. K. Belsky and E. N. Guryanova, *J. Organomet. Chem.*, **219**, 35 (1981).
5. A. L. Rheingold, D. L. Staley and M. E. Fountain *J. Organomet. Chem.*, **365**, 123 (1989).
6. P. Gugger, A. C. Willis and S. B. Wild, *J. Chem. Soc., Chem. Commun.*, 1169 (1990).
7. R. Appel, T. Gaitzsch and F. Knoch, *Angew. Chem. Int. Ed. Engl.*, **24**, 419 (1985).
8. W. Tumas, J. A. Suriano and R. L. Harlow, *Angew. Chem., Int. Ed. Engl.*, **29**, 75 (1990).
9. H. Grützmacher and H. Pritzkow, *Chem. Ber.*, **122**, 1417 (1989).
10. G. Sennyey, F. Mathey, J. Fischer and A. Mitschler, *Organometallics*, **2**, 298 (1983).
11. H. Schmidbaur, P. Nusstein and G. Müller, *Z. Naturforsch., Teil B*, **39**, 1456 (1984).
12. D. Hellwinkel, A. Wiel, G. Sattler and B. Nuber, *Angew. Chem. Int. Ed. Engl.*, **29**, 689 (1990).
13. I. R. Butler, W. R. Cullen, F. W. B. Einstein, S. J. Rettig and A. J. Willis, *Organometallics*, **2**, 128 (1983).
14. G. Becker, B. Becker, M. Birkhahn, O. Mundt and R. E. Schmidt, *Z. Anorg. Allg. Chem.*, **529**, 97 (1985).
15. J. R. Cannon, J. S. Edmonds, K. A. Francesconi, C. L. Raston, J. B. Saunders, B. W. Skelton and A. H. White, *Aust. J. Chem.*, **34**, 787 (1981).

16. P. de Meester, S. S. C. Chu, M. V. Jovanovic and E. R. Biehl, *J. Heterocycl. Chem.*, **22**, 1237 (1985).
17. P. de Meester, S. S. C. Chu and T. L. Chu, *Acta Crystallogr., Sect. C*, **42**, 753 (1986).
18. P. De Meester, S. S. C. Chu, M. V. Jovanovic and E. R. Biehl, *J. Heterocycl. Chem.*, **23**, 1131 (1986).
19. U. Siriwardane, A. Razzuk, S. P. Khanapure, E. R. Biehl and S. S. C. Chu., *J. Heterocycl. Chem.*, **25**, 1555 (1988).
20. W. T. Pennington, A. W. Cordes, J. C. Graham and Y. W. Jung, *Acta Crystallogr., Sect. C*, **39**, 1010 (1983).
21. R. K. Chadha, J. M. Cheayber and J. E. Drake, *J. Cryst. Spectrosc.*, **15**, 53 (1985).
22. G. Baum, A. Greiling, W. Massa, B. C. Hui and J. Lorbeth, *Z. Naturforsch., Teil B*, **44**, 560 (1989).
23. C. A. McAuliffe, B. Beagley, G. A. Gott, A. G. Mackie, P. P. MacRory and R. G. Pritchard, *Angew. Chem. Int. Ed. Engl.*, **26**, 264 (1987).
24. B. Beagley, C. B. Colburn, O. El-Sayrafi, G. A. Gott, D. G. Kelly, A. G. Mackie, C. A. MacAuliffe, P. P. MacRory and R. G. Pritchard, *Acta Crystallogr., Sect. C*, **44**, 38 (1988).
25. H. Schumann, S. Jurgis, E. Hahn, J. Pickardt, J. Blum and M. Eisen, *Chem. Ber.*, **118**, 2738 (1985).
26. G. Gremaud, H. Jungbluth, H. Syoeckli-Evans and G. Süss-Fink, *J. Organomet. Chem.*, **388**, 351 (1990).
27. L. R. Martin, F. W. B. Einstein and R. K. Pomeroy, *Inorg. Chem.*, **22**, 1959 (1983).
28. L. R. Martin, F. W. B. Einstein and R. K. Pomeroy, *Inorg. Chem.*, **24**, 2777 (1985).
29. C. C. Hinkley, M. Matusz and P. D. Robinson, *Acta Crystallogr., Sect. C*, **44**, 371 (1988).
30. M. I. Bruce, M. J. Liddell, O. bin Shawkataly, C. A. Hughes, B. W. Skelton and A. H. White, *J. Organomet. Chem.*, **347**, 207 (1988).
31. H. Lang and L. Zsolnai, *Z. Naturforsch., Teil B*, **45**, 1529 (1990).
32. R. J. H. Clark, A. J. Hempleman, H. M. Dawes, M. B. Hursthouse and C. D. Flint, *J.Chem. Soc., Dalton Trans.* 1775 (1985).
33. R. A. Cipriano, W. Levason, D. Pletcher, N. A. Powell and M. Webster, *J. Chem. Soc., Dalton Trans.*, 1901 (1987).
34. E. C. Alyea, S. A. Dias, G. Ferguson and P. Y. Siew, *Can. J. Chem.*, **61**, 257 (1983).
35. E. C. Alyea, G. Ferguson, J. Malito and B. L. Ruhl, *Can. J. Chem.*, **66**, 3162 (1988).
36. M. Nardelli, C. Pelizzi, G. Pelizzi and P.Tarasconi, *J. Chem. Soc., Dalton Trans.*, 321 (1985).
37. C. Pelizzi, G. Pelizzi and P. Tarasconi, *J. Organomet. Chem.*, **281**, 403 (1985).
38. A. Baiada, F. H. Jardine and R. D. Willet, *Inorg. Chem.*, **29**, 3042 (1990).
39. P. A. Jackson, B. F. G. Johnson, J. Lewis, A. D. Massey, D. Braga, C. Gradella and F. Grepioni, *J. Organomet. Chem.*, **391**, 225 (1990).
40. A. H. Cowley, J. G. Lasch, N. C. Norman and M. Pakulski, *J. Am. Chem. Soc.*, **105**, 5506 (1983).
41. A. H. Cowley, J, E. Kilduff, J. G. Lasch, S. K. Mehrotra, N. C. Norman, M. Pakulski, B. R. Whittlesey, J. L. Atwood and W. E. Hunter, *Inorg. Chem.*, **23**, 2582 (1984).
42. A. H. Cowley, N. C. Norman and M. Pakulski, *J. Chem. Soc., Dalton Trans.*, 383 (1985).
43. H. Lang, G. Huttner, B. Sigwarth, U. Weber, L. Zsolnai, I. Jibril and O. Orama, *Z. Naturforsch., Teil B*, **41**, 191 (1986).
44. A. H. Cowley, J. G. Lasch, N. C. Norman, M. Pakulski and B. R. Whittlesey, *J. Chem. Soc., Chem. Commun.*, 881 (1983).
45. A. H. Cowley, J. E. Kilduff, J. G. Lasch, N.C. Norman, M. Pakulski, F. Ando and T. C. Wright, *Organometallics*, **3**, 1044 (1984).
46. A. H. Cowley, J. G. Lasch, N. C. Norman and M. Pakulski, *Angew. Chem., Int. Ed. Engl.*, **22**, 978 (1983).
47. G. Huttner and I. Jibril, *Angew. Chem., Int. Ed. Engl.*, **23**, 740 (1984).
48. D. Fenske and K. Merzweiler, *Angew. Chem., Int. Ed. Engl.*, **23**, 635 (1984).
49. L.Weber, D. Bungardt, A. Müller and H. Bögge, *Organometallics*, **8**, 2008 (1989).
50. A. L. Rheingold, M. J. Foley and P. J. Sullivan, *Organometallics*, **1**, 1429 (1982).
51. R. Richter, J. Sieler, A. Richter, J. Heinicke, A. Tzschach and O. Lindqvist, *Z. Anorg. Allg. Chem.*, **501**, 146 (1983).
52. I. A. Litvinov, Yu. T. Struchkov, B. A. Arbuzov, E. N. Dianova and E. Ya. Zabotina, *Dokl. Akad. Nauk SSSR*, **268**, 885 (1983).
53. L. Weber, G. Meine and R. Boese, *Angew. Chem. Int. Ed. Engl.*, **25**, 469 (1986).
54. L. Weber, G. Meine, R. Boese and D. Bunghardt., *Z. Anorg. Allg. Chem.*, **549**, 73 (1987).
55. G. Becker and G. Gutekunst, *Z. Anorg. Allg. Chem.*, **470**, 157 (1980).

56. G. Märkl, S. Dietl, M. L. Ziegler and B. Nuber, *Angew. Chem., Int. Ed. Engl.*, **27**, 709 (1988).
57. H. Werner, W. Paul and R. Zolk, *Angew. Chem. Int. Ed. Engl.*, **23**, 626 (1984).
58. O. Mundt, H. Riffel, G. Becker and A. Simon, *Z. Naturforsch., Teil B*, **43**, 952 (1988).
59. Hong Chen, M. M. Olmstead, D. C. Pestana and P. P. Power, *Inorg. Chem.*, **30**, 1783 (1991).
60. G. Becker, G. Gutekunst and C. Witthauer, *Z. Anorg. Allg. Chem.*, **486**, 90 (1982).
61. C. Ni, Z. Zhang, Z. Xie, C. Quian and Y. Huang, *Jiegou Huaxue*, **5**, 181 (1986); *Chem. Abs.*, **107**, 145210 (1987).
62. A. J. Ashe III, W. M. Butler and T. R. Diephouse, *Organometallics*, **2**, 1005 (1983).
63. M. Baudler, Y. Aktalay, T. Heinlein and K.-F. Tebbe, *Z. Naturforsch., Teil B*, **37**, 299 (1982).
64. P. Jutzi, U. Meyer, S. Opiela, M. M. Olmstead and P. P. Power, *Organometallics*, **9**, 1459 (1990).
65. A. Boardman, R. W. H. Small and I. J. Worall, *Inorg. Chim. Acta*, **121**, L35 (1986).
66. R. Borowski, D. Rehder and K. von Deuten, *J. Organomet. Chem.*, **220**, 45 (1981).
67. F. Calderazzo, R. Poli, D. Vitali, J. D. Korp, I. Bernal, G. Pelizzi, J. L. Atwood and W. E. Hunter, *Gazz. Chim. Ital.*, **113**, 761 (1983).
68. G. Thiele, H. W. Rotter, M. Lietz and J. Ellermann, *Z. Naturforsch., Teil B*, **39**, 1344 (1984).
69. J. Ellermann, E. Köck and H. Burzlaff, *Acta Crystallogr., Sect. C*, **42**, 727 1986).
70. O. Mundt, G. Becker, H.-J. Wessely, H. J. Breunig and H. Kischkel, *Z. Anorg. Allg. Chem.*, **486**, 70 (1982).
71. R. L. Wells, Chong-yun Kwag, A. P. Purdy, A. T. McPhail and C. G. Pitt, *Polyhedron*, **9**, 319 (1990).
72. A. L. Rheingold and P.J. Sullivan, *Organometallics*, **2**, 327 (1983).
73. A. L. Rheingold and M. E. Fountain, *Organometallics*, **5**, 2410 (1986).
74. A. L. Rheingold, M. E. Fountain and A. J. DiMaio, *J. Am. Chem. Soc.*, **109**, 141 (1987).
75. A. L. Rheingold, and M. E. Fountain, *Nouv. J. Chem.*, **12**, 565 (1988).
76. P. Mercando, A. J. DiMaio and A. L. Rheingold, *Angew. Chem., Int. Ed. Engl.*, **26**, 244 (1987).
77. H. Werner, W. Paul, J. Wolf, M. Steinmetz, R. Zolk, G. Müller, O. Steigelmann and J. Riede, *Chem. Ber.*, **122**, 1061 (1989).
78. J. R. Harper, M. E. Fountain and A. L. Rheingold, *Organometallics*, **8**, 2316 (1989).
79. A. Barth, G. Huttner, M. Fritz and L. Zsolnai, *Angew. Chem., Int. Ed. Engl.*, **29**, 929 (1990).
80. M. Di Vaira, L. Niccolai, M. Peruzzini and P. Stoppioni, *Organometallics*, **4**, 1888 (1985).
81. R. A. Jones and B. R. Whittlesey, *Organometallics*, **3**, 469 (1984).
82. N. C. Payne and D. W. Stephan, *Inorg. Chem.*, **21**, 182 (1982).
83. R. R. Guimerans, F. E. Wood and A. L. Balch, *Inorg. Chem.*, **23**, 1307 (1984).
84. S. S. D. Brown, I. D. Salter, D. B. Dyson, R. V. Parish, P. A. Bates and M. B. Hursthouse, *J. Chem. Soc., Dalton Trans.*, 1795 (1988).
85. S. S. D. Brown, P. J. McCarthy, I. D. Salter, P. A. Bates, M. B. Hursthouse, I. J. Colquhoun, W. McFarlane and M. Murray, *J. Chem. Soc., Dalton Trans.*, 2787 (1988).
86. O. M. Ni Dhubhgaill, P. J. Sadler and R. Kuroda, *J. Chem. Soc., Dalton Trans.*, 2913 (1990).
87. M. T. Costello, P. E. Fanwick, M. A. Green and R. A. Walton, *Inorg. Chem.*, **30**, 861 (1991).
88. B. F. Hoskins and R. J. Steen, *Aust. J. Chem.*, **36**, 683 (1983).
89. G. Lavigne, N. Lugan and J.-J. Bonnet, *Organometallics*, **1**, 1040 (1982).
90. C. J. Janke, L. J. Tortorelli, J. L. E. Burn, C. A. Tucker and C. Woods, *Inorg. Chem.*, **25**, 4597 (1986).
91. K. Simonsen, M. Hamada, N. Suzuki, M. Kojima, S. Ohba, F. Galsbol, Y. Saito and J. Fujita, *Bull. Chem. Soc. Jpn*, **63**, 2904 (1990).
92. L. Mihichuk, M. Pizzey, B. Robertson and R. Barton, *Can. J. Chem.*, **64**, 991 (1986).
93. S. K. Manocha, L. M. Mihichuk, R. J. Barton and B. E. Robertson, *Acta Crystallogr. Sect. C*, **47**, 722 (1991).
94. R. J. Barton, S. Hu, L. M. Mihichuk and B. E. Robertson, *Inorg. Chem.*, **21**, 3731 (1982).
95. G. L. Roberts, B. W. Skelton, A. H. White and S. B. Wild, *Aust. J. Chem.*, **35**, 2193 (1982).
96. R. Colton, J. Ebner and B. F. Hoskins, *Inorg. Chem.*, **27**, 1993 (1988).
97. M. Wiederhold and U. Behrens, *J. Organomet. Chem.*, **384**, 48 (1990).
98. J. C. Calabrese, T. Herskovitz and J. B. Kinney, *J. Am. Chem. Soc.*, **105**, 5914 (1983).
99. H. C. Jewiss, W. Levason and M. Webster, *Inorg. Chem.*, **25**, 1997 (1986).
100. L. R. Hanton, W. Levason and N. A. Powell, *Inorg. Chim. Acta*, **160**, 205 (1989).
101. S. R. Hall, B. W. Skelton and A. H. White, *Aust. J. Chem.*, **36**, 271 (1983).
102. H. A. Brune, R. Klotzbücher, K. Berhalter and T. Dabaerdemaeker, *J. Organomet. Chem.*, **369**, 321 (1989).
103. D. L. Keppert, J. M. Patrick and A. H. White, *Aust. J. Chem.*, **36**, 469 (1983).

82   D. B. Sowerby

104. F. Cecconi, S. Midollini and A. Orlandini, *J. Chem. Soc., Dalton Trans.*, 2263 (1983).
105. A. L. Balch, L. A. Fossett, M. M. Olmstead, D. E. Oram and P. E. Reedy jun., *J. Am. Chem. Soc.*, **107**, 5272 (1985).
106. A. L. Balch, M. Ghedini, D. E. Oram and P. E. Reedy jun., *Inorg. Chem.*, **26**, 1223 (1987).
107. D. A. Bailey, A. L. Balch, L. A. Fossett, M. M. Olmstead and P. E. Reedy jun., *Inorg. Chem.*, **26**, 2413 (1987).
108. A. L. Balch, M. M. Olmstead and S. P. Rowley, *Inorg. Chem.*, **27**, 2275 (1988).
109. A. L. Balch, M. M. Olmstead, P. E. Reedy jun. and S. P. Rowley, *Inorg. Chem.*, **27**, 4289 (1988).
110. A. L. Balch, E. Y. Fung and M. M. Olmstead, *Inorg. Chem.*, **29**, 3203 (1990).
111. A. L. Balch, B. J. Davis, and M. M. Olmstead, *Inorg. Chem.*, **28**, 3148 (1989).
112. A. L. Balch, V. J. Catalano, M. A. Chatfield, J. K. Nagle, M. M. Olmstead and P. E. Reedy jun., *J. Am. Chem. Soc.*, **113**, 1252 (1991).
113. A. L. Balch. J. K. Nagle, D. E. Oram and P. E. Reedy jun, *J. Am. Chem. Soc.*, **110**, 454 (1988).
114. R. G. Cunninghame, A. J. Downard, L. R. Hanton, S. D. Jensen, B. H. Robinson and J.ʹSimpson, *Organometallics*, **3**, 180 (1984).
115. R. G. Cunninghame, L. R. Hanton, S. D. Jinsen, B. H. Robinson and J. Simpson, *Organometallics*, **6**, 1470 (1987).
116. A. J. DiMaio, T. E. Bitterwolf and A. L. Rheingold, *Organometallics*, **9**, 551 (1990).
117. B. Sigwarth, L. Zsolnai, O. Scheidsteger and G. Huttner, *J. Organomet. Chem.*, **235**, 43 (1982).
118. W. Deck and H. Vahrenkamp, *Z. Anorg. Allg. Chem.*, **598/599**, 83 (1991).
119. A. Winter, L. Zsolnai and G. Huttner, *J. Organomet. Chem.*, **234**, 337 (1982).
120. A. Winter, L. Zsolnai and G. Huttner, *J. Organomet. Chem.*, **250**, 409 (1983).
121. E. Gross, C. Burschka and W. Malisch, *Chem. Ber.*, **119**, 378 (1986).
122. A. M. Arif, R. A. Jones, M. H. Seeberger, B. R. Whittlesey and T. C. Wright, *Inorg. Chem.*, **25**, 3943 (1986).
123. A. J. DiMaio, S. J. Geib and A. L. Rheingold, *J. Organomet. Chem.*, **335**, 97 (1987).
124. A. M. Arif, R. A. Jones, S. T. Schwab and B. R. Whittlesey, *J. Am. Chem. Soc.*, **108**, 1703 (1986).
125. L. R. Frank, I. Jibril, L. Zsolnai and G. Huttner, *J. Organomet. Chem.*, **336**, 337 (1987).
126. R. A. Jones, S. T. Schwab, A. L. Stuart, B. R. Whittlesey and T. C. Wright, *Polyhedron*, **4**, 1689 (1985).
127. H. Schumann, E. Palamidis, J. Loebel and J. Pickard, *Organometallics*, **7**, 1008 (1988).
128. P. B. Hitchcock, M. F. Lappert, A. K. Rai and H. D. Williams, *J. Chem. Soc., Chem. Commun.*, 1633 (1986).
129. J. L. Atwood, A. H. Cowley, W. E. Hunter and S. K. Mehrotra, *Inorg. Chem.*, **21**, 1354 (1982).
130. K. von Deuten, H. Müller and G. Klar, *Cryst. Struct. Commun.*, **9**, 1081 (1980).
131. R. Bohra, H. W. Roesky, M. Noltemeyer and G. M. Sheldrick, *Acta Crystallogr., Sect. C*, **40**, 1150 (1984).
132. G. I. Kokorev, I. A. Litvinov, V. A. Naumov, S. K. Badrutdinov and F. D. Yambushev, *J. Gen. Chem.*, **59**, 1381 (1989).
133. T. Chivers, K. S. Dhathathreyan, C. Lensink and J. F. Richardson, *Inorg. Chem.*, **27**, 1570 (1988).
134. A. Gieren, H. Betz, T. Hübner, V. Lamm, M. Herberhold and K. Guldner, *Z. Anorg. Allg. Chem.*, **513**, 160 (1984).
135. M. Herberhold, W. Buhlmeyer, A. Gieren and T. Hübner, *J. Organomet. Chem.*, **321**, 37 (1987).
136. A. Gieren, T. Hübner, M. Herberhold, K. Guldner and G. Süss-Fink, *Z. Anorg. Allg. Chem.*, **544**, 137 (1987).
137. M. Herberhold, K. Schamel, G. Herrmann, A. Gieren, C. Ruiz-Perez and T. Hübner, *Z. Anorg, Allg. Chem.*, **562**, 49 (1988).
138. F. Edelmann, C. Spang, N. Noltemeyer, G. M. Sheldrick, N. Keweloh and H. W. Roesky, *Z. Naturforsch., Teil B*, **42**, 1107 (1987).
139. A. M. Arif, A. H. Cowley and M. Pakulski, *J. Chem. Soc., Chem. Commun.*, 165 (1987).
140. A. J. DiMaio and A. L. Rheingold, *Organometallics*, **10**, 3764 (1991).
141. A. L. Rheingold and A. J. DiMaio, *Organometallics*, **5**, 393 (1986).
142. R. Bohra, H. W. Roesky, M. Noltemeyer and G. M. Sheldrick, *J. Chem. Soc., Dalton Trans.*, 2011 (1984).
143. J. Ellermann, L. Brehne, E. Lindner, W. Hiller, R. Fawzi, F. L. Dickert and M. Waidhas, *J. Chem. Soc., Dalton Trans.*, 997 (1986).
144. A. A. Gazikasheva, I. A. Litvinov and V. A. Naumov, *J. Struct. Chem.*, **28**, 137 (1987).
145. E. A. Meyers, C. A. Applegate and R. A. Zingaro, *Phosphorus and Sulfur*, **29**, 317 (1987).

146. D. S. Brown, A. G. Massey and T. K. Mistry, *J. Fluorine Chem.*, **16**, 483 (1980).,
147. J. Ellermann, L. Brehm, E. Köck, E. Lidner, W. Hiller and R. Fawzi, *J. Organomet. Chem.*, **336**, 323 (1987).
148. L. M. Mihichuk, M. J. Mombourquette, F. W. B. Einstein and A. C. Willis, *Inorg. Chim. Acta*, **63**, 189 (1982).
149. M. Pizzey, L. Mihichuk, R. J. Barton and B. E. Robertson, *Can. J. Chem.*, **62**, 285 (1984).
150. A. J. DiMaio and A. L. Rheingold, *Inorg. Chem.*, **29**, 798 (1990).
151. J. T. Shore, W. T. Pennington and A. W. Cordes, *Acta Crystallogr., Sect. C*, **44**, 1831 (1988).
152. E. Adams, D. Jeter, A. W. Cordes and J. W. Kolis, *Inorg. Chem.*, **29**, 1500 (1990).
153. A. G. Osborne, R. E. Hollands, R. F. Bryan and S. Lockhart, *J. Organomet. Chem.*, **288**, 207 (1985).
154. R. K. Gupta, A. K. Rai, R. C. Mehotra, V. K. Jain, B. F. Hoskins and E. R. Tiekink, *Inorg. Chem.*, **24**, 3280 (1985).
155. E. W. Abel, M. A. Beckett, P. A. Bates and M. B. Hursthouse, *J. Organomet. Chem.*, **325**, 261 (1987).
156. A. M. Arif, R. A. Jones and K. B. Kidd, *J. Chem. Soc., Chem. Commun.*, 1440 (1986).
157. G. Becker and C. Witthauer, *Z. Anorg. Allg. Chem.*, **492**, 28 (1982).
158. R. A. Bartlett, H. V. R. Diaz, H. Hope, B. D. Murray, M. M. Olmstead and P. P. Power, *J. Am. Chem. Soc.*, **108**, 6921 (1986).
159. K. H. van Bonn, P. Schreyer, P. Paetzold and R. Boese, *Chem. Ber.*, **121**, 1045 (1988).
160. M. A. Petrie, S. C. Shoner, H. V. R. Dias and P. P. Power, *Angew. Chem. Int. Ed. Engl.*, **29**, 1033 (1990).
161. D. E. Heaton, R. A. Jones, K. B. Kidd, A. H. Cowley and C. M. Nunn, *Polyhedron*, **7**, 1901 (1988).
162. A. M. Arif, B. L. Benac, A. H. Cowley, R. Geerts, R. A. Jones, K. B. Kidd, J. M. Power and S. T. Schwab, *J. Chem. Soc., Chem. Commun.*, 1543 (1986).
163. R. L. Wells, A. P. Purdy, A. T. McPhail and C. G. Pitt, *J. Organomet. Chem.*, **308**, 281 (1986).
164. A. P. Purdy, R. L. Wells, A. T. McPhail and C. G. Pitt, *Organometallics*, **6**, 2099 (1987).
165. J. C. Pazik, C. George and A. Berry, *Inorg. Chim. Acta*, **187**, 207 (1991).
166. R. L. Wells, A. P. Purdy, K. T. Higa, A. T. McPhail and C. G. Pitt, *J. Organomet. Chem.*, **325**, C7 (1987).
167. R. L. Wells, J. W. Pasterczyk, A. T. McPhail, J. D. Johansen and A. Alvanipour, *J. Organomet. Chem.*, **407**, 17 (1991).
168. C. G. Pitt, K. T. Higa, A. T. McPhail and R. L. Wells, *Inorg. Chem.*, **25**, 2483 (1986).
169. K. T. Higa, and C. George, *Organometallics*, **9**, 275, (1990).
170. C. J. Carrano, A. H. Cowley, D. M. Giolando, R. A. Jones and C. M. Nunn, *Inorg. Chem.*, **27**, 2709 (1988).
171. R. L. Wells, A. P. Purdy, A. T. McPhail and C. G. Pitt, *J. Organomet. Chem.*, **354**, 287 (1988).
172. A. H. Cowley, R. A. Jones, M. A. Mardones and C. M. Nunn, *Organometallics*, **10**, 1635 (1991).
173. R. L. Wells, L. J. Jones, A. T. McPhail and A. Alvanipour, *Organometallics*, **10**, 2345 (1991).
174. A. H. Cowley, R. A. Jones, K. B. Kidd, C. M. Nunn and D. L. Westmoreland, *J. Organomet. Chem.*, **341**, C1 (1988).
175. A. H. Cowley, D. M. Giolando, R. A. Jones, C. M. Nunn, J. M. Power and W. W. DuMont, *Polyhedron*, **7**, 1317 (1988).
176. L. R. Frank, K. Everetz, L. Zsolnai and G.Huttner, *J. Organomet. Chem.*, **335**, 179 (1987).
177. L. Weber, D. Bunghardt and R. Boese, *Z. Anorg. Allg. Chem.*, **578**, 205 (1989).
178. I. A. Litvinov, V. A. Naumov and N. A. Chedaeva, *J. Struct. Chem.*, **27**, 174 (1986).
179. H. Hope, M. M. Olmstead, P. P. Power and X. Xu *J. Am. Chem. Soc.*, **106**, 819 (1984).
180. A. Belforte, F. Calderazzo, A. Morvillo, G. Pelizzi and D. Vitali, *Inorg. Chem.*, **23**, 1504 (1984).
181. S. C. Sendlinger, B. S. Haggerty, A. L. Rheingold and K. H. Theopold, *Chem. Ber.*, **124**, 2453 (1991).
182. P. Jutzi, T. Wippermann, C. Krüger and H.-J. Kraus, *Angew. Chem., Int. Ed. Engl.*, **22**, 250 (1983).
183. H. Schmidbaur, W. Bublak, B. Huber and G. Müller, *Angew. Chem., Int. Ed. Engl.*, **26**, 234 (1987).
184. E. Hough, D. G. Nicholson and A. K. Vasudevan, *J. Chem. Soc., Dalton, Trans.*, 427 (1987).
185. T. Probst, O. Steigelmann, J. Riede and H. Schmidbaur, *Chem. Ber.*, **124**, 1089 (1991).
186. H. Preut, R. Kasemann and D. Naumann, *Acta Crystallogr., Sect. C*, **42**, 1875 (1986).
187. U. Müller and H.-D. Dörner, *Z. Naturforsch., Teil B*, **37**, 198 (1982).
188. M. P. Bogaard, J. Petersson and A. D. Rae, *Acta Crystallogr., Sect. B*, **37**, 1357 (1981).

189. J. A. Campbell, R. Larsen, C. Campana, S. E. Cremer and A. Gamliel, *Acta Crystallogr., Sect. C*, **43**, 1912 (1987).
190. D. G. Allen, C. L. Raston, B. W. Skelton, A. H. White and S. B. Wild, *Aust. J. Chem.*, **37**, 1171 (1984).
191. A. Kostick, A. S. Secco, M. Billinghurst, D. Abrams and S. Cantor, *Acta Crystallogr., Sect. C*, **45**, 1306 (1989).
192. M. Shao, X. Jin, Y. Tang, Q. Huang and Y. Huang, *Tetrahedron Lett.*, **23**, 5343 (1982).
193. Shen Yangchang, Fan Zhaochang and Qiu Weiming, *J. Organomet. Chem.*, **320**, 21 (1987).
194. G. Ferguson, I. Gosney, D. Lloyd and B. L. Ruhl, *J. Chem. Res. (S)*, 260 (1987).
195. Fan Zhaochang and Shen Yangchang, *Acta Chim. Sinica*, **42**, 759 (1984).
196. G. Ferguson, C. Glidewell, I. Gosney, D. Lloyd, S. Metcalfe and H. Lumbroso, *J. Chem. Soc., Perkin Trans. 2*, 1829 (1988).
197. Xia Zongxiang and Zhang Zhiming, *Acta Chem. Sinica*, **41**, 577 (1983).
198. W. Keim, A. Behr, B. Limbäcker and C. Krüger, *Angew. Chem., Int. Ed. Engl.*, **22**, 503 (1983).
199. L. Weber, R. Boese and W. Meyer, *Angew. Chem., Int. Ed. Engl.*, **21**, 926 (1982).
200. L. Weber, D. Wewers, W. Meyer and R. Boese, *Chem. Ber.*, **117**, 732 (1984).
201. K. Bailey, I. Gosney, R. O. Gould, D. Lloyd and P. Taylor, *J. Chem. Res. (S)*, 386 (1988).
202. H. W. Roesky, M. Witt, W. Clegg, W. Isenberg, M. Noltemeyer and G. M. Sheldrick, *Angew. Chem., Int. Ed. Engl.*, **19**, 943 (1980).
203. M. Witt, H. W. Roesky, M. Noltemeyer, W. Clegg, M. Schmidt and G. M. Sheldrick, *Angew. Chem., Int. Ed. Engl.*, **20**, 974 (1981).
204. H. W. Roesky, N. Bertel, F. Edelmann, M. Noltemeyer and G. M. Sheldrick, *Z. Naturforsch., Teil B*, **43**, 72 (1988).
205. H. W. Roesky, R. Bohra and W. S. Sheldrick, *J. Fluorine Chem.*, **22**, 199 (1983).
206. R. Bohra, H. W. Roesky, J. Lucas, M. Noltemeyer and G. M. Sheldrick, *J. Chem. Soc., Dalton, Trans.*, 1011 (1983).
207. E. Irmer, G. M. Sheldrick, S. S. Parmar and H. K. Saluja, *Acta. Crystallogr., Sect. C*, **44**, 2024 (1988).
208. J. D. Korp, I. Bernal, L. Avens and J. L. Mills, *J. Cryst. Spectrosc.*, **13**, 263 (1983).
209. V. K. Belsky and V. E. Zavodnik, *J. Organomet. Chem.*, **265**, 159 (1984).
210. V. K. Belsky, *J. Organomet. Chem.*, **213**, 435 (1981).
211. C. Lariucci, R. H. de Almeida Santos and J. R. Lechat, *Acta Crystallogr., Sect. C*, **42**, 731 (1986).
212. G. Ferguson, A. J. Lough and C. Glidewell, *J. Chem. Soc., Perkin Trans. 2*, 2065 (1989).
213. G. Ferguson and C. Glidewell, *J. Chem. Soc., Perkin Trans. 2*, 2129 (1988).
214. C. Glidewell, G. S. Harris, H. D. Holden, D. C. Liles and J. S. McKechnie, *J. Fluorine Chem.*, **18**, 143 (1981).
215. B. Beagley, O. El-Sayrafi, G. A. Gott, D. G. Kelly, C. A. McAuliffe, A. G. Mackie, P. P. MacRory and R. G. Pritchard, *J. Chem. Soc., Dalton Trans.*, 1095 (1988).
216. P. G. Jones, A. Olbrich, R. Schelbach and E. Schwartzmann, *Acta Crystallogr., Sect. C*, **44**, 2201 (1988).
217. G. I. Kokorev, I. A. Litvinov, V. A. Naumov and F. D. Yambushev, *Zh. Obshch. Khim.*, **57**, 354 (1987).
218. B. Beagley, D. G. Kelly, P. P. McRory, C. A. McAuliffe and R. G. Pritchard, *J. Chem. Soc., Dalton, Trans.*, 2657 (1990).
219. J. W. Pasterczyk and A. R. Barron, *J. Cryst. Spectrosc.*, **20**, 85 (1990).
220. J. W. Pasterczyk, A. M. Arif and A. R. Barron, *J. Chem. Soc., Chem. Commun.*, 829 (1989).
221. J. S. Edmonds, K. A. Francesconi, P. C. Healy and A. H. White, *J. Chem. Soc., Perkin Trans., 1*, 2989 (1982).
222. K. A. Francesconi, J. S. Edmonds, R. V. Stick, B. W. Skelton and A. H. White, *J. Chem. Soc., Perkin Trans., 1*, 2707 (1991).
223. H. K. Bathla, S. S. Parmar, H. K. Saluja, A. M. Z. Slavin and D. J. Williams, *J. Chem. Soc., Chem. Commun.*, 685 (1987).
224. N. Burford, R. E. v. H. Spence, A. Linden and T. S. Cameron, *Acta Crystallogr., Sect. C*, **46**, 92 (1990).
225. A. G. Groves, W. T. Robinson and C. J. Wilkins, *Inorg. Chim. Acta*, **114**, L29 (1986).
226. W. T. Robinson, C. J. Wilkins and Zhang Zeying, *J. Chem. Soc., Dalton Trans.*, 219 (1990).
227. C. Pelizzi, G. Pelizzi and P. Tarasconi, *J. Chem. Soc., Dalton Trans.*, 2689 (1983).
228. Ng Wee Kong, Chen Wei, V. G. K. Das and R. J. Butcher, *J. Organomet. Chem.*, **361**, 53 (1989).

229. G. Oliva, E. E. Castellano, J. Zukerman-Schpector and A. C. Massabni, *Inorg. Chim. Acta*, **89**, 9 (1984).
230. C. M. de Paula Marques and K. Tomita, *J. Coord. Chem.*, **21**, 367 (1990).
231. L. R. Falvello, M. Gerlock and P. R. Raithby, *Acta Crystallogr., Sect. C*, **43**, 2029 (1987).
232. J. H. Binks, G. J. Dorward, R. A. Howie and G. P. McQuillan, *Inorg. Chim. Acta*, **49**, 251 (1981).
233. R. H. P. Francisco, R. H. de Almeida Santos, J. R. Lechat and A. C. Massabni, *Acta Crystallogr., Sect. B*, **37**, 232 (1981).
234. R. R. Ryan, E. M. Larson, G. F. Payne and J. R. Peterson, *Inorg. Chim. Acta*, **131**, 267 (1987).
235. U. Cassellato, R. Graziani, U. Russo and B. Zarli, *Inorg. Chim. Acta*, **166**, 9 (1989).
236. J. F. de Wet and M. R. Caira, *J. Chem. Soc., Dalton. Trans.*, 2043 (1986).
237. C. Pelizzi and G. Pelizzi, *J. Chem. Soc., Dalton Trans.*, 847 (1983).
238. S. Dondi, M. Nardelli, C. Pelizzi, G. Pelizzi and G. Predieri, *J. Organomet. Chem.*, **308**, 195 (1986).
239. G. Pelizzi, P. Tarasconi, F. Vitali and C. Pelizzi, *Acta Crystallogr., Sect. C*, **43**, 1505 (1987).
240. R. R. Holmes, *Prog. Inorg. Chem.*, **32**, 119 (1984).
241. C. A. Poutasse, R. O. Day, J. M. Holmes and R. R. Holmes, *Organometallics*, **4**, 708 (1985).
242. R. R. Holmes, R. O. Day and A. C. Sau, *Organometallics*, **4**, 714 (1985).
243. R. O. Day, J. M. Holmes, A. C. Sau, J. R. Devillers, R. R. Holmes and J. A. Deiters, *J. Am. Chem. Soc.*, **104**, 2127 (1982).
244. R. H. Fish and R. S. Tannous, *Organometallics*, **1**, 1238 (1982).
245. S. V. L. Narayan and H. N. Shrivastava, *Acta Crystallogr., Sect. B*, **37**, 1186 (1981).
246. V. E. Zavodnik, V. K. Belsky and Yu. G. Galyametdinov, *J. Organomet. Chem.*, **226**, 41 (1982).
247. V. V. Tkachev, V. A. Perov and Yu. F. Gatilov, *Zh. Strukt. Khim.*, **25**, 163 (1984).
248. D. H. Brown, A. F. Cameron, R. J. Cross and M. McLaren, *J. Chem. Soc., Dalton Trans.*, 1459 (1981).
249. W. T. Robinson, C. J. Wilkins and Zhang Zeying, *J. Chem. Soc., Dalton Trans.*, 2187 (1988).
250. P. C. Tellinghuisen, W. T. Robinson and C. J. Wilkins, *J. Chem. Soc., Dalton Trans.*, 1289 (1985).
251. W. Malisch, M. Luksza and W. S. Sheldrick, *Z. Naturforsch., Teil B*, **36**, 1580 (1981).
252. L. Silaghi-Dumitrescu, I. Haiduc and J. Weiss, *J. Organomet. Chem.*, **263**, 159 (1984).
253. C. Silvestru, L. Silaghi-Dumitrescu, I. Haiduc, M. J. Begley, M. Nunn and D. B. Sowerby, *J. Chem. Soc., Dalton Trans.*, 1031 (1986).
254. E. W. Abel, M. A. Beckett, P. A. Bates and M. B. Hursthouse, *Polyhedron*, **7**, 1855 (1990).
255. E. A. Adams, J. W. Kolis and W. T. Pennington, *Acta Crystallogr., Sect. C*, **46**, 917 (1990).
256. A. N. Sobolev, I. P. Romm, V. K. Belsky, O. P. Syutkina and E. N. Guryanova, *J. Organomet. Chem.*, **179**, 153 (1979).
257. A. N. Sobolev, I. P. Romm, V. K. Belsky, O. P. Syutkina and E. N. Guryanova, *J. Organomet. Chem.*, **209**, 49 (1981).
258. H. Schmidbaur, B. Milewski-Mahrla, A. G. Müller and C. Krüger, *Organometallics*, **3**, 38 (1984).
259. A. H. Cowley, R. A. Jones, C. M. Nunn, and D. L. Westmoreland, *Angew. Chem., Int. Ed. Engl.*, **28**, 1018 (1989).
260. M. Birkhahn, P. Krommes, W. Masa and J. Lorbeth, *J. Organomet. Chem.*, **208**, 161 (1981).
261. W. Frank, *J. Organomet. Chem.*, **406**, 331 (1991).
262. A. J. Ashe III, T. R. Diephouse, J. W. Kampf, and S. M. Al-Taweel, *Organometallics*, **10**, 2068 (1991).
263. S. L. Buchwald, R. A. Fisher and B. M. Foxman, *Angew. Chem., Int. Ed. Engl.*, **29**, 771 (1990).
264. M. T. Costello, P. E. Fanwick, K. E. Meyer and R. A. Walton, *Inorg. Chem.*, **29**, 4437 (1990).
265. U. Flörke, M. Woyciechowski and H-J. Haupt, *Acta Crystallogr., Sect. C*, **44**, 2101 (1988).
266. D. Fenske, N. Mronga and K. Dehnicke, *Z. Anorg. Allg. Chem.*, **498**, 131 (1983).
267. C. C. Hinkley, M. Matusz and P. D. Robinson, *Acta Crystallogr., Sect. C*, **44**, 1829 (1988).
268. R. Cini, G. Giorgi and E. Periccioli, *Acta Crystallogr., Sect. C*, **47**, 716 (1991).
269. A. R. Chakravarty, F. A. Cotton, D.A. Tocher and J. H. Tocher, *Inorg. Chim. Acta*, **101**, 185 (1985).
270. D. H. R. Barton, J. Khamsi, N. Ozbalik and J. Reibenspies, *Tetrahedron*, **46**, 3111 (1990).
271. A. L. Rheingold and M. E. Fountain, *J. Cryst. Spectrosc.*, **14**, 549 (1984).
272. P. G. Jones, *Z. Naturforsch., Teil B*, **37**, 937 (1982).
273. J. Vicente, A. Arcas, P. G. Jones and J. Lautner, *J. Chem. Soc., Dalton Trans.*, 451 (1990).
274. A. Baiada, F. H. Jardine, R. D. Willett and K. Emerson, *Inorg. Chem.*, **30**, 1365 (1991).
275. A. L. Rheingold and M. E. Fountain, *Acta Crystallogr., Sect. C*, **41**, 1162 (1985).
276. H. C. Jewis, W. Levason, M. D. Spicer and M. Webster, *Inorg. Chem.*, **26**, 2102 (1987).

277. G. Huttner, U. Weber, B. Sigwarth and O. Scheidsteger, *Angew. Chem., Int. Ed. Engl.*, **21**, 215 (1982).
278. U. Weber, G. Huttner, O. Scheidsteger and L. Zsolnai, *J. Organomet. Chem.*, **289**, 357 (1985).
279. A. H. Cowley, N. C. Norman and M. Pakulski, *J. Am. Chem. Soc.*, **106**, 6844 (1984).
280. A. H. Cowley, N. C. Norman, M. Pakulski, D. L. Bricker and D. H. Russell, *J. Am. Chem. Soc.*, **107**, 8211 (1985).
281. R. E. Desenfants II, J. A. Gaveney jun., R. K. Hayashi, A. D. Rae, L. F. Dahl and A. Bjarnson, *J. Organomet. Chem.*, **383**, 543 (1990).
282. B. F. G. Johnson, J. Lewis, A. J. Whitton and S. G. Bott, *J. Organomet. Chem.*, **389**, 129 (1990).
283. A. J. Ashe III, E. G. Ludwig jun., J. Oleksyszyn and J. C. Huffman, *Organometallics*, **3**, 337 (1984).
284. O. Mundt, H. Riffel, G. Becker and A. Simon, *Z. Naturforsch., Teil B*, **39**, 317 (1984).
285. G. Becker, H. Freudenblum and C. Witthauer, *Z. Anorg. Allg. Chem.*, **492**. 37 (1982).
286. G. Becker, M. Meiser, O. Mundt and J. Weidlein, *Z. Anorg. Allg. Chem.*, **569**, 62 (1989).
287. S. Roller, M. Dräger, H. J. Breunig, M. Ates and S. Gülec, *J. Organomet. Chem.*, **378**, 327 (1989).
288. S. Roller, M. Dräger, H. J. Breunig, M. Ates and S. Gülec., *J. Organomet. Chem.*, **329**, 319 (1987).
289. A. J. Ashe III, W. Butler and T. R. Diephouse, *J. Am. Chem. Soc.*, **103**, 207 (1981).
290. K. von Deuten and D. Rehder, *Cryst. Struct. Commun.*, **9**, 167 (1980).
291. A. H. Cowley, C. M. Nunn and D. L. Westmoreland, *Acta Crystallogr., Sect. C*, **46**, 774 (1990).
292. J. Ellermann, E. Köck and H. Burzlaff, *Acta Crystallogr., Sect. C*, **41**, 1437 (1985).
293. M. Ates, H. J. Breunig, S. Gülec, W. Offermann, K. Häberle and M. Dräger, *Chem. Ber.*, **122**, 473 (1989).
294. H. J. Breunig, K. Häberle, M. Dräger and T. Severengiz, *Angew. Chem., Int. Ed. Engl.*, **24**, 72 (1985).
295. H. J. Breunig, A. Solani-Neshan, K. Häberle and M. Dräger, *Z. Naturforsch., Teil B*, **41**, 327 (1986).
296. I. Bernal, J. D. Korp, F. Calderazzo, R. Poli and I. D. Vitali, *J. Chem. Soc., Dalton Trans.*, 1945 (1984).
297. M. Hall, D. B. Sowerby and C. P. Falshaw, *J. Organomet. Chem.*, **315**, 321 (1986).
298. M. Wieber, D. Wirth and C. Burchka, *Z. Anorg. Allg. Chem.*, **505**, 141 (1983).
299. M. Wieber, D. Wirth, J. Metter and C. Burschka, *Z. Anorg. Allg. Chem.*, **520**, 65 (1985).
300. M. J. Begley, D. B. Sowerby, D. M. Wesolek, C. Silvestru and I. Haiduc, *J. Organomet. Chem.*, **316**, 281 (1986).
301. H. M. Hoffmann and M. Dräger, *J. Organomet. Chem.*, **295**, 33 (1985).
302. H. M. Hoffmann and M. Dräger, *J. Organomet. Chem.*, **320**, 273 (1987).
303. H. M. Hoffmann and M. Dräger, *J. Organomet. Chem.*, **329**, 51 (1987).
304. H. Preut, U. Praekel and F. Huber, *Acta Crystallogr., Sect. C*, **42**, 1138 (1986).
305. H. Preut, F. Huber and K. H. Hengstmann, *Acta Crystallogr., Sect. C*, **44**, 468 (1988).
306. K.-H. Hengstmann, F. Huber and H. Preut, *Acta Crystallogr., Sect. C*, **47**, 2029 (1991).
307. R. A. Fisher, R. B. Nielsen, W. M. Davis and S. L. Buchwald, *J. Am. Chem. Soc.*, **113**, 165 (1991).
308. M. Wieber, H. Höhl and C. Burschka, *Z. Anorg. Allg. Chem.*, **583**, 113 (1990).
309. C. Jones, L. M. Engelhard, P. C. Junk, D. S. Hutchings, W. C. Patalinghug, C. L. Raston and A. H. White, *J. Chem. Soc., Chem. Commun.*, 1560 (1991).
310. M. Hall and D. B. Sowerby, *J. Organomet. Chem.*, **347**, 59 (1988).
311. H. Preut, F. Huber and O. Alonzo, *Acta Crystallogr., Sect. C*, **43**, 46 (1987).
312. F. Calderazzo, F. Marchetti, F. Ungari and M. Wieber, *Gazz. Chim. Ital.*, **121**, 93 (1991).
313. W. S. Sheldrick and C. Martin, *Z. Naturforsch., Teil B*, **46**, 639 (1991).
314. J. von Seyerl, O. Scheidsteger, H. Berke and G. Huttner, *J. Organomet. Chem.*, **311**, 85 (1986).
315. G. Becker, A. Münch and C. Witthauer, *Z. Anorg. Allg. Chem.*, **492**, 15 (1982).
316. A. R. Barron, A. H. Cowley, R. A. Jones, C. M. Nunn and D. L. Westmoreland, *Polyhedron*, **7**, 77 (1988).
317. J. Müller, U. Müller, A. Loss, J. Lorbeth, H. Donath and W. Massa, *Z. Naturforsch., Teil B*, **40**, 1320 (1985).
318. G. E. Forster, I. G. Southerington, M. J. Begley and D. B. Sowerby, *J. Chem. Soc., Chem. Commun.*, 54 (1991).
319. B. Ross, J. Belz and M. Nieger, *Chem. Ber.*, **123**, 975 (1990).
320. M. Wieber, D. Wirth and C. Burschka, *Z. Naturforsch., Teil B*, **39**, 600 (1984).
321. D. Mootz and V. Händler, *Z. Anorg. Allg. Chem.*, **533**, 23 (1986).

322. H. Schmidbaur, R. Nowak, A. Schier, J. M. Wallis, B. Huber and G. Müller, *Chem. Ber.*, **120**, 1829 (1987).
323. H. Schmidbaur, R. Novak, O. Steigelmann and G. Müller, *Chem. Ber.*, **123**, 1221 (1990).
324. H. Schmidbaur, J. M. Wallis, R. Novak, B. Huber and G. Müller, *Chem. Ber.*, **120**, 1837 (1987).
325. H. Schmidbaur, R. Novak, B. Huber and G. Müller, *Organometallics*, **6**, 2266 (1987).
326. A. Lipka and D. Mootz, *Z. Naturforsch., Teil B*, **37**, 695 (1982).
327. L. Korte, A. Lipka and D. Mootz, *Z. Anorg. Allg. Chem.*, **524**, 157 (1985).
328. H. Schmidbaur, R. Nowak, O. Steigelmann and G. Müller, *Chem. Ber.*, **123**, 19 (1990).
329. D. Mootz and V. Händler, *Z. Anorg. Allg. Chem.*, **521**, 122 (1985).
330. E. Hough, D. G. Nicholson and A. K. Vasudevan, *J. Chem. Soc. Dalton Trans.*, 427 (1987).
331. M. Schäfer, J. Pebler, B. Borgsen, F. Weller and K. Dehnicke, *Z. Naturforsch., Teil B*, **45**, 1243 (1990).
332. N. W. Alcock, M. Ravindran, S. M. Roe and G. R. Willey, *Inorg. Chim. Acta*, **167**, 115 (1990).
333. W. Tempel, W. Schwarz and J. Weidlein, *Z. Anorg. Allg. Chem.*, **474**, 157 (1981).
334. P. L. Millington and D. B. Sowerby, *J. Chem. Soc., Dalton Trans.*, 2011 (1981).
335. V. A. Lebedev, R. I. Bochkova, E. A. Kuz'min, V. V. Sharutin and N. V. Belov, *Sov. Phys. Dokl.*, **26**, 920 (1981).
336. K. N. Akatova, R. I. Bochkova, V. A. Lebedev, V. V. Sharutin and N. V. Belov, *Sov. Phys. Dokl.*, **28**, 97 (1983).
337. O. Knop, B. R. Vincent and T. S. Cameron, *Can. J. Chem.*, **67**, 63 (1989).
338. G. Ferguson, C. Glidewell, I. Gosney, D. Lloyd, S. Metcalfe and H. Lumroso, *J. Chem. Soc., Perkin Trans.*, 2, 731 (1988).
339. R. Rüther, F. Huber and H. Preut, *J. Organomet. Chem.*, **295**, 21 (1985).
340. S. P. Bone and D. B. Sowerby, *Phosphorus, Sulphur and Silicon*, **45**, 23 (1989).
341. H. Preut, R. Rüther, F. Huber and A. Blaschette, *Acta Crystallogr., Sect. C*, **45**, 1006 (1989).
342. P. K. Burkert, M. Grommelt, T. T. Pietrass, J. Lachmann and G. Müller, *Z. Naturforsch., Teil B*, **45**, 725 (1990).
343. B. Milewski-Mahrla and H. Schmidbaur, *Z. Naturforsch., Teil B*, **37**, 1393 (1982).
344. H. Schmidbaur, B. Milewski-Marla and F. E. Wagner, *Z. Naturforsch., Teil B*, **38**,1477 (1983).
345. S. Arnold, V. Mansel and G. Klar, *Z. Naturforsch., Teil B*, **45**, 369 (1990).
346. W. Kolondra, W. Schwarz and J. Weidlein, *Z. Anorg. Allg. Chem.*, **501**, 137 (1983).
347. W. Schwarz, W. Kolondra and J. Weidlein, *J. Organomet. Chem.*, **260**, C1 (1984).
348. A. L. Spek, G. Roelofsen, H. A. Meinema and J. G. Noltes, *Acta Crystallogr., Sect. C*, **43**, 1688 (1987).
349. J. Werner, W. Schwarz and A. Schmidt, *Z. Naturforsch., Teil B*, **36**, 556 (1981).
350. M. Hall and D. B. Sowerby, *J. Chem. Soc., Dalton Trans.*, 1095 (1983).
351. T. T. Bamgboye, M. J. Begley and D. B. Sowerby, *J. Organomet. Chem.*, **362**, 77 (1989).
352. M. A. G. M. Tinga, M. K. Groeneveld, O. S. Akkerman, F. Blickelhaupt, W. J. J. Smeets and A. L. Spec, *Recl. Trav. Chim. Pays-Bas*, **110**, 290 (1991).
353. T. Westhoff, F. Huber, R. Rüther and H. Preut, *J. Organomet. Chem.*, **352**, 107 (1988).
354. F. Huber, T. Westhoff and H. Preut, *J. Organomet. Chem.*, **323**, 173 (1987).
355. H. Preut, T. Westhoff and F. Huber, *Acta Crystallogr., Sect. C*, **45**, 49 (1989).
356. T. Westhoff, F. Huber and H. Preut, *J. Organomet. Chem.*, **348**, 185 (1988).
357. G. Ferguson, G. S. Harris and A. Khan, *Acta Crystallogr., Sect. C*, **43**, 2078 (1987).
358. V. A. Lebedev, R. I. Bochkova, L. F. Kuzubova, E. A. Kuz'min, V. V. Sharutin and N. V. Belov, *Sov. Phys. Dokl.*, **27**, 519 (1983).
359. M. Domagala, F. Huber and H. Preut, *Z. Anorg. Allg. Chem.*, **574**, 130 (1989).
360. H. Preut, M. Domagala and F. Huber, *Acta Crystallogr., Sect. C*, **43**, 416 (1987).
361. R. Rüther, F. Huber and H. Preut, *Z. Anorg. Allg. Chem.*, **539**, 110 (1986).
362. M. Domagala, F. Huber and H. Preut, *Z. Anorg. Allg. Chem.*, **582**, 37 (1990).
363. M. Wieber, I. Fetwer-Kremling, H. Reith and C. Burschka, *Z. Naturforsch., Teil B*, **42**, 815 (1987).
364. M. Hall and D. B. Sowerby, *J. Am. Chem. Soc.*, **102**, 628 (1980).
365. R. R. Holmes, R. O. Day, V. Chandrasekhar and J. M. Holmes, *Inorg. Chem.*, **26**, 157 (1987).
366. R. R. Holmes, R. O. Day, V. Chandrasekhar and J. M. Holmes, *Inorg. Chem.*, **26**, 163 (1987).
367. E. R. T. Tiekink, *J. Organomet. Chem.*, **333**, 199 (1987).
368. A. Ouchi and S. Sato, *Bull. Chem. Soc. Jpn.*, **61**, 1806 (1988).
369. H. Preut, R. Rüther and F. Huber, *Acta Crystallogr., Sect. C*, **42**, 1154 (1986).

370. H. Preut, R. Rüther and F. Huber, *Acta Crystallogr., Sect. C*, **41**, 358 (1985).
371. R. Rüther, F. Huber and H. Preut, *J. Organomet. Chem.*, **342**, 185 (1988).
372. D. M. Wesolek, D. B. Sowerby and M. J. Begley, *J. Organomet. Chem.*, **293**, C5 (1985).
373. J. Bordner, G. O. Doak and T. S. Everett, *J. Am. Chem. Soc.*, **108**, 4206 (1986).
374. Ben-Yao Liu, Yih-Tong Ku, Ming Wang, Bo-Yi Wang and Pei-Ju Zheng, *J. Chem. Soc., Chem. Commun.*, 651 (1989).
375. M. Wieber, U. Simonis and D. Kraft, *Z. Naturforsch., Teil B*, **46**, 139 (1991).
376. R. Rüther, F. Huber and H. Preut, *Angew. Chem., Int. Ed. Engl.*, **26**, 906 (1987).
377. M. Wieber, J. Walz and C. Burschka, *Z. Anorg. Allg. Chem.*, **585**, 65 (1990).
378. D. B. Sowerby, M. J. Begley and P. L. Millington, *J. Chem. Soc., Commun.*, 896 (1984).
379. I. G. Southerington, M. J. Begley and D. B. Sowerby, *J. Chem. Soc., Chem. Commun.*, 1555 (1991).
380. G. Ferguson, C. Glidewell, D. Lloyd and S. Metcalfe, *J. Chem. Res. (S)*, 32 (1987).
381. W. Kolondra, W. Schwarz and J. Weidlein, *Z. Naturforsch., Teil B*, **40**, 872 (1985).
382. F. Weller, J. Pebler, K. Dehnicke, K. Hartke and H-M. Wolff, *Z. Anorg. Allg. Chem.*, **486**, 61 (1982).
383. J. Pebler, F. Weller and K. Dehnicke, *Z. Anorg. Allg. Chem.*, **492**, 139 (1982).
384. E. G. Zaitseva, S. V. Medvedev and L. A. Aslanov, *Zh. Strukt. Khim.*, **31**, 104 (1990).
385. E. G. Zaitseva, S. V. Medvedev and L. A. Aslanov, *Zh. Strukt. Khim.*, **31**, 110 (1990).
386. W. Malisch, H-A Kaul, E. Gross and U. Thewalt, *Angew. Chem., Int. Ed. Engl.*, **21**, 549 (1982).
387. B. Murray, J. Hvoslef, H. Hope and P. P. Power, *Inorg. Chem.*, **22**, 3421 (1983).
388. K. H. Whitmire, D. Labahn, H. W. Roesky, M. Noltemeyer and G. M. Sheldrick, *J. Organomet. Chem.*, **402**, 55 (1991).
389. A. N. Sobolev, V. K. Belsky and I. P. Romm, *Koord. Khim.*, **9**, 262 (1983).
390. H. Schumann and L. Eguren, *J. Organomet. Chem.*, **403**, 183 (1991).
391. F. Calderazzo, A. Morvillo, G. Pelizzi and R. Poli, *J. Chem. Soc., Chem. Commun.*, 507 (1983).
392. F. Calderazzo, R. Poli and G. Pelizzi, *J. Chem. Soc., Dalton Trans.*, 2365 (1984).
393. F. Calderazzo, R. Poli and G. Pelizzi, *J. Chem. Soc., Dalton Trans.*, 2535 (1984).
394. O. Mundt, G. Becker, M. Rössler and C. Witthauer, *Z. Anorg. Allg. Chem.*, **506**, 42 (1983).
395. F. Huber, M. Domagala and H. Preut, *Acta Crystallogr., Sect. C*, **44**, 828 (1988).
396. M. Dräger and B. M. Schmidt, *J. Organomet. Chem.*, **290**, 133 (1985).
397. C. Burschka, *Z. Anorg. Allg. Chem.*, **485**, 217 (1982).
398. M. Ali, W. R. McWhinnie, A. A. West and T. A. Harmor, *J. Chem. Soc., Dalton Trans.*, 899 (1990).
399. F. Calderazzo, A. Morvillo, G. Pelizzi, R. Poli and F. Ungari, *Inorg. Chem.*, **27**, 3730 (1988).
400. A. Schier, J. M. Wallis, G. Müller and H. Schmidbaur, *Angew. Chem., Int. Ed. Engl.*, **25**, 757 (1986).
401. W. Frank, J. Weber and E. Fuchs, *Angew. Chem., Int. Ed. Engl.*, **26**, 74 (1987).
402. I. M. Vezzosi, A. F. Zanoli, L. P. Battaglia and A. B. Corradi, *J. Chem. Soc., Dalton Trans.*, 191 (1988).
403. M. G. B. Drew, D. G. Nicholson, I. Sylte and A. Vasudevan, *Inorg. Chim. Acta*, **171**, 11 (1990).
404. N. W. Alcock, M. Ravindran and G. R. Willey, *J. Chem. Soc., Chem. Commun.*, 1063 (1989).
405. A. Schmuck, J. Buschmann, J. Fuchs and K. Seppelt, *Angew. Chem., Int. Ed. Engl.*, **26**, 1180 (1987).
406. A. Schmuck and K. Seppelt, *Chem. Ber.*, **122**, 803 (1989).
407. A. Schmuck, P. Pyykkö and K. Seppelt, *Angew. Chem., Int, Ed. Engl.*, **29**, 213 (1990).
408. A. Schmuck, D. Leopold, S. Wallenhauer and K. Seppelt, *Chem. Ber.*, **123**, 761 (1990).
409. M. Domagala, H. Preut and F. Huber, *Acta Crystallogr., Sect. C*, **44**, 830 (1988).
410. D. H. R. Barton, B. Charpiot, E. T. H. Dau, W. B. Motherwell, C. Pascard and C. Pinchon, *Helv. Chim. Acta*, **67**, 586 (1984).
411. G. Faraglia, R. Graziani L. Volponi and U. Casellato, *J. Organomet. Chem.*, **253**, 317 (1983).

CHAPTER 3

Optically active arsines: preparation, uses and chiroptical properties

S. B. WILD

Research School of Chemistry, Institute of Advanced Studies, Australian National University, Canberra, A.C.T. 0200, Australia

The chemistry of organic arsenic, antimony and bismuth compounds
Edited by S. Patai © 1994 John Wiley & Sons Ltd

I. INTRODUCTION

Following the resolution in 1899 of a simple ammonium ion of the type $[NR^1R^2R^3R^4]^{+1}$, the question of optical activity in arsonium and phosphonium ions of the type $[ER^1R^2R^3R^4]^+$ became paramount[2]. Attempted resolutions over many years of a variety of simple non-cyclic arsonium ions containing four different aryl or alkyl groups, however, were either unsuccessful, discouraging or at least perplexing, and the situation remained that way until 1962 when, following similar work on phosphonium ions and phosphines beginning in 1959, the first simple arsonium ions were resolved and, more significantly perhaps, converted into stable optically active tertiary arsines of the type $AsR^1R^2R^3$. Due to improvements in instrumental techniques, especially as applied to NMR spectroscopy and X-ray crystallography, the field has now developed to a stage where it can be considered routine to resolve a tertiary arsine chiral at arsenic, determine the absolute configurations of the enantiomers and to investigate the stereochemistry, stability and properties of organic and inorganic derivatives. Some observations of historical significance concerning the synthesis and resolution of arsonium ions and arsines are the following: evidence of optical activity in arsonium salts of the type $[AsR^1R^2R^3R^4]X$, including the iodide (1921)[3]; first resolution of a 4-covalent arsenic compound, an arsine sulphide of the type $S=AsR^1R^2R^3$ (1925)[4]; first resolution of a tertiary arsine, a heterocyclic phenoxarsine (1934)[5]; biological synthesis of an arsine of the type $AsR^1R^2R^3$ (1936)[6]; separation of diastereomers of a simple non-cyclic diarsonium picrate and determination of configurational stability at arsenic (1939)[7]; first resolutions of non-cyclic arsonium ions of the type $[AsR^1R^2R^3R^4]ClO_4$ and their conversion into non-cyclic optically active arsines of the type $AsR^1R^2R^3$ (1962)[8]; first resolution of an arsine in a metal complex (1969)[9]; and observation of stereoselectivity in the asymmetric biotransformation of arsenic (1990)[10]. The application of optically active *ortho*-metallated palladium(II) complexes to the resolution of tertiary arsines in 1979[11] also marked a turning point in the field because it provided ready access to separable internal diastereomers (of either hand) that were also suitable for the determination of absolute configurations and optical purities. Nevertheless, the field of organoarsenic chemistry is small by comparison with that of organophosphorus chemistry; the reason for this is unclear. Apart from the lack of an NMR detectable spin-half nucleus, which can be compensated for to a large extent by the attachment of a methyl group to the arsenic, arsenic has many advantages over phosphorus when it comes to the synthesis of organic derivatives. For arsines of the type $AsR^1R^2R^3$ these include: reduced air-sensitivity, increased pyramidal stability and easier recovery from arsonium salts and metal complexes.

The compounds discussed here, in general, will owe their optical activity to configurational stability at arsenic. The meaning of configuration with respect to chirality is that of long usage—the fixed three-dimensional ordering of atoms or groups—and the expressions 'chiral at arsenic' or 'chiral centre' will be used in this work. The stereochemical descriptors R or S based on the Cahn–Ingold–Prelog system[12] of ranking ligands attached to a chiral centre (or axis or plane) according to a set of sequence rules will be employed where absolute configurations are known. The reader should be aware that a reversal of the descriptor R or S occurs when a chiral tertiary arsine coordinates stereospecifically to an element of higher atomic number than carbon. In keeping with IUPAC recommendations and current *Chemical Abstracts* indexing practice, the relative stereochemical descriptors R^*,R^* and R^*,S^* will be applied to diastereomers containing two chiral centres[13], although the older terms racemic (R^*,R^*) and *meso* (R^*,S^*) for symmetrical two-centre systems and *threo* (R^*,R^*) and *erythro* (R^*,S^*) for unsymmetrical two-centre systems will be used occassionally in conjunction with the modern descriptors. The descriptor ($\pm$) has been ommitted from the names of racemates, their formulae and numbers, except where the ommission could lead to confusion. The use of the terms stereoselective and stereospecific will conform with accepted practice[14].

Relatively few applications of optically active tertiary arsines to asymmetric synthesis have been reported by comparison with the extensive work with phosphines[15]. Authoritative accounts of the synthesis and stereochemistry of compounds of Group V elements are available[2,16]; other reviews cover the subject up until 1979[17]. For general treatments of organoarsenic chemistry up until 1976, including optically active compounds, two important works are available[18]. Of related interest is an article on stereochemical aspects of phosphorus chemistry[19] and another published in this series on optically active phosphines: preparation, uses and chiroptical properties[20]. On matters concerning the intricacies of resolutions work, the reader should consult Reference 21, especially Chapter 7, which is entitled *Experimental Aspects and Art of Resolutions*.

In the pages that follow, structural diagrams for tetrahedral molecules do not imply a particular configuration unless the bonds above and below the plane of the paper are indicated as a wedge and as a dashed-line, respectively.

II. ARSONIUM IONS AND ARSINES

In 1902, the unsuccessful fractional crystallization of ($-$)-aspartate and ($+$)-tartrate salts of the ethylmethyl(α-naphthyl)phenylarsonium ion was reported[22] and, in 1912, the failure to resolve allylbenzylmethylphenylarsonium iodide by seeding solutions of the racemate with crystals of the corresponding optically active ammonium salt was reported[23]. The first evidence of optical activity in arsonium compounds was published in 1921[3]. One diastereomer of **1** (where $BCS^- = (+)$-α-bromocamphor-π-sulphonate) was obtained in optically enriched form and was subsequently converted into the $(+)$-iodide **2**, which exhibited a fleeting rotation in organic solvents, especially chloroform. Similar behaviour was exhibited by $(+)$-benzylethyl(n-propyl)(p-tolyl)arsonium bromide[24]. The difficulty attending the resolution of such salts and the ease with which many similar ammonium compounds had been resolved was striking, and difficult to explain at the time. Because quaternary ammonium halides often racemize comparatively rapidly in solvents such as chloroform, whereas nitrates and sulphates possess greater stability, it was suggested that the racemization was due to an equilibrium dissociation of the quaternary halides into the tertiary amine and an alkyl or aryl halide[25]:

$$[NR^1R^2R^3R^4]X \rightleftharpoons NR^1R^2R^3 + R^4X$$

(**1**) $X^- = BCS^-$
(**2**) $X^- = I^-$

R^*,R^* R^*,S^*

(**3**) X = picrate

There is considerable evidence for this type of reaction in organoarsenic chemistry and, indeed 'dissociation–equilibrium' was believed[3] to be the cause of racemization of arsonium halides which, for synthetic reasons, usually contained at least one alkyl substituent (often the benzyl group) on the arsenic atom. In 1939, fractional crystallization was achieved of the diastereomers of the diarsonium picrate **3**[7]. The individual diastereomers of the salt, racemic (R^*,R^*) and *meso* (R^*,S^*), were stable in boiling ethanol, a fact that should have dispelled concerns of dissociation equilibria in halide-free arsonium salts. In 1940, the tetrahedral structure of $[AsPh_4]I$ was established by X-ray crystallography[26].

In the years following, various stable *heterocyclic* arsonium salts were synthesized and resolved, including the *iso*arsinolinium iodide **4**[27] and the spirocyclic arsonium iodides **5**[28] and **6**[29], in which dissociation was unlikely to occur; the optically active iodides in each case were stable in chloroform. The attempted resolutions of a variety of tetra-arylarsonium salts, however, failed, which was attributed to manipulative difficulties, particularly that of crystallization[30]. In 1961, $(+)$-amylbenzylethylphenylarsonium bromide having $[\alpha]_D + 16.5°$ was isolated, but the salt was reported to racemize rapidly in solution[31].

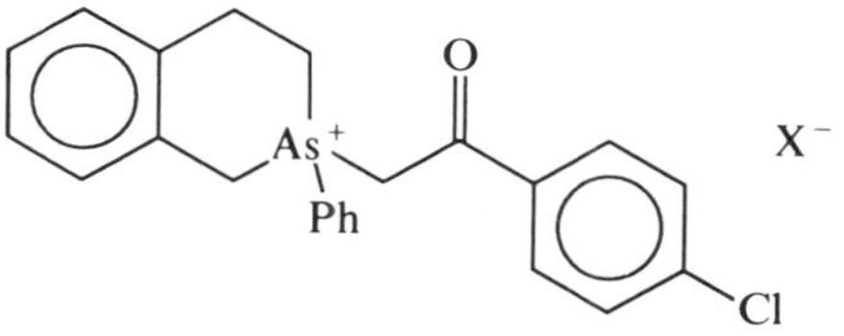

(4)

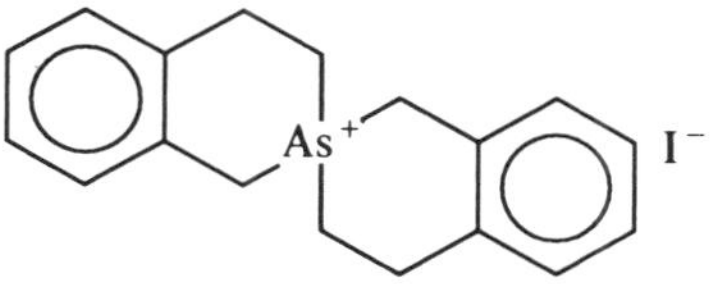

(5)

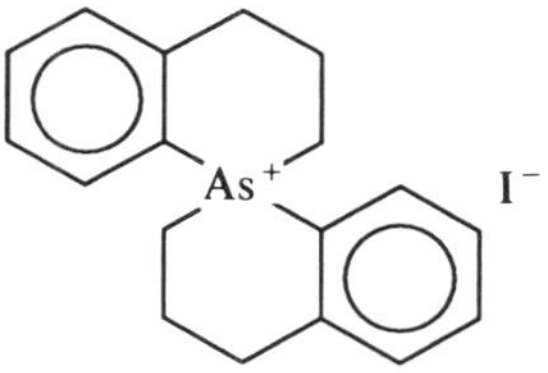

(6)

Impetus for the resolution of simple non-cyclic arsines was provided by the striking results of early calculations concerning the heights of the pyramidal molecules ammonia, phosphine, arsine and stibine, and the times of inversion, which for arsine was determined to be 1.4 years[32]. A necessary condition for resolution is that the intramolecular first-order rate constant for racemization be less than $10^{-5} \sec^{-1}$ which, by substitution into the Arrhenius equation, corresponds to $\Delta E > 100 \, \mathrm{kJ \, mol}^{-1}$. Predictions based on calculations of activation energies from spectroscopic data for PMe_3 indicated that a molecule of this type could be resolved if chemical difficulties could be overcome[33]. Some years later, however, it was concluded from calculations based upon potential energy functions derived from known vibrational frequencies and molecular dimensions that a molecule like $AsMe_3$ or $SbMe_3$ would be stable to racemization at room temperature, but that an analogous phosphine would require work to be done at a low temperature[34]. Thus, it was calculated that the temperature at which the half-time for racemization of $AsMe_3$ would be 2 h was 117 °C; for NMe_3, PMe_3 and $SbMe_3$, the values obtained were -168 °C, 7 °C and 67 °C, respectively. In the light of these calculations, it is interesting to note that a tertiary arsine was resolved before an arsonium ion, although the resolution in 1925 of the tertiary arsine sulphide **7** must be acknowledged as the first unequivocal resolution of an arsenic compound[4]. In 1934, the *heterocyclic* tertiary arsine **8** was resolved via the carboxylic acid group[5], but the source of the optical activity in the phenoxarsine, molecular atropisomerism or configurational stability at arsenic was in dispute for many years[35], despite elegant work in the meantime on the synthesis and resolution of the corresponding phenoxstibine[36], various stibiafluorenes[37], a triarylstibine[38] and a pair of arsafluorenes (see later)[39].

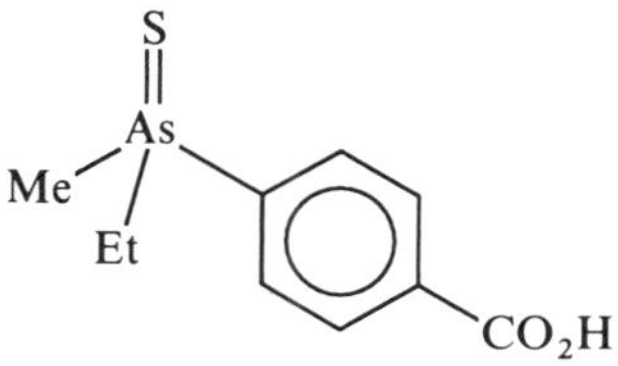

(7)

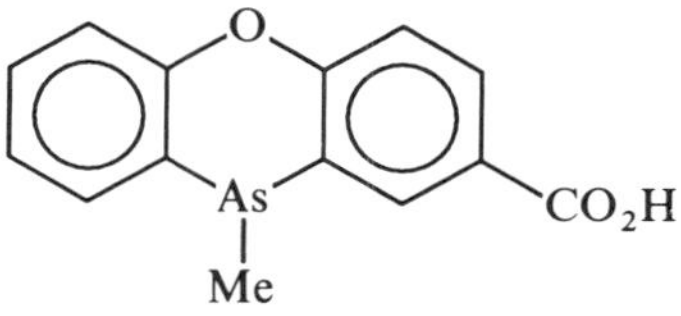

(8)

In a series of papers beginning in 1959[40], the resolutions were reported of a series of phosphonium salts of the type $[PR^1R^2R^3R^4]X$ containing one aryl and three alkyl groups with use of the resolving agent sodium $(-)$-dibenzoylhydrogentartrate, as well as various transformations that enabled a connection to be made between the configurations of the resolved phosphonium ions, phosphine oxides and tertiary phosphines. In 1962, by employing the same resolving agent, $(+)$- and $(-)$-benzylethylmethylphenyl-arsonium perchlorate (9) and $(+)$-benzyl-n-butylmethylphenylarsonium perchlorate (10) were isolated[8]. The electrochemical reduction of $(+)$- and $(-)$-9 and $(+)$-10 at a mercury electrode gave, by elimination of toluene, $(+)$- and $(-)$-ethylmethylphenylarsine (11) and $(+)$-n-butylmethylphenylarsine (12), respectively; quaternization of $(+)$-12 with p-toluenesulphonic acid benzyl ester regenerated the arsonium ion of $(+)$-10 with partial racemization[8]. Interconversions between various optically active arsonium salts and arsines showed that both cathodic cleavage of a benzyl group from an arsonium ion and the quaternization of an arsine with an alkyl halide occurred with retention of configuration at arsenic[41]. The n-propylarsonium salt $(+)$-13 was isolated[42] and shown to have the S absolute configuration by the quasi-racemate method in conjunction with the corresponding (S)-$(+)$-phosphonium bromide[43]. In 1976, with use of a similar method, the absolute configuration of (R)-$(+)$-14 was established[44]. The first arsonium salts to have had their absolute configurations determined by X-ray crystallography, however, where the salts (R)-$(+)$-15 and -16, which were obtained by resolution of the racemic cation with $(-)$-dibenzoylhydrogentartrate and subsequent exchange of the

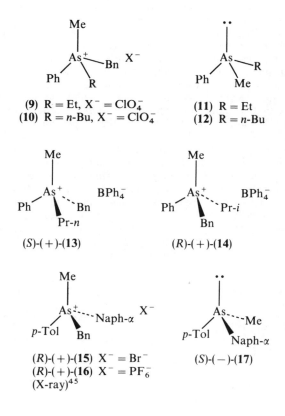

(9) R = Et, $X^- = ClO_4^-$
(10) R = n-Bu, $X^- = ClO_4^-$

(11) R = Et
(12) R = n-Bu

(S)-$(+)$-(13)

(R)-$(+)$-(14)

(R)-$(+)$-(15) $X^- = Br^-$
(R)-$(+)$-(16) $X^- = PF_6^-$
(X-ray)[45]

(S)-$(-)$-(17)

tartrate for bromide by ion-exchange chromatography or hexafluorophosphate by precipitation[45]. The stereospecific removal of the benzyl group from (R)-$(+)$-**15** by a Wittig reaction involving the corresponding arsonium ylide and benzaldehyde is perhaps the most convenient method of cleanly and stereospecifically removing a benzyl group from an arsonium ion[45,46]. The reaction gave (S)-$(-)$-**17** in high yield with complete retention of configuration at arsenic, along with (S,S)-$(+)$-2,3-diphenyloxirane in 8% enantiomeric excess (ee).

The opening up of the various synthetic routes between resolved arsonium ions and arsines, particularly for simple non-cyclic compounds, and the laying down of firm stereochemical foundations for interconversions between them, marked the beginning of an era in which chiral tertiary arsines could be designed, synthesized and resolved for a variety of applications in organic synthesis and coordination chemistry.

III. RESOLUTIONS OF TERTIARY ARSINES CHIRAL AT ARSENIC

A. Direct Resolutions via Salt-forming Groups

1. Heterocyclic arsines

The first resolution of a tertiary arsine chiral at arsenic was reported in 1934[5]. The heterocycle 10-methylphenoxarsine-2-carboxylic acid (**8**) was resolved by the fractional crystallization of diastereomeric $(-)$-strychninium salts which, once separated, were individually converted back into the resolved arsines with dilute hydrochloric acid. In later work, the 10-ethyl[47] and 10-phenyl[48] compounds, **18** and **19**, and the 8-chloro-10-phenyl compound **20**[49] were resolved (Table 1). Each compound had high optical satability, but oxidation of $(-)$-**20** was accompanied by complete loss of activity. It is now known that optically active tertiary arsine oxides racemize instantly in the presence of traces of water[50]. Some years later, in order to dispel a suggestion that the p-carboxyphenyl group had a destabilizing effect on the optical activity of the enantiomers of 10-p-carboxyphenylphenoxstibine[36], 2-chloro-10-p-carboxyphenylphenoxarsine (**21**) was synthesized and resolved (Table 1)[51]. The optical stability of the 10-phenoxarsine was unaffected by the p-carboxyphenyl group in the 10-position. Moreover, the various phenoxarsines were optically stable in boiling ethanol and could be recovered unchanged

TABLE 1. Direct resolutions of heterocyclic arsines

Arsine	$[\alpha]_D{}^a$		Solvent	Resolving agent	Reference
	$(+)$	$(-)$			
Phenoxarsine **8**	95.8	96	EtOH	$(-)$-strychnine	5
Phenoxarsine **18**	119.2	118.8	EtOH	$(-)$-strychnine	47
Phenoxarsine **19**	223.1	223.2	EtOH	$(-)$-PhCHMeNH$_2$	48
Phenoxarsine **20**	69.0	68.7	Me$_2$CO	$(+)$-PhCHMeNH$_2$	49
Phenoxarsine **21**		161.3	CHCl$_3$	$(-)$-cinchonine	51
Arsacridine **30**	84.1	83.9	MeOH	$(+)$-amphetamine	35
Arsafluorene **32**	156	160	C$_5$H$_5$N	$(+)$-PhCHMeNH$_2$	61
Arsafluorene **33**	255	251	EtOH	$(+)$-tartaric acid	61
Arsafluorene **34**	278	280	EtOH	b	61

^{a} Specific rotations $[\alpha]_D$ to nearest 0.1° measured at specified concentrations and temperatures (ca 20 °C).
^{b} Acetamido derivatives prepared from respective enantiomers of amines **33**.

from N sodium hydroxide after several hours heating at $100\,°C$, but a solution of $(+)$-**8** in ethanol containing methyl or ethyl iodide lost its optical activity (slowly at first) in about 15 h.

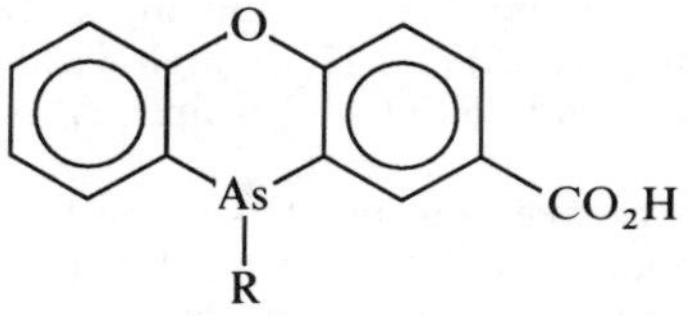

(**18**) R = Et
(**19**) R = Ph   (**20**)

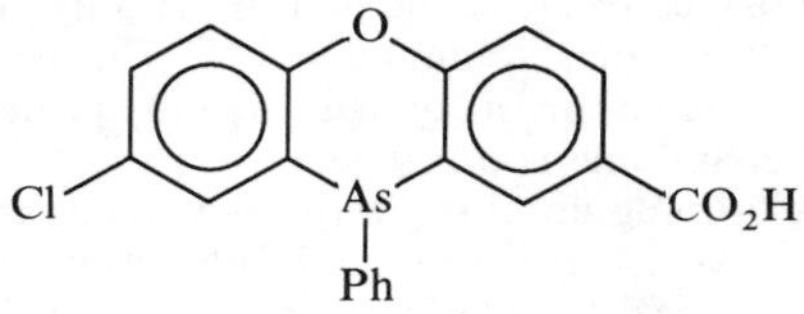

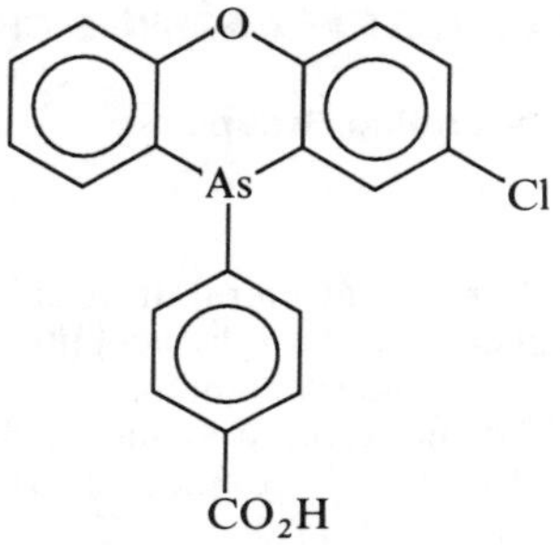

(**21**)

The isolation of the enantiomers of the various phenoxarsines undoubtedly constituted the first resolutions of tertiary arsines, although the optical activity in the molecules was at the time attributed incorrectly to the conformational rigidity of the folded rings, rather than to configurational stability at arsenic, a view supported by the failure to obtain any indication of the separation of the phenoxarsonium salts **22** [where $X^- = (+)$-camphor-10-sulphonate or $(+)$-α-bromocamphor-π-sulphonate][52] and the failure of others to resolve the non-cyclic arsines **23** and **24**[53]. In the latter case, although crystalline diastereomeric $(-)$-strychinium and $(-)$-quininium salts of the carboxylates of **23** and **24** were isolated, the salts in each case exhibited a similar rotation in chloroform. It was not realized at the time that a trace of hydrochloric acid in the chloroform could have racemized the resolved arsines[54]. Moreover, in much of this work dilute hydrochloric acid was used to decompose the diastereomeric ammonium salts of the arsine-carboxylates.

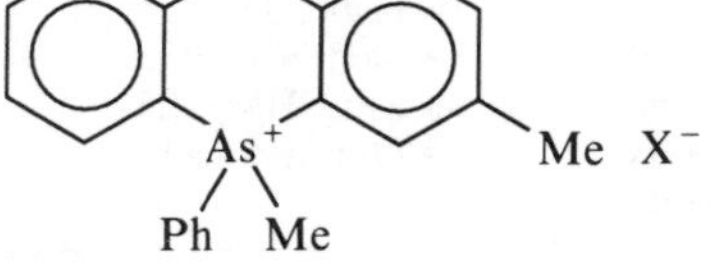

(**22**)

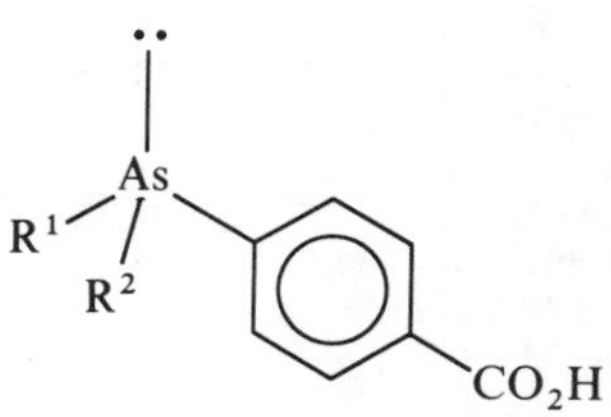

(**23**) R^1 = Et, R^2 = n-Pr
(**24**) R^1 = Me, R^2 = Ph

A report concerning the resolution of the benzophenarsazine **25** [where $X^- = (+)$-α-bromocamphor-π-sulphonate][55] was regarded by workers in the field to be too scant to be meaningful. Crystal structure determinations of the achiral compounds **26** and **27** revealed dihedral angles of 169.3° and 156.3° respectively, between the two halves; the smaller angle in **27** indicated less aromaticity in this tricyclic ring system[56,57].

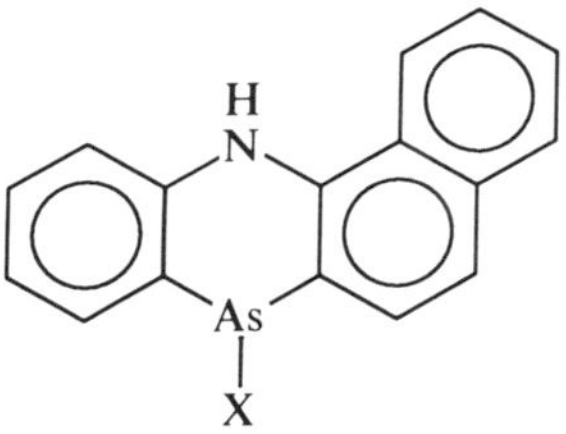

(25)

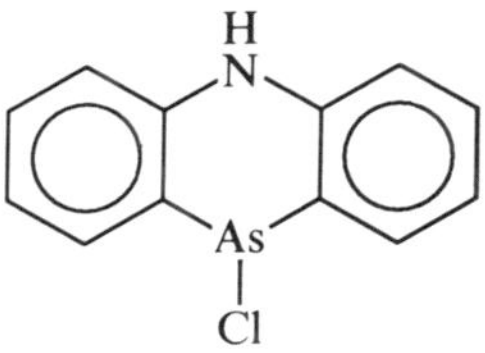

(26)
(X-ray)[56]

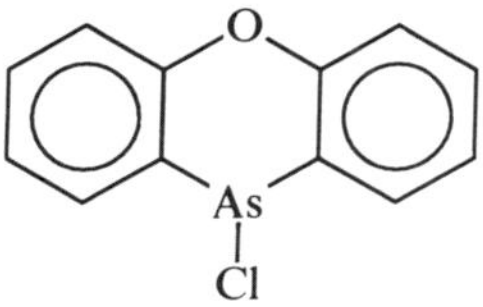

(27)
(X-ray)[57]

In an attempt to avoid any complication in the phenoxarsines due to the cyclic oxygen atoms, achiral 5,10-bis(p-tolyl)-5,10-diarsanthrene (**28**) was synthesized and subjected to fractional crystallization[58]. When folded about the As—As axis to give the correct intervalency angles at arsenic, three diastereomers of **28** are possible, **28a**, **28b** and **28c**. Diastereomers **28a** and **28b** are conformational isomers (atropisomers) related to one another by flexing (equation 1); **28c**, however, is a configurational isomer and interconversion between **28a** and **28b** with **28c** requires inversion at arsenic. (Flexing of **28c** leads to a geometrically indistinguishable molecule). Exhaustive fractional crystallization of **28** separated it into two forms of almost identical melting points, considered to be **28a** and **28c**, together with a form containing equal quantities of the two. Each isomer could be kept in the molten condition at 190 °C for 10 min without any indication of conversion into the other form or decomposition. (This observation was a definite indication of configurational isomerism.) The two pure forms differed in many chemical properties, including giving different methiodides, and were believed to have provided evidence for the stability of 'folded configurations'.

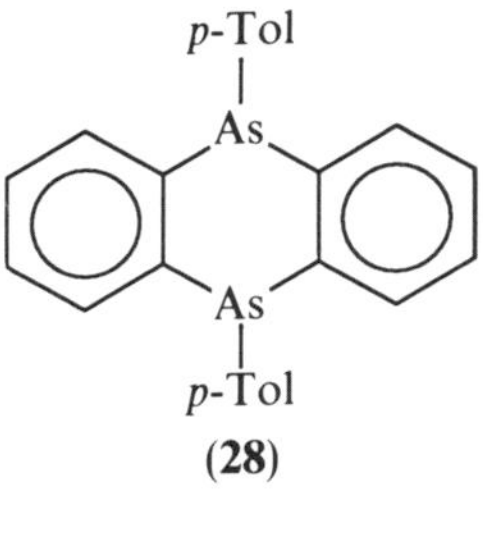

(28)

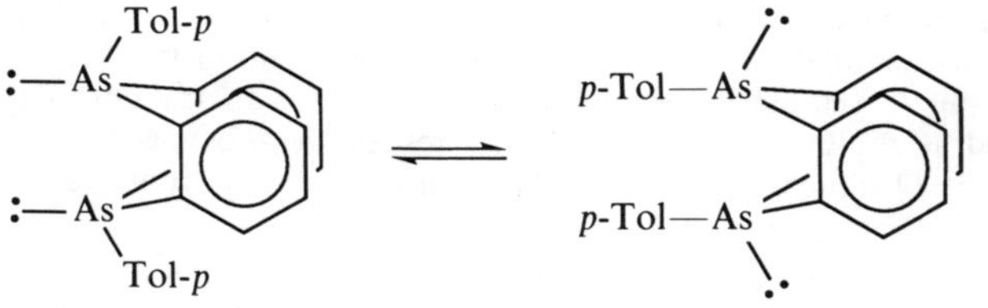

(28a)   (28b)   (1)

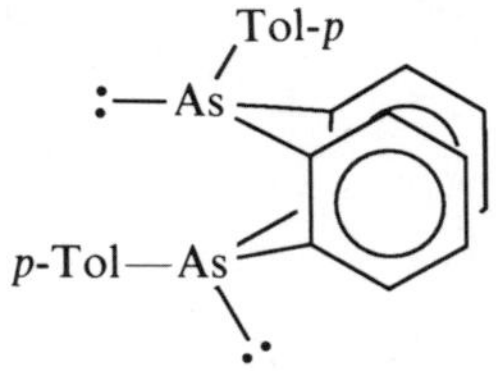

(28c)

This interpretation was subsequently refuted, however, when the 5,10-dihydroars-acridine **30** and the first *non-cyclic* triarylarsine, **31**, were resolved[35]. (It is noteworthy that the diastereomeric salts of **30** and **31** were decomposed with 0.1 N sulphuric acid.) On the basis of the resolutions of **30** and **31**, it was concluded that resolvable phenoxarsines were best pictured as separable and configurationally stable (at arsenic)

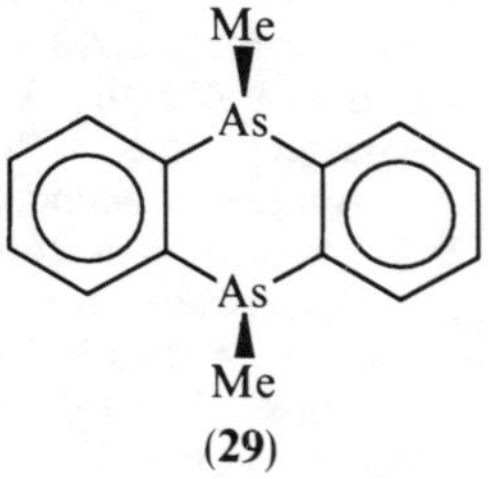

(29)

(X-ray)[60]

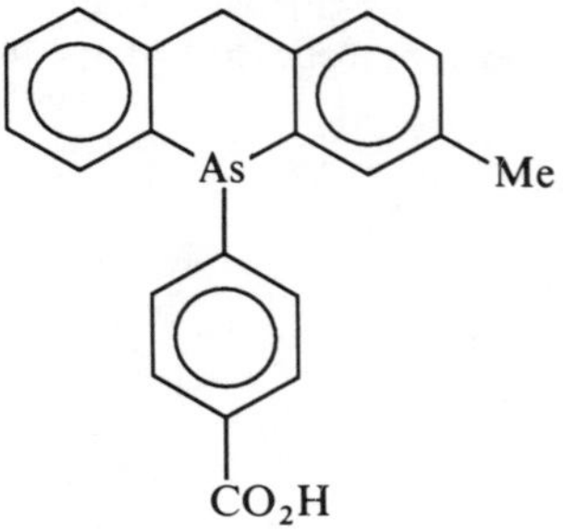

(30) 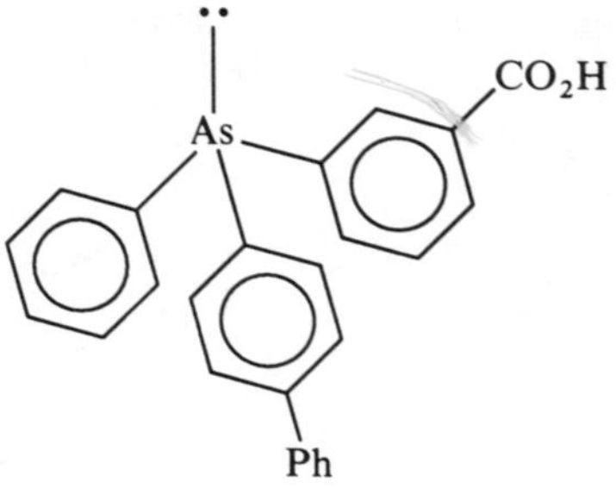   (31)

enantiomers that individually exist in solution as rapidly interconverting folded conformers (equation 2). The crystal structure of a single configurational diastereomer of *cis*-5,10-dihydro-5,10-dimethylarsanthrene (**29**)[59] was determined in 1968[60]. The molecule crystallized in a butterfly conformation. It had been pointed out earlier that the inversion barrier at arsenic in the phenoxarsine **8** should be higher than in a similar acyclic compounds because the C—O—C angle must be distorted in the process of obtaining a planar configuration about arsenic and that a similar consideration applies to inversion about either arsenic in the diarsanthrene **28**[34].

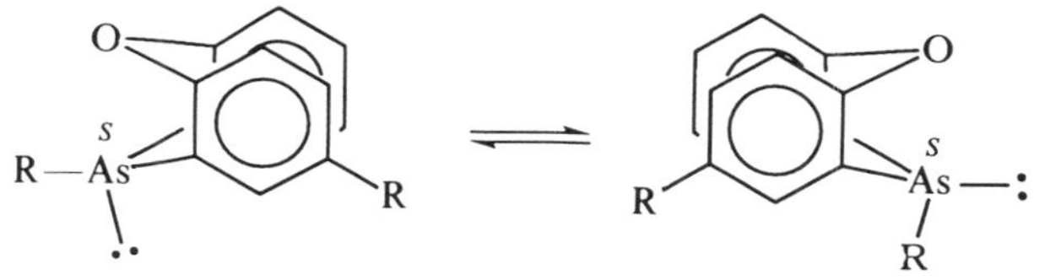

$$\tag{2}$$

Because of earlier doubts concerning the source of the optical activity in resolved phenoxarsines, the substituted 9-arsafluorenes **32–34** were synthesized[61], in which the arsenic is held in a chemically stable five-membered ring. Optical resolution of the acid **32** was achieved with (+)- and (−)-α-methylbenzylamine, giving the stable enantiomers after removal of resolving agent (Table 1). The amino compound **33** was similarly resolved by hydrogentartrate formation with (+)- and (−)-tartaric acid. The free amines **33** solidified as glasses, but gave crystalline acetamido derivatives (**34**) that were completely stable in ethanol, even when heated for 1 h in a sealed tube at 110 °C. The optical stabilities of the arsafluorenes are considerably greater than those of the analogous antimony compounds[37]. An X-ray crystal structure determination established the planarity of the tricyclic ring system in achiral 9-phenyl-9-arsafluorene (**35**), and the pyramidal nature of the arsenic geometry[62].

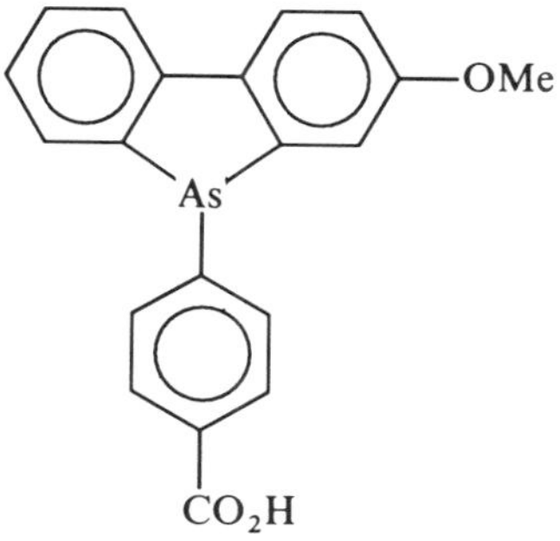

CO₂H

(**32**)

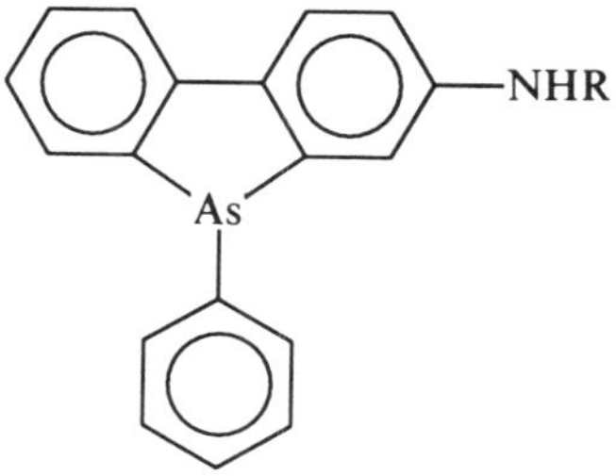

(**33**) R = H
(**34**) R = CH₃CO

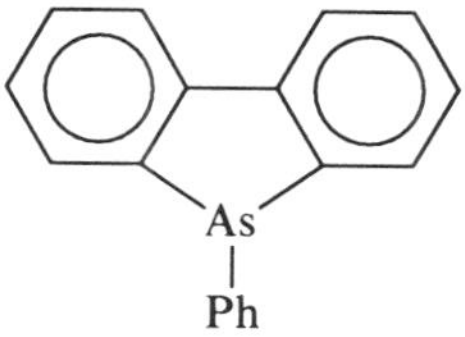

(**35**)
(X-ray)[62]

2. Acyclic arsines

The previously mentioned acyclic triaryl arsine (+)-**31** was resolved in 1962 by the fractional crystallization of (−)-α-phenylethylammonium salts; the (−)-enantiomer was obtained with use of (+)-amphetamine as the resolving agent (Table 2)[35]. This was the first resolution of an acyclic tertiary arsine by a direct method, although the optical purities of the enantiomers were not determined. The early work[53] on the resolution of chiral acyclic p-arsinobenzoic acid has been re-investigated and extended substantially. Thus, the enantiomers of the arsines **36–47** were obtained by the decomposition of separated diastereomeric salts with dilute hydrochloric or sulphuric acid (Table 2)[63-68]. The methyl esters **48–50** of the arsino-benzoic acids **41, 45** and **46** were prepared by treating benzene solutions of the acids with diazomethane[68]. The use of hydrochloric acid for the decompositions of certain diastereomers is of concern in connection with the optical purities of the enantiomers. Cold 0.1–0.2 N sulphuric acid was used in the preparation of the enantiomers of **44, 46, 51** and **52**, however. The magnitudes of the rotations of the enantiomers of the arsines are related to the positions of the substituents on the aromatic rings[69]. When (+)-**36** was heated with allyl chloride in boiling benzene for 5 h, (+)-[As(All)EtPh($C_6H_4CO_2H$-p)]Cl was obtained in high yield[70].

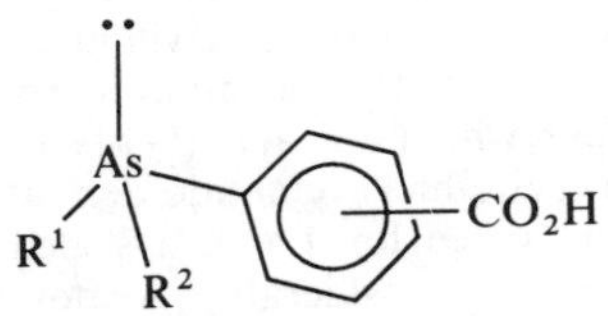

(**36**)–(**47**)

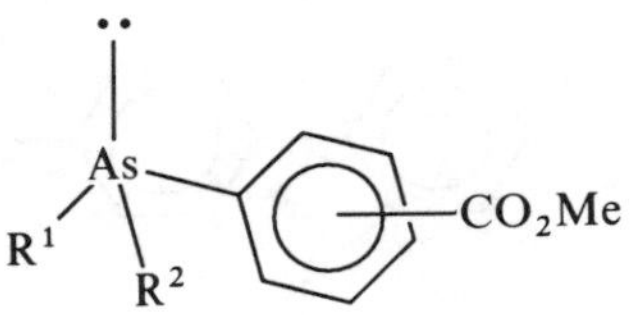

(**48**)–(**50**)

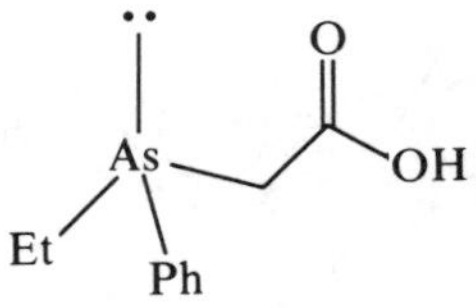

(**51**)

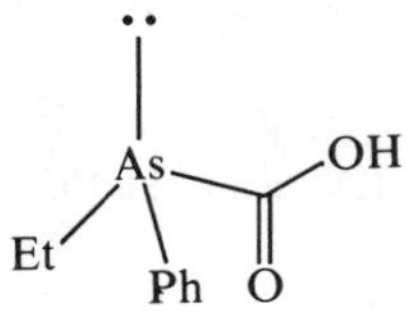

(**52**)

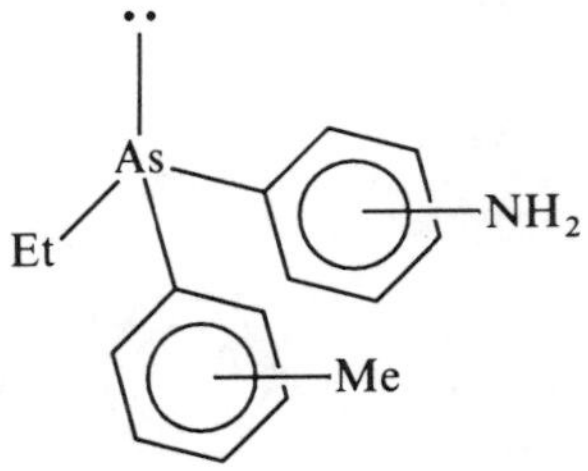

(**53**) − (**61**)

The (+)-enantiomer of the arsino-acetic acid **51** was isolated with use of (−)-α-methylbenzylamine as resolving agent (Table 2)[71]. The racemic arsine was prepared by permanganate oxidation of allylethylphenylarsine, followed by SO_2/HCl reduction of the intermediate arsine oxide.

TABLE 2. Direct resolutions of acyclic arsino-carboxylic acids

Arsino-acid	$[\alpha]_D{}^a$ (+)	(−)	Solvent	Resolving agent	Reference
Arsino-benzoic acids					
AsPh(C$_6$H$_4$Ph-p)(C$_6$H$_4$CO$_2$H-m) (31)	16[b]	24[b,c]	dioxane	(−)-PhCHMeNH$_2$	35
AsEtPh(C$_6$H$_4$CO$_2$H-p) (36)	12.8	12.6[d]	C$_6$H$_6$	(−)-quinine	63, 68
As(β-Naph)Ph(C$_6$H$_4$CO$_2$H-o) (37)	6.0	4.9	C$_6$H$_6$	(−)-quinine	64
AsEt(n-Pr)(C$_6$H$_4$CO$_2$H-p) (38)	3.0	3.0[e]	C$_6$H$_6$	(−)-quinine	64
AsEt(n-Bu)(C$_6$H$_4$CO$_2$H-p) (39)	7.2	6.5	MeNO$_2$	(+)-morphine	65
AsMePh(C$_6$H$_4$CO$_2$H-p) (40)	8.7	6.2	f	(−)-quinine	66
AsEtPh(C$_6$H$_4$CO$_2$H-o) (41)	17.9	18.4	f	(−)-quinine	66, 68
AsMe(C$_6$H$_4$Br-p)(C$_6$H$_4$CO$_2$H-p) (42)	8.6	7.1	MeNO$_2$	(−)-quinine	66
AsEt(C$_6$H$_4$Br-p)(C$_6$H$_4$CO$_2$H-p) (43)	13.5	13.0	MeNO$_2$	(−)-quinine	66
AsEt(C$_6$H$_4$Br-o)(C$_6$H$_4$CO$_2$H-p) (44)	16.8	16.2	C$_6$H$_6$	(−)-quinine	67
AsEtPh(C$_6$H$_4$CO$_2$H-p) (45)	18.4	18.4	C$_6$H$_6$	g	68
AsEtPh(C$_6$H$_4$CO$_2$H-m) (46)	16.6	16.3	C$_6$H$_6$	(−)-quinine	68
AsEt(C$_6$H$_4$Ph-p)(C$_6$H$_4$CO$_2$H-p) (47)	12.8	12.6	CHCl$_3$	(−)-quinine	66
AsEtPh(C$_6$H$_4$CO$_2$Me-o) (48)	25.3	25.3	C$_6$H$_6$	h	68
AsEtPh(C$_6$H$_4$CO$_2$Me-m) (49)	22.6	22.3	C$_6$H$_6$	h	68
AsEtPh(C$_6$H$_4$CO$_2$Me-p) (50)	17.4	17.2	C$_6$H$_6$	h	68
Arsino-acetic acid					
AsEtPh(CH$_2$CO$_2$H) (51)	2.7		EtOH	(−)-PhCHMeNH$_2$	71
Arsino-formic acid					
AsEtPh(CO$_2$H) (52)	9.4		MeOH	(−)-quinine	72

[a] Specific rotations $[\alpha]_D$ to nearest 0.1° measured at specified concentrations and temperatures (*ca* 20 °C).
[b] ORD: $[\phi]_{600}$.
[c] Obtained via (+)-amphetamine.
[d] Values taken from Reference 68.
[e] Obtained via (+)-PhCHMeNH$_2$.
[f] Solvent not specified.
[g] Resolving agent not specified.
[h] Esters prepared from corresponding acids diazomethane.

S. B. Wild

TABLE 3. Direct resolutions of acyclic arsino-anilines with (+)-tartaric acid

	$[\alpha]_D{}^a$		
Arsine	(+)	(−)	Reference
AsEt(C_6H_4Me-p)($C_6H_4NH_2$-o) (53)	2.4	2.9	73
AsEt(C_6H_4Me-p)($C_6H_4NH_2$-m) (54)	2.2	1.9	74
AsEt(C_6H_4Me-p)($C_6H_4NH_2$-p) (55)	2.0	1.5	74
AsEt(C_6H_4Me-o)($C_6H_4NH_2$-o) (56)	2.7	2.3	75
AsEt(C_6H_4Me-o)($C_6H_4NH_2$-m) (57)	2.5	2.0	75
AsEt(C_6H_4Me-o)($C_6H_4NH_2$-p) (58)	2.2	1.7	75
AsEt(C_6H_4Me-m)($C_6H_4NH_2$-o) (59)	2.3	2.1	76
AsEt(C_6H_4Me-m)($C_6H_4NH_2$-m) (60)	2.3	1.9	76
AsEt(C_6H_4Me-m)($C_6H_4NH_2$-p) (61)	2.2	1.7	76

^{a} Specific rotations $[\alpha]_D$ to nearest 0.1° measured at specified concentrations and temperatures in diethyl ether (ca 20 °C).

Sodium ethylphenylarsenide with carbon dioxide yields crystalline ethylphenylarsinoformic acid (52) after treatment with dilute sulphuric acid[72]. The arsinoformic acid was resolved with (−)-quinine. The less-soluble diastereomer, after three recrystallizations from chloroform, was decomposed with dilute sulphuric acid into (−)-52, which required purification by extraction into benzene to separate it from arsine oxide (Table 2).

The arsino-anilines 53–61 were resolved by the fractional crystallization of (+)-hydrogentartrate salts (Table 3)[73–76]. A handicap in evaluating the work relating to the resolution of the compounds 36–61 is the lack of experimental detail, and some inconsistencies in the physical data reported, together with the failure to determine optical purities and absolute configurations.

B. Resolutions via Arsonium Ions

1. Cathodic or alkali metal reduction

An optically active arsonium ion containing an allyl or a benzyl group is quantitatively reduced at a mercury electrode in water or ethanol[8] with retention of configuration at arsenic (and liberation of toluene or propene)[41–43] to give an optically active tertiary arsine, free of functional groups on the organic substituents (equation 3).

$$\begin{array}{c} \text{Ph} \\ | \\ \text{As}^+\cdots\text{R}^3 \\ \text{R}^1\quad\text{R}^2 \end{array} \xrightarrow[\text{H}^+]{2e^-} \begin{array}{c} \cdots \\ | \\ \text{As}\cdots\text{R}^3 \\ \text{R}^1\quad\text{R}^2 \end{array} + \text{PhMe} \qquad (3)$$

In aqueous solution, determinations of half-wave potentials have indicated the following ordering of difficulty of removal of organic substituents from arsonium ions[77]:

$$\text{PhCH}_2(\text{Bn}) < \text{CH}_2{=}\text{CH}{-}\text{CH}_2 \text{ (All)} < \text{CH}_2\text{CO}_2\text{Et} < p\text{-Tol} < \text{Ph} < \text{Et} < \text{Me}$$

Cathodic cleavage of benzyl groups from resolved arsonium salts have been used to produce optically active tertiary arsines (Table 4). Some examples of interconversions

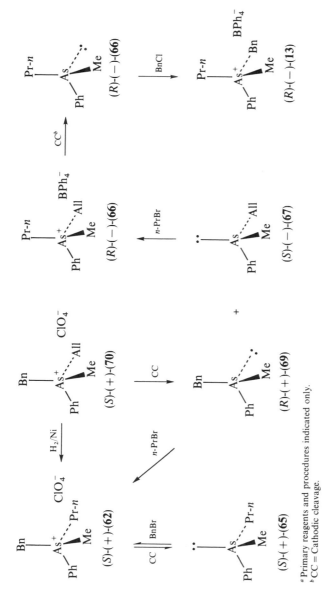

SCHEME 1

[a] Primary reagents and procedures indicated only.

[b] CC = Cathodic cleavage.

between various arsonium salts and tertiary arsines are given in Scheme 1. The absolute configurations indicated in the scheme are based on the assignment for the tetraphenylborate (S)-$(+)$-13[42] and the perchlorate (S)-$(+)$-62[78]. The assignments of the configurations agree with those given in Reference 79. Experimental detail for some of the work indicated in Table 4 is scant and few optical purities have been determined. The allyl group is also readily cleaved at a mercury electrode, but because of the

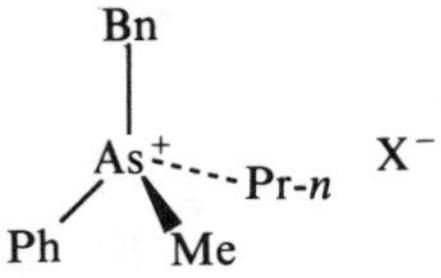

(S)-$(+)$-(62) $X^- = ClO_4^-$ (UV, ORD, CD)[78]
(S)-$(+)$-(63) $X^- = I^-$
(S)-$(+)$-(64) $X^- = Br^-$

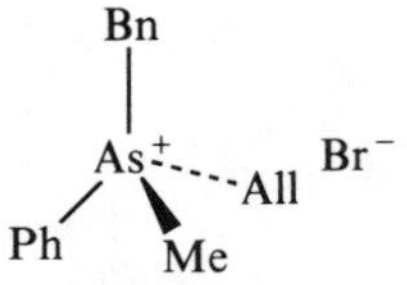

(S)-$(+)$-(71)

similarities of the reduction potentials for benzyl and allyl groups, mixed products are obtained when arsonium ions containing both groups are reduced electrolytically[41]. Potassium or sodium amalgam in methanol can also be used to reduce benzylarsonium salts (Table 4)[80]. The half-wave potentials for arsonium salts (-1.2 to -1.5 V) are *ca* 0.3 V more positive than those of the corresponding phosphonium salts[81]. The distribution of products in the electrolysis of certain arsonium and phosphonium salts is dependent upon the temperature of the reaction and the solvent employed[82]. The irreversible reduction of an arsonium or phosphonium salt at a mercury electrode can proceed by a two-electron or two one-electron steps, as indicated in Scheme 2[81].

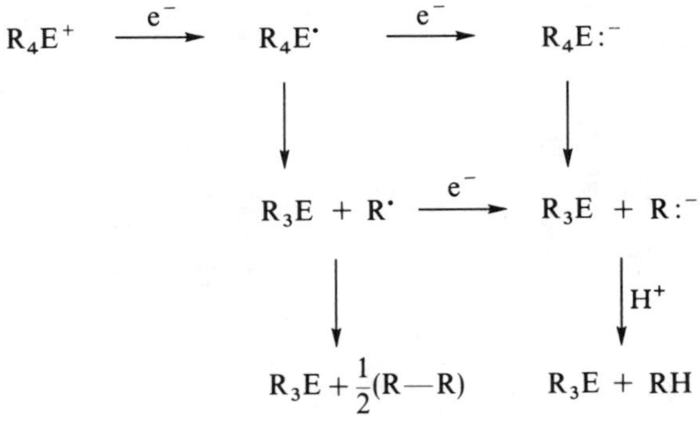

$$R_4E^+ \xrightarrow{e^-} R_4E^{\cdot} \xrightarrow{e^-} R_4E{:}^-$$

$$R_3E + R^{\cdot} \xrightarrow{e^-} R_3E + R{:}^-$$

$$R_3E + \tfrac{1}{2}(R\!-\!R) \qquad R_3E + RH$$

SCHEME 2

2. Hydride reduction

Whereas the lithium aluminium hydride reduction of an optically active benzylphosphonium salt in tetrahydrofuran leads to a racemic[83] or highly racemized[84] tertiary phosphine and toluene, the reduction of $(-)$-benzyl(n-butyl)methylphenylarsonium bromide (68) or (S)-$(+)$-benzylmethylphenyl(n-propyl)arsonium bromide, (S)-$(+)$-64, with this reagent affords $(-)$-n-butylmethylphenylarsine, $(-)$-12, or (S)-$(+)$-methylphenyl(n-propyl)arsine, (S)-$(+)$-65, of similar optical purity to the products obtained by cathodic reduction of the respective arsonium compounds (Table 4)[84]. Moreover,

TABLE. 4. Optically active tertiary arsines via arsonium salts

Arsine	Abs. config.	$[\alpha]_D{}^a$	Solvent	Arsonium salt	Abs. config.	$[\alpha]_D$	Solvent	Ref.
Cathodic cleavage								
AsEtMePh (11)	S	+1.9	MeOH	[AsBnEtMePh]ClO$_4$ (9)	S	+21.4	MeOH	8
	R	−1.2	MeOH		R	−19.3	MeOH	8
As(n-Bu)MePh (12)	S	+12.4	MeOH	[AsBn(n-Bu)MePh]ClO$_4$ (10)	S	+29	MeOH	8
AsMePh(n-Pr) (65)	S	+12.1	b	[AsBnMePh(n-Pr)]ClO$_4$ (62)	S	+33	b	41, 42
AsAllMePh (67)	R	−8.6	b	[AsAllMePh(n-Pr)]BPh$_4$ (66)	R	−2.9	b	41, 42
AsBnMePh (69)	S	−71.4	b	[AsAllBnMePh]ClO$_4$ (70)	S	−14.7	b	41, 42
	R	+46.2	b	[AsAllBnMePh]ClO$_4$ (70)	S	−14.7	b	41, 42
Amalgam reduction								
As(n-Bu)MePh (12)	R	−0.6	MeOH	[AsBn(n-Bu)MePh]ClO$_4$ (10)	R	+1.3	b	79
AsMePh(n-Pr) (65)	R	−8.6	MeOH	[AsBnMePh(n-Pr)]I (63)	R	−16	b	79
Hydride reduction								
As(n-Bu)MePh (12)		−0.8	MeOH	[AsBn(n-Bu)MePh]Br (68)		−2	MeOH	83
AsMePh(n-Pr) (65)	S	+13.5	MeOH	[AsBnMePh(n-Pr)]Br (64)	S	+26	MeOH	83
Cyanolysis								
AsBnMePh (69)	R	+56		[AsAllBnMePh]Br (71)	S	+10.1		85
AsMePh(n-Pr) (65)	R	−10.5		[AsAllMePh(Pr-n)]Br (72)		c		85
Benzylidene transfer								
AsMe(α-Naph)(p-Tol) (17)	S	−116	CHCl$_3$	[AsBnMe(α-Naph)(p-Tol)]Br (15)	R	+54.8		44
AsMePh(8-Quin) (73)	S	+110	Et$_2$O	[AsBnMePh(8-Quin)]Br (74)	R	−3.2		45
1,2-C$_6$H$_4$(AsMePh)$_2$ (75)	S,S	−94.4	Et$_2$O	[AsBnMePh(2-C$_6$H$_4$(AsMePh)]Br (76)	R,S	−108.7		45

^{a} Specific rotations $[\alpha]_D$ to nearest 0.1° measured at specified concentrations and temperatures (*ca* 20 °C).
^{b} Solvent not specified.
^{c} Rotation not measured.

S. B. Wild

quaternization of the optically active tertiary arsines with benzyl bromide gives arsonium salts of the same optical purity as the starting materials, thus establishing that the hydride reduction proceeds with retention of configuration at arsenic in each case (equation 4).

$$
\underset{\substack{R^1 \quad R^2}}{\overset{Ph}{\underset{}{\bigg\uparrow}}}
\underset{R^1 \quad R^2}{As^+\text{---}R^3}
\xrightarrow{\text{LiAlH}_4}
\underset{R^1 \quad R^2}{\overset{\bullet\bullet}{\underset{}{\bigg|}}As\text{---}R^3}
+ \; PhMe
\tag{4}
$$

Lithium aluminium hydride reduction of (R^*,S^*)-$(\pm)$-$(1,2$-$C_6H_4(AsMePh)(AsBnMePh)]$-$ClO_4\cdot H_2O$ or (R^*,R^*)-$(\pm)$-$[1,2$-$C_6H_4(AsMePh)(AsBnMePh)]ClO_4$ in boiling tetrahydrofuran affords (R^*,R^*)-$1,2$-$C_6H_4(AsMePh)_2$ or $(R,*S^*)$-$1,2$-$C_6H_4(AsMePh)_2$, (R^*,R^*)- and (R^*,S^*)-**75**, providing further evidence of the stereospecificity of hydride reduction[85]. The 2-methallyl, 2,2-dimethallyl and propargyl groups are also cleaved from the corresponding triphenylarsonium salts in high yield by lithium aluminium hydride[84].

(R^*,R^*)-(**75**) (R^*,S^*)-(**75**)

3. Cyanolysis

An optically active arsonium ion containing an allyl group attached to the arsenic is cleaved by cyanide in hot aqueous solution to give, by elimination of methacrylonitrile, the optically active tertiary arsine in high yield with retention of configuration at arsenic (equation 5); the reaction proceeds more readily when one (or more) aromatic groups is attached to the arsenic[86]. Thus, (S)-$(+)$-allylbenzylmethylphenylarsonium bromide, (S)-$(+)$-**71**, gives (R)-$(+)$-benzylmethylphenylarsine, (R)-$(+)$-**69**, in 80% yield by cyanolysis (Table 4). The reaction proceeds by an addition–elimination mechanism involving hydrolysis of a β-cyanopropylarsonium intermediate[87].

$$
\underset{\substack{R^1 \quad R^2}}{\overset{}{\underset{}{\bigg\uparrow}}}
As^+\text{---}R^3
\xrightarrow{\text{CN}^-}
\underset{R^1 \quad R^2}{\overset{\bullet\bullet}{\underset{}{\bigg|}}As\text{---}R^3}
+ \; H_2C{=}C\underset{Me}{\overset{CN}{\big\backslash}}
\tag{5}
$$

Chiral arsonium salts can be built up by alternate cyanolysis and quaternization reactions. If the β-position on the allyl substituent is blocked, as in 2-methallyl-triphenylarsonium chloride, the products of the reaction are triphenylarsine (ca 30%) and (2-cyano-2-methylpropyl)triphenylarsonium chloride (ca 60%) (equation 6); the latter compound can be hydrolysed into the corresponding 2-carboxy-2-methylpropylarsonium salt[88].

$$\underset{Ph_3As^{\overset{+}{\cdot}}}{\overset{Me}{\diagdown}}C=CH_2 \quad \xrightarrow[H^+]{CN^-} \quad \underset{Ph_3As^{\overset{+}{\cdot}}}{\overset{NC}{\diagdown}}\overset{Me}{\underset{Me}{\diagup}} \tag{6}$$

4. Hydrolysis

An aqueous solution of (R)-$(-)$-allylmethylphenyl(n-propyl)arsonium bromide, when treated with aqueous sodium hydroxide, decomposes into methylphenyl(n-propyl)-arsine (3) (not isolated) and allyl alcohol[88,89]. Quarternization of the arsine with benzyl bromide regenerates the arsonium bromide of 89.5% optical purity, based upon the rotation of the sample used to prepare the optically active allylarsonium salt. Thus, the hydrolysis of an allylarsonium ion is highly stereoselective giving the tertiary arsine with retention of configuration at arsenic (equation 7).

$$\underset{\underset{Pr\text{-}n}{Ph}}{As^{\overset{+}{\cdots}Me}} \quad \xrightarrow{OH^-} \quad \underset{\underset{Pr\text{-}n}{Ph}}{As_{\cdots Me}} + H_2C\!\!=\!\!\diagup\!\!\diagdown^{OH} \tag{7}$$

As found for the related cyanolysis reaction, aryl groups on the arsenic facilitate the hydrolysis of allylarsonium ions and substituents on the allyl group markedly affect the course of the reaction. At 20 °C, a high yield of (2-hydroxypropyl)triphenylarsonium bromide was isolated from the reaction between allyltriphenylarsonium bromide and aqueous sodium hydroxide[88,89]; upon further heating in aqueous sodium hydroxide the 2-hydroxypropyl compound decomposes into triphenylarsine and allyl alcohol. Sodium ethoxide or methoxide at 20 °C reacts with allyltriphenylarsonium bromide to give the corresponding 2-alkoxypropylarsonium bromide which, when heated at 80 °C with 10% sodium hydroxide, affords triphenylarsine and the respective isopropenyl ethers[88,89]. Allyltriphenylarsonium bromide rearranges partially on basic alumina into the corresponding 1-propenylarsonium compound[90].

Benzyl- or cinnamyl-triphenylarsonium bromides, upon reaction with aqueous sodium hydroxide at 100 °C, decompose into triphenylarsine oxide and toluene or 2-methyl-styrene, respectively[88]. Since tertiary arsine oxides spontaneously racemize in water[50], alkaline hydrolysis of arsonium salts of this type cannot be used for the preparation of optically active arsines.

5. Benzylidene transfer

Equilibrium concentrations of semi-stabilized benzylidene ylides derived from optically pure arsonium salts and alkali metal ethoxides react with aromatic aldehydes to produce high yields of enantiomerically enriched *trans*-2,3-diaryloxiranes and optically pure tertiary arsines with retention of configuration at arsenic (equation 8)[45,46].

$$\underset{\underset{R^2}{R^1}}{As^{\overset{+}{\cdots}R^3}}\!\!\diagup^{Ph} \quad \xrightarrow{PhCHO} \quad \underset{\underset{R^2}{R^1}}{As_{\cdots R^3}} + \underset{O \;\; Ph}{\overset{Ph}{\triangle}} \tag{8}$$

Examples of the reaction are given in Table 4 for the tertiary arsines **17, 73** and **75** (equations 9–11). The absolute configuration of (R)-(+)-**15** was determined by X-ray crystal structure determination[45]; the absolute configurations of (S)-(+)-**73** and (S)-(−)-**75** were determined by X-ray crystal structure analyses of appropriate metal complexes (References 92 and 93, respectively).

$$(9)$$

(R)-(+)-(15)   (S)-(−)-(17)

$$(10)$$

(R)-(−)-(74)   (S)-(+)-(73)

$$(11)$$

(R,S)-(−)-(76)   (S,S)-(−)-(75)

C. Resolutions via Metal Complexes

The formation and separation of a pair of internally diastereomeric metal complexes provides a valuable route of wide applicability to the resolution of tertiary arsines. The method was first applied to the resolution of *trans*-cyclo-alkenes[94] and ethyl-*p*-tolyl sulphoxide[95] in conjunction with (R)-(+)- or (S)-(−)-α-methylbenzylamine on platinum(II), but it was later adapted to the resolution of *t*-butylmethylphenylphosphine in a similar complex containing (+)-deoxyephedrine[96]. An important development occurred some years later, however, when the dimeric complex (R)-(−)-**77** was employed first as a reagent for the generation of a pair of diastereomers by bridge-splitting with a chiral tertiary phosphine[97]. A concern with such reactions is that the bridge-splitting may lack regioselectivity, in which case four diastereomers of the phosphine or arsine complex will be produced. It has been found subsequently that the naphthylamine complexes (R)-$(-)$- or (S)-$(+)$-**78**, although more expensive to prepare, undergo highly regioselective reactions with most unidentates and unsymmetrical bidentates to give diastereomers in which the phosphine or arsine coordinates *trans* to the dimethylamino group (see below).

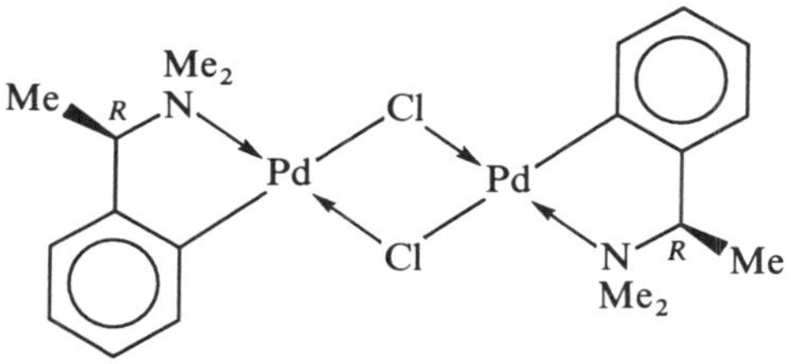

(R)-$(-)$-**(77)**

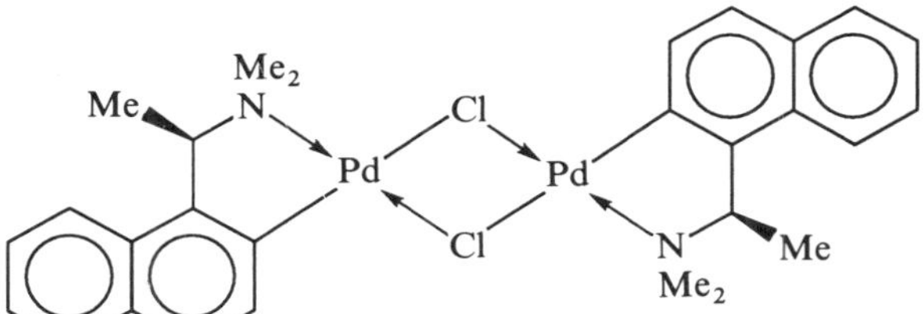

(R)-$(-)$-**(78)**

1. Separation of diastereomers

a. As-Unidentates.   The first resolution of a tertiary arsine by the metal complexation method was reported 1970[9]. The procedure for the resolution of ethylmethylphenylarsine (**11**) is given in Scheme 3. The key step in the sequence is the formation of a pair of diastereomeric salts by a bridge-splitting reaction involving *trans*-$[Pt_2Cl_4(AsEtMePh)_2]$· $C_{10}H_8$ and (R,R)-$(+)$-stilbene diamine. A 10-fold difference in solubility of the salts $[(R,R),R_{As}]$-$(+)$- and $[(R,R),S_{As}]$-$(+)$-**79** in chloroform facilitated the separation. The salts, when individually treated with an excess of cyanide, liberated the resolved arsines (Table 5). Since the replacement of a lone pair by a heavy metal changes the priority of that ligand (or phantom ligand) from 4 to 1, the CIP descriptor must be reversed when the ligand is displaced from the metal and *vice versa*[12]. The rotations for the enantiomers of **11** obtained by the metal complexation route are considerably higher than those reported for samples of the same arsine obtained by cathodic reduction (Table 4).

The crystal and molecular structure of (S)-7-phenyldinaph[2,1-*b*; 1',2'-*d*]arsole (**82**), which was obtained by spontaneous resolution of the racemate from hot methanol, reveals appreciable bending of the distorted naphthyl residues away from each other (Scheme 4)[100]. The molecule is fluctional in solution on the NMR time scale, however, with similar barriers between the conformational isomers (atropisomers) for the 7-phenyl [$\Delta G^{\neq}$ 59 ± 1 kJ mol^{-1} (259 K)] and the 7-methyl [$\Delta G^{\neq}$ 65 ± 1 kJ mol^{-1} (287 K)] compounds. The analogous phospholes are also unsuitable for resolution because of similarly low barriers to inversion of the atropisomers[100,101]. Both arsenic ligands, when coordinated to iron(II) in complexes of the type $[(\eta^5\text{-}C_5H_5)\{1,2\text{-}C_6H_4(PMePh)_2\}FeL]PF_6$,

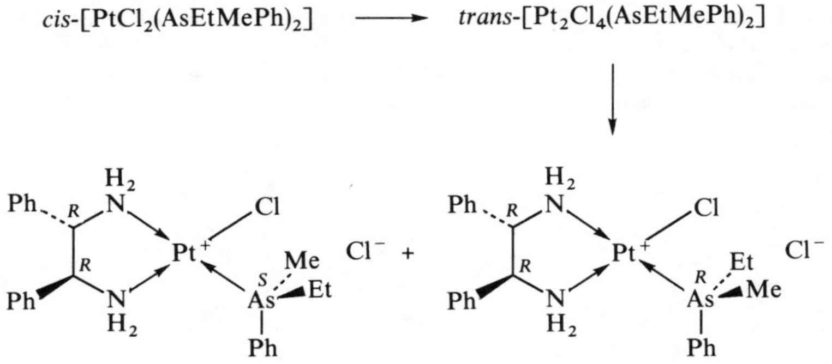

$cis\text{-}[PtCl_2(AsEtMePh)_2]$ $\longrightarrow$ $trans\text{-}[Pt_2Cl_4(AsEtMePh)_2]$

SCHEME 3

are configurationally stable in solution at 20 °C, although it was not possible to separate the diastereomers by fractional crystallization. The 7-methyl substituted arsole is demethylated by bromine giving the corresponding 7-bromo compound which, in turn is converted by additional bromine into configurationally stable $(\pm)$-[2-(2'-bromo-1,1'-binaphthyl)]dibromoarsine.

b. As_2-Bidentates. The chelating bis(tertiary arsine) 1,2-phenylenebis(methylphenyl-arsine) (75) has been synthesized in 89% yield from 1,2-dichlorobenzene and sodium methylphenylarsenide in tetrahydrofuran[85,102]. Crystallization of the equimolar mixture of R^*,R^* and R^*,S^* diastereomers of 75 from diethyl ether–methanol afforded the R^*,R^* diastereomer as air-stable needles, mp 88–88.5 °C; recovery of the more soluble R^*,S^* diastereomer, mp 61–62 °C, was via its sparingly soluble nickle(II) chloride derivative[103].

The diastereomers (R^*,R^*)- and (R^*,S^*)-75 were identified by ^{13}C NMR spectroscopic analysis of the tetracarbonylmolybdenum(0) derivatives in the metal–carbonyl region[104] and by X-ray crystal structure analyses of tricarbonyldiiodomolybdenum(II) derivatives[105]. It is noteworthy that the higher-melting R^*,R^* diastereomer of 75 crystallizes preferenti-ally in a typical second-order asymmetric transformation[21,106] from an equilibrium mixture of the two diastereomers in methanol containing hydrochloric acid and sodium iodide, thereby provding a simple method of converting almost all of the original mixture of 75 into the more desirable R^*,R^* diastereomer[103]:

$$(R^*,S^*)\text{-}75 \rightleftharpoons (R^*,R^*)\text{-}(\pm)\text{-}75 \downarrow$$

TABLE 5. Optically active tertiary arsines via metal complexes

Arsine	Abs. Config.	$[\alpha]_D{}^a$	Solvent	Reference
As-Unidentates				
AsEtMePh (**11**)	R	−3.3	Et$_2$O	9
	S	+3.1	Et$_2$O	9
As$_2$-Bidentates				
1,2-C$_6$H$_4$(AsMePh)$_2$ (**75**)	R	+95	CH$_2$Cl$_2$	103
	S	−95.2	CH$_2$Cl$_2$	103
As,N-Bidentates				
AsMePh(8-Quin) (**73**)	R	−115	Et$_2$O	92
	S	+115	Et$_2$O	92
AsMePh(CH$_2$CH$_2$NH$_2$) (**87**)	R	−13	CH$_2$Cl$_2$	110
	S	+13.1	CH$_2$Cl$_2$	110
AsMe(C$_6$H$_4$CH$_2$OH-*o*)(CH$_2$CH$_2$NH$_2$) (**89**)	R	−65	CH$_2$Cl$_2$	112
	S	+65	CH$_2$Cl$_2$	112
Arsamine **105**	R	−104	CH$_2$Cl$_2$	112
	S	+104	CH$_2$Cl$_2$	112
As,P-Bidentates				
1,2-C$_6$H$_4$(AsMePh)(PMePh)[(*R*,R**)-**93**]	R_{As},R_p	+79.0	CH$_2$Cl$_2$	113
	S_{As},S_p	−79.4	CH$_2$C.$_2$	113
1,2-C$_6$H$_4$(AsMePh)(PMePh) [(*R*,S**)-**93**]	S_{As},R_p	+15.5	CH$_2$Cl$_2$	113
	R_{As},S_p	−15.5	CH$_2$Cl$_2$	113
As,S-Bidentates				
AsMePh(CH$_2$CH$_2$SH) (**96a**)	R	−16.7	CH$_2$Cl$_2$	114
	S	+16.7	CH$_2$Cl$_2$	114
AsMe(C$_6$H$_4$CH$_2$OMe-*o*)(CH$_2$CH$_2$SH) (**96b**)	R	+1.1°	CH$_2$Cl$_2$	116
	S	−1.2°	CH$_2$Cl$_2$	116
Arsathiane **110**	R	−32.5	CH$_2$Cl$_2$	116
	S	+32.5	CH$_2$Cl$_2$	116
As$_4$-Quadridentates				
(*R*,R**)-tetars (**99**)	S	+28.7	CHCl$_3$	117
As$_2$N$_2$-Macrocycles				
Diarsadiimine **102**	R	+99.8	CH$_2$Cl$_2$	119
	S	−99.8	CH$_2$Cl$_2$	119
Diarsadiamine **104**	R	+190	CH$_2$Cl$_2$	119
	S	−190	CH$_2$Cl$_2$	119
As$_2$S$_2$-Macrocycles				
Diarsadithiane **107**	R	−180	CH$_2$Cl$_2$	116
	S	+181	CH$_2$Cl$_2$	116

^{a} Specific rotations $[\alpha]_D$ to nearest 0.1° measured at specified concentrations and temperatures (*ca* 20 °C).

In a bridge-splitting reaction, (*R*,R**)-**75** reacts with (*S*)-(+)-**77** in methanol to give a solution of the *chloride* salts [*S*,(R_{As},R_{As})]- and [*S*,(S_{As},S_{As})]-**82** which, when treated with 1 equiv. NH$_4$PF$_6$, yields 95% of [*S*,(S_{As},S_{As})]-(+)-**83** having $[\alpha]_D$ + 297° (acetone) after recrystallization from acetone–benzene (Scheme 5)[103]. Cyanide cleanly liberated (*R,R*)-(+)-**75** from [*S*,(S_{As},S_{As})]-(+)-**83** in a dichloromethane–water mixture; the arsine was recovered from the organic phase and recrystallized from aqueous methanol (Table 5). It was found later that 1,2-diaminoethane displaced the optically active bis(tertiary

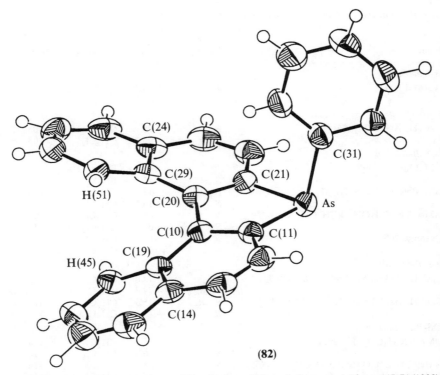

(82)

SCHEME 4. Reprinted by permission of Elsevier Sequoia from *J. Organomet. Chem.*, **445**, 71 (1993)

arsine) from $[S,(S_{As},S_{As})]-(+)$-**83**, giving $(S)-(+)$-**84**. The stereospecificity of the displacement and optical purity of $(R,R)-(+)$-**75** was confirmed by re-preparing $[S,(S_{As}, S_{As})]-(+)$-**83** from $(R,R)-(+)$-**75** and $(S)-(+)$-**77**/NH_4PF_6 and checking the rotation and the ^1H NMR spectrum of the complex. Optically pure $(S,S)-(-)$-**75** was obtained by a similar procedure involving $(R)-(-)$-**77**. In connection with work where solvents for rotations are not specified, it is instructive to note the following values of $[\alpha]_D$ for $10\,g\,l^{-1}$ solutions of (S,S)-**75**: $-99.7°$ (Me_2CO), $-95°$ (CH_2Cl_2), $-57.5°$ (C_6H_6), $-57.4°$ (Et_2O) and $-55.3°$ ($CHCl_3$)[103].

The crystal and molecular structure of $[S,(S_{As},S_{As})]-(+)$-**83** has been determined[93]. The structural analysis provided the absolute chirality of the cation and of the three stereogenic centres present, including the carbon of the *ortho*-metalated $(S)-(-)$-dimethyl-(α-methylbenzyl)amine, which had been established previously on the basis of CD and ORD measurements[107].

The phosphorus analogue of **75** has been synthesized, separated into diastereomers and resolved by a similar strategy to that employed for the arsine[108].

c. As,N-Bidentates. Several chelating asymmetric *As,N*-bidentates have been resolved with use of the palladium(II) resolving agents $(R)-(+)$- and $(S)-(+)$-**78**. Methylphenyl(8-quinolyl)arsine (**73**) was resolved by the fractional crystallization of the diastereomers $(R,R_{As})-(-)$- and $(R,S_{As})-(+)$-**86**[92]. Arsenic-containing ligands are readily displaced from palladium(II) by 1,2-diaminoethane. Thus treatment of $(R,R_{As})-(+)$- or $(R,S_{As})-(-)$-**86** with the diamine liberated the optically pure arsines (Table 5) with formation of

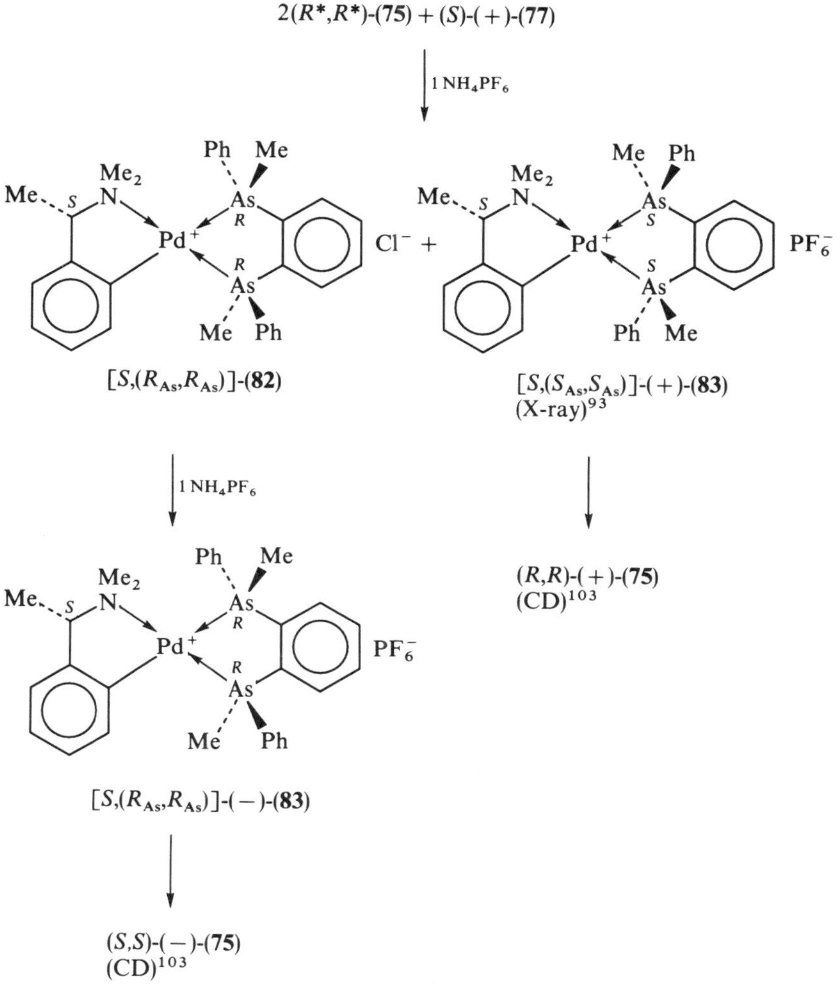

$2(R^*,R^*)$-(**75**) + (S)-(+)-(**77**)

$\downarrow$ 1 NH$_4$PF$_6$

$[S,(R_{As},R_{As})]$-(**82**)

$[S,(S_{As},S_{As})]$-(+)-(**83**)
(X-ray)[93]

$\downarrow$ 1 NH$_4$PF$_6$

$\downarrow$

(R,R)-(+)-(**75**)
(CD)[103]

$[S,(R_{As},R_{As})]$-(−)-(**83**)

$\downarrow$

(S,S)-(−)-(**75**)
(CD)[103]

SCHEME 5

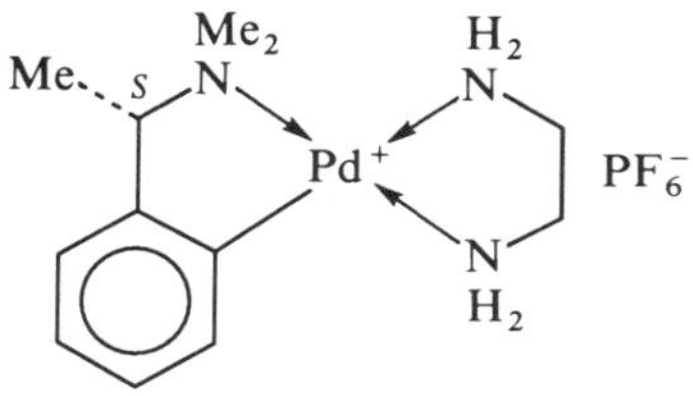

(S)-(+)-(**84**)

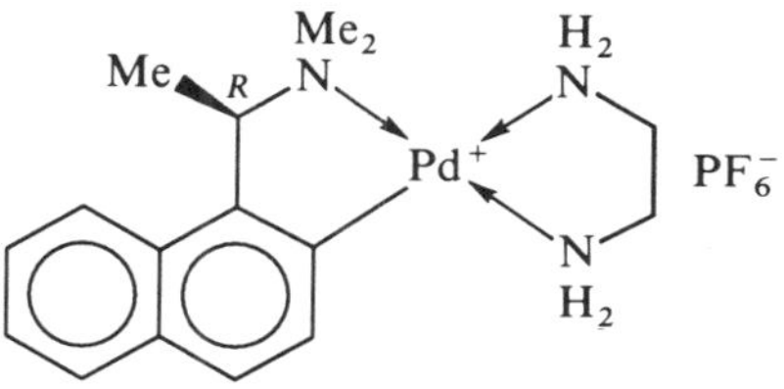

(R)-(−)-(**85**)

(R)-(−)-**85**; the amine complex was converted back into the chloro-bridged resolving
agent (R)-(+)-**78** with hydrochloric acid.

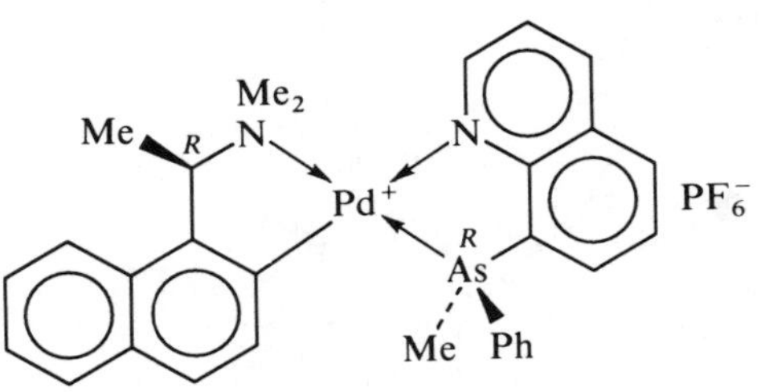

(R,R_{As})-(−)-(**86**) 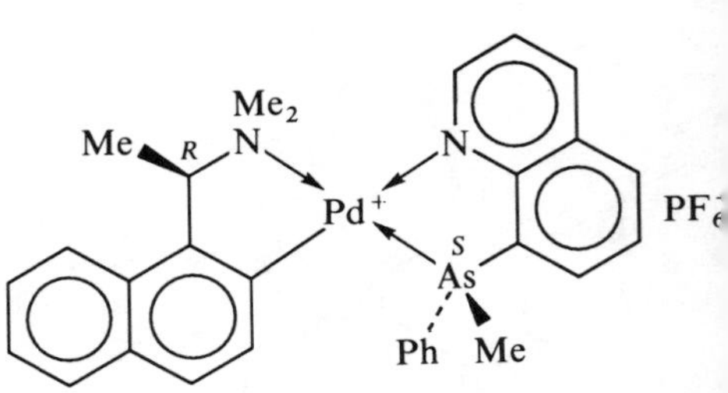 (R,S_{As})-(−)-(**86**)

The absolute configuration of (R)-(−)-**73** was deduced from the ¹H NMR spectrum
of (R,S_{As})-(−)-**86** by comparison with the spectrum of the corresponding phosphorus
compounds for which the crystal and molecular structure had been determined. In
complexes of the type (R,S_{As})-(−)-**86** the γ-naphthalene aromatic proton adjacent to the
Pd—C bond is strongly shielded when the arsenic stereocentre has the S configuration
because of the proximity of the arsenic–phenyl group. Because of the puckering of the
organometallic ring[109], the phenyl group shields the γ-naphthalene proton in the R,S_{As}
diastereomer, but not in the R,R_{As} diastereomer. The upfield shift of the resonance for
the γ-proton is observed generally when the adjacent chiral arsenic stereocentre is
contained in a chelate ring. Consistent results were obtained when the corresponding
complexes of the enantiomers of (R^*,R^*)-**75** were investigated similarly.

The bidentate 1-amino-2-(methylphenylarsino)ethane (**87**) was resolved by separation
of the diastereomers (R,R_{As})-(+)- and (R,S_{As})-(−)-**90**[110]. The absolute configurations of
the pure enantiomers of **87** were deduced from the ¹H NMR spectra of (R,R_{As})-(+)- and
(R,S_{As})-(−)-**90** where, once again, the upfield shift of the γ-naphthalene proton in the
R,S_{As} diastereomer was observed (Table 5). The complex (+)-cis-[Pt((R)-**87**)₂](PF₆)₂,
when stirred for 16 h in acetone at 25 °C in the presence of a trace of (S)-(+)-**87**, produced

(**87**) R = Me
(**88**) R = n-Bu
(**89**) R = 2-$C_6H_4CH_2OH$

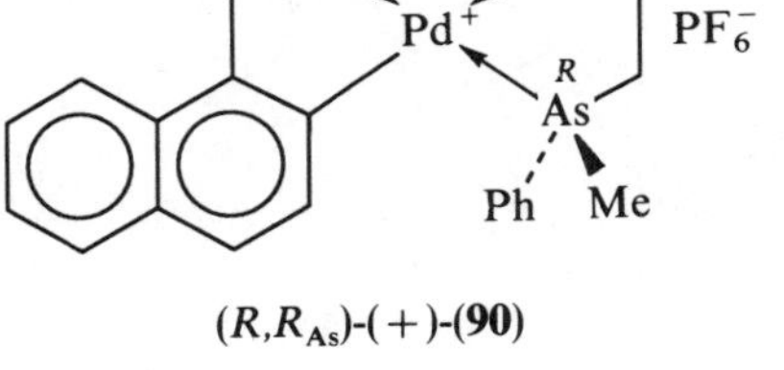

(R,R_{As})-(+)-(**90**) 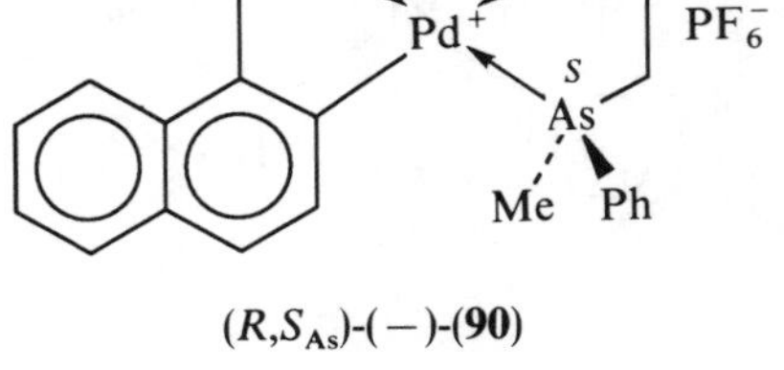 (R,S_{As})-(−)-(**90**)

a mixture of the epimers **91a** and **91b** in high yield. In another laboratory, the resolutions of **87** and the *n*-butyl congener **88** were similarly performed, but neither arsine was isolated[111]. The CD spectra of the diastereomers (S,S_{As})- and (S,R_{As})-**90** (chloride salts) and related complexes were recorded and compared with those of the analogous phosphine complexes[111].

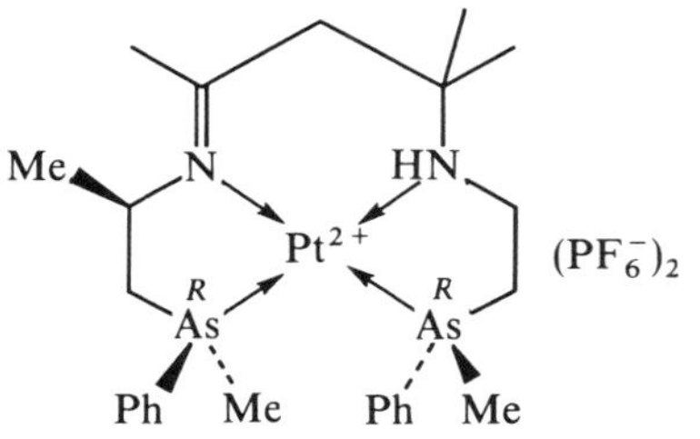

(**91a**)

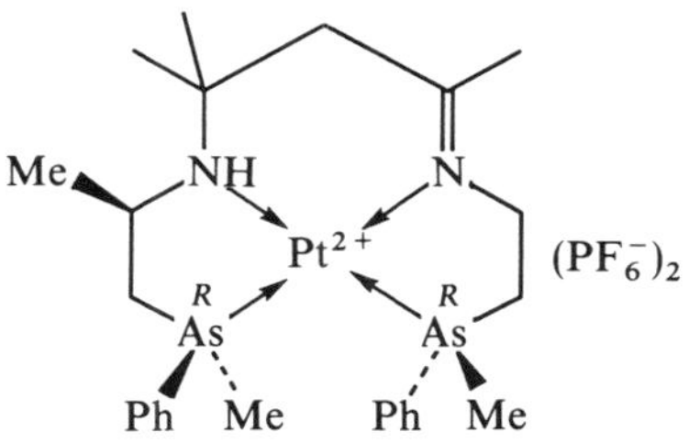

(**91b**)

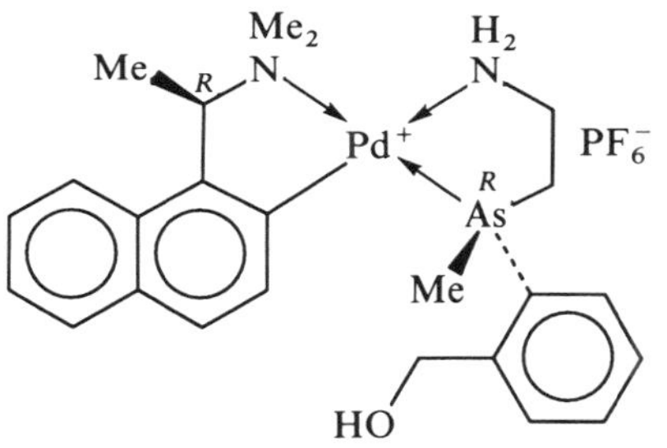

(R,R_{As})-(+)-(**92a**)

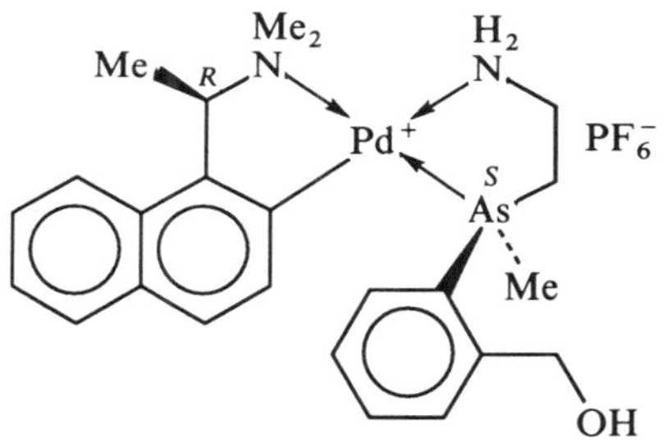

(R,R_{As})-(−)-(**92a**)

The highly functionalized arsine **89** was resolved by the fractional crystallization of the diastereomers (R,R_{As})-(+)- and (R,S_{As})-(−)-**92a**[117]. Oxidation of the R,S_{As} diastereomer with $BaMnO_4$ produced the corresponding aldehyde complex that was converted into a complex of an optically active *trans*-As_2N_2 macrocycle (see below).

d. As,P-Bidentates. The (R^*,R^*)-($\pm$) and (R^*,S^*)-($\pm$) forms of the asymmetric bidentate **93** have been separated and individually resolved by the metal complexation method (Table 5)[113]. The diastereomers of the *As,P* bidentate were identified by examining the ^{13}C NMR spectra of the dicarbonylnickel(0) derivatives in the metal–carbonyl region. The (R^*,R^*)-($\pm$) form of the ligand was resolved by the separation of $[R,(R_{As},R_P)]$-(−)- and $[R,(S_{As},S_P)]$-(+)-**94** and the (R^*,S^*)-($\pm$) form by the separation of $[R,(R_{As},S_P)]$-(+)- and $[R,(S_{As},R_P)]$-(−)-**95**. Absolute configurations were assigned to the four enantiomers of **93** on the basis of the shielding of the γ-naphthalene protons in the ^1H NMR spectra of $[R,(R_{As},S_P)]$-(+)- and $[R,(S_{As},R_P)]$-(−)-**95**, and the corresponding diastereomers of (R^*,R^*)-**93**, namely $[R,(R_{As},R_P)]$-(+)- and $[R,(S_{As},S_P)]$-(−)-**95**.

Although (R^*,R^*)-($\pm$)- and (R^*,S^*)-($\pm$)-**93** epimerize at phosphorus when heated at 140 °C for *ca* 2 h, epimerization at arsenic was not effected by the usual racemizing conditions for tertiary arsines (methanolic HCl); indeed, optically active salts protonated at phosphorus were isolated.

(R_{As},R_P)-$(+)$-(**93**) (S_{As},S_P)-$(-)$-(**93**)

(R_{As},S_P)-$(+)$-(**93**) (S_{As},R_P)-$(-)$-(**93**)

$[R,(R_{As},R_P)]$-$(-)$-(**94**) $[R,(S_{As},S_P)]$-$(+)$-(**94**)

$[R,(R_{As},S_P)]$-$(+)$-(**95**) $[R,(S_{As},R_P)]$-$(-)$-(**95**)

$[R,(R_{As},R_P)]\text{-}(+)\text{-}(95)$ $[R,(S_{As},S_P)]\text{-}(-)\text{-}(95)$

e. As,S-Bidentates. The asymmetric chelating agent (2-mercaptoethyl)methylphenyl-arsine (**96a**) was resolved by the crystallization of the unusual μ-thiolato diastereomer $(R,R,S_{As},S_S)\text{-}(-)\text{-}$**97**, which was obtained from 2 equiv. $(R)\text{-}(-)\text{-}$**78** and 1 equiv. each of **96a** and triethylamine[114]. The crystal and molecular structure of highly crystalline $(R,R,S_{As},S_S)\text{-}(-)\text{-}$**97** was determined. Treatment of the pure diastereomer with 1,2-diaminoethane split off $(R)\text{-}(-)\text{-}$**85** leaving $(R,S_{As})\text{-}(-)\text{-}$**98** from which aqueous KCN liberated $(R)\text{-}(-)\text{-}$**96a** (equation 12), which could be distilled without racemization. The

$(R,R,S_{As},S_S)\text{-}(-)\text{-}$**(97)** $(R,S_{As})\text{-}(-)\text{-}$**(98)**
(X-ray)[114]

optical purity of the arsine was confirmed by preparing the kinetically labile nickel(II) complex of the deprotonated ligand, which showed no evidence of the thermodynamically stable *meso* (R^*,S^*) diastereomer in the ^{1}H NMR spectrum in chloroform[115]. Decomposition of the μ-thiolato complex remaining in the mother liquor after removal of $(R,R,S_{As},S_S)\text{-}(-)\text{-}$**97** with 1,2-diaminoethane and cyanide afforded $(S)\text{-}(+)\text{-}$**96a** of 72% optical purity. The impure material was brought to complete purity by reacting it with 2 equiv. $(S)\text{-}(+)\text{-}$**78** in the presence of base and crystallizing to purity $(S,S,R_{As},R_S)\text{-}(+)\text{-}$**97** from which pure $(S)\text{-}(+)\text{-}$**96a** was liberated with cyanide (Table 5). The analogous light-sensitive 2-mercaptoethylphosphine has also been resolved by the metal complexation method, but it required alkylation before it could be displaced from the metal[121].

(**96a**) Ar = Ph
(**96b**) Ar = 2-$C_6H_4CH_2OMe$

By following an almost identical procedure, the enantiomers of the corresponding 2-methoxymethylaryl compound **96b** were obtained (Table 5)[116]. The enantiomers of **96b** were used for the metal-template synthesis of optically active *trans*-As$_2$S$_2$ macrocycles (see below).

f. As$_4$-Linear quadridentates. The linear tetra(tertiary arsine) **99** (tetars) has been separated into R^*,R^* and R^*,S^* diastereomers by the fractional crystallization of dichlorocobalt(III) complexes in which the ligand acts as a quadridentate[117]. The reaction of ligand with cobalt(II) chloride in methanolic HCl in air gave an *ca* 80% yield of the complex [CoCl$_2$(**99**)]Cl, which consisted of *ca* 60% of the blue *cis* diastereomer (R^*,R^*)-**100** and *ca* 40% of the green *trans* diastereomer (R^*,S^*)-**100**. Decomposition of the two forms of the complex with cyanide afforded the pure R^*,R^* and R^*,S^* diastereomers of **99**. Detailed investigations of the stereochemistry and interconversions of the five possible diastereomers of the octahedral cobalt(III) complexes of the ligand were undertaken[118]. The reaction of the complex (R^*,R^*)-**100** with (−)-dibenzoyl-hydrogentartrate afforded the corresponding pair of diastereomeric *cis*-α-complexes that were separated and individually converted into the perchlorate salts from which the respective enantiomers of (R^*,R^*)-**99** were displaced with cyanide (Table 5). The absolute configurations of the enantiomers were inferred from analyses of the CD spectra of the dichlorocobalt(III) complexes and their relationship to the spectrum of the peroxo complex *cis*-β-[CoO$_2${(R,R)-**99**}]ClO$_4$, for which the crystal and molecular structure had been determined.

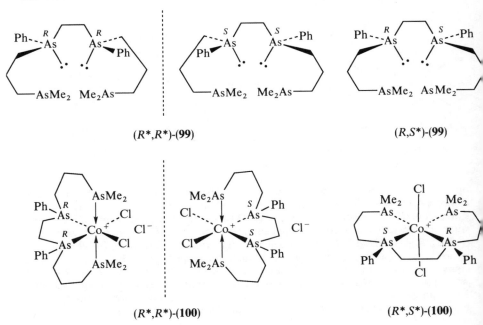

(R^*,R^*)-(**99**) (R,S^*)-(**99**)

(R^*,R^*)-(**100**) (R^*,S^*)-(**100**)

g. As$_2$N$_2$-Macrocyclic quadridentates. When **101** was heated at 80 °C in vacuo, the 14-membered *trans*-As$_2$N$_2$macrocycle (R^*,S^*)-**102** was formed quantitatively and with complete stereoselectivity by elimination of 1,2-bis(methylamino)ethane (Scheme 6)[112]. In chloroform, (R^*,S^*)-**102** rearranges slowly and quantitatively into the chiral seven-

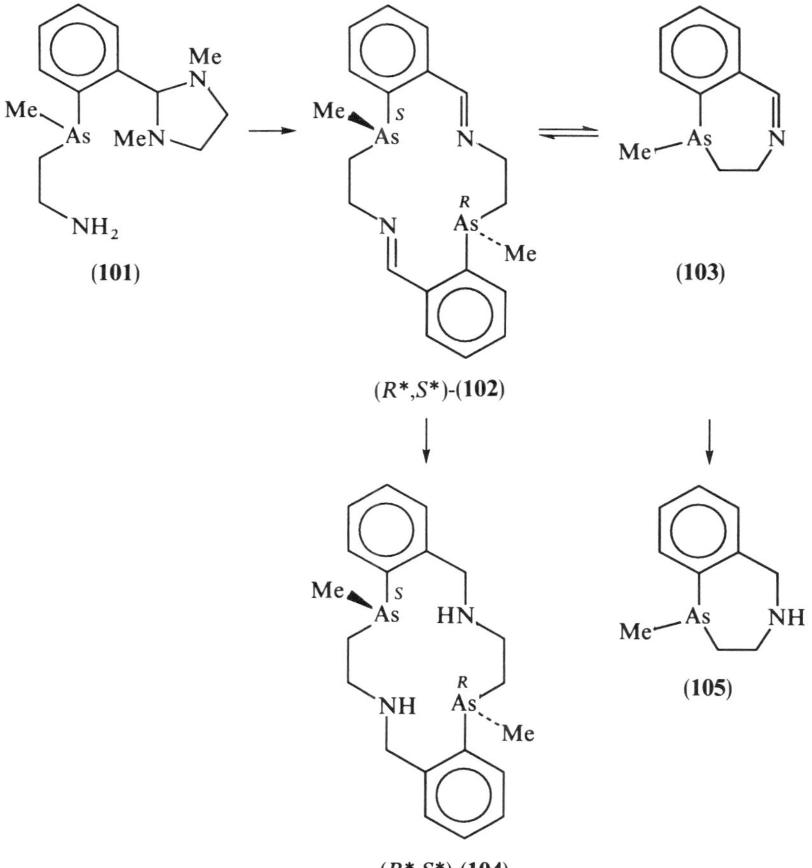

(101)

(R^*,S^*)-(102)

(103)

(R^*,S^*)-(104)

(105)

SCHEME 6

membered As,N heterocycle 103; when the solvent is removed from the solution of the monomer, however, achiral (R^*,S^*)-102 reforms quantitatively. The rearrangement into 103 is catalysed by acid. The behaviour of (R^*,S^*)-102 and 103 parallels that of 4,5-dihydro-3H-2-benzazepine and its dimer, for which a mechanism involving a 1,3-diazetidine intermediate has been proposed[119]. Lithium aluminium hydride reduction of (R^*,S^*)-102 before or after rearrangement yielded the trans-$As_2(NH)_2$ macrocycle (R^*,S^*)-104 or the monomer 105, respectively (Scheme 6). The benzaldehyde complex (R,S_{As})-92b was prepared by $BaMnO_4$ oxidation of the benzyl alcohol complex (R,S_{As})-(−)-92a; the former with 1,2-bis(methylamino)ethane gave the diastereomers of 106 (Scheme 7). Alternatively, 106 was produced in an asymmetric synthesis involving the reaction of (R)-(+)-78 with a solution of monomer 103 (Section III.C.2). Air-stable (R,R)-(−)-102 was isolated by displacement from yellow $[(R,(S_{As},R_{As})]$-(−)- or orange $[(R,(S_{As},S_{As})]$-(−)-106, which are four- and five-coordinate forms of the same complex. The crystal-structures of both complexes were determined. The enantiomers of the R^*,R^*

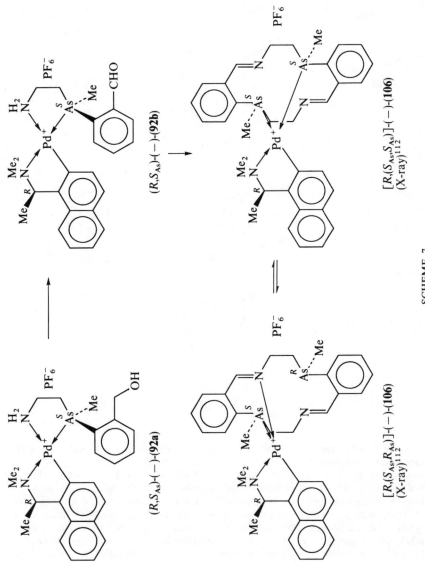

(R,S_{As})-$(-)$-(92a)

(R,S_{As})-$(-)$-(92b)

$[R,(S_{As},S_{As})]$-$(-)$-(106)
(X-ray)^{112}

$[R,(S_{As},R_{As})]$-$(-)$-(106)
(X-ray)^{112}

SCHEME 7

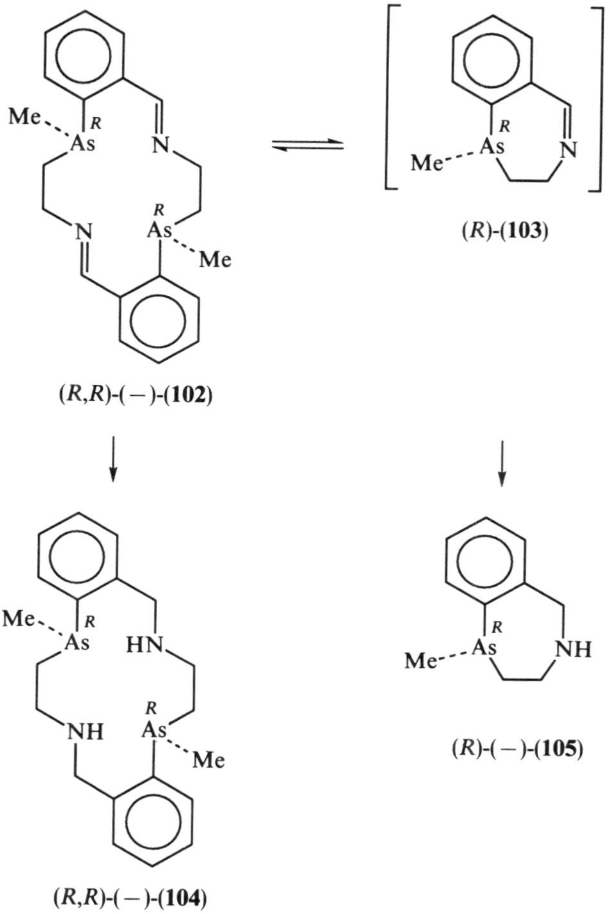

(R,R)-(−)-(102)

(R)-(103)

(R,R)-(−)-(104)

(R)-(−)-(105)

SCHEME 8

form of the diimine macrocycle are individually stable, but when the two are together achiral (R^*,S^*)-**102** crystallizes. The *S,S* enantiomer of (R^*,R^*)-**102** was obtained by similar reactions involving (*S*)-(−)-**78**. Lithium aluminium hydride reductions of the optically active forms of the diimine macrocycle yielded the corresponding enantiomers of the *trans*-$As_2(NH)_2$ diamine macrocycle **104**, although if acid-catalysed rearrangements into the optically active mono-imine were first performed, the enantiomers of the corresponding seven-membered As(NH) hetereocycle **105** were isolated (Scheme 8). The enantiomers of the *trans*-$As_2(NH)_2$ macrocycles crystallize from chloroform as 2-chloroform solvates (Table 5).

h. As_2S_2-Macrocyclic quadridentates. The air-stable enantiomers of (R^*,R^*)-**107** have been isolated by treatment of chloroform solutions of the respective enantiomers of the

(R^*,R^*)-(**107**)
(X-ray)[121]

(R^*,S^*)-(**107**)

palladium(II) complex of deprotonated **96b**, (S,S)-$(-)$- and (R,R)-$(+)$-**108**, with boron tribromide and subsequent displacement of the optically active forms of the trans-As_2S_2 ligand from the intermediate complexes with cyanide[116]. The reaction of the thermo-dynamically preferred trans diastereomer of the (S,S)-$(-)$ precursor complex **108** led streoselectively to the most stable one of six diastereomers of the product complex, namely (S_{As},S_{As},R_S,R_S)-**109**, as shown in Scheme 9. The cyclization proceeds in 79% yield to give 66% (R,R)-$(-)$-**107** and 13% (R)-$(-)$-**110** (Table 5). Use of the racemic form of the palladium complex afforded in a highly stereoselective reaction 50% (R^*,R^*)-**107** and 38% $(\pm)$-**110**. In turn, (R^*,R^*)-**107** was quantitatively converted by a second-order asymmetric transformation into less soluble (R^*,S^*)-**107** by crystallization from chloroform containing a trace of hydrochloric acid.

$$(R^*,R^*)\text{-}\mathbf{107} \rightleftharpoons (R^*,S^*)\text{-}\mathbf{107} \downarrow$$

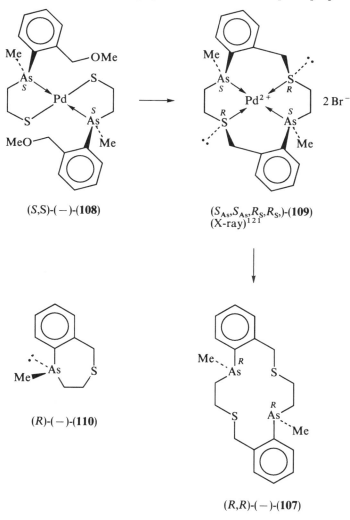

(S,S)-$(-)$-(**108**)

$(S_{As},S_{As},R_S,R_S,)$-(**109**)
(X-ray)[121]

(R)-$(-)$-(**110**)

(R,R)-$(-)$-(**107**)

SCHEME 9

When coordinated to palladium(II), however, $(R*S*)$-**107** quantitatively transforms into $(R_{As}*,R_{As}*,S_S*,S_S*)$-**107** when heated for 30 min in dimethyl sulphoxide. The crystal structures of $(R*,R*)$-**107**, $(R*,S*)$-**108**, $(R_{As}*,R_{As}*,S_S*,S_S*)$-, $(R_{As}*,S_{As}*,R_S*S_S*)$- and $(R_{As}*, S_{As}*,S_S*,R_S*)$-**109** have been determined[120].

2. Asymmetric synthesis

The asymmetric synthesis of coordinated (R)-$(-)$-**11** has been effected by the stereo-selective deprotonation and ethylation at $-65\,°C$ of the diastereomer $[(R_P,R_P),R_{As}]$-$(+)$-**80** (equation 13)[98]. The pure diastereomer of the secondary arsine complex was obtained

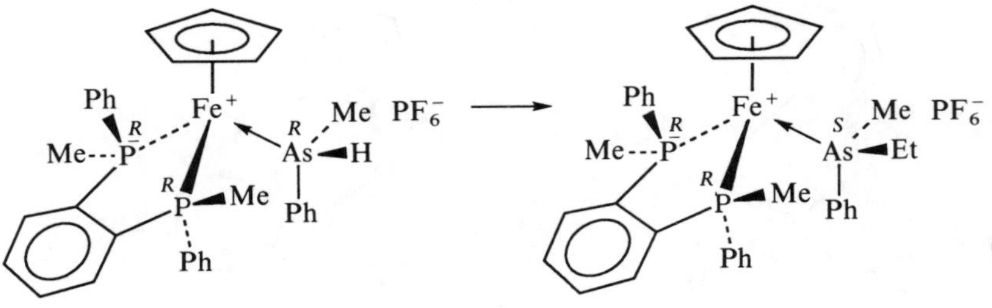

$$[(R_P,R_P),R_{As}]\text{-}(+)\text{-}(\mathbf{80})$$

$$[(R_P,R_P),S_{As}]\text{-}(+)\text{-}(\mathbf{81})$$

(13)

by a second-order asymmetric transformation in *ca* 90% yield in a highly stereoselective synthesis involving $(R_P,R_P)\text{-}(+)\text{-}[(\eta^5\text{-}C_5H_5)\{1,2\text{-}C_6H_4(PMePh)_2\}Fe(MeCN)]PF_6$ and methylphenylarsine. The isolation of $[(R_P,R_P),R_{As}]\text{-}(+)\text{-}\mathbf{80}$ constitutes the first resolution of secondary arsine, at least in coordinated form. The conversion of the resolved secondary arsine into the resolved tertiary arsine complex $[(R_P,R_P),S_{As}]\text{-}(+)\text{-}\mathbf{81}$ at low temperature, which is necessary because of the low barrier to inversion of the intermediate tertiary arsenido complex, is completely stereoselective with retention of configuration at arsenic. Unfortunately, however, due to inertness of the tertiary arsine complex, prolonged heating of the complex with cyanide in methanol was required to liberate the resolved arsine, which had $[\alpha]_D - 2.0°$ (Et$_2$O). Nevertheless, the conversion of methyl-

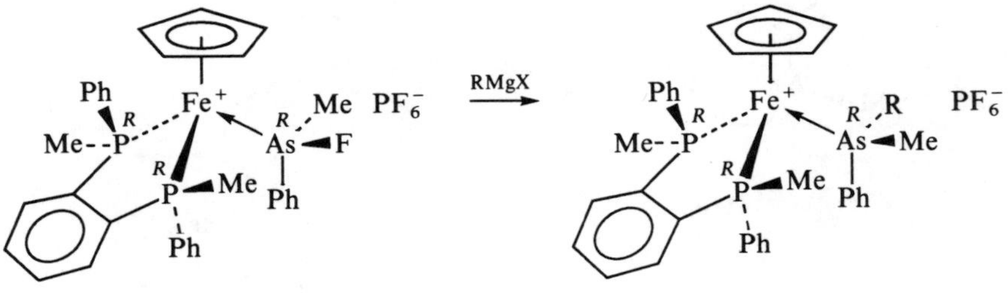

$$[(R^*_P,R^*_P),R^*_{As}]\text{-}(\mathbf{111})$$

$$[(R^*_P,R^*_P),R^*_{As}]\text{-}(\mathbf{112})$$

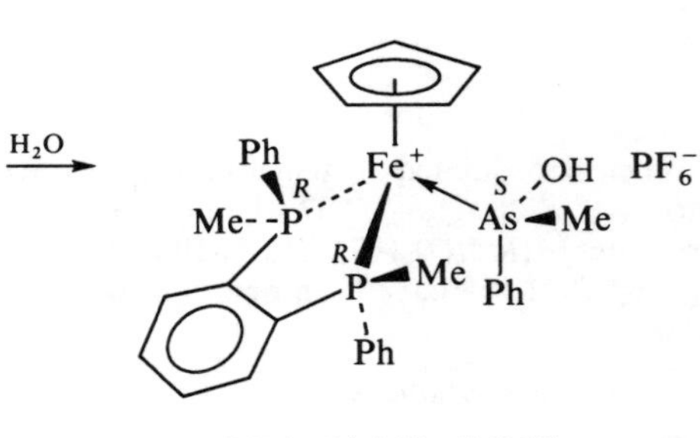

$$[(R^*_P,R^*_P),S^*_{As}]\text{-}(\mathbf{113})$$

SCHEME 10

phenylarsine into (R)-$(-)$-**11** is the first asymmetric synthesis of a tertiary arsine to be facilitated by a metal complex. In a related development, fluoromethylphenylarsine was resolved by the separation of the diastereomers $[(R_P^*,R_P^*),R_{As}^*]$-$(\pm)$- and $[(R_P^*,R_P^*),S_{As}^*]$-$(\pm)$-**111**[122]. The individual diastereomers, when treated with benzyl- or ethyl-magnesium bromide, gave, by substitution of fluoride, the respective tertiary arsine complexes $[(R_P^*,R_P),R_{As}]$-$(\pm)$-**112** with predominant inversion at arsenic. Hydrolysis of either diastereomer of the fluoroarsine complex led to the corresponding hydroxyarsine (arsinous acid) complex $[(R_P^*,R_P^*),S_{As}^*]$-$(\pm)$-**113** with complete inversion at arsenic (Scheme 10 shows the reactions for one enantiomer in each case).

The achiral 14-membered *trans*-diimine macrocycle (R^*,S^*)-**102**, in the presence of trifluoroacetic acid, rearranges quantitatively into the chiral seven-membered monoimine $(\pm)$-**103** (Section III.C.1.h)[112]. If the rearrangement of (R^*,S^*)-**102** is carried out in methanol containing a suspension of (R)-$(-)$-**78**, an orange solution is obtained from which pure $[R,(S_{As},R_{As})]$-$(-)$-**106** can be isolated by the addition of ammonium hexafluorophosphate. The yield of the complex was ca 50%. The addition of more acid and halide in an attempt to facilitate racemization of the free arsine and thereby promote the further crystallization of the complex by second-order asymmetric transformation was unsuccessful. Nevertheless, this highly stereoselective synthesis of $[R,(S_{As},R_{As})]$-$(-)$-**106** is a more expedient route to (R,R)-$(-)$-**102** than the one involving resolution of the benzyl alcohol complex (R,S_{As})-$(-)$-**92a**.

D. Stereoselective Syntheses via Arsenic(V) Acids and Esters and Arsine Sulphides

1. Reductive alkylation of esters of thioarsinic acids and arsine sulphides

A stereoselective synthesis of tertiary arsines of the type AsRMePh is available via the displacement of $(-)$-menthoxide from diastereomerically enriched thioarsinate menthyl esters at $-78\,°C$ with 2 equiv. of an alkyl- or aryl-lithium reagent in diethyl ether (equations 14 and 15)[123]. The CD curves of (R)-$(-)$- and (S)-**114** are enantiomeric in type as found for the analogous menthyl methylphenylphosphinates with which they can be structurally correlated by ^{1}H NMR spectroscopy. The reaction to the arsine proceeds in two steps: the first involves displacement of the menthoxide with 1 equiv. of the lithium reagent to give the tertiary arsine sulphide with a high degree of inversion

(R)-$(-)$-**(114)** (R)-$(+)$-**(115)** (S)-$(+)$-**(65)**   (14)

(S)-**(114)** (R)-$(+)$-**(116)** (S)-$(+)$-**(117)**   (15)

$(R)-(-)-$**(114)**
$(CD)^{123}$

$(S)-$**(114)**
$(CD)^{123}$

at arsenic; the second step is the completely stereoselective desulphurization of the intermediate arsine sulphide by the second equivalent of the organolithium reagent (equation 14).
Thus, $(R)-(-)-$**114** of 76% diastereomeric excess (de), when treated with 2 equiv. n-propyl-
lithium, gave $(S)-(+)-$**65** in 51% yield of 67% ee (Table 6). The optical purity of the arsine was
determined by preparing the arsine sulphide $(R)-(+)-$**115** from the arsine and sulphur
in boiling ethanol and recording the ^{1}H NMR spectrum of the sulphide in the presence
of a chiral europium(III) complex, which caused a doubling of the AsMe resonances.
The arsine itself caused no such doubling. This appears to be the only reported case of
the use of an NMR shift reagent to determine the optical purity of an arsine, albeit via
the sulphide. In the reaction of $(S)-$**114** with β-naphthyllithium, the intermediate arsine
sulphide $(R)-(+)-$**116** was isolated and separately reduced to $(S)-(+)-$**117** with n-BuLi
to confirm the stereoretentive nature of the latter step (equation 15). The first use of
hexachlorodisilane for the reduction of an arsine sulphide to an arsine was also reported
in this work. Thus, $(R)-(+)-$**116** with Si_2Cl_6 after 6 h in boiling benzene gave $(S)-(+)-$**117**
in unspecified yield with retention of configuration at arsenic.

Similar work has been carried out with the menthyl ethylphenylarsinothioates
(**118**)^{124,125} and menthyl phenyl-n-propylarsinothioates^{126}. The diastereomers $(R)-(-)-$
and $(S)-(-)-$**118** exhibit different chemical shifts and coupling constants in their 100-MHz
^{1}H NMR spectra in $(-)-\alpha$-methylbenzylamine, which can be used to assign configurations.
By following similar strategies to those indicated in equations 14 and 15, $(R)-(-)-$ and
$(S)-(-)-$**118** were converted by reaction with n-propylmagnesium bromide into the
$(R)-(+)-$ and $(S)-(-)-$ethylphenyl(n-propyl)arsine sulphides (**119**), respectively, which,
in turn, were individually converted into the respective $(S)-(+)-$ and $(R)-(-)-$ethylphenyl-

$(R)-(-)-$**(118)** $(R)-(+)-$**(119)** $(S)-(+)-$**(120)**   (16)

$(R)-$**(114)** $(R)-(+)-$**(121)** $(S)-(-)-$**(122)**   (17)

TABLE 6. Tertiary arsines via esters of arsenic(V) acids and arsine sulphides

Arsine	Abs. config.	$[\alpha]_D{}^a$	Solvent	Precursor	Abs. config.	$[\alpha]_D$	Solvent	Ref.
Via thioarsinic acid esters								
AsMePh(n-Pr) (65)	S	+10.7	MeOH	As(S)(OMen)MePh (114)	R	*b*		123
AsMe(β-Naph)Ph (117)	S	+1.7	CHCl₃	As(S)(OMen)MePh (114)	R			123
AsEtPh(n-Pr) (120)	S	+14.7	C₆H₆	As(S)(OMen)EtPh (118)	R	−50.4	CCl₄	124,125
Via arsine sulphides								
AsMe(β-Naph)Ph (117)	S	+1.4	CHCl₃	As(S)Me(β-Naph)Ph (116)	R	+7.8	MeOH	123
AsEtPh(n-Pr) (120)	R	−13.2	C₆H₆	As(S)EtPh(n-Pr) (119)	S	−17.7	MeOH	124, 125
	S	+14.4	C₆H₆	As(S)EtPh(n-Pr) (119)	R	+18.3	MeOH	124, 125
AsAllMePh (122)	S	−80.1	MeOH	As(S)AllMePh (121)	R	+28.8	MeOH	127
Via arsinic acid esters								
AsEtPh(n-Pr) (120)	S	+1.2	C₆H₆	As(O)(OMen)EtPh (125)	R	−8.7	MeNO₂	125

^{a} Specific rotations $[\alpha]_D$ to nearest 0.1° measured at specified concentrations and temperatures (*ca* 20 °C).
^{b} Reference 127 gives $[\alpha]_D$ −45.7° (CHCl₃) for the less soluble diastereomer of this compound.

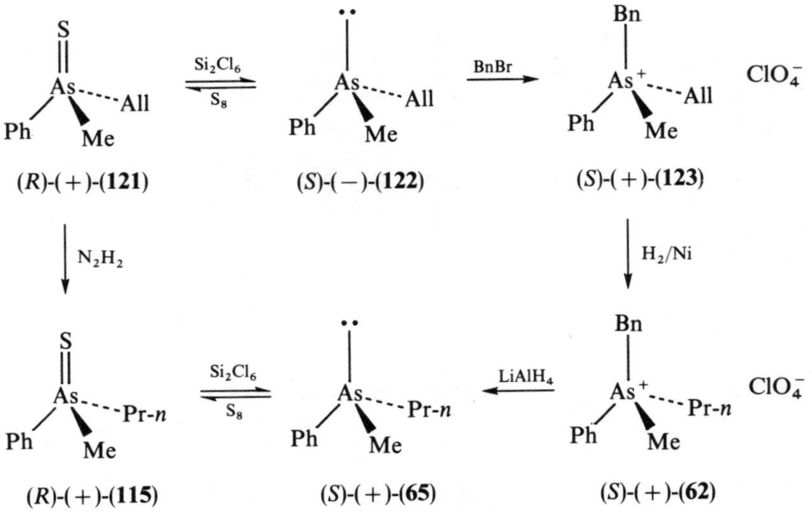

S. B. Wild

SCHEME 11[a]

[a]Primary reagents indicated only.

(n-propyl)arsines (120) with 1 equiv. n-propyllithium [equation 16 shows the reaction of (R)-(−)-118] (Table 6).

In the later work, (S)-(−)-allylmethylphenylarsine, (S)-(−)-122, was obtained by reduction of the corresponding arsine sulphide (R)-(+)-121 with allyllithium which, in turn, had been obtained from (R)-(−)-114 with the same reagent (equation 17)[127]. By chemical correlation and the aid of ORD spectroscopy, the configurations at arsenic in the enantiomers of the arsine 122 and its sulphide were determined, as indicated in Scheme 11.

Phenyl-p-tolylarsinthioic acid itself, 124, has been resolved with use of (−)-quinine[128]. The enantiomers of 124 have $[\alpha]_D$ values of +27.2° and −23.3° in toluene.

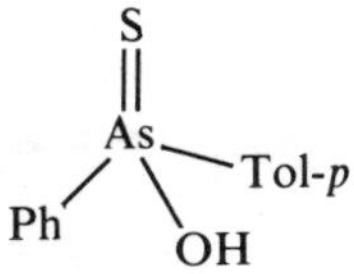

(124)

2. Reductive alkylation of arsinic acids and esters

a. By Grignard reagents. The moisture-sensitive arsinic acid menthyl ester diastereomers (R)-(−)- and (S)-(−)-125, $[\alpha]_D$ − 8.7° and − 2.6° (MeNO$_2$), respectively, were prepared by treatment of the respective thioarsinic acid esters 118 with dinitrogen tetroxide in nitromethane (Scheme 12)[125]. Ester (−)-125, having $[\alpha]_D$ − 8.7°, gave, however, with n-propylmagnesium bromide in THF the tertiary arsine (S)-(+)-120 with low stereoselectivity (Table 7).

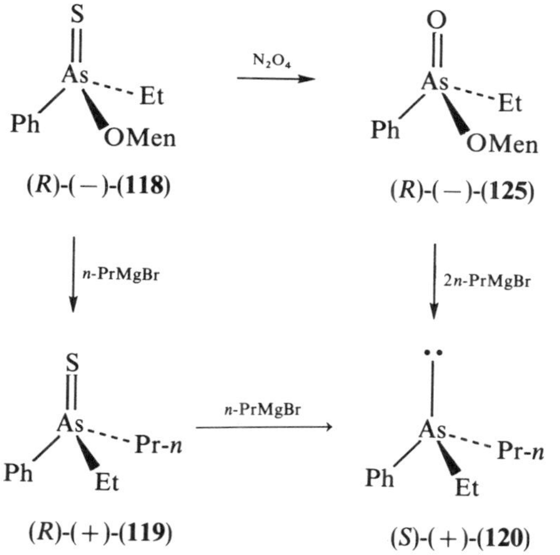

SCHEME 12

b. By microorganisms. The mould *Scopulariopsis brevicaulis*, when growing on bread treated with a 1% solution of ethyl-*n*-propylarsinic acid, produces (*R*)-ethylmethyl-*n*-propylarsine, (*R*)-**126**, in 60% ee[10]. The arsine was aspirated in a stream of air into a solution of the palladium(II) dimer (*S*)-(+)-**78** in toluene, whereupon a mixture of the diastereomers (*S*,*S*$_{As}$)- and (*S*,*R*$_{As}$)-**127** was produced (Scheme 13). The arsine was evolved

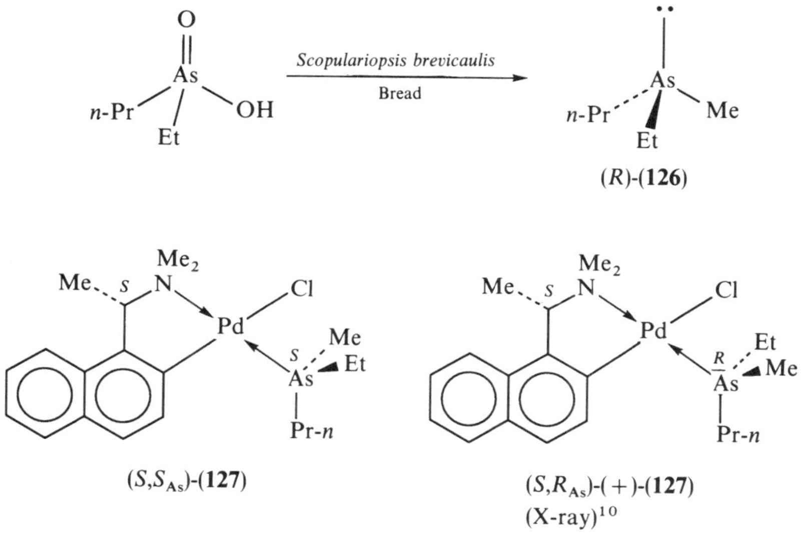

SCHEME 13

TABLE 7. Tertiary arsines via esters of arsenic(III) acids

Arsine	Abs. config.	$[\alpha]_D{}^a$	Solvent	Precursor	Abs. config.	$[\alpha]_D$	Solvent	Ref.
Via arsinous acid esters								
AsMePh(n-Pr) (65)	R	−4.0	MeOH	As(OEt)MePh (128)	R	+6.4	Neat	133
AsMePh(i-Pr) (125)	R	−2.3	MeOH	As(OEt)Ph(i-Pr) (129)	R	+5.2		133
AsEtPh(n-Pr) (120)	R	−5.7	MeOH	As(OC$_6$H$_{11}$-c)EtPh (130)	R	+7.4	Neat	133
AsMePh(n-Pr) (65)	R	−15.3	MeOH	As(OCinch)MePh (131a)		+164.2	CHCl$_3$	134
	S	+16.1	MeOH	As(OCinch)MePh (131b)		+189.6	CHCl$_3$	134
AsBuEt(i-Pr) (138a)		−11.3	MeOH	As(OCinch)Et(i-Pr) (132a)		+147.7	CH$_2$Cl$_2$	134
		+12.0	MeOH	As(OCinch)Et(i-Pr) (132b)		+169.4	CH$_2$Cl$_2$	134
AsMePh(n-Pr) (65)	R	−16.0	C$_6$H$_5$Me	As(OMen)MePh (133)	R	−44.3	Et$_2$O	135
AsMe(i-Bu)(o-Tol) (138b)	S	+26.6	C$_6$H$_5$Me	As(OMen)Me(o-Tol) (134)	S	−26.8	Et$_2$O	135
As(i-Bu)MePh (139)	R	−21.8	C$_6$H$_5$Me	As(OMen)MePh (133)	R	−44.3	Et$_2$O	135
As(i-Bu)Me(o-Tol) (140)	S	+19.1	C$_6$H$_5$Me	As(OMen)Me(o-Tol) (134)	S	−26.8	Et$_2$O	135
Via arsinothious esters								
AsMePh(n-Pr) (65)	R	−3.0	MeOH	As(SPentyl-n)MePh (135)	R	+5.7	Neat	131

^{a} Specific rotations $[\alpha]_D$ to nearest 0.1° measured at specified concentrations and temperatures (*ca* 20 °C).

over 14 days during rapid growth of the mould. After removal of toluene, the 500-MHz ^{1}H NMR spectrum of $(S,S_{As})/(S,R_{As})$-**127** indicated a 60% excess of the S,S_{As} diastereomer. Thus, because of the reversal of the CIP descriptor for the arsine upon coordination to palladium, the arsine produced by the mould was predominantly of the R configuration. The identity of the diastereomeric complexes was established by carrying out a crystal and molecular structure determination on the complex (S,R_{As})-$(+)$-**127**, which was obtained pure after nine recrystallizations of the less soluble component of a synthetic mixture of the two diastereomers.

The discovery of the stereoselective biomethylation of a prochiral arsinic acid by a microorganism opens up an exciting new route to optically active tertiary arsines. Numerous microorganisms reductively methylate arsenic(V) compounds[129]. The biological synthesis of $(\pm)$-**126** by *Scopulariopsis brevicaulis* was described in 1936[6]. At that time, however, it was not recognized that simple tertiary arsines chiral at arsenic were configurationally stable and amenable to optical resolution.

E. Stereoselective Syntheses via Esters of Arsinous and Arsinthious Acids

The optically active arsinous acid esters **128**–**134** and the arsinthious acid ester **135** have been used to prepare optically active tertiary arsines (Table 7). When the arsine sulphide **136** was heated in benzene with *n*-butyl or *n*-propyl bromide, ethyl bromide was eliminated and there was a reversal in the sign of the rotation, which was attributed to the formation of the inverted *n*-butyl and *n*-propyl esters of (*p*-carboxyphenyl)phenyl-arsinthious acid, **137a** and **137b**, respectively (equation 18)[130].

(128)-(134)   (R)-(+)-(135)

(141)

Some years later, stereoselective syntheses of a variety of optically active esters of arsinthious acids were reported[131]. The esters were prepared by treating secondary chloroarsines with thiols in the presence of $(-)$-*N,N*-diethyl-α-methylbenzylamine. The *n*-pentyl ester **135**, when allowed to react with *n*-propylmagnesium bromide, gave (R)-$(-)$-**65** with low stereoselectivity and inversion at arsenic as indicated in equation 19 (Table 7)[131].

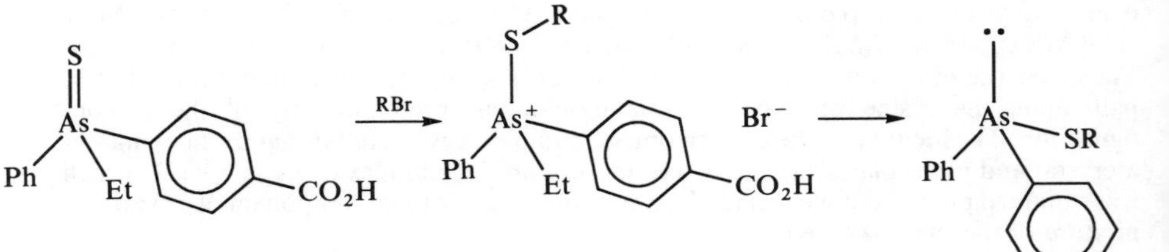

(136)

(137a) R = Bu-*n*
(137b) R = Pr-*n*

(18)

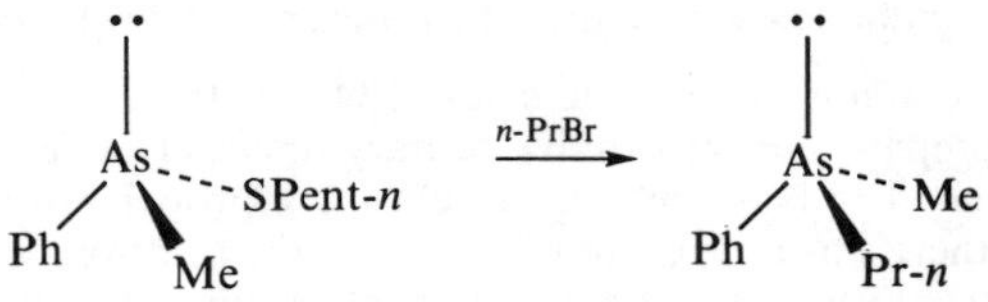

(*R*)-(+)-(135)   (*R*)-(−)-(65)

(19)

Optically active alkylphenylarsinous acid esters can be prepared by treating chloro-alkylphenylarsines with alcohols in the presence of (−)-brucine[132] or (−)-*N*,*N*-diethyl-α-methylbenzylamine[133]. Secondary haloarsines react with the lithium derivative of (−)-cinchonidine to give the crystalline cinchonidine esters of arsinous acids **131** and **132**[134]. Another route to optically active arsinous acid esters is given in Scheme 14[134,135]. Table 7 lists optically active arsines prepared from the esters of arsinous and arsinthious acids by treatment with Grignard reagents.

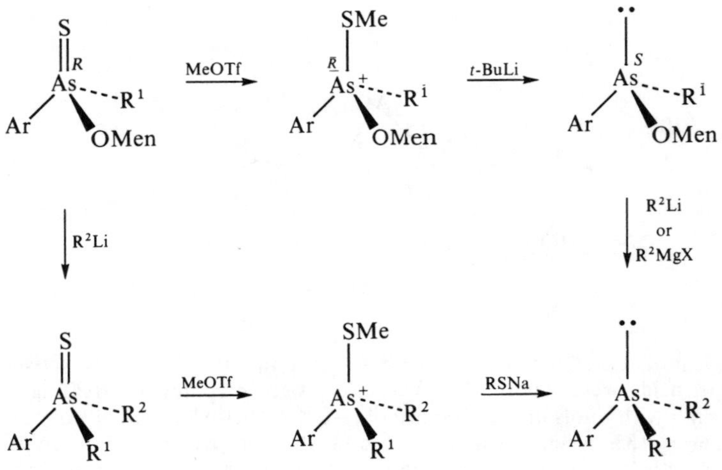

SCHEME 14

In thorough work of related interest, the resolution of the arsonthious ester 2-*p*-carboxyphenyl-5-methyl-1:3-dithia-2-arsindane (141) was described by the fractional crystallization of the (+)-α-methylbenzylammonium and the (−)-quininium salts: the former yielded the (−)-acid, $[\alpha]_D - 8.7°$ (CHCl$_3$); the latter the (+)-acid, $[\alpha]_D + 8.9°$ (CHCl$_3$)[136]. The optical stability of the compound was unaffected by crystallization from boiling ethanol, but in pyridine the rotation of the (−)-acid reached half-value within 6 weeks. The compound racemized more rapidly in 0.1 N sodium hydroxide, which was attributed to fission of the arsenic–sulphur bonds.

F. Tertiary Arsine Oxides and Sulphides

1. Tertiary arsine oxides

Although ethylmethylphenylphosphine oxide was the first tetrahedral phosphorus compound to be resolved[137], only negative results were obtained when attempts were made to resolve methyl(α-naphthyl)phenylarsine oxide (142) by protonation and separation of (+)-α-bromocamphor-π-sulphonate salts, or (*o*-carboxyphenyl)methylphenylarsine oxide (143) or its anhydride 144 after protonation and use of (−)-brucine or (+)-morphine[138]. The reversible combination of a tertiary arsine oxide with water was recognized at the outset[138] and it is this property that is responsible for the racemization[50,139]:

$$R_3As{=}O·H_2O \rightleftharpoons R_3As(OH)_2 \rightleftharpoons [R_3AsOH]OH$$

(142)   (143)   (144)

(145)-(153)

Under anhydrous conditions, however, optically active tertiary arsine oxides can be prepared from optically active tertiary arsines and dinitrogen tetraoxide. Thus, (+)-(*p*-carboxyphenyl)ethylphenylarsine [(+)-36] reacts with N_2O_4 in nitromethane to give a solution of (+)-(*p*-carboxyphenyl)ethylphenylarsine oxide [(+)-147], which lost its optical activity over 1.5 h. The optically active arsine oxides 145, 146, 153 were similarly generated in solution (Table 8), but only racemates or racemic *hydrochlorides* were isolated[139-141]. The oxides can also be prepared from the corresponding arsine sulphides

TABLE 8. Optically active tertiary arsine oxides and sulphides

Compound	$[\alpha]_D{}^a$			Solvent	Synthetic method	Reference
	(+)	(−)				
Arsine oxides						
As(O)MePh(C$_6$H$_4$CO$_2$H-p) (**145**)^b	11.8	9.3	MeNO$_2$	c	139
As(O)EtPh(C$_6$H$_4$CO$_2$H-o) (**146**)^b	21.3	23.6	MeNO$_2$	c	139
As(O)EtPh(C$_6$H$_4$CO$_2$H-p) (**147**)^b	13.5	10.9	MeNO$_2$	c	139
As(O)Et(Pr-n)(C$_6$H$_4$CO$_2$H-p) (**148**)^b	8.6	7.5	MeNO$_2$	c	139
As(O)(Bu-n)Et(C$_6$H$_4$CO$_2$H-p) (**149**)	10.2	9.5	MeNO$_2$	d	140
As(O)Et(C$_6$H$_4$Br-p)(C$_6$H$_4$CO$_2$H-p) (**150**)	16.5	15.1	MeNO$_2$	d	140
As(O)EtPh(C$_6$H$_4$CO$_2$H-m) (**151**)	20.7	19.4	MeNO$_2$	d	140
As(O)Me(C$_6$H$_4$Br-p)(C$_6$H$_4$CO$_2$-p) (**152**)	11.6	10.8	MeNO$_2$	d	140
As(O)Et(C$_6$H$_4$Br-o)(C$_6$H$_4$CO$_2$H-p) (**153**)	21.2	21.0	MeNO$_2$	d	141
Arsine sulphides						
As(S)EtMe(C$_6$H$_4$CO$_2$H-p) (**7**)	19.1	19.1	EtOH	e	4
As(S)EtMePh (**157**)	24.5		MeOH	f	8
As(S)(Bu-n)MePh (**158**)	16		MeOH	f	8
As(S)MePh(n-Pr) (**115**)	13.5		MeOH	f,g	123, 126
As(S)Me(Np-β)Ph (**116**)	9.1		MeOH	f,g	123
As(S)EtPh(n-Pr) (**119**)	18.3	17.7	MeOH	g	124
As(S)AllMePh (**121**)	28.8		MeOH	f,g	127
As(S)EtPh(CO$_2$H) (**159**)	14.2		MeOH	f	72
As(S)EtPh(CH$_2$CO$_2$H-p) (**160**)	4.0		C$_6$H$_6$	f	71
As(S)EtPh(C$_6$H$_4$CO$_2$H-p) (**136**)	19.1	6.9	C$_6$H$_6$	f	130
As(S)Et(C$_6$H$_4$Br-o)(C$_6$H$_4$CO$_2$H-p) (**161**)	23.4	23.1	C$_6$H$_6$	f	67
As(S)Et(C$_6$H$_4$Me-p)(C$_6$H$_4$NH$_2$-o) (**162**)	2.9	2.8	Et$_2$O	f	73
As(S)Et(C$_6$H$_4$Me-p)(C$_6$H$_4$NH$_2$-m) (**163**)	2.5	2.4	Et$_2$O	f	73
As(S)Et(C$_6$H$_4$Me-p)(C$_6$H$_4$NH$_2$-p) (**164**)	2.2	1.9	Et$_2$O	f	73

^{a} Specific rotations $[\alpha]_D$ to nearest 0.1° measured at specified concentrations and temperatures (*ca* 20 °C).
^{b} Arsine oxides not isolated.
^{c} Arsine sulphide plus N$_2$O$_4$ (inversion).
^{d} Arsine plus N$_2$O$_4$.
^{e} (+)-Enantiomer resolved (+)-morphine; (−)-enantiomer resolved (−)-brucine.
^{f} Arsine plus sulphur in boiling benzene.
^{g} Reduction of thioarsinic acid with RMgX or RLi.

by treatment with N$_2$O$_4$ but, unlike direct oxidation, this reaction proceeds with partial racemization and inversion at arsenic, as demonstrated by the transformations indicated in Scheme 15[139]. The absolute configuration given to (+)-**136** in the Scheme is arbitrary. The optically active arsine oxides racemize in the presence of a trace of moisture. The rotation for **153** is the value obtained after two recrystallizations of the compound from ethanol.

When treated with methyl iodide, (+)-**149** in nitromethane yields the crystalline (+)-methoxyarsonium iodide **154**, which decomposes into (−)-methyl n-butyl(p-carboxyphenyl)arsinite (**155**) (equation 20)[142].

2. Tertiary arsine sulphides

The tetrahedral stereochemistry of a tertiary arsine sulphide was first established in 1925, when p-carboxyphenylethylmethylarsine sulphide (**7**) was resolved[4]. Unlike hygroscopic tertiary arsine oxides, arsine sulphides are air-stable non-basic compounds with

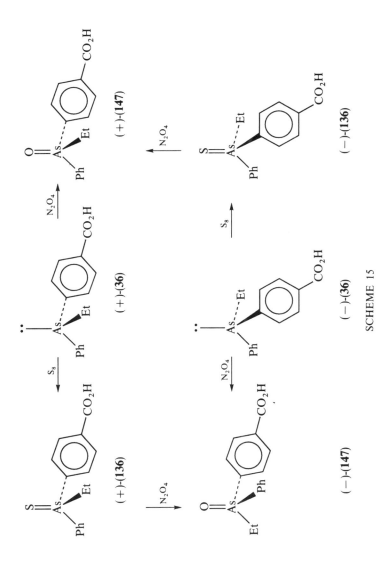

SCHEME 15

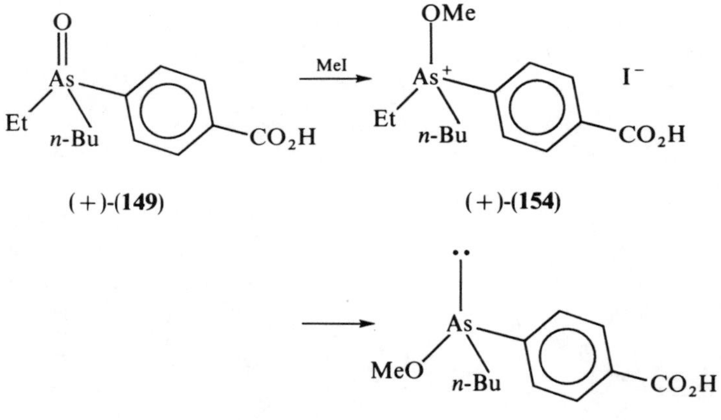

$(+)$-(149) $(+)$-(154)

(20)

$(-)$-(155)

a propensity for crystallization. The arsine sulphide 7, readily formed $(-)$-brucine and $(+)$-morphine salts via the carboxylic acid group that were separated by fractional crystallization and individually decomposed by sodium hydroxide into the crystalline arsine sulphides (Table 8).

Some years later, two isomers of ethylene-1,2-bis(n-butylphenylarsine sulphide) (156) were isolated[7] and were considered to be the racemic (R^*,R^*) and the *meso* (R^*,S^*)

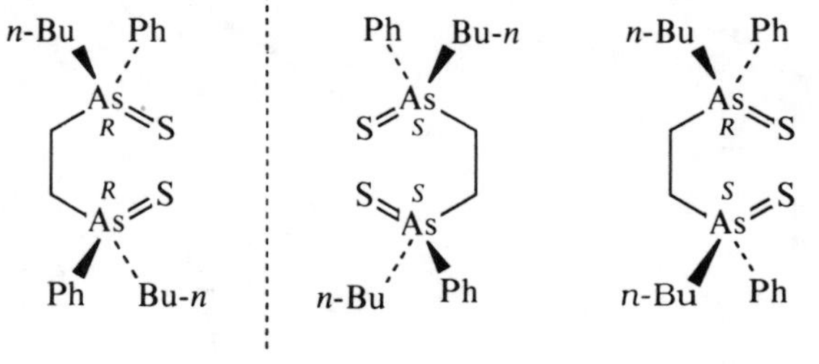

(R^*,R^*)-(156) (R^*,S^*)-(156)

diastereomers. The puzzling observation was made, however, that the lower–melting form, mp 113–116 °C, passed readily (even in hot ethanol) into the higher-melting form, mp 121 °C. The authors surmised that equilibrium of the type

diarsine-disulphide $\rightleftharpoons$ diarsine-monosulphide + sulphur

could occur (although evidence for the dissociation could not be detected by molecular weight measurements) with inversion of the tertiary arsine group facilitating interconversion between the diastereomers.

The first synthesis of an optically active tertiary arsine sulphide from an optically active tertiary arsine was reported in 1962[8], compounds 157 and 158 being prepared by heating the respective optically active tertiary arsines with sulphur in benzene. Along similar lines, arsine sulphides 136, 159–164 were prepared and isolated (Table 8). The sulphurization of the tertiary arsines occurs with retention of configuration at arsenic.

The optically active arsine sulphides **115**, **119** and **121** (Table 8) were prepared by reactions between appropriate menthyl esters of thioarsinic acids and Grignard or alkyllithium reagents, which proceed with inversion of configuration at the arsenic(V) stereocentre.

IV. CONFIGURATIONAL STABILITY AT ARSENIC

A. Tertiary Arsines

1. Pyramidal inversion

The first experimental determination of the inversion barrier of a tertiary arsine was reported in 1971[143]. The kinetics of racemization of (R)-$(-)$- and (S)-$(+)$-**11**, resolved by the metal complexation method[9], at $217.6 \pm 0.3\,°C$ in decalin (sealed tube) was determined polarimetrically in the 310–350 nm region. From the kinetic data, by substitution into the Eyring equation, the free energy of activation, $\Delta G^{\ddagger}$, was calculated to be $175 \pm 2\,kJ\,mol^{-1}$ at $217.6\,°C$. This energy value corresponds to a half-life for racemization of the arsine of ca 740 h at $200\,°C$. It had been reported previously that resolved ethylmethylphenylarsine and methyl(n-propyl)phenylarsine showed no detectable loss of optical activity over 10 h at $200\,°C$[54]. On the basis of photoracemization studies[144], the energy requirement for the thermal racemization of tertiary arsines was calculated to be 244–$251\,kJ\,mol^{-1}$, but this estimate was discounted as invalid because of the irrelevance of the technique to processes involving vibrationally excited states of the electronic ground state[143].

The barrier to pyramidal inversion at arsenic in *iso*propylphenyltrimethylsilylarsine (**165**) is $ca\,74\,kJ\,mol^{-1}$ lower than in **11**, whereas the less electronegative trimethylgermyl substituent reduces the barrier by $ca\,66\,kJ\,mol^{-1}$[1,145]. Thus, the most important influence on pyramidal stability in these systems appears to be the atomic electronegativity of the adjacent heteroatom and not p_π–d_π bonding as had been suggested earlier as the source of barrier lowering in $(R^*,R^*)/(R^*,S^*)$-1,2-dimethyl-1,2-diphenylarsine (**166**)[146].

(165) (R^*,R^*)-(166) (R^*,S^*)-(166)

(167)   (168)   (169)

Cyclic $(4p-2p)\pi$ conjugation contributes to the lowering of the inversion barrier in **167**, however, where the value $\Delta G^{\ddagger} = 145\,kJ\,mol^{-1}$ at 151 °C was determined by an analysis of kinetic data for the equilibration of a 72:28 diastereomeric mixture of the arsindole[147]. The effect will be maximal in the planar transition state. Total line-shape analysis of the temperature-dependent silyl-methyl NMR resonances in **168** indicated a cooperative effect between cyclic $(p-p)\pi$ delocalization and substituent atom electronegativity, since the diminution in barrier height was greater than that afforded by either effect alone.

2. Acid-catalysed inversion

Tertiary arsines are very weak bases; indeed, only one protonated tertiary arsine appears to have been isolated, the salt **169**, which spontaneously decomposes at -50 °C into dimethylarsine and bromotrimethylsilane[148]. (Dimethylarsine and hydrogen chloride react at low temperatures to give dimethylarsonium chloride; the compound decomposes reversibly on warming[149].) Nevertheless, optically active tertiary arsines, especially non-cyclic compounds, racemize rapidly in solutions containing halo-acids[39,144]. The complete racemization of the ($+$)-arsafluorene **32** was observed to occur after about 500 h in chloroform containing 5% ethanol when a trace of hydrogen chloride was added[39]. The racemization did not follow the unimolecular rate law. The effect of the halo-acid is marked in polar solvents such as methanol, were an increasing rate of racemization of (S)-($+$)-methylphenyl-n-propylarsine, (S)-($+$)-**65**, was observed for anions in the order $F^- < Cl^- << Br^- <<< I^{-}$[54]. In solutions containing acetic acid, perchloric acid or sulphuric acid, the racemization of a tertiary arsine in negligible. These results have been corroborated by an investigation of the racemization of (R,R)-($+$)-1,2-phenylenebis(methylphenylarsine), (R,R)-($+$)-**75**, in the presence of acids[150]. The rate of the racemization of the tertiary arsine was found to be proportional to the square of the concentration of the halo-acid. The mechanism proposed for the racemization of a tertiary arsine by an acid HX is given in equation 21.

$$\text{(21)}$$

B. Arsonium Ions

The salts ($+$)-[AsAmylBnEtPh]Br[24], ($-$)-[AsBn(n-Bu)EtPh]Cl and ($+$)-[AsBnEt(n-Pentyl)Ph]Br[151] were reported to completely racemize over *ca* 1 h at 20 °C in chloroform. A dissociation–equilibrium between the salt and the tertiary arsine and alkyl halide was put forward as the mechanism of the racemization, where the alkyl halide was considered to attack the tertiary arsine from the side of the lone-pair (preferred) or from the side of the plane of the three substituents[151]. It was stated that the racemization would be facilitated by raising the temperature due to inversion of the tertiary arsine in line with earlier calculations[34]. This notion of facile dissociation–equilibrium in arsonium salts, first suggested in 1921[3] and carried through until 1964[151], was finally put to rest in 1965, however, when it was shown that the salts ($-$)-[AsBn(n-Bu)EtPh]X (where X = Cl or Br) in chloroform or ethyl acetate were completely stable for days[152]. Moreover, since the enantiomers of AsEtMePh show no evidence of racemization at 200 °C, and can be quaternized with retention of configuration, dissociation–equilibrium in arsonium halides should not lead to racemization. Nevertheless, arsonium halides do racemize when heated in aprotic solvents (iodide > bromide > chloride), the rate of the process increasing along the following sequence: acetone > chloroform > benzyl chloride > acetonitrile > dimethyl formamide > dimethyl sulphoxide. For example, (R)-$(-)$-[AsBn(n-Bu)EtPh]Cl as a 0.155 M solution in chloroform is 32.5% racemized after 70 h at 103 °C. Thus, the choice of iodide as the counterion for arsonium ions in the early resolution work was unfortunate, especially in conjunction with acetone as solvent[3]. In water, methanol, ethanol, acetic acid or N hydrochloric acid, resolved arsonium halides are stable, and arsonium salts containing poorly nucleophilic counterions, such as perchlorate, appear to be stable indefinitely, even in strongly ion-pairing solvents.

The data collected on the racemization of arsonium ions are consistent with a mechanism involving Berry pseudorotation of a transient five-coordinate halogeno-arsenic(V) intermediate (equation 22). The more nucleophilic the halide the faster the racemization.

$$\text{(22)}$$

C. Arsine Complexes

There are several examples of racemization at arsenic in positively charged coordination complexes. When cis-α-[CoCl$_2${(R^*,R^*)-99}]Cl, (R^*,R^*)-100, or $trans$-[CoCl$_2${(R^*,S^*)-99}]Cl, (R^*,S^*)-100, is heated under reflux in acetonitrile at concentrations of greater than 10^{-2} M, equilibrium between the two diastereomers is established within ca 8 h[117,118]. The following mixture of disatereomers prevails at equilibrium: cis-α-(R^*,R^*) (35%), $trans$-(R^*,S^*) (60%) and cis-β-(R^*,R^*) (5%). For the corresponding $bromo$ bromide under similar conditions, the distribution at equilibrium is cis-α-(R^*,R^*) (70%), $trans$-(R^*,S^*) (30%). The optically active complex racemizes completely under similar conditions. The establishment of the equilibrium depends upon the nature of the free and coordinated anion and is a useful method for interconverting diastereomers of tertiary arsines and complexes. The following three points were noted in connection with the above system: (a) no inversion of the free arsine occurred in the presence or absence of chloride over 24 h; (b) no inversion occurred when [Co(MeCN)$_2${(R^*,R^*)-99}](ClO$_4$)$_3$ or [Co(MeCN)$_2$ {(R^*,S^*)-99}](ClO$_4$)$_3$ was boiled in acetonitrile for 24 h; and (c) the rate of approach of the equilibrium appeared to depend upon the concentration of the halide. These observations are reminiscent of those found for the racemization of quaternary arsonium salts. A tertiary arsine coordinated to a metal cation (R$_3$As → M$^+$) is formally analogous to an arsonium ion (R$_3$As$^+$—M) and racemization at arsenic could occur by pseudorota-tion of a transient five-coordinate halogenoarsenic intermediate, as indicated in equation 23. The situation will be more complicated in metal chelates, however, as pseudorotation will be restricted.

There have been two other reports of racemization at arsenic in cationic metal complexes. One concerns the quantitative transformation of [Pd{(R^*,S^*)-107}](ClO$_4$)$_2$ into [Pd{(R^*,R^*)-107}](ClO$_4$)$_2$ when a solution of the former in DMSO was heated at 150 °C for several minutes[116]. The transformation is irreversible and involves unprece-

$$\underset{R^1\overset{|}{\underset{R^2}{}}}{\overset{M}{\underset{}{As^+}}}\text{-}R^3 \quad \underset{X^-}{\overset{X^-}{\rightleftharpoons}} \quad R^1\text{---}\underset{X}{\overset{M}{\underset{|}{As}}}\text{---}\underset{R^2}{R^3} \quad \underset{}{\overset{-X^-}{\rightleftharpoons}} \quad \underset{R^{3\cdots}\overset{|}{\underset{R^2}{}}}{\overset{M}{\underset{}{As^+}}}R^1 \qquad (23)$$

dented inversion at arsenic in a halide-free complex. It was noted also that $(+)$-$[Pt\{(R,R)$-**107**$\}](ClO_4)_2$ racemized more slowly than the *meso* complex was converted into the racemic complex. The observations were rationalized in terms of a relieving of conformational strain in the chelate rings upon transformation of the R^*,S^* complex into the more stable R^*,R^* complex. Crystal structure determinations revealed that the Pd—As and Pd—S bonds in the centrosymmetrical R^*,S^* complex were compressed, but typical in the R^*,R^* complex where the palladium is situated out of the plane of the donors of the macrocycle[120]. In similar fashion, $[Pd\{(R^*,S^*)$-**104**$\}](PF_6)_2$ transforms quantitatively into $[Pd\{(R^*,R^*)$-**104**$\}](PF_6)_2$ when heated for 5 min at 150 °C in DMSO[116]. The transformation occurs under milder conditions for the corresponding chloride salt, however, implying anion participation. Upon being heated for 1 h at 60 °C in water, $[Pd\{(R^*,S^*)$-**104**$\}]Cl_2$ transforms quantitatively into $[Pd\{(R^*,R^*)$-**104**$\}]Cl_2$; the hexafluorophosphate requires 4 h in boiling water for complete conversion. The mechanism for the transformations of the palladium complexes is unclear, but dissociation of the ligand is unlikely because free tertiary arsines are configurationally stable under the non-acidic conditions employed.

V. CHIROPTICAL PROPERTIES, ABSOLUTE CONFIGURATION

The absolute configuration of $(+)$-$[AsBnMePh(n$-Pr$)]BPh_4$, $(+)$-**13**, has been established as S by the quasi-racemate method of Fredga in conjunction with the analogous dextrorotatory phosphorus compound[42], which was structurally characterized by X-ray diffraction and the absolute configuration confirmed by the Bijvoet method[43]. There is also a strong correlation between the CD spectra of (S)-$(+)$-**13** and $(+)$-$[AsBnMePh(n$-Pr$)]X$ (where $X = ClO_4^-$ or $MeSO_3^-$), where a strong positive Cotton effect around 220 mm was observed in each case[78]. The absolute configurations of the salts (R)-$(+)$-**15** and (R)-$(+)$-**16** have been established by X-ray crystal structure analyses[45].

The CD curves of the $[1R, 2S, 5R]$-menthyl methylphenylthioarsinate esters (R)-$(-)$- and (S)-**114** are enantiomeric in type[123], as found for the analogous menthyl methylphenylphosphinates[153]. Moreover, the ^1H NMR spectra of the thioarsinate esters display a characteristic pattern of signals in the isopropyl methyl region that closely resemble a number of menthyl phosphinates, menthyl methylthiophosphinates and menthyl phosphonates, which proved to be a reliable basis for correlating configurations in these systems. Accordingly, based on the large upfield shift of one isopropyl methyl doublet for the diastereomer of **114** having mp 115–117 °C, the configuration at arsenic was assigned as R. The lower melting diastereomer, mp 100–102 °C, therefore has the S configuration at arsenic. As discussed in Section III,D.1, the reaction of (R)-$(-)$- or (S)-**114** with 2 equiv. alkyl- or aryl-lithium at -78 °C leads to optically active tertiary arsines with inversion at arsenic: the first equivalent of the lithium reagent substitutes the menthoxide group with inversion, giving the tertiary arsine sulphide; the second equivalent reduces the arsine sulphide to the arsine with retention. The small degree of racemization that occurs in the overall synthesis takes place during the substitution step, since desulphurization of the intermediate arsine sulphide by n-BuLi is completely stereoselective. A considerable body of related work followed this discovery of the predominantly stereoselective synthesis of tertiary arsines from diastereomers of menthyl thioarsinates (see Table 7).

The absolute configurations of a number of arsonium ions and arsines have been correlated with one another on the basis of the stereospecificity of cathodic cleavage of the benzyl group from an arsonium ion and the quaternization of an arsine by benzyl and other alkyl halides, both of which occur with retention of configuration. Scheme 1 gives the connection between the stereochemistries of various optically active arsonium ions and arsines related to (S)-(+)-**62**.

The configurations of a number of tertiary arsines have been determined by crystal structure determinations of diastereomeric metal complexes containing the arsine and a reference ligand of known configuration. Of particular use in this connection are the palladium(II) complexes (R)-(−)- or (S)-(+)-**77**, and (R)-(+)- or (S)-(+)-**78**, which undergo quantitative bridge-splitting reactions with chiral tertiary arsines to give diastereomeric complexes that, once separated, are ideal for the determination of absolute configurations by X-ray crystallography. The absolute configurations of most of the arsines listed in Table 5 have been determined by this method.

VI. TERTIARY ARSINES CHIRAL AT A SUBSTITUENT

Few optically active tertiary arsines are known that owe their optical activity to a chiral substituent. The first compounds of this type to be reported were the enantiomers of (R*,R*)-2,3-O-isopropylidene-2,3-dihydroxy-1,4-bis(diphenylarsino)butane (diarsop)[154]. The enantiomers of the bis(tertiary arsine), (R,R)-(+)-**170** and (S,S)-(−)-**170**, were synthesized from the ditosylates of commercially available (R,R)-(−)- and (S,S)-(+)-2,3-O-isopropylidenethreitol by treatment with potassium diphenylarsenide–2-dioxane. The compounds are air-stable crystalline solids, mp 64–66 °C (Table 9). A similar preparation of (S,S)-(−)-**170** was reported in the same year[155].

Treatment of (S)-(−)-dimethyl(α-methylbenzyl)amine with n-BuLi followed by chlorodiphenylarsine affords the tertiary arsine (S)-(+)-**171** (amars), which was isolated in 31% yield as the crystalline *hydrochloride* and converted into the free amine with sodium hydroxide[156]. The preparation of this compound by the same method is also reported in Reference 157.

The ferrocenyl-amine (S)-**172**, upon treatment with n-BuLi and iododimethylarsine or chlorodiphenylarsine in diethyl ether, affords the arsines (S,R_{Fe})-**173** or (S,R_{Fe})-**174** with

TABLE. 9. Tertiary arsines chiral at a substituent

Arsine	Abs. config.	$[\alpha]_D$ (deg)	Solvent	Reference
Diarsop (**170**)	R,R	+ 31.6	C_6H_6	154
	S,S	− 27.6[a]	C_6H_6	154
Amars (**171**)	S	− 22.3[b]	CH_2Cl_2	156
Arsamine **173**	R_{Fe},S	c		158
Arsamine **174**	R_{Fe},S	c		158
Neomenars (**175**)	S^d	+ 69.6	CH_2Cl_2	46
Menars (**177**)	R^d	− 79.6	CH_2Cl_2	46

[a] The sample obtained via (R)-(+)-diethyl tartrate had $[\alpha]_D - 3.1°$ (CHCl_3)[155].
[b] The value given in Reference 154 is $[\alpha]_D - 37.3°$ (C_6H_6).
[c] No chiroptical data given.
[d] Absolute configuration of epimeric carbon.

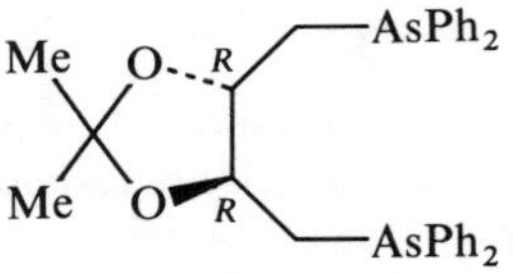

(R,R)-$(+)$-$(\mathbf{170})$ 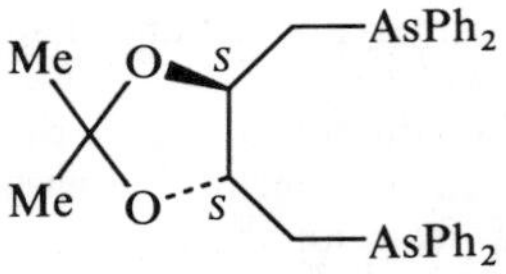 (S,S)-$(-)$-$(\mathbf{170})$

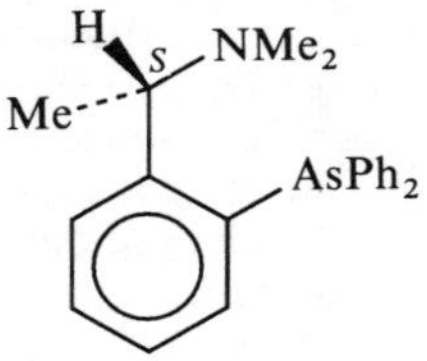

(S)-$(-)$-$(\mathbf{171})$
(X-ray, Rh(I) complex)[156]

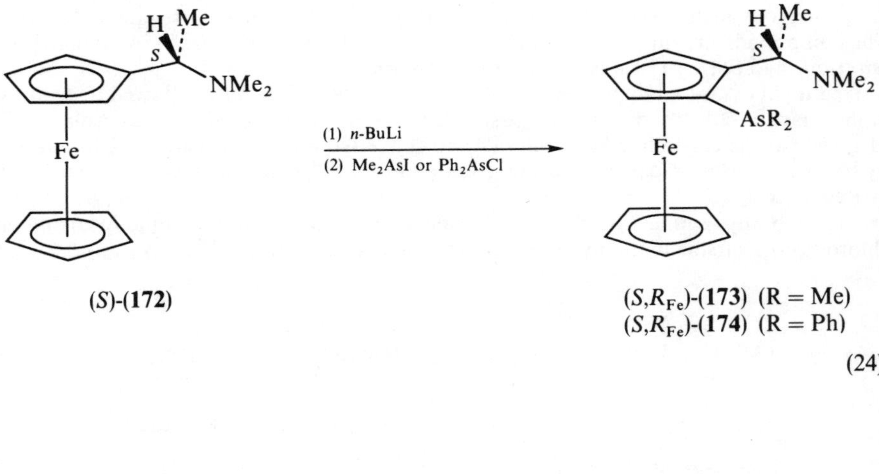

(S)-$(\mathbf{172})$ (S,R_{Fe})-$(\mathbf{173})$ (R = Me)
                                                  (S,R_{Fe})-$(\mathbf{174})$ (R = Ph)

(24)

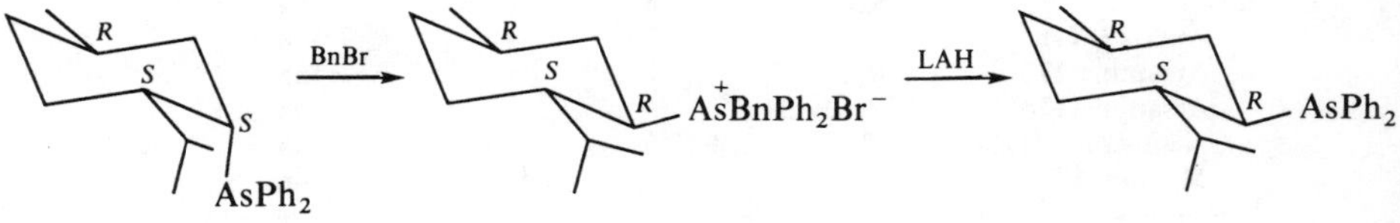

(S)-$(+)$-$(\mathbf{175})$ (R)-$(+)$-$(\mathbf{176})$ (R)-$(-)$-$(\mathbf{177})$

(25)

complete stereoselectivity at planar-chiral-Fe (equation 24)[158]. The yield of (S, R_{Fe})-**174** was 56% after recrystallization from absolute ethanol. No chiroptical data were reported for the compounds.

Potassium diphenylarsenide–2-dioxane reacts with $(-)$-menthyl chloride in boiling tetrahydrofuran to give $(1S,2S,5R)$-$(+)$-neomenthyldiphenylarsine, (S)-$(+)$-**175** (neomenars). The arsine was obtained as colourless needles in ca 40% yield after recrystallization from warm methanol (Table 9). The neomenthylarsine, upon reaction with benzyl bromide in benzene, affords the epimeric $(1R,2S,5R)$-benzylmenthylarsonium bromide, (R)-$(+)$-**176**, which, upon reduction with lithium aluminium hydride, affords $(1R,2S,5R)$-diphenylmenthylarsine, (R)-$(-)$-**177** (menars) (equation 25).

VIII. METHODS OF MEASUREMENT OF ENANTIOMERIC EXCESSES

For a survey of methods available for determining enantiomeric purity, the reader should consult Reference 21, pp. 405–422.

The enantiomeric excesses of chiral tertiary phosphines have been determined by NMR analysis of diastereomeric phosphonium salts formed by stereospecific quaternization of the phosphines with a (R)-$(+)$-2-phenyl-2-methoxyethyl bromide or the corresponding ethyl-1-d_2 compound (the latter for use where the signals for the methylene protons of the reference group in the phosphonium salt interfere with the ^1H NMR analysis)[159]. The conditions employed for the preparation of the phosphonium salts, boiling benzene for 7 h with a two-fold excess of the alkyl bromide, may be harmful if applied to the synthesis of arsonium salts, however, because of the potentially racemizing conditions for the arsonium bromides (see Section IV.B). The use of a chiral europium(III) shift reagent for the determination of an optical purity of a tertiary arsine, at least where there are no polar groups on the substituents, is also inappropriate. A doubling of the AsMe resonance was not observed for (S)-$(+)$-**65** in the presence of tris{3-heptafluoropropylhydroxymethylene-$(+)$-camphorato}europium(III)[123]. The impure arsine sulphide (R)-$(+)$-**115**, however, did give a doubling of the AsMe signal in the presence of the shift reagent. Since the conversion of an arsine into a configurationally stable arsine sulphide is stereospecific and can be effected under non-racemizing conditions, the use of a chiral NMR-shift reagent for the determination of the optical purity of an arsine via the sulphide is a potentially useful method. The stereospecific formation of a diastereomeric pair of metal complexes containing a configurationally homogeneous reference ligand will in general be a more convenient method, however (see below).

The enantiomeric excess of a tertiary arsine can be determined by the analysis of an NMR spectrum of a suitable metal complex containing the arsine and an optically pure reference ligand. Two suitable reference complexes are the palladium(II) dimers **77** and **78**, both of which are readily prepared as either enantiomer from the commercially available optically active amines after methylation. An advantage of this method is that the identities of the diastereomers can be determined by performing an X-ray crystal structure determination on one of them, which will establish the configuration at arsenic with reference to the configuration of the chiral carbon stereocentre of the reference ligand. The dimer (R)-$(+)$-**77** has been employed for the determination of optical purities of certain chelating C_2-diphosphines by NMR spectroscopy[160]. The dimers **77** and **78** undergo bridge-splitting reactions with unidentate tertiary arsines to give neutral complexes in which the arsine can be cis or $trans$ to the dimethylamino group. In all cases investigated in the laboratories of the author, however, the naphthalene dimer **78** has given $trans$ complexes quantitatively and with complete regioselectivity. The chelation of unsymmetrical bidentates also appears to be completely regioselective in the case of (R)-$(-)$-**78** in our experience. It is important that the proportions of the reactants be equivalent because a kinetic resolution of the arsine is possible in such systems. If one

of the groups on the arsenic is methyl and there is good base-line separation for the
AsMe resonances of the two diastereomers of a complex containing a reference ligand
of R configuration, say R,S_{As} and R,R_{As}, the percentage ee of the arsine can be calculated
from equation 26 by substitution of the values of the integrals for the two AsMe signals,
which reflect the concentrations of the enantiomers of the arsine. Equation 26 assumes
a linear relationship between specific rotation and concentration.

$$\text{Percentage ee} = \frac{[R,S_{As}] - [R,R_{As}]}{[R,S_{As}] + [R,R_{As}]} \times 100 = \% R_{As} - \% S_{As} \qquad (26)$$

The accuracy of the determination will depend upon the accuracy of peak areas used
in the calculation, perhaps $\pm 2\%$ [21,161]. For a crystalline complex, a single crystallization
of material of $>95\%$ ee will almost certainly bring it to complete optical purity, *in which
case the absence of NMR signals for the minor diastereomer of the complex to within the
limits of sensitivity of the NMR spectrometer will be the criterion for complete optical
purity of the arsine.* Since the displacement of an arsine from a configurationally
homogeneous complex of this type is stereospecific with retention of configuration at
arsenic and can be carried out under mild conditions, the arsine liberated will also be
optically pure. However, to be certain of the optical purity of the arsine the diastereomer
should be re-prepared on a small scale and checked once again for purity by NMR
spectroscopy.

For bidentates of C_2 symmetry, a pair of diastereomeric salts of the type $[R,(R_{As},R_{As})]$-
and $[R,(S_{As},S_{As})]$-**82** will be obtained in the bridge-splitting reaction (Scheme 5). Substitution of the peak areas of appropriate signals into equation 26 will give the
optical purity of the bidentate. In some cases it will be convenient to exchange the chloride
for hexafluorophosphate or similar anion to increase the solubilities of the diastereomers
and to eliminate complications due to fluctional behaviour of five-coordinate chloro-
complexes, in which case care must be taken to avoid unintentional enrichment of one
of the diastereomers. For unsymmetrical bidentates of C_1 symmetry (asymmetric
bidentates), the regioselectivity of coordination of the donors will again be of concern.
As for the unidentates, however, the regioselectivity of coordination can be quickly
established at the outset by the reaction of the racemic ligand with the reference complex.

A method for determining the optical purity of a (2-mercaptoethyl)phosphine, where
the phosphine serves as its own reference, has been published[162]. The determination
requires the synthesis of a kinetically labile complex of the type $[ML_2]$ that will
equilibrate as a pair of racemic (R^*,R^*) and *meso* (R^*,S^*) diastereomers with minimal
diastereoselection. This will usually be the case for *trans*-$[NiL_2]$ complexes of unsymmetrical bidentates containing a tertiary phosphine or arsine donor and one other
donor that has a propensity for bridging, for example, thiolato-S, which will facilitate
equilibration[115]. Bivalent palladium and platinum complexes of the type $[ML_2]$ with
deprotonated **96a** give equilibrium mixtures of *cis* and *trans* complexes in chloroform
that exhibit minimal $R^*,R^*/R^*,S^*$ diastereoselection for the theremodynamically favoured
trans diastereomers, but considerable diastereoselection for the less stable *cis* diastereomers where there are relatively strong interactions between the groups on the
adjacent chiral As-stereocentres[115]. Under the favourable conditions described above, the
intensities of appropriate NMR signals for the pair of diastereomers produced can be
substituted into equation 26 to obtain an ee for the arsine. If an excess of the ligand is
used, allowance must be made for the free ligand concentration because the ratio of
enantiomers in the excess ligand may differ from the ratio of the diastereomers of the
complex[162]. In the example given in Reference 162, however, allowance for free ligand
concentration barely alters ($<0.2\%$) the outcome of the direct calculation. This method
may be of use when simple complexes of the type (R,R_{As})- and (R,S_{As})-**127** (Scheme 13)

cannot be prepared. As for the *ortho*-metalated palladium complexes, however, the *absence* of NMR signals for a thermodynamically stable *meso* diastereomer of a complex of the type [ML_2] will be the more reliable criterion of optical purity of a ligand.

VIII. APPLICATIONS

A. Asymmetric Synthesis

1. Catalytic hydrogenation

Early work on the catalytic hydrogenation of conjugated dienes by cationic rhodium(I) complexes containing tertiary phosphines and arsines indicated that 1,2-bis(diphenyl-arsino)ethane was an effective adjunct to the metal in such reactions, the arsine-based catalyst giving comparable chemical yields of mono-olefins at somewhat reduced rates to those obtained for the corresponding phosphine-based catalyst but with greater regioselectivity in the products[163]. Similar work with optically active chelating arsines has also given interesting differences in behaviour between the two types of ligands. Thus, the catalytic asymmetric hydrogenation of Z-α-acetamidocinnamic acid in methanol by a rhodium(I) catalyst containing (S,S)-$(-)$-**170** afforded (S)-$(+)$-N-acetylphenylalanine in up to 39% ee when a large excess of triethylamine was present (equation 27)[154],

$$(27)$$

whereas the analogous phosphine, (S)-$(-)$-diop gave the R enantiomer of the product in 88% ee[164]. Hydrogenation of 2-acetamidoacrylic acid using the diarsop catalyst gave (S)-$(-)$-N-acetylalanine in good chemical yield but with low stereoselectivity (8% ee)[154]. Once again, the arsine gave the amino acid of opposite configuration to that obtained with the analogous phosphine. The ferrocenyl arsines (S,R_{Fe})-**173** and (S,R_{Fe})-**174**, however, did not give catalytically active solutions for this reaction when the arsines were mixed with [RhCl(COD)]$_2$ (where COD = cycloocta-1,5-diene)[158].

Detailed work has been carried out on the relative merits of the enantiomers of the *As*-chiral bidentate (R^*,R^*)-**75** and the phosphorus analogue in rhodium(I) complexes for the catalytic asymmetric hydrogenation of a variety of prochiral Z-substituted enamide esters and acids in ethanol[165]. The enantioselectivity of the reaction was remarkably dependent upon the nature of the β-substituent on the enamide–olefin bond and the presence of triethylamine in the reaction mixture, but the arsine out-performed the phosphine for several substrates. For example, the arsine-catalyst gave (S)-$(+)$-N-benzoylvaline in 89% ee from α-benzamido-β,β-dimethylacrylic acid in the presence of triethylamine, whereas the phosphine catalyst failed to effect the reduction at all. This optical yield for the valine derivative is significantly higher than that obtained with use of any chelating phosphine. The enantiomers of the rigid-backboned ligand (R^*,R^*)-**75**, which presents a dissymmetric array of methyl and phenyl groups to the reactants, was much more effective for the reaction than (S,S)-$(-)$-**170**, wherein the non-rigid backbone dissymmetry is transmitted to the reactants via edge-face arrays of diastereotopic phenyl groups. Catalysts containing the enantiomers of (R^*,R^*)-**75** (or the phosphine) also appear to be considerably more active for these reactions than catalysts containing arsines (or phosphines) with flexible alicyclic chelate rings.

2. Catalytic hydrosilylation

The catalytic hydrosilylation of acetophenone or *t*-butyl methyl ketone with diethyl-, methylphenyl- or diphenylsilane in the presence of rhodium(I) catalysts containing (R,R)-(+)- or (S,S)-(−)-**170**, followed by acid cleavage of the intermediate silyl ethers, affords the respective alcohols with optical yields of 10–42%[154]. The synthesis of (R)-(+)-1-phenylethanol from acetophenone and diethylsilane in conjunction with the catalyst derived from (S,S)-(−)-**170** was the most effective reaction (equation 28). In

$$\underset{Ph}{\overset{Me}{\diagdown}}\!\!=\!\!O \;+\; Et_2SiH_2 \quad\xrightarrow{\text{cat.}}\quad \underset{Ph}{\overset{Me}{\diagup}}\!\!\underset{OSiHEt_2}{\overset{R}{\diagdown}}\!\!\cdots\!\!H \quad\xrightarrow{H^+}\quad \underset{Ph}{\overset{Me}{\diagup}}\!\!\underset{OH}{\overset{R}{\diagdown}}\!\!\cdots\!\!H \qquad (28)$$

similar reactions involving the ligand (S)-(−)-**171**, the best result was an 8% optical yield for the reduction of *t*-butyl methyl ketone by diphenylsilane[156]. An X-ray crystal structure analysis of $[Rh\{(S)\text{-}(-)\text{-}\mathbf{171}\}NBD]ClO_4$ (where NBD = bicyclo-2,2,1-norborna-2,5-diene) revealed that the unit cell contained two independent cations with enantiomorphic conformations of the amars chelate rings and associated arsenic–phenyl groups. If this similarity in energy of the conformational δ,λ diastereomers persists in solution, this could account for the low stereoselectivities observed in the products[156].

A comparison between the effectiveness of rhodium(I) complexes of the arsenic ligand (S,S)-(−)-**75** and the phosphorus isostere for the catalytic hydrosilylation of acetophenone and *t*-butyl methyl ketone with diphenylsilane indicated that the two complexes were equally efficient[166]. The conversions were almost quantitative at 25 °C, although the optical yields (4–41%) were improved by lowering the reaction temperatures. The rates of hydrosilylation in these reactions appear to be amongst the highest reported. For example, a standard solution of acetophenone was reduced in the presence of either catalyst within 5 min at 25 °C, whereas reduction of the same substrate with a similar diop catalyst took 20 h at this temperature[167]. Interestingly, the percentage asymmetric synthesis of product under standard ambient conditions was greater for the arsine-containing catalyst (32% ee) than for the phosphine-based catalyst (18% ee), although the performance of the latter was improved (to 36% ee) by carrying out the initial addition of diphenylsilane at −78 °C, and then allowing the reaction mixture to warm to 25 °C.

3. Benzylidene transfer

Chiral arsonium ylides containing the benzylidene group react with aromatic aldehydes to produce *trans*-2,3-diaryloxiranes with optical purities of up to 41%[46]. The degree of asymmetric induction depends upon the nature of the substituents on the ylide, the substrate, and upon the reaction conditions. For the variety of arsonium ylides investigated, however, the yields of recovered optically active arsines were almost quantitative with complete retention of configuration at arsenic in each case. Thus, apart from the value of stereoselective benzylidene transfer for the asymmetric synthesis of *trans*-2,3-diaryl oxiranes, the reaction provides the cleanest route to the recovery of optically active arsines from resolved benzylarsonium salts (see Section III.B.5). The mechanism of the reaction, which must take into account the exclusive formation of *trans*-diaryloxiranes, is believed to occur via the stereoselective decomposition of a pair of (R^*,S^*)-betaines (Scheme 16).

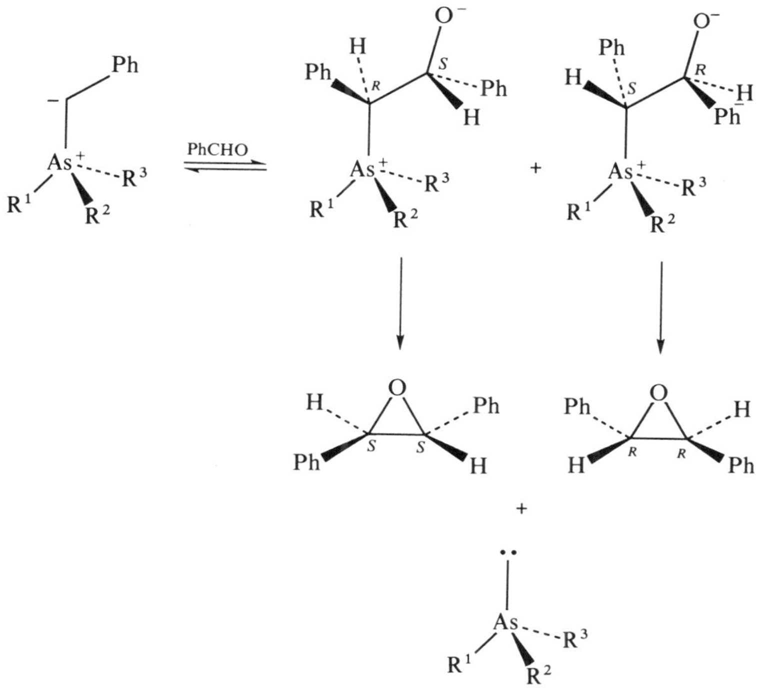

SCHEME 16

B. Stereochemical Probes

Perhaps the most important use of configurationally homogeneous enantiomers and diastereomers of tertiary arsines for the coordination chemist is the opportunity to synthesize particular metal complexes for investigations of stereochemistry, internal rearrangement and stability, especially with the aid of ^{1}H NMR spectroscopy. With diagnostic applications in mind it is important to place strategically methyl groups in the molecule, particularly on the arsenic. Investigations of a variety of metal complexes by NMR spectroscopy have provided unambiguous evidence of frequently unrecognized intramolecular dynamic behaviour and intermolecular ligand exchange in coordination complexes. Detailed knowledge of this kind is essential for the design of metal complexes for the resolution of arsines and for the development of arsine complexes for use in asymmetric synthesis. This is not the place to enlarge upon this topic, but there follows a list of useful references concerning the stereochemistry and stabilities of arsine complexes for the interested reader: nickel(II) complexes[160-170]; palladium(II) and platinum(II) complexes[170-172]; ruthenium(II) complexes[173]; gold(I) complexes[174]; copper(I) and silver(I) complexes[175].

IX. CONCLUSION

Despite the uncertain beginning, a variety of optically active tertiary arsines is now available for application in inorganic and organic synthesis. Tertiary arsines are powerful ligands for metals in a range of oxidation states, a feature that makes them particularly suitable for the design of complexes for stereochemical investigations and for use in asymmetric synthesis. Nevertheless, the field is small by comparison with the extensive one involving similar phosphines. As stated at the outset, the reason for the disparity is unclear, for arsines are frequently superior to the analogous and isosteric phosphines when it comes to their synthesis, handling, stability, recovery from arsonium ions and metal complexes, not to mention their more pleasent smell! Perhaps this review will stimulate further work. The paucity of data concerning the optical purities and absolute configurations for many of the arsines, however, needs to be addressed before the full potential of many of the interesting molecules described above can be realized.

X. REFERENCES

1. W. J. Pope and S. J. Peachey, *J. Chem. Soc.*, 1127 (1989).
2. (a). F. G. Mann, *Prog. Stereochem.*, **2**, 196 (1957);
    (b) W. E. McEwen, in *Topics in Phosphorus Chemistry*, Vol. 2 (Eds. M. Grayson and E. J. Griffith), Interscience, New York, 1965, p. 1.
    (c) L. Homer, *Helv. Chim. Acta* (Alfred Werner commemorative volume), 93 (1967).
3. G. J. Burrows and E. E. Turner, *J. Chem. Soc.*, 426 (1921).
4. W. H. Mills and R. Raper, *J. Chem. Soc.*, 2479 (1925).
5. M. S. Lesslie and E. E. Turner, *J. Chem. Soc.*, 1170 (1934).
6. F. Challenger and A. A. Rawlings, *J. Chem. Soc.*, 264 (1936).
7. J. Chatt and F. G. Mann, *J. Chem. Soc.*, 610 (1939).
8. L. Horner and H. Fuchs, *Tetrahedron Lett.*, 203 (1962).
9. B. Bosnich and S. B. Wild, *J. Am. Chem. Soc.*, **92**, 459 (1970).
10. P. Gugger, A. C. Willis and S. B. Wild, *J. Chem. Soc., Chem. Commun.*, 1169 (1990).
11. N. K. Roberts and S. B. Wild, *J. Chem. Soc., Dalton Trans.*, 2015 (1979).
12. (a) R. S. Cahn, C. K. Ingold and V. Prelog, *Angew. Chem., Int. Ed. Engl.*, **5**, 385 (1966).
    (b) V. Prelog and G. Helmchen, *Angew. Chem., Int. Ed. Engl.*, **21**, 567 (1982).
13. R. S. Cahn and O. C. Dermer, *Introduction to Chemical Nomenclature*, 5th ed., Butterworths, London, 1979, pp. 140–150.
14. J. March, *Advanced Organic Chemistry*, 3rd ed., Wiley, New York, 1985, p. 119.
15. See, for example, H. Brunner, in *The Chemistry of the Metal–Carbon Bond*, Vol. 5 (Ed. F. R. Hartley), Chap. 4, Wiley, Chichester, 1989, p. 109.
16. F. G. Mann, *The Heterocyclic Derivatives of Phosphorus, Arsenic, Antimony and Bismuth*, 2nd ed., Part II, Wiley-Interscience, New York, 1970, pp. 357–584.
17. (a) L. Horner, *Pure Appl. Chem.*, **9**, 225 (1964);
    (b) V. I. Sokolov and O. A. Reutov, *Russ. Chem. Rev.*, **34**, 1 (1965);
    (c) G. K. Kamai and G. M. Usacheva, *Russ. Chem. Rev.*, **35**, 601 (1966);
    (d) F. D. Yambushev and V. I. Savin, *Russ. Chem. Rev.*, **48**, 582 (1979).
18. (a) G. O. Doak and L. D. Freedman, *Organometallic Compounds of Arsenic, Antimony, and Bismuth*, Wiley-Interscience, New York, 1970;
    (b) S. Samaan, in *Methoden der Organischen Chemie (Houben Weyl)* (Ed. E. Müller), Vol. XIII/8, *Metallorganische Verbindungen As, Sb, Bi*, Georg Thieme Verlag, Stuttgart, 1978, pp. 16–441.
19. M. J. Gallagher and I. D. Jenkins, *Top. Stereochem.*, **3**, 1 (1968).
20. H. B. Kagan and M. Sasaki, in *The Chemistry of Organophosphorus Compounds*, Vol. 1 (Ed. F. R. Hartley), Chap. 3, Wiley, Chichester, 1990, p. 51.
21. J. Jacques, A. Collet and H. Wilen, *Enantiomers, Racemates and Resolutions*, Wiley-Interscience, New York, 1981.
22. A. Michaelis, *Annalen*, **321**, 159 (1902).
23. T. F. Winmill, *J. Chem. Soc.*, 718 (1912).
24. G. K. Kamai, *Chem. Ber.*, **66**, 1779 (1933); *Zh. Obshch. Khim.*, **4**, 184 (1934).
25. W. J. Pope and A. W. Harvey, *J. Chem. Soc.*, **79**, 828 (1901).
26. R. C. L. Mooney, *J. Am. Chem. Soc.*, **62**, 2955 (1940).
27. F. G. Holliman and F. G. Mann, *J. Chem. Soc.*, 547, 550 (1943).
28. F. G. Holliman and F. G. Mann, *J. Chem. Soc.*, 45 (1945).
29. F. G. Holliman, F. G. Mann and D. A. Thornton, *J. Chem. Soc.*, 9 (1960).

30. F. G. Mann and J. Watson, *J. Chem. Soc.*, 505 (1947).
31. G. K. Kamai and Y. F. Gatilov, *Dokl. Akad. Nauk SSSR*, **137**, 91 (1961); *Chem. Abstr.*, **55**, 19840 (1961).
32. (a) G. B. B. M. Sutherland, E. Lee and C. K. Wu, *Trans. Faraday Soc.*, **35**, 1373 (1939);
    (b) C. C. Costain and G. B. B. M. Sutherland, *J. Phys. Chem.*, **56**, 321 (1952).
33. J. F. Kincaid and F. G. Henriques, *J. Am. Chem. Soc.*, **62**, 1474 (1940).
34. R. E. Weston, *J. Am. Chem. Soc.*, **76**, 2645 (1954).
35. K. Mislow, A. Zimmerman and J. L. Melillo, *J. Am. Chem. Soc.*, **85**, 594 (1963).
36. I. G. M. Campbell, *J. Chem. Soc.*, 4 (1947).
37. (a) I. G. M. Campbell, *J. Chem. Soc.*, 3109 (1950);
    (b) I. G. M. Campbell, *J. Chem. Soc.*, 4448 (1952);
    (c) I. G.M. Campbell and D. J. Morrill, *J. Chem. Soc.*, 1662 (1955).
38. I. G. M. Campbell, *J. Chem. Soc.*, 3116 (1955).
39. I. G. M. Campbell and R. C. Poller, *J. Chem. Soc.*, 1195 (1956).
40. (a) K. F. Kumli, W. E. McEwen and C. A. VanderWerf, *J. Am. Chem. Soc.*, **81**, 248 (1959);
    (b) M. Zanger, C. A. VanderWerf and W. E. McEwen, *J. Am. Chem. Soc.*, **81**, 3806 (1959);
    (c) A. Bladé-Font, C. A. VanderWerf and W. E. McEwen, *J. Am. Chem. Soc.*, **82**, 2396 (1960).
41. L. Horner and H. Fuchs, *Tetrahedron Lett.*, 1573 (1963).
42. L. Horner, H. Winkler and E. Meyer, *Tetrahedron Lett.*, 789 (1965).
43. A. F. Peerdeman, J. P. C. Holst, L. Horner and H. Winkler, *Tetrahedron Lett.*, 811 (1965).
44. L. B. Ionov, L. A. Kunitskaya, I. P. Mukanov and Y. F. Gatilov, *Zh. Obshch. Khim.*, **46**, 68 (1976); *J. Gen. Chem. USSR (Engl. Transl.)*, **46**, 68 (1976).
45. D. G. Allen, C. L. Raston, B. W. Skelton, A. H. White and S. B. Wild, *Aust. J. Chem.*, **37**, 1171 (1984).
46. (a) D. G. Allen, N. K. Roberts and S. B. Wild, *J. Chem. Soc., Chem. Commun.*, 346 (1978);
    (b) D. G. Allen and S. B. Wild, *Organometallics*, **2**, 394 (1983).
47. M. S. Lesslie and E. E. Turner, *J. Chem. Soc.*, 1268 (1935).
48. M. S. Lesslie and E. E. Turner, *J. Chem. Soc.*, 730 (1936).
49. M. S. Lesslie, *J. Chem. Soc.*, 1001 (1938).
50. L. Horner and H. Winkler, *Tetrahedron Lett.*, 3271 (1964).
51. M. S. Lesslie, *J. Chem. Soc.*, 1183 (1949).
52. M. S. Lesslie and E. E. Turner, *J. Chem. Soc.*, 1051 (1935).
53. G. K. Kamai, *Chem. Ber.*, **68**, 960, 1893 (1935); *J. Gen. Chem. USSR (Engl. Transl)*, **10**, 683 (1940); **12**, 104 (1942); **17**, 2178 (1947).
54. L. Horner and W. Hofer, *Tetrahedron Lett.*, 4091 (1965).
55. C. F. H. Allen, F. B. Wells and C. V. Wilson, *J. Am. Chem. Soc.*, **56**, 233 (1934).
56. A. Camerman and J. Trotter, *J. Chem. Soc.*, 730 (1965).
57. J. E. Stuckey, A. W. Cordes, L. B. Handy, R. W. Perry and C. K. Fair, *Inorg. Chem.*, **11**, 1846 (1972).
58. J. Chatt and F. G. Mann, *J. Chem. Soc.*, 1184 (1940).
59. (a) E. R. H. Jones and F. G. Mann, *J. Chem. Soc.*, 411 (1955);
    (b) F. G. Mann, *J. Chem. Soc.*, 4266 (1963).
60. O. Kennard, F. G. Mann, D. G. Watson, J. K. Fawcett and K. A. Kerr, *J. Chem. Soc., Chem. Commun.*, 269 (1968).
61. I. G. M. Campbell and R. C. Poller, *J. Chem. Soc.*, 1195 (1956).
62. D. Sartain and M. R. Truter, *J. Chem. Soc.*, 4414 (1963).
63. Y. Gatilov, G. KF. Kamai and L. B. Ionov, *Zh. Obshch. Khim.*, **38**, 370 (1968); *J. Gen. Chem. USSR (Engl. Transl.)*, **38**, 368 (1968).
64. Y. F. Gatilov and L. B. Ionov, *Zh. Obshch. Khim.*, **38**, 2561 (1968); *J. Gen. Chem. USSR (Engl. Transl.)*, **38**, 2476 (1968).
65. Y. F. Gatilov, L. B. Ionov and S. S. Molodtsov, *Zh. Obshch. Khim.*, **42**, 1535 (1972); *J. Gen. Chem. USSR (Engl. Transl.)*, **42**, 1527 (1972).
66. Y. F. Gatilov and L. B. Ionov, *Zh. Obshch. Khim.*, **39**, 1064 (1969); *J. Gen. Chem. USSR (Engl. Transl.)*, **39**, 1036 (1969).
67. Y. F. Gatilov and F. D. Yambushev, *Zh. Obshch. Khim.*, **43**, 1132 (1973); *J. Gen. Chem. USSR (Engl. Transl.)*, **43**, 1123 (1973).
68. Y. F. Gatilov, L. B. Ionov and F. D. Yambushev, *Zh. Obshch. Khim.*, **41**, 570 (1971); *J. Gen. Chem. USSR (Engl. Transl.)*, **41**, 565 (1971).

69. Y. F. Gatilov, F. D. Yambushev and N. K. Tenisheva, *Zh. Obshch. Khim.*, **43**, 2273 (1973); *J. Gen. Chem. USSR (Engl. Transl.)*, **43**, 2263 (1973).
70. F. D. Yambushev, Y. F. Gatilov and N. K. Tenisheva, *Uch. Zap. Kazan, Gos. Pedagog. Inst.*, **123**, 45 (1973); *Chem. Abstr.*, **86**, 190114 m (1977).
71. Y. F. Gatilov, L. B. Ionov and F. D. Yambushev, *Zh. Obshch. Khim.*, **40**, 2250 (1970); *J. Gen. Chem. USSR (Engl. Transl.)*, **40**, 2237 (1970).
72. L. B. Ionov, Y. F. Gatilov, I. P. Mukanov and L. G. Kokorina, *Zh. Obshch. Khim.*, **44**, 1874 (1974); *J. Gen. Chem. USSR (Engl. Transl.)*, **44**, 1837 (1974).
73. F. D. Yambushev, Y. F. Gatilov, N. K. Tenisheva and V. I. Savin, *Zh. Obshch. Khim.*, **44**, 1734 (1974); *J. Gen. Chem. USSR (Engl. Transl.)*, **44**, 1701 (1974).
74. Y. F. Gatilov, F. D. Yambushev and N. K. Tenisheva, *Zh. Obshch. Khim.*, **43**, 2681 (1973); *J. Gen. Chem. USSR (Engl. Transl.)*, **43**, 2659 (1973).
75. F. D. Yambushev, Y. F. Gatilov, N. K. Tenisheva and V. I. Savin, *Zh. Obshch. Khim.*, **44**, 2499 (1974); *J. Gen. Chem. USSR (Engl. Transl.)*, **44**, 2458 (1974).
76. F. D. Yambushev, Y. F. Gatilov, N. K. Tenisheva and V. I. Savin, *Zh. Obshch. Khim.*, **45**, 2527 (1975); *J. Gen. Chem. USSR (Engl. Transl.)*, **45**, 2481 (1975).
77. (a) L. Horner, F. Röttger and H. Fuchs, *Chem. Ber.*, **96**, 3141 (1963);
    (b) L. Horner and J. Haufe, *Chem. Ber.*, **101**, 2903 (1968).
78. L. Horner and W.-D. Balzer, *Chem. Ber.*, **102**, 3542 (1969).
79. W. Klyne and J. Buckingham, *Atlas of Stereochemistry*, 2nd ed., Vol. 1, Oxford University Press, New York, 1978, p. 231.
80. L. Horner and K. Dickerhof, *Phosphorus and Sulfur*, **15**, 213 (1983).
81. L. Horner and J. Haufe, *J. Electroanal. Chem.*, **20**, 245 (1969).
82. L. Horner, J. Röder and D. Gammel, *Phosphorus*, **3**, 175 (1973).
83. (a) W. E. McEwen, K. F. Kumli, A. Blade-Font, M. Zanger and C. A. VanderWerf, *J. Am. Chem. Soc.*, **86**, 2378 (1964);
    (b) T. J. Katz, C. R. Nicholson and C. A. Reilly, *J. Am. Chem. Soc.*, **88**, 3832 (1966);
    (c) P. D. Henson, K. Naumann and K. Mislow, *J. Am. Chem. Soc.*, **91**, 5645 (1969).
84. L. Horner and M. Ernst, *Chem. Ber.*, **103**, 318 (1970).
85. K. Henrick and S. B. Wild, *J. Chem. Soc., Dalton Trans.*, 1506 (1975).
86. L. Horner and W. Hofer, *Tetrahedron Lett.*, 3321 (1966).
87. L. Horner, W. Hofer, I. Ertel and H. Kunz, *Chem. Ber.*, **103**, 2718 (1970).
88. L. Horner and S. Samaan, *Phosphorus*, **3**, 153 (1973).
89. L. Horner and S. Samaan, *Phosphorus*, **1**, 207 (1971).
90. L. Horner and I. Ertel, H.-D. Ruprecht and O. Belovsky, *Chem. Ber.*, **103**, 1582 (1970).
91. I. Gosney, T. J. Lillie and D. Lloyd, *Angew. Chem., Int. Ed. Engl.*, **16**, 487 (1977).
92. D. G. Allen, G. McLaughlin, G. B. Robertson, W. L. Steffen, G. Salem and S. B. Wild, *Inorg. Chem.*, **21**, 1007 (1982).
93. B. Skelton and A. H. White, *J. Chem. Soc., Dalton Trans.*, 1556 (1980).
94. (a) A. C. Cope, C. R. Ganellin, H. W. Johnson, T. V. Van Auken and H. S. Winkler, *J. Am. Chem. Soc.*, **85**, 3276 (1963);
    (b) A. C. Cope, K. Banholzer, H. Keller, B. A. Pawson, J. J. Whang and H. J. S. Winkler, *J. Am. Chem. Soc.*, **87**, 3644 (1965).
95. A. C. Cope and E. A. Caress, *J. Am. Chem. Soc.*, **88**, 1711 (1966).
96. T. H. Chan, *J. Chem. Soc., Chem. Commun.*, 895 (1968).
97. (a) S. Otsuka, A. Nakamura, T. Kano and K. Tani, *J. Am. Chem. Soc.*, **93**, 4301 (1971);
    (b) K. Tani, L. D. Brown, J. Ahmed, J. A. Ibers, M. Yokota, A. Nakamura and S. Otsuka, *J. Am. Chem. Soc.*, **99**, 7876 (1977).
98. G. Salem and S. B. Wild, *J. Organomet. Chem.*, **370**, 33 (1989).
99. (a) G. T. Crisp, G. Salem, F. S. Stephens and S. B. Wild, *J. Chem. Soc., Chem. Commun.*, 600 (1987);
    (b) G. T. Crisp, G. Salem, F. S. Stephens and S. B. Wild, *Organometallics*, **8**, 2360 (1989).
100. A. A. Watson, A. C. Willis and S. B. Wild, *J. Organometal. Chem.*, **445**, 71 (1993).
101. A. Dore, D. Fabbri, S. Gladiali and O. De Lucchi, *J. Chem. Soc., Chem. Commun.*, 1124 (1993).
102. S. C. Grocott and S. B. Wild, *Inorg. Chem.*, **21**, 3526 (1982).
103. N. K. Roberts and S. B. Wild, *J. Chem. Soc., Dalton Trans.*, 2015 (1979).
104. K. Henrick and S. B. Wild, *J. Chem. Soc., Dalton Trans.*, 2500 (1974).

105. J. C. Dewan, K. Henrick, D. L. Kepert, K. R. Trigwell, A. H. White and S. B. Wild, *J. Chem. Soc., Dalton Trans.*, 546 (1975).
106. E. E. Turner and M. M. Harris, *Quart. Rev.*, **1**, 299 (1947).
107. J. Cymerman Craig, R. P. K. Chan and S. K. Roy, *Tetrahedron*, **23**, 3573 (1967).
108. N. K. Roberts and S. B. Wild, *J. Am. Chem. Soc.*, **101**, 6254 (1979).
109. (a) C. J. Hawkins, *Absolute Configuration of Metal Complexes*, Chap. 3, Wiley-Interscience, New York, 1971, pp. 63–86;
    (b) C. J. Hawkins and J. Palmer, *Coord. Chem. Rev.*, **44**, 1 (1982).
110. J. W. L. Martin, J. A. L. Palmer and S. B. Wild, *Inorg. Chem.*, **23**, 2664 (1984).
111. Y. Shigetomi, M. Kojima and J. Fujita, *Bull. Chem. Soc. Jpn.*, **58**, 258 (1985).
112. J. W. L. Martin, F. S. Stephens, K. D. V. Weerasuria and S. B. Wild, *J. Am. Chem. Soc.*, **110**, 4346 (1988).
113. G. Salem and S. B. Wild, *Inorg. Chem.*, **22**, 4049 (1983).
114. P.-H. Leung, G. M. McLaughlin, J. W. L. Martin and S. B. Wild, *Inorg. Chem.*, **25**, 3392 (1986).
115. P.-H. Leung, J. W. L. Martin and S. B. Wild, *Inorg. Chem.*, **25**, 3396 (1986).
116. P. G. Kerr, P.-H. Leung and S. B. Wild, *J. Am. Chem. Soc.*, **109**, 4321 (1987).
117. B. Bosnich, W. G. Jackson and S. B. Wild, *J. Am. Chem. Soc.*, **95**, 8269 (1973).
118. B. Bosnich, W. G. Jackson and S. B. Wild, *Inorg. Chem.*, **13**, 1121 (1974).
119. I. M. Goldman, J. K. Larson, J. R. Tretter and E. G. Andrews, *J. Am. Chem. Soc.*, **91**, 4941 (1969).
120. J. MacB. Harrowfield, J. M. Patrick, B. W. Skelton and A. H. White, *Aust. J. Chem.*, **41**, 159 (1988).
121. (a) P.-H. Leung, A. C. Willis and S. B. Wild, *Inorg. Chem.*, **31**, 1406 (1992);
    (b) P. H. Leung and S. B. Wild, *Bull. Sing. N. I. Chem.*, **17**, 37 (1989).
122. G. Salem, G. B. Shaw, A. C. Willis and S. B. Wild, *J. Organomet. Chem.*, **455**, 185 (1993).
123. J. Stackhouse, R. J. Cook and K. Mislow, *J. Am. Chem. Soc.*, **95**, 953 (1973).
124. I. B. Ionov, Y. Y. Samitov, L. G. Kokorina, A. P. Korovyakov and B. D. Chernokal'skii, *Zh. Obshch. Khim.*, **47**, 1276 (1977); *J. Gen. Chem. USSR (Engl. Transl.)*, **47**, 1176 (1977).
125. L. B. Ionov, A. P. Korovyakov, L. G. Kokorina and B. D. Chernokal'skii, *Zh. Obshch. Khim.*, **47**, 1561 (1977); *J. Gen. Chem. USSR (Engl. Transl.)* **47**, 1433 (1977).
126. L. B. Ionov, A. P. Korovyakov and B. D. Chernokal'skii, *Zh. Obshch. Khim.*, **47**, 408 (1977); *J. Gen. Chem. USSR (Engl. Transl.)*, **47**, 376 (1977).
127. L. B. Ionov, V. A. Shchuklin, S. I. Zobnin and G. B. Zamost'yanova, *Zh. Obshch. Khim.*, **52**, 342 (1982); *J. Gen. Chem. USSR (Engl. Transl.)*, **52**, 295 (1982).
128. L. B. Ionov, L. A. Kunitskaya and V. I. Korner, *Zh. Obshch. Khim.*, **45**, 1508 (1975); *J. Gen. Chem. USSR (Engl. Transl.)*, **45**, 1476 (1975).
129. (a) F. Challenger, *Quart. Rev.*, **9**, 255 (1955);
    (b) W. R. Cullen and K. J. Reimer, *Chem. Rev.*, **89**, 713 (1989).
130. Y. F. Gatilov, L. B. Ionov and G. K. Kamai, *Zh. Obshch. Khim.*, **38**, 372 (1968); *J. Gen. Chem. USSR (Engl. Transl.)*, **38**, 370 (1968).
131. L. B. Ionov, A. P. Korovyakov and B. D. Chernokal'skii, *Zh. Obshch. Khim.*, **49**, 2514 (1979); *J. Gen. Chem. USSR (Engl. Transl.)*, **49**, 2219 (1979).
132. L. B. Ionov, A. P. Korovyakov and B. D. Chernokal'skii, *Zh. Obshch. Khim.*, **48**, 940 (1978); *J. Gen. Chem. USSR (Engl. Transl.)*, **48**, 860 (1978).
133. L. B. Ionov, A. P. Korovyakov and B. D. Chernokal'skii, *Zh. Obshch. Khim.*, **49**, 184 (1979); *J. Gen. Chem. USSR (Engl. Transl.)*, **49**, 162 (1979).
134. L. B. Ionov, S. M. Reshetnikov and I. G. Kornienko, *Zh. Obshch. Khim.*, **53**, 2712 (1983), *J. Gen. Chem. USSR (Engl. Transl.)*, **53**, 2443 (1983).
135. L. B. Ionov, S. M. Reshetnikov, L. L. Makarova and T. M. Flegontova, *Zh. Obshch. Khim.*, **55**, 862 (1985); *J. Gen. Chem. USSR (Engl. Transl.)*, **55**, 769 (1985).
136. I. G. M. Campbell, *J. Chem. Soc.*, 1976 (1956).
137. (a) J. Meisenheimer and L. Lichtenstadt, *Chem. Ber.*, **44**, 356 (1911);
    (b) J. Meisenheimer, J. Casper, M. Höring, W. Lauter, L. Lichtenstadt and W. Samuel, *Justus Liebigs Ann. Chem.*, **449**, 224 (1926).
138. (a) J. A. Aeschlimann and N. P. McCleland, *J. Chem. Soc.*, 2025 (1924);
    (b) J. A. Aeschlimann, *J. Chem. Soc.*, 811 (1925).

139. Y. F. Gatilov, L. B. Ionov and V. M. Gavrilov, *Zh. Obshch. Khim.*, **42**, 540 (1972); *J. Gen. Chem. USSR (Engl. Transl.)*, **42**, 538 (1972).
140. Y. F. Gatilov, L. B. Ionov and S. S. Molodtsov, *Zh. Obshch. Khim.*, **42**, 1535 (1972); *J. Gen. Chem. USSR (Engl. Transl.)*, **42**, 1527 (1972).
141. Y. F. Gatilov and F. D. Yambushev, *Zh. Obshch. Khim.*, **43**, 1132 (1973); *J. Gen. Chem. USSR (Engl. Transl.)*, **43**, 1123 (1973).
142. Y. F. Gatilov, L. B. Ionov, S. S. Molodtsov and V. P. Kovyrzina, *Zh. Obshch. Khim.*, **42**, 1959 (1972); *J. Gen. Chem. USSR (Engl. Transl.)*, **42**, 1952 (1972).
143. G. H. Senkler and K. Mislow, *J. Am. Chem. Soc.*, **94**, 291 (1972).
144. L. Horner and W. Hofer, *Tetrahedron Lett.*, 3323 (1966).
145. R. D. Baechler, J. P. Casey, R. J. Cook, G. H. Senkler and K. Mislow, *J. Am. Chem. Soc.*, **94**, 2859 (1972).
146. (a) J. B. Lambert, G. F. Jackson and D. C. Mueller, *J. Am. Chem. Soc.*, **90**, 6401 (1968);
     (b) See also: J. B. Lambert, *Top. Stereochem.*, **6**, 19 (1971).
147. R. H. Bowman and K. Mislow, *J. Am. Chem. Soc.*, **94**, 2861 (1972).
148. C. R. Russ and A. G. MacDiarmid, *Angew. Chem., Int. Ed. Engl.*, **5**, 418 (1966).
149. W. R. Cullen, *Can. J. Chem.*, **41**, 322 (1963).
150. J. MacB. Harrowfield, G. Salem and S. B. Wild, unpublished work.
151. G. K. Kamai and Y. F. Gatilov, *Zh. Obshch. Khim.*, **34**, 782 (1964); *J. Gen. Chem. USSR (Engl. Transl.)*, **34**, 781 (1964).
152. L. Horner and W. Hofer, *Tetrahedron Lett.*, 3281 (1965).
153. (a) R. A. Lewis, O. Korpium and K. Mislow, *J. Am. Chem. Soc.*, **89**, 4786 (1967).
     (b) W. B. Farnham, R. K. Murray and K. Mislow, *J. Am. Chem. Soc.*, **92**, 5809 (1970).
154. A. D. Calhoun, W. J. Kobos, T. A. Nile and C. A. Smith, *J. Organometal. Chem.*, **170**, 175 (1979).
155. B. A. Murrer, J. M. Brown, P. A. Chaloner, P. N. Nicholson and D. Parker, *Synthesis*, 350 (1979).
156. N. C. Payne and D. W. Stephan, *Inorg. Chem.*, **21**, 182 (1982).
157. L. Horner and G. Simons, *Phosphorus and Sulfur*, **14**, 253 (1983).
158. W. R. Cullen and J. D. Woollins, *Can. J. Chem.*, **60**, 1793 (1982).
159. J. P. Casey, R. A. Lewis and K. Mislow, *J. Am. Chem. Soc.*, **91**, 2789 (1969).
160. E. P. Kyba and S. P. Rines, *J. Org. Chem.*, **47**, 4800 (1982).
161. (a) W. H. Pirkle and S. D. Beare, *J. Am. Chem. Soc.*, **91**, 5150 (1969);
     (b) W. H. Pirkle and P. L. Rinaldi, *J. Org. Chem.*, **42**, 3217 (1977).
162. M. L. Pasquier and W. Marty, *Angew. Chem., Int. Ed. Engl.*, **24**, 315 (1985).
163. R. R. Schrock and J. A. Osborn, *J. Am. Chem. Soc.*, **98**, 4450 (1976).
164. (a) H. B. Kagan and T. P. Dang, *J. Am. Chem. Soc.*, **94**, 6429 (1972);
     (b) G. Gelbard, H. B. Kagan and R. Stern, *Tetrahedron*, **32**, 233 (1976).
165. D. G. Allen, S. B. Wild and D. L. Wood, *Organometallics*, **5**, 1009 (1986).
166. A. F. M. Mokhlesur Rahman and S. B. Wild, *J. Mol. Catal.*, **39**, 155 (1987).
167. W. Dumont, J.-C. Poulin, T.-P. Dang and H. B. Kagan, *J. Am. Chem. Soc.*, **95**, 8295 (1973).
168. N. K. Roberts and S. B. Wild, *Inorg. Chem.*, **20**, 1892 (1981).
169. G. Salem and S. B. Wild, *Inorg. Chem.*, **23**, 2655 (1984).
170. P.-H. Leung, J. W. L. Martin and S. B. Wild, *Inorg. Chem.*, **25**, 3396 (1986).
171. N. K. Roberts and S. B. Wild, *Inorg. Chem.*, **20**, 1900 (1981).
172. G. Salem and S. B. Wild, *Inorg. Chem.*, **31**, 581 (1992).
173. (a) S. C. Grocott and S. B. Wild, *Inorg. Chem.*, **21**, 3526 (1982);
     (b) S. C. Grocott and S. B. Wild, *Inorg. Chem.*, **21**, 3535 (1982).
174. J. A. L. Palmer and S. B. Wild, *Inorg. Chem.*, **22**, 4054 (1983).
175. G. Salem, A. Schier and S. B. Wild, *Inorg. Chem.*, **27**, 3029 (1988).

CHAPTER **4**

Thermochemistry of organo-arsenic, antimony and bismuth compounds

JOEL F. LIEBMAN

Department of Chemistry and Biochemistry, University of Maryland, Baltimore County Campus, 5401 Wilkens Avenue, Baltimore, Maryland 21228-5398, USA

JOSÉ ARTUR MARTINHO SIMÕES

Departmento de Engenharia Quimica, Instituto Superior Técnico, 1096 Lisboa Codex, Portugal

and

SUZANNE W. SLAYDEN

Department of Chemistry, George Mason University, 4400 University Drive, Fairfax, Virginia 22030-4444, USA

The chemistry of organic arsenic, antimony and bismuth compounds
Edited by S. Patai © 1994 John Wiley & Sons Ltd

I. INTRODUCTION: SCOPE AND DEFINITIONS

In the previous volumes of the Patai series 'The Chemistry of Functional Groups' the various authors of the thermochemistry chapters used varying definitions of the discipline to define the scope of their chapters. In our chapter, we will limit our attention only to enthalpies of formation (also called heats of formation) with some accompanying discussion of bond enthalpies (a concept related to, but not identical to, the more customary but more ambiguously defined bond energies and bond strengths[1]). We do not include entropies, phase change enthalpies and heat capacities *per se*[2] because of the paucity of data. Neither have we attempted to derive the now standard Benson increments[3] for enthalpies of formation because of the diversity of structural types and since we perceive general inaccuracy, as well as inadequacy, of the data.

More precisely, except for the admittedly disingenuous case of nitrogen, thermochemical data for the organic compounds containing the group 15 elements (i.e. N, P, As, Sb and Bi) are rather scarce and usually of poor quality. Regretfully, we consider the respectably understood organic thermochemistry of nitrogen to be all but irrelevant for the current chapter because:

(a) The majority of simple, isolable and easily purified compounds that characterize the organic chemistry of nitrogen have few or no thermochemically convenient valence isoelectronic counterparts among the heavier group 15 elements. These include the counterparts of nitriles and isocyanides; and azo, azoxy and nitro compounds which are either highly hindered or have not been isolated as bulk, condensed phase samples.

(b) The majority of simple, isolable and easily purified compounds that characterize the organic chemistry of the heavier group 15 elements have few or no thermochemical counterparts among the compounds of nitrogen. These include phosphine oxides and arsenites. The hygroscopicity of amine oxides complicates their combustion calorimetry because of purity problems and organic arsenites and nitrites generally have different stoichiometry and hence structure: the former are $(RO)_3As$ and the latter are $RONO$.

Both of these differences between nitrogen and its heavier congeners relate to the relative dissociation enthalpies of single and multiple bonds connecting these elements. These differences are not unique to the group 15 elements: recent thermochemical reviews involving the lightest two group 16 elements, O and S, document that most oxygen and sulfur-containing functional groups likewise show rather few structural parallels[4,5].

Relatedly, the study of compounds containing the elements adjacent to group 15 are not particularly helpful. Compounds of the group 14 elements (C, Si, Ge, Sn and Pb) generally lack lone pairs on the 'central' atom and so lack

(a) resonance stabilization in their phenyl derivatives,
(b) lone pair–lone pair repulsion with adjacent hetero atoms, and
(c) higher valence oxides with dative bonding to oxygen.

The chemistry of the group 16 elements that adjoin the elements of interest on the other side of the periodic table likewise provides little insight: while the chemistry of oxygen- and sulfur-containing species can provide guidance for those of nitrogen, we have already discounted the relevance of nitrogen chemistry. In principle, sulfur chemistry should help us with phosphorus chemistry, but we admit now that in this study we will only consider the particular cases of organic molecules containing arsenic, antimony and bismuth because the recent reviews[6,7] of the thermochemistry of organophosphorus compounds show the data are generally of insufficient quality. Organo-selenium[8] and -tellurium thermochemistry[9] is seemingly at least as primitive as that of arsenic and antimony. Finally, we note that our understanding of organobismuth chemistry is not aided by that of polonium—quite enigmatically, all polonium isotopes have short half-lives and so polonium is highly radioactive while bismuth (at least as its naturally occurring [209]Bi isotope) is effectively infinitely stable. As such, organopolonium chemistry is essentially unknown.

II. DEFINITIONS AND CONVENTIONS

In what follows, the elements arsenic, antimony and bismuth will be generically written as E and we will refer to their compounds that also contain carbon sometimes as organometallics and sometimes as organometalloids. It is to be noted that the metallic/metalloid/nonmetallic boundaries are not universally agreed upon[10], although it is generally acknowledged that the metallic character of these three elements increases as one proceeds down the periodic table: As < Sb < Bi. After all, aqueous As(III) occurs as a thermochemically characterized anion while aqueous Bi(III) occurs in a plethora of thermochemically characterized mononuclear and polynuclear cations[11]. Much less ambiguous than the assignment of metallic character, and for that matter much less contentious, is our selection of units for this chapter. Following orthodox thermochemical practice, all enthalpies in this chapter will be given in $kJ\,mol^{-1}$ where the conversion factor between this unit and the admittedly more commonly used 'chemists' $kcal\,mol^{-1}$ is $4.184\,kJ\,mol^{-1} = 1\,kcal\,mol^{-1}$; $1\,kJ\,mol^{-1} = 0.2390\,kJ\,mol^{-1}$.

For organometalloid/organometallic compounds of arsenic, antimony and bismuth, all of the available directly determined enthalpies of formation are shown in Table 1, together with an indication of the experimental method used to obtain them and the appropriate literature references. Also included in Table 1 are the available enthalpy of formation data for the homoleptic hydrides, alkoxides and thiolates, respectively, because:

(a) in a thermochemical context, hydrogen is very naturally compared with hydrocarbyl (alkyl and aryl) groups,

(b) these compounds also are trivalent, tricoordinate and carbon-containing even if carbon is not directly bound to the central element E, and

(c) we recall that sulfur is both valence isoelectronic to oxygen, and of comparable electronegativity to carbon.

How can we assess the thermochemical data presented in Table 1? Many were obtained by static-bomb calorimetry. Regrettably, this technique is clearly unsuitable to deal with these substances due to the ill-defined composition of the combustion products (see the discussion in References 13 and 28). The formation of nonstoichiometric oxides upon combustion all but precludes the experimental rigor demanded of the combustion calorimetrist. This fact, by itself, allows us to question the reliability of the values shown for all of the trialkyl compounds and for triphenylantimony. Although the results for triphenylbismuth found by static and rotating-bomb calorimetry overlap within their error bars, this has been suggested to be fortuitous[27].

III. ESTIMATION METHODOLOGIES AND ALIPHATIC DERIVATIVES

A. Trialkyl and Trihydride Species

Knowledge of trends of thermochemical data found for other elements may help to show that, in fact, the values for the trialkyl species in Table 1 are unreasonable. A method that seems to be a generally valid approach is to assess the data for families of compounds, say ML_n, by plotting $\Delta H_f^\circ(ML_n)$ versus $\Delta H_f^\circ(LH)$, either with ML_n and LH in their standard reference states (i.e. in their stable physical states at 298.15 K

TABLE 1. Standard enthalpies of formation of As, Sb and Bi organometalloid/organometallic compounds (data in $kJ\,mol^{-1}$)

Compound	ΔH_f°(l/c) [method]^a	$\Delta H_v^\circ/\Delta H_s^\circ$	ΔH_f°(g)
AsH_3, g			66.4^b [DBK]
As_2H_4, g			147.3^c [MS]
$AsMe_3$, l	$-16.2 \pm 10.2^{d,e}$ [SB]	28.8 ± 1.3^d	12.6 ± 10.1
$AsEt_3$, l	$13.1 \pm 16.8^{d,f}$ [SB]	43.1 ± 4.2^d	56.2 ± 17.3
$AsPh_3$, c	$309.8 \pm 6.7^{d,g}$ [RB]	98.3 ± 8.4^d	408.1 ± 10.7
$AsPh_3O$, c	$112 \quad \pm 17^h$ [RS]	149.0 ± 5.4^i	$261 \quad \pm 18$
	78.6 ± 14.6^i [RB]	149.0 ± 5.4^i	227.6 ± 15.6
$As(OMe)_3$, l	$-592.5 \pm 1.0^{j,k}$ [RS]	42.3 ± 1.3^j	-550.2 ± 1.6
$As(OEt)_3$, l	$-704.4 \pm 0.8^{j,k}$ [RS]	50.6 ± 4.2^j	-653.8 ± 4.3
$As(OPr)_3$, l	$-783.3 \pm 3.5^{j,k}$ [RS]	58.6 ± 8.4^j	-724.7 ± 9.1
$As(S_2CNEt_2)_3$, c	-104.5 ± 7.6^l [RS]	$124 \quad \pm 3^l$	$20 \quad \pm 8$
$As(S_2CNBu_2)_3$, c	-443.1 ± 5.4^m [RS]	$128 \quad \pm 3^m$	-315.1 ± 6.2
SbH_3, g			145.1^b [DBK]
Sb_2H_4, g			239.3^c [MS]
$SbMe_3$, l	$0.8 \pm 25.2^{d,n}$ [SB]	31.4 ± 1.3^d	32.2 ± 25.2
$SbEt_3$, l	$5.2 \pm 10.7^{d,f}$ [SB]	43.5 ± 4.2^d	48.7 ± 11.5
$SbPh_3$, c	$329.2 \pm 17.9^{d,o}$ [SB]	106.3 ± 8.4^d	435.4 ± 19.0
$SbPh_3O$, c	$136 \quad \pm 17^h$ [RS]		
$Sb(S_2CNEt_2)_3$, c	-161.8 ± 7.3^p [RS]	$160 \quad \pm 2^p$	$-2 \quad \pm 8$
$Sb(S_2CNBu_2)_3$, c	-526.2 ± 2.5^m [RS]	$179 \quad \pm 3^m$	-347.2 ± 3.9
BiH_3, g			278^c [MS]
$BiMe_3$, l	$158.4 \pm 14.2^{d,g}$ [SB]	36.0 ± 1.3^d	194.4 ± 14.3
$BiEt_3$, l	$169.7 \pm 16.8^{d,f}$ [SB]	46.0 ± 4.2^d	215.7 ± 17.3
$BiPh_3$, s	$469.2 \pm 16.9^{o,j}$ [SB]	110.9 ± 8.4^j	580.1 ± 18.9
	$489.7 \pm 5.2^{d,r}$ [RB]	110.9 ± 8.4^j	600.6 ± 9.9
$Bi(S_2CNEt_2)_3$, c	-183.6 ± 7.3^p [RS]	$213 \quad \pm 3^p$	$29 \quad \pm 8$
$Bi(S_2CNBu_2)_3$, c	-530.7 ± 5.5^n [RS]	$202 \quad \pm 3^m$	-328.7 ± 6.3

^{a}Methods: DBK = 'Databook: MS = mass spectrometry; RB = rotating-bomb combustion calorimetry; RS = reaction-solution calorimetry; SB = static-bomb combustion calorimetry.

^{b} See Reference 11.
^{c} See Reference 12.
^{d} See Reference 13.
^{e} See Reference 14.
^{f} See Reference 15.
^{g} See Reference 16.
^{h} See Reference 17 as amended by the discussion in the text (cf. Section IV).
^{i} See Reference 18.
^{j} See Reference 19.
^{k} See Reference 20.
^{l} See Reference 21.
^{m} See Reference 22.
^{n} See Reference 23.
^{o} See Reference 24.
^{p} See Reference 25.
^{q} See Reference 26.
^{r} See Reference 27.

and 1 bar, *ca* the more 'conventional' 25 °C and 1 atm) or with both species in the gas phase. For the sake of clarity, a brief description of this method is given below.

It has been observed that many, if not all, of the above plots which involve reliable thermochemical data define excellent straight lines[29]. The meaning of this observation can be obtained by considering Scheme 1, $\Delta H_f^\circ(1)$ and $\Delta H_f^\circ(2)$ being the enthalpies of hypothetical reactions for a 'family' of compounds ML_n, where reactants and products are in the standard reference states (rs) and in the gas phase (g), respectively. The ΔH_v° are vaporization or sublimation enthalpies for species that are liquids or solids under standard conditions, respectively. The empirical linear relation (equation 3) implies that $\Delta H^\circ(1)$ is a constant for the series of compounds ML_n. $\Delta H_r^\circ(1)$, on the other hand, can be expressed in terms of the bond dissociation enthalpies (equation 4) by using Scheme 1.

$$ML_n(\text{rs}) + nHY(\text{rs}) \xrightarrow{\Delta H^\circ(1)} MY_n(\text{rs}) + nLH(\text{rs}) \tag{1}$$

$$\big\downarrow \Delta H_{v1}^\circ \qquad \big\downarrow n\Delta H_{v2}^\circ \qquad\qquad \big\downarrow \Delta H_{v3}^\circ \qquad \big\downarrow n\Delta H_{v4}^\circ$$

$$ML_n(\text{g}) + nHY(\text{g}) \xrightarrow{\Delta H^\circ(2)} MY_n(\text{g}) + nLH(\text{g}) \tag{2}$$

SCHEME 1

$$\Delta H_f^\circ(ML_n, \text{rs}) = a\Delta H_f^\circ(LH, \text{rs}) + b \tag{3}$$

$$\Delta H^\circ(1) = n[\bar{D}(M\!-\!L) - D(L\!-\!H)] + [\Delta H_{v1}^\circ - n\Delta H_{v4}^\circ]$$
$$+ n\Delta H_{v2}^\circ + [nD(H\!-\!Y) - n\bar{D}(M\!-\!Y) - \Delta H_{v3}^\circ] \tag{4}$$

[It is important to note that we are talking about $\bar{D}(M\!-\!L)$, defined as $1/n$ times the enthalpy to disrupt (i.e. dismember) ML_n into $M + nL$—this quantity is generally not equal to $D(ML_{n-1}\!-\!L)$ which corresponds to cleaving only one $M\!-\!L$ bond in ML_n and so forms $ML_{n-1} + L$.] At least for liquid species, it is highly plausible that the second bracketed term in equation 4 is nearly a constant. Elsewhere it was shown[30] that for numerous substituents X, to within a 'few' $kJ\,mol^{-1}$, the enthalpy of vaporization of arbitrary mono functional hydrocarbon derivatives, RX, may be approximated by

$$\Delta H_v^\circ(RX) = 4.7\tilde{n}_C + 1.3n_Q + 3.0 + b(X) \tag{5}$$

where $\tilde{n}_C$ is the total number of nonquaternary carbons, n_Q is the total number of quaternary carbons, and $b(X)$ is a substituent-dependent parameter. It has been shown that X need not be univalent. That is, the enthalpy of vaporization of the divalent oxygen containing ethers (ROR'), and polyvalent sulfur-containing species, sulfides, sulfoxides and sulfones (RSO$_x$R', $x = 0, 1$ and 2, respectively) are all reliably predicted by use of the appropriate 'b' for X = $-$O$-$, $-$S$-$, $-$SO$-$ and $-$SO$_2-$ (with n_O and $\tilde{n}_C$ found by summing the relevant carbon count of R and R'). As such, although in fact untested for organometallics and organometalloids except for boron-containing species[31], we have no reason to believe that for the homoleptic R_3E the enthalpy of vaporization cannot be reliably estimated as

$$\Delta H_v^\circ(R_3E) = 3 \times 4.7\tilde{n}_C + 3 \times 1.3n_Q + 3.0 + b(E) \tag{6}$$

where $\tilde{n}_C$ and n_O refer to the carbon count for each individual R group. Finally, we note that the third bracketed term in equation 4 is constant, implying that $n[\bar{D}(M\!-\!L) - D(L\!-\!H)]$ is also constant for the series of compounds that obey the linear correlation. It is therefore very likely that $\bar{D}(M\!-\!L)$ and $D(L\!-\!H)$ follow nearly parallel trends. Obviously, if the linear correlation holds for reactants and products in the gas phase, then $\Delta H^\circ(2)$ is constant for a given Y and so will be $\bar{D}(M\!-\!L) - D(L\!-\!H)$.

The above mentioned method has been tested for a variety of organic, inorganic and organometallic thermochemical data and there is no apparent reason why it should not work—even approximately—for arsenic-, antimony- and bismuth-containing compounds. For example, we find for the trialkyl arsenites the following relation (also see Figure 1):

$$\Delta H_f^{\circ}(\text{As(OR)}_3, 1) = (2.997 \pm 0.062)\Delta H_f^{\circ}(\text{ROH}, 1) + 125.0(\pm 17.1) \qquad (7)$$

Relatedly for gaseous species, for equation 5 we find a value of $b(\text{AsO}_3)$ of 25.2 ± 1.3, 19.4 ± 4.2 and $13.3 \pm 8.5 \,\text{kJ mol}^{-1}$ for the three cases of R = Me, Et and Pr. Within rather large error bars, these three values are all compatible with ca $22 \,\text{kJ mol}^{-1}$ for the enthalpy of vaporization contribution of an arsenite grouping[32].

But what about the trialkyl arsines, stibines and bismuthines? The logic above suggests that if we had thermochemical data on, say, tripropylarsine, we could reliably predict the desired quantity for tributylarsine. While two points are not enough to confirm consistency of data, they are enough to confirm inconsistency. More precisely, the slopes of the lines in Figure 2 are seen to be negative while the above 'derivation' suggests that they may reasonably be expected to approximately equal n, the number of affixed groups to the central atom. The disconcerting sign of the slope arises from the finding that the enthalpies of formation of the triethyl compounds are more positive than those of the trimethyl compounds.

While the plots of enthalpies of formation indicate that the arsenic, antimony and bismuth data are problematic, they cannot be used to recommend alternative values since $both$ the methyl and ethyl are thought to be affected by large errors. If one of these were known to be reliable, it could be used to predict the second one. The difference between the enthalpies of formation of methyl and ethyl derivatives relates to the electronegativity of the affixed atom or group, wherefore the more electronegative it is, the less relatively stable is the methyl compared to the ethyl derivative[33]. This stability reasoning extends to the relative stability of methyl and ethyl cation, and antithetically to methyl and ethyl anion, hence we conclude that only for extremely electropositive elements such as lithium is it at all conceivable that the methyl compound has the more

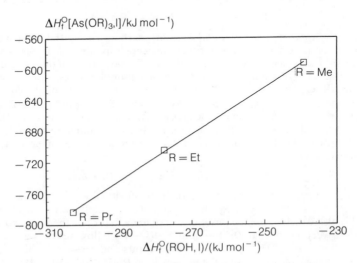

FIGURE. 1. Standard enthalpies of formation of liquid arsenites vs the standard enthalpies of formation of the alcohols

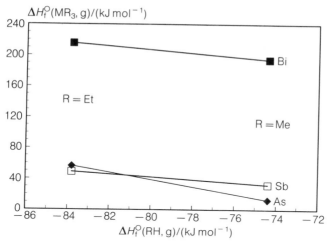

FIGURE. 2. Standard enthalpies of formation of gaseous arsenic, antimony and bismuth trialkyls *vs* the standard enthalpies of formation of the alkanes

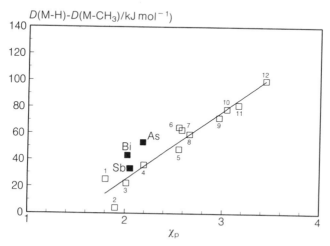

FIGURE. 3. Differences between M—H and M—Me mean bond enthalpies for elements of Groups 14–17 *vs* Pauling electronegativity of M. The key elements of this chapter—As, Sb and Bi—are marked explicitly while $1 \equiv Sn$, $2 \equiv Si$, $3 \equiv Ge$, $4 \equiv P$, $5 \equiv C$, $6 \equiv Se$, $7 \equiv S$, $8 \equiv I$, $9 \equiv Br$, $10 \equiv N$, $11 \equiv Cl$ and $12 \equiv O$

negative enthalpy of formation[34]. The electronegativities of arsenic, antimony and bismuth are comparable to that of the boron[10], for which the triethyl compound is more stable than the trimethyl by ca 42 kJ mol^{-1} in the liquid phase and 26 kJ mol^{-1} in the gas. (We note that the $42 - 26 = 16$ kJ mol^{-1} difference of the two phases is almost exactly the predicted difference of enthalpies of vaporization contribution of the three 'extra' carbons found in the triethyl compound over that in the trimethyl compound, namely 14 kJ mol^{-1}; see Reference 35.)

Crude estimates, say within ca ± 30 kJ mol^{-1}, for the enthalpies of formation of the Me$_3$E, i.e. trimethylarsine, trimethylstibine and trimethylbismuthine, can be obtained. A well-established feature of transition metal thermochemistry asserts that the differences between metal–hydrogen and metal–methyl bond dissociation enthalpies are lower for more electropositive, early elements, than for late transition elements[36]. It can be shown that the differences between $mean$ M—H and M—CH$_3$ bond dissociation enthalpies for groups 14–17 [e.g. $\bar{D}$(M—H) $- \bar{D}$(M—Me) in CH$_4$ and CMe$_4$, NH$_3$ and NMe$_3$, H$_2$O and Me$_2$O, HCl and MeCl] are approximately proportional to the Pauling electronegativity of M: deviation of the experimental data from the 'predicted' rarely exceeds 20 kJ mol^{-1} (see Figure 3). If it is assumed that the reported enthalpies of formation of the gaseous trihydrides AsH$_3$, SbH$_3$ and BiH$_3$ are reliable[37], then we deduce $\bar{D}$(As—Me) = 262, $\bar{D}$(Sb—Me) = 230 and $\bar{D}$(Bi—Me) = 168 kJ mol^{-1}. From these data we find ΔH_f°(AsMe$_3$, g) = -44, ΔH_f°(SbMe$_3$, g) = 14 and ΔH_f°(BiMe$_3$, g) = 142 kJ mol^{-1}, with uncertainties assumed to be ± 35 [i.e. a cautious, conservative $\pm (20^2 \times 3)^{1/2}$] kJ mol^{-1}. If we take the difference of enthalpies of formation of methyl and ethyl derivatives to be that of the tricoordinate, trivalent boron species[13,19], we deduce that ΔH_f°(AsEt$_3$, g) = -70, ΔH_f°(SbEt$_3$, g) = -12 and ΔH_f°(BiEt$_3$, g) = 69 kJ mol^{-1}. Agreement between experiment and our model can only be called horrible. But we already admitted to major problems seemingly inherent in the experimental results for the enthalpies of formation of the trialkyl derivatives of arsenic, antimony and bismuth.

B. Tris(dithiocarbamates)

As intimated earlier, perhaps it was unwise to calibrate our thinking on the trimethyl species and also (admittedly) a $posteriori$ ill-advised for the literature experiments to have been on these species. However, the ca 350 ± 30 kJ mol^{-1} differences in gaseous enthalpies of formation of the ethyl and butyl dithiocarbamates [E(S$_2$CNR$_2$)$_3$ for R = Et and Bu] from References 21, 22 and 25 are in marked deviation from the predictions from use of the 'universal' methylene enthalpy of formation increment[38], 20.6 kJ mol^{-1}. That is, because the butyl species have 12 more methylene groups than the ethyl, we would have expected the difference in enthalpies of formation to be $12 \times 20.6 = 247.2$ kJ mol^{-1}. How do we explain the discrepancy of some $100 + 30$ kJ mol^{-1}? After all, the enthalpies of formation of the dithiocarbamates were determined by reaction calorimetry involving the well-characterized ECl$_3$ and so ill-defined oxides should not be a problem at all.

To be more precise, the enthalpies of formation of the condensed phase species were determined by reaction calorimetry. By definition, the enthalpy of sublimation equals the sum of the enthalpies of fusion and vaporization. The fusion quantities were measured for all of the dithiocarbamates, the vaporization enthalpies were not. How does our earlier experience help us? For simple, i.e. unsubstituted, hydrocarbons, by summing either experimentally measured[39] or estimated enthalpies of fusion[40] and calculated enthalpies of vaporization[41], one can derive[42] encouragingly accurate enthalpies of sublimation[43]. Though we have no such systematic experience with sublimation enthalpies of substituted species, we are nonetheless optimistic that enthalpies of sublimation can also be predicted for these species as well—at least if we use experimentally measured enthalpies of fusion. Admittedly, we have no knowledge of the requisite b for dithiocarbamates to use in equation 6. Recall that it was not needed to compare the enthalpies of vaporization of trimethyl and triethyl borane. In the current case, we again note that the butyl dithiocarbamate species have 12 carbons more than the ethyl and so should have enthalpies of vaporization higher by 12×4.7 or ca $56 \, \mathrm{kJ \, mol^{-1}}$, while taking the requisite differences from the reported values (cf. References 21, 22, and 25) results in the butyl compounds having enthalpies of vaporization some $10–20 \, \mathrm{kJ \, mol^{-1}}$ lower. The discrepancy is ca $70 \, \mathrm{kJ \, mol^{-1}}$. From this we conclude that the differences of the enthalpies of formation of the gaseous dithiocarbamates may well be reduced by this $70 \, \mathrm{kJ \, mol^{-1}}$, and so doing makes the values more in line with those predicted from the 'universal' methylene increment.

IV. THE TRIPHENYL DERIVATIVES AND THEIR OXIDES

We now turn to the three Ph_3E species. The directly measured, combustion-derived, enthalpies of formation of triphenylarsine, triphenylstibine and triphenylbismuthine increase in the order $As < Sb < Bi$ as do the parent trihydrides and our derived values for the trialkyl species. However, we will not attempt to make any further comparison between the enthalpies of formation of the three triphenyl species or between them and the trihydrides and/or trialkyls. In the absence of rotating bomb calorimetric results for triphenylstibine and for all of the trialkyl species, values are too inexact to warrant any further discussion[44].

Relatedly, despite the synthesis and structural characterization of numerous arsine and stibine oxides, bomb calorimetry measurements have only been reported on triphenylarsine oxide[18]. While corresponding measurements have been made on triphenylarsine, it is clearly premature to make general observations as to E—O bond enthalpies in the absence of additional data. In principle, reaction calorimetry should prove useful. Indeed, we note a solution phase (benzene) enthalpy of reaction study[17] of triphenylarsine and t-butyl hydroperoxide according to the reaction

$$Ph_3As \, (soln) + t\text{-}BuOOH \, (soln) \longrightarrow Ph_3AsO \, (soln) + t\text{-}BuOH \, (soln) \qquad (8)$$

The so-derived[45] enthalpy of formation of gaseous triphenylarsine oxide, $261 \pm 18 \, \mathrm{kcal \, mol^{-1}}$, just overlaps the value obtained from rotating bomb calorimetry, $227.6 \pm 15.6 \, \mathrm{kJ \, mol^{-1}}$. While related success arises in the comparison of the values for triphenylphosphine oxide[46], we admit our discomfort with the large error bars accompanying all of these values. As such, while we derive an enthalpy of formation of triphenylstibine oxide using the static bomb results for stibine—and thus enter this value into Table 1—we regrettably consider it pointless even to compare the E—O bond enthalpies for the three triphenyl compounds.

V. IF ONLY SOME NEW MEASUREMENTS WERE MADE

We have essentially exhausted all of the directly measured enthalpies of formation of compounds containing carbon–arsenic, –antimony and –bismuth bonds. However, let us now make use of other thermochemical data and see what can be derived using some plausible estimates. And barring that, let us see what new enthalpies of formation would become available if only some new measurement were made.

A. Chlorophenylarsines and Redistribution Reactions

So far, all of our attention on the trivalent compounds has been related to those species with identical groups on the central atom. There is no reason why this must be so. We now turn to some redistribution reactions of the generic types that shuffle groups on the central E atom, e.g.

$$3EX_2Y \longrightarrow 2EX_3 + EY_3 \tag{9}$$

$$2EX_2Y \longrightarrow EX_3 + EXY_2 \tag{10}$$

$$3EXYZ \longrightarrow EX_3 + EY_3 + EZ_3 \tag{11}$$

The energetics of two such reactions have been reported, namely

$$3PhAsCl_2 \ (l) \longrightarrow Ph_3As \ (l) + 2AsCl_3 \ (l) \tag{12}$$

$$2PhAsCl_2 \ (l) \longrightarrow AsCl_3 \ (l) + Ph_2AsCl \ (l) \tag{13}$$

for which exothermicities of 84.1 and 20.1 kJ mol^{-1} have been determined[47] at ca 250 °C, scarcely gentle reaction conditions[48]. Knowledge of the enthalpy of formation of any two of the four phenyl and/or chloro arsenic species in reactions 12 and 13 would allow the determination of the enthalpies of formation of the other two. While at 250 °C all four arsenic compounds are liquids, it is still necessary to make thermal (heat capacity) corrections to correct these data to those for the standard temperature of 25 °C. We can be optimistic that the net thermal correction will be rather small because there are the same number and type of bonds on both sides of the reaction. However, it is necessary to recall that our enthalpy of formation data (admittedly unreliable) on triphenylarsine is only for the solid and the gas. Determination of the enthalpy of fusion of triphenylarsine is thus a desirable goal. This should not be an excessive obstacle[49], although the toxicity and foul odor associated with most arsenic compounds has no doubt discouraged experimentalists from numerous otherwise reasonable studies of these species. What would we learn if the thermochemical data were available? At the very least we would learn about the energetics consequence of *gem*-halogen, and thus anomeric, interactions[50] in group 15 compounds, and of interactions of 2p with 4p and 4d orbitals in the formal reasonance structures

$$C{=}C{-}\ddot{A}s{\diagdown} \quad\longleftrightarrow\quad C^-{-}C{=}As^+{\diagdown} \quad\longleftrightarrow\quad C^+{-}C{=}\ddot{A}s^-{\diagdown} \tag{14}$$

B. Iodomethylarsines and Iodination Reactions

Consider now the condensed phase iodination reaction

$$Me_2As{-}AsMe_2 \ (l) + I_2 \ (s) \longrightarrow 2Me_2AsI \ (l) \tag{15}$$

that has an accompanying exothermicity[51] of 83.2 kJ. Disappointingly, we fail to gain useful information about the enthalpy of formation of any diarsine or iodoarsine from this result. Lacking this information, one could estimate the enthalpy of formation of $Me_2As{-}AsMe_2$ by assuming thermoneutrality for the admittedly mixed condensed phase reaction

$$\tfrac{4}{3}(Me_3As \ (l)) + \tfrac{2}{3} As \ (s) \longrightarrow Me_2As{-}AsMe_2 \ (l) \tag{16}$$

After all, the same number of C—As and As—As bonds appear on both sides of the reaction. Likewise we should think that reaction 17 should be nearly thermoneutral because of conservation of bond types and anticipated small steric repulsion (because As is a large atom) and small anomeric effects (because the As—I bond is so nonpolar, at least by Pauling electronegativity based reasoning).

$$2Me_3As \ (l) + AsI_3 \ (s) \longrightarrow 3Me_2AsI \ (l) \tag{17}$$

The reader may be skeptical because we are 'mixing' thermochemical data on liquids and solids, and because of the nonmolecular (i.e. infinite lattice) nature of the most stable allotropic form of (solid) arsenic. However, the lack of convincing thermochemical data for the 'conventional' trimethylarsine suggests that none of these reactions is particularly useful for the conceptual understanding and thus utilization of calorimetric results. Besides, from the thermoneutrality assumed for reactions 16 and 17 we would predict that the exothermicity of reaction 18 should be 3/2 of that of reaction 15.

$$As\ (s) + \tfrac{3}{2}\ I_2\ (s) \longrightarrow AsI_3\ (s) \tag{18}$$

we recognize the exothermicity of iodination reaction 18 to equal, by definition, the enthalpy of formation of solid AsI_3. From the NBS tables[11] we find this last quantity equals $-58.2\ \text{kJ mol}^{-1}$, and so (regardless of the enthalpy of formation of trimethylarsine) we would predict the enthalpy of reaction 15 to equal $\tfrac{2}{3}(-58.2) = -38.8\ \text{kJ mol}^{-1}$. There is a discrepancy of some $-83.2 - (-38.8) \cong -45\ \text{kJ mol}^{-1}$. How is this to be explained[52,53]?

C. From Diarsines to Monomeric and Dimeric Antimonin: As—S and Sb—Sb Bond Dissociation Enthalpies

Let us assume for a moment, however, that we had reliable (and direct) thermochemical information about trimethylarsine and tetramethyldiarsine. How valid is it to assume all As—As bonds have comparable dissociation enthalpies[53]? Clearly we know enough not to assume the As—As bond in As_4 is the same as in tetramethyldiarsine. The former is immediately identified as strained and so we are to use the more stable allotrope with an infinite network of nonplanar hexagons. What about Sb—Sb bonds? If we assume constancy of the dissociation enthalpies of these bonds (excepting the strained Sb_4), how about using the bond enthalpy from elemental solid antimony? The enthalpy of dimerization of antimonin. C_5H_5Sb, to form the tricyclic Diels-Alder or [4+2] cycloaddition product has been determined[54] to be $-30.5 \pm 1.3\ \text{kJ mol}^{-1}$ by a careful study of the monomer–dimer equilibrium (equation 19).

$$\tag{19}$$

Interestingly, the product is 'head-to-head', i.e. it has one new C—C and Sb—Sb bond, as opposed to two new C—Sb bonds. Assume we were convinced of the 'more or less' invariance of the Sb—Sb bond enthalpy[55] and that we had a reliable value for the C_{sp^2}—Sb bond enthalpy (say from a more trustworthy measurement of the enthalpy of formation of triphenylstibine[56]). With judicious use of Benson increments and strain corrections[3], one could thus estimate the enthalpy of formation of the dimer, and accordingly derive a trustworthy value for the monomer. We would then have a handle on the aromaticity of antimonin. The enthalpy of reaction of bismin, C_5H_5Bi, to form the analogous dimer has likewise been derived[54] to be $-50\ \text{kJ mol}^{-1}$. However, this does not particularly help us since elemental bismuth is metallic, i.e. it is illogical to consider that Bi (s) is held together by covalent Bi—Bi bonds.

D. Cleaving the First Bond in Trihydrocarbyl Neutrals and Related Ions

There are direct measurements of the dissociation enthalpy for cleaving the first bond in trimethylarsine, -stibine and -bismuthine, and triethyl-stibine and convincing evidence[57] that these $R—ER_2$ bounds have significantly different dissociation enthalpies from the subsequent RE—R and R—E bonds. Reliable measurement of the enthalpy of formation of the various Me_3E species and the E—E bond enthalpies in the corresponding Me_2E—EMe_2 species would give us the enthalpy of formation of both the 'E—E compounds' and by inference of 'mixed' species such as $(CH_3)_2AsI$. While we lack data as to the related phenyl–E bond enthalpies in the neutral species, Ph_nE, there are relevant data for the related cations. More precisely, we know the enthalpies for Ph_nE^+ formation that presumably correspond[58a] to reaction 20

$$Ph_3E \longrightarrow Ph_nE^+ + (3\text{-}n)Ph + e^- \qquad (20)$$

for $n = 0$, 1 and 2 and $E = Sb$ and Bi. However, in no case are there any data as to the threshold energy or appearance potential for the formation of E^+, i.e. $n = 3$, that would be simply formed by the reaction

$$Ph_nE \longrightarrow E^+ + nPh + e^- \qquad (21)$$

Such information for $n = 3$ would give reliable values for the enthalpy of formation of the triphenyl organometalloids and organometallics since the enthalpy of formation of the atomic cations are all available; e.g. see Reference 58b.

E. Thermochemistry of Rearrangements

The energetics of rearrangement of As(III) and As(V) sulfur compounds have been studied[59]. It has been shown that the transformations of Me_2As—S—S—$AsPh_2$ into $Me_2As(S)$—S—$AsPh_2$ and into Me_2As—S—$As(S)Ph_2$ are exothermic by 18.0 and 18.5 $kJ\,mol^{-1}$, respectively. From these experimental findings, and the corollary conclusion that $Me_2As(S)$—S—$AsPh_2$ is more stable than Me_2As—S—$As(S)Ph_2$ by only $0.5\,kJ\,mol^{-1}$, we conclude that the two chemically nonequivalent arsenics in Me_2As—S —$AsPh_2$ are essentially equivalent with regard to the energetics of their sulfiding. Dare we conclude that the sulfiding reaction

$$R_3As + [S] \longrightarrow R_3AsS \qquad (22)$$

is essentially independent of the groups affixed to the arsenic? We know of no enthalpy of formation determination of any R_3As species and the corresponding sulfide, never mind two such pairs of species, to draw any conclusion about the validity of our supposition. And should the reader recall the apparently one paper[60] that deals with such a related pair of stibines, Ph_3Sb and Ph_3SbS, the absence of any and all details save the enthalpies of combustion (not even suggested products, much less the enthalpy of formation) makes this study without much interest except in a historical sense[61].

We close now with the case of the interesting Arbuzov rearrangement of trimethyl-phosphite (reaction 22):

$$(MeO)_3P \xrightarrow{} MeP(O)(OMe)_2 \qquad (22)$$

This reaction is catalyzed by 'Me^+' (e.g. methyl triflate) and has been shown[62] to be exothermic by $100 \pm 10\,kJ\,mol^{-1}$. We know of no such related study for trimethyl arsenite or any other $(RO)_3E$ species. Such information would allow us to interrelate the seemingly well understood E—O single-bonded species with those containing the much more poorly understood E—C bonds.

VI. ACKNOWLEDGMENTS

JFL thanks the Chemical Science and Technology Laboratory, (United States) National Institute of Standards and Technology, for partial support of his research. JAMS thanks Junta Nacional de Investigação Científica e Tecnológica, Portugal (Project PMCT/C/ CEN/42/90) for financial support. A travel grant from The Luso-American Foundation for Development, Portugal, is also acknowledged.

VII. REFERENCES

1. For clarification and extensive discussion of the interrelationships of these alternative concepts as applied to the study of transition metal containing species and catalysis, see J. A. Martinho Simões and J. L. Beauchamp, *Chem. Rev.*, **90**, 629 (1990).
2. We note the recent compendia of these physical properties:
   (a) E. S. Domalski, W. H. Evans and E. D. Hearing, 'Heat Capacities and Entropies of Organic Compounds in the Condensed Phase', *J. Phys. Chem. Ref. Data*, 13 (1984), Supplement 1, and its update by E. S. Domalski and E. D. Hearing, *J. Phys. Chem. Ref. Data*, **19**, 881 (1990).
   (b) V. Majer and V. Svoboda, *Vaporization of Organic Compounds*, IUPAC Chemical Data Series No. 32, Blackwell Scientific Publications, Boston, 1985.
   (c) J. S. Chickos, in *Molecular Structure and Energetics: Physical Measurements* (Vol. 2) (Eds. J. F. Liebman and A. Greenberg), VCH Publishers, Inc., New York, 1987 (enthalpy of sublimation).
   (d) W. E. Acree, Jr., *Thermochim. Acta*, **189**, 37 (1991) (enthalpy of fusion).
   From these compendia we find negligible relevant data nor any useful patterns or insight for our study of arsenic-, antimony- and bismuth-containing compounds.
3. (a) S. W. Benson, F. R. Cruickshank, D. M. Golden, G. R. Haugen, J. E. O'Neal, A. S. Rodgers, R. Shaw and R. Walsh, *Chem. Rev.*, **69**, 279 (1969).
   (b) S. W. Benson, *Thermochemical Kinetics*, 2nd ed., Wiley, New York, 1976.
4. S. W. Slayden and J. F. Liebman, in *The Chemistry of Functional Groups Supplement E2: Chemistry of Hydroxyl, Ether and Peroxide Groups* (Ed. S. Patai), Wiley, Chichester, 1993.
5. J. F. Liebman, K. S. K. Crawford and S. W. Slayden, in *The Chemistry of Functional Groups Supplement S: Chemistry of Sulphur Containing Groups* (Eds. S. Patai and Z. Rappoport), Wiley, Chichester (in press).
6. G. Pilcher, in *The Chemistry of Organophosphorus Compounds* (Ed. F. R. Hartley), Wiley, Chichester, 1990.
7. P. B. Dias, M. E. Minas da Piedade and J. A. Martinho Simões, *Coord. Chem. Rev.* (in press).
8. V. I. Tel'noi, V. N. Larina, E. N. Karataev and V. K. Stankevich, *Metaloorg. Khim.*, **1**, 102 (1988).
9. V. I. Tel'noi, V. N. Larina, E. N. Karataev and E. N. Deryagina, *Russ. J. Phys. Chem.*, **62**, 1623 (1988).
10. L. C. Allen, *J. Am. Chem. Soc.*, **111**, 9003 (1989).
11. D. D. Wagman, W. H. Evans, V. B. Parker, R. H. Schumm, I. Halow, S. M. Bailey, K. L. Churney and R. L. Nuttall, *The NBS Tables of Chemical Thermodynamic Properties: Selected Values for Inorganic and C_1 and C_2 Organic Substances in SI Units*, *J. Phys. Chem. Ref. Data*, 11 (1982), Supplement 2.
12. As_2H_4 and Sb_2H_4, F. E. Saalfield and H. J. Svec, *Inorg. Chem.*, **2**, 50 (1963); BiH_3, F. E. Saalfield and H. J. Svec, *Inorg. Chem.*, **2**, 246 (1963).
13. G. Pilcher and H. A. Skinner, in *The Chemistry of the Metal–Carbon Bond*, Part 1 (Eds. R. Hartley and S. Patai), Wiley, New York, 1982.
14. L. H. Long and J. F. Sackman, *Trans. Faraday Soc.*, **52**, 1201 (1956).
15. W. F. Lautsch, A. Tröber, W. Zimmer, L. Mehner, W. Linck, H. M. Lehmann, H. Brandenberger, H. Korner, H.-J. Metschker, K. Wagner, and R. Kaden, *Z. Chem.*, 415 (1963).
16. C. T. Mortimer and P. Sellers, *J. Chem. Soc.*, 1965 (1964).
17. V. R. Tsvetkov, Y. A. Aleksandrov, V. N. Glushakova, N. A. Skorodumova and G. M. Kol'yakova, *Russ. J. Gen. Chem.*, **40**, 198 (1980).
18. S. Barnes, P. M. Burkinshaw and C. T. Mortimer, *Thermochim. Acta*, **131**, 107 (1986).
19. J. B. Pedley and J. Rylance, *Sussex-N.P.L. Computer Analysed Thermochemical Data: Organic and Organometallic Compounds*, University of Sussex, Brighton, 1977. We note that the successor of this volume, J. B. Pedley, R. D. Naylor and S. P. Kirby, *Thermochemical Data of Organic Compounds*, 2nd ed., *Chapman & Hall, London, 1986*, lacks any data on organometallic/organometalloid species.
20. T. Charnley, C. T. Mortimer and H. A. Skinner, *J. Chem. Soc.*, 1181 (1953).
21. C. Airoldi and A. G. de Sousa, *J. Chem. Soc., Dalton Trans.*, 2955 (1987).
22. C. Airoldi and A. G. de Sousa, *J. Chem. Thermodyn.*, **21**, 283 (1989).
23. L. H. Long and J. F. Sackman, *Trans. Faraday Soc.*, **51**, 1062 (1955).
24. K. H. Birr, *Z. Anorg. Allg. Chem.*, **306**, 21 (1960).
25. C. Airoldi and A. G. de Sousa, *Thermochim. Acta*, **130**, 95 (1988).

26. L. H. Long and J. F. Sackman, *Trans. Faraday Soc.*, **50**, 1177 (1954).
27. W. V. Steele, *J. Chem. Thermodyn.*, **11**, 187 (1979).
28. (a) G. Pilcher, in *International Review of Science: Physical Chemistry and Thermodynamics* (Ed. H. A. Skinner), Butterworths, London, 1975.
(b) G. Pilcher, in *Energetics of Organic Species* (Ed. J. A. Martinho Simões), NATO ASI Series, Kluwer, Dordrecht, 1992.
29. (a) A. R. Dias, J. A. Martinho Simões, C. Teixeira, C. Airoldi and A. P. Chagas, *J. Organomet. Chem.*, **335**, 71 (1987).
(b) A. R. Dias, J. A. Martinho Simões, C. Teixeira, C. Airoldi and A. P. Chagas, *J. Organomet. Chem.*, **361**, 319 (1989).
(c) A. R. Dias, J. A. Martinho Simões, C. Teixeira, C. Airoldi and A. P. Chagas, *Polyhedron*, **10**, 1433 (1991).
(d) J. P. Leal, A. Pires de Matos and J. A. Martinho Simões, *J. Organomet. Chem.*, **403**, 1 (1991).
(e) D. Griller, J. A. Martinho Simões and D. D. M. Wayner, in *Sulfur-Centered Reactive Intermediates in Chemistry and Biology* (Eds. C. Chatgilialoglu and K.-D. Asmus), NATO ASI Series, Plenum, New York, 1991.
(f) J. A. Martinho Simões in *Energetics of Organic Species* (Ed. J. A. Martinho Simões), NATO ASI Series, Kluwer, Dordrecht, 1992.
30. J. S. Chickos, D. G. Hesse, J. F. Liebman and S. Y. Panshin, *J. Org. Chem.*, **53**, 3424 (1988). For the extension to multiply substituted species, see J. S. Chickos, D. G. Hesse and J. F. Liebman, *J. Org. Chem.*, **54**, 5250 (1989).
31. J. F. Liebman, J. S. Chickos and J. Simons, in *Advances in Boron and the Boranes: A volume in honor of Anton B. Burg* (Eds. J. F. Liebman, A. Greenberg and R. E. Williams), VCH Publishers, Inc., New York, 1988.
32. By analogy to simple carboxylic acid derivatives such as esters [cf. J. F. Liebman and J. S. Chickos, *Struct. Chem.*, **1**, 501 (1990)], it is not unreasonable to assume that the vaporization substituent parameter for arsenite esters, $b(AsO_3)$, should approximately equal b (As) + 3b (O). From the somewhat more refined observation that simple summation generally overestimates the b of the composite group by some 5kJ mol^{-1}, we conclude that $b(AsO_3) \cong b$ (As) + 3b (O) − 5 and so b (As) should equal *ca* 12kJ mol^{-1}. Indeed, this value reproduces to within 1kJ mol^{-1} the experimentally measured enthalpies of vaporization of both trimethyl and triethylarsine.
33. R. L. Montgomery and F. D. Rossini, *J. Chem. Thermodynam.*, **10**, 471 (1978). Also see Y.-R. Luo and S. W. Benson, *J. Phys. Chem.*, **92**, 5255 (1988) and *Acc. Chem. Res.*, **25**, 375 (1992).
34. J. F. Liebman, J. A. Martinho Simões and S. W. Slayden, in *Lithium Chemistry: Principles and Applications* (Eds A.-M. Sapse and P. v. R. Schleyer), Wiley, New York (forthcoming).
35. Despite this success and that enunciated in Reference 32, it is to be noted that experience has shown methyl derivatives are generally poorly treated by our enthalpy of vaporization equations (cf. Reference 30. Supplementary material).
36. See Reference 29f. Also see S. W. Benson, *Chem. Rev.*, **78**, 23 (1978) and S. W. Benson, J. T. Francis and T. T. Tsotsis, *J. Phys. Chem.*, **92**, 4515 (1988).
37. This should not be taken for granted even for measurements on the energetics of compounds of arsenic, antimony and bismuth *not* determined by combustion calorimetry. In particular, the quoted mass spectrometric/appearance potential measurement from Reference 12 for the enthalpy of formation of gaseous BiH_3 is higher by 48kJ mol^{-1} than that estimated by analogy to other binary hydrides, cf. S. R. Gunn, *J. Phys. Chem.*, **68**, 949 (1964). Significant errors are not uncommon in such use of electron impact measurements to derive enthalpies of formation: see the discussions in H. M. Rosenstock, K. Draxl, B. M. Steiner and J. T. Herron, *Energetics of Gaseous Ions, J. Phys. Chem. Ref. Data*, 6 (1977), Supplement 1, and S. G. Lias, J. E. Bartmess, J. F. Liebman, J. L. Holmes, R. D. Levin and W. G. Mallard, *Gas-Phase Ion and Neutral Thermochemistry, J. Phys. Chem. Ref. Data*, **17**, (1988), Supplement 1.
38. J. D. Cox and G. Pilcher, *Thermochemistry of Organic and Organometallic Compounds*, Academic Press, London and New York, 1970, pp. 518 ff.
39. J. S. Chickos, R. Annunziata, L. H. Ladon, A. S. Hyman and J. F. Liebman, *J. Org. Chem.*, **51**, 4311 (1986).
40. J. S. Chickos, D. G. Hesse and J. F. Liebman, *J. Org. Chem.*, **55**, 3833 (1990).

41. J. S. Chickos, A. S. Hyman, L. H. Ladon and J. F. Liebman, *J. Org. Chem.*, **46**, 4294 (1981). The basic equation from this paper was used to derive the equations in Reference 30, i.e. the approach for hydrocarbons corresponds to the case $b = 0$.

42. J. S. Chickos, D. G. Hesse and J. F. Liebman, in *Energetics of Organic Species* (Ed. J. A. Martinho Simões), NATO ASI Series, Kluwer, Dordrecht, 1992.

43. The standard errors, ± 7 and $\pm 11 \, \text{kJ mol}^{-1}$ respectively, are within twice the standard error of the experimental measured quantities.

44. What do we know about the normalized difference of enthalpies of formation of phenyl and methyl derivatives, i.e. $(1/n)[\Delta H_f^\circ(\text{Ph}_n X, g) - \Delta H_f^\circ(\text{Me}_n X, g)]$? In Reference 5 it was shown that this difference for X taken as π-withdrawing substituents generally 'hovered' around $130 \pm 5 \, \text{kJ mol}^{-1}$. From Pedley, Naylor and Kirby, *op. cit.* (*cf.* Reference 19), we find the difference for the unsubstituted hydrocarbons (X = H) is $157 \, \text{kJ mol}^{-1}$ and for the halo species the differences are Cl, 134; Br, 141; and I, $150 \, \text{kJ mol}^{-1}$. The differences would seem to roughly track electronegativity or the electron-withdrawing power of the various substituents. However, the difference of the enthalpies of formation of ethane and toluene, i.e. X = Me, is $134.2 \pm 0.7 \, \text{kJ mol}^{-1}$. It would appear that simple patterns are lacking for the comparison of the cases of phenyl and methyl derivatives, even without the additional burden of uncertain enthalpies of formation for the X = As, Sb and Bi containing species.

45. The number presented in Table 1 resulted from accepting the direct thermochemical measurements reported in Reference 17 but using the most current values for the enthalpies of formation of triphenylarsine[13,16] and *t*-butyl hydroperoxide[19] and alcohol[19].

46. D. R. Kirklin and E. S. Domalski, *J. Chem. Thermodyn.*, **20**, 743 (1988).

47. J. D. N. Fitzpatrick, S. R. C. Hughes and E. A. Moellyn-Hughes, *J. Chem. Soc.*, 3542 (1950).

48. Nonetheless, these conditions are milder (and the analysis more complete) than those earlier employed by A. G. Evans and E. Warhurst, *Trans. Farad. Soc.*, **44**, 189 (1948) and so we trust the results in Reference 47 more than those in this earlier paper.

49. In principle, one should be able to obtain the desired number by taking the difference of the available enthalpy of sublimation, $98.3 \pm 8.4 \, \text{kJ mol}^{-1}$, and an estimated enthalpy of vaporization (using equation 6 and $b = 12 \, \text{kJ mol}^{-1}$ as suggested in Reference 32). So doing results in a value of $-2 \pm 9 \, \text{kJ mol}^{-1}$. Even disregarding the physically implausible negative values, it is clear that the value derived is too small. This may be explained by suggesting equation 6 is inappropriate for species such as triphenylarsine wherein there are inadequate intermolecular interactions, i.e, there are effectively more 'quaternary-like' carbons because both sides of the phenyl groups cannot participate. However, this reasoning is currently only an *a posteriori* comment, where we recall in Reference 39 the analogous discussion of the otherwise anomalously low phase change enthalpies of hexaethylbenzene.

50. A. E. Reed and P. v. R. Schleyer , *J. Am. Chem. Soc.*, **109**, 7362 (1987).

51. C. T. Mortimer and H. A. Skinner, *J. Chem. Soc.*, 4331 (1952).

52. Part of the error may be not using liquid I_2 in equation 15 so that all of the species therein are the same phase, liquid, much as all of the species in equation 18 are the same phase, solid. So doing reduces the discrepancy by the heat of fusion of I_2, a quantity estimated to be $13 \, \text{kJ mol}^{-1}$ in M. W. Chase, Jr., C. A. Davies, J. R. Downey, Jr., R. A. McDonald and A. N. Syverud, *The JANAF Tables*, 3rd ed., *J. Phys. Chem. Ref. Data*, **14** (1985), Supplement 1.

53. Another part of the error may arise by equating the As—As bond enthalpy in the solid element with that in Me_2As—$AsMe_2$. We have no way of currently evaluating this error: the As—As bond enthalpy in As_2H_4 is questionable if for no other reason than we suspect the enthalpy of formation of this species (see Reference 12 for the data and the caveats in Reference 37). However, we note that the most current evaluated value for the S—S bond enthalpy in dimethyl disulfide (see Reference 29e) is $272 \, \text{kJ mol}^{-1}$, only $6 \, \text{kJ mol}^{-1}$ larger than that found in gaseous S_8, a quantity naturally defined $\frac{1}{8}[\Delta H_f^\circ(g, S_8) - 8(\Delta H_f^\circ(g, S)]$ (using data from Reference 11).

54. A. J. Ashe, III, T. R. Diephonse and M. Y. El-Sheikh, *J. Am. Chem. Soc.*, **104**, 5693 (1982).

55. One cannot argue that the formation of new C—C and Sb—Sb bonds instead of two C—Sb bonds must be kinetically, and not thermodynamically, controlled because Pauling's electronegativity equations demand heteronuclear bonds to have a higher dissociation enthalpy than the arithmetic (or geometric) mean of the homonuclear components (eg. see L. Pauling, *The Nature of the Chemical Bond*, 2nd ed., Cornell University Press, Ithaca, 1944, pp. 47–52). We note it is not uncommon that heterometallic bonds are weaker than the average of the

corresponding homometallic bonds [see J. A. Connor, in *Energetics of Organic Species* (Ed. J. A. Martinho Simões), NATO ASI Series, Kluwer, Dordrecht, 1992). We merely acknowledge that no reliable data exist for determining Sb—Sb bond dissociation enthalpies—elemental Sb and Sb_2H_4 (see Reference 53) are the sole species for which one can derive any such data.

56. The use of triphenylstibine instead of the trivinyl species implicitly assumes the near-constancy found for the difference of enthalpies of formation of gaseous vinyl and phenyl derivatives. See:

(a) J. F. Liebman, in *Molecular Structure and Energetics: Studies of Organic Molecules* (Vol. 3) (Eds. J. F. Liebman and A. Greenberg) VCH, Deerfield Beach, 1986.

(b) P. George, C. W. Bock and M. Trachtman, in *Molecular Structure and Energetics: Biophysical Aspects* (Vol. 4), (Eds. J. F. Liebman and A. Greenberg), VCH, New York, 1987.

(c) Y.-R. Luo and J. L. Holmes, *J. Phys. Chem.*, **96**, 9568 (1992).

57. See D. F. McMillen and D. M. Golden, *Ann. Rev. Phys. Chem.*, **33**, 493 (1982) and G. P. Smith and R. Patrick, *Int. J. Chem. Kinet.*, **15**, 167 (1983) and relevant references cited therein.

58. (a) V. K. Potapov, A. N. Rodionov, T. L. Evlasheva and K. L. Rogoshin, *High Energy Chem. (USSR)*, **8**, 486 (1974). These fragmentations, unlike those reported for BiH_3, As_2H_4 and Sb_2H_4 (cf. References 12 and the caveats in Reference 37), are more likely to be accurate because they were done by photoionization as opposed to electron impact. However, it is unproven that the ions formed by loss of one or two phenyls from the Ph_3E species has the Ph_nE structure as opposed to some rearranged product, e.g. a heterotropylium ion. Such alternative product formation would affect the appearance potential as well as suggesting structure/energetics ambiguity should some other $(C_6H_5E)^+$ be formed by some fragmentation process, say from Ph_3EO or $PhECl_2$.

(b) See, for example, the compendium by Lias and her coworkers, cited in Reference 37.

59. L. Silaghi-Dumitrescu, I. Silaghi-Dumitrescu and I. Haiduc, *Rev. Roum. Chim.*, **34**, 305 (1989).

60. W. H. Charch, E. Mack, Jr. and C. E. Boord, *Ind. Eng. Chem.*, **18**, 334 (1926).

61. It is to be recalled that not only is there ambiguity as to the products of combustion of any antimony-containing species (cf. Reference 28), thermochemical studies of sulfur compounds show related complications. (See for example, the discussion in W. N. Hubbard, D. W. Scott and G. Waddington, in *Experimental Thermochemistry: Measurements of Heats of Reaction* (Ed. F. D. Rossini), Interscience, New York, 1956.)

62. E. S. Lewis and K. Colle, *J. Org. Chem.*, **46**, 4369 (1981).

CHAPTER **5**

Detection, identification and determination*

T. R. CROMPTON

Hill Cottage, Pen Marian, Nr Llangoed, Beaumaris, Anglesey, Gwynedd, Wales LL588SU, UK

* The literature covered in this summary is up to 1991. World output of papers on these organometallic compounds peaked in 1971 to 1978 (139 papers) declining to 42 papers published during the period 1979 to 1986. A further dramatic decline has occurred since 1986 (9 papers).

The chemistry of organic arsenic, antimony and bismuth compounds
Edited by S. Patai © 1994 John Wiley & Sons Ltd

ABBREVIATIONS

AF	aeration feed
AgDDC	silver diethyldithiocarbamate
DMAA	dimethylarsinic acid
DPP	differential pulse polarography
DSMA	disodium methanearsonate
GC	gas chromatography
MAF	ferric methanearsonate
MMAA	(mono)methylarsonic acid
MSMA	monosodium methanearsonate
PAO	phenylarsine oxide
SBE	settling basis effluent
TLC	thin-layer chromatography

I. ORGANOARSENIC COMPOUNDS

A. Occurrence in Nature

Organoarsenic species are known to vary considerably in their toxicity to humans and animals[1,2]. Large fluxes of inorganic arsenic into the aquatic environment can be traced to geothermal systems[3], base metal smelter emissions and localized arsenite treatments for aquatic weed control. Methylated arsenicals enter the environment either directly as pesticides or by the biological transformation of inorganic species[4,5].

Large amounts of arsenic enter the environment each year owing to the use of arsenic compounds in agriculture and industry as pesticides and wood preservatives. The main amount is used as inorganic arsenic (arsenite, arsenate) and about 30% as organoarsenicals such as mono- and dimethylarsinate[6]. Arsenic is known to be relatively easily transformed between organic and inorganic forms in different oxidation states by biological and chemical action[7,8]. Until recently, most of the analytical work has been concerned only with the total content of arsenic. However, since the toxicity and biological activity of the different species vary considerably, information about the chemical form is of great importance in environmental analysis. Arsenic is an ubiquitous element on the earth, and the presence of inorganic arsenic and several methylated forms in the environment has been well documented[9]. The occurrence of biomethylation of arsenic in microorganisms[10], soil[11], animals and humans[12] has also been demonstrated. Therefore, investigation of the fate of arsenicals in the environment and in living organisms requires analytical methods for the complete speciation of these compounds.

It has also been shown that arsenic is incorporated into marine and freshwater organisms in the form of both water-soluble and lipid-soluble compounds[13]. Recent studies have shown the presence of arsenite [As(III)], arsenate [As(V)], methylarsonic acid, dimethylarsinic acid and arsenobetaine (AB)[4]. Methylated arsenicals also appear in the urine and plasma of mammals, including man, by biotransformation of inorganic arsenic compounds[12]. Several methods have been devised to characterize these arsenicals.

Information concerning the methylated form of arsenic in the environment has been scant until recently. Braman and Foreback[14] detected traces of the following arsenicals in natural water, bird eggshells, seashells and human urine: dimethylarsinic acid, methylarsonic acid, arsenate and arsenite. They suggested that dimethylarsinic acid was the major and ubiquitous form of arsenic and that methylarsonic acid was also present, but in lower concentration. Edmonds and Francesconi[15] reported the presence of di- and trimethylated arsenic in marine fauna.

In microorganisms, methylation of arsenic has been generally known[16-18]. However, there has been little evidence that this phenomenon occurs in higher animals. Lasko and Peoples[12] observed that the methylated form of arsenic was excreted in the urine of cows and dogs treated with inorganic forms of arsenic. Organoarsenic compounds were measured by Creselius[19] in human urine after drinking inorganic arsenic-rich wine.

B. Determination of Arsenic

1. Spectrophotometric methods

Haywood and Riley[20] showed that tetraphenylarsonium chloride, 1-(o-arsonophenyl-azo)-2-naphthol-3,6-disulphuric acid and o-arsonophenylazo-p-dimethylaminobenzene are quantitatively decomposed in seawater by ultraviolet radiation, and 98% of the As can be recovered.

Organic arsenic species can be rendered reactive either by UV photolysis or by oxidation with potassium permanganate or a mixture of nitric and sulphuric acids. Arsenic(V) can be determined separately from total inorganic arsenic after extracting As^{III} as its pyrrolidine dithiocarbamate into chloroform.

In the method[20] for inorganic arsenic the sample is treated with sodium borohydride and the arsine evolved is absorbed in a solution of iodine. The resultant arsenate ion is determined photometrically by a molybdenum blue method. For seawater the range, standard deviation and detection limit are $1-4 \mu g l^{-1}$, 1.4% and $0.14 \mu g l^{-1}$, respectively; for potable waters they are $0-800 \mu g l^{-1}$, about 1% (at $2 \mu g l^{-1}$ level) and $0.5 \mu g l^{-1}$, respectively. Silver and copper cause serious interference at concentrations of a few tens of $mg l^{-1}$; however, these elements can be removed by preliminary extraction with a solution of dithizone in chloroform or by ion exchange.

In replicate analyses on seawater samples, mean ($\pm SD$) arsenic concentrations of 2.63 ± 0.05 and $2.49 \pm 0.05 \mu g l^{-1}$ were found. The recovery of arsenic was checked by analysing arsenic-free seawater which had been spiked with known amounts of arsenic(V). The results showed a linear relationship between absorbance and arsenic concentration and that arsenic could be recovered from seawater with an average efficiency of 98.0% at levels of 1.3–6.6 $\mu g\,l^{-1}$. Analogous experiments in which arsenic(III) was used gave similar recoveries.

Although purely thermodynamic considerations suggest that arsenic should exist in oxic seawaters almost entirely in the pentavalent state, equilibrium rarely appears to be attained, probably because of the existence of biologically mediated reduction processes and arsenic in most of these waters exists to an appreciable extent in the trivalent state; As^{III}: As^{V} ratios as high as 1:1 have been found in a number of instances.

Haywood and Riley[20] found that As(III) can be separated from As(V) even at levels of $2\,\mu g\,l^{-1}$, by extracting it as the pyrrolidine dithiocarbamate complex with chloroform. They applied this technique to samples of seawater spiked with As(V) and As(III) and found that arsenic(V) could be satisfactorily determined in the presence of arsenic(III).

In a further method the sample is heated in a sealed tube for 5 min with a mixture (3 + 1) of magnesium and magnesium oxide, which converts all the arsenic into magnesium arsenide[21–23]. On decomposition by dilute sulphuric acid, arsine is evolved and is absorbed in a 0.5% solution of silver diethyldithiocarbamate (AgDDC) in pyridine. The colour produced has an absorption maximum at 560 nm and is proportional to concentration up to 20 μg of arsenic in 3 ml of solution. Alternatively, the arsine is oxidized by bromine and determined iodimetrically.

Sandhu and Nelson[24] have studied the interference effects of several metals on the determination of organically bound arsenic at the 0–100 $\mu g\,l^{-1}$ range in waste water by the AgDDC spectrophotometric method. Antimony and mercury interfere specifically, forming complexes with AgDDC at absorbance maxima at 510 and 425 nm, respectively. Recovery of arsenic released by digesting solutions was tested and shown to give about 90% recovery of organic arsenic.

The differentiation of 'inorganic' As(III) and As(V) can be achieved by exploitation of the pH sensitivity of the reduction of arsenic compounds by sodium tetrahydroborate(III), as adapted to analyses by the AgDDC spectrophotometric procedure[25].

White and Englar[26] have described a procedure for the determination of inorganic and organic arsenic in marine brown algae. Inorganic arsenic was removed by distillation as the corresponding trichloride and assessed by absorption of the arsine–AgDDC complex. Severe digestion conditions for organic arsenic were found necessary to determine total arsenic analysed subsequently by the AgDDC–pyridine reagent. Arsenic in the inorganic and organic forms ranged from 0.5 to 2.7 and 40.3 to 89.7 $\mu g\,g^{-1}$ dry weight, respectively, in Laminariaceae, Alariaceae and Lessoniaceae collected in British Columbia, providing levels of inclusion similar to commercially available brown seaweed products. By contrast Sargassum muticum contained 20.8 $\mu g\,g^{-1}$ dry weight of inorganic arsenic, about 38% of the total concentration, and a commercial specimen of Hizikia fusiforme contained 71.8 $\mu g\,g^{-1}$ inorganic arsenic, some 58% of the total concentration, suggesting the propensity of members of the Sargassaceae family to accumulate the inorganic form of arsenic.

a. Hydride generation atomic absorption spectrometry. Arsenic being present in inorganic form and as a variety of organic compounds implies the use of analytical strategies which use physical and chemical separations, as well as general or specific detectors. Since the lower organic arsenic compounds, as well as some inorganic arsenic compounds, are easily vaporized, both gas chromatography and direct vapour generation atomic absorption are favoured analytical methods.

A review of the analytical chemistry of arsenic in the sea, including occurrence, analytical methods and the establishment of analytical standards, has been published[27].

The major organic arsenic compound in the environment is dimethyl arsinate. Specific and sensitive methods for the determination of this compound are needed, since the direct atomic spectrometric method does not distinguish between different organoarsenic species present. Hydride generation and selective vaporization of cold-trapped arsines in combination with various detection systems seem to be the methods most frequently used[4,28,28] and are applicable to As^{III}, As^V, monomethylarsinates, dimethyl arsinate and trimethylarsine oxide. The optimum conditions for generation of the various arsines are, however, different for each species with regard to the pH of the generating solution[30,31]. These methods also suffer from interferences from numerous inorganic ions[32]. Preconcentration in a toluene cold-trap following arsine generation was successful for monomethyl and dimethyl arsinate but non-quantitative recoveries of dimethyl arsenite were reported, probably because of the problem mentioned above with the arsine generation step. Molecular rearrangements occurring during arsine generation are, however, reported to be minimized if sodium tetrahydroborate is introduced as a pellet[33]. To simplify the determination of the different species, improved separation of the arsines using gas chromatography may be necessary. Whilst the hydride generation atomic absorption spectrometric method is slightly more complicated than direct atomic absorption spectrometry, conversion of the various forms of arsenic with sodium borohydride to substituted arsines permits differentiation of the various forms of arsenic.

The arsines are collected in a cold trap (liquid nitrogen), then vaporized separately by slow warming, and the arsenic measured by monitoring the intensity of an arsenic spectral line, as produced by a direct-current electrical discharge[28,34,35]. A similar method was used by Talmi and Bostick[33] who used a mass spectrometer as the detector. Their method had a sensitivity of $0.25\,\mu g\,l^{1-}$ for water samples.

Another variation on the method[4] with slightly higher sensitivity (several nanograms per litre) used a liquid nitrogen cold trap and gas chromatography separation, but used the standard gas chromatography detectors or atomic absorption, for the final measurement, detecting four arsenic species in natural waters.

The differentiation of inorganic As(III) and As(V) can be achieved by exploitation of the pH sensitivity of the reduction of arsenic compounds by sodium borohydride, as adapted to analysis by atomic absorption spectrometry[36]. An alternative to pH control of arsine production involves suppression of As(V) reduction by the addition of DMF[37].

Numerous papers[38-65] have been published discussing the use of sodium borohydride for reducing inorganic arsenic compounds to arsine preparatory to its determination by atomic absorption spectrometry[4,28,66] and other means such as inductively coupled plasma emission spectrometry, microwave emission spectrometry, electron capture and flame ionization detection[4], discharge emission[7,33] and neutron activation analysis[67,68]. Separation of the compounds is achieved by gas chromatography[4], or sequential volatilization[28,66]. Braman and Foreback[7] and others have reported molecular rearrangements[33] and incomplete recoveries at low concentrations[70,71], while discussing application of these techniques to organoarsenic compounds.

Kolthoff and Belcher[69] used a digestion procedure employing sulphuric acid and hydrogen peroxide for the decomposition of organoarsenic compounds in natural water. This procedure was applied to three organoarsenicals containing $5\,\mu g$ of arsenic using 125 ml arsine generators. After reaction with concentrated sulphuric acid and 30% hydrogen peroxide followed by reduction with hydrochloric acid, potassium iodide and stannous chloride, arsine was generated into a silver diethyldithiocarbamate reagent and then determined spectrophotometrically. Alternatively, following hydrogen peroxide–sulphuric acid decomposition inorganic arsenic was determined by atomic absorption spectroscopy.

The percentage recoveries obtained by the above digestion followed by atomic absorption spectrometric determination are given in Table 1. The recoveries obtained by the same method when applied to arsenic spiked water samples are shown in Table 2. Figure 1 shows the effect of ultra-violet irradiation as a function of time for triphenylarsine oxide, disodium methanearsonate and dimethylarsinic acid. The extent of arsenic recovery using

TABLE 1. Recovery of arsenic employing wet digestion with 5 ml of 30% hydrogen peroxide and 5 ml of sulphuric acid with analyses by silver diethyldithiocarbamate and atomic absorption[69]

Compound	Recovery (%)[a]	
	AgDDC	Atomic absorption
Triphenylarsine oxide	101.0 ± 5.2 (4)	100.8 (2)
Disodium methanearsonate	103.5 ± 1.2 (4)	100.8 (2)
Dimethylarsinic acid	99.9 ± 2.3 (3)	99.2 (2)

[a] Each sample contained 5 μg arsenic. Number in parentheses indicates the number of samples run.

TABLE 2. Recovery of arsenic from triphenylarsine oxide, disodium methanearsonate and dimethylarsinic acid spiked into a water samples A and B with wet digestion employing 15 ml of 30% hydrogen peroxide and 5 ml of sulphuric acid with analysis by silver diethyldithiocarbamate[a69]

Sample	Recovery (%)[b]	
	Sample 1	Sample 2
100 ml A	($9.7\,\mu g\,As\,l^{-1}$)	($6.9\,\mu g\,As\,l^{-1}$)
100 ml A + triphenylarsine oxide	92.4	94.1
100 ml A + disodium methanearsonate	90.4	89.1
100 ml AF + dimethylarsinic acid	96.0	90.2
100 ml B	($22.6\,\mu g\,As\,l^{-1}$)	($18.5\,\mu g\,As\,l^{-1}$)
100 ml B + triphenylarsine oxide	101.1	100.6
100 ml B + disodium methanearsonate	98.1	104.4
100 ml B + dimethylarsinic acid	89.4	96.6

[a] All 5 μg weights are as arsenic.
[b] Corrected for amount of arsenic initially present.

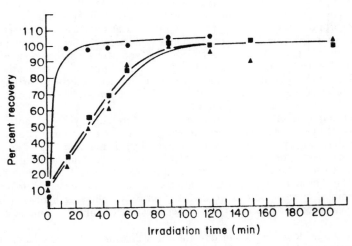

FIGURE 1. Recovery of arsenic after ultra-violet exposure as a function of time: ● triphenylarsine oxide; ■ disodium methanearsonate; ▲ dimethylarsinic acid. Reproduced from Reference 69 by permission of Interscience Publishers

TABLE 3. Recovery of arsenic from samples spiked into water with digestion by UV and analysis by high-sensitivity atomic absorption[69]

Compound	Recovery (%)[a]		
	Sample 1	Sample 2	Sample 3
Triphenylarsine oxide	110.0	100.0	102.8
Disodium methanearsonate	100.0	102.4	102.8
Dimethylarsinic acid	100.0	109.8	88.6

[a] Corrections were made for arsenic present in the samples.

photooxidation in conjunction with high sensitivity analysis when applied to water samples is shown in Table 3.

The method described above gives good results with arsenic recoveries ranging from 98.5 to 104.9%. The same method when applied to primary settled raw sewage gave arsenic recoveries ranging from 89.1 to 96.0%.

Arsenic recoveries of 89.4 to 104.4% were experienced from an activated sludge effluent sample. Employing the high-sensitivity arsenic analysis by atomic absorption resulted in arsenic recoveries of 99.2–100.8%.

The hydrogen peroxide–sulphuric acid digestion seemed to provide the most consistent and complete recoveries of any of the wet digestive procedures examined, and coupled with the silver diethyldithiocarbamate colorimetric analysis resulted in quantitative arsenic recoveries from waste water samples.

Kolthoff and Belcher[69] showed that a 15-min exposure to UV of triphenylarsine oxide resulted in greater than 99% photodecomposition of the compound. The monoalkylated arsenic compound reacted much more slowly, requiring 2 h for complete decomposition so, obviously, this method has to be used with caution. Four-hour decomposition produced 100–110% conversion of three other organoarsenic compounds.

Edmonds and Francesconi[72] have reported that methylarsonic acid (MMAA) and dimethylarsinic acid (DMAA) occurring universally in the environment may be estimated directly by vapour generation atomic absorption spectrometry without prior digestion. Sodium borohydride treatment produces methylarsine and dimethylarsine, respectively, which are swept directly into a hydrogen–nitrogen entrained air flame by the excess hydrogen generated by hydrolysis of the sodium borohydride. These methylated arsines are estimated in a manner identical to the arsine produced following acid digestion or dry ashing. The calibration curves and instrument response are directly comparable and are dependent only on the quantity of arsenic entering the flame. Estimations were carried out as described by Duncan and Parker[75] with similar calibration curves and instrumental response. Detection limits were 500 ng for inorganic arsenic and 1 μg for MMAA and DMAA. Aqueous solutions of methylarsonic acid and dimethylarsinic acid were adjusted to 2% in hydrochloric acid before addition of the sodium borohydride solution.

Standard mixtures of sodium arsenate, MMAA and DMAA were treated with sodium borohydride and the mixed arsines generated trapped in a glass-bead-packed tube at -180 °C. The cooling agent was removed and the trap allowed to warm slowly in the laboratory atmosphere. A valve assembly kept the trap sealed and was released periodically with a simultaneous flow of nitrogen through the trap into the flame. The valve was opened for 5 s each min. The instrument response was recorded on a chart moving at 2 mm min^{-1} and resembled, in outline, a gas chromatographic trace with arsine peaking at 3 min, methylarsine at 8 min and dimethylarsine at 13 min.

Arsine itself and methyl arsines have different responses in colorimetric versus atomic absorption methods of analysis. The colorimetric AgDDC method is much more responsive to arsine than to methyl arsines, whilst vapour generation atomic absorption spectrometry is equally responsive to both forms of arsenic. This explains the difference in the results obtained by the groups of Uthe[13] and Penrose[74] in the determination of organoarsenic compounds. These authors each compared the efficiency of wet ashing (oxidizing acid) and dry ashing (magnesium oxide/magnesium nitrate). Whereas Uthe[73] found wet ashing gave a slightly higher recovery than dry ashing (both approach 100%), Penrose[74] obtained good recovery for dry ashing but less than 5% for wet ashing. The difference lies in their methods of estimating the arsine generated after the reductive step. Penrose used a colorimetric method whereas Uthe used vapour generation atomic absorption spectrometry. Therefore, around 100% recovery would have been achieved even if no oxidation had occurred under the digestion conditions.

Fishman and Spencer[76] used UV radiation or an acid persulphate digestion procedure to decompose organoarsenic compounds. The automated method of Agemian and Cheam[77] uses hydrogen peroxide and sulphuric acid for the destruction of organic matter, combined with permanganate–persulphate oxidation for the complete recovery of organoarsenic compounds from fish. An automated system based on sodium borohydride reduction with atomization in a quartz tube is used for the determination of the inorganic arsenic thus produced.

Agemian and Cheam[77] found that in the sodium borohydride reduction of inorganic arsenic to arsenic, concentrations from 0.5 to 1.5 M of hydrochloric acid gave the highest sensitivity; both As(III) and As(V) were equivalently detected. When the hydrochloric acid concentration was increased from 2 to 6 M, the sensitivity for both species decreased, particularly for As(V). Replacement of the hydrochloric acid line with a sulphuric acid line reduced the sensitivity for As(III) by about 30% and As(V) gave a sensitivity of about 50% of AS(III).

Stringer and Attrep[78] compared hydrogen peroxide–sulphuric acid digestion and U.V. photodecomposition methods for the decomposition in water samples of triphenylarsine oxide, disodium methane–arsonate and DMAA to inorganic arsenic prior to reduction to arsine and determination by atomic absorption spectroscopy or by the AgDDC spectrophotometric method[79].

Stringer and Attrep[78] carried out their UV photodecompositions using a medium-pressure 450-W mercury arc lamp mounted vertically in the middle of ten 1-inch-diameter silica sample tubes. Prior to irradiation, the samples were acidified with nitric acid and some 30% hydrogen peroxide was added. Following irradiation the samples were made up to a known volume, and an aliquot was transferred to an Erlenmeyer flask, which also served as the arsine generator. Prior to determination by atomic absorption spectrometry the inorganic arsenic content was determined spectrophotometrically with AgDDC.

Fleming and Taylor[80] described a method for the determination of total arsenic in organoarsenic compounds by arsine generation and atomic absorption spectrophotometry using a flame-heated silica furnace. Denyszyn and coworkers[81] collected arsine at the $2 \mu g/m^3$ level produced from organoarsine compounds, on charcoal, then desorbed the arsine in acid and analysed it by electrothermal atomic absorption spectrophotometry. Mean percentage recovery and standard deviation were, respectively, 89.1% and $\pm$ 0.10.

Howard and Arbab-Zavar[29] have described a technique for the determination of 'inorganic' As(III) and As(V), 'methylarsenic' and 'dimethylarsenic' species which is based on the trapping of arsines and selective volatilization into a heated quartz atomizer tube situated in the optical path of an atomic-absorption spectrometer. Improved reproducibility is obtained by the use of a continuous flow reduction stage and detection limits are approximately 0.25 ng. For a typical sample volume of 10 ml this corresponds to a detection limit of 0.025 ng ml^{-1} of arsenic. Interference effects encountered by earlier workers[82,83] were investigated and depression of results was observed in the presence of silver(I), gold(III), chromium(VI), iron(II), iron(III), germanium(IV), molybdenum(VI), antimony(III), antimony(V), tin(V), manganese(VII) and nitrate. Various approaches to overcoming such interferences were investigated, and for general use masking with EDTA is advocated. The choice of extraction procedures of speciation analysis is discussed.

The apparatus consists of an arsine generator, the trapping section and the detector. Sample solution is mixed with acid or buffer and is then reduced with sodium tetrahydroborate solution during passage through a mixing coil. The resultant mixture is transported by nitrogen carrier gas to a gas liquid separator from where the gas stream passes through a lead acetate scrubber, to remove residual hydrogen sulphide, and then through a drying agent to the trap. The addition of EDTA to the sample solution is achieved by the insertion of an additional 14-turn mixing coil and reagent line into the system prior to the acidification of the sample. Design and preliminary silanization of the trap components with trimethylchlorosilane are crucial to the sensitivity and resolution of the system.

Following condensation of arsines at $-196\,°C$, the trap is allowed to warm to room temperature when the arsines volatilize in order of increasing boiling point and are swept into the atomization cell. This is a quartz tube heated by a conventional 10-cm-path-length air–acetylene flame aligned in the light path of a background-corrected atomic absorption spectrometer. The following conditions were used throughout the work: light source, arsenic hollow-cathode lamp; wavelength, 193.7 nm; spectral band pass, 1 nm; damping time constant, 1.5 s. The spectrometer output was monitored on a chart recorder (Tekman TE200).

During the evolution of the trapped arsines it is possible to stop the pump to conserve reductant solution, but then the sensitivity of the analysis is reduced as poor atomization results from the absence of air–hydrogen flames at the ends of the atomization tube.

To prepare the sample approximately 1 g was weighed into a boiling tube and 10 ml concentrated hydrochloric acid added. The mixture is heated at 65–70 °C, filtered and made up to 25 ml with concentrated hydrochloric acid.

Using a pumping rate of 2.5 ml min^{-1} reagents, effective reduction was achieved with a 2% m/V sodium borohydride solution in the presence of hydrochloric acid[57].

The reduction of 'inorganic' As(III) and As(V) species can be controlled by adjustment of the pH under which the reduction is performed[30]. At pH 5 arsenic is produced solely from As(III), whilst in the presence of 1 M hydrochloric acid arsenic is produced from both As(III) and As(V). By performing the reduction at two different pH values it is therefore possible to determine 'inorganic' As(III) and 'total inorganic arsenic', and hence 'inorganic' As(V) the difference. Analysis of 12 synthetic mixtures containing 50 ng of As(III) and As(V) gave recoveries of 98.0 ± 1.5% and 99.0%, respectively.

The carrier-gas flow rate influences both the sensitivity and resolution of the method affecting the steady-state arsenic concentration in the atomizer cell and the rate of utilization. The effect on sensitivity of varying carrier-gas flow rate is demonstrated in Figure 2. The peak height for As(III) or As(V), 'methyl' and 'dimethylarsenic' decreased with increasing flow rate whereas instrument stability and resolution improved with increasing gas flow rates. The optimum flow rate is 150 ml min^{-1}.

Calibration is linear in this method for all four species over the range 0–50 ng of arsenic (corresponding to concentrations of 0–10 ng ml^{-1} of arsenic for a 5-ml sample); see Figure 3. Above 50 ng, however, an unexplained transition is evident to a region in which the peak area is still linearly related to arsenic concentration, but with reduced sensitivity. Reproducibility and detection limits data are reported in Table 4.

Without a complexing agent depression of results was observed in the presence of silver(I), gold(III), chromium(VI), iron(II), iron(III), germanium(IV), molybdenum(VI), antimony(III), antimony(V), tin(II), manganese(VII) and nitrite. In the presence of EDTA complexing agent full recoveries were obtained with all elements except gold(III), antimony(V) and tin(II). No effects were observed for up to 10 μg ml^{-1} of aluminium(III),

T. R. Crompton

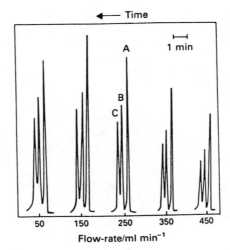

FIGURE 2. Typical arsine signals under conditions of varying carrier-gas flow rates: (A) arsenic(III) or arsenic(V), (B) methylarsenic, (C) dimethylarsenic. Reproduced from Reference 29 by permission of The Royal Society of Chemistry

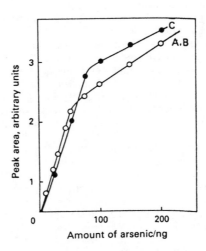

FIGURE 3. Typical calibration graphs for (A) arsenic(III) or arsenic(V), (B) methylarsenic, (C) dimethylarsenic. Reproduced from Reference 29 by permission of The Royal Society of Chemistry

TABLE 4. Detection limit and reproducibility[29]

| Arsenic species | Detection limit/ng | Relative standard deviation (%) | |
		10 ng of arsenic	50 ng of arsenic
Arsenic(III) (pH 5)	0.62	7.3	3.2
Arsenic(V)	0.23	6.9	2.8
'Methylarsenic'	0.26	3.5	2.7
'Dimethylarsenic'	0.25	3.8	1.9

TABLE 5. Effect of 5000-fold excess of iron on the determination of 10 ng of arsenic[29]

| | Relative arsenic response (%) | | | |
| | Unmasked | | EDTA masking | |
	Fe(II)	(Fe(III)	Fe(II)	Fe(III)
As(III)	91	102	95	98
As(V)	93	98	97	97
'Methylarsenic'	81	15	99	98
'Dimethylarsenic'	0	0	103	71

bismuth(III), calcium(II), cadmium(II), cobalt(II), copper(II), mercury(II), manganese(II), magnesium(II), sodium(I), lead(II), selenium(IV), selenium(VI), vanadium(IV) and tellurium(IV); up to 1000 μg ml^{-1} of perchlorate, bromide, iodide, nitrate, sulphate or cysteine hydrochloride did not interfere.

All observed interferences, except those due to antimony, gold(III) and nitrite, were masked by the pumped addition of 0.02 M EDTA (the pH adjusted to 3 with dilute hydrochloric acid) to the sample prior to the reduction step. In the presence of a 5000-fold excess of iron(III), however, depression of the dimethylarsenic signal cannot be completely restored by EDTA (Table 5). Extraction of the sample solution (at pH 2.0) with a 0.005 M solution of dithizone (diphenylthiocarbazone) in dichloromethane overcomes interferences due to silver(I), gold(III), chromium(VI), molybdenum(VI) and tin(II).

An attractive approach to the removal of interferents involves the use of the chelating ion-exchange resin Chelex 100. With the exception of iron(II) and iron(III), all identified cationic interferents can be overcome by passing the sample solution (at pH 4.0) through a column of this resin. The ubiquitous nature of iron, however, precludes the application of this procedure in most practical circumstances.

This procedure was applied to the specification analysis of arsenic in estuarine samples, following the hydrochloric acid digestion procedure. Some typical results are given in Table 6 and clearly demonstrate the marked distinction of arsenic species present in environmental samples. It is to be noted that in molluscs and algae the predominant species is dimethylarsenic due to biological methylation reactions occurring in the food chain. The water samples, however, reflect the contrasting geological surroundings of the sampling sites together with a small concentration of dimethylarsenic derived from planktonic conversion and decaying organic material.

Persson and Irgum[84] have described a method for the determination of DMMA in sea water in the sub-ppm range by electrothermal atomic absorption spectrometry after preconcentration on a strong cation exchange column and elution with ammonia.

TABLE 6. Analysis of estuarine samples[29a]

Sample type	Description	Source	Total arsenic(III) and arsenic(V)	As(III)	As(V)	Methyl arsenic	Dimethyl arsenic
Mollusc	Nucella lapillus	Brean Down, Somerset	0.99 ± 0.04	—	—	0.26 ± 0.01	1.26 ± 0.04
	Patella vulgata	Brean Down, Somerset	0.17 ± 0.01	—	—	0.03 ± 0.01	0.27 ± 0.01
	Littorina littorea	Southsea, Hampshire	1.59 ± 0.07	—	—	0.29 ± 0.01	1.74 ± 0.04
Algae	Fucus vesiculosis	Hurst Spit, Dorset	0.20 ± 0.01	—	—	0.05 ± 0.01	2.79 ± 0.08
River water		Beaulieu River, Hampshire	—	$0.10\ \mu\text{g l}^{-1}$	$1.02\ \mu\text{g l}^{-1}$	$0.06\ \mu\text{g l}^{-1}$	$0.23\ \mu\text{g l}^{-1}$
		Restronguet Creek, Cornwall	—	$1.6\ \mu\text{g l}^{-1}$	$19.4\ \mu\text{g l}^{-1}$	$<0.2\ \mu\text{g l}^{-1}$	$<0.2\ \mu\text{g l}^{-1}$

[a] Results are in micrograms per gram unless otherwise stated.

b. Direct atomic absorption spectrometry. This technique has been used for the determination of methylated arsenicals in water[72], seawater[84,85] and marine sediments[86].

Persson and Irgum[84] succeeded in determining sub-ppm levels of dimethyl arsinate by preconcentrating the organoarsenic compound on a strong cation-exchange resin (Dowex AG 50 W-XB). By optimizing the elution parameters, dimethyl arsinate can be separated from other arsenicals and from group I and II metals, which can interfere in the final determination. Graphite furnace atomic absorption spectrometry was used as a sensitive and specific detector for arsenic. The described technique allows dimethyl arsinate to be determined in a sample (20 ml) containing a 10^5-fold excess of inorganic arsenic with a detection limit of 0.02 ng As per ml.

A Perkin-Elmer 372 atomic absorption spectrometer with a HGA74 graphite furnace was used with temperature-controlled heating of the graphite tube[85]. Good recoveries were obtained from the artificial seawaters, even at the $0.05 \mu g \, l^{-1}$ level, but for natural seawater samples the recoveries were lower (74–85%). This effect could be attributed to organic sample components that eluted from the column together with dimethyl arsinate.

Maher[86] has described ion-exchange chromatographic and hydride generation atomic absorption spectrometric methods for the determination of inorganic arsenic, mono- and dimethylarsenic in marine sediments. Sediments were initially digested with 6 N hydrochloric acid, followed by further digestion of undissolved solids with boiling sodium hydroxide–sodium chloride. Both extracts were analysed.

Krynitsky[231] has studied the preparation of biological tissue for the determination of arsenic by graphite furnace atomic absorption spectrometry. Zeeman graphite furnace atomic absorption spectrometry has been used to determine organoarsenic compounds in horse urine[232].

c. Solvent extraction atomic absorption spectrometry. Masui and coworkers[87] have described a method for the separation of As(III), As(V) and organic arsenic from water samples by solvent extraction. The hydride generation atomic absorption spectrometric method[86] has been applied to the analysis of algae. The organoarsenic compounds were extracted from the algae by digestion with 0.1 M sodium hydroxide, followed by filtration, concentration to dryness and dissolution of the residue in 8.5 M hydrochloric acid. A toluene extract of the acidic phase was prepared and arsenic back-extracted from this phase with hydrochloric acid–potassium dichromate reagent. Arsenic species were separated on an ion-exchange column. Arsenic in each of the fractions was reduced to arsine on a zinc reductor column and evaluated by carbon tube atomic absorption spectrometry at 193.7 nm.

Typical results obtained in a study of arsenic in several species of macro algae, tissues of *Mercenaria mercenaria* and estuarine sediments collected from the southern coast of England were inorganic arsenic ($0.1–3.2 \, \text{mg kg}^{-1}$, monomethylarsenic ($0.2–0.6 \, \text{mg kg}^{-1}$), dimethy larsenic ($7.6–15.6 \, \text{mg kg}^{-1}$) and total arsenic ($20–49 \, \text{mg kg}^{-1}$).

2. Titration procedures

A digestion procedure for arsenic which is not subject to interference by magnesium, calcium, strontinum barium, cobalt, nickel, zinc, manganese cadmium, copper and halogens involves heating the sample to fuming in a Kjeldahl flask containing concentrated sulphuric acid, copper sulphate and concentrated nitric acid[88]. After cooling, further portions of nitric acid are added and the sample is heated to fuming until colourless or pale blue. It is then diluted, heated to boiling point and an aqueous solution of barium nitrate is added. The solution is then cooled, filtered, and neutralized to methyl orange with sodium hydroxide solution. Aqueous 25% nitric acid and an excess of 0.1 N silver nitrate solution is added, followed by dropwise addition of a concentrated solution of sodium acetate until precipitation is complete. The solution is then made up to volume, filtered and the excess of silver nitrate back-titrated with 0.1 N ammonium thiocyanate solution.

A micro Carius procedure for arsenic in which the sample is digested with fuming nitric acid to form As(V), which is then determined by iodine titration, yields low results[89,90]. However, when the digestion temperature and time were increased and potassium chloride was added, the end-points became sharp[91] and the results were in close agreement with theory.

Tuckerman and collaborators[91] state that chloric acid is to be preferred to the more widely used sulphuric acid or sulphuric-nitric acid digestions or alkaline fusions recommended for the determination of arsenic in organic compounds. Excess chloric acid is easily removed by boiling to leave a perchloric acid solution of inorganic As(V). Rapid micro and semi-micro methods for the determination of arsenic based on chloric acid digestion are described.

Samples of p-arsanilic acid ranging from 1.6–4.0 mg in weight were run by the micro method and gave excellent reproducibility of the absorbance values. A sample of p-arsanilic acid (theoretical arsenic content 34.5%) gave an average arsenic content of 34.6% with a 99% confidence limit of 0.6%[92]. Using the average absorbance per mg of arsenic found for p-arsanilic acid, samples of 4-)2-hydroxyethylureido)-phenylarsonic acid were found to contain the theoretical 24.6% of arsenic by the micro method.

Chloride does not interfere in the procedure. Phosphorus interferes in the micro method by formation of a heteropoly blue similar to that formed by pentavalent arsenic.

TABLE 7. Comparison of percentage arsenic found by the micro Carius method at different temperatures, with and without potassium chloride[89]

Digestion temp: Digestion time : Sample	Theory	I 250 °C 8 h, KCl absent	II 300 °C 10 h, KCl present
As_2O_3	75.73	71.99 72.01	76.04 75.66 75.59 75.66
$C_6H_5CH_2AsO_3H_2$	34.67	31.12 31.63	34.50 34.49
$C_6H_5AsO_3H_2$	37.08	36.62 34.06 35.26 35.21	36.91 36.88 36.97
$HOC_6H_4AsO(OH)_2$	34.36	30.76 31.70	34.46 34.42
$(C_6H_5)_3As$	24.46	turbid	24.40 24.46 24.51 24.49
$((C_6H_5)_3AsCH_3)I$	16.72	turbid	16.83 16.72

Phosphorus and arsenic, if present together, may be separated and determined, after digestion of the sample with chloric acid, by the method given by Jean[93].

Organic arsenicals may be analysed by heating them in a Kjeldahl flask with a mixture of concentrated nitric and sulphuric acids and hydrogen peroxide[94]. After cooling, the mixture is diluted with water and treated with concentrated sulphuric acid and zinc. The arsine evolved passes into a solution in pyridine of silver diethyldithiooccarbamate, and the molar absorptivity of the resulting solution is measured at 540 nm. Halogens and sulphur do not interfere. Arsenic in 10-mg amounts of organic compounds may be determined iodimetrically[45]. The method is suitable for the determination of all types of organic arsenic compounds, including those which give low results by the classical wet oxidation methods using sulphuric and nitric acids.

DiPietro and Sassaman[89] modified the micro Carius procedure for arsenic originally described by Steyermark[90] in which the sample is digested with fuming nitric acid to form As(V). The arsenic is then determined by iodine titration. DiPietro and Sassaman[89] evaluated this procedure by checking it against several pure organic arsenic compounds. None of these compounds yielded correct values for arsenic using the Steyermark[90] procedure (Table 7). Because incomplete destruction of the arsenoorganic compound was suspected, the digestion temperature and time were increased to 300 °C for 10 hours. Following this severe treatment, all solutions remained clear upon the addition of potassium iodide and the end-points became sharp. Results obtained by the modified procedure[89] are shown in Table 7.

3. Digestion procedures for arsenic

Armstrong and coworkers[96] observed that organic matter in seawater could be oxidized to carbon dioxide on exposure to U.V. radiation. This approach has also been used to decompose for o-arsanilic acid and for sodium cacodylate giving recoveries 97 and 108% respectively.

Most of the classical procedures for decomposing organoarsenic compounds present in samples prior to the determination of total inorganic arsenic incorporate some mode of wet or dry digestion to destroy the organically bound arsenic, in addition to all other organic constituents present in the sample.

Probably the most frequently used method of digestion incorporates the use of nitric and sulphuric acids. Kopp and Bandemar[97] used this method and experienced 91 to 114% recovery of arsenic trioxide added to de-ionized water and 86 to 100% recovery of the compound added to river water. Evens and Bandemar[98] recovered 87% of the arsenic trioxide added to eggs. By modifying the method by the addition of perchloric acid, and Caldwell and coworkers[99] observed 80 to 90% arsenic recovery with o-nitrobenzene-arsenic acid, 85 to 94% with o-arsanilic acid and 76.7% with disodium methylarsenate.

Two uncertainties seem to arise when reviewing digestive methods using nitric and sulphuric acid. First, the addition of inorganic arsenic to an organic matrix and subsequent recovery of all the inorganic arsenic added is not definite proof of total recovery of any organoarsenicals present. Secondly, the choice of o-nitrobenzenearsonic acid and o-arsanilic acid seems unfortunate since both compounds represent arsenic attached to an aromatic ring which is atypical of cacodylic acid and disodium methanearsonate (DSMA), two widely used organoarsenicals.

A relatively simple digestive method employing 30% hydrogen peroxide in the presence of sulphuric acid was reported by Kolthoff and Belcher[69] and subsequently used by Dean and Rues[100] to determine arsenic in triphenylarsine.

Stringer and Attrep[101] applied the Dean and Rues[100] and the ultra-violet[96] decomposition methods to the determination of organoarsenic compounds in waste water, using either the AgDDC colorimetric procedure described by Kopp[102] or the arsine–atomic absorption method described by Manning[103]. The organoarsenicals investigated were disodium methanearsonate, DMAA and triphenylarsineoxide. The digestive methods gave quantitative arsenic recoveries for the three organoarsenic compounds. The U.V. photodecomposition proved to be an effective digestive technique, requiring a 4 h irradiation to decompose a primary settled raw wastewater sample containing spiked quantities of the three organoarsenicals.

The percentage recoveries obtained by the digestion procedure with AgDDC analysis and atomic absorption analysis are given in Table 8. Results obtained by this digestion method applied to the aeration feed (AF) and settling basis effluent (SBE) samples spiked with each of the three organoarsenicals equivalent to 5 μg or arsenic are presented in Table 9.

The effects of ultra-violet irradiation as a function of time for triphenylarsine oxide, DSMA and DMAA are illustrated in Figure 4. The extent of arsenic recovery using photooxidation in conjunction with high-sensitivity arsenic analysis when applied to the AF and SBE samples is shown in Table 10.

A chemical analysis of the aerator feeding and settling basin effluent samples is provided in Table 11 in order to illustrate the overall water quality of the samples used in this study.

TABLE 8. Recovery of arsenic from triphenylarsine oxide, disodium methane arsonate and dimethylarsinic acid employing wet digestion with 5 ml of 30% hydrogen peroxide and 5 ml of sulphuric acid with analyses by AgDDC and atomic absorption[101]

	Recovery (%)[a]	
Compound	AgDDC	Atomic absorption
Triphenylarsine oxide	101.0 ± 5.2 (4)	100.8 (2)
Disodium methanearsonate	103.5 ± 1.2 (4)	100.8 (2)
Dimethylarsinic acid	99.9 ± 2.3 (3)	99.2 (2)

[a] Each sample contained 5 μg As. Number in parentheses indicates the number of samples run.

TABLE 9. Recovery of arsenic from triphenylarsine oxide, disodium methane-arsonate and dimethylarsinic acid spiked into AF and SBE sample with wet digestion employing 15 ml of 30% hydrogen peroxide and 5 ml of sulphuric acid with analysis by AgDDC[a101]

	Recovery (%)[b]	
Sample	Sample 1	Sample 2
100 ml AF	(9.7 μg As/l)	(6.9 μg As/l)
100 ml AF + triphenylarsine	92.4	94.1
100 ml AF + disodium methanearsonate	90.4	89.1
100 ml AF + dimethylarsinic acid	96.0	90.2
100 ml SBE	(22.7 μg As/l)	(18.5 μg As/l)
100 ml SBE + triphenylarsine oxide	101.1	100.6
100 ml SBE + disodium methanearsonate	98.1	104.4
100 ml SBE + dimethylarsinic acid	89.4	96.6

[a] All 5 μg weights are as arsenic.
[b] Corrected for amount of arsenic initially present.

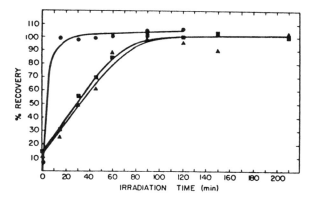

FIGURE 4. Recovery of arsenic after ultra-violet exposure as a function of time: ● triphenylarsine oxide; ■ disodium methanearsenate; ▲ dimethylarsinic acid. Reprinted with permission from Reference 101. Copyright (1979) American Chemical Society

TABLE 10. Recovery of arsenic from triphenylarsine oxide, disodium methanearsonate and dimethylarsinic acid spiked into AF and SBE with digestion by U.V. and analysis by high-sensitivity atomic absorption[101]

Compound	Recovery (%)		
	Aerator feed Sample 1	Aerator feed Sample 2	Settling basin effluent
Triphenylarsine oxide	110.0	100.0	102.8
Disodium methanearsonate	100.0	102.4	102.8
Dimethylarsinic acid	100.0	109.8	88.6

TABLE 11. Chemical analysis of the aerator feed sample and the settling basin effluent sample taken at Dallas[101]

Parameter[a]	Samples	
	Aerator feed	Settling basin
pH at 25 °C	7.3	7.7
Total alkalinity as $CaCO_3$	216.0	160.0
Ammonia nitrogen	12.7	1.93
Organic nitrogen	12.8	3.70
Nitrite–nitrate nitrogen	0.2	1.7
Nitrite nitrogen	0.01	0.08
Total phosphorus as P	5.5	6.5
Chemical oxygen demand	222.2	42.65
Total organic carbon as C	57.0	12.0
Total inorganic carbon as C	50.9	33.0
Total carbon as C	107.0	45.0
Suspended solids	132	21

[a] All values reported in mg/l units.

The digestive method employing hydrogen peroxide and sulphuric acid with analysis by AgDDC gave extremely good reduction in both organic and total carbon content.

The SBE sample showed 102.8% recoveries for both DSMA and triphenylarsine oxide. An 88.6% recovery of DMAA added to SBE indicated that 2 h irradiation was insufficient for all of this compound to photodecompose in the presence of other organic matter.

There were several observations made concerning the arsenic analysis by both the AgDDC technique and the high-sensitivity atomic absorption procedure during the course of this investigation. Turbidity problems were initially encountered in the AgDDC colorimetric procedure.

In conclusion, the data support the premise that the recovery of arsenic from triphenylarsine oxide, DSMA and DMAA is complete by wet digestion using hydrogen peroxide–sulphuric acid and by photodecomposition using ultraviolet light, when the arsenic recoveries ranged from 98.5 to 104.9%. This same method with a primary settled raw sewage sample gave arsenic recoveries ranging from 89.1 to 96.0%, and 98.4 to 104.4% with an activated sludge effluent sample. The same digestive method employing arsenic analysis by atomic absorption resulted in arsenic recoveries of 99.2 to 100.8%.

The hydrogen peroxide–sulphuric acid digestion seemed to provide the most consistent and complete recoveries of any of the wet digestive procedures examined. Previous studies using nitric–sulphuric acid, with and without ammonium oxalate, gave arsenic recoveries ranging from 80.2 to 105.9% and from 100.5 to 110.5%, respectively. This digestion procedure coupled with the AgDDC colorimetric analysis resulted in quantitative arsenic recoveries from wastewater samples. The hydrogen peroxide–sulphuric acid digestion combined with the high-sensitivity arsenic analysis gave acceptable recoveries of arsenic and should be considered a viable technique.

The most encouraging digestive technique examined during the course of this investigation was the photodecomposition approach using UV light. The time-dependent study of this method with the three organoarsenicals showed that 15 min exposure of triphenylarsine oxide resulted in greater than 99% photodecomposition of the compound, while the monoalkylated arsenic compound reacted much slower, requiring 2 h for complete decomposition.

4. Oxygen flask combustion procedures

Corner[104] was the first to recommend the determination of arsenic in organic materials by oxygen flask combustion. Arsenic attacks the platinum sample holder, hence many workers recommend a silica spiral, but the quartz devitrified during even a single combustion. Spectrographic analysis has shown that arsenic forms alloys with platinum even under the strongly oxidizing conditions obtained when paper impregnated with potassium nitrate is burned[105]. The following is a fairly conventional procedure for the determination of arsenic. After combustion the arsenite and arsenate formed are absorbed in a suitable solution, usually sodium hydroxide. Following conversion to arsenic trichloride, this is distilled into sodium bicarbonate solution and is estimated iodometrically. Merz[106] pointed out that 'phosphorous-resistant' platinum can be used as the sample holder for certain materials, but in general a silica spiral is recommended[107], but with this combustion is never as satisfactory as with platinum. For example, Corner[104] was unable to achieve proper and immediate combustion of resinous materials. Some workers[105,108] are of the opinion that wet combustion methods are preferable to the oxygen flask method for the determination of organic arsenic.

The method described by Merz[106] involves burning the sample in an oxygen-filled flask, and the combustion products are absorbed in dilute iodine solution in which As(III) is quantitatively oxidized to As(V). The arsenic is determined by the molybdoarsenate blue reaction, with hydrazine sulphate as reductant; the excess of iodine from the absorber does not interfere. Concentrations of arsenic greater than 10 μg are measured on a filter photometer and those of less than 10 μg with a spectrophotometer. The mean error for a single determination is ± 0.2%.

Puschel and Stefanac[109] use alkaline hydrogen peroxide in the oxygen flask method to oxidize arsenic to arsenate. The arsenate is titrated directly with standard lead nitrate solution with 4-(2-pyridylazo) resorcinol or 8-hydroxy-7-(4-sulpho-1-naphthylazo) quinoline-5-sulphonic acid as indicator. Phosphorus interferes in this method. The precision at the 99% confidence limit is within ± 0.67% for a 3-mg sample. In another variation, Stefanac[110] used sodium acetate as the absorbing liquid, and arsenite and arsenate are precipitated with silver nitrate. The precipitate is dissolved in potassium nickel cyanide ($K_2Ni(CN)_4$) solution and the displaced nickel is titrated with EDTA solution, with murexide as indicator. The average error is within ± 0.19% for a 3-mg sample. Halogens and phosphate interfere in the procedure.

Belcher and collaborators[111] described a method for arsenic in which the sample (30–60 mg) is burnt by a modified oxygen flask procedure and arsenic is determined by the precipitation as quinoline molybdoarsenate, which is then reduced with hydrazine sulphate and determined spectrophotometrically at 840 nm. The absolute accuracy is within ± 1% for arsenic. Phosphorus interferes in this procedure.

Wilson and Lewis[112] developed a method for arsenic, which includes precipitation as ammonium uranyl arsenate and subsequent ignition under controlled conditions to triuranium octaoxide. Phosphate and vanadate form similar insoluble ammonium uranyl salts and, in their presence the method is not applicable.

Various sample supports have been tried in the hope that they might prove more satisfactory than platinum[112], and results obtained with steel and aluminium supports gave satisfactory recoveries of arsenic from o-arsanilic acid, arsenious oxide and acetarsol.

5. Determination of carbon

The determination of carbon and arsenic in organic compounds using magnesium fusion and the elemental analysis of organoarsines and organobromoarsines have been reviewed[113,114].

6. Determination of fluorine

A volumetric microdetermination based on oxygen flask combustion for 5–40% of fluorine in organic compounds containing arsenic has been described[115]. The sample (containing 0.2–0.6 mg of fluorine) is burnt in an oxygen combustion flask and the products are absorbed in 5 ml of water. Hexamine and murexide-naphthol green B (C.I. Acid Green 1) are added and the flask contents are titrated with 0.005 M cerium(IV) sulphate to a green end-point. Methods have been reported for the determination of arsenic in organoarsenic compounds with halogen attached either to the arsenic or to the carbon, and in organoarsenic compounds with halogen in the anion[230].

7. Determination of sulphur

Compounds containing arsenic which are unstable in oxygen or moist air can be stabilized by exposure to sulphur for 8–48 h in vacuo. The resulting compounds may be analysed by pyrolysis for sulphur and arsenic[116].

C. Titration Procedures

Potentiometric titration and colorimetric methods have been described for the determination of β-chlorovinyldichloroarsine (Lewisite)[117,118]. In the latter the organic arsenic is mineralized by refluxing with aqueous sodium hydrogen carbonate and then oxidized to

As(V) by the addition of aqueous iodine, followed by conversion to molybdoarsenate and reduction to molybdenum blue by boiling under reflux with a sulphuric acid solution of ammonium vanadate and hydrazine sulphate. The method is sufficiently sensitive to determine down to 3 μg of Lewisite per millilitre of sample.

The pharmaceutical sodium methyl arsinate has been determined by non-aqueous titration with mercury(II) acetate[119.]

D. Spectrophotometric Procedures

Phenarsazine derivatives can be determined spectrophotometrically as the disodium salt of dinitrophenarsazinic acid at 520 nm[120]. The sample is dissolved in glacial acetic acid and converted by nitric acid to dinitrophenarsazinic acid. Addition of excess of sodium hydroxide yields a violet disodium salt suitable for photometric evaluation. From 1 to 8 μg/ml of phenarsazine can be determined by this method with an error of $\pm 4\%$. A spectrophotometric method has been described for the determination of 4-hydroxy-3-nitrophenylarsenic acid in animal feeds[121].

Two digestion methods have been compared for the determination of AS in wastewater samples[122], namely a wet method employing hydrogen peroxide–sulphuric acid and UV photodecomposition. The organoarsenicals investigated were DSMA, DMAA, and triphenylarsine oxide. All the digestive methods gave quantitative arsenic recoveries and the UV photodecomposition also proved to be effective, requiring 4 h irradiation to decompose samples containing the three organoarsenicals. Arsenic was determined in the digests by the AgDDC spectrophotometric method. Spectrophotometric procedures have also been used to obtain speciation data on monomethyl arsonate and dimethyl arsinate[123,126].

E. Polarographic Procedures

Substituted diarsines (R_2As–AsR_2) can be determined polarographically[127], the $E_{1/2}$ of the anodic wave being independent of the nature of the substituent. The analysis must be carried out in the absence of oxygen, which oxidizes the As—As bond. Concentrations of diarsines down to about 10^{-4} M can be determined by this procedure.

Various workers have reported on the polarographic reduction of organoarsenic compounds[128–132]. In the methods described for the determination of arsine in air and other gases, various absorbing solutions have been used, including mixtures of potassium permanganate, concentrated sulphuric acid and bromine[133], a mixture of silver nitrate and AgDDC[134] and a solution of ammonium nitrate in ethanol[135]. The last solution is also a suitable medium for the polarographic determination of arsine and is sensitive enough to determine down to 5×10^{-4} M of arsine in gas mixtures: phosphine interferes in this determination.

Arsenic(III) and arsenic(V), monomethylarsonate and dimethylarsinate have been determined by differential pulse polarography after separation by ion-exchange chromatography[136]; detection limits for the latter two are 18 and 8 ppb, respectively. Diphenylarsenic acid has been studied polarographically[137].

Anodic stripping voltammetry has been used to determine total arsenic species[138]. Pulse polarographic methods have been applied to aqueous and non-aqueous solutions of methyl- and dimethylarsenic acids at concentration levels down to 0.1 μg/ml[139]. These arsenicals are electroactive in aqueous buffers and in non-aqueous media in which the acidic supporting electrolyte, guanidinium perchlorate, is employed. A direct method of analysis, based on differential pulse polarography, is reported. Detection limits of roughly 0.1 μg/ml (for MMAA) and 0.3 μg/ml (for DMAA) are achieved with non-aqueous electrolytes, and working curves are linear over at least 3 orders of magnitude change in concentration. A procedure for separately analysing MMAA and DMAA in solutions containing both acids is given, based on prior separation by ion-exchange chromatography. The mechanism of the reduction of DMAA in pH 4 buffer was studied, and dimethylarsine was the major product identified.

The determination of inorganic arsenic by differential pulse polarography and differential pulse anodic stripping voltammetry has been reported[140,141]. The polarographic reduction of dimethylarsinic acid and methylarsonic acid has also been reported[130]. Bess and coworkers[131] studied the differential pulse polarography of a series of alkylarsonic and dialkylarsinic acids below pH 2. The peak potentials were pH dependent, shifting to more anodic values at lower pH values. Recently, Bess and coworkers[132] reported on the differential pulse polarography of aromatic arsonic and arsinic acids. They found that peak potentials as well as peak currents were pH dependent below pH 2.

Phenylarsonic acid and phenylarsonous acid were studied extensively by Watson and Svehla[128,129] using dc polarography at a dropping mercury electrode. They suggested that phenylarsine oxide exists in aqueous solution as phenylarsonous acid and the diffusion current was pH independent. Unfortunately, above pH 2 their waves were poorly formed and exhibited broad maxima. Most of the work was carried out in 0.1 M hydrochloric acid at concentrations below 1×10^{-4} M, where phenylarsine oxide showed two main reduction processes. The half-wave potentials were shown to be pH dependent as the reduction processes became increasingly irreversible with increasing pH. A reaction scheme was proposed, according to which the reduction product reacts with the electroactive species with the formation of an insoluble polymeric product. The first wave was attributed to the reduction of phenylarsine oxide to phenylarsine, where each mole of phenylarsine combines with an additional two moles of phenylarsine oxide to form the insoluble polymeric product. The second wave was due to an increase of phenylarsine oxide molecules that undergo reduction to phenylarsine and a decrease of phenylarsine oxide (PAO) molecules that react with the phenylarsine. At this wave there is a net increase in the average number of electrons consumed per molecule of PAO.

Phenylarsine oxide is used as a titrant for the direct and indirect determination of residual chlorine and ozone in water and wastewater. Preliminary investigations on the direct measurement of PAO by differential pulse polarography (DPP) indicate that this technique is a promising method for lowering the detection limits in the indirect measurement of these oxidants[142]. The control of pH is a necessary consideration in free and combined chlorine analysis with PAO[143,146] as well as the stability and measurement of ozone[144,146]. According to a study conducted by Lowry and collaborators[145], pH was shown to exert a strong influence on the DPP behaviour of PAO. In general, PAO exhibited three reduction peaks, A, B and C, with varying pH dependencies, as shown in Figure 5. Two unresolved peaks, B and C, occurred at the more cathodic potential. At low pH, peak B predominates while at high pH value, peak C predominates. At about pH 4.7, both peaks B and C approach equal height. Better resolution could not be obtained with decreases in scan rate (0.5 mV/s) or smaller modulation amplitudes (5, 10, 25 mV). The pH dependence of the current is inconsistent with the results reported by previous investigators. It is apparent that the pH dependence is not an artifact of differential pulse polarography. To determine if this behaviour is characteristic of a dropping mercury electrode only, an investigation of the mechanism of reduction of PAO by stationary electrode voltammetry and coulometry was carried out. The optimum sensitivity is obtained by using peak A in the lower pH range, over a PAO range of 1.23×10^{-6} M to 1.23×10^{-5} M.

Gifford and Bruckenstein[147] generated the hydrides of As(III), Sn(II) and Sb(III) by sodium borohydride reduction and separated them on a column of Poropak Q. Detection was at a gold gas-porous electrode by measurement of the respective electrooxidation

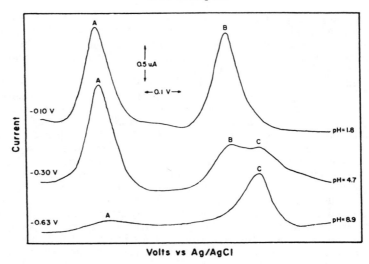

FIGURE 5. DPP of PAO at various pH values: (PAO) 3.08×10^{-6} M, scan
rate 2 mV/s, drop time 2 s, pulse amplitude 199 mV, ionic strength (μ) 0.025.
Reprinted with permission from Reference 145. Copyright (1978) American
Chemical Society

currents. Detection limits were (5 ml samples) As(III), 0.2 ppb; Sn(II), 0.8 ppb; and Sb(III),
0.2 ppb.

Watson has discussed the polarographic behaviour of phenyl arsonic acid[128], phenyl
arsenooxide[129], triphenylarsenooxide[148] and diphenylarsinic acid[149].

Diphenylarsinic acid gives rise to a single cathodic wave at -0.8 V below pH 6,
displaying an adsorption pre-wave, and some limited inhibition effects. The latter are
removed by addition of a surface-active agent, thus yielding a well-formed, diffusion-
controlled wave, the height of which is proportional to concentration (up to 1×10^{-3} M)
and independent of pH. An irreversible reduction to tetraphenyldiarsine has been found.
In mixtures with phenylarsonic acid and/or triphenylarsine oxide, the wave at pH 1 gives
the total concentration of all three, while that at pH 5.3 gives the concentration of diphenyl
and triphenyl species and that at pH 7 triphenylarsine oxide alone.

Watson[149] investigated the effect of a surface-active agent by recording the current–
potential curve for 2×10^{-4} M diphenylarsinic acid in 0.1 M hydrochloric acid in the
presence of an increasing concentration of Triton X-100. The small discontinuity moved
rapidly to more positive potentials to merge with the main wave while the pre-wave and
maximum were suppressed. By a concentration of 0.005% of Triton X-100 a single,
well-formed wave resulted (Figure 6). Above this concentration no further change was
detectable. The half-wave potential and the over-all height underwent no major change,
indicating that there had been no change in the basic reduction process.

In the presence of 0.01% of Triton X-100 good proportionality has been found between
the wave height and the concentration of diphenylarsinic acid, with much improved
standard errors of estimate of the graph (2–3% of the average wave height). Calibration
graphs were prepared from several solid samples of diphenylarsinic acid. Within the limits
of experimental error the slopes for various samples are equal, confirming that solid
diphenylarsinic acid yields a reproducible concentration of the electroactive form. In this

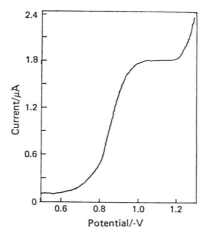

FIGURE 6. Current–potential curve for a 0.0002 M solution of diphenylarsenic acid in 0.1 M hydrochloric acid containing 0.01% of Triton X-100. Reproduced from Reference 149 by permission of The Royal Society of Chemistry

process the potential-determining step yields the diphenyldihydroxyarsine radical **II**, which then adds on a second electron–proton pair to form the hydroxide **III** and then the diphenylarsine **IV**. Further reactions may then occur, such as the reaction of **IV** with **III** to produce the tetraphenyl diarsine **V**. This reduction mechanism requires six electrons in order to reduce two molecules of diphenylarsinic acid, that is, three electrons per molecule. The experimentally obtained value of 2.95 ± 0.16 electrons per molecule confirms that this is the actual reduction mechanism.

$$\text{(C}_6\text{H}_5\text{)}_2\text{As—H} + \text{(C}_6\text{H}_5\text{)}_2\text{As—OH} \longrightarrow \text{(C}_6\text{H}_5\text{)}_2\text{As—As(C}_6\text{H}_5\text{)}_2$$

      (IV)   **(III)**   **(V)**

In 0.1 M hydrochloric acid, the waves of diphenylarsinic acid and phenylarsonic acid are not resolved but give a combined wave height. By increasing the pH above 3, the phenylarsonic acid becomes electroinactive. The practical working range for diphenyl-arsinic acid above pH 3 is considerably narrowed by partial overlap with decomposition of the hydrogen ion. By working at pH 5.3 in phosphate buffer containing 0.01% of Triton X-100, specific determination of diphenylarsinic acid is possible in up to a 100-fold excess of phenylarsonic acid. Accuracies of about 3% have been obtained by using the standard additions technique with five additions. The phenylarsonic acid content is obtained from the combined wave height in solution in 0.1 M hydrochloric acid.

Similarly, in 0.1 M hydrochloric acid or at pH 5.3 resolution from the wave of triphenylarsine oxide is not possible. At pH $\geqslant$ 7 only the latter species is active. Thus, by choice of a suitable pH, specific determination of phenylarsonic acid, diphenylarsinic acid and triphenylarsineoxide in mixtures is possible. Inorganic arsenic(V) is electroinactive while the tetraphenylarsonium ion is reduced at much more negative potentials. Arsenic(III) species are reduced at more positive potentials. Polarography is thus of value for the quantitative speciation of organoarsenic samples.

Henry and coworkers[150] reported a method for the determination of As(III) and As(V) and total inorganic arsenic by differential pulse polarography. Arsenic(III) was measured indirectly in 1 M perchloric acid or 1 M hydrochloric acid[140]. Total inorganic arsenic was determined in either of these supporting electrolytes after the reduction of electroinactive As(V) with aequeous sulphur dioxide. As(V) was evaluated by difference. Sulphur dioxide was selected because it reduced As(V) rapidly and quantitatively, and its excess was readily removed from the reaction mixture.

F. Gas Chromatography

1. Alkyl, aryl and vinyl arsines

Gudzinowicz and Martin[151] applied gas chromatography to the separation of eight substituted organoarsines and substituted organobromoarsines of the type RAsR'R", where R is an alkyl or aryl group, R' is an alkyl group, CF_3 or C_3F_7 and R" is CF_3 or C_3F_7, ranging in molecular weight from 156 to 306. They found that an almost linear relationship existed between log retention time and either the boiling points or the molecular weights of each component of a homologous series. Chromatography was carried out at 290 °C with argon at 40 ml per minute as carrier gas; an argon ionization detector was used.

Gudzinowicz and Martin[151] used the same method for quantitative studies of organoarsenic (triphenylarsine and trivinylarsine) and organobromo-arsenic (methyl, ethyl and butyl bromoarsines) compound mixtures.

Table 12 shows the relative retention time date for the separation of a five-component mixture containing dissimilar structures.

TABLE 12. Retention times of specific organo- and organo-bromoarsines relative to methylbutylbromoarsine[a][151]

Compound	Relative retention time
Dimethylbromoarsine	0.43
Methylethylbromoarsine	0.70
Methylbutylbromoarsine	1.00
Vinyldibromoarsine	1.53
Tributylarsine	2.13

[a] Operating conditions: column temperature, 95 °C; flash heater, 165 °C; detector temperature, 195 °C; cell voltage, 1250 v.; argon flow rate, 40 cc/min.

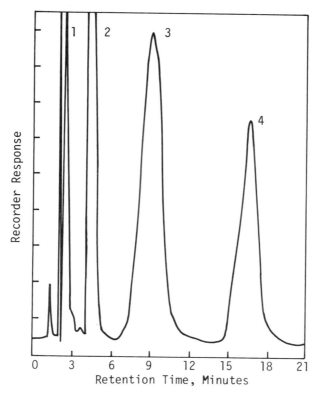

FIGURE 7. Gas chromatogram of (1) benzene, (2) trivinylarsine, (3) divinylbromoarsine, (4) vinyldibromoarsine. Operating conditions: column temperature 80 °C, flash heater 167 °C, detector temperature 235 °C, cell voltage 125 °V, argon flow rate 33 cc/min. Reprinted with permission from Reference 151. Copyright (1962) American Chemical Society

Figures 7 and 8 are chromatograms of homologous compounds where either or both the organo- and/or bromo-contents in each series are varied.

Gudzinowicz and Martin[151] emphasize the possibility of thermally induced molecular rearrangements, particularly for vinyl compounds, under otherwise normal gas chromatographic conditions. Such molecular changes impose a definite limitation upon the usefulness of this method, and their absence under actual separation conditions should first be established.

Figure 8 shows the effect of the alkyl chain length on the retention of organobromoarsines. A plot of the logarithm of the net retention time of these compounds vs the boiling point of each component gives a nearly linear relationship (Figure 9). However, in the same conditions methyldibromoarsine could not be resolved from methylbutylbromoarsine.

The observed behaviour fits the relationship (Grant and Vaughan[152], Pecsok[153]) below,

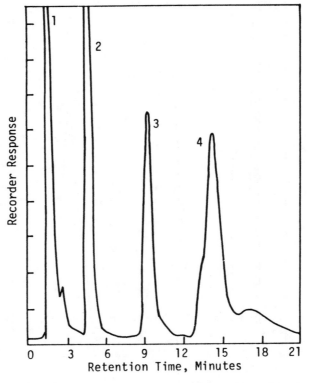

FIGURE 8. Gas chromatogram of (1) methylene chloride, (2) dimethylbromoarsine, (3) methylethylbromoarsine, (4) methyl-butylbromoarsine. Operating conditions: column temperature 68 °C, flash heater 132 °C, detector temperature 200 °C, cell voltage 1250 V, argon flow rate 44 cc/min. Reprinted with permission from Reference 151. Copyright (1962) American Chemical Society

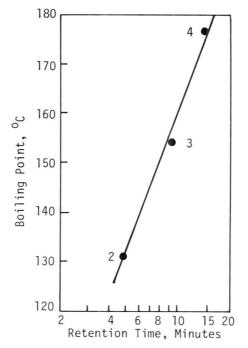

FIGURE 9. Plot of logarithm of net retention time vs boiling point: (2) dimethylbromoarsine, (3) methylethylbromoarsine, (4) methylbutylbromoarsine. Reprinted with permission from Reference 151. Copyright (1962) American Chemical Society

when the net retention time is the abnormal retention time corrected for air time:

$$\ln t_{\mathrm{R}} = \frac{-K_{\mathrm{T}} B_{\mathrm{p}}}{RT} + \ln c$$

where t_{R} = retention time, c = column constant, K_{T} = Trouton's constant, B_{p} = boiling point, R = gas constant and T = absolute temperature.

Figure 10 which relates molecular weight to retention time, shows that the following equation is applicable to these arsines:

$$\ln t_{\mathrm{R}} = \frac{-K_{\mathrm{i}}(M - 170)^{1/3}}{RT} + \ln c$$

where K_{i} = constant relating molecular weight increments to heats of vaporization, M = molecular weight of the compound and 170 = molecular weight of the constant atomic grouping CH_3AsBr.

Gudzinowicz and Martin[151] also applied their technique to the quantitative determination of substituted arsines. For this study they chose a close boiling binary arsine mixture,

T. R. Crompton

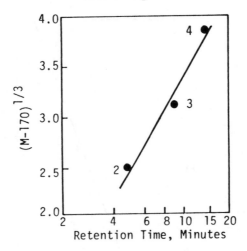

FIGURE 10. Plot of logarithm of net retention time vs $(M - 170)^{1/3}$ relating retention time to molecular weight; (2) dimethylbromarsine, (3) methylethylbromarsine, (4) methylbutylbromarsine. Reprinted with permission from Reference 151. Copyright (1962) American Chemical Society

a 2:1 ratio of dimethylbromoarsine (b.p. 125–130 °C/720 mm) to trivinylarsine (b.p. 130 °C). The mixture was dissolved in carbon tetrachloride and chromatographed with a copper column operated at 66 °C and packed with squalene on Chromosorb W. With a helium inlet pressure of 25 p.s.i. and a flow rate of 350 to 355 cc per minute, dimethyl-bromoarsine and trivinylarsine were eluted in 4.1 and 5.2 min, respectively.

Because of the high toxicity and water sensitivity of these compounds, a calibrated microliter capillary pipet was used in conjunction with the liquid sampling introduction system. This choice of sampling was made recognizing that peak *height* is highly dependent on flow rate, temperature, sample size and method of injection—factors which affect efficiency and resolution. Peak *area* is much less dependent on these operating parameters. The linear relationship obtained between the peak height measurement and the volume of arsines injected is noted in Figure 11. The reproducibility of the peak height measurements for seven injections of trivinylarsine from the same pipet was $\pm$ 3%.

A linear plot relating peak height to milligrams of dimethylbromoarsine present in the sample was also obtained. With this plot and a knowledge of the sample density, it is possible to make direct concentration determinations on a weight per cent basis. This peak height method yields more accurate results than peak area integrations when there is significant band overlapping.

Schwedt and Russel[154] have described a method for the gas chromatographic determination of arsenic (as triphenylarsine) in biological material. Down to 2 mgl of arsenic in the sample could be determined by their procedure.

Andreae[4] described a method for the sequential determination of arsenate, arsenite, mino-, di- and trimethylarsine, MMAA and DMAA and trimethylarsine oxide in natural waters with detection limits of several ngl. The arsines are volatilized from the sample by

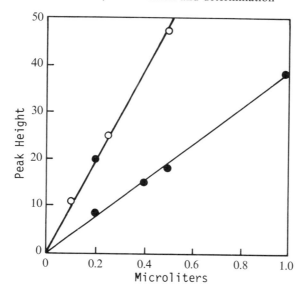

FIGURE 11. Gas chromatography of substituted arsines. Peak height vs sample volume: ○——○ trivinylarsine, ●——● dimethyl-bromarsine. Reprinted with permission from Reference 151. Copyright (1962) American Chemical Society

gas stripping; the other species are then selectively reduced to the corresponding arsines and volatilized. The arsines are collected in a cold trap cooled with liquid nitrogen. They are then separated by slow warming of the trap or by gas chromatography and measured with atomic absorption, electron capture and -or flame ionization detectors. He found that these arsenic species all occurred in natural water samples.

The gas chromatograph is equipped with an electron capture detector or with an atomic absorption detection system using samples of 1–50 ml.

Volatile arsines in the sample are stripped out by bubbling a helium stream through the sample, which is buffered to pH of about 6. The As(III) is converted to arsine by sodium borohydride and stripped from the solution with helium. Then hydrochloric acid is added, to bring the pH to about 1, and the addition of more sodium borohydride reduces MMAA, DMAA and trimethylarsine oxide to the corresponding arsines, which are also swept out of the solution by the helium stream.

Table 13 presents some values obtained for inorganic arsenic, MMAA and DMAA in natural water samples.

Andreae[4] commented that if stored in air-tight containers, the free arsines are stable in solutions for a few days, but they are lost depending on the temperature, concentration and acidity of the samples. The above technique has been discussed by other workers[155–157].

Lussi-Shlatter and Brandenberger[158,159] have reported a method for inorganic arsenic and phenylarsenic compounds based upon gas chromatography. However, methylarsenic species were not examined.

Methods involving reduction to produce hydrides followed by separation and detection by an emission-type detector have been used for organoarsenic compounds[28]. The design of a glow discharge tube proposed earlier[160] as an element-specific detector for gas

TABLE 13. Arsenic species concentrations in natural waters (μg/l As)[4]

Locality and sample type	As(III)	As(V)	MMAA	DMAA
Seawater, Scripps Piper, La Jolla, Calif.				
5 Nov. 1976	0.019	1.75	0.017	0.12
11 Nov. 1976	0.034	1.70	0.019	0.12
Seawater, San Diego Trough				
Surface	0.017	1.49	0.005	0.21
25 m below surface	0.016	1.32	0.003	0.14
50 m below surface	0.016	1.67	0.003	0.004
75 m below surface	0.021	1.52	0.004	0.002
100 m below surface	0.060	1.59	0.003	0.002
Sacramento River, Red Bluff, Calif.	0.040	1.08	0.021	0.004
Owens River, Bishop, Calif.	0.085	2.5	0.062	0.22
Colorado River, Parker, Ariz.	0.114	1.95	0.063	0.051
Colorado River, Slough Near Topcock Calif.	0.085	2.25	0.13	0.31
Saddleback Lake, Calif.	0.053	0.020	0.002	0.006
Rain, La Jolla, Calif.				
10 Sept. 1976	0.002	0.180	0.002	0.024
11 Sept. 1976	0.002	0.094	0.002	0.002

chromatography has been modified[161] to overcome its principal drawback, namely that it appeared to be subject to coating of the tube walls with decomposition products thus attenuating the light signal as chromatographic peaks passed through the discharge.

Parris and collaborators[162] have described a procedure utilizing a commercial atomic absorption spectrophotometer with a heated graphite tube furnace atomizer linked to a

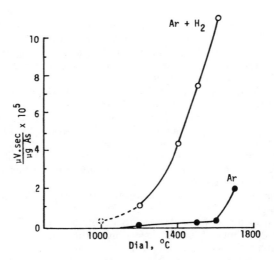

FIGURE 12. Comparison of atomization efficiency curves for trimethylarsine in a silica-lined graphite furnace with and without hydrogen added to the carrier gas. Reprinted with permission from Reference 162. Copyright (1977) American Chemical Society

199

TABLE 14. Calibration curves for GC-AA analyses of trace trimethylarsine in nitrogen[162]

Organometal[a]	Concentration range (ng)	N	r	(± % std error)[b]	a (± % std error)[c]	RSD (%)[d]	Sensitivity (g^{-1})	Detection limit (ng)	Conditions[e]
Me$_3$As	0–320	30	0.965	606,888 (5.13)	−6,465 (77.2)	33.2	516	5	1800 °C, G, H$_2$
Me$_3$As	0–512	28	0.975	757,800 (4.48)	−20,279 (3.30)	44.3	197	14	1800 °C, P, H$_2$
Me$_3$As	0–384	34	0.995	798,021 (1.83)	+13,317 (19.6)	11.5	327	8	1500 °C, S, H$_2$
Me$_3$As	128–1152	13	0.930	218,456 (12.0)	−18,669 (94.4)	910	7.6	288	1700 °C, S
Me$_3$As	128–1024	8	0.979	130,726 (8.56)	+5,776 (134)	50.9	13.6	180	1700 °C, S
Me$_3$As	64–1284	25	0.949	296,958 (6.90)	−56,537 (26.9)	122	9.1	228	Pulsed, 2700 °C, G
Me$_3$As	64–1284	25	0.936	176,777 (7.84)	−39,485 (26.1)	102	8.0	259	Pulsed, 2700 °C, G

[a] Ca 200–400 p.p.m. gaseous organometal in ultra-pure N$_2$; typically 0.05 to 1.0 cm^3 samples taken for GC-AA run.
[b] In units of V s^{-1} g^{-1}.
[c] In units of v.s.
[d] Calculated at 0.1 g metal.
[e] Ar gas flow from GC at 20 ml min^{-1}, with ca 1–2 ml min^{-1} H$_2$ added as noted; G = graphite furnace tube alone; S = graphite tube fitted with SiO$_2$, linear; P = graphite tube coated with pyrolitic carbon.

gas chromatograph for the determination of down to $5\,\mu gl^{-1}$ (as arsenic) of trimethyl-arsine in respirant gases produced in microbiological reactions. The authors[162] investigated the efficiency of atomization for trimethylarsine on two different tubes (bare graphite and silica-lined graphite) using inert (pure argon) and reactive (argon with 10% hydrogen) carrier gases. The results are shown in Figure 12 and in Table 14. With inert carrier gas, very little atomization occurs at less than 1400 °C on either surface. At the upper operating range on silica (1600 °C) atomization is observed, but the sensitivity is low. On bare graphite, higher temperatures can be routinely used, and even with the inert carrier gas atomization begins at a lower temperature than on silica.

Parris and coworkers[162] calculated figures of merit for the various modes of operation studied with their atomic absorption detector utilizing the procedure of Mandel and Stiehler[163]. The latter devised a simple quantitative measure of merit for comparing proportionality quantities such as these calibration curves. They define sensitivity as a criterion which takes into account not only the reproducibility of precision of the test procedure, but which is also diagnostic of small variations in the property measured. Thus,

$$\text{sensitivity} = \frac{\text{slope of the calibration curve}}{\text{standard deviation of measurement(s)}} = \frac{m}{\sigma_m}\,\text{ng}^{-1}$$

where m is the slope determined for the measured range of interest and σ_m is the standard deviation for some measurement taken within the range.

It is common practice to compare lowest limits of detection or response to various analytes. A number of schemes have been introduced in order to provide such a threshold figure of merit, but perhaps the most reliable is that of Burrell[164] which employs an extension of the sensitivity concept. If measurement of σ_m is restricted to a series of n replicate analyses at a very low concentration, that is, where the precision becomes low, then a practical confidence interval can be imposed which will permit objective evaluation of a conservative detection limit, based on the t statistic appropriate to the σ_m determined for n observations. Skogerboe and Grant[165] have demonstrated the application of δ in the form, where $(1 - \alpha)$ is the confidence interval required.

In view of the problems inherent in the pulsed mode of operation including serious detector-induced tailing of the chromatogram, Parris and coworkers[162] adopted a continuous mode of operation (Figure 13) in which quasi-static thermal and chemical regimes could be maintained. This method offers advantages over the pulsed mode in terms of control of decomposition, volatilization and atomization of the analyte. These workers showed that the maximum temperature which could be maintained indefinitely by the furnace was represented by a power supply dial setting of about 2000 °C. Above this setting, the furnace tended to overheat. To avoid contact of the analyte with the porous and chemically reactive graphite furnace tube, the authors[162] also carried out experiments with vitreous silica furnace liners. The exact conditions employed and the observed sensitivity and detection limits are included in Table 14.

Investigations of the retention of trimethylarsine on graphite and lined tubes as a function of furnace temperature showed that, below 400 °C, molecular trimethylarsine adsorbs on the surfaces; between 400 and 1000 °C, trimethylarsine decomposes and elemental arsenic is retained on the surface; above 1000 °C, trimethylarsine is decomposed and elemental arsenic is volatilized. Porous graphite or new alumina-lined tubes were very retentive, but silica-lined tubes retained relatively little arsenic. Alumina liners were found to undergo considerable physical and chemical change when heated over 1500 °C and their sorptive properties changed with use (Figure 14).

Parris and coworkers[162] applied this technique to the examination of simulated respirant atmospheres that might be found over a culture of microorganisms methylating arsenic, also tin and selenium as trimethylarsine, dimethylselenium and tetramethyltin.

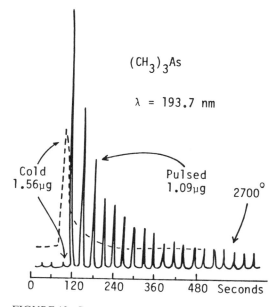

FIGURE 13. Comparison of pulsed GC-AA output continuous UV detection output. The pulsed output was obtained using a 0.85 ml injection at 382-ppm trimethylarsine. The power supply dial was set at 2000 °C and the pulse program was set to give a 7-s pulse on a 30-s interval. The background corrector of the detector was not employed and, as indicated by the arrow, molecular absorption of the 193.7-nm EDL-As can be detected. The continuous output was obtained using 0.20 ml of 2536-p.p.m. trimethylarsine with the furnace at ambient temperature and very low attenuation of signal to the recorder

$$\delta = \frac{t(n-1, 1-\alpha)}{\psi}(ng)$$

EDL = electrodeless discharge lamp. Reprinted with permission from Reference 162. Copyright (1977) American Chemical Society

Samples of the mixture were analysed in the molecular-U.V. detection mode. For chromatogram A in Figure 15 the arsenic lamp (193.7 nm) without background correction was used as an ultra-violet source to detect volatile molecular species. Two peaks are observed for this mixture with retention time 95 ± 4 s and 110 ± 4 s. It was established independently that under these GC conditions (i.e. column 42 ± 2 °C, isothermal) the retention times for the components of the mixture are: $(CH_3)_3As$, 93 ± 3 s; $(CH_3)_2Se$, 92 ± 2 s; $(CH_3)_4Sn$, 109 ± 2 s.

Of the three chromatograms (B, C, D) determined under atomization conditions (1800 °C dial, D_2 background corrector on) with the respective EDL sources (As, Se, Sn) operating in their recommended modes, only the selenium chromatogram C looks like the chromatograms obtained without other compounds present. In the cases of arsenic (B),

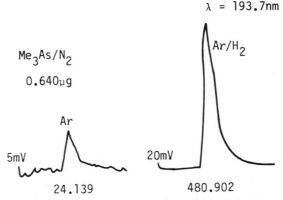

FIGURE 14. Comparison of atomic absorption of arsenic
(0.64 μg) in silica-lined graphite furnace at 1500 °C (dial) with
and without hydrogen in the carrier gas. The chromatogram on
the right is attenuated by a factor of four relative to the one on
the left (note the difference in noise). Reprinted with permission
from Reference 162. Copyright (1977) American Chemical
Society

and especially tin (D), element-specific absorption is observed, not only at the retention
time at which the respective compounds are expected but also at the times associated with
retention of the other metal components of the mixture. Apparently a pre-existing deposit
of an element (e.g. tin) is carried into the analytical volume during the thermal decomposition of mixtures of compounds containing other elements. In the case of selenium where
the volatility of the element precludes formation of deposits, this effect is minimized.
Experiments which support the above interpretation of the chromatograms in Figure 15
include the observation that with a *new* graphite tube the only major tin-specific peak
observed when the mixture is first injected is the 110 ± 4 s peak. However, on subsequent
injections the 92 ± 2 s peak is introduced and grows in magnitude. Injections of methyl
bromide into the GC-AA furnace at 1800 °C failed to remove completely accumulated tin
deposits. Large amounts of tin were indeed detected (tin lamp with D_2 background
correction) as the metal was chemically transported off the surface by methyl, bromine and
hydrogen radicals, but the technique failed to remove efficiently all the deposited tin.

Odanaka and collaborators[166] have reported that the combination of gas chromatography with a multiple ion detection system and hydride generation technique is useful for
the quantitative determination of arsine, monomethyl-, dimethyl- and trimethylarsenic
compounds and this approach is applicable to the analysis of environmental and
biological samples including waters, crops, plants, biological materials and river and
marine sediments.

In this method arsine and methylarsines produced by sodium borohydride reduction
are collected in n-heptane (-80 °C) and then determined. The limit of detection for a 50 ml
sample was $0.2–0.4\,\mu g\,l^{-1}$ of arsenic. Relative standard deviations ranged from 2% to 5%
for distilled water replicates spiked at the $10\,\mu g\,l^{-1}$ level. Recoveries of all four arsenic
species ranged from 85% to 100% (river water), 92–103% (crops and plants), 87–103%
(biological materials) and 72–102% (sediments).

DMAA yields with $NaBH_4$ predominantly dimethylarsine; MMAA yields predominantly methylarsine and trimethylarsine oxide yields mainly trimethylarsine. The volatile

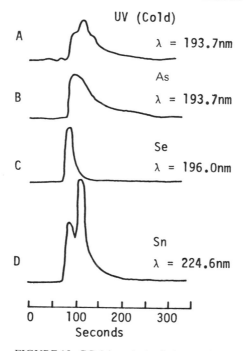

FIGURE 15. GC-AA analysis of mixture of tetramethyltin, trimethylarsine and dimethylselenium. In this series of experiments, the simulated respirant atmosphere described in the text was analysed first (A) by molecular absorption of the 193.7-nm light from the As-EDL source (no background correction) (EDL = electrodeless discharge lamp). Two peaks are observed in the chromatogram which correspond to the retention times of $(CH_3)_3As$ and $(CH_3)_2Se$ (95 ± 4 s) and $(CH_3)_4Sn$ (110 ± 4 s). Using atomization at 1800 °C, with H_2 in bare graphite furnace, the chromatograms B, C and D were obtained using the arsenic (193.7 nm), selenium (196.0 nm) and tin (224.6 nm) lamps with background correction. Note that the injection used for the molecular detection in A was twice as large as those in B, C and D and that B, C and D are attenuated by a factor of 10 relative to A. Reprinted with permission from Reference 162. Copyright (1977) American Chemical Society

arsines were collected in a *n*-heptane cold trap (-80 °C). The helium flow was continued for 1 min to ensure complete generation and trapping of the arsines. After the collection of the arsines, the *n*-heptane solution was injected into the gas chromatograph/mass spectrometer.

The following ions were characteristic and intense ions in the mass spectra: arsine m/z

TABLE 15. Organo- and organobromo-arsines[a][167]

Compound	Molecular weight	Boiling point (°C)
Dimethylbromoarsine	185	128–130/720 mm
Methylethylbromoarsine	199	152–155
Methylbutylbromoarsine	227	172–178/720 mm
Tributylarsine	246	114/10 mm
Trivinylarsine	156	130
Triphenylarsine	306	—
Methyldibromoarsine	250	179–181/720 mm
Vinyldibromoarsine	262	74–76/14 mm

[a] Triphenylarsine (b.p. 360 °C) was eluted in 4.2 min using the following higher-temperature operating conditions: column temperature, 290 °C; flash heater, 340 °C; detector temperature, 365 °C; argon pressure, 30 p.s.i.; flow rate, 40 cc per min.

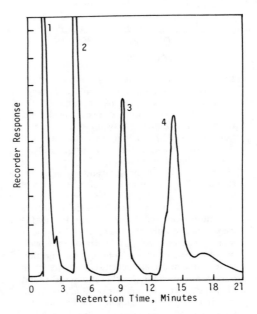

FIGURE 16. Gas chromatogram of (1) methylene chloride, (2) dimethylbromoarsine, (3) methylethylbromoarsine, (4) methylbutylbromoarsine. Operating conditions: column temperature 68 °C, flash heater 132 °C, detector temperature 200 °C, cell voltage 1250 V, argon flow rate 44 cc/min. Reprinted with permission from Reference 151. Copyright (1962) American Chemical Society

$78 M^+$, 76 $(M-2)^+$; methylarsine m/z 92 M^+, 90 $(M-2)^+$, 76 $((M-CH_3)-1)^+$; dimethylarsine m/z 106 M^+, 90 $((M-CH_3)-1)^+$; trimethylarsine m/z 120 M^+, 105 $((M-CH_3)-1)^+$, 103 $((M+CH_3)^+ +2)^+$. Simultaneous determination of all four arsenicals could not be conducted at one injection because of the limited range of detectable mass spectra in the system used.

2. Bromoarsines

Gudzinowicz and Martin[167] applied gas chromatography to the separation of eight substituted organoarsines and substituted organobromarsines of the type RAsR'R" where R is alkyl or aryl, R' is alkyl, CF_3 or C_3F_7, ranging in molecular weight from 156 to 306. They found that an almost linear relationship existed between log retention times and either the boiling point or the molecular weights of each component of a homologous series. Chromatography was carried out on silicone gum rubber on Chromosorb W, at 290 °C with argon as carrier gas; an argon ionization detector was used. Some data for the arsenic derivatives investigated are shown in Table 15.

Figure 16 shows the effect of the aklyl chain length on the retention of organobromoarsines. A plot of the logarithm of the net retention time of these compounds vs the boiling point of each component gives a nearly linear relationship (Figure 17). However, when the same operating conditions were used, methyldibromoarsine could not be resolved from methylbutylbromoarsine.

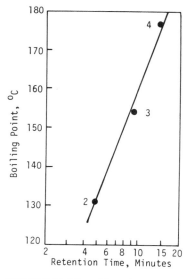

FIGURE 17. Plot of logarithm of net retention time vs boiling point: (2) dimethylbromoarsine, (3) methylethylbromoarsine, (4) methylbutylbromoarsine. Reprinted with permission from Reference 151. Copyright (1962) American Chemical Society

3. Perfluoroorganoarsenic compounds

The separation of perfluorinated organoarsenic compounds has been studied by Gudzinowicz and Driscoll[168]. In this work they used a Perkin-Elmer Model 154C Vapour Fractometer with helium as the carrier gas.

The data for the arsenic compounds investigated are shown in Table 16. The results show relative retention volumes corrected for void volume and pressure drop obtained for various arsines using different chromatographic columns and operating conditions. A comparison of retention volumes for several aryl arsenic derivatives at 200 °C on both columns (conditions 'C' and 'D' in Table 16) indicates that these compounds can be separated at markedly decreased retention times with the shorter SE-30 column provided, however, that diphenylperfluoromethyl and diphenylperfluoropropyl arsines are not present in the same mixture (Figure 18). Figure 19 shows the separation of several lower-boiling lower-molecular-weight compounds.

Examination of the molecular structures and boiling points of the organoarsenic compounds lead Gudzinowicz and Driscoll[168] to the conclusion that the introduction of either a perfluoromethyl or -propyl group into a molecule in place of its alkyl analogue lowers the boiling point of the arsenic compounds investigated even though a larger, higher-molecular-weight grouping has been incorporated, and also decreases the retention times.

In the diphenylarsenic series, the substitution of C_3F_7 or CF_3 only increases the retention time slightly, even though there is a difference of 100 in molecular weight.

TABLE 16. Relative retention volumes for several alkyl/aryl and perfluorinated organoarsines at various operating conditions[a][168]

			A	B	C	D
Compound	Mol. wt.	b.p. (°C)	Rel. ret. vol.	Rel. ret. vol.	Rel. ret. vol.	Rel. ret. vol.
$C_2H_5As(CF_3)_2$	242	77	0.51			
$(C_2H_5)_2AsCF_3$	201	112	1.06			0.33
$C_4H_9As(CF_3)_2$	270	118	1.00[b]			0.23
$(C_4H_9)_2AsCF_3$	258	186	1.64			
◯—$As(C_3F_7)_2$	490	128/68 mm		0.65	0.66	0.84
◯—$As \begin{smallmatrix} CH_3 \\ C_3F_7 \end{smallmatrix}$	336	123/69 mm		1.00[c]	1.00[d]	1.00[e]
$(◯—)_2AsCF_3$	298				4.91	4.58
$(◯—)_2AsC_3F_7$	398				5.16	

[a] Operating conditions:
A. $5\frac{1}{2}$ ft column with 15% by weight SE-30 on Fluoropak 80, 100 °C. Column temperature, 35 cc/min helium flow rate at 1 atm and 25 °C.
B. $5\frac{1}{2}$ ft column with 15% by weight SE-30 on Fluoropak 80, 150 °C. Column temperature, 25.4 cc/min helium flow rate at 1 atm and 25 °C.
C. $5\frac{1}{2}$ ft column with 15% by weight SE-30 on Fluoropak 80, 200 °C. Column temperature, 52 cc/min helium flow rate at 1 atm and 25 °C.
D. 25 ft column with 33% by weight Kel-F Wax 400 on Chromosorb W, 200 °C. Column temperature, 88 cc/min helium flow rate at 1 atm and 25 °C.
[b] Retention time = 9.75 min.
[c] Retention time = 22.40 min.
[d] Retention time = 4.15 min.
[e] Retention time = 14.42 min.

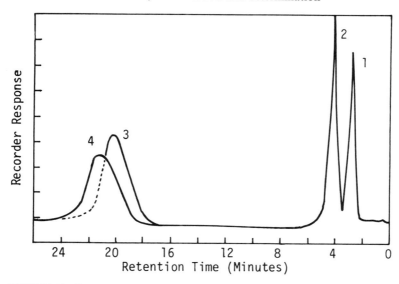

FIGURE 18. Gas chromatogram of (1) $C_6H_5As(C_3F_7)_2$ (2) $C_6H_5AsCH_3C_3F_7$, (3) $(C_6H_5)_2AsCF_3$, (4) $(C_6H_5)_2AsC_3F_7$. Operating conditions: $5\frac{1}{2}$ ft column packed with 15% w/w SE-30 on Fluoropak 80, column temperature 200 °C, helium gas carrier pressure 20 psi., helium flow rate 52 cc/m. Reproduced from Reference 168 by permission of Preston Publications Inc.

4. Methylarsenic and dimethylarsinic acids

Gas-phase methods have been described for the separation and identification of inorganic species as well as for MMAA and DMMA[169,170].

Braman and Foreback succeeded in separating and quantitatively determining As(III), As(V), MMAA and DMAA by a procedure based on sodium borohydride reduction to the corresponding arsines with a dc helium plasma emission spectrometer as the detector[7].

Talmi and Bostick[33] and Talmi and Norvell[186] improved the method by using gas chromatography. Although the method is highly sensitive, it has some disadvantages. It cannot discriminate all arsenic compounds in the biological samples because the same arsines can be produced by reduction from different organoarsenicals. Furthermore, the collection efficiency is incomplete for the very volatile arsine (b.p. -55 °C), and two reduction steps are necessary to discriminate between As(III) and As(V).

Andreae[4] described a method for the sequential determination of arsenate, arsenite, mono-, di- and trimethyl arsine, MMAA, DMAA and trimethylarsine oxide in natural waters with detection limits of several ng/l. The arsines are volatilized from the sample by gas stripping; the other species are then selectively reduced to the corresponding arsines and volatilized. The arsines are collected in a cold trap cooled with liquid nitrogen. They are then separated by slow warming of the trap or by gas chromatography, and measured with atomic absorption, electron capture and/or flame ionization detectors. He found that these four arsenic species all occurred in natural water samples.

The gas chromatograph is equipped with a ^{63}Ni electron capture detector mounted in parallel with a flame ionization detector[177].

Any volatile arsines in the sample are stripped out by bubbling a helium stream through

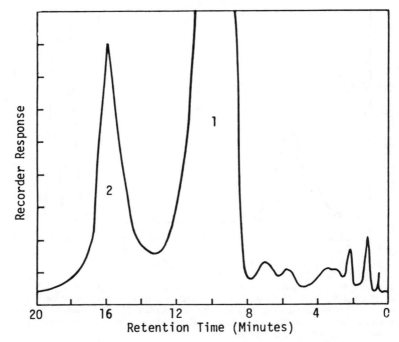

FIGURE 19. Gas chromatogram of (1) $C_4H_9As(CF_3)_2$, (2) $(C_4H_9)_2AsCF_3$. Operating conditions: $5\frac{1}{2}$ ft column packed with 15% w/w SE-30 on Fluoropak 80; column temperature 100 °C, helium carrier gas pressure 10 p.s.i., helium flow rate 35 cc/m. Reproduced from Reference 168 by permission of Preston Publications, Inc.

the sample. Then by successive additions of sodium borohydride As(V), MMAA, DMMA and trimethylarsine oxide are converted to the corresponding arsines and swept out of the solution.

For direct detection by the atomic absorption technique, the stripping gas stream can be used to carry the sample into the burner. The arsines are condensed by immersing the gas trap in liquid nitrogen. By slowly warming the traps to room temperature, the arsines are released in the sequence of their boiling points.

Using gas chromatography, the arsines are rapidly evaporated by immersing the trap in hot water, and detected by a flame ionization detector, or an electron capture detector, or an atomic absorption detector.

Inorganic forms of arsenic could always be determined by the relatively rapid boiling point–separation–atomic absorption detection technique [about 20 min per sample for As(III) and As(V)]; the slower and more cumbersome gas chromatographic method with electron capture detection allows the determination of the organic arsenic species in the low parts per trillion range often found in natural waters. For samples with relatively high concentrations of the organic forms, e.g. urine, a configuration using gas chromatographic separation and detection by atomic absorption or flame ionization detection allows working in the high nanogram to microgram range.

Odanaka and coworkers[178] detected dimethylarsinate as a metabolite in the blood, urine and feces of rats, which had been administered orally a 100 mg/kg dose of ferric

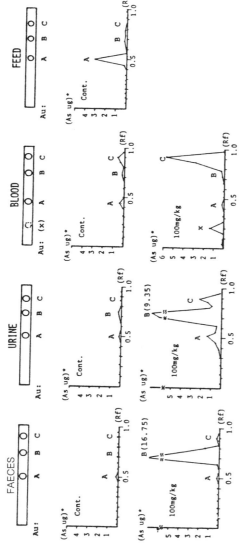

FIGURE 20. Thin-layer chromatograms of arsenic in urine, faeces, blood and feed extract: Au, authentic compounds; A, arsenic acid; B, DSMA; C, DMAA; (X, unknown). Faeces and urine are the second-day samples and blood is the fifth-day sample after treatment of MAF. The asterisk indicates the values in the faeces and blood are As μg/g of sample weight and in the urine and feed are As μg/10 g of sample weight. Reprinted with permission from Reference 178. Copyright (1978) American Chemical Society

methanearsonate, MAF, $(CH_3AsO_3)_3Fe_2$, a compound used as a fungicide for controlling sheath blight in rice in Japan. Following separation of inorganic arsenic and methanearsonate by TLC, dimethylarsinate was converted to dimethylarsine by reduction and identified by GC-MS.

Determination of the arsenic content of each sample was performed by the AgDDC method following wet ashing. Reductive volatilization of arsenic was carried out[33] with sodium borohydride and subsequent analysis by GC-MS.

TLC analysis of the extracts of rat blood, urine feces and feed samples indicated the presence of three major components, as shown in Figure 20. Components A, B and C had

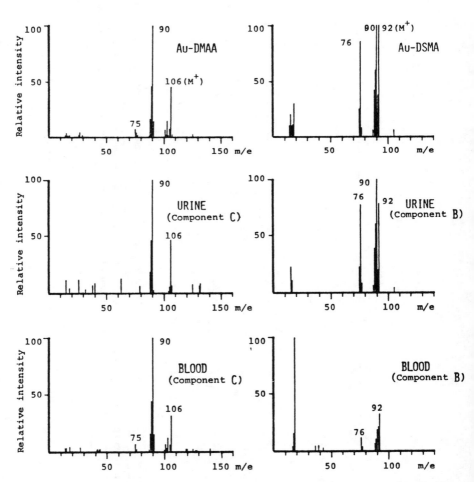

FIGURE 21. Mss spectrum of reduction derivation of authentic arsenic and arsenic obtained from TLC development of urine and blood extract: left, corresponds to compound C (DMAA) on the TLC plate; right, corresponds to compound B (DSMA) on the TLC plate; upper, authentic compounds; middle, urine of the second day; lower, blood of the fifth day. Reprinted with permission from Reference 178. Copyright (1978) American Chemical Society

mobilities similar to the authentic compounds: arsenic acid, DSMA and DMAA respectively. The component A in the urine was identified as arsenate by the AgDDC method. By their comparative GC-MS spectra (Figure 21) the components B and C in the blood and urine were distinctly identified as methanearsonate and dimethylarsinate. The component X which was observed in the blood could not be identified. These results indicated that the major portion of arsenic in feces was methanearsonate, with dimethylarsinate as a minor portion.

Odanaka and collaborators[166] applied GC with multiple ion detection after hydride generation to the determination of mono- and dimethyl arsenic compounds, trimethylarsenic oxide and inorganic arsenic in soil and sediments. Recoveries were 87 to 103%. They also studied the fate of arsenicals in some laboratory animals.

Arsenate was orally administered (1 mg As/kg to rate, mice and hamsters). Urine samples were collected at 24 h intervals and were analysed. Inorganic, mono-, di- and trimethylated arsenic compounds were detected as urinary metabolites among which the lalst one was detected as a new metabolite. These data indicate that inorganic arsenic may be biomethylated up to the trimethylated stage in these animals.

5. Methyl arsine oxides

Soderquist and coworkers[179] determined hydroxydimethyl arsine oxide in water and soil by converting it to iododimethylarsine by hydrogen iodide followed by chromatography and electron capture detection. The recovery of hydroxydimethylarsine oxide (0.15 ppm) added to pure water was 92.3% $\pm$ 7.4%. The corresponding results from soil were 91.3 $\pm$ 5.1%.

Odanaka and coworkers[166] have reported that the combination of gas chromatography with a multiple ion detection system is useful for the quantitative determination of inorganic, mono-, di- and trimethylarsenic oxide compounds, and this approach is applicable to the analysis of environmental and biological samples. The limit of detection for a 50-ml sample was 0.2–0.4 ng ml^{-1} of arsenic. Recoveries of all four arsenic species from river water ranged from 85% to 100% (Table 17).

The following ions were characteristic as intense ions in the mass spectra of arsines: arsine, m/z 78 M$^+$, 76 (M $-$ 2)$^+$; methylarsine: m/z 92 M$^+$. 90 (M $-$ 2)$^+$, 76 ((M $-$ CH$_3$) $-$ 1)$^+$; dimethylarsine m/z 106 M$^+$, 90 ((M $-$ CH$_3$) $-$ 1)$^+$; trimethylarsine, m/z

TABLE 17. Analysis of standard arsenic[a][166]

Standard	Generated arsines	Arsine species (%)
DSMA	arsine	0.08
	methylarsine	99.8
	dimethylarsine	0.05
	trimethylarsine	0.03
DMAA	arsine	0.07
	methylarsine	0.05
	dimethylarsine	99.8
	trimethylarsine	0.06
Trimethyl arsine oxide	arsine	0.05
	methylarsine	0.05
	dimethylarsine	0.08
	trimethylarsine	99.8

[a] Reduction samples containing 20 μg of standard arsenic.

$120\,M^+$, $105\,((M - CH_3) - 1)^+$, $103\,((M - CH_3) - 2)^+$. Simultaneous determination of all four arsenicals could not be done at one injection because of the limited range of detectable mass spectra in the multiple ion detection system used.

The calibration plots for each arsenic compound are linear from 0.2 ng of As ml^{-1} to 2000 ng of As nl^{-1} in aqueous solutions (50 nl). The absolute detection limit per injected sample is 30 pg of As for arsine (m/z 78) and trimethylarsine (m/z 120) and 20 pg of As for methylarsine and dimethylarsine.

The reproducibility of the method for distilled water samples containing 0.5 µg of arsenic for each arsenic species was quite good. The relative standard deviation values were 5% for arsenate, 2% for DSMA and DMAA and 5% for trimethylarsine oxide. Recoveries of inorganic arsenic and mono-, di- and trimethylarsenic compounds were in the range 83 to 110% in spiking experiments on river water.

6. Miscellaneous organoarsenic compounds

Ives and Guiffrida[180] investigated the applicability of the potassium chloride thermionic detector and the flame ionization detector to the determination of organoarsenic compounds in the presence of organophosphorus and nitrogen-containing compounds.

Weston and coworkers[181] determined arsanilic acid and carbarsone in animal feeding stuffs. The additives were extracted from the food with water; the carbarsone present was converted to arsanilic acid that was then reduced to aniline, which can be separated by steam distillation and determined by chromatography and a flame ionization detector.

Schwedt and Ruessel[182] have described a method for the gas chromatographic determination of arsenic as triphenylarsine in biological material. The sample is treated with diphenylmagnesium followed by mercaptoacetic acid. The solution is chromatographed and determined by flame ionization. Down to 2 p.p.m. of arsenic in the sample could be determined by this procedure.

Inorganic and methylated organoarsenic compounds have been converted to their diethyldithiocarbamate[183] or trimethylsilyl derivatives[171,184] prior to determination by gas chromatography.

Hanamura and collaborators[185] applied thermal vaporization and plasma emission spectrometry to the determination of organoarsenic compounds in fish.

7. Alternate detectors

The gas chromatography–microwave plasma detector (GC–MPD) technique has been applid to the analysis of alkyl arsenic acids in environmental samples[33,186].

Braman and coworkers[28] investigated methods involving reduction to produce hydrides followed by separation and detection by an emission-type detector for the analysis of organoarsenic compounds.

Feldman and Batistoni[161] modified the design of a glow discharge tube proposed earlier as an element-specific detector for GC by Bramen and Dynako[160] to overcome its principal drawback, namely the fact that it appeared to be subject to coating of the tube walls by decomposition products of the sample, thus attenuating the light signal. Feldman and Batistoni[161] also comment that several of the earlier workers who have developed optical emission detectors have remarked that spectral background correction would be beneficial, but did not do this. Even when used, the possible advantages of background correction have not always been fully enjoyed, because most of the devices proposed for this purpose have had a fixed wavelength interval between the line of interest and the position at which background intensity was measured. Inevitably, occasions arise on which these positions are occupied by other lines or bands, so that the simple background correction hoped for cannot be performed. This difficulty can often be circumvented by making the separation of the two observation points variable as suggested by Defreese and Malmstadt[187]. To prevent mechanical interference between the two photomultipliers, the beam is split inside the monochromator, with the normal fixed slit at one exit and the moveable slit at the other. Less UV energy is lost if one replaces the existing mirror by a quartz-substrate mirror with a latticework coating pattern, or if one cuts away the upper half of the existing deflecting mirror, as was done in the present case.

Based on these principles, Feldman and Batistoni[161] developed a simple helium glow discharge detector with a stable but inexpensive power supply to detect various metals (Al, As, Cr, Cu), as well as Pi, Si, C and S in gas chromatographic effluents. Improved glow chamber design prevents degradation products from coating the observation window. The monochromator is provided with an internal beam-splitter and a side-exit port. A moveable exit slit mounted on the latter permits background corrections to be made at the most suitable distance from the elemental line detected. Selectivity and versatility are greatly improved by this type of background detection.

Table 18 tabulates limits of detection for arsenic and various other elements studied by Feldman and Batistoni[161]. The authors point out that the detection limit for a given element achieved by their system is governed by a number of factors. These included the characteristics of the glow discharge and its chamber, characteristics of the optical system outside and inside the monochromator and those of the electrical detection/amplification system.

The selectivity of the technique was measured by Feldman and Batistoni[161] by chromatography of a silylized mixture of phenylarsonic acid, nonanoic acid, undecanoic acid and three aliphatic hydrocarbons. In each case a flame ionization detection trace was obtained simultaneously with the glow discharge detector trace. The selectivity of a given Si or As line was defined as the ratio of the peak height obtained per gram atom of carbon in the form of the interfering compound tested.

Figure 22 shows results obtained in the gas chromatography of a silylated mixture of aliphatic acids, phenylarsenic acid and hydrocarbons using a flame ionization detector and a glow discharge detector set at the silicon and arsenic wavelengths.

Ives and Guiffrida[180] investigated the applicability of the potassium chloride thermionic detector and the flame ionization detector to the determination of organoarsenic compounds in the presence of organophosphorus and nitrogen-containing compounds.

TABLE 18. Limits of detection for various elements using the helium glow discharge detector (preliminary values)[161]

Atom or molecule observed	Compound used	Wavelength of line or bandhead line or bandhead (nm)	Detection limit of element[b] without background correction	
			(ng)	(ng/cm³)
P	Tributyl phosphite	213.62	6.4	0.13
P	Tributyl phosphate	213.62	18	0.4
Si	Silylized ester of undecanoic acid	251.61	10	0.2
Al	Al trifluoroacetylacetonate	396.15	12	0.2
As	AsH₃	228.81	0.3[a]	0.001
Cr	Cr acetylacetonate	425.43	1.4	0.028
Cu	Cu hexafluoroacetylacetonate	324.75	110	2.2

[a] Observed by release of AsH_3, from a liquid nitrogen-cooled trap, not by gas chromatograph. Carrier gas velocity, 600 cm³/min.
[b] Quantity required to give a peak twice as high as the peak-to-peak base line noise.

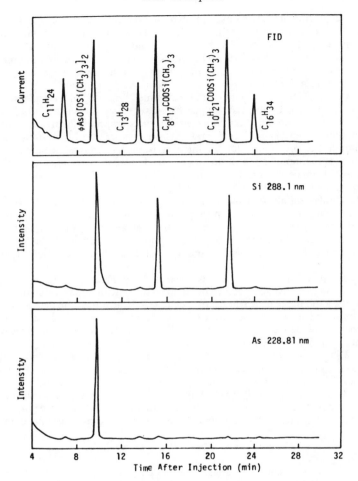

FIGURE 22. Gas chromatograms of a silylated mixture of aliphatic acids, phenylarsonic acid and hydrocarbons. Top curve: flame ionization detector (FID). Centre and bottom curves: glow discharge detector (GDD), with monochromator set as indicated. Reprinted with permission from Reference 161. Copyright (1977) American Chemical Society

8. Arsine

Arsine is included here as it often occurs as an impurity in organoarsenic preparations.

Gas chromatography has been used[188] to determine arsine in hydrogen-rich mixtures. The arsine was detected on a column containing dioctyl phthalate on polyoxyethylene-glycol as adsorbent with hydrogen as the carrier gas. The limit of detection as arsenious oxide was 0.001 mg. In addition, determinations of down to 4.2×10^{-4} g l of arsine in silane on various columns have been developed[190], using dry nitrogen as carrier gas and a katharometer detector after passage of the dry gas through a furnace at 1000 °C to decompose the arsine to hydrogen.

Gifford and Bruckenstein[189] separated and determined arsine, stannane and stibene by gas chromatography with a gold gas-porous electrode detector. Detection limits for 5-ml samples were: As(III) 0.2 p.p.b., Sn(II) 0.8 p.p.b. and Sb(III) 0.2 p.p.b.

Covello et al.[191], have described a procedure in which arsenic in toxicological samples is reduced to arsine, which is then swept out of the reaction flask with helium to a column of silica gel. Detection is achieved by thermal conductivity at levels down to 2 μg arsenic.

Arsine, present as an impurity in acetylene, has been determined[192] on a column packed with tritolyl phosphate on Chromosorb 102 at 30 °C with hydrogen or helium as carrier gas, and thermal conductivity detection. Peaks were obtained for nitrogen, methane, arsine, hydrogen sulphide and phosphine.

Molodyk and collaborators[193] have described a gas chromatographic procedure for the determination of parts per million of methane, ethylene, acetylene and ethane in arsine.

G. Other Chromatographic Techniques

1. Ion-exchange chromatography

Various workers have discussed the application of ion-exchange chromatography to the separation and determination of organoarsenic compounds[4,84,86,175,194-210].

Ion chromatography can separate these arsenicals in the liquid phase, using ion-exchange resins for the separation of As(III), As(V), methylarsonic acid and dimethylarsinic acid in biological, water and soil samples[194,195,198-200,202,203].

Because a selective and sensitive detection is necessary after the separation, atomic absorption spectrometry has been used for this purpose[175]. DC plasma and microwave helium plasma atomic emission spectrometry have been employed for gas-phase detection[4,196,201]. For the liquid phase, a graphite furnace Zeeman effect atomic absorption method has been used with the automated sampler[195]. Inductively coupled argon plasma emission spectrometry seems another choice as a high performance liquid chromatography detector because it has high sensitivity for arsenic, low chemical interference and wide dynamic range. Dimethyl arsinite has a strong affinity for acid-charged cation exchange resins[194,196]. Yamamoto[194] reduced the separated compounds to arsines for spectrometric evaluation. Elton and Geiger[197] used this fact to separate monomethyl arsonate and dimethyl arsinite prior to determinations of the organoarsenicals by differential pulse polarography. The authors reported detection limits of 100 μg l^{-1} and 300 μg l^{-1}, respectively.

Dietz and Perez[196] used ion-exchange chromatography to achieve separations of monomethylarsonate, dimethylarsenite and tri- and pentavalent arsenic. Here, further inorganic speciation relies on redox-based colorimetry[204]. Both the accuracy and precision suffer from the low As(III)/As(IV) and As(total)/P ratios normally encountered in the environment.

Iverson, Anderson and coworkers[199] separated various arsenic species, such as arsenite (AsO$_3$$^{3-}$), arsenate (AsO$_3$$^{3-}$), monomethyl arsonate and dimethyl arsinate by liquid column chromatography and detected them by use of differential pulse polarography[197,205] or flameless atomic absorption. The separations are done by using a gravity column chromatography and by collecting fractions as they elute. Such procedures are very time-consuming.

Stockton and Irgolic[195] separated As(III), As(V), arsenobetaine and arsenocholine by high performance liquid chromatography using a reversed-phase ion suppression technique. This method seemed applicable to the separation of these arsenicals with better resolution than the conventional ion-exchange chromatography using resins.

Stockton and Irgolic[95] developed a low-cost automated interface for the Hitachi Zeeman GF-AA which makes possible the use of this instrument as an element-specific detector for HPLC. The interface consists of an Altex slider injection valve with pneumatic actuators and a 40-μl sample loop, a linear actuator and a sequence control circuit.

Arsenocholine, arsenobetaine and inorganic arsenic were separated on a microparticulate reverse-phase column with the sodium salts of heptanesulphonic acid (Figure 23) or dodecylbenzenesulphonic acid (Figure 24) as counter ion for the arsonium salts. The solvent is made 1.0 M with respect to acetic acid, to suppress the dissociation of the carboxylic acid group in arsenobetaine and achieve greater retention on the reverse-phase column. By varying the acetonitrile/water ratio in the eluent, the retention volume of arsenocholine can be changed from total retention to co-elution with arsenobetaine. Total retention of arsenobetaine on the top of the column was achieved with dodecylbenzenesulphonate. The procedure is useful for the preconcentration of arsenobetaine from very dilute solutions. Arsenite and arsenate were not retained at all with either of the sulphonates and eluted with the solvent front (Figure 23). The graphite furnace atomic absorption detector (GF-AA) responds to the arsenic compounds even though there is insufficient response by refractive index and ultraviolet detectors. The GF-AA detector shows signals corresponding only to the arsenic-containing compounds. The retention times for and the quantities of these compounds present in the mixture can be obtained from the signals.

Henry and Thorpe[205] determined monomethylarsonate, dimethylarsinite and tri- and pentavalent inorganic arsenic by coupling a digestion and reduction scheme with ion-exchange chromatography. However, the utility of this technique for routine environmental analysis is limited, since the implementation time is substantial. This method also relies on estimating As(V) by arithmetic difference.

Ricci and coworkers[206] have described a highly sensitive, automated technique for the determination of MMAA, DMAA, p-aminophenyl arsonate, arsenite and arsenate. This procedure is based on ion-chromatography on a Dionex column, with 0.0024 M $NaHCO_3$/0.0019 M Na_2CO_3/0.001 M $Na_2B_4O_7$ eluent, when all the compounds except arsenite and dimethyl arsinite are separated effectively. For separation of the last two, a lower ionic strength eluent (0.005 M $Na_2B_4O_7$) can be used in a separate analysis. The detection system utilizes a continuous arsine generation system followed by heated quartz furnace atomization and atomic absorption spectrometry. Detection limits of less than 10 ng/ml were obtained for each species.

A typical gradient elution of the five components is shown in Figure 25.

Precision was approximately 11% at the 20 ng/ml concentration. Using high performance liquid chromatography with inductively coupled argon plasma–atomic emission spectrometric detection, high performance liquid chromatography has been used to separate mixtures of arsenic compounds on anion and cation exchange columns using phosphate buffer[207]. The inductively coupled argon plasma–atomic emission spectrometric detection is used as a selective detector by observing arsenic emissions at 193.6 nm. The detection limit was at the nanogram level (Figure 26).

The ionic characteristics of the arsenicals may explain the elution sequence of arsenobetaine, arsenite and arsenate: arsenobetaine → As(III) → As(V) in anion-exchange chromatography and As(III) → As(V) → arsenobetaine in cation-exchange chromatography. DMAA eluted later than MMAA on both columns, indicating the affinity of methyl groups of arsenic to alkyl groups of the column packing. In ion-suppression reversed-phase chromatography, DMAA was eluted later than MMAA.

As(III) was oxidized to As(V) during sample storage. This tendency was especially marked in dilute solution. When a mixed solution of As(III) and As(V) (7 μg/ml As each) was stored at room temperature for 4 weeks, As(III) disappeared and, instead, twice the amount of As(V) appeared on the chromatogram. A mixed solution at 70 μg/ml As each retained As(III) at the 20% level during the same period. However, stock solutions which

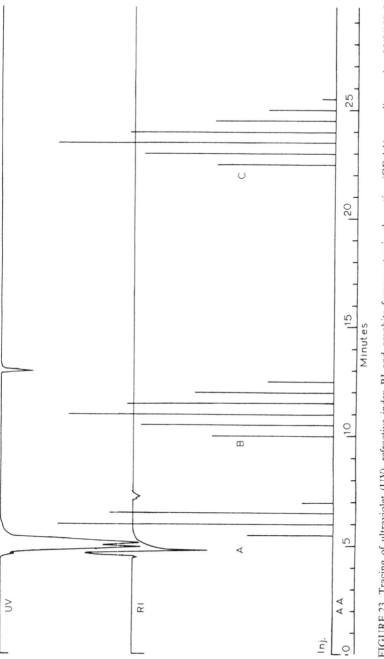

FIGURE 23. Tracing of ultraviolet (UV), refractive index RI and graphite furnace atomic absorption (GF-AA) recordings using 95/5/6 H_2O/acetonitrile/acetic acid and 0.005 M heptanesulphonic acid as solvent A = 1 μg arsenate/arsenite mixture; B = 1 μg arsenobetaine; C = 1 μg arsenocholine. Flow rate was 0.5 ml/min on a μ-Bondapak (C_{18}) reverse-phase column. R.I. attenuation was 4X and the UV was 0.1 A.U.F.S. Reproduced from Reference 195 by permission of Gordon and Breach

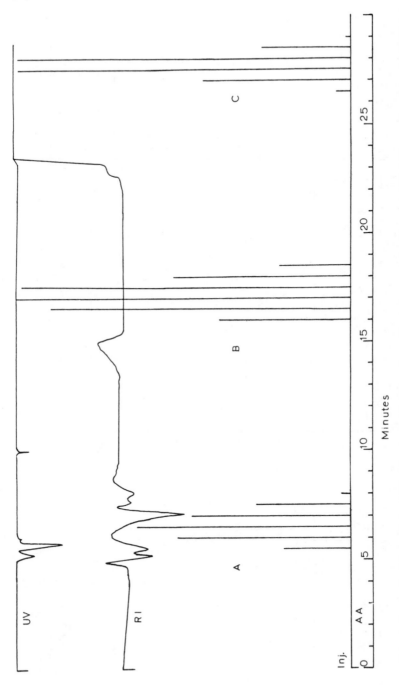

FIGURE 24. Tracing of ultraviolet (UV), refractive index (RI) and graphite furnace atomic absorption (GF-AA) recordings using $2/3/0.6\,H_2O/$ acetonitrile/acetic acid and $0.005\,M$ sodium dodecylbenzenesulfonate. The solvent was changed to acetonitrile at minute 17. $A = 1\,\mu g$ mixture of arsenite/ arsenate; $B = 1\,\mu g$ arsenobetaine; $C = 1\,\mu g$ arsenocholine. Flow rate was $0.5\,ml/min$. R.I. attenuation was 4X on a Water's model 202 differential refractometer and the UV sensitivity was $0.1\,A.U.F.S.$ on a Model 25 Beckman spectrophotometer. Reproduced from Reference 195 by permission of Gordon and Breach

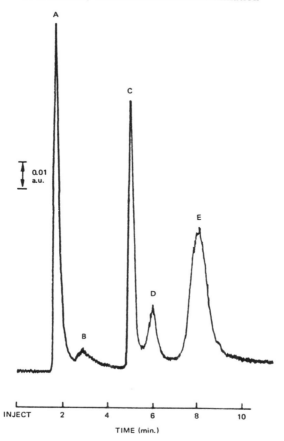

FIGURE 25. Ion chromatography-atomic absorption spectroscopy chromatogram of trace inorganic and organic arsenic species separated by gradient elution. Peak identification is as follows: peak A, DMA (20 ng/ml); peak B, AsO_3^{3-} (20 ng/ml); peak C, MMA (20 ng/ml); peak D, p-aminophenyl arsonate, p-APA (20 ng/ml); peak E, AsO_4^{3-} (60 ng/ml). Reprinted with permission from Reference 206. Copyright (1981) American Chemical Society

contained 0.7% As(III) did not show deterioration during 4 weeks. Therefore, standard mixture solutions should be prepared just before high performance liquid chromatographic analysis.

Morita and coworkers[207] pointed out that with the argon/hydrogen flame and 193.7 nm line, the sensitivity of atomic absorption spectrometry was about one-twentieth of that obtained using inductively coupled argon plasma–atomic emission spectrometry with high performance liquid chromatography detector. Therefore, atomic absorption spectrometry can be applied only for relatively concentrated samples. Arsenic monitored with dc plasma atomic emission spectrometry gave a sensitivity of about one-fifth of that obtained using the inductively coupled plasma techniques.

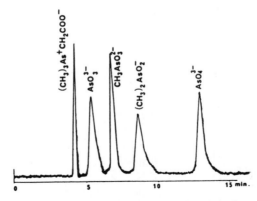

FIGURE 26. Separation of five arsenicals by anion-exchange chromatography. Each arsenical containing 350 ng of As was injected. Reprinted with permission from Reference 206. Copyright (1981) American Chemical Society

Maher[86] used ion-exchange chromatography to separate inorganic arsenic and methylated arsenic species in marine organisms and sediments. This procedure involves the use of solvent extraction to isolate the arsenic species, which are then separated by ion-exchange chromatography and determined by arsine generation.

Grabinski[208] has described an ion-exchange method for the separation of four arsenic species on a single column containing both cation and anion exchange resins. Flameless atomic absorption spectrometry with a deuterium arc background correction is used as the detection system because of its linear response and lack of specificity for these compounds combined with its resistance to matrix bias in this type of analysis. Arsenic recoveries ranged from 97% to 104% for typical lake water samples while 96% to 107% were obtained from arsenic-contaminated sediment interstitial water. The detection limit was 10 p.p.b. for each individual arsenic species.

The practical utility of Grabinski's[208] technique is shown in a sample chromatogram of Lake Mendota (Madison, WI) water (Figure 27).

Interstitial water from sediment sampled at the mouth of the Menominee River near Marinette, WI, was also analyzed for each arsenic species. The water was contaminated with arsenic MMAA and had considerable air exposure. The results of this analysis are presented in Table 19.

Persson and Irgum[84] determined sub-p.p.m. concentrations of DMAA in seawater by electrothermal atomic absorption spectrometry. Graphite-furnace atomic absorption spectrometry was used as a sensitive and specific detector for arsenic. The technique allowed DMAA to be determined in a sample (20 ml) containing a 10^5-fold excess of inorganic arsenic with a detection limit of 0.02 ng As ml^{-1}.

Good recoveries (82–98%) were obtained from spiked artificial seawaters, even at the 0.05–p.p.b. level, but for natural seawater samples the recoveries were lower (74–85%). Higher recoveries (100–102%) are obtained when the standard addition technique is used.

Hanamura and collaborators[209] applied thermal vaporization and plasma emission spectrometry to the determination of organoarsenic compounds in fish.

Aggett and Kadwani[210] point out that hydride generation is of course limited to the determination of those species which can be converted into volatile hydrides. Fortunately, the four species most commonly considered to be of environmental importance at the

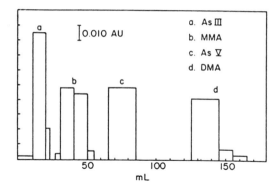

FIGURE 27. Sample chromatogram of spiked Lake Mendota Water. Reprinted with permission from Reference 208. Copyright (1981) American Chemical Society

TABLE 19. Arsenic speciation from spiked and unspiked contaminated sediment interstitial water[208]

	Species			
	As(III)	MMAA	As(V)	DMAA
Interstitial water (ng)	0	2278	300	18
Spike (ng)	300	300	300	300
Interstitial water + spike (ng)	310	2576	620	330
Recovery of spike (ng)	310	298	320	312
Recovery of spike (%)	103	99	107	104

present time, i.e. arsenate, arsenite, MMAA and DMAA, are amenable to this form of analysis. However, conventional ion-exchange and ion-chromatographic methods appear to possess the potential advantage that it should be possible to extend or modify them to include the analysis of additional, environmentally important arsenic species should that become necessary.

Since the four arsenic species mentioned above are weak acids, the dissociation constants of which are quite different (Table 20), it seemed that separation by anion-exchange chromatography was both logical and possible. The authors[210] employed a two-stage single-column anion-exchange method using hydrogen carbonate and ammonium chloride as eluate anions. These species appear to have no adverse effects in subsequent analytical procedures. The analyses were performed by hydride generation atomic-absorption spectroscopy, although any extension to more general speciation

TABLE 20. Dissociation constants of arsenic species[210]

Acid	pk_{a1}	pk_{a2}	pk_{a3}
Arsenic acid	2.25	7.25	12.30
Arsenious acid	9.23		
Monomethylarsonic acid	4.26	8.25	
Dimethylarsinic acid	6.25		

would require the use of a more general technique for analysis, such as graphite furnace atomic absorption spectroscopy or inductively coupled plasma atomic emission spectroscopy.

Preliminary experiments were conducted with various combinations of resins and eluates in the pH range 5–7. In these, the only promising separations were obtained with carbon dioxide–hydrogen carbonate as the eluate.

The behaviour of arsenic(III) and DMAA on SRA 70 resin as a function of the pH is shown in Figure 28. Neither MMAA nor arsenic(V) was eluted the pH range 4.8–6.4. This is understandable in terms of the dissociation constants of the respective acids and the nature of the buffer solution used for elution.

Arsenic(III) exists as an undissociated molecule over the pH range studied and, as a consequence, is eluted rapidly in a manner independent of pH. At lower pH the elution of DMAA overlaps that of arsenic(III), but as the pH is raised there is greater tendency for DMAA to be retained by the resin, presumably a consequence of dissociation. This is reversed somewhat in the region of pH 6.0–6.4, probably owing to the increase in hydrogen carbonate concentration in the eluent in this pH region.

Elution of MMAA was accelerated and complete separation of the four species was obtained by eluting first with carbon dioxide–hydrogen carbonate buffer at pH 5.5 ± 0.3 followed by elution with carbon dioxide–ammonium chloride buffer solution at pH 4.0–4.2.

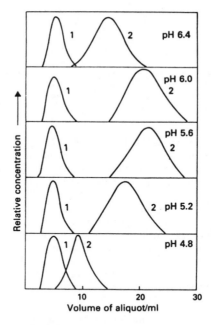

FIGURE 28. Separation of arsenic(III) and dimethylarsinic acid on SRA70 by elution with carbon dioxide–hydrogen carbon as a function of pH: 1, arsenic(III): 2, dimethylarsinic acid. Reproduced from Reference 210 by permission of The Royal Society of Chemistry

A problem that arose was that arsenic(III) was oxidized to arsenic(V) during elution, as indicated by the 70–80% recoveries for arsenic(III) and by the elution of low concentrations of arsenic when arsenic(III) was eluted on its own. This problem was removed when resins were treated with nitric acid (1 mol l^{-1}) and EDTA (0.1 mol l^{-1}, pH 5) before use. Recoveries obtained with resin treated in this way are shown in Table 21 and a chromatogram of a synthetic mixture of the four species is shown in Figure 29.

The application of these methods to interstitial waters of lake sediments and also to species of lake-weed revealed no methylated arsenic species, only As(III) and As(V) in the 0.5 μg l^{-1} range.

Ebden and coworkers[233] used coupled chromatography–atomic spectroscopy for arsenic speciation. Beauchemin and collaborators[234] identified and determined arsenic species in dogfish muscle. The arsenic species were identified using liquid chromatography–inductively coupled plasma–MS, TLC and electron impact–MS. Results indicate that arsenobetaine constitutes about 84% of the arsenic present in the sample analysed.

TABLE 21. Recoveries of arsenic species on treated SRA 70[210]

Species	Mean recovery (%)[a]
Arsenic(III)	97.0
Arsenic(V)	98.1
Monomethylarsonate	98.8
Dimethylarsinate	94.8

[a] Four samples analysed.

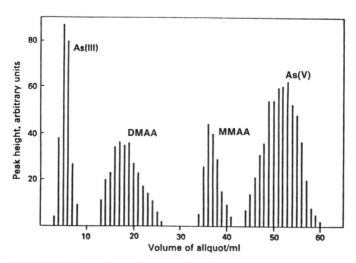

FIGURE 29. Separation of a synthetic mixture of arsenic(III) (0.5 μg), dimethylarsinic acid (0.5 μg), monomethylarsonic acid (0.5 μg) and arsenic(V) (1 μg) on SRA 70. Reproduced from Reference 210 by permission of The Royal Society of Chemistry

2. Thin-layer chromatography

The application of this technique has been discussed by various workers[178,211-215].

TLC on alumina using light petroleum as the solvent has been used to separate Ph_3M where M is a phosphorus, arsenic, antimony or bismuth[212]. R_F values have been determined on silica gel G and neutral alumina for mixtures of labelled triphenylarsine, triphenyl phosphate, triphenyl phosphite, triphenylphosphine oxide, triphenylphosphine and tritolylphosphate using acetone–light petroleum, benzene–acetone and chloroform–acetone as solvents[214]. The chromatograms were automatically examined radiometrically. A range of aromatic compounds containing arsenic or phosphorus has been separated on silica gel, aluminium oxide and magnesium silicate[213]. Twenty-four solvents were studied and R_D values and their standard deviations tabulated.

Von Endt and coworkers[215] used monosodium methane arsonate as model compound in a study of organic arsenicals in soil and used TLC on silica gel G coated plates to separate monosodium methane arsonate (MSMA), arsenate and arsonite in soil extracts.

Silica gels G and H, and cellulose were examined as the solid phases for chromatography of methanearsonate, arsenite and arsenate. Several sprays for the visualization of the arsenicals on plates were tested. Three of the more successful reagents and the colour produced with the final product are shown in Table 22.

TABLE 22. Spray reagents used to detect arsenite arsenate and MSMA on thin-layer chromatograms[215]

	Colour and detection limit		
Spray	Na arsenite	Na arsenate	MSMA
2N HCl–12% (NH$_4$)$_2$S (1 to 1)	yellow 1.5 μg		
1% (NH$_4$)$_6$Mo$_7$O$_{24}$ 1% SnCl$_2$ in 10% HCL 12% (NH$_4$)$_2$S–H$_2$O (1 to 1)		blue 1.5 μg	blue 1.5 μg
1% Et$_2$NCS$_2$ in 50% H$_2$O–acetone 0.1% dithizone in benzene		Orange 2.5–3.0 μg	Orange 2.5–3.0 μg

TABLE 23. Separation of sodium arsenite sodium arsenate and MMSA using several thin-layer chromatographic systems

		R_f		
Support	Solvent	Na arsenite	Na arsenate	MSMA
HN cellulose 300 G	MeOH–BuOH–H$_2$O (3:1:1)	0.80	0.55	0.60
	Pyridine–ethyl acetate–H$_2$O (2:5:5) upper layer	0.43		
Silica gel H	MeOH–NH$_4$OH–10% TCA–H$_2$O (50:15:5:30)	0.45	0.27–0.60	0.67
Silica gel G	As above	0.57	0.00–0.53	0

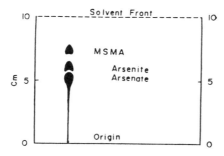

FIGURE 30. Thin-layer chromatography of three arsenic compounds. Reprinted with permission from Reference 215. Copyright (1968) American Chemical Society

Several acidic solvent systems and types of supports work with varying efficiency (Table 23). Cellulose and silica gel H (with organic binder) allow the movement of arsenite, but arsenate and disodium methanearsonate (DSMA) remain on the origin. DSMA, arsenate and arsenite separate best on thin-layer plates coated with silica gel G (calcium sulphate binder).

Figure 30 illustrates the separation which may be expected when microgram quantities of three arsenic compounds are cochromatographed. Silica gel G-coated glass plates were used and developed in a methanol–ammonia–trichloroacetate–water solvent system. Arsenate (valence $+5$) migrates 6/10 of the distance to the front with tailing. Arsenite has an R_f of 0.8 and MSMA has an R_f of 0.85. Arsenite was visualized by using the ammonium sulphide spray, arsenate and MSMA by using the ammonium molybdate–stannous chloride spray.

3. Paper chromatography

This technique has had very limited application to the analysis of organoarsenic compounds[216–218].

Six organoarsenic compounds (arsanilic acid, arsenosobenzene, arsphenamine, 4-hydroxy-3-nitrophenylarsonic acid, 4-nitrophenylarsonic acid and p-ureidophenylarsonic acid) were separated by two-dimensional chromatography on sheets of Whatman No. 1 paper with a solvent consisting of water and nitric acid diluted with methyl cyanide[218]. The compounds were located on the chromatogram and identified by their quenched or fluorescent areas in U.V. light. Final identification was made by spraying with ethanolic ammoniacal silver nitrate or ethanolic pyrogallol, with air drying between the sprayings. The identification limit is 1 μg for each compound.

II. ANALYSIS OF ORGANIC ANTIMONY COMPOUNDS

A. Elements

1. Spectrophotometric methods

Several methods for the determination of antimony in organoantimony compounds have been described[113,116,219–221]. In one, the organoantimony compound is burnt with metallic magnesium to convert antimony to magnesium antimonide[113,219]. This is then decomposed with dilute sulphuric acid to produce stibine, which is absorbed in 6 N hydrochloric acid containing sodium nitrite. The resulting hexachloroantimonic acid can be determined colorimetrically after having been converted to a blue-coloured compound by reaction with methyl violet. The blue colour is extracted with toluene for spectrophotometric evaluation. Nitrogen, phosphorus and arsenic do not interfere.

In an alternative procedure the decomposition is effected by the use of concentrated sulphuric acid and potassium sulphate, followed by the addition of 30% hydrogen peroxide. The solution is adjusted to 6 N with respect to hydrochloric acid and antimony is extracted with diisopropyl ether. Colour is developed by the addition to the ether phase of 0.02% Rhodamine B in 1 N hydrochloric acid[220]. Compounds containing trivalent antimony which are unstable in oxygen or moist air may be stabilized by exposure to sulphur for 8–48 h in vacuo and the resulting compounds are examined by pyrolysis[116].

2. Atomic absorption spectrometry

Marr and coworkers[222] have described a procedure for the microdetermination of antimony in organoantimony compounds by atomic absorption spectrophotometry. They compared air–acetylene and air–hydrogen flames and prefer the latter on account of the lower noise. The effects of varying instrumental and chemical parameters were also studied.

A number of relatively involatile organoantimony compounds and inorganic antimony salts were analysed by the proposed method and also by an alternative spectrophotometric method.

In the atomic absorption procedure 5–10 mg of organoantimony compound is dissolved in 5 ml of butan-2-one, acetone, dimethoxyethane or THF depending on the solubility of the organoantimony compound. Concentrated hydrochloric acid and ethanol are added and the mixture diluted with water. The solution is aspirated into a fuel-rich air–hydrogen or air–acetylene flame and the absorbance measured at 217.6 nm in the flame just above the burner head. The calibration graph is slightly curved over the range

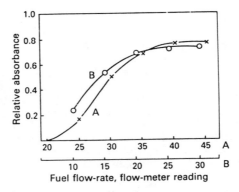

FIGURE 31. Effect of fuel flow-rate on antimony absorption: A, air–hydrogen flame; B, air–acetylene flame. Antimony concentration, $100\,\mu\text{g ml}^{-1}$ in 5% V/V hydrochloric acid; air pressure, 35 (flow-meter reading); height above burner, 2 mm; wavelength, 217.6 nm. Reproduced from Reference 222 by permission of The Royal Society of Chemistry

$20–100\,\mu g\,ml^{-1}$, but linear below $20\,\mu g\,ml^{-1}$. The method is simple, accurate and rapid, and equally useful for inorganic and organic antimony compounds.

In the spectrophotometric procedure the sample is decomposed by boiling with concentrated sulphuric acid and hydrogen peroxide in a micro-Kjeldahl flask. To the solution is added sulphuric acid, ascorbic acid and potassium iodide. The absorbance is measured at 425 nm against water. A linear calibration graph was obtained for the range $4–20\,\mu g\,ml^{-1}$ of antimony.

In the atomic absorption method maximum absorption was obtained using fuel-rich flames and a height just above the burner (Figures 31 and 32). The results shown in Figure 33 indicate that ethanol enhances the absorption signal in both air–hydrogen and air–acetylene. The different degree of enhancement suggests that ethanol not only changes the nebulization process but also plays a role in the atomization of antimony in the flame.

Many organoantimony compounds are not particularly soluble in ethanol and it is necessary to use a second organic solvent to assist dissolution of the samples. The effect of

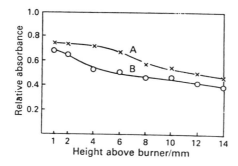

FIGURE 32. Effect of height above burner on antimony absorption: A, air–hydrogen flame, fuel-to-oxidant ratio 1.2: B, air–acetylene flame, fuel-to-oxidant ratio 0.7. Other conditions as for Figure 31. Reproduced from Reference 222 by permission of The Royal Society of Chemistry

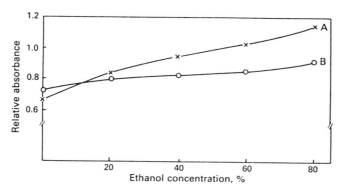

FIGURE 33. Effect of ethanol concentration on antimony absorption. Conditions as for Figure 31. Reproduced from Reference 222 by permission of The Royal Society of Chemistry

TABLE 24. Results for the analysis of organoantimony compounds[222]

| Compound | Antimony, % | | | | Solvent[a] used for atomic absorption |
| | Theory | Spectro-photometry | Atomic absorption | | |
			Air–acetylene	Air–hydrogen	
$(C_6H_5)_4SbCl$	26.1	25.9	26.3	26.2	Butan-2-one
$(C_6H_5)_3SbCl_2$	28.7	27.9	27.8	28.0	Tetrahydrofuran
$(C_6H_5)_3SbI_2$	20.0	19.6	19.9	19.8	Dimethoxyethane
$(C_6H_5)_3Sb^b$	34.5	34.6	34.4	34.5	Dimethoxyethane
$(C_6H_5)_5Sb$	24.0	23.6	24.0	23.8	Butan-2-one (warm)
$(C_6H_5)_4SbS(p\text{-}CH_3C_6H_4)$	22.0	21.6	21.8	21.8	Butan-2-one
$(C_6H_5)_3SbBr_2$	23.7	23.6	23.4	23.5	Butan-2-one
$\{[(C_6H_5)_3Sb]_2O\}(NO_3)_2$	28.7	28.5	28.5	28.5	Butan-2-one
$(C_6H_5)_4SbS(p\text{-}NH_2C_6H_4)$	21.9	21.6	22.1	21.8	Butan-2-one
$(C_6H_5)_3CH_3SbBF_4$	26.7	26.7	26.6	26.8	Butan-2-one

[a] In addition to ethanol
[b] Commercial compound.

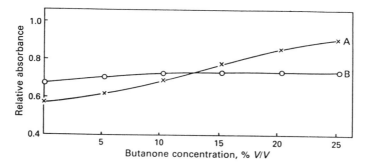

FIGURE 34. Effect of butanone concentration of antimony absorption in A, air–hydrogen and B, air–acetylene flames with fuel-to-oxidant ratios of 1.6 and 0.7, respectively. Solution composition: 20% V/V water, 5% V/V hydrochloric acid, 75–50% V/V ethanol, 0–25% V/V butan-2-one. Other conditions as for Figure 31. Reproduce from Reference 222 by permission of The Royal Society of Chemistry

this additional solvent on the antimony atomic-absorption signal was investigated for solutions also containing 50% of ethanol in order to discover suitable flame conditions under which a reasonable volume of additional solvent could be tolerated. Butan-2-one, THF and dimethoxyethane all exhibited qualitatively the same depression in fuel-rich flames and enhancement in fuel-depleted flames, typified by the curves in Figure 34 for butan-2-one. In practice, a moderately fuel-rich flame was found to be most suitable.

The results obtained for the analysis of a number of organoantimony compounds by the atomic absorption and the spectrophotometric methods are summarized in Table 24. The results obtained by the two methods are generally in good agreement.

The relative standard deviations and the noise levels for different flames are compared in Table 25. The air–hydrogen and air–acetylene flames gave almost equal sensitivities for the determination of antimony at 217.6 nm. However, the air–hydrogen flame was less noisy than the air–acetylene flame and can be confidently recommended for work in the range $10–50\,\mu g\,ml^{-1}$ of antimony.

Andreae and collaborators[223] have described a method for the determination of methylantimony species, Sb(III) and Sb(V) in natural waters by atomic absorption spectrometry with hydride generation. Some results are also reported for estuary and sea waters. The limit of detection was $0.3–0.6\,\mu g\,l^{-1}$ for a 100-ml water sample.

TABLE 25. Comparison of the performance of the spectrophotometric and atomic-absorption methods[222]

Method	Compound	No. of determinations	Relative standard deviation (%)	Noise level[a] (%)
Spectrophotometry	$(C_6H_5)_3SbCl_2$	6	1.8	—
	$(C_6H_5)_5Sb$	6	1.5	—
Atomic absorption				
Air–acetylene	$(C_6H_5)_3SbCl_2$	5	0.73	3
	$(C_6H_5)_5Sb$	6	0.75	
Air–hydrogen	$(C_6H_5)_3SbCl_2$	6	0.54	1
	$(C_6H_5)_5Sb$	5	0.50	

[a]Noise level is the percentage of signal from the most concentrated working standard used to construct the calibration graph.

B. Polarographic Procedures

Besada and Ibrahim[224] have described a polarographic method for the determination of antibilharzial organoantimony compounds in river waters.

C. Gas Chromatography

A gas chromatographic method for the analysis of organoantimony compounds uses a special sample injector to avoid oxidation of the sample[225]. Separation was achieved on a 1-m column of Chromosorb W containing 7.5% of paraffin wax (m.p. 63–64 °C)–triphenylamine (17:3), and using dry purified helium as the carrier gas and a thermistor detector. The column temperature ranged from 73 to 165 °C, depending on the type of compound being determined. The gas chromatography–microwave plasma detector technique has been applied to the analysis of organoantimony compounds in environmental samples[186].

D. Other Chromatographic Techniques

Thin-layer chromatography has been used to separate triphenylantimony from its phosphorus, arsenic and bismuth analogues. R_F values have been reported for these compounds on alumina plates, using light petroleum as developing solvent[226].

III. ORGANOBISMUTH COMPOUNDS

A. Elements

Bismuth in pharmaceuticals such as bismuth tribromophenoxide, bismuth salicylate and bismuth subgallate has been determined complexiometrically[227]. The sample is shaken with dilute nitric acid for 5 min and then diethyl ether added with gentle swirling followed by water, and the solution is heated to 60 °C. The solution is then cooled to 50 °C, ignoring the precipitate, and a 1% solution of methyl thymol blue in potassium nitrate is added and the solution titrated immediately with 0.1 M EDTA–disodium salt to a pale red-violet colour. Then 10% aqueous ammonia is added and this solution titrated further until it becomes yellow.

B. Spectrophotometric Methods

In a spectrophotometric method for the assay of glycobiarsol tablets (bismuth glycolylarsanilate) the sample is allowed to react with aqueous EDTA–disodium salt solution and the absorbance of the acidified solution is measured at 258 nm[228].

C. Thin-layer Chromatography

TLC on alumina has been applied to the separation of compounds of the type Ph_3M where M is bismuth, phosphorus, arsenic or antimony[212]. Light petroleum is the recommended elution solvent.

D. Paper Chromatography

Basic bismuth gallate and free tribromophenol can be detected in small samples of bismuth tribromophenoxide by the simultaneous ascending chromatography of a neutral ethanolic suspension of one portion of the sample and an acidified ethanolic solution of another portion[229]. The two spots are placed side by side and developed with butanol saturated with water. After 2 h the chromatogram is treated with aqueous iron(II) ammonium sulphate solution at about halfway between the origin and the front, and silver nitrate solution is applied close to the front. The presence of tribromophenol is shown by an orange spot with silver nitrate on the neutral chromatogram and basic bismuth gallate by a blue-black spot with iron(II) ammonium sulphate solution on the acid chromatogram. An orange fleck on the acid chromatogram is due to the acid decomposition of the bismuth tribromophenoxide.

IV. REFERENCES

1. J. L. Webb, in *Enzyme and Metabolic Indicators*, Vol. 3, Chap. 6, Academic Press, New York, 1966.
2. D. Lisk, *N.Y. State J. Med.*, **71**, 2541 (1971).
3. R. E. Stanffer, J. W. Ball and E. A. Jenne, Geological Survey Professional Paper 1044F, US Govt. Printing Office, Washington D.C., 1980.
4. M. O. Andreae, *Anal. Chem.*, **49**, 820 (1977).
5. P. T. S. Wong, Y. K. Chau, L. Luton, G. A. Bengut and D. J. Swaine, *Methylation of Arsenic in the Aquatic Environment*, Conference Proceedings on Trace Substances in Environmental Health, XI, Hemphill, University of Missouri, U.A.A., 1977.
6. Pesticide Use Report, California Department of Agriculture, 1972.
7. R. S. Braman and C. C. Foreback, *Science*, **182**, 1247 (1973).
8. E. A. Woolson (Ed.) *Arsenical Pesticides* ACS Symp. Ser. No. 7, 1975.
9. R. S. Braman, in *Arsenical Pesticides* (Ed: E. A. Woolson), ACS Symp. Ser. No 7, 1975, p. 108.
10. D. P. Cox, in *Arsenical Pesticides* (Ed. E. A. Woolson), ACS Symp. Ser. No. 7, 1975.
11. D. W. Von Endt, P. C. Kearney, and D. D. Kanfman, *J. Agric. Food Chem.*, **16**, 17 (1968).
12. J. H. Lasko and S. A. Peoples, *J. Agric. Food Chem.*, **23**, 674 (1975).
13. A. C. Chapman, *Analyst (London)*, **51**, 548 (1926).
14. R. S. Braman and C. C. Foreback, *Science*, **182**, 1249 (1973).
15. J. S. Edmonds and K. A. Francesconi, *Nature (London)*, 3 346 (1977).
16. B. C. McBride and R. S. Wolf, *Biochemistry*, **10**, 4312 (1971).
17. D. B. Cox and M. Alexander, *Microb. Ecol.*, **1**, (1974).
18. D. B. Cox M. Alexander, *Bull. Environ. Contam. Toxicol.*, **9**, 84 (1973).
19. E. A. Creselius, in *Chemical Changes in Arsenic Following Digestion by Man*, U.S. Energy Research and Development Adminstration, Contract No. E (45–1):1830 (1975).
20. M. G. Haywood and J. P. Riley, *Anal. Chim. Acta*, **85**, 219 (1976).
21. V. J. Neulenhoff, *Pharm. Weekbl.*, **100**, 409 (1968).
22. M. Juracek and J. Jenik, *Collect. Czech. Chem. Commun.*, **20**, 550 (1955).
23. M. Jutacek and J. Jenik, *Collect. Czech. Chem. Commun.*, **21**, 890 (1956).
24. S. S. Sandhu and P. Nelson, *Anal. Chem.*, **50**, 322 (1978).
25. A. G. Howard and M. H. Arbab-Zavar, *Analyst (London)*, **105**, 338 (1980).
26. J. N. C. White and J. R. Englar, *Botanica Marina*, **159**, 26 (1983).
27. W. R. Penrose, *Crit. Rev. Environ. Control*, 4, 465 (1974).
28. R. S. Braman, D. L. Johnson and C. C. Foreback, *Anal. Chem.*, **49**, 621 (1977).
29. A. G. Howard and M. H. Arbab-Zavar, *Analyst (London)*, **106**, 213 (1981).
30. M. H. Arbab-Zavar and A. G. Howard, *Analyst (London)*, **105**, 744 (1980).
31. T. A. Hinners, *Analyst (London)*, **105**, 751 (1980).
32. F. D. Pierce and H. R. Brown, *Anal. Chem.*, **49**., 1417 (1977).
33. Y. Talmi, and D. T. Bostick, *Anal. Chem.*, **47**, 2145 (1975).
34. D. L. Johnson and R. S. Braman, *Deep Sea Research*, **22**, 503 (1975).
35. R. S. Braman and G. C. Foreback, *Science*, **182** 1247 (1973).
36. J. Aggett and A. C. Aspell, *Analyst (London)*, **101**, 341 (1976).
37. W. H. Clement and S. D. Faust, *Environ. Lett.*, **5**, 155 (1973).
38. D. G. Iveson, M. A. Anderson, R. R. Holm and R. R. Stanforth, *Environ. Sci. Technol.*, **13**, 1491 (1979).
39. M. Yamamoto, *Proc. Soil Sci. Soc. Am.* **39**, 859 (1975).

40. G. E. Pacey and J. A. Ford, *Talanta*, **28**, 935 (1981).
41. A. A. Grabinski, *Anal. Chem.*, **108**, 1495 (1983).
42. W. L. Jolly, *J. Am. Chem. Soc.*, **83**, 335 (1961).
43. W. Holak, *Anal. Chem.*, **31**, 1713 (1969).
44. E. F. Dalton and A. J. Malanoski, *At. Absorpt. Newsl.*, **10**, 92 (1971).
45. F. J. Fernandez and D. C. Manning, *At. Absorpt. Newsl.*, **10**, 86 (1971).
46. D. C. Manning, *At. Absorpt. Newsl.*, **10**, 123 (1971).
47. F. J. Fernandez, *At. Absorpt. Newsl.*, **12**, 93 (1973).
48. F. J. Schmidt and J. L. Royer, *Anal. Lett.*, **6**, 17 (1973).
49. Kan Kwok-Tai, *Anal. Lett.*, **6**, 603 (1973).
50. P. N. Vijan and G. R. Wood, *At. Absorpt. Newsl.*, **13**, 33 (1974).
51. R. K. Skogerboe and A. P. Bejnink, *Anal. Chim. Acta*, **94**, 297 (1977).
52. R. D. Kadag and G. D. Christian, *Anal.Chim. Acta*, **88**, 117 (1977).
53. M. Thompson, S. J. Paklavanpour and G. F. Welton, *Analyst*, **103**, 568 (1978).
54. R. C. Chu, G. P. Barron and P. A. W. Baumgarner, *Anal. Chem.* **44**, 1476 (1972).
55. M. Lansford, E. M. McPerson and M. J. Fishman, *At. Absorpt. Newsl.*, **13**, 103 (1974).
56. R. S. Braman, L. L. Justan and C. C. Foreback, *Anal. Chem.*, **44**, 2195 (1972).
57. K. C. Thompson and D. R. Thomerson, *Analyst*, **99**, 595 (1974).
58. A. E. Smith, *Analyst*, **100**, 300 (1975).
59. M. Bedard and J. D. Kerbyson, *Can. J. Spectrosc.*, **21**, 64 (1976).
60. H. D. Fleming and R. G. Ide, *Anal. Chim. Acta*, **83**, 67 (1976).
61. P. N. Vijan, A. C. Rayner, D. Sturgiss and G. R. Wood, *Anal. Chim. Acta*, **82**, 329 (1976).
62. P. N. Vijan and G. R. Wood, *Talanta*, **23**, 89 (1976).
63. S. Greenfield, I. L. Jones, H. McD. McGeachin and P. B. Smith, *Anal. Chim. Acta*, **74**, 225 (1975).
64. R. H. Wendt and V. A. Fassel, *Anal. Chem.*, **37**, 920 (1965).
65. J. Aggett and A. C. Aspell, *Analyst*, **101**, 341 (1976).
66. E. A. Crecelius, *Anal. Chem.*, **50**, 826 (1978).
67. A. U. Shalkh and D. E. Tallman, *Anal. Chim. Acta*, **98**, 251 (1978).
68. S. Gohda, *Bull. Chem. Sec. Jpn.*, **48**, 1213 (1975).
69. M. M. Kolthoff and R. Belcher, *Volumetric Analysis*, Vol. 3, Interscience Publishers, New York, (1967) pp. 511–513.
70. J. E. Portman and J. P. Riley, *Anal. Chim. Acta*, **31** 509 (1964).
71. M. B. Casvalho and D. M. Hercules, *Anal. Chem.*, **50**, 2030 (1978).
72. J. S. Edmonds and K. A. Francesconi, *Anal. Chem.*, **48**, 2019 (1976).
73. J. F. Uthe, H. C. Freeman, J. R. Johnston and P. Michalik, *J. Assoc. Off. Anal. Chem.*, **57**, 1363 (1974).
74. W. R. Penrose, R. Black and M. J. Hayward, *J. Fish Res. Bd. Can.*, **32**, 1275 (1975).
75 L. Duncan and C. R. Parker, Varian Techtron, Palo Alto, Calif., 'Technical Topics' (1974).
76. M. Fishman and R. Spencer, *Anal. Chem.*, **49**, 1599 (1977).
77. H. Agemian and V. Cheam, *Anal. Chim. Acta*, **101**, 193 (1978).
78. C. E. Stringer and M. Attrep, *Anal. Chem.*, **51**, 731 (1979).
79. D. C. Manning, Perkin Elmer Corp., *At. Absorpt. Newsl.*, **10**, 6 (1971).
80. D. E. Fleming and G. A. Taylor, *Analyst*, **102**, 101 (1977).
81. R. B. Denyszyn, P. M. Grohe and D. E. Wagoner, *Anal. Chem.*, **50**, 1094 (1978).
82. F. D. Pierce and H. R. Brown, *Anal. Chem.*, **48**, 693 (1976).
83. P. B. Vijan and G. R. Wood, Perkin Elmer Corp. *At. Absorpt. Newsl.*, **13**, 33 (1974).
84. J. A. Persson and K. Irgum, *Anal. Chim. Acta.*, **138**, 111 (1982).
85. G. Lundgren, L. Lundwork and G. Johansson, *Anal. Chem.*, **46**, 1928 (1974).
86. W. A. Maher, *Anal. Chim. Acta.*, **126**, 157 (1981).
87. A. Masui, C. Tsutsumi and A. Teda, *Agric. Biol. Chem.*, **42**, 2139 (1978).
88. R. Pietsch, *Z. Anal. Chem.*, **144**, 353 (1955).
89. C. DiPietro and Sassaman, W. A. *Anal. Chem.*, **36**, 2213 (1964).
90. A. Steyermark, in *Quantitative Organic Microanalysis*, 2nd ed., Academic Press, New York, London, (1961), p. 367.
91. M. M. Tuckerman, J. H. Hodecker, B. C. Southworth and K. D. Fliescher, *Anal. Chim. Acta.*, **21**, 463 (1959).
92. R. D. Strickland and C. M. Maloney, *Anal. Chem.*, **29**, 1870 (1957).
93. M. Jean, *Anal. Chim. Acta*, **14**, 172 (1956).

94. H. Kashiwagi, Y. Tukamoto and M. Kan, *Ann. Rep. Takeda Res. Lab.*, **22**, 69 (1962).
95. G. Bahr, H. Bieling and K. H. Thiele, *Z. Anal. Chem.*, **143**, 103 (1954).
96. P. A. J. Armstrong, P. M. Williams and J. D. H. Strickland, *Nature (London)*, **211**, 418 (1966).
97. R. J. Kopp and S. C. Bandemar, *Anal. Chem.*, **45**, 1789 (1973).
98. R. J. Evans and S. C. Bandemar, *Anal. Chem.*, **26**, 595 (1954).
99. J. S. Caldwell, R. L. Lishka and E. F. McFarren, *J. Am. Water Works Assoc.*, **65**, 731 (1973).
100. J. A. Dean and R. E. Rues *Anal. Lett.*, **2**, 105 (1969).
101. C. E. Stringer and M. Attrep, *Anal. Chem.* **51**, 731 (1979).
102. J. F. Kopp, *Anal. Chem.*, **45**, 1789 (1973).
103. D. C. Manning, *At. Absorpt. Newsl.*, **10**, 6 (1971).
104. M. Corner, *Analyst (London)*, **84** 41 (1959).
105. R. Belcher, A. M. G. Macdonald and T. S. West, *Talanta*, **1**, 408 (1958).
106. W. Merz, *Mikrochim. Acta*, 640 (1959).
107. K. Eder, *Mikrochim. Acta*, 471 (1960).
108. W. H. Guttermann, S. F. John, J. E. Barry, D. L. Jones and E. D. Lisk, *J. Agric. Food Chem.*, **9**, 50 (1961).
109. R. Puschel and Z. Stefanac, *Mikrochim. Acta*, **6**, 1108 (1962).
110. Z. Stefanac, *Mikrochim. Acta.*, **6**, 1115 (1962).
111. R. Belcher, A. M. G. Macdonald, S. E. Phang and T. S. West, *J. Chem. Sec.*, 2044 (1965).
112. A. D. Wilson and D. T. Lewis, *Analyst (London)*, **88**, 510 (1963).
113. M. Juracek and J. Jenik, *Sb. Ved. Pr. Vys. Sk. Chem. Technol. Pardubice*, 105 (1959).
114. L. Maier, D. Seyferth, F. G. A. Stove and B. G. Rochow, *J. Am. Chem. Soc.*, **79**, 5884 (1957).
115. H. Trutnovsky, *Mikrochim. Acta*, **7**, 499 (1963).
116. A. P. Terent'ev, M. A. Volodina and E. G. Fursova, *Dokl. Akad Nauk SSSR.*, **169**, 851 (1966).
117. P. Malatesta and A. Lorenzini, *Ric. Sci.*, **28**, 1874 (1958).
118. R. M. Fournier, *Mem. Poudres*, **40**, 385 (1958).
119. N. Z. Bruja, *Rev. Chim. (Bucharest)*, **17**, 359 (1966).
120. G. Bruekner and M. Parkany, *Magy. Kem. Foly.*, **68**, 164 (1962).
121. J. W. Cavett, *J. Assoc. Off. Agric. Chem.*, **39**, 857 (1956).
122. C. E. Stringer and M. Attrep, *Anal. Chem.*, **51**, 731 (1979).
123. S. A. Peoples J. Lakso and T. Lais, *Proc. West. Pharmacol. Soc.*, **14**, 178 (1971)..
124. M. G. Haywood and J. P. Riley, *Anal. Chim. Acta.*, **85**, 219 (1975).
125. T. Kamada, *Talanta*, **23**, 835 (1976).
126. S. S. Sandhu, *Analyst (London)*, **101**, 856 (1976).
127. H. Matschiner, *Tzschack Z. Chem.*, **5**, 144 (1965).
128. A. Watson and G. Svehla, *Analyst (London)*, **100**, 489 (1975).
129. A. Watson and G. Svehla, *Analyst (London)*, **100**, 573 (1975).
130. R. Elton and W. E. Geiger, *Anal. Lett.*, **9**, 665 (1976).
131. R. C. Bess, K. J. Irgolic, J. E. Flannery and T. H. Ridgway, *Anal. Lett.*, **9**, 1091 (1976).
132. R. C. Bess, K. J. Irgolic, J. E. Flannery, and T. H. Ridgway, *Anal. Lett.*, **9**, 1091 (1976).
133. M. V. Alekseeva, *Inform. Meted. Materily Gas Nauch. Issledovatel Sanit. Inst.*, **5**, 16; *Ref. Zh. Khim.*, **40**, 375 (1955).
134. H. M. Factory Inspectorate, Ministry of Labour, *Methods for the Detection of Toxic Substances in Air*, No. 9, *Arsine*, 2nd ed., H. M. Stationery Office, London.
135. V. Vasak, *Collect. Czech. Chem. Commun.*, **24** 3500 (1959).
136. F. T. Henry and T. M. Thorpe, *Anal. Chem.*, **52**, 80 (1980).
137. A. Watson, *Analyst (London)*, **103**, 332 (1978).
138. S. W. Lee and T. C. Meranger, *Anal. Chem.*, **53**, 130 (1981).
139. R. K. Elton and W. E. Geiger, *Anal. Chem.*, **50**, 712 (1978).
140. D. J. Myers and J. Osteryoung, *Anal. Chem.*, **45**, 267 (1973).
141. G. Forsberg, J. W. O'Laughlin and R. C. Megargle, *Anal. Chem.*, **47**, 1586 (1975).
142. J. H. Lowry, R. B. Smart and K. H. Mancy, *Anal. Lett.*, **10**, 979 (1977).
143. H. C. Marks and J. R. Glass, *J. Am. Water Works Assoc.*, **34**, 1227 (1942).
144. W. Stumm, *J. Boston Soc. Civ. Eng.*, **45**, 68 (1958).
145. J. H. Lowry, R. B. Smart and K. H. Mancy, *Anal. Chem.*, **50**, 1303 (1978).
146. *Standard Methods for the Examination of Water and Wastewater*, 13th ed., American Public Health Association, 1971.
147. P. R. Gifford and S. Bruckenstein, *Anal. Chem.*, **52**, 1028 (1980).

148. A. Watson and G. Svehla, *Analyst (London)*, **100**, 584 (1975).
149. A. Watson, *Analyst (London)*, **103**, 332 (1978).
150. F. T. Henry, T. O. Kirch and T. M. Thorpe, *Anal. Chem.*, **51**, 21 (1979).
151. B. J. Gudzinowicz and H. F. Martin, *Anal. Chem.*, **34**, 648 (1962).
152. D. W. Grant and G. A. Vaughan, *J. Appl. Chem.*, **6**, 145 (1956).
153. R. L. Pecsok, in *Principles and Practice of Gas Chromatography*, Wiley, New York, 1959, p. 31.
154. E. G. Schwedt and H. A. Russel, *Chromatographia*, **5**, 242 (1972).
155. A. W. Fickett, E. H. Daughtrey and P. Mishak, *Anal. Chim. Acta.*, **79**, 93 (1975).
156. J. D. Lodmell, Ph.D. Thesis, University of Tennessee, Knoxville, Tenn. (1973).
157. L. D. Johnson, K. O. Gerhart and W. A. Aue, *Sci. Total Environ.*, **1**, 108 (1972).
158. B. Lussi-Schlatter and H. Brandenberger, *Advances in Mass Spectrometry in Biochemistry and Medicine*, Vol. 2, Spectrum Publications, New York, 1976, pp. 231–248.
159. B. Lussi-Schlatter and H. Brandenberger, *Z. Klin. Chem. Biochem.*, **12**, 224 (1974).
160. R. S. Braman and A. Dynako, *Anal. Chem.*, **40**, 95 (1968).
161. C. Feldman and D. A. Batistoni, *Anal. Chem.*, **49**, 2215 (1977).
162. G. E. Parris, W. R. Blair and T. E. Brinkman, *Anal. Chem.*, **49**, 378 (1977).
163. J. Mandel and R. D. Stiehler, *J. Res. Natl. Bur. Stand.*, **53**, 155 (1954).
164. D. C. Burrell, *Atomic Spectrometric Analysis of Heavy Metal Pollutants in Water*, Ann Arbor Science Publishers Inc., Ann Arbor, Mich., 1974.
165. R. K. Skogerboe and C. L. Grant, *Spectrosc. Lett.*, **3**, 215 (1970).
166. Y. Odanaka, N. Tsuchiya, O. Matano and S. Goto, *Anal. Chem.*, **55**, 929 (1983).
167. B. Z. Gudzinowicz and H. F. Martin, *Anal. Chem.*, **34**, 648 (1962).
168. B. J. Gudzinowicz and J. L. Driscoll, *J. Gas Chromatoger.*, **1**, 25 (1963).
169. J. Tadmor, *J. Gas Chromatogr.*, **2**, 385 (1964).
170. D. Vranti-Piscou, J. Kontoyannakos and G. Parissakis, *J. Chromatogr. Sci.*, **9**, 499 (1971).
171. W. C. Butto and W. T. Rainey, *Anal. Chem.*, **43**, 538 (1971).
172. F. Roy, *J. Gas Chromatogr.*, **6**, 245 (1968).
173. B. J. Gudzinowicz and J. L. Driscoll, *J. Gas Chromatogr.*, **1** 108 (1972).
174. L. Johnson, K. Gerhardt and W. Aue, *Sci. Tot. Environ.*, **1**, 108 (1972).
175. J. S. Edmonds and K. A. Francesconi, *Anal. Chem.*, **48**, 2019 (1976).
176. L. R. Overby, S. F. Bocchieri and R. C. Frederickson, *J. Assoc. Off. Anal. Chem.*, **48**, 17 (1965).
177. Y. K. Chau, P. T. S. Wong and P. D. Goulden, *Anal. Chem.*, **47**, 2279 (1975).
178. Y. Odanaka, O. Matano and S. Goto, *J. Agric. Food Chem.*, **26**, 505 (1978).
179. C. J. Soderquist, D. G. Crosby and J. B. Bowers, *Anal. Chem.*, **46**, 155 (1974).
180. N. F. Ives and L. Guiffrida, *J. Assoc. Off. Anal. Chem.*, **53**, 973 (1970).
181. E. Weston, B. B. Wheals and M. J. Kensett, *Analyst*, **96**, 601 (1970).
182. G. Schwedt and H. A. Ruessel, *Chromatographia*, **5**, 242 (1972).
183. E. H. Daughtrey, A. W. Fitchett and P. Mushak, *Anal. Chim. Acta*, **79**, 199 (1975).
184. F. T. Henry and T. M. Thorpe, *J. Chromatogr.*, **166**, 577 (1978).
185. S. Hanamura, B. W. Smith and J. D. Winefordner, *Anal. Chem.* **55**, 2026 (1983).
186. Y. Talmi and V. E. Norvell, *Anal. Chem.*, **47**, 1510 (1975).
187. J. D. Defreese and H. V. Malmstadt, Paper No. 26 presented before The Federation of Analytical Chemistry and Spectroscopic Societies, Indianapolis, Ind., November, 1975.
188. M. Iguchi, N. Nishiyama and Y. Nagese, *J. Pharm. Soc. Jpn.*, **86**, 1408 (1960).
189. P. R. Gifford and S. Bruckenstein, *Anal. Chem.*, **52**, 1028 (1980).
190. A. D. Zorin, G. G. Devyatykh, V. Ya. Dudorov and A. M. Amel'cheuko, *Zh. Anorg. Khim.*, **9**, 2525 (1964).
191. M. Covello, G. Ciampa and E. Ciamillo, *Farmaco, Ed. Prat.*, **22**, 218 (1967).
192. L. Chelmu, *Chim. Anal.*, **2**, 212 (1972).
193. A. D. Molodyk, G. V. Bondar and L. N. Morozova, *Zavod Lab.*, **38**, 129 (1972).
194. M. Yamamoto, *Soil Sci. Soc. Am. Proc.*, **39**, 859 (1975).
195. R. A. Stockton and K. J. Irgolic, *Int. J. Environ. Anal. Chem.*, (1979).
196. E. A. Dietz and M. E. Perez, *Anal. Chem.*, **48**, 1088 (1976).
197. R. K. Elton and W. E. Geiger, Jr., *Anal. Chem.*, **50**, 712 (1978).
198. P. F. Reay and C. J. Asher, *Anal. Biochem.*, **78**, 557 (1977).
199. D. G. Iverson, M. A. Anderson, T. R. Holm and R. R. Stanforth, *Environ. Sci. Technol.* **13**, 1491 (1979).
200. G. E. Pacey and J. A. Ford, *Talanta*, **28**, 935 (1981).

201. L. R. Overby, S. F. Bocchieri and R. C. Frederickson, *J. Assoc. Off. Anal. Chem.*, **48**, 17 (1965).
202. E. A. Woolson and N. Ahronson, *J. Assoc. Off. Anal. Chem.*, **63**, 523 (1980).
203. R. Iadevaia, N. Ahronson and E. A. Woolson, *J. Assoc. Off. Anal. Chem.*, **63**, 742 (1980).
204. D. L. Johnson and M. E. Q. Pilson, *Anal. Chim. Acta*, **58**, 289 (1972).
205. F. T. Henry and T. M. Thorpe, *Anal. Chem.*, **52**, 80 (1980).
206. G. R. Ricci, L. S. Shepard, G. Colovos and N. H. Hester, *Anal. Chem.*, **53**, 610 (1981).
207. M. Morita, T. Uchiro and K. Fuwa, *Anal. Chem.*, **53**, 1806 (1981).
208. A. A. Grabinski, *Anal. Chem.*, **53**, 966 (1981).
209. S. Hanamura, B. W. Smith and J. D. Winefordner, *Anal. Chem.*, **55**, 2026 (1983).
210. J. Aggett and R. Kadwani, *Analyst (London)*, **108**, 1495 (1983).
211. H. Abe, K. Anma, K. Ishikawa and K. Akasaki, *Bunseki Kagaku*, **29**, 44 (1980).
212. J. M. Vobetsky, V. D. Nefedov and E. N. Sinotova, *Zh. Ubshch. Khim.*, **33**, 4023 (1963).
213. K. Berei and L. Vasana, *Magy. Kem. Foly.*, **73**, 313 (1967).
214. K. Berei, *J. Chromatogr.*, **20**, 406 (1965).
215. D. W. Von Endt, P. C. Kearney and D. D. Kaufmann, *J. Agric. Food Chem.*, **16**, 17 (1968).
216. V. Mikelukova, *J. Chromatogr.*, **34**, 284 (1968).
217. R. M. Sach, F. B. Anastasia and W. A. Walls, *Proc. Northeast. Weed Control Conf.*, **24**, 316 (1970).
218. L. C. Mitchell, *J. Assoc. Off. Agric. Chem.*, **42**, 684 (1959).
219. J. Jenik, *Chem. Listy*, **55**, 509 (1961).
220. Y. Kinoshita, *J. Pharm. Soc. Jpn.*, **78**, 315 (1958).
221. H. Bieling and K. H. Thiele, *Z. Anal. Chem.*, **145**, 105 (1955).
222. I. L. Marr, J. Anwar and B. B. Sithole, *Analyst*, **107**, 1212 (1982).
223. M. D. Andreae, J. F. Asmode and P. Foster, *Anal. Chem.*, **53**, 1766 (1981).
224. T. A. Besada and L. F. Ibrahim, *Analyst (London)*, **112**, 549 (1987).
225. P. Longi and R. Mazzocchi, *Chim. Ind. (Milan)*, **48**, 718 (1966).
226. J. M. Vobetsky, V. D. Nefedov and E. N. Sinotova, *Zh. Obshch. Khim.*, **33**, 4023 (1963).
227. R. Schmitz, *Dtsch. Apoth-Ztg.*, **100**, 693 (1960).
228. M. M. Auerbach and W. W. Haughtaling, *Drug Stand.*, **28**, 115 (1960).
229. A. Z. Cartiglioni, *Anal. Chem.*, **161**, 40 (1958).
230. G. Bohr, H. Bieling and K. H. Thiele, *Z. Anal Chem.*, **145**, 105 (1955).
231. A. J. Krynitsky, *Anal. Chem.*, **59**, 1884 (1987).
232. Perkin Eluer Ltd., *Atomic Spectroscopy*, No. 9, April (1986).
233. L. Ebdon, S. Hill, A. P. Walton and R. W. Ward, *Analyst (London)*, **113**, 1159 (1988).
234. D. Beauchemin, M. E. Bednas, S. S. Berman, J. W. Mcharen, K. W. M. Siu and R. E. Sturgeon, *Anal. Chem.*, **60**, 2209 (1988).

CHAPTER **6**

Mass spectra of the organometallic compounds of As, Sb, and Bi

YURII S. NEKRASOV and DMITRI V. ZAGOREVSKII*

Institute of Organo-Element Compounds, Russian Academy of Sciences, Moscow, Russia, CIS, 117334

I. INTRODUCTION

The aim of this work was to discuss the monomolecular ion decomposition reactions of transition metal complexes containing As, Sb or Bi. It is useful, however, to start with a short preview on the general tendencies in mass spectral behaviour of the metal-free derivatives of these VA Group elements.

* Present address: Chemistry Department, University of Ottawa, Ottawa, Ontario, Canada K1N 6N5.

The chemistry of organic arsenic, antimony and bismuth compounds
Edited by S. Patai © 1994 John Wiley & Sons Ltd

The organic derivatives of P, As, Sb and Bi have been the subject of a large number of mass spectrometric investigations. References for these works can be found in the periodical *Mass Spectrometry*[1-10]. The mass spectral characteristics of these compounds, the most studied of which are the phosphorus derivatives, have also been reviewed[11-14].

The main factor affecting the fragmentation of both the tri- and the pentavalent derivatives of VA Group elements is the strength of the E—C or E–heteroatom bonds. The following trends in the mass spectral fragmentation have been observed:

(1) The molecular ion ($M^{+\cdot}$) peak abundance decrease in the order $E = P > As > Sb > Bi$. This was exemplified by the Ph_3E compounds[15]. Among them, Ph_3P displays the most abundant molecular ion (42% of the total ion current). The corresponding values for Ph_3As and Ph_3Sb are 25.5 and 14.9% respectively, whereas the mass spectrum of Ph_3Bi contains only a trace amount of the molecular ion.

(2) An increase in the size of E results in an increase in the intensity of the peaks corresponding to low-mass ions. For example, the molecular ion is the most abundant in the mass spectrum of Ph_3P. The mass spectra of Ph_3As and Ph_3Sb are dominated, rather, by PhE^+ ions, and the $E^{+\cdot}$ ion is the major peak for Ph_3Sb[16].

(3) Fragmentations involving rearrangements are more prevalent in the derivatives of the elements from the beginning of the group. Some examples illustrating this tendency will follow.

The decomposition of the molecular ion of ethyl phosphinate, $(EtO)_3PO$, involves a number of rearrangements, including single (C_2H_4 and C_2H_4O loss) and double (C_2H_3 elimination) H-atom migration[17-19]. Arsenic derivatives, unlike their phosphorus analogues, decompose mainly by simple bond cleavage, resulting in the elimination of the H·, Me·, Et· and EtO· radicals. The weakness of the As—O bond compared with the P—O bond leads to the unique fragment ions, $(M - OEt - O)^+$ and $As(OH)_2{}^+$, due to cleavage of the As—O bond[20,21].

Another example is hydrogen elimination from Ph_2E^+, leading to ions, **I**, containing a new C—C bonds. As mentioned above, the abundance of this rearrangement ion decreases as the size of E increases. Hence, the mass spectrum of Ph_3Bi does not contain a peak corresponding to this ion[16].

The mass spectrum of Ph_3PS also displays some rearrangement ions. One of them corresponds to hydrogen loss from the molecular ion with the formation of the onium ion, **II**[16]. This ion is the most abundant observed in the mass spectrum. The As and Sb containing analogues have not been observed[16].

(I)   (II)

The E—C and E–heteroatom bond strengths are not the only factors governing the decomposition of ions with similar structures. The ionization energy of E is also important. For example, the decomposition of Ph_2E^+ leads to formation of $C_{12}H_{10}{}^{+\cdot}$ or E^+, depending on the ionization energy of E. When E = Sb, the organic ions are formed since the ionization energy of biphenyl (8.27 eV) is lower than that of Sb (Table 1). The ionization energy of Bi is lower than that of Ph_2 and hence Bi^+ ion is the exclusive product of Ph_2Bi^+ fragmentation[15].

The behaviour of most transition metal containing gas-phase ions can easily be understood in terms of charge localization on the central metal atom[22]. This is displayed by the preference of so-called 'organometallic'-type fragmentation, which includes the metal–ligand bond cleavage and some specific, metal controlled ('catalysed') transformations in the ligands.

Fragmentation pathways of the organometallic derivatives of As, Sb and Bi can be conveniently separated into two classes: the 'organometallic', mentioned above, and the 'organoelement', which includes processes characteristic to the organoelement part of the molecule, namely, ER_n. In addition, the presence of various heteroelements, M and E, in the molecule can lead to new processes which are unusual for both the 'simple' organometallic compounds and the VA Group element organic derivatives. The competition between these processes is controlled by the nature of the transition and main group elements, and by the ligands at M and substituents at E.

Ionization energies of transition metals (e.g. 6.76 eV for Cr, 7.43 eV for Mn, 7.88 eV for Fe[23]) are usually lower than those of As and Sb and comparable with that of Bi (Table 1). For this reason, according to the Stevenson's rule[24], the positive charge is preferably localized on the transition metal atom, and 'organometallic'-type fragmentation might be predicted. According to the above data an increase in the abundance of the 'organoelement'-type processes can be expected in going down the group from As to Bi. In general, fragmentation will depend on a number of electronic structural factors and, unfortunately, cannot easily be predicted. This is illustrated by the absence of regularity in the ionization potentials of the inorganic and organic derivatives of P, As, Sb and Bi listed in Table 1. The values change over a wide range depending on the nature of the substituents. This absence of regularity does, however, introduce the possibility of changing the properties of the transition metal derivatives of Group 15 elements by varying their composition.

TABLE 1. Ionization potentials (in eV) of ERR'_2[a]

R	R'	E = P	As	Sb	Bi
—	—	10.486	9.7883 ± 0.002	8.641	7.298
H	H	9.869 ± 0.002	9.89	9.54 ± 0.03	(10.1)
F	F	11.44	12.84 ± 0.05	(12.1)	
Cl	Cl	9.91	10.55 ± 0.025	10.1 ± 0.1	10.4
Br	Br	(9.7)	(10.0)		
Me	Me	8.06 ± 0.05	(8.2)	(7.7)	
Et	Et	8.15 ± 0.11	7.9[b]		
Pr^n	Pr^n		7.8[b]		
Bu^n	Bu^n	(7.5)	8.3[b]		
Me	H	9.12 ± 0.07	8.5		
H	Me	8.47 ± 0.07	8.1		
OMe	OMe	8.50	7.93		
Ph	Ph	7.39 ± 0.03	7.32 ± 0.05	7.26 ± 0.05	7.45 ± 0.05
Et_2As	Et		7.4[b]		
$Pr_2{}^nAs$	Pr^n		7.0[b]		
$Bu_2{}^nAs$	Bu^n		8.2[b]		
H_3Si	H		10.1 ± 0.1[c]		
—	C_5H_5[d]	9.2[e]	8.8[e]	8.3[e]	7.9[f]

[a] All values were taken from Reference 23 unless otherwise stated. Estimated values are in parentheses.
[b] G. M. Bogolyubov, N. N. Grishin and A. A. Petrov, Zh. Obshch. Khim., **41**, 1710 (1971).
[c] F. E. Saafield and M. V. McDowell, Inorg. Chem., **6**, 96 (1967).
[d] Elementabenzenes.
[e] Vertical Ip's C. Batich, E. Heilbronner, V. Hornung, A. J. Ashe, III, D. T. Clark, U. T. Cobley, D. Kilcast and I. Scanlan, J. Am. Chem. Soc., **95**, 928 (1973).
[f] Vertical Ip. J. Bustide, E. Heilbronner and J. P. Mayer, Tetrahedron Lett., 411 (1976).

The organometallic derivatives of As, Sb, and Bi, studied by mass spectrometry, are listed in the tables in the corresponding sections of this chapter. References in these Tables marked by an asterisk contain a discussion of the mass spectral data. The others have reported only selected ion peaks. For convenience, all references containing information about the mass spectra of the organometallic derivatives of As, Sb and Bi[25-92] are listed in alphabetic order of the names of the first authors. In addition to the common abbreviations and designations we have used the following:

E   main group element
L   ligand at the transition metal atom
R   radical at the main group element
Cp′   methylcyclopentadienyl
Cp*   pentamethylcyclopentadienyl

All mass spectra, unless otherwise stated, were produced by electron impact ionization.

II. MASS SPECTRA OF THE ORGANOMETALLIC DERIVATIVES OF As, Sb AND Bi

A. Complexes of the Type $(LM)_nER_{3-n}$

The mass spectra of 16 such complexes have been reported (Table 2).
The molecular ions of $CpM(CO)_3EMe_2$ (M = Mo, W) complexes are quite abundant. They exhibit four main decomposition pathways, typified by the decomposition of $CpW(CO)_3SbMe_2$, 3 (Scheme 1)[69,70].

TABLE 2. Organometallic derivatives of type LM—ERR′

No.	E	LM	R	R′	Reference
1	As	$CpMo(CO)_3$	Me	Me	72
2	As	$CpW(CO)_3$	Me	Me	72
3	Sb	$CpW(CO)_3$	Me	Me	69
4	Sb	$CpFe(CO)_2$	$CpFe(CO)_2$	$CpFe(CO)_2$	28
5	Sb	$CpFe(CO)_2$	$CpFe(CO)_2$	Cl	28
6	Sb	$CpW(CO)_3$	$CpW(CO)_3$	$CpW(CO)_3$	28
7	Sb	$CpW(CO)_3$	$CpW(CO)_3$	Cl	28
8	Sb	$CpMo(CO)_3$	$CpMo(CO)_3$	$CpMo(CO)_3$	28
9	Sb	$CpMo(CO)_3$	$CpMo(CO)_3$	Cl	28
10	Sb	$CpCr(CO)_3$	$CpCr(CO)_3$	Me	70
11	Sb	$CpMo(CO)_3$	$CpMo(CO)_3$	Me	70
12	Sb	$CpW(CO)_3$	$CpW(CO)_3$	Me	70
13	Sb	$CpCr(CO)_3$	$CpMo(CO)_3$	Me	70
14	As	$CpW(CO)_3$	Bu^t	NSO	56
15	As	$LFe(CO)_2{}^a$	Br	Br	61
16	Sb	$LFe(CO)_2{}^a$	Br	Br	61

^{a}

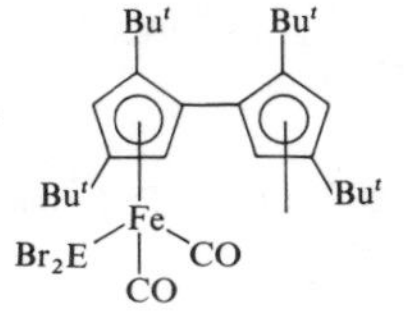

According to the classification of the organometallic ion decomposition reactions given earlier, 'organometallic'-type dissociations in Scheme 1 include CO and $SbMe_2$ loss. These simple metal ligand bond dissociations lead to $CpW(CO)_nSbMe_2^{+\cdot}$ and $CpW(CO)_n^+$ ($n = 1-3$). Examples of the 'organoelement'-type dissociations include cleavage of the M—E bond giving $SbMe_2^+$ ions as well as the consecutive elimination of methyl groups or transition metal containing species from $CpW(CO)_nSbMe_2^{+\cdot}$. The latter gives rise to $CpW(CO)_nSbMe_m^+$ ($m = 0, 1$) ions.

In complexes of the type $CpM(CO)_nEMe_2$, $SbMe_2$ loss occurs directly from the molecular ion. This differs from what is observed for the other one-electron ligand containing complexes $CpM(CO)_nL$, where L = alkyl, halogen, phenyl etc., in which L is lost from $CpML^+$, the ion formed by the loss of all carbonyl ligands. This suggests that the Sb—M bond is significantly weaker than the M—Me, M—Ph and M—Hal bonds. Fragmentation of the ER_2 ligand in these complexes usually proceeds after the loss of all carbonyl groups. Complex 1 is the exception to this 'rule', methyl group elimination is observed from the molecular ions, and hence competes with the loss of the weakly bound CO ligands.

In addition to the simple bond cleavage reactions described above, fragmentation of $CpM(CO)_3EMe_2$ (M = Mo, W) complexes also involves some rearrangement. For example, a rearrangement process observed for As-containing complexes of this type (E = As) is the migration of the cyclopentadienyl ligand from the transition metal atom, M, to the As atom with the formation of $CpAsMe_2^{+\cdot}$ and $CpAsMe^+$ ions[72]. Similar processes, leading to cyclopentadienyl nido-clusters, CpE^+, have been observed in the mass spectra of IV main group element organometallic compounds, $CpM(CO)_nER_3$ (E = C, Si, Ge, Sn, Pb; M = Mo, W, $n = 3$; M = Fe, $n = 2$)[93].

Another rearrangement observed for $CpM(CO)_3AsMe_2$ (M = Mo, W) is methane elimination from $(M - nCO)^+$ ions leading to abundant $CpM(CO)_nAsCH_2^+$ [or $CH_3C_5H_4M(CO)_nAs^+$]. Note that methane molecule loss is also characteristic of $PhAsMe_2$ under electron impact[94].

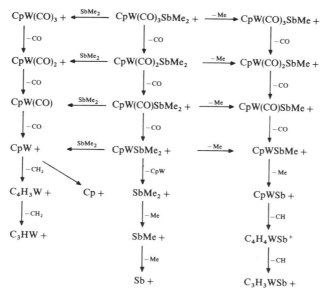

SCHEME 1

The predominant dissociation pathways of the bimetallic complex $[CpW(CO)_3]_2SbMe$ (**12**) are similar to those for $CpW(CO)_3SbMe_2$ (**3**). Fragmentation of **4** consists of the elimination of 6 CO groups, whereas in **12**, the most abundant reactions involve W—Sb bond cleavage forming $CpW(CO)_3SbMe^+$ and $CpW(CO)_3^+$. Methyl group loss is observed only from $(M - nCO)^+$, where $n = 2$–6. $(M - 2CO)^{+\cdot}$ also loses SbMe, giving rise to the metal–metal bond containing product ion, $[Cp(CO)_2W—W(CO)_2Cp]^{+70}$.

The mass spectrum of the heterobimetallic complex $CpCr(CO)_3SbMeMo(CO)_3Cp$ exhibits a series of peaks corresponding to $L_nMSbMe_m^+$ ($m = 0,1$) ions, which contain only one transition metal atom. The chromium containing fragment ions are more abundant than their molybdenum analogues reflecting the higher Cr—Sb compared with Mo—Sb bond strength[70]. The comparison of relative abundances of fragment ions containing the M_3Sb unit, in mass spectra of the trimetallic complexes $[CpM(CO)_3]_3Sb$ (**6** M = W; **8** M = Mo) shows that the W—Sb bond is stronger than Mo—Sb bond[28]. The high abundance of $Cp_3W_3(CO)_6Sb^{+\cdot}$ ions in the mass spectrum of **6** allowed the authors[28] to propose the possibility of the synthesis of the corresponding neutral compound, an analogue for the corresponding As containing complex, which has been synthesized before.

Elimination of $Bu^{t\cdot}$ and $NSO^\cdot$ from the complex $CpW(CO)_3As(Bu^t)NSO$ (**14**) occurs only after the loss of all carbonyl ligands and also involves hydrogen rearrangement leading to elimination of neutral C_4H_8 and HNSO molecules[56].

A competitive elimination of carbonyl ligands and $AsBr_2$ occurs from **15**. Bromine atom loss is also observed but only after the elimination of 4 CO groups, i.e. from $(M - 4CO)^{+\cdot}$ ions. The mass spectrum of this compound exhibits $(M - 4CO - 2As - Br)^+$ ions resulting from the migration of three Br atoms from As to Fe followed by the loss of an As atom and a Br atom[61].

B. Arsilidene (Dimetallaarsacumulene) Complexes, LM---AsR---ML

Mass spectra of arsilidene complexes, $[Cp'Mn(CO)_2]_2AsR$, **17**–**24** (Table 3), contain molecular ion peaks. The most common fragment ions correspond to elimination of 2CO, R and $Cp'Mn(CO)_2$. The abundances of these ions depend strongly on the R substituent at the As atom (Table 4). Substituents R such as F, I, OCS, NCS and N_3 are easily eliminated directly from the molecular ion. Loss of Ph, however, occurs only after elimination of all carbonyl ligands, and R = H, R = c-Hex are not lost at all. Only compounds with R = Ph, H and c-Hex have mass spectra which display peaks corresponding to $(M - 2CO)^{+\cdot}$ and $(M - 4CO)^{+\cdot}$. The formation of $(M - R - nCO)^+$ ($n = 2,4$) ions is characteristic of complexes containing R = I, OCS, NCS and N_3. The complex having R = F is intermedi-

TABLE 3. Arsilidene complexes, LM---AsR---ML

No.	E	LM	R	Reference
17	As	$Cp'Mn(CO)_2$	Ph	53
18	As	$Cp'Mn(CO)_2$	c-Hex	53
19	As	$Cp'Mn(CO)_2$	H	87
20	As	$Cp'Mn(CO)_2$	I	87
21	As	$Cp'Mn(CO)_2$	OSC	87
22	As	$Cp'Mn(CO)_2$	NSC	87
23	As	$Cp'Mn(CO)_2$	N_3	87
24	As	$Cp'Mn(CO)_2$	F	86
25	As	$W(CO)_5$	Cp*	60
26	As	$Cr(CO)_5$	Cp*	60

TABLE 4. The relative abundances of ions in the mass spectra of some arsilidene complexes $Cp'Mn(CO)\cdots AsR\cdots Mn(CO)_2Cp'$

R	P^+	$(P-R)^+$	$(P-2CO)^+$	$(P-R-2CO)^+$	$(P-4CO)^+$	$(P-R-4CO)^+$	$Cp'Mn(CO)As^+$	$Cp'Mn^+$	Mn^+	Reference
H	22	—	30	—	100	—	13	[a]	[a]	87
I	21	97	—	12	—	18	100	[a]	[a]	87
OCS	2	43	—	20	—	42	100	[a]	[a]	87
NCS	9	12	—	6	—	14	100	[a]	[a]	87
N_3	0.3	0.3	—	0.1	—	0.2	0.5	100	[a]	87
F	0.2	1.8	1.8	1.4	—	2.4	36	[a]	81	86
Ph	4	—	6	—	16	8	[a]	100	60	53
c-Hex	6	—	7	—	48	—	[a]	100	56	53

[a] Abundances of ions have not been reported.

ate in that its mass spectrum contains peaks due to all of the ions mentioned above, except $(M - 4CO)^{+\cdot}$.

Molecular ion peaks were not found in the mass spectra of $[(CO)_5M]_2AsCp^*$, **25** and **26**. The peak with highest m/z value corresponds to the $(M - Cp^*)^+$ ion, which further eliminates consecutively 10 CO molecules producing ions of composition $AsM_2(CO)_n^+$ ($n = 0-9$). The base peak in their mass spectra is due to the $C_5Me_5^+$ ion[60].

C. Complexes of the Type LM—ER₃

Complexes of the form $(CO)_5MEPh_3$ (Table 5) are characterized by sufficiently high stability under electron impact ionization to form molecular ions of intermediate abundance[30]. Their decomposition involves consecutive decarbonylation forming $M(CO)_nEPh_3^{+\cdot}$ ions ($n = 0-4$). Cleavage of the M—E bond is observed only in $Ph_3EM^{+\cdot}$ with both $M^{+\cdot}$ and $EPh_3^{+\cdot}$ ions being produced.

TABLE 5. Organometallic derivatives of the type LM—ERR'R"

No.	E	R	R'	R"	ML	Reference
27	As	Ph	Ph	Ph	$W(CO)_5$	5*
28	Sb	Ph	Ph	Ph	$W(CO)_5$	5*, 7
29	As	Ph	Ph	Ph	$Mo(CO)_5$	5*
30	Sb	Ph	Ph	Ph	$Mo(CO)_5$	5*
31	Sb	Ph	Ph	Ph	$Cr(CO)_5$	7
32	As	Ph	Ph	Ph	$Cr(CO)_5$	7
33	As	Ph	Ph	Ph	$Mn(CO)(CS)Cp'$	40*
34	Sb	Ph	Ph	Ph	$Mn(CO)(CS)Cp'$	40*
35	As	Ph	Ph	Ph	$Mn(CO)_2Cp'$	40*
36	Sb	Ph	Ph	Ph	$Mn(CO)_2Cp'$	40*
37	As	Ph	Ph	Ph	$Mn(CO)_2C_4H_4N^a$	64*
38	Sb	Ph	Ph	Ph	$Mn(CO)_2C_4H_4N^a$	64*
39	As	Ph	Ph	Ph	$Mn(CO)_2C_9H_7{}^b$	64*
40	As	Ph	Ph	Ph	$Mn(CO)_2C_9H_7{}^b$	64*
41	As	Ph	Ph	Ph	$Fe(CO)_2C_8H_6{}^c$	39*
42	Sb	Ph	Ph	Ph	$Fe(CO)_2C_8H_6{}^c$	39*
43	As	H	H	H	$Cr(CO)_5$	50
44	As	H	H	H	$Mo(CO)_5$	50
45	As	H	H	H	$W(CO)_5$	50
46	Sb	H	H	H	$Cr(CO)_5$	50, 51*
47	Sb	H	H	H	$Mo(CO)_5$	51*
48	Sb	H	H	H	$W(CO)_5$	51*
49	As	H	H	H	$Fe(CO)_4$	50
50	As	H	H	H	$Mn(CO)_2Cp$	49
51	As	H	H	H	$Mn(CO)_2Cp$	76
52	As	Me	Me	Me	$Cr(CO)_3(=CMe)Br$	52
53	Bi	Me	Me	Me	$Fe^+(CO)_2Cp \cdot BF_4{}^-$	81
54	As	NMe_2	NMe_2	NMe_2	$Cr(CO)_5$	65*
55	As	NMe_2	NMe_2	NMe_2	$Mo(CO)_5$	65*
56	As	NMe_2	NMe_2	NMe_2	$W(CO)_5$	65
57	As	NMe_2	NMe_2	NMe_2	$Fe(CO)_4$	65*
58	As	NMe_2	NMe_2	NMe_2	$Cr(CO)_4As(NMe_2)_3$	65*
59	Sb	$SiMe_3$	$SiMe_3$	$SiMe_3$	$Cr(CO)_5$	31
60	Sb	$SnMe_3$	$SnMe_3$	$SnMe_3$	$Cr(CO)_5$	31
61	Sb	$SiMe_3$	$SiMe_3$	$SiMe_3$	$Mo(CO)_5$	31

TABLE 5. (continued)

No.	E	R	R'	R''	ML	Reference
62	Sb	SnMe₃	SnMe₃	SnMe₃	Mo(CO)₅	31
63	Sb	SiMe₃	SiMe₃	SiMe₃	W(CO)₅	31
64	Sb	SnMe₃	SnMe₃	SnMe₃	W(CO)₅	31
65	As	Me	Me	NMe₂	Cr(CO)₅	66
66	As	Me	Me	NMe₂	Mo(CO)₅	66
67	As	Me	Me	NMe₂	W(CO)₅	66
68	As	Me	Me	Cl	Cr(CO)₅	66
69	As	Me	Me	OEt	Cr(CO)₅	66
70	As	Me	Me	SEt	Cr(CO)₅	66
71	As	Me	Me	Cl	Mo(CO)₅	66
72	As	Me	Me	OEt	Mo(CO)₅	66
73	As	Me	Me	SEt	Mo(CO)₅	66
74	As	Me	Me	Cl	W(CO)₅	66
75	As	Me	Me	OEt	W(CO)₅	66
76	As	Me	Me	SEt	W(CO)₅	66
77	As	Me	Me	OEt	Fe(CO)₄	66
78	As	Me	Me	SEt	Fe(CO)₄	66
79	As	Me	Me	OH	Cr(CO)₅	66
80	As	Me	Me	OH	Mo(CO)₅	66
81	As	Me	Me	OH	W(CO)₅	66
82	As	Me	NMe₂	NMe₂	Cr(CO)₅	66
83	As	Me	NMe₂	NMe₂	Mo(CO)₅	66
84	As	Me	NMe₂	NMe₂	W(CO)₅	66
85	As	Me	NMe₂	NMe₂	Fe(CO)₄	66
86	As	Me	OEt	OEt	Cr(CO)₅	66
87	As	Me	SEt	SEt	Cr(CO)₅	66
88	As	Me	OEt	OEt	Mo(CO)₅	66
89	As	Me	OEt	OEt	W(CO)₅	66
90	As	Me	SEt	SEt	W(CO)₅	66
91	As	Me	OEt	OEt	Fe(CO)₄	66
92	As	Mn(CO)₅	Me	Me	Fe(CO)₅	41
93	Sb	Mo(CO)₃Cp	Me	Me	Ni(CO)₃	69
94	Sb	W(CO)₃Cp	W(CO)₃Cp	Me	Cr(CO)₅	71
95	Sb	W(CO)₃Cp	W(CO)₃Cp	Me	Mn(CO)₂Cp	71
96	As	Mn(CO)₅	Me	Me	Cr(CO)₄Lᵈ	43
97	As	ᵉ	Me	Me	W(CO)₅	91
98	As	AsH(Ph)Rᶠ	H	Ph	Cr(CO)₅	59
99	As	AsMe₂Rᵍ	Me	Me	Fe(CO)₄	37
100	As	Me	ʰ	ʰ	Fe(CO)₄	37
101	Sb	W(CO)₃Cp	W(CO)₃Cp	Me	Ni(CO)₃	77
102	Sb	W(CO)₃Cp	Me	Me	Ni(CO)₃	77

ᵃ C₄H₄N—pyrrolyl.
ᵇ C₉H₇—indenyl.

ᶜ C₈H₆ =

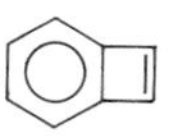

ᵈ L = AsMe₂Mn(CO)₅.
ᵉ R = N(Me)AsMe₂W(CO)₅.
ᶠ R = Cr(CO)₅.
ᵍ R = o-C₆H₄AsMe₂Fe(CO)₄.

ʰ R' + R'' =

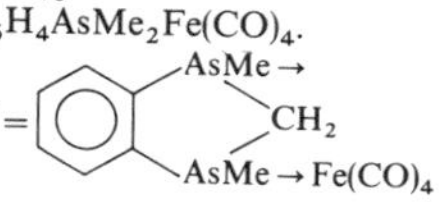

The fragmentation of $Cp'Mn(CO)_2EPh_3$ (E = P, As **35**, Sb **36**) is virtually identical to that observed for the above complexes. It starts with the loss of two CO groups and only then eliminates the EPh_3 ligand. The following order in Mn—L bond strengths was deduced from the ionization and appearance energy measurements for $Cp'Mn(CO)_2EPh_3$ and $Cp'Mn(CO)(CS)EPh_3$: L = CO < CS $\ll$ SbPh$_3$ < AsPh$_3$ < PPh$_3$ < Cp'[40]. The Mn—E bond dissociation energies in $Cp'MnEPh_3^+$ ions are 7.43 ± 0.1, 6.29 ± 0.04 and 6.13 ± 0.05 eV for E = P, As and Sb, respectively. The substitution of one CO group by the σ-donating EPh_3 ligand in the $CP'Mn(CO)_3$ molecule leads to an increase in the Mn—CO bond strength and to a decrease in the ionization energies of the complexes in the following order[40]:

$$Cp'Mn(CO)_3 > CP'Mn(CO)_2PPh_3 > CP'Mn(CO)_2AsPh_3 > Cp'Mn(CO)_2SbPh_3$$

Ip (eV)   7.86   6.55   6.38   6.37

Phenyl radical loss from the complexes **36–40**, $LMn(CO)_2EPh_3$ (L = Cp', pyrrolyl, indenyl)[40], is observed only from $MnEPh_3^+$ ions, i.e. after all other ligands are lost. This process can be accompanied by the loss of a hydrogen atom leading to $C_{10}H_8EMn^+$, which contains a 9-elementafluorenyl species coordinated to the metal atom. Elimination of C_6H_6 (or $C_6H_5^{\cdot} + H^{\cdot}$) has also been observed for ionized EPh_3 molecules[15,16]. Fragmentation of the metal containing ions $MnEPh_3^+$ also includes migration of the Ph group to the metal atom, producing $MnPh^+$ ions. All of these processes are also observed for $FeEPh_3^+$ ions originating from $C_8H_8Fe(CO)_2EPh_3$ (**41** E = As, **42** E = Sb)[39].

Arsine and stibine containing metal carbonyl complexes, $M(CO)_5EH_3$ (**43–48**; M = Cr, Mo, W), decompose by three competitive routes, CO, H and EH_3 loss with the formation of $M(CO)_nEH_m^+$ and $M(CO)_n^{+\cdot}$ ions ($n = 0$–4; $m = 0$–3). The ratio of the total abundances of ions from these two species, ($[M(CO)_nEH_m^+]/[M(CO)_n^{+\cdot}]$), increases going from Cr to W, and from As to Sb, reflecting a change in M—E bond strengths in the order Sb > As, and W > Mo > Cr[50,51].

The fragmentation of $CpMn(CO)_2AsH_3$ (**50**) consists of the consecutive elimination of a CO and an AsH_3 ligand leading to $CpMn(CO)_nAsH_3^{+\cdot}$ and $CpMn(CO)_n^{+\cdot}$ ions. The elimination of a hydrogen atoms proceeds only after the loss of all carbonyl groups, $CpMnAsH_m^+$ ions ($m = 0$–2) being produced. The decarbonylated ion, $CpMnAsH_3^{+\cdot}$, also loses the hydrocarbon ligand, forming $MnAsH_3^+$. The latter eliminates from one to three hydrogen atoms. Hydrogen atom migration from As to Mn in $MnAsH_3^+$ leads to the manganese hydride cation, MnH^+[49].

The ionization energy of $CpMn(CO)_2AsH_3$ (7.16 ± 0.1 eV[49]) is substantially lower than that of $CpMn(CO)_3$ (8.06 eV[23]) and close to the ionization energies of $CpMn(CO)_2L$ complexes, which contain a cycloolefin ligand, L (7.0–7.3 eV[23]). Thus, the donor–acceptor properties of AsH_3 as a ligand are similar to that of olefines[49].

Substitution of the hydrogen atoms in the arsine ligand by fluorine atoms results both in a substantial increase in ionization energy [12.8 eV for the $CpMn(CO)_2AsF_3$[76]] and in the weakening of the Mn—AsF_3 bond strength relative to the Mn—AsH_3 bond. This is supported by the observation of a higher abundance of $CpMn(CO)_n^{+\cdot}$ ions (7.8% for $n = 2$ and 18% for $n = 1$) compared with $(M - CO)^{+\cdot}$ (1.7%) and $(M - 2CO)^{+\cdot}$ (5.7%) ions[76].

Unlike $CpMn(CO)_2AsH_3$, for which dehydrogenation occurs only after the loss of all CO ligands, $CpMn(CO)_2AsF_3$ is characterized by the successful competition of F atom loss with carbonyl elimination even in the molecular ions. This shows that the As—H bond is stronger than the As—F bond. Similar to their arsine containing analogues, the dissociation of the Mn—As bond in the fluoroarsine complex can proceed via migration of an F atom to Mn giving rise to MnF^+ ions. The abundance of MnF^+ in the mass spectrum of $CpMn(CO)_2AsF_3$ (5.4%) is higher than of MnH^+ in the mass spectrum of $CpMn(CO)_2AsH_3$ (2.6%)[49].

Molecular ions of tris(dimethylamino)arsine containing complexes, $(Me_2N)_3$-$AsM(CO)_m$ (**54–57**; $n = 5$, M = Cr, Mo, W; $n = 4$, M = Fe), decompose by two major routes: (1) with consecutive decarbonylation leading to $(Me_2N)_3AsM(CO)_n^{+\cdot}$ ions ($n = 0$–4), and (2) by Me_2N loss with further carbonyl group elimination giving rise to $(Me_2N)_2AsM(CO)_n^+$ ions ($n = 0$–5). The comparison of $Z = [M - Me_2N]^+/\{[M - Me_2N]^+ + [M - CO]^{+\cdot}\}$ values, which characterizes the relative E—N and M—CO bond strengths in the molecular ions of tris(dimethylamino)phosphine and tris(dimethylamino)arsine complexes, shows that this ratio is always higher for the arsines. Thus, the As—N bond is less stable than the P—N bond. The mass spectrum of the tungsten containing complex, **56**, exhibits peaks due to ions in the series $(Me_2N)_mAsW(CO)_n^+$ ($n = 0$–5, $m = 1, 2$; $n = 0$–2, $m = 0$) which correspond to the loss of two and three Me_2N groups, whereas only one Me_2N radical is eliminated for the chromium containing analogue[65]. This indicates that the W—CO bond is stronger than the Cr—CO bond.

The As—N bond dissociation in $(Me_2N)_3AsM^{+\cdot}$ (M = Cr, Mo, W) includes a hydrogen rearrangement resulting in $MeN=CH_2$ loss and $(Me_2N)_2AsMH^+$ formation. This ion (M = Cr) decomposes by the migration of the Me_2N group from As to Cr giving rise to the Me_2NAsH^+ cation.

The fragmentation of $[(Me_2N)_3As]_2Cr(CO)_4$, **58**, leads to $(Me_2N)_3AsCr^{+\cdot}$, which itself undergoes a unique migration of three Me_2N groups from As to Cr forming $(Me_2N)_3Cr^+$[65].

Another example of complex rearrangement is the decomposition of $(Me_2N)_2AsFeH^+$, originating from the molecular ion of **57**, which eliminates neutral arsine, AsH_3, forming $(MeNCH_2)_2Fe^+$[65].

Thus, the presence of three types of heteroatoms (N, As and transition metal atom) in molecules of the type $(Me_2N)_3AsM(CO)_5$ (M = Cr, Mo, W) results in specific rearrangements of their ions in the gas phase.

The fragmentation of $(Me_3E)_3SbM(CO)_5$ (**59–64**; M = Cr, Wo, E = Si, Sn) includes the consecutive loss of carbonyl ligands forming $(Me_3E)_3SbM(CO)_n^{+\cdot}$ ($n = 0$–4). Elimination of Me_3Si groups also occurs after the loss of two, three and five carbonyl ligands for W, Mo and Cr containing complexes, respectively. Unlike the tris(dimethylamino)arsine complexes described above, the molecular ions of the tris(trimethylsilyl) derivatives of Mo and W are characterized by Me group(s) elimination leading to $(M - nMe)^+$ ions ($n = 1, 3, 5$). The consecutive demethylation also occurs for carbonyl free $(Me_3E)_2SbM^{+\cdot}$ ions (M = Mo, W; E = Si, Sn), forming ions of the series $MSbE_2Me_n^+$ ($n = 0$–5 for M = W and Cr, E = Sn; $n = 3$–5 for M = Mo, E = Si). Dissociation of the M—Sb bond leads to $(Me_3E)_3Sb^{+\cdot}$. The most abundant ion in the mass spectra of these compounds is Me_3E^+[31].

The mass spectra of complexes of the type $(CO)_nMAsMeRR'$ (**65–91**; M = Fe, Cr, Mo, W; R, R' = Me, Me_2N, EtO, EtS) have been reported without discussion[66].

The mass spectra of As and Sb derivatives of the type $LM—ER_2—ML'$, where ER_2 is a one-electron ligand with respect to one of the metal atoms and a two-electron ligand for the other, have been reported. For example, $(CO)_5MnAsMe_2[Fe(CO)_5]$ (**92**) has both σ-Mn—As and donor–acceptor Fe—As bonds. The mass spectrum of this complex exhibits peaks due to ions with three and two heteroatoms $[(CO)_nFeMnAsMe_2^{+\cdot}$, $n = 0$–9; $FeMnAsMe_m^{+\cdot}$, $FeAsMe_m^{+\cdot}$ and $MnAsMe_m^+$, $m = 0$–2; $FeMn^{+\cdot}$, $FeAs^+$, $MnAs^+]$, as well as monoheteroatomic ions $(Fe^+, Mn^{+\cdot}, As^+)$.

Both decarbonylation and methyl groups loss are characteristic of the next bimetallic complex, $Cp(CO)_3MoSbMe_2Ni(CO)_3$ (**93**). Its mass spectrum shows peaks due to $Cp(CO)_nMoSbMe_m^+$ ($n = 0$–9, $m = 1, 2$). Sb–N bond cleavage leads to ions of two series, $CpMo(CO)_nSbMe_m^+$ ($n = 0$–3, $m = 1, 2$) and $Ni(CO)_n^{+\cdot}$ ($n = 0$–3). Elimination of species containing two metal atoms gives rise to $SbMe_m^+$ ($m = 0$–2)[69].

The fragmentation of $[CpW(CO)_3]_2Sb(Me)Cr(CO)_5$ (**94**) and $[CpW(CO)_3]_2Sb(Me)Mn(CO)_2Cp$ (**95**) is similar to that of **93**. The mass spectra of these compounds exhibit ions derived from competitive carbonyl and methyl loss. These are $Cp_2(CO)_nW_2CrSbMe^+$ ($n = 0-11$) and $Cp_2(CO)_nW_2CrSb^+$ ($n = 0-8$) for **94**, and $Cp_3(CO)_nW_2MnSbMe^+$ ($n = 0-8$) and $Cp_2(CO)_nW_2MnSb^+$ ($n = 0-4$) for **95**[71].

An important series of fragment ions for the trimetallic complex $(CO)_5MnAs(Me)_2$-$Cr(CO)_4As(Me)_2Mn(CO)_5$ (**96**) corresponds to the elimination of carbonyl ligands. Methyl loss occurs only from the fully decarbonylated ion, $MnAs(Me)_2CrAs(Me)_2Mn^+$, and leads to $MnAsMeCrAsMeMn^+$ and $Mn_2CrAs_2^+$. Other fragments, containing from one to four heteroatomic elements, such as $Cr(AsMe_2)_2^+$, $MnCrAs_2^+$, $MnCrAsMe_n^+$ ($n = 0-2$), $MnAsMe_n^+$ ($n = 0-2$), $CrAsMe_n^+$ ($n = 0-2$), $AsMe_n^+$ ($n = 0-2$), $Mn^{+\cdot}$ and $Cr^{+\cdot}$, are also present in the mass spectrum of **96**.

Dissociation of $(CO)_5WAs(Me)_2N(Me)As(Me)_2W(CO)_5$ (**97**) under electron impact proceeds by carbonyl ligand loss and As—W bond dissociation leading to two principal series of ions—$(CO)_nWAs_2NMe_5^+$ and $W(CO)_n^{+\cdot}$[91].

Molecular ions of $(CO)_5CrAs(H)PhAs(H)PhCr(CO)_5$ (**98**) decompose by consecutive elimination of 10 CO groups and two chromium atoms, forming abundant $Ph_2As_2CH_2^+$ ions; the latter dissociates to $PhAs_2^+$ and $PhAs^{+\cdot}$[59].

The dissociation pathway particular to the bimetallic complexes **99** and **100** is methane elimination from the decarbonylated ions[37].

D. Complexes of the Type

$$X \underset{E'R_2}{\overset{ER_2}{\diagup\diagdown}} ML$$

In this class of organometallic complexes (Table 6), two ER_2 groups comprise a four-electron donor ligand. The fragmentation of o-phenylenebisdimethylarsine (das) carbonyl compounds of VIB group metals, $(das)M(CO)_4$ (**103–105**; M = Cr, Mo, W), and consists of the competitive elimination of CO and Me groups, forming $(M - nCO - mMe)^+$ ($n = 0-3$; $m = 0-2$) ions. The fragmentation pathways of free and M coordinated o-phenylenebisdimethylarsines are considerably different. Consecutive methyl radical elimination is characteristic of the $C_6H_4(AsMe_2)_2^{+\cdot}$ molecular ion, whereas As—Me bond dissociation in $(das)M^+$ proceeds via hydrogen rearrangement resulting in methane elimination[33,74].

The molecular ions of $(das)Fe(NO)_2$ (**106**) and $(das)Co(CO)NO$ (**107**) decompose by the elimination of CO, NO and methyl groups[74]. As in the previous case $(das)M^+$ ions lose CH_4.

The strongest bond in complex **110** is the Pt—As bond. Hence its major decomposition pathways are associated with Pt—Cl, Pt—Cp and As—Et bond cleavage. Fragmentation includes the consecutive elimination of either Cp, Et_2, Cl and Et_2 or Cl, Et_2, Et_2 and Cp, both producing $PtAs_2C_6H_4^{+\cdot}$[36].

Peaks due to the elimination of five neutral species with mass 28 amu are present in the mass spectra of $(Ph_2AsCH_2CH_2AsPPh_2)M(CO)_4$ (**111–113**; M = Cr, Mo, W) suggesting that all carbonyl ligands as well as the bridging ethylene group are lost. Phenyl loss is also observed, proceeding only after the loss of all CO ligands and of the C_2H_4 bridge[33].

The fragmentation of 1,2-bis(dimethylarsino)ethylene (edas) complexes, $(edas)M(CO)_4$ (**114–117**; M = Cr, Mo, W, $n = 4$; M = Ni, $n = 2$) can be rationalized by competitive decarbonylation and demethylation leading to ions of the series $(M - nCO-mMe)^+$ ($n = 0-4$, $m = 0,1$ for M = Cr; $n = 0-4$, $m = 0$ for M = Mo; $n = 1,2$, $m = 1$ and $n = 0-4$, $m = 0$ for M = W; $n = 1,2$, $m = 0$ and $n = 1$, $m = 2$ for M = Ni). Ions $(M - 3CO)^{+\cdot}$ and $(M - 4CO)^{+\cdot}$ in the mass spectra of **115** (M = Mo) and **116** (M = W) eliminate methane.

Carbonyl ligand loss from 1,2-bis(dimethylarsino)-3,3',4,4'-tetrafluorobutane (**118**) is accompanied by F atom(s) elimination, forming ions of the series $(M - nCO)^{+\cdot}$ ($n = 1-5$,

TABLE 6. Compounds of the type

$$X \diagup \begin{matrix} ER_2 \\ E'R_2 \end{matrix} \diagdown ML$$

No.	E	E′	R	X	ML	Reference
103	As	As	Me	—C_6H_4—	$Cr(CO)_4$	33*
104	As	As	Me	—C_6H_4—	$Mo(CO)_4$	33*, 74
105	As	As	Me	—C_6H_4—	$W(CO)_4$	33*
106	As	As	Me	—C_6H_4—	$Fe(NO)_2$	74
107	As	As	Me	—C_6H_4—	$Co(CO)(NO)$	74
108	As	As	Me	—C_6H_4—	$Ni(CO)_2$	74
109	As	As	Me	—C_6H_4—	$Fe(CO)_3$	37
110	As	As	Et	—C_6H_4—	$Pt(Cl)Cp$	36
111	As	As	Ph	—C_2H_4—	$Cr(CO)_4$	33*
112	As	As	Ph	—C_2H_4—	$Mo(CO)_4$	33*
113	As	As	Ph	—C_2H_4—	$W(CO)_4$	33*
114	As	As	Me	—C_2H_2—	$Cr(CO)_4$	74
115	As	As	Me	—C_2H_2—	$Mo(CO)_4$	74
116	As	As	Me	—C_2H_2—	$W(CO)_4$	74
117	As	As	Me	—C_2H_2—	$Ni(CO)_2$	74
118	As	As	Me	—C_4F_4—[a]	$Fe_3(CO)_{12}$	38
119	As	P	Me	—$C_6H_4SiMe_2$—	$Cr(CO)_4$	27
120	As	P	Me	—$C_6H_4SiMe_2$—	$Mo(CO)_4$	27
121	As	P	Me	—$C_6H_4SiMe_2$—	$W(CO)_4$	27
122	P	As	Me	—$C_6H_4SiMe_2$—	$Cr(CO)_4$	27
123	P	As	Me	—$C_6H_4SiMe_2$—	$Mo(CO)_4$	27
124	P	As	Me	—$C_6H_4SiMe_2$—	$W(CO)_4$	27
125	As	As	Me	—$C_6H_4SiMe_2$—	$Cr(CO)_4$	27
126	As	As	Me	—$C_6H_4SiMe_2$—	$Mo(CO)_4$	27
127	As	As	Me	—$C_6H_4SiMe_2$—	$W(CO)_4$	27
128	N	As	Me	—$C_6H_4SiMe_2$—	$Cr(CO)_4$	27
129	N	As	Me	—$C_6H_4SiMe_2$—	$Mo(CO)_4$	27
130	N	As	Me	—$C_6H_4SiMe_2$—	$W(CO)_4$	27
131	Sb	Sb	Me	$CpMo(CO)_3$	$CpMo(CO)_3$	77
132	Sb	Sb	Me	$CpW(CO)_3$	$CpW(CO)_3$	77
133	As	As	Me	$CpW(CO)_3$	$CpW(CO)_3$	77
134	As	As	Ph	$Fe(CO)_3$	$Fe(CO)_3$	45

[a]
$$\begin{matrix} F_2 \\ F_2 \end{matrix} \diagup \begin{matrix} Me_2 \\ As \rightarrow Fe(CO)_2 \\ OC \quad CO \quad Fe(CO)_4 \\ As \rightarrow Fe(CO)_2 \\ Me_2 \end{matrix}$$

10) and $(M - nCO - mF)^+$ ($n = 1–10$, $m = 1, 2$). Migration of fluorine atoms to the iron atom is observed after the loss of all carbonyl ligands and leads to the elimination of neutral FeF_2 [38].

The mass spectra of compounds 119–130 ($X = —C_6H_4SiMe_2—$) exhibit molecular ion peaks as well as peaks due to $(M - nCO)^{+\cdot}$ ($n = 1–4$)[27].

Some information is available concerning the bimetallic complexes, 131–134, which contain bridging ER_2 groups [45,77]. The mass spectrum of 133 [$LM = CpW(CO)_3$, $E = As$] exhibits peaks due to $Cp_2W_2As_2Me_m(CO)_n^+$ ($n = 0–4$, $m = 3, 4$) ions corresponding to all possible combinations of CO and methyl loss. Cleavage of the As—W bond results in the formation of $CpW(CO)_3{}^+$. The decarbonylated ions decompose by consecutive elimination of all Me groups to form $Cp_2W_2As_2{}^{+\cdot}$, and by loss of $AsMe_2{}^{77}$.

E. Complexes with Bridging ER_n Groups

The compounds of this type and their structures are listed in Tables 7 and 8. In the bimetallic compounds **135–138**, the ER_n group is σ-bonded to the Fe atom and involved in a donor–acceptor bond with the other metal atom, M. The molecular ions of these compounds decompose by the loss of their CO ligands. Methyl radical elimination also occurs, but only after all CO groups are lost, giving rise to $CpMAsMe^+$ ions for compounds **135** and **136** and $FeAsM^+$ ions for compounds **137** and **138**. A second dissociation route for $LMFeAsMe_2{}^+$ is iron atom loss leading to $CpMAsMe_2{}^+$ in the mass spectra of **135** and **136** and $MAsMe_2{}^+$ in the mass spectra of **137** and **138**. These latter ions fragment by the loss of two methyl radicals. $CpMFeAsMe_2{}^+$ can also eliminate the metal atom M. The tri-, di- and monoatomic fragment ions, $MFeAs^+$ (M = W, Mn, Co), MAs^+ (M = Mn, Co), $FeAs^+$, MFe^+ (M = Mo, Co), $Fe^{+\cdot}$, $M^{+\cdot}$ and As^+, which have no carbon containing groups, are present in high abundances in the mass spectra of these compounds[42].

The mass spectra of bimetallic complexes with a bridging $AsMe_2$ group exhibit abundant molecular ions. Their fragmentation starts with carbonyl ligands loss. The other dissociation routes depend on the nature of the ligand L[55]. The abundant peaks in the mass spectra of **143** and **144** [L = SCF_3 and $P(CF_3)_2$, respectively] are due to ions of the type $Mn_2F(CO)_nAsMe_2{}^{+\cdot}$ ($n = 0, 1, 3, 4$), which result from F atom migration to Mn[55].

The bridged complex $Me_2AsRe(CO)_4S_2$ (**145**, X = S_2) undergoes consecutive loss of four CO ligands. Methyl elimination occurs from $(M - 2CO)^{+\cdot}$ and $(M - 3CO)^{+\cdot}$ Ions $(M - nCO)^{+\cdot}$ ($n = 2$–4) also decompose by cleavage of the As—Re bond to form $S_2Re(CO)_n{}^+$ ($n = 2$–4) and $AsMe_2{}^+$. Double methyl radical loss from the decarbonylated ion, $S_2ReAsMe_2{}^+$, leads to S_2ReAs^{+67}.

The molecular ions of **146** (X = $-PPh_2CH_2-$) lose four CO ligands and the bridging X group to yield Cr^+ ions. The decarbonylated ion, $(M - 4CO)^{+\cdot}$, can eliminate its Cr atom thereby forming the metal-free $PPh_2CH_2{}^+$ organic ion. The mass spectrum of this compound also shows a peak due to $PPh_3{}^{+\cdot}$ formed by Ph migration from As to P[92].

TABLE 7. Complex of the type $(CO)_4M \xleftarrow{} AsR_2$ with bridging X

No.	R	M	—X—	Reference
135	Me	Fe	$CpMo(CO)_2$	42*
136	Me	Fe	$CpW(CO)_2$	42*
137	Me	Fe	$Mn(CO)_4$	42*
138	Me	Fe	$Co(CO)_3$	42*
139	Me	Mn	$-H-Mn(CO)_4-$	55
140	Me	Mn	$-Fe-Mn(CO)_4-$	55
141	Me	Mn	$-I-Mn(CO)_4-$	55
142	Me	Mn	$-SMe-Mn(CO)_4-$	55
143	Me	Mn	$-S(CF_3)-Mn(CO)_4-$	55
144	Me	Mn	$-P(CF_3)_2-Mn(CO)_4-$	55
145	Me	Re	$-S_2-$	67
146	Ph	Cr	$-PPh_2CH_2-$	92
147	Ph	Cr	$-CH_2P(Ph)_2CH_2-$	92

TABLE 8. Complexes with bridging ER_n groups

No.	Compound	Reference
148		35
149		84
150		57
151 152	E = As E = Sb	58 58
153		54
154 155	R = Me R = Ph	82 82

Decarbonylation of the molecular ions of 147 ($X = -CH_2PPh_2CH_2-$) is accompanied by methylene loss leading to $(M - nCO - CH_2)^{+\cdot}$ ($n = 0-4$)[92].

The mass spectrum of the chelate complex 148, with bridging $AsPh_2$ group, exhibits three abundant peaks corresponding to $M^{+\cdot}$, $(M - 2CO)^{+\cdot}$ and $Cr^{+\cdot}$. The bisarene structure has been proposed for the $(M - 2CO)^{+\cdot}$ ion.

(148) $M^{+\cdot}$

The prominant ions in the mass spectrum of 149 are due to CO loss, $(M - nCO)^{+\cdot}$ ($n = 2-12$). Also present are peaks due to the inorganic ions $Fe_2As_2^{+\cdot}$, $Fe_2^{+\cdot}$ and As^+. The organic fragment Bu^{t+} is very abundant, whereas others, such as $MeAs^{+\cdot}$ and Bu^tS^+, are minor components.

The molecular ions of the bimetallic complex $[Cr(CO)_5]_2AsPh$ (150) lose consecutively 10 CO groups and the Cr atom leading to $CrAsPh^{+}$[57]. Its trimetallic analogue 151 also loses consecutively all carbonyl ligands and two iron atoms with the formation of

$(CO)_n CrFe_2 EPh^+$ ($n = 1$–13), $CrFeEPh^+$ and $CrEPh^+$ ions. The mass spectrum of the antimony containing complex, **152**, exhibits peaks due to the inorganic ion $CrFe_2 Sb^{+58}$.

The mass spectrum of $[CpCr(CO)_2]_2 As_2$ (**153**), containing an As_2 bridging group, displays peaks due to $Cp_2 Cr_2(CO)_n As_2^{+\cdot}$ ($n = 0$, 2–4), $Cp_2 Cr_2(CO)_n As^+$ ($n = 0, 1$), $CpCr_2 As_2^+$, $CpCrAs_2^+$, $CpCrAs^{+\cdot}$, $Cp_2 Cr^{+\cdot}$ and $CpCr^+$ produced by the loss of CO, Cp and As. Chromocenium ions, $Cp_2 Cr^{+\cdot}$, are formed by the migration of the cyclopenta-dienyl ligand from one Cr atom to the other [28].

The molecular ions of the arsaacetylene complexes **154** and **155** lose consecutively 6 carbon monoxide molecules, a Co atom and the $RC{\equiv}As$ ligand forming $RAsCo_2(CO)_n^+$ (R = Me, Ph), $AsCo^{+\cdot}$ (R = Ph) and $Co^{+\cdot 82}$.

F. Organometallic Derivatives of Arsenic-containing Heterocycles

Arsenic-containing heterocyclic derivatives are multifunctional ligands. For example, arsole can form several types of complexes: (a) one involving the lone electron pair on As (**156**), (b) as a two-electron donor to one metal atom and, at the same time, as a one-electron donor to the other transition element (**157**), (c) as a four-electron ligand of the π-butadiene type due to the presence of two double $C{=}C$ bonds (**158**), and (d) as a five-electron elementacyclopentadienyl ligand (**159–162**). Each of these types of com-pounds have been characterized by mass spectrometry (Tables 9 and 10). Unfortunately a detailed discussion of their mass spectral behaviour is unavailable.

The mass spectrum of **156** exhibits peaks due to the series $(M - nCO)^{+\cdot}$ ($n = 0$–4). The base peak in the mass spectrum is due to the metal-free arsole ion[89].

The molecular ion of the bimetallic complex **157** is unstable, but peaks due to $Cp_2 Fe_2 C_4 H_2 Me_2 As^+$ and $CpFeC_4 H_2 Me_2 As^+$ are present to confirm the assigned struc-ture of this compound[88].

Elimination of 7CO groups and two iron atoms from the molecular ions of **158** yields metal-free $PhAsC_4 H_2 Me_2^+$, which is the most abundant ion in the mass spectrum[89].

Complex **159** is unstable under mass spectrometric conditions. In addition to the molecular ion and ions corresponding to carbonyl ligands loss, its mass spectrum exhibits peaks due to the dimer ion, $[LMn(CO)_4]_2$, where L = tetraphenylarsole[25].

The volatility of compound **160** is very low, but its chemical ionization mass spectrum exhibits a molecular ion peak[63].

Information concerning the mass spectra of the ferrocene-like complexes **161** and **162** is limited and focused on the presence of molecular ion peaks[88].

The fragmentation characteristics of VI group transition metal complexes containing arsabenzene derivatives as a π-ligand depend on both the nature of R and the central metal atom[73]. The common decomposition pathways of their molecular ions, like benzenetri-carbonyl chromium derivatives, are carbonyl ligand elimination with the formation of $(M - nCO)^{+\cdot}$ ($n = 1$–3). The $(M - 3CO)^{+\cdot}$ ions of the cyclohexyl substituted complexes **166–168** undergo the loss of up to three H_2 molecules forming a diphenylarsabenzene ligand coordinated to the Cr ion (see below). Elimination of $2H_2$, in the case of Mo containing complex, occurs from $(M - 2CO)^{+\cdot}$. The same process occurs in the W analogue, **168**, proceeding even from $(M - CO)^{+\cdot}$.

The $(M - 3CO)^{+\cdot}$ ions of **163–165**, which contains a t-butyl group as a substituent in the heterocyclic ligand, decompose with the elimination of either $CH_3^{\cdot}$ or CH_4. The resulting ions, $(M - 3CO - Me)^+$ and $(M - 3CO - CH_4)^{+\cdot}$, undergo further dehydrogenation[73].

TABLE 9. Organometallic derivatives of arsenic-containing heterocycles

No.	Compound	Reference

156

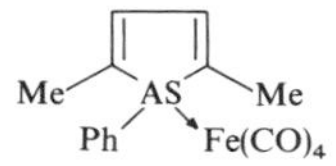

89

157

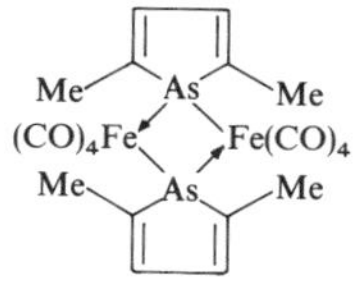

88

158

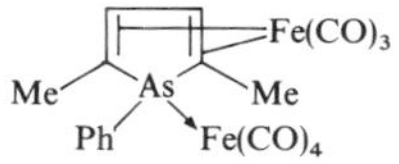

89

	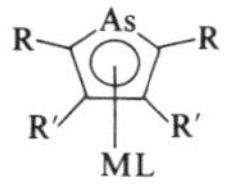 R	R	ML	
159	Ph	Ph	$Mn(CO)_3$	25
160	H	Me	$Mn(CO)_2PPh_3$	63
161	H	Me	FeCp	88
162	H	Me	$(C_4H_2Me_2As)Fe$	88

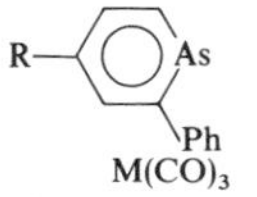

	M	R	
163	Cr	Bu^t	73
164	Mo	Bu^t	73
165	W	Bu^t	73
166	Cr	c-Hex	73
167	Mo	c-Hex	73
168	W	c-Hex	73

$(CO)_5Cr \longleftarrow PhAs \longrightarrow AsPb \longrightarrow Cr(CO)_5$
                              $\diagdown X \diagup$

169   X =   59

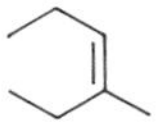

170   X =   59

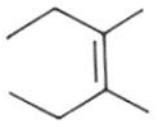

171   X =   59

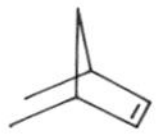

172   X =   59

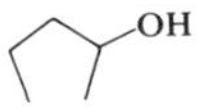

173   X =   59

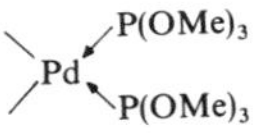

(continued)

TABLE 9. (*continued*)

No.	Compound	Reference
174		59
175	$n = 1$	44
176	$n = 2$	44
177	$n = 3$	44
178		79
179		80
180		78
181		80
182		80
183		80
184	$CpCoB_9H_9Sb_2$	68
185	$CpCoB_9H_9AsSb$	69

TABLE 10. Organometallic derivatives of cyclic arsines

No.	Compound	Reference

$$\begin{array}{ccc} R & & M(CO)_5 \\ & \diagdown As \diagup & \\ R-As & & As-R \\ & As-As \rightarrow M(CO)_5 & \\ R \diagup & & \diagdown R \end{array}$$

	M	R	
186	Cr	Me	47
187	Mo	Me	47
188	W	Me	47
189	Cr	Ph	59

190

$$\begin{array}{ccc} & Ph & \\ & | & \\ & As & \\ Ph-As & \diagup \diagdown & As-Ph \\ & As-As \rightarrow Cr(CO)_5 & \\ Ph \diagup & & \diagdown Ph \end{array}$$

59

191

$$\begin{array}{ccc} Ph & & Ph \\ \diagdown & & \diagup \\ As & - & As \\ | & & | \\ As & - & As \rightarrow Cr(CO)_5 \\ Ph \diagup & & \diagdown Ph \end{array}$$

59

Mass spectra of the bimetallic diphenylarsene derivatives, **169–172**, have been reported[59]. They yield both molecular and $(M - nCO)^{+\cdot}$ ($n = 5, 8, 9, 10$) ions of low abundance. The dominant peaks correspond to the metal-free ions $Ph_2As_2^{+\cdot}$, $PhAs_2^+$ and $PhAs^{+\cdot}$, the latter being the base peak.

The mass spectra of **173** and **174** do not exhibit either molecular or other palladium containing ions. The chromium containing ions, $Cr_2As_2Ph_2^+$ and $CrAs_2Ph^+$, are present, but the major peaks correspond to the metal-free ion $Ph_2As_2^{+\cdot}$ and its dissociation products, As_2Ph^+ and $AsPh^{+\cdot}$[59].

The most abundant dissociation product of the mono-, bi- and trimetallic chromium carbonyl derivatives, **175–177**, and base peak in their mass spectra, is the metal-free arsaadamantane ligand ion, $C(CH_2AsO)_3^+$ (Scheme 2)[44].

Mass spectra of the pentaarsacyclopentadienyl analogue of ferrocene, **178**, and the corresponding triple-decker 'sandwich' compound **179** have been reported[79]. The mass spectrum of **178** exhibits a molecular ion and fragment ions formed by the loss of two or four As atoms. No molecular ion peak was found for **179**, but the presence of $(M - CpFe)^+$, $C_5Me_5FeAs_3^+$, $CpFe^+$ and As_n^+ was in agreement with the prescribed structure of this compound.

The triple-decker dimolybdenium complex, **180**, unlike the Fe containing analogue, **179**, displays a molecular ion as a prominent peak in the mass spectrum[78].

Mass spectra of complexes **181–183**, containing As_nCo_m cluster units, exhibit abundant molecular ions[80]. Consecutive loss of Cp* ligands from $M^{+\cdot}$ of **182** leads to ligand-free $Co_3As_6^+$ ions. Arsenic atom elimination from the inorganic skeleton gives rise to $Cp^*Co_2As_2^+$ and $Cp^*CoAs_2^+$ among others.

Mass spectrometry was useful in determining the molecular masses of the stibaborane complexes $B_9H_9AsSbCoCp$, **185**, and $B_9H_9Sb_2CoCp$, **184**[68].

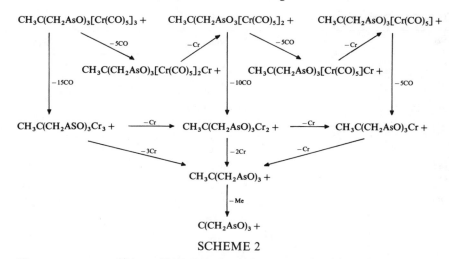

SCHEME 2

The mass spectra of bimetallic carbonyl metal compounds with cyclic arsine ligands have been discussed[47,59] (Table 10). Molecular ion peaks are present for the pentamethyl-cyclopentaarsine containing complexes of chromium and tungsten, $(AsMe)_5[M(CO)]_2$ (**186, 188**). Their decomposition includes CO and/or $M(CO)_5$ loss giving rise to the ions $(AsMe)_5M_2(CO)_n^{+\cdot}$ $(n = 0-9)$, $(AsMe)_5M(CO)_n^+$ $(n = 0-5)$, $M_2As_mMe^+$ and MAs_mMe^+ $(m = 2-4)$, $M_2As_2Me_2^+$, $M_2AsMe_m^+$ $(m = 2-4)$, MAs_m^+ $(m = 2-5)$, MAs_3CH^+ and $AsMe_2^+$. The latter is the most abundant in the mass spectra[47].

Fragmentation of $(AsPh)_5[Cr(CO)_5]_2$ (**189**) leads to a small yield of $(AsPh)_5Cr_2^+$, $(AsPh)_5Cr^+$ and $(AsPh)_5^{+\cdot}$ ions. The latter ion decomposes giving rise to $(AsPh)_m^+$ $(m = 1-4)$ and $PhAs_2^+$, of which $PhAs_2^+$ is the most abundant. All of the above ions, except $(AsPh)_5Cr_2^+$, are also formed upon ionization and fragmentation of the mono-metallic complex $(AsPh)_5Cr(CO)_5$[59].

The tetraphenylcyclotetraarsine compound, $(AsPh)_4Cr(CO)_5$ (**191**), produces a mass spectrum which exhibits a molecular ion, the decarbonylated ions, $(AsPh)_4Cr(CO)_n^+$ $(n = 0-4)$, and the metal-free ligand ion $As_4Ph_4^+$. Further fragmentation of this ligand ion yields $(AsPh)_n^+$ $(n = 1-3)$, of which $AsPh^{+\cdot}$ is predominant[59].

G. Complexes of the Type $(LM)_n(AsR)_m$

Information concerning these compounds is limited (Table 11).

The predominant fragment ions displayed by compounds **192** and **193** are $(C_6R_5As)_2$-$Fe(CO)_n^{+\cdot}$ $(R = H, F; n = 0-4)$, $As_2Fe^{+\cdot}$ and $Fe^{+\cdot}$, the latter corresponding to the consecutive elimination of four CO, two C_6R_5 groups and two arsenic atoms[46,62].

Ionization of the diphenyldiarsine palladium complex, **194**, yields a molecular ion, a ligand-free $Pd^{+\cdot}$ ion, the metal-free $As_2Ph_2^{+\cdot}$ ion and fragmentation products of the latter[59].

The mass spectra of $(AsR)_4Fe_2(CO)_6$ (**195**, R = Me; **196**, R = Et), $(AsMe)_5Mn_2(CO)_8$ (**197**) and $(AsMe)_8Mn_2(CO)_6$ (**198**), containing the metal atom in the arsenic cycles, exhibit molecular ion peaks, peaks due to ions produced by their decarbonylation and ions of composition $Mn_nAs_mMe_k^+$ $(n = 1, 2, m = 0-4, k = 0-4$ for **195**; $n = 2, m = 4, k = 0-4$ for **196**; $n = 1, 2, m = 1-5, k = 0-4$ for **197**; $n = 1, 2, m = 3-8, k = 0-6$ for **194**). The most abundant ion in the mass spectra of **197** and **198** is $MnMe^+$, originating from the migration of a methyl group from As to the metal atom[45].

TABLE 11. Complexes of the type $(LM)_n(AsR)_m$

No.	Compound	Reference

$$C_6R_5As \overset{\diagdown}{\underset{ML}{\diagup}} AsC_6R_5$$

No.		Reference
192	$ML = Fe(CO)_4$ $R = H$	62
193	$ML = Fe(CO)_4$ $R = F$	46
194	$ML = Pd(PPh_3)_2$ $R = H$	59

| **195** | $R = Me$ | 45 |
| **196** | $R = Et$ | 45 |

| **197** | | 45 |

| **198** | | 45 |

199	$M = Cr$	47
200	$M = Mo$	47
201	$M = W$	47

| **202** | $R = Et$ | 47 |
| **203** | $R = Pr$ | 47 |

Neither molecular nor $(M - nCO)^{+\cdot}$ ions are present in the mass spectra of $[(AsMe)_5M(CO)_4]_2$ (**199**, M = Cr; **200**, M = Mo; **201**, M = W). As well, $(AsMe)_5^{+\cdot}$ and its decomposition products are absent. The peaks of highest m/z value in the mass spectrum of **199** correspond to $Cr_2As_9Me_9^{+\cdot}$; other ions of composition $Cr_2As_nMe_m^{+}$ are present[47].

The mass spectrum of $[(CO)_3Mo(AsEt)_4]_2$ (**202**) exhibits peaks assigned to the ions $Mo_2As_8Et_n^{+}$ ($n = 0–3$), $Mo_2As_nEt^{+}$ ($n = 2, 6$), $Mo_2As_n^{+}$ ($n = 2, 5–8$), $MoAs_4^{+}$ and $AsEt^{+}$. In addition to $MoAs_nPr_m^{+}$, the mass spectrum of its isopropyl analogue, **203**, shows a molecular ion[47].

H. Ionic Complexes

The electron impact mass spectrum of the cluster $\{[CpFe(CO)_2]_3SbCl\}_2[FeCl_4]\cdot CH_2Cl_2$ (**204**) has been recorded at $200\,°C^{90}$. It exhibits neither a molecular ion nor ions containing the Fe_3Sb group. At the same time, however, ions containing Fe_2Sb $[ClSbFe_2(CO)_nCp^{+}]$ and FeSb $[ClSbFeCp^{+}$ and $SbFe(CO)_nCp^{+}$, where $n = 2–0]$ have been identified.

Several non-volatile ionic complexes of As and Sb have been characterized by fast atom bombardment mass spectrometry (Table 12). The mass spectrum of **53** (a mixture of 3-nitrobenzyl alcohol with CH_2Cl_2 was used as the matrix) exhibits $(M + BF_2)^{+}$, $M^{+\cdot}$, $(M - 2CO)^{+\cdot}$, $CpFe(CO)_2^{+}$, $Cp_2Fe^{+\cdot}$ and Bi^{+} ions[81].

Fragmentation of **205–208** (Table 12) under fast atom bombardment (matrices—glycerol and thioglycerol) include the consecutive loss of 2As and X atoms from the inorganic cycle leading to $(M - 2As - X)^{+}$, which further eliminates $Ph^{\cdot}$, $PPh_2^{\cdot}$ or $Ph_2^{\cdot} + PPh_2^{\cdot}$[34].

The formation of Bi_nNa_m ($n, m = 1, 0;\ 2, 3;\ 3, 4;\ 4, 5;\ 5, 5;\ 6, 5;\ 7, 4;\ 8, 5;\ 9, 6;$ and $14, 7$) clusters has been observed by laser ionization mass spectrometry. The intensity distribution for these ions shows a maximum corresponding to their most stable electronic and geometric configurations according to the 'magic numbers' of Zintl[95]. For example, $Bi_3Na_4^{+}$, $Bi_4Na_3^{+}$, $Bi_5Na_4^{+}$ and $Bi_7Na_4^{+}$ ions correspond to Zintl's Bi_3^{3-}, Bi_4^{2-}, Bi_5^{3-} and Bi_7^{3-} ions, respectively.

I. Organometallic Compounds of As, Sb and Bi Without M—E Bonds

Examples of this type of compounds can be found in Table 13 (**209–214**). Fragmentation of the molecular ion of $CpMn(CO)_2CNAsMe_2$ (**209**) starts with the elimination of two

TABLE 12. Ionic clusters

No.	Compound	Method	Reference
204	$\{[CpFe(CO)_2]_3SbCl\}_2[FeCl_4]CH_2Cl_2$	EI	90
205	X = S	FAB[a]	34
206	X = Se	FAB[a]	34
207	X = Te	FAB[a]	34
208	Bi_nNa_m clusters	LI[b]	48*

[a]Fast atom bombardment.
[b]Laser ionization.

TABLE 13. Complexes without M—E bonds

No.	Compound	Reference
209	M = Mn	29
210	M = Re	29
211		83
212		83
213		92
214		58

carbonyl groups giving rise to $(M - 2CO)^{+\cdot}$, which itself forms $CpMn^{+\cdot}$ by loss of $CNAsMe_2$ or $MnCNAsMe_2{}^{+\cdot}$ by loss of Cp. Demethylation can occur from both $(M - 2CO)^{+\cdot}$ and $(M - 2CO - Cp)^{+}$ forming $CpMnCNAsMe^{+}$ and $MnCNAsMe^{+}$. For the isostructural complex $CpRe(CO)_2CNAsMe_2$ (**210**), methyl loss ('organoelement' type of fragmentation) competes with carbonyl ligand loss ('organometallic' type of fragmentation) and leads to $CpRe(CO)_nCNAsMe^{+}$ ions $(n = 2, 1)$. The $(M - Me)^{+}$ ion eliminates the AsMe group giving rise to $CpRe(CO)_2CN^{+\,29}$.

Mass spectrometry has been successfully used for the identification of ferrocenophane, which contains a bridging AsPh group (**211**), and its pentacarbonyltungsten complex, **212**[83]. Molecular ion peaks have been observed for the two compounds.

Complex **213**, like its isomer with Cr—As bond, **147**, decomposes with the elimination of both CH_2 and CO groups leading to $(M - CH_2 - nCO)^{+\cdot}$ ions $(n = 2, 4)$. Elimination of the Cr atom from $(M - CH_2 - 4CO)^{+\cdot}$ yields $Ph_2AsCH_2PPh_2{}^{+\cdot\,92}$.

Decomposition involving methyl and BiMe loss competes with the decarbonylation process in **214**, leading to various metal containing ions, $(M - Me - nCO)^{+}$ $(n = 0-4)$, $(M - 2Me - nCO)^{+\cdot}$ $(n = 2, 3)$, $(M - BiMe - nCO)^{+\cdot}$ $(n = 0-4)$, $(M - Me - BiMe - nCO)^{+}$ $(n = 1-3)$ and $(M - 2Me - BiMe - 4CO)^{+\cdot}$. The ions $Bi_2{}^{+\cdot}$ and $BiMe_n{}^{+}$ $(n = 0-3)$ are abundant, and Bi^{+} ion is the base peak in the mass spectrum[85].

III. CONCLUSION

The data presented above showed that the decomposition of ions containing a transition metal atom and As, Sb or Bi proceeds by various pathways. All of these processes can be separated into two main reaction types, 'organometallic' and 'organoelement'. Reactions of the first type are associated with charge localization on the transition metal atom, whereas the second are characterized by significant participation of the main group element in charge delocalization. The ratio between these two general reaction types as well as the relative abundances of processes within each of them are a complex function of the transition metal, main group element, and ligands and radicals connected to these atoms.

The majority of fragment ions in the mass spectra of the compounds discussed contains the transition metal atom, reflecting the predominant charge localization on the organometallic part of their molecular ions. Both simple bond cleavages and rearrangement reactions have been observed. The most common reaction type is the metal–ligand bond dissociation. Again, the dissociation characteristics of the metal–ligand bonds depend upon the entire ligand arrangement on the central metal atom, but the following tendencies can be noted. Firstly, the ER_3-type ligands are connected to the metal atom significantly stronger than some other two–electron ligands, e.g. carbonyl and thiocarbonyl groups. At the same time, M—ER_3 bonds are weaker than the metal–cyclopentadienyl bond. Secondly, the M—ER_3 bond strengths decrease as the atomic number of E increases.

Localization of the positive charge on the transition metal atom causes some specific rearrangements in the E-containing ligands, such as dehydrogenation of the cyclohexyl substituent yielding new π-bond formation with the metal atom. Another example is methane loss from $1,2\text{-}(Me_2E)_2C_6H_4$ coordinated to the transition metal ion, which is not observed in the ions of the corresponding metal-free derivatives.

A most interesting reaction associated with the positive charge localization on the transition metal atom is radical migration from the main group element to M. The extent of this reaction depends upon the migrating radical. Electronegative groups like F, Br and Me_2N display the highest tendency for migration and the simultaneous migration of two and even three of these groups sometimes occurs. Migration of other radicals is rarer. The migratory aptitude of the phenyl group is only moderate, but still more than that of the H atom.

The abundance of 'organoelement'-type fragmentation reactions relative to 'organometallic' ones has a tendency to increase as the atomic number of the E element increases. Such behaviour can be explained by a more successful competition of ER_n groups with a heavier E atom for the positive charge and/or by decreasing of the E—R bond strength in going from As to Sb and Bi. The first conclusion does not obviously follow from a comparison of Ip values of the metal-free ER_3 molecules (E = Me, Et, Ph; Table 1) and their organometallic derivatives. For example, the substitution of a CO ligand in $Cp'Mn(CO)_3$ by ER_3 leads to a substantial, but almost identical decrease in the Ip values relative to the Ip values of the initial molecules. At the same time, the dissociation characteristics of $L_mMER_n^+$ ions show that the combined influence of the organometalic and organoelement parts increases the abundance of E-containing ions. Formation of these ions, including the metal-free cations, can be associated with charge localization on the E atom.

The second important factor is the E—R bond strength, which intuitively should be weaker in ER_n^+ species with the heavier E atom. For example, Me loss from dimethylarsinometalcarbonyl ions occurs only after the loss of the carbonyl ligands, whereas it competes with CO elimination from the molecular ions of its Sb-containing analogue. Loss of all organic radicals connected to the E atom together with the elimination of ligands from the transition metal atom(s) leads very often to the formation of the ligand-free inorganic ions, $M_m E_n^+$. Their future investigation may provide useful information about the properties of the smallest semiconductor species doped by the metal atom.

IV. REFERENCES

1. M. I. Bruce, in *Mass Spectrometry (Specialist Periodic Report)*, Vol. 1 (Ed. D. H. Williams), Chap. 5, The Chemical Society, London, 1971, pp. 182–252.
2. M. I. Bruce, in *Mass Spectrometry (Specialist Periodic Report)*, Vol. 2 (Ed. D. H. Williams), Chap. 5, The Chemical Society, London, 1973, pp. 193–263.
3. T. R. Spalding, in *Mass Spectrometry (Specialist Periodic Report)*, Vol. 3 (Ed. R. A. W. Johnstone), Chap. 5, The Chemical Society, London, 1975, pp. 143–223.
4. T. R. Spalding, in *Mass Spectrometry (Specialist Periodic Report)*, Vol. 4 (Ed. R. A. W. Johnstone), Chap. 12, The Chemical Society, London, 1977, pp. 268–330.
5. T. R. Spalding, in *Mass Spectrometry (Specialist Periodic Report)*, Vol. 5 (Ed. R. A. W. Johnstone), Chap. 13, The Chemical Society, London, 1979, pp. 312–346.
6. R. G. Cragg, in *Mass Spectrometry (Specialist Periodic Report)*, Vol. 6 (Ed. R. A. W. Johnstone), Chap. 12, The Chemical Society, London, 1981, pp. 294–328.
7. R. G. Cragg, in *Mass Spectrometry (Specialist Periodic Report)*, Vol. 7 (Ed. R. A. W. Johnstone), Chap. 10, The Chemical Society, London, 1983, pp. 388–427.
8. J. Charalambous, in *Mass Spectrometry (Specialist Periodic Report)*, Vol. 8 (Ed. M. E. Rose), Chap. 11, The Chemical Society, London, 1985, pp. 333–360.
9. J. Charalambous, in *Mass Spectrometry (Specialist Periodic Report)*, Vol. 9 (Ed. M. E. Rose), Chap. 11, The Chemical Society, London, 1987, pp. 373–406.
10. J. Charalambous, in *Mass Spectrometry (Specialist Periodic Report)*, Vol. 10 (Ed. M. E. Rose), Chap. 10, The Chemical Society, London, 1989, pp. 323–356.
11. R. G. Gillis and J. L. Occolowitz, in *Analytical Chemistry of Phosphorus Compounds*, (Ed. M. Halmann) Interscience, New York, 1970, p. 295–331.
12. I. Granoth in *Topics in Phosphorus Chemistry*, Vol. 8. (Eds. E. J. Griffits and M. Grayson), Wiley-Interscience, New York, 1976, pp. 41–98.
13. M. R. Litzow and T. R. Spalding, *Mass Spectra of Inorganic and Organometallic Compounds*, Elsevier, Amsterdam, 1973.
14. T. Nishiwaki, *Heterocycles*, **2**, 473 (1974).
15. T. R. Spalding, *Org. Mass Spectrom.*, **11**, 1019 (1976).
16. C. Glidewell, *J. Organomet. Chem.*, **116**, 199 (1976).
17. S. Theppendahl, P. Jacobsen and J. Wieczorkowski, *Acta Chem. Scand.*, **28B**, 657 (1974).
18. G. Pietter and E. M. Gaydon, *Org. Mass Spectrom.*, **10**, 122 (1975).
19. S. Sass and T. L. Fisher, *Org. Mass Spectrom.*, **14**, 257 (1979).
20. P. Frøyen and J. Müller, *Org. Mass Spectrom.*, **7**, 73 (1973).
21. P. Frøyen and J. Müller, *Org. Mass Spectrom.*, **9**, 132 (1974).
22. See, for example, J. Charalambous, *Mass Spectrometry of Metal Compounds*, Butterworths, London, 1975.
23. S. G. Lias, J. E. Bartmess, J. F. Liebman, J. L. Holmes, R. D. Levin and W. G. Mallard, 'Gas Phase Ion and Neutral Thermochemistry', *J. Phys. Chem. Ref. Data*, **17**, Suppl. 1 (1988).
24. D. P. Stevenson, *Discuss. Faraday Soc.*, **10**, 35 (1951).
25. E. W. Abel, *J. Chem. Soc., Chem. Commun.*, 258 (1973).
26. A. J. Ashe III, J. W. Kampf and D. B. Puranik, *Organometallics*, **11**, 2743 (1992).
27. P. Aslanidis and J. Grobe, *J. Organomet. Chem.*, **249**, 103 (1983).
28. A. M. Barr, M. D. Kerlogue, N. C. Norman and P. M. Webster, *Polyhedron*, **8**, 2495 (1989).
29. H. Behrens, G. Landgraf, P. Merbach, M. Moll and K.-H. Trummer, *J. Organomet. Chem.*, **253**, 217, (1983).
30. S. T. Bond and N. V. Duffy, *J. Inorg. Nucl. Chem.*, **35**, 3241 (1973).
31. H. J. Breunig and W. Fichtner, *J. Organomet. Chem.*, **222**, 97 (1981).
32. R. A. Brown and G. R. Dobson, *J. Inorg. Nucl. Chem.*, **33**, 892 (1971).
33. R. A. Brown, J. R. Paxson and G. R. Dobson, *Org. Mass Spectrom.*, **7**, 1059 (1973).
34. G. Cetini, L. Operti, G. A. Vaglio, M. Peruzzini and P. Stoppioni, *Polyhedron*, **6**, 1491 (1987).

35. R. Colton and C. J. Rix, *Aust. J. Chem.*, **24**, 2461 (1971).
36. R. J. Cross and R. Wardle, *J. Chem. Soc. (A)*, 2000 (1971).
37. W. R. Cullen and D. A. Harbourne, *Can. J. Chem.*, **47**, 3371 (1969).
38. W. R. Cullen, D. A. Harbourne, B. V. Liengme and J. R. Sams, *Inorg. Chem.*, **9**, 702 (1971).
39. A. Efraty, M. H. A. Huang and C. A. Weston, *J. Organomet. Chem.*, **91**, 327 (1971).
40. A. Efraty, D. Liebman, M. H. A. Huang and C. A. Weston, *Inorg. Chim. Acta*, **39**, 105 (1980).
41. W. Ehrl and H. Vahrenkamp, *Chem. Ber.*, **106**, 2556 (1973).
42. W. Ehrl and H. Vahrenkamp, *Chem. Ber.*, **106**, 2563 (1973).
43. W. Ehrl, R. Rinck and H. Vahrenkamp, *J. Organomet. Chem.*, **56**, 285 (1973).
44. J. Ellermann, Su Ping Ang, M. Leitz and M. Moll, *J. Organomet. Chem.*, **222**, 105 (1981).
45. P. S. Elmes and B. O. West, *J. Organomet. Chem.*, **32**, 365 (1971).
46. P. S. Elmes, P. Leverett and B. O. West, *J. Chem. Soc., Chem. Commun.*, 747 (1971).
47. P. S. Elmer and B. O. West, *Aust. J. Chem.*, **23**, 2247 (1970).
48. R. W. Farley and A. W. Castleman, Jr., *J. Am. Chem. Soc.*, **111**, 2734 (1989).
49. E. O. Fischer, W. Bathelt, M. Herberhold and J. Müller, *Angew. Chem., Int. Ed. Engl.*, **7**, 634 (1968).
50. E. O. Fisher, W. Bathelt and J. Müller, *Chem. Ber.*, **103**, 1815 (1970).
51. E. O. Fisher, W. Bathelt and J. Müller, *Chem. Ber.*, **104**, 986 (1971).
52. E. O. Fisher and K. Richter, *Chem. Ber.*, **109**, 2547 (1976).
53. L.-R. Frank, K. Evertz, L. Zsolnai and G. Huttner, *J. Organomet. Chem.*, **335**, 179 (1987).
54. L. Y. Goh, R. C. S. Wong, W.-H. Yip and T. C. W. Mak, *Organometallics*, **10**, 875 (1991).
55. J. Grobe and F. Kober, *J. Organomet. Chem.*, **29**, 295 (1971).
56. M. Herberhold, T. Treibner, T. Chivers and S. S. Kumaravel, *Z. Naturforsch.*, **46b**, 169 (1991).
57. G. Huttner and H. G. Schmid, *Angew. Chem., Int. Ed. Engl.*, **14**, 433 (1975).
58. G. Huttner, G. Mohr, P. Friedrich and H. G. Schmid, *J. Organomet. Chem.*, **160**, 59 (1978).
59. I. Jibril, L.-R. Frank, L. Zholnai, K. Evertz and G. Huttner, *J. Organomet. Chem.*, **393**, 213 (1990).
60. P. Jutzi and R. Kross, *J. Organomet. Chem.*, **390**, 317 (1990).
61. P. Jutzi, J. Schnitter, J. Dahlhaus, D. Gestmann and H.C. Leue, *J. Organomet. Chem.*, **415**, 117 (1991).
62. M. Jakob and E. Weiss, *J. Organomet. Chem.*, **153**, 33 (1978).
63. D. L. Kershner and F. Basolo, *J. Am. Chem. Soc.*, **109**, 7396 (1989).
64. R. B. King and A. Efraty, *Org. Mass Spectrom.*, **3**, 1227 (1970).
65. R. B. King and T. F. Korenowski, *Org. Mass Spectrom.*, **5**, 939 (1971).
66. P. Köber and M. Kerber, *Z. Anorg. Allg. Chem.*, **507**, 119 (1983).
67. E. Lindner and H.-M. Ebinger, *J. Organomet. Chem.*, **66**, 103 (1974).
68. J. L. Little, *Inorg. Chem.*, **18**, 1598 (1979).
69. W. Malisch and P. Panster, *J. Organomet. Chem.*, **99**, 421 (1975).
70. W. Malisch and P. Panster, *Chem. Ber.*, **108**, 700 (1975).
71. W. Malisch and P. Panster, *Chem. Ber.*, **108**, 716 (1975).
72. W. Malisch, M. Kuhn, W. Albert and H. Rößner, *Chem. Ber.*, **113**, 3318 (1980).
73. G. Märkl, H. Baier, R. Liebl and K. K. Mayer, *J. Organomet. Chem.*, **217**, 333 (1981).
74. H. G. Metzger and R. D. Feltham, *Inorg. Chem.*, **10**, 951 (1971).
75. J. Müller, *Chem. Ber.*, **102**, 152 (1969).
76. J. Müller and K. Fenderl, *Angew. Chem., Int. Ed. Engl.*, **10**, 418 (1971).
77. P. Panster and W. Malisch, *Chem. Ber.*, **109**, 3842 (1976).
78. A. L. Rheingold, M. J. Foley and P. J. Sallivan, *J. Am. Chem. Soc.*, **104**, 4727 (1982).
79. O. J. Scherer, C. Blath and G. Wolmershäuser, *J. Organomet. Chem.*, **387**, C21 (1990).
80. O. J. Scherer, K. Pfeiffer, G. Heckmann and G. Wolmershäuser, *J. Organomet. Chem.*, **425**, 141 (1992).
81. H. Schumann and L. Eguren, *J. Organomet. Chem.*, **403**, 183 (1991).
82. D. Seyferth and J. S. Merola, *J. Am. Chem. Soc.*, **100**, 6783 (1978).
83. D. Seyferth and H. P. Withers, Jr., *Organometallics*, **1**, 1275 (1981).
84. L.-C. Song and Q.-M. Hu, *J. Organomet. Chem.*, **414**, 219 (1991).
85. O. Stelzer, E. Unger and V. Wray, *Chem. Ber.*, **110**, 3430 (1977).
86. A. Strube, G. Huttner and L. Zholnai, *Z. Anorg. Allg. Chem.*, **577**, 263 (1989).
87. A. Strube, G. Hunter, L. Zholnai and W. Imhof, *J. Organomet. Chem.*, **399**, 281 (1990).
88. G. Thiollet, F. Mathey and R. Poilbank, *Inorg. Chim. Acta*, **32**, L67 (1979).

89. G. Thiollet and F. Mathey, *Inorg. Chim. Acta*, **35**, L331 (1979).
90. Trinh-Toan and L. F. Dahl, *J. Am. Chem. Soc.*, **93**, 2654 (1971).
91. H. Vahrenkamp, *Chem. Ber.*, **105**, 3574 (1972).
92. L. Weber and D. Wewers, *Chem. Ber.*, **117**, 1103 (1984).
93. See, for example, R. Pikver, E. Suurmaa, E. Lippmaa, D. V. Zagorevskii, Yu. S. Nekrasov and V. F. Sizoi, *Izv. Akad. Nauk SSSR, Ser. Khim.*, 1670 (1983) and references cited therein.
94. K. Henrick, M. Mickewicz, N. Roberts, E. Shewchuk and S. B. Wild. *Aust. J. Chem.*, **28**, 1473 (1975).
95. J. D. Corbett, *Chem. Rev.*, **85**, 383 (1985).

CHAPTER **7**

PES of organic derivatives of As, Sb and Bi

L. SZEPES, A. NAGY and L. ZANATHY

General and Inorganic Chemistry Department, Eötvös Loránd University, H-1518 Budapest, P.O. Box 32, Hungary

ABBREVIATIONS

AO	atomic orbital
BE	binding energy
das	*o*-phenylenebis(dimethylarsine)
dam	bis(diphenylarsino)methane

The chemistry of organic arsenic, antimony and bismuth compounds
Edited by S. Patai © 1994 John Wiley & Sons Ltd

dpm	bis(diphenylphosphino)methane
E	Group 15 element
E_{kin}	kinetic energy
EN	electronegativity
FWHM	full width at half maximum
HeI	HeI radiation source ($hv = 21.218$ eV)
HeII	HeII radiation source ($hv = 40.814$ eV)
$+/-$ I	$+/-$ inductive effect
IE	ionization energy
INDO	intermediate neglect of differential overlap
JT	Jahn–Teller
LCBO	linear combination of bonding orbitals
lp	lone pair
MO	molecular orbital
OS	oxidation state
PE	photoelectron
PES	photoelectron spectroscopy
PEPICO	photoelectron–photoion coincidence
QAS	tris(o-diphenylarsinophenyl)arsine
QP	tris(o-diphenylphosphinophenyl)phosphine
SCF	self-consistent field
SO	spin–orbit
sp	square planar
tbp	trigonal bipyramid
TPEPICO	threshold photoelectron–photoion coincidence
TPES	threshold photoelectron spectroscopy
UV	ultraviolet
UPS	ultraviolet photoelectron spectroscopy
X	halogen
XPS	X-ray photoelectron spectroscopy
ZEKE	zero kinetic energy

I. INTRODUCTION

Since its advent photoelectron spectroscopy has provided a great wealth of information about molecular electronic structure. Contributing to many fields of chemistry the technique has proved an invaluable tool in such principal areas of interest as substituent effects, or different electronic interactions[1,2]. High-temperature photoelectron spectroscopy has made it possible to study isolated inorganic molecules with ionic bonding[3] and investigations of species adsorbed on surfaces leads to a better understanding of catalytically induced reactions[4,5]. Coupled with pyrolysis or microwave discharge techniques PE spectra of free radicals and transient species can be observed, from which important structural and energetic data can be drawn[6,7].

Organometallic chemistry displays a great variety of metal–ligand interactions which have broadened our knowledge about chemical bonding. Photoelectron spectroscopy, providing the most direct experimental approach to electronic structure, plays an essential role in understanding the nature of these interactions[8-10].

This chapter is based on the photoelectron spectroscopy of As-, Sb- and Bi- organic derivatives. The general remark made on the chemistry of the organyls of Group 15 also holds for their UPS studies; 'Limited practical utility, combined with a large variety of structural and bonding peculiarities have developed the study of Group 15 organometallics into what one may call 'chemistry for chemists'[11].

TABLE 1. Organic derivatives of arsenic, antimony and bismuth investigated by HeI UPS

Compound	References	Other investigations
$MeAsH_2$	39	
Me_2AsH	39	
Me_3As	26	
Me_3Sb	26	
Me_3Bi	41, 54	TPEPICO
$t-Bu_3Sb$	70	
CF_3AsH_2	39	
$(CF_3)_2AsH$	39	
Me_2AsCF_3	40	
Me_2SbCF_3	40	
$MeAs(CF_3)_2$	40	
$MeSb(CF_3)_2$	40	
$(CF_3)_3As$	39	
$(CF_3)_3Sb$	39, 49	Vacuum pyrolysis
Ph_2AsH	56	
Ph_3As	55, 56	
Ph_3Sb	55, 56	
Me_2AsCl	63	
Me_2AsBr	63	
Me_2AsI	63	
Me_2SbBr	63	
Me_2SbI	63	
$MeAsCl_2$	63	
$MeAsBr_2$	63	
$MeAsI_2$	63	
$MeSbI_2$	63	
Me_2AsCN	64	
$(SiH_3)_3As$	69	
$(Me_3Si)_3As$	70	
$(Me_3Si)_3Sb$	70	
$(Me_3Ge)_3Sb$	70	
$(Me_3Sn)_3Sb$	70	
$Ph_2AsSiMe_3$	72	
$Ph_2SbSiMe_3$	72	
F_3SiAsH_2	71	
$F_3SiAsMe_2$	71	
$Cl_3SiAsMe_2$	71	
Me_4As_2	74	
Me_4Sb_2	75	
$(CF_3)_4As_2$	74	
Me_2AsSPh	76	
Me_2AsSMe	78	
$Me_2AsSeMe$	78	
$(PhS)_3As$	76	
Me_5Sb	65, 66	
Me_4SbF	66	
Me_3AsF_2	66	Vacuum pyrolysis
Me_3AsCl_2	66, 68	Vacuum pyrolysis
Me_3SbF_2	66	Vacuum pyrolysis
Me_3SbCl_2	65, 66	Vacuum pyrolysis
Me_3SbBr_2	65	
Me_3SbI_2	65	

(*continued*)

TABLE 1. (*continued*)

Compound	References	Other investigations
	82	
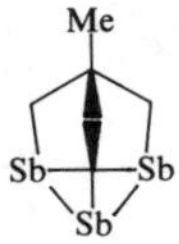	82	
$t\text{-Bu}_3(\text{AsP}_2)$	80, 81	
$t\text{-Bu}_3(\text{As}_3)$	81	
$t\text{-Bu}_4(\text{As}_4)$	49	
$(\text{CF}_3)_4(\text{As}_4)$	49	HeII UPS
$\text{Me}_5(\text{As}_5)$	49	
$\text{Me}_3\text{As}{=}\text{CH}_2$	84	
$\text{Me}_3\text{As}{=}\text{CHSiMe}_3$	84	
$\text{Me}_3\text{As}{=}\text{C(SiMe}_3)_2$	84	
$\text{C}_5\text{H}_5\text{As}$	85, 89	HeII UPS Angular distribution of photoelectrons
$\text{C}_5\text{H}_5\text{Sb}$	85	
$\text{C}_5\text{H}_5\text{Bi}$	86	
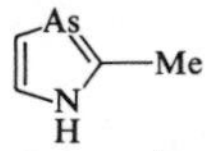	91	
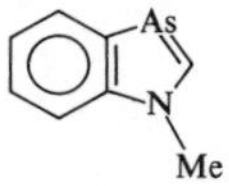	94	
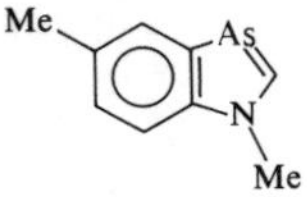	94	
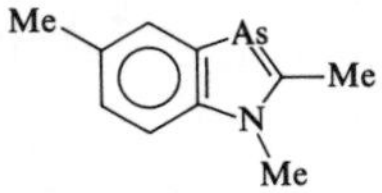	94	
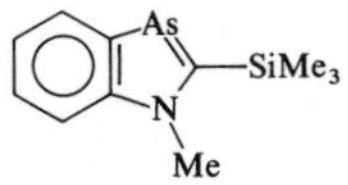	94	
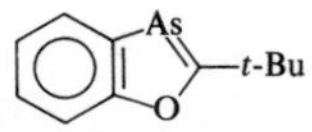	94	
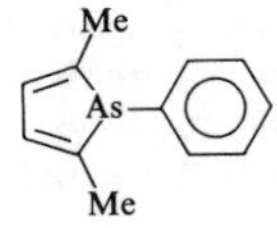	95	

(*continued*)

TABLE 1. (*continued*)

Compound	References	Other investigations
(structure: benzo ring with As–Me, =N, –t-Bu)	97	
cis-[PtMe$_2$(AsMe$_3$)$_2$]	45	
cis-[Pt(CF$_3$)$_2$(AsMe$_3$)$_2$]	59	
trans-[PdCl$_2$(AsEt$_3$)$_2$]	62	
trans-[PtCl$_2$(AsEt$_3$)$_2$]	62	
[Fe(CO)$_4$(AsPh$_3$)]	60	
[W(CO)$_5$(AsPh$_3$)]	61	
[Cr(CO)$_5$(Me$_2$AsSMe)]	79	
[Mo(CO)$_5$(Me$_2$AsSMe)]	79	
[W(CO)$_5$(Me$_2$AsSMe)]	79	
[Cr(CO)$_5$(Me$_2$AsSeMe)]	79	
[Mo(CO)$_5$(Me$_2$AsSeMe)]	79	
[W(CO)$_5$(Me$_2$AsSeMe)]	79	
[Cr(CO)$_5$(Me$_2$AsAsMe$_2$)]	79	
[Mo(CO)$_5$(Me$_2$AsAsMe$_2$)]	79	
[W(CO)$_5$(Me$_2$AsAsMe$_2$)]	79	

After an introduction to the basic principles and recently developed experimental techniques, the PE spectra of the compounds listed in Table 1 are discussed.

A. The Photoelectron Experiment

Photoelectron spectroscopy is a sort of molecular spectroscopy which is based on the photoionization process. Several outstanding books have been published on the subject[12-14], so basic principles are only briefly outlined here.

The sample is bombarded by monochromatic vacuum ultraviolet photons. As a result, a positive ion in its *i*th electronic state and a free electron are formed, which is represented by equation 1.

$$M_0 + h\nu \longrightarrow M_i^+ + e \qquad (1)$$

The energy balance of the reaction can be written as

$$h\nu = IE_i + E_{kin} \qquad (2)$$

The energy of the bombarding photons is known, the kinetic energy of the emitted electron (E_{kin}) is measured, so the energy required to remove an electron from the molecule can be determined. This amount of energy is called ionization energy, IE_i, or electron binding energy, BE_i. In other words, the *i*th ionization energy is the total energy difference between the neutral molecule and the *i*th particular state of the ion. In the case of molecules, vibrational and rotational excitations may occur upon ionization, so the energy balance can be rewritten as

$$h\nu = IE_i + E_{kin} + \Delta E_{vib} + \Delta E_{rot} \qquad (3)$$

By using low-energy photons (energy less than 50 eV) valence shell ionizations occur, and the technique is called ultraviolet photoelectron spectroscopy (UPS). When X-rays are the impinging radiation, core-level ionizations take place providing the basis for X-ray photoelectron spectroscopy (XPS).

B. The Photoelectron Spectrum

The photoelectron spectrum consists of a series of bands where the number of electrons (N) is related to their energy (E_{kin}). The energy of the photoelectron depends on the investigated molecule and on the photon energy. Band intensities are determined by the probability of the ionization events which is again a function of photon energy. The number of photoionization events, N_i, is given by

$$N_i = I_v \cdot \sigma_i \cdot n \cdot l \tag{4}$$

where n is the number density of the target, l is the path length traversed by the radiation of flux I_v and σ_i is the photoionization cross section. The latter is the sum of cross sections for individual ionizations (of different orbitals) called partial cross sections, σ_j. The partial cross section is defined by the relation

$$\sigma_j = \left(N_j / \sum_j N_j \right) \sigma_i \tag{5}$$

where N_j represents the number of ions in state j; it can also be equated to the number of electrons formed in a specific process. The summation over all ionic states gives the total number of ions produced. The number of ions in state j related to the total number of ions is known as branching ratio or relative abundance.

Ionization cross sections depend on the photon energy; its change has a pronounced effect on relative band intensities. These intensity shifts help to assign bands in the spectra, especially in the case of d- and f-block organometallic compounds[8, 9].

Another factor influencing band intensities in photoelectron spectroscopy is the angular distribution of photoelectrons. This means that band areas in the photoelectron spectrum depend on the angle (α) between the directions of photon and electron propagation. These photoelectron distributions depend on the energy of the electrons and on the nature of the orbital from which they are ejected. The number of electrons N_j ejected per unit solid angle in a specific direction by unpolarized radiation is given by

$$N_j \propto (\sigma_j / 4\pi) [1 - \beta/4(3 \cos^2\alpha - 1)] \tag{6}$$

where β is termed the asymmetry parameter ranging in value from -1 to $+2$, while α is the angle between the ejected electron and the photon beam.

Values of β determined from angular distributions provide important information about the photoionization process and may help in the orbital assignment of bands in a PE spectrum. The problem of photoionization cross sections and photoelectron angular distributions is thoroughly discussed by Samson[15, 16].

C. Interpretation of Photoelectron Spectra

The interpretation of a PE spectrum includes the assignment of bands to appropriate ionic states and the determination of which particular orbital of the neutral system is involved in a given ionization.

The primary experimental data obtained from PE spectra are ionization energies and it is the well-known Koopmans' approximation which relates ionization energies to orbital energies. It states that the ith ionization energy is given by the negative of the ith SCF orbital energy[17]:

$$IE_i = - \varepsilon_i^{SCF} \tag{7}$$

By neglecting the reorganization energy in the ion and the correlation energy in both the ionic and neutral states, its approximate nature is obvious, however it has proved to be very useful in the spectrum assignment of many molecules because trends are reflected correctly.

On the theoretical basis of the Koopmans' theorem *ab initio* and semi-empirical calculations can be used as important aids in the analysis of PE spectra[13]. Currently, the MNDO method enjoys great popularity because ionization energies of even large and complex molecules can be calculated with an accuracy of few tenths of eV[18]. This level of accuracy has been achieved by re-parametrization of the procedure. Further improvements in the calculation of ionization energies can be achieved by the separate SCF and configuration interaction (CI) calculations for the neutral molecule and ion.

In cases where Koopmans' approximation fails badly, the direct way of calculating ionization energies, the method of Green's functions, has been applied successfully[19], although we are not aware of its use in case of organometallic compounds. Applications of a simplified Green's function approach to transition metal complexes, however, have been reported [20-22]. The scattered-wave X_α method[23], on the contrary, has wide applications in describing metal–metal bonds and the nature of multicentre interactions within metal-lacycles[24].

Apart from quantum chemical calculations there are many experimental guides which help to interpret PE spectra. Their outlook feature provides much help in the spectral analysis. Vibrational fine structure of PE bands, for instance, contains useful information about the bonding character of the ionized orbital and about the ion structure. In the absence of fine structure, bandwidth can also be associated with orbital character: the sharper the band, the more non-bonding is the ionized orbital. Furthermore, ionizations from degenerate orbitals can give rise to additional bands: spin–orbit coupling and the Jahn–Teller effect may complicate the spectrum on one hand; on the other, they show characteristic splittings which may serve as orienting guidance to assignment.

Relative band intensities give a rough estimate of orbital degeneracies, since those are proportional to the number of equivalent electrons available for ionization from the appropriate orbitals. This consideration is especially useful in the case of large polyatomic molecules, where spectral features such as vibrational structure and spin–orbit splitting play a limited role in the assignment procedure. Ionization of electrons localized mainly within a certain part of the molecule can be identified on the basis of their ionization energy and band shape. PE bands originating from the ionization of metal d electrons confined to transition-metal centres in organometallic complexes can be generally found in the lowest-ionization energy part of the spectra. Removal of electrons from metal–carbon or metal–metal bonds also results in low-energy ionic states. For instance, in the HeI photoelectron spectra of compounds Me_3M—$M'Me_3$ (M, M' = Si, Ge and Sn) the first and second ionization energies can easily be attributed to the ionizations from the M—M' (a_1) and M(M')C_3 (e) orbitals, respectively, on the basis of their energy and relative band intensity[25]. Heteroatom lone-pair ionization bands have also characteristic low ionization energies and relatively narrow peaks. This is exemplified by the PE spectra of Me_3E (E = P, As and Sb) derivatives showing a weak band at about 8 eV which can be ascribed to ionization from the lone pair of the Group 15 heteroatom[26]. The same is true for isolated double bonds, such as the C=C bond, and for other group orbitals.

Investigation of chemically related series of compounds is also highly informative. Systematic changes made in the structure or composition of the studied molecule result in systematic changes in the PE spectra, thus assignment is facilitated by their comparison. The comparative measurement of isostructural main group compounds contributes substantially to the interpretation of their spectra and to the better understanding of their valence electron structure. As will be shown later, this type of approach to assignment has been widely used in the photoelectron spectroscopy of As-, Sb- and Bi-organic derivatives.

As discussed earlier, partial ionization cross sections depend on the energy of the exciting radiation. Different relative PE band intensities are obtained when we use a different excitation energy. The most widely used experimental technique is to run spectra with HeI (21.218 eV) and HeII (40.814 eV) excitation sources and compare the corresponding cross-section areas. This empirical band assignment has proved to be especially effective in identifying molecular orbitals of predominantly metallic character (metal d orbitals) in the case of organometallic complexes[8, 9], because these bands have the largest HeII relative intensity increase. An experimental observation particularly relevant to our present topic is that—considering the elements of a given group—molecular orbitals composed in great extent of atomic s and p orbitals have generally smaller relative HeII cross sections for heavier elements than for lighter ones. New perspectives of cross-sectional studies are given by the application of synchrotron radiation sources which are becoming more and more available[27].

Angular distribution studies are also promising tools for orbital assignment. In these experiments electron signals are measured at different angles and, from the experimental angular distribution, the β asymmetry parameter can be obtained, which is a function of the nature of the vacated orbital at a given photon energy. Because of their complexity from both theoretical and experimental points of view, angular distribution measurements are not performed routinely. In cases where band overlapping complicates the evaluation of spectra, angular distribution experiments may be of great help as shown by the example of benzene, where ionization from the $3e_{1u}$ orbital was identified on the basis of angular distribution data[28].

The investigation of Rydberg states in the far UV spectra of molecules may. also contribute to the interpretation of PE spectra. From the quantum defects and the number of Rydberg series that converge on a given limit, the ionized molecular orbital can be deduced. The convergence limit of a series corresponds to an ionization energy in the PE spectrum.

Obviously, there are various theoretical and experimental approaches to the assignment problem, and none of them can be considered as an absolute method. The use of as many independent ways as possible leads to correct and reliable assignments.

D. Experimental Aspects

Basic instrumentation is not discussed here since detailed descriptions are given in the literature[12–14]. New experimental developments with special relevance to the covered topics will be outlined shortly.

In a conventional PE apparatus the spectrum is obtained by scanning the electron kinetic energy at a fixed wavelength of photons. Most frequently applied photon sources are HeI and HeII radiation produced by low-pressure discharge through helium.

If a continuum light source is available, the wavelength may be scanned and electrons of fixed kinetic energy can be detected. Zero- (or near-zero-) energy threshold electrons can be detected with high efficiency by a simple electron collection system[29,30]. The technique relies on the fact that initially zero-energy electrons may be easily directed toward the electron detector by very weak electric fields, while electrons formed with an appreciable kinetic energy are scattered in all directions. This is the basis of threshold photoelectron spectroscopy (TPES) where zero- (or near-zero-) energy electrons are detected while the photon energy is scanned.

The spectrum obtained by TPES is similar to the PE spectrum produced by an apparatus operating at fixed photon energy; however, threshold electron analysis has several advantages as listed below:

(1) large PE signals as a consequence of the fact that the analyser collects electrons from all directions;

(2) the PE intensity does not depend on the β asymmetry parameter;

(3) there is no need for energy calibration by external standards (e.g. Ar, Xe) because an absolute calibration of energy is available.

TPES has been applied to study fine details of photoionization[31-33], but this technique is also limited as far as spectral resolution is concerned because of its inability to eliminate spurious peaks originating from low-energy electrons.

Further developments, namely zero-field ionization and delayed pulsed-field extraction, have made it possible to detect electrons of exactly zero kinetic energy[34]. This forms the basis of high-resolution zero kinetic energy (ZEKE) photoelectron spectroscopy which produces rotationally resolved spectra of molecular ions[35] and provides information about low-frequency vibronic states in the ions of molecular complexes. For instance, in the ZEKE spectrum of the phenol–H_2O (1:1) molecular complex, harmonic progression related to the rotation of the phenyl ring around the hydrogen bond (16 cm^{-1}) has been observed[36]. The approximately three-order improvement in energy resolution offers profoundly new possibilities also in the photoionization studies of organometallic systems.

Combination of PES with mass spectrometry in a coincidence system leads to an extremely powerful technique called photoelectron–photoion coincidence (PEPICO) spectroscopy, which renders it possible to examine dissociations of energy-selected molecular ions.

In the ionization process

$$M + h\nu \longrightarrow M^+ + e \qquad (8)$$

the molecular ion M^+ can be formed in different internal energy states expressed by

$$E_{ion} = h\nu - IE - E_{kin}(e) \qquad (9)$$

where E_{ion} is the ion internal energy transferred upon ionization.

If ions are detected in delayed coincidence with electrons of energy E_{kin}, ions with internal energy E_{ion} are selected from all the other ions formed. This technique permits one to study the fate of a molecular ion in a specified internal-energy state. Like in PES there are two types of coincidence experiments. In one case, the apparatus operates at fixed photon energy, usually at 21.218 eV (HeI resonance line), and ions are detected in coincidence with electrons of selected energies[13]. In the other experimental arrangement a continuum light source is applied together with a vacuum UV monochromator, and zero kinetic energy electrons are collected in coincidence with ions, so that the ion internal energy is given by $h\nu - IE$[32, 37, 38].

A typical PEPICO apparatus is shown in Figure 1, where an acceleration region followed by a drift tube is used to determine the ionic species. The coincidence condition is achieved by using electron and ion signals as start and stop inputs, respectively, to a time-to-pulse height converter whose output is sent to a multichannel pulse height analyser.

By comparing threshold and energetic electron detection in coincidence studies, one can conclude that in the former case a 50–500 times higher signals-to-noise ratio is possible, while in the latter case the photon flux of a HeI radiation source is greater by several orders of magnitude. As a final result, coincidence rates are comparable in the two types of experiment. In practice, the threshold approach seems to be preferred because a wider variety of states is available since autoionization aids to excite levels not available in conventional photoelectron spectroscopy.

By using this PEPICO apparatus a great amount of dynamical information can be

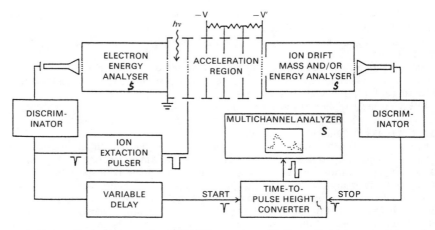

FIGURE 1. Block diagram of a PEPICO experiment[32]. Reproduced by permission of Academic Press

drawn from the ion time-of-flight distribution. For further details the reader is referred to References 32, 37 and 38.

The result of another type of PEPICO experiment is the breakdown diagram, which is a plot of the fractional abundance of the ions formed by the dissociation of energy-selected parent ions as a function of the photon energy. It is also represented as a plot of branching ratios to the various dissociation products as a function of the precursor ion internal energy. The breakdown diagrams are obtained directly from a series of PEPICO mass spectra at several photon energies.

II. PHOTOELECTRON SPECTROSCOPY OF As, Sb AND Bi ORGANYLS

A. Simple Trivalent Derivatives

The HeI PE spectroscopy of the title compounds focuses on the characterization of the substituent effects influencing the lone-pair ionization energies of the Group 15 elements, which can be assigned in all cases to the first band in the spectra.

The simplest organoarsenic compounds investigated are the members of the series Me_nAsH_{3-n} ($n = 1–3$) and their perfluoromethyl analogues[39].

As can be seen in Figure 2, the spectra of methylarsines are very similar showing up systematic trends in ionization energies and relative band intensities. It is evident from the spectra that the IE_1 values and the intensity of the first band relative to those found in the energy range of 10.5–16 eV decrease as the number of methyl groups increases. Supposing that the HeI ionization cross section of the MO attributed to the arsenic lone pair (n_{As}) does not change significantly in the series, IE_1 can be assigned to n_{As} followed by ionizations from the σ_{As-C} and σ_{As-H} orbitals (Table 2). If we consider the change in the IE_1 values along the series Me_nAsH_{3-n} as well as $(CF_3)_nAsH_{3-n}$ ($n = 0–3$) and compare them to IE_1 of AsH_3, the $+I$ effect of the methyl group(s) and the $-I$ effect of the CF_3 substituent(s) are clearly demonstrated.

The 'composite molecule' approach to the molecular orbitals of $MeAsH_2$ (Figure 3) provides further aid to the interpretation of its PE spectrum, namely:

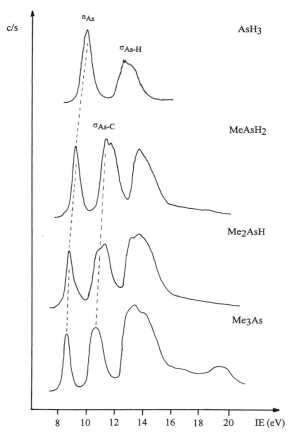

FIGURE 2. HeI PE spectra of Me_nAsH_{3-n} $(n = 0-3)^{39}$. Reproduced by permission of Elsevier Sequoia SA

(1) The MO corresponding to the first band in the spectrum of $MeAsH_2$ can be composed by the antisymmetric combination of CH_4 $(1t_2)$ and AsH_3 $(2a_1)$ orbitals. The constructed MO has substantial σ_{C-H} contribution, that is why in adition to the $+$ I effect of the Me group, hyperconjugation has to be taken into account as a destabilizing factor.

Because of energetic reasons, a similar effect is not expected to play an important role in the case of CF_3AsH_2. This idea is supported also by the qualitative MO diagram of $CF_3PH_2{}^{39}$.

(2) The orbital diagram of Figure 3 shows that the second band in the PE spectrum can be related to ionization from the σ_{As-C} MO, composed from the CH_4 $(1t_2)$ and AsH_3 $(2e)$ orbitals.

The data in Table 2 also reveal that CF_3 substitution of methylarsines does not change the order of orbitals. However, stabilization of all orbitals occurs as a result of the strong $-$ I effect of the perfluoromethyl group(s).

The same sort of substituent effect can be traced in Table 3, where the first and second ionization energies of pure and mixed methyl/perfluoromethyl triorganyls are sum-

TABLE 2. Vertical ionization energies (in eV) and orbital characters of
R_nAsH_{3-n}[39,42]

Compound	IE_1	IE_2		IE_3
AsH_3	10.58 (n)	12.7	(σ_{As-H})	
$MeAsH_2$	9.50 (n)	11.2–12.6^a	(σ_{As-C})	12.02 (σ_{As-H})
Me_2AsH	8.87 (n)	10.5–11.9^a	(σ_{As-C})	11.35 (σ_{As-H})
Me_3As	8.65 (n)	10.70	(σ_{As-C})	
CF_3AsH_2	11.00 (n)	12.73	(σ_{As-C})	13.39 (σ_{As-H})
$(CF_3)_2AsH$	11.26 (n)	12.63	(σ_{As-C})	13.31 (σ_{As-H})
$(CF_3)_3As$	11.41 (n)	12.77	(σ_{As-C})	

^{a} FWHM.

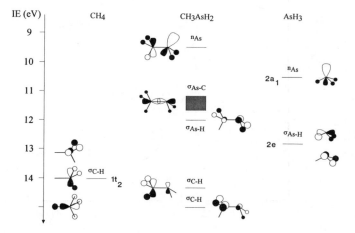

FIGURE 3. Qualitative MO assignments of $MeAsH_2$ based on the 'composite molecule' ($CH_4 + AsH_3 \longrightarrow CH_3AsH_2$) approach

marized. The corresponding values of the analogous N and P compounds and those of Group 15 hydrides are also included for the sake of comparison. Band assignments are facilitated by chemically related comparisons[39] and by CNDO calculations carried out for the $Me_nE(CF_3)_{3-n}$ (E = P and As, $n = 0$–3) series[40].

The PE spectra of the Me_3E molecules (Figure 4) show a new feature: the first ionization energies, assigned to the lone pairs of the Group 15 elements, are nearly constant and range between 8.44–8.65 eV[26]. (A similar phenomenon has been found in the case of the corresponding hydrides[41-43].) This is in contrast with the behaviour of other series where the ionization energy of a given orbital, localized mainly in the vicinity of the heteroatom, changes parallel with the electronegativity of the heteroatom. For example, IE_1 values in the series Me_2Y (Y = O, S, Se and Te) decrease gradually on going from O to Te[44].

Crucial to the understanding of this anomalous constancy of the first ionization energies are the SCF-MF-X_α calculations performed for the EH_3 molecules[45]. In agreement with the decreasing HEH bond angle (E = N: 107.8°; P: 93.6°; As: 91.8° and Sb: 91.3°[46]) the s

TABLE 3 Vertical ionization energies (in eV) and orbital characters of EH_3 and $Me_nE(CF_3)_{3-n}$[26,39-41]

Compound	IE_1 (n_E)	IE_2	
NH_3	10.83	16.0	(σ_{N-H})
PH_3	10.59	13.6	(σ_{P-H})
AsH_3	10.58	13.0	(σ_{As-H})
SbH_3	10.02	11.9	(σ_{Sb-H})
Me_3N	8.54	12.8	(σ_{N-E-C})
Me_3P	8.60	11.34	(σ_{P-C})
Me_3As	8.65	10.70	(σ_{As-C})
Me_3Sb	8.48	10.3^a	(σ_{Sb-C})
Me_3Bi	8.44	$9.39, 10.24^a$	(σ_{Bi-C})
CF_3PMe_2	9.7	12.3	(σ_{P-C})
CF_3AsMe_2	9.8	11.6	(σ_{As-C})
CF_3SbMe_2	9.4	10.9	(σ_{Sb-C})
$(CF_3)_2PMe$	10.8	13.1	(σ_{P-C})
$(CF_3)_2AsMe$	10.7	12.2	(σ_{As-C})
$(CF_3)_2SbMe$	10.2	11.5	(σ_{Sb-C})
$(CF_3)_3P$	11.57	13.44	(σ_{P-C})
$(CF_3)_3As$	11.41	12.77	(σ_{As-C})
$(CF_3)_3Sb$	10.74	11.97	(σ_{Sb-C})

^{a} SO and/or JT splitting.

character of the lone pair is 7% in NH_3, 19% in PH_3 and 20% in AsH_3. (The small difference between the last two values agrees well with the experimentally determined slight difference in bond angles.) The increased s character stabilizes the HOMO of the hydrides (n_E)[47] while the lowering of electronegativity on going from nitrogen to antimony is a destabilizing factor. The sum of these two counteracting effects may be considered responsible for the small variation of the IE_1 values (see Table 3). This reasoning is quite acceptable as far as the IE_1 difference between ammonia and phosphine is concerned, but is hard to apply to the heavier analogues where differences in the HEH bond angles are small. A similar situation is found in the case of the Me_3E (E = N, P, As, Sb and Bi) series. Substantial change of the CEC bond angle is only experienced between the nitrogen and phosphorus trimethyls; the values are: $110.9°$ in Me_3N, $98.8°$ in Me_3P, $96.1°$ in Me_3As, $94.1°$ in Me_3Sb and $97.1°$ in Me_3Bi[48]. Not surprisingly, the first ionization energies are destabilized compared to the hydrides as a consequence of the $\Sigma + I$ effect of the methyl groups. The anomalous constancy of the first ionization energy values has been interpreted in terms of geometry, hybridization and symmetry[26]. On going down the series, the growing s character of the lone pair stabilizes the first ionization energies while the decreasing electronegativity of the central atom and the hyperconjugative effect of the methyl groups shift them to lower values. As an overall result, the lone-pair ionization energies become nearly constant.

The fact that in the Me_3E-type compounds methyl groups have considerable hyperconjugative effect in addition to their $+ I$ effect is clearly demonstrated in Figure 5, where first ionization energies of the $Me_nE(CF_3)_{3-n}$ compounds (E = P, As, Sb; $n = 0-3$) vs. E and n are plotted[40]. The near constancy of the first ionization energies in the series $n = 3$ disappears by the gradual substitution of CF_3 groups for CH_3. In the case of the $(CF_3)_3E$ compounds, the first ionization energies show the same trend as the central atom

L. Szepes, A. Nagy and L. Zanathy

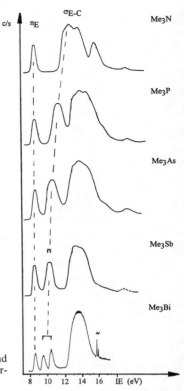

FIGURE 4. HeI PE spectra of Me_3E ($E = N, P, As, Sb$ and Bi) with qualitative assignments[26,41]. Reproduced by permission of Verlag der Zeitschrift der Naturforschung

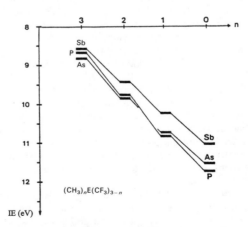

FIGURE 5. The IE_1 values of $Me_nE(CF_3)_{3-n}$ plotted as a function of E and n[40]. Reproduced by permission of VCH Verlagsgesellschaft mbH

electronegativities. The range of the IE_1 values at given n is inversely proportional to n. This correlation also implies that CF_3 groups compared to methyls do not have a substantial hyperconjugative effect. Due to the lack of bond angles, it is hard to decide whether the geometric factors discussed above play an important role in the ΔIE_1 patterns. Even if their operation cannot be ruled out, basically substituent electron-withdrawing/donating effects seem to be influential. This idea is supported by Figure 6 where the plots of the sum of substituent group electronegativities ($EN_{Me} = 2.35$, $EN_{CF_3} = 3.3$)[40] versus IE_1 show reasonable linearity.

The second band in the spectra can be assigned to electron removal from the σ_{E-C} orbitals which are spin–orbit (and Jahn–Teller) split in the cases of Me_3Sb and Me_3Bi. The variation of the IE_2 values in a given series reflects the same substituent and atomic properties as experienced in the case of lone-pair ionizations.

Special techniques in addition to the UPS measurements, namely vacuum pyrolysis and TPEPICO, have been applied in the cases of $(CF_3)_3Sb$ and Me_3Bi.

The vacuum pyrolysis of $(CF_3)_3Sb$ in the temperature range of 298–1100 K was aimed to prepare $CF_3^{\cdot}$ radicals[49]. The PE spectra of pyrolysis products formed at different temperatures are shown in Figure 7. From these spectra there is no evidence for the formation of $CF_3^{\cdot}$ whose IE_1 is expected between 9.25 and 10.10 eV[50]. However, production of CF_2 species seemed to be likely at 1100 K on the basis of some characteristic IE values. Another pyrolysis product, C_2F_4, was also identified by the analysis of vibrational fine structure and other spectral features.

Organometallic compounds, especially As-, Sb- and Bi-organic derivatives, have not been frequently studied by PEPICO spectroscopy. Only few publications are centred on PEPICO studies applied to the investigation of ion dissociation dynamics of organometallic systems and metalcarbonyls[51–53]. Recently Nagaoka and coworkers investigated ionic fragmentation processes in organometallic molecules of Group 12–15 elements following $(n-1)d$ core photoionization[54]. The methyl derivatives of Zn, Ga, Ge, Sn, Pb and Bi were studied by use of the TPEPICO technique. Beside the interesting PE spectral feature of

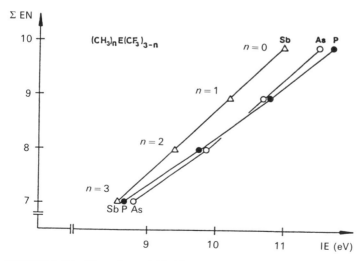

FIGURE 6. The IE_1 values of $Me_nE(CF_3)_{3-n}$ (E = P, As and Sb; $n = 0$–3) versus Σ EN of the ligands[40]. Reproduced by permission of VCH Verlagsgesellschaft mbH

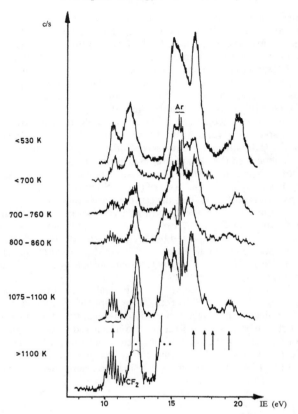

FIGURE 7. HeI PE spectra of $(CF_3)_3Sb$ at different temperatures using a heated Al_2O_3 inlet tube. The bands marked by arrows refer to C_2F_4, those marked by asterisks to an unidentified species[49]. Reproduced by permission of The Royal Society of Chemistry

Group 14 tetramethyls (quintet splitting) two questions have been studied in detail: first, the reason for the predominant production of the metal ion; second, whether this metal ion formation process is characteristic in all cases of the investigated molecules. This last problem has special relevance to the photochemical vapour deposition of metals in the processes of semiconductor production. By analysing TPE spectra, photoionization efficiency curves and TPEPICO breakdown curves it can be shown that in all cases, except $BiMe_3$, metal ions are predominantly formed in a non-statistical process from the $[(n-1)d]^9$ hole state before electronic relaxation. The monomethyl–metal ions are likely to be produced from both core and valence ionized states while di- and trimethyl–metal ion formations occur from valence ionized states in statistical fragmentation processes. The decomposition processes of $BiMe_3$ following the 5d ionization seem to be statistical in contrast to the other investigated molecules. The exceptional behaviour of $BiMe_3$ is related to the fact that, while its molecule ion in the d-core ionized state belongs to the same point group as the neutral molecule in its ground state, the other ground state parent molecules have different geometries from those of their $(n-1)$d-core ionized states.

This example points to the importance of organometallic molecules as models for testing the theory of ionic unimolecular processes.

The low-energy part of the PE spectra of the Ph_2EH (E = N, P and As) and Ph_3E (E = N, P, As and Sb) derivatives has been interpreted by the analysis of all symmetry allowed interactions that can take place among the various orbitals (Figure 8, Table 4)[55-57]. The propeller-like conformation of the molecules allows some interaction between the appropriate π_{ring} orbitals and the resulting combinations may further interact with some other orbitals like the n_E lone pair or the d orbitals of the heteroatom as well as the molecular orbitals localized on the EC_3 group. In phenylarsines and -stibine this mixing of orbitals is not substantial, as indicated by the shoulders of the second band in the PE spectra of arsine derivatives and the regular shape of the same band in the spectrum of Ph_3Sb. The weakness or absence of interaction can be explained by the unfavourable geometrical and $n-\pi$ overlap factors. Consequently, the three ionization regions between 7 and 11 eV can be assigned to electron removal from the heteroatom lone pair, the uppermost π_{ring} and the σ_{E-C} orbitals, respectively.

It is worthwhile to mention that $\pi-\pi$ and $n-\pi$ interactions are so strong in NPh_3 that it

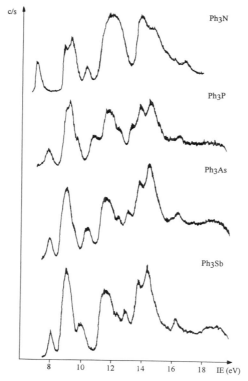

FIGURE 8. HeI PE spectra of Ph_3E (E = N, P, As and Sb)[57]

TABLE 4. Vertical ionization energies (in eV) and orbital characters of Ph_3E and Ph_2EH; $\Delta IE_1 = IE_{1(Me_3E)} - IE_{1(Ph_3E)}$[39,55,56,63]

Compound	IE_1	IE_2	IE_3	IE_4	IE_5	ΔIE_1
Ph_3N	7.00 $(\pi - n)$	8.81 (π)	9.25 (π)	10.29 $(\pi + n)$		1.54
Ph_3P	7.85 $(\pi - n)$	9.02 $(\pi)^a$	9.18 (π)	9.68 $(\pi + n)^a$	10.81 (σ_{P-C})	0.75
Ph_3As	8.03 (n)		9.10 (π)	9.58 $(\pi)^a$	10.47 (σ_{As-C})	0.62
Ph_3Sb	8.08 (n)		9.06 (π)		10.04 (σ_{Sb-C})	0.40
Ph_2NH	7.35 $(\pi - n)$	9.00 (π)	9.3 (π)	10.51 $(\pi + n)$	11.7 (σ_{N-C})	1.60
Ph_2PH	8.29 $(\pi - n)$	9.08 (π)		9.84 $(\pi + n)$	10.9 (σ_{P-C})	0.79
Ph_2AsH	8.43 (n)	9.06 (π)	9.78 (π)		10.56 (σ_{As-C})	0.49

^{a} Shoulder.

is no longer possible to speak about separate n and π orbitals. This mixing decreases on going down the Group 15, which is further supported by the following facts:

(a) the shape of the first PE band becomes sharper and sharper on going towards $SbPh_3$, indicating an increasing non-bonding character for the lone pair;

(b) the IE_1 shows a downward shift with respect to the corresponding EMe_3 and this shift decreases in the order $PPh_3 > AsPh_3 > SbPh_3$ (see Table 4).

Less pyramidalization of the Ph_3E compounds—due to the greater spatial requirement of the phenyl groups—may contribute to the decrease of the first IE on going from the EMe_3 to the EPh_3 compounds (E = P, As and Sb). The experimentally determined CAsC angle in Ph_3As is $102°$[58], larger by $6°$ than that of Me_3As[48].

Some organometallic complexes containing organoarsine ligand(s) have been studied by UPS. The ligands concerned are (see also Table 1): trimethylarsine (Pt complexes)[45,59], triphenylarsine (Fe and W complexes)[60,61] and also triethylarsine (Pd and Pt complexes)[62] whose individual PE spectrum—to the best of our knowledge—has not been recorded so far. These investigations provide information about various aspects of the ligand–metal dative bond. From the spectra of fourteen cis-$[Me_2PtL_2]$ complexes[45], for example, the relative σ-donor and π-acceptor strength of the L ligands has been evaluated, assuming that the former is mainly reflected in the first (metal d-type) IE of the complex while the latter is proportional to the splitting between two metal d bands of appropriate symmetries. Virtually no difference has been found between the behaviour of Me_3P and Me_3As. The band arising from the coordinated As lone-pair orbital has not been located, hence no information has been given about its stabilization in the complex. In the tetracarbonyliron[60] and pentacarbonyltungsten[61] complexes of triphenylarsine the ionization from n_{As} occurs at 9.15 and 8.98 eV, respectively, showing a nearly 1-eV stabilization relative to free Ph_3As. Analysis of the metal d-orbital ionizations has showed that Ph_3As is a weaker σ-donor and a weaker π-acceptor than Ph_3P[60] and thus the net donating ability of the two compounds is very similar[61]. In the spectra of the two triethylarsine complexes, $trans$-$[MCl_2(Et_3As)_2]$ (M = Pd or Pt), the band at 9.79 (Pd) or 9.95 (Pt) eV has been attributed to the orbital having predominant n_{As} character. The lone-pair IE of free Et_3As is expected to be lower than 8.65 eV (IE_1 of Me_3As); the sizeable stabilization on complexation has been explained with significant through-space interaction between the lone pairs of As and Cl[62].

The first several ionization energies in the PE spectra[63] of some organohaloarsines and -stibines Me_2EX and $MeEX_2$ are summarized in Table 5, together with those of some reference compounds. These molecules have C_s symmetry allowing no degenerate molecular orbitals. Thus, in the absence of accidental degeneracy, one ionization from n_E and two

TABLE 5. Vertical ionization energies (in eV) and orbital characters of Me_nEX_{3-n}[63]

Compound	n_E	n_X	$\sigma_{E-C}, \sigma_{E-X}$
$MeAsCl_2$	10.02	11.50, 11.59[a], 12.55	13.36
$MeAsBr_2$	9.58	10.29, 10.67, 10.80, 11.65	12.48
$MeAsI_2$	9.08	9.40, 9.83, 10.25, 10.93	11.79
$MeSbI_2$	8.7	9.05, 9.59, 9.87, 10.48	11.27
Me_2NCl	9.17	11.22, 11.81	13.2
Me_2PCl	9.15	11.0, 11.74	12.72
Me_2AsCl	9.45	10.74, 11.49	12.08–12.47
Me_2PBr	9.24	10.47, 11.06	12.20
Me_2AsBr	9.29	10.18, 10.82	11.80
Me_2SbBr	9.02	10.01, 10.61	11.2–11.5
Me_2AsI	8.85	9.41, 10.13	11.21
Me_2SbI	8.81	9.39, 10.03	10.88–11.09

[a] Band of double intensity.

n_X-type ionizations per halogen atom are expected, followed by bands corresponding to E—C and E—X bonding σ orbitals. The spectra have been interpreted in accord with these expectations, placing the n_E band at the lowest IE in all cases. The assignment is somewhat ambiguous in the case of the diiodides where the highest-energy iodine lone-pair combinations are energetically very close to the lone pair of the Group 15 atom.

The strong -I effect of halogen atoms results in characteristic trends in the IE values of the compounds in question, as expected. All orbitals are stabilized (1) with increasing electronegativity of the halogen atom(s) and (2) for given E and X, with increasing 'n' in a $Me_{3-n}EX_n$ series. An exception is the Me_2PX series where substitution of Cl for Br actually destabilizes the phosphorus lone pair. This apparent contradiction might point to the importance of conjugative destabilization of n_P, i.e. the stronger mesomeric effect of the chlorine substituent. For the Me_2AsX series the trend is 'normal' implying a less significant mesomeric interaction. This idea is further substantiated by the comparison of the spectra of Me_2NCl, Me_2PCl and Me_2AsCl (Figure 9). The trend of n_E ionization energies is quite strange at first sight: virtually no difference between N and P and, further, a marked stabilization on going over to As. The contrast with the EN sequence is even more obvious than in the case of trimethyls, and the arguments used there can be applied for Me_2ECl as well, but here the key role probably belongs to mesomeric interaction between the lone pairs of E and Cl. The significance of this interaction should be greatly reduced when N is replaced by P and further by As, which is nicely reflected by the variation in the shape of the spectral band belonging to the lone pair: the well-observable narrowing is an indication of increasing non-bonding character of the corresponding orbital.

Very similar is the trend of the lone-pair ionization energies of Me_2NCN, Me_2PCN and Me_2AsCN (9.44, 9.80 and 9.82 eV, respectively)[64]. $n_E-\pi_{CN}$ conjugation is only significant for E = N, where the HOMO is actually an (n–π) out-of-phase combination. For E = P and particularly for E = As the interaction is strongly reduced, as shown by the increased IE values and by the narrowing of the lone-pair band. These findings are in good accord with the experimental geometries of these compounds (established by microwave spectroscopy): dimethyl cyanamide is nearly planar while Me_2PCN and Me_2AsCN have pyramidal structures.

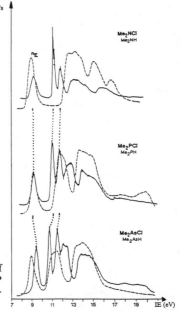

FIGURE 9. HeI PE spectra of Me_2ECl. The spectra of Me_2EH are shown for reference as a broken line (E = N, P and As)[63]. Reproduced by permission of Verlag der Zeitschrift der Naturforschung

B. Pentavalent Derivatives

A qualitative picture of bonding in these compounds, having a regular or distorted trigonal bipyramidal structure, is given by the results of Extended Hückel[65] or SCC-X_α[66] MO calculations performed on the hypothetical EH_5 molecules. The MO diagram for AsH_5 (symmetry group D_{3h}) is illustrated in Figure 10. In the absence of the most typical feature of trivalent Group 15 compounds—the lone pair, the HOMO (a_1') is an essentially non-bonding (slightly antibonding) combination of predominant ligand character. The following two molecular orbitals, e and a_2'', are metal–ligand bonding in relation to the equatorial and axial ligands, respectively. Finally, the deepest-lying a_1' MO is predicted to be out of the range of HeI PE spectroscopy and contain large s_E contributions.

This simple MO scheme is fully suitable for the interpretation of the PE spectrum of Me_5Sb[65] (Figure 11). Three band regions are clearly distinguished in the spectrum with intensity maxima at 7.20, 10.5 and 13.2 eV. In this order, they are assigned to the non-bonding ligand-localized orbital a_1', to the Sb—C bonding orbitals (e and a_2'') and to the C—H bonding orbitals. The antibonding nature of the HOMO is demonstrated by the vibrational fine structure of the lowest-IE band: a progression with a spacing of 800 cm^{-1} is observed while the Sb—C stretching frequencies in the neutral molecule are below that value. There is a marked difference between the spectra of Me_5Sb and the structurally related Me_5Ta: in the latter case, the second band located at 9.25 eV IE has been assigned to the a_1' orbital on the basis of HeII/HeI intensity relations[67]. This fact illustrates the different bonding properties of the a_1' orbital in the two compounds: it has a proper symmetry to mix with the d_{z^2} orbital of the central atom, but this mixing results in appreciable metal–ligand bonding in the tantalum compound only. In Me_5Sb the IE of a_1' is lower by 2 eV which—together with the vibrational fine structure—indicates that the involvement of the d orbitals of antimony in the bonding is negligible, even in the

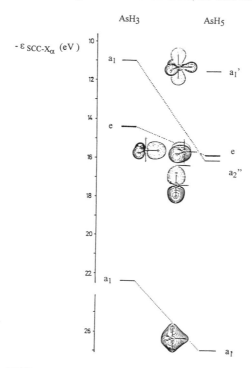

FIGURE 10. SCC-X$_\alpha$ MO diagram for the transition
AsH$_3$ ⟶ AsH$_5$[66]. Reproduced by permission of
The Royal Society of Chemistry

pentavalent state. The σ_{Sb-C} orbitals of Me$_5$Sb are stabilized by 0.2 eV relative to those of Me$_3$Sb. According to Extended Hückel MO calculations, the flattening of the SbH$_3$ molecule ($C_{3v} \longrightarrow D_{3h}$) stabilizes the corresponding (e) MO by 0.7 eV; assuming a similar behaviour for Me$_3$Sb, the addition of the two axial methyl groups has actually a weak destabilizing effect.

The comparison of the PE spectra of Me$_3$AsCl$_2$ and Me$_3$SbCl$_2$ on the one hand, and Me$_3$TaCl$_2$ on the other,[65,68] yields conclusions similar to the case of pentamethyls. In these compounds the a_1' orbital is stabilized by the $-$ I effect of the halogen atoms. It is assigned to the band at 13.85 eV in the spectrum of the tantalum compound; this rather high value indicates large metal d_{z^2} admixture and significant bonding character. In the case of Me$_3$AsCl$_2$ and Me$_3$SbCl$_2$ the assignment is not unambiguous because the bands arising from chlorine lone pairs overlap with the a_1' band. The resulting band system is located between 10–11.5 eV and 9.5–11 eV, respectively. Thus the stabilization of a_1' relative to Me$_5$Sb is nearly 3 eV, which is still less than the stabilization occurring in the case of the analogous tantalum compounds. This again speaks for the absence of d-orbital participation in the bonds around As and Sb and may be a reason for the greater chemical activity of the axial halogens in Me$_3$AsCl$_2$ or Me$_3$SbCl$_2$ than in Me$_3$TaCl$_2$.

There are some other mixed arsenic(V) and antimony(V) organohalides studied by UPS; the spectra of Me$_4$SbF and Me$_3$SbF$_2$ are shown in Figure 11 as examples[66]. The former

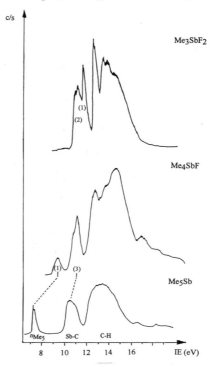

FIGURE 11. HeI PE spectra of Me_nSbF_{5-n} ($n = 3-5$). The numbers in parentheses refer to relative band intensities[66]. Reproduced by permission of The Royal Society of Chemistry

one can be assigned by analogy with Me_5Sb: the non-bonding HOMO (9.15 eV) and the σ_{Sb-C} band (10.84 eV) are clearly separated from each other as well as from the rest of the spectrum containing overlapping σ_{C-H} and n_F ionizations. The interpretation of the spectra of the Me_3EX_2 derivatives, however, is always ambiguous because the presence of two axial halogen atoms stabilizes the a_1' orbital to such an extent that it is shifted into the complex band system of halogen lone-pair ionizations and attempts to locate it have not been really convincing[65,66,68] (SCC-X_α MO calculations[66] on AsH_3F_2 place the a_1' ligand non-bonding orbital second after a doubly degenerate fluorine lone-pair combination). Vacuum pyrolysis of the Me_3EX_2 dihalides (E = As or Sb, X = F or Cl) in the temperature range of 900–1400 K has been performed[66] and the products have been analysed by field ionization mass spectrometry and by UPS. The general conclusion (not surprisingly) was that thermal stability increases from Me_3AsX_2 to Me_3SbX_2, and from Me_3ECl_2 to Me_3EF_2. PE spectra of the pyrolysis products indicated that the expected MeCl elimination from the dichlorides did not take place; HCl loss was observed instead and both UPS and MS results were consistent with the formation of $Me_2E(CH_2Cl)$ as the final product. Me_3AsF_2 lost HF as well but Me_3SbF_2 and Me_4SbF both decomposed with MeF elimination yielding Me_3Sb and, in the former case, some Me_2SbF as well.

C. Arsines and Stibines Substituted with Si-, Ge- or Sn-containing Functional Groups

The electronegativity of silicon is lower than that of carbon, therefore—strictly in terms of the inductive effect only—the silyl group is expected to possess stronger electron-releasing properties than the methyl group, and the trimethylsilyl group ought to be more electron-donating than the *t*-butyl group. Ge- and Sn-functional groups, in turn, should be even more electron-releasing than their Si-analogues. Hence the substitution of Si, Ge or Sn for a carbon atom bonded to a Group 15 element should shift the effective charge on the latter in a negative direction and, consequently, result in a lowering of the binding energy of electrons localized largely on the Group 15 atom—first of all the lone pair. UPS investigation of a great number of substituted amines and phosphines, however, has revealed the absence of such a straightforward relationship. Silicon in particular appears to stabilize the lone pair relative to carbon whenever it is directly bonded to the nitrogen or phosphorus atom. Thus, the first IE of trisilylamine (9.7 eV) is significantly higher than that of trimethylamine (8.5 eV); somewhat smaller stabilization is observed for the trisilylphosphine–trimethylphosphine pair (9.3 and 8.6 eV, respectively)[69]. Group 14 heavier elements have a similar but usually smaller stabilizing effect. On the other hand, if silicon is in a β position relative to the central atom, the destabilization expected on the basis of EN trends occurs. The factors most commonly held responsible for the anomalous stabilization are: (1) delocalization of the lone pair into the vacant 3d orbitals of the Si atom; (2) delocalization into the vacant antibonding σ orbitals of the SiH_3 or SiC_3 moiety; (3) reduced ability of the SiH_3 group to destabilizing hyperconjugative interaction with the lone pair. None of these concepts is able to offer a satisfactory rationalization of all experimental facts on its own, and there is no general agreement on the relative importance of each factor.

The PE spectrum of trisilylarsine[69] is extremely similar to that of trisilylphosphine, both in appearance and in the energetic positioning of the bands. The band assigned to the doubly degenerate Si—As bonding orbital (10.2 eV) is somewhat destabilized with respect to the phosphorus analogue (10.6 eV), but the band corresponding to the lone pair is found exactly at the same IE for both compounds. Hence the lone pairs of P and As are equally stabilized by *ca* 0.7 eV relative to the trimethyls. The authors' explanation invokes the concept of electron donation from the lone-pair orbital (supposed to be mainly p_z in character) to the available d orbitals of silicon, although elements below the 2nd period are not generally considered good donors for this kind of interaction. The stabilization is more than half as large as for the corresponding amines; if one applies the concept of (p–d)-bonding consistently, this would imply quite significant delocalizing power for both phosphorus and arsenic which conflicts with a lot of experimental evidence of various kinds. The bond angles around P and As are noticeably sharper in the trisilyl than in the trimethyl compounds, the angle in trisilylarsine (94.1°) being half-way between arsine (92.0°) and trimethylarsine (96.1°)[48]. It might be argued that a corresponding change in the hybridization of the lone pair is at least partly responsible for the observed stabilization.

The central valence angle in trisilylstibine is by 5° narrower than in trimethylstibine[48], so that IE of the lone pair of the former would represent valuable information on the validity of these hypotheses. The PE spectrum of $(SiH_3)_3Sb$ is not available as yet, but those of the $(Me_3Si)_3E$ derivatives, E = N—Sb, have been recorded, together with some *t*-butyl, trimethylgermyl and trimethylstannyl analogues[70]. The first band in all spectra has been assigned to ionization from the lone-pair orbital of the Group 15 atom and the measured IE values are summarized in Table 6. The lone-pair IE of $(Me_3Si)_3Sb$ is by 0.73 eV lower than that of $(Me_3Si)_3As$. This is the largest difference known for any pair of analogous As- and Sb-compounds, though it is not far from the range of commonly occurring values. What is more, the difference between the first ionization energies of

TABLE 6. Lone-pair vertical ionization energies of the $(Me_3M)_3E$ molecules (in eV)[70]

M =	C	Si	Ge	Sn
E = N		8.60		7.57
E = P	7.68	8.21		7.81
E = As			8.30	
E = Sb	7.62	7.57	7.58	7.64

Me_3E and $(Me_3Si)_3E$ steadily increases down the group: E = N, $\Delta = -0.06$ eV; E = P, $\Delta = 0.39$ eV; E = As, $\Delta = 0.35$ eV; E = Sb, $\Delta = 0.91$ eV. These results seem to argue for the importance of lone-pair delocalization from phosphorus and arsenic and the relative absence of such an interaction in the case of antimony. It is difficult to judge the possibility of alternative explanations, since direct structural information on the trimethylsilyl derivatives of As and Sb is not available.

In striking contrast to the trimethylsilyl compounds, the variation of the central atom in the Me_3E, $(Me_3C)_3E$ or $(Me_3Sn)_3E$ series results in only very slight changes in the lone-pair IE; the values for the N, P, As and Sb compounds are within a range of 0.2–0.25 eV in each series [data for $(Me_3C)_3As$ and $(Me_3Sn)_3As$ have not been reported]. This is explained by generalization of the concept used for the trimethyl series, i.e. increasing involvement of the valence s orbital of the Group 15 atom in the hybrid orbital of lone-pair character counteracts the decrease of atomic ionization energies down the group. The anomaly shown by the trimethylsilyl series supports the idea that bonding between silicon and Group 15 elements is in some way special, and if the key is delocalization then silicon as an electron pair acceptor is far superior to the other elements in its group. Still, the whole picture is far from being clear, e.g. the striking constancy of the first ionization energies of all four antimony compounds in Table 6 is difficult to rationalize unless one expects the inductive effect of all Me_3M (M = Group 14 element) groups to be quite the same.

The rest of the features in the spectra of $(Me_3M)_3E$ molecules allows a simple and consistent interpretation. In the order of increasing IE one can observe bands attributable to M—E, M—C and C—H bonding orbitals. The positions of all of these bands are in good agreement with values found for simpler compounds containing the same bond types. The orbitals show the expected destabilization with increasing atomic number of the elements on which they are mainly localized. This is in line with the general tendency that the energy of bonding σ orbitals is shifted stronger than that of the lone pair when analogous compounds of N, P, As and Sb are compared.

The study of some trihalogensilylarsines[71], namely F_3SiAsH_2, $F_3SiAsMe_2$ and $Cl_3SiAsMe_2$, on the other hand, provides a quite different picture of the As—Si bond. The PE spectra of the above compounds have been compared in several aspects to those of analogues containing lighter elements. The results fail to reveal any effect particular to silicon and absent from carbon chemistry. Thus the replacement of the hydrogen atom with a —PH_2 or —AsH_2 group in both F_3CH and F_3SiH causes—in all four cases—a very similar destabilization (0.4–0.5 eV) of the levels assigned to lone-pair combinations of the fluorine atoms. This leads the authors to conclude that the interaction between the Group 15 element and the trifluorosilyl group is best described as basically a pure inductive effect. Not unexpectedly, the substitution of the —$AsMe_2$ group results in a markedly stronger destabilization of the n_F orbitals (almost 1 eV). In the trichloro derivatives the n_{Cl} levels are less affected by the introduction of the —PMe_2 or —$AsMe_2$ group (in both cases 0.4–0.5 eV destabilization relative to Cl_3SiH), which is in accord with the weaker electron-withdrawing ability of the Cl_3Si— group and also speaks for the inductive nature of charge transfer.

Considering the lone-pair ionization energies, the effect of varying substitution on N, P or As is remarkably similar or, better to say, identical within experimental error, as shown by the compilation of the available relevant data in Table 7 (the values are taken from Reference 71 unless the source is indicated). It is also seen that the stabilizing effect of the trifluorosilyl group is in all cases inferior to that of the trifluoromethyl group, which is just the reverse of the relationship between —SiH_3 and —CH_3. These results seem to rule out any appreciable delocalization of the lone pair of N, P or As to the —SiF_3 group. They do not necessarily mean, however, the uselessness of the whole concept of lone-pair delocalization; it may be argued, e.g., that the trifluorosilyl group is a poor electron pair acceptor because the vacant orbitals of Si are less available as a consequence of competition offered by the lone pairs of the F atoms.

The effect of silicon substitution on the electronic structure of aromatic amines and their heavier analogues has been investigated on the model series Ph_2ESiMe_3, E = N—Sb[72]. In the PE spectra of all four compounds the first band lies at fairly low IE (around 7.5 eV) and is well separated from the rest of the spectrum. By analogy with the diphenyls and triphenyls, it has been assigned to the out-of-phase $\pi_{Ph}-n_N$ combination for E = N and to the lone pair for E = P—Sb. Ionization from aromatic π orbitals occurs at about 9 eV followed by a fairly crowded spectral region containing the σ_{E-C} and σ_{E-Si} bands. The first IE is nearly constant throughout the series (E = N, 7.44; E = P, 7.74; E = As, 7.72; E = Sb, 7.66 eV), in marked contrast to the diphenyls (E = N, 7.35; E = P, 8.29; E = As, 8.43 eV) and triphenyls where the first IE of the nitrogen derivative is by about 1 eV below those of other members of the series. The substitution of a trimethylsilyl group for the H atom noticeably destabilizes the HOMO of Ph_2PH and Ph_2AsH but results in a slight stabilization in the case of diphenylamine. Effects like this very often evoke the concept of $(p-d)\pi$ bonding for explanation; in this particular case it is more likely that, as a consequence of silylation, a significant geometrical change (pyramidalization) leading to a reduced $\pi_{Ph}-n_N$ overlap occurs at nitrogen but not at phosphorus or arsenic. As a result, the variation of bond angles in the Ph_2ESiMe_3 series is much smaller than in the Ph_2EH series. Obviously this does not necessarily mean that delocalization to silicon is unimportant; the observed effects might well result from operation of both factors.

TABLE 7. Lone-pair vertical ionization energies of some CF_3- and SiF_3-substituted amines, phosphines and arsines (in eV); $\Delta IE_{(C-Si)} = IE_{F_3CER_2} - IE_{F_3SiER_2}$

Compound	IE (n_E)	$\Delta IE_{(C-Si)}$	Reference
F_3CPH_2	11.15		
F_3SiPH_2	11.06	0.09	
F_3CAsH_2	11.00		
F_3SiAsH_2	10.90	0.10	
F_3CNMe_2	9.99		110
F_3SiNMe_2	9.60	0.39	
F_3CAsMe_2	9.8		40
$F_3SiAsMe_2$	9.04	0.4	

D. Open-chain Molecules with As—S, As—Se, As—As and Sb—Sb Bonds

If several atoms in a molecule possess non-bonding electron pairs, the possibility of lone pair–lone pair interaction arises[73]. In the compounds to be discussed below, one or several immediate neighbours of the As atom have lone pairs. The extent of lp–lp interaction in these species is a function of molecular conformation; the overlap depends strongly on the torsional angle around the bond in question. On the other hand, the split between the energy levels of the in-phase and out-of-phase combinations of the lone pairs can be determined from the PE spectrum, providing a direct measure of the strength of the interaction. Thus the PE spectra of such molecules give information about their conformational characteristics or, the other way round, geometric parameters known from other sources may be used to facilitate the interpretation of spectra.

The PE spectra[74] of $Me_2AsAsMe_2$ and $(CF_3)_2AsAs(CF_3)_2$ feature three and two bands, respectively, in the energy region where ionization from the lone pairs is expected. In the case of the tetramethyl derivative the first and third bands are of nearly equal intensity while the second one is significantly smaller and its position is approximately half-way between the other two (7.91, 8.85 and 9.50 eV). Ionizations from the bonding σ orbitals cannot be sensibly placed into the energy region below 9.5 eV; furthermore, the second band is far too small to be attributed to an ionization of the same molecule. Consequently, one must assign all three bands to lone-pair combinations, assuming the presence of two rotational isomers of this compound in the gas phase. The rotamer with *trans* position of the lone pairs is more abundant and gives rise to the two intense bands at 7.91 and 9.50 eV. The splitting over 1.5 eV is in good accord with the expectation that interaction of the lone pairs is maximized at *trans* conformation. The second rotamer is supposed to have *gauche* conformation; at this geometry lp–lp interaction is diminished and the energy levels of the two lp orbitals remain nearly degenerate. Ionization from these orbitals results in a single PE band at 8.85 eV. This value is by 0.15 eV higher than the average lp IE of the *trans* rotamer thanks to the net destabilizing effect of the *trans* lone-pair interaction in the latter. The average IE for both rotamers is also reasonably close to the first IE of dimethylarsine, 9.14 eV, which supports the correctness of the above interpretation. The next ionization appears as a shoulder on the high-IE side of the third band at 9.82 eV and has been attributed to the As—As bonding σ orbital.

The composition of the $Me_2AsAsMe_2$ rotameric mixture was determined by band area measurement as 88% *trans* and 12% *gauche*. $(CF_3)_2AsAs(CF_3)_2$, on the other hand, was found to exist in the gas phase exclusively as the *trans* conformer. The split between the two bands assigned to lone-pair ionizations (10.39 and 11.94 eV) is very close to that for the *trans* rotamer of tetramethyldiarsine; the average lp IE is just between those for $(CF_3)_2AsMe$ and $(CF_3)_3As$, also in accord with what one would expect on the basis of electronegativity considerations. Similar results were obtained for Me_2PPMe_2 and $(CF_3)_2PP(CF_3)_2$, both of which were found to exist as rotameric mixtures with predominance of the *trans* conformer. The existence of rotamers of Me_4P_2, $(CF_3)_4P_2$, and Me_4As_2, as well as the absence of the *gauche* form of $(CF_3)_4As_2$, has also been established by vibrational spectroscopy, although the per-cent compositions obtained there differ from the results of the UPS investigation.

Recently, the spectrum of $Me_2SbSbMe_2$ has been recorded[75] and it has been found to be very similar to that of tetramethyldiarsine. Lone-pair ionizations from the *trans* rotamer occur at 7.72 and 9.4 eV while the small peak corresponding to the *gauche* rotamer appears at 8.65 eV. The rotameric composition was determined on the basis of band areas as 88% *trans* and 12% *gauche*, just as in the case of $Me_2AsAsMe_2$. The band arising from the Sb—Sb bonding σ orbital has been tentatively placed at 9.08 eV.

The effect of conformational equilibrium on the PE spectrum is not always so clearly manifested. The low-IE region of the spectrum[76] of Me_2AsSPh contains three distinct bands at 8.21, 8.78 and 9.22 eV, and a shoulder at 9.75 eV. The third band, which is the sharpest and most intense one, can be safely assigned to the antisymmetric phenyl π orbital on the basis of its IE which is very similar to that in benzene or thioanisole. Further assignment, supported with simple empirical LCBO calculations[77], implicitly assumes that no interaction between the lone pairs of sulphur and arsenic occurs. Thus a bonding, a non-bonding and an antibonding MO are formed by combining the lone-pair orbital of sulphur (which is postulated to be of predominant p character) with the two symmetric phenyl π orbitals. Interaction with the highest-lying phenyl σ orbital is also accounted for. As a result, the first ionization is assigned to the antibonding π–n combination while the shoulder at 9.75 eV corresponds to the non-bonding combination. The second peak is attributed to ionization of the arsenic lone pair. Its position, at 8.78 eV, is just slightly stabilized with respect to trimethylarsine, in agreement with expectation. Further support for the validity of the above assignment is provided by the spectrum of $(PhS)_3As$. The equilibrium conformation of this compound (Figure 12), known from NMR measurements, prohibits any interaction between the lone pairs of As and S. The PE bands at 8.12, 9.23 and 10.0 eV have nearly the same position and relative intensity as the 1., 3., and 4. bands[76] in the spectrum of Me_2AsSPh but the second band of the latter is almost absent; just a noisy plateau appears in the region about 8.5 eV. This is in accord with the lesser relative number of As lp electrons in $(PhS)_3As$ and confirms the assignment given for Me_2AsSPh.

Thus the assumption about the absence of lp–lp interaction in dimethyl(phenyl-thio)arsine seems to be justified. One of the possible equilibrium conformations where the interaction is at minimum is shown in Figure 13. The plane of the paper contains the $C_{Ph}SAs$ moiety and bisects the CAsC angle; the phenyl—S—As—methyl dihedral angle (Φ) is 50.2°, and the p-type lone pair of sulphur and the arsenic lone pair are perpendicular to each other. The dipole moment of the molecule is not consistent with this geometry, but the contradiction is resolved by the suggestion that the compound is an equilibrium mixture of two rotamers with $\Phi = 50.2°$ and $\Phi = 230.2°$. There is no lp–lp interaction in

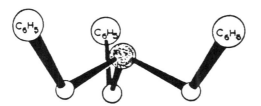

FIGURE 12. The equilibrium conformation of $(PhS)_3As$[76]. Reproduced by permission of Elsevier Sequoia S.A.

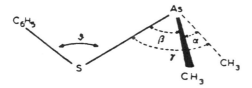

FIGURE 13. Geometrical arrangement of $PhSAsMe_2$[76]. Reproduced by permission of Elsevier Sequoia S.A.

either of these conformers, and a mixture of 45% of the first and 55% of the second one is calculated to have a dipole moment equal to the experimental value. One must keep in mind, however, that it may not be necessary to find a universal geometric model that simultaneously fits the results of solution and gas-phase experiments.

In contrast to the phenylthio-arsines, the PE spectra of dimethyl(methylthio)arsine and dimethyl(methylseleno)arsine[78] bear evidence of significant lp–lp interaction. In both spectra, five individual bands can be identified, the first two being separated from the rest by a 1.5-eV-wide valley. These two bands are found at 8.5 and 9.2 eV in the case of Me_2AsSMe, and at 8.2 and 9.0 eV in the case of $Me_2AsSeMe$. They are assigned to the out-of-phase and in-phase combination of the lone pairs, respectively. Qualitatively, the assignment follows from the fact that the first IE of both compounds is slightly lower than the IE of the lone pair in either Me_2S/Me_2Se or Me_3As (8.71, 8.40 and 8.65 eV) while the splitting between the first two bands is far larger than the IE difference between the corresponding methyl compounds. Three further ionizations are attributed to As—S (As—Se) σ-, C—As σ- and S (Se) sp^n-type non-bonding orbitals, in order of increasing IE. An INDO calculation on Me_2PSMe, whose PE spectrum is very similar to that of the arsenic analogue, has been performed to show the rotational-angle dependence of orbital separation. As seen in Figure 14, the lp–lp interaction is minimized when the S–methyl bond is in the plane bisecting the CPC angle ($\theta = 0°$ or $\theta = 180°$); maximum interaction corresponds to the conformation rotated by 90°. This result confirms that the lone pair of sulphur occupies an orbital of predominant p character, in contrast to the sp^n-type lone pair of P or As. CNDO calculations on Me_2AsSMe and $Me_2AsSeMe$, both assuming $\theta = 150°$, reproduce the trends in experimental ionization energies reasonably well but the separation between the two lp combinations is somewhat too small, implying that the methyl group is still further from the *trans* position in the equilibrium conformation.

There are also PE spectra available of all the six complexes containing one of the above two molecules coordinated as a ligand to a Group 6 metal pentacarbonyl moiety[79]. In these quite similar spectra, the lowest-energy ionization, occurring about 7.5–8 eV, is attributable to six electrons in orbitals of predominant metal d character. The following two bands in the 8.5–10 eV region correspond to the lone pairs of the As-functional ligand. The splitting between these two bands is 1.2–1.3 eV for the complexes of Me_2AsSMe and 1.4 eV in the case of $Me_2AsSeMe$, nearly double that in the spectra of the free ligands. The authors' interpretation ignores lp–lp interaction in the coordinated ligand and, since coordination always occurs through the As atom, considers the lower-IE orbital as a slightly perturbed sulphur or selenium p_π lone pair and the higher-IE orbital as the arsenic sp^n-type lone pair significantly stabilized by interaction with empty metal d orbitals. This approach seems somewhat arbitrary because neither the spectra nor the published results of CNDO calculations (performed for the Cr complexes) contain any direct evidence on the extent of lp–lp interaction in the complexed ligand. The stabilization of the coordinated As lone pair is almost 1.5 eV relative to trimethylarsine; the lone pair of S or Se is at just a little higher IE than in the dimethyl derivatives.

In the course of the same study, the spectra of the three Group 6 metal pentacarbonyl tetramethyldiarsine complexes have also been recorded and interpreted in a similar manner. The situation is common because only one of the arsenic atoms is involved in coordination with the metal centre. Accordingly, the two lp ionizations are attributed to the uncomplexed and to the complexed As atom in order of increasing energy. Disturbing is, however, their separation of only 0.8–0.9 eV which is very moderate relative to the expected complexation effect. In addition, the bands are rather broad and not well resolved. The authors suggest that different conformers of $Me_2AsAsMe_2$ may be present in the complex, just as in the case of the free molecule; this obviously implies significant lone-pair interaction.

In all the nine complexes so far mentioned, the metal d levels are destabilized relative to

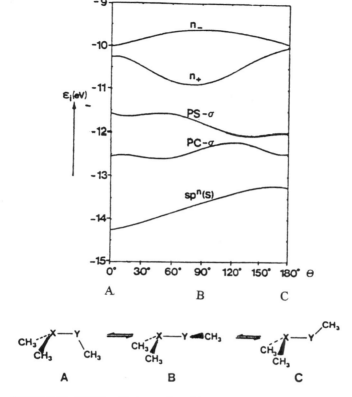

FIGURE 14. INDO orbital energies of Me_2PSMe as functions of the θ rotational angle[78]. Reproduced by permission of VCH Verlagsgesellschaft mbH

the hexacarbonyls by 0.6–0.8 eV demonstrating that the As-functional ligands are better σ donors and weaker π acceptors than CO. There is little difference between the behaviour of the As compounds and their phosphorus analogues, the As ligands being the marginally stronger donors. The metal d band usually has hardly any structure, i.e. the d orbitals remain close to degenerate in these compounds. The small splitting observed has a tendency to increase with growing atomic number of both the metal and the coordinated atom. This effect is attributed by the authors to greater diffuse overlap, leading to a larger covalent interaction in the complexes of heavier elements.

E. Rings of Group 15 Heavy Atoms

Three-, four- and five-membered rings consisting exclusively of Group 15 heavier atoms have been studied. Apart from the spectrum of a molecule containing an Sb_3 ring moiety, only compounds of arsenic, namely the As_3, AsP_2, As_4 and As_5 ring systems, are concerned. The several highest occupied molecular orbitals are, in all cases, divided into two classes: lone-pair combinations on the one hand, and ring bonding orbitals usually described in the framework of the Walsh model[80,81] on the other. The HOMO always corresponds to a lone-pair combination. The Walsh treatment of ring orbitals can strictly be applied only if the molecule is symmetric with respect to the ring plane. This is obviously impossible for rings of P, As or Sb atoms (in addition, the four- and five-membered rings are not planar at all), therefore the Walsh orbitals are not isolated by symmetry from the σ orbitals located outside the ring plane and can mix with the latter. Still, the qualitative picture provided by the Walsh model seems to be adequate for description of the As—P, As—As or Sb—Sb bonding orbitals in these rings. For the purpose of illustration the three lp combinations and the two highest-lying Walsh-type orbitals for the *cis* and *trans* configurations of a P_3 ring are presented schematically, together with their energy levels, in Figure 15 as provided by an MNDO calculation[81].

The most complete investigation of effects of variation of the Group 15 atom is a study on the three analogous cage compounds with nortricyclane skeleton whose PE spectra[82] are demonstrated in Figure 16. Systematic changes may be traced down the P–As–Sb row, the most obvious ones being the decreasing IE and the growing split in the first band (or rather band pair in the case of Sb). This band is assigned to two lone-pair combinations which are degenerate at C_{3v} symmetry (see Figure 15). Splitting of the band may be caused

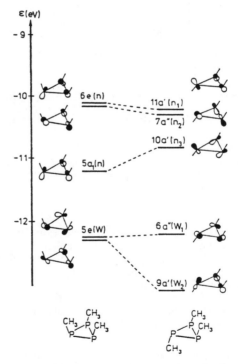

FIGURE 15. Correlation between the molecular orbitals in *cis* and *trans* configurations of trimethylcyclotriphosphane according to an MNDO calculation[81]. Reprinted with permission from Reference 81. Copyright (1985) American Chemical Society

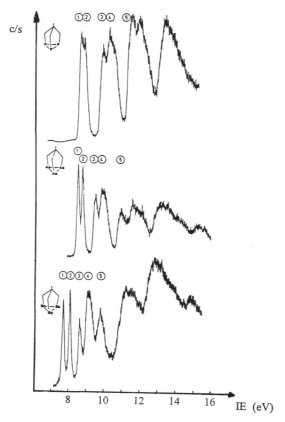

FIGURE 16. HeI PE spectra of P_3-, As_3- and Sb_3-substituted nortricyclanes. Reprinted with permission from Reference 82. Copyright (1985) American Chemical Society

by operation of the Jahn–Teller effect and/or spin–orbit coupling both of which remove degeneracy in the ion[83]. The authors have performed model calculations in an attempt to reproduce the observed splittings as a function of atomic number (SO splitting is taken to be proportional to Z^2) and atomic mass (which determines the parameters for JT splitting) and thus to assess the relative importance of the two mechanisms for each of the compounds. The results of the calculations fit the experiment very well and lead to the conclusion that both effects are simultaneously operative in all three cases, although they are of similar magnitude in the As compound only. The large splitting observed for the Sb derivative is mainly due to spin–orbit coupling while in the P compound the latter is negligible relative to Jahn–Teller splitting.

Further bands have been assigned in accord with the general model: band No.3 (see spectra)—to the third, least repulsive, lone–pair combination; band No. 4—to the highest-lying, doubly degenerate, Walsh orbitals; band No.5 probably arises from an E—C bonding orbital although it may be mixed with C—C and C—H σ orbitals, especially in the case of phosphorus. The IE of all orbitals decreases markedly in the

P–As–Sb sequence; the average difference between the As and Sb compound is 0.7 eV concerning either lp or Walsh orbitals. This value is conspicuously, although not anomalously, high with respect to other pairs of congeners and one can only speculate about the rigid geometry preventing change of hybridization on transition from As to Sb and thus allowing the effect of decreasing atomic ionization energies to come across.

The spectra of simple three-membered rings containing arsenic, namely $(t\text{-Bu})_3P_2As$ and $(t\text{-Bu})_3As_3$, have also been recorded[81] along with $(t\text{-Bu})_3P_3$ and a number of other phosphorus-containing cycles[80]. They bear some resemblance to those of the cage compounds in the sense that five maxima can be identified in the low-energy region but the bands are much wider and not so well resolved, even below 11 eV where surely no ionization from the t-butyl σ orbitals occurs (Figure 17). All these compounds exist in *trans* configuration, consequently the symmetry is lowered to C_s and the orbitals doubly degenerate in the cage compounds are split in the monocycles (cf. Figure 15). Calculations show that lower symmetry also leads to a stronger mixing between the lone pairs, the Walsh orbitals and other σ orbitals of the molecule. Accordingly, the first five ionizations are assigned to five non-degenerate orbitals, the first three to lone-pair combinations, and the fourth and fifth to the two highest-lying Walsh-type orbitals. Correlation of energy levels of the three related compounds is shown in Figure 18. The striking feature of this diagram is the very close similarity of all energy levels of the triphosphane and the triarsane while the orbitals of the mixed derivative are markedly stabilized. The origin of this phenomenon is thought to be the reduced overlap between the atomic orbitals of P and As with respect to the P—P or As—As interaction in which atomic orbitals of the same diffuseness are involved. Repulsion between the lone pairs causes a net destabiliz-

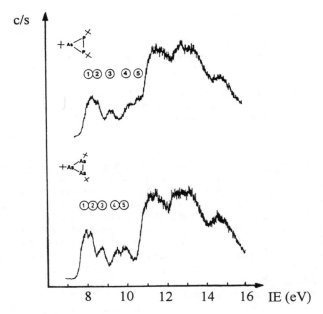

FIGURE 17. HeI PE spectra of $(t\text{-Bu})_3AsP_2$ and $(t\text{-Bu})_3As_3$. Reprinted with permission from Reference 81. Copyright (1985) American Chemical Society

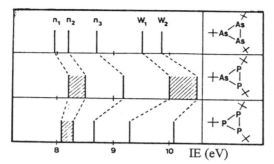

FIGURE 18. Correlation of the first five bands in the PE spectra of $(t\text{-Bu})_3As_3$, $(t\text{-Bu})_3AsP_2$ and $(t\text{-Bu})_3P_3$. Reprinted with permission from Reference 81. Copyright (1985) American Chemical Society

ation of the corresponding orbitals, hence the trend in the lp ionization energies is adequately explained. However, one observes the escalation of the same trend in the case of the Walsh bonding orbitals which renders the above explanation problematic. Given the poor separation of the bands, perhaps an extra unidentified band might be hidden somewhere in the spectrum of the diphosphaarsirane although this is not obvious from intensity relationships. Figure 19 shows the comparison of energy levels of tri-$(t\text{-butyl})$cyclotriarsane and of the tricyclic molecule containing the As_3 ring. The general destabilization of orbitals in the former compound is certainly due to the strong positive inductive effect of three t-butyl groups. The shift of the first two lone-pair combinations and of the second Walsh orbital is significantly less, because these orbitals are stabilized relative to the others when the ring conformation changes from cis to trans (Figure 15).

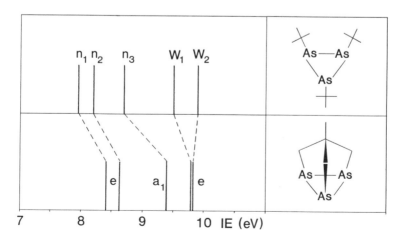

FIGURE 19. Correlation between the first bands in the PE spectra of $(t\text{-Bu})_3As_3$ and $MeC(CH_2)_3As_3$ [81,82]

Larger ring systems have been investigated by Elbel and coworkers[49]. The spectra of $(t\text{-Bu})_4As_4$ and $(CF_3)_4As_4$ are shown in Figure 20 together with the proposed assignment for the t-butyl-substituted compound. The MO diagrams are drawn according to results of SCC-X_α calculations on As_4H_4, however the ordering of orbitals is not based on these calculations but first of all on comparison with spectra of the phosphorus analogues. The assignment follows the generally observed trend that on transition from P to As, bonding σ levels are in most cases shifted more than lone-pair orbital energies. Thus lone-pair combinations belonging to the a_1, e and b_2 species have been assigned to the first, second and fifth band, respectively, while the third and fourth bands have been attributed to ionizations from Walsh-type, mainly As—As bonding orbitals, e and b_1. The assignment for $(CF_3)_4As_4$ is similar, albeit less certain, due to poorer spectrum quality [the sample was known to be contaminated with $(CF_3)_5As_5$]: the first wide band arises from three lone-pair combinations, the second corresponds to Walsh-type orbitals and the third band is supposed to include ionization from the fourth, deepest-lying lp-combination orbital. The latter two bands are shifted by the strong negative inductive effect of trifluoromethyl groups into the $11-13$ eV region, hence it is reasonable to assume that they also contain ionizations from the As—C bonding σ orbitals, more so because little further structure is seen up to 15 eV where ionization from fluorine lone-pair combinations begins. The PE spectrum of Me_5As_5 is also presented, but its detailed assignment has been hindered by severe complications like pseudorotation of the ring and an equilibrium reaction leading

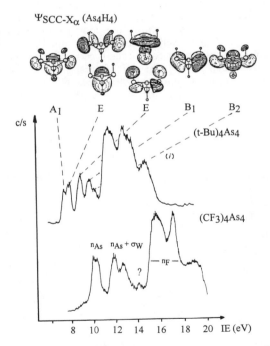

FIGURE 20. HeI PE spectra of $(t\text{-Bu})_4As_4$ and $(CF_3)_4As_4$ and plots of SCC-X_α calculated wavefunctions of As_4H_4[49]. Reproduced by permission of The Royal Society of Chemistry

to unavoidable presence of Me_3As_3. It is only suggested that the bands roughly follow the order $n_{As}-\sigma_{As-As}-\sigma_{As-C}$. It is interesting to compare the IE of the HOMO of the R_nAs_n species: 7.95 eV for $n = 3$, $R = t$-Bu[81]; 7.32 eV for $n = 4$, $R = t$-Bu; and 7.51 eV for $n = 5$, $R = Me$. These values might imply a growing repulsive interaction between As lone pairs with increasing ring size; the trend in average lp ionization energies is, unfortunately, not so easy to follow because of difficulty of identification of spectral bands corresponding to deeper-lying lone-pair combinations (which, by the way, may have a significant degree of mixing with σ-bonding orbitals).

The assignments outlined above rest predominantly on qualitative considerations based on not too obvious energy differences; given the complexity of the electronic structure of the molecules in question, one must treat them with a reasonable degree of caution. It should be remembered, for example, that the first assignment for $(t$-Bu$)_3P_2As$, just like the present assignments for the four-membered rings, placed two ring bonding Walsh orbitals at an intermediate energy between lone-pair combinations[80]. It was supported by semiempirical calculations; nevertheless, it has been modified consequently[81]. Direct analogy should not be drawn between the two cases, but the multitude of factors having a smaller or greater influence on energy levels of various orbitals must be recognized.

F. Compounds with E=C Double Bonds and Pyrrole Analogues

Investigations in this field can be divided into four classes dealing with: (1) arsenic ylides, (2) heavy analogues of pyridine, (3) molecules containing a five-membered azarsole or oxarsole ring with a double bond at As and (4) arsole and azarsole rings with three single bonds at As (pyrrole analogues).

1. Arsenic ylides

The first band in the PE spectra[84] of $Me_3As=CH_2$, $Me_3As=CHSiMe_3$ and $Me_3As=C(SiMe_3)_2$ occurs at 6.72, 6.56 and 6.66 eV, respectively, and is extremely far separated from the rest of the spectrum. It can be assigned without doubt to the special ylidic π bond. The electron pair forming this bond is largely localized on the carbon atom, hence the structure is best described as

$$Me_3As=CH_2 \quad \longleftrightarrow \quad Me_3As^+ - {}^-CH_2$$

The assignment of some further bands is also quite obvious. The second band at 11.4 eV in the spectrum of the methylide and its analogues in the spectra of the silylated compounds clearly correspond to As—C bonding σ orbitals. This band is preceded by another one centred at 9.90 eV in the spectrum of $Me_3As=CHSiMe_3$ and at 9.88 eV in the spectrum of $Me_3As=C(SiMe_3)_2$; the attribution of these ionizations to Si—C σ orbitals is fairly evident as well. The As—C orbitals are stabilized by 0.7 eV relative to their position in trimethylarsine, the IE of the Si—C orbitals, on the other hand, is lowered by the same amount with respect to tetramethylsilane as a result of charge separation according to the zwitterionic canonical form.

The HOMO of $Me_3As=CH_2$ is destabilized by only 0.09 eV relative to the phosphorus analogue while the difference is 0.4 eV when the E—C σ orbitals are compared. (π-Ionization energies of $Me_3P=O$ and $Me_3As=O$ differ by 0.8 eV.) The origin of this phenomenon is a geometric change at the carbanion centre: in contrast to the planar arrangement in the phosphorus ylide, the $AsCH_2$ moiety is shown to be pyramidal and close to sp^3 hybridization by NMR. Thus the HOMO of the arsenic ylide has an increased s_C character which is responsible for its relatively weaker destabilization. The π-bond character of the HOMO is reduced, the electron pair is localized on the carbon atom to a higher degree; in other words, the importance of the zwitterionic canonical structure is greater in the case of the As compound. This is in accord with the lesser ability of the As atom to form multiple bonds, with increased reactivity of arsenic ylides relative to their phosphorus congeners, as well as with the conclusions drawn from Extended Hückel MO calculations. Stabilization of the E—C σ orbitals relative to their position in Me_3E is also slightly greater in the case of As.

Substitution of trimethylsilyl groups for the methylene hydrogens of $Me_3As=CH_2$ has little effect on the IE of the compound's HOMO. Considering the sizeable positive inductive effect of the —$SiMe_3$ group and also the possibility of hyperconjugative destabilization, the near-constancy of the first ionization energies can be interpreted as evidence of compensation of the above factors by the stabilizing effect of delocalization from the carbanion centre to the Si-substituent(s). CNDO/2 calculations on the $Me_3P=CH_2 - Me_3P=CHSiMe_3 - Me_3P=C(SiMe_3)_2$ series have demonstrated a steady decrease of the HOMO p_C coefficient in the above order together with a practically constant total effective charge on the ylidic C atom. Silicon substitution increases the thermal and chemical stability of methylenearsoranes, not through stabilization of the energy level of the HOMO but rather by delocalization of the ylidic π electrons over a more extensive conjugated system and thus decreasing the population of the HOMO at the carbanion centre.

2. Pyridine analogues

Apart from the trimethyl derivatives, pyridine and its four heavy analogues C_5H_5E constitute—to the best of our knowledge—the only other complete series of Group 15 organoelement compounds investigated by UPS[85,86]. The spectra are shown in Figure 21 and the vertical ionization energies of the bands are summarized in Table 8 together with the assignments proposed by the authors. Bismabenzene[86] is thermally very labile, so only the first band stands out clearly; the rest of the spectrum is partly obscured by the high background caused by the presence of decomposition products. Still, bands 2–4 are thought to arise from ionization of C_5H_5Bi itself. The diagram in Figure 22 points out the correlation between orbitals of similar character of benzene, pyridine and the heavier congeners. Assignments for pyridine and phosphabenzene have been supported with high-level ab initio calculations[87] while those for the compounds with a heavier hetero-atom are based on essentially qualitative considerations. An almost perfectly linear correlation, including all five compounds, has been demonstrated between the experimental IE of bands assigned to π-type orbitals and the first (p-) IE of the free heteroatoms. The two π orbitals belonging to the b_1 irreducible representation (symmetry group C_{2v}) have been analysed in this way and the slope of the regression in both cases almost perfectly equals the square of the heteroatom AO coefficient in the given MO according to simple Hückel MO treatment (1/3 and 1/6).

For all five compounds the assignment proposes benzene-like orbitals implying aromaticity of the molecules. The higher-lying b_1 orbital is similar to the symmetric component of the benzene HOMO and has significant heteroatom character; a_2 is analogous to the antisymmetric component and has a node at the heteroatom. For further correlations see Figure 22. The lone pair of the heteroatom is expected to lie in the plane of the molecule and belongs to the a_1 irreducible representation. The orbital sequence is identical in all cases except pyridine, where the ordering of the two highest π orbitals is reversed. Nitrogen has a net electron-withdrawing effect with respect to the carbon atoms and thus stabilizes b_1 relative to a_2 while the other Group 15 atoms are electron-releasing and have just the opposite effect. Both π and σ ring bonding orbitals with significant heteroatom participation have a strong tendency to destabilization on going down the group. As we have seen, this trend can be quantitatively correlated with atomic

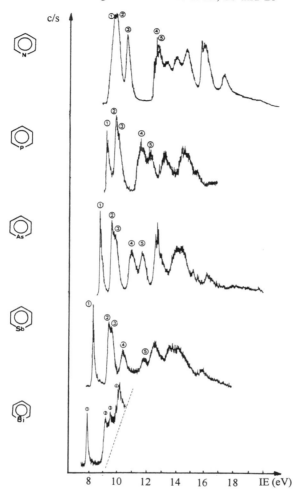

FIGURE 21. HeI PE spectra of C_5H_5E (E = N, P, As, Sb and Bi)[85,86]

TABLE 8. Vertical ionization energies (in eV) and assignment of bands in the PE spectra of the C_5H_5E molecules. The relative sequence of bracketed pair of orbitals is uncertain (see text)[86]

Band No.	C_5H_5N	C_5H_5P	C_5H_5As	C_5H_5Sb	C_5H_5Bi
1.	$\begin{cases} 9.7(a_2,\pi) \\ 9.8(a_1,n) \end{cases}$	$9.2(b_1,\pi)$	$8.8(b_1,\pi)$	$8.3(b_1,\pi)$	$7.9(b_1,\pi)$
2.		$\begin{cases} 9.8(a_2,\pi) \\ 10.0(a_1,n) \end{cases}$	$\begin{cases} 9.6(a_2,\pi) \\ 9.9(a_1,n) \end{cases}$	$\begin{cases} 9.4(a_2,\pi) \\ 9.6(a_1,n) \end{cases}$	$\begin{cases} 9.2(a_2,\pi) \\ 9.6(a_1,n) \end{cases}$
3.	$10.5(b_1,\pi)$				
4.	$12.5(b_2,\sigma)$	$11.5(b_2,\sigma)$	$11.0(b_2,\sigma)$	$10.4(b_2,\sigma)$	$10.2(b_2,\sigma)$
5.	$12.6(b_1,\pi)$	$12.1(b_1,\pi)$	$11.8(b_1,\pi)$	$11.7(b_1,\pi)$	

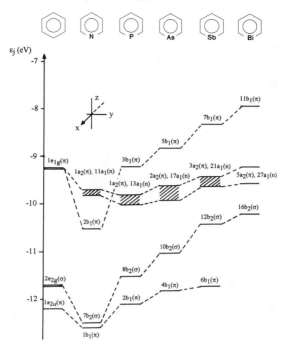

FIGURE 22. Orbital correlation diagram for C_6H_6 and C_5H_5E (E = N, P, As, Sb and Bi). Reprinted from Reference 86, Copyright (1976), with permission from Pergamon Press Ltd, Headengton Hill Hall, Oxford OX3 OBW, UK

p-ionization energies in the case of π orbitals, most likely thanks to a negligible contribution from atomic s orbitals. The steep fall of the IE of the $b_2(\sigma)$ orbital is probably assisted by changing hybridization of the heteroatom, i.e. decreasing s character of the σ bonds. The energy of the a_1 and a_2 orbitals, on the other hand, changes little throughout the whole series. This constancy is self-evident in the case of a_2 and it might also be expected for the lone pair (a_1) showing similar behaviour in other series of compounds (e.g., trimethyl derivatives). The explanation is again the increasing reluctance of s electrons to take part in chemical bonding with increasing atomic number and consequently compensation of lower atomic ionization energies by increasing s character of the lone pair. This concept (at least on its own) is not really satisfactory for the interpretation of properties of simple R_3E derivatives in view of the observed rather moderate variation of valence angles accompanying the expected change of hybridization in the P–As–Sb–Bi sequence. In the C_5H_5E series, however, the CEC angle decreases rather sharply[88]: E = N, 117°; E = P, 101°; E = As, 97°; E = Sb, 93°; E = Bi, 90°—the range corresponds to a change from nearly perfect sp^2 hybridization at nitrogen to virtually pure p character of the σ bonds at bismuth, giving a boost to the credibility of the above explanation.

The spectra do not provide appreciable information with respect to the relative ordering of a_1 and a_2. The corresponding bands (1. and 2. in the spectrum of pyridine, and 2. and 3. in all other cases) are very poorly resolved and the ordering proposed in Table 8 is 'mainly inspired guesswork' (supported by later *ab initio* calculations in the case of pyridine and phosphabenzene). In order to clarify the situation, additional UPS investigation has been performed on C_5H_5P and C_5H_5As: spectra obtained with HeII exciting radiation (Figure 23) and angle-dependent HeI spectra have been recorded[89]. The most unambiguous conclusions can be drawn concerning the first spectrum band: it has markedly lower relative intensity at higher photon energy and it is associated with a relatively high value of the β asymmetry parameter, as follows from the angular distribution of photoelectrons. Both these results argue for significant phosphorus 3p or arsenic 4p participation in the HOMO, in accord with the original assignment. The question about individual assignments for the second and third band still remains somewhat obscure. The relative intensity of band No.3—in fact, rather a shoulder on the high-IE side of band No.2—seems to be reduced in the HeII spectra, implying greater P or As character, but the phenomenon is not very obvious: the composite band (2. + 3.) does have different shapes in the HeI and HeII spectra but its half-width is approximately equal in both (even a little greater in the latter case). The angular dependence of bands 2. and 3. is nearly the same, indicating very similar values of β for the a_1 and a_2 orbitals and, of course, giving no information about their sequence.

Gas-phase XPS studies of arsabenzene and phosphabenzene were performed by Ashe and coworkers[90] in order to interpret their anomalously low basicity relative to pyridine. Correlation between core ionization energies and proton affinities has been invoked for the explanation. Good correlation can be expected if 'there is no change in geometry during protonation, or if all molecules of a given series of compounds undergo similar geometrical changes on protonation'. Well-established linear correlations exist for oxygen and nitrogen compounds (including pyridine) and phosphines as well. A comparison of the proton affinity and core-ionization data for arsabenzene (and phosphabenzene) with those

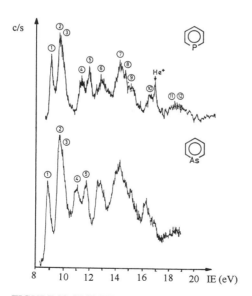

FIGURE 23 HeII PE spectra of C_5H_5E (E = P and As). Reproduced from Reference 89 by permission of Helvetica Chemica Acta

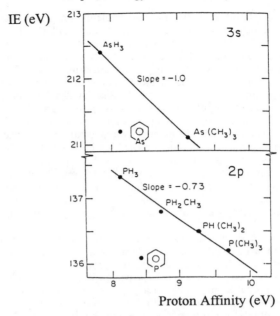

FIGURE 24. Correlation between core ionization energy
and proton affinity for arsenic and phosphorus compounds.
Reprinted with permission from Reference 90. Copyright
(1979) American Chemical Society

for arsines (and phosphines) shows a large deviation from the correlation line established
by the reference compounds (see Figure 24). These experimental findings are consistent
with the suggestion that, while for the aliphatic arsenic (and phosphorus) compounds
bond angles change significantly on protonation, the ring structure of arsabenzene (and
phosphabenzene) prevents the necessary rearrangement on protonation which would
decrease the diffuseness (s character) of the lone pair through rehybridization (increased p
character). For the aliphatic and aromatic nitrogen compounds no geometric rearrangement takes place upon protonation; that is why the corresponding points are on the same
line (higher p character of the nitrogen lone pair).

3. Five-membered rings with an As=C double bond

The simplest such heterocycle investigated by UPS is 2-methyl-1,3-azarsole, studied
together with 1,3-azaphosphole and 2-methyl-1, 3-azaphosphole[91]. The spectra of the
three compounds are qualitatively very similar, containing three relatively low-energy
ionizations ahead of a broader band system ascribed to σ-bonding orbitals. The first two
bands (7.76 and 8.80 eV in the case of the As compound) arise from the two highest π
orbitals of the molecule while the third band corresponds to the lone pair of the heavy
heteroatom (9.79 eV for As). The assignment has been aided by *ab initio* calculations on the
unsubstituted rings (this has been so far the only instance of assignment of the PE spectrum
of an organoarsenic compound with the help of non-empirical calculations) and it is in
accord with qualitative expectations, too. The HeII PE spectrum of 2-methyl-1,3-azaphos-

phole was also recorded where the relative intensity of the third band is greatly reduced, thus confirming its assignment to an orbital of predominant phosphorus character. The IE difference between the analogous P and As compounds is the greatest in the case of the first π band, less for the second π band, and the lone pairs differ insignificantly. The energy shifts of these orbitals follow the same trends established for the pyridine analogues, especially if one considers that, according to the calculations, the HOMO of $1H$-1,3-azaphosphole and -azarsole is largely localized on the C=E double bond. The calculations also indicate a similar narrowing of the valence angle at E in C_3H_3NE: 106.3° for E=N, 88.9° for E=P and 83.6° for E=As. (It should be noted that experimental valence angles in C_5H_5P and C_5H_5As were reproduced by the same method within an error of 1°.)

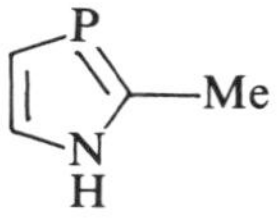

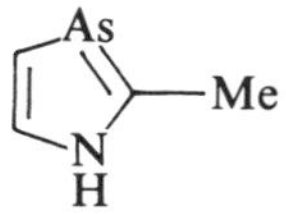

1H-1,3-azaphosphole   2-methyl-1H-1,3-azaphosphole   2-methyl-1H-1,3-azarsole

Both the calculated molecular geometry and calculated ΔE of bond separation reactions of the C_3H_3NE rings point to their aromaticity regardless of E being nitrogen, phosphorus or arsenic; some calculations even attribute the greatest aromatic character to the As compound. UPS confirms the conclusions of the calculations concerning the electronic structure of these molecules: they all possess three π orbitals of aromatic type, two of which have quite low IE while the third one lies significantly deeper. The HOMO is more concentrated on the E=C bond, while the other two π orbitals have more C and N character but extensive conjugation is well illustrated by comparing the HOMO of C_3H_3NP (8.44 eV) with isolated P=C double bonds in HP=CH$_2$ (10.30 eV)[92], HP=CH—CH$_3$ (9.75 eV)[92] or CH$_3$—P=CH$_2$ (9.69 eV)[93]. Similar conclusions have been drawn from the study of the benzazarsoles and benzoxarsole shown below[94]. The orbital diagram in Figure 25, based on CNDO/S calculations, offers a general qualitative picture of the electronic structure of these molecules. They possess altogether five π orbitals delocalized over the whole molecule, and three of these are expected to lie high enough to give rise to distinct bands in the PE spectrum outside the region of σ ionizations. Indeed, the more or less resolved band systems in the spectral region below 10 eV IE are attributable to four ionizations in all cases: three from π orbitals and one, the one with the highest IE (9.3–9.7 eV for the benzazarsoles and 9.9 eV in the case of the benzoxarsole), from the orbital hosting the As lone pair. Like for all other

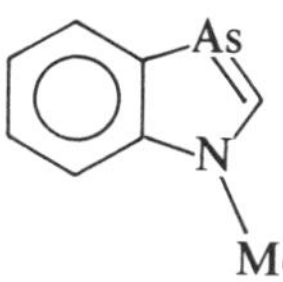

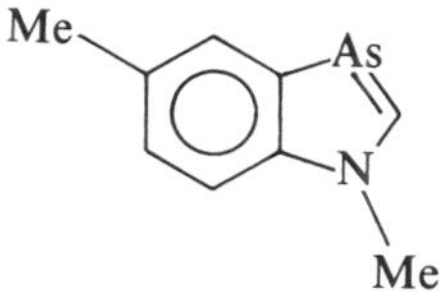

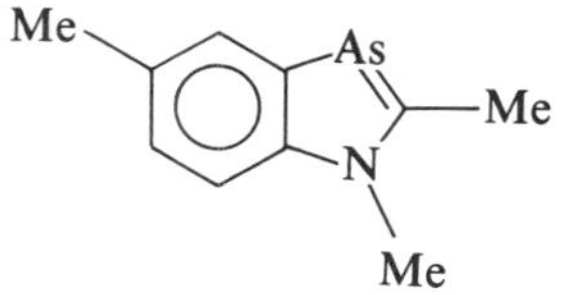

1-methyl-1,3-benzazarsole 1,6-dimethyl-1,3-benzazarsole 1,2,6-trimethyl-1,3-benzazarsole

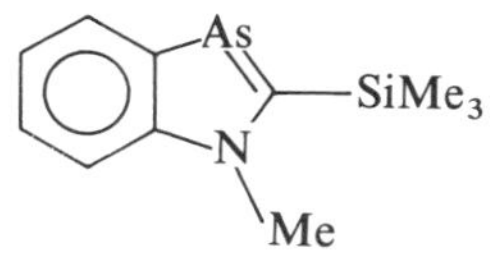

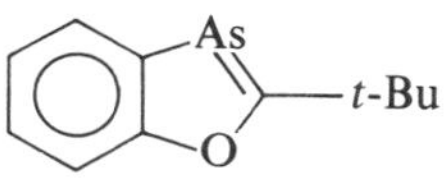

1-methyl-2-trimethylsilyl-1,3-benzazarsole   2-t-butyl-1,3-benzoxarsole

As-aromatic compounds, the calculations indicate a significant share of the As=C π bond in the HOMO. Conjugation with the π system of the C_6 ring lifts the HOMO of the benzazarsoles to 7.1–7.5 eV, i.e. noticeably above its energy level in 2-methyl-1,3-azarsole. In the orbital diagram in Figure 25 one sees that the first, third and fifth π orbitals possess more or less significant As character, consequently their IE must shift when P is substituted for As; the second and fourth π orbitals, on the other hand, should be quite insensitive to such an exchange. Among the compounds investigated there are two pairs of related molecules suitable for direct comparison: 1-methyl-1,3-benzazaphosphole and -benzazarsole, and 2-t-butyl-1,3-benzoxaphosphole and -benzoxarsole. Unfortunately, the large conjugated π system of these molecules has a strong levelling effect so the experimentally observed IE shifts are too small to identify reliable trends, although the shift of IE_1 and IE_3 is indeed marginally greater than others in the spectra of the first pair of compounds. Insignificant IE changes following chemical substitution leave little ground for more detailed assignment on a qualitative basis; thus further assignment (according to the orbital sequence in Figure 25) rests on the results of CNDO/S calculations. The significance of the latter is difficult to estimate, since all orbital energies calculated for the benzazarsoles match the experimental ionization energies within 0.2 eV (most are within 0.1 eV) which is really a degree of accuracy unprecedented not only for semiempirical but also for non-empirical Hartree–Fock calculations and quite unexpected in view of the approximative nature of Koopmans' theorem.

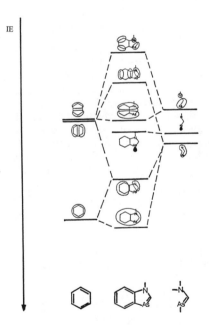

FIGURE 25. Orbital diagram for 1H-1,3-benzazarsole according to CNDO/S calculations[94]. Reproduced by permission of Elsevier Sequoia S.A

4. Pyrrole analogues

Non-aromaticity of As-analogues of pyrroles has been proved beyond all doubt by various chemical and spectroscopic methods as well as by quantum chemical calculations[91]. Two such compounds have been studied by UPS: 1-phenyl-2,5-dimethylarsole and 2-*t*-butyl-3-methyl-1,3-benzazarsole, and both spectra have been interpreted accordingly. The low-IE region of the spectrum of the former compound[95] features three bands at

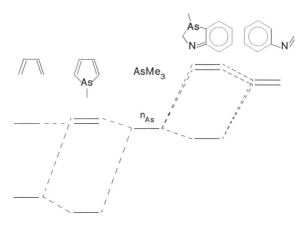

1-phenyl-2,5-dimethylarsole   2-*t*-butyl-3-methyl-1,3-benzazarsole

8.0, 8.6 and 9.15 eV. The third band is attributed doubtlessly to phenyl π orbitals on the basis of its position and double intensity. Its absence of splitting implies that interaction between the phenyl π system and the As lone pair is insignificant. The IE of the highest π orbital is 8.57 eV in cyclopentadiene and about 8.28 eV in the methylcyclopentadienes[96], therefore a value close to 8 eV can be expected for a doubly methylated cyclopentadiene. The HOMO of 1-phenyl-2,5-dimethylarsole has just that IE indicating a very small degree of conjugation between the butadiene fragment of the five-membered ring and the As lone pair. Accordingly, the latter is found at 8.6 eV, i.e. only 0.05 eV above its energy level in trimethylarsine. 1-phenyl-2,5-dimethylphosphole has an almost identical PE spectrum; saturation of its five-membered ring shifts the energy of the phosphorus lone pair only by 0.15 eV, once more confirming that very little conjugation takes place. Similarly, the first two (strongly overlapping) bands in the PE spectrum of 2-*t*-butyl-3-methyl-1,3-benzazarsole[97] (8.25 and 8.45 eV) have been assigned to π orbitals delocalized over all ring atoms but arsenic, the band at 9.66 eV to the pyridine-like lone pair of the N atom and the band at

FIGURE 26. Interaction of the As lone pair with the conjugated π-system in 1*H*-arsole and in 3*H*-1,3-benzazarsole[97]

8.96 eV to the As lone pair. The latter is somewhat stabilized relative to its energy in the previous compound, which is easily explained allowing that there is yet a slight interaction between the lone pair and the π system of the rings (Figure 26). In the arsole ring the orbital hosting the lone pair has zero overlap with the π-HOMO for symmetry reasons; the π orbital of proper symmetry lies much deeper, hence the combination of predominant n_{As} character is destabilized. In the benzazarsole ring system, however, interaction of the lone pair with the π orbitals lying above it is not forbidden by symmetry and the stabilized combination carries the largest contribution from the former. The shift is not large in either case, signifying again the small extent of conjugation.

G. X-ray Photoelectron Spectroscopy of As-, Sb- and Bi-Organic Derivatives

As mentioned in the introduction, X-ray photoelectron spectroscopy (XPS) is based on core-electron ionizations. Core electrons retain their atomic character to a great extent, therefore XPS is basically an analytical technique, which is able to provide an elemental analysis of the investigated sample. Core electrons, however, suffer small changes in ionization energy as their chemical environment is changed. As a result, atoms in different environments and in different oxidation states may be identified in the spectrum. The chemical structure dependence of carbon 1s ionization energy, for example, has been nicely demonstrated in the case of several organic compounds[98]; chemical shifts for atoms other than carbon have also been reported in great number[99].

By considering the above facts it is not surprising that XPS has only limited application in the investigation of organic As-, Sb- and Bi-derivatives. On the basis of the studied molecular systems and the obtained experimental data, papers published in this field can be grouped as follows:

(1) XPS studies of Group 15 organometallic compounds with the intention of structure and bonding elucidation, reporting certain core-electron ionization energy data of the metal atom (e.g. As 3s or Sb 3d).

(2) Investigation of transition metal complexes which contain—among others— organoarsenic ligand(s). Core binding energies of the central atom and that of ligands (including As) are determined and occasionally compared.

(3) The same kind of compounds as in (2), but no attention is paid to the As core ionizations. In this case examinations are centred on structural problems and dependence of transition metal nd binding energies upon various factors. This type of research has no close relation to the subject of this chapter.

A wide variety of antimony compounds, including organoantimony oxides, halides and simple organoantimony compounds of both typical oxidation states (i.e. III and V), where examined by XPS[100,101]. The antimony 3d(3/2, 5/2) binding energies range from 538.4, 529.3 eV for Ph_3SbS to 540.8, 531.2 eV for Me_3SbBr_2. It is clearly demonstrated that correlation of Sb 3d binding energies with formal oxidation states and gross structural differences is rather poor. For example, Sb 3d binding energies are lower by 0.2 eV for Ph_3SbS than for Ph_3Sb, or complexation of Me_3SbCl_2 with $SbCl_3$ yields shifts only within the range of standard deviations. It is evident from both publications that, on this level of measurements (FWHM = 1.8–2.3 eV and 1.5–1.8 eV), XPS is not a powerful technique for studying structure and bonding in organoantimony compounds.

By use of a monochromatized synchrotron radiation source it is possible to obtain better instrumental resolution, and the variable photon energy enables one to find the optimum cross section for valence and core ionization studies. These instrumental improvements are demonstrated by the XPS studies of a number of solid Sn, In, Sb and Pb organometallic and inorganic compounds[102]. Sb 4d(3/2, 5/2) binding energies were determined for Ph_4SbCl, the only Sb-organic derivative in the series, with improved instrumental resolution (0.43 eV) and line width (1.01 eV). The corresponding values are 36.12 and

34.88 eV, respectively, showing approximately 2 eV destabilization relative to the free atom 4d binding energies (38 and 37 eV, respectively[103]).

Transition metal complexes are ideally suited for study by XPS, because the electronic environment of the transition metal atom or ion can be probed directly.

Brant and coworkers investigated a series of cobalt and iron complexes with o-phenylenebis(dimethylarsine) (das) ligand[104]. The N 1s, As 3d and Fe, Co 2p binding energies of these $[MLL' (das)_2]^{m+}$ (L = L' = Cl, Br, NCS; L = NO, L' = Cl, Br, I, NCS; L = Cl, L' = CO, CH_3CN) have been determined. While N 1s and metal 2p binding energies have been found to be sensitive to electronic features and chemical environment, As 3d binding energies are unaffected by the other ligands and are independent of m ionic charge. The average As 3d binding energies are 43.72, 43.70 and 44.10 eV for m = 0, 1 and 2, respectively. These experimental findings led the authors to the conclusion that the effects of the Madelung potential on the core binding energies are small.

The same constancy of As 3d binding energies of the attendant ligand of bis(diphenylarsino)methane (dam) has been found in the dipalladium complex $Pd_2(dam)_2Cl_2$ upon insertion of small molecules into the Pd—Pd bond[105]. The As 3d binding energy in $Pd_2(dam)_2Cl_2$ is 43.8 eV and that of the insertion products $Pd_2(dpm)_2 (\mu - X)Cl_2$ (X = CO or SO_2) are 44.0 eV in both cases. Also, small variations in Pd 3d binding energies, resulting from the insertion reactions, were experienced. All these facts indicate that the whole process occurs with minimal changes in the atomic charges.

In another detailed study, about 70 nickel compounds were studied by X-ray photoelectron spectroscopy[106]. Ni 2p(3/2, 1/2) binding energies and shake-up satellites[99] associated with them were determined. In some cases N 1s or As 3p(3/2) binding energies of ligands were also measured. Basically, the dependence of Ni 2p binding energies and shake-up satellites on stereochemistry, magnetic properties and ligand surroundings was investigated.

Ni 2p and As 3p binding energies, nickel oxidation states (OS) and geometry of complexes containing das and QAS, tris(o-diphenylarsinophenyl)arsine ligands are summarized in Table 9. The organophosphorus analogue of the last complex is also included for the sake of comparison [QP = tris(o-diphenylphosphinophenyl)-phosphine].

In the case of compounds 1,2 and 3 it is clearly demonstrated that Ni 2p binding energies can be correlated with Ni oxidation states, i.e. Ni 2p binding energies increase in the order II < III < IV. This does not hold for the whole series, partly because arsenic atoms donate efficiently to the metal.

Another factor influencing Ni 2p binding energies is stereochemistry. In agreement with the concluding remarks of the publication[106], it is reflected also by the above series that Ni 2p binding energies increase in the order planar < (tetrahedral) < octahedral when nickel is bonded to the same ligand.

Ni 2p ionization energies are very similar in both complexes 4 and 5. This fact points to the same electronic effect of organoarsenic and organophosphorus ligands on the central

TABLE 9. Oxidation state, geometry and binding energies (eV) of compounds of nickel[106]

Compounds of nickel	No	OS of Ni	Geometry	Ni 2p(3/2)	Ni 2p(1/2)	As 3p(3/2)
$Ni(das)_2Cl_2$	1	II	sp	853.7	870.8	—
$[Ni(das)_2Cl_2][Cl]$	2	III	Octahedral	853.9	871.6	141.0
$[Ni(das)_2Cl_2][ClO_4]_2$	3	IV	Octahedral	854.7	872.6	—
$[Ni(QAS)Cl][ClO_4]$	4	II	tbp	853.3	871.0	—
$[Ni(QP)Cl][ClO_4]$	5	II	tbp	853.2	870.6	—

atom. Similarity between the chemical behaviour of P- and As-organic derivatives has been frequently experienced in the framework of this chapter.

Several other transition metal complexes containing organoarsenic ligands have been investigated by the XPS technique. In these studies structural problems[107,108] and dependence of central atom binding energies upon the nature of the ligands and the metal oxidation state[109] have been investigated. These studies do not add further information to the PES of As-, Sb- and Bi-organic compounds and lie outside the scope of this chapter.

III. CONCLUDING REMARKS

There are relatively few publications dealing with the PES of As-, Sb- and Bi-organyls. Systematic investigation of chemically related series of compounds is rare and is limited to the simplest trivalent derivatives. Apart from a few exceptions, calculations on the As, Sb or Bi compounds themselves are absent or performed on a rather low empirical or semi-empirical level. The interpretation of the spectra is usually based on qualitative considerations and on calculations performed for analogous compounds of phosphorus. As a result, in most cases only the low-ionization energy part of the spectra is interpreted and the assignments are often somewhat ambiguous, bearing a certain degree of arbitrariness.

If one compares the results obtained for related compounds of the Group 15 elements, the properties of the nitrogen derivative are usually clearly out of line with those of the heavier congeners. Much smaller differences are experienced in the P–As–Sb–Bi sequence, the compounds of phosphorus and arsenic usually showing the greatest similarity. Substitution of Sb for As has a general destabilizing effect which is most pronounced for E—C (and seemingly also for E—E) bonding σ orbitals; the ionization energy of the lone pair is influenced to a lesser extent. Data on Bi compounds are too scarce to allow any generalized statements.

The lone pair of formally sp^3 hybridized As and Sb apparently lacks conjugation with carbon-based π systems. Quite the opposite, the E=C double bond has a marked tendency to take part in conjugative interaction and to be incorporated into aromatic systems.

IV. REFERENCES

1. E. Heilbronner and J. P. Maier, in *Electron Spectroscopy*, Vol. 1 (Eds. C. R. Brundle and A. D. Baker), Chap. 5, Academic Press, London, New York, San Francisco, 1977, pp. 205–292.
2. M. Thompson, P. A. Hewitt and D. S. Wooliscroft, in *Handbook of X-Ray and Ultraviolet Photoelectron Spectroscopy* (Ed. D. Briggs), Chap. 10, Heyden, London, Philadelphia, Rheine, 1977, pp. 341–379.
3. J. Berkowitz, in *Electron Spectroscopy*, Vol. 1 (Eds. C. R. Brundle and A. D. Baker), Chap. 7, Academic Press, London, New York, San Francisco, 1977, pp. 355–433.
4. A. M. Bradshaw, L. S. Cederbaum and W. Domcke, in *Structure and Bonding*, Vol. 24, Springer-Verlag, New York, Heidelberg, Berlin, 1975, pp. 133–169.
5. C. T. Au, A. F. Carley and M. W. Roberts, *Int. Rev. Phys. Chem.*, **5**, 57 (1986).
6. H. Bock and B. Solouki, *Angew. Chem.*, **93**, 425 (1981).
7. N. P. C. Westwood in *Molecular Ions* (Eds. J. Berkowitz and K. Groeneveld), NATO ASI Series, Series B: Physics, Vol. 90, Plenum Press, New York and London, 1983, pp. 275–278.
8. J. C. Green, *Structure and Bonding*, Vol. 43, Springer-Verlag, New York, Heidelberg, Berlin, 1981, pp. 37–112.
9. C. Furlani and C. Cauletti, *Structure and Bonding*, Vol. 35, Springer-Verlag, Berlin, Heidelberg, New York, 1978, pp. 119–169.
10. D. L. Lichtenberger and G. E. Kellogg, in *Gas Phase Inorganic Chemistry* (Ed. D. Russel), Plenum Press, New York and London, 1989, pp. 245–277.
11. Ch. Elschenbroich and A. Salzer, *Organometallics*, VCH, Weinheim, 1989, p. 147.
12. J. W. Rabalais, *Principles of Ultraviolet Photoelectron Spectroscopy*, Wiley, New York, 1977.

13. J. H. D. Eland, *Photoelectron Spectroscopy*, Butterworths, London, 1974.
14. C. R. Brundle and A. D. Baker (Eds.), *Electron Spectroscopy: Theory, Techniques and Applications*, Vols. 1–4, Academic Press, London, New York, San Francisco, 1977–81.
15. J. A. R. Samson, in *Electron Spectroscopy*, Vol. 4 (Eds. C. R. Brundle and A. D. Baker), Chap. 6, Academic Press, London, New York, Toronto, Sidney, San Francisco, 1981, pp. 361–396.
16. J. A. R. Samson, *Phys. Rep.*, **28**, 303 (1976).
17. T. Koopmans, *Physica*, **1**, 104 (1933).
18. J. P. Stewart, *J. Computer-Aided Molecular Design*, **4**, 1 (1990).
19. W. von Niessen, in *Molecular Ions* (Eds. J. Berkowitz and K. Groeneveld), NATO ASI Series, Series B: Physics, Vol. 90, Plenum Press, New York and London, 1983, pp. 355–406.
20. F. Ecker and G. Hohlneicher, *Theor. Chim. Acta*, **25**, 189 (1972).
21. L. S. Cederbaum and W. Domcke, *Adv. Chem. Phys.*, **36**, 205 (1977).
22. W. von Niessen, L. S. Cederbaum, W. Domcke and J. Schirmer, in *Computational Methods in Chemistry* (Ed. J. Bargon), Plenum Press, New York, 1980.
23. J. W. D. Conolly, in *Modern Theoretical Chemistry*, Vol. VII (Ed. G. A. Segal), Plenum, New York, 1977, p. 105.
24. G. Granozzi, *Invited Lecture at XVIIIth EUCMOS*, Amsterdam, 1987.
25. L. Szepes, T. Korányi, G. Náray-Szabó, A. Modelli and G. Distefano, *J. Organomet. Chem.*, **217**, 35 (1981).
26. S. Elbel, H. Bergmann and W. Enßlin, *J. Chem. Soc., Faraday Trans. 2*, **70**, 555 (1974).
27. I. Nenner and J. A. Beswick, in *Handbook on Synchrotron Radiation*, Vol. 2 (Ed. G. V. Marr), Chap. 6, North-Holland, Amsterdam, 1987, pp. 355–466.
28. T. A. Carlson and C. P. Anderson, *Chem. Phys. Lett.*, **10**, 561 (1971).
29. R. Spohr, P. M. Guyon, W. A. Chupka and J. Berkowitz, *Rev. Sci. Instrum.*, **42**, 1872 (1971).
30. T. Baer, W. B. Peatman and E. W. Schlag, *Chem. Phys. Lett.*, **4**, 243 (1969).
31. M. J. Hubin-Franskin, J. Delwich, P. M. Guyon and I. Nenner, in *Molecular Ions* (Eds. J. Berkowitz and K. Groeneveld), NATO ASI Series, Series B: Physics, Vol. 90, Plenum Press, New York and London, 1983, pp. 279–282.
32. T. Baer, in *Gas Phase Ion Chemistry*, Vol. 1 (Ed. M. T. Bowers), Chap. 5, Academic Press, 1979, pp. 153–196.
33. T. Baer, P. M. Guyon, I. Nenner, A. Tabché-Fouhaille, R. Botter, L. F. A. Ferreira and T. Govers, *J. Chem. Phys.*, **70**, 1585 (1979).
34. K. Müller-Dethlefs and E. W. Schlag, *Annu. Rev. Phys. Chem.*, **42**, 109 (1991).
35. L. A. Chewter, M. Sander, K. Müller-Dethlefs and E. W. Schlag, *J. Chem. Phys.*, **86**, 4737 (1987).
36. Reiser and coworkers, quoted in Reference 34, in the legend to Figure 12 on p. 129.
37. T. Baer, in *Advances in Chemical Physics*, Vol. LXIV (Eds. I. Prigogine and S. A. Rice), Wiley, New York, 1986, pp. 111–202.
38. T. Baer, J. Booze and K. M. Weitzel, in *Vacuum Ultraviolet Photoionization and Photodissociation of Molecules and Clusters* (Ed. C. Y. Ng), Chap. 5, World Scientific, Singapore, 1991, pp. 259–296.
39. S. Elbel, H. tom Dieck and R. Demuth, *J. Fluorine Chem.*, **19**, 349 (1982).
40. R. Gleiter, W. D. Goodmann, W. Schäfer, J. Grobe and J. Apel, *Chem. Ber.*, **116**, 3745 (1983).
41. M. Grodzicki, H. Walther and S. Elbel, *Z. Naturforsch.*, **39b**, 1319 (1984).
42. A. W. Potts and W. C. Price, *Proc. R. Soc. London, Ser A*, **326**, 181 (1972).
43. G. R. Branton, D. C. Frost, C. A. McDowell and I. A. Stenhouse, *Chem. Phys. Lett.*, **5**, 1 (1970).
44. S. Cradock and R. A. Whiteford, *J. Chem. Soc., Faraday Trans. 2*, **68**, 281 (1972).
45. D.-S. Yang, G. M. Bancroft, L. Dignard-Bailey, R. J. Puddephatt and J. S. Tse, *Inorg. Chem.*, **29**, 2487 (1990).
46. N. N. Greenwood and A. Earnshaw, *Chemistry of the Elements*, Chap. 13, Pergamon Press Ltd., 1984 (reprinted with corrections in 1986 by A. Wheaton & Co. Ltd., Exeter), p. 650.
47. A. D. Walsh, *J. Chem. Soc.*, 2296 (1953).
48. Landolt-Börnstein, *Numerical Data and Functional Relationships in Science and Technology*, New Series, Group II: *Atomic and Molecular Physics*, Vol. 15 (Supplement to Vol. II/7): *Structure Data of Free Polyatomic Molecules* (Editor in chief: O. Madelung), Springer-Verlag, Berlin–Heidelberg, 1987.
49. S. Elbel, H. Egsgaard and L. Carlsen, *J. Chem. Soc., Dalton Trans.*, 195 (1988).
50. J. J. Kaufman and W. S. Koski, *J. Am. Chem. Soc.*, **82**, 3262 (1960).
51. L. Szepes and T. Baer, *J. Am. Chem. Soc.*, **106**, 273 (1984).

52. P. R. Das, T. Nishimura and G. G. Meisels, *J. Phys. Chem.*, **89**, 2808 (1985).
53. K. Norwood, A. Ali, G. D. Flesch, and C. Y. Ng, *J. Am. Chem. Soc.*, **112**, 7502 (1990).
54. S. Nagaoka, S. Suzuki, U. Nagashima, T. Imamura and I. Koyano, *J. Phys. Chem.*, **94**, 2283 (1990).
55. G. Distefano, S. Pignataro, L. Szepes and J. Borossay, *J. Organomet. Chem.*, **102**, 313 (1975).
56. T. P. Debies and J. W. Rabalais, *Inorg. Chem.*, **13**, 308 (1974).
57. L. Szepes, Dissertation, L. Eötvös University, Budapest, 1976, p. 55.
58. N. N. Greenwood and A. Earnshaw, *Chemistry of the Elements*, Chap. 13, Pergamon Press Ltd., 1984 (reprinted with corrections in 1986 by A. Wheaton & Co. Ltd., Exeter), p. 694.
59. D.-S. Yang, G. M. Bancroft, R. J. Puddephatt and J. S. Tse, *Inorg. Chem.*, **29**, 2496 (1990).
60. A. Flamini, E. Semprini, F. Stefani, G. Cardaci, G. Bellachioma and M. Andreocci, *J. Chem. Soc., Dalton Trans.*, 695 (1978).
61. H. Daamen, A. Oskam and D. J. Stufkens, *Inorg. Chim. Acta*, **38**, 71 (1980).
62. J. N. Louwen, R. Hengelmolen, D. M. Grove, D. J. Stufkens and Ad Oskam, *J. Chem. Soc., Dalton Trans.*, 141 (1986).
63. S. Elbel and H. tom Dieck, *Z. Naturforsch.*, **31b**, 178 (1976).
64. S. Elbel, H. tom Dieck and R. Demuth, *Z. Naturforsch.*, **31b**, 1472 (1976) and references cited therein.
65. S. Elbel and H. tom Dieck, *Z. Anorg. Allg. Chem.*, **483**, 33 (1981).
66. S. Elbel, H. Egsgaard and L. Carlsen, *J. Chem. Soc., Dalton Trans.*, 481 (1987).
67. J. C. Green, D. R. Lloyd, L. Galyer, K. Mertis and G. Wilkinson, *J. Chem. Soc., Dalton Trans. 2*, 1403 (1978).
68. S. Elbel, M. Grodzicki, L. Pille and G. Rünger, *J. Mol. Struct.*, **175**, 441 (1988).
69. S. Cradock, E. A. V. Ebsworth, W. J. Savage and R. A. Whiteford, *J. Chem. Soc., Faraday Trans. 2*, **68**, 934 (1972).
70. W.-W. Du Mont, H. J. Breunig, H. Schumann, H. Götz, H. Juds and F. Marschner, *J. Organomet. Chem.*, **96**, 49 (1975).
71. R. Demuth, *Z. Naturforsch.*, **32b**, 1252 (1977).
72. G. Distefano, L. Zanathy, L. Szepes and H. J. Breunig, *J. Organomet. Chem.*, **338**, 181 (1988).
73. M. Thompson, P. A. Hewitt and D. S. Wooliscroft, in *Handbook of X-Ray and Ultraviolet Photoelectron Spectroscopy* (Ed. D. Briggs), Chap. 10, Heyden, London, Philadelphia, Rheine, 1977, pp. 351–355.
74. A. H. Cowley, M. J. S. Dewar, D. W. Goodman and M. C. Padolina, *J. Am. Chem. Soc.*, **96**, 2648 (1974).
75. H. J. Breunig, L. Szepes and A. Nagy, unpublished results (1992).
76. G. Distefano, A. Modelli, A. Grassi, G. C. Pappalardo, K. J. Irgolic and R. A. Pyles, *J. Organomet. Chem.*, **220**, 31 (1981).
77. F. Brogli, E. Heilbronner, J. Wirz and E. Kloster-Jensen, *Helv. Chim. Acta*, **58**, 2620 (1975).
78. M. C. Böhm, M. Eckert-Maksic, R. Gleiter, J. Grobe and Duc Le Van, *Chem. Ber.*, **114**, 2300 (1981).
79. M. C. Böhm, R. Gleiter, J. Grobe and Duc Le Van, *J. Organomet. Chem.*, **247**, 203 (1983).
80. R. Gleiter, M.C. Böhm and M. Baudler, *Chem. Ber.*, **114**, 1004 (1981).
81. R. Gleiter, W. Schäfer and M. Baudler, *J. Am. Chem. Soc.*, **107**, 8043 (1985).
82. R. Gleiter, H. Köppel, P. Hofmann, H. R. Schmidt and J. Ellermann, *Inorg. Chem.*, **24**, 4020 (1985).
83. J. H. D. Eland, *Photoelectron Spectroscopy*, Chap. 6, Butterworths, London, 1974, pp. 131–155.
84. K.-H. A. O. Starzewski, W. Richter and H. Schmidbaur, *Chem. Ber.*, **109**, 473 (1976).
85. C. Batich, E. Heilbronner, V. Hornung, A. J. Ashe III, D. T. Clark, U. T. Cobley, D. Kilcast and I. Scanlan, *J. Am. Chem. Soc.*, **95**, 928 (1973).
86. J. Bastide, E. Heilbronner, J. P. Maier and A. J. Ashe III, *Tetrahedron Lett.*, 411 (1976).
87. W. von Niessen, G. H. F. Diercksen and L. S. Cederbaum, *Chem. Phys.*, **10**, 345 (1975).
88. Ch. Elschenbroich and A. Salzer, *Organometallics*, VCH Verlags-GmbH, Weinheim (Germany), 1989, p. 162.
89. A. J. Ashe III, F. Burger, M. Y. El-Sheik, E. Heilbronner, J. P. Maier and J.-F. Muller, *Helv. Chim. Acta*, **59**, 1944 (1976).
90. A. J. Ashe III, M. K. Bahl, K. D. Bomben, W.-T. Chan, J. K. Gimzewski, P. G. Sitton and T. D. Thomas, *J. Am. Chem. Soc.*, **101**, 1764 (1979) and references cited therein.
91. T. Veszprémi, L. Nyulászi, J. Réffy and J. Heinicke, *J. Phys. Chem.*, **96**, 623 (1992).

92. S. Lacombe, D. Gonbeau, J.-L. Cabioch, B. Pellerin, J.-M. Denis and G. Pfister-Guillouzo, *J. Am. Chem. Soc.*, **110**, 6964 (1988).
93. H. Bock and M. Bankmann, *Angew. Chem., Int. Ed. Engl.*, **25**, 265 (1986).
94. L. Nyulászi, G. Csonka, J. Réffy, T. Veszprémi and J. Heinicke, *J. Organomet. Chem.*, **373**, 49 (1989).
95. W. Schäfer, A. Schweig, G. Märkl, H. Hauptmann and F. Mathey, *Angew. Chem.*, **85**, 140 (1973).
96. S. Evans, M. L. H. Green, B. Jewitt, A. F. Orchard and C. F. Pygall, *J. Chem. Soc., Faraday Trans. 2*, **68**, 1847 (1972).
97. L. Nyulászi, G. Csonka, J. Réffy, T. Veszprémi and J. Heinicke, *J. Organomet. Chem.*, **373**, 57 (1989).
98. D. T. Clark, in *Handbook of X-Ray and Ultraviolet Photoelectron Spectroscopy* (Ed. D. Briggs), Chap. 6, Heyden, London, Philadelphia, Rheine, 1977, pp. 211–247.
99. J. A. Connor, in *Handbook of X-Ray and Ultraviolet Photoelectron Spectroscopy* (Ed. D. Briggs), Chap. 5, Heyden, London, Philadelphia, Rheine, 1977, pp. 183–209.
100. T. Birchall, J. A. Connor and I. H. Hillier, *J. Chem. Soc., Dalton Trans.*, 2003 (1975).
101. W. E. De Bock, D. W. Wambeke, D. F. Van De Vondel and G. P. Van Der Kelen, *J. Mol. Struc.*, **140**, 303 (1986).
102. G. M. Bancroft, T. K. Sham, D. E. Eastman and W. Gudat, *J. Am. Chem. Soc.*, **99**, 1752 (1977).
103. *Free Atom Subshell Binding Energies*, in *Handbook of X-ray and Ultraviolet Photoelectron Spectroscopy* (Ed. D. Briggs), Heyden, London, Philadelphia, Rheine, 1977.
104. P. Brant and R. D. Feltham, *Inorg. Chem.*, **19**, 2673 (1980).
105. P. Brant, L. S. Benner and A. L. Balch, *Inorg. Chem.*, **18**, 3422 (1979).
106. L. J. Matienzo, L. I. Yin, S. O. Grim and W. E. Swartz, *Inorg. Chem.*, **12**, 2762 (1973).
107. J. H. Enemark, R. D. Feltham, J. Riker-Nappier and K. F. Bizot, *Inorg. Chem.*, **14**, 624 (1975).
108. Chan-Cheng Su and J. W. Faller, *J. Organomet. Chem.*, **84**, 53 (1975).
109. B. J. Bridson, W. S. Mialki and R. A. Walton, *J. Organomet. Chem.*, **187**, 341 (1980).
110. R. H. Staley, M. Taagepera, W. G. Henderson, I. Koppel, J. L. Beauchamp and R. W. Taft, *J. Am. Chem. Soc.*, **99**, 328 (1977).

CHAPTER 8

Lewis acidity, basicity, H-bonding and complexing of organic arsenic, antimony and bismuth compounds*

KATHARINA C. H. LANGE and THOMAS M. KLAPÖTKE[†]

Institut für Anorganische und Analytische Chemie, Technische Universität Berlin, Strasse des 17. Juni 135, D-10623 Berlin, Germany

*In this chapter, full lines are used both for covalent chemical bonds as well as for partial bonds and for coordination. Hence in many cases in the structures and equations *two* solid lines originate from Li and *four* solid lines from As, Sb or Bi.
[†] Author to whom correspondence should be addressed.

The chemistry of organic arsenic, antimony and bismuth compounds
Edited by S. Patai © 1994 John Wiley & Sons Ltd

ABBREVIATIONS

The following abbreviations, arranged alphabetically below, are used besides of the well known abbreviations which are listed in each volume:

B.E.	Bond Energy
DADPE	$Ph_2As(CH_2)_2PPh_2$
DIARS	$1,2\text{-}(Me_2As)_2C_6H_4$
DPAE	$Ph_2As(CH_2)_2AsPh_2$
EDTA	ethylenediamine tetraacetate
HNAP-H	N-(2-hydroxy-1-naphthalidene-2-aminopyridine)
HNPD-H	N-(2-hydroxy-1-naphthalidene-o-phenylenediamine)
Mes	mesityl; $1,3,5\text{-}(CH_3)_3C_6H_3$
Ox	oxalate
PA	proton affinity
Q-H	quinoline
QSH	8-mercaptoquinoline
SAP-H	N-salicylidene-2-aminopyridine

SPD-H   *N*-salicylidene-*o*-phenylenediamine
tripod   tripodal ligand $[CH_3C(CH_2PPH_2)_3]$

I. OUTLINES

The aim of this review on the organoelement chemistry of arsenic, antimony and bismuth is to focus on the hydrogen bonding, the complexing and, closely related, the acidity and basicity of such compounds. This chapter is not exhaustive in scope, but rather consists of surveys of the most recent decade of work in this still developing area. This chapter emphasizes the synthesis, reactions and molecular structures of the class of compounds outlined above (less attention is paid to mechanism, bond theory, spectroscopic properties and applications which can be found in other specialized chapters). Especially the single-crystal X-ray diffraction technique has elucidated many novel and unusual structures of molecules and of the solid state in general. Not unexpectedly, certain organoelement compounds present problems concerning their classification as *n*-coordinated species, since it is sometimes difficult to distinguish between a weak long-range interaction in the solid state and the fact that two atoms can be forced a little bit closer together by crystal lattice effects.

As organoelement chemistry is the discipline dealing with compounds containing at least one direct element–carbon bond, in this chapter we discuss arsenic, antimony and bismuth species in which at least one organic group is attached through carbon. Classical species containing E—C σ-covalent bonds (E = As, Sb, Bi) as well as π-complexes involving dative bonds will also be considered.

In the section on hydrogen bonding both simple σ-bound covalent compounds (e.g. R_2EH) as well as complexes containing coordinative bonds (for definition see below) will be discussed (e.g. $R_2EH_2^-$).

Following the old definition by Sidgwick (from 1927)[1], that the coordinate bond is a covalent bond formed by the donation of one pair of electrons from the donor atom (e.g. H^-, Cl^-) or molecule (C_6H_6, ER_3, etc.) to a complex centre, in the section on complexing (naturally including acidity and basicity) only species are considered that contain at least one dative-covalently bound ligand. This implies that neutral three- and five-coordinated species of the type R_3E and R_5E (E = As, Sb, Bi) have not been included.

Into each of the two main sections on hydrogen bonding and complexing, a short, more theoretical paragraph has been inserted in order to understand the interrelations more fully. In the H-bonding section there is a thermodynamic discussion on the stability of ammonium vs. arsonium salts. In the complexing/acidity/basicity section there is a subsection on s-orbital contraction due to the post transition metal effect, the lanthanoid contraction and relativistic effects in order to understand (for example) the relatively high basicity of PR_3 and SbR_3 compounds compared with their AsR_3 and BiR_3 analogues.

II. H-BONDING

A. Introduction

The group 15 organoelement hydrides of the type R_nEH_{3-n} (E = As, Sb, Bi) are highly reactive and very sensitive towards oxidation by air (see below). The methyl and ethyl derivatives are pyrophoric. However, due to the low polarity of the E—H bond these species are *relatively* moisture-stable (cf. χ (Pauling) = H, 2.20; As, 2.18; Sb, 2.05; Bi, 2.02)[2]. In general, the (thermal) stability decreases for the central atoms in the order As > Sb > Bi. The more H atoms are present the lower is the stability of their hydrides.

Moreover, from spectroscopic data (IR) the dissociation energies for As—H bonds in arsines show progressive weakening of bonds with methyl substitution which is in agreement with the experimental observation for the group 15 hydrides in general that the bond strength of the E—H diminishes with methylation[3,4]. The organic arsenic, antimony and bismuth hydrides are best prepared by the reduction of the corresponding organoelement halides usually with zinc powder or zinc amalgam (for arsenic only), alkali metal borohydrides or lithium alanate at low temperatures (equations 1–3)[5].

$$MeAsCl_2 \xrightarrow{Zn/Cu,\ HCl} MeAsH_2 \qquad (1)$$

$$Me_2SbBr \xrightarrow[thf,\ -60\,°C]{LiBH(OMe)_3} Me_2SbH \qquad (2)$$

$$MeBiCl_2 \xrightarrow[Me_2O,\ -110\,°C]{LiAlH_4} MeBiH_2 \qquad (3)$$

Although no organoelement(V) hydrides of arsenic, antimony and bismuth are known, primary and secondary arsines (oxidation state III) may be obtained by the reduction of organoarsenic(V) compounds (equations 4 and 5)[6]

$$Me_2AsO(OH) \xrightarrow[-ZnCl_2]{Zn,\ HCl} Me_2AsH \qquad (4)$$

$$PhAsCl_4 \xrightarrow[-Cl_2]{LiBH_4,\ Et_2O} PhAsH_2 \qquad (5)$$

As discussed above, the electronegativities of As and H are essentially identical. However, to discuss the O_2 oxidation (e.g. by air) it might be useful to consider arsenic in the primary arsine as As(+III) (although this procedure could be regarded as being too formalistic). In the first step O_2 oxidizes the hydride H to give water and cyclopolyarsines (equation 6).

$$MeAs^{(+III)}H_2 + 1/2\,O_2 \longrightarrow H_2O + 1/n\,(MeAs^{+1})_n \qquad (6)$$

This in fact would correspond to a reduction of arsenic yielding the intermediate compound with an As—As bond. As this homocyclic species is even less stable towards oxygen, it readily takes up O_2 to generate $(MeAsO)_n$ (arsonousacid anhydride, also called arsoxane or arsenoxomethyl) which finally, in the presence of moisture, gives $MeAsO(OH)_2$ (arsonic acid) (equations 7 and 8)[6].

$$(MeAs)_n + n/2\,O_2 \longrightarrow (MeAsO)_n \qquad (7)$$
$$(MeAsO)_n + n\,H_2O + n/2\,O_2 \longrightarrow n\,MeAsO(OH)_2 \qquad (8)$$

Secondary arsines are usually oxidized to yield the corresponding diarsines (equation 9); they themselves are often extremely O_2 sensitive and can easily be converted to give arsinous acid anhydrides (diarsoxanes) (equation 10)[5] and finally (see above) arsinic acid (equation 11).

$$2\,Me_2AsH + 1/2\,O_2 \xrightarrow[-H_2O]{} Me_2As—AsMe_2 \text{ (cacodyl)} \qquad (9)$$

$$Me_2As—AsMe_2 + 1/2\,O_2 \longrightarrow Me_2As—O—AsMe_2 \text{ (cacodyl oxide)} \qquad (10)$$

$$Me_2As—O—AsMe_2 + O_2 + H_2O \longrightarrow 2\,Me_2As(O)OH \qquad (11)$$

$(Me_2As)_2$ and $(Me_2As)_2O$ were the first organoarsenic (and organometallic!) compounds obtained by L. C. Cadet in 1760 (cf equations 9 and 10).

B. Reactions

The polarity of the M—C and of the M—H bonds increases descendingly in the group from arsenic to bismuth. Thus the organobismuth hydrides are the least stable and the most reactive. Several books and monographs are available describing the reactivity of group 15 organohydrides[5-8]. In this context we like to focus on the more recent studies of the hydrides and summarize the highlights of the last years.

1. Arsenic

a. Alkyl arsines. The addition of arsines to carbon–carbon double and triple bonds is well reported (see above). A nice example is the addition of R_2AsH to acetylenes or unsaturated aldehydes (equation 12)[9].

$$R_2AsH + MeCH=CHCHO \longrightarrow MeCH(AsR_2)CH_2CHO \qquad (12)$$

Heterocyclic five-membered arsoles can similarly be prepared by the addition of primary arsines to diacetylenes (equation 13)[5].

$$PhAsH_2 + RC\equiv C-C\equiv CR \longrightarrow R\underset{\underset{Ph}{As}}{\overset{}{\diagdown}}R \qquad (13)$$

Many of the primary and secondary alkylarsines (especially the methyl derivatives Me_nAsH_{3-n}) are of particular interest as arsenic precursors in organometallic vapour-phase epitaxy[10] (cf weakening of AsH bonds with methyl substitution; see above and Reference 3).

Although it has been shown that PhHAs—AsHPh is non-existent (at least very unstable)[11], the tetraphenyl substituted derivative $(Ph_2As)_2$ is generated from the reaction of Ph_2AsH and F_2 according to equation 14[12]. However, F_2 readily cleaves the As—As bond to oxidize the diarsine, yielding the stable diphenyltrifluoroarsorane (equation 15). This is a further example demonstrating the weakness of homonuclear E—E single bonds between the group 15 elements [B.E. (E—E): E = P, 48; E = As, 35; E = Sb, 29 kcal mol^{-1}][6].

$$2 Ph_2AsH + F_2 \xrightarrow[-2HF]{} Ph_2As—AsPh_2 \qquad (14)$$

$$Ph_2As—AsPh_2 + 3 F_2 \longrightarrow 2 Ph_2AsF_3 \qquad (15)$$

The alkyl hydrides $MeAsH_2$ and Me_2AsH were shown to react with acetonitrile to give the highly reactive organoarsine species $MeAs(CH_2CH_2CN)_2$ and $Me_2AsCH_2CH_2CN$, respectively[13]. A new selective route yielding silicon-bound organoarsenic(III) compounds has recently been published for mono- and disilylated arsines (equations 16 and 17)[14].

$$PhMeAsH + Me_3SiOSO_2CF_3 \longrightarrow PhMeAsSiMe_3 + CF_3SO_3H \qquad (16)$$

$$PhAsH_2 + 2 Me_3SiOSO_2CF_3 \longrightarrow PhAs(SiMe_3)_2 + 2 CF_3SO_3H \qquad (17)$$

b. Cyclic compounds. Whereas the primary methylarsine $MeAsH_2$ is in its oxidation pathway initially converted into cyclopentamethylarsine, $(MeAs)_5$ (see above), this puckered ring is best prepared from the reaction of $MeAsH_2$ and dibenzylmercury[5]. A new and interesting derivatization for the relatively easily available $(MeAs)_5$ in high yields provides the reaction with other primary arsines (equation 18)[15].

$$(MeAs)_5 \xrightarrow{\text{RAsH}_2} (RAs)_n \tag{18}$$

$$R = Et, Pr, \qquad n = 5$$
$$R = Ph, \text{ } p\text{-Tol}, \quad n = 6$$

A three-membered As_3 ring is formed as part of a polycyclic cage according to equation 19[5].

$$MeC(CH_2AsI_2)_3 \xrightarrow{\text{Na(thf)}} \tag{19}$$

A similar reaction starting from the tetrasubstituted $C(CH_2AsI_2)_4$ (see equation 20) formed a new primary organoarsenic hydride[16].

$$C(CH_2AsI_2)_4 \xrightarrow[\text{thf}]{\text{LiAlH}_4} \tag{20}$$

A six-membered As—Li heterocycle was obtained as a trimer of $LiAs[CH(SiMe_3)_2]_2$ prepared according to equation 21[17].

$$R_2AsCl \xrightarrow{\text{Li}} \tag{21}$$

$$R = CH(SiMe_3)_2$$

This metallacycle is itself a strong reducing agent and generates R_2AsH or R_2AsMe on being treated with HCl or MeCl, respectively[17].

A planar centrosymmetric four-membered Ga—As ring (essentially a R_2As—GaR'_2 dimer) was obtained from a binary arsine and Ph_3Ga (equation 22) and was characterized by single-crystal X-ray diffraction[18].

$$2(Me_3SiCH_2)_2AsH + 2Ph_3Ga \xrightarrow{-PhH} \tag{22}$$

Whereas the ring distances are essentially identical (251.8, 253.0 pm) the angle at the As (94.9°) is substantially larger than the Ga angle with 85.1°[18].

c. Arsine species coordinated to transition metal fragments. In general there are only few examples of well characterized complexes containing primary arsines coordinated to transition metals.

The iron(II) facilitated synthesis of the first isolated 1-phenylarsetane complex and its characterization by X-ray diffraction[19] (1) describes one of the few examples where an As heterocycle behaves as a complex ligand. The arsine ligand is with a Fe—As bond length of 232.6 pm just weakly coordinated to the iron centre (sum of van der Waals radii, 228 pm)[20] and can therefore really be regarded as an organoarsenic(III) arsetane.

Cp
│
Ph   Fe —— As
  \   ╱   │
Me — P │   Ph
        │
        P
        │ `Me
        Ph

(1) Structure of the first isolated arsetane complex

A related cationic iron complex contains the coordinated AsHMePh ligand and therefore possesses a chiral centre at the As atom (2)[21]. The group 6 arsinidene complexes $(CO)_5M(AsH_2Ph)$ (M = Cr, Mo, W) were prepared in a straightforward condensation reaction from the free ligand base $PhAsH_2$ and $(CO)_5M(THF)$[22]. Analogous examples of group 7 were also prepared and characterized[22]: (i) $(CO)_5MnAsH_2Ph$ (IR, ^1H NMR), (ii) $Cp(CO)_2Mn(PhAsH—AsHPh)Mn(CO)_2Cp$ (X-ray) (3).

Cp
│
Ph   Fe —— As —— Me
  \   ╱   │   `H
Me — P │   Ph
        │
        P
        │ `Me
        Ph

(2) Structure of the cation containing a chiral four-coordinated AsMePhH ligand

          Cp
          │
Ph   H   Mn(CO)₂
│   │   ╱
As —— As
╱   `H   `Ph
(CO)₂Mn
│
Cp

(3) Structure of $Cp(CO)_2Mn(AsHPh—AsHPh)Mn(CO)_2Cp$

The As—As bond distance in the dinuclear complex was determined to be equal to 246.0 pm and is therefore nicely in agreement with an As—As single bond (cf Me_5As_5, 242.8 pm; Ph_6As_6, 245.6 pm). This example shows that although free $(AsPhH)_2$ is unstable, it can be stabilized in complex bonding and represents a good back-bonding ligand (d: Mn–As 231.9 pm; covalent radii: As, 123 pm; Mn, 146 pm)[22].

The first example of an organolanthanide compound with a lanthanide–arsenic bond was obtained from the reaction according to equation 23 and characterized by an X-ray diffraction study[23]. Although in the next example the arsine PhAsH$_2$ represents only the starting material, the preparation of the first phenyl azide analogue PhAs$_3$ (coordinated to Co) is an interesting and recent example from organoarsenic chemistry[24]: [(tripod)Co-μ_2-(η^3-PhAs$_3$)Co(tripod)]$^{2+}$.

$$\text{Cp}_2\text{Lu}(\mu\text{-CH}_3)_2\text{Li(TMEDA)} + 2\,\text{Ph}_2\text{AsH} \xrightarrow[-2\,\text{CH}_4]{} \underset{\substack{\text{As} \\ \text{Ph}_2}}{\overset{\substack{\text{Ph}_2 \\ \text{As}}}{\text{Cp}_2\text{Lu}}} \diagdown \text{Li} \underset{\substack{\text{N} \\ \text{Me}_2}}{\overset{\substack{\text{Me}_2 \\ \text{N}}}{\diagdown}} \qquad (23)$$

Finally, the NMR spectra of Ph$_n$AsH$_{3-n}$ and Ph$_n$AsH$_{2-n}$Na in THF have been studied. The results are consistent with the existence of a pπ–pπ interaction between As and the phenyl ring in the case of Ph$_n$AsH$_{2-n}$ anions but not in the case of the neutral species[25].

2. Antimony

A new preparation of primary and secondary stibines by hydrolysis (methanolysis) of the corresponding silyl compounds was recently reported (equations 24 and 25)[26]. This seems to be a useful method as the Si precursors are readily available from the corresponding chlorides and Me$_3$SiCl/Mg in THF[26].

$$\text{PhSb(SiMe}_3)_2 + 2\,\text{MeOH} \xrightarrow[-30\,°\text{C}]{\text{Et}_2\text{O}} \text{PhSbH}_2 + 2\,\text{Me}_3\text{SiOMe} \qquad (24)$$

$$\text{Ph}_2\text{Sb(SiMe}_3) + \text{MeOH} \xrightarrow[-30\,°\text{C}]{} \text{Ph}_2\text{SbH} + \text{Me}_3\text{SiOMe} \qquad (25)$$

The first secondary stibine, the structure of which was elucidated by X-ray diffraction, was Mes$_2$SbH[27]. This hydride is exceptionally stable and possesses (as expected) a trigonal pyramidal structure (CSbC, 101.7°). Reaction according to equation 26 formed the planar (X-ray) Cu$_2$Sb$_2$ heterocycle containing essentially identical Cu–Sb distances of 270 pm in agreement with a real Cu–Sb single bond (sum of covalent radii = 268 pm)[27].

$$\text{Mes}_2\text{SbH} \xrightarrow[\text{2. CuCl, Me}_3\text{P}]{\text{1. } n\text{-BuLi}} [\text{Mes}_2\text{SbCu(PMe}_3)_2]_2 \qquad (26)$$

The diphenylstibine anion SbPh$_2^-$ (V-shape, X-ray) was found in a complex prepared according to equation 27[28].

$$\text{Ph}_2\text{SbH} + \text{BuLi} + 2[12\text{-crown-4}] \xrightarrow{\text{thf}} [\text{Li(12-crown-4)}_2][\text{SbPh}_2]\ 1/3\ \text{thf} \qquad (27)$$

C. Proton Affinity and Basicity

It is a generally accepted tenet that the Lewis basicity of trivalent group 15 compounds decreases with the heavier congeners. As the triorgano substituted complexes are generally easier to investigate than the hydrides, numerous studies have been concerned with the Lewis base interaction of R$_3$E compounds (E = P, As, Sb, Bi) with boron Lewis acids[29] and with the proton affinity (PA) of R$_3$E and EH$_3$ species[30-33]. The proton affinities of a series of Ph$_3$E compounds (E = P, As, Sb) have been determined by bracketing using reactant ion monitoring: Ph_3P, 232 ± 1; Ph_3As, 216 ± 2; Ph_3Sb, 202 ± 2 kcal mol^{-1} [30]. For AsH_3 the gas-phase acidity, $PA(AsH_2^-)=360\pm10$ kcal mol^{-1}, and basicity $PA(AsH_3)=175\pm5$ kcal mol^{-1} were determined.

In order to understand the difference in basicity, proton affinity and stability of complexes of N, on the one hand, and As, Sb and Bi on the other, we discuss the stability of ammonium vs arsonium chloride. Using a simple Born–Haber approach as described in Reference 34 we can estimate the heat of reactions 28–31 (ΔH_r°) as summarized in Table 1. All data were estimated as illustrated in Schemes 1–5.

$$AsH_4Cl(s) \longrightarrow H_2(g) + AsH_2Cl(g) \qquad (28)$$

$$NH_4Cl(s) \longrightarrow H_2(g) + H_2NCl(g) \qquad (29)$$

$$AsH_3(g) + HCl(g) \longrightarrow AsH_4^+Cl^-(s) \qquad (30)$$

$$NH_3(g) + HCl(g) \longrightarrow NH_4^+Cl^-(s) \qquad (31)$$

The energy level diagram in Scheme 6 clearly shows that NH_4Cl is thermodynamically stable with respect to both, decomposition to NH_3/HCl and H_2/NH_2Cl. AsH_4Cl,

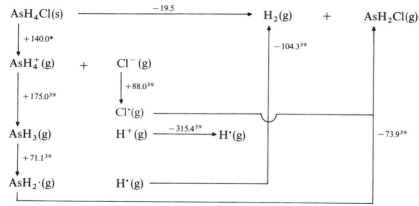

*$V_m(AsH_4^+)$ was taken equal to $V_m(GeH_4)=84\text{ Å}^3(d=1.52\text{ g/cm}^3)$; $(Cl^-)=1.80\text{ Å}$, this gives $V_m=33\text{ Å}^3$ and $U_L(AsH_4^+Cl^-)=140$ kcal mol$^{-1})$
$V_m=$ molecular volume (in Å^3)
$U_L=$ crystal lattice energy

SCHEME 1. Born–Haber cycle to estimate the heat of reaction 28

TABLE 1. ΔH_r° values for equations 28–32 (all data in kcal mol^{-1})

Equation	ΔH_r°
28	-19.5
29	$+102.5$
30	$+15.7$
31	-44.7
32	-3.8

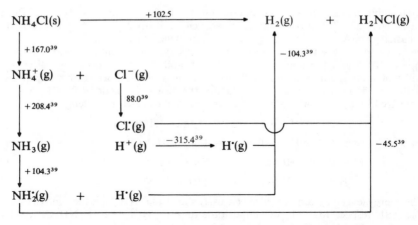

SCHEME 2. Born–Haber cycle to estimate the heat of reaction 29

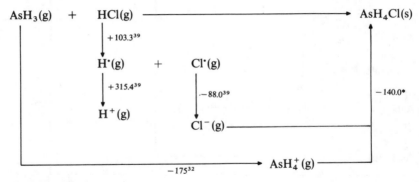

*See footnote to Scheme 1.

SCHEME 3. Born–Haber cycle to estimate the heat of reaction 30

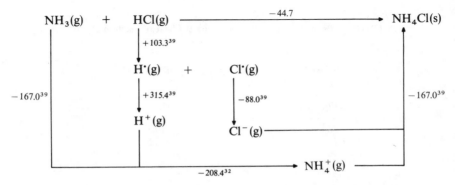

SCHEME 4. Born–Haber cycle to estimate the heat of reaction 31

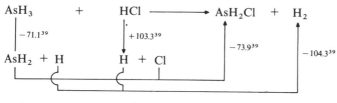

$$\Delta H = -3.8 \, \text{kcal mol}^{-1}$$

SCHEME 5. Born–Haber cycle to estimate the heat of reaction 32

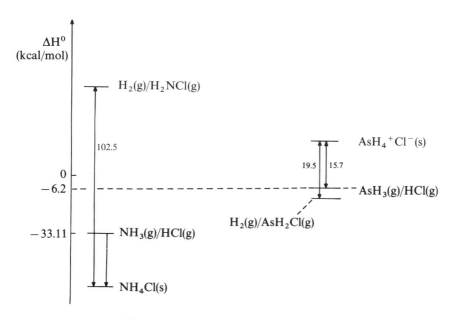

SCHEME 6. Stability of NH_4Cl vs AsH_4Cl

however, is unstable with respect to the dissociation into AsH_3/HCl. Moreover, the formation of H_2/AsH_2Cl should thermodynamically be even slightly more favourable (Scheme 5, equation 32, Table 1). The two major differences concerning the stability of

$$AsH_3(g) + HCl(g) \longrightarrow AsH_2Cl(g) + H_2(g) \tag{32}$$

NH_4Cl vs AsH_4Cl are the lattice energy and the proton affinity, because:

- $U_L(NH_4Cl) > U_L(AsH_4Cl)$
  (167 > 140 kcal mol^{-1})
- $PA(NH_3) > PA(AsH_3)$
  (208 > 175 kcal mol^{-1})

moreover

- BE(NH) > BE(As—H)
  (104   > 71 kcal mol^{-1})
- BE(NCl) < BE(AsCl)
  (45   < 74 kcal mol^{-1}).

III. COMPLEXING, ACIDITY AND BASICITY

A. Introduction

The first part of this section deals with complex compounds with arsenic, antimony or bismuth acting as central atoms. They have been ordered according to the coordination number of the element and, within these sections, according to the donor properties of the ligands and to an ionic or covalent type of the complex. In the second part we report on organoelement compounds coordinated to transition metals or main group elements.

Organoelement(III) and organoelement(V) compounds [rarely organoelement(I) species] are able to form ionic complexes. However, antimony in the oxidation state $+I$ is to be found in the anion $[SbPh_2]^-$, which itself is a very good reducing agent and may attack the alkali metal cation. This problem of stability was solved by sequestering the Li cation with a bulky crown ether. Another possibility of stabilizing the anion is its coordination to transition metal carbonyls.

Organoelement(III) anions are four-substituted and singly-charged. Usually the kind of substitution varies from different halogen substitution to donor ligands such as oxygen or sulphur. These anions are generally four-coordinated but their coordination number may increase to five by substitution with multidentate ligands. Dimeric element(III) anions may also realize a coordination number of five, containing bridging atoms like chlorine, e.g. $[PhSbCl_2(\mu\text{-}Cl)]_2^{2-}$. Doubly-charged element(III) anions are five-substituted (and also five-coordinated), while element(V) ions can either be four-substituted cations or six-substituted anions, respectively.

Adduct complexes are formed by organoelement(III) species with halogens where the central atom is four-coordinated, e.g. Ph_3AsI_2. Moreover, organoelement(III) compounds can yield adducts with oxygen donor molecules where the four-coordinated central atom exists in a Ψ-trigonal bipyramidal coordination sphere. Also, intermolecular interactions can exist in those adducts with formation of polymer structures and increase of the element's coordination number to five. Furthermore, adduct complexes are formed by intermolecular interactions between organoelement(III) or organoelement(V) units, respectively. Special kind of adducts are the arene complexes, which are formed by interactions between trihalogenoelement(III) species and aromatic ring systems.

Moreover, organoelement(III) and -element(V) compounds may be coordinated by multidentate ligands with donor atoms like oxygen, sulphur or nitrogen. These mostly bidentate ligands have bridging or chelating ability. There are only few examples of potential chelating ligands behaving as monodentate ligands, e.g. $CF_3CO_2^-$ in $[BiPh_2(O_2CCF_3)_2]^-$. Carbonyl groups may behave as μ^2-oxygen ligands; moreover, thiocarbamate and dithiocarbamate species as well as ligands containing oxygen and nitrogen together or nitrogen and sulphur are known as donor molecules. Depending on their mode of action (chelating and/or bridging) these multidentate ligands increase the coordination number of the element concerned. Thus three-substituted element(III) species may be five- or six-coordinated and five-substituted element(V) compounds might show coordination numbers of six or seven.

B. Phosphorus and Antimony vs. Arsenic and Bismuth— The Post Transition Metal Effect, Lanthanoid Contraction and Relativistic Effects

Many physical and chemical properties down group 15 (and not only there) exhibit a saw-tooth behaviour, superimposed on the regular trend down a column. For example, one can quote the tendency of arsenic and bismuth to be trivalent whereas phosphorus and antimony are both tri- and pentavalent. For arsenic and bismuth it is difficult to achieve the highest oxidation state while P(V) and Sb(V) compounds are well established. For example, PCl_5 and $SbCl_5$ are stable whereas $AsCl_5$ is stable at low temperatures only (it decomposes to give $AsCl_3$ and Cl_2) and $BiCl_5$ is hitherto unknown. Moreover, arsenic and especially bismuth exhibit a low basicity compared with phosphorus and antimony. Therefore arsines and BiR_3 compounds are much weaker complex ligands than phosphine and SbR_3 derivatives. It was proposed that the first anomaly at row four (As) is caused by the *post transition metal effect* (d contraction), caused by an increase of the effective nuclear charge for 4s electrons due to filling the first d-shell (3d). A similar interpretation is possible for the second anomaly at row 6 (Bi). This effect is commonly called *lanthanoid contraction* due to the effect of filling the 4f shell. However, more recent calculations show that for the second minimum (Bi), relativistic effects and the lanthanoid contraction were found to be equally important. The main relativistic effect on atomic orbitals responsible for the resistance of bismuth to achieve its highest oxidation state and which explains its low basicity ('inert s-pair') is the relativistic radial contraction and (energetic) stabilization of the s (or p) shells in heavy-element chemistry (cf relativistic effects increase, for valence shells, roughly like Z^2). Let us finally compare the radial 1s shrinkage for the elements As, Sb and Bi that are the subjects of this chapter. The non-relativistic limit for a 1s electron is Z au. Thus the 1s electron of arsenic has a v/c of $33/137 \approx 0.24$, whereas these values are 0.37 for antimony and roughly 0.61 for bismuth ($c = 137$ a.u.) [c is the finite speed of light and v is the average radial velocity of the electrons in the 1s shell (v is roughly Z au)]. As the relativistic mass increase is given by $m = m_0/[1-(v/c)^2]^{0.5}$ with the effective Bohr radius $a_0 = (4\pi\varepsilon_0)/(\hbar^2/me^2)$ the radial 1s shrinkage for arsenic and antimony is 3 or 8%, respectively, whereas it is already 26% (!) for bismuth. A detailed discussion of relativistic effects in structural chemistry is given in the literature[35-37].

1. Examples

If one compares the E(III)/E(V) (E = P, As) bond energies and the ionization energies of oxidation state + III P and As compounds, both the diminished stability of the As(V)–element bond [vs P(V)-element] and the unusual high I_P of As(III) compounds [vs P(III) compounds] can best be explained by the post transition metal effect (see above). Two examples from inorganic chemistry (just due to the better accessibility of data) are given in Table 2[38,39].

The proton affinities of Ph_3E (E = N, P, As, Sb) compounds also reflect the trend shown in the I_P values and the P.A. of PPh_3 is greater than that of Ph_3N and Ph_3As (cf also the section on H-bonding)[30]. As expected there is a general trend of decreasing thermal stability of the E—C bond in EPh_3 compounds (E = P, As, Sb, Bi) in the order: P > As > Sb > Bi. However, due to the post transition metal effect $SbPh_5$ is more stable than $AsPh_5$ (the instability of PPh_5 is mostly due to steric effects) and the stability of EPh_5 species is in the order: Sb > As > P > Bi[40,41]. The instability of the purple $BiPh_5$ and its chromophoric behaviour (formation of $BiPh_4^+ Ph^-$, charge transfer; all other EPh_5 compounds are colourless) can both be explained by the poor shielding ability of

TABLE 2. Bond energies and ionization potentials of As(III) vs P(III) compounds

Compound	B.E (E-Hal)/kcal mol^{-1}	I_p/eV
AsF_3	116.5	
AsF_5	97	
PF_3	117	
PF_5	110	
$AsCl_3$	74	11.7
PCl_3	77	10.5

the 4f electrons and relativistic contraction of the 6s orbital[42]. Thus not only the lower basicity of Bi(III) organyles but also the higher acidity of Bi(V) compounds can be understood by these effects: (i) with PhLi BiPh$_5$ forms (Ph$_6$Bi)$^-$; (ii) coordinative Bi—O bonds tend to be stronger than Sb—O bonds[41,43].

Since the ionization potentials I_P (see above: AsCl$_3$ > PCl$_3$) are directly related to the absolute electronegativity χ by

$$\chi = (I_P + E_A)/2 \qquad (E_A = \text{electron affinity})$$

the extended HSAB principle (hard soft acid base principle) can also help to determine the behaviour of ER$_3$ bases. Therefore, not only the knowledge of χ but also the identification of absolute hardness η

$$\eta = (I_P - E_A)/2$$

with the energy gap between HOMO and LUMO is very important (Scheme 7). Soft molecules with a small gap are more polarizable than hard ones. Theory shows that electron density changes in a reaction depend on the size of the gap, thus soft molecules should be more reactive than hard molecules.

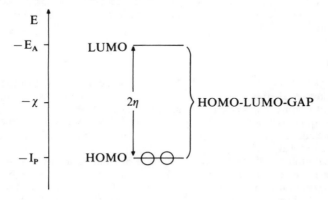

SCHEME 7. Orbital energy diagram

Let us consider, for example, the coordination of the bases Me$_3$P and Me$_3$As to the (hypothetical) complex centre Ni. Experimental values for χ and η (determined from exp. I_P and E_A data) are given in Table 3.

The ΔE data of Table 3 (gain in energy due to electron transfer) clearly show that

TABLE 3. HSAB parameters

	χ/eV	η/eV	$\Delta E/eV^a$	$\Delta E/\text{kcal mol}^{-1}$
Ni	4.4	3.3	—	—
PMe$_3$	2.8	5.9	−0.0696	−1.6
AsMe$_3$	2.8	5.5	−0.0727	−1.7

^{a}Calculated from ($\Delta E = -\Delta\chi^2/4\Sigma\eta$) for the coordination of PMe$_3$ and AsMe$_3$ to Ni.

PMe$_3$ and AsMe$_3$ are essentially identical in their coordination behaviour towards the Lewis acid Ni (AsMe$_3$ is the slightly better base). Although AsMe$_3$ and PMe$_3$ do not differ in their absolute electronegativities (post-transition metal effect, see above), AsMe$_3$ is slightly softer than PMe$_3$ which favours the coordination to Ni. An extensive discussion of the extended HSAB principle is given in the literature[44-49].

C. Complexing

1. Bi-coordinated complexes

There are few examples of bi-coordinated complexes reported in the recent literature. One of them is the species [Li(12-crown-4)$_2$][SbPh$_2$]1/3 THF. The diphenylstibine anion is bi-coordinated and the Sb—C bond distances range from 212.9 to 215.7 pm. The CSbC angles are 96.8°, 97.6° and 102.4°, respectively. This compound was prepared by reaction of Ph$_2$SbH with BuLi and 12-crown-4.

The [SbPh$_2$]$^-$ anion can also be generated in THF by electrochemical reduction of (Ph$_2$Sb)$_2$O, (Ph$_2$Sb)$_2$, Ph$_2$Sb(^{n}Bu), Ph$_3$Sb under the presence of the electrolyte salt ^{n}Bu$_4$NPF$_6$. However, the anion is unstable and decomposes to give the neutral Ph$_2$Sb(^{n}Bu)[50].

Reaction of the tertiary phenylstibine with BuLi/crown ether yielded the complex [Li(12-crown-4)$_2$][Sb$_3$Ph$_4$]·THF containing the bent trinuclear anion [Ph$_2$SbSbSbPh$_2$]$^-$ with a SbSbSb angle of 88.8° and CSbC angle of 92.7°. The Sb—Sb bond distances are to 276.1 pm, whereas the Sb—C bond lengths are 216.6 pm and 219.0 pm[28].

A bis(pentamethylcyclopentadienyl)element cation Cp$_2$*E$^+$ (Cp* = C$_5$Me$_5$; E = As, Sb) was prepared from Cp$_2$*EF and BF$_3$ as its BF$_4$$^-$ salt. The X-ray structure analysis revealed a non-centrosymmetric E—Cp* bond situation. Therefore the E—Cp* bond may best be described by a η^2 or η^3 rather than a η^5 coordination. As expected, the Cp* ligands are not in a coplanar position, they form an angle of 36.5°. The BF$_4$$^-$ anion seems to be important in stabilizing the cation; attempts with AlCl$_4$$^-$, BCl$_4$$^-$, BBr$_4$$^-$ and SbCl$_6$$^-$ were not successful[51]. This may very well be explained by the exceptionally weak basicity of the BF$_4$$^-$ anion and one could predict that AsF$_6$$^-$ and SbF$_6$$^-$ salts should also be stable.

2. Tri-coordinated complexes

The preparation, isolation and structural characterization of salts containing the first arsolidinium cations have recently been reported. Reactions between equimolar amounts of 2-chloro-1,3-dimethyl-1,3-diaza-2-arsolidine or 2-chloro-1,3-dithia-2-arsolidine with AlCl$_3$ or GaCl$_3$ result in quantitative chloride ion abstraction to give the corresponding cations [N̄(Me)CH$_2$CH$_2$N(Me)Ās]$^+$ and [S̄CH$_2$CH$_2$SĀs]$^+$ as tetrachloroaluminate and gallate salts. X-ray crystallography reveals a centrosymmetric dimeric arrangement for three derivatives in which the cations are bound together by relatively weak As—N or As—S interactions. A complex between the respective chloroarsolidine and arsolidinium cation is formed in reaction mixtures which provide appropriate stoichiometry. The X-ray structure of the thia derivative, $[\overline{SCH_2CH_2SAs}\dot{S}As(Cl)SCH_2\dot{C}H_2][GaCl_4]$, shows a single, noticeably weaker inter-ring connection, but otherwise the compound exhibits similar structural features. The cations are viewed as examples of stable arsenium (carbene analogue) units, unique in the absence of a shield or Hückel π-delocalization[52].

3. Tetra-coordinated complexes

a. Tetra-coordinated ionic complexes. The biarsonium salt $[(Ph_2MeAs)_2CH_2]^{2+}$ $2I^-$ was prepared by successive ionization of $(Ph_2As)_2CH_2$. By treatment with CH_3I only one As atom becomes quaternary, while in the next step methylfluorosulphate is used for the quaternization of the second As atom[53].

An example for a very unusual and interesting steric conformation is given in figure 4[54].

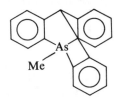

(4) Structure of a propeller-like tetracoordinated As(V) cationic complex

The salt $[Me_4Sb]I$ contains tetrahedral tetramethylstibonium cations which show rather long SbI contacts of 406 pm. The $[Me_4Sb]^+I^-$ unit also forms even longer contacts of 435 pm towards the adjacent molecule. Thus six of these $[Me_4Sb]^+I^-$ units form a six-membered ring-like structure[55].

A tetraphenyl borate salt containing the tetraphenylstibonium cation of the type $[Ph_4E]^+[Ph_4B]^-$ (E = As, Sb) was prepared in 1988[56].

In the binuclear arsonium salt $[Me_2AsAsMe_2I]^+[GaI_4]^-$ the As atoms of the cation show a bond distance of 242.7 pm[57]. Arsenic in the $+V$ oxidation state is to be found in a series of cations of the type $[Me_2AsX_2]^+$ (X = F, I) and $[MeAsI_3]^+$; the anion of these complexes is $[AsF_6]^-$[58]. Stabilized by the same anion are the cations $[Ph_3EI]^+$ with E = As, Sb and Bi in the oxidation state $+V$[59]. The complex $[Ph_2AsF_2]^+[AsF_6]^-$ containing the diphenyldifluoroarsonium cation was prepared from benzene and arsenic pentafluoride[60].

A Sb(III) containing anion is represented in the salt $[NMe_4]^+[Ph_2SbCl_2]^-$. The anion shows Ψ-trigonal bipyramidal configuration. The chlorine atoms are in axial positions while the phenyl groups and the lone-pair of electrons are situated equatorially, and therefore the C(Ph)SbC(Ph) angle is decreased to 97.5°. This anion is monomeric in contrast to the dimeric trichlorophenylbismuth(III) anion $[PhBiCl_3]_2^{2-}$[61]. Anions of the type $[Ph_2ECl_2]^-$ (E = Sb, Bi) can easily be prepared by reduction of diphenylantimony(III)chloride or diphenylbismuth(III)bromide with $CoCp_2$ or $CoCpCl$. The cation in these compounds is $[CoCp_2]^+$. In $[PPh_4]^+[Ph_2BiBr_2]^-$ the anions are monomeric and Ψ-trigonal bipyramidal configurated. The bromine atoms occupy the axial positions as chlorine in the analogous $[Ph_2SbCl_2]^-$ anion (see above and **5**)[62].

The bismuth(III) anion of the species $[NEt_4]^+[BiBr_2I(2\text{-}(2\text{-pyridyl})phenyl)]^-$ is monomeric and only tetra-coordinated. There is no interaction between the metal and the nitrogen donor atom of the 2-(2-pyridyl)phenyl[63]. Also, in $[BiPh_4]^+[Ph_2Bi(CF_3CO_2)_2]^-$

(5) Structure of the tetracoordinated anion in [PPh$_4$][Ph$_2$BiBr$_2$]

the anion is tetra-coordinated though it is bound to two phenyl groups and two carboxylate ligands. In this complex the carboxylate units act via oxygen (η^1) as monodentate ligands and not bidentate as usual. The bismuth anion shows distorted Ψ-trigonal bipyramidal configuration; the phenyl ligands and the stereochemical active lone-pair of electrons are situated equatorially. The PhBiPh angle has a value of 94.8° [64]. The X-ray structure analysis of (2,6-Me$_2$C$_6$H$_3$)$_3$Sb(OH)I indicates partly ionic character for this complex by formation of (2,6-Me$_2$C$_6$H$_3$)$_3$Sb$^+$OH $\cdots$ I[65]. The cationic As(V) diol species [H$_2$C(CH$_2$)$_3$CH$_2$As(OH)$_2$]Cl can also be regarded as dialkylarsonic acid; moreover, the molecules are connected by intermolecular OH $\cdots$ Cl $\cdots$ HO$^-$ bonds[66].

b. Adduct species. The charge transfer complex [Ph$_3$As·I$_2$] is formed by thermal decomposition of [Mn(OAsPh$_3$)$_3$I$_2$SO$_2$]. This species can be regarded as a complex containing an end-on bound I$_2$ unit to the As centre. Thus the lone-pair of electrons of As interacts with the LUMO σ* orbital of I$_2$. The two-coordinated iodine (formally negatively charged) may be considered as being sp^{3}d hybridized and therefore one would expect a linear AsII moiety. This in fact was established by X-ray diffraction (AsII, 174.8°). Moreover, as the electron density in the antibonding σ*(I$_2$) orbital increases by coordination to Ph$_3$As, the rather long I—I distance of 300.5 pm (of I$_2$, 266.0 pm) agrees well with this concept (AsI, 265.3 pm)[67,68].

The first step of the reaction of Me$_2$AsNEt$_2$ with BH$_3$·THF at −90 °C yields two differently coordinated adducts. BH$_3$ is coordinated either to the Me$_2$As unit or to the NEt$_2$ group. The latter complex rearranges to form an intermediate with the Me$_2$As coordinated adduct and finally cleaves to give Me$_2$AsH·BH$_3$, Et$_2$NBH$_2$ and Et$_2$NB$_2$H$_5$[69].

An organobismuth halide adduct [BiPh$_2$Br(THF)] was obtained by redistribution between BiPh$_3$ and BiBr$_3$ in Et$_2$O and subsequent crystallization from THF/hexane solution. This adduct is monomeric (Ψ-trigonal bipyramidal) with bromine and THF in the axial positions. The Br—Bi bond distance of 274.1 pm is longer than the one in the polymeric bismuthine [BiPhBr$_2$(THF)]. However, the Bi—O(THF) distance of 258.9 pm is shorter than in [BiPhBr$_2$(THF)]. For this reason the authors discuss the Bi—Br σ* orbital to be the acceptor of bismuth for oxygen from THF. The BrBiO angle was determined to be 173.5°[62].

Tetracyanoethylene (TCNE) coordinates weakly to Ph$_3$E (E = As, Sb, Bi) in CH$_2$Cl$_2$ solution to form Ph$_3$E–TCNE complexes[70]. Reaction of phenyldichlorarsine PhAsCl$_2$ with K$_2$SN$_2$ led to the formation of the diarsine Ph$_2$As—N=S=N—AsPh$_2$ with weak intramolecular As$\cdots$As interactions (337.9 pm) (see **6**)[71].

$$\begin{array}{c}
\quad\; S \\
N{=}\!\!\diagdown\!\!\diagup\!\!{=}N \\
| \qquad\qquad | \\
Ph_2As\text{------}AsPh_2
\end{array}$$

(6) Structure of the diarsine Ph$_2$As—N=S=N—AsPh$_2$

Similarly, the related tetramethyldistibine Me_4Sb_2 shows intermolecular interactions of the antimony atoms (367 pm). Within the Me_4Sb_2 unit the Sb–Sb distance is 286 pm. The terminal CSbC angle is $179.2°$[72]. The azides Me_2SbN_3 as well as Me_2BiN_3 are polymers in the solid state with strong metal–αN interactions and zig-zag configuration. The Sb–N distances of 232.2 pm and 243.4 pm are not significantly different. The angles deviate in comparison to the ideal values of a trigonal bipyramidal configuration, e.g. the NSbN angle is $169.6°$, the CSbC angle is $93.36°$ and the Sb—N—Sb angle is $126°$[73]. The adduct $[(3-Y—C_6H_3O_2-1,2)CpSb]\cdot(Me_2SO)$ (Y = H, OH) contains a π bound Cp ring. In this complex dimethylsulphoxide acts as oxygen donor[74].

c. Tetra-coordinated complexes with polydentate N, O, S-containing donor ligands. $Ph_2SbOCOR$ or $Ph_2BiOCOR$ are known containing R = Et, Pr, Me_2CH, Bu^t, $MeCOCH_2CH_2Ph$. In these complexes the bidentate carboxylate ligands do not act by chelating but by bridging under formation of polymers, which were investigated by IR spectroscopy[75]. The polymer diphenylacetatostibine shows distorted Ψ-trigonal bipyramidal configuration with the phenyls and the stereochemical active lone-pair of electrons being in equatorial positions while the bridging O atom as well as the closer bound oxygen are situated axially. Whereas the shorter Sb—O bond distance is 213.7 pm, the Sb distance toward the bridging oxygen atom has a mean value of 255 pm. The OSbO angle was determined to be $168°$, the CSbC angle is $95°$[76]. $Ph_2Sb(O_2PPh_2)$ as well as $Ph_2Sb(OSPPh_2)$ contain Ψ-trigonal bipyramidal coordinated antimony with the stereochemically active lone-pair of electrons in the equatorial position. The bidentate diphenylphosphinate and the diphenylmonothiophosphinate ligands again act by bridging and form chain-like structures. The Sb—O bond distances are similar with 223 pm and 229 pm, respectively, the Sb–S distance is 275.3 pm (OSbO, $166.7°$)[77]. Other ligands containing oxygen as well as sulphur donor atoms are derivatives of the monothio-β-ketone $NaSC(Ph)CH(CO)$—R with R = Ph, $4-MeC_6H_4$, $4-FC_6H_4$, $4-ClC_6H_4$, $4-BrC_6H_4$, $4-MeOC_6H_4$. Treatment of these species with Ph_2SbCl in THF yielded the product $Ph_2SbSC(Ph)CH(CO)R$. IR spectroscopy revealed that antimony is tetra-coordinated by two phenyl groups and one S and one O atom of the ketone[78]. The molecules of 2-phenyl-1,3-dithia-2-stibolane show intermolecular Sb $\cdots$ S interactions under formation of helical chain-like structures. The intermolecular Sb–S distances have values of 324 pm while the intramolecular Sb–S distances are 243 pm and 246 pm, respectively (cf. **7**)[79].

(7) Structure of 2-phenyl-1,3-dithia-2-stibolane

A nice example for a complex containing a S,S' coordinated chelate ligand is represented by the species $Ph_2SbS_2P(OR)_2$ (R = Me, Et, Pr, Me_2CH, Bu, Me_2CHCH_2). IR and multinuclear NMR spectroscopy (^1H, ^{13}C, ^{31}P) indicate a Ψ-trigonal bipyramidal structure. The species is monomeric in solution and shown in **8**[80].

(8) Structure of $Ph_2SbS_2P(OR)_2$

Analogous conditions are to be found in $Ph_2SbS_2P(O_2R)$, $R = CMe_2CMe_2$, $CH_2CMe_2CH_2$, $CH_2CEt_2CH_2$, CMe_2CH_2CHMe; $ORO = (OPh)_2$, where phosphorus itself is a member of a heterocycle[81]. By refluxing or vacuum pyrolysis of $[(CF_3)_2AsCINSiMe_3]_2$ an As and N containing tetrameric heterocycle of the composition $[(CF_3)_2AsN]_n$ $(n = 3, 4)$ was prepared. This eight-membered $(AsN)_4$ ring is slightly puckered. The As atoms are substituted with two CF_3 units each. The As—N bond distances (171.6 pm, 173.2 pm) are shorter than those of single bonds (see **9**)[82] (sum of covalent radii: 191 pm)[20] and indicate a bond order of greater than one.

(9) Structure of $[(CF_3)_2AsN]_4$

d. Tetra-coordinated arene complexes. Arene complexes of the type $(EX_3)_2C_6R_6$ and $(EX_3)C_6R_6$ may be regarded as containing tetra- *or* hexa-coordinated E (E = As, Sb, Bi) and will be discussed in section C, 5, b.

4. Penta-coordinated complexes

μ-Oxobistrimethylantimony(V)phenylsulphonate[83] $(Me_3SbO_3SPh)_2O \cdot 2H_2O$ exists in the ionic form $[(Me_3SbOH_2)_2O]^{2+}[PhSO_3]_2^-$. Antimony is coordinated in a distorted trigonal bipyramidal configuration, H_2O and the bridging O atoms are in axial positions. The Sb atoms and the bridging oxygen form an angle of 153.2°, while the angle of bridging O, Sb and O(H_2O) varies between 171° and 172°. The H_2O molecule is only weakly coordinated to antimony and forms hydrogen bridges to the oxygen of the sulphonate anions. The Sb$\cdots$O (anion) distance is in the order of magnitude of the sum of the van der Waals radii[20], and therefore a weak cation–anion interaction can be considered[83]. Reaction of $SbCl_3$ with KCN yielded the Sb(III) containing anion $[SbCl_3(CN)_2]^{2-}$[84]. Trihalogenomethylantimonate(III) $[MeSbClBr_2]^-$ exists as a dimer in which antimony is penta-coordinated. It was prepared from $MeSbBr_2$ and R_4ECl ($R_4E = Bu_4N$, Ph_4P, Ph_4Sb) in CH_2Cl_2 solution. The complex was extensively studied by IR spectroscopy[85].

The dimeric Bi(III) anion in $[NEt_4]_2[(BiPhI_3)_2]$ is penta-coordinated with tetragonal pyramidal configuration (C_{2v}). The phenyl ligands occupy the *cis* apical positions on the same side of the Bi_2I_6 plane where two bridging I atoms are located between the two Bi atoms. The structure is slightly distorted and the two bismuth atoms lie 16.2 pm and 9.8 pm above the I_4 planes. The bismuth distances to the bridging iodine (328.8 pm, 328.5 pm, 325.7 pm, 332.7 pm) are longer than to the terminal iodine with mean values of 295 pm and 296 pm[62].

Organobismuth halides of the type $[BiPhX_2(THF)]$ (X = Br, I) show intermolecular halogen–bismuth interaction under formation of polymer chains. Bi again has tetragonal pyramidal configuration with an apical phenyl group. THF is weakly coordinated to Bi via oxygen, the Bi—O bond distance corresponds to 267.1 pm in the bromine compound and to 280.8 pm in the iodine species. In the bromine complex Bi lies 4 pm out of the plane while the iodine compound is not distorted. The bismuth distances to the bridging halides differ from those to terminal atoms. The distances to the bridging halides have mean values of 303.8 pm for Bi—Br and 322.7 pm for Bi—I[62]. The structure of the chlorine-substituted Sb(III) anion [(PhSbCl$_3$)$_2$]$^{2-}$ is analogous to the [(PhBiI$_3$)$_2$]$^{2-}$ anion. The Sb—Cl(bridging) distances with 300.7 pm and 310.3 pm are also longer than the Sb—Cl(terminal) distances of 243.2 pm and 244.4 pm. Although the compound is basically five-coordinated, there are very weak intermolecular Sb···Cl interactions (375.6 pm) and therefore in the solid state one could also classify this complex as being six-coordinated[86]. On the contrary, the tetragonal pyramidal [PhSbCl$_4$]$^{2-}$ has a monomeric structure. The phenyl group again is apical; the Sb–Cl distances range from 253.7 pm to 277.0 pm[61]. The analogous Sb(III) anion [MeSbI$_4$]$^{2-}$ is monomeric and has the same structure. As usual there are two longer Sb—I bond distances of 305 pm and 320 pm and two shorter values of 299 pm and 291 pm. There are very weak Sb···I cation–anion interactions in [Me$_4$Sb]$_2$$^+$[MeSbI$_4$]$^{2-}$. The hypervalent Sb(III) anion possesses a 4-electron 3-centre bond[55].

The organo- and halogen-substituted Bi(III) anion in the complex Na[BiBr$_2$(O$_2$CEt)(2-(2-pyridyl)phenyl)] contains one monodentate pyridylphenyl ligand and one bidentate carboxylato ligand. IR spectroscopy could not reveal whether the carboxylate group is a chelate or a bridging ligand. Since there is no Bi–N interaction with the pyridylphenyl ligand, bismuth is five-coordinated[63]. Thiocarboxylato ligands are bidentate in organoantimony(III) compounds. If thioacetate is coordinated to antimony [e.g. in PhSb(SOCMe)$_2$, 10] the Sb–S distances are significantly shorter (245 pm) than the Sb–O distances (281 pm). This may very well be explained by both the relatively strong Sb—S bond according to the HSAB principle (Sb and S are soft, O is hard) and the fact that oxygen prefers to form π bonds (to carbon) rather than two σ bonds (one to C, one to Sb)[87].

(10) Structure of PhSb(SOCMe)$_2$

Methyldiiodostibine [MeSbI$_2$] has a tetragonal pyramidal structure, caused by intermolecular Sb···I interactions (340 and 347 pm). Whereas the iodine atoms are in the plane, the methyl groups are apical. The MeSbI$_2$ molecules associate via iodine bridges to form linear chains with alternating short and long Sb–I distances[55] (11).

(11) Structure of MeSbI$_2$

The complex PhBi(S$_2$COMe)$_2$ contains two bidentate xanthogenato chelate ligands which lie equatorially in the coordination octahedron of bismuth, while the phenyl group and the stereochemically active lone-pair of electrons occupy the axial positions[88].

Acting as tridentate ligands in Sb(III) complexes are diphenyldithiophosphinate and -arsinate. They form intra- as well as intermolecular interactions which lead to the formation of weakly bound dimers. Two molecules of Ph$_2$SbS$_2$EPh$_2$ (E = P, As) react to form a dimer containing a Sb$_2$S$_4$E$_2$ heterocycle; the structure is shown as 12[89].

(12) The dimeric structure of $[Ph_2SbS_2EPh_2]_2$ (E = P, As)

The tridentate sulphur containing ligand in $Ph_2SbS_2P(OPr^i)_2$ also favours the formation of polymer chains. However, the structure differs from the complex mentioned above and contains one strong Sb—S bond of 254 pm and two weaker intermolecular Sb—S bonds (see 13)[90].

(13) The chain-like structure of $[Ph_2SbS_2P(OPr^i)_2]$

In complexes of the type $PhE(III)(S_2PO_2R)_2$ (E = As, Sb; R = $CH_2CMe_2CH_2$, CMe_2CMe_2, CMe_2CH_2CHMe, CHMeCHMe) the sulphur-containing ligands behave as bidentate ligands and lead to five-coordinated As and Sb, which was proved by IR, ^{1}H and ^{31}P NMR spectroscopy. The ligands contain heterocycles of phosphorus, oxygen and carbon[91]. In the complex $PhSb[(SCH_2CH_2)_2O]$antimony is also five-coordinated. There are transannular Sb—O bonds and Sb—S contacts toward the adjacent molecules[92]. The species $[Ph_2Sb(SQ)]$ (QSH = 8-mercaptoquinoline] contains an Sb—S bond of 244.4 pm. Moreover, the sulphur–nitrogen donor ligand coordinates one nitrogen atom to antimony with an Sb—N distance of 267.7 pm. In addition to this there are weak intermolecular Sb–Sb interactions of 388.4 pm. The configuration is best described by a distorted tetragonal pyramid, where a phenyl group and the lone-pair of electrons possess the axial positions[93]. In $[PhSb(SC_5H_4N)_2]$ two sulphur- and nitrogen-containing donor ligands (2-pyridinethiolate) coordinate to antimony yielding a Ψ-octahedral configuration with two S and two N atoms equatorial and one phenyl group and the lone-pair of electrons in the axial positions. Sb···Sb interactions are quite weak (415 pm)[94]. The complex $[(3-Y-C_6H_3O_2-1,2)SbCp]\cdot 2L$ (Y = H, OH; L = formamide, dimethylformamide, carbamide) contains, beside O donor ligands, a σ-bonded Cp ring[74].

5. Hexa-coordinated complexes

a. Hexa-coordinated ionic complexes. In anionic Sb(V) complexes of the type $[R_4E][R'SbCl_{5-n}Br_n]$ (R' = Me, R_4E = Ph_4P, Ph_4Sb; R' = Ph, R_4E = Ph_4P, Et_4N; n = 0–5) the stability decreases in the order $[MeSbCl_5]^- \approx [MeSbCl_4Br]^- \gg$

$[MeSbCl_3Br_2]^- \approx [MeSbCl_2Br_3]^- > [MeSbClBr_4]^- \approx [MeSbBr_5]^-$; the salts become more instable by increasing replacement of chlorine by bromine. This may be explained by both, the decreasing polarity of the Sb—X (X = Cl, Br) bond and the increasing steric hindrance with increasing bromine substitution[85]. Reaction of $(C_6F_5)_3SbCl_2$ with CsF in MeOH or of $(C_6F_5)_3SbF_2$ with CsF in MeOH yielded the complex $Cs[(C_6F_5SbF_3]$. ^{19}F NMR spectroscopy indicates a meridional configuration for the anion[95]. In $[PPh_4]_2[Sb_2I_8]$ weak interactions exist between the phenyl groups of the cation and the antimony atoms in the anion. Antimony is surrounded by iodine in a tetragonal pyramidal configuration, two iodine atoms act bridging between two antimony centres, the Sb—I bond distances range between 276.7 and 335.3 pm. One phenyl group is η^2 bond to antimony, a second phenyl group of the same cation is connected with antimony of an adjacent anion, thus forming a chain-like polymer structure (see **14**)[96].

(14) The chain-like structure of $[PPh_4]_2[Sb_2I_8]$

The Sb(V) anion in $[SbPh_4][SbPh_2(Ox)_2]$ has a highly distorted octahedral configuration; the oxalate ligands are bidentate and situated in *cis*-positions to each other. The slightly unsymmetrical Sb—O bond distances have values of 202.9, 203.2, 207.2 and 207.7 pm[97]. $[Et_4N][Ph_3SbCl(O_2C_{10}H_6)]$ has been prepared from Ph_3SbCl_2 with 2,3-naphthalenediol and Et_3N[98]. The coordination compound $Li[(Tol)_3MeE-2-\{OC(CF_3)_2C_6H_4\}]$ (E = Sb, Bi) exists in different isomers; the structure is given as **15**[99,100]. Another stable six-coordinated bismuth containing complex is $Li[(Tol)_2Bi-2-\{OC(CF_3)_2\}C_6H_4]$, which was prepared from a five-coordinated bismuth educt and a dilithium reagent[101].

Tol$_3$Me

(15) Structure of $Li-[(Tol)_3MeE-2-\{OC(CF_3)_2C_6H_4\}]$ (E = Sb, Bi) (Tol = p-CH$_3$C$_6$H$_4$)

b. Hexa-coordinated arene complexes. Arene complexes containing arsenic, antimony or bismuth halides are formed by element (As, Sb, Bi)–arene η^6-interaction. The metal atoms lie above the plane of the coordinated arene ligands and interact additionally with halides of adjacent molecules to gain Ψ-octahedral configuration. There

TABLE 4. Arene complexes $(EX_3)_n \cdot$ arene $(E = As, Sb$ Bi; $X = Cl$, Br, $n = 1,2$)

EX_3	n	Arene	Reference
$AsCl_3$	1	Me_6C_6	102
$AsCl_3$	2	Et_6C_6	102
$SbCl_3$	2	$Me_4C_6H_2$	102
$SbCl_3$	2	pyrene	103
$SbBr_3$	1	Me_6C_6	104
$SbBr_3$	2	phenanthrene	103
$SbBr_3$	2	9,10-dihydroanthracene	105
$BiCl_3$	1	Me_6C_6	104
$BiCl_3$	2	pyrene	106

are two classes of EX_3 arene compounds: (i) $(EX_3) \cdot$ arene and (ii) $(EX_3)_2 \cdot$ arene $(E = As,$ Sb, Bi; $X = Cl$, Br); examples for both types are given in Table 4.

The 2:1 complexes possess an inverse sandwich structure. On both sides of the aromatic ring system there is interaction with an element(III) halide. Tetrameric EX_3 units form three-dimensional organometallic polymers with the halogens in the bridging positions. Arene complexes of the composition 1:1 form 'half-sandwich' ('piano chair') structures. The metal atom lies above the arene ring centre, it has three strong and two weaker bonds to X (X = halogen) and exhibits (the arene coordinations included) an octahedral configuration. Moreover, there are bridging halides between the metal centres causing a two-dimensional polymeric structure. For example, the Bi—Cl bond distances in $(BiCl_3) \cdot$ Mes are 248.9, 248.3, 246.5 pm and 336.8, 330.2 pm; the structure is illustrated in **16**.

(**16**) Coordination of Bismuth in $(BiCl_3) \cdot$ Mes (Mes $= 1,3,5\text{-}Me_3C_6H_3$)

In the related antimony and bismuth complexes $(SbBr_3) \cdot (C_6Me_6)$ and $(BiCl_3) \cdot (C_6Me_6)$ there is an anomaly of the η^6 bond which is more significant for Sb than for Bi. The Sb distances to the arene centre in $(SbBr_3) \cdot (C_6Me_6)$ range from 322 to 333 pm (cf section III.c.6). In the η^6-arene–$BiCl_2^+$ complex $[(Me_6C_6) \cdot BiCl_2][AlCl_4]$ two terminal chlorine atoms have strong bonds to bismuth (243.3 pm, 244.5 pm) while three bridging chlorine atoms are weakly coordinated (326.3, 336.9, 294.6 pm). Two of them belong to one $[AlCl_4]^-$ anion, the third one belongs to another $[AlCl_4]^-$ anion. The arene centre coordinates closely to bismuth with a distance of 272 pm. A distance of 328.4 pm between antimony and the π system is to be found in $(SbCl_3)_2 \cdot$ pyrene[103]. Hepta- and octa-coordinated arene complexes are discussed in section C, 6 (see below).

c. Hexa-coordinated complexes containing N, O, S donoratoms. Arsenic, antimony and bismuth in their $+ V$ oxidation state form octahedral configurations if coordinated by three carboxylato groups. The carboxylato ligands occur usually bidentately chelating or bridging, sometimes they may also act as monodentate oxygen donor ligands. The bidentate carboxylato ligands form asymmetrical bonds between the two oxygen atoms and the coordinated metal. β-Diketones are known as classic examples of chelate ligands.

The monomeric complex of the type R_3SbBrL (R = Me, Ph) can contain various ligands, e.g. the moieties of ethyl acetoacetate, salicylaldehyde, o-hydroxyacetophenone or 2-hydroxy-1-naphthaldehyde. The six-coordinated antimony possesses an octahedral structure[107]. Acetylacetone, when reacted with $(Ph_3SbCl)_2O$ and Ph_3SbO, led to cleavage of the Sb—O bond and yielded $Ph_3Sb(acac)X$ (X = OH, Cl), with acac acting as bidentate ligand. IR and ^1H NMR spectroscopy indicate the existence of complex isomers[108]. The dimer $(Ph_3SbO_2C_6H_4)_2 \cdot H_2O$ contains two differently coordinated Sb atoms. One of them is five-coordinated and forms a trigonal bipyramid, while the other is six-coordinated due to the coordination of an additional H_2O molecule[109]. The adduct species $Ph_2SbCl_3 \cdot H_2O$ and $Ph_2SbBr_3 \cdot MeCN$ possess distorted octahedral structure. The first contains a long Sb—O bond distance; H_2O is coordinated in *trans* position to one of the Cl atoms, which shows a shorter Sb–Cl distance compared with the other Cl atoms. Moreover, there is an additional intermolecular interaction of Cl atoms with the H atoms of H_2O. $Ph_2SbCl_3 \cdot H_2O$ was prepared by crystallization of the anhydrous precursor compound from MeCN containing approximately 5% H_2O. The same technique starting with Ph_2SbBr_3 did not lead to an analogous product but yielded $Ph_2SbBr_3 \cdot MeCN$. This adduct contains a longer Sb–N distance of 253 pm and a Br—Sb—N angle of 82.8°[110]. Reaction of $MeSb(OEt)_2$ with phthalic acid yielded a dimer containing a fourteen-membered heterocycle. By additional coordination of one molecule of ethanol, each of the antimony atoms gained octahedral coordination (see **17**). The carboxylato groups form asymmetric bonds with Sb—O bond distances of 211.3 and 265.3 pm or 208.8 and 298.5 pm, respectively. The Sb–O distance to the coordinated EtOH molecule was determined to be 264.7 pm[111].

(**17**) Structure of the dimer $[MeSb\{(O_2C)_2C_6H_4-1,2\} \cdot EtOH]_2$

In $PhSb(OAc)_4$ one of the acetyl ligands acts bidentately chelating while three are monodentate ligands. Two of the latter ones occupy the axial positions of the octahedron while the bidentate acetyl and the remaining ligands lie equatorially[112]. Treatment of 1-trimethylsiloxyl-1-alkoxycyclopropane with $SbCl_5$ causes ring cleavage and substitution of one Cl atom by a metal–carbon bond and an additional coordination of carboxyl oxygen to antimony. Thus antimony reaches a six-coordinated structure as shown in **18**[113].

(**18**) Structure of [Cl$_4$Sb(CH$_2$)$_2$COOCHMe$_2$]

Ph$_3$Sb(O$_2$C$_6$H$_4$NO$_2$) exists as a weakly bound dimer in the solid state. 4-Nitrocatecholate acts as a bidentate chelate ligand. The metal atom interacts with oxygen of the alkoxy group of an adjacent molecule thus reaching octahedral configuration. The intermolecular Sb–O distance of 334.1 pm is smaller than the sum of the van der Waals radii[20] (360 pm). The Sb—O bonds within the molecule are stronger with distance values of 207.8 and 203.9 pm. The dimer forms a four-membered Sb$_2$O$_2$ heterocycle (cf **19**)[114].

(**19**) Structure of the dimer [Ph$_3$Sb(O$_2$C$_6$H$_4$NO$_2$)]$_2$

In dimeric cyclodiarsazane, arsenic becomes six-coordinated due to the formation of N—As donor bonds. The structure is given as **20**[115].

(**20**) The structure of the dimeric cyclodiarsazane

Schiff's bases containing oxygen as well as nitrogen donor atoms can act as bi- or tridentate ligands. In the complexes R$_3$SbBr(SAP) (R = Me, Ph; SAP-H = N-salicylidene-2-aminopyridene) and R$_3$SbBr(HNAP) [R = Me, Ph; HNAP-H = N-(2-hydroxy-1-naphthalidene)-2-aminopyridene] the ligands behave as bidentate ligands, effecting a six-coordinated structure for antimony[116]. 8-Quinolinato ligands can be mono- or bidentate O, N donor ligands. When reacted with triorgano–Sb(V) compounds they form complexes of the type [R$_3$SbQ$_2$] (R = Me, Et, Ph; Q = 8-quinolinato ligand) where the central metal is five-coordinated. Only for [Ph$_3$SbQ$_2$] ^{1}H and ^{15}N NMR, UV–VIS and IR spectroscopy indicate an equilibrium of a mono- and bidentate bond situation concerning the quinolinato ligands. Compounds of the type [R$_3$SbClQ] (R = Me, Et, Ph) and [Me$_3$SbBrQ] with only one quinolinato ligand usually possess six-coordinated antimony atoms[117]. In the tetraorgano Sb(V) quinolinato complex [R$_4$SbQ] (R = Me, Ph) antimony is coordinated in a distorted octahedral structure with Sb—O and Sb—N bond distances of 218.7 pm and 246.3 pm, respectively (cf **21**)[118].

(21) Structure of R_4SbQ (R = Me, Ph; Q = 8-quinolinato ligand)

Some Bi(V) compounds contain 2-methyl-8-quinolinato ligands in bidentate function, e.g. $[Ph_3BiClQ]^{119}$ or $[Ph_2BiX_2Q]$ (X = Cl, Br)[120]. N, N'-Dimethylbenzamidine is used as a bidentate N,N donor ligand for organo Sb(V) complexes, e.g. $[LPh_nSbCl_{4-n}]$ (LH = N,N'-dimethylbenzamidine). The ligand coordinates to antimony with formation of a four-membered SbN_2C heterocycle. The compound Ph_2SbCl_2L contains six-coordinated antimony in a distorted octahedral coordination sphere. The NSbN angle of 60.6° and the ClSbCl angle of 172.4° indicate a significant rate of distortion. The Sb—Cl bond distances are 242.6 and 245.4 pm, the Sb–N lengths are 212.5 and 216.7 pm (22)[121].

(22) Structure of Ph_2SbCl_2L (L = N,N'-dimethylbenzamidine)

In general, to stabilize a nitrogen–antimony(V) bond, the basicity of nitrogen must be reduced by electron withdrawing groups or by a mesomeric effect, and finally the nitrogen containing agent must be stable towards oxidation. The complex $(MeNSbCl_3)_4$ was prepared from $SbCl_5$ and $(Me_3Si)_2NMe$. The product is a tetramer and contains six-coordinated antimony; its cubic molecular structure is formed by Sb and N atoms. The Sb—N bond distances range from 215.1 to 218.9 pm and are comparable to those in S_2N_2 compounds; the angles have values of 99.6–100.8° for SbNSb and 78.2–79.1° for NSbN. The cubane-like structure is shown as 23[122].

(23) Structure of $[MeNSbCl_3]_4$

Oxadithia- or trithiastibocanes or -bismocanes of the type $RE[(SCH_2CH_2)_2X]$ (R = Ph, substituted Ph; X = O, S) contain six-coordinated antimony or bismuth in the +III oxidation state. The metal atom realizes its coordination number of six by transannular interaction with oxygen or sulphur and additional intermolecular interactions.

The trithiastibocane $RSb[(SCH_2CH_2)_2S]$ (R = Ph, 4-Tol, 4-nitrophenyl) was prepared from dichloroarsine, dithiol and triethylamine. It contains an eight-membered metallacycle with boat-chair conformation[92,123]. In the analogue with R = 4-nitrophenyl the transannular $Sb \cdots S$ 1,5-interaction shows a distance of 319 pm while the intermolecular $Sb \cdots S$ interactions have values of 353 pm. An analogous bismuth complex is the oxadithiabismocane $PhBi[(SCH_2CH_2)_2O]$ in chair–chair conformation (see **24**). The eight-membered heterocycle contains two rather short Bi–S distances of 256.0 and 260.2 pm while the intermolecular Bi–S distances extend to 344.0 and 350.9 pm. The transannular $Bi \cdots O$ distance is 297 pm[124].

(**24**) Structure of the oxadithiabismocane $PhBi[(SCH_2CH_2)_2O]$

$PhAs(S_2CNEt_2)_2$ is monomeric in the solid state while the analogous methylbis(dithio-carbamato)stibane $MeSb(S_2CNEt_2)_2$ exists as weakly bound dimers. The chelating S donor ligand forms asymmetrical Sb—S bonds (see **25**); their distances are 255.4, 296.0 pm and 253.8, 290.4 pm. The intermolecular $Sb \cdots S$ distance is quite long (384.7 pm) but ranges within the sum of van der Waals radii[20] (Sb 228 + S 185 = 413 pm)[125].

(**25**) Structure of the stibane $MeSb(S_2CNEt_2)_2$ with intermolecular contacts

The analogous bismuth compound $PhBi(S_2CNEt_2)_2$ is also a dimer containing six-coordinated bismuth. However, the conformation is Ψ pentagonal bipyramidal because of the stereochemical activity of the lone-pair of electrons. The four S atoms of the coordinated ligands and one S atom of the adjacent molecule (in interaction with bismuth) occupy the equatorial positions. The phenyl group and the lone-pair of electrons lie in the apical positions. The Bi—S bond distances range from 267.1 and 292.6 pm to 267.6 and 294.2 pm. The Bi–S intermolecular distance is naturally longer with 342.1 pm. The angles between two ligand sulphur atoms and bismuth are 63.9° and 63.5°, while the angles at the bismuth between ligand sulphur and intermolecular bound sulphur are 144.0° and 143.6°[63]. In case of methyl substituted bismuth the analogous complex shows the following structural parameters; Bi—S bond distances: 267, 295 pm and 270, 298 pm; Bi—S intermolecular distances: 327 and 336 pm[126]. Substitution of phenyl or methyl in $LBi(S_2CNEt_2)_2$ with L = 2-(2-pyridyl)phenyl yielded a monomeric compound. However, it has Ψ-pentagonal bipyramidal coordination due to the strong Bi–N interactions and the stereochemically active lone-pair of electrons. Four S atoms of the thiocarbamate ligands and the N atom of the pyridylphenyl ligand occupy the equatorial places while the σ-bond carbon of the pyridylphenyl and the lone-pair of electrons are in the apical positions. All angles, however, differ from the ideal values by approximately 10°. The Bi—S bond distances are not so significantly different from each other (279.8, 287.8, 276.0, 289.5 pm); the Bi–N bond distance is 255.3 pm[63].

 d. Further hexa-coordinated complexes. The adduct species $Ph_3SbCl_2 \cdot SbCl_3$ forms polymeric chains due to Sb···Cl interactions with distances of 326.2 pm[127]. Two diorganoantimony(V)trihalides Ph_2SbBr_2F and Ph_2SbBr_3 form an adduct by fluorine bridging. One antimony atom has trigonal bipyramidal configuration, the other is six-coordinated with formation of a distorted octahedral structure (see **26**), which is caused by the formation of a rather strong Sb···F···Sb bridge[128].

(**26**) Structure of the adduct species $Ph_2SbBr_2F \cdot Ph_2SbBr_3$

 The axial fluorine atom increases the Lewis acidity of this antimony atom and thus effects stronger bonding to the bromine in the axial position. The Sb—$Br_{ax.}$ bond distance is 251.0 pm and is substantially shorter than the Sb—$Br_{eq.}$ bonds of 262.6 and 260.0 pm. Moreover, an angular stretching is observed. A similar structural situation exists in $(Ph_2SbBrF)_2O$ (see **27**)[128].

(**27**) Structure of $(Ph_2SbBrF)_2O$

 The $Ph_2SbBrFO$ unit exhibits a distorted octahedron due to the F donor atom of the second unit. The mutual Sb–Sb repulsion ($d_{Sb-Sb} > 550 \, pm$) results in a quite long Sb–F contact of 330 pm.

6. Hepta- and octa-coordinated complexes

 a. Arene complexes. For six-coordinated arene complexes, see section III.C.5.b above. Haloarsines, -stibines, and -bismuthines of the elements in their +III oxidation state form arene complexes with aromatic cycles by π bonding. Coordination of neutral arenes to metal atoms (d elements) is discussed on the base of simple models of valence theory[129]. The species EX_3 (E = As, Sb, Bi; X = Cl, Br) act as Lewis acids and form donor–acceptor complexes with aromatic compounds. The different metal atoms show different stereoscopic pictures in their coordination. So bismuth is centrally situated above the aromatic ring forming an η^6 bond, while antimony prefers an asymmetric position toward the ridge of the ring system, nearly with formation of an η^2 bond.

These highly coordinated arene complexes are composed in the type $(E(III)X_3)_2 \cdot (arene)$. The metal atoms are connected by halide bridges to reach higher coordination states and form polymers. The complex $(SbCl_3)_2 \cdot (C_6H_6)$ has pentagonal bipyramidal coordination with benzene in the axial position. The antimony atoms form intermolecular interactions to chlorine of two adjacent $SbCl_3$ molecules, thus reaching seven-coordination. The distances to bridging Cl range between 340.1 and 396.8 pm and are shorter than the sum of the van der Waals radii[20] with 400 pm. The stronger intramolecular Sb—Cl bond distances come to 234.7 and 238.0 pm. The chlorine bridging effects $[SbCl_3]_n$ layers with inserted benzene molecules (see **28**)[130]. The distances between Sb atoms and benzene ring centres are 330 and 322 pm.

(**28**) Coordination in the arene complex $(SbCl_3)_2 \cdot (C_6H_6)$

Bismuth also is seven-coordinated in the 2:1 complex $(BiCl_3)_2 \cdot (C_6Me_6)$. To the metal atom two close-lying chlorine atoms (240.4 pm, 243.8 pm) coordinate as well as two slightly distant bridging Cl atoms (288.7 pm), two weakly bond Cl atoms (307 pm, 371.6 pm) and finally the arene by η^6 bond. The compound consists of tetrameric Bi_4Cl_{12} units and forms a three-dimensioned network. The polymeric compound has a sandwich structure (**29**)[131,129].

(**29**) Structure of $(BiCl_3)_2(C_6Me_6)$

For both of these complexes the lone-pair electrons are not stereochemically active with respect to the polyhedral coordination, but stabilize the $6s^2$ and $5s^2$ level, respectively (inert pair effect). The interactions between metals and arenes are rather weak but significantly influence the coordination sphere in comparison to $BiCl_3$ or $SbCl_3$.

b. Crown ether complexes. These complexes are composed from EX_3 or REX_2 ($E = Sb$, Bi; $X = $ halide, pseudohalide; $R = $ organic rest) and a crown ether. The metal atom is coordinated to its ligands and to the oxygen of the crown ether and lies in a sandwich-like conformation between the ether and the other ligands. Organic ligands of metal(III) compounds influence their Lewis acidity. This anisotropic effect is of significance in the complex $[SbBr_2R(15\text{-crown-}5)]$ ($R = Ph$, Me) $(\mathbf{30})^{132}$.

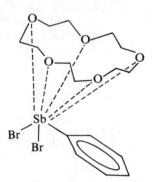

(30) Structure of $[SbBr_2R(15\text{-crown-}5)](R = Ph$, Me)

In both compounds all Sb—O bond distances are different ($R = $ Me: Sb–O $= 283–327$ pm; $R = $ Ph: Sb–O $= 286–335$ pm). The longest Sb—O bond is in *trans* position to the ligand R; the dihedral angle ($R = $ Ph) is $17.1°$. So the ligand R causes an anisotropic decrease in the antimony atom's Lewis acidity. The Sb ligand angles are approximately $90°$, ($R = $ Me: BrSbBr $= 91.7°$, $BrSbC_{Me}$: $90.6°$, $92.7°$; $R = $ Ph: BrSbBr $= 91.5°$, $BrSbC_{Ph} = 90.1°$, $90.9°$). The lone-pair of electrons of antimony probably extends to the configuration space of the crown ether. However, ^{121}Sb Mössbauer spectroscopy indicates an increasing s character of the lone-pair electrons and they may not be stereochemically active.

c. Chelate complexes containing N, O, S donor atoms. In E(V) compounds of the type R_3EL_2 carboxylato ligands act chelating, and intramolecular interactions between the metal and the oxygen of the carboxyl groups cause an increase in the coordination number and frequently distorted pentagonal bipyramidal configurations are formed ($E = $ Sb, Bi; $R = $ organic ligand; $L = $ bidentate carboxylato ligand). Examples for Sb and

TABLE 5. Compounds of the type R_3EL_2 ($L = O_2CR'$)

E	R	R'	Reference
Sb	Me, Ph, C_6H_{11},	Me, Ph, heterocycles	
	4-An, 4-FC_6H_4,	2-furyl, 2-thienyl, 2-pyrryl,	
	2,4,6-$Me_3C_6H_2$	2-(N-methyl)pyrryl	133
	Ph	Me	134
	Ph	CF_3	43
Bi	Ph	CF_3	43
	Ph	2-furoato(1-)	135
	Ph, 4-Tol	2-furyl, 2-thienyl, 2-pyrryl,	
		2-(N-methyl)pyrryl	133

Bi complexes are summarized in Table 5. The lack of As(V) complexes of the type R_3AsL_2 may be explained by the resistance of As to realize high coordination numbers and oxidation states, which are usually discussed on the basis of the post-transition metal effect[38]. Element(V) species of the type R_3EL possessing only one carboxylato chelate ligand which coordinates symmetrically show trigonal bipyramidal structures.

Comparing $Ph_3Bi(O_2CCF_3)_2$ with $Ph_3Sb(O_2CCF_3)_2$, the difference in the coordination spheres becomes evident and it can be stated that the Bi compound shows distorted pentagonal bipyramidal conformation while the Sb complex is in a distorted trigonal bipyramidal arrangement. The $C_{Ph}BiC_{Ph}$ angles are 140.6°, 110.1° and 109.1° (31)[43].

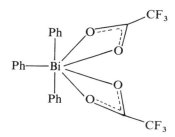

(31) Structure of $Ph_3Bi(O_2CCF_3)_2$

The Bi compound has asymmetrical Bi—O bond distances of 230.8 or 230.9 pm and 298.0 or 298.1 pm. The Sb—O bond distances differ even more with values of 211.6 or 215.3 pm and 320.9 or 323.1 pm. Nevertheless, these long Sb–O distances do not exceed the sum of the van der Waals radii[20] of 360 pm. The authors, however, conclude that these long bond lengths indicate that there is no interaction between the antimony atoms and the oxygen of the carboxyl group. The comparable stronger intramolecular interaction in case of Bi compounds points to higher Lewis acidity of the Bi(V) species.

However, for heterocyclic substituted carboxylate ligands the situation is different. IR spectroscopy (v_{C-O} and $v_{C=O}$) reveals that in this case the Sb compounds possess stronger Sb···O (carboxyl) interactions than the analogous Bi complex. For example, in $Ph_3Sb[O_2C-(2-thienyl)]_2$ the antimony atom exists in a distorted pentagonal bipyramidal structure with C_{Ph}—Sb—C_{Ph} angles of 145.9°, 104.4° and 109.5°. The apical positions are occupied by one phenyl group each. In the equatorial positions there are one phenyl group and four atoms of the carboxylato ligand.

For complexes of the type $R_3E(O_2CR')_2$ [E = Sb, Bi; R = aryl (Ph, 4-Tol); R' = heterocycles as like 2-furyl, 2-thienyl, 2-pyrryl, 2-N-methylpyrryl] it can be concluded from IR spectroscopic data (v_{CO} in solution as well as in the solid state) that Bi shows less interaction with the carboxylato oxygen than does Sb. The authors discuss this effect on the basis of the higher mass of bismuth compared with antimony[133].

Organic substituents introduce an inductive effect to the Lewis acidity of Sb in the series of Me < 4-Tol < Ph < 4-FC$_6$H$_4$. With increased Lewis acidity the Sb—O bond becomes stronger. However, in the case of bulky substituents like cyclohexyl or mesityl the Sb—O(carboxyl) contacts are decreased, due to the steric hindrance[133]. Bis-[2-furoato(1-)]triphenylbismuth also possesses pentagonal bipyramidal coordination. The Bi—O(carboxyl) bond distances reveal a rather ionic character of the carboxylato ligand[135]. An oxygen-bridged Sb(V) dimer which contains seven-coordinated antimony and asymmetric chelate ligands was obtained by hydrolysis of $Ph_2Sb(OAc)_3$ (32). The Sb—O bond distances have mean values of 216 and 247 pm[136].

(32) Structure of $[Ph_2Sb(OAc)_2]_2O$

Some further examples for seven-coordinated Sb(V) atoms are: R_3SbBrL (R = Me, Ph), L is a tridentate monobasic Schiff's base ligand containing O, N, N donor groups. In $R_3SbBr(SPD)$ and in $R_3SbBr(HNPD)$ antimony is seven-coordinated due to the tridentate ligand (33) (SPD-H and HNPD-H, see abbreviations). The compounds were characterized by IR and ^1H and ^{13}C NMR spectroscopy[116].

(33) Structures of the ligands SPD-H and HNPD-H

The antimony in the adducts $SbF_3 \cdot L$ and $(C_6H_4O_2)SbF \cdot L$ is pseudo-seven-coordinated, as proved by IR and ^{121}Sb Mössbauer spectroscopy. L represents a tridentate nitrogen-containing donor ligand such as pyridine-2-carboxaldehyde-2-pyridylhydrazone, 2,2':6',2''-terpyridine or 2,4,6-tris(2-pyridyl)-1,3,5-triazine[137].

7. Organoelement compounds coordinated to main group elements or transition metals

a. Main group element complexes. Reaction of organoarsenic(III) compounds $[R_2As(III)X]$ with Ga(III) salts led to cyclic complexes with arsenic behaving as a doubly bridging ligand, 34 denotes the structure of this type of complexes while Table 6 gives some examples for gallium complexes containing various organoarsenic ligands.

(34) Structures of four-membered heterocycles $[R_2As\text{-}GaR'_2]_2$
(R = CH_2SiMe_3; R' = Cl, Ph, $As(CH_2SiMe_3)_2$)

With R = Me_3SiCH_2 and R'' = Ph the X-ray data of the complex in 34 reveal a planar, centrosymmetric As_2Ga_2 ring with angles of As—Ga—As 85.08° and Ga—As—Ga 94.92°. The Ga—As bond lengths are 251.8 pm and 253.0 pm, thus being approximately 10 pm longer than the sum of the covalent radii with 246 pm (Pauling)[20]. In the Ga complex $\{[(Me_3SiCH_2)_2As]_2GaCl\}_2$ NMR spectroscopy proves the dynamic behaviour

TABLE 6. Complexes of the type $[R_2As—GaR'_2]_2$

Arsenic educt	Gallium educt	As—Ga compound	Reference
Bu^t_2AsLi, MeLi	$GaCl_3$	$[Bu^t_2AsGaMe]_2$	138
$(Me_3SiCH_2)_2AsLi$	$GaCl_3$	$[\{(Me_3SiCH_2)_2As\}_2GaAs(CH_2SiMe_3)_2]_2$	139
$(Me_3SiCH_2)_2As_2SiMe_3$	$GaCl_3$	$\{[(Me_3SiCH_2)_2As]_2\text{-}GaCl\}_2$	140
$(Me_3SiCH_2)_2AsH$	Ph_3Ga	$[(Me_3SiCH_2)_2AsGaPh_2]_2$	18

of the As ligand with a rapid exchange of the terminal and bridging positions. The cluster-like compound $[(PhAsH)(R_2Ga)(PhAs)_6(RGa)_4]$ with $R = CH_2SiMe_3$ containing an As_7Ga_5 centre is obtained by reaction of $Ga(III)(CH_2SiMe_3)_3$ with the primary phenylarsine $PhAsH_2$[141]. The phenyl-substituted arsenic acts as a doubly or triply bridging ligand. An interesting gallium complex with an antimony ligand which forms a six-membered Sb_3Ga_3 ring (35) can be obtained from $GaCl_3$ and $Bu^t_2SbSiMe_3$[142].

(35) Structure of $[R_2SbGaCl_2]_3$ ($R = Bu^t$)

This compound in boat conformation contains a Ga_2Sb_2 plane; the mean Ga—Sb bond distances are 266.1 pm, in agreement with the sum of the Pauling covalent radii[20] (266 pm).

As far as we know, the first indium(III) complex with a dialkylstibine ligand was prepared according to equation 33; the structure is given as 36.

$$4(Me_3C)_2SbSiMe_3 + 2InCl_3 \xrightarrow{-78\,°C} [(Me_3C)_2SbInClSb(CMe_3)_2]_2 + 4Me_3SiCl \quad (33)$$

(36) Structure of $[R_2SbInClSbR_2]_2$ ($R = CMe_3$)

This antimony ligand shows dynamic behaviour, proved by NMR spectroscopy, and its position rapidly varies from terminal to bridging[143]. The In—Sb bond distances of 284.4 pm are within the sum of the van der Waals radii[20] (284 pm).

The group 1 element lithium is proved to form four- or six-membered rings with arsenic ligands like As_2Li_2 and As_2LiLu (Lu = Lutetium) or As_3Li_3 rings; the bisubstituted As(III) ligands are situated bridging between the Li atoms. Metallation of $[(Me_3Si)_2CH]_2AsCl$ with lithium yields the trimeric $[LiAs\{CH(SiMe_3)_2\}_2]_3$ which possesses a six-membered $(LiAs)_3$ ring[17]. This compound is a strong reducing agent and, by treatment with HCl or MeCl, it decomposes to give R_2AsH or R_2AsMe which demonstrates a rather weak As—Li bond in the six-membered heterocycle. A square planar Li_2As_2 unit was found in the dimeric $[Li(THF)\{As(Bu^t)As(Bu^t)_2\}]_2$ (see **37**) with Li—As bond distances of 258 pm. The transannular Li–Li distance of 329.0 pm is too long to expect a significant bonding interaction.

(**37**) Structure of $[Li(THF)\{As(Bu^t)As(Bu^t)_2\}]_2$

The first known complex with a direct lanthanide arsenic bond was prepared from $Cp_2Lu(\mu\text{-}CH_3)_2Li(TMEDA)$ in benzene solution with successive replacement of $\mu\text{-}CH_3$ by $\mu\text{-}AsPh_2$ (generated from Ph_2AsH) acting as a soft donor. The structure of this complex is given as **38**; the Lu—As bond distances are 289.6 and 287.0 pm, respectively[23].

(**38**) Structure of $Cp_2Lu(\mu\text{-}AsPh_2)_2Li(TMEDA)$

However, using the secondary arsine Bu^t_2AsH no substitution occurs because of the higher Lewis basicity of Ph_2AsH compared with Bu^t_2AsH.

b. Transition metal complexes.   Organic compounds of arsenic, antimony and bismuth, especially in their $+III$ oxidation state, are well known as ligands in transition metal complexes due to their donor properties as Lewis bases (soft donors). They are frequently used as two-electron donors. These ligands may occupy terminal positions or may act by multidentate bridging or chelating. The donor properties depend on the element itself and on the kind of its substituents. The donor ability is reduced by strong electronegative substituents. The Lewis acidity decreases in the sequence $M = As > Sb > Bi$. Silver complexes of the type $[Ag(Ph_3E)][ClO_4]$ ($E = As$, Sb, Bi) show by means of thermodynamical data that the ability to form complexes of Ph_3E with $AgClO_4$ decreases in the order $E = (P)$, As $> Sb > Bi > (N)$[144]. The reason may be an increase of the E—C bond polarization as the electronegativity of E decreases within the group[145]. A further reason for the weak Lewis basicity of the triply coordinated Bi(III) might be the fact that the lone-pair of electrons is located in an orbital with predominant s-character[146]. In the same way R_3Bi compounds (R = alkyl, aryl) are less used as ligands in transition metal complexes[147]. The reactivity of Bi_2Ph_4 towards cobalt carbonyls in comparison to the P, As and Sb analogues is weak[148].

Transition metal substitued arsenic(III) species of the type R_2As—$M(CO)_3Cp$

(M = group 6 metal) possess a substantially greater Lewis basicity than their triorgano-arsine counterparts[149].

Concerning ligands of organoarsenic, -antimony and -bismuth compounds in main group and transition metal complexes, some contributions are cited in the list of references[150,151,7]. Thus we fixed our attention to the literature within the last ten years to deal with topics of current interest.

i. Group 4 Elements.   Using the classical arsenic–phosphorus four-electron ligand $[Ph_2P(CH_2)_2AsPh_2]$ the binuclear η^5-cyclopentadienyl-titanium(IV) complex $[(Cp_2TiCl)_2$-μ-$(Ph_2P(CH_2)AsPh_2)]$ was prepared in 1981 and ten years later its structure was elucidated by X-ray analysis and is shown as **39**[152,153].

(**39**) Structure of $[(Cp_2TiCl)_2\mu$-$(Ph_2P(CH_2)AsPh_2)]$ (X = P and As)

Another example for a new titanium complex $[Cp_2Ti(As_2Me_4)]_2[AsF_6]_4$ contains a six-membered metallacycle. The temperature-dependent ^1H NMR spectra indicate dynamic behaviour and a chair conformation (**40**).

(**40**) Structure of $[Cp_2Ti(As_2Me_4)]_2[AsF_6]_4$

Moreover, the cationic complex $[Cp_2Ti(AsMe_3)_2][AsF_6]_2$ was prepared in 1991 from $Cp_2Ti(AsF_6)_2$ and the Lewis base $AsMe_3$ in SO_2 solution[154].

(**41**) Structure of $Cp(CO)_3V(Ph_2AsAsPh_2)$

ii. Group 5 Elements.   Treatment of $CpV(CO)_4$ with Ph_4As_2 yielded in a photochemical reaction the substitution of one CO by tetraphenyldiarsine (see **41**). The As—As bond distance in the complex does not significantly differ from that in the diarsine[155].

iii. Group 6 Elements.   Organoarsenic, antimony and bismuth compounds are frequently used as ligands for group 6 metals.

Terminal ligands.   Compounds of $R_3E(III)(E = As, Sb, Bi; R = alkyl, aryl)$ are suitable for terminal π ligands; the complex formation and donor capacity decrease in the sequence $As > Sb > Bi$, e.g. $BiPh_3$ shows weak donor properties (see Section II.B).

As and Sb trialkyl substituted molybdenum and tungsten complexes can be prepared from those which are As and Sb dialkyl substituted by means of transmethylation (equation 34)[156].

$$Cp(CO)_3M - E(CH_3)_2 \xrightarrow{P(OCH_3)_3} \quad \underset{\underset{CO}{\underset{|}{Me_2E}}{\overset{Cp}{\overset{|}{M}}}\diagdown^{CO}_{P(OCH_3)_3} \quad \longrightarrow \quad \underset{\underset{CO}{\underset{|}{Me_3E}}{\overset{Cp}{\overset{|}{M}}}\diagdown^{CO}_{\underset{O}{\overset{\|}{P(OCH_3)_2}}}$$

$$(34)$$

The intermediate of the irreversible isomerization can be isolated. The EMe_2 unit is σ bonded in the educt; it is a centre of basic activity, showing tendency for reactions with electrophiles. Thus the electron-rich state of the atoms of group 15 elements leads to irreversible isomerization with shifting of one methyl group from the PR_3 ligand to arsenic or antimony, respectively. The donor capacity of the transition metal is enlarging this ability.

A cationic Sb(III) ligand is part of the tungsten complex $[Cp(CO)_2(Me_3P)WSb(Me)-(Cl)Bu^t]I$, which was prepared from $trans\text{-}Cp(CO)_2(MeP)WSb(Cl)Bu^t$ and MeI in Et_2O[157]. The first transition metal complexes containing a metal–arsenic double bond were synthesized in 1983[149]. Transition metal substituted arsines of the type $Cp(CO)_3M—AsBu^t_2$ (M = Mo, W) show increased Lewis basicity compared with triorganoarsines R_3As. From the above-mentioned complex one carbonyl is cleavable already at 60 °C (equation 35).

$$\underset{\underset{OC}{\diagup}\underset{CO}{\overset{|}{M}}\underset{CO}{\diagdown}\underset{Bu^t}{\overset{As\diagdown}{\underset{|}{Bu^t}}}}{\overset{Cp}{\overset{|}{OC—M—As}}} \quad \underset{\underset{RT}{+CO}}{\overset{+60\,°C}{\underset{\xrightleftharpoons{\hspace{1cm}}}{-CO}}} \quad \underset{\underset{OC}{\diagup}\underset{CO}{\overset{}{}}}{\overset{Cp}{\overset{|}{M=As}}}\diagup^{Bu^t}_{\diagdown Bu^t} \qquad (35)$$

The *t*-butyl group shows rotation around the transition metal–arsenic axis at room temperature. Cleavage of the carbonyl is favoured since the sp^2 hybridized arsenic atom allows more space for the *t*-butyl group than the sp^3 hybridized one. The double bond

$$\underset{\underset{CO}{\underset{|}{S}}{\underset{OC}{\diagup}\overset{Cp}{\overset{|}{W}}\diagdown}}{\quad} \text{—As(Bu}^t)_2 \qquad\qquad \underset{\underset{CO}{\underset{|}{CH_2}}{\underset{OC}{\diagup}\overset{Cp}{\overset{|}{W}}\diagdown}}{\quad} \text{—As(Bu}^t)_2$$

$$\text{(a)} \qquad\qquad\qquad\qquad\qquad \text{(b)}$$

(42) Structure of the threefold metallacycles $Cp(CO)_2W(SAsBu^t_2)$ and $Cp(CO_2)W(CH_2AsBu^t_2)$

is stable towards MeI but reacts with HCl and nucleophiles like Me_3P or t-BuCN. Reaction with S_8 or CH_2N_2 yielded three-membered metallacycles which contain the arsenic as an η^2 ligand (**42a,b**)[158].

Arsanylketene is stabilized as ligand in the tungsten complex $[Cp(CO)\{PMe_3\}$-$ClWAs(Ph)_2C(Me)=C=O]$ (equation 36)[159].

$$(36)$$

By thermal cleavage, this complex was transformed into an ionic arsenic–tungsten heterocycle (**43**)[160,161].

(**43**) Structure of the cation in $[Cp(PMe_3)(CO)\overline{W-As(Me)_2C(Tol)=C(PMe_3)-O}]$ I

Treatment of the chromium complex $(CO)_5Cr\text{-}SbCl_2Bu^t$ with K_2SN_2 in MeCN yielded another heterocyclic compound, a sulphur diimid derivative which contains As or Sb as a terminal monodentate ligand: $[(CO)_5CrE(Bu^t)(NSN)_2EBu^t]$, (E = As, Sb)[162]. Quite unusual terminal ligands are organocyclotriarsanes and -tristibanes containing three-membered carbon bridged As or Sb rings as monodentate ligands. These compounds are coordinated to group 6 metal carbonyls like $MeC(CH_2Sb)_3M(CO)_5$ (M = Cr, Mo, W) as shown by **44**[163,164].

(**44**) Structure of $MeC(CH_2E)_3Cr(CO)_5$ (E = As, Sb)

The Sb—Sb bond distances are 280 pm; the Sb $\cdots$ Sb contacts extend from 396 to 406 pm, thus being shorter than the sum of van der Waals radii[20] with 440 pm. The analogous arsenic complex shows As—As$_{Cr}$ mean distances of 240.7 pm; the As—As bond distances normally are 246.2 pm[165].

Moreover, there are some reactions of interest which show complex formations with the $BiPh_3$ ligand, although the lone-pair of electrons is quite inert. $BiPh_3$ was coordinated to metal carbonyl complexes of the type $[Cr(CO)_5EPh_3]$ (E = P, As, Sb, Bi). The CBiC angle in $BiPh_3$ is 94° and increases to a value of 98.7° in the complex molecule, which indicates a higher s-component of the Bi—Cr bond[147]. Other examples with EPh_3 as a ligand are the complexes of the type $[MI_2(CO)_3L(EPh_3)]$ (E = As, Sb, Bi; M = Mo, W; L = MeCN, PR_3, AsR_3, $P(OR)_3$]. These compounds were prepared by exchange reactions of ligands (see equation 37)[166,167].

$$MI_2(CO)_3(MeCN)_2 + EPh_3 \xrightarrow[-\text{MeCN}]{\text{CH}_2\text{Cl}_2} [MI_2(CO)_3(MeCN)(EPh_3)] \qquad (37)$$

EPh_3 (E = P, As, Sb, Bi) was also coordinated to a molybdenum carbonyl. This ionic complex was formed by treatment of the $[CpMo(CO)_3]_2$ dimer with ferricenium tetrafluoroborate in the presence of EPh_3 (E = P, As, Sb, Bi) and $Ph_2P(CH_2)PPh_2$, where the Mo—Mo bond of the dimer is cleaved in a photochemical reaction[168]. Similarly, the highly reactive tributyl bismuth can be coordinated to group 6 metal carbonyls in a photochemical reaction (equation 38)[169].

$$M(CO)_6 \xrightarrow[-\text{CO}]{hv,\text{THF},\text{BiBu}_3} (CO)_5MBiBu_3 \qquad (38)$$

M = Cr, Mo, W

Bridging and chelating ligands. An example for a bidentate chelate ligand is an antimony sulphur diimid[170] which coordinates to group 6 transition metals to form complexes of the type $[Bu^tSb(N=S=N)]_2M(CO)_4$ (M = Cr, Mo, W) **(45)**.

(45) Structure of $[Bu^tSb(N=S=N)]_2M(CO)_4$ (M = Cr, Mo, W)

The binuclear complex $\{[Mo(CO)_4]_2(\mu\text{-}Ph_2AsAsPh)(\mu\text{-}Ph_2As)\}$ was prepared by thermolysis of $[\eta^6\text{-cyclo-}(PhAs)_9][Mo_2(CO)_9]$ or $[\eta^4\text{-catena-}(PhAs)_8]\text{-}[Mo(CO)_6]$[171]. The first complex mentioned above contains a Mo_2As_2 rhombus with a Mo—Mo bond distance of 310.8 pm and an As—As bond distance of 242.7 pm. The As—As bond length is relatively short and therefore one can consider a strong σ bond where the repulsion of the lone-pair of electrons (cf free, uncoordinated R_2As—AsR_2) is diminished due to the coordination to the Mo atoms **(46)**.

(46) Structure of $\{[-Mo(CO)_4]_2(\mu\text{-}Ph_2AsAsPh)(\mu\text{-}Ph_2As)\}$

Tetranuclear complexes with bridging μ-AsR_2 groups and one metal–metal bond such as $\{[(CO)_4Cr]_2(\mu\text{-}AsMe_2)_2\}$ or $\{[(CO)_4Cr](\mu\text{-}AsMe_2)_2Fe(CO)_3\}$ contain planar As_2—M_2 rings[172].

Treatment of As bridged binuclear compounds containing a MM′As metallacycle (M, M′ = group 6 and 8 transition metals) with nucleophiles, e.g. organometallic dimethylarsenides, causes opening of the rings and addition of a terminal As ligand[173,174]. As-bridged trinuclear complexes were reacted with Lewis bases like Me_2As—$FeCp(CO)_2$, Me_2As—$MCp(CO)_3$, Me_2As—$FeCp(CO)PMe_3$ and Me_2As—$MCp(CO)_2PMe_3$ (M = Cr, Mo, W) and afforded either complexes with three differently coordinated metal atoms and two bridging $AsMe_2$ groups of a chain-like type or three-membered metallacycles with one new terminal organometallic $AsMe_2$ ligand (see equation 39).

$$\begin{array}{c}
\overset{\displaystyle AsMe_2}{\underset{\displaystyle Met\text{——}Met'}{\diagup\diagdown}} \quad + \quad Me_2AsMet'' \quad\longrightarrow\quad \overset{\displaystyle AsMe_2 \quad AsMe_2}{\underset{\displaystyle Met \quad\; Met' \quad\; Met''}{\diagup\diagdown\diagup\diagdown}}
\end{array}$$

$$\begin{aligned}
Met,\ Met',\ Met'' &= FeCp(CO)_2,\ MCp(CO)_3,\ M = Cr,\ Mo,\ W;\\
&\quad FeCp(CO)_2PMe_3,\ MCp(CO)_2PMe_3,\ M = Cr,\ Mo,\ W.
\end{aligned}$$

(39)

Examples for bridging antimony ligands μ-R_2Sb (R = Me, Ph) are the complexes $[Cp(CO)_2MP(OCH_3)_3]_2[\mu\text{-}SbMe_2]_2$ (M = Mo, W)[156] and the μ-$SbPh_2$ bridged chromium tungsten complex $[(CO)_2(Et_2NC)L_2W(\mu\text{-}SbPh_2)Cr(CO)_5]$ (L_2 = 2,2′-bipyridine, 1,10-phenanthroline)[175]. The bidentate four-electron donors of the type $R_2E(CH_2)_nE'R_2$ (E = E′ = As, R = CH_3, Ph, n = 1,2; E = As, E′ = P, R = CH_3, Ph, n = 1,2; E = Sb, E′ = P, R = CH_3, Ph, n = 1,2)[176] are commonly used ligands. The dimeric compound fac,fac-$Mo_2(CO)_6(DADPE)_3$ was prepared from $Mo(CO)_3(C_7H_8)$ and DADPE in CH_2Cl_2 solution. Here the DADPE ligands show both bridging and chelating properties (**47**). In the chelate the Mo–As and Mo–P distances are 256.8 pm and 261.8 pm, respectively. The distance from Mo to the bridging atom is 254.9 pm[177].

(**47**) Structure of fac, fac-$Mo_2(CO)_6(DADPE)_3$ (X = As and P)

The chelate ligand $Ph_2SbCH_2SbPh_2$ was treated with the sulphur ylide $(CO)_5Cr[CH_2\text{-}S(O)Me_2]$ and yielded the four-membered metallacycle $(CO)_4Cr[Ph_2SbCH_2SbPh_2]$[178]. In contrast, the chromium complex $(CO)_4Cr[Ph_2SbCH_2PPh_2CH_2]$ contains a five-membered metallacycle with one Cr—Sb and one Cr—CH_2 bond. A planar $[CrAs_4]$ heterocycle was found in the mononuclear chromium complex $(CO)_4\overset{\shortmid}{Cr}$—$(AsClBu^t)(AsBu^t)_2$-$(AsClBu^t)$, where the arsenic coordinated to chromium is acting as a two-electron donor. The As–As distances of the cycle vary from long (247.8 pm) via short (242.7 pm) to long (247.9 pm). While the short distance corresponds to an As—As single bond, the others are longer due to the coordination to the Lewis acidic metal atom (**48**).

(48) Structure of $(CO)_4Cr[ClAsBu^tAsBu^t]_2$

A six-membered $[Cr_2As_4]$ cycle exists in the complex $(CO)_4Cr(\mu\text{-}As_2Me_4)_2Cr(CO)_4$. The bridging As ligands show As–As distances of 244.2 pm corresponding to the value of a single bond[180]. In comparison to arsenic, compounds with Sb—Sb bonds rarely exist. In 1985 the complex $[(CO)_5Cr]_2[\mu\text{-}(SbClBu^t)_2]$ was reported[181]. It contains a bridging moiety with a Sb–Sb distance of 290.1 pm, similar to that in uncoordinated Sb_2R_4 compounds.

Treating cyclopolyarsines cyclo-$(RAs)_n$ with chromium or molybdenum carbonyl compounds yielded dimers bridged by a polyarsenic cycle. The complexes $[Cp_2Mo_2(CO)_4][\mu,\eta^2\text{-}catena\text{-}(CH_3As)_5]^{182}$ and also $\{[CH_2Cp_2Mo_2(CO)_4]\text{-}catena\text{-}(\mu,\eta^2\text{-}(CH_3As)_4\}^{183}$ do not have any metal–metal bond. The first complex mentioned above shows a Mo–Mo distance of 416.9 pm; the As—As bond lengths within the cycle are consistent with common values. In the second complex the mean value for the As—As bond lengths is 244.0 pm. The bridging As atoms are at a distance of 283.5 pm and 274.8 pm, respectively, but obviously not bound to each other **(49)**.

(49) Structure of $[Cp_2Mo_2(CO)_4][\mu, \eta^2\text{-}catena\text{-}(CH_3As)_5]$

With increasing number of atoms in the polyarsines, their bridging structure becomes analogous to that of cubanes. The type of $[cyclo\text{-}(RAs)_nM_2(CO)_6]$ $(M = Cr, n = 9, R = Me^{184}; M = Mo, n = 10, R = Me; M = Mo, n = 9, R = Ph^{185})$ can be prepared from the homologous metal–carbonyl compound and a corresponding cyclo polyarsine. The As–As– and the As–metal distances are consistent with the common values. A triple bridging polyarsine ligand is included in the molybdenum complex $[Mo(CO)_3][\eta^3\text{-}cyclo\text{-}(CH_3As)_6S_3]^{186}$. However, polyarsines may be coordinated chelating to one metal centre without cleavage of their cyclic structure. Thus the complex $[cyclo\text{-}(PhAs)_6Mo(CO)_4]$ contains an arsenic cycle of twist-boat conformation with As–As mean distances of 245 pm and 98.6° angles, while the shortest transannular As–As distance is 319.6 pm[187].

Diarsine and distibine ligands. The π bond in ligands of the type $RE = ER$ $(E = P,$ As, Sb) is naturally very reactive but may be protected by a side-on coordination to a transition metal atom. The ligand is η^2-bound with a bond order of approximately 1.5. Therefore, arsenic and antimony coordinate either via the lone-pair of electrons as σ

ligands or via the double bond as π ligands (**50**).

(**50**) Possible η^2 coordination of the RE=ER ligand to a $M(CO)_5$ fragment

E=As, M=Cr, R=Ph; E=Sb, M=Cr, R=Bu^t, Ph

The antimony complexes $[(CO)_5MSbR]_2[M(CO)_5]$; (M=Cr, W, R = Bu^t; M=W, R = Ph) were obtained from the appropriate dinuclear carbonyl compound and $RSbCl_2$[188,189]. Values of 272.0 pm and 270.6 pm indicate that the Sb–Sb distance is shorter than that of a single bond. The analogous arsenic complex $L(CO)_4Cr[(CO)_5$-CrAsPh]_2$ [L = P(OMe)_3, bipyridine] was prepared from the reaction of $[(CO)_5$-CrAsHPh]_2$ with Pt/C and L at room temperature. In this complex the As—As bond distance is 234 pm and thus approximately 10 pm shorter than a double bond[190]. π-Bound RAs=AsR' can coordinate as a monodentate ligand to a metal carbonyl while the double bond remains uncoordinated. In the complex $[(CO)_5CrAs(R)=AsR']$ (R' = CH_2(SiMe_3), R' = 2,4,6-Bu$^t_3$C_6H_2) the As–As distance is only 224.6 pm. This is substantially shorter than the value for a single bond. The complex cited above shows that chromium, the arsenic coordinated to chromium and the arsenic-bound carbon are located in one plane which indicates that only one As is directly coordinated to chromium.[191]

Arsinidene and stibinidene complexes. Arsenic and antimony in the oxidation state +I can also be coordinated to transition metal carbonyls. For example, the ER (E = As, Sb) fragment is situated in a bridging position between the metal atoms in species like $RE[M(CO)_5]_2$ (E = Sb, M = Cr, R = Bu^t[192]; M = W, R = CH(SiMe_3)_2[193], Bu^t[188]; E = As, M = Cr, R = Ph[194]] (**51**).

(**51**) Structure of $RE[M(CO)_5]_2$

E is coordinated trigonal planar and sp^2 hybridized. The MEM unit can be regarded as a 3-centre 4-electron π-system. The complex acts as a Lewis acid which forms adducts to compensate the electron deficiency.

Other π-ligands. Polymolybdato complexes often contain cross-linked organoelement oxygen groups (element = As, Sb, Bi). The element atoms are differently coordinated, e.g. in the guanidine salt $(CN_3H_6)[(PhAs)_2Mo_6O_{25}H]\cdot H_2O$ there are tetrahedral $RAsO_3$ units bridged to molybdenum atoms through oxygen bonds[195]. The antimony ligand in $[Bu_4N][Ph_2SbO(MoO_4)]_2$, however, is octahedral coordinated in Ph_2SbO_4 units. Two of these units and two tetrahedral MoO_4 groups form a planar eight-membered ring[196]. In the compound $[Bu_4N]_2[Ph_3Bi(MoO_4)_2]\cdot 3H_2O$ bismuth is coordinated trigonal bipyramidal, where the phenyl groups lie in equatorial and two oxygen atoms in axial positions[197].

iv. Group 7 elements

Bridging and chelating ligands. Mononuclear cationic rhenium complexes containing the well-known ligand-type $R_2E(CH_2)_nER_2$ are, for example, the two air-stable compounds $[Re(V)N(Ph_2As(CH_2)_2AsPh_2)_2Cl]^+[ClO_4]^-$ and $[Re(V)O_2(Ph_2P(CH_2)_2$ $AsPh_2)_2]^+[ClO_4]^-$. In these species rhenium is coordinated distorted octahedral, forming five-membered metallacycles of the type ReE_2C_2 (E = As, P)[198]. The Re–As distances average to mean values of 254 pm in the above complexes. Diorganoarsine and -diarsine as well as diorganostibine and -distibine may act as bridging ligands under formation of multiple membered heterocycles. The complex $[(Ph_3Sb(CO)_3Re(\mu_2\text{-}PPh_2)(\mu_2\text{-}SbPh_2]Re(CO)_4]^{199}$ contains a Re_2SbP rhombus and also one terminal Ph_3Sb group and one ReRe bridging Ph_2Sb unit. The bridging Re–Sb distance of 274.0 pm is longer than the 267.1 pm Re–Sb distance of the terminal group.

An interesting example for a manganese complex is $Mn_2(CO)_7X(AsMe_2)_3$ (X = Cl, Br)[200]. This species was prepared from $KMn(CO)_5$ and Me_2AsCl via $Me_2AsAsMe_2\text{-}Mn(CO)_4Br$. The latter was thermally treated, yielding a binuclear complex containing a Mn—As—As—Mn—As heterocycle. The Mn–As distances of 246.2 pm and 250.2 pm of the Mn—As—Mn moiety correspond to the value of a single bond while the Mn–As distances in the Mn—As—As—Mn unit are slightly longer (239.2 pm and 238.6 pm). The As–As distance of 242.2 pm corresponds to a strong σ bond. This may indicate a bond contraction within the Mn_2As_3 five-membered ring possessing a distorted envelope conformation. The bidentate As_2Me_4 unit acts as a four-electron donor while the $AsMe_2$ fragment has three-electron donor properties (52).

$$\begin{array}{c} Me_2 \quad Me_2 \\ As\text{——}As \\ \diagup \qquad \diagdown \\ (CO)_4Mn \qquad\qquad Mn(CO)_3Cl \\ \diagdown \qquad \diagup \\ As \\ Me_2 \end{array}$$

(52) Structure of $Mn_2(CO)_7Cl(AsX)_3$ (X = Me)

$[(CO)_3Mn(\mu\text{-}Br)_2(\mu\text{-}Ph_4As_2)Mn(CO)_3]$ was directly prepared from $(CO)_5MnBr$ and Ph_4As_2. The compound shows an increased As–As distance of 248.9 pm, which may partly be due to the steric influence of the bridging Br atoms (53)[201].

$$\begin{array}{c} Br \\ \diagup \quad \diagdown \\ (CO)_3Mn \qquad Mn(CO)_3 \\ \diagdown \quad \diagup \\ Br \\ \diagdown \quad \diagup \\ Ph_2As\text{——}AsPh_2 \end{array}$$

(53) Structure of $[(CO)_3Mn(\mu\text{-}Br)_2(\mu\text{-}Ph_4As_2)Mn(CO)_3]$

$(RXE)_2$ units are able to bridge dimeric manganese complexes. For example, the compound $[Cp(CO)_2Mn(PhHAs—AsHPh)Mn(CO)_2Cp]$ contains an As—As moiety with an As–As distance of 246 pm corresponding to a single bond. The Mn–As distance is 231.9 pm, thus being significantly shorter than the sum of the covalent radii, which indicates a good coordinative bond ability of the $Ph_2As_2H_2$ ligand[22]. In a related complex the $R_2Cl_2Sb_2$ ligand represents the bridging unit in the dimer [Cp(CO)$_2$Mn(RClSb—SbClR)Mn(CO)$_2$Cp][181]. The AsR$_2$ unit can act either in bridging two metal carbonyl fragments, e.g. L(CO)$_4$MnAs(Me)$_2$Fe(CO)$_4$ [L = CO, PMe$_3$, P(OMe)$_3$], or by formation of three-membered MAsM' cycles like L(CO)$_4$Mn(μ-AsMe$_2$)Fe(CO)$_4$ [L = CO, PMe$_3$, P(OMe)$_3$][202,203].

Polymethylarsine (MeAs)$_n$ reacts with the rhenium carbonyl Re$_2$(CO)$_{10}$, yielding a coordination compound of very complex structure and electron arrangement, {[cyclo-(MeAs)$_7$(As)Re(CO)$_4$]Re(CO)$_6$}. This rhenium complex contains a nine-membered As$_8$Re ring, which contains one As atom without any methyl substitution and is coordinated to two Re(CO)$_3$ units.

Arsinidene and stibinidene complexes. Arsenic and antimony in their +I oxidation state can be stabilized by a metal–arsenic (or antimony) π bond with a trigonal planar coordination. PhE[Mn(CO)$_2$Cp]$_2$ was prepared as its arsenic[204] and antimony derivative[205,206].

v. Group 8 elements

Terminal ligands. Triorganoarsines, -stibines and -bismuthines ER$_3$ (E = As, Sb, Bi; R = alkyl, aryl) in variable composition are known to coordinate as terminal π ligands to group 8 metals. Species of the type [(CO)$_4$Fe(ER$_3$)] (R = alkyl) were prepared from Fe(CO)$_5$ and the quaternary salt (Bu$_4^t$Sb)$^{+207}$ in a sealed tube at 403 K. The iron atom is coordinated trigonal bipyramidal; the Fe–Sb distance is 254.7 pm. Alkyl-substituted bismuth ligands (R = C$_2$H$_5$, C$_3$H$_7$, C$_4$H$_9$) were reacted with Fe$_2$(CO)$_9$ and yielded the corresponding mononuclear complexes[169]. Aryl-substituted ligands being coordinated to metal complexes have also been reported. [(CO)$_2$Fe(PPh$_3$)$_2$EPh$_3$] (E = As, Sb) was prepared by reductive elimination of Br$_2$ from [(CO)$_2$Fe(PPh$_3$)$_2$Br$_2$] (E = As, Sb)[208]. {(CO)$_2$Fe[P(OPh)$_3$]$_2$SbR$_3$} (R = C$_6$H$_5$, MeC$_6$H$_4$, cyclo-C$_6$H$_{11}$) was obtained from SbR$_3$ and {(CO)$_2$Fe[P(OPh)$_3$]$_2$} in a photochemical reaction[209]. There are also cationic complexes containing these terminal ligands. An example is [CpFe{P(OMe)$_3$}$_2$EPh$_3$] [PF$_6$] (E = As, Sb) with a Fe–Sb distance of 248.1 pm[210]. In the series [CpFe(CO)$_2$(ER$_3$)][BF$_4$] (E = As, Sb, R = Me, Et, Pr^i, Ph; E = Bi, R = Me, Pr^i, Ph) the ligand BiEt$_3$ is missing[145]. NMR data indicate a decrease of the FeE bond strength in the series As > Sb > Bi. The Fe–Bi distance is 257.0 pm in the case of the BiPh$_3$ ligand. This value is a somewhat longer than the sum of the covalent radii[20] with 253 pm. AsPh$_3$ acts as a ruthenium ligand in [Ru$_2$(η^4-μ_2-o-C$_6$H$_4$O$_2$)(CO)$_4$(AsPh$_3$)]$_2$ where the Ru–As distance was found to be 249.5 pm[211]. Interesting compounds were prepared by coordination of optically active secondary arsines to cationic iron complexes. So AsHMePh[21], AsH(C$_3$H$_6$Cl)Ph and AsH(μ-CH$_2$CH$_2$CH$_2$)Ph[19] were used as ligands L in the coordination compound [CpFe{C$_6$H$_4$(PMePh)-1,2}L][PF$_6$]. The species with the ligand AsH(μ-CH$_2$CH$_2$CH$_2$)Ph was the first isolated 'arsetan' complex and shows a Fe–As distance of 232.6 pm (sum of the van der Waals radii[20]: 228 pm); see Section II.B.1.c.

Coordination of {2,4,6-(Bu$_3^t$)-C$_6$H$_2$}As=PCH(SiMe$_3$)$_2$ to Fe$_2$(CO)$_9$ led to a complex containing an As—P double bond and an As—Fe bond: (CO)$_4$Fe{2,4,6-(Bu$_3^t$)-C$_6$H$_2$}-As=PCH(SiMe$_3$)$_2$ (**54**)[212].

The cyclic triarsine species MeC(CH$_2$As)$_3$ was coordinated to ruthenium under formation of Ru(CO)$_4$L[213].

Multidentate ligands. An arsenic species of the type R$_2$As(CH$_2$)$_n$AsR$_2$ was found to be able to coordinate as a classical bidentate chelating ligand to ruthenium under formation of the Ru(III)–chelate complex [Ru(III)(HEDTA)(DPAE)][214]. Some trinuclear metal complexes with various metal carbonyl centres, bridged by AsR$_2$ units forming As metal ring systems, have been reported with the group 6 elements.

(54) Structure of $(CO)_4Fe\{2,4,6\text{-}(Bu_3^t)\text{-}C_6H_2\}As = PCH(SiMe_3)_2$

The cationic complex $\{[Cp(CO)_2Fe]_2(\mu\text{-}AsMe_2\}^+Cl^-$ contains an $AsMe_2$ unit bridging the iron centres[215]. The same As ligand was also used for bridging two ruthenium carbonyls under the formation of a four-membered Ru_2As_2 cycle: $[Cl(CO)_3Ru(\mu\text{-}AsMe_2)]_2$[216]. The complex $[(CO)_3Ru(\mu\text{-}AsMe_2)]_2$ contains an additional Ru—Ru bond. A five-membered Fe—Co—As—Fe—As ring system is part of the multinuclear complex $Fe_2CoCp_3(CO)_2(AsMe_2)_2$[200]. Vibrational spectra as well as X-ray analyses indicate fluctional behaviour of the molecules with variable metal positions **(55)**.

In $[(\mu\text{-}Bu^tS)(\mu\text{-}MeAs)Fe_2(CO)_6]_2$ the two Fe—Fe units are each bridged by one arsenic atom. The internuclear As–As distance is 243.5 pm, corresponding to the common single-bond value **(56)**[217].

(55) Structure of $Fe_2CoCp_3(CO)_2(AsMe_2)_2$

(56) Structure of $[(\mu\text{-}Bu^tS)(\mu\text{-}MeAs)Fe_2(CO)_6]_2$

Reaction of $Fe(CO)_5$ with cyclo-$(AsMe)_5$ yielded the dimer $[Fe(CO)_3]_2[\mu,\eta^2\text{-catena}(MeAs)_4]$ with a four-membered AsMe chain[218]. The trinuclear iron cluster $[Fe(CO)_3]_3[SbCH(SiMe_3)_2]_2$ contains an antimony ligand. The Sb atoms are bound to each other, secondly to the organo group and thirdly coordinated to three Fe atoms each[219]. μ_3-SbR ligands are rare. However, recently a species of the type $[HFe_3(CO)_9(\mu_3\text{-}SbBu^t)][PPN]$ has been reported[220].

Diarsine ligands. The double bond in As_2R_2 can be stabilized via a π bond by coordination to iron carbonyls. In $[(CO)_4Fe(\eta^2\text{-}PhAsAsPh)]$ the As–As distance is 236.5 pm[221]; in the complex $[(CO)_4Fe(\eta^2\text{-}C_6F_5AsAsC_6F_5)]$ it comes to 238.8 pm[222].

vi. Group 9 elements

Terminal ligands. The ionic complex $[CoCp_2]^+[EPh_2Cl_2]^-$ (E = Sb, Bi) was prepared from $CoCpCl_2$ or $CoCp_2$ and EPh_2Cl. The anion is Ψ-trigonal bipyramidal coordinated[223]. Tributylarsine coordinates to rhodium with Rh–As distances of 248.3 and 248.5 pm under formation of the binuclear complex $[(AsBu^t_3)(CO)Rh]_2(\mu\text{-}Cl)(\mu\text{-}SR)$ $[R = 5\beta\text{-}Me\text{-}2\alpha(1\text{-methylethyl})cyclohexyl]$[224]. In $Co_2(CO)_6[MeC(CH_2As)_3]$ the cage-like triarsine $MeC(CH_2As)_3$ is acting as terminal ligand[213].

Multidentate ligands. E_2Ph_4 (E = As, Sb) was reacted with $Co_2(CO)_8$ under formation of a bridged polymer complex of the type $[Co(\mu\text{-}EPh_2)(CO)_3]_n$. The corresponding Bi_2Ph_4 shows only weak reactivity toward the cobalt carbonyl, without complex formation[148]. Treatment of $Co_2(CO)_8$ with the polyarsenic species $(PhAs)_6$ led to an As bridged cluster: $[Co_8(\mu_6\text{-}As)(\mu_4\text{-}As)(\mu_4\text{-}AsPh)_2(CO)_{16}]_2{}^{225}$. The multinuclear complex $Fe_2CoCp_3(CO)_3(AsMe_2)_2$ has already been considered in Section III.C.7.b.v.

vii. Group 10 elements

Terminal ligands. The platinum(II) complex $[LL'Pt(\mu\text{-}O)_2E(O)R]$ contains $AsPh_3$ as a terminal ligand (E = P, L = $AsPh_3$, L' = PR_3^-, R = Ph; E = P, L = L' = $AsPh_3$, R = Ph, Me, OPh; E = As, L = L' = $AsPh_3$, R = Ph). Some of these species contain phosphorus ligands as well[226]. $Pt(AsBu_3)_4{}^{227}$ is a nice example of a homoleptic completely with arsenic ligands substituted complex.

Multidentate ligands. The dinickel complex $[(PMe_3)Ni(\mu\text{-}AsBu_2^t)]_2{}^{228}$ was prepared from $AsBu_2^tLi$ and $NiCl_2(PMe_3)_2$ under formation of a Ni—Ni bond and $AsBu_2^t$ bridges. Reaction of Na_2PdCl_4 with $AsBu_3^t$ led to the air-stable palladium complex $PdHCl[AsBu_3^t]_2{}^{229}$. Bis(dimethylarsenic) sulphide is bridging two $PtMe_3X$ units to give the platinum complex $[Me_3Pt(\mu\text{-}X)]_2[(\mu\text{-}AsMe_2)_2S]^{230}$. A type of platinum and palladium complexes containing As ligands of special steric demand is that of $[M(o\text{-}C_6H_4CH_2AsR_2)_2]$ (M = Pd, Pt; R = Me, Ph). Some species were prepared from $1,2\text{-}Li(CH_2AsR_2)C_6H_4$ and $MCl_2 \cdot 2Et_2S$ (57)[231].

(57) Structure of $[M(o\text{-}C_6H_4CH_2AsR_2)_2]$ (M = Pd, Pt; R = Me, Ph)

In $Me_2Pt(Ph_2AsCH_2AsPh_2)PtMe_2{}^{232}$ the two $PtMe_2$ units are bridged by the $Ph_2AsCH_2AsPh_2$ ligand. Nickel in the oxidation state + II can be stabilized by bidentate arsenic donor ligands, e.g. in five-coordinated bis triarsine nickel complexes $[Ni(mtas)_2]^{2+}$, $[Ni(ptas)_2]^{2+}$, $[Ni(diars)(mtas)]^{2+}$ (ligands, see 58)[233].

Arsinidene complexes. The As—As double bond of arsenic species can be stabilized by coordination to the palladium complex $[(CO)_5Cr(Ph)As=As(Ph)Cr(CO)_5]PdL_2$ [L = $P(OMe)_3$, bipyr]. Here the As—As distance is 236.3 ± 0.5 pm, while the Pd—As distances are 246.0 pm and 243.9 pm, respectively[190]. Reaction of a $NiCl_2$ complex with the ligand $PhAs(SiMe_3)_2$ led to the formation of a species with an $\eta^2\text{-}PhAsAsPh$ ligand; the As—As distance was determined to be 237.2 pm (59)[234].

viii. Group 11 and 12 elements. The first antimony complex of a late transition metal was $[MesSbCu(PMe_3)_2]_2$. The corresponding dimer was prepared from Mes_2SbH by lithiation and reaction with CuCl and $PMe_3{}^{27}$. ER_3 can be used as a ligand coordinating to silver under complex formation of the type $[Ag(ER_3)_n]^+X^-$ (E = As, R = Ph, n = 1–3, X = $NO_3{}^{235}$; R = Ph, n = 4, X = $SnPh_2(NO_3)_2Cl$, $(d_{As-Ag} = 264.3\text{–}270.0\,pm)$; R = Ph, n = 3, X = Cl^{236}; R = C_6F_5, n = 2, X = $ClO_4{}^{237}$]. $BiPh_3$ is also able to coordinate to $AgClO_4$ but considerably less strongly[238]. The chelate ligand $CH_2(AsPh_2)_2$ was found

(58) Structure of the ligands in the nickel complexes: $[Ni(mtas)_2]^{2+}$, $[Ni(ptas)_2]^{2+}$, $[Ni(diars)(mtas)]^{2+}$, (diars = $1,2-(Me_2As)_2C_6H_4$), mtas, R = Me; ptas, R = Ph)

(59) Structure of a nickel complex containing an $\eta^2PhAsAsPh$ ligand

to bridge two $Hg(C_6F_5)_2$ units under formation of $[Hg(C_6F_5)_2AsPh_2]_2CH_2$ (As–Hg distance: 340 pm)[239].

IV. ACKNOWLEDGEMENTS

The authors wish to thank Mrs C. B. F. Klapötke for improving the manuscript. We also thank the Deutsche Forschungsgemeinschaft (DFG KI -636/1–2) and the Fonds der Chemischen Industrie for financial support.

V. REFERENCES

A. General Literature

Annual surveys

G. O. Doak and L. D. Freedman, *Bismuth. Annual Survey, J. Organomet. Chem.*

L. D. Freedman and G. O. Doak, *Antimony. Annual Survey, J. Organomet. Chem.*

D. B. Sowerby *et al., Elements of Group 5, Coord. Chem. Rev.*

J. L. Wardell, *Group V: Arsenic, Antimony, and Bismuth, Organomet. Chem.*

G. O. Doak and L. D. Freedman, *Organometallic Compounds of Arsenic, Antimony and Bismuth*, Wiley, New York, 1970.

C. Elschenbroich and A. Salzer, *Organometallchemie*, 3rd ed., Teubner, Stuttgart, 1990.

N. N. Grennwood and A. Earnshaw, *Chemistry of the Elements*, Pergamon, Oxford, 1984.

I. Haiduc and J. J. Zuckerman, *Basic Organometallic Chemistry*, W. de Gruyter, Berlin, New York, 1985.

D. Hellwinkel, *Penta- and Hexaorganyl Derivatives of the Group Five Elements*, in *Topics in Current Chemistry*, Vol. 109 (Ed. F. L. Boschke), Springer-Verlag, Berlin, 1983.

J. E. Huheey, *Anorganische Chemie*, W. de Gruyter, Berlin, New York, 1988.

C. A. McAuliffe, *Transition Metal Complexes of Phosphorous, Arsenic and Antimony Ligands*, Macmillan Press, London, 1973.

C. A. McAuliffe and W. Levason, *Phoshine, Arsine and Stibine Complexes of the Transition Elements*, Elsevier, Amsterdam, 1979.

P. Powell, *Organometallic Chemistry*, 2nd ed., Chapman & Hall, London, New York, 1988.

D. F. Shriver, P. W. Atkins and C. H. Langford, *Anorganische Chemie*, VCH Verlagsgesellschaft, Weinheim, 1992.

J. L. Wardell, *Arsenic, Antimony and Bismuth*, Vol. 2 (Eds. G. Wilkinson, R. D. Gillard and J. A. McCleverty), Pergamon, Oxford, 1987.

M. Wieber, *Bismut-organische Verbindungen*, in *Gmelins Handbuch der Anorganischen Chemie*, Vol. 47, 8th ed., Springer-Verlag Berlin, 1977.

B. Literature Cited in the Text

1. N. N. Greenwood and A. Earnshaw, *Chemistry of the Elements*, Pergamon, Oxford, 1984, p. 1082.
2. D. F. Shriver, P. W. Atkins and C. H. Langford, *Anorganische Chemie*, VCH Verlagsgesellschaft, Weinheim, 1992, p. 703.
3. D. C. McKean, I. Torto and A. R. Morrison, *J. Phys. Chem.*, **86**, 307 (1982).
4. D. C. McKean, M. W. Mackenzie and I. Torto, *Spectrochim. Acta*, **38A**, 113 (1982).
5. I. Haiduc and J. J. Zuckerman, *Basic Organometallic Chemistry*, W. de Gruyter, Berline, New York, 1985.
6. C. Elschenbroich and A. Salzer, *Organometallchemie*, 3rd ed., Teubner, Stuttgart, 1990.
7. M. Wieber, *Bismut-organische Verbindungen*, in *Gmelins Handbuch der Anorganischen Chemie*, Vol. 47, 8th ed., Springer-Verlag, Berlin, 1977.
8. G. O. Doak and L. D. Freedman, *Organometallic Compounds of Arsenic, Antimony and Bismuth*, Wiley, New York, 1970.
9. R. D. Gigauri, N. T. Gurgenidze, B. D. Chernokal'skü, M. A. Indzhiiya and G. N. Chachava, *Izv. Akad. Nauk SSR, SER. Khim.*, **7**, 142 (1981).
10. R. M. Lum and J. K. Klingert, *J. Appl. Phys.*, **66**, 3820 (1989).
11. G. Wittig, M. A. Jesatis and M. Goz, *Justus Liebigs Ann. Chem.*, **577**, 1 (1952).
12. I. Ruppert and V. Bastian, *Angew. Chem.*, **90**, 226 (1978); *Angew. Chem., Int. Ed.Engl.*, **17**, 214 (1978).
13. P. B. Chi and F. Kober, *Z. Anorg. Allg. Chem.*, **513**, 35 (1984).
14. W. Uhlig and A. Tzschach, *Z. Anorg. Allg. Chem.*, **576**, 281 (1989).
15. V. K. Gupta, L. K. Krannich and C. L. Watkins, *Inorg. Chem.*, **26**, 1638 (1987).
16. J. Ellermann and M. Lietz, *Z. Naturforsch.*, **37B**, 845 (1982).
17. P. B. Hitchcock, M. F. Lappert and S. J. Smith, *J. Organomet. Chem.*, **320**, C27 (1987).
18. R. L. Wells, A. P. Purdy, A. T. McPhail and C. G. Pitt, *J. Organomet. Chem.*, **308**, 281 (1986).
19. A. Bader, D. D. Pathak, S. B. Wild and A. C. Willis, *J. Chem. Soc., Dalton Trans.*, 1751 (1992).
20. E. Fluck and K. Heumann, *Periodensystem der Elemente*, VCH Verlagsgesellschaft, Weinheim, 1989.
21. G. Salem and S. G. Wild, *J. Organomet. Chem.*, **370**, 33 (1989).
22. G. Huttner, H. G. Schmid and H. Lorenz, *Chem. Ber.*, **109**, 3741 (1976).
23. H. Schumann, E. Palamides, J. Löbel and J. Pickardt, *Organometallics*, **7**, 1008 (1988).
24. A. Barth, G. Huttner, M. Fritz and L. Zsolnai, *Angew. Chem.*, **102**, 956 (1990); *Angew. Chem., Int. Ed. Engl.*, **29**, 929 (1990).
25. R. Batchelor and T. Birchall, *J. Am. Chem. Soc.*, **104**, 674 (1982).
26. M. Ates, H. J. Breunig and S. Gulec, *Phosphorus Sulfur Silicon Relat. Elem.*, **44**, 129 (1989).
27. A. H. Cowley, R. A. Jones, C. M. Nunn and D. L. Westmoreland, *Angew. Chem.*, **101**, 1089 (1989); *Angew. Chem., Int. Ed. Engl.*, **28**, 1018 (1989).
28. R. A. Bartlett, H. V. R. Dias, H. Hope, B. D. Murray, M. M. Olmstead and P. P. Power, *J. Am. Chem. Soc.*, **108**, 6921 (1986).
29. D. C. Mente, J. L. Mills and R. E. Mitchell, *Inorg. Chem.*, **14**, 123 (1975).
30. V. T. Tran and B. Munson, *Org. Mass Spectrom.*, **21**, 41 (1986).
31. O. W. Kolling and E. A. Mawdsley, *Inorg. Chem.*, **9**, 408 (1970).
32. K. H. Wyatt, D. Holtz, T. B. McMahon and J. L. Beauchamp, *Inorg. Chem.*, **13**, 1511 (1974).
33. S. Ahrland and F. Hultén, *Inorg. Chem.*, **26**, 1796 (1987).
34. N. Burford, J. Passmore and J. C. P. Sanders, *From Atoms to Polymers* (Eds. J. F. Liebman and A. Greenberg), VCH Verlagsgesellschaft, New York, 1989.
35. P. Pyykkö, *Chem. Rev.*, **88**, 563 (1988).
36. K. S. Pitzer, *Acc. Chem. Res.*, **12**, 271 (1979).
37. P. Pyykkö and J. P. Desclaux, *Acc. Chem. Res.*, **12**, 276 (1979).
38. J. E. Huheey, *Anorganische Chemie*, W. de Gruyter, Berlin, New York, 1988.
39. D. A. Johnson, *Some Thermodynamic Aspects of Inorganic Chemistry*, Cambridge University Press, Cambridge, 1982.
40. W. Levason, B. Sheik and F. P. McCullough, *J. Coord. Chem.*, **12**, 5693 (1982).
41. V. V. Sharutin, *Zh. Obshch. Khim.*, **58**, 2305 (1988).
42. A. Schmuck, J. Buschmann, J. Fuchs and K. Seppelt, *Angew. Chem.*, **99**, 1206 (1987); *Angew. Chem., Int. Ed. Engl.*, 1180 (1987).

43. G. Ferguson, B. Kaitner, C. Glidewell and S. Smith, *J. Organomet. Chem.*, **419**, 283 (1991).
44. R. G. Parr and R. G. Pearson, *J. Am. Chem. Soc.*, **105**, 7512 (1983).
45. R. G. Parr, R. A. Donnelly, M. Levy and W. E. Palke, *J. Chem. Phys.*, **68**, 3801 (1978).
46. R. G. Pearson, *Inorg. Chem.*, **27**, 734 (1988).
47. R. G. Pearson, *Chem. Ber.*, **27**, 444 (1991).
48. R. G. Pearson, *J. Am. Chem. Soc.*, **110** 7684 (1988).
49. W. B. Jensen, *Chem. Rev.*, **78**, 1 (1978).
50. Y. Mourad, Y. Munier, H. J. Breunig and M. Ates, *J. Organomet. Chem.*, **398**, 85 (1990).
51. P. Jutzi, T. Wippermann, C. Krüger and H. J. Kraus, *Angew. Chem.*, **95**, 244 (1983); *Angew. Chem., Int. Ed. Engl.*, **22**, 250 (1983).
52. N. Burford, T. M. Parks, B. W. Royan, B. Borecka, T. S. Cameron, J. F. Richardson, E. J. Gabe and R. Hynes, *J. Am. Chem. Soc.*, **114**, 8147 (1992).
53. H. Schmidbaur and P. Nusstein, *Organometallics*, **4**, 344 (1985).
54. F. Smit and C. H. Stam, *Acta Crystallogr.*, **B36**, 455 (1980).
55. H. J. Breunig, K. H. Ebert, S. Gülec, M. Dräger, D. B. Sowerby, M. J. Begley and U. Behrens, *J. Organomet. Chem.*, **427**, 39 (1992).
56. J. L. Aubagnac, F. H. Cano, R. Claramunt, J. Elguero, R. Faure, C. Foces-Foces and P. Raj, *Bull. Soc. Chim. Fr.*, 905 (1988).
57. A. Boardman, R. W. H. Small and I. J. Worrall, *Inorg. Chim. Acta*, **121**, L35 (1986).
58. I. Tornieporth-Oetting and T. Klapötke, *Chem. Ber.*, **123**, 1657 (1990).
59. I. Tornieporth-Oetting and T. Klapötke, *J. Organomet. Chem.*, **379**, 251 (1989).
60. F. L. Tanzella and N. Bartlett, *Z. Naturforsch.*, **36B**, 1461 (1981).
61. M. Hall and D. B. Sowerby, *J. Organomet. Chem.*, **347**, 59 (1988).
62. W. Clegg, R. J. Errington, G. A. Fisher, D. C. R. Hockless, N. C. Norman, A. G. Orpen and S. E. Stratford, *J. Chem. Soc., Dalton Trans.*, 1967 (1962).
63. M. Ali, W. R. McWhinnie, A. A. West and T. A. Hamor, *J. Chem. Soc., Dalton Trans.*, 899 (1990).
64. D. H. R. Barton, B. Chapriot, E. T. H. Dau, W. B. Motherwell, C. Pascard and C. Pichon, *Helv. Chim. Acta*, **67**, 586 (1984).
65. G. Ferguson, G. S. Harris and A. Khan, *Acta Crystallogr.*, **C43**, 2078 (1987).
66. J. W. Pasterczyk, A. M. Arif and A. R. Barron, *J. Chem. Soc., Chem. Commun.*, 829 (1989).
67. C. A. McAuliffe, B. Beagley, G. A. Gott, A. G. Mackie, P. P. MacRory and R. G. Pritchard, *Angew. Chem.*, **99**, 237 (1987); *Angew. Chem., Int. Ed. Engl.*, **26**, 264 (1987).
68. B. Beagley, C. B. Colburn, O. E. Sayrafi, A. G. Gott, D. G. Kelly, A. G. Mackie, C. A. McAuliffe, P. P. MacRory and R. G. Pritchard, *Acta Crystallogr.*, **C44**, 38 (1988).
69. R. K. Kanjolia, L. K. Krannick and C. L. Watkins, *Inorg. Chem.*, **24**, 445 (1985).
70. J. E. Frey, R. D. Cole, E. C. Kitchen, L. M. Suprenant and M. S. Sylwestrzak, *J. Am. Chem. Soc.*, **107**, 748 (1985).
71. A. Gieren, H. Betz, T. Hübner, V. Lamm, M. Herberhold and K. Guldner, *Z. Allg. Anorg. Chem.*, **513**, 160 (1984).
72. O. Mundt, H. Riffel, G. Becker and A. Simon, *Z. Naturforsch.*, **39B**, 317 (1984)
73. J. Müller, U. Müller, A. Loss, J. Lorberth, H. Donath and W. Massa, *Z. Naturforsch.*, **40B**, 1320 (1985).
74. M. K. Rastogi, *Synth. React. Inorg. Met. Org. Chem.*, **14**, 799 (1984).
75. F. Huber and S. Bock, *Z. Naturforsch.*, **37B**, 815 (1982).
76. S. P. Bone and D. B. Sowerby, *J. Organomet. Chem.*, **184**, 181 (1980').
77. M. J. Begley, D. B. Sowerby, D. Weseolek, C. Silvestru and I. Haiduc, *J. Organomet. Chem.*, **316**, 281 (1986).
78. R. Gupta, Y. P. Singh and A. K. Rai, *Main Group Met. Chem.*, **12**, 117 (1989)
79. H. M. Hoffmann and M. Dräger, *J. Organomet. Chem.*, **329**, 51 (1987).
80. R. Karra, Y. P. Singh and A. K. Rai, *Phosphorous, Sulfur Silicon Relat. Elem.*, **45**, 145 (1989).
81. S. K. Pandey, G. Srivastava and R. C. Mehrotra, *J. Indian Chem. Soc.*, **66**, 558 (1989).
82. R. Bohra, H. W. Roesky, J. Lucas, M. Noltemeyer and G. M. Sheldrick, *J. Chem. Soc., Dalton Trans.*, 1011 (1983).
83. R. Rüther, F. Huber and H. Preut, *J. Organomet. Chem.*, **342**, 185 (1988).
84. M. K. Rastogi, *Synth. React. Inorg. Met.-Org. Chem.*, **17**, 525 (1987).
85. M. Wieber and J. Walz, *Z. Anorg. Allg. Chem.*, **583**, 102 (1990).
86. H. Preut, F. Huber and G. Alonzo, *Acta Crystallogr.*, **C43**, 46 (1987).
87. M. Hall, D. B. Sowerby and C. P. Falshaw, *J. Organomet. Chem.*, **315**, 321 (1986).

88. C. Burschka, Z. Allg. Anorg. Chem., 485, 217 (1982).
89. C. Silvestru, L. Silaghi-Dumitrescu, I. Haiduc, M. J. Begley, M. Nunn and D. B. Sowerby, J. Chem. Soc., Dalton Trans., 1031 (1986).
90. C. Silvestru, M. Curtui, I. Haiduc, M. J. Begley and D. B. Sowerby, J. Organomet. Chem., 426, 49 (1992).
91. R. K. Gupta, A. K. Rai, R. C. Mehrotra and V. K. Jain, Inorg. Chim. Acta, 88, 201 (1984).
92. H. M. Hoffmann and M. Dräger, J. Organomet. Chem., 295, 33 (1985).
93. H. Preut, U. Praeckel and F. Huber, Acta Crystallogr., C42, 1138 (1986).
94. H. Preut, F. Huber and K. H. Hengstmann, Acta Crystallogr., C44, 468 (1988).
95. R. Kasemann and D. Naumann, J. Fluorine Chem., 41, 321 (1988).
96. S. Pohl, W. Saak and D. Haase, Angew. Chem., 99, 462 (1987); Angew. Chem., Int. Ed. Engl., 26, 467 (1987).
97. P. L. Millington and D. B. Sowerby, J. Chem. Soc., Dalton Trans., 1199 (1992).
98. R. R. Holmes, R. O. Day, V. Chandrasekhar and J. M. Holmes, Inorg. Chem., 26, 157 (1987).
99. K. Y. Akiba, H. Fujikawa, Y. Sunaguchi and Y. Yamamoto, J. Am. Chem. Soc., 109, 1245 (1987).
100. Y. Yamamoto, H. Fujiikaura, H. Fujishima and K. Y. Akiba, J. Am. Chem. Soc., 111, 2276 (1989).
101. K. Y. Akiba, K. Ohdoi and Y. Yamamoto, Tetrahedron Lett., 30, 953 (1989).
102. H. Schmidbaur, R. Nowak, O. Steigelmann and G. Müller, Chem. Ber., 123, 1221 (1990).
103. D. Mootz and V. Händler, Z. Anorg. Allg. Chem., 521, 122 (1985).
104. H. Schmidbaur, R. Nowak, A. Schier, B. Wallis, B. Huber and G. Müller, Chem. Ber. 120, 1837 (1987).
105. H. Schmidbaur, R. Nowak, O. Steigelmann and G. Müller, Chem. Ber., 123, 19 (1990).
106. J. M. Vezzosi, A. F. Zanoli, L. P. Battaglia and A. B. Corradi, J. Chem. Soc., Dalton Trans., 191 (1988).
107. D. M. Joshi and N. K. Jha, Synth. React. Inorg. Met.-Org. Chem., 17, 961 (1987).
108. R. G. Goel and D. R. Ridley, J. Organomet. Chem., 182, 207 (1979).
109. M. Hall and D. B. Sowerby, J. Am. Chem. Soc., 102, 628 (1980).
110. T. T. Bamgboye, M. J. Begley and D. B. Sowerby, J. Organomet. Chem., 362, 77 (1989).
111. M. Wieber, D. Wirth and C. Burschka, Z. Naturforsch., 39B, 600 (1984).
112. M. Wieber, I. Fetzer-Kremeling and H. Reith, Z. Naturforsch., 42B, 815 (1987).
113. E. Nakamura, J. Shimada and I. Kuwajima, Organometallics, 4, 641 (1985).
114. R. R. Holmes, R. O. Day, V. Chandrasekhar and J. M. Holmes, Inorg. Chem., 26, 163 (1987).
115. P. Maroni, Y. Madaule and T. Seminario, Can. J. Chem., 63, 636 (1985).
116. N. K. Jha and D. M. Joshi, Synth. React. Met.-Org. Chem., 16, 947 (1986).
117. V. K. Jain, J. Mason and R. C. Mehrotra, J. Organomet. Chem., 309, 45 (1986).
118. H. Schmidbaur, B. Milewski-Mahria and F. E. Wagner, Z. Naturforsch., 38B, 1477 (1983).
119. P. Bras, A. van der Gen and J. Wolters, J. Organomet. Chem., 256, C1 (1983).
120. G. Faraglia, R. Grazani and L. Volponi, J. Organomet. Chem., 253, 317 (1983).
121. F. Weller, J. Pebler, K. Dehnicke, K. Hartke and H. M. Wolff, Z. Anorg. Allg. Chem., 486, 61 (1982).
122. W. Neubert, H. Pritzkow and H. P. Latscha, Angew. Chem., 100, 298 (1988); Angew. Chem., Int. Ed. Engl., 27, 287 (1988).
123. H. M. Hoffmann and M. Dräger, J. Organomet. Chem., 320, 273 (1987).
124. M. Dräger and B. M. Schmidt, J. Organomet. Chem., 290, 133 (1985).
125. M. Wieber, D. Wirth, J. Metter and C. Burschka, Z. Anorg. Allg. Chem., 520, 65 (1985).
126. C. Burschka and M. Wieber, Z. Naturforsch., 34B, 1037 (1979).
127. M. Hall and D. B. Sowerby, J. Chem. Soc., Dalton Trans., 1095 (1983).
128. S. P. Bone, M. J. Begley and D. B. Sowerby, J. Chem. Soc., Dalton Trans., 2085 (1992).
129. A. Schier, J. M. Wallis, G. Müller and H. Schmidbaur, Angew. Chem., 98, 742 (1986); Angew. Chem., Int. Ed. Engl., 25, 757 (1986).
130. D. Mootz and V. Händler, Z. Anorg. Allg. Chem., 533, 23 (1986).
131. H. Schmidbaur, R. Nowak, A. Schier, J. M. Wallis, B. Huber and G. Müller, Chem. Ber., 120, 1829 (1987).
132. M. Schäfer, J. Pebler and K. Dehnicke, Z. Anorg. Allg. Chem., 611, 149 (1992).
133. M. Domagala, F. Huber and H. Preut, Z. Anorg. Allg. Chem., 574, 130 (1989).
134. S. P. Bone and D. B. Sowerby, Phosphorus, Sulfur Silicon Relat. Elem., 45, 23 (1989).
135. M. Domagala, H. Preut and F. Huber, Acta Crystallogr., C44, 830 (1988).

136. D. B. Sowerby, M. J. Begley and P. J. Mulligan, *J. Chem. Soc., Chem. Commun.*, **896** (1984).
137. G. Alonzo, M. Consiglio, F. Maggio and N. Bertrazzi, *Inorg. Chim. Acta*, **147**, 217 (1988).
138. A. M. Arif, B. L. Benac, A. H. Cowley, R. Geerts, R. A. Jones, K. B. Kidd, J. M. Power and S. T. Schwab, *J. Chem. Soc., Chem. Commun.*, 1543 (1986).
139. R. L. Wells, A. P. Purdy, K. T. Higa, A. T. McPhail and C. G. Pitt, *J. Organomet. Chem.*, **325**, C7 (1986).
140. C. G. Pitt, A. P. Purdy and K. T. Higa, *Organometallics*, **5**, 1266 (1986).
141. R. L. Wells, A. P. Purdy, A. T. McPhail and G. C. Pitt, *J. Chem. Soc., Chem. Commun.*, 487 (1986).
142. A. H. Cowley, R. A. Jones, K. B. Kidd, C. M. Nunn and D. L. Westmoreland, *J. Organomet. Chem.*, **341**, C1 (1988).
143. A. R. Barron, A. H. Cowley, R. A. Jones, C. M. Nunn and D. L. Westmoreland, *Polyhedron*, **7**, 77 (1988).
144. S. Ahrland, T. Berg and P. Trinderup, *Acta Chem. Scand.*, **A31**, 775 (1977).
145. H. Schumann and L. Eguren, *J. Organomet. Chem.*, **403**, 183 (1991).
146. N. C. Norman, *Chem. Soc. Rev.*, **17**, 269 (1988).
147. A. J. Carty, N. J. Taylor, A. W. Coleman and M. F. Lappert, *J. Chem. Soc., Chem. Commun.*, 639 (1979).
148. F. Calderazzo, R. Poli and G. Pelizzi, *J. Chem. Soc., Dalton Trans.*, 2535 (1984).
149. M. Luksza, S. Himmel and W. Malisch, *Angew. Chem.*, **95**, 418 (1983); *Angew. Chem., Int Ed. Engl.*, **22**, 416 (1983).
150. C. A. McAuliffe and W. A. Levason, *Phosphine, Arsine and Stibine Complexes of the Transition Elements*, Elsevier, Amsterdam, 1979.
151. G. Wilkinson, F. G. A. Stone and E. W. Abel (Eds.), *Comprehensive Organometallic Chemistry*, 9 Vols., Pergamon Press, Oxford, 1982.
152. L. P. Battaglia, M. Nardelli, L. Pelizzi, G. Brederi and G. P. Chinsoldi, *J. Organomet. Chem.*, **259**, C7 (1981).
153. L. P. Battaglia and A. B. Corradi, *Inorg. Chim. Acta*, **30**, 125 (1991).
154. P. Gowik and T. Klapötke, *J. Organomet. Chem.*, **204**, 349 (1991).
155. R. Borowski, D. Rehder and K. J. v. Deuten, *J. Organomet. Chem.*, **220**, 45 (1981).
156. W. Malisch and R. Janta, *Angew. Chem.*, **90**, 221 (1978); *Angew. Chem., Int. Ed. Engl.*, **17**, 211 (1978).
157. R. Schemm and W. Malisch, *J. Organomet. Chem.*, **288**, C9 (1985).
158. A. Meyer, A. Hartl and W. Malisch, *Chem. Ber.*, **116**, 348 (1983).
159. F. R. Kreißl, M. Wolfgruber, W. Sieber and H. G. Alt, *Angew. Chem.*, **95**, 159 (1983); *Angew. Chem., Int, Ed. Engl.*, **22**, 149 (1983).
160. F. R. Kreißl, M. Wolfgruber and W. Sieber, *Angew. Chem.*, **95**, 1002 (1983); *Angew. Chem., Int. Ed. Engl.*, **22**, 1001 (1983).
161. F. R. Kreißl, M. Wolfgruber and W. Sieber, *Organometallics*, **2**, 1266 (1983).
162. M. Herberhold, K. Schamel, A. Gieren and T. Hübner, *Phosphorus, Sulfur Silicon Relat. Elem.* **41**, 355 (1989).
163. J. Ellermann and A. Veit, *Angew. Chem.*, **94**, 377 (1982); *Angew. Chem., Int. Ed. Engl.*, **21**, 375 (1982).
164. J. Ellermann and A. Veit, *J. Organomet. Chem.*, **290**, 307 (1985).
165. J. Ellermann, H. A. Lindner, H. Schnössner, G. Thiele and G. Zoubek, *Z. Naturforsch.*, **33B**, 1386 (1978).
166. P. K. Baker and S. G. Fraser, *J. Coord. Chem.*, **16**, 97 (1987).
167. P. K. Baker, S. G. Fraser and T. M. Matthews, *Inorg. Chim. Acta*, **150**, 217 (1988).
168. H. Schumann, *J. Organomet. Chem.*, **323**, 97 (1987).
169. H. J. Breunig and U. Gräfe, *Z. Anorg. Allg. Chem.*, **510**, 104 (1984).
170. M. Herberhold and K. Schamel, *Z. Naturforsch.*, **43B** 998 (1988).
171. A. L. Rheingold and M. E. Fountain, *New J. Chem.*, **12**, 565 (1988).
172. H. Vahrenkamp and E. Keller, *Chem. Ber.*, **112** (1979) 1991.
173. H. J. Langenbach and H. Vahrenkamp, *Chem. Ber.*, **113**, 2189 (1980).
174. H. J. Langenbach and H. Vahrenkamp, *Chem. Ber.*, **113**, 2200 (1980).
175. A. C. Filippou, E. O. Fischer, H. G. Alt and U. Thewalt, *J. Organomet. Chem.*, **326**, 59 (1987).
176. Further types of bidentate ligands are described in: C. A. McAuliffe and W. Levason, *Phosphine, Arsine and Stibine Complexes of the Transition Elements*, Elsevier, Amsterdam, 1979.

177. B. F. Abrahams, R. Colton, B. F. Hoskins and K. McGregor, *Aust. J. Chem.*, **45**, 941 (1992).
178. L. Weber, D. Wewers and E. Lücke, *Z. Naturforsch.*, **40B**, 968 (1985).
179. R. A. Jones and B. R. Whittlesey, *Organometallics*, **3**, 469 (1984).
180. F. A. Cotton and T. R. Wedd, *Inorg. Chim. Acta*, **10**, 127 (1974).
181. U. Weber, L. Zsolnai and G. Huttner, *Z. Naturforsch.*, **40B**, 872 (1985).
181. U. Weber, L. Zsolnai and G. Huttner, *Z. Naturforsch.*, **40b**, 872 (1985).
182. A. L. Rheingold and M. R. Churchill, *J. Organomet. Chem.*, **243**, 165 (1983).
183. A. J. DiMaio, T. Bitterwoff amd A. L. Rheingold, *Organometallics*, **9**, 551 (1990).
184. P. S. Elmes, B. M. Gatehouse, D. J. Lloyd and B. O. West, *J. Chem. Soc, Chem. Commun.*, **953** (1974).
185. A. L. Rheingold, M. L. Fountain and A. J. DiMaio, *J. Am. Chem. Soc.*, **109**, 141 (1987).
186. A. J. DiMaio and A. L. Rheingold, *Inorg. Chem.*, **29**, 798 (1990).
187. A. L. Rheingold and M. L. Fountain, *Organometallics*, **5**, 2410 (1986).
188. U. Weber, G. Huttner, O. Schneidsteger and L. Zsolnai, *J. Organomet. Chem.*, **289**, 357 (1985).
189. G. Huttner, U. Weber, B. Sigwarth and O. Schneidsteger, *Angew. Chem.*, **94**, 210 (1982); *Angew. Chem., Int. Ed. Engl.*, **21**, 215 (1982).
190. G. Huttner and I. Jibril, *Angew. Chem.*, **96**, 709 (1984); *Angew. Chem., Int. Ed. Engl.*, **23**, 740 (1984).
191. A. H. Cowley, J. G. Lasch, N. C. Norman and M. Pakulski, *Angew. Chem.*, **95**, 1019 (1983); *Angew. Chem., Int. Ed. Engl.*, **22**, 978 (1983).
192. U. Weber, L. Zsolnai and G. Huttner, *J. Organomet. Chem.*, **260**, 281 (1984).
193. A. M. Arif, A. H. Cowley, N. C. Norman and M. Pakulski, *Inorg. Chem.*, **25**, 4836 (1986).
194. G. Huttner, and H. G. Schmid, *Angew. Chem.*, **87**, 454 (1975); *Angew. Chem., Int. Ed. Engl.*, **14**, 433 (1975); G. Huttner, J. v. Seyeri, M. Marsili and H. G. Schmid, *Angew. Chem.*, **87**, 455 (1975); *Angew. Chem., Int. Ed. Engl.*, **14**, 434 (1975).
195. B. Y. Liu, Y. T. Ku, M. Wang and P. J. Zheng, *Chin. J. Chem.*, 121 (1990).
196. B. Y. Liu, Y. T. Ku, M. Wang, B. Wang and P. J. Zheng, *J. Chem. Soc., Chem. Commun.*, 651 (1989).
197. W. C. Klemperer and R. S. Liu, *Inorg. Chem.*, **19**, 3863 (1980).
198. V. W. W. Yam, K. K. Tam, M. C. Cheng, S. M. Peng and Y. Wang, *J. Chem. Soc., Dalton Trans.*, 1717 (1992).
199. U. Flörke, M. Woyciechowski and H. J. Haupt, *Acta Crystallogr.*, **C44**, 2101 (1988).
200. E. Röttinger, A. Trenkle, R. Müller and H. Vahrenkamp, *Chem. Ber.*, **113**, 1280 (1980).
201. F. Calderazzo, R. Poli, D. Vitale, J. D. Korp, I. Bernal, G. Pelizzi, J. L. Atwood and W. E. Hunter, *Gazz. Chim. Ital.*, **113**, 761 (1983).
202. H. J. Langenbach and H. Vahrenkamp, *Chem. Ber.*, **113**, 3773 (1979).
203. H. J. Langenbach and H. Vahrenkamp, *J. Organomet. Chem.*, **191**, 391 (1980).
204. J. v. Seyerl, B. Sigwarth and G. Huttner, *Chem. Ber.*, **114**, 727 (1981); J. v. Seyerl, B. Sigwarth, H. G. Schmid, H. Mohr, A. Frank, M. Marsili and G. Huttner, *Chem. Ber.*, **114**, 1392 (1981); J. v. Seyerl, B. Sigwarth and G. Huttner, *Chem. Ber.*, **114**, 1407 (1981).
205. J. V. Seyerl and G. Huttner, *Angew. Chem.*, **90**, 911 (1978); *Angew. Chem., Int. Ed. Engl.*, **17**, 843 (1978).
206. U. Kirchgässner and U. Schubert, *Chem. Ber.*, **122**, 1481 (1989).
207. A. L. Rheingold and M. E. Fountain, *Acta Crystallogr.*, **C41**, 1162 (1985).
208. S. Vancheessan, *Ind. J. Chem.*, **22A**, 54 (1983).
209. M. Wieber and H. Höhl, *Z. Naturforsch.*, **44B**, 1149 (1989).
210. H. Schumann, L. Eguren and J. W. Ziller, *J. Organomet Chem.*, **408**, 361 (1991).
211. D. S. Bohle and P. A. Goodsen, *J. Chem. Soc., Chem. Commun.*, 1205 (1992).
212. A. H. Cowley, J. E. Kilduff, J. G. Lasch, N. C. Norman, M. Pakulski, F. Ando and T. C. Wright, *Organometallics*, **3**, 1044 (1984).
213. J. Ellemann and L. Mader, *Z. Anorg. Allg. Chem.*, **485**, 36 (1982).
214. M. M. T. Khan, K. S. Venkatasubramanian, Z. Shirin and M. M. Bhadbhade, *J. Chem. Soc., Dalton Trans.*, 885 (1992).
215. W. Malisch, H. Blau, H. Rößner and G. Jäth, *Chem. Ber.*, **113**, 1180 (1980).
216. E. Roland and H. Vahrenkamp, *Chem. Ber.*, **113**, 1799 (1980).
217. L. C. Song and Q. M. Hu, *J. Organomet. Chem.*, **414**, 219 (1991).
218. P. S. Elmes and B. O. West, *J. Organomet Chem.*, **32**, 365 (1971).

219. A. H. Cowley, N. C. Norman, M. Pakulski, D. L. Bricker and D. H. Russel, *J. Am. Chem. Soc.*, **107**, 8211 (1985).
220. W. Deck and H. Vahrenkamp, Z. *Anorg. Allg. Chem.*, **598**, 83 (1991).
221. M. Jacob and E. J. Weiss, *J. Organomet. Chem.*, **153**, 31 (1978).
222. P. S. Elmes, P. Leverett and B. O. West, *J. Chem. Soc., Chem. Commun.*, 747 (1971).
223. F. Calderazzo, F. Marchetti, F. Ungari and M. Wieber, *Gazz. Chim. Ital.*, **121**, 93 (1991).
224. H. Schumann, B. Gorella, M. Eisen and J. Blum, *J. Organomet. Chem.*, **412**, 251 (1991).
225. A. L. Rheingold and P. J. Sullivan, *J. Chem. Soc., Chem. Commun.*, **243**, 165 (1983).
226. R. D. W. Kemmitt, S. Mason, J. Fawcett and D. R. Russel, *J. Chem. Soc., Dalton Trans.*, 851 (1992).
227. D. H. Goldsworthy, F. R. Hartley, G. L. Marshall and S. G. Murray, *Inorg. Chem.*, **24**, 2849 (1985).
228. R. A. Jones and B. R. Whittlesey, *Inorg. Chem.*, **25**, 825 (1986).
229. R. G. Goel and W. O. Ogini, *Inorg. Chim. Acta*, **44**, L165 (1980).
230. R. W. Abel, M. A. Beckett, P. A. Bates and M. B. Hursthouse, *J. Organomet. Chem.*, **325**, 261 (1987).
231. H. P. Abicht and K. Issleib, *J. Organomet. Chem.*, **289**, 201 (1985).
232. G. B. Jacobson and B. L. Shaw, *J. Chem. Soc., Chem. Commun.*, 692 (1985).
233. A. J. Downard, L. R. Hanton and R. L. Paul, *J. Chem. Soc., Chem. Commun.*, 235 (1992).
234. D. Fenske and K. Merzweiler, *Angew. Chem.*, **96**, 600 (1984); *Angew. Chem., Int. Ed. Engl.*, **23**, 635 (1984).
235. M. Nardelli, C. Pelizzi, G. Pelizzi and P. Tarasconi, *J. Chem. Soc., Dalton Trans.*, 321 (1985).
236. C. Pelizzi, G. Pelizzi and P. Tarasconi, *J. Organomet. Chem.*, **281**, 403 (1985).
237. M. Baudler, D. Koch and B. Carlson, *Chem. Ber.*, **11**, 1217 (1978).
238. F. Hultèn and I. Persson, *Inorg. Chim. Acta*, **128**, 43 (1987).
239. A. J. Canty and B. M. Gatehouse, *J. Chem. Soc., Dalton Trans.*, 511 (1972).

Notes added in proof

1. Quite recently Norman et al.[240] reported on structural studies of the complexes [BiPhCl$_2$(THF)], [N^{n}Bu$_4$]$_2$[Bi$_2$Ph$_2$Br$_6$] and [NEt$_4$][BiPh$_2$I$_2$]. The compound [BiPhCl$_2$(THF)] comprises a chloride-bridged polymeric chain with each Bi centre in a square-based pyramidal coordination environment. A phenyl group occupies the apical site whilst the four basal positions are occupied by three Cl atoms, one terminal and two bridging, and the O of a coordinated THF molecule. The structure of the dianion [Bi$_2$Ph$_2$Br$_6$]$^{2-}$ consists of a planar Bi$_2$Br$_6$ unit and *trans* phenyl groups with each Bi centre also adopting a square-based pyramidal coordination geometry with apical phenyls. The structure of the anion [BiPh$_2$I$_2$] can be described as disphenoidal or equatorially vacant, trigonal bipyramidal with axial iodides and equatorial phenyls.
2. [Bi(C$_6$H$_6$)Cl$_3$], obtained from a benzene solution of BiCl$_3$, has been shown by X-ray diffraction to be a layer polymer with very weak bismuth–benzene π-bonding[241].
3. The Sb centre in the structurally characterized complex [H(py)$_2$][SbI$_4$(dmpe)] (py = pyridine, dmpe = 1,2-bis(dimethylphosphino)ethane] shows significant distortions from octahedral geometry which are discussed in terms of an arrested double S_N2 transition state for the nucleophilic substitution of two iodine anions by dmpe[242].

240. W. Clegg, J. Errington, G. A. Fisher, R. J. Flynn, and N. C. Norman, *J. Chem. Soc., Dalton Trans.*, 637 (1993).
241. W. Frank, J. Schneider, and S. Müller-Becker, *J. Chem. Soc., Chem. Commun.*, 799 (1993).
242. W. Clegg, M. R. J. Elsegood, V. Graham, N. C. Norman, and N. L. Pickett, *J. Chem. Soc., Dalton Trans.*, 997 (1993).

CHAPTER **9**

Substituent effects of arsenic, antimony and bismuth groups*,†,‡

MARVIN CHARTON

Chemistry Department, School of Liberal Arts and Sciences, Pratt Institute, Brooklyn, New York 11205, USA

---

*For definitions of various terms, symbols and abbreviations used in this chapter, see Section VI (glossary).
†In all Tables, references in boldface refer to the numbers of equations in this chapter used to estimate the values of the constants reported in the Table.
‡In all Tables, values labelled E are estimates.

The chemistry of organic arsenic, antimony and bismuth compounds
Edited by S. Patai © 1994 John Wiley & Sons Ltd

I. STRUCTURAL EFFECTS

A. Introduction

The objective of this work is to present models for the quantitative description of structural effects of substituents involving Group 15 elements. It is useful for this purpose to summarize the types of substituents we shall consider here. They include:

1. Dicoordinate substituents $-M = M - Z$ in which one of the bonding orbitals involving M is a π orbital. M may be N, P, As or Sb; Z may be any group. Examples are $N = NMe$, $P = PPh$.

2. Tricoordinate substituents $-MZ^1Z^2$ in which all the bonding orbitals involving M are σ. M may be N, P, As, Sb or Bi; Z may be any group. Z^1 may equal Z^2. Examples are NF_2, PH_2, $AsMe_2$, $SbEt_2$ and $BiPh_2$.

3. Tetracoordinate substituents $-M^1(M^2)Z^1Z^2$. M^1 may be N, P, As, Sb or Bi; M^2 may be O or, less frequently, S and Z may be any group. Z^1 may equal Z^2. Examples are $P(S)Me_2$, $As(O)Me_2$ and $Sb(O)Ph_2$. For all Group 15 elements other than nitrogen the $M^1 - M^2$ bond involves some degree of π bonding.
Tetracoordinate substituents $[MZ^1Z^2Z^3Z^4]^+$. M may be N, P or As; Z may be any group.

4. Pentacoordinate substituents $-MZ^1Z^2Z^3Z^4$. M may be P, As, Sb or Bi; Z may be any group. All the Z^i may be the same. Examples are PF_4, $AsMe_4$, $SbPh_4$ and $BiPh_4$.

Inherent in the concept of molecular structure is the notion that properties of all kinds, chemical, physical and biological, must vary with structural change. At first the structure – property relationships (SPR) reported were qualitative. As quantitative measurements of these properties accumulated, attempts were made to formulate quantitative descriptions of the dependence of properties on structure. It is to these methods for the quantitative description of structural effects that we now turn our attention.

B. Structure–Property Quantitative Relationships (SPQR)

The quantitative descriptions of the dependence of properties on structure, referred to above, are termed structure–property quantitative relationships (SPQR). These relationships can be classified according to the type of property:

1. Quantitative structure–chemical reactivity relationships (QSRR). These properties involve the formation and/or cleavage of chemical bonds. Equilibrium constants, rate constants and oxidation–reduction potentials are typical examples of this type of property.

2. Quantitative structure–chemical property relationships (QSCR). These properties involve the difference in intermolecular forces between an initial and a final state. Examples are equilibrium constants for hydrogen bonding, and partition; chromatographic properties such as capacity factors in high performance liquid chromatography, retention times in gas chromatography and R_F values in thin layer chromatography; melting and boiling points, solvent effects and solubilities.

3. Quantitative structure–physical property relationships (QSPR). These involve infrared, ultraviolet, nuclear magnetic resonance and other types of spectra, bond lengths and bond angles, dipole moments, ionization potentials and electron affinities.

4. Quantitative structure–activity relationships (QSAR). These involve any type of property associated with biological activities. The data range from rate and equilibrium constants for pure enzymes to toxicities in large multicellular organisms.

1. The nature of SPQR

There are three different types of chemical species (molecules, ions or radicals) for which SPQR can be determined:

1. Species with the structure XGY, where X is a variable substituent, Y an active group (an atom or group of atoms at which a measurable phenomenon takes place) and G is a skeletal group to which X and Y are bonded. In a given data set G and Y are held constant and only X varies.

2. Species with the structure XY in which the variable substituent X is directly attached to the constant active site Y.

3. Species in which no distinction between substituent and active site is possible; the entire species is the active site and it varies.

The function of SPQR is to provide a quantitative description of the exchange in some measurable quantity Q with a corresponding change in the structure of the substituent X, all other pertinent variables such as the conditions of the measurement being held constant. Thus:

$$(\partial Q/\partial X)_{G,Y,T,P,Sv,I,\ldots} = Q_X \tag{1}$$

where G is the skeletal group, Y the active site, T the temperature, P the pressure, Sv the solvent and I the ionic strength, all of which are constant throughout the data set.

We assume that Q_X will be a linear function of some number of parameters which represent the effects of structural variation of X. Then:

$$Q_X = a_1 p_{1X} + a_2 p_{2X} + a_3 p_{3X} + \cdots + a_0$$

$$= \sum_{i-1}^{n} a_i p_{iX} + a_0 \tag{2}$$

where the p_i are parameters which account for the structural effects of X on Q. These parameters can be obtained in four ways:

1. From quantum chemical calculations.
2. From molecular mechanics calculations.
3. From a reference set by definition. This method assumes that structural effects on the data set to be studies are a linear function of those which occur in the reference set.
4. From comparative molecular field analysis (COMFA).

Once suitable parameters are available, the values of Q can be correlated with them by means of simple linear regression analysis if the model requires only a single variable, or with multiple linear regression analysis if it requires two or more variables. We will consider here only those parameters which are defined directly or indirectly from suitable reference sets.

2. Uses of SPQR

SPQR can be used to provide mechanistic information about chemical and enzymatic reactions. They are useful in the prediction of chemical reactivities and properties, of physical properties and of biological activities. This has led to their very extensive use in the design of medicinal drugs and pesticides. This involves not only maximization of activity and minimization of side effects, but desirable properties such as improved solubility, longer shelf life and controlled release. They have also become an important method in environmental science, where they can be used to predict toxicities and other properties of environmental interest. Finally, SPQR provide an effective, efficient and convenient method for storing the experimental results of studies on the variation of measured properties as a function of structural change.

C. The Nature of Structural Effects

It is helpful to divide structural effects into three categories:

1. Electrical effects. These effects cause a variation in the electron density at the active site. They account for the ability of a substituent to stabilize or destabilize a cation, anion or radical.
2. Steric effects. These effects result from the repulsion between valence electrons in orbitals on atoms which are in close proximity but not bonded to each other.
3. Inter- and intramolecular force effects. These effects account for the interactions between the substituent and its immediate surroundings, such as the medium, a surface or a receptor site.

Electrical effects are the preponderant factor in chemical reactivities and physical properties. Intermolecular forces are usually the preponderant effect in bioactivities. Either electrical effects or intermolecular forces may be the prrdominant factor in chemical properties. Steric effects only occur when the substituent and the active site are in close proximity to each other and even then rarely account for more than 20% of the overall substituent effect.

II. ELECTRICAL EFFECTS

A. Introduction

The earliest successful parameterization of electrical effects is that of Hammett[1-3]. He defined the σ_m and σ_p constants using the ionization constants of 3- and 4-substituted benzoic acids in water at 25 °C as the reference set and hydrogen as the reference substituent to which all others are compared. For hydrogen, the values of the σ_m and σ_p constants were defined as zero. Thus:

$$\sigma_X \equiv \log \frac{K_X}{K_H} \tag{3}$$

These parameters were intended to apply to XGY systems in which the skeletal group is phenylene. Hammett found it necessary to define an additional set of parameters, σ_p^-, in order to account for substituent effects in systems with an active site which has a lone pair on the atom adjacent to the skeletal group. The reference set in this case was the ionization constants of 4-substituted phenols in water at 25 °C. Brown and Okamoto[4,5] later defined another set of constants, σ_p^+, to account for substituent effects in benzene derivatives with electronically deficient active sites. In this case the reference set was the rate constants for the solvolysis of 4-substituted cumyl chlorides in 90% aqueous acetone at 25 °C. Finally, Wepster and coworkers[6] and Taft[7] both independently proposed constants intended to represent substituent effects in benzene derivatives with minimal delocalized effect. Using the Taft notation these constants are written as σ_p^0. The reference systems were of the type 4-XCH$_4$CH$_2$Y, as it was argued that the methylene group intervening between the phenylene group and the active site acted as an insulator preventing conjugation between X and Y. These parameters differ in electronic demand. They are used in the Hammett equation, which may be written in the form:

$$Q_X = \rho\sigma_X + h \tag{4}$$

where Q_X is the value of the quantity of interest when the substituent is X, and σ_X is either σ_{mX}, σ_{pX}, σ_{pX}^0, σ_{pX}^+ or σ_{pX}^-; ρ and h are the slope and intercept of the line. The choice of parameters depends on the location of the substituent and the electronic demand of the active site.

Taft and his coworkers[8-10] went on to develop a diparametric model which separated the electrical effect into contributions from the 'inductive' and resonance effects. This separation actually depends on the difference in the extent of electron delocalization when a substituent is bonded to an sp^3-hybridized carbon atom on the one hand or to an sp^2- or sp-hybridized carbon atom on the other. As the first case represents minimal delocalization and the second extensive delocalization, we have referred to them as the localized and delocalized electrical effects. The diparametric model of electrical effects can be written in the form:

$$Q_X = L\sigma_{lX} + D\sigma_{DX} + h \tag{5}$$

where σ_l and σ_D are the localized and delocalized electrical effect parameters, respectively. Taft and coworkers[10] suggested that four σ_D constants are required. They are σ_{RX}, σ_{RX}^0, σ_{RX}^+ and σ_{RX}^-, and they correspond to the σ_p constants described above. Charton noted that in cases of very large electron demand two additional σ_D constants were required, $\sigma_R^\oplus$ for highly electron-deficient (positive) active sites[11] and $\sigma_R^\ominus$ for active sites that are very electron-rich (negative)[12].

An alternative diparametric model was proposed by Yukawa and Tsuno[13] for use with electron-poor active sites. The equation was originally written as:

$$Q_X = \rho\sigma_X + \rho r(\sigma_X^+ - \sigma_X) \tag{6}$$

A later version has the form:

$$Q_X = \rho\sigma_X + \rho r(\sigma_X^+ - \sigma_X^0) \tag{7}$$

A similar relationship:

$$Q_X = \rho\sigma_X + \rho r(\sigma_X^- - \sigma_X) \tag{8}$$

has been proposed for electron-rich active sites[13]. We will refer to these relationships as the

YT equations. They have the advantage that both *meta-* and *para*-substituted compounds may be included in the same data set on the assumption that ρ_m is equal to ρ_p. This assumption is usually, though not always, justified.

It became clear that there was a need for a more general model of electrical effects. The use of the LD equation for the description of chemical reactivities required either an *a priori* knowledge of the type of σ_D substituent constant required or a comparison of the results obtained using each of the available σ_D constants. The use of the YT equation has generally been restricted to electronically deficient active sites.

In 1987 a triparametric model of the electrical effect was introduced[14] that can account for the complete range of electrical effects on chemical reactivities of closed-shell species (carbenium and carbanions), that is, reactions which do not involve radical intermediates. The basis of this model was the observation that the σ_D constants differ in their electronic demand. If it is assumed that they are generally separated by an order of magnitude in this variable, then it is possible to assign to each σ_D type a corresponding value of the electronic demand, η. Thus the equation:

$$\sigma_{DX} = a_1\eta + a_0 = \sigma_e\eta + \sigma_d \tag{9}$$

is obeyed. The intercept of this linear relationship represents the intrinsic delocalized (resonance) effect, σ_{dX}; the slope represents the sensitivity of the X group to the electronic demand of the active site. On substituting equation 6 into the LD equation we obtain the triparametric relationship:

$$Q_X = L\sigma_{lX} + D\sigma_{dX} + R\sigma_{eX} + h \tag{10}$$

The σ_l values are identical to σ_I. The symbol was changed in order to be consistent with the other symbols used in the equation.

When the composition of the electrical effect, P_D, is held constant, the LDR equation simplifies to the CR equation:

$$Q_X = C\sigma_{ldX} + R\sigma_{eX} + h \tag{11}$$

where σ_{ld} is a composite parameter. It is defined by the relationship:

$$\sigma_{ldX} = l\sigma_{eX} + d\sigma_{dX} \tag{12}$$

The distinction between pure and composite parameters is that the former represent a single effect while the latter represent a mixture of two or more. The percent composition of these parameters is given by:

$$P_D = \frac{100d}{l+d} \tag{13}$$

If we let the constant value of P_D be k', then the σ_{ldX} parameter for a given value of k' can be written as:

$$\sigma_{ldXk'} = \sigma_{lX} + [k'/100 - k']\sigma_{DX} \tag{14}$$

If we set:

$$k^* = k'/(100 - k') \tag{15}$$

equation 14 gives:

$$\sigma_{ldXk'} = \sigma_{lX} + k^*\sigma_{dX} \tag{16}$$

It has been shown that the Yukawa–Tsuno equation for 4-substituted benzene derivatives is approximately equaivalent to the CR equation[15]. This has led to the development of a modified Yukawa–Tsuno (MYT) equation, which has the form:

$$Q_X = \rho\sigma_X + R\sigma_{eX} + h \tag{17}$$

with σ taking the value σ_m for 3-substituted benzene derivatives and $\sigma_{50.0}$ for 4-substituted benzene derivatives, while σ_{ex} for substituents in the *meta* position is 0.

When the sensitivity to electronic demand is held constant, the LDR equation reverts to the LD equation (equation 5). It is possible to combine 3- and 4-substituted benzene derivatives into a single data set using a modified form of the LD equation (the MLD equation):

$$Q_X = \rho'\sigma_X + D\sigma_{DX} + h \qquad (18)$$

where σ is σ_m for 3-substituted and σ_l for 4-substituted ones, while σ_D is 0 for 3-substituents.

When both the electronic demand and the composition of the electrical effect are held constant, a set of composite parameters is obtained, having the form:

$$\sigma_{k'/kX} = l\sigma_{1X} + d\sigma_{dX} + r\sigma_{eX} \qquad (19)$$

with:

$$k' = P_D = \frac{100d}{(l+d)}; k = \eta = r/d \qquad (19a)$$

The Hammett substituent constants are special cases of these parameters.

The $\sigma_{k'/k}$ values describe the overall electrical effect of the X group. They are obtained from the expression:

$$\sigma_{k'/kx} = \sigma_{1X} + [P_D/100 - P_D](\sigma_{dX} + \eta\sigma_{eX}) \qquad (20)$$

$$= \sigma_{1X} + k^*(\sigma_{dX} + k\sigma_{eX}) \qquad (20a)$$

A plot of the $\sigma_{k'/kX}$ values for a group with $X = P_D$, $Y = \eta$ and $Z = \sigma_{k'/k}$ produces a surface that characterizes the electrical effect of the X group.

1. Estimation of electrical-effect parameters

It is very often necessary to estimate values of electrical-effect parameters for groups for which no measured values are available. This is of particular importance with regard to substituents in which arsenic, antimony or bismuth is the central atom. It has long been known the electrical-effect parameters of substituents X whose structure can be written as MZ_n^i are a function of the electrical effect of the Z^i when M is held constant[16,23]. In later work it was shown that when Z^i is held constant and M is allowed to vary, the substituent constant is a function of the Allred–Rochow electronegativity of M, χ_M and the number of Z groups, n_Z. It has been shown that when both Z and M vary, σ_1 values for all groups of the type $X = MZ^1Z^2Z^3$ are given by an equation of the form:

$$\sigma_{1X} = a\chi_M + L\Sigma\sigma_{1Z} + D\Sigma\sigma_{dZ} + R\Sigma\sigma_{eZ} + h \qquad (21)$$

σ_{dX} and σ_{ex} for $M(lp)_n Z^1Z^2Z^3$ groups can be calculated from an equation of this type. In an extension of this work based on results obtained for the prediction of pK_a values of oxyacids, we have used a correlation equation of the form:

$$\sigma_X = a\chi_M + L\sum\sigma_{1Z} + D\sum\sigma_{dZ} + R\sum\sigma_{eZ} + a_H n_H + a_{1p} n_{1p} + a_{1d} n_{Od1}$$
$$+ a_{2d} n_{Od2} + a_p n_{Op} + a_q q + h \qquad (22)$$

Values of σ_1, σ_d and σ_e for the Group 15 substituents of the type $M(lp)_n Z^1Z^2O$ may be calculated from the equations:

$$\sigma_{1X} = 0.341(\pm 0.0101)\Sigma\sigma_{1Z} + 0.128(\pm 0.0194)\Sigma\sigma_{dZ} + 0.314(\pm 0.0813)\Sigma\sigma_{eZ}$$
$$+ 0.0329(\pm 0.0128)\chi + 0.348(\pm 0.0182)n_{Od1} + 0.205(\pm 0.0239)n_{Od2}$$
$$+ 0.296(\pm 0.0143)n_{Op} + 0.149(0.00707)n_{1p} - 0.0559(\pm 0.0366) \qquad (23)$$

$100R^2$, 94.88; $A100R^2$, 94.33; F, 148.2; S_{est}, 0.0400; S^0, 0.242; n, 73; P_D, 27.3(± 4.42); η, 2.45(± 0.514).

$$\sigma_{dX} = 0.442(\pm 0.0443)\Sigma\sigma_{IZ} + 0.242(\pm 0.0381)\Sigma\sigma_{dZ} - 0.0529(\pm 0.0228)n_H$$
$$+ 0.673(\pm 0.0459)n_{Od1} + 0.753(\pm 0.0483)n_{Op} + 0.0828(\pm 0.0203)n_{lp} \tag{24}$$
$$- 0.228(\pm 0.0830)$$

$100R_2$, 95.85; $A100R^2$, 95.38; F, 174.7; S_{est}, 0.0694; S^0, 0.219; n, 61; P_D, 35.3(± 6.33).

$$\sigma_{eX} = 0.132(\pm 0.0149)\Sigma\sigma_{IZ} + 0.0152(\pm 0.00737)n_H + 0.0584(\pm 0.0670)\chi_M$$
$$+ 0.317(\pm 0.0184)n_{Od1} - 0.0662(\pm 0.0151)n_{Od2} + 0.208(\pm 0.0171)n_{Op}$$
$$+ 0.0825(\pm 0.00699)n_{lp} - 0.423(\pm 0.0258) \tag{25}$$

$100R^2$, 92.93; $A100R^2$, 91.64; F, 62.76; S_{est}, 0.0195; S^0, 0.295; n, 53; P_D, 37.5(± 7.34);η, 4.02(± 0.457).

M is the atom of the group which is bonded to either the skeletal group or the active site, where χ_M is the Allred–Rochow electronegativity[24], n_H is the number of hydrogen atoms attached to M, n_{Od1} and n_{Od2} the first and second oxygen atoms involved in pdπ bonding, n_{Op} the number of oxygen atoms involved in ppπ bonding and n_{lp} the number of lone pairs on M.

Again, in an extension of previous work we have used the correlation equation:

$$\sigma_X = L_2\sigma_{IZ2} + D_2\sigma_{dZ2} + R_2\sigma_{eZ2} + a_1\chi_{M^1}$$
$$+ a_2\chi_{M^2} + L_1\sigma_{IZ^1} + a_\pi n_\pi + h \tag{26}$$

for the estimation of σ_1, σ_d and σ_e constants of groups of the types $M^1Z^1 = M^2Z^2$ and $M^1 \equiv Z^2$.

$$\sigma_{1M^1Z^1M^2Z^2} = 0.292(\pm 0.0337)\sigma_{RZ2} + 0.175(\pm 0.0382)\sigma_{dZ2}$$
$$+ 0.0814(\pm 0.0249)\chi_{M^1} + 0.205(\pm 0.164)\chi_{M^2}$$
$$+ 0.394(\pm 0.0462)\sigma_{IZ^1}\chi + 0.206(\pm 0.0338)\sigma_{dZ^1} \tag{27}$$
$$+ 0.201(\pm 0.0166)n_\pi - 0.803(\pm 0.0932)$$

$100R^2$, 93.68; $A100R^2$, 92.50; F, 65.66; S_{est}, 0.0297; S^0, 0.282; n, 39; P_{D1}, 34.3(± 6.51); P_{D2}, 37.5(± 9.14).

$$\sigma_{d,M^1Z^1M^2Z^2} = 0.473(\pm 0.0717)\sigma_{IZ2} + 0.272(\pm 0.0869)\sigma_{dZ2}$$
$$+ 2.19(\pm 0.330)\sigma_{eZ2} + 0.299(\pm 0.0574)\chi_{M^1}$$
$$+ 0.432(\pm 0.0334)\chi_{M^2} + 0.148(\pm 0.0640)\sigma_{dZ^1}$$
$$+ 0.877(\pm 0.381)\sigma_{eZ^1} - 1.77(\pm 0.172) \tag{28}$$

$100R^2$, 94.95; $A100R^2$, 93.94; F, 77.93; S_{est}, 0.0586; S^0, 0.254; n, 37; P_{D2}, 36.5($\pm(12.9)$); η_2, 8.04.

$$\sigma_{e,M^1Z^1M^2Z^2} = 0.169(\pm 0.0297)\sigma_{IZ2} - 0.0540(\pm 0.0238)\sigma_{dZ2}$$
$$+ 0.422(\pm 0.0735)\sigma_{eZ2} + 0.0694(\pm 0.0134)\chi_{M^1}$$
$$+ 0.0878(\pm 0.0183)\sigma_{IZ^1} - 0.269(\pm 0.0351) \tag{29}$$

$100R^2$, 74.78; $A100R^2$, 71.30; F, 16.61; S_{est}, 0.0149; S^0, 0.553; P_{D2}, 24.2(± 11.4); η_2, -7.82.

Values of σ_1, σ_d and σ_e for Group 15 substituents are given in Tables 1, 2 and 3. Estimated values of σ_D parameters may be calculated from the equations[14]:

$$\sigma_R = 0.934\sigma_{dX} + 0.308\sigma_{eX} - 0.0129 \tag{30}$$
$$\sigma_{RX}^0 = 0.770\sigma_{dX} - 0.288\sigma_{eX} - 0.0394 \tag{31}$$

TABLE 1. Values of σ_I. Group 15 substituents

Z^1, Z^2	M=N	Ref.	P	Ref.	As	Ref.	Sb	Ref.	Bi	Ref.
Dicoordinate	**M=MZ**									
H	0.31E	17	−0.01E	27	0.03E	27	−0.08E	27	−0.12E	27
CF$_3$	0.44E	17	0.13E	27	0.17E	27	0.06E	27	0.02E	27
CN	0.46E	27	0.18E	27	0.22E	27	0.11E	27	0.06E	27
Me	0.25E	27	−0.04E	27	0.00E	27	−0.11E	27	−0.15	27
OMe	0.27E	27	−0.02E	27	0.02E	27	−0.09E	27	−0.13E	27
NMe$_2$	0.21E	27	−0.08E	27	−0.04E	27	−0.15E	27	−0.19E	27
t-Bu	0.21	17	−0.04E	27	0.00E	27	−0.11E	27	−0.15E	27
Ph	0.27	17	0.00E	27	0.04E	27	−0.07E	27	−0.11E	27
F	0.35E	27	0.06E	27	0.10E	27	−0.01E	27	−0.05E	27
NH$_2$	0.21E	27	−0.08E	27	−0.04E	27	−0.15E	27	−0.19E	27
OH	0.28E	27	−0.01E	27	0.03E	27	−0.08E	27	−0.12E	27
M^1=CH	0.20E	17	0.08E	27	0.09E	27	0.06E	27	0.05E	27
M^1=CMe	0.13E	27	0.05E	27	0.06E	27	0.04E	27	0.02E	27
M^1=CPh	0.13E	17	0.09E	27	0.10E	27	0.08E	27	0.06E	27
M^1=PH	0.13E				0.00E	27	0.03E	27	−0.04E	27
M^1=AsH			0.02E	27			0.00E	27	−0.02E	27
M^1=SbH			−0.06E	27	−0.05E	27			−0.09E	27
M^1=BiH			−0.09E	27	−0.08E	27	−0.11E	27		
M=O	0.37	27								
M=C=O	0.34E	17								
M=C=S	0.54E	17								
M$_3$	0.43	14								
M≡C	0.63	16								
Tricoordinate	**MZ^{1}Z^2**									
H, H	0.17	14	0.18	14	0.14E	19	0.15E	23	0.15E	23
H, CF$_3$	0.34E	14	0.28E	14						
H, CN	0.45E	14	0.39E	14						
H, CONH$_2$	0.23									
H, CHO	0.33	14								
H, Me	0.13	14	0.12E	19						
H, SO$_2$Me	0.42E	19								

(continued)

TABLE 1. (continued)

Z^1, Z^2	M=N	Ref.	P	Ref.	As	Ref.	Sb	Ref.	Bi	Ref.
$(CF_3)_2$	0.49E	14	0.45E	23	0.46E	23	0.44E	23	0.44E	23
$(CN)_2$	0.71E	14	0.55E	23	0.55E	23	0.54E	23	0.53E	23
H, $COCF_3$	0.46	19								
H, Ac	0.28	14	0.21E	14						
H, CO_2Me	0.28E	19								
H, Et	0.17E	18								
Me_2	0.17	14	0.10E	14	0.10E	23	0.09E	23	0.09E	23
$(OMe)_2$	0.22E	23	0.18E	23	0.19E	23	0.18E	23	0.17E	23
$(SMe)_2$	0.22E	23	0.19E	23	0.19E	23	0.18E	23	0.17E	23
$(C_2H)_2$	0.32E	23	0.29E	23	0.30E	23	0.28E	23	0.28E	23
$(CH{=}CH_2)_2$	0.17E	23	0.14E	23	0.14E	23	0.13E	23	0.13E	23
Et_2	0.15E	14	0.10E	23	0.11E	23	0.09E	23	0.09E	23
$(OEt)_2$	0.20E	23	0.17E	23	0.17E	23	0.16E	23	0.15E	23
$(SEt)_2$	0.20E	23	0.16E	23	0.17E	23	0.16E	23	0.17E	23
$(NMe_2)_2$	-0.01E	23	-0.04E	23	-0.04E	23	-0.05E	23	-0.06	23
$(CH_2CH{=}CH_2)_2$	0.14E	23	0.11E	23	0.11E	23	0.10E	23	0.09E	23
H, Ph	0.20	14								
H, Bz	0.28	19								
$(C_6F_5)_2$	0.38E	23	0.35E	23	0.35E	23	0.34E	23	0.34E	23
Ph_2	0.21E	14	0.13E	14	0.14E	23	0.33E	23	0.12	23
$(OPh)_2$	0.28E	23	0.25E	23	0.26E	23	0.24E	23	0.24	23
$(SPh)_2$	0.21E	23	0.18E	23	0.18E	23	0.17E	23	0.17	23
$(Bzl)_2$	0.15E	23	0.11E	23	0.12E	23	0.10E	23	0.10E	23
Cl_2	0.44E	23	0.40E	23	0.41E	23	0.39E	23	0.39E	23
F_2	0.47E	23	0.43E	23	0.44E	23	0.42E	23	0.42E	23
H, NH_2	0.11E	14								
$(NH_2)_2$	0.05E	23	0.02E	23	0.03E	23	0.01E	23	0.01E	23
H, OH	0.15E	19								
$(OH)_2$	0.26E	23	0.23E	23	0.23E	23	0.22E	23	0.21E	23
O_2	0.67	14								

377

Tetracoordinate	M(O)Z^{1}Z^2							
(O)H$_2$	0.36E	23	0.36E	23	0.35E	23	0.35E	23
O(CF$_3$)$_2$	0.65E	23	0.65E	23	0.64E	23	0.64E	23
O(CN)$_2$	0.74E	23	0.75E	23	0.74E	23	0.73E	23
(O)Me$_2$	0.30	16	0.30E	23	0.29E	23	0.29E	23
O(OMe)$_2$	0.36	14	0.39E	23	0.38E	23	0.37E	23
O(SMe)$_2$	0.39E	23	0.39E	23	0.38E	23	0.37E	23
O(C$_2$H)$_2$	0.49E	23	0.49E	23	0.48E	23	0.48E	23
O(CH=CH$_2$)$_2$	0.34E	23	0.34E	23	0.33E	23	0.33E	23
(O)Et$_2$	0.28	16	0.30E	23	0.29E	23	0.29E	23
O(OEt)$_2$	0.32E	14	0.37E	23	0.36E	23	0.35E	23
O(SEt)$_2$	0.36E	23	0.37E	23	0.35E	23	0.35E	23
O(NMe$_2$)$_2$	0.16E	23	0.16E	23	0.15E	23	0.14E	23
O(CH$_2$CH=CH$_2$)$_2$	0.31E	23	0.31E	23	0.30E	23	0.30E	23
(O)Pr$_2$	0.26	16						
(O)Bu$_2$	0.25	16						
O(C$_6$F$_5$)$_2$	0.55E	23	0.55E	23	0.54E	23	0.54E	23
(O)Ph$_2$	0.26	16	0.34E	23	0.33E	23	0.32E	23
O(OPh)$_2$	0.45E	23	0.45E	23	0.44E	23	0.44E	23
O(SPh)$_2$	0.38E	23	0.38E	23	0.37E	23	0.36E	23
(O)Bzl$_2$	0.31E	23	0.32E	23	0.30E	23	0.30E	23
(O)Cl$_2$	0.60E	23	0.61E	23	0.59E	23	0.59E	23
(O)F$_2$	0.63E	23	0.64E	23	0.62E	23	0.62E	23
O(NH$_2$)$_2$	0.22E	23	0.22E	23	0.21E	23	0.21E	23
O(OH)$_2$	0.43E	23	0.43E	23	0.42E	23	0.41E	23
(S)Ph$_2$	0.28	19						
Pentacoordinate	**MZ$^1_2$Z$^2_2$**							
Ph$_4$	−0.04E	23	−0.03E	23	−0.04E	23	−0.05E	23
F$_4$	0.55E	23	0.56E	23	0.55E	23	0.54E	23

TABLE 2. Values of σ_d.

Z^1, Z^2	M=N	Ref.	P	Ref.	As	Ref.	Sb	Ref.	Bi	Ref.
Dicoordinate										
	M=MZ									
H	0.27E	17	-0.41E	28	-0.32E	28	-0.57E	28	-0.67E	28
CF_3	0.24E	17	-0.24E	28	-0.15E	28	-0.40E	28	-0.50E	28
CN	0.44E	28	-0.23E	28	-0.13E	28	-0.38E	28	-0.48E	28
Me	0.15E	28	-0.52E	28	-0.42E	28	-0.68E	28	-0.78E	28
OMe	0.11E	28	-0.56E	28	-0.46E	28	-0.72E	28	-0.81E	28
NMe_2	-0.37E	28	-1.03E	28	-0.94E	28	-1.19E	28	-1.29E	28
t-Bu	0.07	17	-0.53E	28	-0.44E	28	-0.69E	28	-0.79E	28
Ph	0.12	17	-0.41E	28	-0.55E	28	-0.81E	28	-0.90E	28
F	0.47E	28					-0.35E	28	-0.45E	28
NH_2	-0.13E	28	-0.80E	28	-0.70E	28	-0.96E	28	-1.06E	28
OH	0.17E	28	-0.49E	28	-0.40E	28	-0.65E	28	-0.75E	28
M^1=CH	-0.02E	17	-0.22E	28	-0.19E	28	-0.26E	28	-0.31E	28
M^1=CMe	-0.10E	28	-0.33E	28	-0.29E	28	-0.37E	28	-0.42E	28
M^1=CPh	-0.07E	17	-0.46E	28	-0.42E	28	-0.50E	28	-0.55E	28
M^1=PH					-0.38E	28	-0.46E	28	-0.50E	28
M^1=AsH			-0.35E		-0.48E	28	-0.40E	28	-0.44E	28
M^1=SbH			-0.51E		-0.48E	28			-0.60E	28
M^1=BiH			-0.58E		-0.54E	28	-0.63E	28		
M=C=O	-0.17E	17								
M=C=S	-0.11E	17								
M_3	-0.27E	14								
M=O	0.45E	28								
M≡C		16								
Tricoordinate										
	MZ^1Z^2									
H, H	-0.68	14	-0.53E	14	-0.53E	19	-0.48E	24	-0.46E	24
H, CF_3	-0.26E	14	-0.11E	14						
H, CN	-0.41E	14	-0.26E	14						
H, $CONH_2$	-0.45E	19								
H, CHO	-0.33	14								
H, Me	-0.67	14								
H, SO_2Me	-0.21E	19								
$(CF_3)_2$	-0.13E	14	0.01E	24	-0.01E	24	0.04E	24	0.06E	24

Z^1Z^2										
			0.15E	24	0.14E	24	0.18E	24	0.20E	24
$(CN)_2$	−0.17E	14								
H, $COCF_3$	−0.19E	19	0.21E	14						
H, Ac	−0.35	14								
H, CO_2Me	−0.42E	19								
H, Et	−0.65E	18								
Me_2	−0.66	14	−0.49E	14	−0.50E	14	−0.45E	24	−0.44E	24
$(OMe)_2$	−0.54E	24	−0.41E	24	−0.43E	24	−0.38E	24	−0.36E	24
$(SMe)_2$	−0.46E	24	−0.33E	24	−0.35E	24	−0.30E	24	−0.28E	24
$(C_2H)_2$	−0.29E	24	−0.16E	24	−0.18E	24	−0.13E	24	−0.11E	24
$(CH{=}CH_2)_2$	−0.48E	24	−0.35E	24	−0.37E	24	−0.32E	24	−0.30E	24
Et_2	−0.65E	14	−0.48E	24	−0.49E	24	−0.45E	24	−0.43E	24
$(OEt)_2$	−0.56E	24	−0.34E	24	−0.45E	24	−0.40E	24	−0.38E	24
$(SEt)_2$	−0.50E	24	−0.37E	24	−0.39E	24	−0.34E	24	−0.32E	24
$(NMe_2)_2$	−0.71E	24	−0.58E	24	−0.60E	24	−0.55E	24	−0.53E	24
$(CH_2CH{=}CH_2)_2$	−0.60E	24	−0.47E	24	−0.49E	24	−0.44E	24	−0.42E	24
H, Ph	−0.62E	14	−0.01E	24	−0.11E	24	−0.07E	24	−0.05E	24
H, Bz	−0.39E	19	−0.40E	14	−0.38E	14	−0.33E	24	−0.31E	24
$(C_6F_5)_2$	−0.23E	24	−0.30E	24	−0.32E	24	−0.27E	24	−0.25E	24
Ph_2	−0.56E	14	−0.30E	24	−0.32E	24	−0.27E	24	−0.28E	24
$(OPh)_2$	−0.43E	24	−0.44E	24	−0.46E	24	−0.41E	24	−0.40E	24
$(SPh)_2$	−0.43E	24	−0.13E	24	−0.15E	24	−0.10E	24	−0.08E	24
$(Bzl)_2$	−0.57E	24	−0.16E	24	−0.18E	24	−0.13E	24	−0.11E	24
Cl_2	−0.26E	24								
F_2	−0.29	14								
H, NH_2	−0.64E	24								
$(NH_2)_2$	−0.72E	19								
H, OH	−0.39E	24	−0.59E	24	−0.61E	24	−0.56E	24	−0.54E	24
$(OH)_2$	−0.50E	14	−0.38E	24	−0.39E	24	−0.34E	24	−0.33E	24
O_2	0.18	14								
Tetracoordinate $M(O)Z^1Z^2$										
$(O)H_2$	0.08E	24	0.08E	24	0.06E	24	0.10E	24	0.13E	24
$O(CF_3)_2$	0.60E	24	0.60E	24	0.58E	24	0.63E	24	0.65E	24
$O(CN)_2$	0.74E	24	0.74E	24	0.73E	24	0.77E	24	0.79E	24
$(O)Me_2$	0.14	17	0.14	17	0.09E	17	0.14E	17	0.15E	17
$O(OMe)_2$	0.24E	14	0.24E	14	0.16E	14	0.21E	14	0.23E	14

(continued)

TABLE 2. (*continued*)

Z^1, Z^2	M = N	Ref.	P	Ref.	As	Ref.	Sb	Ref.	Bi	Ref.
$O(SMe)_2$			0.26E	**24**	0.24E	**24**	0.29E	**24**	0.31E	**24**
$O(C_2H)_2$			0.43E	**24**	0.41E	**24**	0.46E	**24**	0.48E	**24**
$O(CH{=}CH_2)_2$			0.24E	**24**	0.22E	**24**	0.27E	**24**	0.29E	**24**
$(O)Et_2$			0.11E	**24**	0.10E	**24**	0.15E	**24**	0.16E	**24**
$O(OEt)_2$			0.24	14	0.14E	**24**	0.19E	**24**	0.21E	**24**
$O(SEt_2)$			0.22E	**24**	0.20E	**24**	0.25E	**24**	0.27E	**24**
$O(NMe_2)_2$			0.01E	**24**	−0.01E	**24**	0.04E	**24**	0.06E	**24**
$O(CH_2CH{=}CH_2)_2$			0.12E	**24**	0.10E	**24**	0.15E	**24**	0.17E	**24**
$(O)Pr_2$				16						
$(O)Bu_2$				16						
$O(C_6F_5)_2$			0.49E	**24**	0.48E	**24**	0.52E	**24**	0.54E	**24**
$(O)Ph_2$			0.23E	**24**	0.21E	**24**	0.26E	**24**	0.28E	**24**
$O(OPh)_2$			0.29E	**24**	0.27E	**24**	0.32E	**24**	0.34E	**24**
$O(SPh)_2$			0.29E	**24**	0.27E	**24**	0.32E	**24**	0.34E	**24**
$(O)Bzl_2$			0.15E	**24**	0.13E	**24**	0.18E	**24**	0.19E	**24**
$(O)Cl_2$			0.46E	**24**	0.44E	**24**	0.49R	**24**	0.51E	**24**
$(O)F_2$			0.43E	**24**	0.41E	**24**	0.46E	**24**	0.48E	**24**
$O(NH_2)_2$			0.00E	**24**	−0.02E	**24**	0.03E	**24**	0.05E	**24**
$O(OH)_2$			0.21E	**24**	0.20E	**24**	0.25E	**24**	0.26E	**24**
$(S)Ph_2$			0.29E	19						
Pentacoordinate	**MZ_4**									
Ph			−0.40E	**24**	−0.41E	**24**	−0.36E	**24**	−0.35E	**24**
F			0.00E	**24**	−0.02E	**24**	0.03E	**24**	0.05E	**24**

TABLE 3. Values of σ_e Group 15 substituents

Z^1, Z^2	$M=N$	Ref.	P	Ref.	As	Ref.	Sb	Ref.	Bi	Ref.
Dicoordinate	M=MZ									
H	−0.080	17	−0.13E	29	−0.12E	29	−0.14E	29	−0.15E	29
CF_3	0.060	17	−0.076E	29	−0.067E	29	−0.093E	29	−0.10E	29
CN	0.011E	29	−0.059E	29	−0.050E	29	−0.076E	29	−0.086E	29
Me	−0.063E	29	−0.13E	29	−0.12E	29	−0.15E	29	−0.16E	29
OMe	−0.003E	29	−0.073E	29	−0.063E	29	−0.089E	29	−0.10E	29
NMe_2	−0.093E	29	−0.16E	29	−0.15E	29	−0.18E	29	−0.19E	29
t-Bu	−0.075	17	−0.13E	29	−0.13E	29	−0.15E	29	−0.16E	29
Ph	−0.059	17	−0.15E	29	−0.14E	29	−0.17E	29	−0.18E	29
F	0.079E	29					−0.008E	29	−0.019E	29
NH_2	−0.045E	29	−0.12E	29	−0.11E	29	−0.13E	29	−0.14E	29
OH	0.015E	29	−0.055E	29	−0.045E	29	−0.071E	29	−0.082E	29
$M^1=CH$	−0.060E	17	−0.13E	29	−0.12E	29	−0.14E	29	−0.15E	29
$M^1=CMe$	−0.063E	29	−0.13E	29	−0.12E	29	−0.15E	29	−0.15E	29
$M^1=CPh$	0.063	17	−0.13E	29	−0.13E	29	−0.15E	29	−0.16E	29
$M^1=PH$					−0.12E	29	−0.14E	29	−0.15E	29
$M^1=AsH$			−0.13E	29	−0.12E	29	−0.14E	29	−0.15E	29
$M^1=SbH$			−0.13E	29	−0.12E	29	−0.14E	29	−0.15E	29
$M^1=BiH$			−0.13E	29						
$M=C=O$	0.070E	17								
$M=C=S$	−0.090E	17								
M_3	−0.12E	14								
$M=O$	−0.056	29								
$M=C$		16								
Tricoordinate	MZ^1Z^2									
H, H	−0.13	14	−0.18E	14	−0.17E	19	−0.20E	25	−0.21E	25
H, CF_3	−0.12E	14	−0.17E	14						
H, CN	−0.12E	14	−0.17E	14						
H,$CONH_2$	−0.13E	19								
H, CHO	−0.23E	14								
H, Me	−0.18	14	−0.23E	19						
H, SO_2Me	−0.18E	19								
$(CF_3)_2$	−0.014E	14	−0.11E	14	−0.10E	25	−0.12E	25	−0.13E	25

(continued)

TABLE 3. (*continued*)

Z^1, Z^2	$M=N$	Ref.	P	Ref.	As	Ref.	Sb	Ref.	Bi	Ref.
$(CN)_2$	-0.014E	14	-0.086E	25	-0.078E	25	-0.10E	25	-0.11E	25
H, COCF$_3$	-0.10E	19								
H, Ac	-0.088	14	-0.14E	14						
H, CO$_2$Me	-0.13E	19								
H, Et	-0.15E	18								
Me$_2$	-0.24	14	-0.27E	14	-0.27E	25	-0.28E	25	-0.29E	25
(OMe)$_2$	-0.21E	25	-0.27E	25	-0.26E	25	-0.28E	25	-0.29E	25
(SMe)$_2$	-0.22E	25	-0.28E	25	-0.28E	25	-0.30E	25	-0.31E	25
(C$_2$H)$_2$	-0.15E	25	-0.21E	25	-0.20E	25	-0.22E	25	-0.23E	25
(CH=CH$_2$)$_2$	-0.22E	25	-0.28E	25	-0.27E	25	-0.29E	25	-0.30E	25
Et$_2$	-0.18E	14	-0.26E	25	-0.22E	25	-0.28E	25	-0.29E	25
(OEt)$_2$	-0.22E	25	-0.28E	25	-0.27E	25	-0.29E	25	-0.30E	25
(SEt)$_2$	-0.23E	25	-0.29E	25	-0.28E	25	-0.30E	25	-0.31E	25
(NMe$_2$)$_2$	-0.37E	25	-0.43E	25	-0.42E	25	-0.45E	25	-0.46E	25
(CH$_2$CH=CH$_2$)$_2$	-0.21E	25	-0.27E	25	-0.26E	25	-0.28E	25	-0.29E	25
H, Ph	-0.18E	14								
H, Bz	-0.15E	19								
(C$_6$F$_5$)$_2$	-0.11E	25	-0.17E	25	-0.16E	25	-0.18E	25	-0.19E	25
Ph$_2$	-0.24E	14	-0.29E	14	-0.28E	25	-0.30E	25	-0.31E	25
(OPh)$_2$	-0.19E	25	-0.25E	25	-0.24E	25	-0.26E	25	-0.27E	25
(SPh)$_2$	-0.24E	25	-0.30E	25	-0.29E	25	-0.31E	25	-0.32E	25
(Bzl)$_2$	-0.21E	25	-0.27E	25	-0.26E	25	-0.28E	25	-0.29E	25
Cl$_2$	-0.088E	25	-0.15E	25	-0.14E	25	-0.16E	25	-0.17E	25
F$_2$	-0.069E	25	-0.13E	25	-0.12E	25	-0.14E	25	-0.15E	25
H, NH$_2$	-0.11E	14								
(NH$_2$)$_2$	-0.31E	25	-0.37E	25	-0.36E	25	-0.38E	25	-0.39E	25
H, OH	-0.10E	19								

	−0.19E	n	−0.25E	n	−0.24E	n	−0.26E	n	−0.27E	n
$(OH)_2$	−0.19E	25								
O_2	−0.088	14								
Tetracoordinate	$M(O)Z^1Z^2$									
$(O)H_2$			0.045E	25	0.053E	25	0.031E	25	0.022E	25
$O(CF_3)_2$			0.12E	25	0.13E	25	0.11E	25	0.10E	25
$O(CN)_2$			0.15E	25	0.16E	25	0.13E	25	0.13E	25
$(O)Me_2$			−0.036	14	−0.021E	25	−0.044E	25	−0.052E	25
$O(OMe)_2$			−0.033E	14	−0.026E	25	−0.048E	25	−0.057E	25
$O(SMe)_2$			−0.49E	25	−0.041E	25	−0.063E	25	−0.072E	25
$O(C_2H)_2$			0.024E	25	0.032E	25	0.010E	25	0.001E	25
$O(CH{=}CH_2)_2$			−0.046E	25	−0.037E	25	−0.060E	25	−0.068E	25
$(O)Et_2$			−0.036E	14	−0.022E	25	−0.044E	25	−0.053E	25
$O(OEt)_2$			−0.033	14	−0.035E	25	−0.057E	25	−0.066E	25
$O(SEt)_2$			−0.055E	25	−0.047E	25	−0.069E	25	−0.078E	25
$O(NMe_2)_2$			0.20E	25	−0.19E	25	−0.21E	25	−0.22E	25
$O(CH_2CH{=}CH_2)_2$			−0.031E	25	−0.022E	25	−0.045E	25	−0.053E	25
$(O)Pr_2$			−0.036E	16						
$(O)Bu_2$			−0.036E	16						
$O(C_6F_5)_2$			0.066E	25	0.074E	25	0.052E	25	0.043E	25
$(O)Ph_2$			−0.21	14	−0.041E	25	−0.063E	25	−0.072E	25
$O(OPh)_2$			−0.014E	25	−0.005E	25	−0.028E	25	−0.036E	25
$O(SPh)_2$			−0.066E	25	−0.058E	25	−0.080E	25	−0.089E	25
$(O)Bzl_2$			−0.036E	25	−0.026E	25	−0.049E	25	−0.057E	25
$(O)Cl_2$			0.087E	25	0.095E	25	0.073E	25	0.064E	25
$(O)F_2$			0.11E	25	0.12E	25	0.093E	25	0.084E	25
$O(NH_2)_2$			−0.13E	25	−0.12E	25	−0.15E	25	−0.15E	25
$O(OH)_2$			−0.011	25	−0.003E	25	−0.025E	25	−0.034E	25
$(S)Ph_2$			−0.12	19						
Pentacoordinate	MZ_4									
Ph			−0.43E	25	−0.42E	25	−0.40E	25	−0.45E	25
F			−0.12E	25	−0.11E	25	−0.13E	25	−0.14E	25

TABLE 4. Values of σ_R. Group 15 substituents

Z^1, Z^2	$M=N$	Ref.	P	Ref.	As	Ref.	Sb	Ref.	Bi	Ref.
Dicoordinate	$M=MZ$									
H	0.23E	30	−0.42E	30	−0.34E	30	−0.58E	30	−0.67E	30
CF_3	0.24E	30	−0.25E	30	−0.16E	30	−0.40E	30	−0.50E	30
CN	0.41E	30	−0.23E	30	−0.14E	30	−0.38E	30	−0.48E	30
Me	0.12E	30	−0.53E	30	−0.43E	30	−0.68E	30	−0.78E	30
OMe	0.10E	30	−0.55E	30	−0.45E	30	−0.70E	30	−0.79E	30
NMe_2	−0.38E	30	−1.01E	30	−0.93E	30	−1.17E	30	−1.26E	30
t-Bu	0.04E	30	−0.54E	30	−0.45E	30	−0.69E	30	−0.79E	30
Ph	0.09E	30	−0.43E	30	−0.56E	30	−0.81E	30	−0.90E	30
F	0.46E	30	−0.18E	30	−0.09E	30	−0.33E	30	−0.43E	30
NH_2	−0.14E	30	−0.79E	30	−0.69E	30	−0.94E	30	−1.03E	30
OH	0.16E	30	−0.55E	30	−0.39E	30	−0.63E	30	−0.73E	30
$M^1{=}CH$	−0.04E	30	−0.25E	30	−0.22E	30	−0.29E	30	−0.34E	30
$M^1{=}CMe$	−0.11E	30	−0.35E	30	−0.31E	30	−0.39E	30	−0.44E	30
$M^1{=}CPh$	−0.09E	30	−0.47E	30	−0.43E	30	−0.51E	30	−0.56E	30
$M^1{=}PH$					−0.02E	30	−0.52E	30	−0.58E	30
$M^1{=}AsH$		16	−0.39E	30	−0.42E	30	−0.49E	30	−0.54E	30
$M^1{=}SbH$		16	−0.47E	30	−0.46E	30			−0.63E	30
$M^1{=}BiH$		16	−0.51E	30			−0.61E	30		
M_3	−0.31	16								
$M{=}O$	0.40	30								
Tricoordinate	MZ^1Z^2									
H, H	−0.80	16	−0.53E	30	−0.55E	30	−0.51E	30	−0.50E	30
H, $CONH_2$	−0.47	16								
H, CHO	−0.40	16								
$(CF_3)_2$	−0.13E	30	−0.03E	30	−0.04E	30	0.00E	30	0.01E	30
$(CN)_2$	0.17E	30	0.11E	30	0.11E	30	0.14E	30	0.15E	30
H, Ac	−0.35	16								
Me_2	−0.88	16	−0.54E	30	−0.55E	30	−0.51E	30	−0.50E	30
$(OMe)_2$	−0.57E	30	−0.47E	30	−0.48E	30	−0.44E	30	−0.43E	30
$(SMe)_2$	−0.50E	30	−0.40E	30	−0.41E	30	−0.37E	30	−0.36E	30
$(C_2H)_2$	−0.32E	30	−0.22E	30	−0.23E	30	−0.11E	30	−0.18E	30
$(CH{=}CH_2)_2$	−0.52E	30	−0.41E	30	−0.43E	30	−0.24E	30	−0.24E	30
Et_2	−0.66E	30	−0.53E	30	−0.53E	30	−0.51E	30	−0.49E	30
$(OEt)_2$	−0.59E	30	−0.49E	30	−0.50E	30	−0.46E	30	−0.45E	30
$(SEt)_2$	−0.54E	30	−0.40E	30	−0.45E	30	−0.41E	30		
$(NMe_2)_2$	−0.78E	30	−0.68E	30	−0.69E	30	−0.65E	30	−0.64E	30
$(CH_2CH{=}CH_2)_2$	−0.63E	30	−0.52E	30	−0.54E	30	−0.50E	30	−0.48E	30
H, Ph	−0.86	16								

H, Bz	−0.47	16	−0.06E	16	−0.15E	30	−0.12E	30	−0.11E	30
(C₆F₅)₂	−0.25E	30	−0.43E	30	−0.44E	30	−0.32E	30	−0.32E	30
Ph₂	−0.60E	30	−0.36E	30	−0.37E	30	−0.33E	30	−0.32E	30
(OPh)₂	−0.46E	30	−0.37E	30	−0.39E	30	−0.35E	30	−0.36E	30
(SPh)₂	−0.48E	30	−0.50E	30	−0.51E	30	−0.47E	30	−0.47E	30
(Bzl)₂	−0.60E	30	−0.17E	30	−0.18E	30	−0.14E	30	−0.13E	30
Cl₂	−0.27E	30	−0.19E	30	−0.21E	30	−0.17E	30	−0.15E	30
F₂	−0.29E	30								
H, NH₂										
(NH₂)₂	−0.76E	30	−0.67E	30	−0.68E	30	−0.64E	30	−0.63E	30
H, OH										
(OH)₂	−0.53E	30	−0.43E	30	−0.44E	30	−0.40E	30	−0.39E	30
O₂	0.10	16								
M(O)Z¹Z²										
Tetracoordinate										
(OH)₂			0.08E	30	0.07E	30	0.10E	30	0.13E	30
O(CF₃)₂			0.60E	30	0.58E	30	0.62E	30	0.64E	30
O(CN)₂			0.74E	30	0.73E	30	0.76E	30	0.78E	30
(O)Me₂			0.08E	30	0.08E	30	0.12E	30	0.12E	30
O(OMe)₂			0.21	16	0.14E	30	0.18E	30	0.20E	30
O(SMe)₂			0.23E	30	0.39E	30	0.25E	30	0.27E	30
O(C₂H)₂			0.41E	30	0.19E	30	0.43E	30	0.45E	30
O(CH=CH₂)₂			0.21E	30	0.09E	30	0.23E	30	0.25E	30
(O)Et₂			0.09E	16	0.12E	30	0.13E	30	0.13E	30
O(OEt)₂			0.24	30	0.17E	30	0.16E	30	0.18E	30
O(SEt)₂			0.19E	30	−0.07E	30	0.21E	30	0.23E	30
O(NMe)₂			−0.05E	30	0.09E	30	−0.03E	30	−0.01E	30
O(CH₂CH=CH₂)₂			0.10E	30			0.12E	30	0.14E	30
(O)Pr₂			0.28	16						
(O)Bu₂			0.48E	30						
O(C₆F₅)₂			0.34	16	0.47E	30	0.50E	30	0.52E	30
O(OPh)₂			0.27E	30	0.18E	30	0.22E	30	0.24E	30
O(SPh)₂			0.25E	30	0.25E	30	0.29E	30	0.31E	30
(O)Bzl₂			0.13E	30	0.23E	30	0.27E	30	0.29E	30
(O)Cl₂			0.46E	30	0.11E	30	0.15E	30	0.16E	30
(O)F₂			0.43E	30	0.44E	30	0.48E	30	0.50E	30
O(NH₂)₂			−0.04E	30	0.42E	30	0.46E	30	0.47E	30
O(OH)₂			0.19E	30	−0.06E	30	−0.02E	30	0.00E	30
(S)Ph₂			0.22	16	0.18E	30	0.22E	30	0.23E	30
MZ₄										
Pentacoordinate										
Ph			−0.23E	30	−0.34E	30	−0.57E	30	−0.47E	30
F			−0.04E	30	−0.10E	30	−0.01E	30	0.00E	30

TABLE 5. Values of σ_R^0. Group 15 substituents

Z^1, Z^2	$M=N$	Ref.	P	Ref.	As	Ref.	Sb	Ref.	Bi	Ref.
Dicoordinate	M = MZ									
H	0.19E	31	-0.32E	31	-0.25E	31	-0.44E	31	-0.51E	31
CF$_3$	0.13E	31	-0.20E	31	-0.14E	31	-0.32E	31	-0.40E	31
CN	0.30E	31	-0.20E	31	-0.13E	31	-0.31E	31	-0.38E	31
Me	0.09E	31	-0.40E	31	-0.33E	31	-0.52E	31	-0.59E	31
OMe	0.05E	31	-0.45E	31	-0.38E	31	-0.57E	31	-0.63E	31
NMe$_2$	-0.30E	31	-0.79E	31	-0.72E	31	-0.90E	31	-0.98E	31
t-Bu	0.04E	31	-0.41E	31	-0.34E	31	-0.53E	31	-0.60E	31
Ph	0.07E	31	-0.31E	31	-0.42E	31	-0.61E	31	-0.68E	31
F	0.30E	31	-0.19E	31	-0.12E	31	-0.31E	31	-0.38E	31
NH$_2$	-0.13E	31	-0.62E	31	-0.55E	31	-0.74E	31	-0.82E	31
OH	0.09E	31	-0.40E	31	-0.33E	31	-0.52E	31	-0.59E	31
M^1=CH	-0.04	31	-0.17E	31	-0.15E	31	-0.20E	31	-0.23E	31
M^1=CMe	-0.10	31	-0.26E	31	-0.23E	31	-0.28E	31	-0.32E	31
M^1=CPh	-0.08E	31	-0.36E	31	-0.33E	31	-0.38E	31	-0.42E	31
M^1=PH					-0.30E		-0.35E		-0.38E	
M^1=AsH			-0.01E	31		31	-0.31E		-0.34E	
M^1=SbH			-0.39E	31	-0.37E	31			-0.46E	
M^1=BiH			-0.45E	31	-0.42E		-0.48E			
M=C=O	-0.17	17								
M=C=S	-0.07	17								
M$_3$	-0.21E	17								
M=O	0.32	31								
Tricoordinate	MZ^{1}Z^2									
H, H	-0.42	16	-0.38E	16	-0.40E	32	-0.35E	32	-0.33E	32
H, CHO	-0.24	16								
(CF$_3$)$_2$	-0.14E	32	-0.00E	32	-0.02E	32	0.03E	32	0.04E	32
(CN)$_2$	0.19E	32	0.10E	32	0.09E	32	0.13E	32	0.15E	32
H, Ac	-0.25	16	-0.35	17						
Me$_2$	-0.44	16			-0.35E	32	-0.31E	32	-0.29E	32
(OMe)$_2$	-0.39E	32	-0.28E	32	-0.30E	32	-0.25E	32	-0.23E	32
(SMe)$_2$	-0.33E	32	-0.21E	32	-0.23E	32	-0.18E	32	-0.17E	32
(C$_2$H)$_2$	-0.22E	32	-0.10E	32	-0.12E	32	-0.01E	32	-0.06E	32
(CH=CH$_2$)$_2$	-0.35E	32	-0.23E	32	-0.25E	32	-0.08E	32	-0.08E	32
Et$_2$	-0.49E	32	-0.33E	32	-0.35E	32	-0.31E	32	-0.29E	32
(OEt)	-0.41E	33	-0.29E	33	-0.31E	32	-0.26E	32	-0.25E	32

$(SEt)_2$	−0.30E	**32**	−0.24E	**32**	−0.26E	**32**	−0.21E	**32**	−0.20E	**32**
$(NMe_2)_2$	−0.48E	**32**	−0.36E	**32**	−0.38E	**32**	−0.30E	**32**	−0.32E	**32**
$(CH_2CH{=}CH_2)_2$	−0.44E	**32**	−0.32E	**32**	−0.34E	**32**	−0.33E	**32**	−0.28E	**32**
$(C_6F_5)_2$	−0.18E	**32**	−0.00E	**32**	−0.08E	**32**	−0.04E	**32**	−0.02E	**32**
Ph_2	−0.40E	**32**	−0.23E	**32**	−0.25E	**32**	−0.14E	**32**	−0.13E	**32**
$(OPh)_2$	−0.32E	**32**	−0.20E	**32**	−0.22E	**32**	−0.17E	**32**	−0.15E	**32**
$(SPh)_2$	−0.30E	**32**	−0.18E	**32**	−0.20E	**32**	−0.16E	**32**	−0.16E	**32**
$(Bzl)_2$	−0.42E	**32**	−0.30E	**32**	−0.32E	**32**	−0.27E	**32**	−0.26E	**32**
Cl_2	−0.21E	**32**	−0.10E	**32**	−0.11E	**32**	−0.07E	**32**	−0.05E	**32**
F_2	−0.24E	**32**	−0.13E	**32**	−0.14E	**32**	−0.10E	**32**	−0.08E	**32**
$(NH_2)_2$	−0.50E	**32**	−0.39E	**32**	−0.41E	**32**	−0.36E	**32**	−0.34E	**32**
$(OH)_2$	−0.37E	**32**	−0.26E	**32**	−0.27E	**32**	−0.23E	**32**	−0.22E	**32**
O_2	0.10	16								
Tetracoordinate	**$M(O)Z^1Z^2$**									
$(O)H_2$			0.01E	**32**	−0.01E	**32**	0.03E	**32**	0.05E	**32**
$O(CF_3)_2$			0.39E	**32**	0.37E	**32**	0.41E	**32**	0.43E	**32**
$O(CN)_2$			0.49E	**32**	0.48E	**32**	0.52E	**32**	0.53E	**32**
$(O)Me_2$			0.08E	17	0.04E	**32**	0.08E	**32**	0.09E	**32**
$O(OMe)_2$			0.15E	17	0.09E	**32**	0.14E	**32**	0.15E	**32**
$O(SMe)_2$			0.17E	**32**	0.16E	**32**	0.20E	**32**	0.22E	**32**
$O(C_2H)_2$			0.28E	**32**	0.27E	**32**	0.31E	**32**	0.33E	**32**
$O(CH{=}CH_2)_2$			0.16E	**32**	0.14E	**32**	0.19E	**32**	0.20E	**32**
$(O)Et_2$			0.06E	**32**	0.04E	**32**	0.09E	**32**	0.10E	**32**
$O(OEt)_2$			0.09E	**32**	0.08E	**32**	0.12E	**32**	0.14E	**32**
$O(SEt)_2$			0.15E	**32**	0.13E	**32**	0.17E	**32**	0.19E	**32**
$O(NMe_2)_2$			0.03E	**32**	0.01E	**32**	0.05E	**32**	0.07E	**32**
$O(CH_2CH{=}CH_2)_2$			0.06E	**32**	0.04E	**32**	0.09E	**32**	0.11E	**32**
$O(C_6F_5)_2$			0.32E	**32**	0.31E	**32**	0.35E	**32**	0.36E	**32**
$(O)Ph_2$			0.14E	**32**	0.13E	**32**	0.18E	**32**	0.20E	**32**
$O(OPh)_2$			0.19E	**32**	0.17E	**32**	0.22E	**32**	0.23E	**32**
$O(SPh)_2$			0.20	**32**	0.19E	**32**	0.23E	**32**	0.25E	**32**
$(O)Bzl_2$			0.09E	**32**	0.07E	**32**	0.11E	**32**	0.12E	**32**
$(O)Cl_2$			0.29E	**32**	0.27E	**32**	0.32E	**32**	0.33E	**32**
$(O)F_2$			0.26E	**32**	0.24E	**32**	0.29E	**32**	0.31E	**32**
$O(NH_2)_2$			0.00E	**32**	−0.02E	**32**	0.03E	**32**	0.04E	**32**
$O(OH)_2$			0.13E	**32**	0.12E	**32**	0.16E	**32**	0.17E	**32**
Pentacoordinate	**MZ_4**									
Ph			0.01E	**32**	−0.09E	**32**	−0.27E	**32**	−0.18E	**32**
F			−0.01E	**32**	−0.06E	**32**	0.02E	**32**	0.04E	**32**

TABLE 6. Values of σ_R^+. Group 15 substituents

Z^1, Z^2	$M{=}N$	Ref.	P	Ref.	As	Ref.	Sb	Ref.	Bi	Ref.
Dicoordinate	$M{=}MZ$									
H	0.19E	30	−0.64E	32	−0.52E	32	−0.83E	32	−0.85E	32
CF_3	0.45E	32	−0.34E	32	−0.23E	32	−0.55E	32	−0.67E	32
CN	0.56E	32	−0.29E	32	−0.17E	32	−0.49E	32	−0.61E	32
Me	0.10E	32	−0.75E	32	−0.62E	32	−0.96E	32	−1.09E	32
OMe	0.18E	32	−0.67E	32	−0.54E	32	−0.87E	32	−0.99E	32
NMe_2	−0.51E	32	−1.35E	32	−1.23E	32	−1.56E	32	−1.69E	32
t-Bu	−0.01E	32	−0.76E	32	−0.67E	32	−0.97E	32	−1.10E	32
Ph	0.07E	32	−0.68E	32	−0.80E	32	−1.14E	32	−1.26E	32
F	0.74E	32	−0.11E	32	−0.01E	32	−0.31E	32	−0.44E	32
NH_2	−0.16E	32	−1.02E	32	−0.90E	32	−1.21E	32	−1.34E	32
OH	0.28E	32	−0.56E	32	−0.44E	32	−0.76E	32	−0.89E	32
$M^1{=}CH$	−0.08E	32	−0.44E	32	−0.38E	32	−0.50E	32	−0.57E	32
$M^1{=}CMe$	−0.17E	32	−0.55E	32	−0.49E	32	−0.64E	32	−0.71E	32
$M^1{=}CPh$	−0.14E	32	−0.69E	32	−0.65E	32	−0.77E	32	−0.85E	32
$M^1{=}PH$					−0.58E	32	−0.71E	32	−0.77E	32
$M^1{=}AsH$			−0.18E	32	−0.69E	32	−0.65E	32	−0.71E	32
$M^1{=}SbH$			−0.74E	32	−0.75E	32			−0.88E	32
$M^1{=}BiH$			−0.81E	32			−0.89E	32		
$M{=}C{=}O$	0.04E	17								
$M{=}C{=}S$	−0.24E	17								
M_3	−0.47E	17								
$M{=}O$	0.43E	32								
Tricoordinate	MZ^1Z^2									
H, H	−1.10	16	−0.85E	16	−0.85E	32	−0.86E	32	−0.86E	32
H, Me	−1.16	16								
$(CF_3)_2$	−0.09E	32	−0.15E	32	−0.15E	32	−0.14E	32	−0.14E	32
$(CN)_2$	−0.04E	32	0.05E	32	0.05E	32	0.05E	32	0.05E	32
H, Ac	−0.47	16								
Me_2	−1.22	16	−1.02E	16	−1.03E	32	−1.00E	32	−1.01E	32
$(OMe)_2$	−0.94E	32	−0.94E	32	−0.93E	32	−0.93E	32	−0.93E	32
$(SMe)_2$	−0.88E	32	−0.87E	32	−0.89E	32	−0.88E	32	−0.88E	32
$(C_2H)_2$	−0.55E	32	−0.54E	32	−0.54E	32	−0.44E	32	−0.53E	32
$(CH{=}CH_2)_2$	−0.90E	32	−0.89E	32	−0.89E	32	−0.72E	32	−0.74E	32
Et_2	−0.99E	32	−0.99E	32	−0.91E	32	−1.00E	32	−1.00E	32
$(OEt)_2$	−0.99E	32	−0.98E	32	−0.98E	32	−0.97E	32	−0.97E	32

$(CH_2CH=CH_2)_2$	−1.01E	32	−1.00E	32	−1.00E	32	−0.99E	32	−0.99E	32
H, Bz	−0.34	16								
$(C_6F_5)_2$	−0.40E	32	−0.30E	32	−0.38E	32	−0.39E	32	−0.39E	32
Ph_2	−1.03E	32	−0.93E	32	−0.93E	32	−0.82E	32	−0.84E	32
$(OPh)_2$	−0.79E	32	−0.78E	32	−0.78E	32	−0.77E	32	−0.77E	32
$(SPh)_2$	−0.89E	32	−0.88E	32	−0.88E	32	−0.87E	32	−0.91E	32
$(Bzl)_2$	−0.97E	32	−0.97E	32	−0.97E	32	−0.96E	32	−0.97E	32
Cl_2	−0.39E	32	−0.38E	32	−0.38E	32	−0.37E	32	−0.37E	32
F_2	−0.38E	32	−0.37E	32	−0.37E	32	−0.36E	32	−0.36E	32
$(NH_2)_2$	−1.34E	32	−1.34E	32	−1.34E	32	−1.33E	32	−1.33E	32
$(OH)_2$	−0.86E	32	−0.86E	32	−0.85E	32	−0.84E	32	−0.85E	32
O_2	0.10	16								
Tetracoordinate	$M(O)Z^1Z^2$									
$(O)H_2$	0.25E		0.25E	32	0.24E	32	0.26E	32	0.26E	32
$O(CF_3)_2$	0.96E		0.96E	32	0.97E	32	0.97E	32	0.97E	32
$O(CN)_2$	1.17E		1.18E	32	1.16E	32	1.18E	32	1.18E	32
$(O)Me_2$	0.10E		0.12E	32	0.13E	32	0.12E	32	0.12E	32
$O(OMe)_2$	0.19E		0.19E	32	0.19E	32	0.19E	32	0.19E	32
$O(SMe)_2$	0.24E		0.24E	32	0.24E	32	0.24E	32	0.24E	32
$O(C_2H)_2$	0.58E		0.57E	32	0.58E	32	0.58E	32	0.58E	32
$O(CH=CH_2)_2$	0.23E		0.22E	32	0.23E	32	0.23E	32	0.23E	32
$(O)Et_2$	0.11E		0.13E	32	0.14E	32	0.13E	32	0.13E	32
$O(OEt)_2$	0.17E		0.15E	32	0.18E	32	0.15E	32	0.15E	32
$O(SEt)_2$	0.19E		0.18E	16	0.19E	32	0.19E	32	0.19E	32
$O(NMe_2)_2$	−0.34E		−0.34E	32	−0.33E	32	−0.33E	32	−0.33E	32
$O(CH_2CH=CH_2)_2$	0.13E		0.13E	32	0.13E	32	0.14E	32	0.14E	32
$O(C_6F_5)_2$	0.73E		0.74E	32	0.73E	32	0.73E	32	0.73E	32
$(O)Ph_2$	0.27		0.21E	32	0.21E	32	0.21E	32	0.21E	32
$O(OPh)_2$	0.35E		0.35E	32	0.35E	32	0.35E	32	0.35E	32
$O(SPh)_2$	0.24E		0.23E	32	0.24E	32	0.24E	32	0.24E	32
$(O)Bzl_2$	0.15E		0.10E	32	0.16E	32	0.15E	32	0.15E	32
$(O)Cl_2$	0.74E		0.74E	32	0.74E	32	0.75E	32	0.75E	32
$(O)F_2$	0.76E		0.76E	32	0.76E	32	0.76E	32	0.76E	32
$O(NH_2)_2$	−0.21E		−0.20E	32	−0.22E	32	−0.20E	32	0.20E	32
$O(OH)_2$	0.27E		0.28E	32	0.28E	32	0.27E	32	0.27E	32
Pentacoordinate	MZ_4									
Ph	−0.95E		−1.06E	32	−1.35E	32	−1.26E	32		
F	−0.18E		−0.24E	32	−0.17E	32	−0.17E	32		

TABLE 7. Values of σ_R^-. Group 15 substituents

Z^1, Z^2	M=N	Ref.	P	Ref.	As	Ref.	Sb	Ref.	Bi	Ref.
Dicoordinate	M=MZ									
H	0.43E	33	−0.26E	33	−0.17E	33	−0.42E	33	−0.52E	33
CF₃	0.18E	33	−0.15E	33	−0.06E	33	−0.30E	33	−0.40E	33
CN	0.48E	33	−0.16E	33	−0.07E	33	−0.31E	33	−0.40E	33
Me	0.27E	33	−0.38E	33	−0.28E	33	−0.53E	33	−0.63E	33
OMe	0.13E	33	−0.51E	33	−0.42E	33	−0.67E	33	−0.75E	33
NMe₂	−0.27E	33	−0.91E	33	−0.82E	33	−1.06E	33	−1.15E	33
t-Bu	0.20E	33	−0.39E	33	−0.29E	33	−0.54E	33	−0.64E	33
Ph	0.23E	33	−0.22E	33	−0.40E	33	−0.64E	33	−0.73E	33
F	0.41E	33	−0.22E	33	−0.14E	33	−0.38E	33	−0.48E	33
NH₂	−0.07E	33	−0.71E	33	−0.61E	33	−0.88E	33	−0.97E	33
OH	0.17E	33	−0.46E	33	−0.38E	33	−0.62E	33	−0.72E	33
M¹=CH	0.07E	33	−0.04E	33	−0.02E	33	−0.07E	33	−0.11E	33
M¹=CMe	−0.01E	33	−0.16E	33	−0.14E	33	−0.18E	33	−0.22E	33
M¹=CPh	0.02E	33	−0.31E	33	−0.27E	33	−0.33E	33	−0.37E	33
M¹=PH					−0.24E	33	−0.30E	33	−0.33E	33
M¹=AsH			0.23E	33			−0.23E	33	−0.26E	33
M¹=SbH			−0.37E	33	−0.35E	33			−0.44E	33
M¹=BiH			−0.45E	33	−0.42E	33	−0.49E	33		
M=C=O	−0.30E	17								
M=C=S	0.02E	17								
M₃	−0.11E	17								
M=O	0.60E	33								
Tricoordinate	MZ¹Z²									
H, H	−0.55	16	−0.30E	33	−0.33E	33	−0.22E	33	−0.19E	33
(CF₃)₂	−0.12E	33	−0.19E	33	0.15E	33	0.24E	33	0.28E	33
(CN)₂	0.54E	33	0.31E	33	0.28E	33	0.36E	33	0.40E	33
(C₂H)₂	−0.09E	33	0.15E	33	0.12E	33	0.31E	33	0.24E	33
(CH=CH₂)₂	−0.19E	33	0.05E	33	0.01E	33	0.28E	33	0.30E	33
Me₂	−0.30	16	−0.12E	33	−0.14E	33	−0.06E	33	−0.04E	33
(OMe)₂	−0.28E	33	−0.03E	33	−0.07E	33	0.02E	33	0.05E	33
(SMe)₂	−0.17E	33	0.07E	33	0.05E	33	0.14E	33	0.18E	33
Et₂	−0.45E	33	−0.13E	33	−0.20E	33	−0.06E	33	−0.02E	33
(OEt)₂	−0.28E	33	−0.04E	33	−0.08E	33	0.01E	33	0.05E	33
(SEt)₂	−0.20E	33	0.04E	33	0.00E	33	0.09E	33	0.13E	33
(CH₂CH=CH₂)₂	−0.21E	33	0.04E							

$(NMe_2)_2$	−0.21E	33	0.03E	33	−0.01E	33	0.09E	33	0.15E	33
H, Ph	−0.49	16								
$(C_6F_5)_2$	−0.08E	33	0.26E	33	0.13E	33	0.21E	33	0.25E	33
Ph_2	−0.25E	33	0.05E	33	0.02E	33	0.21E	33	0.22E	33
$(OPh)_2$	−0.18E	33	0.06E	33	0.02E	33	0.11E	33	0.15E	33
$(SPh)_2$	−0.10E	33	0.14E	33	0.10E	33	0.19E	33	0.19E	33
$(Bzl)_2$	−0.31E	33	−0.07E	33	−0.11E	33	−0.02E	33	0.01E	33
Cl_2	−0.15E	33	0.09E	33	0.05E	33	0.14E	33	0.18E	33
F_2	−0.22E	33	0.03E	33	−0.01E	33	0.08E	33	0.12E	33
$(NH_2)_2$	−0.31E	33	−0.08E	33	−0.12E	33	−0.03E	33	0.01E	33
$(OH)_2$	−0.26E	33	−0.03E	33	−0.06E	33	0.03E	33	0.06E	33
O_2	0.37E	16								
Tetracoordinate	**$M(O)Z^1Z^2$**									
$(O)H_2$			0.02E	33	−0.01E	33	0.07E	33	0.11E	33
$O(CF_3)_2$			0.49E	33	0.45E	33	0.54E	33	0.58E	33
$O(CN)_2$			0.60E	33	0.57E	33	0.67E	33	0.69E	33
$(O)Me_2$			0.17E	33	0.14E	33	0.23E	33	0.25E	33
$O(OMe)_2$			0.26E	33	0.22E	33	0.32E	33	0.35E	33
$O(SMe)_2$			0.37E	33	0.34E	33	0.43E	33	0.47E	33
$O(C_2H)_2$			0.45E	33	0.42E	33	0.51E	33	0.54E	33
$O(CH{=}CH_2)_2$			0.35E	33	0.31E	33	0.40E	33	0.44E	33
$(O)Et_2$			0.18E	33	0.15E	33	0.24E	33	0.27E	33
$O(OEt)_2$			0.24E	33	0.22E	33	0.29E	33	0.34E	33
$O(SEt)_2$			0.34E	33	0.30E	33	0.39E	33	0.43E	33
$O(NMe_2)_2$			0.33E	33	0.29E	33	0.38E	33	0.42E	33
$O(CH_2CH{=}CH_2)_2$			0.19E	33	0.15E	33	0.24E	33	0.28E	33
$O(C_6F_5)_2$			0.45E	33	0.43E	33	0.51E	33	0.54E	33
$(O)Ph_2$			0.30E	32	0.30E	33	0.40E	33	0.43E	33
$O(OPh)_2$			0.35E	33	0.32E	33	0.41E	33	0.44E	33
$O(SPh)_2$			0.43E	33	0.40E	33	0.49E	33	0.53E	33
$(O)Bzl_2$			0.23E	33	0.23E	33	0.28E	33	0.31E	33
$(O)Cl_2$			0.39E	33	0.35E	33	0.44E	33	0.48E	33
$(O)F_2$			0.31E	33	0.28E	33	0.38E	33	0.41E	33
$O(NH_2)_2$			0.21E	33	0.17E	33	0.27E	33	0.30E	33
$O(OH)_2$			0.26E	33	0.23E	33	0.32E	33	0.35E	33
Pentacoordinate	**MZ_4**									
Ph			0.57E	33	0.42E	33	0.18E	33	0.32E	33
F			0.19E	33	0.10E	33	0.24E	33	0.28E	33

TABLE 8. Values of $\sigma_R^{\oplus}$. Group 15 substituents

Z^1, Z^2	M = N	Ref.	P	Ref.	As	Ref.	Sb	Ref.	Bi	Ref.
Dicoordinate	M = MZ									
H	−0.02E	34	−0.99E	34	−0.85E	34	−1.22E	34	−1.37E	34
CF$_3$	0.48E	34	−0.59E	34	−0.45E	34	−0.84E	34	−0.98E	34
CN	0.52E	34	−0.52E	34	−0.37E	34	−0.75E	34	−0.91E	34
Me	−0.09E	34	−1.12E	34	−0.97E	34	−1.38E	34	−1.53E	34
OMe	0.09E	34	−0.95E	34	−0.80E	34	−1.19E	34	−1.34E	34
NMe$_2$	−0.81E	34	−1.82E	34	−1.68E	34	−2.08E	34	−2.23E	34
t-Bu	−0.23E	34	−1.13E	34	−1.03E	34	−1.39E	34	−1.54E	34
Ph	−0.11E	34	−1.07E	34	−1.19E	34	−1.31E	34	−1.75E	34
F	0.82E	34	−0.21E	34	−0.07E	34	−0.46E	34	−0.62E	34
NH$_2$	−0.35E	34	−1.40E	34	−1.25E	34	−1.63E	34	−1.78E	34
OH	0.23E	34	−0.80E	34	−0.66E	34	−1.04E	34	−1.20E	34
M^1=CH	−0.28E	34	−0.77E	34	−0.70E	34	−0.86E	34	−0.95E	34
M^1=CMe	−0.38E	34	−0.90E	34	−0.82E	34	−1.02E	34	−1.12E	34
M^1=CPh	−0.35E	34	−1.05E	34	−1.00E	34	−1.17E	34	−1.27E	34
M^1=PH					−0.92E	34	−1.09E	34	−1.17E	34
M^1=AsH			−0.50E	34	−1.04E	34	−1.02E	34	−1.10E	34
M^1=SbH			−1.11E	34	−1.10E	34			−1.29E	34
M^1=BiH			−1.19E	34			−1.28E	34		
M=C=O	0.05E	17								
M=C=S	−0.50E	17								
M$_3$	−0.67	11								
M=O	0.28E	34								
M≡C	−0.30	11								
Tricoordinate	MZ^1Z^2									
H, H	−1.05	11	−1.15	11	−1.28E	34	−1.34E	34	−1.36E	34
H, Me	−1.23	11	−1.31	11						
(CF$_3$)$_2$	−0.23E	34	−0.43E	34	−0.42E	34	−0.44E	34	−0.45E	34
(CN)$_2$	−0.40E	34	−0.18E	34	−0.16E	34	−0.20E	34	−0.22E	34
H, Ac	−0.75	11								
Me$_2$	−1.38	11	−1.47E	11	−1.63E	34	−1.61E	34	−1.64E	34
(OMe)$_2$	−1.45E	34	−1.53E	34	−1.51E	34	−1.53E	34	−1.55E	34
(SMe)$_2$	−1.39E	34	−1.47E	34	−1.50E	34	−1.51E	34	−1.53E	34
(C$_2$H)$_2$	−0.93E	34	−1.01E	34	−1.00E	34	−0.91E	34	−1.03E	34
(CH=CH$_2$)$_2$	−1.42E	34	−1.50E	34	−1.48E	34	−1.32E	34	−1.35E	34
Et$_2$	−1.46E	34	−1.57E	34	−1.43E	34	−1.61E	34	−1.63E	34

(NMe₂)₂	−2.23E	34	−2.33E	34	−2.32E	34	−2.37E	34	−2.39E	34
(CH₂CH=CH₂)₂	−1.52E	34	−1.60E	34	−1.58E	34	−1.60E	34	−1.61E	34
H, Ph	−1.40	11	−1.41	11						
(C₆F₅)₂	−0.71E	34	−0.69E	34	−0.76E	34	−0.79E	34	−0.81E	34
Ph₂	−1.58E	34	−1.55E	34	−1.53E	34	−1.45E	34	−1.48E	34
(OPh)₂	−1.24E	34	−1.32E	34	−1.31E	34	−1.33E	34	−1.34E	34
(SPh)₂	−1.44E	34	−1.51E	34	−1.50E	34	−1.52E	34	−1.57E	34
(Bzl)₂	−1.48E	34	−1.56E	34	−1.55E	34	−1.56E	34	−1.59E	34
Cl₂	−0.66E	34	−0.75E	34	−0.73E	34	−0.75E	34	−0.77E	34
F₂	−0.62E	34	−0.71E	34	−0.69E	34	−0.71E	34	−0.72E	34
(NH₂)₂	−2.02E	34	−2.11E	34	−2.10E	34	−2.12E	34	−2.13E	34
(OH)₂	−1.33E	34	−1.42E	34	−1.39	34	−1.41E	34	−1.43E	34
O₂	0.08	11								
Tetracoordinate M(O)Z¹Z²										
(O)H₂			0.24E	34	0.24E	34	0.21E	34	0.21E	34
O(CF₃)₂			1.12E	34	1.14E	34	1.12E	34	1.10E	34
O(CN)₂			1.40E	34	1.42E	34	1.35E	34	1.38E	34
(O)Me₂			−0.05E	34	0.00E	11	−0.03E	34	−0.05E	34
O(OMe)₂			0.06E	11	0.06E	34	0.03E	34	0.02E	34
O(SMe)₂			0.09E	34	0.09E	34	0.07E	34	0.06E	34
O(C₂H)₂			0.56E	34	0.57E	34	0.54E	34	0.53E	34
O(CH=CH₂)₂			0.07E	34	0.09E	34	0.06E	34	0.05E	34
(O)Et₂			−0.04E	34	0.00E	11	−0.02E	34	−0.04E	34
O(OEt)₂			0.03E	11	0.00E	34	0.02E	34	−0.04E	34
O(SEt)₂			0.02E	34	0.02E	34	0.00E	34	−0.01E	34
O(NMe₂)₂			−0.78E	34	−0.76E	34	−0.78E	34	−0.80E	34
O(CH₂CH=CH₂)₂			−0.01E	34	0.00E	34	−0.03E	34	−0.03E	34
O(C₆F₅)₂			0.79E	34	0.81E	34	0.77E	34	0.76E	34
(O)Ph₂			0.16E	34	0.06E	34	0.03E	34	0.02E	34
O(OPh)₂			0.25E	34	0.27E	34	0.24E	34	0.23E	34
O(SPh)₂			0.06E	34	0.06E	34	0.04E	34	0.03E	34
(O)Bzl₂			0.01E	34	−0.06E	34	−0.01E	34	−0.02E	34
(O)Cl₂			0.83E	34	0.84E	34	0.82E	34	0.80E	34
(O)F₂			0.89E	34	0.90E	34	0.86E	34	0.85E	34
O(NH₂)₂			−0.52E	34	−0.51E	34	−0.56E	34	−0.54E	34
O(OH)₂			0.17E	34	0.19E	34	−0.17E	34	0.14E	34
Pentacoordinate MZ₄										
Ph			1.19E	34	1.04E	34	0.85E	34	1.00E	34
F			0.36E	34	0.26E	34	0.42E	34	0.47E	34

TABLE 9. Values of $\sigma_R^\ominus$. Group 15 substituents

Z^1, Z^2	M=N	Ref.	P	Ref.	As	Ref.	Sb	Ref.	Bi	Ref.
Dicoordinate	M=MZ									
H	0.51E	35	−0.03E	35	0.03E	35	−0.16E	35	−0.23E	35
CF_3	0.06E	35	−0.02E	35	0.05E	35	−0.13E	35	−0.21E	35
CN	0.41E	35	−0.06E	35	0.01E	35	−0.16E	35	−0.23E	35
Me	0.34E	35	−0.14E	35	−0.07E	35	−0.24E	35	−0.31E	35
OMe	0.12E	35	−0.35E	35	−0.28E	35	−0.46E	35	−0.52E	35
NMe_2	−0.10E	35	−0.56E	35	−0.50E	35	−0.67E	35	−0.74E	35
t-Bu	0.29E	35	−0.15E	35	−0.06E	35	−0.25E	35	−0.32E	35
Ph	0.29E	35	0.03E	35	−0.14E	35	−0.31E	35	−0.37E	35
F	0.23E	35	−0.22E	35	−0.16E	35	−0.33E	35	−0.40E	35
NH_2	0.00E	35	−0.33E	35	−0.38E	35	−0.58E	35	−0.65E	35
OH	0.12E	35	−0.45E	35	−0.27E	35	−0.45E	35	−0.52E	35
$M^1=CH$	0.16E	35	0.16E	35	0.16E	35	0.15E	35	0.13E	35
$M^1=CMe$	0.08E	35	0.05E	35	0.06E	35	0.07E	35	0.05E	35
$M^1=CPh$	0.11E	35	−0.08E	35	−0.04E	35	−0.06E	35	−0.08E	35
$M^1=PH$					−0.03E	35	−0.05E	35	−0.06E	35
$M^1=AsH$			0.41E	35			0.01E	35	0.00E	35
$M^1=SbH$			−0.13E	35	−0.13E	35			−0.16E	35
$M^1=BiH$			−0.20E	35	−0.19E	35	−0.22E	35		
M=C=O	−0.39	17								
M=C=S	0.15	17								
M_3	0.08	17								
M≡C	0.62E	35								
Tricoordinate	MZ^1Z^2									
H, H	−0.30E	35	0.02E	35	0.03E	35	0.11E	35	0.16E	35
H, $CONH_2$		16								
H, CHO		16								
$(CF_3)_2$	−0.09E	35	0.34E	35	0.29E	35	0.40E	35	0.45E	35
$(CN)_2$	0.75E	35	0.41E	35	0.37E	35	0.48E	35	0.53E	35
H, Ac	−0.09E	17								
H, Et	−0.22E	17								
Me_2	0.05E	17	0.31E	35	0.30E	35	0.38E	35	0.42E	35
$(OMe)_2$	0.08E	35	0.39E	35	0.34E	35	0.45E	35	0.50E	35
$(SMe)_2$	0.19E	35	0.50E	35	0.48E	35	0.60E	35	0.65E	35
$(C_2H)_2$	0.15E	35	0.47E	35	0.41E	35	0.62E	35	0.58E	35
$(CH=CH_2)_2$	0.17E	35	0.48E	35	0.43E	35	0.71E	35	0.74E	35
Et_2	−0.12E	35	0.29E	35	0.16E	35	0.38E	35	0.43E	35
$(OEt)_2$	0.09E	35	0.40E	35	0.35E	35	0.46E	35	0.51E	35
$(SEt)_2$	0.18E	35	0.49E	35	0.44E	35	0.55E	35	0.60E	35

$(NMe_2)_2$	0.39E	35	0.70E	35	0.65E	35	0.79E	35	0.84E	35
$(CH_2CH{=}CH_2)_2$	0.02E	35	0.33E	35	0.28E	35	0.39E	35	0.44E	35
H, Ph		16								
H, Bz		16								
$(C_6F_5)_2$	0.09E	35	0.50E	35	0.37E	35	0.47E	35	0.52E	35
Ph_2	0.15E	35	0.50E	35	0.45E	35	0.66E	35	0.69E	35
$(OPh)_2$	0.13E	35	0.44E	35	0.39E	35	0.50E	35	0.56E	35
$(SPh)_2$	0.28E	35	0.60E	35	0.54E	35	0.66E	35	0.68E	35
$(Bzl)_2$	0.05E	35	0.36E	35	0.31E	35	0.42E	35	0.46E	35
Cl_2	0.00E	35	0.32E	35	0.26E	35	0.38E	35	0.43E	35
F_2	−0.09E	35	0.22E	35	0.17E	35	0.29E	35	0.34E	35
$(NH_2)_2$	0.21E	35	0.51E	35	0.46E	35	0.57E	35	0.62E	35
$(OH)_2$	0.06E	35	0.36E	35	0.32E	35	0.43E	35	0.47E	35
O_2	0.41	17								
Tetracoordinate $M(O)Z^1Z^2$										
$(O)H_2$			−0.06E	35	−0.10E	35	0.00E	35	0.06E	35
$O(CF_3)_2$			0.24E	35	0.19E	35	0.30E	35	0.35E	35
$O(CN)_2$			0.29E	35	0.25E	35	0.38E	35	0.40E	35
$O(Me)_2$			0.20E	35	0.15E	35	0.27E	35	0.30E	35
$O(OMe)_2$			0.28E	16	0.23E	35	0.35E	35	0.40E	35
$O(SMe)_2$			0.41E	35	0.36E	35	0.48E	35	0.52E	35
$O(C_2H)_2$			0.36E	35	0.31E	35	0.43E	35	0.48E	35
$O(CH{=}CH_2)_2$			0.38E	35	0.33E	35	0.45E	35	0.49E	35
$(O)Et_2$			0.21E	35	0.16E	35	0.28E	35	0.32E	35
$O(OEt)_2$			0.26E	35	0.24E	35	0.32E	35	0.41E	35
$O(SEt)_2$			0.38E	35	0.34E	35	0.46E	35	0.60E	35
$O(NMe_2)_2$			0.61E	35	0.56E	35	0.67E	35	0.72E	35
$O(CH_2CH{=}CH_2)_2$			0.21E	16	0.16E	35	0.28E	35	0.33E	35
$(O)Bu_2$										
$O(C_6F_5)_2$			0.29E	35	0.26E	35	0.36E	35	0.41E	35
$O(Ph)_2$			0.29E	16	0.33E	35	0.45E	35	0.49E	35
$O(OPh)_2$			0.33E	35	0.28E	35	0.40E	35	0.45E	35
$O(SPh)_2$			0.49E	35	0.44E	35	0.56E	35	0.61E	35
$(O)Bzl_2$			0.25E	35	0.27E	35	0.32E	35	0.36E	35
$O(Cl)_2$			0.20E	35	0.15E	35	0.27E	35	0.32E	35
$(O)F_2$			0.10E	35	0.05E	35	0.18E	35	0.23E	35
$O(NH_2)_2$			0.39E	35	0.34E	35	0.48E	35	0.50E	35
$O(OH)_2$			0.24E	16	0.21E	35	0.32E	35	0.36E	35
$(S)Ph_2$										
Pentacoordinate MZ_4										
Ph			1.19E	35	1.04E	35	0.85E	35	1.00	35
F			0.36E	35	0.26E	35	0.42E	35	0.47E	35

TABLE 10. Values of σ_m. Group 15 substituents

Z^1, Z^2	M=N	Ref.	P	Ref.	As	Ref.	Sb	Ref.	Bi	Ref.
Dicoordinate										
	M = MZ									
H	0.38E	36	−0.24E	36	−0.16E	36	−0.38E	36	−0.46E	36
CF$_3$	0.60E	36	0.01E	36	0.09E	36	−0.14E	36	−0.22E	36
CN	0.66E	36	0.07E	36	0.16E	36	−0.07E	36	−0.17E	36
Me	0.29E	36	−0.31E	36	−0.23E	36	−0.46E	36	−0.54E	36
OMe	0.33E	36	−0.27E	36	−0.18E	36	−0.41E	36	−0.50E	36
NMe$_2$	0.03E	36	−0.57E	36	−0.49E	36	−0.71E	36	−0.80E	36
t-Bu	0.21E	36	−0.32E	36	−0.24E	36	−0.46E	36	−0.55E	36
Ph	0.30E	36	−0.24E	36	−0.25E	36	−0.48E	36	−0.56E	36
F	0.61E	36	0.01E	36	0.09E	36	−0.01E	36	−0.22E	36
NH$_2$	0.15E	36	−0.45E	36	−0.37E	36	−0.59E	36	−0.68E	36
OH	0.38E	36	−0.22E	36	−0.14E	36	−0.36E	36	−0.45E	36
M^1=CH	0.17E	36	−0.07E	36	−0.05E	36	−0.12E	36	−0.15E	36
M^1=CMe	0.07E	36	−0.15E	36	−0.11E	36	−0.19E	36	−0.23E	36
M^1=CPh	0.08E	36	−0.16E	36	−0.13E	36	−0.19E	36	−0.24E	36
M^1=PH					−0.21E	36	−0.29E		−0.32E	
M^1=AsH			−0.19E	36	−0.30E	36	−0.23E		−0.27E	36
M^1=SbH			−0.33E	36	−0.35E	36	−0.43E	36	−0.41E	36
M^1=BiH			−0.39E	36						
M=C=O	0.34	17								
M=C=S	0.46	17								
M$_3$	0.27	17								
M=O	0.53E	36								
Tricoordinate										
	MZ^{1}Z^2									
H, H	−0.21E	16	−0.12E	36	−0.16E	36	−0.15E	36	−0.15E	36
H, CONH$_2$	0.06E	16								
H, CHO	0.19E	16								
(CF$_3$)$_2$	0.46E	36	0.41E	36	0.41E	36	0.40E	36	0.40E	36
(CN)$_2$	0.72E	36	0.58E	36	0.58E	36	0.57E	36	0.56E	36
H, Ac	0.16E	16								
Me$_2$	−0.16E	16	−0.25E	36	−0.25E	36	−0.25E	36	−0.25E	36
(OMe)$_2$	−0.11E	36	−0.14E	36	−0.13E	36	−0.13E	36	−0.14E	36
(SMe)$_2$	−0.08E	36	−0.10E	36	−0.11E	36	−0.12E	36	−0.12E	36
(C$_2$H)$_2$	0.13E	36	0.11E	36	0.12E	36	0.14E	36	0.11E	36
(CH=CH$_2$)$_2$	−0.14E	36	−0.16E	36	−0.16E	36	−0.11E	36	−0.11E	36
Et$_2$	−0.20E	36	−0.24E	36	−0.21E	36	−0.25E	36	−0.25E	36
(OEt)$_2$	−0.14E	36	−0.16E	36	−0.16E	36	−0.17E	36	−0.18E	36
(SEt)$_2$	−0.13E	36	−0.16E	36	−0.15E	36	−0.15E	36	−0.14E	36
(NMe$_2$)$_2$	−0.51E	36	−0.53E	36	−0.53E	36	−0.55E	36	−0.55E	36

Substituent	Val	N	Val	N	Val	N	Val	N	Val	N
$(CH_2CH=CH_2)_2$	-0.21E	36	-0.23E	36	-0.23E	36	-0.24E	36	-0.24E	36
H, Ph	-0.02E	16								
H, Bz	0.11E	16								
$(C_6F_5)_2$	0.24E	36	0.26E	36	0.22E	36	0.22E	36	0.22E	36
Ph_2	-0.14E	36	-0.18E	36	-0.17E	36	-0.14E	36	-0.16E	36
$(OPh)_2$	0.01E	36	-0.01E	36	0.00E	36	-0.02E	36	-0.01E	36
$(SPh)_2$	-0.09E	36	-0.12E	36	-0.12E	36	-0.12E	36	-0.13E	36
$(Bzl)_2$	-0.19E	36	-0.22E	36	-0.21E	36	-0.23E	36	-0.23E	36
Cl_2	0.31E	36	0.27E	36	0.28E	36	0.27E	36	0.27E	36
F_2	0.34E	36	0.31E	36	0.32E	36	0.30E	36	0.30E	36
$(NH_2)_2$	-0.41E	36	-0.44E	36	-0.43E	36	-0.44E	36	-0.44E	36
$(OH)_2$	-0.04E	36	-0.06E	36	-0.06E	36	-0.06E	36	-0.08E	36
O_2	0.74	16								
Tetracoordinate $M(O)Z^1Z^2$										
$(O)H_2$			0.44E	36	0.44E	36	0.43E	36	0.44E	36
$(O)(CF_3)_2$			0.99E	36	0.99E	36	0.98E	36	0.98E	36
$(O)(CN)_2$			1.15E	36	1.17E	36	1.15E	36	1.15E	36
$(O)Me_2$			0.34E	36	0.34E	36	0.34E	36	0.33E	36
$(O)(OMe)_2$			0.46	16	0.46E	36	0.45E	36	0.44E	36
$(O)(SMe)_2$			0.48E	36	0.48E	36	0.47E	36	0.46E	36
$(O)(C_2H)_2$			0.70E	36	0.69E	36	0.69E	36	0.69E	36
$(O)(CH=CH_2)_2$			0.43E	36	0.42E	36	0.42E	36	0.42E	36
$(O)Et_2$			0.32E	36	0.35E	36	0.34E	36	0.34E	36
$(O)(OEt)_2$			0.47	16	0.42E	36	0.43E	36	0.41E	36
$(O)(SEt)_2$			0.43E	36	0.44E	36	0.42E	36	0.42E	36
$(O)(NMe_2)_2$			0.05E	36	0.05E	36	0.04E	36	0.04E	36
$(O)(CH_2CH=CH_2)_2$			0.36E	36	0.36E	36	0.35E	36	0.35E	36
$(O)Bu_2$			0.37E	16						
$(O)(C_6F_5)_2$			0.81E	36	0.81E	36	0.80E	36	0.80E	36
$(O)Ph_2$			0.40E	16	0.42E	36	0.41E	36	0.40E	36
$(O)(OPh)_2$			0.58E	36	0.57E	36	0.57E	36	0.57E	36
$(O)(SPh)_2$			0.47E	36	0.47E	36	0.46E	36	0.45E	36
$(O)Bzl_2$			0.37E	36	0.36E	36	0.36E	36	0.36E	36
$(O)Cl_2$			0.86E	36	0.87E	36	0.85E	36	0.86E	36
$(O)F_2$			0.90E	36	0.91E	36	0.89E	36	0.89E	36
$(O)(NH_2)_2$			0.15E	36	0.15E	36	0.14E	36	0.15E	36
$(O)(OH)_2$			0.53E	36	0.53E	36	0.52E	36	0.51E	36
$(S)Ph_2$			0.38	16						
Pentacoordinate MZ_4										
Ph			-0.72E	36	-0.77E	36	-1.07E	36	-0.47E	36
F			0.50E	36	-0.10E	36	0.50E	36	0.49E	36

TABLE 11. Values of σ_p. Group 15 substituents

Z^1, Z^2	M=N	Ref.	P	Ref.	As	Ref.	Sb	Ref.	Bi	Ref.
Dicoordinate	M = MZ									
H	0.53E	37	−0.51E	37	−0.37E	37	−0.75E	37	−0.90E	37
CF₃	0.75E	37	−0.16E	37	−0.02E	37	−0.40E	37	−0.54E	37
CN	0.93E	37	−0.08E	37	0.07E	37	−0.31E	37	−0.47E	37
Me	0.36E	37	−0.65E	37	−0.50E	37	−0.90E	37	−1.05E	37
OMe	0.39E	37	−0.62E	37	−0.47E	37	−0.87E	37	−1.00E	37
NMe₂	−0.22E	37	−1.22E	37	−1.08E	37	−1.47E	37	−1.62E	37
t-Bu	0.23E	37	−0.66E	37	−0.53E	37	−0.91E	37	−1.06E	37
Ph	0.36E	37	−0.52E	37	−0.61E	37	−1.00E	37	−1.14E	37
F	0.90E	37	−0.11E	37	0.03E	37	−0.35E	37	−0.50E	37
NH₂	0.06E	37	−0.96E	37	−0.81E	37	−1.20E	37	−1.35E	37
OH	0.48E	37	−0.53E	37	−0.39E	37	−0.77E	37	−0.92E	37
M¹=CH	0.15E	37	−0.23E	37	−0.18E	37	−0.30E	37	−0.37E	37
M¹=CMe	−0.01E	37	−0.37E	37	−0.31E	37	−0.44E	37	−0.52E	37
M¹=CPh	0.02E	37	−0.46E	37	−0.41E	37	−0.53E	37	−0.60E	37
M¹=PH					−0.46E	37	−0.59E	37	−0.65E	37
M¹=AsH			−0.42E	37			−0.50E	37	−0.57E	37
M¹=SbH			−0.66E	37	−0.61E	37			−0.80E	37
M¹=BiH			−0.76E	37	−0.70E	37				
M=C=O	0.25E	17					−0.84E	37		
M=C=S	0.38E	17								
M₃	0.08E	17								
M=O	0.79E	37								
Tricoordinate	MZ¹Z²									
H, H	−0.63	16	−0.46E	37	−0.51E	37	−0.48E	37	−0.46E	37
H, CONH₂	−0.24E	16								
H, CHO	−0.07E	16								
(CF₃)₂	0.37E	37	0.39E	37	0.39E	37	0.40E	37	0.41E	37
(CN)₂	0.83E	37	0.65E	37	0.65E	37	0.66E	37	0.66E	37
H, Ac	−0.07E	16								
Me₂	−0.71E	16	−0.60E	37	−0.61E	37	−0.57E	37	−0.57E	37
(OMe)₂	−0.47E	37	−0.44E	37	−0.44E	37	−0.41E	37	−0.41E	37
(SMe)₂	−0.40E	37	−0.35E	37	−0.37E	37	−0.35E	37	−0.35E	37
(C₂H)₂	−0.07E	37	−0.03E	37	−0.03E	37	0.08E	37	0.00E	37
(CH=CH₂)₂	−0.47E	37	−0.42E	37	−0.44E	37	−0.26E	37	−0.26E	37
Et₂	−0.63E	37	−0.58E	37	−0.54E	37	−0.57E	37	−0.56E	37
(OEt)₂	−0.52E	37	−0.47E	37	−0.48E	37	−0.46E	37	−0.46E	37
(SEt)₂	−0.47E	37	−0.43E	37	−0.43E	37	−0.41E	37	−0.39E	37
(NMe₂)₂	−1.01E	37	−0.96E	37	−0.97E	37	−0.96E	37	−0.96E	37

Ligand		Ref		Ref		Ref		Ref		Ref
$(CH_2CH=CH_2)_2$	-0.61E	37	-0.57E	37	-0.58E	37	-0.55E	37	-0.54E	37
H, Ph	-0.56E	16								
H, Bz	-0.19E	16								
$(C_6F_5)_2$	0.08E	37	0.22E	37	0.13E	37	0.14E	37	0.15E	37
Ph_2	-0.53E	37	-0.45E	37	-0.45E	37	-0.34E	37	-0.36E	37
$(OPh)_2$	-0.29E	37	-0.24E	37	-0.24E	37	-0.23E	37	-0.22E	37
$(SPh)_2$	-0.40E	37	-0.35E	37	-0.36E	37	-0.34E	37	-0.36E	37
$(Bz)_2$	-0.57E	37	-0.54E	37	-0.54E	37	-0.52E	37	-0.52E	37
Cl_2	0.13E	37	0.17E	37	0.17E	37	0.18E	37	0.19E	37
F_2	0.15E	37	0.18E	37	0.18E	37	0.20E	37	0.21E	37
$(NH_2)_2$	-0.90E	37	-0.86E	37	-0.86E	37	-0.85E	37	-0.84E	37
$(OH)_2$	-0.38E	37	-0.34E	37	-0.34E	37	-0.032	37	-0.33E	37
O_2	0.77	16								
Tetracoordinate $M(O)Z^1Z^2$										
$(O)H_2$			0.50E	37	0.48E	37	0.50E	37	0.52E	37
$O(CF_3)_2$			1.37E	37	1.36E	37	1.38E	37	1.39E	37
$O(CN)_2$			1.63E	37	1.63E	37	1.64E	37	1.65E	37
$(O)Me_2$			0.39E	37	0.39E	37	0.41E	37	0.41E	37
$O(OMe)_2$			0.59	16	0.55E	37	0.57E	37	0.57E	37
$O(SMe)_2$			0.63E	37	0.61E	37	0.63E	37	0.64E	37
$O(C_2H)_2$			0.96E	37	0.95E	37	0.97E	37	0.98E	37
$O(CH=CH_2)_2$			0.56E	37	0.55E	37	0.57E	37	0.58E	37
$(O)Et_2$			0.38E	37	0.40E	37	0.42E	37	0.42E	37
$O(OEt)_2$			0.57	16	0.50E	37	0.53E	37	0.52E	37
$O(SEt)_2$			0.55E	37	0.55E	37	0.56E	37	0.57E	37
$O(NMe_2)_2$			0.02E	37	0.01E	37	0.03E	37	0.03E	37
$O(CH_2CH=CH_2)_2$			0.42E	37	0.41E	37	0.43E	37	0.44E	37
$(O)Bu_2$			0.53E	16						
$O(C_6F_5)_2$			1.11E	37	1.11E	37	1.12E	37	1.13E	37
$(O)Ph_2$			0.60E	16	0.53E	37	0.55E	37	0.56E	37
$O(OPh)_2$			0.75E	37	0.74E	37	0.76E	37	0.77E	37
$O(SPh)_2$			0.63E	37	0.62E	37	0.64E	37	0.64E	37
$(O)Bz_2$			0.45E	37	0.43E	37	0.46E	37	0.46E	37
$(O)Cl_2$			1.15E	37	1.15E	37	1.16E	37	1.17E	37
$(O)F_2$			1.17E	37	1.17E	37	1.18E	37	1.19E	37
$O(NH_2)_2$			0.13E	37	0.12E	37	0.13E	37	0.15E	37
$O(OH)_2$			0.65E	37	0.65E	37	0.67E	37	0.66E	37
$(S)Ph_2$			0.50E	16						
Pentacoordinate MZ_4										
Ph			-0.85E	37	-0.97E	37	-1.43E	37	-0.76E	37
F			0.47E	37	-0.17E	37	0.50E	37	0.50E	37

TABLE 12. Values of σ_p^0. Group 15 substituents

Z^1, Z^2	M=N	Ref.	P	Ref.	As	Ref.	Sb	Ref.	Bi	Ref.
Dicoordinate	M=MZ									
H	0.52E	**38**	−0.38E	**38**	−0.26E	**38**	−0.58E	**38**	−0.71E	**38**
CF_3	0.67E	**38**	−0.08E	**38**	−0.04E	**38**	−0.28E	**38**	−0.41E	**38**
CN	0.84E	**38**	−0.01E	**38**	0.11E	**38**	−0.21E	**38**	−0.35E	**38**
Me	0.36E	**38**	−0.50E	**38**	−0.37E	**38**	−0.70E	**38**	−0.83E	**38**
OMe	0.37E	**38**	−0.49E	**38**	−0.37E	**38**	−0.70E	**38**	−0.81E	**38**
NMe_2	−0.10E	**38**	−0.95E	**38**	−0.84E	**38**	−1.16E	**38**	−1.28E	**38**
t-Bu	0.25E	**38**	−0.50E	**38**	−0.39E	**38**	−0.71E	**38**	−0.84E	**38**
Ph	0.36E	**38**	−0.37E	**38**	−0.44E	**38**	−0.77E	**38**	−0.89E	**38**
F	0.76E	**38**	−0.09E	**38**	0.03E	**38**	−0.29E	**38**	−0.42E	**38**
NH_2	0.10E	**38**	−0.76E	**38**	−0.63E	**38**	−0.96E	**38**	−1.09E	**38**
OH	0.43E	**38**	−0.42E	**38**	−0.30E	**38**	−0.62E	**38**	−0.75E	**38**
M^1=CH	0.18E	**38**	−0.13E	**38**	−0.09E	**38**	−0.19E	**38**	−0.24E	**38**
M^1=CMe	0.04E	**38**	−0.25E	**38**	−0.20E	**38**	−0.30E	**38**	−0.36E	**38**
M^1=CPh	0.06E	**38**	−0.31E	**38**	−0.27E	**38**	−0.36E	**38**	−0.42E	**38**
M^1=PH					−0.34E	**38**	−0.44E	**38**	−0.48E	**38**
M^1=AsH			−0.30E	**38**			−0.36E	**38**	−0.42E	**38**
M^1=SbH			−0.51E	**38**	−0.47E	**38**			−0.62E	**38**
M^1=BiH			−0.60E	**38**	−0.55E	**38**	−0.66E	**38**		
M=C=O	0.24E	17								
M=C=S	0.46E	17								
M_3	0.20E	17								
M=O	0.73E	**38**								
Tricoordinate	$\mathbf{MZ^1Z^2}$									
H, H	−0.40E	**38**	−0.27E	**38**	−0.32E	**38**	−0.28E	**38**	−0.27E	**38**
$(CF_3)_2$	0.40E	**38**	0.45E	**38**	0.45E	**38**	0.46E	**38**	0.48E	**38**
$(CN)_2$	0.89E	**38**	0.68E	**38**	0.67E	**38**	0.68E	**38**	0.69E	**38**
H, Ac	0.00E	17								
H, Et	−0.39E	17								
Me_2	−0.44	20	−0.36E	**38**	−0.37E	**38**	−0.34E	**38**	−0.34E	**38**
$(OMe)_2$	−0.26E	**38**	−0.21E	**38**	−0.22E	**38**	−0.19E	**38**	−0.19E	**38**
$(SMe)_2$	−0.20E	**38**	−0.14E	**38**	−0.16E	**38**	−0.13E	**38**	−0.13E	**38**
$(C_2H)_2$	0.06E	**38**	0.12E	**38**	0.12E	**38**	0.20E	**38**	0.14E	**38**
$(CH=CH_2)_2$	−0.27E	**38**	−0.21E	**38**	−0.22E	**38**	−0.07E	**38**	−0.08E	**38**
Et_2	−0.41E	**38**	−0.35E	**38**	−0.34E	**38**	−0.34E	**38**	−0.33E	**38**
$(OEt)_2$	[illegible]	[illegible]	[illegible]	[illegible]	[illegible]	[illegible]	[illegible]	[illegible]	[illegible]	[illegible]

(NMe₂)₂	−0.68E	38	−0.63E	38	−0.64E	38	−0.62E	38	−0.62E	38
(CH₂CH=CH₂)₂	−0.39E	38	−0.34E	38	−0.35E	38	−0.32E	38	−0.31E	38
(C₆F₅)₂	0.19E	38	0.31E	38	0.24E	38	0.25E	38	0.26E	38
Ph₂	−0.29E	38	−0.23E	38	−0.23E	38	−0.14E	38	−0.15E	38
(OPh)₂	−0.10E	38	−0.05E	38	−0.05E	38	−0.04E	38	−0.02E	38
(SPh)₂	−0.19E	38	−0.13E	38	−0.15E	38	−0.12E	38	−0.13E	38
(Bzl)₂	−0.36E	38	−0.31E	38	−0.31E	38	−0.30E	38	−0.30E	38
Cl₂	0.23E	38	0.28E	38	0.27E	38	0.28E	38	0.30E	38
F₂	0.25E	38	0.29E	38	0.29E	38	0.30E	38	0.31E	38
(NH₂)₂	−0.60E	38	−0.55E	38	−0.56E	38	−0.54E	38	−0.53E	38
(OH)₂	−0.18E	38	−0.13E	38	−0.14E	38	−0.11E	38	−0.12E	38

Tetracoordinate $M(O)Z^1Z^2$

(O)H₂	0.45E	38	0.44E	38	0.46E	38	0.48E	38
O(CF₃)₂	1.20E	38	1.18E	38	1.21E	38	1.22E	38
O(CN)₂	1.41E	38	1.42E	38	1.43E	38	1.44E	38
(O)Me₂	0.38E	38	0.38E	38	0.40E	38	0.41E	38
O(OMe)₂	0.51E	38	0.53E	38	0.55E	38	0.56E	38
O(SMe)₂	0.60E	38	0.59E	38	0.61E	38	0.62E	38
O(C₂H)₂	0.87E	38	0.85E	38	0.87E	38	0.89E	38
O(CH=CH₂)₂	0.54E	38	0.52E	38	0.55E	38	0.56E	38
(O)Et₂	0.37E	38	0.39E	38	0.41E	38	0.42E	38
O(OEt)₂	0.43E	38	0.49E	38	0.52E	38	0.52E	38
O(SEt)₂	0.54E	38	0.54E	38	0.55E	38	0.56E	38
O(NMe₂)₂	0.12E	38	0.11E	38	0.13E	38	0.13E	38
O(CH₂CH=CH₂)₂	0.41E	38	0.40E	38	0.42E	38	0.44E	38
O(C₆F₅)₂	0.99E	38	0.98E	38	1.00E	38	1.01E	38
(O)Ph₂	0.45E	38	0.51E	38	0.54E	38	0.54E	38
O(OPh)₂	0.70E	38	0.69E	38	0.71E	38	0.72E	38
O(SPh)₂	0.61E	38	0.60E	38	0.62E	38	0.62E	38
(O)Bzl₂	0.44E	38	0.43E	38	0.44E	38	0.45E	38
(O)Cl₂	1.02E	38	1.02E	38	1.03E	38	1.05E	38
(O)F₂	1.04E	38	1.04E	38	1.05E	38	1.06E	38
O(NH₂)₂	0.19E	38	0.18E	38	0.20E	38	0.22E	38
O(OH)₂	0.62E	38	0.61E	38	0.63E	38	0.63E	38

Pentacoordinate MZ_4

Ph	−0.63E	38	−0.73E	38	−1.14E	38	−0.46E	38
F	0.55E	38	0.11E	38	0.57E	38	0.57E	38

TABLE 13. Values of σ_p^+. Group 15 substituents

Z^1, Z^2	M=N	Ref.	P	Ref.	As	Ref.	Sb	Ref.	Bi	Ref.
Dicoordinate	M=MZ									
H	0.59E	39	−0.99E	39	−0.77E	39	−1.35E	39	−1.59E	39
CF$_3$	1.08E	39	−0.41E	39	−0.20E	39	−0.80E	39	−1.02E	39
CN	1.28E	39	−0.30E	39	0.07E	39	−0.66E	39	−0.90E	39
Me	0.38E	39	−1.20E	39	−0.97E	39	−1.59E	39	−1.82E	39
OMe	0.51E	39	−1.09E	39	−0.85E	39	−1.46E	39	−1.68E	39
NMe$_2$	−0.58E	39	−2.15E	39	−1.93E	39	−2.54E	39	−2.77E	39
t-Bu	0.18E	39	−1.22E	39	−1.03E	39	−1.61E	39	−1.84E	39
Ph	0.37E	39	−1.03E	39	−1.19E	39	−1.81E	39	−2.03E	39
F	1.40E	39	−0.18E	39	0.04E	39	−0.56E	39	−0.79E	39
NH$_2$	−0.06E	39	−1.67E	39	−1.44E	39	−2.03E	39	−2.26E	39
OH	0.66E	39	−0.91E	39	−0.70E	39	−1.29E	39	−1.53E	39
M^1=CH	0.06E	39	−0.59E	39	−0.50E	39	−0.70E	39	−0.82E	39
M^1=CMe	−0.15E	39	−0.80E	39	−0.69E	39	−0.93E	39	−1.06E	39
M^1=CPh	−0.10E	39	−0.96E	39	−0.89E	39	−1.09E	39	−1.22E	39
M^1=PH					−0.90E	39	−1.12E		−1.22E	
M^1=AsH			−0.86E				−0.99E	39	−1.11E	39
M^1=SbH			−1.20E		−1.12E	39			−1.44E	39
M^1=BiH			−0.35E		−1.25E	39	−1.48E	39		
M=C=O	0.33E	17								
M=C=S	0.21E	17								
M$_3$	−0.25E	17								
M=O	1.02E	39								
Tricoordinate	MZ^{1}Z^2									
H, H	−1.3	20	−1.08E	39	−1.12E	39	−1.12E	39	−1.12E	39
(CF$_3$)$_2$	0.33E	39	0.25E	39	0.25E	39	0.26E	39	0.26E	39
(CN)$_2$	0.74E	39	0.65E	39	0.65E	39	0.65E	39	0.64E	39
H, Ac	−0.6	20								
Me$_2$	−1.7	20	−1.38E	39	−1.40E	39	−1.36E	39	−1.37E	39
(OMe)$_2$	−1.17E	39	−1.17E	39	−1.16E	39	−1.15E	39	−1.15E	39
(SMe)$_2$	−1.07E	39	−1.06E	39	−1.09E	39	−1.07E	39	−1.08E	39
(C$_2$H)$_2$	−0.49E	39	−0.48E	39	−0.47E	39	−0.32E	39	−0.46E	39
(CH=CH$_2$)$_2$	−1.15E	39	−1.14E	39	−1.15E	39	−0.88E	39	−0.90E	39
Et$_2$	−1.34E	39	−1.24E	39	−1.24E	39	−1.36E	39	−1.35E	39
(OEt)$_2$	−1.25E	39	−1.24E	39	−1.24E	39	−1.23E	39	−1.24E	39
(?)$_2$	−1.1?E	20	−1.1?E		−1.17E		−1.16E		−1.1?E	39

	Col 0	Ref	Col 1	Ref	Col 2	Ref	Col 3	Ref	Col 4	Ref
,	1.4	20								
H, Bz	−0.6	20								
$(C_6F_5)_2$	−0.22E	39	−0.06E	20	−0.19E	39	−0.20E	39	−0.19E	39
Ph_2	−1.29E	39	−1.20E	20	−1.19E	39	−1.03E	39	−1.07E	39
$(OPh)_2$	−0.87E	39	−0.86E	39	−0.85E	39	−0.85E	39	−0.84E	39
$(SPh)_2$	−1.08E	39	−1.07E	39	−1.08E	39	−1.06E	39	−1.11E	39
$(Bzl)_2$	−1.29E	39	−1.29E	39	−1.29E	39	−1.28E	39	−1.30E	39
Cl_2	−0.14E	39	−0.14E	39	−0.14E	39	−0.13E	39	−0.13E	39
F_2	−0.10E	39	−0.10E	39	−0.10E	39	−0.09E	39	−0.09E	39
$(NH_2)_2$	−1.90E	39	−1.91E	39	−1.90E	39	−1.90E	39	−1.90E	39
$(OH)_2$	−1.00E	39	−1.01E	39	−1.00E	39	−0.98E	39	−1.01E	39
O_2	0.79	20								
Tetracoordinate $M(O)Z^1Z^2$										
$(O)H_2$			0.69E	39	0.68E	39	0.67E	39	0.69E	39
$O(CF_3)_2$			2.05E	39	2.05E	39	2.06E	39	2.07E	39
$O(CN)_2$			2.46E	39	2.48E	39	2.45E	39	2.47E	39
$(O)Me_2$			0.43E	39	0.46E	39	0.46E	39	0.46E	39
$O(OMe)_2$			0.63E	16	0.65E	39	0.66E	39	0.66E	39
$O(SMe)_2$			0.75E	39	0.74E	39	0.75E	39	0.75E	39
$O(C_2H)_2$			1.34E	39	1.33E	39	1.34E	39	1.34E	39
$O(CH{=}CH_2)_2$			0.67E	39	0.67E	39	0.67E	39	0.68E	39
$(O)Et_2$			0.43E	39	0.47E	39	0.48E	39	0.47E	39
$O(OEt)_2$			0.54E	39	0.58E	39	0.62E	39	0.58E	39
$O(SEt)_2$			0.64E	16	0.64E	39	0.64E	39	0.64E	39
$O(NMe_2)_2$			−0.32E	39	−0.33E	39	−0.31E	39	−0.32E	39
$O(CH_2CH{=}CH_2)_2$			0.49E	16	0.48E	39	0.49E	39	0.50E	39
$(O)Bu_2$				16						
$O(C_6F_5)_2$			1.62E	39	1.62E	39	1.61E	39	1.62E	39
$O(Ph)_2$			0.64E	16	0.64E	16	0.65E	39	0.64E	39
$O(OPh)_2$			0.96E	39	0.96E	39	0.96E	39	0.97E	39
$O(SPh)_2$			0.74E	39	0.73E	39	0.74E	39	0.74E	39
$(O)Bzl_2$			0.52E	39	0.47E	39	0.52E	39	0.52E	39
$(O)Cl_2$			1.68E	39	1.68E	39	1.68E	39	1.69E	39
$(O)F_2$			1.73E	39	1.73E	39	1.72E	39	1.73E	39
$O(NH_2)_2$			−0.08E	39	−0.08E	39	−0.10E	39	−0.06E	39
$O(OH)_2$			0.82E	39	0.83E	39	0.83E	39	0.82E	39
Pentacoordinate MZ_4										
Ph			−1.75E	39	−1.93E	39	−2.59E	39	−1.82E	39
F			0.31E	39	−0.40E	39	0.33E	39	0.33E	39

TABLE 14. Values of σ_p^-. Group 15 substituents

Z^1, Z^2	M=N	Ref.	P	Ref.	As	Ref.	Sb	Ref.	Bi	Ref.
Dicoordinate	M=MZ									
H	0.92E	40	−0.37E	40	−0.20E	40	−0.66E	40	−0.84E	40
CF$_3$	0.87E	40	−0.02E	40	0.15E	40	−0.31E	40	−0.49E	40
CN	1.24E	40	0.04E	40	0.22E	40	−0.23E	40	−0.42E	40
Me	0.66E	40	−0.55E	40	−0.38E	40	−0.84E	40	−1.02E	40
OMe	0.55E	40	−0.66E	40	−0.48E	40	−0.95E	40	−1.11E	40
NMe$_2$	−0.07E	40	−1.26E	40	−1.10E	40	−1.55E	40	−1.72E	40
t-Bu	0.51E	40	−0.57E	40	−0.39E	40	−0.85E	40	−1.03E	40
Ph	0.64E	40	−0.33E	40	−0.48E	40	−0.94E	40	−1.10E	40
F	1.04E	40	−0.15E	40	0.01E	40	−0.44E	40	−0.49E	40
NH$_2$	0.20E	40	−1.00E	40	−0.83E	40	−1.30E	40	−1.48E	40
OH	0.62E	40	−0.57E	40	−0.41E	40	−0.86E	40	−1.04E	40
M^1=CH	0.36E	40	0.01E	40	0.05E	40	−0.06E	40	−0.12E	40
M^1=CMe	0.16E	40	−0.18E	40	−0.12E	40	−0.22E	40	−0.30E	40
M^1=CPh	0.20E	40	−0.30E	40	−0.23E	40	−0.34E	40	−0.42E	40
M^1=PH					−0.32E	40	−0.45E	40	−0.50E	40
M^1=AsH	0.15E	17	−0.24E	40	−0.53E	40	−0.33E	40	−0.39E	40
M^1=SbH	0.34E	20	−0.57E	40	−0.65E	40			−0.70E	40
M^1=BiH	0.08E	20	−0.70E	40			−0.78E	40		
M=C=O	1.46E	20								
M=C=S										
M$_3$										
M=O										
Tricoordinate	MZ^{1}Z^2									
H, H	−0.51	40	−0.21E	40	−0.30E	40	−0.18E	40	−0.14E	40
(CF$_3$)$_2$	0.51E	40	0.77E	40	0.74E	40	0.81E	40	0.85E	40
(CN)$_2$	1.51E	40	1.06E	40	1.04E	40	1.11E	40	1.14E	40
Me$_2$	−0.35E	20	−0.17E	20	−0.18E	20	−0.12E	20	−0.09E	20
(OMe)$_2$	−0.16E	40	0.05E	40	0.02E	40	0.10E	40	0.13E	40
(SMe)$_2$	−0.03E	40	0.18E	40	0.15E	40	0.23E	40	0.26E	40
(C$_2$H)$_2$	0.24E	40	0.45E	40	0.43E	40	0.62E	40	0.53E	40
(CH=CH$_2$)$_2$	−0.13E	40	0.09E	40	0.05E	40	0.34E	40	0.36E	40
Et$_2$	−0.44E	40	−0.17E	40	−0.22E	40	−0.12E	40	−0.08E	40
(OEt)$_2$	−0.20E	40	0.02E	40	−0.02E	40	0.06E	40	0.08E	40
(SEt)$_2$	−0.10E	40	0.10E	40	0.07E	40	0.15E	40	0.21E	40

Compound										
(…CH₂)₂	0.34E	40	—	—	—	—	—	—	—	—
(C₆F₅)₂	0.35E	40	0.69E	40	0.54E	40	0.61E	40	0.65E	40
Ph₂	−0.16E	40	0.43	20	0.05E	40	0.25E	40	0.25E	40
(OPh)₂	0.05E	40	0.26E	40	0.24E	40	0.30E	40	0.34E	40
(SPh)₂	0.02E	40	0.23E	40	0.19E	40	0.27E	40	0.27E	40
(Bzl)₂	−0.29E	40	−0.09E	40	−0.12E	40	−0.05E	40	−0.02E	40
Cl₂	0.36E	40	0.56E	40	0.54E	40	0.61E	40	0.65E	40
F₂	0.34E	40	0.54E	40	0.51E	40	0.58E	40	0.62E	40
(NH₂)₂	−0.48E	40	−0.28E	40	−0.31E	40	−0.24E	40	−0.20E	40
(OH)₂	−0.07E	40	0.13E	40	0.10E	40	0.18E	40	0.19E	40
O₂	1.29	20								

Tetracoordinate — M(O)Z¹Z²

Compound										
(O)H₂			0.55E	40	0.51E	40	0.58E	40	0.63E	40
O(CF₃)₂			1.54E	40	1.50E	40	1.58E	40	1.62E	40
O(CN)₂			1.82E	40	1.80E	40	1.88E	40	1.90E	40
(O)Me₂			0.74	20	0.57E	20	0.65E	40	0.67E	40
O(OMe)₂			0.78E	40	0.79E	40	0.87E	40	0.90E	40
O(SMe)₂			0.95E	40	0.92E	40	1.00E	40	1.02E	40
O(C₂H)₂			1.22E	40	1.19E	40	1.27E	40	1.31E	40
O(CH=CH₂)₂			0.86E	40	0.82E	40	0.90E	40	0.94E	40
(O)Et₂			0.59E	40	0.58E	40	0.66E	40	0.69E	40
O(OEt)₂			0.84	20	0.74E	40	0.81E	40	0.85E	40
O(SEt)₂			0.87E	40	0.84E	40	0.91E	40	0.95E	40
O(NMe₂)₂			0.50E	40	0.46E	40	0.54E	40	0.57E	40
O(CH₂CH=CH₂)₂			0.63E	40	0.59E	40	0.68E	40	0.72E	40
O(C₆F₅)₂			1.33E	40	1.31E	20	1.38E	40	1.42E	40
(O)Ph₂			0.88	20	0.81E	40	0.89E	40	0.92E	40
O(OPh)₂			1.03E	40	0.99E	40	1.07E	40	1.11E	40
O(SPh)₂			1.00E	40	0.96E	40	1.05E	40	1.07E	40
(O)Bzl₂			0.68E	40	0.68E	40	0.72E	40	0.75E	40
(O)Cl₂			1.33E	40	1.31E	40	1.38E	40	1.41E	40
(O)F₂			1.30E	40	1.27E	40	1.35E	40	1.39E	40
O(NH₂)₂			0.48E	40	0.44E	40	0.53E	40	0.56E	40
O(OH)₂			0.89E	40	0.87E	20	0.95E	40	0.96E	40
(S)Ph₂			0.63	20						

Pentacoordinate — MZ₄

Compound										
Ph			−0.10E	40	−0.29E	40	−0.86E	40	0.05E	40
F			0.90E	40	0.04E	40	0.96E	40	0.98E	40

$$\sigma_R^+ = 1.05\sigma_{dX} + 2.14\sigma_{eX} - 0.0731 \tag{32}$$

$$\sigma_R^- = 1.13\sigma_{dX} - 1.58\sigma_{dX} + 0.00272 \tag{33}$$

$$\sigma_R^\oplus = 1.15\sigma_{dX} + 3.81\sigma_{eX} - 0.0262 \tag{34}$$

$$\sigma_R^\ominus = 1.01\sigma_{dX} - 3.01\sigma_{dX} - 0.00491 \tag{35}$$

Values of these parameters are reported in Tables 4–9. Estimated values of Hammett σ constants can be calculated from the relationships[17]:

$$\sigma_{mX} = 1.02\sigma_{1X} + 0.385\sigma_{dX} + 0.661\sigma_{eX} + 0.0152 \tag{36}$$

$$\sigma_{pX} = 1.02\sigma_{1X} + 0.989\sigma_{dX} + 0.837\sigma_{eX} + 0.0132 \tag{37}$$

$$\sigma_{pX}^0 = 1.06\sigma_{1X} + 0.796\sigma_{dX} + 0.278\sigma_{eX} + 0.0289 \tag{38}$$

$$\sigma_{pX}^+ = 1.10\sigma_{1X} + 0.61\sigma_{dX} + 2.76\sigma_{eX} + 0.0394 \tag{39}$$

$$\sigma_{pX}^- = 1.35\sigma_{1X} + 1.36\sigma_{dX} - 1.28\sigma_{eX} + 0.0176 \tag{40}$$

Tables 10–14 present values of the Hammett substituent constants.

2. Ionic groups

In earlier work[21] we proposed that ionic substituent constants could be described by the relationship:

$$\sigma_{X^q} = \sigma_{1X^q} + \sigma_{DX^q} + \sigma_{qX^q} \tag{41}$$

where X^q is an ionic substituent and σ_q is a parameter representing the effect of the charge q. We have shown that the pK_a values of oxyacids of the type HOX, where X is $[M(OH)_m O_n]^q$, are well described by equation 23 with q representing the charge on the X group. This result is in accord with equation 41. We conclude that, in general, ionic substituents can be described by simply adding the term $a_q q$ to the equation used to estimate parameter values for uncharged groups. Thus:

$$\sigma_{X^q} = a_\chi \chi_M + \Sigma \sigma_{1X^q} + \Sigma \sigma_{dX^q} + \Sigma \sigma_{eX^q} + a_q q_{X^q} + h \tag{42}$$

B. Electrical Effects of Group 15 Substituents

1. Classification of substituent electrical effects

It has long been the custom to classify substituents as either electron acceptor (electron withdrawing, electron sink), EA, or electron donor (electron releasing, electron source), ED. Actually there is a third category to consider, groups whose electrical effect is not significantly different from zero (NS groups). Most groups vary in the nature of their electrical effect to a greater or lesser extent depending on the electronic demand of the phenomenon being studied, the skeletal group, if any, to which they are bonded and the experimental conditions. Only a few groups are in the same category throughout the range of P_D and η normally encountered. What is needed is a method for characterizing the type of electrical effect of which a group is capable. We have noted above that a plot of the $\sigma_{k'/k,X}$ values for a group with $X = P_D$, $Y = \eta$ and $Z = \sigma_{k'/k}$ produces a surface that characterizes the electrical effect of the X group. We may generate a matrix of these values by calculating them for values of P_D in the range 10 to 90 in increments of 10 and values of η in the range -6 to 6 in increments of 1. The resulting 9 by 13 group matrix has 117 values. We define $\sigma_{k'/k,X}$ values greater than 0.05 as EA, $\sigma_{k'/k,X}$ values less than -0.05 as ED and $\sigma_{k'/k,X}$ values between 0.05 and -0.05 as NS. The variability of the electrical effect of a group can be quantitatively described by the percent of the matrix area in the $P_D-\eta$ plane in which

TABLE 15. Substituent electrical-effect matrices for typical Group 15 substituents[a]

AsH$_2$

η	10	20	30	40	50	60	70	80	90
+6	−0.03	−0.25	−0.52	−0.89	−1.41	−2.18	−3.48	−6.06	−13.81
+5	−0.01	−0.21	−0.45	−0.78	−1.24	−1.93	−3.08	−5.38	−12.28
+4	0.01	−0.16	−0.38	−0.67	−1.07	−1.68	−2.68	−4.70	−10.75
+3	0.02	−0.12	−0.31	−0.55	−0.90	−1.42	−2.29	−4.02	−9.22
+2	0.04	−0.08	−0.23	−0.44	−0.73	−1.17	−1.89	−3.34	−7.69
+1	0.06	−0.03	−0.16	−0.33	−0.56	−0.91	−1.49	−2.66	−6.16
0	0.08	0.01	−0.09	−0.21	−0.39	−0.66	−1.10	−1.98	−4.63
−1	0.10	0.05	−0.01	−0.10	−0.22	−0.40	−0.70	−1.30	−3.10
−2	0.12	0.09	0.06	0.01	−0.05	−0.14	−0.30	−0.62	−1.57
−3	0.14	0.14	0.13	0.13	0.12	0.11	0.09	0.06	−0.04
−4	0.16	0.18	0.20	0.24	0.29	0.37	0.49	0.74	1.49
−5	0.18	0.22	0.28	0.35	0.46	0.62	0.89	1.42	3.02
−6	0.19	0.26	0.35	0.47	0.63	0.88	1.28	2.10	4.55

As(CN)$_2$

η	10	20	30	40	50	60	70	80	90
+6	0.51	0.47	0.41	0.33	0.22	0.06	−0.22	−0.76	−2.40
+5	0.52	0.49	0.44	0.38	0.30	0.18	−0.03	−0.45	−1.70
+4	0.53	0.51	0.48	0.44	0.38	0.29	0.15	−0.14	−1.00
+3	0.54	0.53	0.51	0.49	0.46	0.41	0.33	0.17	−0.30
+2	0.55	0.55	0.54	0.54	0.53	0.53	0.51	0.49	0.41
+1	0.56	0.57	0.58	0.59	0.61	0.64	0.69	0.80	1.11
0	0.57	0.59	0.61	0.64	0.69	0.76	0.88	1.11	1.81
−1	0.57	0.60	0.64	0.70	0.77	0.88	1.06	1.42	2.51
−2	0.58	0.62	0.68	0.75	0.85	0.99	1.24	1.73	3.21
−3	0.59	0.64	0.71	0.80	0.92	1.11	1.42	2.05	3.92
−4	0.60	0.66	0.74	0.85	1.00	1.23	1.60	2.36	4.62
−5	0.61	0.68	0.78	0.90	1.08	1.35	1.79	2.67	5.32
−6	0.62	0.70	0.81	0.96	1.16	1.46	1.97	2.98	6.02

AsMe$_2$

η	10	20	30	40	50	60	70	80	90
+6	−0.14	−0.43	−0.81	−1.31	−2.02	−3.08	−4.85	−8.38	−18.98
+5	−0.11	−0.36	−0.69	−1.13	−1.75	−2.68	−4.22	−7.30	−16.55
+4	−0.08	−0.30	−0.58	−0.95	−1.48	−2.27	−3.59	−6.22	−14.12
+3	−0.05	−0.23	−0.46	−0.77	−1.21	−1.87	−2.96	−5.14	−11.69
+2	−0.02	−0.16	−0.35	−0.59	−0.94	−1.46	−2.33	−4.06	−9.26
+1	0.01	−0.09	−0.23	−0.41	−0.67	−1.06	−1.70	−2.98	−6.83
0	0.04	−0.02	−0.11	−0.23	−0.40	−0.65	−1.07	−1.90	−4.40
−1	0.07	0.04	0.00	−0.05	−0.13	−0.24	−0.44	−0.82	−1.97
−2	0.10	0.11	0.12	0.13	0.14	0.16	0.19	0.26	0.46
−3	0.13	0.18	0.23	0.31	0.41	0.57	0.82	1.34	2.89
−4	0.16	0.25	0.35	0.49	0.68	0.97	1.45	2.42	5.32
−5	0.19	0.31	0.46	0.67	0.95	1.38	2.08	3.50	7.75
−6	0.22	0.38	0.58	0.85	1.22	1.78	2.71	4.58	10.18

As(OMe)$_2$

η	10	20	30	40	50	60	70	80	90
+6	−0.03	−0.31	−0.66	−1.14	−1.80	−2.80	−4.45	−7.77	−17.72
+5	0.00	−0.24	−0.55	−0.96	−1.54	−2.41	−3.85	−6.73	−15.38
+4	0.03	−0.18	−0.44	−0.79	−1.28	−2.01	−3.24	−5.69	−13.04
+3	0.06	−0.11	−0.33	−0.62	−1.02	−1.63	−2.63	−4.65	−10.70
+2	0.08	−0.05	−0.22	−0.44	−0.76	−1.23	−2.03	−3.61	−8.36

(continued)

TABLE 15. (continued)

+1	**0.11**	0.02	*−0.11*	*−0.27*	*−0.50*	*−0.85*	*−1.42*	*−2.57*	*−6.02*
0	**0.14**	**0.08**	0.01	*−0.10*	*−0.24*	*−0.46*	*−0.81*	*1.53*	*−3.68*
−1	**0.17**	**0.15**	**0.12**	**0.08**	0.02	*−0.07*	*−0.21*	*−0.49*	*−1.34*
−2	**0.20**	**0.21**	**0.23**	**0.25**	**0.28**	**0.32**	**0.40**	**0.55**	**1.00**
−3	**0.23**	**0.28**	**0.34**	**0.42**	**0.54**	**0.72**	**1.01**	**1.59**	**3.34**
−4	**0.26**	**0.34**	**0.45**	**0.60**	**0.80**	**1.11**	**1.61**	**2.63**	**5.68**
−5	**0.29**	**0.41**	**0.56**	**0.77**	**1.06**	**1.49**	**2.22**	**3.67**	**8.02**
−6	**0.32**	**0.47**	**0.67**	**0.94**	**1.32**	**1.88**	**2.83**	**4.71**	**10.36**

[a] Values in boldface indicate that the groups behaves as an electron acceptor, values in italics indicate that it acts as an electron donor and values in ordinary typeface indicate that the group exerts no significant electrical effect.

the group is in each category (P_{EA}, P_{ED} and P_0). Approximate measures of these quantities are given by the relationships.

$$P_{EA} = \frac{100 n_{EA}}{n_T}, \; P_0 = \frac{100 n_{NS}}{n_T}, \; P_{ED} = \frac{100 n_{ED}}{n_T} \tag{43}$$

where n_{EA}, n_{NS}, n_{ED} and n_T are the number of EA, the number of NS, the number of ED and the total number of values in the matrix. Matrices for a number of substituents are given in Table 15, values of P_{EA}, P_{ED} and P_0 for many substituents are reported in Table 16. We may now classify groups into seven types:

1. Entirely EA ($P_{EA} = 100$).
2. Predominantly EA ($100 > P_{EA} \geqslant 75$).
3. Largely EA ($75 > P_{EA} \geqslant 50$).
4. Ambielectronic ($50 > P_{EA}$ or P_{ED}).

5. Largely ED ($75 > P_{ED} \geqslant 50$).
6. Predominantly ED ($100 > P_{ED} \geqslant 75$).
7. Entirely ED ($P_{ED} = 100$).

C. The Nature of Group 15 Substituent Electrical Effects

The overall electrical effect of a substituent, as noted above, is a function of its σ_l σ_d and σ_e values. It depends on the nature of the skeletal group G, the active site Y, the type of phenomenon studied and the medium. These are the determinants of the values of P_D and η, which in turn control the mix of σ_l, σ_d and σ_e.

1. Dicoordinate substituents

The results reported in Table 16 show that our model predicts $N=NZ$ and $N=CHZ$ groups should generally fall into the category of predominantly electron-acceptor substituents. The overall electrical effect of doubly bonded phosphorus and arsenic substituents seems to be more dependent on the nature of the Z group than is the case for nitrogen substituents. Electron-acceptor Z groups result in ambielectronic or largely electron-acceptor substituents. Other Z groups result in electron-donor substituents. For antimony and bismuth substituents, our model predicts an overall electron-donor effect in almost all cases. It is interesting to compare these results with those for a number of substituted vinyl (Vi) groups also reported in Table 16. These groups are more likely to have an overall electron-acceptor effect.

2. Tricoordinate substituents

Based on the results set forth in Table 17, all of the tricoordinate substituents of Group 15 elements seem to have an overall electrical effect which is determined by the nature of

TABLE 16. Values of P_{EA}, P_0 and P_{ED} for Group 15 dicoordinate substituents

X	P_{EA}	P_0	P_{ED}	X	P_{EA}	P_0	P_{ED}	X	P_{EA}	P_0	P_{ED}
N=NH	93	2	5	P=PH	18	13	69	As=AsH	27	14	59
N=NCF$_3$	97	1	2	P=PCF$_3$	44	12	44	As=AsCF$_3$	55	11	34
N=NMe	89	3	9	P=PMe	10	13	77	As=AsMe	17	13	70
N=NOMe	100	0	0	P=POMe	0	3	97	As=AsOMe	0	14	86
N=NNMe$_2$	44	9	48	P=PNMe$_2$	0	0	100	As=AsNMe$_2$	0	2	98
N=NPh	88	3	9	P=PPh	24	11	65	As=AsPh	19	12	69
N=CH$_2$	73	6	21	P=CH$_2$	41	11	48	As=CH$_2$	43	13	44
Sb=SbH	9	6	85	Bi=BiH	7	3	90	ViH	53	9	38
Sb=SbCF$_3$	17	19	64	Bi=BiCF$_3$	6	18	76	ViCF$_3$	84	5	11
Sb=SbMe	7	3	90	Bi=BiMe	3	3	94	ViCN	73	5	22
Sb=SbOMe	0	0	100	Bi=BiOMe	0	0	100	ViMe	39	15	46
Sb=SbNMe$_2$	0	0	100	Bi=BiNMe$_2$	0	0	100	ViPh	39	10	50
Sb=SbPh	7	6	87	Bi=BiPh	4	2	94				
Sb=CH$_2$	38	10	51	Bi=CH$_2$	31	15	55				

M. Charton

TABLE 17. Values of P_{EA}, P_0 and P_{ED} for Group 15 tricoordinate substituents

X	P_{EA}	P_0	P_{ED}	X	P_{EA}	P_0	P_{ED}	X	P_{EA}	P_0	P_{ED}
NH$_2$	25	9	66	PH$_2$	40	9	51	AsH$_2$	36	9	55
N(CF$_3$)$_2$	81	4	15	P(CF$_3$)$_2$	79	4	17	As(CF$_3$)$_2$	80	3	16
N(CN)$_2$	87	0	13	P(CN)$_2$	91	1	8	As(CN)$_2$	92	1	7
NMe$_2$	37	7	56	PMe$_2$	39	7	54	AsMe$_2$	39	6	55
N(OMe)$_2$	43	6	51	P(OMe)$_2$	45	7	48	As(OMe)$_2$	46	6	48
N(NMe$_2$)$_2$	32	7	61	P(NMe$_2$)$_2$	36	3	62	As(NMe$_2$)$_2$	36	3	62
NPh$_2$	43	7	50	PPh$_2$	44	7	49	AsPh$_2$	44	7	50
NCl$_2$	67	3	30	PCl$_2$	68	4	27	AsCl$_2$	68	4	27
NO$_2$	95	1	4								
SbH$_2$	39	9	52	BiH$_2$	41	9	50	SH	47	7	45
Sb(CF$_3$)$_2$	79	3	17	Bi(CF$_3$)$_2$	79	3	18	SCF$_3$	77	3	20
Sb(CN)$_2$	91	1	9	Bi(CN)$_2$	89	3	9	SCN	80	3	16
SbMe$_2$	39	6	55	BiMe$_2$	40	5	55	SMe	52	6	42
Sb(OMe)$_2$	46	6	48	Bi(OMe)$_2$	47	5	48	SOMe	52	4	44
Sb(NMe$_2$)$_2$	36	3	62	Bi(NMe$_2$)$_2$	36	3	62				
SbPh$_2$	49	5	46	BiPh$_2$	49	5	46	SPh	56	4	39
SbCl$_2$	68	4	27	BiCl$_2$	69	3	27	SCl	64	3	32

TABLE 18. Values of P_{EA}, P_O and P_{ED} for Group 15 tetracoordinate substituents

X	P_{EA}	P_O	P_{ED}	X	P_{EA}	P_O	P_{ED}	X	P_{EA}	P_O	P_{ED}
$P(O)H_2$	92	2	6	$As(O)H_2$	89	3	9	$Sb(O)H_2$	97	1	2
$P(O)(CF_3)_2$	99	0	1	$As(O)(CF_3)_2$	97	1	2	$Sb(O)(CF_3)_2$	100	0	0
$P(O)(CN)_2$	99	0	1	$As(O)(CN)_2$	98	0	2	$Sb(O)(CN)_2$	100	0	0
$P(O)Me_2$	95	2	3	$As(O)Me_2$	99	1	0	$Sb(O)Me_2$	95	3	3
$P(O)(OMe)_2$	100	0	0	$As(O)(OMe)_2$	100	0	0	$Sb(O)(OMe)_2$	99	0	1
$P(O)(NMe_2)_2$	60	6	34	$As(O)(NMe_2)_2$	61	5	34	$Sb(O)(NMe_2)_2$	60	6	34
$P(O)Ph_2$	100	0	0	$As(O)Ph_2$	99	1	0	$Sb(O)Ph_2$	97	0	3
$P(O)Cl_2$	99	1	0	$As(O)Cl_2$	99	0	1	$Sb(O)Cl_2$	100	0	0
$Bi(O)H_2$	100	0	0	$S(O)CF_3$	89	2	9	$S(O)_2CF_3$	100	0	0
$Bi(O)(CF_3)_2$	100	0	0								
$Bi(O)(CN)_2$	100	0	0	$S(O)Me$	92	1	7	$S(O)_2Me$	97	0	3
$Bi(O)Me_2$	93	2	5								
$Bi(O)(OMe)_2$	97	0	3	$S(O)Et$	92	1	7	$S(O)_2Et$	97	0	3
$Bi(O)(NMe_2)_2$	58	7	35	$S(O)Ph$	87	3	10	$S(O)_2Ph$	89	1	10
$Bi(O)Ph_2$	96	2	3					$S(O)_2NH_2$	92	3	5
$Bi(O)Cl_2$	100	0	0								

the substituent Z bonded to the central atom. Electron-acceptor Z groups in MZ_2 generally result in an overall electron-acceptor effect; the reverse is the case for electron-donor Z groups. We have included, for purposes of comparison, values for SZ groups in Table 17. The overall electrical effect of these groups (and OZ and SeZ groups as well) tends to be electron acceptor.

3. Tetracoordinate substituents

As is shown by the values in Table 18, our model predicts that all Group 15 substituents having the structure $M(O)Z_2$ are overall electron acceptors. We have reported values of P_{EA}, P_0 and P_{ED} for S(O)Z and S(O)$_2$Z in Table 18. They too show an overall predominant electron-acceptor effect. We have been unable to extend our estimating model to substituents in which the oxygen atom is replaced by sulfur, due to a lack of sufficient experimental data for these groups.

The second type of tetracoordinate Group 15 substituent is the ionic group $[MZ^1Z^2Z^3]^q$. Unfortunately we are not yet able to estimate the electrical-effect substituent constants of ionic groups and have been therefore unable to discuss them.

4. Pentacoordinate substituents

We have estimated values of electrical-effect substituent constants for only two examples of MZ_4 groups: those where Z is fluorine or phenyl. We believe that the estimated σ_1 values for these groups are reasonably correct. The values of σ_d and σ_e, however, must be regarded as very uncertain because we have no direct evidence that our model is suitable for their estimation. We are therefore unable to discuss the electrical effects of these groups at this time.

D. The Validity of the Estimation Model

It is remarkable how little is available in the literature on the determination of substituent constants for a wide structural range of Group 15 substituents. The estimation models reported in this work have been used successfully with other types of groups in the past. We have also shown that a similar model can be successfully used for the quantitative description of the pK_a values of oxyacids. This data set included phosphorus and arsenic derivatives. We believe that the estimated parameter values reported here are reliable for nitrogen and phosphorus, and probably reasonable for arsenic substituents as well. Regrettably, we have no evidence available which permits us to determine their reliability for antimony and bismuth substituents. This is particularly the case for dicoordinate groups of these elements. It seems highly unlikely that the Sb$=$Sb and Bi$=$Bi bonding is comparable to that in the analogous nitrogen, phosphorus or arsenic substituents. It is to be hoped that our predictions will challenge experimenters to undertake the investigation of the substituent effects of these groups so that, at some later time, the nature of their substituent effects may be clarified.

III. STERIC EFFECTS

A. Introduction

The steric effect was introduced qualitatively as a concept by Kehrmann[25]. By the end of the 19th century V. Meyer[26] and J. Sudborough[27] were accumulating quantitative results supporting the steric-effects explanation of rate retardation in the estification of suitably substituted benzoic and acrylic acids. Early reviews of steric effects are given by Stewart[28a], Wittig[29] and somewhat later by Wheland[30].

B. The Nature of Steric Effects

1. Primary steric effects

These effects result from repulsions between electrons in valence orbitals on atoms which are not bonded to each other. They are thought to arise from the interpenetration of occupied orbitals on one atom by electrons on the other, resulting in a violation of the Pauli exclusion principle. *All steric interactions raise the energy of the system in which they occur.* In terms of their effect on chemical reactivity, they may either decrease or increase a rate or equilibrium constant, depending on whether steric interactions are greater in the reactant or in the product (equilibria) or transition state (rate).

2. Secondary steric effects

These effects may result from the shielding of an active site from attack by a reagent or, alternatively, from solvation. They may also be due to a steric effect on the concentration of the reacting conformation of a chemical species.

3. Direct steric effects

These effects occur when the active site at which a measurable phenomenon occurs is in close proximity to the substituent. Examples of systems exhibiting direct steric effects are *ortho*-substituted benzenes, **1**, cis-substituted ethylenes, **2**, and the *ortho* (2, 2-, 2, 1- and 2, 3-) and peri (1, 8-) substituted naphthalenes, **3, 4, 5** and **6**, respectively.

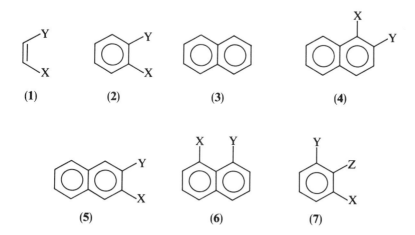

| (1) | (2) | (3) | (4) |

| (5) | (6) | (7) |

4. Indirect steric effects

These effects are observed when the steric effects of the variable substituent is relayed by a constant substituent between it and the active site, as in **7**, where Y is the active site, Z is the constant substituent and X is the variable substituent.

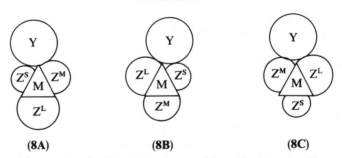

FIGURE 1. Possible conformations of a spherical active site Y adjacent to a substituent $MZ^LZ^MZ^S$. Conformation **8A** has the lowest energy, conformation **8C** the highest

5. The directed nature of steric effects

Steric effects are vector quantities. This is easily shown by considering what happens when a nonsymmetric substituent is in contact with an active site. Consider, for example, the simple case of a spherical active site Y, in contact with a carbon substituent, $CZ^LZ^MZ^S$, where superscripts L, M and S represent the largest, medium-sized and smallest Z groups, respectively. There are three possible conformations of this system which are shown in top views in Figure 1. As all steric interactions raise the energy of the system, the preferred conformation will be the one that results in the lowest energy increase. This is the conformation which presents the smallest face to the active site, conformation **8A**. This is the basis of the minimum steric interaction (MSI) principle which states: *a nonsymmetric substituent prefers that conformation which minimizes steric interactions.* The directed nature of steric effects permits us to draw a conclusion of vital importance: that the substituent volume is not an acceptable measure of its steric effect. Thus, for example, the pentyl and the 1, 1-dimethylpropyl groups have the same volume, but the steric effect of the former is less than half that of the latter. Although there are still some workers who are unable to comprehend this point, it is nevertheless true that group volumes are not appropriate as steric parameters. They are actually measures of group polarizability.

C. The Monoparametric Model of Steric Effects

Stewart, in the second edition of his book[28b], suggested a parallel between rate of esterification of 2-substituted benzoic acids and the molecular weights of the substituents, the nitro group deviating strongly from this relationship. The first attempt to define a set of steric parameters was due to Kindler[31]. Unfortunately, these parameters were later shown to be a function of electrical effects. The first successful parameterization of the steric effect was due to Taft[32], who defined the steric parameter E_s by the expression

$$E_{sx} \equiv \delta \log(k_X/k_{Me}) \tag{44}$$

where k_X and k_{Me} are the rate constants for the acid catalyzed hydrolysis of the corresponding esters XCO_2Ak and $MeCO_2Ak$, respectively. The value of δ is taken as 1.000 for this purpose.

A few $E_{So,X}$ parameters were defined from the rates of acid catalyzed hydrolysis of 2-substituted alkyl benzoates. These parameters, however, have been shown to be a mix of electrical and steric effects with the former predominating. The Taft $E_{s,X}$ values suffered from several deficiencies:

(1) Their validity as measures of steric effects was unproven.

(2) They were determined from average values of the rate constant.

(3) They were available only for derivatives of sp^3-hybridized carbon groups and for hydrogen.

(4) The use of the methyl group as the reference substituent meant that they were not compatible with electrical-effect substituent constants for which the reference substituent is hydrogen.

The first problem was resolved when it was shown that the E_s values are a linear function of van der Waals radii. The latter have long been held to be an effective measure of atomic size. The second and third problems were solved by Charton, who proposed the use of the van der Waals radius as a steric parameter[33] and developed a method for the calculation of group van der Waals radii for tetracoordinate symmetric top substituents MZ_3 such as the methyl and trifluoromethyl groups[34]. In later work the hydrogen atom was chosen as the reference substituent and the steric parameter υ was defined as

$$\upsilon_X \equiv r_{VX} - r_{VH} = r_{VX} - 1.20 \tag{45}$$

where r_{VX} and r_{VH} are the van der Waals radii of the X and H groups in Ångstrom units[35]. Expressing r_V in these units is preferable to the use of picometers, because the coefficient of the steric parameter is then comparable in magnitude to the coefficients of the electrical-effect parameters. Whenever possible, υ parameters are obtained directly from van der Waals radii or calculated from them. These are considered to be primary values. For the greater number of substituents, however, υ parameters must be calculated from the regression equations obtained for correlations of rate constants with primary values. The values obtained from such correlations are considered to be secondary υ values. Available values of υ for Group 15 substituents are reported in Table 19. It has been shown that all other measures of atomic size are a linear function of van der Waals radii. There is, then, no reason for preferring one measure of atomic size over another. As values of υ were developed for a wide range of substituent types with central atoms including oxygen, nitrogen, sulfur and phosphorus as well as carbon, these parameters provide the widest structural range of substituents for which a measure of the steric effect is available.

1. Steric classication of substituents

We may conveniently classify substituents into three categories based on the degree of conformational dependence of their steric effects:

1. No conformational dependence (NCD). Groups of this type include monatomic substituents such as hydrogen and the halogens, cylindrical substituents such as the ethynyl and cyano groups, and tetracoordinate symmetric top substituents such as the methyl, trifluoromethyl and silyl groups.

2. Minimal conformational dependence (MCD). Among the groups of this type are nonsymmetric substituents with the structure $MH_n(lp)_{3-n}$ such as the hydroxyl and amino groups and nonsymmetric substituents with the structure $MZ_2^S Z^L$, where S stands for small and L for large.

3. Strong conformational dependence (SCD). Such groups are those with the structures $MZ_2^L Z^S$ and $MZ^L Z^M Z^S$, where the superscript M indicates medium. Other groups in this category are planar π-bonded groups such as Ph, carboxy and nitro, and quasi-planar π-bonded groups such as dimethylamino and cyclopropyl.

The steric parameter for NCD groups can be obtained directly from van der Waals radii or calculated from them. The values for SCD groups are often obtainable from van der Waals radii, although in some cases they must be derived as secondary values from regression equations obtained by correlating rate constants with known values of the steric parameter. Steric parameters for SCD groups of the nonsymmetric type are only obtainable from regression equations. In the case of planar π-bonded groups the maxi-

TABLE 19. Values of υ, υ_1, υ_2 and υ_3

Z^1, Z^2	υ M=N	υ_1	υ_2	υ_3	υ P	υ_1	υ_2	υ_3	υ As	υ_1	υ_2	υ_3
Dicoordinate	M=MZ											
H		0.35	0.35	0		0.60	0.60	0		0.65	0.65	0
CF$_3$		0.35	0.35	0.90		0.50	0.60	0.90		0.65	0.65	0.90
CN		0.35	0.35	0.40		.60	0.60	0.40		0.65	0.65	0.40
Me		0.35	0.35	0.52		0.60	0.60	0.52		0.65	0.65	0.52
OMe		0.35	0.35	0.32		0.60	0.60	0.32		0.65	0.65	0.32
NMe$_2$		0.35	0.35	0.39		0.60	0.60	0.39		0.65	0.65	0.39
t-Bu		0.35	0.35	1.24		0.60	0.60	1.24		0.65	0.65	1.24
Ph		0.35	0.35	0.57		0.60	0.60	0.57		0.65	0.65	0.57
F		0.35	0.35	0.27		0.60	0.60	0.27		0.65	0.65	0.27
NH$_2$		0.35	0.35	0.35		0.60	0.60	0.35		0.65	0.65	0.35
OH		0.35	0.35	0.32		0.60	0.60	0.32		0.65	0.65	0.32
M^1=CH		0.35	0.57	0		0.60	0.57	0		0.65	0.57	0
M^1=CMe		0.35	0.57	0.52		0.60	0.57	0.52		0.65	0.57	0.52
M^1=CPh		0.35	0.57	0.57		0.60	0.57	0.57		0.65	0.57	0.57
N=C=O		0.35	0.57	0.32								
N=C=S		0.35	0.57	0.60								
N$_3$		0.35	0.35	0.35								
Tricoordinate	MZ'Z^2											
H, H	0.35	0.35	0	0	0.60	0.60	0	0		0.65	0	0
H, CF$_3$		0.35	0.90	0.27		0.60	0.90	0.27				
H, CN		0.35	0.40	0.40		0.60	0.40	0.40				
H, CONH$_2$		0.35	0.50	0.32		0.60	0.50	0.32				
H, CHO		0.35	0.50	0		0.60	0.50	0				
H, Me	0.39	0.35	0.52	0		0.60	0.52	0				
H, SO$_2$Me		0.35	1.03	0.52		0.60	1.03	0.52				
(CF$_3$)$_2$		0.77	0.90	0.27		1.00	0.90	0.27		1.07	0.90	0.27
(CN)$_2$		0.60	0.40	0.40		0.60	0.40	0.40				

Me$_2$	0.39	0.52	0	0.84	0.52	0	0.93	0.52	0
(OMe)$_2$	0.58	0.32	0.52	0.75	0.32	0.52	0.83	0.32	0.52
(SMe)$_2$	0.80	0.60	0.52	0.88	0.60	0.52	1.00	0.60	0.52
H, COCF$_3$	0.35	0.50	0.32	0.60	0.50	0.32			
H, Ac	0.35	0.50	0.32	0.60	0.50	0.32			
H, CO$_2$Me	0.35	0.50	0.32	0.60	0.50	0.32			
H, OH	0.39	0.32	0	0.75	0.32	0	0.83	0.32	0
(OH)$_2$	0.52	0.32	0	0.75	0.32	0	0.83	0.32	0

0.63

Tetracoordinate

M(E)Z^{1}Z^2

(OH)$_2$	0.60	0.32	0	0.65	0.32	0
O(CF$_3$)$_3$	1.29	0.90	0.27	0.88	0.90	0.27
O(CN)$_2$	1.10	0.40	0.40		0.40	0.40
O(OMe)$_2$	1.17	0.52	0.52	1.29	0.52	0.52
O(OMe)$_2$	1.04	0.52	0	1.19	0.52	0
O(SMe)$_2$	1.24	0.52	0	1.36	0.52	0
O(C$_2$H)$_2$	1.21	0.58	0.58		0.58	0.58
O(CH=CH$_2$)$_2$	1.20	0.57	0.57	1.29	0.57	0.57
O(Et)$_2$	1.17	0.52	0.52	1.19	0.52	0.52
O(OEt)$_2$	1.04	0.52	0.52	1.36	0.52	0.52
O(SEt)$_2$	1.24	0.52	0.52		0.52	0.52
O(NMe$_2$)$_2$	1.27	0.52	0	1.29	0.52	0
O(CH$_2$CH=CH$_2$)$_2$	1.17	0.52	0.57		0.52	0.57
O(C$_6$F$_5$)$_2$	1.20	0.57	0.57	1.19	0.57	0.57
O(Ph)$_2$	1.20	0.57	0.57	1.36	0.57	0.57
O(OPh)$_2$	1.04	0.57	0.57	1.26	0.57	0.57
O(SPh)$_2$	1.24	0.57	0.57	1.33	0.57	0.57
O(Bzl)$_2$	1.17	0.57	0.57	1.11	0.57	0.57
O(Cl)$_2$	1.21	0.55	0	1.25	0.55	0
O(F)$_2$	1.04	0.27	0	1.19	0.27	0
O(NH$_2$)$_2$	1.07	0.35	0	1.25	0.35	0
O(OH)$_2$	1.04	0.32	0	1.19	0.32	0

(continued)

TABLE 19. (*continued*)

Z^1, Z^2	$M = Sb$			Bi		
	v_1	v_2	v_3	v_1	v_2	v_3
Dicoordinate	$M = MZ$					
H	0.84	0.84	0	0.90	0.90	0
CF_3	0.84	0.84	0.90	0.90	0.90	0.90
CN	0.84	0.84	0.40	0.90	0.90	0.40
Me	0.84	0.84	0.52	0.90	0.90	0.52
OMe	0.84	0.84	0.32	0.90	0.90	0.32
NMe_2	0.84	0.84	0.39	0.90	0.90	0.39
t-Bu	0.84	0.84	1.24	0.90	0.90	1.24
Ph	0.84	0.84	0.57	0.90	0.90	0.57
F	0.84	0.84	0.27	0.90	0.90	0.27
NH_2	0.84	0.84	0.35	0.90	0.90	0.35
OH	0.84	0.84	0.32	0.90	0.90	0.32
$M^1{=}CH$	0.84	0.57	0	0.90	0.57	0
$M^1{=}CMe$	0.84	0.57	0.52	0.90	0.57	0.52
$M^1{=}CPh$	0.84	0.57	0.57	0.90	0.57	0.57
Tricoordinate	MZ^1Z^2					
H, H	0.84	0	0	0.90	0	0
$(CF_3)_2$	1.22	0.90	0.27	1.27	0.90	0.27
$(CN)_2$		0.40	0.40		0.40	0.40
Me_2	1.09	0.52	0	1.08	0.52	0
$(OMe)_2$	0.99	0.32	0.52	1.05	0.32	0.52
$(SMe)_2$		0.60	0.52		0.60	0.52
$(C_2H)_2$		0.58	0.58		0.58	0.58
$(CH{=}CH_2)_2$		0.57	0.57		0.57	0.57
Et_2	1.09	0.52	0.52	1.08	0.52	0.52
$(OEt)_2$	0.99	0.32	0.32	1.05	0.32	0.32
$(SEt)_2$		0.60	0.52		0.60	0.52
$(NMe_2)_2$		0.39	0.52		0.39	0.52

$(CH_2CH{=}CH_2)_2$	1.09	0.52	0.57	1.08	0.52	0.57
$(C_6F_5)_2$		0.57	0.57		0.57	0.57
Ph_2		0.57	0.57		0.57	0.57
$(OPh)_2$	0.99	0.32	0.57	1.05	0.32	0.57
$(SPh)_2$		0.60	0.57		0.60	0.57
$(Bzl)_2$	1.09	0.52	0.57	1.08	0.52	0.57
Cl_2	1.12	0.55	0	1.17	0.55	0
F_2	0.96	0.27	0	1.00	0.27	0
$(NH_2)_2$		0.35	0		0.35	0
$(OH)_2$	1.29	0.32	0	1.34	0.32	0
Tetracoordinate $\quad$ $M(E)Z^1Z^2$						
$(O)H_2$	0.84	0.32	0	0.90	0.32	0
$O(CF_3)_2$	1.52	0.90	0.27	1.56	0.90	0.27
$O(CN)_2$		0.40	0.40		0.40	0.40
$(O)Me_2$	1.39	0.52	0	1.44	0.52	0
$O(OMe)_2$	1.29	0.32	0.52	1.34	0.32	0.52
$O(SMe)_2$		0.60	0.52		0.60	0.52
$O(C_2H)_2$		0.58	0.58		0.58	0.58
$O(CH{=}CH_2)_2$		0.57	0.57		0.57	0.57
$(O)Et_2$	1.39	0.52	0.52	1.44	0.52	0.52
$O(OEt)_2$	1.29	0.32	0.52	1.34	0.32	0.52
$O(SEt)_2$		0.60	0.52		0.60	0.52
$O(NMe_2)_2$		0.39	0.52		0.39	0.52
$O(CH_2CH{=}CH_2)_2$	1.39	0.52	0.52	1.44	0.52	0.52
$O(C_6F_5)_2$		0.57	0.57		0.57	0.57
$(O)Ph_2$		0.57	0.57		0.57	0.57
$O(OPh)_2$	1.29	0.32	0.57	1.34	0.32	0.57
$O(SPh)_2$		0.60	0.57		0.60	0.57
$(O)Bzl_2$	1.39	0.52	0.57	1.44	0.52	0.57
$(O)Cl_2$	1.42	0.55	0.57	1.48	0.55	0.57
$(O)F_2$	1.22	0.27	0	1.25	0.27	0
$O(NH_2)_2$		0.35	0		0.35	0
$O(OH)_2$	1.29	0.32	0	1.34	0.32	0

mum and minimum values of the steric parameter are available from the van der Waals radii. These groups are sufficiently common and important to be worthy of a more detailed discussion.

2. Planar π-bonded groups

These $(X_{p\pi})$ groups represent an especially difficult problem because their delocalized electrical effect depends on the steric effect when they are bonded to planar π-bonded skeletal groups, $G_{p\pi}$. The σ_d and σ_e electrical-effect parameters are a function of the dihedral angle formed by $X_{p\pi}$ and $G_{p\pi}$. Thus:

$$\sigma_{dX,\theta} = \sigma_{dX,0} \cos^2\theta \tag{46}$$

and

$$\sigma_{e,X,\theta} = \sigma_{e,X,0} \cos^2\theta \tag{47}$$

where θ is the dihedral angle of interest, and $\sigma_{dX,0}$ and $\sigma_{eX,0}$ are the values of σ_d and σ_e when the substituent and skeletal group are coplanar ($\theta = 0$). The effective value of υ is given by the expression:

$$\upsilon = d\cos\theta + r_{vZS} - 1.20 \tag{48}$$

where Z^S is the smaller of the two Z groups attached to the central atom, M, of the $X_{p\pi}$ group. There is no simple a priori way to determine θ. It could conceivably be estimated by molecular mechanics calculations, but there is some reason to believe that θ is a function of the medium. Alternatively, the $X_{p\pi}$ group can be included in the data set by means of an iteration procedure. The method requires an initial correlation of the data set with all $X_{p\pi}$ groups excluded. This constitutes the basis set. The correlation equation used for this purpose is the LDRS equation:

$$Q_X = L\sigma_{lX} + D\sigma_{dX} + R\sigma_{eX} + S_\upsilon + h \tag{49}$$

The correlation is then repeated for each $X_{p\pi}$ group using v values increasing incrementally by some convenient amount from the minimum which represents the half-thickness of the group to the maximum which occurs when $X_{p\pi}$ is nearly perpendicular to $G_{p\pi}$. The proper value of θ is that which results in the best fit of the data to the correlation equation, and for which the L, D, R and S values obtained are in best agreement with those of the basis set.

D. Multiparametric Models of Steric Effects

When the active site is itself large or, alternatively, when the phenomenon studied is some form of biactivity in which binding to a receptor is the key step, a simple monoparametric model of the steric effect will often be insufficient. It is then necessary to make use of a multiparametric model of steric effects. Four multiparametric models are available: that of Verloop[36], the simple branching model, the expanded branching model and the segmental model. The Verloop model suffers from the fact that its parameters measure maximum and minimum distances perpendicular to the group axis. These maxima and minima may occur at any point in the group skeleton (the longest chain in the group). The steric effect, however, may be very large at one segment of the chain and negligible at others. If a data set is large, the likelihood that the maximum and minimum distances of all groups are located at the same segment, and that it is this segment at which the steric effect is important, is very small. The Verloop model will therefore not be discussed further.

1. The branching equations

The simple branching model[36,37] for the steric effect is given by the expression:

$$S\psi = \sum_{i-1}^{m} a_i n_i + a_b n_b \tag{50}$$

where the a_i and a_b are coefficients, n_i is the number of branches attached to the i-th atom and n_b is the number of bonds between the first and last atoms of the group skeleton. So, n_b is a measure of group length. For saturated cyclic substituents it is necessary to determine values of n_i from an appropriate regression equation. For planar π-bonded groups n_i is taken to be 1 for each atom in the group skeleton. For other groups n_i is obtained simply by counting branches. The model makes the assumption that all of the branches attached to a skeleton atom are equivalent. This is only a rough approximation. Distinguishing between branches results in an improved model called the *expanded branching equation*:

$$S\psi = \sum_{i-1}^{m} \sum_{j-1}^{3} a_{ij} n_{ij} + a_b n_b \tag{51}$$

which allows for the difference in steric effect that results from the order of branching[36,37]. This difference is a natural result of the MSI principle. The first branch has much the smallest effect, because a conformation in which it is rotated out of the way of the active site is possible. This rotation becomes more difficult with the second branch and impossible with the third. The problem with the expanded branching method is that it requires a large number of parameters. Rarely does one encounter a data set large enough to permit its use.

2. The segmental model

As both branching methods have problems associated with them, the segmental method[37] is often the simplest and most effective. In this model each atom of the group skeleton together with the atoms attached to it constitutes a segment of the substituent. Applying the MSI principle, the segment is considered to have that conformation which presents its smallest face to the active site. The segment is assigned the υ value of the group which it most resembles. Values of the segmental steric parameters υ_i, where i designates the segment number, are given in Table 19. Numbering starts from the first atom of the group skeleton, namely the atom which is attached to the rest of the species. The expression for the steric effect using the segmental model is:

$$S\psi = \sum_{i-1}^{m} S_i \upsilon_i \tag{52}$$

when only steric effects are present:

$$Q_X = S\psi_X \tag{53}$$

In the general case electrical effects are present, and the LDRS equation in the form

$$Q_X = l\sigma_{DX} + d\sigma_{dX} + R\sigma_{eX} + S\psi_X + h \tag{54}$$

is required.

IV. INTERMOLECULAR FORCES

A. Introduction

Inter- and intramolecular forces (imf) are of vital importance in the quantitative description of structural effects on bioactivities and chemical properties. They may make a

TABLE 20. Intermolecular forces and their parameterization[a]

Interaction	Parameterization
molecule – molecule	
Hydrogen bonding (hb)	n_H, n_n $(\sigma_l, \sigma_d, \sigma_e)$
Dipole – dipole (dd)	μ_{MeX} or μ_{PhX} $(\sigma_l, \sigma_d, \sigma_e)$
Dipole – induced dipole (di)	μ_{MeX} or μ_{PhX}, $\alpha(\sigma_l, \sigma_d, \sigma_e)$
Induced dipole – induced dipole (ii)	α
charge transfer (ct)	n_A, n_D $(\sigma_l, \sigma_d, \sigma_e)$
ion – molecule	
Ion – dipole (Id)	i
Ion – induced dipole (Ii)	i

[a] Parameters in parentheses may contribute to representing the interaction. The dd, di and ii interactions are known collectively as Van der Waals forces. The dd interactions are also called Keesom forces, the di interactions are also called Debye forces and the ii interactions are also known as London or dispersion forces. The ct interactions are also known as donor – acceptor forces.

significant contribution to chemical reactivities and some physical properties as well. Types of intermolecular forces and their present parameterization are listed in Table 20.

B. Parameterization of Intermolecular Forces

1. Hydrogen bonding

For hydrogen-bonding parameters we have used n_H, the number of OH and/or NH bonds in the substituent, and n_n, the number of lone pairs on oxygen and/or nitrogen atoms[40-42]. The use of these parameters is based on the argument that if one of the phases involved in the phenomenon studied includes a protonic solvent, particularly water, then all of the hydrogen bonds that the substituent is capable of forming will indeed form. For such a system, hydrogen-bond parameters defined from equilibria in highly dilute solution in an (inert) solvent are unlikely to be a suitable model. A more sophisticated parameterization than that described above would be the use of the hydrogen-bond energy for each type of hydrogen bond formed. Thus for each substituent the parameter E_{hbX}, would be given by the equation:

$$E_{hbX} = \sum_{i=1}^{m} n_{hbi} E_{hbi} \tag{55}$$

where E_{hbX} is the hydrogen-bonding parameter, E_{hbi} is the energy of the i-th type of hydrogen bond formed by the substituent X and n_{hbi} is the number of such hydrogen bonds. The validity of this parameterization is as yet untested.

2. van der Waals interactions

These interactions (dd, di, ii) are a function of dipole moment and polarizability. It has been shown that the dipole moment cannot be replaced entirely by the use of electrical-effect substituent constants as parameters. This is because the dipole moment has no sign. Either an overall electron-donor group or an overall electron-acceptor group may have the same value of μ. We have therefore made use of the dipole moments of MeX and PhX as parameters for substituents bonded to sp^3- and sp^2-hybridized carbon atoms of a skeletal group. Application to substituents bonded to sp-hybridized carbon atoms should require a set of dipole moments for $HC{\equiv}CX$.

We have chosen as the polarizability parameter the quantity α, which is given by the expression:

$$\alpha \equiv \frac{MR_X - MR_H}{100} = \frac{MR_X}{100} - 0.0103 \tag{56}$$

where MR_X and MR_H are the group molar refractivities of X and H, respectively[40-42]. The factor 1/100 is introduced to scale the α parameter so that its coefficients in the regression equation are roughly comparable to those obtained for the other parameters used. There are many other polarizability parameters including parachor, group molar volumes of various kinds, van der Waals volumes and accessible surface areas any of which would do as well, because they are all highly collinear in each other[43-45]. The proposal of other polarizability parameters seems to be a popular avocation of many.

Values of α can be estimated by additivity from the values for fragments. They may also be estimated from group molar refractivities calculated from the equation:

$$MR_X = 0.320n_c + 0.682n_b - 0.0825n_n + 0.991 \tag{57}$$

where n_c, n_b and n_n are the number of core, bonding and nonbonding electrons, respectively, in the group X[43].

3. Charge transfer interactions

These interactions can be roughly parameterized by the indicator variables n_A and n_D; n_A takes the value 1 when X is a charge transfer acceptor and 0 when it is not, while n_D takes the value 1 when X is a charge transfer donor and 0 when it is not. An alternative parameterization makes use of the first ionization potential of MeX (ip_{MeX}) as the electron-donor parameter and the electron affinity of MeX as the electron-acceptor parameter. We have generally found the indicator variables n_A and n_D to be sufficient. This parameterization accounts for charge transfer interactions directly involving the substituent. If the substituent is attached to a π-bonded skeletal group, then the skeletal group is capable of charge transfer interaction the extent of which is modified by the substituent. This is accounted for by the electrical-effect parameters.

4. The intermolecular force (IMF) equation

We may now write a general relationship for the quantitative description of intermolecular forces:

$$Q_X = L\sigma_{lX} + D\sigma_{dX} + R\sigma_{eX} + A\alpha_X + H_1 n_{HX}$$
$$+ H_2 n_{nX} + I i_X + B_{DX} n_{DX} + B_{AX} n_{AX} + S\psi_X + B^0 \tag{58}$$

Values of the IMF parameters for Group 15 substituents are set forth in Table 21.

V. APPLICATIONS

A. Introduction

We have applied the methods described above to a number of data sets involving chemical reactivities, chemical properties and physical properties. In several cases, the data sets were chosen because they provided an opportunity to test the validity of some of the parameters estimated in this work. All of the data sets studied are reported in Table 22.

TABLE 21. Values of μ, α, n_H and n_n for Group 15 substituents[a]

Z^1, Z^2	μ M=N	α	n_H	n_n	μ P	α	n_H	n_n	μ As	α	n_H	n_n
Dicoordinate	M = MZ											
H	0, —	0.070	0	2		0.192	0	0		0.298	0	0
CF_3		0.100	0	2		0.222	0	0		0.328	0	0
CN		0.113	0	3		0.235	0	1		0.341	0	1
Me		0.116	0	2		0.228	0	0		0.334	0	0
OMe		0.138	0	4		0.250	0	2		0.356	0	2
NMe_2		0.205	0	3		0.327	0	1		0.433	0	0
t-Bu		0.246	0	2		0.368	0	0		0.474	0	0
Ph		0.303	0	2		0.425	0	0		0.531	0	0
F		0.069	0	2		0.181	0	0		0.287	0	0
NH_2		0.104	2	3		0.226	2	1		0.332	2	1
OH		0.078	1	4		0.190	1	2		0.306	1	2
M^1=CH		0.080	0	1		0.232	0	0		0.338	0	0
M^1=CMe		0.116	0	1		0.278	0	0		0.384	0	0
M^1=CPh		0.313	0	1		0.475	0	0		0.584	0	0
N=C=O		0.078	0	3								
N=C=S		0.162	0	3								
N_3	1.56, 2.17	0.092	0	1								
NO		0.042	0	3								
Tricoordinate	MZ^1Z^2											
H, H	1.49, 1.296	0.044	2	1	—, 1.100	0.112	0	0		0.146	0	0
H, CF_3		0.095	1	1		0.150	0	0				
H, CN		0.091	1	2		0.163	0	0				
H, $CONH_2$	4.31, 4.34	0.122	3	4		0.190	0	0				
H, CHO		0.093	1	3		0.161	0	0				
H, Me	1.77, —	0.093	1	1		0.157	0	0				
H, SO_2Me	4.60, —	0.159	1	5		0.227	0	4				
$(CF_3)_2$		0.133	0	1		0.177	0	0				
$(CN)_2$		0.144	0	3		0.197	0	2				
Me_2	1.60, 0.612	0.145	0	1	1.31, 1.192	0.202	0	0	—, 1.10	0.236	0	0
$(OMe)_2$		0.159	0	5	—, 3.29	0.233	0	4		0.280	0	4
$(SMe)_2$		0.279	0	0		0.353	0	0		0.400	0	0
H, $COCF_3$		0.151	1	3		0.197	0	2				
H, Ac	3.75, 3.71	0.139	1	3		0.212	0	2				
H, CO_2Me	3.69, —	0.152	1	5		0.220	0	4				
H, Et	1.71, —	0.140	1	1		0.195	0	0				
$(C_2H)_2$		0.193	0	1		0.261	0	0				
$(CH=CH_2)_2$		0.223	0	0		0.291	0	0		0.314	0	0
Et	1.81	0.232	0	0	1.42 —	0.279	0	0	1.32 —	0.330	0	0

Substituent		δ				δ				δ		
$(OEt)_2$		0.251	0	5		0.319	0	4		0.372	0	4
$(SEt)_2$		0.373	0	1		0.441	0	0		0.492	0	0
$(NMe_2)_2$		0.313	0	3		0.381	0	2		0.434	0	2
$(CH_2CH=CH_2)_2$	1.14,—	0.293	1	1		0.361	0	0		0.414	0	0
H, Ph	3.66,—	0.290	1	1		0.345	0	0				
H, Bz		0.327	1	3		0.395	0	0				
$(C_6F_5)_2$		0.483	0	1		0.521	0	0				
Ph_2	0.71,—	0.539	0	1		0.598	0	0	1.15,—	0.604	0	0
$(OPh)_2$		0.557	0	5		0.625	0	4		0.630	0	4
$(SPh)_2$		0.689	0	1		0.757	0	0		0.678	0	4
$(Bzl)_2$		0.603	0	1		0.671	0	0		0.810	0	0
Cl_2		0.123	0	1		0.191	0	0		0.724	0	0
F_2		0.021	3	1		0.089	3	2		0.244	0	0
H, NH_2	1.67,—	0.074	3	2		0.146	3	2		0.142	0	0
$(NH_2)_2$		0.111	4	3		0.179	4	2				
NO_2	4.26, 3.56	0.063	2	4		0.119	1	2		0.332	4	2
H, OH		0.052	2	3		0.127	2	2				
$(OH)_2$		0.059	2	5			2	4		0.180	2	4

Tetracoordinate $M(E)Z'Z^2$

Substituent		δ				δ				δ		
$(OH)_2$		0.118	0	2						0.205	0	2
$O(CF_3)_2$		0.177	0	2						0.264	0	2
$O(CN)_2$		0.203	0	4						0.290	0	4
$O(Me)_2$	4.39, 4.29	0.189	0	2						0.276	0	2
$O(OMe)_2$		0.208	0	6	—, 5.12					0.320	0	6
$O(SMe)_2$		0.353	0	2						0.440	0	2
$O(C_2H)_2$		0.267	0	2						0.354	0	2
$O(CH=CH_2)_2$		0.297	0	2						0.384	0	2
$O(Et)_2$		0.284	0	2						0.370	0	2
$O(OEt)_2$		0.301	0	6						0.412	0	6
$O(SEt)_2$		0.447	0	2						0.530	0	2
$O(NMe_2)_2$		0.387	0	2						0.474	0	2
$O(CH_2CH=CH_2)_2$		0.367	0	2						0.454	0	2
$O(C_6F_5)_2$		0.557	0	2	5.41,—					0.644	0	2
$(O)Ph_2$	4.44,—	0.593	0	2						0.670	0	2
$O(OPh)_2$		0.631	0	6						0.718	0	6
$O(SPh)_2$		0.757	0	2						0.850	0	2
$(O)Bzl_2$		0.677	0	2						0.764	0	2
$(O)Cl_2$		0.197	0	2						0.284	0	2
$(O)F_2$		0.095	4	4						0.182	4	4
$O(NH_2)_2$		0.185	2	2						0.272	2	2
$O(OH)_2$		0.133	4	4						0.220	4	4
$S(Ph)_2$	4.87,—	0.595	0	0							2	6

(continued)

TABLE 21. *(continued)*

Z^1, Z^2	α	n_H	n_n	α	n_H	n_n
	Bi	μ		**M = Sb** (μ), **M = MZ**		
Dicoordinate						
H	0.316	0	0	0.522	0	0
CF_3	0.356	0	0	0.562	0	0
CN	0.369	0	1	0.575	0	1
Me	0.362	0	0	0.568	0	0
OMe	0.384	0	2	0.590	0	2
NMe_2	0.461	0	1	0.667	0	1
t-Bu	0.502	0	0	0.708	0	0
Ph	0.559	0	0	0.765	0	0
F	0.315	0	0	0.521	0	0
NH_2	0.360	2	1	0.566	2	1
OH	0.352	1	2	0.558	1	2
M^1=CH	0.208	0	0	0.311	0	0
M^1=CMe	0.254	0	0	0.357	0	0
M^1=CPh	0.451	0	0	0.554	0	0
Tricoordinate						
				MZ^1Z^2		
H, H	0.179	0	0	0.282	0	0
$(CF_3)_2$	0.238	0	0	0.341	0	0
$(CN)_2$	0.264	0	2	0.367	0	2
Me_2	0.250	0	0	0.353	0	0
$(OMe)_2$	0.294	0	4	0.397	0	4
$(SMe)_2$	0.414	0	0	0.517	0	0
$(C_2H)_2$	0.328	0	0	0.431	0	0
$(CH=CH_2)_2$	0.358	0	0	0.461	0	0
Et_2	0.344	0	0	0.447	0	0
$(OEt)_2$	0.386	0	4	0.489	0	4
$(SEt)_2$	0.506	0	0	0.609	0	0
$(NMe_2)_2$	0.448	0	2	0.551	0	2
$(CH_2CH=CH_2)_2$	0.428	0	0	0.531	0	0

(C₆F₅)₂		0.618	0	0	0.721	0	0
Ph₂	0.77, —	0.644	0	0	0.747	0	0
(OPh)₂		0.692	0	4	0.795	0	4
(SPh)₂		0.824	0	0	0.927	0	0
(Bzl)₂		0.738	0	0	0.841	0	0
Cl₂		0.258	0	0	0.361	0	0
F₂		0.156	0	0	0.259	0	0
(NH₂)₂		0.246	4	2	0.349	4	2
(OH)₂		0.194	2	4	0.297	2	4
Tetracoordinate							
(O)H₂		0.219	0	2	0.322	0	2
O(CF₃)₂		0.278	0	2	0.381	0	2
O(CN)₂		0.304	0	4	0.407	0	4
(O)Me₂		0.290	0	2	0.393	0	2
O(OMe)₂		0.334	0	6	0.437	0	6
O(SMe)₂		0.454	0	2	0.557	0	2
O(C₂H)₂		0.368	0	2	0.471	0	2
O(CH=CH₂)₂		0.398	0	2	0.501	0	2
(O)Et₂		0.390	0	2	0.493	0	2
O(OEt)₂		0.426	0	2	0.529	0	2
O(SEt)₂		0.546	0	2	0.649	0	2
O(NMe₂)₂		0.588	0	4	0.591	0	4
O(CH₂CH=CH₂)₂		0.468	0	2	0.571	0	2
O(C₆F₅)₂		0.658	0	2	0.761	0	2
(O)Ph₂		0.684	0	2	0.787	0	2
O(OPh)₂		0.732	0	6	0.835	0	6
O(SPh)₂		0.864	0	2	0.967	0	2
(O)Bzl₂		0.778	0	2	0.881	0	2
(O)Cl₂		0.298	0	2	0.361	0	2
(O)F₂		0.196	0	2	0.299	0	2
O(NH₂)₂		0.286	4	2	0.389	4	2
O(OH)₂		0.234	2	4	0.337	2	4

[a] Values of μ given are for the substituent bonded to sp²- and sp³-hybridized carbon, respectively. Values of α were estimated assuming additivity when MR values were unavailable.

TABLE 22. Data used in correlations

1. pKa, XOH, H2O, 25 °C[a]
H, 16.04; Me, 15.54; Et, 15.93; Ph, 10.02; Cl, 7.537; Br, 8.66; I, 10.64; OH, 11.95; OMe, 11.08; C_6F_5, 5.53; Bz, 3.903; CN, 3.57; Ac, 4.456; HCO, 3.451; COCF3, 0.22; 1-Nh, 9.40; 2-Nh, 9.57; NO2, −1.37; OEt, 11.35; CH2CF3, 12.43; CH2OH, 13.57; H2P(O), 0.93; Me2As(O), 5.85; (CF3)2As(O), 1.12; Ph2P(O), 1.42; (OMe)2P(O), 0.99; (OH)2AsO, 2.399; As(OH)2, 9.771; Me2P(O), 2.78; Et2P(O), 2.99; PhS(O), 2.46; PhSe(O), 4.40; MeSO2, −0.72; (OPh)2P(O), 1.55.

2. pKa, H2O, 3-XC6H4CO2H[a]
H, 4.1998; Me, 4.252; t-Bu, 4.199; Cl, 3.822; Br, 3.810; I, 3.851; OH, 4.076; OMe, 4.093; OPh, 3.951; OAc, 4.01; NH2, 4.598; NMe2, 5.10; CHO, 3.951; Ac, 3.825; NHAc, 4.09; CN, 3.598; NO, 3.960; NO2, 3.56; SiMe3, 4.24; i-Pr, 4.28; CF3, 3.79; P(O)(OMe)2, 3.65; P(O)Me2, 3.82; P(O)Et2, 3.81; F, 3.865; CH2Vi, 4.266; 1-MeVn, 4.248.

3. pKa, H2O, 4-XC6H4CO2H[a]
H, 4.1998; Me, 4.370; Et, 4.353; i-Pr, 4.354; t-Bu, 4.389; Cl, 3.9863; Br, 4.0020; I, 3.98; OH, 4.580; OMe, 4.478; OAc, 4.37; NH2, 4.853; NMe2, 5.03; CHO, 3.75; Ac, 3.700; COEt, 3.72; NHAc, 4.29; CN, 3.551; NO2, 3.442; NO, 3.27; SOMe, 3.66; SiMe3, 4.27; i-Pr, 4.42; P(O)(OMe)2, 3.61; P(O)(OEt)2, 3.60; P(O)Me2, 3.70; (O)Et2, 3.71; F, 4.141; SO2NH2, 3.97; CH2Vi, 4.326; CH2OH, 4.16.

4. pKa, 3-XC6H4CO2H, 23.6 mole% aq. EtOH, 25 °C[a]
POMe2, 4.92; POEt2, 4.99; PO-i-Pr2, 5.06; POBu2, 5.15; PO(OMe)2, 5.04; PO(OEt)2, 5.03; PO(OPr)2, 5.10; PO(NMe2)2, 5.16; PO(Ph)2, 5.11; PSEt2, 5.03; PSPh2, 5.24; AsPh2, 5.62 AsOPh2, 4.93; PPh2, 5.51; H, 5.72; Me, 5.88; Cl, 5.17; Br, 5.15; I, 5.26; OMe, 5.59; OEt, 5.63; OCF3, 5.15; OPh, 5.32; Ac, 5.21; OAc, 5.16; CO2Me, 5.16; CO2Et, 5.20; NH2, 5.79; NMe2, 5.92; N(CF3)2, 5.14; NPh2, 5.77; NHAc, 5.52; CN, 4.85; NO2, 4.51; N3, 5.28; N=CHPh, 4.53; SMe, 5.53; SAc, 5.17; SOF3, 5.13; SOCF3, 4.74; SO2Me, 4.78; SF5, 4.82; CF3, 5.11; SO2CF3, 4.54 CH2Ph, 5.84; c-pr, 5.85; Ph, 5.29; SiMe3, 6.00; CH2Cl, 5.59; CH2Br, 5.56; SO2F, 4.53; NHCHO, 5.38; CH2OPh, 5.67; CH2OMe, 5.69; c-Bu, 5.89; cPe, 5.93; c-Hx, 5.99, CH2CN, 5.49; NHCOCF3, 5.19; NHCO2Me, 5.40; NHCO2Et, 5.59; OSO2Me, 5.12; 2-furyl, 5.65; 2-thienyl, 5.61; 3-thienyl, 5.70; OH, 5.72.

5. pKa, 4-XC6H4CO2H, 23.6 mole% sq. EtOH, 25 °C[a]
POMe2, 4.91; POEt2, 4.92; PO-i-Pr2, 5.00; POBu2, 4.94; PO(OMe)2, 4.88; PO(OEt)2, 4.90; PO(OPr)2, 4.92; PO(OBu)2, 4.87; PO(NMe2)2, 5.02; PO(Ph)2, 4.88; PSEt2, 4.92; PSPh2, 4.97; AsPh2, 5.54; AsOPh2, 4.77; PPh2, 5.39; H, 5.72; Me, 5.96; Cl, 5.32; Br, 5.27; OMe, 6.03; OEt, 6.04; OCF3, 5.19; OPh, 5.50; Ac, 5.10; OAc, 5.29; CO2Me, 5.07; CO2Et, 5.08; NH2, 6.47; NMe2, 6.75; N(CF3)2, 4.94; NPh2, 6.10; NHAc, 5.81; CN, 4.70; NO2, 4.47; N3, 5.45; N=CHPh, 5.74; SMe, 5.74; SAc, 5.09; SOF3, 4.98; SOCF3, 4.65; SO2Me, 4.68; SF5, 4.70; CF3, 4.94; SO2CF3, 4.24; CH2Ph, 5.88; c-Pr, 5.91; Ph, 5.35; SiMe3, 5.80; CH2Cl, 5.54; CH2Br, 5.54; SO2F, 4.33; NHCHO, 5.65; CH2OPh, 5.61; CH2OMe, 5.68; c-Bu, 5.89, c-Pe, 5.89; c-Hx, 5.89; CH2CN, 5.46; NHCOCF3, 5.51; NHCO2Me, 5.65; NHCO2Et, 5.91; OSO2Me, 5.16; 2-furyl, 5.71; 2-thienyl, 5.67; 3-thienyl, 5.77; OH, 6.25.

6. pKa, 3-XC6H4CO2H, 48.1 mole% aq. EtOH, 25 °C[a]
H, 6.66; Cl, 6.12; Br, 6.05; NPh2, 6.77; NO2, 5.52; PPh2, 6.52; P(O)Ph2, 6.07; P(S)Ph2, 6.17; AsPh2, 6.60; As(O)Ph2, 5.90.

pKa, 4-XPnCO2H, 48.1 mole% aq. EtOH, 25 °C[a]
H, 6.66; Me, 6.85; Cl, 6.23; Br, 6.19; OMe, 7.08; NPh2, 7.10; NO2, 5.41; PPh2, 6.34; P(O)Ph2, 5.82; P(S)Ph2, 5.95; AsPh2, 6.53; As(O)Ph2, 5.76.

7. pK_a 3-XC_6H_4OH, H_2O, 25 °C[a]

H, 10.020; Me, 10.098; Et, 10.07; *t*-Bu, 10.119; F, 9.29; Cl, 9.023; Br, 9.009; I, 9.06; OH, 9.34; OMe, 9.649; OEt, 9.54; CHO, 8.989; Ac, 9.246; COEt, 9.21; CO$_2$Et, 9.10; CONH$_2$, 9.30; CN, 8.61; NO$_2$, 8.346; SMe, 8.75; SO$_2$Me, 9.53; SOMe, 8.61; CH$_2$OH, 9.33; CH$_2$OH, 9.83; OCF$_3$, 9.02; SCF$_3$, 8.97; SO$_2$CF$_3$, 7.87; SF$_5$, 8.57; Ph, 9.50(26 °C); PMe$_2$, 9.66; P(O)(OEt)$_2$, 8.89; P(O)Me$_2$, 8.66(23 °C); P(S)Me$_2$, 8.88; CF$_3$, 8.952.

8. pK_a 4-XC_6H_4OH, H_2O, 25 °C[a]

H, 10.020; Me, 10.276; Et, 10.21; *t*-Bu, 10.232; F, 9.810; Cl, 9.378; Br, 9.36; I, 9.305; OMe, 10.209; OEt, 10.13; CHO, 7.625; Ac, 8.047; COEt, 8.05; Bz, 7.95; CO$_2$Me, 8.47; CO$_2$Et, 8.50; CONH$_2$, 8.56; CN, 7.97; NO$_2$, 6.36; NO$_2$, 7.140; SMe, 9.53; SAc, 8.88; SOMe, 8.28; SO$_2$Me, 7.83; CH$_2$OH, 9.82; OCF$_3$, 9.35; SCF$_3$, 8.66; SO$_2$CF$_3$, 6.79; SF$_5$, 8.37; SCN, 8.57; Ph, 9.40(26 °C); PMe$_2$, 9.41; P(O)Me$_2$, 8.45; P(O)(OEt)$_2$, 8.28(23 °C); P(S)Me$_2$, 8.45; CF$_3$, 8.675; N=CHPh, 9.20; N=NPh, 8.37.

9. pK_a, 3-substituted XC_6H_4OH, 23.6 mole% aq. EtOH, 25 °C[a]

H, 11.02; Me, 11.35; *t*-Bu, 11.78; Br, 11.50; Ac, 10.36; OAc, 9.08; OAc, 10.05(20 °C); CO$_2$Et, 9.1; CN, 9.54; NO$_2$, 9.35; PMe$_2$, 10.90; P(O)Ph$_2$, 10.20; P(S)Ph$_2$, 10.14; PPh$_2$, 10.82; Ph, 11.10(22 °C).

10. pK_a, 4-substituted XC_6H_4OH, 23.6 mole% aq. EtOH, 25 °C[a]

H, 11.01; Me, 11.49; *t*-Bu, 11.85; Br, 10.60; I, 10.42; OMe, 11.50; CHO, 8.40; Ac, 9.06; OAc, 9.49; CO$_2$Et, 8.3; CONH$_2$, 9.65; CN, 8.80; NO$_2$, 7.95; PMe$_2$, 10.67; P(O)Me$_2$, 9.58; P(O)Ph$_2$, 9.48; P(S)Ph$_2$, 9.49; PPh$_2$, 10.46; Ph, 10.92(20 °C); N=NPh, 9.26; NO 6.90.

11. ΔHNP, Me$_2$NC$_6$H$_4$X-4, CHCl$_3$-AcOH (2:1), 20 °C[b]

Br, −604; PPh$_2$, −574; CONH$_2$, −540; CO$_2$Me, −520; POMe$_2$, −516; Ac, −515; PSPh$_2$, −512; POPh$_2$, −490; CHO, −465; CN, −458; NO$_2$, −403.

12. $10^{-5}k$ 3-/4-$XC_6H_4CH_2CHOTsMe$, 50% aq. EtOH, 55 °C[c]

4-OMe, 74.83; 4-Me, 18.09; 4-*t*-Bu, 15.68; 4-SMe, 16.71; 3-Me, 8.43; H, 7.056; 3-OMe, 5.01; 4-F, 4.99; 4-Cl, 3.41; 3-Cl, 2.78; 3-CF$_3$, 2.388; 3-CN, 2.050; 4-NO$_2$, 2.100.

13. $10^{-5}k$ 3-/4-$XC_6H_4CH_2CHOTsMe$, 30% aq. EtOH, 55 °C[c]

4-OMe, 270.1; 4-Me, 83.77; 4-*t*-Bu, 67.41; 4-SMe, 62.13; 3-Me, 34.74; H, 30.44; 3-OMe, 17.71; 4-F, 20.62; 4-Cl, 11.37; 3-Cl, 8.515; 3-CF$_3$, 7.383; 3-CN, 6.552; 4-NO$_2$, 5.525.

14. $10^{-5}k$ 3-/4-$XC_6H_4CH_2CHOTsMe$, 50% aq. TFE, 55 °C[c]

4-OMe, 236.4; 4-Me, 53.78; 4-*t*-Bu, 57.61; 4-SMe, 37.17; 3-Me, 14.77; H, 9.47; 3-OMe, 4.580; 4-F, 5.274; 4-Cl, 1.87; 3-Cl, 1.012; 3-CF$_3$, 0.6917; 3-CN, 0.576; 4-NO$_2$, 0.431.

15. k, 2-$XC_6H_4CO_2H$ + Ph$_2$CN$_2$, in Ph(CH$_2$)$_2$OH at 30 °C[d]

H, 2.87; Me, 2.35; Et, 2.55; *i*-Pr, 3.39; *t*-Bu, 6.40; F, 7.96; Cl, 14.5; Br, 17.3; I, 16.9; CF$_3$, 20.4; OMe, 2.73; OEt, 2.55; OPh, 6.77; NH$_2$, 1.48; NHMe, 1.28; NHPh, 3.25; NHAc, 12.9; SH, 7.40; SMe, 5.27; SO$_2$Me, 56.2; CN, 23.95; OAc, 6.66; CH$_2$CO$_2$Me, 3.65; CH$_2$Ph, 3.72; CH$_2$CH$_2$Ph, 3.07.

(*continued*)

TABLE 22. (continued)

16. $\Delta\nu C(sp)$— H, $XC\equiv CH + Ph_3PO$ in CCl_4, 20 °C[e]

Pe, 118; Et_3Si, 128; OEt, 133; Ph, 145; SEt, 149; $4-BrC_6H_4$, 153; CH_2Cl, 150; CH_2Br, 148; $4-NO_2C_6H_4$, 169; Ac, 188; CO_2Et, 188; CF_3, 216; CN, 258.

17. log K. $XC\equiv CH + Ph_3PO$ in CCl_4, 20 °C[e]

Pe, -0.19; Et_3Si, -0.11; OEt, -0.12; Ph, 0; SEt, 0.15; $4-BrC_6H_4$, 0.18; CH_2Cl, 0.11; CH_2Br, 0.10; $4-NO_2C_6H_4$, 0.21; Ac, 0.47; CO_2Et, 0.42; CN, 1.32.

18. Peak oxidation potentials E_p(v vs. Ag/AgCl) in MeCN; $(4-XC_6H_4)_2$Se, MeCN[f]

NO_2, 1.76; CO_2H, 1.54 Cl, 1.44; H, 1.38; F, 1.38; Me, 1.32; NHAc, 1.25; OMe, 1.22; NH_2, 0.80; NMe_2, 0.68.

19. Peak oxidation potentials E_p(v vs Ag/AgCl) in MeCN; $(4-XC_6H_4)_2$Te, MeCN[f]

NO_2, 1.14; Cl, 0.98; H, 0.95; F, 0.98; Me, 0.89; OMe, 0.80; NH_2, 0.56; NMe_2, 0.50; CF_3, 1.12; CO_2Me, 1.02; Br, 0.89; Ph, 0.80; OH, 0.80; NHPh, 0.66.

20. δ, ^{19}F NMR, $3-FC_6H_4X$[g]

$P(NMe_2)_2$, 0.63; PMe_2, 0.05; $P(OMe)_2$, -0.03; PPh_2, -0.64; PF_2, -2.11; PCl_2, -2.62; $P(CF_3)_2$, -3.12; $P(CN)_2$, -4.65; PF_4, -2.62; SPh, -0.88; SOPh, -3.05; SO_2Ph, -3.14; $N(CN)_2$, -5.75; $N(CF_3)_2$, -2.86; OMe, -0.98; OPh, -1.88; NMe_2, -0.08; SMe, -0.23; F, -3.03; Br, -2.43; Me, 1.23; t-Bu, 0.45; $SiMe_3$, 0.85; Vi, 0.63; Ph, 0, H, 0; CF_3, -2.10; CN, -2.73; NO_2, -3.43; CO_2Et, -0.15; Ac, -0.60.

21. δ, ^{19}F NMR, $4-FC_6H_4X$[g]

$P(NMe_2)_2$, 2.33; PMe_2, 1.40; $P(OMe)_2$, -1.94; PPh_2, -0.45; PF_2, -8.30; PCl_2, -7.37; $P(CF_3)_2$, -8.82; $P(CN)_2$, -9.33; PF_4, -12.09; SPh, 1.40; SOPh, -3.34; SO_2Ph, -7.23; $N(CN)_2$, 0.64; $N(CF_3)_2$, -3.19; OMe, 11.70; OPh, 7.45; NH_2, 14.40; NMe_2, 15.90; SMe, 4.40; F, 6.80; Cl, 3.20; Br, 2.60; I, 1.70; Me, 5.40; Et, 5.05; CH_2Cl, 0.50; CH_2CN, 1.30; Vi, 1.45; Ph, 3.00; C_2H, 2.35; H, 0; CF_3, -5.05; CN, -8.95; NO_2, -9.20; CO_2Et, -5.90; Ac, -6.10.

22. δ ^{13}C NMR, $LNi(CO)_3$[h]

PCl_3, -190.23; $P(OPh)_3$, -193.33; $P(OMe)_3$, -194.82; $P(OEt)_3$, -195.25; $P(O-i-Pr)_3$, -195.54; $AsPh_3$, -195.80; PPh_3, -195.94; $SbPh_3$, -196.50; PMe_3, -196.69; PEt_3, -197.18; PBu_3, -197.33.

23. δ ^{125}Te NMR, $(4-XC_6H_4)_2Te$[f]

NO_2, 753; Cl, 697; H, 689; F, 687; Me, 661; OMe, 649; NH_2, 637; NMe_2, 612; CF_3, 727; Br, 699; Ph, 670; OH, 657; NHPh, 649.

Abbreviations: Vi, vinyl; C_2, ethynylene; Pe, pentyl.

[a] Reference 46. [b] Reference 50. [c] Reference 51. [d] Reference 52. [e] Reference 53. [f] Reference 54. [g] Reference 55–57. [h] Reference 58.

B. Chemical Reactivity

1. Equilibria

The pK_a values of substituted hydroxyl compounds[46], XOH, in water at 25 °C were correlated with the LDR equation. Best results were obtained on the exclusion of the values for $H_2P(O)OH$ and $Ph_2P(O)OH$. The best regression equation is:

$$pK_{a_X} = -16.4(\pm 1.93)\sigma_{lX} - 10.1(\pm 1.25)\sigma_{dX} + 22.9(\pm 5.83)\sigma_{eX} + 12.835(\pm 0.812) \quad (59)$$

$100R^2$, 87.36; A$100R^2$, 86.57; F, 71.43; S_{est}, 1.90; S^0, 0.378; n, 35; P_D, 38.2($\pm$ 5.75); η, $-2.26(0.504)$.

The range of the data encompassed more than 17 orders of magnitude in this set. The value of P_D shows that the localized electrical effect is predominant. The value of η shows that the electronic demand is comparable to that observed for the ionization of 4-substituted phenols. The statistics reported above are described in the Glossary. In view of the many different original sources of the data and the enormous range of structural type, the goodness of fit is quite reasonable.

We have also examined the ionization of 3- or 4-substituted benzoic acids in water and in aqueous ethanol[46]. Wepster and coworkers[47] have noted a medium effect on the ionization of benzoic acids in mixed solvents. We have applied the IMF equation to this problem in order to determine the nature of this effect[48,49]. The correlation equation used has the form:

$$Q_X = L\sigma_{lX} + D\sigma_{dX} + R\sigma_{eX} + A\alpha_X + H_1 n_{HX} + H_2 n_{nX} + h \quad (60)$$

This relationship was found useful in accounting for medium effects on carboxylic acid ionization. For the 3-substituted compounds the best regression equation was obtained after the exclusion of the dimethylamino data point. It is:

$$pK_{a_X} = -1.06(\pm 0.0668)\sigma_{lX} - 0.449(\pm 0.0440)\sigma_{dX} - 1.11(\pm 0.365)\sigma_{eX}$$
$$- 0.431(\pm 0.158)\alpha_X + 4.216(\pm 0.0368) \quad (61)$$

$100R^2$, 94.83; A$100R^2$, 94.15; F, 100.8; S_{est}, 0.0660; S^0, 0.252; n, 27; P_D, 29.8 ($\pm$ 3.32).

For the 4-substituted acids in water, the best regression equation was obtained on the exclusion of the data points for the sulfonamido and acetoxy groups; it is:

$$pK_{a_X} = -1.01(\pm 0.0464)\sigma_{lX} - 1.09(\pm 0.0323)\sigma_{dX}$$
$$- 1.20(\pm 0.200)\sigma_{eX} + 4.162(\pm 0.0192) \quad (62)$$

$100R^2$, 98.75; A$100R^2$, 98.65; F, 658.7; S_{est}, 0.0508; S^0, 0.120; n, 29; P_D, 51.9 ($\pm$ 2.08); η. 1.10 ($\pm$ 0.180).

In 23.6 mole% aqueous ethanol, the best regression equation obtained for the 3-sustituted benzoic acids is:

$$pK_{a_X} = -1.57(\pm 0.0763)\sigma_{lX} - 0.631(\pm 0.0683)\sigma_{dX} + 0.493(\pm 0.242)\sigma_{eX}$$
$$- 0.221(\pm 0.103)\alpha_X + 5.822(\pm 0.0378) \quad (63)$$

$100R^2$, 92.52; A$100R^2$, 92.13; F, 176.2; S_{est}, 0.107; S^0, 0.285; n, 62; P_D, 28.6($\pm$ 3.37).

For the 4-substituted benzoic acids in the same solvent, best results were obtained on the exclusion of the dimethylamino data point giving the regression equation:

$$pK_{a_X} = -1.69(\pm 0.117)\sigma_{lX} - 1.30(\pm 0.0958)\sigma_{dX} + 0.895(\pm 0.350)\sigma_{eX} - 0.399(\pm 0.145)\alpha_X$$
$$+ 0.0710(\pm 0.0334)n_{HX} + 0.0223(\pm 0.0127)n_{nX} + 5.82(\pm 0.0522) \quad (64)$$

$100R^2$, 92.19; A$100R^2$, 91.47; F, 104.3; S_{est}, 0.147; S^0, 0.297; n, 60; P_D, 43.5($\pm$ 3.90); η, $-0.690(\pm 2.65)$.

As the number of data points available for the 3-substituted benzoic acids in 48.1 mole% aqueous ethanol was insufficient for the use of an LDR-type equation, the 3- and 4-substituted benzoic acids were combined into a single data set using the correlation equation:

$$Q_X = \rho\sigma_X + R\sigma_{eX} + A\alpha_X + H_2 n_{nX} + h \tag{65}$$

This relationship is based on the modified Yukawa–Tsuno equation (MYT). No term in n_H was included, because only one substituent had an NH or OH bond. The best regression equation obtained is:

$$pK_{a_X} = -1.31(\pm 0.101)\sigma_X + 0.864(\pm 0.404)\sigma_{eX} - 0.222(\pm 0.126)\alpha_X + 6.621(\pm 0.0625) \tag{66}$$

$100R^2$, 91.70; $A100R^2$, 90.77; F, 62.58; S_{est}, 0.148; S^0, 0.320; n, 21; P_D, 50; η, -0.657 (± 0.303).

We have also examined the ionization of phenols[46]. The correlation equation used was the same as that for the benzoic acids. Best results for the 3-substituted phenols were obtained on exclusion of the methylsulfonyl data point; the regression equation is:

$$pK_{a_X} = -2.37(\pm 0.118)\sigma_{1X} - 0.553(\pm 0.0873)\sigma_{dX} - 1.58(\pm 0.344)\sigma_{eX} + 10.108(\pm 0.0659) \tag{67}$$

$100R^2$, 95.28; $A100R^2$, 94.93; F, 174.8; S_{est}, 0.123; S^0, 0.233; n, 30; P_D, 18.9(± 3.14).

Best results for the 4-substituted phenols were again obtained on exclusion of the methylsulfonyl data point. The regression equation is:

$$pK_{a_X} = -2.77(\pm 0.141)\sigma_{1X} - 2.54(\pm 0.108)\sigma_{dX} + 5.69(\pm 0.535)\sigma_{eX} + 10.005(\pm 0.0683) \tag{68}$$

$100R^2$, 97.29; $A100R^2$, 97.13; F, 383.4; S_{set}, 0.157; S^0, 0.175; n, 36; P_D, 47.8(± 2.59); η, $-2.24(\pm 0.188)$.

As was noted above, the η values for the 4-substituted phenols and for substituted hydroxyl compounds are the same.

Also studied were the ionization constants of phenols in 23.6 mole% aqueous ethanol. Exclusion of data point for acetyl gave best results; the regression equation for the 3-substituted phenols is:

$$pK_{a_X} = -2.73(\pm 0.357)\sigma_{1X} - 0.537(\pm 0.324)\sigma_{dX} + 11.167(\pm 0.123) \tag{69}$$

$100R^2$, 89.28; $A100R^2$, 88.31; F, 41.64; S_{est}, 0.2555; S^0, 0.373; n, 13; P_D, 16.4(± 10.2).

For the 4-substituted phenols, best results are obtained on the exclusion of the values for the acetoxy and nitroso groups. The best regression equation is:

$$pK_{a_X} = -3.22(\pm 0.317)\sigma_{1X} - 3.34(\pm 0.254)\sigma_{dX} + 6.33(\pm 0.818)\sigma_{eX} + 11.253(\pm 0.120) \tag{70}$$

$100R^2$, 96.37; $A100R^2$, 95.89; F, 123.9; S_{est}, 0.236; S^0, 0.216; n, 18; P_D, 50.9(± 5.00); η, $-1.89(\pm 0.198)$.

Finally, we have considered the half-neutralization potentials (HNP) of 4-substituted N,N-dimethylanilines in 2:1 chloroform–acetic acid[50]. The best regression equation is:

$$\Delta HNP_X = 235(\pm 46.7)\sigma_{1X} + 235(\pm 36.4)\sigma_{dX} - 437(\pm 113)\sigma_{eX} + 648(\pm 23.5) \tag{71}$$

$100R^2$, 90.03; $A100R^2$, 87.54; F, 21.08; S_{est}, 20.9; S^0, 0.396; n, 11; P_D, 49.9(± 9.97); η, $-1.86(\pm 0.385)$.

2. Reaction rates

We have studied three data sets for the solvolysis of 1-(3'- or 4'-substituted phenyl)-2-propyl tosylates[51]. We have used the MYT equation in the form:

$$Q_X = \rho\sigma_X + R\sigma_{eX} + A\alpha_X + h \tag{72}$$

The MYT equation uses for 3-substituted compounds σ_m values for σ and 0 for σ_e, while for 4-substituted compounds it uses σ_{C50} values for σ. The best regression equation obtained for 50% aqueous ethanol is:

$$\log 10^{-5}k = -1.15(\pm 0.162)\sigma_X - 3.74(\pm 1.21)\sigma_{eX} + 0.908(\pm 0.0647) \tag{73}$$

$100R^2$, 86.67; $A100R^2$, 85.46; F, 32.51; S_{est}, 0.185; S^0, 0.416; n, 13; P_D, 50; η, 3.25(± 0.948).
The large positive value of η obtained suggests the formation of a carbenium ion. The best regression equation obtained for the results in 30% aqueous ethanol is:

$$\log 10^{-5}k = -1.34(\pm 0.158)\sigma_X - 3.34(\pm 1.18)\sigma_{eX} + 1.505(\pm 0.0631) \tag{74}$$

$100R^2$, 89.94; $A100R^2$, 87.93; F, 26.83; S_{est}, 0.187; S^0, 0.381; n, 13; P_D, 50; η, 2.84(± 1.06).
The results for solvolysis in 50% aqueous trifluoroethanol are:

$$\log 10^{-5}k = -2.39(\pm 0.232)\sigma_X - 5.31(\pm 2.10)\sigma_{eX} + 1.034(\pm 0.133) \tag{75}$$

$100R^2$, 93.54; $A100R^2$, 92.25; F, 43.47; S_{est}, 0.259; S^0, 0.305; n, 13; P_D, 50; η, 2.22(± 0.852).
In order to provide an example involving steric effects, we have examined the correlation of rate constants for the reaction of 2- substituted benzoic acids with diazodiphenylmethane in 2- phenylethanol as the solvent[52]. The correlation equation used has the form:

$$Q_X = L\sigma_{1X} + D\sigma_{dX} + R\sigma_{eX} + A\alpha_X + H_1 n_{HX} + H_2 n_{nX} + S_1 \upsilon_{1X} + S_2 \upsilon_{2X} + S_3 \upsilon_{3X} + h \tag{76}$$

The regression obtained is

$$\log k_X = 1.63(\pm 0.0925)\sigma_{1X} + 0.720(\pm 0.0727)\sigma_{dX} + 0.350(\pm 0.0663)\upsilon_{1X}$$
$$- 0.174(\pm 0.0905)\upsilon_{2X} + 0.724(\pm 0.235)\alpha_X - 0.0271(\pm 0.0149)n_{nX} \tag{77}$$
$$+ 0.335(\pm 0.0570)$$

$100R^2$, 97.08; $A100R^2$, 96.31; F, 99.65; S_{est}, 0.0785; S^0, 0.201; n, 25; P_D, 30.7(± 3.46); η, 0.
There is a significant dependence on the steric effect of the first segment which is adjacent to the active site. The positive sign indicates steric acceleration of the rate. There may possibly be a small steric effect due to the second segment which has a negative sign indicating rate deceleration.

C. Chemical Properties

As our first example of the application of this methodology to chemical properties, we have chosen two sets of measurements of hydrogen-bonding activity in terminal acetylenes[53]. The hydrogen-bond acceptor used was triphenylphosphine oxide in CCl_4. Both equilibrium constants for hydrogen-bond formation and the differences Δv in the infrared frequency of the hydrogen-bonded and the free C(sp)—bond have been correlated with the LDRA equation. The best regression equations obtained are for the Δv values:

$$\Delta v_X = 185(\pm 15.8)\sigma_{1X} + 76.4(\pm 11.9)\sigma_{dX} + 128(\pm 4.40) \tag{78}$$

$100R^2$, 95.23; $A100R^2$, 94.79; F, 99.7; S_{est}, 0.933; S^0, 0.249; n, 13; P_D, 29.2(± 5.08); η, 0.
The corresponding equation for the $\log K$ values is:

$$\log K_X = 1.93(\pm 0.251)\sigma_{1X} + 0.768(\pm 0.184)\sigma_{dX} - 0.14(\pm 0.066) \tag{79}$$

$100R^2$, 89.99; $A100R^2$, 88.98; F, 40.44; S_{est}, 0.141; S^0, 0.365; n, 12; P_D, 28.4(± 7.56); η, 0.

As a second example of chemical properties we have considered peak oxidation potentials of 4, 4′-disubstituted diphenyl selenides and diphenyl tellurides[54]. We consider these quantities to be chemical properties because no covalent bonds have been formed or broken. The best regression equations obtained were for the selenides:

$$E_{pX} = 0.489(\pm 0.108)\sigma_{lX} + 0.635(\pm 0.0801)\sigma_{dX} + 1.54(\pm 0.332)\sigma_{eX} + 1.41(\pm 0.0483) \quad (80)$$

$100R^2$, 96.88; $A100R^2$, 95.99; F, 62.16; S_{est}, 0.0699; S^0, 0.228; n, 10; P_D, 56.5(± 9.80); η, 2.42(± 0.424).

For the tellurides:

$$E_{pX} = 0.275(\pm 0.0353)\sigma_{lX} + 0.372(\pm 0.0237)\sigma_{dX} + 1.34(\pm 0.143)\sigma_{eX}$$
$$+ 0.409(\pm 0.114)\alpha_X + 0.954(\pm 0.0167) \quad (81)$$

$100R^2$, 98.91; $A100R^2$, 98.51; F, 204.1; S_{est}, 0.0242; S^0, 0.130; n, 14; P_D, 57.5(± 5.27); η, 3.60(± 0.309).

The η values obtained are in agreement with oxidation to the radical cation.

D. Physical Properties

We have examined several data sets of NMR chemical shifts as examples of physical properties. Thus, we have correlated ^{19}F chemical shifts of 3- or 4-substituted fluorobenzenes with the LDRA equation[55-57]. Best results for the 4-substituted fluorobenzenes were obtained on exclusion of the value for the dimethylamino group. The regression equation is:

$$\delta^{19}F_X = -12.9(\pm 2.01)\sigma_{lX} - 21.3(\pm 1.49)\sigma_{dX} + 37.6(\pm 4.58)\sigma_{eX} + 3.56(0.932) \quad (82)$$

$100R^2$, 87.91; $A100R^2$, 87.29; F, 92.12; S_{est}, 2.23; S^0, 0.366; n, 42; P_D, 62.4(± 6.32); η, $-1.76(\pm 0.176)$.

For the 3-substituted compounds, the best regression equation is:

$$\delta^{19}F = -7.49(\pm 0.502)\sigma_{lX} - 1.68(\pm 0.704)\alpha_X + 1.230(\pm 0.255) \quad (83)$$

$100R^2$, 86.41; $A100R^2$, 86.04; F, 111.3; S_{est}, 0.631; S^0, 0.384; n, 38; P_D, 0; η, 0.

Also examined were the ^{13}C chemical shifts of substituted nickel carbonyls, $X—Ni(CO)_3^{58}$. The ligands, X, had the form MZ_3 where M is a group 15 element, usually phosphorus. It can be seen on inspection of the equations we have used to estimate electrical-effect parameters for Group 15 tricoordinate substituents that we can use these parameters as a measure of the electrical effect of the X groups in this data set. The best regression equation obtained is:

$$\delta^{13}C_X = 22.9(\pm 0.933)\sigma_{lX} - 199.22(\pm 0.177) \quad (84)$$

$100R^2$, 98.52; F, 600.9; S_{est}, 0.261; S^0, 0.134; n, 11.

Due to collinearities between σ_l and σ_d and σ_e, the nature of the electrical effect in this set is uncertain.

Finally, we have correlated the ^{125}Te chemical shifts of the 4, 4′-disubstituted diphenyl tellurides[54], obtaining

$$\delta^{125}Te_X = 86.0(\pm 12.0)\sigma_{lX} + 88.5(\pm 8.70)\sigma_{dX} + 61.7(\pm 34.9)\sigma_{eX} + 682(\pm 5.56) \quad (85)$$

$100R^2$, 96.54; $A100R^2$, 95.85; F, 83.72; S_{est}, 8.20; S^0, 0.224; n, 13; P_D, 50.7(± 6.58); η, 0.698(± 0.389).

E. Conclusions

We have described a number of examples of the application of the LDR, MYT and IMF equations to the quantitative description of chemical reactivities, chemical properties and physical properties. In many of the data sets studied, we have used parameter values for Group 15 substituents estimated in this work. These substituents include many tricoordinate and tetracoordinate phosphorus groups and some tricoordinate and tetracoordinate arsenic groups. The estimated electrical-effect substituent constants for arsenic groups appear to be reasonable. Unfortunately, we have unable to find data sets which include antimony and bismuth substituents. Our results for these groups are therefore purely speculative. We have found no data which permit a test of the estimated steric parameters for Group 15 substituents. Clearly, this area requires experimental study. Quantitative determinations of suitable chemical reactivities, chemical properties and physical properties are required to fill in this enormous gap in our understanding of the structural effects of Group 15 substituents.

VI. GLOSSARY

General

X A variable substituent.
Y An active site.The atom or group of atoms at which a measurable phenomenon occurs.
G A skeletal group to which X and Y may be attached.
Parameter   An independent variable.
Pure parameter   A parameter which represents a single effect.
Composite parameter   A parameter which represents two or more effects.
Monoparametric equation   A relationship in which the effect of structure on a property is represented by a single, generally composite parameter. Examples are the Hammett and Taft equations.
Diparametric equation   A relationship in which the effect of structure on a property is represented by two parameters, one of which is generally composite. Examples discussed in this work include the LD, CR and MYT equations. Other examples are the Taft, Eherenson and Brownlee DSP (dual substituent parameter), Yukawa–Tsuno (YT) and the Swain, Unger, Rosenquist and Swain (SURS) equations. The DSP equation is a special case of the LDR equation with the intercept set equal to zero. It is inconvenient to use and has no advantages. The SURS equation uses composite parameters which are of poorer quality than those used with the LDR and DSP equations. The MYT equation has all the advantages of the YT equation and gives results which are easier to interpret.
Multiparametric equation   An equation which uses three or more parameters all of which may be either pure or composite.

Electrical-effect parameterization

σ_l The localized (field and/or inductive) electrical-effect parameter. It is identical to σ_1. Though other localized electrical-effect parameters such as σ_l^q and σ_F have been proposed, there is no advantage to their use. The σ^* parameter has sometimes been used as a localized electrical-effect parameter; such use is generally incorrect.
σ_d The intrinsic delocalized (resonance) electrical-effect parameter. It represents the delocalized electrical effect in a system with zero electronic demand.
σ_e The electronic demand sensitivity parameter. It adjusts the delocalized effect of a group to meet the electronic demand of the system.
σ_D A composite delocalized electrical-effect parameter which is a function of σ_d and σ_e.Examples of σ_D constants are the σ_R^+ and σ_R^- constants.The $\sigma_{R,k}$ constants, where k designates the value of the electronic demand η, are also examples of σ_D constants.
σ_R A composite delocalized electrical-effect parameter of the σ_D type with η equal to 0.380. It is derived from 4-substituted benzoic acid pK_a values.

σ_R^0 A composite delocalized electrical-effect parameter of the σ_D type with η equal to -0.376. It is derived from 4-substituted phenylacetic acid pK_a values.

σ_R^+ A composite delocalized electrical-effect parameter of the σ_D type with η equal to 2.04. It is derived from rate constants for the solvolysis of 4-substituted cumyl chlorides.

$\sigma_R^\oplus$ A composite delocalized electrical-effect parameter of the σ_D type with η equal to 3.31. It is derived from ionization potentials of the lowest-energy π orbital in substituted benzenes.

$\sigma_R^\ominus$ A composite delocalized electrical-effect parameter of the σ_D type with η equal to -2.98. It is derived from pK_a values of substituted nitriles.

σ_R^- A composite delocalized electrical-effect parameter of the σ_D type with η equal to -1.40. It is derived from pK_a values of substituted anilinium ions.

$\sigma_{k/k}$ A composite parameter which is a function of σ_1, σ_l and σ_e. Its composition is determined by the values of k and k'. The Hammett σ_m and σ_p constants are of this type.

$\sigma_{Ck'}$ A composite constant that is a function of σ_1 and σ_d, its composition is determined by the value of k'.

σ Any electrical-effect parameter.

η The electronic demand of a system or of a composite electrical-effect parameter that is a function of both σ_d and σ_e. It is represented in subscripts as k. It is a descriptor of the nature of the electrical effect. It is given by R/D, where R and D are the coefficients of σ_e and σ_d, respectively.

P_D The percent delocalized effect. It too is a descriptor of the nature of the electrical effect. It is represented in subscripts as k'.

LDR equation   A triparametric model of the electrical effect.

P_{EA} The percent of the $\sigma_{k'/k}$ values in a substituent matrix which exhibit an electron-acceptor electrical effect.

P_{ED} The percent of the $\sigma_{k'/k}$ values in a substituent matrix which exhibit an electron-donor electrical effect.

P_0 The percent of the $\sigma_{k'/k}$ values in a substituent matrix which do not exhibit a significant electrical effect.

Steric-effect parameterization

r_V The van der Waals radius. A useful measure of group size. The internuclear distance of two nonbonded atoms in contact is equal to the sum of their van der Waals radii.

υ A composite steric parameter based on van der Waals radii. For groups whose steric effect is at most minimally dependent on conformation, it represents the steric effect due to the first atom of the longest chain in the group and the branches attached to that atom. The only alternative monoparametric method for describing steric effects is that of Taft which uses the E_S parameter. This was originally developed only for alkyl and substituted alkyl groups and for hydrogen. Hansch and Kutter[50] have estimated E_S values for other groups from the υ values using a method which, in many cases, disregards the MSI principle. It is best to avoid their use.

Simple branching equation (SB)   A topological method for describing steric effects which takes into account the order of branching by using as parameters n_i, the number of atoms other than H that are bonded to the i-th atoms of the substituent.

Expanded branching equation (XB)   A topological method for describing steric effects which takes into account the order of branching by using as parameters n_{ij} the number of j-th branching atoms bonded to the ith atoms of the substituent.

n_b The number of bonds in the longest chain of a substituent. It is a steric parameter which serves as a measure of the length of a group along the group axis.

MSI principle   The principle of minimal steric interaction which states that the preferred conformation of a group is that which results in the smallest possible steric effect.

Intermolecular force parametrization

α A polarizability parameter defined as the difference between the group molar refractivities for the group X and for H divided by 100. Many other polarizability parameters, such as the van der Waals volume, the group molar volume and the parachor, can be used in its place. All of these polarizability parameters are very highly linear in each other.

n_H A hydrogen-bonding parameter which represents the long-pair acceptor (proton donor) capability of a group. It is defined as the number of OH and/or NH bonds in the group.

n_n A hydrogen-bonding parameter which represents the lone-pair donor (proton acceptor) capability of the group. It is defined as the number of lone pairs on O and/or N atoms in the group.

i A parameter which represents ion–dipole and ion–induced dipole interactions. It is defined as 1 for ionic groups and 0 for nonionic groups.

n_D A charge transfer donor parameter which takes the values 1 when the substituent can act as a charge transfer donor and 0 when it cannot.

n_A a charge transfer acceptor parameter which takes the values 1 when the substituent can act as a charge transfer acceptor and 0 when it cannot.

IMF equation   A multiparametric equation, which models phenomena that are a function of the difference in intermolecular forces between an initial and a final state.

Statistics

Correlation equation   An equation with a data set correlated by simple (one parameter) or multiple (two or more parameters) linear regression analysis.

Regression equation   The equation obtained by the correlation of a data set with a correlation equation.

n The number of data points in a data set.

Degrees of freedom (DF)   Defined as the number of data points (n), minus the number of parameters (N_p) plus 1 [$DF = n - (N_p + 1)$].

F statistic   A statistic which is used as a measure of the goodness of fit of a data set to a correlation equation. The larger the value of F, the better the fit. Confidence levels can be assigned by comparing the F value calculated with the values in an F table for the N_p and DF values of the data set.

$100R^2$ A statistic which represents the percent of the variance of the data accounted for by the regression equation. It is a measure of the goodness of fit.

$A100R^2$ The adjusted value of $100R^2$, which takes into account the number of independent variables. The difference Δ between $A100R^2$ and $100R^2$ is a measure of the extent of collinearity among the independent variables. The smaller the values of Δ, the less the collinearity and the more reliable the results.

S_{est} The standard error of the estimate. It is a measure of the error to be expected in predicting a value of the dependent variable from the appropriate parameter values.

S^0 Defined as the ratio of S_{est} to the root-mean-square of the data. It is a measure of the goodness of fit. The smaller the value of S^0, the better the fit.

VII. REFERENCES

1. L. P. Hammett, *J. Am. Chem. Soc.*, **59**, 96 (1937).
2. L. P. Hammett, *Trans. Faraday Soc.*, **34**, 156 (1938).
3. L. P. Hammett, *Physical Organic Chemistry*, 1st ed., McGraw-Hill, New York, 1940, p. 184.
4. H.C. Brown and Y. Okamoto, *J. Am. Chem. Soc.*, **79**, 1913 (1957).
5. L. M. Stock and H. C. Brown, *Adv. Phys. Org. Chem.*, **1**, 35 (1963).

6. H. van Bekkum, P. E. Verkade and B. M. Wepster, *Recl. Trav. Chim. Pays-Bas*, **78**, 815 (1959).
7. R. W. Taft, *J. Phys. Chem.*, **64**, 1805 (1960).
8. R. W. Taft, *J. Am. Chem. Soc.*, **79**, 1045 (1957).
9. R. W. Taft and I. C. Lewis, *J. Am. Chem. Soc.*, **80**, 2436 (1958).
10. S. Ehrenson, R. T. C. Brownlee and R. W. Taft, *Prog. Phys. Org. Chem.*, **10**, 1 (1973).
11. M. Charton, in *Molecular Structures and Energetics* **4** (Eds. A Greenberg and J. F. Liebman), VCH Publ., Weinheim, 1987, pp. 261–317.
12. M. Charton, *Bull Soc. Chim. Belg.*, **91**, 374 (1982).
13. Y. Yukawa and Y. Tsuno, *Bull. Chem. Soc. Jpn.*, **32**, 965, 971 (1959).
14. M. Charton, *Prog. Phys. Org. Chem.*, **16**, 287 (1987).
15. M. Charton and B. I. Charton, *Abstr. 10th Int. Conf. Phys. Org. Chem.*, Haifa, 1990, p. 24.
16. M. Charton, *Prog. Phys. Org. Chem.*, **13**, 119 (1981).
17. M. Charton, in *The Chemistry of the Functional Groups. Supplement A, The Chemistry of Double Bonded Functional Groups*, Vol. 2, 1 (Eds. S. Patai and Z. Rappoport), Wiley, Chichester, 1989, pp. 239–298.
18. M. Charton, in *The Chemistry of Sulfenic Acids, Esters and Derivatives* (Ed. S. Patai), Wiley, Chichester, 1990, pp. 657–700.
19. M. Charton, unpublished results.
20. O. Exner, in *Correlation Analysis in Chemistry:Recent Advances* (Eds. N. B. Chapman and J. Shorter), Plenum, New York, 1978, pp. 439–540.
21. M. Charton, in *Chemistry of the Functional Groups; Supplement C. Triply Bonded Groups* (Ed. S. Patai), Wiley, Chichester, 1983, pp. 269–323.
22. M. Charton, *J. Org. Chem.*, **28**, 3121 (1963).
23. M. Charton, *J. Org. Chem.*, **49**, 1997 (1984).
24. A. Allred and E. G. Rochow, *J. Inorg. Nucl. Chem.*, **5**, 264 (1958).
25. F. Kehrmann, *Chem. Ber.*, **21**, 3315 (1888); **23**, 130 (1890); *J. prakt. chem.*, [2] **40**, 188, 257 (1889); [2] **42**, 134 (1890).
26. V. Meyer, *Chem. Ber.*, **27**, 510 (1894); **28**, 1254, 2773, 3197 (1885); V. Meyer and J. J. Sudborough, *Chem. Ber.*, **27**, 1580, 3146 (1894); V. Meyer and A. M. Kellas, *Z. phys. chem.*, **24**, 219 (1897).
27. J. J. Sudborough and L. L. Lloyd, *Trans Faraday Soc.*, **73**, 81 (1898); J. J. Sudborough and L. L. Lloyd, *Trans. Chem. Soc.*, **75**, 407 (1899).
28. (a) A. W. Stewart, *Stereochemistry*, Longmans Green, London, 1907, pp. 314–443.
    (b) A. W. Stewart, *Stereochemistry*, 2nd ed., Longmans Green, London, 1919, pp. 184–202.
29. G. Wittig, *Stereochemie*, Akademische Verlagsgesellschaft, Leipzig, 1930, pp. 333–361.
30. G. W. Wheland, *Advanced Organic Chemistry*, 3rd edn., Wiley, New York, 1960, pp. 498–504.
31. K. Kindler, *Ann. Chem.*, **464**, 278 (1928).
32. R. W. Taft, in *Steric Effects in Organic Chemistry*, (Ed. M. S. Newman), Wiley, New York, 1956, pp. 556–675.
33. M. Charton, *J. Am. Chem. Soc.*, **91**, 615 (1969).
34. M. Charton, *Prog. Phys. Org. Chem.*, **8**, 235 (1971).
35. M. Charton, *Prog. Phys. Chem.*, **10**, 81 (1973).
36. M. Charton, *Top. Curr. Chem.*, **114**, 57 (1983).
37. M. Charton, *Stud. Org. Chem.* **42**, 629 (1992).
38. M. Charton, *J. Org. Chem.*, **48**, 1011 (1983); M. Charton, *J. Org. Chem.*, **48**, 1016 (1983).
39. A. Verloop, W. Hoogenstraaten and J. Tipker, *Drug Design*, **7**, 165 (1976).
40. M. Charton and B. I. Charton, *J. Theoret. Biol.*, **99**, 629 (1982).
41. M. Charton, in *Trends in Medicinal Chemistry'88* (Eds. H. van der Goot, G. Domany, L. Pallos and H. Timmerman), Elsevier, Amsterdam, 1989, pp. 89–108.
42. M. Charton, *Prog. Phys. Org. Chem.*, **18**, 163 (1990).
43. M. Charton and B. I. Charton, *J. Org. Chem.*, **44**, 2284 (1979).
44. M. Charton, *Top. Curr. Chem.*, **114**, 107 (1983).
45. M. Charton, in *Rational Approaches to the Synthesis of Pesticides* (Eds. P. S. Magee, J. J. Menn and G. K. Koan), American Chemical Society, Washington, D.C., 1984, pp. 247–278.
46. V. A. Palm (Ed.), *Tables of Rate and Equilibrium Constants of Heterolytic Organic Reactions*, Vol. I, Moscow, 1975; Supplementary Vol. I, Tartu State University, Tartu, 1984.
47. A. J. Hoefnagel and B. M. Wepster, *J. Chem. Soc., Perkin Trans.*, 977 (1989); A. J. Hoefnagel and B. M. Wepster, *Collect. Czech. Chem. Commun.*, **55**, 119 (1990).

48. M. Charton and B. I. Charton, *Abstr. 3rd Eur. Symp. Org. Reactivity*, Göteborg, 1991, p. 102; M. Charton, *Abstr. Fourth Kyushu International Symposium on Physical Organic Chemistry*, Fukuoka/Ube, 1991, pp. 42–48.
49. M. Charton and J. Shorter, *Abstr. 11th Int. Conf. Phys. Org. Chem.*, Ithaca, New York 1992, p. 148; M. Charton and B. I. Charton, *Abstr. 11th Int. Conf. Phys. Org. Chem.*, Ithaca, New York, 1992, p. 149.
50. G. P. Schiemenz, *Angew. Chem.*, **78**, 145 (1966).
51. M. Goto, K. Funatsu, N. Arita, M. Mishima, M. Fujio and Y. Tsuno, *Mem. Fac. Sci. Kyushu Univ.*, **17**, 123 (1989).
52. M. H. Aslam, A. G. Burden, N. B. Chapman, J. Shorter and M. Charton, *J. Chem. Soc., Perkin Trans. 2*, 500 (1981).
53. C. Laurence and R. Queignac, *J. Chem. Soc., Perking Trans. 2*, 1915 (1992).
54. L. Engman, J. Persson, C. M. Andersson and M. Berglund, *J. Chem. Soc., Perkin Trans. 2*, 1309 (1992).
55. R. W. Taft, E. Price, I. R. Fox, I. C. Lewis, K. K. Anderson and G. T. Davis, *J. Am. Chem. Soc.*, **85**, 3146 (1963); J. W. Rakshys, R. W. Taft and W. A. Sheppard, *J. Am. Chem. Soc.*, **90**, 5236 (1968).
56. H. Schindlbauer and W. Prikoszovich, *Chem. Ber.*, **102**, 2914 (1969); W. Prikoszovich and H. Schindlbauer, *Chem. Ber.*, **102**, 2922 (1969).
57. L. J. Kaplan and J. C. Martin, *J. Am. Chem. Soc.*, **95**, 793 (1973).
58. G. M. Bodner, *Inorg. Chem.*, **14**, 1932 (1975).
59. E. Kutter and C. Hansch, *J. Med. Chem.*, **12**, 647 (1969).

CHAPTER 10

Thermochromism of organometallic derivatives containing As, Sb or Bi

H. J. BREUNIG

Universität Bremen, Fachbereich 2 (Chemie), Postfach 330440, D-28334 Bremen Germany

I. INTRODUCTION

Some organometallic derivatives containing Sb—Sb or Bi—Bi bonds exhibit drastic colour changes on melting or dissolution in organic solvents. A representative example is

---

The chemistry of organic arsenic, antimony and bismuth compounds
Edited by S. Patai © 1994 John Wiley & Sons Ltd

tetramethyldistibane,

$$
\begin{array}{ccc}
\text{Me} & & \text{Me} \\
\diagdown & & \diagup \\
& \text{Sb} \!\!-\!\!-\!\!-\!\! \text{Sb} & \\
\diagup & & \diagdown \\
\text{Me} & & \text{Me}
\end{array}
$$

a compound that is yellow as a liquid and in solution. The crystals of this distibane are bright red at temperatures close to the melting point of 18 °C. On further cooling a blue shine is observed. At liquid nitrogen temperature the solid becomes yellow again. Similar colour changes have been observed for several distibanes, dibismuthanes and related compounds.

The first syntheses of tetramethyldistibane, and hence the first observations of the unusual colour changes, date back to the classical work of Paneth who obtained tetramethyldistibane by the reaction of methyl radicals with elemental antimony[1,2]. Fifty years later the experimental basis for the understanding of the colour phenomena of Me_4Sb_2 was laid independently by the groups of Ashe[3] and Becker[4]. It was shown that in the crystal the Me_4Sb_2 molecules are aligned to linear chains with close intermolecular contacts between antimony atoms.

Delocalization of the electrons along the antimony chains was assumed to be responsible for the bathochromic shift of the absorption from the yellow liquid to the red solid[5]. Similar colour phenomena have also been observed for Me_4Bi_2[1,6,7], a compound that is red or orange in solution in organic solvents and green in the crystalline state.

In sharp contrast to their heavier congeners, Me_4As_2 and Me_4P_2 do not exhibit colour changes, since they are colourless both as liquids and as solids. Also, many distibanes and dibismuthanes do not exhibit colour changes. Compounds like Ph_4Sb_2, Mes_4Sb_2 or Ph_4Bi_2 remain yellow or orange on phase transitions.

The term 'thermochromism' is often used for the colour changes of Me_4Sb_2 and related compounds, although it does not cover all the phenomena. In many cases the most drastic bathochromic changes occur reversibly on transitions from fluid to solid phases. A common structural feature of thermochromic derivatives of antimony and bismuth is the formation of polymeric chains or layers in the deeply coloured crystals. These extended structures are formed by association of molecules through secondary bonds. As criteria for secondary bonding, a considerable shortening of intermolecular distances compared to the sum of van der Waals radii or the ratio between contact distances (E...E) and bond distances (E–E) are used[25]. Secondary bonding may be discussed for (E...E)/(E–E) < 1.4. If the formation of extended structures through considerable secondary bonding results in a remarkable bathochromic shift of the area of absorption in the visible part of the UV–VIS spectra, the compound appears to be thermochromic.

Progress in the field of thermochromic distibanes and dibismuthanes has been summarized in the excellent article[8] of Ashe summarizing the literature until mid–1988. Overviews of their recent work in the field have also been published by the group of Becker and by our group[9,10].

In this chapter the topic of colour phenomena that are related to secondary bonding is extended beyond the thermochromic distibanes and comprises also cyclostibanes and organometallic derivatives with bonds of antimony or bismuth with selenium, tellurium or iodine.

II. DIARSANES, DISTIBANES AND DIBISMUTHANES

Distibanes and dibismuthanes have received most of the attention with respect to colour changes. Diarsanes do not show similar phenomena but have been investigated for a better understanding of the thermochromic properties of the heavier congeners. Not all the distibanes and dibismuthanes exhibit colour changes. There is a strong influence of the organic substituents. Red or orange solid phases are formed of distibanes, R_4Sb_2 with $R = Me$, Et, $CH_2=CH$, Me_3M ($M = Si$, Ge, Sn) and various distibanes where the antimony atoms are incorporated in five-membered unsaturated rings (distiboles). Often the substituents that cause colour changes in the case of distibanes have similar effects on dibismuthanes and therefore groups of compounds with the same substituents are discussed together.

A. Tetramethyl Derivatives

1. Syntheses

Many methods have been developed for the synthesis of Me_4As_2 since the early experiments of Cadet de Gassincourt, who obtained tetramethyldiarsane among other products of the reaction of As_2O_3 and potassium acetate[11]. The modern syntheses often use Me_2AsH as starting material. For example, good yields of Me_4As_2 are obtained by reaction of Me_2AsH with Me_2AsNMe_2 or Me_2AsNMe_2, BH_3[12]. For a recent summary of the methods for the preparation of Me_4As_2 and other diarsanes, see References 13 and 14.

Tetramethyldistibane has been obtained by various reductive methods. A simple but effective method is the reduction[15,16] of Me_2SbBr with Mg in THF (equation 1).

$$2Me_2SbBr + Mg \longrightarrow Me_2SbSbMe_2 + MgBr_2 \tag{1}$$

The synthetic methods for distibanes have been summarized by Wieber[17].

Tetramethyldibismuthane is obtained by a reaction sequence starting from Me_3Bi and Na in liquid ammonia. The intermediate, Me_2BiNa, is not isolated but oxidized with $ClCH_2CH_2Cl$ to give the dibismuthane (equations 2 and 3) in 70% yield[7,18].

$$Me_3Bi + 2Na \longrightarrow Me_2BiNa + MeNa \tag{2}$$
$$2Me_2BiNa + ClCH_2CH_2Cl \longrightarrow Me_2BiBiMe_2 + 2NaCl \tag{3}$$

Tetramethyldibismuthane is also obtained by reduction of dimethylbismuth bromide with sodium in liquid ammonia in 78% yield[19].

2. Colours and structures

Tetramethyldiarsane, -distibane and -dibismuthane are bad-smelling, air-sensitive liquids. The thermal stability of Me_4Bi_2 is very low. As was mentioned above, only the distibane and the disbismuthane show colour changes whereas the diarsane is colourless in all the phases. The colours and the UV–VIS spectral data in the different states of aggregation are given in Table 1. Spectra have been recorded using absorption spectroscopy of highly diluted solutions or reflexion spectroscopy of concentrated solutions, pure liquids or solid phases. Spectra of Me_4Sb_2 are shown in Figure 1. The spectra show that, with growing concentration, the edge of absorption is moved to longer wavelengths. This results in a broadening of the area of absorption from a band of absorption to a structured

TABLE 1. Properties of tetramethyl distibane and dibismuthane

Compound	Solid colour	λ_{max} (nm)	Liquid colour	λ_{max} (nm)	Solution colour	λ_{max} (nm)	m.p. (°C)	dec. (°C)	References
$Me_2SbSbMe_2$	red	540	yellow	450	yellow	215	17.5	160	1, 2, 7, 21, 22
$Me_2BiBiMe_2$	green	665	red		red	264	−12.5	25	1, 6, 7

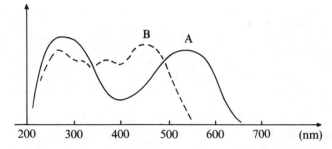

FIGURE 1. Diffuse reflection spectra of solid (A) and liquid (B) Me_4Sb_2

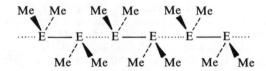

FIGURE 2. Linear chains in crystals of tetramethyl diarsane, distibane and dibismuthane

plateau at high concentrations. The most important spectral change is between the liquid and the solid phases. This bathochromic shift of the edge of absorption is characteristic for thermochromic distibanes and related compounds. Despite the remarkable differences with respect to the colour changes, the solid state structures of tetramethyldiarsane, -stibane and -bismuthane are similar. The three compounds form chains of E_2 units where the methyl groups adopt the *trans* (antiperiplanar) conformation as shown for Me_4Sb_2 (Figure 2). Selected structural data are given in Table 2.

With respect to the origin of thermochromism, the intermolecular contact distances are important. These distances decrease from As to Bi. If the ratio of inter- and intramolecular distances is used as a measure for possible intermolecular interactions with $E \ldots E/E-E = 1.4$ as a borderline between van der Waals (v.d.w.) interactions and secondary bonding[25], there is a clear distinction between the non-thermochromic diarsane and the thermochromic distibane and dibismuthane.

A simple qualitative explanation of the colour effects of Me_4Sb_2 and Me_4Bi_2 is based on the assumption of p orbital overlap in the bismuth or antimony chains (Figure 3). The transition energy between the orbitals in the monomers is larger than the band gap in the polymer and thus the bathochromic shift of the edge of absorption between monomers in

TABLE 2. Selected structural data of tetramethyl diarsane, distibane and dibismuthane

Compound	E–E (pm)	E...E (pm)	2r (v.d.W.) (pm)	E...E/E–E	E–E...E angle (deg)	References
As_2Me_4 solid	243	370	400	1.52	179	20
Sb_2Me_4 solid	284	368	440	1.30	179	3
solid	286	365		1.28		4
gas	282					23
$Bi_2Me_4{}^a$	312	358	460	1.15	178	20

[a] Preliminary results.

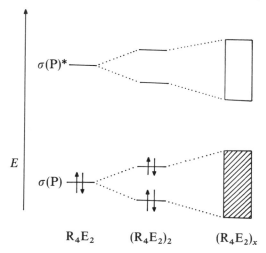

FIGURE 3. Energy levels of distibane (E = Sb) or
dibismuthane (E = Bi) chains

dilute solution and the polymers in the chain is qualitatively explained. This model has
also been used[10] for the interpretation of the UV/VIS spectra in medium concentrations. It
is assumed that equilibria between monomers and oligomers of growing chain length exist
(equation 4).

$$R_4Sb_2 + (R_4Sb_2)_n \rightleftharpoons (R_4Sb_2)_{n+1} \tag{4}$$

An increase in the chain lengths and in the relative amount of oligomers in the equilibria
may be responsible for the observed broadening of the area of absorption with growing
distibane concentration.

The extent of delocalization of electrons, and hence the colours of the antimony chains,
may also be influenced by the presence of different conformers[24] and interaction of lone
pairs. A more extended conjugation should be possible in all *trans* chains of the solid
whereas *gauche* interactions in the liquid might interrupt the delocalization. IR and
Raman spectroscopic investigations of liquid and solid tetramethyldistibane have been
carried out. There are considerable spectral changes between the solid state spectra and
the liquid. The possibility of the presence of both *gauche* and *trans* conformers in liquid
tetramethyldistibane has, however, been discussed controversely[8,26,27].

More theoretical work seems to be necessary for a better understanding also of the
quantitative aspects of the colour changes.

B. Other Tetraalkyl Distibanes and Dibismuthanes

The distibanes and dibismuthanes with longer *n*-alkyl or branched alkyl substituents
are interesting, because in this family of compounds the boderline between thermochromism and non-thermochromic behaviour is reached and some of the colour effects are
different than those observed in the case of the methyl derivatives. For instance, tetraethyl-
distibane does not change colour on melting but reversibly becomes orange on cooling of
the solid. The distibanes, R_4Sb_2, with longer *n*-alkyl or branched alkyl groups (R = Pr^n,
Pr^i, Bu^n) are not thermochromic. They are yellow in concentrated solutions in organic
solvents, in the liquid and in the solid.

TABLE 3. Colours of tetraalkyldibismuthanes and distibanes

Compound	Solid	Liquid	Solution[c]	References
Et_4Sb_2	orange[a], yellow[b]	yellow	yellow	1, 2, 15, 21
Et_4Bi_2	red, violet	red	red, yellow[d]	28, 29
$Pr_4^nSb_2$	yellow	yellow	yellow, colourless[d]	30, 31
$Pr_4^nBi_2$	red, violet	red	red, yellow[d]	29
$Pr_4^iSb_2$	yellow	yellow	yellow, colourless[d]	32
$Pr_4^iBi_2$	red, violet	red	red, yellow[d]	29
$Bu_4^nSb_2$	yellow	yellow	yellow, colourless	32

[a] -110 °C.
[b] -60 °C.
[c] In hydrocarbons.
[d] Dilute solution.

Tetraalkyldibismuthanes, R_4Bi_2 with R = Et, Pr^n, Pr^i, Bu^n, are red liquids that freeze to red violet glasses on cooling. The colour of the solutions in organic solvents depends on the concentration. With decreasing concentration the colour changes from red to yellow. Yellow phases are also observed at liquid nitrogen temperatures. With respect to thermochromism these bismuth derivatives may be considered as borderline cases.

A comparison of the colours of analogous compounds of Sb and Bi is given in Table 3.

C. Tetraaryldistibanes and -dibismuthanes

The tetraaryldistibanes are typical non-thermochromic compounds preserving the yellow colour on phase transitions. The study of the structures and spectra of these derivatives is, however, useful for the sake of comparison. Tetraaryldibismuthanes have been characterized[35] as 'slightly thermochromic' because they show only minor colour changes between more orange or more yellow on cooling from RT to liquid N_2 temperature. Colours and some structural data are given in Table 4. The comparison of the distances and angles of the non-thermochromic derivatives Ph_4Sb_2 and Mes_4Sb_2 with the data of Me_4Sb_2 and other thermochromic distibanes is convincing evidence for the relation between thermochromism and intermolecular association. Ph_4Sb_2 and Mes_4Sb_2 are good examples for crystals built of molecules that have only van der Waals contacts. Closer intermolecular contacts between Sb atoms are hindered because the aryl groups on

TABLE 4. Properties of tetraaryl distibanes and dibismuthanes

		distances and angles			
	Colours	E–E (pm)	E...E (pm)	E–E...E	References
Ph_4Sb_2	yellow	283.7	429	108	33
Ph_4Bi_2	yellow, orange	299.0	[a]		34, 35
p-Tol_4Sb_2	yellow				36
p-Tol_4Bi_2	yellow, orange				37
Mes_4Sb_2	yellow	284.9	638		38

[a] Molecules packed by van der Waals interactions.

each antimony are in a perpendicular position to each other. In accordance with the non-thermochromic behaviour, the UV-VIS spectra of Ph_4Sb_2 in saturated solutions and in the solid state are very similar[55]. There is, however, a hypsochromic shift of the absorption on dilution.

The small colour changes of the tetraaryldibismuthanes do not result from considerable secondary bonding phenomena. This was proven by X-ray structure analysis of crystals of Ph_4Bi_2 in which the molecules approach to van der Waals distances.

D. Heterocyclic and Vinylic Derivatives

Distibanes and dibismuthanes in which the antimony or bismuth atoms are incorporated in five-membered heterocycles are named bistiboles and bibismoles. This important class of thermochromic compounds has been studied intensively by the group of Ashe. The synthesis and X-ray structure determination of tetramethylbistibole,

has provided the major impetus to progress in the field of thermochromic compounds of antimony and bismuth. The purple blue crystals of tetramethylbistibole contain chains of molecules in the all-*trans* conformation with very short intermolecular Sb...Sb contacts. Structural parameters of tetramethylbistibole and the corresponding colourless arsenic compound are listed in Table 5. The corresponding arsole and stibole differ not only with respect to the distances between the molecules, but also with respect to the conformations. Both parameters may influence the solid state colours of these derivatives. The colours of tetramethylbistibole and of related compounds are given in Table 6. Thermochromism is found in analogous compounds of antimony and bismuth. Not all the heterocyclic

TABLE 5. Structural data of tetramethyl biarsole and bistibole

	Conformation	Distances (pm)	(pm)	Angle (deg)	References
	gauche	As—As 244	As...As >400		43
	trans	Sb—Sb 284	Sb...Sb 363	Sb—Sb...Sb 173.5	42

TABLE 6. Colours of bistiboles and bibismoles

| | E = Sb Colours | | | | E = Bi Colours | | | |
	solution	liquid	solid	refs	solution	liquid	solid	refs
HC=C(CH₃)–C(CH₃)=CH ring, Sb)₂	yellow	yellow	blue	42	red		black	41
H₂C–C(=CH₂)–C(=CH₂)–CH₂ ring, E)₂	yellow	yellow	orange	40	red		violet	40
H₂C–CH₂–H₂C–CH₂ ring, E)₂	yellow	yellow	orange	7, 39	red	red	violet	7

distibanes proved to be thermochromic. The derivatives

$$(H_2C \underset{CH_2-CH_2}{\overset{CH_2-CH_2}{\diagup}} Sb)_2 \quad \text{and} \quad (\underset{HC=CEt}{\overset{HC=CEt}{|}} Sb)_2$$

do not exhibit colour changes[39,44]. Like in the case of the tetraalkyl derivatives, the larger substituents hinder the formation of deeply coloured phases and the compounds are yellow.

The thermochromic effect of heterocyclic bistiboles has been the subject of several theoretical publications[5,40,45] with focus on the unsaturated compounds. Band structures have been calculated for the antimony chain of

$$[(\underset{HC=CH}{\overset{HC=CH}{|}} Sb)_2]$$

with extended Hückel methods. The HOMO of the monomeric distibane, and hence the valence band of the polymer, results from a mixing of $Sb(p_z)$ and $Sb(n)$ orbitals whereas the LUMO and the conduction band is made from the π^* orbitals of the diene. For an instructive and critical summary of the theoretical work, see Reference 8.

A detailed study of a series of tetravinyl-distibanes and -dibismuthanes showed that the

TABLE 7. Colours of tetravinyl distibanes and dibismuthanes

Compound	Solution/solid	Solid	References
$(CH_2=CH)_4Sb_2$	yellow	violet	46
$[CH_2=C(CH_3)]_4Sb_2$	yellow	orange	46
$[CH_2=C(CH_3)]_4Bi_2$	red	purple	7
$[E-CH_3(H)CH=CH]_4Sb_2$	yellow	yellow	46
$[Z-CH_3(H)CH=CH]_4Sb_2$	yellow	yellow	46
$[(CH_3)_2C=CH]_4Sb_2$	yellow	yellow	46
$[(CH_3)_2C=CH]_4Bi_2$	red	red	7

presence of two $C=C$ units adjacent to antimony is not a sufficient condition for thermochromism. The colours of the solid phases of these related compounds appeared to be very sensitive to the steric situation on the organic substituents. A rationale could be that the steric requirements for the chain formation in the crystal must be fulfilled. Example of the colours of vinylic distibanes and dibismuthanes are given in Table 7. Colour changes are only observed for $(CH_2=CH)_4Sb_2$, $[CH_2=C(CH_3)]_4Sb_2$ and $[CH_2=C(CH_3)]_4Bi_2$ whereas other vinylic derivatives preserve their colour on phase transitions[46].

E. Arsinostibanes and Stibinobismuthanes

Attempts have been made to study the properties of arsinostibanes or stibinobismuthanes with respect to thermochromism. These heterobinuclear compounds have, however, only been obtained in equilibria with the homobinuclear derivatives (equations 5 and 6).

$$R_2Sb-SbR_2 + R'_2BiBiR'_2 \rightleftharpoons 2R'_2BiSbR_2 \tag{5}$$
$$R, R' = Me^{47}; \quad R = Me, \quad R' = Pr^{n48}$$

$$R_2SbSbR_2 + R_2AsAsR_2 \rightleftharpoons 2R_2AsSbR_2 \tag{6}$$
$$R, R' = Me^{47}$$

In the case of the reaction of a distibole with a diarsole, however, a sufficiently stable arsinostibine[47]

was isolated, which proved to be thermochromic, being yellow as a liquid but violet in the solid state.

F. Other Derivatives of Distibanes in Equilibria

Distibanes of the type R_2SbSbR_2'[49] bearing different substituents and also catena-tristibanes $R_2SbSbRSbR_2$[50] are of interest with respect to thermochromism. Until the present these species were only obtained in equilibrium mixtures. Examples are given in equations 7 and 8.

TABLE 8. Properties and structural data of organometaldistibanes and dibismuthanes

Compound	Colours		Distances		Angles (deg)	Temp.	References
	solution	solid	(pm) Sb–Sb	(pm) Sb...Sb	Sb–Sb...Sb	(°C)	
$(Me_3Si)_4Sb_2$	yellow	red	287	399	165.8	20	51, 52
			286	389	168.4	–120	53
$(Me_3Ge)_4Sb_2$	yellow	red	285	390	170.5	22	54, 55
				386		–110	55
$(Me_3Sn)_4Sb_2^a$	yellow	red	288	388	173	20	53
			287	381	173	–120	53
$(Me_3Sn)_4Sb_2^b$	yellow	red	288	389	174	20	56, 57, 58
			Bi–Bi	Bi...Bi	Bi–Bi...Bi		
$(Me_3Si)_4Bi_2$	red	green	304	380	169	20	59, 60

[a,b] Different modifications.

$$R_2SbSbR_2 + R_2'SbSbR_2' \rightleftharpoons 2R_2SbSbR_2' \qquad (7)$$

$$R_2SbSbR_2 + 1/x\,(R'Sb)_x \rightleftharpoons R2SbSbR'SbR_2 \qquad (8)$$

$$R = Me, Et;\ R' = Et, Pr^n, Bu^n$$

Colour changes are only observed when a thermochromic distibane takes part in the equilibrium and therefore it is difficult to decide if the mixed compound is thermochromic. In samples containing $EtSb(SbMe_2)_2$ and Me_4Sb_2, the area of absorption was extended to longer wavelengths compared with the solid state UV–VIS spectra of the distibane[50].

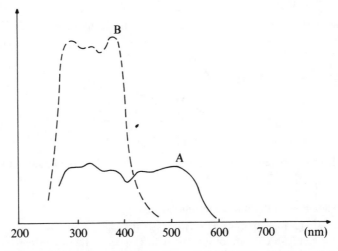

FIGURE 4. Diffuse reflection spectra of $(Me_3Sn)_4Sb_2$:(A) red solid, (B) yellow saturated solution in C_6H_{12} at RT

G. Tetrakis(trimethylsilyl, -germyl and -stannyl)distibanes and Dibismuthanes

The distibanes and dibismuthanes with trimethylsilyl, -germyl and -stannyl substituents are another important group of compounds with respect to colour phenomena, because they exhibit clear colour changes and are well characterized by X-ray diffraction and other methods. The colours and structural parameters are given in Table 8. The colours of the melt are not always known because of decomposition. For the demonstration of the colour effect, therefore, the UV-VIS spectra of the solid are compared with the spectra of the saturated solutions (Figure 4). The crystal structures of all the derivatives have the same basic features that have been observed in the structures of the other deeply coloured distibanes as well. There are antimony or bismuth zig-zag chains with alternate long and short distances of the heavy elements built of molecules in the all-*trans* conformation. In the distibane series there are considerable differences in the intermolecular distances at RT. These variations, however, do not influence very much the colours of the crystals or the UV-VIS spectra. On cooling, the intermolecular Sb ... Sb distances become shorter. A dramatic deepening of the colour is, however, not observed. For $(Me_3Sn)_4Sb_2$ two modifications of red crystals have been prepared with interesting differences in the chain structure[53].

III. ORGANOARSENIC AND ORGANOANTIMONY HOMOCYCLES

Dramatic colour changes are sometimes observed in systems of compounds with the empirical formula $(RE)_n$ ($E = As, Sb$) when ring/chain equilibria are involved and the number of element–element bonds in rings or chains is changed. The chemistry of the cycles[61] and the related chains[62] has been reviewed and is also dealt with in a chapter of this book. There are two principal motives for colour changes: reversible formation of intermolecular contacts or ring/chain transformations.

A. Organoantimony Homocycles with Intermolecular Contacts

Remarkable colour changes that are related to secondary bonding phenomena are rare in the chemistry of organoantimony homocycles. This is not unexpected, because bulky organic substituents are often used to protect the antimony rings and to avoid inter-molecular contacts. The yellow tetramer $(Bu^tSb)_4$[63] is an example of such an isolated ring. The organoantimony homocycles where colour changes and association play a role are listed in Table 9. Only the phenylantimony hexamer and the cage compound $MeC(CH_2Sb)_3$ show some colour changes between the crystalline state and the solution. These molecules are associated to stacks or layers through Sb ... Sb contact distances that indicate only little secondary bonding. The interdependence of intermolecular association and a deep colour of the solid is, however, not observed in the case of the yellow

TABLE 9. Colours and structural data of some associated organoantimony homocycles

| Compound | Colours | | Distances | | |
	solution	solid	Sb–Sb (pm)	Sb ... Sb (pm)	References
$(MesSb)_4$	yellow	yellow	285	390	64
$(PhSb)_6$	yellow	orange	284	419	65
$MeC(CH_2Sb)_3$	yellow	orange	281	397–401	66

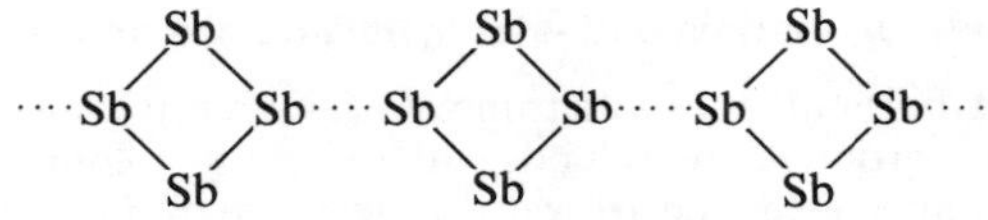

FIGURE 5. Association of Sb atoms in (MesSb)$_4$

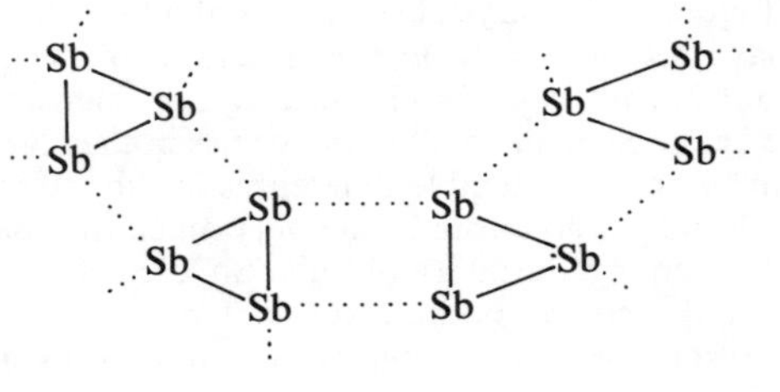

FIGURE 6. Intermolecular association of the Sb skeleton in MeC(CH$_2$Sb)$_3$

mesitylantimony tetramer. The reason for this violation of the rule is not known. The association of (MesSb)$_4$ and MeC(CH$_2$Sb)$_3$ is shown in Figures 5 and 6.

B. Colour Changes Caused by Ring–Chain Interconversions

A prominent example for colour changes due to ring/chain interconversions is the methylarsenic system[62]. (MeAs)$_5$ is yellow in all phases but polymerizes with formation of a purple-black ladder polymer (equation 9).

$$1/5 \,(\text{MeAs})_5 \rightleftharpoons 1/2x \,[(\text{MeAs})_2]_x \tag{9}$$

The structure of this methylarsenic polymer (Figure 7) is very important with respect to colour phenomena and intermolecular interactions. The structure contains two types of As–As distances. The shorter distances correspond to 'normal' covalent arsenic–arsenic bonds and the longer distances result from semibonding interactions.

There are close relations to the chain structures of the thermochromic distibanes. The analogue would be a distibane chain with equal distances between the antimony atoms. Such regular chains should, however, be unstable with respect to Peierls distortion[5] but have been discussed as intermediates between the two modifications of (Me$_3$Sn)$_4$Sb$_2$[53].

Equilibria between rings and chains are common in the alkylantimony ring and chain

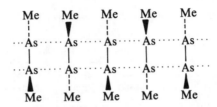

FIGURE 7. Bonding and semibonding interactions of the purple–black methylarsenic polymer

TABLE 10. Colours of seleno- and telluro-stibanes and bismuthanes

	Solution	Liquid	Solid	References
$(Me_2Sb)_2Se$	yellow	orange	red	67
$(Me_2Sb)_2Te$	yellow	brown	blue	67
$Me_2SbTeMe$	red		orange	72, 73, 74
$Et_2SbTeMe$	yellow		orange	73, 74
$p\text{-}Tol_2BiTePh$			yellow, red[a]	76
$Pr_2^nBiTe\ (p\text{---}Tol)$	orange		red	75
$EtSb(TePh)_2$	yellow		red	80

[a] Two modifications.

chemistry (equation 10). The rings are stable only in solution. When the solvent is

$$1/n\,(RSb)_n \underset{\text{yellow}}{\overset{}{\longleftrightarrow}} \quad 1/x\,(RSb)_x \atop \text{black} \tag{10}$$

$$R = Et, Pr^n, Bu^{n\,64}; \ n = 4, 5; \ x \gg 5$$

removed, unstable red oils are formed that polymerize to give the black solid polymers. The structure of the polymeric material is not known. On concentration of the solutions, red oils are formed that slowly give the polymers.

IV. CHALCOGENOSTIBANES AND -BISMUTHANES

The isolobal relations $R_2E^V \leftrightarrow RE^{VI}$ and $RE^V \leftrightarrow E^{VI}$ ($E^V = Sb, Bi; E^{VI} = Se, Te$) that are supported by the thermochromic properties of some ditellurides[60] have stimulated the search for colour effects of compounds with bonds between Sb, Bi and Se or Te. As a result of these efforts, knowledge in synthetic chemistry in the field has expanded considerably and methods have been developed for the preparation of the following types of novel compounds[67,68,72–80]:

$$R_2SbESbR_2 \qquad R_2SbER' \qquad RSb(ER')_2 \qquad Sb(SeR)_3$$
$$R_2BiEBiR_2 \qquad R_2BiER' \qquad RBi(ER')_2$$
$$(E = Se, Te; R, R' = Alkyl, Aryl)$$

Some of the novel compounds, mainly tellurostibanes and tellurobismuthanes, showed considerable colour changes. These derivatives are listed in Table 10. Little is known, however, of the solid structures of these derivatives. One of the few novel selenostibanes, $(MeSe)_3Sb$, has been characterized by X-ray diffraction and indeed considerable association via intermolecular Sb . . . Se contacts between 355 pm and 366 pm has been observed. These contacts are considerably shorter than the sum of van der Waals radii of Sb (220 pm) and Se (190 pm), but the ratio Sb...Se/Sb–Se is close to 1.4 and therefore only weak secondary bonding may be assumed. Consequently, the compound does not change the yellow colour on phase transitions. A section of the structure is given in Figure 8.

This overview shows that more structural and spectroscopic work is needed after the preparative screening of the field. Most promising with respect to colour changes are the antimony or bismuth ditellurides, $RE^V(TeR)_2$ ($E^V = Sb, Bi$).

V. HALOGENOSTIBANES AND -BISMUTHANES

Halogens have also been considered with respect to isolobal relations with organoantimony or organobismuth fragments. The relation $R_2Sb \leftrightarrow I$ is confirmed by some analogies

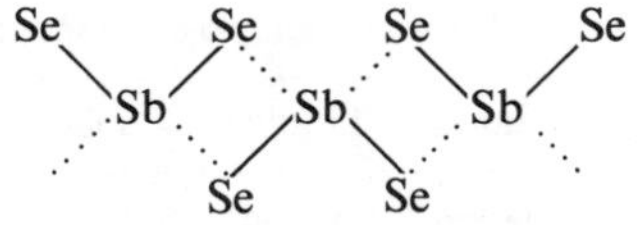

H. J. Breunig

Sb–Se 258 pm; Sb...Se 355–366 pm

FIGURE 8. Intermolecular contacts in a section of the crystal structure of $(MeSe)_3Sb$

TABLE 11. Distances in crystals of $PhSbX_2$ (X = Cl, Br, I) and $MeSbI_2$

	Sb–X (pm)	Sb ... X (pm)	Sb–C (pm)	References
$PhSbCl_2$	238	386	215	9, 70
	241	344		
$PhSbBr_2$	253	406	216	9, 70
	253	362		
$PhSbI_2$	274	408	214	9, 70
	275	382		
$MeSbI_2$	280	347	230	69
	276	340		

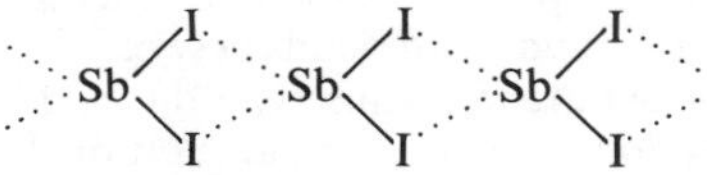

Sb–I 276–280 ppm; Sb...I 340–347 pm

FIGURE 9. Association of $MeSbI_2$ in the crystal

between $Me_2SbSbMe_2$ and I_2. Iodine is a classical example for colour changes on phase transitions and intermolecular association. The mixed compound, Me_2SbI, does not display different colours. It is an unstable yellow oil that forms a yellow glassy solid on cooling. A better candidate for structural investigations is the stable crystalline compound $MeSbI_2$. Indeed, there has been some interest in the structures and intermolecular association of organoantimony and organobismuth dihalides. Despite the considerable degree of association, the colour changes of the yellow organoantimony diiodides are not dramatic. More drastic effects are probable for analogous organobismuth iodides. Early reports on the violet colour of $PhBiI_2$ could not, however, be confirmed[9].

Structural data of phenyl and methylantimony diiodide are listed in Table 11.

The intermolecular association leads to the formation of coordination polymers. The crystal structures of the phenylantimony dihalides have been determined at − 120 °C. The three dihalides crystallize in the same type of crystal structure, i.e. isotypically. There is intermolecular association not only through antimony/halogen contacts but also through η^3-arene interaction. These interactions lead to a two-dimensional network. The intermolecular antimony/halogen contacts are shorter than the sum of van der Waals radii[71] of Sb (220 pm), and Cl (170–190), Br (180–200) or I (195–212) respectively. The intermolecular interactions of phenylantimony dichloride, dibromide and diiodide are not very strong and do not vary significantly with the nature of the halogen. Shorter intermolecular distances have been observed for $MeSbI_2$. The chain structure of this diiodide is shown in Figure 9. UV-VIS spectra of a saturated solution of $MeSbI_2$ in C_6H_{12} ($\lambda_{max} = 360\,nm$) and for the solid ($\lambda_{max} = 495\,nm$) show that there is a considerable bathochromic shift on crystallization of the diiodide from solution. These changes occur, however, on the edge of the visible region of the spectra and are not spectacular.

VI. REFERENCES

1. F. A. Paneth, *Trans. Faraday Soc.*, **30**, 179 (1934).
2. F. A. Paneth and H. Loleit, *J. Chem. Soc.*, 366 (1935).
3. A. J. Ashe III, E. G. Ludwig, J. Oleksyszyn and J. C. Huffman, *Organometallics*, **3**, 337 (1984).
4. O. Mundt, H. Riffel, G. Becker and A. Simon, *Z. Naturforsch.*, **39b**, 317 (1984).
5. T. Hughbanks, R. Hoffman, M.-Hwan Whangbo, K. R. Stewart, O. Eisenstein and E. Canadell, *J. Am. Chem. Soc.*, **104**, 3876 (1982).
6. A. J. Ashe III and E. G. Ludwig, *Organometallics*, **I**, 1408 (1982).
7. A. J. Ashe III, E. G. Ludwig and J. Oleksyszyn, *Organometallics*, **2**, 1859 (1983).
8. A. J. Ashe III, *Adv. Organomet. Chem.*, **30**, 77 (1990).
9. G. Becker and O. Mundt, in *Unkonventionelle Wechselwirkungen in der Chemie metallischer Elemente* (Ed. B. Krebs), DFG Forschungsbericht, VCH Verlagsgesellschaft, Weinheim, 1992, p. 199.
10. H. J. Breunig and S. Gülec, in *Unkonventionelle Wechselwirkungen in der Chemie metallischer Elemente* (Ed. B. Krebs), DFG Forschungsbericht, VCH Verlagsgesellschaft, Weinheim, 1992, p. 218.
11. L. Cadet de Gassincourt, *Mémoires de mathematique et physique des savants étrangers*, **3**, 623 (1760).
12. V. K. Gupta, L. K. Krannich and C. L. Watkins, *Inorg. Chem.*, **25**, 2553 (1986).
13. H. H. Sisler, in *Inorganic Reactions and Methods*, Volume 7 (Ed. J. J. Zuckerman) VCH Publishers, New York, 1988, p. 31.
14. S. Samaan, in *Methoden der organischen Chemie* (ouben-Weyl), 4. Aufl. Vol XIII/8 (Ed. E. Müller), Thieme, Stuttgart, 1978, p. 144.
15. H. J. Breunig, V. Breunig-Lyriti and T. P. Knobloch, *Chemiker Zeitung*, **101**, 399 (1977).
16. H. J. Breunig, in *Organometallic Syntheses*, Vol. 3 (Eds. R. B. King and J. J. Eisch), Elsevier, Amsterdam, 1986, p. 625.
17. M. Wieber, in *Gmelin Handbook of Inorganic Chemistry, Sb Organoantimony Compounds*, Springer, Berlin, Heidelberg, New York, 1981, p. 74.
18. A. J. Ashe, III and E. G. Ludwig, Jr., *Organometallics*, **1**, 1408 (1982).
19. M. Wieber and I. Sauer, *Z. Naturforsch.*, **39b**, 887 (1984).
20. O. Mundt, H. Riffel, G. Becker and A. Simon, *Z. Naturforsch.*, **43b**, 952 (1988).
21. S. Roller, M. Dräger, H. J. Breunig, M. Ates and S. Gülec, *J. Organomet. Chem.*, **378**, 327 (1989).
22. S. Gülec, Ph.D. Thesis, Bremen, 1990.
23. A. G. Csaszar, L. Hedberg, K. Hedberg, E. G. Ludwig, Jr. and A. J. Ashe III, *Organometallics*, **5**, 2251 (1986).
24. V. Breunig-Lyriti, Ph.D. Thesis, Bremen, 1979.
25. A. F. Trotman-Dickenson (Ed.), *Comprehensive Inorganic Chemistry*, Vol. 2, Pergamon Press, Oxford, 1973, p. 1175.
26. H. J. Breunig, V. Breunig-Lyriti and W. Fichtner, *Z. Anorg. Allg. Chem.*, **478**, 111 (1982).
27. H. Bürger, R. Eujen, G. Becker, O. Mundt, M. Westerhausen and C. Witthauer, *J. Mol. Struct.*, **98**, 265 (1983).
28. H. J. Breunig and D. Müller, *Angew. Chem.*, **94**, 448 (1982).
29. H. J. Breunig and D. Müller, *Z. Naturforsch.*, **38b**, 125 (1983).
30. H. A. Meinema, H. F. Martens and J. G. Noltes, *J. Organomet. Chem.*, **51** (1973) 223.
31. H. J. Breunig and H. Jawad, *Z. Naturforsch.*, **37b**, 1104 (1982).
32. H. J. Breunig and W. Kanig, *J. Organomet. Chem.*, **186**, C5 (1980).
33. K. V. Deuten and D. Rehder, *Cryst. Struct. Commun.*, **9**, 167 (1980).
34. F. Calderazzo, A. Morvillo, G. Pelizzi and R. Poli, *J. Chem. Soc., Chem. Commun.*, 507 (1983).
35. F. Calderazzo, R. Poli and G. Pelizzi, *J. Chem. Soc., Dalton Trans.*, 2365 (1984).

36. F. F. Blicke and U. O. Oakdale, *J. Am. Chem. Soc.*, **55**, 1198 (1933).
37. M. Wieber and I. Sauer, *Z. Naturforsch.*, **42b**, 695 (1987).
38. A. H. Cowley, C. M. Nunn and D. L. Westmoreland, *Acta Crystallogr., Sect. C*, **46**, 774 (1990); M. Ates, H. J. Breunig, K. H. Ebert, R. Kaller, M. Dräger and U. Behrens, *Z. Naturforsch.*, **47b**, 503 (1992).
39. H. A. Meinema, H. F. Martens, J. G. Noltes, N. Bertazzi and R. Barbieri, *J. Organomet. Chem.*, **136**, 173 (1977).
40. A. J. Ashe III, C. M. Kausch and O. Eisenstein, *Organometallics*, **6**, 1185 (1987).
41. A. J. Ashe III and F. J. Drone, *Organometallics*, **3**, 495 (1984).
42. A. J. Ashe III, W. Butler and T. R. Diephouse, *J. Am. Chem. Soc.*, **103**, 207 (1981).
43. A. J. Ashe III, W. M. Butler and T. R. Diephouse, *Organometallics*, **2**, 1005 (1983).
44. F. J. Drone, *Diss. Abstr. Int. B*, **47**, 1059 (1986) cited in Reference 8.
45. E. Canadell and S. S. Shaik, *Inorg. Chem.*, **26**, 3797 (1987).
46. A. J. Ashe III, E. G. Ludwig Jr. and H. Pommering, *Organometallics*, **2**, 1573 (1983).
47. A. J. Ashe III and E. G. Ludwig Jr., *J. Organomet. Chem.*, **303**, 197 (1986).
48. H. J. Breunig and D. Müller, *Z. Naturforsch.*, **41b**, 1129 (1986).
49. H. J. Breunig and S. Gülec, *Polyhedron*, **7**, 2601 (1988).
50. M. Ates, H. J. Breunig, K. Ebert, S. Gülec, R. Kaller and M. Dragër, *Organometallics*, **11**, 145 (1992).
51. H. J. Breunig and V. Breunig-Lyriti, *Z. Naturforsch.*, **34b**, 926 (1979).
52. G. Becker, H. Freudenblum and C. Witthauer, *Z. Anorg. Allg. Chem.*, **506**, 42 (1983).
53. G. Becker, M. Meiser, O. Mundt and J. Weidlein, *Z. Anorg. Allg. Chem.*, **569**, 62 (1989).
54. H. J. Breunig, *Z. Naturforsch.*, **33b**, 244 (1978).
55. S. Roller, M. Dräger, H. J. Breunig, M. Ates and S. Gülec, *J. Organomet. Chem.*, **378**, 327 (1989).
56. S. Roller, M. Dräger, H. J. Breunig, M. Ates and S. Gülec, *J. Organomet. Chem.*, **329**, 319 (1987).
57. H. J. Breunig, *Z. Naturforsch.*, **33b**, 990 (1978).
58. H. J. Breunig, *Z. Naturforsch.*, **39b**, 111 (1984).
59. G. Becker and M. Rößler, *Z. Naturforsch.*, **37b**, 91 (1982).
60. O. Mundt, G. Becker, M. Rössler and C. Witthauer, *Z. Anorg. Allg. Chem.*, **506**, 42 (1983).
61. D. B. Sowerby, in *The Chemistry of Inorganic Homo- and Heterocycles*, Vol. 2, (Eds. I. Haiduc and D. B. Sowerby), Academic Press, London, 1987, p. 729.
62. A. L. Rheingold, in *Homoatomic Rings, Chains and Macromolecules of Main-Group Elements* (Ed. A. L. Rheingold), Elsevier, Amsterdam, 1977, p. 385.
63. O. Mundt, G. Becker, H.-J. Wessely, H. J. Breunig and H. Kischkel, *Z. Anorg. Allg. Chem.*, **486**, 70 (1982).
64. M. Ates, H. J. Breunig, S. Gülec, W. Offermann, K. Häberle and M. Dräger, *Chem. Ber.*, **122**, 473 (1989).
65. H. J. Breunig, A. Soltani-Neshan, K. Häberle and M. Dräger, *Z. Naturforsch.*, **41b**, 327 (1986).
66. J. Ellermann, E. Köck and H. Burzlaff, *Acta Crystallogr.*, **C41**, 1437 (1985).
67. H. J. Breunig and H. Jawad, *J. Organomet. Chem.*, **277**, 257 (1984).
68. M. Wieber and I. Sauer, *Z. Naturforsch.*, **39b**, 1668 (1984).
69. H. J. Breunig, K. H. Ebert, S. Gülec, M. Dräger, D. B. Sowerby, M. J. Begley and U. Behrens, *J. Organomet. Chem.*, **427**, 39 (1992).
70. G. Becker, O. Mundt, H. Stadelmann and H. Thurn, *Z. Anorg. Allg. Chem.*, **617**, 59 (1992).
71. J. E. Huheey, *Anorganische Chemie*, Walter de Gruyter, Berlin, 1988, p. 278.
72. A. J. Ashe III and E. G. Ludwig Jr., *J. Organomet. Chem.*, **303**, 197 (1986).
73. H. J. Breunig and S. Gülec, *Z. Naturforsch.*, **41b**, 1387 (1986).
74. H. J. Breunig, *Phosphorus and Sulfur*, **38**, 97 (1988).
75. W. W. du Mont, T. Severengiz, H. J. Breunig and D. Müller, *Z. Naturforsch.*, **40b**, 848 (1985).
76. M. Wieber and I. Sauer, *Z. Naturforsch.*, **42b**, 695 (1987).
77. M. Wieber and I. Sauer, *Z. Naturforsch.*, **39b**, 1668 (1984).
78. H. J. Breunig and S. Gülec, *Z. Naturforsch.*, **43b**, 998 (1988).
79. H. J. Breunig, S. Gülec, B. Krebs and M. Dartmann, *Z. Naturforsch.*, **44b**, 1351 (1989).
80. H. J. Breunig, S. Gülec and R. Kaller, *Phosphorus, Sulfur, and Silicon*, **67**, 33 (1992).

CHAPTER **11**

Electrochemistry

MERETE FOLMER NIELSEN

Department of Chemistry, University of Copenhagen, Denmark

The chemistry of organic arsenic, antimony and bismuth compounds
Edited by S. Patai © 1994 John Wiley & Sons Ltd

I. INTRODUCTION

Of organic As, Sb and Bi compounds, the electrochemistry of organic As compounds has been most explored, the electrochemistry of organic Sb compounds to a lesser extent and the electrochemistry of Bi compounds only very little[1]. Much of the electrochemical work on organic As compounds has been carried out in aqueous solution for analytical purposes due to the earlier widespread industrial applications of organic As compounds and their toxicity to humans. However, the analytical aspects of some of the studies have been reviewed previously[2,3] and will not be treated here, but the general and mechanistic aspects of these studies are included.

Application of catalytic amounts of inorganic salts of the three elements in preparative organic electrochemistry (see, e.g., References 4 and 5) probably involves organometallic intermediates, but this area is not included in the chapter due to the uncharacterized nature of these intermediates.

The electrochemical studies of organic As, Sb and Bi compounds have mainly been conducted by application of classical polarographic methods and cyclic voltammetry at low scan rates combined with preparative and coulometric experiments. Modern high-precision techniques operating at short time scales have not been applied but may in the future advantageously be applied in the unraveling of several of the yet unanswered questions regarding mechanistic and kinetic details of the reactions.

Much of the electrochemistry of organic As compounds closely parallels that of phophorous analogues (which have been explored in much greater detail), whereas the electrochemistry of other organic As, Sb and Bi compounds parallels the electrochemistry of other organometallics. Comparisons are made where relevant.

The chapter is organized in sections according to the type of the electrochemical process (reductions and oxidations) and the oxidation state of the metal atom with further division into classes of compounds. Finally, a brief section is included dealing with the influence of organic As and Sb compounds as ligands on the electrochemistry of transition metal complexes.

II. REDUCTION OF PENTAVALENT As, Sb AND Bi COMPOUNDS

A. Introduction

Reduction of pentavalent compounds has been the most active area of electrochemistry of organic derivatives of arsenic due to the previously widespread application of pentavalent arsenic compounds in medicine, agriculture and chemical warfare. The number of electrochemical studies including organic derivatives of pentavalent antimony and bismuth is, in comparison, rather limited.

The pentavalent compounds may conveniently be divided into five groups: quaternary salts, R_4M^+, tertiary dihalides, R_3MX_2, -onic acids, $RMO(OH)_2$ -inic acids, $R_2MO(OH)$ and tertiary oxides, R_3MO. The five groups will be treated separately in the following sections.

B. Reduction of Quaternary Arsonium and Stibonium Salts, R_4M^+

The quaternary organic derivatives of As, Sb and Bi, like the corresponding derivatives of P, are strong electrolytes in polar solvents. The nature of the anions, normally halides, has no influence on the electrochemistry of the cations. However, no systematic studies of the electrochemistry of bismonium salts have been reported.

Electrochemical studies of quaternary arsonium and stibonium salts have mostly been carried out in aqueous or buffered aqueous solution using polarographic methods, i.e. reduction at a dropping mercury electrode[6]. A few studies in polar aprotic solvents have been carried out.

1. The effect of structure on the reduction potential

All of the quarternary arsonium salts are reduced in a chemically irreversible two-electron process at mercury electrodes in aqueous solution and thereby resemble the analogous phosphonium salts[7-9]. In contrast, the stibonium salts, with the exception of the tetramethylstibonium ion, are reduced in two distinct one-electron steps at mercury electrodes[10]. The arsonium salts are in aqueous solution reduced at potentials which are shifted approximately 300 mV in the positive direction compared with those of the corresponding phosphonium salts[9], whereas the first electron transfer to the stibonium salts takes place at a further 0.5–1.0 V in the positive direction; cf Table 1.

The half-wave potentials are dependent on substrate concentration as well as on the drop time of the mercury electrode and the height of the mercury column, and frequently the current (i) versus potential (E) curves exhibit large polarographic maxima due to adsorption processes. These effects can be minimized by addition of maximum suppressors like gelatine or Triton X-100. A consequence of this non-ideal polarographic behaviour is that results obtained under even slightly different conditions only can be compared in a semi-quantitative way.

The effect on the polarographic half-wave potential of variation of the four organic substituents of the arsonium ions has been studied in aqueous solution[7,9]. The two-electron reduction process is chemically irreversible and, consequently, the effect of a change in substituents on the value of the half-wave potential ($E_{1/2}$) reflects the combined effects of the substituent on the electron-accepting properties of the ion and the kinetics of the follow-up reaction. Taking the tetraphenyl arsonium ion as reference it appears (Table 1) that substitution in one of the phenyl groups has a minor influence on the value of $E_{1/2}$, whereas substitution of one of the phenyl groups by an alkyl group shifts the potential in the negative direction to the extent of approximately 120 mV per alkyl group[9]. This effect is smaller than the effect found for the corresponding phosphonium salts in which substitution of a phenyl group by an alkyl group shifts the value of $E_{1/2}$ approximately 180 mV in the negative direction[9]. Substitution of one of the phenyl groups in Ph_4As^+ by a more electron-withdrawing group such as an alkyl acetate group, —CH_2COOR, or by a benzyl or an allyl group shifts the value of $E_{1/2}$ in the positive direction.

For the stibonium ions the successive substitution of all four phenyl groups in Ph_4Sb^+ by methyl indicates a tremendous difference between As and Sb in sensitivity to

TABLE 1. Half-wave potentials, $E_{1/2}$, for reduction of arsonium, stibonium and bismonium salts, R_4M^{+a}

Compound		$E_{1/2}/V$ vs NHE^b	Reference
Ph_4As^+		-1.14	7
		-1.180	9
	in DME^c	-2.1	11
	in DMF^d	-1.36	12
Ph_4Sb^+	pH 7	-0.394 (1), -0.959 (2)	10
		-0.6 (1), -1.0 (2)	13
Ph_4Bi^+	in DME^c	-2.2	11
$Ph_3(PhCH_2)As^+$		-0.89	7
		-1.016	9
$Ph_3(o\text{-}Tol)As^+$		-1.192	9
$Ph_3(m\text{-}Tol)As^+$		-1.187	9
$Ph_3(p\text{-}Tol)As^+$		-1.185	9
$Ph_3(p\text{-}Cumyl)As^+$		-1.167	9
$Ph_3(p\text{-}PhPh)As^+$		-1.027	9
$(p\text{-}Tol)_4Sb^+$	pH 7	-0.42 (1), -0.98 (2)	14
Ph_3AllAs^+		-0.96	7
Ph_3EtAs^+		-1.36	7
		-1.292	9
Ph_3MeAs^+		-1.20	7
		-1.301	9
Ph_3MeSb^+		-0.47 (1), -1.26 (2)	10
$Ph_3(CH_2CO_2menthyl)As^+$		-0.87	7
$Ph_3(CH_2CO_2Et)As^+$		-1.07	7
$Ph_3(CH_2COPh)As^+$		-1.26	7
$Ph_2MeAllAs^+$		-1.32	7
$Ph_2Et_2As^+$		-1.89	7
$Ph_2Me_2Sb^+$		-0.66 (1), -1.28 (2)	10
$PhMe(PhCH_2)(CH_2CO_2menthyl)As^+$		-0.86	7
$Ph(PhCH_2)_2MeAs^+$		-1.00	7
$PhMe_3As^+$		-2.05	7
$PhMe_3Sb^+$		-0.955 (1), -1.275 (2)	10
$PhEtMe_2As^+$		-2.07	7
$(PhCH_2)_3MeAs^+$		-1.08	7
$(p\text{-}Tol)_2(PhCH_2)MeAs^+$		-1.19	7
Me_4Sb^+		-1.250	10

[a] In water unless otherwise noted.
[b] All potentials are given relative to the NHE although they may have been measured relative to another reference electrode. This conversion of potentials has been made throughout the chapter.
[c] DME = 1,2-dimethoxyethane = glyme.
[d] DMF = N,N-dimethylformamide.

substitution. The potential of the first reduction process is shifted 560 mV in the negative direction from Ph_4Sb^+ to $PhMe_3Sb^{+10}$, but the corresponding shift for As is only 90 mV.

2. Mechanistic studies and the effect of mercury electrodes

In classic polarographic analysis a plot of $\log[i/(i_d-i)]$ vs $-E$ (i is the current at the potential E, and i_d is the limiting, diffusion-controlled current at the plateau of the polarogram) yields a straight line from which the number of electrons, n, can be determined. This is a consequence of equation 1 which, however, is valid only for reversible electron transfer and in the absence of fast chemical follow-up reactions[6].

$$E = E_{1/2} + \frac{RT}{nF} \ln \left[\frac{i_d - i}{i} \right] \tag{1}$$

Application of equation 1 to data for the irreversible reduction of the arsonium salts therefore gives inconclusive results with respect to the number of electrons involved[9]. However, by coulometry the amount of current necessary for complete reduction of a known amount of substrate is measured, and a value of $n = 2$ has been so determined for a number of the arsonium salts[9]. Together with the results of product analysis it may be concluded that the overall reaction in aqueous solution is described by equation 2. A detailed mechanism has not been established but several pathways may be envisaged[9] as outlined in Scheme 1.

$$R_4As^+ + 2e^- + H^+ \longrightarrow R_3As + RH \tag{2}$$

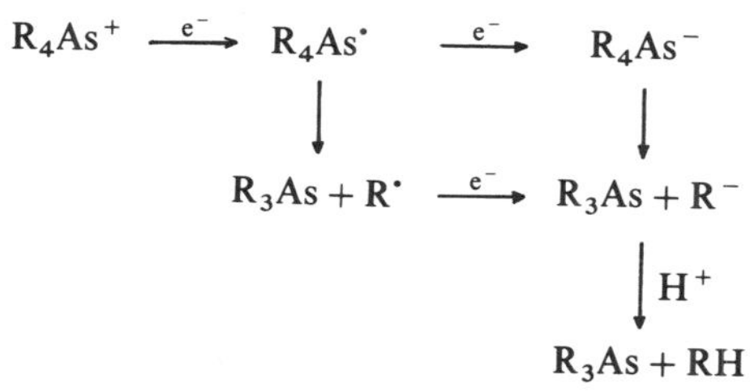

$$R_4As^+ \xrightarrow{e^-} R_4As^{\cdot} \xrightarrow{e^-} R_4As^-$$

$$\downarrow \qquad\qquad\qquad \downarrow$$

$$R_3As + R^{\cdot} \xrightarrow{e^-} R_3As + R^-$$

$$\downarrow H^+$$

$$R_3As + RH$$

SCHEME 1

From preparative studies where the relative amounts of the possible hydrocarbon products (RH) in the unsymmetrically substituted arsonium salts were determined, the following order of cleavage was found[7,8]:

$$CH_2COPh, CH_2COOEt, CH_2Ph, All > Ph > n\text{-}Bu > p\text{-}Tol > Et > Me$$

The results have been statistically corrected for differences in the number of identical substituents. The order of cleavage roughly reflects the relative stabilities of the anionic leaving groups and is almost the same order of cleavage found for reduction of the analogous phosphonium compounds[8]. This may be taken as evidence that the alkyl group is cleaved as an anion and not as a radical and this conclusion is further supported by the absence in the products of dimers of the type R—R. However, the experimentally detected adsorption indicates that products, e.g. $R_4As^{\cdot}$ or $R^{\cdot}$, are adsorbed to the mercury electrode surface, and fast cleavage of the radical can therefore not be ruled out.

The mechanism of reduction at mercury apparently changes dramatically when As is replaced by Sb. This is seen in the shift from one two-electron process to two one-electron processes, where the potential of the first reduction is shifted at least 0.5 V in the positive direction relative to the single reduction process for the analogous arsonium salt. The potential of the second reduction process remains close to that for the arsonium compound; cf Table 1. The main difference between the arsonium and the stibonium salts is their tendency to adsorb on the mercury electrode. From electrocapillary measurements, in which the drop time (t) of the mercury electrode is measured as a function of potential (E) in the absence and in the presence of the stibonium ions, it appears that the t/E relationship is severely effected by the presence of the stibonium ions; the effect is the greater, the larger the number of phenyl groups attached to Sb^{10}. The change in drop time is related to the surface tension of the mercury drop which again reflects the immediate surroundings of the surface, and changes arising from addition of substrate thereby reflect any specific adsorption of substrate or reduction product[6]. The results of the electrocapillary measurements for the stibonium ions, $Ph_mMe_{4-m}Sb^+$, show that the ions as well as the products of the first electrode process adsorb on the electrode[10], whereas the products of the second electrode process do not adsorb because all the electrocapillary curves in this potential range coincide with the curve of the background. For Me_4Sb^+ and for Ph_4As^+ the electrocapillary curves are identical to the background curve in the entire potential range[10,15], and similar reduction behaviour was found for the two ions. The origin of the differences in adsorption behaviour between Ph_4As^+ and Ph_4Sb^+ is not clear, but it has been suggested[15] that the difference could be attributed to the ability of Sb compounds to adopt a distorted trigonal bipyramidal configuration with the mercury at one of the apices, whereas the shorter bond lengths in the As compounds should prevent a change from the tetrahedral configuration.

The direct involvement of the mercury in the reduction process is further indicated by the different behaviour exhibited by Ph_4Sb^+ on stationary carbon electrodes[16]. In that case reduction of Ph_4Sb^+ takes place in a single two-electron process, whereas when the carbon electrode is covered by 1–20 monolayers of mercury, the first scan in cyclic voltammetry shows the same characteristics as voltammetry on a stationary hanging mercury drop electrode, i.e. two distinct reduction peaks[16].

In preparative electrochemical reductions of $Ph_mMe_{4-m}Sb^+$ it has been shown that the only Sb-containing product is a tertiary stibine irrespective of whether the electrolysis is carried out at a potential between the first and the second reduction potential or beyond the second reduction potential[15].

For the reduction of stibonium salts at mercury electrodes two slightly differing mechanisms have been proposed:

Mechanism 1 (Reference 15):

$$R_4Sb^+ \cdots Hg + e^- \longrightarrow R_3Sb + [RHg^\bullet] \quad (3)$$

$$2[RHg^\bullet] \longrightarrow R_2Hg + Hg \quad (4)$$

$$[RHg^\bullet] + e^- + H^+ \longrightarrow RH + Hg \quad (5)$$

Mechanism 1 is based on the results of studies of Ph_4Sb^+ in buffered aqueous solution[15]. In cyclic voltammetry (CV) the peak corresponding to the second electron transfer (equation 5) is considerably smaller than the peak corresponding to the first reduction (equation 3), which according to Mechanism 1 could be explained by the influence of the disproportionation reaction 4 on the concentration of $[RHg^\bullet]$. Involvement of H^+ in the second electron transfer (equation 5) is indicated by the dependency of $E_{1/2}$ on pH[15]—the wave shifts in the positive direction when pH is decreased. A similar variation with pH was observed for reduction of $(p\text{-Tol})_4Sb^{+14}$. Further support for Mechanism 1 comes from the fact that it is Ph_2Hg and not benzene which is isolated after controlled potential electrolysis at the potential of the first reduction process, for which $n = 1$[15], whereas only benzene was isolated after electrolysis at the potential of the second reduction process, for which $n = 2$[15]. Additionally, a polarogram of an equimolar mixture of Ph_4Sb^+ and Ph_2Hg showed only three reduction processes: the reduction of Ph_2Hg to $PhHg^\bullet$, the first reduction of Ph_4Sb^+ and a third wave of double height attributed to the reduction of $PhHg^\bullet$ resulting from both the prior processes[15].

The second mechanism was proposed based on a study of the series of stibonium salts of the composition $Ph_mMe_{4-m}Sb^+$ mentioned above[10].

Mechanism 2 (Reference 10):

$$Ph_mMe_{4-m}Sb^+ + e^- \longrightarrow Ph_mMe_{4-m}Sb^\cdot \tag{6}$$

$$Ph_mMe_{4-m}Sb^\cdot + Hg \longrightarrow Ph_{m-1}Me_{4-m}Sb + [PhHg^\cdot] \tag{7}$$

$$2[PhHg^\cdot] \longrightarrow Ph_2Hg + Hg \tag{8}$$

$$Ph_mMe_{4-m}Sb^\cdot + e^- \longrightarrow Ph_mMe_{4-m}Sb^- \tag{9}$$

$$Ph_mMe_{4-m}Sb^- + H^+ \longrightarrow Ph_{m-1}Me_{4-m}Sb + PhH \tag{10}$$

In Mechanism 2 the second electron transfer (equation 9) is a direct reduction of the initially formed radical and not of the phenylmercury radical (cf reaction 5 in Mechanism 1). In this study no pH dependence of the potential of the second reduction process was observed[10]; the reason for this discrepancy is not clear. It was argued in References 1 and 10 that the presence of only three reduction peaks in the experiment where equimolar amounts of Ph_2Hg and Ph_4Sb^+ were studied was due to a fortuitous coincidence of two separate peaks. The alternative intermediate proposed in Mechanism 2, $Ph_mMe_{4-m}Sb^\cdot$, which should give rise to the second peak by reaction 9[10], is, however, not a likely candidate, because the lifetime as measured by cyclic voltammetry is so short that no trace of reoxidation was found even when the sweep rate was increased to $100\,V\,s^{-1}$. The experimental evidence on which Mechanism 2 was mainly based comes from the apparent difference in product distribution. The ratio of phenyl to methyl cleavage differs, when the preparative electrolysis of the three salts with a mixture of methyl and phenyl substituents is carried out at the potential of the first and second reduction, respectively. No difference in product distribution was found for $PhMe_3Sb^+$ (only Me_3Sb found) but for $Ph_2Me_2Sb^+$ as well as for Ph_3MeSb^+ the amount of product due to cleavage of a phenyl group increased when the reduction was carried out at the potential of the second reduction[1,10] (NB mistake in Table II in Reference 10). These results were interpreted as preferential cleavage of a radical, reaction 7, after the first reduction, with the ratio of phenyl to methyl cleavage being determined by the thermodynamic stability of the resulting products. The roughly equal stabilities of $Ph^\cdot$ and $Me^\cdot$ make the relative stability of the product tertiary stibine the factor determining which group is cleaved. Consequently, the cleavage should favour symmetrically substituted over unsymmetrically substituted stibines, and Ph_2MeSb over $PhMe_2Sb$. On the other hand, after two-electron reduction, i.e. reaction via equations 6, 9 and 10, the relative stability of the Sb-containing products is identical to the previous situation but anion stability which favours phenyl cleavage is claimed to be responsible for the shift in observed product distribution in favour of products due to cleavage of the phenyl group[1,10].

No systematic studies of the reduction of R_4M^+ have been carried out in aprotic solvents. Reduction of Ph_4As^+ in DMF on mercury has been reported[12] to take place in one two-electron process, while the reduction of Ph_4Bi^+ in MeCN has been reported to take place in two one-electron steps[1]. No mention was made of the number of electrons involved in the reduction processes in DME[11].

There are indications, however, that the reduction mechanism in aprotic solvents may be different for cations in which one of the substituents possesses α-hydrogens. The reduction of a series of 10-methyl-10-arylphenoxarsinium salts of the general structure 1 has been studied in DMF by polarography in the absence and in the presence of added proton donors[17].

In the absence of proton donors the reduction of 1 proceeds in two one-electron steps in contrast to the two-electron reduction found for Ph_4As^+ under the same conditions[17]. Upon addition of phenol, the first reduction wave grows to a two-electron wave at the expense of the second reduction wave[17] and thereby parallels the behaviour observed for the reduction in aprotic solvents of phosphonium salts bearing α-hydrogens[18]. In both

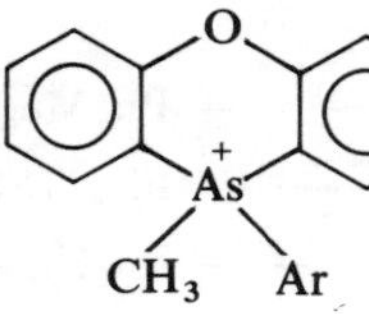

(1)

cases the suggested mechanism (Scheme 2) involves protonation of the expelled carbanion (for **1** the aryl substituent is expected to be cleaved[17]) by the α-hydrogen of the substrate cation which is converted to the corresponding ylide. This ylide may then be reduced at a lower potential corresponding to the second reduction wave observed in the absence of proton donors.

$$R_3(CH_2R')M^+ + 2e^- \longrightarrow R_2(CH_2R')M + R^-$$

$$R_3(CH_2R')M^+ + R^- \longrightarrow R_3M{=}CHR' + RH$$

Total reaction: $2R_3(CH_2R')M^+ + 2e^- \longrightarrow R_2(CH_2R')M + R_3M{=}CHR' + RH$

SCHEME 2

When a proton donor is added, the mechanism changes from two-electron reduction of half the amount of substrate ($n = 1$) to two-electron reduction of all the substrate ($n = 2$), because the added proton donor successfully competes with the substrate in proton transfer to the carbanion. In the case of the phosphonium ions the presence of the ylide was confirmed by the reduction potential of independently prepared ylide and by the isolation of the ylide after preparative electrolysis[18]. The ylide corresponding to structure **2** was not isolated[17] but from the height of the second polarographic wave the reduction of the supposed ylide is a two-electron process[17].

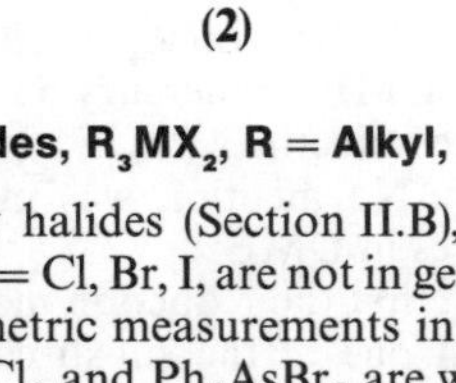

(2)

C. Reduction of Tertiary Dihalides, R_3MX_2, R = Alkyl, Aryl, X = Cl, Br, I

In contrast to the quaternary halides (Section II.B), the tertiary dihalides, R_3MX_2, M = As, Sb, Bi, R = alkyl, aryl, X = Cl, Br, I, are not in general strong electrolytes but have a covalent structure. Conductometric measurements in the polar aprotic solvent acetonitrile have shown[19] that Ph_3AsCl_2 and Ph_3AsBr_2 are weak electrolytes with molar conductances (Λ_m) smaller than 20 ohm^{-1} cm^2 mol^{-1}, while Ph_3AsI_2 and Ph_3AsBrI have $\Lambda_m = 40$–70 ohm^{-1} cm^2 mol^{-1} (for strong electrolytes Λ_m is of the order of 100–200 ohm^{-1} cm^2 mol^{-1}). The analogous Ph_3PX_2 compounds are all strong electrolytes.

The actual course of the partial ionization reaction of Ph_3AsX_2 has been shown to depend on the nature of X^{19}. When X = Br the ionization equilibrium is a simple dissociation reaction:

$$Ph_3AsBr_2 \rightleftharpoons Ph_3AsBr^+ + Br^- \tag{11}$$

For the dichloride, $X = Cl$, the ionization takes place as a disproportionation reaction:

$$2Ph_3AsCl_2 \rightleftharpoons Ph_3AsCl^+ + Ph_3AsCl_3^- \tag{12}$$

The diiodide and the mixed bromide–iodide ionize according to equation 13, where in the mixed compound the halide forming the strongest bond to As, i.e. the bromide, remains in the complex:

$$2Ph_3AsIX \rightleftharpoons Ph_3AsX^+ + Ph_3As + I_3^- \tag{13}$$

The electrochemical reduction of R_3MX_2 has received very little attention, and the influence of the partial ionizations described above for As has not been discussed. The data in Table 2 summarize the measured potentials for reduction of R_3MX_2. The most interesting observation is that the reduction of tertiary dihalides such as Ph_3AsX_2 apparently takes place at *higher* potentials than the reduction of the corresponding quaternary salt, Ph_4As^+; cf Table 1.

In one of the few studies made it is concluded that the reduction of Ph_3MX_2 at mercury in glyme (DME) takes place in a single two-electron process according to equation 14, i.e. by formation of the corresponding tertiary arsine, stibine or bismuthine, R_3M,

$$R_3MX_2 + 2e^- \longrightarrow R_3M + 2X^- \tag{14}$$

which can be further reduced at lower potentials[11]; cf Section III.A. For M = As or Bi, the polarographic waves showed maxima which could not be suppressed[11], and involvement of mercury is likely. The differences in $E_{1/2}$ values found for Ph_3MBr_2, M = As, Sb, Bi[11] (cf Table 2) may therefore be caused by differences in adsorption rather than by the changes in electronic structure. The reduction of Ph_3BiCl_2 in CH_2Cl_2 at mercury electrodes also takes place in a single process which was shown by coulometry to require two electrons[21]. The reaction is also in this case associated with cleavage of two Cl^-, but the Bi-containing product was suggested to be a mixed Bi–Hg complex, $Ph_2BiHgPh$[21]. The mixed Bi–Hg product was not isolated but suggested on the basis of its reduction potential, which coincides with the potential of reduction of a mercury containing product formed during oxidation of Ph_3Bi[21]; cf Section V.E.

TABLE 2. Half-wave potentials, $E_{1/2}$, for reduction of tertiary dihalides of As, Sb and Bi, R_3MX_2

Compound	Conditions	$E_{1/2}$/V vs NHE[a]	Reference
Me_3SbCl_2	1 M HCl	−0.54	20
	1 M NH_4OH	−1.00	20
Me_3SbBr_2	1 M NH_4OH	−1.00	20
Ph_3AsBr_2	DME	−0.0 (1), −2.5 (2)[b]	11
Ph_3SbCl_2	1 M HCl	−0.67	20
	DME	−1.1 (1), −2.4 (2)[b]	11
Ph_3SbBr_2	DME	−0.5 (1), −2.4 (2)[b]	11
Ph_3BiCl_2	CH_2Cl_2	−0.2 (1), −1.3 (2)[c]	21
Ph_3BiBr_2	DME	−0.2 (1), −2.2 (2)[b]	11

[a] All potentials have been recalculated to be given against NHE.
[b] Reduction of the product, R_3M.
[c] Reduction of a product, possibly $Ph_2BiHgPh$[21].

D. Reduction of Arsonic Acids, RAsO(OH)₂, R = Alkyl, Aryl

Studies of the electrochemical reduction of $RMO(OH)_2$ are scarce for $M = Sb$ and Bi, while the reduction of arsonic acids, $RAsO(OH)_2$, especially the arylarsonic acids, $ArAsO(OH)_2$, has received considerable attention since 1930[22-24]. The electrochemical reductions of arsonic acids have been carried out almost exclusively by polarographic methods in buffered aqueous solution, and the work has focused on three main aspects: (1) the effects of substituents on $E_{1/2}$ and correlation of $E_{1/2}$ with pK_a, (2) mechanistic investigations and (3) analytical applications.

1. The effect of structure on the reduction potential

Like most other organic arsenic containing compounds, the arsonic acids are more or less susceptible to adsorption, and because the influence of adsorption on the half-wave potential is dependent on substituents, concentration and the presence of surfactants, it is

TABLE 3. Half-wave potentials, $E_{1/2}$, for reduction of arylarsonic acids in buffered aqueous solution

Substituent	pH	$E_{1/2}$/V vs NHE	Reference
—	2.97	−0.97	22
	2.2	−0.820	25
	1.7	−0.705	26
	1.0	−0.737	27
	1.10	−0.640[a]	28
2-Cl	1.0	−0.873	27
2-MeO	1.0	−0.835	27
2-NH₂	2.2	−0.831	25
	1.7	−0.697	26
	1.0	−0.654	27
2-NO₂	1.7	−0.620 (2)[b]	26
2-OH	1.0	−0.600	27
3-NH₂	2.2	−0.840	25
3-NO₂	1.7	−0.710 (2)[b]	26
	1.0	−0.68 (2)[b]	29
4-Cl	1.7	−0.660	26
4-Me	2.97	−0.99	22
	1.7	−0.715	26
4-MeCONH	2.97	−1.00	22
4-MeO	2.97	−1.06	22
4-NH₂	2.2	−0.857	25
	1.7	−0.740	26
4-NO₂	1.7	−0.760 (2)[b]	26
4-OH	1.7	−0.722	26
	1.3	−0.70	29
2,4-di-OH	1.7	−0.750	26
2,4-di-OH-5-NO₂	1.7	−0.790	26
3-NH₂-4-OH	1.7	−0.735	26
3-NO₂-4-OH	1.7	−0.775 (2)[b]	26
α-Naphthyl AsO(OH)₂	1.3	−0.35	30

[a] Determined by differential pulse polarography.
[b] Reduction potential for the arsonic acid group reduction; reduction of the nitro group takes place at higher potentials.

difficult to compare $E_{1/2}$ values determined under even slightly differing conditions. Any discussion of the influence of structure on the reduction potential can therefore only be qualitative except in cases where the potentials have been determined under exactly identical conditions.

It is a general observation that reduction of arsonic acids in aqueous solution only takes place under acidic conditions (otherwise, the reduction of water is the first electrochemical process taking place when the potential is scanned in the negative direction), and that the value of the half-wave potential is pH-dependent. In Table 3 the half-wave potentials for reduction of a number of arylarsonic acids are given. For several of the acids, values of $E_{1/2}$ are given at more than one pH value in order to facilitate direct comparison within as large a number of substituted acids as possible. Data from older studies obtained at large concentrations of the substrate ($> 10\,mM$) and under unbuffered conditions[22,23] are not included.

Nitro-substituted phenylarsonic acids are reduced in two chemically irreversible processes in acidic aqueous solution corresponding to initial reduction of the nitro group to a hydroxylamine[26] or an amine[27,31] followed by reduction of the arsonic acid group. However, when a nitro group—or a naphthylazo group—is present *ortho* to the arsonic acid group, the reduction of the nitro or azo group has been reported to be severely influenced by adsorption[32]. In polar aprotic solvents like DMF or MeCN without added acids, the product of the initial one-electron reduction of nitro-substituted phenylarsonic acid is relatively stable, and ESR spectra confirm that the spin density primarily is located in the nitro group[33], i.e. the initial reduction process can be regarded as reduction of a substituted nitrobenzene.

Correlation of the half-wave potentials of *meta*- and *para*-substituted phenylarsonic acids with the Hammett substituent constants, σ, has been tried[26]. The equation $E_{1/2} = (0.053 \pm 0.010)\sigma - 0.705\,V$ vs NHE ($r = 0.9946$) was obtained for a series of five compounds (unsubstituted, 4-OH, 4-NH$_2$, 4-Me, 3-NH$_2$-4-OH) studied in 0.02 M HCl, while under the same conditions the equation $E_{1/2} = 0.173\sigma - 0.703\,V$ vs NHE ($r = 0.9991$) was obtained for the second reduction of a series of three NO$_2$-substituted phenylarsonic acids using the σ value for —NHOH for the reduced nitro group[26], i.e. apparently two different correlations exist. However, the possible protonation of the amino or hydroxylamino groups under the experimental conditions was not taken into account, and the validity of assigning the σ value of the reduced substituent to the combined substituent/arsonic acid reduction taking place at the potential of the second wave was not discussed. It is therefore not surprising that the data cannot be described by a simple Hammett relation. For the simple *para*-substituted phenylarsonic acids a linear relation (equation 15) was found between the half-wave potential and the pK_a value of the arsonic acid.

$$E_{1/2} = -0.438\,pK_a - 0.547\,V \text{ vs NHE} \tag{15}$$

The influence of *ortho* substituents on the polarographic half-wave potential shows, when compared to the influence of *para* substituents, that intramolecular hydrogen-bonding between the OH in the arsonic acid group and *ortho* substituents like OH, NH$_2$ and NHOH probably takes place, resulting in larger positive shifts of $E_{1/2}$ than expected from purely electronic effects. The *ortho* chloro substituent, probably for steric reasons (twisting of the arsonic acid group), results in a negative shift of $E_{1/2}$[27]; cf Table 3. The *ortho*-substituted phenylarsonic acids do not obey the linear relation between $E_{1/2}$ and pK_a found for the *para*-substituted acids (equation 15).

Aliphatic arsonic acids have been investigated to a much smaller extent than the aromatic arsonic acids. A series of alkylarsonic acids with pK_a values in the range 3.6–4.2 have been studied in buffered aqueous solution[34] and, like the arylarsonic acids, the reduction wave vanishes at higher pH values (> 2.0). The alkylarsonic acids have $E_{1/2}$ values which are in general lower than those found for aromatic arsonic acids, but which

TABLE 4. Half-wave potentials, $E_{1/2}$, for the reduction of alkylarsonic acids

Compound	pH[a]	$E_{1/2}$/V vs NHE	Reference
MeAsO(OH)$_2$	MeOH[b]	−1.43	35
	MeCN, 4% MeOH[b]	−1.39	35
	1.78	−1.09	34
EtAsO(OH)$_2$	1.78	−0.96	34
PrAsO(OH)$_2$	1.7	−0.87	26
	1.78	−0.91	34
BuAsO(OH)$_2$	1.78	−0.86	34
PenAsO(OH)$_2$	1.78	−0.80	34
HexAsO(OH)$_2$	1.78	−0.72	34
HeptAsO(OH)$_2$	1.78	−0.65	34
OctAsO(OH)$_2$	1.78	−0.60	34
PhCH$_2$AsO(OH)$_2$	1.10	−0.471[c]	28
PhCH$_2$CH$_2$AsO(OH)$_2$	1.10	−0.567[c]	28

[a] In aqueous solution unless otherwise noted.
[b] Using the acidic guanidinium perchlorate (0.1 M) as supporting electrolyte; with neutral supporting electrolytes no reduction takes place.
[c] Determined by differential pulse polarography.

steadily increase with the size of the alkyl group (cf Table 4). For the largest alkyl groups (hexyl to octyl) the arsonic acids exhibit strong adsorption effects[34], which only in the case of hexylarsonic acid could be suppressed by addition of Triton X-100. Also, in aprotic solvents acidic conditions are necessary for the reduction to take place; in MeCN this involves use of a supporting electrolyte (0.1 M) with an acidic cation such as guanidinium[35]. Addition of weaker acids like phenol or benzoic acid is not effective in promoting the reduction of the alkylarsonic acids in aprotic solvents[35].

2. Mechanistic studies

The arylarsonic acids have pK_a values in the range 3.5–4.5, and they can be reduced only within a narrow pH range well below their pK_a values. For phenylarsonic acid the peak height in differential pulse polarography (DPP) was observed to double, when pH was changed from 2.0 to 1.1[28]. The limiting polarographic currents, i_l, are diffusion-controlled[25,26,30], $i_l = i_d$, except for reduction of α-naphthylarsonic acid, where the limiting current seems to be partially controlled by adsorption as judged from the effect of the height of the mercury column, h, on the limiting current[30], i_l. For diffusion-controlled processes the limiting current is proportional to the square root of the (corrected) height of the mercury column[6].

In an early study of three arylarsonic acids[29] the number of electrons consumed in the reduction process was estimated to be equal to six and, consequently, that the product was the corresponding arylarsine, ArAsH$_2$. The estimation of the number of electrons was made by comparison of the polarographic wave height for the arsonic acid process with the wave heights of the two first processes observed for 3-nitrophenylarsonic acid at pH 0.9, which were attributed to the four-electron reduction of nitro to hydroxylamine and the two-electron reduction of hydroxylamine to amine, respectively[29].

Values of n have been based on a combination of the polarographic wave heights and on the results of microcoulometry, in which a small volume of solution containing a known concentration of the substrate is partially (15–20%) reduced by application of the dropping mercury electrode as cathode, while the height of the limiting current is followed as a function of time[36]. The number of electrons (n) consumed in the reduction of arylarsonic acids was in this way determined to be equal to four[25,26], and the following reaction sequence was suggested for the polarographic reduction[26]:

$$ArAsO(OH)_2 + 2e^- + 2H^+ \longrightarrow ArAs(OH)_2 + H_2O \qquad (16)$$

$$ArAs(OH)_2 + 4e^- + 4H^+ \longrightarrow ArAsH_2 + 2H_2O \qquad (17)$$

$$ArAs(OH)_2 + ArAsH_2 \longrightarrow 2/x(ArAs <)_x + 2H_2O \qquad (18)$$

The intermediate, arylarsonous acid [$ArAs(OH)_2$] is identical with the species obtained when the trivalent arsenic compound arylarsine oxide, $ArAs=O$, is dissolved in water and it is *easier* to reduce than the corresponding pentavalent arylarsonic acid; cf Section III.D. The reaction sequence 16–18 requires an average of four electrons per molecule of the arsonic acid, because the six-electron product, arylarsine ($ArAsH_2$), reacts with the two-electron intermediate, $ArAs(OH)_2$, with formation of a polymeric arsenobenzene, $(ArAs <)_x$. The polymer has been isolated after preparative reduction of $PhAs=O$; cf Section III.D. The same number of electrons, $n = 4$, was obtained in a different study[25] and the structure $ArAs=AsAr$ was assigned to the product. The product was not isolated but a similar structure was shown to be the main product of preparative reduction of 4-aminophenylarsine oxide, $4\text{-}NH_2C_6H_4As=O$[37], cf Section III.D. However, it was shown in the same, very early study[37] that preparative reduction of 4-aminophenylarsonic acid in 2 M HCl using an amalgamated Zn electrode resulted in 4-aminophenylarsine which could be isolated as the hydrochloride (80%) (the 4-aminophenylarsine is oxidized in air). Preparative reduction of a number of *ortho*-substituted phenylarsonic acids at mercury electrodes in 2 M HCl also leads to the corresponding *ortho*-substituted phenylarsines in good yields (65–80%)[27]. If the *ortho* substituent is NO_2, the expected 2-aminophenylarsine is obtained in a mixture with other products, because the product of the reduction of the nitro group is very dependent on the exact conditions. Prior chemical reduction of the nitro group to the amine gave a much cleaner arsine product as a result of the electro-chemical reduction of the arsonic acid group[27,31].

In macroscale reduction of unsubstituted phenylarsonic acid at a mercury pool electrode, the final product is an oxygen-containing polymer, which contains As—As as well as As—O—As bonds[30]. Precipitation of $PhAsH_2$ was also observed during the electrolysis[30], and it was suggested that products were formed by the following reactions:

$$PhAsO(OH)_2 + 2e^- + 2H^+ \longrightarrow PhAs(OH)_2 + H_2O \qquad (19)$$

$$PhAs(OH)_2 \rightleftharpoons 1/y(PhAs{-}O)_y + H_2O \qquad (20)$$

$$PhAs(OH)_2 + 4e^- + 4H^+ \longrightarrow PhAsH_2 \qquad (21)$$

The difference between the products obtained when $PhAsO(OH)_2$ is reduced under polarographic and macroscale conditions, respectively, has been explained by the differences in the relative concentrations of the different species[30]. Under polarographic conditions the concentration of $PhAs(OH)_2$ is so low that polymerization (reaction 20) does not take place, and $PhAs(OH)_2$ reacts preferentially with $PhAsH_2$ (reaction 18). Precipitation of $PhAsH_2$ is also of minor importance under polarographic conditions due to the lower concentrations.

Polarographic analysis of a series of substituted phenylarsonic acids[26,29,30] shows that the reduction process is electrochemically irreversible giving linear plots of log $[(i_d - i)/i]$ vs E from which a value of $\alpha n = 0.49$ was determined according to equation 22[26] for all members of the series except for the *p*-Cl derivative[26]. The αn value was independent of pH in the range 1–2 and independent of the concentration of the arsonic acid[26].

$$E = E_{1/2} + \frac{0.0542}{\alpha n F} \log\left(\frac{i_d - i}{i}\right) \qquad (22)$$

A smaller pH-independent value of αn (0.30) was found for the unsubstituted phenylarsonic acid by application of cyclic voltammetry (CV)[38], and the value of αn increased slightly by increasing concentration of the substrate, probably due to adsorption at the highest concentrations (> 1 mM).

Both wave heights and half-wave potentials are dependent on pH in reduction of the arylarsonic acids, and it was found[26] that a plot of $E_{1/2}$ vs pH in the pH range 1.0–2.0 was linear for all of the derivatives except the nitro-substituted one with slopes equal to the slopes of the log $[(i_d - i)/i]$ vs E plots. This behaviour corresponds to the involvement of one proton prior to the irreversible electron transfer step[26,29]. The initial step, equations 16 and 19, in the reduction process therefore includes the microscopic steps described by equations 23 and 24.

$$ArAsO(OH)_2 + H^+ \rightleftharpoons ArAs(OH)_3{}^+ \qquad (23)$$

$$ArAs(OH)_3{}^+ + e^- \longrightarrow ArAs(OH)_3{}^{\cdot} \qquad (24)$$

The values of the reaction constant, ρ, found in the Hammett correlations of $E_{1/2}$ discussed above, are relatively small compared to other polarographic reaction constants[39]. This has also been taken as an indication of electron transfer to the protonated substrate in which case a positively charged species is reduced to a neutral one with the consequence that stabilization of the reduction product by electron-withdrawing groups is of limited importance[26].

All of the investigated alkylarsonic acids have pH-dependent half-wave potentials in buffered aqueous solution[34], and the reduction process is chemically irreversible. The reduction of methylarsonic acid, $MeAsO(OH)_2$, in buffered aqueous solution takes place at such negative potentials (-1.09 V vs NHE) that interference from the background current excludes mechanistic measurements[35,40]. In 0.1 M H_2SO_4 no reduction peak was observed for $MeAsO(OH)_2$ by CV on a mercury electrode before the background[41], but if the cathodic sweep was continued into the background, reduction took place as evidenced by a sharp oxidation peak in the following anodic sweep[41] attributed to the oxidation of $MeAsH_2$ produced in the cathodic scan and adsorbed on the mercury electrode. The same oxidation peak was observed when $MeAsO(OH)_2$ was chemically reduced prior to the electrochemical experiment[41]. In MeCN, using guanidinium perchlorate as supporting electrolyte, the wave is separated from the background, and it was shown that the limiting current was diffusion-controlled[35]. Contrary to the pH dependence found in aqueous solution, $E_{1/2}$ in MeCN was independent of concentration of the acidic supporting electrolyte in the concentration range 0.01–0.5 M[35]. Coulometric experiments on $MeAsO(OH)_2$ in 96% MeCN/4% MeOH gave an n value close to one[35], which is in disagreement with all of the mechanisms discussed above for the arylarsonic acids in aqueous solution. However, no products were isolated and no mechanism suggested to explain the coulometric n value. Preparative experiments with product isolation have not been reported for any of the alkylarsonic acids either in aqueous or in non-aqueous solutions.

E. Reduction of Arsinic Acids, $R_2AsO(OH)$, R = Alkyl, Aryl

The electrochemistry of the mono acidic analogues of arsonic acids, the arsinic acids, has been explored to a much smaller extent than that of the arsonic acids (Section II.D). This is despite the fact that dimethylarsinic acid (cacodylic acid), and especially its reduction product tetramethyldiarsine (cacodyl), have been known for more than a hundred years. Salts of dimethylarsinic acid have, like the salts of methylarsonic acid, been widely used as herbicides and pesticides, and the major part of the electrochemical studies are aimed at analysis.

1. The effect of structure on the reduction potential

The dialkylarsinic acids have larger pK_a values (in the range 6.0–6.5) than the corresponding alkylarsonic acids, and they are electrochemically reducible in buffered aqueous solution in a wider pH range (pH < 4)[34]. For alkyl groups larger than ethyl, all the dialkylarsinic acids exhibit adsorption effects which can be minimized by addition of Triton X-100[34]. Similarly to alkylarsonic acids, the half-wave potential for reduction of dialkylarsinic acids shifts in the positive direction when the size of the alkyl group increases (cf Table 5) and when the pH of the solution is decreased[34]. In general, the dialkylarsinic acids are slightly easier to reduce than the corresponding monoalkylarsonic acid, although quantitative comparisons are difficult to make because the half-wave potentials for the two series have been obtained at different pH. The same tendency was found for the reduction potentials of diphenyl-, dibenzyl- and bis(2-phenylethyl)arsinic acid compared to phenyl-, benzyl- and 2-phenylethylarsonic acid[28] (cf Tables 3, 4 and 5); also for these compounds the reduction potentials are pH-dependent, shifting in the positive direction when pH is lowered[28].

A qualitative study of the effect of the position of the nitro groups in bis(nitrophenyl)arsinic acids on the potential of the second wave has been reported[43]. Like the nitrophenylarsonic acids, the reduction of nitro-substituted diphenylarsinic acids takes place in two distinct steps. The first reduction wave corresponds to reduction of the nitro

TABLE 5. Half-wave potentials, $E_{1/2}$, for the reduction of arsinic acids[a]

Compound	pH[a]	$E_{1/2}$/V vs NHE	Reference
Me$_2$AsO(OH)	4.3	−1.13	35
	MeOH[b]	−1.56	35
	MeCN, 4% MeOH[b]	−1.50	35
	2.3	−0.86	40
Et$_2$AsO(OH)	2.66	−0.88	34
Pr$_2$AsO(OH)	2.66	−0.81	34
Bu$_2$AsO(OH)	2.66	−0.76	34
(c-Hex)$_2$AsO(OH)	2.66	−0.71	34
(PhCH$_2$)$_2$AsO(OH)	1.80	−0.480[c]	28
(PhCH$_2$CH$_2$)$_2$AsO(OH)	1.80	−0.540[c]	28
Ph$_2$AsO(OH)	1.0	−0.6	42
	1.80	−0.430[c]	28
2, 2′-bis(NH$_2$C$_6$H$_4$)AsO(OH)	1.04	−0.632	25
3, 3′-bis(NH$_2$C$_6$H$_4$)AsO(OH)	1.04	−0.559	25
4, 4′-bis(NH$_2$C$_6$H$_4$)AsO(OH)	1.04	−0.534	25
3, 4′-bis(NH$_2$C$_6$H$_4$)AsO(OH)	1.04	−0.541	25

[a] In aqueous solution unless otherwise noted. All potentials converted to potentials vs NHE.
[b] 0.1 M guanidinium perchlorate used as supporting electrolyte.
[c] Determined by application of differential pulse polarography (DPP).

group to hydroxylamine and the second, non-ideal, wave was interpreted as a combination of further reduction of the hydroxylamine to amine and reduction of the arsinic acid group[43]; in all cases the second wave of the substituted compounds appeared at higher potentials than the reduction wave of the unsubstituted diphenylarsinic acid[43].

Dimethylarsinic acid appears to be the only arsinic acid studied in non-aqueous solvents[35]. In agreement with the findings for methylarsonic acid, reduction only takes place in MeOH or in MeCN in the presence of an acidic supporting electrolyte (guanidinium perchlorate). In contrast to the results obtained in buffered aqueous solution[34], the half-wave potential for reduction of $Me_2AsO(OH)$ in MeOH or in MeCN is *lower* than the half-wave potential for reduction of $MeAsO(OH)_2$ under the same conditions[35].

2. Mechanistic studies

Of the diarylarsinic acids only the unsubstituted diphenylarsinic acid has been studied in detail by polarography in buffered aqueous solution[42]. The reduction is associated with adsorption as indicated by a prewave and a polarographic maximum on the main wave which, however, could be suppressed by addition of Triton X-100[42]. The height of the reduction wave is directly proportional to the concentration of $Ph_2AsO(OH)$ in the range 10^{-4}–10^{-3} M, and the limiting current (i_l) is almost entirely diffusion-controlled[42]. Similar conclusions were drawn for a series of bis(aminophenyl)arsinic acids based on measurements of i_l as a function of the concentration and the height of the mercury column[25].

For the unsubstituted diphenylarsinic acid in 0.1 M HCl, $n = 3$ was determined by microcoulometry[42]. The same n value was found for $Ph_2AsO(OH)$ at pH 2.2[43], and for the series of bis(aminophenyl)arsinic acids[25] by microcoulometry at pH 2.2. Similarly, a value of $n = 3$ was found for reduction of the arsinic acid group in a series of bis(nitrophenyl)-arsinic acids at pH 2.2. The value was based on the relative wave heights of the first (four-electron reduction of each of the two nitro groups) and the second (two-electron reduction of each of the two hydroxylamine groups plus reduction of the arsinic acid group) reductions[43]. In all three cases the product was suggested to be the dimer $(Ar_2AsAsAr_2)$[25,42,43], although no products were isolated. It was proposed that the product was formed via reactions 25–27[42].

$$Ar_2AsO(OH) + 2e^- + 2H^+ \longrightarrow Ar_2AsOH + H_2O \qquad (25)$$

$$Ar_2AsOH + 2e^- + 2H^+ \longrightarrow Ar_2AsH + H_2O \qquad (26)$$

$$Ar_2AsH + Ar_2AsOH \longrightarrow Ar_2AsAsAr_2 + H_2O \qquad (27)$$

The intermediate two-electron product, arsinous acid (Ar_2AsOH), is expected to be reduced further to the arsine (Ar_2AsH) at *higher* potentials than the initial reduction of the starting material (cf Section III.C.), but the four-electron product appears from the coulometry not to be the final product due to a favourably competing dimerization (reaction 27). The dimerization reaction is analogous to the oligomerization reaction 18 of $PhAs(OH)_2$ and $PhAsH_2$, which was suggested to take place during the reduction of $PhAsO(OH)_2$[26] (cf Section II.D.2).

Polarographic analysis of the reduction wave for $Ph_2AsO(OH)$ shows that the reduction process is electrochemically irreversible like the reduction of the corresponding arsonic acid. From equation 22 the pH-independent value $\alpha n = 0.66$ was determined[42], which is larger than the value found for reduction of $PhAsO(OH)_2$ ($\alpha n = 0.49$[26], Section II.D.2). Like with arylarsonic acids, a linear dependence of $E_{1/2}$ on pH was found for the reduction of $Ph_2AsO(OH)$ in the pH range 1.0–2.4. In combination with the value of αn, the slope of the $E_{1/2}$ vs pH relationship indicates by analogy with the arylarsonic acids that one proton is involved before the irreversible electron transfer[42]. The initial reaction 25 is therefore likely to include the microscopic steps described by equations 28 and 29.

$$Ph_2AsO(OH) + H^+ \rightleftharpoons Ph_2As(OH)_2^+ \tag{28}$$

$$Ph_2As(OH)_2^+ + e^- \longrightarrow Ph_2As(OH)_2^{\cdot} \tag{29}$$

Detailed electrochemical studies of the reduction of dialkylarsinic acids have only been carried out for $Me_2AsO(OH)$[35,40]. The limiting polarographic current is diffusion-controlled and proportional to the concentration of $Me_2AsO(OH)$ (2×10^{-5}–3×10^{-3} M) in buffered aqueous solution in the pH range 1.9–5.4[35,40].

The number of electrons required for reduction of $Me_2AsO(OH)$ has been determined by bulk electrolysis, $n \approx 5$[35], but the approach was complicated by passivation of the mercury pool electrode, probably due to coating by the product. Mass spectrometric analysis (MS) of the gaseous products showed peaks corresponding to AsH_3^+, $MeAsH_2^+$, Me_2AsH^+ and Me_3As^+ as well as traces of $Me_2AsAsMe_2^+$[35,40]. The major product detected by gas chromatography (GC) was the four-electron product Me_2AsH (30%) together with trace amounts of $MeAsH_2$, whereas no trace of the dimer ($Me_2AsAsMe_2$) or of Me_3As was detected by GC[35]. The additional species found by MS may arise from fragmentation processes[35,40], and it was concluded[35] that the main reaction taking place can be described by the overall equation 30.

$$Me_2AsO(OH) + 4e^- + 4H^+ \longrightarrow Me_2AsH + 2H_2O \tag{30}$$

In preparative electrolysis at constant current in 1 M H_2SO_4 using an amalgamated zinc electrode, the main product was, however, shown to be the dimer ($Me_2AsAsMe_2$)[37] ($n = 3$) in agreement with the findings for $Ph_2AsO(OH)$, while only small amounts of Me_2AsH were found[37]. The differences in the main product formed during preparative electrolysis may be due to differences in concentrations and in cell geometry. If $Me_2AsAsMe_2$ is formed as an intermediate product in concentrated solutions, its insolubility may prevent the further reduction to the dimethylarsine if it precipitates, and the electrode is not a mercury pool. The reduction potential of $Me_2AsAsMe_2$ has not been determined.

Polarographic analysis of the reduction of $Me_2AsO(OH)$ shows the electron transfer to be irreversible, as observed for $Ph_2AsO(OH)$, and a value of $\alpha n = 0.56$ was determined at pH 4.0[35]. The half-wave potential ($E_{1/2}$) was also found to vary linearly with pH in the pH range 1.5–5.0 with a slope corresponding to the involvement of one proton prior to the rate-determining step[35] as found for $Ph_2AsO(OH)$. The initial steps in the reduction of $Me_2AsO(OH)$ can therefore also be described by reactions 28 and 29 in accord with the known amphoteric nature of $Me_2AsO(OH)$ [pK_a for $Me_2As(OH)_2^+$ is ca 1.6][35]. Above pH 5 the slope of $E_{1/2}$ vs pH increases noticeably due to the ionization of $Me_2AsO(OH)$ [pK_a for $Me_2AsO(OH) = 6.2$] and the consequent requirement for more than one proton to form the electroactive species[35].

In non-aqueous solution (MeOH or MeCN) using the acidic guanidinium perchlorate as supporting electrolyte, $E_{1/2}$ was independent of the concentration of the guanidinium ion in the range 0.01–0.5 M[35]. Bulk electrolysis in 96% MeCN, 4% MeOH gave $n = 1.1$, despite the fact that both the height of the polarographic wave and the value of αn determined from equation 22 were similar to those found in buffered aqueous solution[35]. No products were isolated or identified. The reason for the discrepancy between polar-ographic and coulometric results in not clear but parallels the findings for reduction of $MeAsO(OH)_2$ under the same conditions[35] (cf Section II.D.2).

F. Reduction of $R_3M{=}O$ and Related Compounds

In contrast to the electrochemical studies of arsonic and arsinic acids (Sections II.D and II.E), most of the electrochemical studies of trisubstituted arsine and stibine oxides and their derivatives have been made in non-aqueous solution. The largest number of studies concern the reduction of triphenylarsine oxide (Ph_3AsO) which, independently of the solvent, is reduced in a chemically irreversible process. The only R_3MO species known to give a relatively stable one-electron reduction product is $(4\text{-}NO_2C_6H_4)Et_2AsO^{33}$, which like 4-nitrophenylarsonic acid in aprotic solvents gives an anion radical with the spin density located primarily in the nitro group. No systematic studies have been made of the influence of changing the alkyl or aryl substituents on the reduction properties of the arsine or stibine oxides.

1. Reduction of Ph₃AsO in buffered aqueous solution and in water/methanol mixtures

In contrast to phenylarsonic acid and diphenylarsinic acid, triphenylarsine oxide has been reported to be electroactive in a wide pH range, pH $1-12^{44-46}$. The electrochemical behaviour of Ph_3AsO in buffered aqueous solution is severly distorted by adsorption[45]. In addition to a polarographic maximum, the plateau current shows a number of sharp discontinuities. The discontinuities have been explained by potential-dependent breakdowns of Ph_3AsO (and its reduction product) molecules adsorbed in orientationally different, well-defined states[45]. In 0.1 M HCl well-defined polarograms could be obtained for 2×10^{-4} M solutions of Ph_3AsO if relatively large amounts of surface active agent ($> 0.005\%$ Triton X-100) were added; under those conditions a single well-defined polarographic wave was observed, $E_{1/2} = -0.56$ V vs NHE^{45}. The adsorption problems are also diminished by changing the solvent from water to water/methanol mixtures; in buffered 40/60 (v/v) $H_2O/MeOH$ the adsorption maximum completely disappeared[46]. The limiting current, i_l, was shown to be diffusion-controlled in buffered aqueous solution based on the effect of the height of the mercury column on the limiting current, and proportional to the substrate concentration, whereas the foot of the wave still showed partial adsorption-controlled behaviour[45]. Diffusion control was also found in buffered $H_2O/MeOH$ mixtures[46].

A value of $n = 2$ was determined by microcoulometry in 0.1 M HCl, and after microcoulometry Ph_3As was identified by TLC as the product[45]. In 10/90 (v/v) $H_2O/MeOH$ mixtures $n = 2$ was determined from the height of the polarographic waves and an estimate of the diffusion coefficient for Ph_3AsO in this medium ($D = 0.47 \times 10^{-5}$ cm^2 s^{-1})[46].

The half-wave potential for reduction of Ph_3AsO in buffered aqueous solution is pH-dependent like the half-wave potential for the reduction of arsonic and arsinic acids. The potential shifts in the negative direction when pH is increased[44-46], and it was concluded[45] that one proton was involved prior to the rate-determining step, although due to the adsorption problems the polarographic analysis gave less reliable data than for the reduction of arsonic and arsinic acids. The same conclusion was reached for reduction in buffered $H_2O/MeOH$ mixtures[46], where it additionally was shown that at pH > 6 the diffusion current decreases and completely vanishes at pH 12. Polarographic investigations of the reduction of Ph_3AsO in pure MeOH in the presence of small amounts of benzoic acid showed that when the concentration of the benzoic acid was smaller than twice the substrate concentration, the wave height was directly proportional to the concentration of the benzoic acid, indicating that the overall reduction process consumes two electrons and two protons[46].

Based on the above observations, the suggested mechanism for reduction of Ph_3AsO under acidic conditions in water or water/methanol mixtures is described by equations $31-33^{44,45}$:

$$Ph_3AsO + H^+ \rightleftharpoons Ph_3AsOH^+ \tag{31}$$

$$Ph_3AsOH^+ + e^- \longrightarrow Ph_3AsOH^{\cdot} \tag{32}$$

$$Ph_3AsOH^{\cdot} + e^- + H^+ \longrightarrow Ph_3As + H_2O \tag{33}$$

2. Reduction of R_3MO and R_3MS in non-aqueous solvents

In the absence of benzoic acid the half-wave potential for reduction of Ph_3AsO in MeOH is very close to the background reduction $(E_{1/2} \approx -1.9$ V vs NHE$)^{46}$. In polar aprotic solvents (DMSO, DMF and MeCN) Ph_3AsO is reduced in the same potential range (cf Table 6). Reduction of the corresponding Ph_3SbO in MeCN takes place at only a slightly more positive potential[47] (cf Table 6), whereas the corresponding Ph_3PO in DMF is reduced at a potential approximately 150 mV lower than Ph_3AsO^{48}. Replacing O in Ph_3AsO with S shifts the $E_{1/2}$ value more than 200 mV in the positive direction in DMF[48].

A series of aryldiethylarsine oxides has been studied in DMF[49], and from Table 6 it is seen that exchange of two of the phenyl groups in Ph_3AsO with two ethyl (or two propyl)

TABLE 6. Half-wave potentials, $(E_{1/2})$ for Ph_3MO and derivatives in non-aqueous solution[a]

Compound	Solvent	$E_{1/2}$/V vs NHE	References
Ph_3AsO	DMSO	-2.1	50
	DMF	$-2.1, -2.090, -2.09$	12, 48–50
	MeCN	-2.1	51
	MeOH	-1.9	46
	H_2O/MeOH, 8:2[b]	-0.7	46
$PhEt_2AsO$	DMF	-2.315	49
$(4\text{-}BrC_6H_4)Et_2AsO$	DMF	-1.91 (1), -2.310 (2)	49
$(4\text{-}ClC_6H_4)Et_2AsO$	DMF	-2.00 (1), -2.300 (2)	49
$(4\text{-}MeOC_6H_4)Et_2AsO$	DMF	-2.435	49
$(4\text{-}Me_2NC_6H_4)Et_2AsO$	DMF	-2.54	49
$(4\text{-}MeC_6H_4)Et_2AsO$	DMF	-2.365	49
$(4\text{-}MeC_6H_4)Pr_2AsO$	DMF	-2.390	49
$PhPr_2AsO$	DMF	-2.320	49
Ph_3SbO	MeCN	-2.02	47
Ph_3AsOH^+	MeCN	-0.55^c	51
	MeCN	-0.85	51
$Ph_3AsOHCl$	MeCN	-1.2 (1), -2.40 (2)	52
$(Ph_3AsO)_2H^+$	MeCN	-0.94^c	51
	MeCN	-1.3	51
Ph_3AsS	DMF	-1.845 (1), -2.540 (2)	48
$\begin{array}{c} S \\ \parallel \\ Ph_2As \\ \diagdown(CH_2)_2 \\ Ph_2As \\ \parallel \\ S \end{array}$	DMF	-1.990	48
$\begin{array}{c} S \\ \parallel \\ Ph_2As \\ \diagdown(CH_2)_3 \\ Ph_2As \\ \parallel \\ S \end{array}$	DMF	-1.972	48

[a] At Hg electrodes unless otherwise noted.
[b] At pH 5.1, reduction of Ph_3AsOH^+, see text.
[c] At a Pt electrode.

groups shifts the half-wave potential approximately 0.2 V in the negative direction. Halogen substitution in the *para* position of the phenyl group gives rise to two polarographic waves when $ArEt_2AsO$ is reduced in DMF; the first two-electron reduction process is attributed to the cleavage of the halogen as a halide ion followed by protonation of the phenyl group and formation of $PhEt_2AsO$, which is subsequently reduced in the second electrochemical process[49]. Based on the reduction potentials for four members of the series $ArEt_2AsO$ (cf Table 6), a Hammett reaction constant $\rho = 0.38$ V was determined ($r = 0.990$)[49], i.e. a much larger dependence on substituents than found for the half-wave potential for reduction of arylarsonic acids in acidic aqueous solution (cf Section II.D.1). Due to the limited number of compounds all of which are electron-donating (H, Me, MeO, Me_2N) the correlation with σ^+ was almost as good ($r = 0.987$) giving $\rho^+ = 0.12$ V[49]. In the same study[49] the σ_p values corresponding to Et_2As and Et_2AsO as substituents were determined from the Hammett correlation of $E_{1/2}$ for the reversible one-electron reduction of substituted nitrobenzenes in DMF, $\rho = 0.45$ V; $\sigma_p(Et_2As) = 0.09$ and $\sigma_p(Et_2AsO) = 0.33$ were found[49].

The reduction of Ph_3AsO in MeOH and in the aprotic solvents is diffusion-controlled[46,50], and from the heights of the limiting currents in MeOH, DMF and DMSO and the estimated diffusion coefficients in these solvents, the reduction process was shown to be a two-electron process[46,50]. In DMF, $n = 2$ was also found for the series $ArEt_2AsO$ based on the relative heights of their polarographic reduction waves and the height of the wave corresponding to one-electron reduction of benzophenone in DMF[49]. However, the mechanism and the products of the reduction of Ph_3AsO in aprotic solvents (and in MeOH) are apparently quite different from what was found in acidic aqueous solution. The suggested overall process in non-aqueous solution can be formulated as equation 34[12,46,50] (and in Reference 49 also for ArR_2AsO, R = Et, Pr):

$$Ph_3AsO + 2e^- \longrightarrow Ph_2AsO^- + Ph^- \qquad (34)$$

i.e. one of the As—C bonds and not the As—O bond is cleaved in the process. The products were not isolated but the absence of a reduction wave at lower potentials, corresponding to the reduction of Ph_3As (cf Section III.A) in aprotic solvents, supports the view that the final product is different from Ph_3As. For the sulphur analogue, Ph_3AsS, reduction in DMF follows a different path[48]. In this case the product is Ph_3As, as concluded from the appearance of a reduction wave corresponding to reduction of Ph_3As, and the following path has been suggested[48]:

$$Ph_3AsS + 2e^- \longrightarrow Ph_3As + S^{2-} \qquad (35)$$

Also, in case of the two reported bisulphides, $Ph_2As(S)(CH_2)_nAs(S)Ph_2$, $n = 2, 3$, cleavage of the As—S bonds leading to $Ph_2As(CH_2)_nAsPh_2$, $n = 2, 3$, has been reported[48]. Equation 35 is in agreement with results obtained in reduction of Ph_3PS (the reduction of Ph_3PO is a chemically reversible one-electron process in DMF)[48].

Addition of water (10–40%) to DMSO does not change the mechanism of reduction of Ph_3AsO; only a small potential shift in the positive direction is observed, which can be explained by a solvent effect[50]. Addition of benzoic acid to MeOH changed the path from equation 34 to equations 31–33 (cf Section II.F.1) but this is not the case when benzoic acid is added to aprotic solvents. In the case of DMSO and DMF a new polarographic wave is observed at higher potentials ($E_{1/2} \approx -1.9$ V vs NHE) when benzoic acid is added, with a height independent of the concentration of Ph_3AsO but proportional to the benzoic acid concentration; the original wave corresponding to reduction of Ph_3AsO is unaffected[50]. The new reduction process was shown to be the direct reduction of protons, the wave being present even in the absence of Ph_3AsO[50]. This observation suggests that the protonation equilibrium 31 in DMSO and DMF is pulled to the left by the direct reduction of PhCOOH. This is in contrast to the effect observed when phenol (PhOH) was added to a DMF solution of Ph_3AsO, in which case the reduction wave was shifted in the positive direction, splitting into two closely spaced waves with a total height corresponding to an overall four-electron process[49]. The mechanism suggested to account for these observations is given by equations 31, 36 and 37[49].

$$Ph_3AsOH^+ + 2e^- + H^+ \longrightarrow PhH + Ph_2AsOH \qquad (36)$$

$$Ph_2AsOH + 2e^- + 2H^+ \longrightarrow Ph_2AsH + H_2O \qquad (37)$$

This latter mechanism with initial two-electron reduction of the protonated oxide is, however, not in accord with the detailed study of the reduction of genuine samples of Ph_3AsOH^+ in MeCN (cf Section II.F.3).

3. Reduction of Ph_3AsOH^+ in MeCN

Although Ph_3AsOH^+ is apparently not the electroactive species when benzoic acid is added to Ph_3AsO in aprotic solvents (cf Section II.F.2), Ph_3AsOH^+ can be obtained as a stable perchlorate salt by oxidation of Ph_3As in 'wet' MeCN[52] (cf Section V.B), and its reduction in MeCN has been studied separately[51]. The mechanism of reduction of Ph_3AsOH^+ in MeCN depends on the electrode material. The 'polarogram' of Ph_3AsOH^+ on a Pt electrode with periodic renewal of the diffusion layer[53] shows three waves[51]. The reduction at the lowest potential was identified as being due to reduction of Ph_3AsO, which apparently was produced in the processes taking place at higher potential[51]. The two first waves were of equal heights and both resulted in formation of H_2, as indicated by the presence of two corresponding oxidation peaks when CV on a stationary Pt electrode was made on the solution. The two oxidation peaks were identical to the two oxidation peaks observed in CV on a solution containing Ph_3AsO and H_2. Furthermore, the two reduction waves were shifted in the positive direction and the size of the oxidation peaks increased, when a platinized Pt electrode was used[51].

The value of n for the first reduction process was determined as 0.5 by coulometry on a Pt electrode. After exhaustive electrolysis a hydrogen-bonded dimeric species, $[Ph_3AsO \cdots H \cdots OAsPh_3]^+$, was isolated (as the perchlorate salt). However, if the electrolysed solution was treated with anhydrous $HClO_4$, the original polarogram was restored[51]. This is in agreement with reactions 38–40:

$$Ph_3AsOH^+ + e^- \longrightarrow Ph_3AsO + \tfrac{1}{2}H_2(ads.) \qquad (38)$$

$$Ph_3AsO + Ph_3AsOH^+ \rightleftharpoons [Ph_3AsO \cdots H \cdots OAsPh_3]^+ \qquad (39)$$

$$[Ph_3AsO \cdots H \cdots OAsPh_3]^+ + H^+ \longrightarrow 2Ph_3AsOH^+ \qquad (40)$$

The relevance of the equilibrium 39 was further confirmed by addition of Ph_3AsO to a solution of Ph_3AsOH^+, which was accompanied by disappearance of the first reduction wave and doubling of the second, demonstrating that the second reduction process corresponds to the reduction of $[Ph_3AsO \cdots H \cdots OAsPh_3]^+$. In accord with this, electrolysis carried out at the potential of the second reduction wave gave $n = 1$ and the only As product isolated was Ph_3AsO[51] reactions 38, 39 and 41.

$$[Ph_3AsO \cdots H \cdots OAsPh_3]^+ + e^- \longrightarrow 2Ph_3AsO + \tfrac{1}{2}H_2(ads.) \qquad (41)$$

Again, the original polarogram could be restored by addition of anhydrous $HClO_4$ to the electrolysed solution[51] confirming that both reduction processes are in fact catalytic reductions of protons in different states of activity. Reactions 38 and 41 probably occur via adsorbed intermediates like $Ph_3AsOH^\cdot$ and $(Ph_3AsO)_2H^{\cdot}$[51].

When a mercury or a glassy carbon electrode is used instead of Pt, three reduction waves are again present but the two first reductions take place at lower potentials than at Pt, and the first wave is in this case twice as high as the second. The third wave again corresponded to reduction of Ph_3AsO^{51}. Electrolysis at the potential of the first wave gave $n = 0.66$, and equal amounts of Ph_3As and $(Ph_3AsO)_2HClO_4$ were isolated. Electrolysis at the potential of the second wave (on a new solution) gave $n = 1$, and equal amounts of Ph_3As and Ph_3AsO were isolated. The potential of the second wave was shown to correspond to the reduction of the hydrogen-bonded dimer[51]. These observations were rationalized by equations 42–45[51]:

$$2Ph_3AsOH^+ + 2e^- \longrightarrow 2Ph_3AsOH^{\bullet} \tag{42}$$

$$2Ph_3AsOH^{\bullet} \longrightarrow Ph_3AsO + Ph_3As + H_2O \tag{43}$$

$$Ph_3AsO + Ph_3AsOH^+ \rightleftharpoons [Ph_3AsO\cdots H\cdots OAsPh_3]^+ \tag{44}$$

$$2[Ph_3AsO\cdots H\cdots OAsPh_3]^+ + 2e^- \longrightarrow 2Ph_3AsO + 2Ph_3AsOH^{\bullet} \tag{45}$$

Reactions 42–44 account for the reduction at the potential of the first wave giving a total of 2 electrons per 3 molecules of substrate, resulting in formation of equal amounts of Ph_3As and hydrogen-bonded dimer, while reaction 45 followed by reaction 43 accounts for the further reduction at the potential at the second wave giving additionally 2 electrons per 6 molecules of substrate, resulting in additional formation of 1 molecule of Ph_3As and 3 molecules of Ph_3AsO. Hence reduction at the potential of the second wave is an overall one-electron process resulting in equimolar amounts of Ph_3As and Ph_3AsO. The scheme and the influence of the acid–base behaviour of the arsenic species is further confirmed by electrolyses of Ph_3AsOH^+ and $HClO_4$ in equimolar amounts, which result in an overall two-electron process with Ph_3As as the only product[51].

The two compounds $Ph_3AsOHCl$ and $Ph_3AsOHBr^{54}$ are, in contrast to $Ph_3AsOHClO_4^{52}$, only weak electrolytes in MeCN with molar conductances in the range 20–24 $ohm^{-1} cm^2 mol^{-1}$ [54]. Preparative electrolysis of MeCN solutions of the two compounds showed that the partial ionization takes place according to equation 46 and not according to equation 47, as demonstrated by formation of halogen (Cl_2 or Br_2) at the anode and H_2 at the cathode.

$$Ph_3AsOHX \rightleftharpoons Ph_3AsOH^+ + X^- \tag{46}$$

$$Ph_3AsOHX \rightleftharpoons Ph_3AsX^+ + OH^- \tag{47}$$

The arsenic-containing product was shown to be Ph_3AsO by infrared (IR) spectroscopy of the catholyte[54]. The reduction of $Ph_3AsOHCl$ on Pt takes place according to equation 38 as confirmed by the presence of only two waves in the polarogram of which the one at the lowest potential corresponds to reduction of Ph_3AsO^{52}. Reactions 39 and 41 interfere negligibly, because the small amounts of free Ph_3AsOH^+ prevent formation of the dimeric hydrogen-bonded species[52].

4. Reduction of $[Ph_3SbOSbPh_3]^{2+}$ in MeCN

The stable product of oxidation of Ph_3Sb in wet MeCN is *not* Ph_3SbOH^+, which would be expected from the formation of Ph_3AsOH^+ by oxidation of Ph_3As^{52}. Instead, the dicationic dimer $[Ph_3SbOSbPh_3]^{2+}$ is formed[47] (cf Section V.B) which is formally the product of condensation of two Ph_3SbOH^+. The oxybis(triphenylantimony) dication can be isolated as the diperchlorate and its electrochemical reduction has been studied by polarography[47]. In dry MeCN $[Ph_3SbOSbPh_3]^{2+}$ shows three well-defined reduction waves (0.05 V, -2.02 V and -2.69 V vs NHE) in addition to an ill-defined wave associated with a large minimum around -0.3 V vs NHE. The waves at -2.02 V and -2.69 V were identified as corresponding to the reduction of Ph_3SbO and Ph_3Sb (cf Section III.A), respectively[47]. Addition of increasing amounts of water to the solution decreased the height of the first reduction wave while the ill-defined wave developed into two well-resolved ones at -0.36 V and -0.60 V vs NHE[47]. This complicated cathodic behaviour was attributed to the reduction of $[Ph_3SbOSbPh_3]^{2+}$ in different states of activity determined by the concentration of Ph_3SbO, which behaves as a ligand toward the dicationic complex[47].

The simplest cathodic behaviour is observed in the presence of two equivalents of anhydrous $HClO_4$ in dry MeCN. In that case clean two-electron reduction of each Sb unit takes place at the potential of the first wave (0.05 V vs NHE) upon formation of Ph_3Sb, and equations 48–50 describe the reaction sequence:

$$2[Ph_3SbOSbPh_3]^{2+} + 4e^- \longrightarrow 2Ph_3SbO + 2Ph_3Sb \qquad (48)$$

$$2Ph_3SbO + 2H^+ \longrightarrow 2Ph_3SbOH^+ \qquad (49)$$

$$2Ph_3SbOH^+ \rightleftharpoons [Ph_3SbOSbPh_3]^{2+} + H_2O \qquad (50)$$

In the absence of protons coulometry shows that the first wave changes from a four-electron to a one-electron wave (per dication of substrate) and only half a mole of Ph_3Sb per mole of substrate dication was found. This is due to the fact that the Ph_3SbO formed in equation 48 cannot regenerate the substrate dication in the absence of protons but coordinates to the starting dication with formation of a new complex (reaction 51) which is electrochemically inactive at the potential of the first wave. The new complex can be isolated after exhaustive electrolysis at the potential of the first wave[47].

$$[Ph_3SbOSbPh_3]^{2+} + Ph_3SbO \longrightarrow [Ph_3SbOSbPh_3(Ph_3SbO)]^{2+} \qquad (51)$$

The isolated complex $[Ph_3SbOSbPh_3(Ph_3SbO)]^{2+}$ is reduced at -0.36 V vs NHE in a process requiring approximately 0.66 electrons per complex dication, and Ph_3Sb and a new complex dication with two Ph_3SbO ligands could be isolated, as accounted for by equations 52 and 53[47].

$$[Ph_3SbOSbPh_3(Ph_3SbO)]^{2+} + 2e^- \longrightarrow 2Ph_3SbO + Ph_3Sb \qquad (52)$$

$$2[Ph_3SbOSbPh_3(Ph_3SbO)]^{2+} + 2Ph_3SbO \longrightarrow 2[Ph_3SbOSbPh_3(Ph_3SbO)_2]^{2+} \qquad (53)$$

Finally, $[Ph_3SbOSbPh_3(Ph_3SbO)_2]^{2+}$ can be reduced at -0.6 V vs NHE in a two-electron process (equation 54):

$$[Ph_3SbOSbPh_3(Ph_3SbO)_2]^{2+} + 2e^- \longrightarrow 3Ph_3SbO + Ph_3Sb \qquad (54)$$

The three-step reduction of $[Ph_3SbOSbPh_3]^{2+}$ in different states of activity has an overall stoichiometry of two electrons per substrate molecule and results in formation of equal amounts of Ph_3SbO and Ph_3Sb. This is exactly what is found, if the original, dry solution is reduced at -1.0 V vs NHE corresponding to the potential of reduction of $[Ph_3SbOSbPh_3(Ph_3SbO)_2]^{2+}$[47].

The effect of water on the polarographic behaviour of $[Ph_3SbOSbPh_3]^{2+}$ has been attributed to partial hydrolysis according to equation 55 which explains the concommitant decrease of the first wave and development of a second well-defined wave at -0.36 V.

$$3[Ph_3SbOSbPh_3]^{2+} + H_2O \rightleftharpoons 2[Ph_3SbOSbPh_3(Ph_3SbO)]^{2+} + 2H^+ \qquad (55)$$

However, the number of electrons and the amount of Ph_3Sb formed when the water containing solution was electrolysed at the potential of the second wave (-0.45 V) were higher than expected from reactions 52 and 53[47]. This behaviour has been explained by participation of the protons liberated by the hydrolysis (reaction 55) in the overall reaction according to equation 56[47].

$$2[Ph_3SbOSbPh_3(Ph_3SbO)]^{2+} + 2H^+ + 4e^- \longrightarrow$$
$$[Ph_3SbOSbPh_3(Ph_3SbO)_2]^{2+} + 2Ph_3Sb + H_2O \quad (56)$$

III. REDUCTION OF TRIVALENT As, Sb AND Bi COMPOUNDS

The electrochemical reductions of trivalent As, Sb and Bi compounds have mainly been carried out in non-aqueous solution, in contrast to the electrochemical studies of pentavalent As, Sb and Bi compounds which have almost exclusively been carried out in aqueous solution, and have been much less focused on analytical applications. The trivalent compounds are divided into five groups based on the identity of the substituents: compounds with three organic ligands, R_3M, compounds with two organic ligands, R_2MX, dimeric compounds with an oxygen bridge, Ph_2MOMPh_2, compounds with one organic ligand, RMX_2, and finally PhAsO which is the only trivalent compound studied mainly in aqueous solution. The five groups are treated separately in the following sections.

A. Reduction of R_3M, R = Alkyl, Aryl

1. Reduction of non-cyclic compounds

Trisubstituted phosphines, arsines, stibines and bismuthines are electrochemically inert within a large potential range spanning approximately $-2.5\,V$ to $1.3\,V$ vs NHE (cf Section V.A). This property is one of the main reasons for the widespread application of, in particular, the phosphines and arsines as 'innocent' ligands for transition metal complexes in a variety of oxidation states (cf Section VII). Due to the rather low potentials necessary to reduce compounds of the type R_3M, aprotic solvents are required in order to avoid discharge of protons in the medium prior to reduction of the substrate. Furthermore, Ph_3M (M = As, Sb, Bi) is known to adsorb strongly to Hg electrodes in aqueous or hydroxylic solutions as, e.g., demonstrated by electrocapillary measurements in MeOH for Ph_3As and Ph_3Sb^{55} or by the inhibiting effect of Ph_3As, Ph_3Sb and Ph_3Bi on ion transfer in the reduction of Cd^{2+} to $Cd(Hg)$ in $MeOH^{56}$.

Triphenylphosphine (Ph_3P) forms stable anion radicals on a voltammetric or polarographic time scale in $DMF^{12,48}$, whereas Ph_3M (M = As, Sb, Bi) in DMF is reduced in a chemically irreversible two-electron process[12]. The only trisubstituted arsine reported to undergo reversible one-electron reduction is $(4-NO_2C_6H_4)Et_2As$, and in this case ESR measurements confirm that the reduction may be regarded as one-electron reduction of a substituted nitrobenzene with the unpaired electron located mainly in the nitro group[33]. In Table 7 half-wave potentials ($E_{1/2}$) are collected for the chemically irreversible reduction of R_3M (M = As, Sb, Bi) in different aprotic solvents, and it appears that $E_{1/2}$ is almost completely independent of the choice of solvent. From Table 7 it is also seen that the value of $E_{1/2}$ for reduction of Ph_3M increases, when M is changed from As to Sb to Bi. The half-wave potential for the reversible reduction of Ph_3P does not belong to this series; Ph_3P is approximately $0.1\,V$ easier to reduce than Ph_3As.

When one of the phenyl groups in Ph_3As is substituted by an α-naphthyl group the reduction potential increases by $450\,mV$, whereas further substitution of the remaining phenyl groups by α-naphthyl groups has a relatively small effect (cf Table 7). The same pattern is followed when phenyl groups are substituted by α-naphthyl groups in Ph_3P^{48}, and in both cases the presence of a naphthyl group gives rise to an additional reduction process at lower potentials attributed to reduction of the naphthyl substituent[48].

On a preparative time scale, two-electron reduction of Ph_3As in DMF leads to

TABLE 7. Half-wave potentials, $(E_{1/2})$ for the chemically irreversible reduction of R_3M in aprotic solvents[a]

Compound	Solvent	$E_{1/2}$/V vs NHE	Reference
Ph_3As	DMF	-2.550	48
	DMF	-2.54	12
	DME	-2.5	11
$Ph_2(\alpha\text{-Naph})As$	DMF	-2.100 (1)	48
		-2.320 (2)[b]	48
$Ph(\alpha\text{-Naph})_2As$	DMF	-2.050 (1)	48
		-2.290 (2)[b]	48
$(\alpha\text{-Naph})_3As$	DMF	-2.030 (1)	48
		-2.280 (2)[b]	48
Ph_3Sb	MeCN	-2.69	47
	DME	-2.4	11
	DMF	-2.38	12
Ph_2BuSb	THF	-2.61	57
Ph_3Bi	DME	-2.2	11
	DMF	-2.25	12

[a] At Hg electrodes.
[b] Reduction of the naphthalene substituent, see text.

formation of Ph_2AsH and benzene according to equation 57[48], where the protons are supplied by the medium, probably by residual water. Preparative reduction of the mixed phenyl-α-naphthylarsines leads exclusively to cleavage of the As—phenyl bond in analogy with observations made for the corresponding phosphines[48], leading to the disubstituted arsine and benzene.

$$Ph_3As + 2e^- + 2H^+ \longrightarrow Ph_2AsH + PhH \qquad (57)$$

No mechanistic studies have revealed the details of the cleavage reaction. The cleavage is assumed to take place after the first electron transfer[12,48,57] either as $Ph^{\cdot}$[48,57] or as Ph^-[57]. Irrespective of which of the two cleavage reactions actually takes place, the radical product ($Ph^{\cdot}$ or $Ar_2M^{\cdot}$) can be assumed to be further reduced to the corresponding anion in a fast reaction either in solution or at the electrode surface, due to their much higher reduction potential.

In scrupulously dried aprotic solvents, Ph_2M^- is relatively stable even on the time scale of preparative electrolysis (cf Section IV). On a voltammetric time scale the two-electron reduction of Ph_3Sb in THF gives rise to a new, chemically irreversible one-electron oxidation on the reverse scan corresponding to oxidation of Ph_2Sb^- to $Ph_2Sb^{\cdot}$[57] (cf Section IV.A). The peak potential of this oxidation is in the range -0.25 V to -1.25 V vs NHE, dependent on the activity of the Pt electrode[57]. The radical $Ph_2Sb^{\cdot}$ rapidly dimerizes, and the reduction of the dimer (cf Section IV.A) can be observed if the scan is reversed again[57]. After preparative two-electron reduction of Ph_3Sb in THF, a red solution of Ph_2Sb^- is obtained but the colour fades and the height of the peak corresponding to oxidation of Ph_2Sb^- gradually decreases due to reaction between Ph_2Sb^- and the tetrabutylammonium cation (Bu_4N^+) of the supporting electrolyte (reaction 59). The mixed tertiary stibine (Ph_2BuSb) and Bu_3N were isolated[57]. The reactions are summarized in equations 58 and 59.

$$Ph_3Sb + 2e^- \longrightarrow Ph_2Sb^- + Ph^- \qquad (58)$$
$$Ph_2Sb^- + Bu_4N^+ \longrightarrow Ph_2BuSb + Bu_3N \qquad (59)$$

Reaction 59 also takes place when Bu_4N^+ is added to a THF solution of Ph_2Sb^- generated by reaction of Ph_3Sb with alkali metal[57]. If the preparative electrolysis is instead carried out in DMF, the reactivity of Ph_2Sb^- is diminished due to improved solvation, but in addition to Ph_2BuSb small amounts of Ph_2SbH are formed due to trace amounts of water in DMF[57]. Reduction of Ph_2BuSb in THF takes place at -2.61 V vs NHE and the voltammogram closely resembles that of Ph_3Sb. The group cleaved upon two-electron reduction of Ph_2BuSb is Bu^-[57]. The extra product suggested in DMF (Ph_2SbH) is very unstable and was not isolated, but a new reduction peak was observed at -1.5 V vs NHE and attributed to this species[57].

2. Reduction of heterocycles containing As

The electrochemical reduction of five-membered rings containing one As atom (arsoles) has only been reported in a few cases, and no electrochemical studies of the corresponding stiboles and bismoles have been reported.

The polarographic reduction of 1,2,3,4,5-pentaphenylarsole (structure **3**) in 1,2-dimethoxyethane (DME) shows two waves (cf Table 8), which are chemically irreversible[58].

(3)

The first reduction wave was reported to be a two-electron wave, but if electrolysis at the potential of this wave was stopped after 1 F mol^{-1} a blue solution of the anion radical was obtained[58]. The half-life of the anion radical was about 1 minute under these conditions[58]. Identical behaviour was observed for other five-membered ring systems in which the As atom was exchanged with B, Si, Ge, Sn or P[58], and suggested to be due to a comproportionation mechanism (reactions 60 and 61).

$$3 + 2e^- \longrightarrow 3^{2-} \tag{60}$$

$$3^{2-} + 3 \rightleftharpoons 2 3^{-\cdot} \tag{61}$$

TABLE 8. Half-wave potentials ($E_{1/2}$) for the reduction of substituted arsoles

Compound	Solvent	$E_{1/2}$ V vs NHE	Reference
1,2,5-triphenylarsole (**4**)	MeCN	-1.53 (1)^a -1.79 (2)^b	59
1-methyl-2,5-diphenylarsole (**6**)	MeCN	-1.71 (1)^a	59
1,2,3,4,5-pentaphenylarsole (**3**)	DME	-1.7 (1)^b -2.1 (2)^b	58

^{a} Chemically reversible, $E_{1/2} = E^0$ determined by CV.
^{b} Chemically irreversible.

The equilibrium constant for reaction 61 must be large in order to obtain a solution of $3^{-\cdot}$, but further reduction at the potential of the first wave was suggested to lead to quantitative formation of the unstable dianion giving unidentified products[58]. The reduction process taking place at lower potential was not commented on.

The mechanism proposed above is in disagreement with a later, more detailed study of the related 1,2,5-triphenylarsole (4) in MeCN[59]. Cyclic voltammetry of 4 shows two reduction peaks (cf Table 8), of which the first is a chemically reversible one-electron reduction on the time scale of the voltammetric experiment ($v = 0.05$–$0.5\ V\ s^{-1}$), whereas the second is chemically irreversible[59]. Changing the As substituent from Ph to Me shifts the reduction potentials in the negative direction (cf Table 8) and diminishes the chemical stability of the anion radical (the reduction is not completely reversible at $v = 1\ V\ s^{-1}$)[59].

Exhaustive electrolysis of 4 in MeCN at the potential of the first reduction requires $2\ F\ mol^{-1}$, and not $1\ F\ mol^{-1}$ as suggested from the CV experiments; the final product is 1-H-2,5-diphenylarsole (5)[59]. In contrast, chemical reduction by Li or K in DME gives a stable solution of the anion radical ($4^{-\cdot}$) as confirmed by ESR spectroscopy[59]. The difference between the electrochemical and the chemical reduction has been explained by the difference in the media; due to small amounts of residual water in MeCN during the electrochemical reduction, the disproportionation equilibrium (equation 62), which is strongly displaced to the left, is pulled to the right by protonation of the dianionic product (equation 63) in accord with the observation of a chemically irreversible second electron transfer in CV[59].

$$2\,4^{-\cdot} \;\rightleftharpoons\; 4 + 4^{2-} \tag{62}$$
$$4^{2-} + 2H^+ \;\longrightarrow\; PhH + 5 \tag{63}$$

However, based on the mechanism for reduction of the non-cyclic R_3M compounds (Section III.A.1) and the studies of the reaction of $4^{-\cdot}$ (formed by chemical reduction) with alkyl halides discussed below, it seems to be more likely that the experimentally found two-electron reduction on a preparative time scale[59] is due to a slow cleavage of the As—Ph bond in the anion radical.

Reaction in DME of chemically generated $4^{-\cdot}$ with MeI, EtI or $PhCH_2Cl$ leads to formation of 1-Me-, 1-Et- and 1-($PhCH_2$)-2,5-diphenylarsole in 63%, 56% and 23% yield, respectively[59]. Three possible mechanisms were discussed: (a) disproportionation (equation 62) followed by cleavage of the As—Ph bond (equation 64) and reaction between the arsolyl anion and RX (equation 65), (b) nucleophilic attack of $4^{-\cdot}$ on RX leading to a tetra-substituted As intermediate (equation 66) followed by cleavage of $Ph^{\cdot}$ (equation 67) and (c) initial, reversible cleavage of $4^{-\cdot}$ into $Ph^{\cdot}$ and an arsolyl anion (equation 68), followed by equation 65.

$$\tag{64}$$

$$\tag{65}$$

$$\text{(structure)} + \text{RX} \longrightarrow \text{(structure)} + \text{X}^- \qquad (66)$$

$$\text{(structure)} \longrightarrow \text{(structure)} + \text{Ph}^\cdot \qquad (67)$$

$$\text{(structure)} \rightleftharpoons \text{(structure)} + \text{Ph}^\cdot \qquad (68)$$

Mechanism (a) was ruled out because the application of stoichiometric amounts of the reducing alkali metal should result in a 50% recovery of the substrate, **4**, and none was found. Mechanism (b) was deemed unlikely because reaction took place (although only in 23% yield) when $RX = PhCH_2Cl$, and according to the order of cleavage found after reduction of tetra-substituted arsonium salts (cf Section II.B.2), $PhCH_2$ and not Ph should be cleaved in reaction 67. In none of the cases was starting material recovered. Mechanism (c), with a small equilibrium constant for the dissociation reaction 68, is therefore the most likely[59]. Formation of the arsolyl anion is favoured by the stability of the cyclic 6e structure analogous to the cyclopentadienyl anion. The existence of equilibrium 68 also accounts for the observed consumption of $2\,F\,mol^{-1}$ in the preparative electrolysis, because the $Ph^\cdot$ formed will immediately be reduced either at the electrode surface or, more likely, by $4^{-\cdot}$ in solution.

Based on the product distribution (partial recovery of the starting material in addition to the substitution product) it was suggested[59] that the anion radical ($6^{-\cdot}$) derived from 1-Me-2,5-diphenylarsole (**6**) due to its lower potential of formation reacts with RX by two parallel reactions—by S_N2 (like $4^{-\cdot}$) and by electron transfer (ET) (reaction 69[59], cf Section IV.B).

$$6^{-\cdot} + RX \longrightarrow 6 + R^\cdot + X^- \qquad (69)$$

It may be concluded that the main difference between the reduction of the arsoles and the tertiary arsines is the potential of reduction (the 1-phenylarsoles are reduced at potentials which are approximately 1 V higher than Ph_3As) and the enhanced stability of the arsole anion radicals.

B. Reduction of R_2MX, R = Alkyl, Aryl, X = Cl, Br, I

Very few studies of the electrochemical reduction of compounds of the type R_2MX have been made. Cacodyl chloride, Me_2AsCl, was earlier[60] shown to be electrochemically reducible, and it was demonstrated that the dimer $Me_2AsAsMe_2$ (cacodyl) was an intermediate on the way to the final product, Me_2AsH (dimethylarsine). The reduction of Me_2AsCl on a Pt electrode in a mixture of EtOH and HCl was studied by measuring the volume and composition of the gaseous products escaping from the cathode compartment (H_2 and Me_2AsH) as a function of the current passed through the cell[60]. It was found that the amount of Me_2AsH (and H_2) produced during the electrolysis could not account for the current passed through the cell, as long as the current passed was smaller than the amount necessary for full reduction of Me_2AsCl to Me_2AsH. On the basis of a separate demonstration that cacodyl could be reduced to dimethylarsine (cf Section IV.A) it was concluded that the intermediate was cacodyl[60].

It was later shown that the series of compounds Ph_2MX, M = As, Sb, Bi, all undergo chemically irreversible one-electron reduction in DME followed by formation of the corresponding dimer as described by equations 70 and 71[11].

$$Ph_2MX + e^- \longrightarrow Ph_2M^{\cdot} + X^- \tag{70}$$

$$2Ph_2M^{\cdot} \longrightarrow Ph_2MMPh_2 \tag{71}$$

The formation of Ph_2MMPh_2 by electrochemical reduction of Ph_2MX is analogous to the formation of R_2MMR_2 (R = i-Pr, n-Bu, M = Sb) by chemical reduction of R_2MX by a reducing agent such as Mg in THF[61].

The half-wave potentials for the chemically irreversible reduction of Ph_2MX are, within experimental error, almost independent of the identity of M but they depend on X as is apparent from the data in Table 9. The half-wave potentials are in all cases very much higher than the half-wave potentials for reduction of the analogous phosphinic halides which are reduced at potentials around -2.4 V vs NHE yielding diphenylphosphine, Ph_2PH, as a result of $H^{\cdot}$ abstraction from the solvent[11]. This change in potential when going from P to As, Sb or Bi is in agreement with the increasing polarization of the M—X bond with increasing metal character of M. That the M–halogen bond is indeed very polar is also indicated by the similarity of half-wave potentials for Ph_2MI and Ph_2MClO_4, M = As, Sb, where the perchlorates are unambiguously ionic. Surprisingly, the half-wave potentials for reduction of Ph_2MX are in general shifted 2 V in the positive direction relative to the half-wave potentials for reduction of the corresponding Ph_4M^+ under the same conditions (cf Section II.B) and thereby resemble the series of pentavalent dihalides (cf Section II.C).

All of these compounds show a second reduction at lower potentials with a half-wave potential characteristic of M and completely independent of X[11]. This second reduction is due to reduction of the dimer formed in reaction 71 as shown by comparison with the reduction of genuine samples of the dimers (Ph_2MMPh_2); cf Section IV.A[11].

TABLE 9. Half-wave potentials ($E_{1/2}$) for the chemically irreversible reduction of R_2MX

Compound	Solvent	$E_{1/2}$/V vs NHE	Reference
Ph_2AsBr^a	DME	0.0	11
Ph_2AsI^a	DME	-0.1	11
$Ph_2AsClO_4^{a,b}$	DME	-0.1	11
$(Naph)_2AsCl$	DME	-0.2	11
Ph_2SbI	DME	-0.1	11
$Ph_2Sb(OAc)$	DME	-0.5	11
$Ph_2SbClO_4^b$	DME	0.0	11
Ph_2BiCl	DME	-0.1	11
Ph_2BiI	1 M HCl	0.075	20

[a] Shows polarographic maximum.
[b] In contrast to the other compounds, Ph_2MClO_4 is ionic.

C. Reduction of (Ph₂M)₂O

Only two compounds in this group have been studied electrochemically, $(Ph_2As)_2O^{11}$ and $(Ph_2Sb)_2O^{11,57}$. In contrast to Ph_2MX, $(Ph_2M)_2O$ compounds are reduced in DME and in THF in two-electron processes[11,57] and at potentials which are approximately 1.5 V lower (cf Table 10).

When $(Ph_2Sb)_2O$ is reduced at Pt electrodes in THF, the cyclic voltammetric peak potential is very dependent on the state of the surface, and the reduction is accompanied by the appearance on the reverse scan of an oxidation peak corresponding to oxidation of Ph_2Sb^- (cf Sections III.A.1 and IV.A). If an activated Pt electrode is employed, a second oxidation peak is observed at higher potentials[57]; the origin of this peak was not investigated. The two-electron process has been described by equation 72[11,57], and the second oxidation peak observed by CV in THF may be due to oxidation of Ph_2SbO^-.

$$(Ph_2M)_2O + 2e^- \longrightarrow Ph_2M^- + Ph_2MO^- \tag{72}$$

After preparative electrolysis of $(Ph_2Sb)_2O$ in THF (2 F mol^{-1}) the polarogram shows a new reduction wave which is at a lower potential than the original reduction process, in addition to the oxidation wave corresponding to Ph_2Sb^-[57]. After the electrolysis the red colour fades the oxidation wave disappears and the new reduction wave increases due to formation of Ph_2BuSb by reaction between Ph_2Sb^- and the Bu_4N^+ of the supporting electrolyte (Equation 59); cf Section III.A.1[57]. The fate of Ph_2SbO^- was not investigated in this study. Preparative electrolysis in DME of $(Ph_2As)_2O$ gave stable anionic products (equation 72) despite the presence of Bu_4N^+[11]. Addition of two equivalents of Ph_2AsBr to the electrolysed solution resulted in disappearance of the yellow colour, and the appearance of a reduction wave at the same potential as that of the original reduction wave but of doubled height[11]. This behaviour has been explained by equation 73 and the coincidence of the reduction waves corresponding to $(Ph_2As)_2O$ and to $(Ph_2As)_2$ (cf Section IV.A)[11]. The mechanisms of the reactions in equation 73 were not discussed (cf Section IV.B).

$$Ph_2As^- + Ph_2AsO^- + 2Ph_2AsBr \longrightarrow (Ph_2As)_2 + (Ph_2As)_2O + 2Br^- \tag{73}$$

D. Reduction of RMX₂, R = Alkyl, Aryl, X = Cl, Br, I

The compounds in the series $RSbBr_2$, R = Et, Pr, Bu are very weak electrolytes in THF as shown by the absence of an oxidation wave for Br^- [62]. The polarogram of $RSbBr_2$ in THF shows several ill-defined waves in the range 0.0–1.8 V vs NHE[62]. Electrolysis at a potential of -1.5 V vs NHE yields a green solution after 2 F mol^{-1}, which shows a single oxidation wave around -0.3 V vs NHE independent of the substituent R. Reoxidation of the electrolysed solution restores the initial polarogram[62]. After two-electron reduction a yellow product could be isolated which, by MS and NMR spectroscopy, was shown for R = Et to be a mixture of $(RSb)_4$ and $(RSb)_5$ and almost exclusively $(RSb)_5$ when R = Pr, Bu; these products are identical to the cyclic Sb(I) products obtained by reaction of Mg

TABLE 10. Half-wave potentials $(E_{1/2})$ for the chemically irreversible reduction of $(Ph_2M)_2O$ at Hg electrodes in aprotic solvents

Compound	Solvent	$E_{1/2}$/V vs NHE	Reference
$(Ph_2As)_2O$	DME	-1.8	11
$(Ph_2Sb)_2O$	DME	-1.6	11
$(Ph_2Sb)_2O$	THF	-1.71	57

with $RSbBr_2$ in THF[62]. The products eventually turned black and insoluble and were identified as polymeric structures with the formula $(RSb)_x, x \gg 5$[62], analogously to the products found after chemical reduction of $RSbBr_2$ by Mg in THF[61].

The cyclic products $(RSb)_4$ and $(RSb)_5$ can both be reduced in well-defined two-electron processes with $E_{1/2}$ in the range -1.4 V to -1.9 V vs NHE. For $(PrSb)_5$ this reduction process was shown by CV to give rise to a new oxidation around -1.3 V vs. NHE[62]. Scheme 3 was proposed to account for the observations[62]. The dianion $(RSb)_n^{2-}$ is probably non-cyclic, and the over-all chemical reversibility is a combination of a two-electron reduction/oxidation process and a chemical reaction[62].

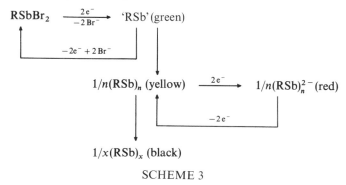

SCHEME 3

Similar reduction behaviour was previsouly observed when $RSbI_2$, $R = Pr$, Bu, were reduced in acidic buffered EtOH/water mixtures or in DMF[63]. However, in these solvents only the insoluble black-violet polymeric species $(RSb)_x$ could be isolated after preparative electrolysis, and the product also contained finely divided Hg (from the electrode) which made further characterization difficult[63]. However, in these solvents the polymeric product could be reoxidized to the original substance[63].

In basic (0.1–1.0 M NaOH) EtOH/water mixtures the reduction of $RSbI_2$ took place at lower potentials (≈ -1.0 V vs NHE) and required 2–3 F mol^{-1}. Only a trace amount of the polymer was formed, whereas the main product in this case appeared to be amalgamated Sb, the formation of which was suggested to take place via reactions 74 and 75, with reaction 74 competing favourably with the oligomerization reaction in Scheme 3.

$$\text{'RSb'} + OH^- \longrightarrow RH + SbO^- \tag{74}$$
$$SbO^- + e^- + H_2O \longrightarrow Sb + 2OH^- \tag{75}$$

In DME $PhAsCl_2$ is reported to be reduced in a two-electron process with $E_{1/2} \approx 0.0$ V vs NHE associated with a polarographic maximum[11]. No products were identified. The reduction of $PhSbCl_2$ has been reported to be at -1.0 V vs NHE in 1 M NaOH[20].

E. Reduction of PhAsO

Phenylarsine oxide (PhAsO) is one of the few trivalent compounds the reduction of which has been studied in aqueous solution, and it is reduced at a potential which is almost 1 V less cathodic than the reduction potential of the pentavalent $PhAsO(OH)_2$. In the solid state PhAsO (arsenosobenzene) exists as a tetrameric oxide (structure 7)[2], whereas in aqueous solution it exists as a monomeric dihydroxy compound $[PhAs(OH)_2$, phenyl-arsonous acid][64] with a pK_a of about 11[3].

(7)

Polarographic studies of PhAsO in unbuffered or in buffered aqueous solution at pH > 2 show two poorly defined waves with maxima[2,64]. However, in 0.1 M HCl at substrate concentrations below 10^{-4} M, the polarogram shows two well-defined waves ($E_{1/2} \approx 0.15$ V and $E_{1/2} \approx -0.20$ V vs NHE)[64]. Both waves appeared to be diffusion-controlled[64], with the wave heights directly proportional to the substrate concentration in the range 10^{-5}–10^{-4} M^2. At higher concentrations the height of the first wave becomes constant, and one or two new waves appear at lower potentials[64]. This observation was explained by adsorption of an insoluble product at the electrode surface which inhibits further reduction of substrate at the potential of the first wave[64]; the new waves could be suppressed by addition of Triton X-100^2. The relative heights of the two waves were found to be approximately 3:2 and independent of pH below pH 12, although the reproducibility decreased with increasing value of pH[64]. This is in contrast to the results obtained in another study[65], in which it was shown by application of DPP and CV on 3×10^{-6} M buffered solutions of PhAsO that the height of the first peak decreased with increasing pH and almost disappeared at pH 9. Furthermore, it appeared that the second peak in fact consists of two peaks, the first dominating at low and the second at high pH values, with equal heights at pH close to 5[65].

In both studies the dependence of the two half-wave potentials on pH was found to be linear: $dE_{1/2}(1)/dpH = -0.059$ V, $dE_{1/2}(2)/dpH = -0.08$ V (pH 1.0–2.2)[64], $dE_p(1)/dpH = 0.068$ V, $dE_p(2)/dpH = -0.073$ V (pH 1.2–7.6)[65], although in the CV study, the change in pH from 1.2 to 7.6 was associated with a decrease in the peak currents (I_p) of 52% for $I_p(1)$ and 20% for $I_p(2)$[65]. The differences in the influence of pH may be due to the differences in substrate concentration (almost an order-of-magnitude larger substrate concentration in Reference 64 than in Reference 65) and the non-ideal behaviour of the polarographic waves above pH 2 in Reference 64.

Polarographic analysis of the first wave in 0.1 M HCl showed that a plot of $\log(i_d - i)$ vs E was linear, which is consistent with a reversible reduction to an insoluble product[64]. Application of DPP showed a change in the width of the first peak when pH was changed from 2 to 8 which suggested that the electrode process changed from a reversible to an irreversible process[65]. At the same time it was concluded that the product of the first reduction process adsorbed to the electrode surface[65], and from CV it appears that this adsorbed product could be oxidized at 0.06 V vs NHE (pH 4.7). When the sweep is reversed after the second double peak, a new oxidation peak at -0.30 V vs NHE arises in addition to the oxidation peak at 0.06 V vs NHE (pH 4.7)[65].

The two studies of the reduction of PhAsO disagree as to the number of electrons transferred as well as the products. The slope of the plot of $\log(i_d - i)$ vs E and microcoulometry indicate $n = 1.25 - 1.30$ for the first reduction process[64], whereas exhaustive electrolysis in 0.1 M HClO$_4$ at the potential of the first process gave $n = 2.02 \pm 0.02$[65]. The solid product formed during microcoulometry gave a fragmentation pattern in MS compatible with a polymer containing three As atoms to one O atom[64], whereas the insoluble product formed by exhaustive electrolysis gave an MS fragmentation pattern corresponding to a polymer consisting exclusively of PhAs units[65]. Reoxidation of the exhaustively electrolysed solution at a potential positive of the observed oxidation peak at 0.06 V vs NHE resulted in reappearance of the original voltammogram[65], i.e. the oxidation peak observed in CV corresponds to a process in which the insoluble product is reoxidized to phenylarsine oxide.

The process taking place at the potential of the first reduction has been described by reaction 76[64,65] followed by reaction 77[64] or reaction 78[65], where in both cases the final oligomers are insoluble. The presence of $PhAsH_2$ during the electrolysis was detected by its characteristic odour[64].

$$PhAs(OH)_2 + 4e^- + 4H^+ \longrightarrow PhAsH_2 + 2H_2O \qquad (76)$$

$$PhAsH_2 + 2PhAs(OH)_2 \longrightarrow 1/x[(PhAs)_3O]_x + H_2O \qquad (77)$$

$$PhAsH_2 + PhAs(OH)_2 \longrightarrow 1/3[(PhAs)]_6 + 2H_2O \qquad (78)$$

Microcoulometry at the potential of the second reduction wave in 0.1 M HCl with substrate concentrations below 10^{-4} M gave $n = 1.98 \pm 0.04$ with precipitation of product[64], and the process taking place at this potential was described by reactions 76 and 78[64], i.e. the same overall electrode process but with a different follow-up reaction than that suggested for reduction at the potential of the first wave. This is in contrast to the results obtained by exhaustive electrolysis in 0.1 M $HClO_4$ with a substrate concentration of 4.7×10^{-5} M, where $n = 3.96 \pm 0.08$ was found[65]. No product precipitated before the electrolysed solution was exposed to air[65], and reoxidation of the electrolysed solution gave no sign of re-formation of $PhAsO$[65]. When the reduction was carried out in two steps, most of the insoluble product formed in the first step precipitated onto the mercury pool electrode and could be further reduced in the second step. Reaction 79 was suggested to account for the second reduction step.

$$1/3[(PhAs)]_6 + 4e^- + 4H^+ \longrightarrow 2PhAsH_2 \qquad (79)$$

The reason for the different number of electrons, and consequently for the differences in products found in the two studies, may be the difference in the applied coulometric methods. The microcoulometric method applied in Reference 64 is not exhaustive as is the coulometric method applied in Reference 65, and the products formed in the former study may therefore be of an intermediate nature. Insoluble products formed in the bulk solution during microcoulometry precipitate and never get in contact with the electrode, whereas precipitation onto the mercury pool electrode in the exhaustive electrolysis allows further reduction of an insoluble intermediate.

In a much earlier study of the reduction of $4\text{-}NH_2C_6H_4AsO$ at a Pb electrode[37] another intermediate (ArAs=AsAr, Ar = $4\text{-}NH_2C_6H_4$) was isolated after preparative electrolysis in basic solution, and reactions 76 and 80 were suggested to account for this product[37].

$$ArAsH_2 + ArAsO \longrightarrow ArAs{=}AsAr + H_2O \qquad (80)$$

At lower pH the main product of the reduction changed from ArAs=AsAr to the arylarsine ($4\text{-}NH_2C_6H_4AsH_2$). The fact that $4\text{-}NH_2C_6H_4AsO$ is easier to reduce than $4\text{-}NH_2C_6H_4AsO(OH)_2$ was used to explain that reduction of $4\text{-}NH_2C_6H_4AsO(OH)_2$ in acidic solution does not lead to the oxide but directly to the arylarsine[37] (cf Section II.D.2).

IV. REDUCTION OF DIMERIC COMPOUNDS OF As, Sb AND Bi

Homobimetallic species of the structure R_2MMR_2, M = As, Sb, Bi, are compounds formally containing M in the oxidation state II and the electrochemical reduction of these dimeric compounds is therefore described in this separate section together with a summary of the typical reactions of the metalloid anions formed in the reduction process.

A. Formation of Ph$_2$M$^-$

In the series M = P, As, Sb, Bi the reduction potential of the homobimetallic species Ph$_2$MMPh$_2$ increases, when M is changed from P to Bi, paralleling the decrease in M—M bond strength from P—P to Bi—Bi[66]. In DMF, Ph$_2$AsAsPh$_2$ is easier to reduce than is Ph$_2$PPPh$_2$ by approximately 150 mV[67], and in Table 11 the data for reduction in DME[11] show that the change from As to Sb and from Sb to Bi are both associated with a positive shift in the half-wave potential of 0.2 V. Substitution of one of the phenyl groups by an aliphatic chain shifts the potential in the negative direction by almost 0.4 V[67] (cf Table 11).

Independently of the choice of the polar aprotic solvent, the reduction of Ph$_2$MMPh$_2$ is a chemically irreversible two-electron reduction[11,66–68] leading to formation of two metalloid anions (equation 71).

$$Ph_2MMPh_2 + 2e^- \longrightarrow 2Ph_2M^- \qquad (81)$$

In an early general study of reductive electrochemical cleavage of bimetallic compounds in DME[66] it was found that as for the As, Sb and Bi dimers, homobimetallics always cleave after the uptake of *two* electrons, whereas heterobimetallic compounds (R$_n$M^{1}M^{2}R$_m$) may cleave after the uptake of only *one* electron giving a metalloid anion [R$_n$M^1]$^-$ and a paramagnetic species [R$_m$M^2]$^{\cdot}$[66]. The products obtained after heterolytic cleavage at the anion radical stage were found to be governed by the relative reduction potentials for the two corresponding homobimetallic compounds in the way that [R$_n$M^1]$^-$, i.e. the anion, corresponds to the homobimetallic compound with the highest oxidation potential[66]. Bimetallic species with bridging ligands may in some cases be electrochemically reduced without complete cleavage of the dimeric structure, probably because the complex is held together by the bridges while the metal–metal bond is cleaved[66]. Whether this is the case when structure **8** is reduced[67] (cf Table 11) was not investigated.

$$Ph—As—As—Ph$$
$$\diagup \qquad \diagdown$$
$$CH_2CH_2CH_2$$

(8)

Reduction of tetraalkyldiarsines has only been reported in the case of Me$_2$AsAsMe$_2$ (cacodyl)[60], and then only preparative aspects in acidic EtOH. The final product was shown to be dimethylarsine (Me$_2$AsH)[60], most likely formed by protonation of Me$_2$As$^-$.

The metalloid anions (Ph$_2$M$^-$, M = As, Sb, Bi) are also formed by reduction of Ph$_3$M (Section III.A.1) or by reduction of (Ph$_2$M)$_2$O (Section III.C) and can be oxidized electrochemically (cf Table 12). In THF Ph$_2$Sb$^-$ is oxidized at about -1.2 V vs NHE as

TABLE 11. Half-wave potentials ($E_{1/2}$) for the chemically irreversible reduction of (Ph$_2$M)$_2$ in aprotic solvents

Compound	Solvent	$E_{1/2}$/V vs NHE	Reference
(Ph$_2$As)$_2$	DME	−1.8	11
	DMF	−1.655	67
(Ph$_2$Sb)$_2$	DME	−1.6	11
	THF	−1.60	68
(Ph$_2$Bi)$_2$	DME	−1.4	11
PhAs-AsPh (8) | | (CH$_2$)$_3$	DMF	−2.085	67

TABLE 12. Half-wave potentials ($E_{1/2}$) for the chemically irreversible reduction of anions Ph_2M^{-} [a]

Compound	Solvent	$E_{1/2}$/V vs NHE[b]	Reference
Ph_2As^-	DME	−1.1	69
Ph_2Sb^-	DME	−1.1	69
	THF	−1.25	57
Ph_2Bi^-	DME	−1.2	69

[a] At a Pt electrode.
[b] The potentials are rather uncertain.

shown by CV[68] (cf Section III.A.1), in a one-electron process to give the corresponding radical ($Ph_2Sb^{\cdot}$) which rapidly dimerizes (reaction 82).

$$2Ph_2Sb^{\cdot} \longrightarrow Ph_2SbSbPh_2 \qquad (82)$$

Identical behaviour was observed for Ph_2As^- in DME[11]. A second oxidation peak at a higher potential, $E_{1/2} \approx -0.4$ V vs NHE, was attributed to either the further oxidation of $Ph_2As^{\cdot}$ or to the oxidation of the dimer formed by a reaction analogous to reaction 82[11] (cf Section VI.A). Oxidation of $Ph_2As^{\cdot}$ or of the dimer gives Ph_2As^+, and reduction of Ph_2As^+ could be observed as a new peak in CV if the scan was reversed after the second oxidation peak[11]. The oxidation of $Ph_2AsAsPh_2$ has, in contrast to the oxidation of $Ph_2SbSbPh_2$, not been studied independently (cf Section VI.A).

B. Reaction of Ph_2M^-

The metalloid anions Ph_2As^- (yellow) and Ph_2Sb^- (orange/red) are relatively stable in aprotic solvents[11,57,68], whereas in DME Ph_2Bi^- (green) gradually disproportionates precipitating free Bi metal[11]. After preparative reduction of Ph_3Sb in THF Ph_2Sb^- is also formed (cf Section III.A.1), and it was reported that Ph_2Sb^- reacted with the cation of the supporting electrolyte (Bu_4N^+) according to reaction 59 (Section III.A.1). However, this substitution process is slow compared to the reaction of Ph_2M^- with good electrophiles like alkyl halides (RX).

The kinetics of the reaction (equation 83) of a series of electrogenerated metalloid anions including Ph_2As^-, Ph_2Sb^- and Ph_2Bi^- with alkyl halides have been studied in DME[69].

$$Ph_2M^- + RX \longrightarrow Ph_2MR + X^- \qquad (83)$$

The anions were generated by exhaustive electrolysis of the dimers, and the electrophile (Me_2CHBr) was subsequently added. The kinetics of reaction 83 was followed by monitoring the decrease with time in the current at a rotating Pt-disk electrode (RDE) held at the potential of the Ph_2M^- oxidation[69]. The oxidation currents were in separate experiments shown to be directly proportional to the concentration of Ph_2M^- in the range 5×10^{-4} M–5×10^{-3} M[69], and the kinetic experiments were carried out under second-order conditions[69]. The measured rate constants are given in Table 13, and for comparison the rate constant for reaction of a 'normal' nucleophile (PhS^-) with Me_2CHBr is included. From the data in Table 13 it appears that the metalloid anions (Ph_2M^-) are excellent nucleophiles, with rate constants more than three orders of magnitude larger than the rate constant found for PhS^-, and apparently the nucleophilicity increases in the series from Ph_2As^- to Ph_2Bi^-, although the rate constant for reaction of Ph_2Bi^- is rather uncertain due to experimental difficulties (cf footnote to Table 13)[69].

TABLE 13. Second-order rate constants, k, for the
reaction between Ph_2M^- and Me_2CHBr (reaction 83)
in DME at 25°C[a]

Anion	$k/M^{-1}s^{-1}$
Ph_2As^-	85
Ph_2Sb^-	140
Ph_2Bi^-	$> 100^b$
PhS^-	0.06

[a] Data from Reference 69.
[b] This rate constant was difficult to obtain due to insolubility
of the precursor complex and instability of the anion.

Including the larger series of metalloid anions studied in Reference 69 (derived from
main group metal and transition metal complexes of very different structures) it was
shown[69] that the logarithm of the rate constant for reaction 83 plotted against the
potential of the chemically irreversible oxidation of the anions gave a roughly linear
correlation, indicating that the anions which are easiest to oxidize are the most reactive.
The relative rate constants covered a range of twelve orders of magnitude with the
compounds Ph_2M^-, $M = As$, Sb, Bi, being the most reactive. The possibility that electron
transfer (ET) is the initial step in the substitution reaction was not discussed, and the
products were not isolated. However, polarography did not detect side-products from
reaction of $Ph_2M^{\cdot}$, such as the dimer or Ph_2MH, and GC failed to detect side-products
from reactions of $Me_2CH^{\cdot}$[69].

The anions Ph_2As^- and Ph_2Sb^- have been prepared $in\ situ$ by alkali metal reduction in
liquid ammonia and, like the corresponding Ph_2P^-, they react with haloarenes (ArX) by a
photostimulated $S_{RN}1$-mechanism[70,71], i.e. by a mechanism involving ET as an important
step. The mechanism may be formulated as in equations 84–87.

$$Ph_2M^- + ArX \xrightarrow{h\nu} Ph_2M^{\cdot} + ArX^{-\cdot} \tag{84}$$

$$ArX^{-\cdot} \longrightarrow Ar^{\cdot} + X^- \tag{85}$$

$$Ar^{\cdot} + Ph_2M^- \rightleftharpoons Ph_2ArM^{-\cdot} \tag{86}$$

$$Ph_2ArM^{-\cdot} + ArX \longrightarrow Ph_2ArM + ArX^{-\cdot} \tag{87}$$

The coupling reaction 86 has been formulated as a reversible reaction which could also
take the form of equation 88, in order to account for the formation of scrambled products,
$PhAr_2M$, Ph_3M and Ar_3M. Dissociation according to equation 88 leads to the formation
of $PhArM^-$ and $Ph^{\cdot}$ which can react like Ph_2M^- and $Ar^{\cdot}$, respectively, in reactions 84 and
86[70,71].

$$Ph_2ArM^{-\cdot} \rightleftharpoons PhArM^- + Ph^{\cdot} \tag{88}$$

The predominance of either ET (equation 87) or cleavage (equations 86 and 88) was
shown[71] to depend on the relative energies of the σ^* and π^* molecular orbitals (MOs) in
the intermediate $Ph_2ArM^{-\cdot}$, i.e. whether the odd electron occupies the σ^* or π^* MO. The
energy of the σ^* can be varied by changing from As to Sb, while the energy of the π^* is
dependent on the aryl moiety[71]. If the odd electron occupies π^*, the unscrambled
substitution product is obtained, whereas location of the odd electron in σ^* may lead to
bond dissociation and consequently to scrambled products. The energy of the π^* MO can
be approximated by the reversible reduction potential of the non-halogen substituted aryl

compound, because the π^* of the intermediate is almost exclusively aryl ligand based[71]. The energy of the σ^* MO can be approximated by (a) the energy of the σ MO, which subsequently can be approximated by the bond dissociation energy of M—Ph, or (b) the reversible reduction potential of Ph_2ArM[71]. When the reversible reduction potentials for Ph_2ArM are inaccessible due to fast follow-up reactions of the anion radicals, the potentials of the chemically irreversible reductions may be used as rough approximations. As discussed in Section III.A.1, the potentials of the chemically irreversible reduction of Ph_3M seem to follow the expected change in bond strength between C and M when M is varied. Consequently, for the same haloaromatic compound, ArX, the ratio of simple substitution product to scrambled products is expected to increase if M is changed from Bi to As (or P), because the energy of σ^* increases from Bi to P. The experimentally observed product ratios are in agreement with these expectations[71].

Addition of Ph_2AsBr to DME solutions of Ph_2As^- or addition of $Ph_2Sb(OAc)$ to DME solutions of Ph_2Sb^- leads to formation of the substitution products $Ph_2AsAsPh_2$ or $Ph_2SbSbPh_2$, respectively[11]. In these cases an ET mechanism (reactions 89 and 90) is likely to be followed, due to the ease of reduction of Ph_2AsBr and $Ph_2Sb(OAc)$ (cf Table 9 in Section III.B).

$$Ph_2M^- + Ph_2MX \longrightarrow Ph_2M^\cdot + Ph_2M^\cdot + X^- \tag{89}$$

$$2\,Ph_2M^\cdot \longrightarrow Ph_2MMPh_2 \tag{90}$$

V. OXIDATION OF TRIVALENT As, Sb AND Bi COMPOUNDS

Studies of the electrochemical oxidation of trivalent organic As, Sb and Bi compounds have mainly been focused on oxidations of trisubstituted arsines and, to a smaller extent, on the oxidations of trisubstituted stibines, whereas only a few preparative studies of oxidations of trisubstituted bismuthines have been reported.

A. Effect of Structure on the Oxidation Potential

With only few exceptions, all of the electrochemical oxidation studies have been carried out in MeCN containing varying amounts of water. The compounds of the structure $R^1R^2R^3M$, R^1, R^2, R^3 = aryl, alkyl, M = As, Sb (Bi), are in most cases (for exceptions see below) oxidized in a chemically irreversible process in MeCN. Half-wave potentials ($E_{1/2}$) for the chemically irreversible oxidations of $R^1R^2R^3M$ are given in Table 14.

Compared to the analogous trisubstituted phosphines, the trisubstituted arsines are oxidized at slightly higher potentials (50–100 mV) as expected from the decrease in electron-donating ability in the series N, P, As[72]. However, the analogous stibines are oxidized at almost the same potentials as the arsines but differences in the rates of the follow-up reactions may be the cause of this apparent coincidence in irreversible potentials. Trimesitylarsine and trimesitylstibine are oxidized in diffusion-controlled, chemically reversible one-electron processes[73,74], and the reversible potentials for these oxidations show (cf Table 14) that the stibine is 60 mV more difficult to oxidize than the arsine.

Electrochemical oxidation of Ph_3Bi has only been carried out in CH_2Cl_2, and $E_{1/2}$ for this chemically irreversible oxidation is higher than $E_{1/2}$ for Ph_3As and Ph_3Sb in MeCN. Although the solvents are different, this finding is in good agreement with the relative rates of chemical oxidation of Ph_3M (M = As, Sb, Bi) using substituted cation radicals of triarylamines as oxidizing agents[79]; Ph_3As and Ph_3Sb are oxidized by $(p\text{-Tol})_3N^{+\cdot}$ in CH_2Cl_2 with comparable rates ($k = 6.6\,M^{-1}s^{-1}$ for Ph_3As and $k = 5.7\,M^{-1}s^{-1}$ for Ph_3Sb), wheras Ph_3Bi requires the stronger oxidant $(p\text{-BrC}_6H_4)_3N^{+\cdot}$ to give a measurable rate of reaction ($k = 49.7\,M^{-1}s^{-1}$), when studied by stopped flow technique[79].

TABLE 14. Half-wave potentials ($E_{1/2}$) for the chemically irreversible oxidation at Pt electrodes of trisubstituted
arsines, stibines and bismuthines in MeCN[a]

Compound	$E_{1/2}$/V vs NHE	References
Triaryl compounds:		
Ph_3As	1.365^b	72
	$1.260^{b,c}$	72
	1.72	73, 75, 76
	1.65^d	77
	0.63^e	78
	1.35	51
$Ph_2As(\alpha\text{-Naph})$	1.335^b	72
	$1.230^{b,c}$	72
$PhAs(\alpha\text{-Naph})_2$	1.345^b	72
$(\alpha\text{-Naph})_3As$	1.430^b	72
	1.70	76
$(p\text{-Tol})_3As$	1.57	73, 76
$(p\text{-An})_3As$	1.47	73, 76
$(p\text{-ClC}_6H_4)_3As$	1.91	76
Mes_3As	1.23^f	73, 76
Ph_3Sb	1.215^b	72
	1.65	74, 76
	1.5	47
	0.66^e	78
$(p\text{-Tol})_3Sb$	1.56	74, 76
$(p\text{-An})_3Sb$	1.481	74, 76
Mes_3Sb	1.29^f	74, 76
$((MeO)_3C_6H_2)_3Sb$	0.98	74, 76
Ph_3Bi	1.42^g	79
Trialkyl compounds:		
Et_3As	1.50	73, 75, 76
Pr_3As	1.48^d	77
	1.53	76
Bu_3As	1.115^b	72
	$1.045^{b,c}$	72
	1.46^d	77
	1.56	76
Pe_3As	1.40^d	77
	1.58	76
Bu_3Sb	1.48	76
Pe_3Sb	1.50	76
Compounds with mixed ligands:		
Ph_2AsEt	1.230^b	72
	$1.145^{b,c}$	72
$Ph_2As(OMe)$	1.83	76
$PhAsEt_2$	1.180^b	72
	$1.050^{b,c}$	72
$PhAs(c\text{-Hex})_2$	1.260^b	72
	$1.150^{b,c}$	72

TABLE 14. (Contd.)

Compound	$E_{1/2}$/V vs NHE	References
PhAs(CH$_2$)$_4$	1.205^b	72
(p-An)AsEt$_2$	1.37	76
EtAs(OMe)$_2$	2.37	76
EtAs(OEt)$_2$	2.32	76
EtAs(OBu)$_2$	2.28	76
Bu$_2$AsC≡CH	1.96	76
Compounds with two or three As atoms:		
Ph$_2$AsCH$_2$AsPh$_2$	0.77^{e,h}	80
	0.78^{e,g}	80
	0.79^e	80
	2.02^i	80
MeAs((CH$_2$)$_3$AsMe$_2$)$_2$ (9)	0.01^{e,j}	81
Bu$_2$AsC≡CAsBu$_2$	1.65	76

^{a} Using neutral supporting electrolytes unless otherwise noted.
^{b} Carbon paste (CP) electrode.
^{c} In a 1:1 MeCN/H$_2$O mixture.
^{d} Using pyridinium tetrafluoroborate (0.1 M) as supporting electrolyte.
^{e} Hg electrode; the process involves reaction with Hg, see Section V.E.
^{f} Chemically reversible.
^{g} In CH$_2$Cl$_2$.
^{h} In acetone.
^{i} E_{peak} determined by CV.
^{j} In a 1:1 benzene/EtOH mixture.

The potentials for the chemically irreversible oxidation of trisubstituted arsines in MeCN at CP electrodes correlate roughly with the sum of the Taft σ^* values for the alkyl or aryl substituents (four compounds)[72], with a slope that is slightly larger than the corresponding correlation for the analogous phosphines.

For the series of symmetrically substituted compounds Ar$_3$P, Ar$_3$As and Ar$_3$Sb the half-wave potentials for the irreversible oxidations were correlated with the $\Sigma\sigma$ values and the three Hammett relations, equations 91–93, in which the potentials are referred to NHE and m is the number of compounds, were found[76].

$$Ar_3P: \quad E_{1/2} = (0.34 \pm 0.06)\Sigma\sigma + (1.37 \pm 0.05) \quad (r = 0.987, m = 7) \quad (91)$$

$$Ar_3As: \quad E_{1/2} = (0.31 \pm 0.03)\Sigma\sigma + (1.71 \pm 0.02) \quad (r = 0.999, m = 5) \quad (92)$$

$$Ar_3Sb: \quad E_{1/2} = (0.22 \pm 0.03)\Sigma\sigma + (1.65 \pm 0.05) \quad (r = 0.998, m = 5) \quad (93)$$

The reaction constant (ρ) in equation 93[76] is different from that reported in Reference 74 based on four of the five potentials ($\rho = 0.62$ in Reference 74), and the ρ value in equation 93 should in fact be 0.28 if the σ value for o-Me and o-MeO are taken to be equal to the σ values for p-Me and p-MeO, respectively, as applied in References 73 and 76 for o-Me. Consequently, the ρ values for the three series are not very different, and interpretation of any differences is dubious due to the varying influence of the homogenous follow-up reactions on the values of $E_{1/2}$. The influence of the homogenous follow-up reaction on $E_{1/2}$ may change when the substituents in the phenyl rings are changed, or when the central atom is changed.

If one of the phenyl groups in Ph$_3$As is exchanged with an alkoxy group (OMe), the potential of the chemically irreversible oxidation is shifted 0.11 V in the positive direction[76] (cf Table 14). If, however, two of the alkyl groups in R_3As are exchanged with alkoxy groups, the $E_{1/2}$ value for the irreversible oxidation is shifted dramatically in the positive direction[76] (cf Table 14) and trialkoxyarsines are not electrochemically oxidizable in the accessible potential range[76].

When water is added to the MeCN solutions of trisubstituted arsines, the oxidation potentials are shifted in the negative direction (cf Table 14) and the shift in potential is larger for the aryl-substituted arsines than for the alkyl-substituted analogues[72]. The shift is probably due to increase in the rate of the follow-up reaction of the initially formed cation radical, whereas the different magnitude of the effect for aryl and alkyl compounds probably reflects differences in the contributions from the heterogeneous charge transfer rate constant to the value of the potential. Partial control of the potential by the heterogeneous charge transfer rate constant is expected to be of more importance for the alkyl than for the aryl compounds, due to smaller values of the heterogeneous rate constant for the alkyl than for the aryl compounds and comparable rate constants for the homogeneous follow-up reaction for the two types of compounds.

The application of a rotating ring-disk electrode (RRDE) in the study of the oxidation of tri-substituted arsines and stibines shows[73,74,76], as mentioned above, that when the substituents are sufficiently electron-donating as in the case of Mes_3As and Mes_3Sb (Mes = mesityl = 2,4,6-trimethylphenyl), the cation radical formed at the disk is sufficiently stable to allow detection at the ring as also found for Mes_3P. However, the stability of the cation radicals decreases in the order P > As > Sb, as reflected in the rotation speed necessary to obtain comparable reduction currents at the ring[74]. The differences in reactivity of the cation radicals have been attributed to the increase in bond lengths, M—C, when M is changed from P to As to Sb, making the central atom M more susceptible to attack from nucleophiles (H_2O) the longer the bond length[74].

On a polarographic time scale in 'wet' MeCN, the oxidation of Ph_3As is a two-electron process[52] affording triphenylarsine oxide (Ph_3AsO) as product. When studied by RDE, oxidation of Mes_3As takes place in two consecutive one-electron steps as evaluated from the heights of the plateau currents by comparison with phenothiazine, and the second oxidation step is chemically irreversible at all the employed rotation speeds[73]. For the other tri-substituted arsines and stibines studied in References 73 and 74, the height of the limiting current of the first oxidation process on the RDE corresponds to the transfer of 1.3–1.7 electrons per molecule for the triarylarsines but 0.94 for Et_3As[73], and to 0.6–1.4 electrons per molecule for the stibines[74]. The number of electrons was almost unaffected by addition of small amounts of water but *increased* and approached $n = 2$ as the rotation speed was increased, and the experimentally found $n < 2$ is therefore not a consequence of a relatively slow follow-up reaction. The scheme suggested to account for these observations includes, in addition to the product-forming reactions 94–96, a deactivation of the starting material via protonation (reaction 97) by the protons liberated in reaction 95. If the forward reaction in equation 97 is fast compared to the time scale of the experiment, and the equilibrium constant is large, only half of the substrate will be oxidized in a two-electron process, and the experimentally determined value of n will be equal to one.

$$R_3M \;\rightleftharpoons\; R_3M^{+\bullet} + e^- \tag{94}$$

$$R_3M^{+\bullet} + 2H_2O \longrightarrow R_3MOH^\bullet + H_3O^+ \tag{95}$$

$$R_3MOH^\bullet + R_3M^{+\bullet} \longrightarrow R_3MOH^+ + R_3M \tag{96}$$

or

$$R_3MOH^\bullet \longrightarrow R_3MOH^+ + e^-$$

$$R_3M + H_3O^+ \rightleftharpoons R_3MH^+ + H_2O \tag{97}$$

At higher rotation speeds, the protonation of the substrate (equation 97) may, in the case of the triaryl-substituted compounds, be out-run thereby increasing the value of n[73,74].

For Et_3As the forward reaction (equation 97) may be faster giving $n = 1$ even at higher rotation speeds[73]. The influence of the acid–base reaction 97 was further supported in the case of $((MeO)_3C_6H_2)_3Sb$ for which $n = 0.63$ was found in MeCN with residual water, whereas $n = 2$ was found upon addition of acetate ions[74].

B. Preparative Studies in 'Wet' Acetonitrile

Preparative electrochemical oxidation of Ph_3As in 'wet' MeCN ('wet' ≈ 0.1 M H_2O) represents an alternative route to the synthesis of triphenylarsine oxide derivatives from Ph_3As. Exhaustive electrolysis at Pt electrodes at the first oxidation potential of Ph_3As gives, in a two-electron process, a cationic species (hydroxytriphenylarsonium, Ph_3AsOH^+) which can be isolated as the perchlorate or chloride[52]. In contrast to the perchlorate, the chloride is not fully dissociated in MeCN[52]. If a tetrafluoroborate salt is used as supporting electrolyte, the isolated product is Ph_3AsOBF_3 and not $Ph_3AsOH^+BF_4^-$ [52] probably because HF is eliminated from the initially formed $Ph_3AsOH^+BF_4^-$ [52]. In contrast to the product isolated after two-electron oxidation of Ph_3As in wet MeCN, preparative two-electron oxidation of Ph_3Sb under the same conditions leads to the oxybis(triphenylantimony) dication $[Ph_3SbOSbPh_3]^{2+}$, which can be isolated as the diperchlorate[47].

The quantitative formation of Ph_3AsOH^+ in a two-electron oxidation is in accord with reactions 94–97, when it is assumed that on the time scale of preparative electrolysis equilibrium 97 is pulled to the left. The oxidation of Ph_3Sb probably follows the same mechanism, but the initially formed Ph_3SbOH^+ dimerizes and eliminates a molecule of water[47] (reaction 98).

$$2Ph_3SbOH^+ \rightleftharpoons [Ph_3SbOSbPh_3]^{2+} + H_2O \tag{98}$$

After two-electron oxidation of Ph_3As, potentiometric titration with Bu_4NOH shows three inflection points in the ratio 2:1:1. The first inflection corresponds to neutralization of the protons formed during the electrolysis, and the second to neutralization of half the amount of Ph_3AsOH^+. The base formed in this neutralization process (Ph_3AsO) reacts with Ph_3AsOH^+ forming a tightly hydrogen-bonded complex $[Ph_3AsO \cdots H \cdots OAsPh_3]^+$ which is neutralized in the last acid–base reaction[52]. The electrochemical reduction of Ph_3AsOH^+ and $[Ph_3AsO \cdots H \cdots OAsPh_3]^+$ has been studied[52] (cf Section II.F.3) Potentiometric titration of the solution with Bu_4NOH after two-electron oxidation of Ph_3Sb shows two inflection points, the first of which again corresponds to the neutralization of the protons formed during the electrolysis, whereas the second is attributed to the neutralization of Ph_3SbOH^+. This demonstrates the reversibility of the condensation reaction 98[47]. The electrochemical reduction of $[Ph_3SbOSbPh_3]^{2+}$ has also been studied[47] (cf Section II.F.4).

When chloride ions are added to the MeCN solution of Ph_3As, the oxidation of Ph_3As on a Pt electrode takes place at a much lower potential (negative shift ≈ 0.8 V) than the direct oxidation of Ph_3As alone, and also at a lower potential than oxidation of Cl^- alone under the same conditions[52]. This behaviour has been described as an electrocatalytic process involving initial oxidation of Cl^- (reaction 99) followed by a fast reaction between $Cl^{\cdot}$ and Ph_3As (reaction 100), where the initial reaction between $Cl^{\cdot}$ and Ph_3As successfully competes with the formation of Cl_2 which is not formed during the electrolysis[52].

$$Cl^- \longrightarrow Cl^{\cdot} + e^- \tag{99}$$

$$2Cl^{\cdot} + Ph_3As + H_2O \longrightarrow Ph_3AsOHCl + HCl \tag{100}$$

The electrochemical reduction of the isolated product ($Ph_3AsOHCl$) has been studied independently[52] (cf Section II.F.3) and, because it is a weaker electrolyte than $Ph_3AsOHClO_4$, its reduction is much simpler.

C. Formation of Quaternary Arsonium and Stibonium Salts

Preparative oxidation of Et_3As and Ph_3As^{75} or Ph_3Sb^{82} in the presence of aromatic or heteroaromatic compounds leads to formation of quaternary arsonium or stibonium salts[75,82]. The oxidations were carried out in nominally dry MeCN in the presence of an excess of the aromatic compound which, in the case of the arsines, were naphthalene, thiophene, pyridine and furan[75], and in the case of Ph_3Sb were thiophene and toluene[82].

Electrophilic aromatic substitution is the mechanism suggested in both cases [75,82], the initial step being attack of the cation radical on the aromatic compound (ArH) (reaction 101), followed by deprotonation of the intermediate (reaction 102) and further oxidation of the radical to the resulting arsonium or stibonium ion (reaction 103).

$$R_3M^{+\cdot} + ArH \longrightarrow R_3MArH^+ \tag{101}$$

$$R_3MArH^+ \longrightarrow R_3MAr^\cdot + H^+ \tag{102}$$

$$R_3MAr^\cdot \longrightarrow R_3MAr^+ + e^-$$

or   (103)

$$R_3MAr^\cdot + R_3M^{+\cdot} \longrightarrow R_3MAr^+ + R_3M$$

In the deprotonation step (reaction 102), the substrate (R_3M) has been suggested to act as base[75] by analogy with the deactivation equilibrium of the substrate (equation 97) believed to operate when R_3As is oxidized in 'wet' MeCN (cf Section V.A). If a base is deliberately added (Na_3PO_4) the yields increase[82]. Whether the second electron transfer (reaction 103) is heterogenous or homogeneous was not discussed, and the reduction potentials for the resulting R_3MAr^+ in MeCN have not been determined. No further attempts to verify the mechanism were made due to severe adsorption and distortion of the voltammograms of R_3M in the presence of the aromatic compounds.

The yields of arsonium salts were reported to be 77–80%, based on the stoichiometry $2R_3As + ArH \rightarrow R_3AsAr^+ + R_3AsH^+ + 2e^-$, R = Ph, Et, whereas the yields of the stibonium salts were lower (46–51%) based on the stoichiometry $Ph_3Sb + ArH + PO_4^{3-} \rightarrow Ph_3SbAr^+ + HPO_4^{2-} + 2e^-$.

The arylation reaction takes place in competition with reaction of the cation radical with residual water (reaction 95; cf Sections V.A and V.B), which may cause the yields of the quaternary ions to be relatively low. Similar competition has been observed when Ph_3P is oxidized in MeCN in the presence of benzene[83], in which case the Ph_2PO as well as Ph_3PH^+ were found as products in addition of Ph_4P^{+} [83].

D. Formation of R_3MF_2, R = Alkyl, Aryl, M = As, Sb

When Ph_3As was oxidized in 'wet' $MeCN/Et_4NBF_4$ (0.1 M), Ph_3AsOBF_3 was obtained whereas preparative electrolysis at constant current in nominally dry $MeCN/Et_4NBF_4$ leads to formation of the pentacoordinated arsorane Ph_3AsF_2 (30%)[84]. The same type of product is obtained when trialkylarsines, trialkylstibines and triarylstibines are oxidized under similar conditions[84,85] (cf Table 15). It appears that the cation radicals derived from the tertiary stibines are fluorinated with a much larger efficiency than are the arsines. However, the yield of R_3AsF_2 is greatly improved if the source of fluoride is changed from BF_4^- to F^- in the form of n-pentylammonium fluoride $(PeNH_3F)^{77}$ (cf Table 15).

Cyclic voltammetry of R_3As in MeCN using 0.1 M pyridinium tetrafluoroborate as supporting electrolyte gives values of n in the range $0.66-1.40^{77}$ and addition of a twofold excess of $PeNH_3F$ slightly reduces the value of n, whereas the half-wave potentials are cathodically shifted by $30-60\,mV^{77}$. The decrease in n was attributed to adsorption of the products, and the values of n determined from coulometric experiments were in all cases close to 2 in the presence of two equivalents of $PeNH_3F^{77}$. Similar behaviour was observed when Et_4NBF_4 in a large excess was added to MeCN solutions containing tertiary stibines[85].

TABLE 15. Yields of R_3MF_2 isolated after constant current oxidation of R_3M in MeCN at a Pt electrode in the presence of different sources of fluorine[a]

Substrate	Source of F	Yield (%)	Reference
Pr_3As	Et_4NBF_4[b]	26	84
	$PeNH_3F$	83	77
Bu_3As	Et_4NBF_4[b]	28	84
	$PeNH_3F$	86	77
Pe_3As	Et_4NBF_4[b]	34	84
	$PeNH_3F$	90	77
Ph_3As	Et_4NBF_4[b]	30	84
	$PeNH_3F$	81	77
Bu_3Sb	Et_4NBF_4	95	85
Pe_3Sb	Et_4NBF_4	93	85
Ph_3Sb	Et_4NBF_4	96	85
Mes_3Sb[c]	Et_4NBF_4	96	85

[a] The concentration of fluorine-containing compound was twice the concentration of the substrate, unless otherwise noted.
[b] Relative concentrations of substrate and fluorine source were not given.
[c] 2:1 mixture of MeCN and benzene.

E. Oxidations Involving Mercury

Oxidation of bis(diphenylarsino)methane (dpam) in MeCN at a Pt electrode is by CV observed as a broad, irreversible peak at 2.02 V vs NHE[80]. When the electrode material is changed to Hg, a diffusion-controlled, chemically reversible one-electron oxidation process is observed with $E_{1/2} = 0.79$ V vs NHE, and the same behaviour is observed (with approximately the same $E_{1/2}$ value) in acetone and in CH_2Cl_2[80] (cf Table 14). Coulometry in acetone gave $n = 1$. In acetone and in CH_2Cl_2 a new second reduction was found for dpam at Hg electrodes close to the potential of oxidation of the mercury. This second oxidation was adsorption-controlled as judged from electrocapillary measurements and the variation of the limiting polarographic current with substrate concentration, but the nature of this oxidation was not studied further[80].

After exhaustive electrochemical oxidation of dpam at Hg in acetone, polarography shows the above-mentioned second oxidation wave as well as a reduction wave at 0.76 V vs NHE[80]. This polarogram was identical to the polarogram obtained if Hg(I) or Hg(II) salts were added to the original solution[80]. The reaction sequence given by equations 104 and 105[80] is consistent with these findings.

$$2\text{dpam} + 2\text{Hg} \rightleftharpoons 2[\text{Hg(I)(dpam)}]^+ + 2e^- \tag{104}$$

$$2[\text{Hg(I)(dpam)}]^+ \rightleftharpoons [\text{Hg(II)(dpam)}_2]^+ + \text{Hg} \tag{105}$$

Reactions 104 and 105 are also in agreement with the results obtained for oxidation of Ph_3As and Ph_3Sb at Hg electrodes in MeCN[78]. In those cases, the oxidation potentials are shifted approximately 0.6 V in the negative direction relative to the oxidation potentials obtained at Pt electrodes (cf Table 14) and the number of electrons involved in the oxidations was equal to one on a polarographic as well as on a coulometric time scale[78]. Furthermore, preparative oxidation of Ph_3As in MeCN at Hg gave $(Ph_3As)_2Hg^{2+}$ in 85% yield, which was isolated as the diperchlorate[78]. Identical behaviour was found for a number of tertiary phosphines[78].

Oxidation of the bismuth analogue (Ph_3Bi) apparently follows a different path. In polarography at Hg in CH_2Cl_2 a single drawn out wave at ≈ 0.7 V vs NHE is observed[21], and the limiting current is proportional to the concentration of Ph_3Bi in the range

5×10^{-6} M to 2×10^{-4} M[21]. The final product of the oxidation is Ph_2Hg but the amount of product formed is not directly proportional to the substrate concentration, and the final Bi-containing product was not identified. Cationic as well as radical Bi intermediates were envisaged in the over-all transfer of Ph from Bi to Hg^{21}.

Methyl-bis(3-propyldimethylarsine)arsine (9) behaves almost as dpam[81]. No electrochemical oxidation is observed at Pt electrodes in 1:1 benzene/EtOH mixtures but in polarography at a Hg electrode a single oxidation wave with a maximum is observed at 0.01 V vs NHE.

(9)

The limiting polarographic current was not proportional to the substrate concentration and a slightly soluble mercury compound was formed[81]. The nature of the product and the number of electrons involved in the process were not determined. Chemical oxidation of **9** by I_2 in an overall six-electron process (corresponding to two-electron oxidation of each As atom) leads to formation of the partly ionic hexaiodide[81].

VI. OXIDATION OF LOWER VALENT As, Sb AND Bi COMPOUNDS

A. Oxidation of As—As and Sb—Sb Compounds

The electrochemical reduction of the tetra-substituted diarsines, distibines and disbismuthines (Ph_2MMPh_2) has been studied in some detail (cf Section IV.A) whereas studies of the electrochemical oxidation of these compounds are rather sparse. Table 16 presents reported chemically irreversible oxidation potentials for these dimers.

In MeCN at a CP electrode, $Ph_2AsAsPh_2$ is oxidized at a potential approximately 50 mV higher than the oxidation of the analogous diphosphine $(Ph_2PPPh_2)^{72}$, i.e. with almost the same potential difference as found for oxidation of Ph_3As and Ph_3P (cf Section V.A). Direct comparison between the oxidation potential for $Ph_2AsAsPh_2$ and $Ph_2SbSbPh_2$ is difficult due to the difference in solvent as well as electrode material.

TABLE 16. Half-wave potentials $(E_{1/2})$ for the chemically irreversible oxidation of R_2MMR_2, M = As, Sb

Compound	Solvent	$E_{1/2}$/V vs NHE	Reference
$Ph_2AsAsPh_2$	MeCN	1.910[a]	72
$Ph_2SbSbPh_2$	THF	0.410[b]	68
PhAs⟩(CH₂)₃ (8) PhAs⟩	MeCN	1.110[a]	72

[a] At a carbon paste electrode.
[b] At a Pt electrode.

No mechanistic studies of the oxidation of the diarsines have been made, but cyclic voltammetry of $Ph_2SbSbPh_2$ in THF at a Pt electrode showed a single oxidation peak associated with a reduction peak at the reverse scan[68] ($E_{1/2} \approx 0.25$ V vs NHE). Exhaustive electrolysis in THF at a Pt electrode at $-30\,°C$ gave, in a two-electron process, a cationic product species attributed to Ph_2Sb^+ (reaction 106)[68] and polarography of the solution showed, like the cyclic voltammogram of the initial solution, a reduction wave attributed to the reduction of Ph_2Sb^+ (reaction 108). Re-reduction of the oxidized solution led to re-formation of the substrate, probably via dimerization of the radicals (reactions 108 and 109; cf Section IV.A) whereas addition of Cl^- led to formation of Ph_2SbCl (reaction 107).

$$Ph_2SbSbPh_2 \longrightarrow 2Ph_2Sb^+ + 2e^- \tag{106}$$

$$Ph_2Sb^+ + Cl^- \longrightarrow Ph_2SbCl \tag{107}$$

$$Ph_2Sb^+ + e^- \longrightarrow Ph_2Sb^\cdot \tag{108}$$

$$2Ph_2Sb^\cdot \longrightarrow Ph_2SbSbPh_2 \tag{109}$$

B. Formation of Organic Sb and Bi Compounds from Sacrificial Anodes

Processes analogous to the well-known electrochemical synthesis of tetraethyllead (Et_4Pb) from sacrificial Pb electrodes have been described for Sb and Bi. Et_3Bi is formed in good yields (94% based on the current passed) when an aqueous solution of $NaBEt_4$ is oxidized at a Bi anode[86]. If the anode material is changed to Sb, Et_3Sb is formed in moderate yields based on the current passed (25–50% dependent on the current density)[86]. The lower yields at the Sb anode were ascribed to passivation of the electrode[86].

The process which can be described by the general equation 110 also takes place with other alkyl groups than ethyl and with $NaAlR_4$ in place of $NaBR_4$[87,88].

$$M + 3NaBR_4 \longrightarrow MR_3 + 3Na^+ + 3BR_3 + 3e^- \tag{110}$$

Application of a mixture of $NaAlEt_4$ and $NaAlEt_3(OR)$ increased the yield of $SbEt_3$ formed by oxidation at an Sb electrode to 96% based on the amount of current passed through the solution[88].

C. Formation of Organic Bi Compounds from Bi Cathodes

Application of a metallic Bi *cathode* for the preparative reduction of $CH_2{=}CHCN$, ICH_2CH_2CN, MeI and EtI has been shown to lead to the unexpected formation of organobismuth compounds[89]. In aqueous solution electrolysis of acrylonitrile ($CH_2{=}CHCN$) and iodopropionitrile (ICH_2CH_2CN) both gave tris(β-cyanoethyl)bismuth [$Bi(CH_2CH_2CN)_3$] as product[89]. In MeCN, electrolysis of ICH_2CH_2CN gave, in addition to $Bi(CH_2CH_2CN)_3$, a smaller amount of bis(β-cyanoethyl)bismuth iodide [$IBi(CH_2CH_2CN)_2$][89]. In MeCN, reduction of MeI and EtI at Bi cathodes gave (upon bromination during work-up) Me_2BiBr and Et_2BiBr in 56–59% yields[89].

The reaction seems to be rather general and takes place for compounds of the structures $RCH{=}CHX$, RCH_2Y, and $RCHYCH_2X$, where R = H, alkyl, X = CN, $CONH_2$, COOH, Y = Br, I[90]. However, no investigations have been made to clarify the mechanism of these oxidations of metallic bismuth to trivalent bismuthines during electrochemical *reduction*. A chemical analogue to this electrochemical reduction leading to an oxidized Bi species has been described[91]. If arsenic is reduced by Na metal in liquid NH_3, a formal 'As^{3-}' species is formed which, upon addition of ArX in a photostimulated $S_{RN}1$ reaction, forms Ar_3As[91]. The analogous reaction of Sb with Na or K is much slower and the yield of Ar_3Sb lower[91]. However, the mechanism for reduction by Na must be quite different from any mechanism which can be envisaged for the above-described electrochemical reactions.

VII. INFLUENCE OF As- AND Sb-CONTAINING LIGANDS ON THE ELECTROCHEMISTRY OF TRANSITION METAL COMPLEXES

Only a brief summary of the influence of As-, and in a few cases Sb-containing ligands on the electrochemistry of their complexes with transition metals will be given in the following. This section deals with general trends which are exemplified by studies where comparison can be made between analogous P-, As- and Sb-containing ligands. Ligands containing Bi compounds are rare, and electrochemical studies including such complexes are virtually non-existent.

A. Classification of Ligands

Trialkyl and triphenyl substituted tertiary arsines and stibines are, like the phosphine analogues, suitable as ligands in transition metal complexes in several oxidation states, due to their low reduction potentials (cf Table 7 in Section III.A.1) and relatively high oxidation potentials (cf Table 14 in Section V.A).

The character of the lone pair of R_3M, R = alkyl, aryl, M = P, As, Sb, changes from a p-type orbital in R_3P towards more s-type orbitals when going to R_3As and R_3Sb, and at the same time the basicity of R_3M decreases down the group[92].

The potentials measured for reduction of Ph_3P, Ph_3As and Ph_3Sb (cf Section III.A.1) are not a very adequate measure of the π-acceptor strength of Ph_3M as a ligand, since the processes are chemically irreversible. The π-acceptor strength is of major importance for the bonding to low-valent transition metals. The π-acceptor strength is expected to decrease from Ph_3P to Ph_3Sb based on the efficiency of the overlap[93], whereas the ease of reduction, as expressed by the potentials for the irreversible reduction in aprotic solvents, shows that Ph_3P is easier to reduce than Ph_3As but Ph_3As is more difficult to reduce than Ph_3Sb. Similarly, the potentials for the chemically irreversible oxidation of Ph_3M may not adequately describe the σ-donor strength of Ph_3M as a ligand, which is the dominant binding form when the ligand is coordinated to transition metals in the higher oxidation states[94]. However, the σ-donor strength decreases from R_3P to R_3Sb[95] and R_3P has been shown to be more easily oxidized than R_3As for a number of substituents (R)[72]. However, there is almost no difference in the potentials for the chemically irreversible oxidation of Ph_3As and Ph_3Sb.

Several approaches have been made (see Reference 96 and references cited therein) to establish a quantitative, empirical scale describing the influence of a large collection of ligands of wide structural diversity on the redox potentials of transition metal complexes. All the approaches are based on the assumption that the relative contribution of each ligand to the redox potential of a transition metal complex is independent of the presence of other ligands, the identity of the transition metal, the oxidation and spin states of this metal, the solvent etc.[96]. Hence the approaches resemble the empirical use of substituent constants in the Hammett relation in organic chemistry.

The electrochemical model reaction applied in the E_L scale established in Reference 96 is the reversible potential for Ru(III)/Ru(II) in MeCN. The electrochemical ligand scale E_L was set up so that for a complex $RuL_x^1L_y^2L_z^3$, where L^1, L^2 and L^3 are different ligands and x, y and z are the numbers of each ligand in the complex, the potential E of the Ru(III)/Ru(II) couple vs NHE in MeCN is given by equation 111, where $E_L(L^i)$ is the parameter corresponding to ligand L^i

$$E = xE_L(L^1) + yE_L(L^2) + zE_L(L^3) \tag{111}$$

In Reference 96 more than 200 different ligands, including five As-containing ligands and one Sb-containing ligand, are associated with an E_L value, and Table 17 gives the E_L values

TABLE 17. E_L parameter values for P-, As- and Sb-containing ligands[a]

Ligand	E_L/V
Ph_3P	0.39
Ph_3As	0.38
Ph_3Sb	0.38
$(Ph_2P)_2CH_2(dppm)$	0.43
$(Ph_2As)_2CH_2(dpam)$	0.35
$1,2\text{-}(Ph_2P)_2C_6H_4$	0.45
$1,2\text{-}(Ph_2As)_2C_6H_4$	0.34
$1,2\text{-}(Me_2P)_2C_6H_4$	0.31
$1,2\text{-}(Me_2As)_2C_6H_4$	0.30
$Ph_2PCH_2CH_2PPh_2(dppe)$	0.36
$Ph_2AsCH_2CH_2AsPh_2(dpae)$	0.44

[a] Data from Reference 96. For details of the definition of E_L see Reference 96. For the bidentate ligands, the E_L value represents the contribution from *each* of the coordinating atoms.

for these As and Sb ligands together with the E_L values for the phosphine analogues. Except for the E_L values for 1,2-bis(diphenylphosphino)ethane (dppe) and 1,2-bis-(diphenylarsino)ethane (dpae), the E_L values for the As ligands are smaller than the E_L values for the analogous P ligands, indicating that, relative to the corresponding P ligands, the As ligands are more efficiently stabilizing the higher oxidation state [Ru(III)] or destabilizing the lower oxidation state [Ru(II)]. For the single Sb ligand included, it appears that it has an effect practically equal to that of the As ligand. The reason for the deviation of dppe relative to dpae in this general trend is not known, but for the related bis(diphenylarsino)methane (dpam) it is known in a single case, $Cr(CO)_2(dpam)$, that the coordination of dpam is special; only one of the two As atoms is coordinated to Cr, but one of the phenyl groups at the other As atom acts as a π-ligand to Cr^{80}. The corresponding dppm complex has the formula $Cr(CO)_2(dppm)_2^{80}$, and the possibility of such structural differences between the complexes leading to the assignment of the E_L values for dppe and dpae cannot be ruled out.

In addition to the ligands included in Table 17, Ph_3AsO and mixed phosphine–arsine bidentate ligands are commonly used. The ligands Ph_3MO coordinate to the transition metals via the oxygen atom, and due to a larger electron density at oxygen in Ph_3AsO than in Ph_3PO, Ph_3AsO form stronger complexes than Ph_3PO^{97}.

B. Examples. Comparison with Phosphor Analogues

Table 18 displays redox potential data for ten, structurally different transition metal complexes for which the potentials have been determined for an arsine ligand as well as for the analogous phosphine ligand (and, in some cases, the stibine ligand). It appears from the data in Table 18 that no common 'rule' determines the relative potentials of the phosphine, the arsine and the stibine analogues within a single type of complex system. For the first four systems in Table 18, data for P, As and Sb ligands are all available but only in two cases is a systematic variation of E found when the ligand is changed from PPh_3 to $AsPh_3$ to $SbPh_3$ [$Rh_2(MeCONH)_4(MPh_3)_2$ and $Re_2H_8(MPh_3)_4$], and in both cases the stabilization of the higher oxidation states decreases from PPh_3 to $SbPh_3$ in contrast to the expectations based on the E_L values in Table 17. For System 2 the same trend is observed,

TABLE 18. Examples of the influence of M (P, As, Sb) in ligands on the redox potentials of transition metal complexes[a]

Complex	E_{red3}	E_{red2}	E_{red1}	E_{ox1}	E_{ox2}	Reference
System 1:						
[Ru(bpy)$_2$Cl(PPh$_3$)]$^+$				1.28		98
			-1.05^b	1.18^b		99
[Ru(bpy)$_2$Cl(AsPh$_3$)]$^+$				1.25		98
			$-1.11^{b,c}$	1.17^b		99
[Ru(bpy)$_2$Cl(SbPh$_3$)]$^+$				1.27		98
			$-1.13^{b,c}$	1.17^b		99
[Ru(bpy)$_2$(H$_2$O)(PPh$_3$)]$^{2+}$				1.57		98
				1.04^d	1.30^d	99
[Ru(bpy)$_2$(H$_2$O)(AsPh$_3$)]$^{2+}$				1.56		98
				1.04^d	1.21^d	99
[Ru(bpy)$_2$(H$_2$O)(SbPh$_3$)]$^{2+}$				1.63^c		98
				1.06^d	1.34^d	99
System 2:						
Rh$_2$(MeCONH)$_4$(PPh$_3$)$_2$				0.50^e		93
Rh$_2$(MeCONH)$_4$(AsPh$_3$)$_2$				0.61^e		93
Rh$_2$(MeCONH)$_4$(SbPh$_3$)$_2$				0.65^e		93
System 3:						
[Fe(CN)$_5$PPh$_3$]$^{3-}$				0.52^f		100
[Fe(CN)$_5$AsPh$_3$]$^{3-}$				0.56^f		100
[Fe(CN)$_5$SbPh$_3$]$^{3-}$				0.56^f		100
System 4:						
Re$_2$H$_8$(PPh$_3$)$_4$				0.00	0.80^c	92
Re$_2$H$_8$(AsPh$_3$)$_4$				0.05	0.90^c	92
Re$_2$H$_8$(SbPh$_3$)$_4$				0.15	1.09^c	92
Re$_2$H$_8$(PMe$_2$Ph)$_2$(PPh$_3$)$_2$				-0.06	0.77^c	92
Re$_2$H$_8$(PMe$_2$Ph)$_2$(AsPh$_3$)$_2$				-0.04	0.86^c	92
Re$_2$H$_8$(PMe$_2$Ph)$_2$(SbPh$_3$)$_2$				0.05	0.74^c	92
Re$_2$H$_8$(PEt$_2$Ph)$_2$(PPh$_3$)$_2$				-0.09	0.74^c	92
Re$_2$H$_8$(PEt$_2$Ph)$_2$(AsPh$_3$)$_2$				-0.09	0.95^c	92
Re$_2$H$_8$(PEt$_2$Ph)$_2$(SbPh$_3$)$_2$				0.04	0.88^c	92
System 5:						
Re$_2$(μ-S)(μ-Br)Br$_3$(CS)(dppm)$_2$		-1.07^c	-0.25	0.66	1.62	101
Re$_2$(μ-S)(μ-Br)Br$_3$(CS)(dpam)$_2$		-1.27	-0.15	0.67	1.59	101

$[Re_2(\mu\text{-}S)(\mu\text{-}Br)Br_2(MeCN)(CS)(dppm)_2]^+$	-1.08	-0.64^c	-0.08	1.04		101
$[Re_2(\mu\text{-}S)(\mu\text{-}Br)Br_2(MeCN)(CS)(dpam)_2]^+$	-1.09	-0.62	0.00	0.99		101
$[Re_2(\mu\text{-}S)(\mu\text{-}Br)Br_2(EtCN)(CS)(dppm)_2]^+$	-1.11	-0.67^c	-0.11	1.01		101
$[Re_2(\mu\text{-}S)(\mu\text{-}Br)Br_2(EtCN)(CS)(dpam)_2]^+$	-1.12	-0.64	0.00	0.98		101
System 6:						
$[Ni(o\text{-}C_6H_4(PMe_2)_2)Cl_2]^+$				0.60^b	1.42^b	102
$[Ni(o\text{-}C_6H_4(AsMe_2)_2)Cl_2]^+$				0.45^b	1.54^b	103
$[Ni(o\text{-}C_6F_4(PMe_2)_2)Cl_2]^+$				0.71^b	1.64^b	102
$[Ni(o\text{-}C_6F_4(AsMe_2)_2)Cl_2]^+$				0.52^b	1.59^b	103
System 7:						
$Fe_4(CO)_{12}BHAu_2(PPh_3)_2$			-0.86^g	0.63^g	0.89^g	104
$Fe_4(CO)_{12}BHAu_2(AsPh_3)_2$			-0.80^g	0.93^g		104
System 8:						
$mer\text{-}Cr(CO)_3(\eta^1\text{-}dppe)(\eta^2\text{-}dppe)$				0.38	1.44^c	105
$mer\text{-}Cr(CO)_3(\eta^1\text{-}dpadpe)(\eta^2\text{-}dpadpe)^h$				0.36	1.45^c	106
$mer\text{-}Cr(CO)_3(\eta^1\text{-}dpae)(\eta^2\text{-}dpae)$				0.35	1.62^c	106
System 9:						
$Re_2Cl_4(dppe)_2$				0.47	1.28	107
$Re_2Cl_4(dpadpe)_2$				0.47	1.30	107
$Re_2I_4(dppe)_2$				0.53	1.16	107
$Re_2I_4(dpadpe)_2$				0.52	1.15	107
System 10:						
$Rh_2(CO)_2(\mu\text{-}PhNC(Me)NPh)_2(PPh_3)_2$				0.16	1.49	108
$Rh_2(CO)_2(\mu\text{-}PhNC(Me)NPh)_2(PPh_3)(AsPh_3)$				0.13	1.45	108

[a] In CH_2Cl_2 unless otherwise noted. The potentials are reversible potentials (taken as the midpoint between the peak potentials for the reduction and oxidation processes) for one-electron exchanges, unless otherwise noted, and the designations E_{red1}, E_{red2}, E_{red3}, E_{ox1}, and E_{ox2} refer to the 1, 2, and 3. reduction and 1. and 2. oxidation process, respectively, relative to the oxidation state of the complex given in the first column. All potentials are referred to NHE.
[b] In MeCN.
[c] $E_{red1} = E_p$ due to chemical irreversibility.
[d] In H_2O, pH 2.
[e] In the presence of an excess of the free ligand (MPh_3) to suppress dissociation.
[f] In H_2O, pH 5.
[g] The product of the electrochemical process reacts in a slow chemical follow-up reaction; see discussion in the text.
[h] dpadpe = $Ph_2AsCH_2CH_2PPh_2$.

although E_{ox1} for the As and Sb complexes are identical within experimental error. For System 4 the potential of the first oxidation (E_{ox1}) follows the same pattern when only two phosphine ligands are exchanged with $AsPh_3$ and $SbPh_3$ ligands, but the potential of the chemically irreversible second oxidation for the $SbPh_3$ complex is in between the values for the P and the As analogues or even lower than the potential for the P analogue.

For the ruthenium complexes in System 1, the potential (E_{ox1}) of the Ru(II)/Ru(III) process (which was the initial model reaction for the definition of the E_L scale) shows very little sensitivity to the nature of MPh_3—however, in all cases E_{ox1} for the complex with the $AsPh_3$ ligand is lower than (or equal to) E_{ox1} for the complex with the PPh_3 ligand as predicted by the E_L values. For System 1 the data obtained in water are for combined electron and proton transfer processes, showing pH dependence[98], and the Ru(IV) oxo complexes resulting from the second oxidation process are potent catalysts for electro-catalytic oxidation of organic substrates[98].

The complex with o-$C_6F_4(AsMe_2)_2$ as ligand is the only one of the four complexes in System 6 which retains a six-coordinate structure in solution, and this fact results in a much larger heterogeneous electron transfer rate constant for the first oxidation of this complex compared to the other three complexes[103]. Independently of whether the ligand is fluorinated or not, the complexes with the bidentate As ligands are more easily oxidized than the corresponding complexes with P ligands.

The dirhenium complexes in System 5 exhibit a peculiar dependence on change from bidentate P ligands to bidentate As ligands. In all cases the As ligands seem to stabilize preferentially the highest oxidation state for the two most extreme electron transfer reactions (the second reduction and the second oxidation for the neutral complexes, third reduction and first oxidation for the positively charged ones), whereas the opposite is the case for the two intermediate electron transfers.

The auraferraborane clusters of System 7 show a marked dependence on the identity of the MPh_3 ligand. The As complex is easier to reduce than the corresponding P complex and the reduction product is more stable ($t_{1/2} \approx 0.1$ s for the As complex, and $t_{1/2} \approx 0.01$ s for the P complex[104]). The major difference, however, is found for the oxidations: while the P complex shows two well-defined electrochemically and chemically quasi-reversible processes in CV, the As complex only gives one well-defined oxidation peak at a potential ca 0.3 V higher than E_{ox1} for the P complex (i.e. close to E_{ox2} for the P complex). A difference of ca 0.3 V in the redox potential upon a change from PPh_3 to $AsPh_3$ seems very unlikely based on the other data in Table 18, and the explanation of the observation is probably that a small pre-peak in the CV of the As complex (at a potential close to E_{ox1} for the P complex) is in fact the first oxidation of the As complex. The unexpected small size may be due to a reorganization process prior to electron transfer, which is much slower for the As than for the P complex.

The two examples of the effect of changing from a bidentate P—P ligand to a bidentate P—As ligand in System 9 show that even for closely related systems the effects may be different. When the additional ligands are chlorides, exchange of P—P ligand with the P—As ligand leads to increases in the oxidation potential (and similar behaviour is found for the analogous bromide complexes[107]) but if the additional ligands are iodides the effect of changing the P—P ligand to the P—As ligand is the reverse.

The chromium carbonyl complexes with one monodentate and one bidentate ligand in System 8 show systematic changes in E_{ox1} and E_{ox2} (but in opposite directions), when the ligands are changed from a P—P to a P—As and an As—As type (for η^1-dpadpe the coordination to Cr is through P^{106}). The first, chemically reversible oxidation is made easier in going from dppe to dpae, while the second, chemically irreversible oxidation becomes increasingly more difficult. The one-electron oxidation products slowly decompose to $trans$-$[Cr(CO)_2(\eta^2$-L-L$)_2]^+$ [105,106].

VIII. REFERENCES

1. M. D. Morris, *Electroanal. Chem.*, **7**, 79 (1974).
2. A. Watson and G. Svehla, *Proc. Soc. Anal. Chem.*, 163 (1974).
3. A. Watson, in *Polarography of Molecules of Biological Significance* (ed. W. F. Smyth), Chap. 10, Academic Press, England, 1979.
4. Y. Ikeda and E. Manda, *Chem. Lett.*, 839 (1989).
5. Y. Ikeda, *Chem. Lett.*, 1719 (1990).
6. A. J. Bard and L. R. Faulkner, *Electrochemical Methods*, Wiley, New York, 1980.
7. L. Horner, F. Röttger and H. Fuchs, *Chem. Ber.*, **96**, 3141 (1963).
8. L. Horner and J. Haufe, *Chem. Ber.*, **101**, 2903 (1968).
9. L. Horner and J. Haufe, *J. Electroanal. Chem.*, **20**, 245 (1969).
10. G. L. Kok and M. D. Morris, *Inorg. Chem.*, **11**, 2146 (1972).
11. R. E. Dessy, T. Chivers and W. Kitching, *J. Am. Chem. Soc.*, **88**, 467 (1966).
12. S. Wawzonek and J. H. Wagenknecht, in *Polarography 1964*, **2** (ed. G. J. Hills), Macmillan, London, 1966, p. 1035.
13. M. Shinagawa, H. Matsuo and H. Okashita, *Japan Analyst*, **7**, 219 (1958).
14. H. E. Affsprung and A. B. Gainer, *Anal. Chim. Acta*, **27**, 578 (1962).
15. M. D. Morris, P. S. McKinney and E. C. Woodbury, *J. Electroanal. Chem.*, **10**, 85 (1965).
16. T. R. Williams and P. S. McKinney, *J. Electroanal. Chem.*, **30**, 131 (1971).
17. G. A. Savicheva, V. N. Nikulin, B. D. Chernokal'skii, V. I. Gavrilov, V. N. Khlebnikov and T. A. Sozinova, *Zh. Obshch. Khim.*, **46**, 2068 (1976).
18. J.-M. Savéant and S. K. Binh, *Bull. Soc. Chim. Fr.*, 3549 (1972).
19. A. D. Beveridge and G. S. Harris, *J. Chem. Soc.*, 6076 (1964).
20. M. K. Saikina, *Uchenye Zapiski Kazan. Gosudarst. Univ. Im. V. I. Ul'yanova–Lenina Khim.*, **116**, 129 (1956); *Chem. Abstr.*, **51**, 7191e (1956).
21. A. M. Bond, R. T. Gettar, N. M. McLachlan and G. B. Deacon, *Inorg. Chim. Acta*, **166**, 279 (1989).
22. B. Breyer, *Chem. Ber.*, **71**, 163 (1938).
23. B. Breyer, *Biochem. Z.*, **301**, 65 (1939).
24. H. Erlenmeyer and E. Willi, *Helv. Chim. Acta*, **18**, 733 (1935).
25. M. Maruyama and T. Furuya, *Bull. Chem. Soc. Jpn.*, **30**, 657 (1957).
26. A. Watson and G. Svehla, *Analyst*, **100**, 489 (1975).
27. A. Tzschach and H. Matschiner and H. Biering, *Z. Anorg. Allg. Chem.*, **436**, 60 (1977).
28. R. C. Bess, K. J. Irgolic, J. E. Flannery and T. H. Ridgway, *Anal. Lett.*, **10**, 415 (1977).
29. I. M. Kolthoff and R. A. Johnson, *J. Electrochem. Soc.*, **98**, 138 (1951).
30. A. Watson, W. F. Smyth, G. Svehla and K. Vadasi, *Z. Anal. Chem.*, **253**, 106 (1971).
31. A. Tzschach and H. Biering, *J. Organomet. Chem.*, **133**, 293 (1977).
32. C. P. Wallis, *J. Electroanal. Chem.*, **1**, 307 (1959/60).
33. A. V. Il'yasov, Y. A. Levin and I. D. Morozova, *J. Mol. Struct.*, **19**, 671 (1973).
34. R. C. Bess, K. J. Irgolic, J. E. Flannery and T. H. Ridgway, *Anal. Lett.*, **9**, 1091 (1976).
35. R. K. Elton and W. E. Geiger, Jr., *Anal. Chem.*, **50**, 712 (1978).
36. G. A. Gilbert and E. K. Rideal, *Trans. Faraday Soc.*, **47**, 396 (1951).
37. F. Fichter and E. Elkind, *Chem. Ber.*, **49**, 239 (1916).
38. A. V. Trubachev, M. A. Pletnev, S. M. Reshetnikov and L. B. Ionov, *Elektrokhimiya*, **23**, 1652 (1987).
39. P. Zuman, *Substituent Effects in Organic Polarography*, Plenum Press, New York, 1967.
40. R. Elton and W. E. Geiger, *Anal. Lett.*, **9**, 665 (1976).
41. G. Spini, A. Profumo and T. Soldi, *Anal. Chim. Acta*, **176**, 291 (1985).
42. A. Watson, *Analyst*, **103**, 332 (1978).
43. M. Maruyama and T. Furuya, *Bull. Chem. Soc. Jpn.*, **30**, 650 (1957).
44. M. R. Jan and W. F. Smyth, *Analyst*, **109**, 1483 (1984).
45. A. Watson and G. Svehla, *Analyst*, **100**, 584 (1975).
46. D. Jannakoudakis, P. G. Mavridis and J. Markopoulos, *Chem. Chron.*, **5**, 249 (1976).
47. G. Schiavon, S. Zecchin, G. Cogoni and G. Bontempelli, *J. Electroanal. Chem.*, **59**, 195 (1975).
48. H. Matschiner, A. Tzschach and A. Steinert, *Z. Anorg. Allg. Chem.*, **373**, 237 (1979).
49. Yu. M. Kargin, N. I. Semakhina, B. D. Chernokal'skii, A. S. Gel'fond and G. K. Kamai, *Izv. Akad. Nauk SSSR, Ser. Khim.*, 2488 (1970).

50. D. Jannakoudakis, J. Markopoulos and P. G. Mavridis, *Chem. Chron.*, **5**, 263 (1976).
51. G. Schiavon, S. Zecchin, G. Cogoni and G. Bontempelli, *J. Electroanal. Chem.*, **52**, 459 (1974).
52. S. Zecchin, G. Schiavon, G. Cogoni and G. Bontempelli, *J. Organomet. Chem.*, **81**, 49 (1974).
53. G. Schiavon, G.-A. Mazzocchin and G. G. Bombi, *J. Electroanal. Chem.*, **29**, 401 (1971).
54. G. S. Harris and F. Inglis, *J. Chem. Soc. (A)*, 497 (1967).
55. P. Nikitas, A. Anastopoulos and D. Jannakoudakis, *J. Electroanal. Chem.*, **145**, 407 (1983).
56. A. Anastopoulos and I. Moumtzis, *Electrochim. Acta*, **35**, 1805 (1990).
57. Y. Mourad, Y. Mugnier, H. J. Breunig and M. Ates, *J. Organomet. Chem.*, **398**, 85 (1990).
58. R. E. Dessy and R. L. Pohl, *J. Am. Chem. Soc.*, **90**, 1995 (1968).
59. G. Märkl, H. Hauptmann and A. Merz, *J. Organomet. Chem.*, **249**, 335 (1983).
60. E. M. Dehn, *Am. Chem. J.*, **33**, 88 (1905).
61. H. J. Breunig and W. Kanig, *J. Organomet. Chem.*, **186**, C5 (1980).
62. Y. Mourad, Y. Mugnier, H. J. Breunig and M. Ates, *J. Organomet. Chem.*, **406**, 323 (1991).
63. G. Chobert and M. Devaud, *Electrochim. Acta*, **25**, 637 (1980).
64. A. Watson and G. Svehla, *Analyst*, **100**, 573 (1975).
65. J. H. Lowry, R. B. Smart and K. H. Mancy, *Anal. Chem.*, **50**, 1303 (1978).
66. R. E. Dessy, P. M. Weissman and R. L. Pohl, *J. Am. Chem. Soc.*, **88**, 5117 (1966).
67. H. Matschiner, F. Krech and A. Steinert, *Z. Anorg. Allg. Chem.*, **371**, 256 (1969).
68. Y. Mourad, Y. Mugnier, H. J. Breunig and M. Ates, *J. Organomet. Chem.*, **388**, C9 (1990).
69. R. E. Dessy, R. L. Pohl and R. B. King, *J. Am. Chem. Soc.*, **88**, 5121 (1966).
70. R. A. Rossi, R. A. Alonso and S. M. Palacios, *J. Org. Chem.*, **46**, 2498 (1981).
71. R. A. Alonso and R. A. Rossi, *J. Org. Chem.*, **47**, 77 (1982).
72. H. Matschiner, L. Krause and F. Krech, *Z. Anorg. Allg. Chem.*, **373**, 1 (1970).
73. Yu. M. Kargin, E. V. Nikitin, O. V. Parakin, A. A. Kazakova, Y. G. Galyametdinov and B. D. Chernokal'skii, *Dokl. Akad. Nauk SSSR*, **240**, 1383 (1978).
74. E. V. Nikitin, A. A. Kazakova, O. V. Parakin, B. S. Mironov and Yu. M. Kargin, *Dokl. Akad. Nauk SSSR*, **251**, 1175 (1980).
75. E. V. Nikitin, A. A. Kazakova, O. V. Parakin, Y. G. Galyametdinov, B. D. Chernokal'skii and Yu. M. Kargin, *Dokl. Akad. Nauk SSSR*, **250**, 899 (1980).
76. A. S. Romakhin, E. V. Nikitin, O. V. Parakin, Yu. A. Ignat'ev, B. S. Mironov and Yu. M. Kargin, *Zh. Obshch. Khim.*, **56**, 2597 (1986).
77. O. V. Parakin, E. V. Nikitin, Y. A. Ignat'ev, A. S. Romakhin and Yu. M. Kargin, *Zh. Obshch. Khim.*, **55**, 1496 (1985).
78. L. Horner and J. Haufe, *Chem. Ber.*, **101**, 2921 (1968).
79. V. Yu. Atamanyuk, V. G. Koshechko and V. D. Pokhodenko, *Zh. Org. Khim.*, **16**, 1901 (1979).
80. A. M. Bond, R. Colton and J. J. Jackowski, *Inorg. Chem.*, **18**, 1977 (1979).
81. J. Masek, *Collect Czech. Chem. Commun.*, **30**, 4117 (1965).
82. E. V. Nikitin, O. V. Parakin and Yu. M. Kargin, *Zh. Obshch. Khim.*, **54**, 1789 (1984).
83. G. Schiavon, S. Zecchin, G. Cogoni and G. Bontempelli, *J. Electroanal. Chem.*, **48**, 425 (1973).
84. E. V. Nikitin, A. A. Kazakova, Y. A. Ignat'ev, O. V. Parakin and Yu. M. Kargin, *Zh. Obshch. Khim.*, **53**, 230 (1983).
85. E. V. Nikitin, A. A. Kazakova, O. V. Parakin and Yu. M. Kargin, *Zh. Obshch. Khim.*, **52**, 2027 (1982).
86. K. Ziegler and O.-W. Steudel, *Justus Liebigs Ann. Chem.*, **652**, 1 (1962).
87. K. Ziegler and H. Lehmkuhl, *Ger. Patent 1,161,562; Chem. Abstr.*, **60**, 11623h (1964).
88. K. Ziegler and H. Lehmkuhl, *Ger. Patent 1,127,900; Chem. Abstr.*, **57**, 11235e (1958).
89. N. Chernykh and A. P. Tomilov, *Elektrokhimiya*, **10**, 1424 (1974).
90. A. P. Tomilov and I. N. Chernykh, *Otkrytiya, Izobret., Prom. Obaztsy, Tovanye Znaki*, **51**, 68 (1974); *Chem. Abstr.*, **80**, 133627h (1974).
91. E. R. Bornancini, R. A. Alonso and R. A. Rossi, *J. Organomet. Chem.*, **270**, 177 (1980).
92. M. T. Costello, G. A. Moehring and R. A. Walton, *Inorg. Chem.*, **29**, 1578 (1990).
93. S. P. Best, P. Chandley, R. J. H. Clark, S. McCarthy, M. B. Hursthouse and P. A. Bates, *J. Chem. Soc., Dalton Trans.* 581 (1989).
94. C. A. McAuliffe and W. Levason, *Phosphine, Arsine and Stibine Complexes of the Transition Elements*, Elsevier, Amsterdam, 1979.
95. C. Elschenbroich and A. Salzer, *Organometallics. A Consise Introduction*, Chap. 9, VCH Publishers, New York, 1989.
96. A. B. P. Lever, *Inorg. Chem.*, **29**, 1271 (1990).

97. G. F. Payne and J. R. Peterson, *Radiochim. Acta*, **39**, 155 (1986).
98. M. E. Marmion and K. J. Takeuchi, *J. Am. Chem. Soc.*, **110**, 1472 (1988).
99. B. P. Sullivan, D. J. Salmon and T. J. Meyer, *Inorg. Chem.* **17**, 3334 (1978).
100. M. M. Monzyk and R. A. Holwerda, *Polyhedron*, **9**, 2433 (1990).
101. J.-S. Qi, P. W. Schrier, P. E. Fanwick and R. A. Walton, *Inorg. Chem.*, **31**, 258 (1992).
102. S. J. Higgins and W. Levason, *Inorg. Chem.*, **24**, 1105 (1985).
103. L. R. Hanton, J. Evans, W. Levason, R. J. Perry and M. Webster, *J. Chem. Soc., Dalton Trans.*, 2039 (1991).
104. C. E. Housecroft, M. S. Shongwe, A. L. Rheingold and P. Zanello, *J Organomet. Chem.*, **408**, 7 (1991).
105. A. M. Bond, R. Colton and K. McGregor, *Inorg. Chem.*, **25**, 2378 (1986).
106. R. N. Bagchi, A. M. Bond, R. Colton, I. Creece, K. McGregor and T. Whyte, *Organometallics*, **10**, 2611 (1991).
107. P. Brant, H. D. Glicksman, D. J. Salmon and R. A. Walton, *Inorg. Chem.*, **17**, 3203 (1978).
108. N. G. Connelly, G. Garcia, M. Gilbert and J. S. Stirling, *J. Chem. Soc., Dalton Trans.*, 1403 (1987).

CHAPTER **12**

Radical intermediates in radiation chemistry of As, Sb and Bi compounds

M. GEOFFROY and T. BERCLAZ

Department of Physical Chemistry, University of Geneva, 30 Quai Ernest Ansermet, 1211 Geneva, Switzerland

I. INTRODUCTION

Exposure of phosphorus-containing compounds to ionizing radiation gives rise to a large variety of paramagnetic species which, most often, have been identified by Electron Spin

The chemistry of organic arsenic, antimony and bismuth compounds
Edited by S. Patai © 1994 John Wiley & Sons Ltd

Resonance (ESR). The effects of radiation on arsenic, antimony and bismuth compounds are, however, considerably less documented; a compilation of the corresponding information has been published in Landolt–Börnstein[1] and this subject is periodically reviewed in Specialist Reports of the Royal Chemical Society[2]. The purpose of this chapter is not to present an exhaustive compilation of all the species produced by radiolysis of organic compounds which contain an As, Sb or Bi atom, but only to deal with the most important types of radiogenic radicals formed from these compounds and, when possible, to compare the structures of these radicals with those of the corresponding phosphorous species.

Each heavy atom of the column V of the Periodic Table can show several valence states, can exhibit different coordination numbers and can accommodate various charges; such a 'versatility' results in a considerable variety of radical species which can be radiolytically generated from organic compounds containing these atoms. Moreover, as illustrated by arsoranyl radicals, each radical can often adopt several structures. Such a richness in reaction intermediates makes the identification of radiogenic species very difficult and the interpretation of ESR spectra, for example, requires one to be aware of the hyperfine parameters previously obtained not only with organic radicals produced by radiolysis but, also, with inorganic radicals produced by photolysis or by chemical reaction. This is why some of these species will also be mentioned in this chapter. A second difficulty arises from the complexity of the spectra: several species are generally trapped in the same sample and for each radical the various tensors are often not aligned. In these conditions the analysis of powder spectra can be ambiguous and valuable structural information can only be obtained from single-crystal ESR measurements. Special attention will therefore be paid to the studies which use this technique.

Arsenic, antimony and bismuth are particularly appropriate for ESR studies since all the naturally abundant isotopes of these nuclei have a nuclear spin different from zero and a rather large gyromagnetic ratio: ^{75}As, natural abundance 100%, $I = 3/2$, $g_n = 0.959$; ^{121}Sb, natural abundance 57.3%, $I = 5/2$, $g_n = 1.343$; ^{123}Sb, natural abundance 42.7%, $I = 7/2$, $g_n = 0.727$; ^{209}Bi, natural abundance 100%, $I = 9/2$, $g_n = 0.938$. A good description of the single occupied molecular orbital (SOMO) is therefore expected to be furnished by the resulting magnetic hyperfine interaction. Moreover, the nuclear spins are superior to 1/2 and the quadrupole moments are sufficiently large to lead to an observable hyperfine quadrupolar interaction; this interaction, which reflects the electric field gradient due to all the electrons of the heteroatom, depends upon the participation of this heteroatom in the various occupied molecular orbitals (MO)[3]. The interpretation of the hyperfine tensors is generally based on a comparison between the experimental interaction and the values expected for an ns or an np electron of the atom under study[4−7]. These atomic constants have been calculated by Morton and Preston[8] and are given in Table 1.

TABLE 1. Atomic constants for phosphorus, arsenic, antimony and bismuth

Atom	A_{iso} (MHz)[8]	$T_{//}$ (MHz)[8]
^{31}P	13306	733
^{75}As	14660	667
^{121}Sb	35100	1257
^{209}Bi	77530	1327

II. DICOORDINATED SPECIES

A. R_2M Radicals

1. R_2As

The first attempts to produce R_2As radicals by γ-irradiation were unsuccessful[9]; whereas radiolysis of PH_3 in dilute solutions in a krypton matrix at 4.2 K generated H_2P radicals as well as H and P atoms, for AsH_3 solutions only H and As atoms were observed. The first ESR study[10] of an arsinyl radical appeared in 1966 and concerned the Ph_2As radical produced by photolysis of gaseous Ph_2AsH and trapped after condensation on a finger dewar. Me_2As and Ph_2As have then been trapped[11,12] in frozen ethanolic solutions of arsines and diarsines after UV- or γ-irradiations. From these studies it was clear that the delocalization of the unpaired electron onto the R groups in R_2As was very small, but the disordered nature of the trapping matrix prevented a detailed description of these arsinyl radicals. Such a description could be obtained[13] with a single crystal of triphenylarsine oxide: X-irradiation of Ph_3AsO generated, indeed, the Ph_2As radical and the angular variation of the resulting ESR spectra led to the determination of the g and the ^{75}As magnetic hyperfine tensors. Moreover, for several orientations corresponding to a small coupling the forbidden $\Delta M_I = 1, 2$ transitions were observed. The positions of these lines not only provide the ^{75}As quadrupole tensor but also yield the relative sign of the magnetic and quadrupole constants. The resulting eigenvalues are shown in Table 2 while the mutual orientation of the eigenvectors is shown in Figure 1 together with the structure of the radical.

TABLE 2. g eigenvalues and ^{75}As hyperfine interactions for Ph_2As trapped in a single crystal[13]

Tensor	Eigenvalues		
g tensor	2.052	1.999	2.012
^{75}As magnetic coupling	629 MHz	−106 MHz	−123 MHz
^{75}As quadrupolar coupling	−19 MHz	10 MHz	9 MHz

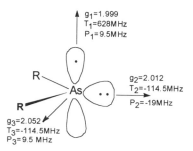

FIGURE 1. Structure of an arsinyl radical together with the mutual orientation of the ESR tensors (T: ^{75}As magnetic hyperfine tensor, P: ^{75}As quadrupolar interaction)

The arsenic spin densities have been estimated by comparing the isotropic and anisotropic components of the magnetic coupling with the constants associated with a 4s and a 4p orbital of atomic arsenic, respectively. The unpaired electron is mainly confined in an arsenic 4p orbital while the isotropic coupling constant is essentially due to inner-shell polarization. In accord with this structure the value of g measured along the direction of the semi-occupied 4p orbital is close to that of the free electron. In a Townes and Dailey analysis[3,14] the three eigenvalues of the quadrupolar tensor can be expressed as a function of the RAsR angle γ, of the occupation number a of the π orbital and of the occupation number b of the σ orbitals. From this analysis the bond angle γ was estimated to be close to 98° which, as expected when going from a second- to a fourth-row central atom, is considerably less than the value of 110° reported by Roncin and coworkers[15] for the HNH angle in $NH_2{}^{\cdot}$.

Although arsinyl radicals can be studied at room temperature when they are trapped in a crystalline matrix, they are generally unstable in solution. An interesting behaviour arises in the case of bulky ligands, like $(Me_3Si)_2CH$ or $(Me_3Si)_2N$, which confer exceptional kinetic stability on the radical. These radicals were prepared by photolysis of degassed toluene solutions of R_2AsCl in the presence of an electron-rich olefin[16,17]. The resulting isotropic ^{75}As constants, measured in the liquid phase, are consistent with the value found for $Ph_2As^{\cdot}$ in a solid matrix[13] and are shown in Table 3.

2. $R_2Sb^{\cdot}$

The diphenylantimony radical has been produced in a single crystal of triphenyl antimony in two different ways: X-irradiation[18] and ultraviolet irradiation[19]. Although both studies conclude that there is a strong localization of the unpaired electron in an antimony 5p orbital, the g and magnetic hyperfine tensors are somewhat different. This difference can tentatively be explained by supposing that the radicals produced by photolysis are located near the surface of the crystal and that the large linewidth is due to some motion and to some disorder. The $Ph_2Sb^{\cdot}$ species produced by radiolysis[18] exhibits narrow EPR lines and the expected signals due to the two isotopes could be easily followed as well as, for many orientations, the forbidden $\Delta M_I = \pm 1$ transitions. Due to the large spin–orbit coupling for antimony, the g anisotropy is large and there is a significant orbital contribution to the magnetic coupling tensor. The corrected 'spin only' values of the ^{121}Sb magnetic hyperfine coupling are $T_1 = 1270\,MHz$, $T_2 = -752\,MHz$ and $T_3 = -699\,MHz$. These values correspond to an antimony 5p spin density equal to unity. From the forbidden transitions it was possible to determine the quadrupolar interaction tensor, and to estimate the interbonding angle γ. This angle, close to 98°, is very similar to that found for the arsenic homologue[13].

Since the diphenylphosphinyl radical could also be trapped in single-crystal matrices

TABLE 3. ^{31}P or ^{75}As isotropic coupling constants for some phosphinyl and arsinyl radicals

Radical	Isotropic coupling constant (MHz)	Reference
$[CH(SiMe_3)_3]_2P^{\cdot}$	270	16
$[CH(SiMe_3)_3]_2As^{\cdot}$	105	16
$[N(SiMe_3)_2]P^{\cdot}$	260	16
$[N(SiMe_3)_2]As^{\cdot}$	89	16
$Ph_2P^{\cdot}$	287	20
$Ph_2As^{\cdot}$	133	13

TABLE 4. Comparison of ESR parameters and spin densities for $Ph_2M^\cdot$ radicals

Central atom	Isotropic coupling (MHz)	Anisotropic coupling (MHz)			s-Character	p-Character
		τ_1	τ_2	τ_3	ρ_s	ρ_p
^{31}P [20]	287	485	−232	−253	0.03	0.66
^{75}As [13]	133	495	−239	−256	0.01	0.75
^{121}Sb [18]	−60	1330	−692	−640	0.00	1.00

(X-irradiated single crystal of Ph_3PO[20], Ph_3PS[20], Ph_3PBH_3[21], Ph_3P[22]), it is worthwhile comparing the members of the series $Ph_2M^\cdot$ (M = P, As, Sb). The three sets of tensors shown in Table 4 indicate very similar structures with a high spin density in the π orbital of the central atom; in contrast to $Ph_2N^\cdot$[23], these three radicals do not exhibit any delocalization onto the phenyl rings and, at least for the As and Sb species, they are strongly bent.

B. (RM=MR)⁻· Radical Anions

(RP=PR)⁻· radical anions could be generated rather easily by electrochemical reduction of diphosphenes[24-27] and the resulting ESR spectra gave evidence of the P=P π^* character of the SOMO. Although a reversible one-electron reduction process could also be observed by cyclic voltametry for a diarsene compound[26]—$(Me_3Si)_3CAs=AsC(SiMe_3)_3$—the ESR spectrum of the corresponding anion could not be obtained.

III. TRICOORDINATED SPECIES

A. $R_3M^{+\cdot}$ Radical Cations

The phosphoniumyl radicals are very well known radicals which have already been reviewed in the present series[28]. The possibility of forming arsoniumyl radicals by photolysis was mentioned twenty years ago by Preer and coworkers[11]. Symons and coworkers[29,30] showed that γ-irradiation of solutions of Ph_3As in sulphuric acid at 77 K produces the radical cation. In this sample the irradiated molecule is in fact the arsonium cation, Ph_3AsH^+, which, after ejection of an electron, loses a proton and forms $R_3As^{+\cdot}$:

$$(Ph_3AsH)^+ \longrightarrow Ph_3As^{+\cdot} + (H^+) + e^-$$

The same radiation process has been observed with Ph_2PMe and $PhPMe_2$. In the case of a solution of triphenylarsine oxide in H_2SO_4, Ph_3AsOH^+ captures an electron and yields also the $Ph_3As^{+\cdot}$ species:

$$(Ph_3AsOH)^+ + e^- + H_2SO_4 \longrightarrow Ph_3As^{+\cdot} + H_2O + HSO_4^-$$

In methanolic solution, however, in contrast to Ph_3PO, Ph_3AsO does not give rise to the radical cation but only to the tetracoordinated arsenic radicals (see below).

Although the ESR spectra of $R_3As^{+\cdot}$ have been obtained from irradiated frozen solutions, the corresponding tensors can be determined without any difficulty since both the g and the ^{75}As tensors exhibit axial symmetry and have their eigenvectors mutually aligned. Some of these tensors are given in Table 5.

The decomposition of the ^{75}As coupling into isotropic and anisotropic constants leads to arsenic 4s and 4p spin densities equal to 0.06 and 0.75, respectively. From the resulting hybridization ratio λ^2, close to 10.8, the out-of-plane angle ζ can be calculated by using the Coulson formula and by assuming a C_{3v} symmetry (see Figure 2). The value of ζ (11°)

TABLE 5. g tensor and ^{31}P or ^{75}As hyperfine interaction for some $R_3M^{+\cdot}$ species

Species	$T_{//}$ (MHz)	$T_\perp$ (MHz)	$g_{//}$	$g_\perp$	Reference
$Ph_3As^{+\cdot}$	1302	646	1.995	2.017	30
$Et_3As^{+\cdot}$	1540	784	2.00	2.03	30
$Ph_3P^{+\cdot}$	1554	884	2.001	2.0055	31

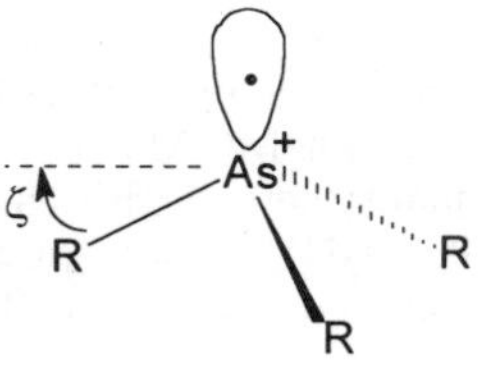

FIGURE 2. Structure of an arsoniumyl radical cation

indicates a pyramidal structure which is quite similar to that obtained for $Ph_3P^{+\cdot}$ after X-irradiation of a single crystal of Ph_3PBCl_3[31] and whose ESR parameters are also shown in Table 5.

B. Radical Anions $R_3As^{-\cdot}$

The anion derived from 1,2-phenylenephosphorochloridite is one of the rare examples of a radiogenic radical anion produced by electron capture by a phosphine[32]. In the case of arsenic compounds, $AsCl_3^{-\cdot}$ and $AsF_3^{-\cdot}$ were produced by γ-irradiation of polycrystalline trihalogenoarsines at 77 K[33]. This last species exhibits hyperfine coupling with ^{75}As ($T_{//} = 666$ MHz), two ^{19}F of one kind ($T_{//} = 240$ MHz) and one ^{19}F of another kind ($T_{//} = 78$ MHz). These arsenic radical anions adopt an approximate planar T-shape structure with the unpaired electron located in an orbital perpendicular to the molecular plane.

C. Neutral Species R_2MY (Y = O, S, Se)

Arsonyl radicals have the general formula $RA\dot{s}(O)R'$. Such a radical was observed after γ-irradiation of n-propylarsonic acid[34] and probably results from the decomposition of the electron excess species $(PrAs(O)(OH)_2)^- \rightarrow PrA\dot{s}O_2^- + H_2O$.

TABLE 6. Arsenic spin densities and hybridization ratio for some arsonyl and oxygen containing arsoranyl radicals

Radical	c_p^2	c_s^2	$\lambda^2 = c_p^2/c_s^2$	Reference
$(OH)_2A\dot{s}O$	0.63	0.10	6.3	35
$A\dot{s}O_3^{2-}$	0.57	0.10	5.7	39
$PrA\dot{s}O_2^-$	0.62	0.12	5.2	34
$(OH)A\dot{s}(O)CH_2As(O)(OH)_2$	0.62	0.1	5.6	35
$Me_2A\dot{s}O_2^{2-}$	0.46	0.18	2.5	38
$PrA\dot{s}O_3^{3-}$	0.58	0.23	2.5	34
$(OH)_2A\dot{s}(O^-)CH_2As(O)(OH)_2$	0.48	0.17	2.8	35

Similarly, X-irradiation of a single crystal of methylenediarsonic acid[35] produced the radical $(OH)(O)A\dot{s}CH_2As(O)(OH)_2$ whose structure is practically the same as that obtained for the phosphorus analogue[36]. The inorganic radicals $\cdot AsO_3^{--}$ [34,37] and $(HO)_2A\dot{s}O$ [34,35] have also been observed. All these arsonyl-type radicals are pyramidal with a hybridization ratio $\lambda^2 = c_p^2/c_s^2$ which lies between 5 and 6.3. As shown in Table 6, these values differentiate them clearly from the trigonal bipyramid structure observed for some oxygen-containing arsoranyl radicals (see Section IV.B).

IV. TETRACOORDINATED SPECIES

A. Dimeric Radical Cations $(R_3M—MR_3)^{+\cdot}$

Whereas γ-irradiation of solutions of tertiary arsines in H_2SO_4 leads to the formation of the monomeric radical cation $R_3As^{+\cdot}$ [30], irradiation of the pure crystalline compound can lead to the arsinyl radical in the case of aryl substituents, or to the dimeric radical cation in the case of alkyl arsines[40]. This formation of a dimeric species has also been observed with tertiary phosphines and the ESR parameters for $Et_3As—AsEt_3^{+\cdot}$, $Me_3As—AsMe_3^{+\cdot}$ and $(R_3P—PR_3)^{+\cdot}$ are given in Table 7 together with the values obtained after optimization of the structure of $(H_3As—AsH_3)^{+\cdot}$ [41] by ab initio calculations.

These results show that the excess electron, produced by radiolysis and captured by two arsine molecules which dimerize, is located in an As—As σ^* orbital. As stressed by Symons[40], the R_3As moiety is more pyramidal in the dimeric cation radical than in the $R_3As^{+\cdot}$ species, in accordance with the pyramidalization observed when passing from $R_3As^{+\cdot}$ to R_3As.

B. $R_4As^{\cdot}$ Radicals and $R_3As...X^{\cdot}$ Adducts

The phosphorus homologues of $R_4As^{\cdot}$ radicals, the phosphoranyl radicals, have been considerably studied during the past 15 years and several review articles have been dedicated to this species[42-44]. This great interest for phosphoranyl radicals is due to the fact that the $R_4P^{\cdot}$ radical often appears as an important intermediate and that its structure frequently governs the reaction mechanism. The structure and reaction properties of arsoranyl and phosphoranyl radicals are very similar; nevertheless, some notable differences exist, due for example to the greater electron affinity of arsenic than of phosphorus[30].

Exposure of a large variety of arsenic-containing organic compounds to ionizing radiation yields arsoranyl radicals; the most representative examples are radiolysis of solutions of arsonium salts or arsine-oxide in methanol[30] and radiolysis of crystalline arsenic chalcogenides[45]. In order to specify the structure of the arsoranyl radicals, we will first describe some results obtained after X-irradiation of single crystals. By analogy with phosphoranyl radicals the structures in Figure 3 can be expected for an arsoranyl radical. In the Trigonal Bipyramid Structure (TBP), the unpaired electron is formally confined in a

TABLE 7. ^{31}P or ^{75}As couplings and spin densities for some $(R_3M—MR_3)^{+\cdot}$ species

Radical	Hyperfine coupling $T_{//}$	Hyperfine coupling $T_{\perp}$	Isotropic coupling	s-Character ρ_s	p-Character ρ_p
$(Et_3As)_2^{+\cdot}$ [40]	1302	1050	1134	0.08	0.25
$(Et_3P)_2^{+\cdot}$ [40]	1512	1153	1272	0.09	0.32
$(H_3As)_2^{+\cdot}$ [41]	1472	1124	1240	0.08	0.34

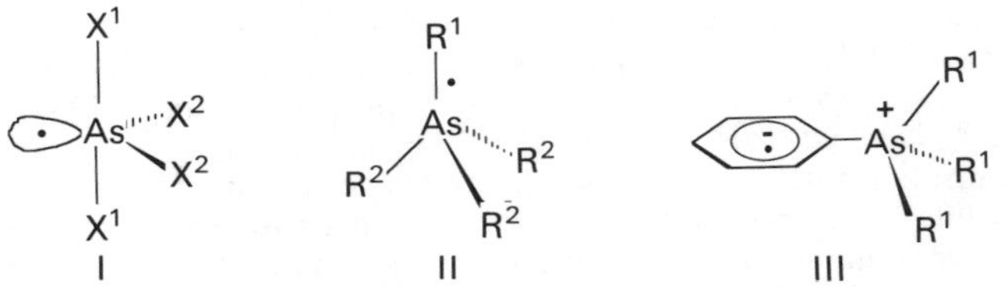

M. Geoffroy and T. Berclaz

I   II   III

[Trigonal Bipyramid Structure] $[C_{3v}$ structure] [arsonium-type structure]

FIGURE 3. Limit structures for arsoranyl radicals

nonbonding sp^2 orbital, whereas in Structure II it is localized in an As—R^1 σ^* orbital. The third structure corresponds, in fact, to an arsonium cation in which one of the ligands is a phenyl radical anion.

$Ph_3AsO^{-\cdot}$, $Ph_3AsS^{-\cdot}$ and $Ph_3AsSe^{-\cdot}$ have been trapped in irradiated single crystals of the corresponding chalcogenide[45]. $Ph_3AsS^{-\cdot}$ is stable even at room temperature, whereas $Ph_3AsO^{-\cdot}$ disappears at 266 K and $Ph_3AsSe^{-\cdot}$ disappears at 226 K. $Ph_3AsBr^{\cdot46}$ has been trapped in an irradiated single crystal of $(Ph_3AsCH_3)^+Br^-$ and is stable at room temperature. Some examples of ^{75}As hyperfine tensors are shown in Table 8.

In the TBP structure, the maximum coupling with the central atom is expected to be oriented perpendicular to the maximum coupling of the apical ligand R^1 and this is indeed the case for $\cdot PF_4$[47] and $\cdot POCl_3^{-}$[48]. For the pyramidal structure, the maximum coupling of the central atom is expected to be oriented along the direction of the maximum coupling of the ligand and this situation has been observed for $Ph_3PCl^{\cdot}$[49]. For $Ph_3AsBr^{\cdot46}$, the maximum bromine-$T_{//}$ direction makes an angle of 35° with the ^{75}As-$T_{//}$ direction. It is known from single-crystal NQR studies that for a bond involving a halogen atom, the electric field gradient is in general aligned along the bond direction and the mutual orientation of the hyperfine tensors for $Ph_3AsBr^{\cdot}$ therefore shows that the As-$T_{//}$ direction makes an angle of 35° with the As—Br bond. A similar result was obtained with $Ph_3AsCl^{\cdot}$[50]; in this case the (^{75}As-$T_{//}$, ^{35}Cl-$T_{//}$) angle was found to be equal to 29°. It therefore appears that the structure of $Ph_3AsX^{\cdot}$ (X = Br, Cl) radicals is intermediate between the TBP and the C_{3v} structure. Some hyperfine tensors obtained with arsoranyl radicals have been decomposed into isotropic and anisotropic coupling constants and the resulting spin densities are shown in Table 9 together with the hybridization ratio $\lambda^2 = c_p^2/c_s^2$.

TABLE 8. ^{75}As hyperfine couplings for $Ph_3AsX^{-\cdot}$ (X = O, S, Se) trapped in single-crystal matrices[45]

X	Eigenvalues (MHz)	Isotropic coupling	Anisotropic coupling		
			τ_1	τ_2	τ_3
O	2202	1966	236	-88	-147
	1878				
	1819				
S	1930	1728	202	-95	-105
	1633				
	1623				
Se	1641	1485	265	-60	-96
	1425				
	1389				

TABLE 9. ^{75}As hyperfine couplings, spin densities and hybridization ratio for $Ph_3AsY^\bullet$ (Y = Cl, Br)

Radical	A_{iso}	τ_{max}	ρ_s	ρ_p	$\lambda^2 = \rho_p/\rho_s$
$Ph_3AsCl^{\bullet}$ [50]	1918	341	0.13	0.51	3.9
$Ph_3AsBr^{\bullet}$ [46]	1688	350	0.11	0.52	4.7

All the triphenyl arsoranyl radicals have similar parameters and therefore adopt a similar structure in which the unpaired electron is mainly localized in an arsenic p orbital. Moreover, for $Ph_3AsX^{-\bullet}$ (X = chalcogen) the total arsenic spin density decreases with the difference of electronegativity ($X_{chalcogen} - X_{arsenic}$) which is consistent with a strong contribution of the C_{3v} structure since, for such a structure, the unpaired electron is located in an arsenic-axial ligand σ^* bond.

It is rather difficult to follow the various radical reactions which occur in irradiated arsenic-containing compounds because the overlap of the EPR signals precludes the identification of the radiogenic species. However, in the case of single-crystal samples X-irradiated at low temperature, it is possible to follow independently the various sets of lines of each radical, to determine without ambiguity the hyperfine tensors of each species and finally to identify the radical present at several temperatures. The temperature dependence of each signal then gives evidence about reorientations and transformations of the radiogenic radicals. Such a study has been performed for sodium dimethyl arseniate[38]. X-irradiation of a single crystal of $Me_2As(O)O^-Na^+$ at 77 K leads to the formation of both the arsoranyl radical $Me_2A\dot{s}O_2{}^{2-}$ and the radical $Me(\dot{C}H_2)A\dot{s}(O)O^-$ resulting from the homolytic scission of a C—H bond. This last radical is stable up to 295 K, whereas the former decreases 165 K and 215 K to give rise to two tricoordinated species $MeAsO_2{}^{-\bullet}$ and $Me_2A\dot{s}O$. In the absence of oxygen these to radicals are stable at room temperature, but in the presence of oxygen they are transformed into $^\bullet AsO_2{}^{2-}$.

The identification of radicals trapped in an X-irradiated single crystal of diarsonic acid[35] has shown that most of the radiation processes observed with dicarboxylic acid were also involved in the arsenic homologues. Malonic acid, for example, has been widely studied[51] and it was shown that the electron-rich and electron-deficient primary species reacted with an undamaged neighboring molecule to yield the observed radicals: $HOOCCH_2\dot{C}(OH)_2$, $HOOCCH_2CO$, $HOO\dot{C}HCOOH$ and $HOO\dot{C}CH_2$. In diarsonic acid the corresponding species were also identified: the arsoranyl-type radical $(OH)_2As(O)CH_2A\dot{s}(OH)_3$, the tricoordinated arsenic species $(HO)_2As(O)CH_2A\dot{s}(O)OH$ and the carbon-centered radical $(HO)_2As(O)\dot{C}HAs(O)(OH)_2$. Although the trapping of $^\bullet AsO_3{}^{2-}$ indicated that a homolytic scission of the C—As bond occurred, the radical $(HO)_2As(O)\dot{C}H_2$ was not observed, in contrast with the study on $(HO)_2P(O)CH_2P(O)(OH)_2$ which showed an abundant formation of $(HO)_2P(O)\dot{C}H_2$. Storing the irradiated crystal of diarsonic acid two days at room temperature leads to the trapping of an additional radical characterized by a very large ^{75}As coupling and which is due to a pentacoordinated arsenic radical anion $[(OH)_3(O^-)AsCH_2As(O)(OH)_2]^{-\bullet}$ (see below).

Most attention has been paid to radiation mechanisms involving phenyl-containing arsoranyl radicals. As pointed out by Eastland and Symons[30], the electron capture by a tetracoordinated pentavalent arsenic bound to an aromatic ring can potentially give rise to two species: (1) a radical, like III (Figure 3), resulting from an electron addition on the phenyl ring and which, after protonation, can yield the cyclohexadienyl radical, (2) a radical, like II (Figure 3), due to the addition of the excess electron on the arsenic atom. This radical can relax either by bond stretching to form the arsoniumyl species $Ar_3As^{+\bullet}$ or by bond bending to yield the arsoranyl radical I. γ-irradiation of $Ph_4As^+Br^{-}$ [52]

TABLE 10. ^{75}As coupling and spin densities for arsoranyl radicals produced in irradiated methanolic solutions

Radical	A_{iso} (MHz)	$\tau_{//}$ (MHz)	$\tau_{\perp}$ (MHz)	ρ_s	ρ_p	Reference
Ph$_4$As$^{\cdot}$	1500	165	-81	0.10	0.25	52
Ph$_4$As$^{\cdot}$	1867	235	-117	0.12	0.35	30

and Ph$_3$AsO^{30} in CD$_3$OD at 77 K has been shown to produce Ph$_4$As$^{\cdot}$ and Ph$_3$AsO$^{-\cdot}$, respectively. The high arsenic spin density found for these species (Table 10) indicates that the electron addition occurs on arsenic in preference to the benzene ring, whereas for Ph$_3$PO in methanol electron capture is mainly on the aromatic ring.

Irradiation of methyltriphenylarsonium iodide illustrates nicely the importance of the arsoranyl radical as a reaction intermediate[53]. X-irradiation of a single crystal of (Ph$_3$As^{13}CH$_3$)$^+$I$^-$, at 77 K, produces a species whose hyperfine interactions with As, ^{13}C and three protons reveal the trapping of the adduct Ph$_3$As...$\dot{C}$H$_3$. This species disappears between $-170\,°$C and $-100\,°$C, whereas the signals of the arsoranyl Ph$_3$AsI$^{\cdot}$[54], trapped in low abundance at 77 K, increase in this temperature range and disappear at $-20\,°$C. In contrast to these results, the irradiation of a methanolic solution of Ph$_3$AsMe$^+$I$^-$[55] indicates that the arsoranyl Ph$_3$AsMe is generated at 77 K and that Me$^{\cdot}$ and Ph$_3$AsCH$_2^{\cdot}$ are formed by annealing (Figure 4). The first event in the radiolysis of the crystalline salt[54] is probably the formation of iodine atoms (the electron-loss species); the capture of the ejected electron by the arsonium cation produces the arsoranyl radical which, after stretching of the As—Me bond, gives rise to the arsine-methyl radical adduct. A slight temperature increase enables a neighboring iodine atom to react with this adduct and, by substitution of the methyl group, to yield the more stable triphenyliodoarsoranyl radical. A similar process was observed in crystalline Ph$_3$AsMe$^+$Br$^-$[46]; in this case, the greater electronegativity of the bromine conferred a greater stability on the bromoarsoranyl radical which could be observed at room temperature. In methanol most of the excess electrons originate from solvent molecules and the probability, for Ph$_3$AsMe, to be attacked by an iodine atom is low enough to allow the radical Ph$_3$A$\dot{s}$Me, perhaps slightly stabilized by bending, to be observed.

Although the following species have not been produced by radiolysis, it is worthwhile mentioning them because their study gave interesting information about the formation, the structure and the reactivity of arsoranyl radicals. The reaction of triphenylarsine[56,57] with RO radicals formed by photolysis of peroxides generates the Ph$_3$AsOR radical which was shown to be more stable than R'OP(OR)$_3$. These arsoranyl radicals decompose by α-scission (R'OA$\dot{s}$Ph$_3 \rightarrow$ R'OAsPh$_2$ + Ph$^{\cdot}$) rather than by β-scission (R'OA$\dot{s}$Ph$_3 \rightarrow$ R'$^{\cdot}$ + OAsPh$_3$). If photolysis occurs in the presence of air, the pentacoordinated species Ph$_3$(OR')AsOO$^{\cdot}$ is formed. When an alkoxydiphenylarsine (Ph$_2$AsOR) is photolyzed in the presence of di-t-butylperoxide, an arsoranyl radical containing two alkoxy groups in

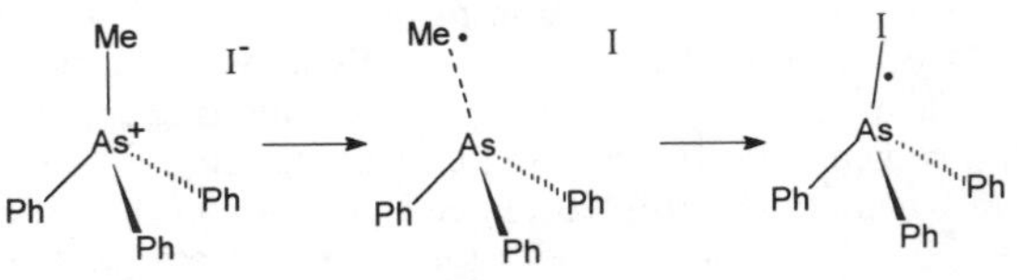

FIGURE 4. Formation of Ph$_3$As...$\dot{C}$H$_3$ and Ph$_3$AsI$^{\cdot}$ in an X-irradiated single crystal of Ph$_3$As$^+$I$^-$

TABLE 11. ESR parameters for some arsoranyl radicals containing RS or RO ligands

Radical	^{75}As-A_{iso} (MHz)	$g_{average}$	Reference
Me_3AsS^tBu	1829	2.016	58
Me_3AsSEt	1824	2.016	58
Ph_3AsO^tBu	1876	2.014	57
$Me_2As(O^tBu)_2$	2146	2.010	58

apical position is formed, and the higher electronegativity of OR′ over Ph causes a large increase in the isotropic ^{75}As coupling constant. Arsoranyl radicals containing a SR ligand have also been formed by photolysis[58] and the corresponding coupling constants are shown in Table 11 for the sake of comparison with the alkoxy analogues.

Although this chapter deals essentially with organic compounds, it may be mentioned that many arsoranyl radicals have been formed by radiolysis of inorganic compounds; the most representative examples are $^{.}As(OH)_4$[59] (trigonal bipyramid), $^{.}AsO_4^{4-}$[60,61] and $^{.}AsF_4$[62,63].

Radicals containing a tetracoordinated antimony atom are very rare. As far as we know only $^{.}SbH_3F$ has been reported: γ-irradiated solutions of PH_3, AsH_3 and SbH_3 in SF_6 at low temperature exhibit ESR spectra due to the corresponding H_3MF radical[64]. The s spin density increases when going from P to As and to Sb, in accordance with the decreasing electronegativity. This variation confirms the antibonding nature of the semi-occupied orbital.

V. PENTACOORDINATED SPECIES: $MR_5^{-.}$ RADICAL ANIONS

The formation of a radical anion centered on a pentacoordinated arsenic has often been observed after irradiation of inorganic compounds. The most investigated species are probably fluorine derivatives of arsenic: $AsF_5^{-.}$[65,66], for example, has been trapped at 300 K in crystalline KPF_6 containing traces of $KAsF_6$[67]. The EPR spectrum exhibits hyperfine coupling with an arsenic nucleus (5060 MHz), four equivalent fluorine nuclei (507 MHz) and a fifth fluorine nucleus (14 MHz). This radical anion therefore adopts a square pyramidal structure similar to the structure observed for $PF_5^{-.}$. The arsenic 4s spin density (0.349) is appreciably higher than that measured for the phosphorus 3s spin density (0.286), whereas the inverse variation is observed for the basal fluorine 2s spin densities [$\rho(F, 2s) = 0.009$ for AsF_5^-, $\rho(F, 2s) = 0.0105$ for $PF_5^{-.}$]. This variation, which accompanies a decrease in electronegativity of the central atom, reflects the antibonding nature of the SOMO. AsF_4O^{2-} has been produced by radiolysis of $KAsF_6$ recrystallized from water[68]. It was shown from the fluorine hyperfine interaction that two structures exist for this anion: in the first one the apical position is occupied by the oxygen atom whereas in the second one this position is occupied by a fluorine atom. Comparison of $AsF_5^{-.}$ and $^{.}AsF_4O^{2-}$[68] shows that replacement of a basal F in $AsF_5^{-.}$ by a less electronegative atom causes an increase in the coupling of the F atom *trans* to the substituent but a decrease in the two *cis* F atoms. A similar result was obtained with $SF_5^{-.}$. The formation of the pentafluoroarsenic radical anion is not limited to radiation processes; this species is also produced by reacting AsF_5 with organic molecules: when AsF_5 is combined in vacuum at 77 K with butadiyne it undergoes a one-electron transfer and the ESR signals due to $AsF_5^{-.}$ can be detected[69].

The formation of pentacoordinated arsenic radical anions by the radiolysis of organic compounds was reported for $R_3AsF_2^{-.}$[70]; this species was trapped together with the

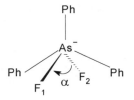

FIGURE 5. Structure of the $Ph_3AsF_2^{-\cdot}$ radical anion

diphenylarsinyl radical in an X-irradiated single crystal of Ph_3AsF_2. The various hyperfine tensors (^{75}As: 2983, 2433, 2424 MHz, ^{19}F: 938, 261, 271 MHz, ^{19}F: 841, 175, 157 MHz) were not drastically different from those expected for an arsoranyl radical, but the mutual orientation of the eigenvectors together with the comparison of the crystallographic bond directions with these eigenvectors led to the identification of a pentacoordinated radical anion adopting the structure shown in Figure 5 with $\alpha = 135°$.

As already mentioned in Section IV.B, the arsoranyl radical trapped in X-irradiated methylenediarsonic acid[35] is replaced, at room temperature, by a radical which exhibits a very strong ^{75}As coupling ($T_1 = 4516$ MHz, $T_2 = 4144$ MHz and $T_3 = 4138$ MHz) and a smaller coupling with a second arsenic nucleus ($T_1 = 129$ MHz, $T_2 = 98$ MHz, $T_3 = 92$ MHz). These values are in accordance with the addition of a hydroxyl ion to the arsoranyl radical to form the $[(OH)_3As(O^-)CH_2As(O)(OH)_2]^{-\cdot}$ species.

It seems that radical anions centered on a pentacoordinated antimony atom have never been detected, and previous observations of SbF_5^- have been assigned later to an impurity species SbF_5O^- [71,72].

VI. HEXACOORDINATED SPECIES: $^\cdot MF_6^{2-}$ AND $MF_6^\cdot$

γ-Irradiation, at 77 K, of polycrystalline salts of $CsAsF_6$ and $CsSbF_6$ yields $^\cdot AsF_6^{2-}$ and $^\cdot SbF_6^{2-}$, respectively[73], while $^\cdot BiF_6^{2-}$ was trapped [74,75] in crystalline $CsAsF_6$ doped with BiF_6. In these ions, the SOMO consists mainly of the central atom s orbital and the resulting isotropic coupling is considerable. The six ligand p-orbitals point toward the central atom and the antibonding character of the resulting molecular orbital manifests itself by the increase of the metal isotropic coupling constant when passing from As to Sb and to Bi.

Irradiation of $MF_4^+AsF_6^-$ and $NF_4^+SbF_6^-$ led to the formation of NF_3^+ together with that of a species exhibiting strong hyperfine interaction with two fluorine nuclei, a small coupling with As or Sb and a weak interaction with additional fluorines. Symons and coworkers[76] tentatively identified this species as being $AsF_6^\cdot$ or $SbF_6^\cdot$.

VII. LIGAND-CENTERED RADICALS

A. Radical Carbon α to an Arsenic Atom

The most abundant radical species trapped in a freshly X-irradiated single crystal of methylendiarsonic acid[35] is the carbon-centered radical $(OH)_2(O)As\dot{C}HAs(O)(OH)_2$ which has been identified from its hyperfine couplings with a proton ($T_1 = 95$ MHz, $T_2 = 61$ MHz, $T_3 = 25$ MHz) and two ^{75}As nuclei ($T_1 = 126$ MHz, $T_2 = 117$ MHz, $T_3 = 109$ MHz). In contrast to previous observations with methylendiphosphonic acid, the homolytic scission of the C—M bond is considerably less effective than the C—H scission. The rupture of a C—H bond located β to an arsenic atom was also reported for $Me_2AsO_2^-$ [38] and for $MeAsO_3^{2-}$ [34].

B. Radical Carbon β to an Arsenic Atom

γ-Irradiation of n-propylarsonic acid[77,78] at 77 K showed that two species were mainly produced: the propyl radical presumably formed by a dissociative electron-capture process and a radical centered on a carbon located in β position from the arsenic atom. Two mechanisms have been proposed by Symons to explain the formation of this last radical either a proton loss from the cation or a radical attack on the undamaged molecule:

$$MeCH_2CH_2As(O)(OH)_2 + R^{\cdot} \longrightarrow MeC^{\cdot}HCH_2As(O)(OH)_2 + RH$$

The preferential scission of a C—H bond β to an arsenic atom was explained by a stabilization effect due to hyperconjugation. This hyperconjugation is revealed by the very high hyperfine coupling with ^{75}As (^{75}As-T: 225 G, 247 G, 225 G) which indicates an arsenic spin density ρ_s close to 6%; similar values were obtained for radicals containing an $AsEt_3^+$ or an $AsEt_2$ moiety linked to a $R\dot{C}HCH_2$ group. These species adopt therefore a preferred conformation which allows a substantial overlap between the carbon π-orbital and the carbon–arsenic σ-bond. Such interaction is not observed when the arsenic atom is replaced by a nitrogen atom, but exists for phosphorus and for heavier atoms like Sn or Ge.

C. Adducts with Organic Molecules

Reacting $MCl_{3-n}Ph_n$ (M = As, Sb), $AsMe_2Br$, SbF_3 with 9,10-phenantraquinone or benzothiophene-4,5-dione in diglyme at $T > 160\,°C$ produces paramagnetic species which exhibit hyperfine coupling with As, Sb or Bi[79,80]. In these species As, Sb or Bi are chelated by the *ortho* quinone. Although N,N'-di-t-butyl-1,4-diaza-1,3-butadiene can form complexes with phosphorus radicals, attempts to generate adducts with heavier atoms of the group V have been unsuccessful[81]. The complexation of trivalent salts of P, As, Sb and Bi has also been reported by Prokof'ev and coworkers[82,83].

VIII. INORGANIC RADICALS

Aqueous arsenious, arsenite and arsenate solutions were studied by pulse radiolysis[84] and showed the formation of $As(OH)_4$, $HAsO_3^-$, $As(OH)_3O^-$. Many inorganic compounds in which the arsenic atom is only bound to oxygen atoms have been exposed to ionizing radiation. The purpose of these studies was often to obtain information about phase transitions (e.g. ferroelectric–antiferroelectric) and the resulting arsonyl-type radicals (AsO_3^{--}, see Reference 37) or arsoranyl-type radicals (AsO_4^{4-}, see Reference 60) were used as a spin probe[60,61,85,86]. Several inorganic arsenic-centered radical ions were also detected by studying the effects of γ-irradiation on arsenic sulfide chalcogenide glasses[87,88] oxide glasses[89] and germane/arsine mixtures[90]. Localized paramagnetic states have also been induced in glassy As_2Se_3 and As_2S_3 by infrared radiation[91]. γ-Radiation induced surface hole centers were observed in silica gels containing deposited arsenic ions[92] and in UV-irradiated As-doped soda-lime-silica glasses[93].

IX. REFERENCES

1. M. Lehnig (Ed.), *Landolt-Börnstein: Numerical Data and Functional Relationships in Science and Technology. Atomic and Molecular Physics. Magnetic Properties of Free Radicals*, Vol. II/9c2 (1979), II/17a (1987), II/17f (1988), Springer-Verlag, Berlin.
2. *Specialist Periodical Reports, Electron Spin Resonance*, Royal Society of Chemistry Cambridge, U.K.
3. E. A. C. Lucken, *Nuclear Quadrupole Constants*, Academic Press, London, 1969.
4. P. W. Atkins and M. C. R. Symons, *The Structure of Inorganic Radicals*, Elsevier, Amsterdam, 1967.

5. J. E. Wetz and J. R. Bolton, *Electron Spin Resonance. Elementary Theory and Practical Applications*, McGraw-Hill, New York, 1972.
6. N. M. Atherton, *Electron Spin Resonance Theory and Applications*, Ellis Horwood, Chichester, 1973.
7. W. Gordy, *Theory and Applications of Electron Spin Resonance*, Wiley, New York, 1980.
8. J. R. Morton and K. F. Preston, *J. Magn. Reson.*, **30**, 577 (1978).
9. R. L. Morehouse, J. J. Christansen and W. Gordy, *J. Chem. Phys.*, **45**, 1747 (1966).
10. U. Schmidt, K. Kabitzke, K. Markau and A. Muller, *Chem. Ber.*, **99**, 1497 (1966).
11. J. Preer, F-D. Tsay and H. B. Gray, *J. Am. Chem. Soc.*, **94**, 1875, (1972).
12. A. R. Lyons and M. C. R. Symons, *J. Am. Chem. Soc.*, **95**, 3483 (1973).
13. M. Geoffory, L. Ginet and E. A. C. Lucken, *J. Chem. Phys.*, **65**, 729 (1976).
14. C. H. Townes and B. P. Dailey, *J. Chem. Phys.*, **17**, 782 (1949).
15. J. P. Michaut, J. Roncin and R. Marx, *Chem. Phys. Lett.*, **36**, 599 (1975).
16. M. J. S. Gynane, A. Hudson, M. F. Lappert, P. P. Power and H. Goldwhite, *J. Chem. Soc., Dalton Trans.*, 2428 (1980).
17. M. J. S. Gynane, A. Hudson, M. F. Lappert, P. P. Power and H. Goldwhite, *J. Chem. Soc., Chem. Commun.*, 623 (1976).
18. M. Geoffroy, L. Ginet and E. A. C. Lucken, *J. Chem. Phys.*, **66**, 5292 (1977).
19. W. T. Cook and J. S. Vincent, *J. Chem. Phys.*, **67**, 5766 (1977).
20. M. Geoffroy, E. A. C. Lucken and M. Mazeline, *Mol. Phys.*, **28**, 839 (1974).
21. T. Berclaz and M. Geoffroy, *Helv. Chim. Acta.*, **61**, 684 (1978).
22. W. T. Cook, J. S. Vincent, I. Bernal and F. Ramirez, *J. Chem. Phys.*, **61**, 3479 (1974).
23. F. A. Neuegebauer and S. Bamberger, *Chem. Ber.*, **107**, 2362 (1974).
24. B. Cetinkaya, A. Hudson, M. F. Lappert, and H. Goldwhite, *J. Chem. Soc., Chem. Commun.*, 609 (1982).
25. M. Geoffroy, G. Terron, A. Jouaiti, P. Tordo and Y. Ellinger, *Bull. Magn. Reson.*, in press.
26. M. Culcasi, G. Gronchi, J. Escudié, C. Couret, L. Pujol and P. Tordo, *J. Am. Chem. Soc.*, **106**, 3130 (1986).
27. B. Cetinkaya, P. B. Hitchcock, M. Lappert, A. J. Thorne and H. Goldwhite, *J. Chem. Soc., Chem. Commun.*, 691 (1982).
28. P. Tordo, in *The Chemistry of Organophosphorus Compounds* (Ed. F. R. Hartley), Wiley, Chichester, 1990, p. 137.
29. I. Marov and M. C. R. Symons, *Russ. J. Inorg. Chem.*, **16**, 1193, (1971).
30. G. W. Eastland and M. C. R. Symons, *J. Chem. Soc., Perkin Trans. 2*, 833 (1977).
31. T. Berclaz and M. Geoffroy, *Mol. Phys.*, **30**, 549 (1975).
32. M. Cattani-Lorente, M. Geoffroy, S. P. Mishra, J. Weber and G. Bernardinelli, *J. Am. Chem. Soc.*, **108**, 7148 (1986).
33. S. Subramanian and M. T. Rogers, *J. Chem. Phys.*, **57**, 4582 (1972).
34. A. R. Lyons and M. C. R. Symons, *J. Chem. Phys.*, **60**, 164 (1974).
35. M. Geoffroy and A. Llinares, *Helv. Chim. Acta*, **64**, 329 (1981).
36. M. Geoffroy, L. Ginet and E. A. C. Lucken, *Mol. Phys.*, **31**, 745 (1976).
37. A. Reuveni and Z. Luz., *J. Magn. Reson.*, **16**, 339 (1974).
38. M. Geoffroy and A. Llinares, *Helv. Chim. Acta*, **62**, 1605 (1979).
39. J. Gaillard, O. Constantinescu and B. Lamotte, *J. Chem. Phys.*, **55**, 5442 (1971).
40. A. R. Lyons and M. C. R. Symons, *J. Chem. Soc., Faraday Trans. 2*, 1589 (1972).
41. T. Berclaz and M. Geoffroy, unpublished results.
42. W. G. Bentrude, in *The Chemistry of Organophosphorous Compounds* (Ed. F. R. Hartley), Wiley, Chichester, 1990, p. 531.
43. B. P. Roberts, in *Advances in Free Radical Chemistry* (Ed. G. H. Williams), **6**, Heyden and Sons, London, 1980, p. 225.
44. W. G. Bentrude, *Acc. Chem. Res.*, **15**, 117 (1982).
45. M. Geoffroy, A. Llinares and E. Krzywanska, *J. Magn. Reson.*, **58**, 389 (1984).
46. M. Geoffroy, and A. Llinares, *Helv. Chim. Acta*, **66**, 76 (1983).
47. A. Hasegawa, K. Ohnishi, K. Sogabe and M. Miura, *Mol. Phys.*, **30**, 1367 (1975).
48. T. Gillbro and F. Williams, *J. Am. Chem. Soc.*, **96**, 5032 (1974).
49. T. Berclaz, M. Geoffroy and E. A. C. Lucken, *Chem. Phys. Lett.*, **36**, 677 (1975).
50. T. Berclaz, M. Geoffroy and E. A. C. Lucken, *J. Magn. Reson.*, **33**, 577 (1979).

51. M. Kikuchi, N. Leray, J. Roncin and B. Joukoff, *Chem. Phys.*, **12**, 169 (1976).
52. S. A. Fieldhouse, H. C. Starkie and M. C. R. Symons, *Chem. Phys. Lett.*, **23**, 508 (1973).
53. M. Geoffroy and A. Llinares, *Mol. Phys.*, **41**, 55 (1980).
54. M. Geoffroy, A. Llinares and S. P. Mishra, *J. Chem. Soc., Faraday Trans. 1*, **82**, 521 (1986).
55. M. C. R. Symons and G. D. G. McConnachie, *J. Chem. Soc., Faraday Trans. 1*, **80**, 211 (1984).
56. E. Furimsky, J. A. Howard and J. R. Morton, *J. Am. Chem. Soc.*, **94**, 5932 (1972).
57. E. Furimsky, J. A. Howard and J. R. Morton, *J. Am. Chem. Soc.*, **95**, 6574 (1973).
58. A. G. Davies, D. Griller and B. P. Roberts, *J. Organomet. Chem.*, **38**, C8 (1972).
59. I. S. Ginns, S. P. Mishra and M. C. R. Symons, *J. Chem. Soc., Dalton Trans.* 2509, (1973).
60. N. S. Dalal, J. A. Hebden, D. E. Kennedy and C. A. McDowell, *J. Chem. Phys.*, **66**, 4425 (1977).
61. J. Gaillard and P. Gloux, *Solid State Commun.*, **17**, 817 (1975).
62. A. J. Colussi, J. R. Morton and K. F. Preston, *Chem. Phys. Lett.*, **30**, 317 (1975).
63. A. R. Boate, J. R. Morton, and K. F. Preston, *J. Magn. Reson.*, **24**, 259 (1976).
64. A. J. Colussi, J. R. Morton and K. F. Preston, *J. Phys. Chem.*, **79**, 1855 (1975).
65. A. R. Boate, A. J. Colussi, J. R. Morton and K. F. Preston, *Chem. Phys. Lett.*, **37**, 135 (1976).
66. J. R. Morton, K. F. Preston and S. J. Strach, *J. Phys. Chem.*, **83**, 2628 (1979).
67. J. R. Morton, K. F. Preston and S. J. Strach, *J. Phys. Chem.*, **83**, 3418 (1979).
68. J. R. Morton, K. F. Preston and S. J. Strach, *J. Magn. Reson.*, **37**, 321 (1980).
69. P. J. Russo, M. M. Labes and G. E. Kemmerer, *J. Chem. Soc., Chem. Commun.*, 701 (1981).
70. M. Geoffroy, J. Hwang and A. Llinares, *J. Chem. Phys.*, **76**, 5191 (1982).
71. F. G. Herring, J. H. Wang and W. C. Lin, *J. Phys. Chem.*, **71**, 2086 (1967).
72. M. C. R. Symons, *J. Chem. Soc.* (A), 2393 (1971).
73. A. R. Boate, J. R. Morton and K. F. Preston, *Chem. Phys. Lett.*, **50**, 65 (1977).
74. A. R. Boate, J. R. Morton and K. F. Preston, *J. Magn. Reson.*, **29**, 274 (1978).
75. A. R. Boate, J. R. Morton and K. F. Preston, *J. Chem. Phys.*, **67**, 4302 (1977).
76. S. P. Mishra, M. C. R. Symons, K. O. Christe, R. D. Wilson and R. I. Wagner, *Inorg. Chem.*, **14**, 1103 (1975).
77. A. R. Lyons and M. C. R. Symons, *J. Chem. Soc., Chem. Commun.*, 1068 (1971).
78. A. R. Lyons and M. C. R. Symons, *J. Chem. Soc., Faraday Trans. 2*, **68**, 622 (1972).
79. A. Alberti and A. Hudson *J. Organomet. Chem.*, **182**, C49 (1979).
80. A. Alberti, A. Hudson, A. Maccioni, G. Podda and G. F. Pedulli, *J. Chem. Soc., Perkin Trans. 2*, 1274 (1982).
81. A. Alberti and A. Hudson, *J. Organomet. Chem.*, **248**, 199 (1983).
82. A. I. Prokof'ev, N. A. Malisheva, B. L. Tumanskii, N. N. Bubnov, S. P. Solodovnikov and M. I. Kabachnik, *Izv. Akad. Nauk, SSSR, Ser. Khim.*, 1976 (1978).
83. A. I. Prokof'ev, N. A. Malisheva, N. N. Bubnov, S. P. Solodovnikov and M. I. Kabachnik, *Izv. Akad. Nauk SSSR, Ser. Khim.*, 169 (1983).
84. U. K. Kläning, B. H. J. Bielski and K. Sehested, *Inorg. Chem.*, **28**, 2717 (1989).
85. H. A. Farach, M. A. Mesa, R. J. Creswick, C. P. Poole Jr and T. Mzoughi, *Ferroelectrics*, **117**, 171 (1991).
86. B. Rakvin, P. K. Kahol and N. S. Dalal, *Mol. Phys.*, **68**, 1185 (1989).
87. T. Budinas, P. Mackus, I. V. Savytsky and O. I. Shpotyuk, *J. Non-Cryst. Solids*, **90**, 521 (1987).
88. B. T. Kolomiets and V. M. Lyubin, *Disordered Semiconductors*, Plenum Press, New York, 1987, p. 151.
89. H. Imagawa, H. Hosono and H. Kawazoe, *J. Phys.* (*Paris*), *Colloq.* C9, **43**, 169 (1982).
90. P. Benzi, M. Castiglioni, L. Operti, G. A. Vaglio and P. Volpe, *Materials Engineering*, **1**, 965 (1990).
91. S. G. Bishop, J. A. Freitas Jr and U. Strom, *AIP Conf. Proc.*, **120**, 86 (1984).
92. V. B. Kazanskii, S. L. Kalyagin, G. A. Kozlov, S. A. Surin and B. N. Shelimov, translated from *Kinetika i Kataliz*, **19**, 1264 (1978).
93. H. Hosono and Y. Abe, *J. Non-Cryst. Solids*, **125**, 98 (1990).

CHAPTER **13**

Thermolysis

C. A. McAULIFFE and A. G. MACKIE

Department of Chemistry, University of Manchester Institute of Science and Technology, Manchester M60 IQD, UK

I. INTRODUCTION

Some years ago the thermolysis of organopnictogens was considered an esoteric subject of limited interest. The topic has been transformed into one of the growth points of hetero-organic chemistry because of the enormous economic potential of metal organic chemical vapour deposition (MOCVD) processes for the production of III–V type of

The chemistry of organic arsenic, antimony and bismuth compounds
Edited by S. Patai © 1994 John Wiley & Sons Ltd

semiconductor materials such as GaAs and SbIn. Bismuth chemistry has also been boosted by the race to produce economically significant superconducting materials. Organobismuth compounds are used as precursors in the preparation of both bulk phase and films of superconducting materials. Thus, the bias of this chapter has been led by contemporary economic needs and not the arcane interests of academia. A broad view is taken of organopnictogens inasmuch as bridged species, i.e. metal-organic (M—E—C) compounds, are included, as are the hydrides. To exclude such compounds would be to elevate taxonomy over the logical development of our chosen themes.

The words *pyrolysis* and *thermolysis* tend to be used as synonyms. We use them in the sense that a thermolysis is a controlled pyrolysis. A thermolytic reaction implies a single substance substrate and a controlled thermal input. Also implied is a vacuum, an inert atmosphere or unreactive solvent. However, we will include in this discussion the occasional example in which the atmosphere may not be inert, e.g. MOCVD processes that use hydrogen as a carrier gas. We draw a distinction between decomposition in oxygen (air) and in other gases. In oxygen, a decomposition is generally regarded as combustion, unless the process is so controlled as to produce incompletely oxidized products. Such processes are normally excluded, with the notable exception of the bismuth-containing superconductors. A pyrolysis nomenclature has been presented by Ericsson and Lattimer (see Appendix) and was developed in a series of informal workshops at recent international conferences on analytical and applied pyrolysis.

Compounds which decompose at ambient temperatures are not usually described in terms of thermolysis. However, there is no logical distinction between compounds that decompose in the range 200 K to 300 K and those in the range 300 K to 400 K. Such compounds as are of interest are also included.

In an effort to keep the references to a minimum, only 1980 and later chemistry is fully referenced. The earlier chemistry is economically referenced. The chemistry asserted *ex cathedra* will be found in one of the works of scholarship listed in the brief bibliography.

The emphasis of this short review is on areas of contemporary commercial interest, although not exclusively. Readers will probably notice the many evasions of our own definitions. This is because we feel empowered by definitions inasmuch as they define a problem to be addressed and solved. We do not seek to teach but to facilitate and/or stimulate useful connections to be made with the areas of expertise of our readers.

Finally, we are writing for laboratory-based workers rather than our 'pencil and paper colleagues' to whom we apologise for the lack of 'figures'.

II. GROUP 13 LIGANDS

A. Boron Ligands

Aminoarsine-borane adducts of the type $R_{3-n}As(NR_2^1)_n(BH_3)_m$ (R = alkyl; $n = 1$–3; $m = 1, 2$) are stable at low temperatures, but on warming to the ambient undergo thermolysis. The thermolytic products are complex because of side reactions, but among the significant primary breakdown products are reduced arsenic species, e.g. R_2AsH. The theoretical interest in these complexes derives from which of the alternative Lewis base sites, arsenic or nitrogen, forms the adduct with the Lewis acid BH_3. Krannich and coworkers[1] have investigated such species mainly by variable-temperature multinuclear NMR spectroscopy.

The antimony/borane adduct dimethylstibino-borane decomposes to a product analysed as SbB and described as a black solid[2]:

$$[Me_2SbBH_2]_n \xrightarrow{\ 210\,°C\ } SbB \qquad (1)$$

B. Aluminium Ligands

Our main discussion of the MOCVD preparation of III–V semiconductors is in Section III.A. Single-source precursors for III–V semiconductors have shown advantages and $[Bu_2^t As.AlEt_2]_2$ has been proposed by Cowley and coworkers[3] as such a source. Other possibilities are available such as $Et_2Al \cdot \overline{AsCHMe \cdot CHMe \cdot CHMe \cdot CHMe}^4$. Aluminium containing III–V semiconductors are often ternary phases. Cowley and coworkers[5] have synthesized a potential ternary precursor containing an aluminium/arsenic moiety.

$$
\begin{array}{c}
Bu_2^t \\
As \\
Me_2Ga \diagup \diagdown AlMe_2 \\
\diagdown \diagup \\
As \\
Bu_2^t
\end{array}
\xrightarrow{\Delta} \quad 2\,Al_{0.5}Ga_{0.5}As \tag{2}
$$

C. Gallium Ligands

The organopnictogen compounds involving Pn—Ga bonds are either simple adducts of the type R_3Pn—GaR_3 (Pn = pnictogen: P, As, As, Sb) or species which can be thought of as substituted gallanes of the type $(R_2Pn)_{3-n}GaR_n$, and which are usually oligomeric.

Group 13 Lewis acids form adducts with Lewis bases of Group 15. These adducts have found some use as single-source precursors in the MOCVD preparation of films of III–V semiconductors, e.g. Me_3In–PEt_3[6]. The adducts were an attractive possibility since they are more stable to the ambient than their component parts, e.g. $Me_3Ga.AsMe_3$ is less susceptible to aerial oxidation than Me_3Ga or Me_3As. More important though is the advantage of the in-built 1:1 arsenic-to-gallium stoichiometry of R_3GaAsR_3 adducts for the formation of GaAs compared to the complex metering of independent supplies of Me_3Ga and Me_3As:

$$
R_3Ga.AsR_3 \xrightarrow{\Delta} GaAs + \text{volatile products} \tag{3}
$$

Unfortunately, the adducts partially dissociate with a consequential loss of stoichiometry before the thermolytic process that produces GaAs, and an excess of R_3As is necessary[7]. This is due to the Ga—As bond being the weakest bond in either Me_3GaAsH_3 or $Me_3GaAsMe_3$. To overcome this problem of the weak Ga—As bond in the gallane–arsine adducts, Cowley and coworkers[8] have proposed the use of arsinogallanes $[R_2GaAsR_2^1]_n$ and, more specifically, $[Me_2GaAsBu_2^t]_2$[9] (1).

The stronger 'respectable' two-electron gallium arsenic σ-bond in 1 insures that As—C and Ga—C bond fission is favoured. The vacuum pyrolysis of 1 affords mainly isobutene and methane with a small amount of isobutane as the volatile products. Cowley and coworkers[8] have proposed a hydrogen transfer from the isobutyl (As) group to the gallium methyl substituent resulting in the elimination of methane. The transfer is thought to be mediated by the gallium moiety, since the crystal structure[9] is consistent with a close Ga...H interaction if 2.62 Å with a CH_3 of the t-butyl group (see Figure 1) providing a plausible low-energy decomposition pathway (equation 4).

$$\tag{4}$$

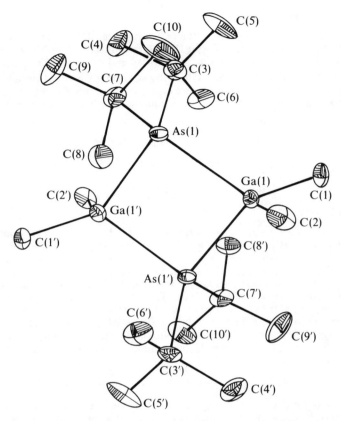

FIGURE 1. ORTEP view of $[Cl_2GaAsBu^t_2]_2$; key bond lengths (Å) and angles (deg): As(1)—Ga(1) 2.541(1), As(1')—Ga(1) 2.558(1), Ga(1)—C(1) 2.016(5), Ga(1)—C(2), 2.028(5), As(1)—C(3), 2.046(5), Ga(1)—As(1)—Ga(1') 95.69(2), As(1)—Ga(1)—As(1') 84.31, As(1)—Ga(1)—C(1) 115.1(2), As(1)— Ga(1)—C(2) 116.2(2), C(1)—Ga(1)—C(2) 109.3(3), C(3)—As(1)—C(7) 110.3(2)

Interestingly, if the substituent on arsenic is the less sterically demanding isopropyl group, the arsinogallane is a trimer[10] $[Me_2GaAsPr^i_2]_3$, (2). This structural change leads to a siginificant change in the mechanism of thermolysis as shown by the emergence of C_4 and C_6 hydrocarbons, presumably $MePr^i$ and Pr^i_2. The significance of the bulk of the substituent on arsenic has been further illustrated by Cowley and coworkers[11] in preparing the trimeric GaAs precursor $[Bu^t_2GaAsH_2]_3$, (3). In the solid state 3 decomposes slowly at 155 °C and more rapidly at 250 °C affording GaAs, but with significant carbon inclusion. Refluxing 3 in toluene for 20 minutes gives a red-brown GaAs containing trapped organic debris. Cowley and coworkers[11] infer that their results support a Langmuir–Hinshelwood type mechanism for the formation of GaAs by surface-bound species:

$$Me_3Ga + PnH_3 \longrightarrow Me_2GaPnH_2 + CH_4 \qquad (5)$$

$$Me_2GaPnH_2 \longrightarrow MeGaPnH + CH_4 \tag{6}$$

$$MeGaPnH \longrightarrow GaPn + CH_4 \tag{7}$$

However, despite careful work it was not possible to isolate the conclusive $[Bu^tGaAsH]_n$.

Some organoarsinogallanes have been characterized as stable crystalline compounds which are unstable in solutions, e.g. $[\{(Me_3SiCH_2)_2As\}_3Ga]$ is thermally unstable in solution, decomposing at room temperature or over to $[(Me_3SiCH_2)_2As]_2$ and other unidentified products[12]. Thermolysis of $[\{(Me_3SiCH_2)_2As\}_2GaX]_2$ affords ligand redistribution products, *inter alia* a trimeric arsinogallane:

$$[\{(Me_3SiCH_2)_2As\}_2GaX]_2 \xrightarrow{\Delta} [(Me_3SiCH_2)_2As]_2 + [(Me_3SiCH_2)_2AsGaX_2]_3$$

$$+ \text{ other products} \tag{8}$$

In contrast, when the substituent on gallium is a phenyl group, a dimeric arsinogallane is obtained[13]:

$$[\{(MeSiCH_2)_2As\}_2GaPh]_2 \xrightarrow{\Delta} [(Me_3SiCH_2)_2As]_2 + \text{grey powder}$$

$$+ [(Me_3SiCH_2)_2AsGaPh_2]_2 \tag{9}$$

D. Indium and Thallium Ligands

Trimethylindium etherate (b.p. $56\,°C/15\,torr$) is more convenient to use than the reactive solid $InMe_3$ (m.p. $88\,°C$) for epitaxial growth. The complex with $SbEt_3$ is decomposed *in situ* to form an InSb layer[14]:

$$Me_3In.OEt_2 + SbEt_3 \longrightarrow Me_3In.SbEt_3 + OEt_2$$

$$\downarrow$$

$$InSb \longleftarrow [MeInSbEt]_n \longleftarrow Me_2In.SbEt_2 \tag{10}$$

Cowley and coworkers[14], in continuing their search for III–V semiconductor precursors, have proposed $[(Bu^t_2Sb)_2InCl]_2$ for InSb band systems.

III. GROUP 14 LIGAND

A. Carbon Ligands

The arsines, AsR_3, are among the most numerous and widely studied organo-arsenic compounds. They include both acyclic (e.g. Me_3As) and heterocyclic (e.g. cyclo-$C_5H_{10}AsCH_3$) structures. Arsines are classified as derivatives of the parent compound AsH_3 by successive substitutions to form the family of compounds $R_{3-n}AsH_n$ ($n = 0-3$). Thus we include in this section arsine itself, AsH_3, together with the primary, secondary and tertiary arsines.

The thermolysis of AsH_3 is the final stage of the famous forensic proof of arsenic poisoning—the Marsh test, in which an arsenic mirror is formed. The same pyrolysis is the basis of the modern spectroscopic determination of arsenic. An example of the contemporary relevance of organoarsenic analytical pyrolysis is given by the analysis of oil shale using the hyphenated techniques HPLC-GFAA (Graphite Furnace Atomic Adsorption Spectroscopy). Arsenic and organoarsenic compounds were shown to be formed or released during the controlled pyrolysis ($500\,°C$) of oil shale kerogen[15]. The main species identified were $MeAs(O)(OH)_2$, $PhAs(O)(OH)_2$ and AsO_4^{3-} together with

TABLE 1. Thermochemical data[a] for group 15 compounds[16]

Compound	$\Delta_f H_e$	$\Delta\Delta$ Values[c]	$\Delta_f H_s$	Error[b]
AsMe$_3$	3 ± 2.5	(Me/H) = −4.5	+5.0	[2]
AsEt$_3$	13.4 ± 4	(Et/Me) = +3.5	−1.0	−14.4
AsPr$_3$			−16.0	
AsBu$_3$			−31.0	
AsPh$_3$	97.6	(Ph/Me) = 33		
SbMe$_3$	7.7 ± 6	(Me/H) = −9	10.0	+2
SbEt$_3$	10.4 ± 2.5	(Et/Me) = +1	6.0	−4
SbPr$_3$			−9.0	
SbBu$_3$			−24.0	
SbPh$_3$	104.1 ± 5	(Ph/Me) = 32.0		
BiMe$_3$	46.1 ± 5	(Et/Me) = +1.8	50.0	[4]
BiEt$_3$	51.6 ± 4		47.00	−5
BiPr$_3$			32.00	
BiBu$_3$			17.0	
BiPh$_3$	138.7 ± 4	(Ph/Me) = 31.0		

[a] Uncertainties not mentioned if less than $2 \, \text{kcal} \, \text{mol}^{-1}$.
$\Delta_f H_e$ = experimental value of the heat of formation.
$\Delta_f H_s$ = Author's suggested value for the heat of formation.
[b] Error = $\Delta_f H_s - \Delta_f H_e$
[c] $\Delta\Delta = \Delta(\Delta_f H_e)$ for the pair of substituents in parentheses.

traces of Me$_2$As(O)OH and AsO$_2^-$. Revised thermochemical data for some of the arsines, stibines and bismuthines are given in Table 1.

The thermolysis of arsine and the arsines has become of great commercial as well as scientific interest due to their role in the formation of gallium arsenide films by MOCVD processes. The now classical MOCVD preparation of gallium arsenide developed by Manasevit[17] (equation 11) is undergoing continuing refinement as new gallium and arsenic precursor compounds are investigated.

$$\text{GaMe}_3 + \text{AsH}_3 \xrightarrow{\Delta} \text{GaAs} + 3\,\text{CH}_4 \tag{11}$$

We confine our discussion to the arsenic precursors and their pyrolysis. The reader is referred to the technical literature for the MOCVD preparation of GaAs[18].

The technical/commercial importance of the arsine thermolysis mechanisms is the role of decomposition intermediates in the deleterious incorporation of carbon into the growth of MOCVD-produced GaAs. Wendt and Speckman[19] in their investigations into the thermal decomposition of arsines show the crucial importance of the production of AsH$_2$ radicals in the formation of low-carbon GaAs.

In the MOCVD environment arsine is commonly diluted with a carrier gas, usually hydrogen. Under these conditions the decomposition of AsH$_3$ is rapid in the range 650–710 °C (Figure 2)[20]. The overall reaction is:

$$4\,\text{AsH}_3 \xrightarrow{\Delta} \text{As}_4 + 6\,\text{H}_2 \tag{12}$$

The decomposition of AsH$_3$ is catalysed by the presence of Me$_3$Ga (Figure 2).

In the heterogeneous decomposition in the rate-limiting step[21] is the removal of the first hydrogen on the surface:

$$\text{AsH}_3(\text{g}) \longrightarrow \text{AsH}_{2(\text{Ad})} + \text{H}_{(\text{Ad.})} \tag{13}$$

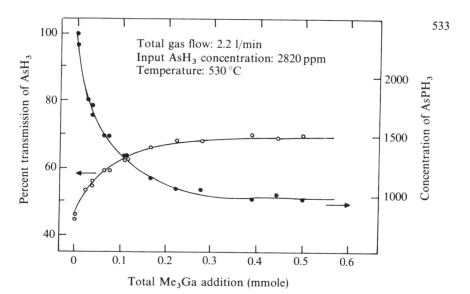

FIGURE 2. AsH₃ decomposition depending on the amount of Ga in the hot region

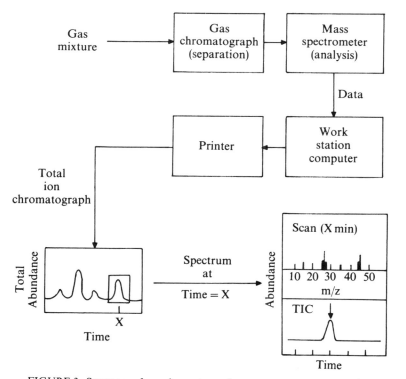

FIGURE 3. Summary of gas chromatograph–mass spectrometer operation

with activation energy of $23.2 \, \text{kcal mol}^{-1}$, the second and third hydrogen being more readily lost. Over GaAs the AsH_3 is known to decompose at $172 \, ^\circ C$.

The homogeneous process from mass spectroscopy is shown to occur:

$$AsH_3 \xrightarrow{\; < 620 \, ^\circ C \;} As + H_2 \tag{14}$$

$$\xrightarrow{\; > 620 \, ^\circ C \;} As_2 + As_4 + H_2 \tag{15}$$

The heterogeneous decomposition mechanism is characterized by easy chemisorption followed by rapid decomposition. Again the presence of adsorbed organogallium species catalyses this process, thus the formation of GaAs is a synergetic, mutually catalytic process.

Triethylarsine pyrolysis under MOCVD-type conditions was investigated by Speckman and Wendt[19,22] using GC–mass spectroscopy. In studies of this type it is essential to be mindful of the apparatus employed. Accordingly the appratus is shown schematically in Figures 3 and 4. Their results are summarized in Figures 5 and 6.

The thermolysis of $AsEt_3$ begins at $400 \, ^\circ C$ as indicated by the formation of the first traces of Et_2AsH. The first hydrocarbon to be detected is ethane at $450 \, ^\circ C$. The

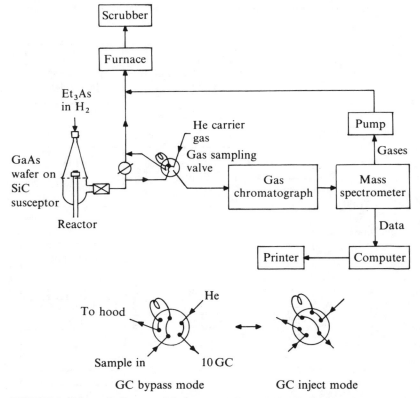

FIGURE 4. Schematic diagram of the integrated GC–MS/OMCVD reactor system used for decomposition studies and sampling valve operation

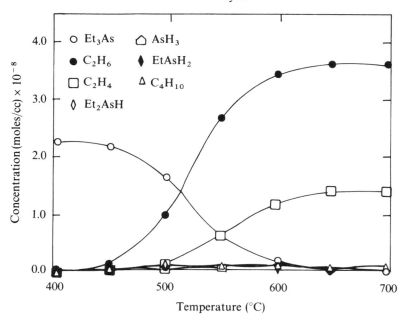

FIGURE 5. Thermal decomposition of triethylarsenic in hydrogen and total reaction mixture composition as a function of temperature

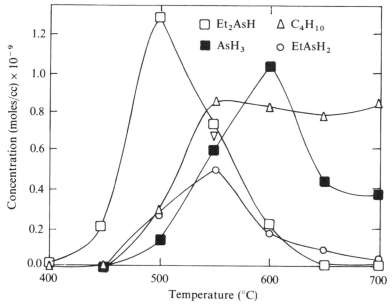

FIGURE 6. Trace reaction products formed during the thermolysis of triethylarsenic at different temperatures

concentration of ethane increases rapidly with temperature and it is always the dominant organic species at all temperatures. Ethylene is only formed above 500 °C and always as a minor product. Under the experimental conditions no Et_3As is detected above 700 °C. Two mechanisms have been proposed for the thermolysis. The first is based on a β-elimination as the first step, with the concerted build-up of an As—H bond, together with the elimination of C_2H_4.

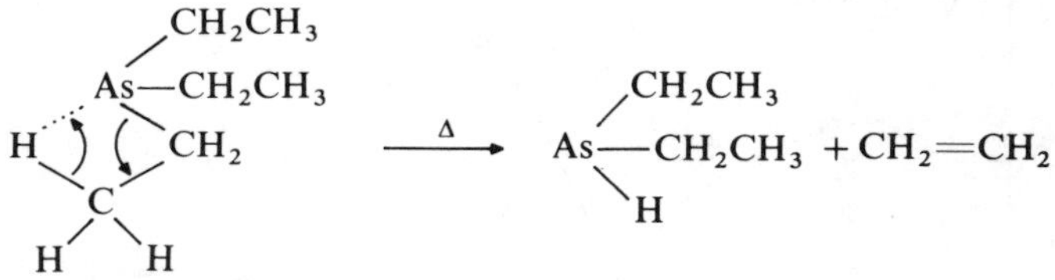

$$
\begin{array}{c}
\text{As—CH}_2\text{CH}_3 \\
\text{CH}_2
\end{array}
\quad \xrightarrow{\Delta} \quad
\text{As—CH}_2\text{CH}_3 + \text{CH}_2\!=\!\text{CH}_2 \qquad (16)
$$

In the Alternative mechanism the first step is a homolytic As—C bond fission forming ethyl (Et˙) and arsenyl (Et_2As˙) radicals. Both these radical species abstract hydrogen forming neutral molecules.

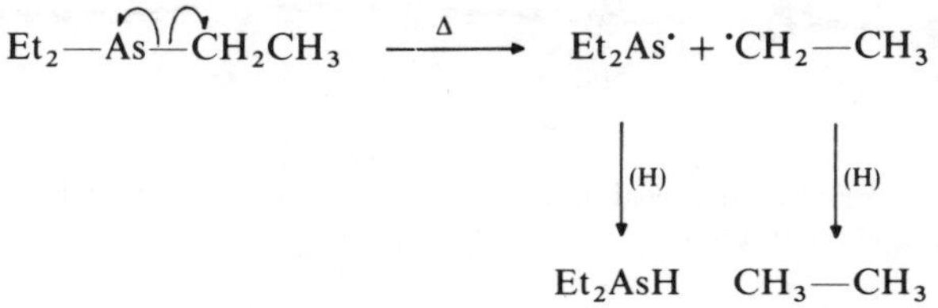

$$
Et_2\!-\!\overset{\frown\frown}{As}\!-\!CH_2CH_3 \quad \xrightarrow{\Delta} \quad Et_2As\cdot + \,\cdot CH_2\!-\!CH_3
$$

$$
\Big\downarrow{}_{(H)} \qquad \Big\downarrow{}_{(H)} \qquad\qquad (17)
$$

$$
Et_2AsH \qquad CH_3\!-\!CH_3
$$

Thus, the crucial difference in the proposed pathways is the emergence of C_2H_4 or C_2H_6 as the mechanistic indicator. In fact ethane, C_2H_6, is the observed product. Based on this and on other data Speckman and Wendt[22] proposed the reaction pathway for the pyrolysis of Et_3As shown in Scheme 1. Their observations are consistent with the data from the low-pressure pyrolysis of Et_3As obtained by Jensen and coworkers[23]. Both these groups' results are in conflict with observations of Melas and collaborators[24], who interpreted their thermolysis study of $AsEt_3$ interms of a β-elimination mechanism.

The pyrolysis mechanisms proposed by Wendt and Speckman[19] for Et_2AsH and for $EtAsH_2$ are given in Schemes 2 and 3. The formation of ethylene requires significantly more thermal energy than that required to produce Et_2AsH. The low concentration of butane is explained by the relatively low concentration of CH_3CH_2 in the gas phase diluted by the hydrogen carrier gas. After a study of the pyrolysis activation energies

TABLE 2. Pyrolysis activation energies of MOCVD arsenic precursors[25]

Arsenic precursor	Pyrolysis activation energy (eV)
Me_3As	2.93
Et_3As	1.58
Et_2AsH	0.82
Bu^tAsH_2	0.82
AsH_3	0.74

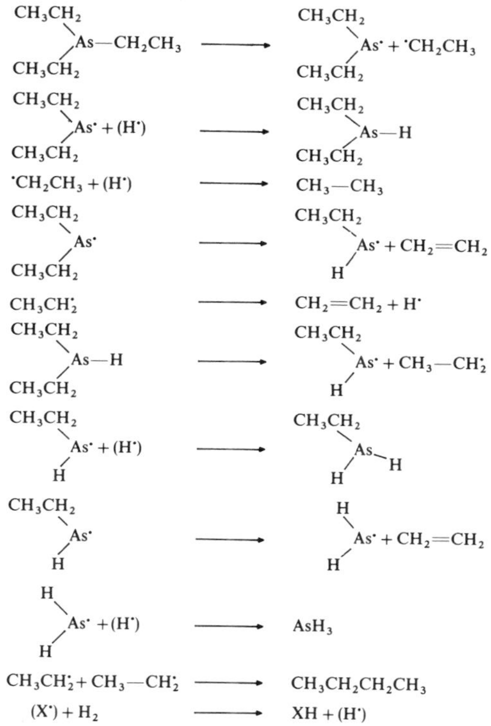

$$\begin{array}{l} CH_3CH_2 \\ \qquad \diagdown \\ \qquad \quad As{-}CH_2CH_3 \\ \diagup \\ CH_3CH_2 \end{array} \longrightarrow \begin{array}{l} CH_3CH_2 \\ \qquad \diagdown \\ \qquad \quad As^{\cdot} + {}^{\cdot}CH_2CH_3 \\ \diagup \\ CH_3CH_2 \end{array}$$

$$\begin{array}{l} CH_3CH_2 \\ \qquad \diagdown \\ \qquad \quad As^{\cdot} + (H^{\cdot}) \\ \diagup \\ CH_3CH_2 \end{array} \longrightarrow \begin{array}{l} CH_3CH_2 \\ \qquad \diagdown \\ \qquad \quad As{-}H \\ \diagup \\ CH_3CH_2 \end{array}$$

$${}^{\cdot}CH_2CH_3 + (H^{\cdot}) \longrightarrow CH_3{-}CH_3$$

$$\begin{array}{l} CH_3CH_2 \\ \qquad \diagdown \\ \qquad \quad As^{\cdot} \\ \diagup \\ CH_3CH_2 \end{array} \longrightarrow \begin{array}{l} CH_3CH_2 \\ \qquad \diagdown \\ \qquad \quad As^{\cdot} + CH_2{=}CH_2 \\ \diagup \\ H \end{array}$$

$$CH_3CH_2^{\cdot} \longrightarrow CH_2{=}CH_2 + H^{\cdot}$$

$$\begin{array}{l} CH_3CH_2 \\ \qquad \diagdown \\ \qquad \quad As{-}H \\ \diagup \\ CH_3CH_2 \end{array} \longrightarrow \begin{array}{l} CH_3CH_2 \\ \qquad \diagdown \\ \qquad \quad As^{\cdot} + CH_3{-}CH_2^{\cdot} \\ \diagup \\ H \end{array}$$

$$\begin{array}{l} CH_3CH_2 \\ \qquad \diagdown \\ \qquad \quad As^{\cdot} + (H^{\cdot}) \\ \diagup \\ H \end{array} \longrightarrow \begin{array}{l} CH_3CH_2 \\ \qquad \diagdown \\ \qquad \quad As{\diagdown} \\ \diagup \qquad \quad H \\ H \end{array}$$

$$\begin{array}{l} CH_3CH_2 \\ \qquad \diagdown \\ \qquad \quad As^{\cdot} \\ \diagup \\ H \end{array} \longrightarrow \begin{array}{l} H \\ \quad \diagdown \\ \qquad As^{\cdot} + CH_2{=}CH_2 \\ \diagup \\ H \end{array}$$

$$\begin{array}{l} H \\ \quad \diagdown \\ \qquad As^{\cdot} + (H^{\cdot}) \\ \diagup \\ H \end{array} \longrightarrow AsH_3$$

$$CH_3CH_2^{\cdot} + CH_3{-}CH_2^{\cdot} \longrightarrow CH_3CH_2CH_2CH_3$$

$$(X^{\cdot}) + H_2 \longrightarrow XH + (H^{\cdot})$$

SCHEME 1. Proposed mechanism for the pyrolysis of triethylarsine[22] (X = organic and/or organometallic species)

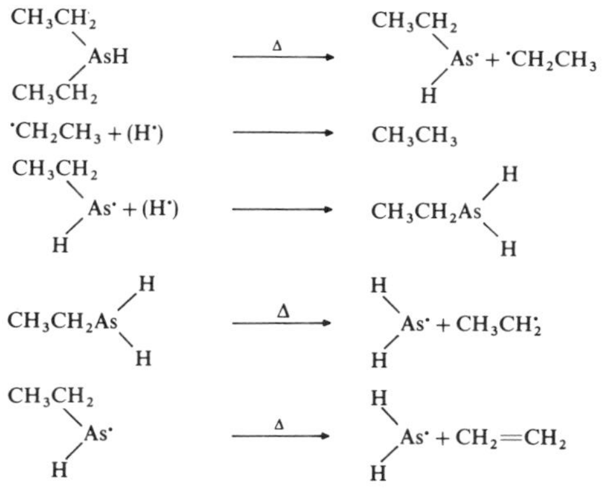

$$\begin{array}{l} CH_3CH_2 \\ \qquad \diagdown \\ \qquad \quad AsH \\ \diagup \\ CH_3CH_2 \end{array} \xrightarrow{\Delta} \begin{array}{l} CH_3CH_2 \\ \qquad \diagdown \\ \qquad \quad As^{\cdot} + {}^{\cdot}CH_2CH_3 \\ \diagup \\ H \end{array}$$

$${}^{\cdot}CH_2CH_3 + (H^{\cdot}) \longrightarrow CH_3CH_3$$

$$\begin{array}{l} CH_3CH_2 \\ \qquad \diagdown \\ \qquad \quad As^{\cdot} + (H^{\cdot}) \\ \diagup \\ H \end{array} \longrightarrow \begin{array}{l} \qquad \qquad \quad H \\ \qquad \qquad \diagup \\ CH_3CH_2As \\ \qquad \qquad \diagdown \\ \qquad \qquad \quad H \end{array}$$

$$\begin{array}{l} \qquad \qquad \quad H \\ \qquad \qquad \diagup \\ CH_3CH_2As \\ \qquad \qquad \diagdown \\ \qquad \qquad \quad H \end{array} \xrightarrow{\Delta} \begin{array}{l} H \\ \quad \diagdown \\ \qquad As^{\cdot} + CH_3CH_2^{\cdot} \\ \diagup \\ H \end{array}$$

$$\begin{array}{l} CH_3CH_2 \\ \qquad \diagdown \\ \qquad \quad As^{\cdot} \\ \diagup \\ H \end{array} \xrightarrow{\Delta} \begin{array}{l} H \\ \quad \diagdown \\ \qquad As^{\cdot} + CH_2{=}CH_2 \\ \diagup \\ H \end{array}$$

SCHEME 2. Proposed mechanism for the thermolysis of Et_2AsH[19]

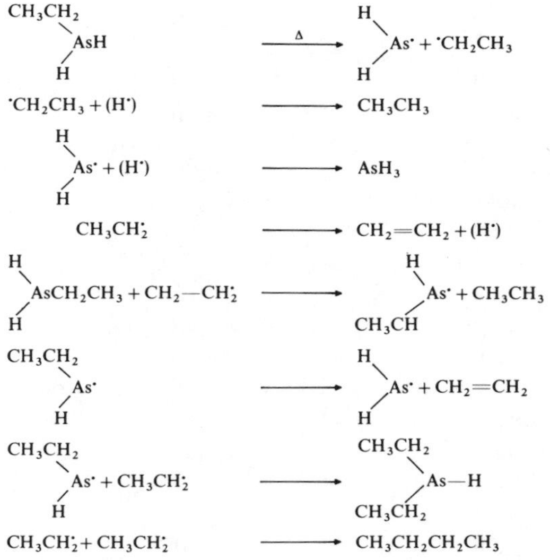

$$
\begin{array}{c}
\overset{\displaystyle CH_3CH_2}{\underset{\displaystyle H}{\diagup}}AsH \xrightarrow{\ \Delta\ } \overset{\displaystyle H}{\underset{\displaystyle H}{\diagup}}As\cdot + \cdot CH_2CH_3
\end{array}
$$

$$\cdot CH_2CH_3 + (H\cdot) \longrightarrow CH_3CH_3$$

$$\overset{\displaystyle H}{\underset{\displaystyle H}{\diagup}}As\cdot + (H\cdot) \longrightarrow AsH_3$$

$$CH_3CH_2^\cdot \longrightarrow CH_2{=}CH_2 + (H\cdot)$$

$$\overset{\displaystyle H}{\underset{\displaystyle H}{\diagup}}AsCH_2CH_3 + CH_2{-}CH_2^\cdot \longrightarrow \overset{\displaystyle H}{\underset{\displaystyle CH_3CH}{\diagup}}As\cdot + CH_3CH_3$$

$$\overset{\displaystyle CH_3CH_2}{\underset{\displaystyle H}{\diagup}}As\cdot \longrightarrow \overset{\displaystyle H}{\underset{\displaystyle H}{\diagup}}As\cdot + CH_2{=}CH_2$$

$$\overset{\displaystyle CH_3CH_2}{\underset{\displaystyle H}{\diagup}}As\cdot + CH_3CH_2^\cdot \longrightarrow \overset{\displaystyle CH_3CH_2}{\underset{\displaystyle CH_3CH_2}{\diagup}}As{-}H$$

$$CH_3CH_2^\cdot + CH_3CH_2^\cdot \longrightarrow CH_3CH_2CH_2CH_3$$

SCHEME 3. Proposed thermolytic mechanism for the decomposition of EtAsH$_2$[19]

of arsenic precursors, Table 2, Kobayashi and coworkers[25] proposed that Bu^{t}AsH and Et$_2$AsH would be particularly efficaceous as arsenic precursors in MOCVD processes.

Stringfellow and coworkers[26] showed that the Bu^{t}AsH$_2$ and Bu$^t_2$AsH pyrolyses were accelerated by an increase in the t-butyl radical concentration. The independent t-butyl radical source was azotertiary-butane (Bu$^t_3$N$_2$). The self-pyrolysis temperatures for 50% decomposition were reduced to 350 °C for Bu$^t_2$AsH and 300 °C for Bu^{t}AsH$_2$ under the gas-flow conditions of MOCVD reactors. On the basis of analyses of the products of the pyrolyses they proposed the plausible mechanisms shown in Schemes 4 and 5.

$$R\cdot + RAsH_2 \longrightarrow RH + R(H)As\cdot$$

$$R\cdot + R(H)As\cdot \longrightarrow R_2AsH$$

$$R(H)As\cdot \longrightarrow R\cdot + HAs\cdot$$

SCHEME 4. Proposed mechanism for the Bu$^t_3$N$_2$ catalysed pyrolysis of Bu^{t}AsH$_2$ (R = Bu^t)[26]

$$R\cdot + R_2AsH \longrightarrow RH + R_2As\cdot$$

$$R\cdot + R_2As\cdot \longrightarrow R_3As$$

$$R_2As\cdot \longrightarrow R\cdot + RAs\cdot$$

$$RAs\cdot \longrightarrow R\cdot + As^\circ$$

SCHEME 5. Proposed mechanism for the Bu$^t_2$N$_2$ catalysed pyrolysis of Bu$^t_2$AsH$_2$ (R = Bu^t)[26]

Zorin and collaborators[27] used highly rectified SbH$_3$·to prepare ultra-pure antimony by pyrolysis at 300–400 °C. The thermolysis of Et$_3$Sb has been investigated by Yablokov's group[28] in the range 305–340 °C. The volatile products are mainly ethane with small amounts of butane, butenes, propene, ethylene and methane. They proposed the following mechanism based on the first-order kinetics:

$$Et_3Sb \longrightarrow Et_2Sb^{\cdot} + Et^{\cdot} \tag{18}$$

$$Et_2Sb^{\cdot} \longrightarrow EtSb^{\cdot\cdot} + Et^{\cdot} \tag{19}$$

$$EtSb^{\cdot\cdot} \longrightarrow Sb^{\circ} + Et^{\cdot} \tag{20}$$

$$2Et^{\cdot} \longrightarrow EtH + C_2H_2 \text{ and other hydrocarbons} \tag{21}$$

Stibines with ligands containing β-OH groups have been used to synthesize olefins by β-elimination on thermolysis[29].

$$Ph_2SbCH_2CR(OH)Ph \xrightarrow{180\,^\circ C} PhRC{=}CH_2 \quad R = H, Ph \tag{22}$$

The stibine $Cp^*_3Sb(Cp^* = cyclo-C_5Me_5)$ is unstable, only existing at low temperature ($-78\,^\circ C$). On warming to room temperature it arranges with ligand loss to form cyclo-$[SbCp^*]_4$[30].

Trimethylstibine[31] has been used to prepare semiconductor films, e.g. InAsSb. However, there are two main drawbacks to using Me_3Sb. Firstly, carbon encapsulation by the growing film can be serious especially with aluminium-containing layers owing to the strength of the Al—C bond[32,33]. The other problem is associated with the relatively high decomposition temperature of Me_3Sb. This is important when, for instance, bismuth is one of the players where low growth temperatures are required[34,35]. At $400\,^\circ C$, films of InSb contain droplets of In owing to the slowness of the Me_3Sb thermolysis. Hence the importance of the search for Sb precursors with low dissociation temperatures. The proposed alternative MOCVD stibine precursors include $(CH_2{:}CH)_3Sb$[36], Pr^i_3Sb[37], $(CH_3CH{:}CH)_3Sb$[37], Me_2SbBu^t [38,39], $Me_3CCH_2SbH_2$[40], $(Me_3CCH_2)_2SbH$[40] and $(CF_3)_3Sb$[41].

Bismuthine and the primary and secondary alkyl bismuthines are thermally unstable.

$$BiH_nMe_{3-n} \longrightarrow n/3\,BiH_3 + (3-n)/3\,BiMe_3 \tag{23}$$

$BiMe_3$ is the only bismuthine that can be distilled at atmospheric pressure. Complete vapour pressure and thermal stability data were determined by Amberger[42]. Over the temperature range $367-409\,^\circ C$ the decomposition of Me_3Bi is a first-order homogeneous process[43]. This is consistent with successive loss of methyl ligands. Me_3Bi is used in an MOCVD process to produce $InSb_{1-x}Bi_x$ semiconductors which are infrared emittors and detectors[44,45].

The advantages of MOCVD processing for high-quality Bi–Sc–Ca–Cu–O (BSCCO) type superconducting films lies in the control over film thickness and composition, the versatility in being able to coat complex shapes and low growth temperature. Finally, MOCVD processes have the commercial advantage of large through-puts.

$BiPh_3$ has a vapour pressure of 1 torr at $98\,^\circ C$ and is carried to the deposition site in an argon carrier gas. At the deposit site the atmosphere is dioxygen. Natori and co-workers[46] used dipivaloylmethanato (DMP) derivatives of the other metals, viz $Sr(DMP)_2$ $Ca(DMP)_2$ and $Cu(DMP)_2$. The deposition site was held at $800\,^\circ C$ and the film had a superconducting onset temperature (T_c) of 110 K. Hamaguchi and collaborators[47] also used Ph_3Bi, but employing the hexafluoroacetylacetonate derivatives of the other metals.

A disadvantage of Ph_3Bi is its relatively high stability, the rate of decomposition being slow at $650\,^\circ C$ judged by the Bi/Cu ratio of the product[48] which was also noted by other workers[49]. It was found that added water reduced the decomposition temperature of Ph_3Bi[50]. These superconducting films have also been prepared using bismuth alkoxides; see Section V.A. Five-coordinate organoarsenic(V) compounds in which all the ligands are carbon commonly undergo reductive thermolysis, e.g. the well-known example:

$$Ph_5As \xrightarrow{ca\ 150\,^{\circ}C} Ph_3As + Ph{-}Ph \qquad (24)$$

This propensity has been harnessed synthetically, e.g. the conversion of 5-substituted-5,5′-spiro-bi[dibenzarsole] to a nine-membered arsenic hetereocycle[51].

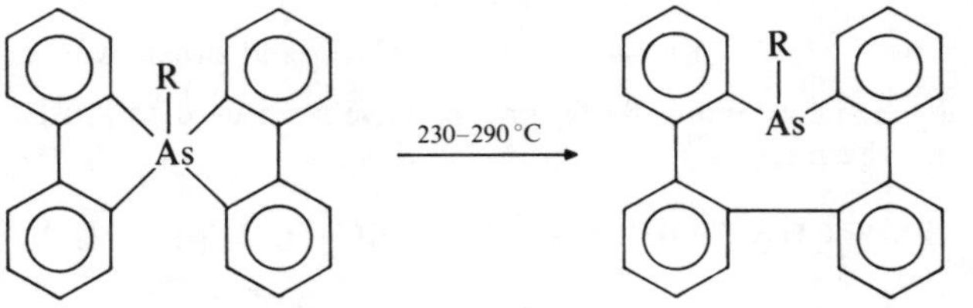

(25)

The antimony analogue $SbPh_5$ is also pyrolysed on heating:

$$SbPh_5 \xrightarrow{160\text{–}200\,^{\circ}C} Ph_3Sb + Ph{-}Ph \qquad (26)$$

However, it was found[52] that the decomposition is catalysed by copper(II) acetate, when it is quantitive at room temperature in toluene solution.

The arsonium ylides are the most synthetically useful of the organoarsenic(V) compounds. Trimethylarsonium methylide[53] and triphenylarsonium benzylide[54] undergo thermolysis under the mild conditions of refluxing benzene, affording the arsine and an olefin.

The proposed mechanism is an initial As=C cleavage generating a carbene which attacks another ylide molecule. The carbene–ylide intermediate (**4**) then decomposes, generating the observed olefin (Scheme 6).

$$R_3As{=}CHR' \xrightarrow{\Delta} R_3As{:} + {:}CHR' \qquad (27)$$

$$R_3As{=}CHR' + {:}CHR' \longrightarrow \left[R_3\overset{+}{A}{-}CHR\;\;|\;\;{-}CHR \right] \longrightarrow R'_3As + CHR'{=}CHR' \qquad (28)$$

SCHEME 6. Proposed mechanism for the thermolysis of arsonium ylides[53]

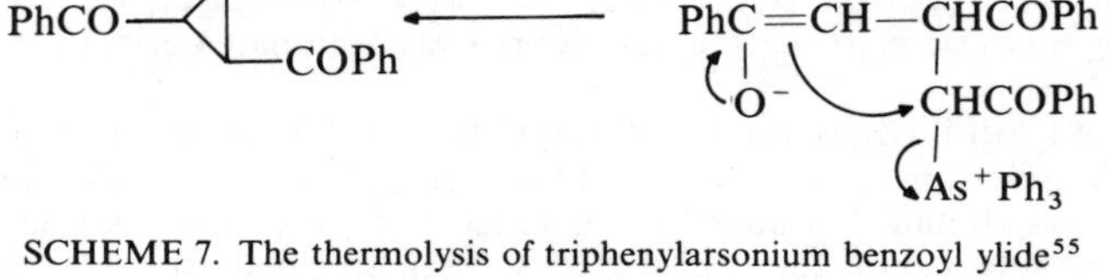

SCHEME 7. The thermolysis of triphenylarsonium benzoyl ylide[55]

If the olefin generated is susceptible to nucleophilic attack, then a cyclopropane may be formed, e.g. the thermolysis of triphenylarsonium benzoylylide[55] in refluxing toluene affords trans-1,2,3-tribenzoylcyclopropane (Scheme 7). The ylide is stable in refluxing benzene.

An interesting exception to the general rule that phosphonium ylides are more stable than arsonium ylides is provided by $Ph_3P=CHNO_2$ which decomposes spontaneously at room temperature, whereas the arsenic analogue is stable. This attributed to the greater strength of the P—O bond in the reaction intermediate compared to the As—O bond[56].

The pyrolysis of quaternary arsonium salts in general affords arsines with the elimination of MeX when possible:

$$[R_3AsMe]^+X^- \xrightarrow{\Delta} R_3As + MeX \qquad (29)$$

This is used synthetically, e.g. by Mann and Jones[57] in preparing a 1,4-diarsenane.

$$(30)$$

B. Silicon and Other Group 14 Ligands

For some applications it is desirable to have Si-dopped GaAs. Kikkawa and co-workers[58] found Bu^tAsH_2 a particularly useful arsenic MOCVD precursor because of the high concentration of the AsH_2 species produced in the initial pyrolysis step. The AsH_2 formed a silyl species as shown in their proposed scheme.

$$Bu^tAsH_2 \longrightarrow AsH_2 + C_4H_9 \qquad (31)$$

$$AsH_2 + AsH_2 \longrightarrow AsH_3 + AsH \qquad (32)$$

$$AsH_2 + SiH_4 \longrightarrow H_2AsSiH_3 + H \qquad (33)$$

$$AsH + SiH_4 \longrightarrow H_2AsSiH_3 \qquad (34)$$

$$AsH_3 + SiH_2 \longrightarrow H_2AsSiH_3 \qquad (35)$$

Tris(triethylsilyl) antimony decomposes at 300 °C affording antimony almost quantitatively together with a mixture of high boiling silanes[59].

Tris(triethylstannyl) antimony on thermolysis (220 °C/7 h) affords elemental antimony, and some metallic tin[59]. Catalytic quantities of $AlBr_3$ accelerate the thermolysis. A similar result was obtained with the bismuth analogue:

$$4(Et_3Sn)_3M \xrightarrow{AlBr_3} 9 Et_4Sn + 3 Sn + 4 M \quad (M = Sb, Bi) \qquad (36)$$

The uncatalysed bismuth reaction, which takes place at a significantly lower temperature, gives a distannane:

$$2(Et_3Sn)_3Bi \xrightarrow{160 °C} 2 Bi + 3 Et_3SnSnEt_3 \qquad (37)$$

IV. GROUP 15 LIGANDS

A. Nitrogen Ligands

Some interest has been shown in using aminoarsines, $As(NR_2)_3$, as MOCVD arsenic precursors, since they are volatile and *relatively* less toxic than AsH_3[60]. $Bi(NR_2)_3$ (sublimes at 10^{-2} torr at $30\,°C$) is also being investigated[61]. The thermolysis of azido arsines is a process to approach with extreme caution since explosions with azides are not unknown! Sowerby and Revitt[62] prepared diarsines by thermolytic nitrogen elimination:

$$R_2AsN_3 \xrightarrow{\;220\,°C\;} R_2As\!\!-\!\!AsR_2 + N_2 \qquad (38)$$
$$R = Me,\ Et,\ Ph$$

The above reaction only occurs where there is one phenyl ligand. In the case of Ph_2AsN_3 an AsN heterocycle, $[As(Ph)N]_4$[63] is obtained.

The mean arsenic–nitrogen bond energies in arsenic trihalide AsX_3 ($X = Cl$, Br, I) complexes with nitrogen ligands range from 121 to $168\,kJ\,mol^{-1}$[64]. Substitution of Ph for one of the halogens gives mean energies of the As—N bond in $PhAsX_2$–nitrogen ligand complexes in the range 113 to $217\,kJ\,mol^{-1}$[65]. The $PhAsCl_2$ complexes have weaker As—N bonds than those of $AsCl_3$, while $PhAsI_2$ complexes are stronger than those of AsI_3. The adducts $PhAsX_2\cdot L$ (L = pyridine, β- or α-picoline; $X = Cl$, Br, I) were all unstable in the liquid phase, the thermograms showing 100% weight loss, usually in one step in the range 350–480 °C.

Arsenic phosphorus heterocycles disproportionate on thermolysis[66,67].

$$\overline{As(Bu^t)P(Bu^t)P}(Bu^t) \xrightarrow{\;150\,°C\;} (PBu^t)_4 + (AsBu^t)_3 \qquad (39)$$

The antimony analogue decomposes completely at room temperature in 2 h.

The thermolysis of cyclopolyarsines has been briefly reviewed[68]. In general $(AsR)_n$ compounds decompose into the arsine AsR_3 and elemental arsenic As_4 or the diarsine R_2As—AsR_2 and As_4.

$$(PhAs)_6 \xrightarrow{\;196\,°C\;} AsPh_3 + As_4 \qquad (40)$$

$$(MeAs)_5 \xrightarrow{\;180\,°C/90\,h\;} AsMe_3 + As_4 \qquad (41)$$

West and coworkers[69] on the basis of $^1H\,NMR$ spectroscopy suggest that $(MeAs)_5$ is in rapid equilibrium with open-chain oligomers, which are in turn in slow equilibrium with higher polymeric species:

$$(MeAs)_5 \underset{rapid}{\overset{100\,°C}{\rightleftharpoons}} MeAs(AsMe)_xAsMe \underset{slow}{\rightleftharpoons} (MeAs)_n \qquad (42)$$
$$x = \text{low integer}$$

The thermolysis of $(Bu^tAs)_5$ affords polycyclic arsines such as $Bu^t_6As_8$[70] (5).

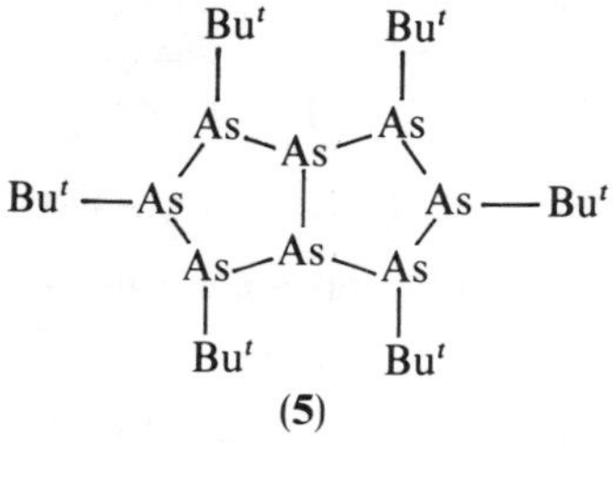

(5)

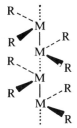

FIGURE 7. Ordered stacking of pnictogen atoms in thermochromic compounds

Antimony and bismuth both form compounds of the type R_2M—MR_2 (M = Sb, Bi) many of which exhibit reversible thermochromic behaviour[71]. This property has been ascribed to the M...M bonding in the solid phase which is lost on melting (Figure 7).

The early examples of these compounds indicated that bismuth and antimony analogues behaved similarly. Recently, however, Buchwald and coworkers[72] provided the first example of antimony and bismuth compounds (6) which behaved differently.

$$M = Sb, Bi$$
$$R = Me, SiMe_3$$

(6)

The bistiboles are orange solids that melt to orange liquids. In contrast, the bibismoles are lustrous deep-green crystals that melt to give, reversibly, orange (R = Me) and red (R = Me$_3$Si) oils.

The thermolysis of $(Me_2Sb)_2$ gives $SbMe_3$ and elemental antimony:

$$3\,Me_2Sb\cdot SbMe_2 \xrightarrow{\;200\,°C/20\,h\;} 4\,Me_3Sb + 2\,Sb \qquad (43)$$

The perfluoromethyl analogue decomposes slowly at room temperature.

The arsenic(V) compounds $R_3As{=}NH$ on thermolysis retain the nitrogen ligand[73]:

$$Bu_3^tAs{=}NH \xrightarrow{\;100\,°C\;} Bu_2^tAsNH_2 + Me_2C{=}CH_2 \qquad (44)$$

The bismuth(V) species $Ph_3Bi(N_3)_2$ decomposes slowly if stored at room temperature. On heating, it is reduced to Bi(III) by loss of PhN_3[74].

$$Ph_3Bi(N_3)_2 \xrightarrow{\;\Delta\;} Ph_2BiN_3 + PhN_3 \qquad (45)$$

Continued heating causes the azide to disproportionate:

$$Ph_2BiN_3 \xrightarrow{\;\Delta\;} Ph_3Bi + Bi + N_2 \qquad (46)$$

V. GROUP 16 LIGANDS

A. Oxygen Ligands

The pyrolysis of organoarsenic compounds containing the arsenyl moiety has some limited preparative applications [arsenyl (As=O) by analogy with phosphoryl (P=O)]. The compounds are based on the arsonic acid $RAs(O)(OH)_2$, the arsinic acid $R_2As(O)OH$ and the arsine oxide $R_3As=O$ structures. The acids are in interesting contrast with the phosphorus series. The phosphonic and phosphinic esters are prepared from the phosphorus(III) precursors via the Arbuzov synthesis. This synthetic route fails with the arsenic analogues, and further, if an alkyl halide or a salt is added in the pyrolysis of arsonic or arsinic acid esters a retro-Arbuzov reaction takes place[75,76].

The thermolysis of the arsonic acids affords initially an anhydride and, on further heating, a polymeric anhydride[77,78]:

$$MeAs(O)(OH)_2 \xrightarrow{\Delta} Me(OH)As(O)O(O)As(OH)Me \xrightarrow{\Delta} (MeAsO_2)_n \quad (47)$$

The pyrolysis of arsonic acid esters results in low yields of arsenic(III) esters.

$$RAs(O)(OR)_2 \xrightarrow{115-282\,^{\circ}C} As(OR)_3\,(2-11\%) + RAs(OR)_2\,(6-40\%) \quad (48)$$

Some examples are given in Table 3. For the esters $RAs(O)(OR')_2$ an empirical stability series for the organic ligand R of the arsenic was established for this pyrolysis.

$$Bu^i > Bu^n > Et > Me > Bz > allyl > Pr^i$$

It has already been noted that the esters of arsinous and arsonous acids disproportionate of heating. The formation of $As(OR)_3$ in the pyrolysis of arsonic acid derivatives is thought to be the product of the secondary disproportionation reaction of the primary product $RAs(OR)_2$.

The esters of arsinic acid are more thermally labile, often resinifying on distillation. However, catalysed pyrolyses using alkyl halides, salts or mineral acids afford useful yields (ca 50%) of arsenic(III) esters[75]. The mechanism is considered to involve the pyrolysis of an alkoxyarsonium salt 7 is formed with the catalyst.

$$R_2As(O)OR' + R''CH_2X \longrightarrow [R_2As(OR')(OCH_2R'')]^+X^- \quad (49)$$

7

There are three main decomposition pathways for 7: elimination of (i) RX, (ii) R'X or (iii) R''CHO shown in Scheme 8. The route via (i) has been termed a retro-Arbuzov reaction. The decomposition pathway is structure-sensitive, route (i) being characteristic for R = alkyl. When R = aryl, route (ii) is favoured especially if the volatile halide is

TABLE 3. The pyrolysis of the esters of arsonic acids

Arsonic acid ester	Pyrolysis temperature (°C)	Pyrolysis product	Reference
$MeAs(O)(OMe)_2$	225	$As(OMe)_3$	79
$MeAs(O)(OEt)_2$	boiling point	$MeOAs(OEt)_2$	79
$MeAs(O)(OBu)_2$	200	$MeAs(OBu)_2$ $BuOH$	80
$EtAs(O)(OBu)_2$	200	$EtOAs(OBu)_2$	81
$PrAs(O)(OEt)_2$	< 200	$PrOAs(OEt)_2$	81
$PrAs(O)(OBu)_2$	200	$PrAs(OBu)_2$ $PrOAs(OBu)_2$	81

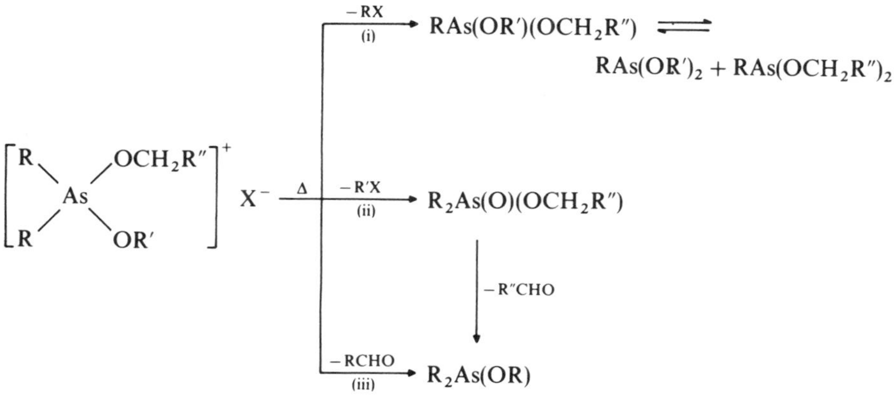

$$
\left[\begin{array}{cc} R & OCH_2R'' \\ & As \\ R & OR' \end{array}\right]^+ X^- \xrightarrow{\Delta}
\begin{array}{l}
\xrightarrow[\text{(i)}]{-RX} RAs(OR')(OCH_2R'') \rightleftharpoons \\
\qquad\qquad RAs(OR')_2 + RAs(OCH_2R'')_2 \\[1em]
\xrightarrow[\text{(ii)}]{-R'X} R_2As(O)(OCH_2R'') \\
\qquad\qquad\Big\downarrow {-R''CHO} \\[0.5em]
\xrightarrow[\text{(iii)}]{-RCHO} R_2As(OR)
\end{array}
$$

R = alkyl, aryl; R' = alkyl; R'' = alkyl, aryl

SCHEME 8. The alternative reaction routes of the intermediate alkoxyarsonium salts in the alkyl halide catalysed decomposition of the esters of arsinic acids[75,76]

removed. Route (iii) is favoured by higher temperatures. The arsenic(III) derivative of perfluoroacetic acid on thermolysis affords an arsine incorporating the trifluoromethyl ligand[82].

$$CF_3C(O)OAsMe_2 \xrightarrow{\Delta} CF_3AsMe_2 \tag{50}$$

The pyrolysis of arsine oxides affords reduction products, mainly arsines with lesser amounts of dialkylarsinite esters. Disproportionation reactions occur in these reactions, shown by the isolation of Bu_3As from the pyrolysis of $Bu_2MeAs{=}O$. Some examples are given in Table 4.

Hydroxytetraarylstiboranes of the type $(XC_6H_4)Ph_3SbOH$ undergo thermolysis at 30–50 °C in p-xylene with the elimination of PhH and C_6H_5X, resulting in a mixture of $(XC_6H_4)Ph_2Sb{=}O$ and $Ph_3Sb{=}O$[87]. McEwen and coworkers[87] proposed the radical mechanism shown in Scheme 9. However, Barton and coworkers[88] showed that radical trapping with $Ph_2C{=}CH_2$ caused no change in the observed yield of C_6H_6 seemingly excluding a radical mechanism.

$$(XC_6H_4)Ph_3SbOH \xrightarrow{\Delta} (XC_6H_4)Ph_3SbO^{\cdot}$$

$$(XC_6H_4)Ph_3SbO^{\cdot} \longrightarrow (XC_6H_4)Ph_2Sb{=}O + Ph^{\cdot}$$

$$(XC_6H_4)Ph_3SbOH + Ph^{\cdot} \longrightarrow (XC_6H_4)Ph_3SbO^{\cdot} + PhH$$

$$(XC_6H_4)Ph_3SbO^{\cdot} \longrightarrow Ph_3Sb{=}O + XC_6H_4^{\cdot}$$

$$(XC_6H_4)Ph_3SbOH + XC_6H_4^{\cdot} \longrightarrow (XC_6H_4)Ph_3SbO^{\cdot} + C_6H_5X$$

SCHEME 9. Proposed mechanism for the decomposition of hydroxytetraarylstiboranes[87]

$[Ph_3Sb(Br)O]_2$ on thermolysis in chlorobenzene at 45 °C affords $[Ph_3SbBr]_2O$ and singlet oxygen, proposed on the basis of trapping experiments[89] with α-terpinene.

TABLE 4. Pyrolysis of arsine oxides $R_3As=O$

Arsine oxide	Main pyrolysis identified	Reference
Me_2BuAsO	Me_2BuAs, $MeBu_2As$	83
Et_3AsO	Et_3As, EtOH	84
Pr_3AsO	Pr_3As, PrOH	85
$MeBu_2O$	$MeBu_2As$, Bu_3As	86
$Ph_2(HCO_2CH_2CH_2)AsO$	$(Ph_2As)_2O$, EtCOOH	86

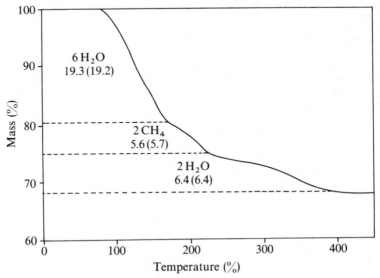

$$(51)$$

Stiboranes with two oxygen ligands of the type $Ph_3Sb(OAr)_2$ undergo a two-stage thermolysis[90]. The first stage is dependent on the structure of the aryl ligand.

$$Ph_3Sb(OAr)_2 \xrightarrow{\Delta} Ph_3Sb + \text{other products} \qquad (52)$$

The second stage, the decomposition of Ph_3Sb, affords elemental Sb in N_2 atmosphere and Sb_2O_3 in air.

Disodium-di-μ-oxo-bis[trihydroxymethylantimonate(V)] was shown to contain the

FIGURE 8. Thermogravimetric analysis of $Na_2[Me(OH)_3SbO]_2 \cdot 6 H_2O$ (theoretical weight loss in parentheses)

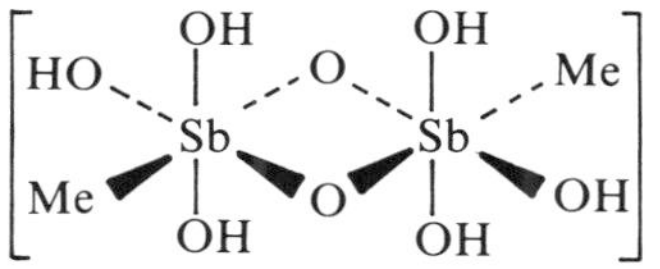

$$\left[\begin{array}{c} \text{HO} \underset{\text{Me}}{\overset{\text{OH}}{\underset{|}{\overset{|}{\text{Sb}}}}} \cdots \text{O} \cdots \underset{\text{OH}}{\overset{\text{OH}}{\underset{|}{\overset{|}{\text{Sb}}}}} \cdots \text{Me} \\ \text{O} \quad \text{OH} \end{array}\right]$$

(8)

anion **8** by X-ray crystallography[91]. Thermogravimetric analysis (Figure 8) is consistent with Scheme 10.

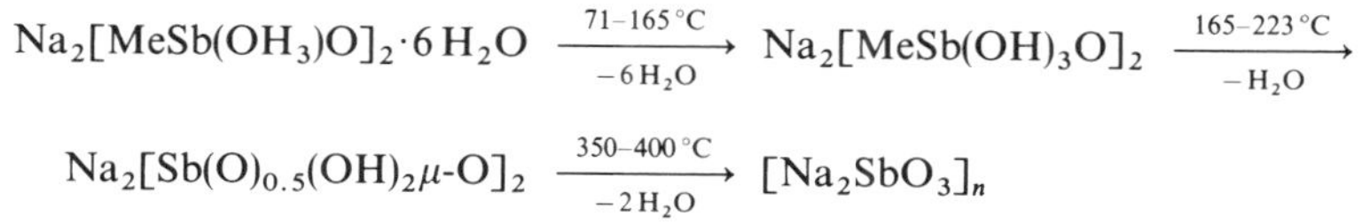

$$Na_2[MeSb(OH_3)O]_2 \cdot 6\,H_2O \xrightarrow[-6\,H_2O]{71-165\,°C} Na_2[MeSb(OH)_3O]_2 \xrightarrow[-H_2O]{165-223\,°C}$$

$$Na_2[Sb(O)_{0.5}(OH)_2\mu\text{-}O]_2 \xrightarrow[-2\,H_2O]{350-400\,°C} [Na_2SbO_3]_n$$

SCHEME 10. The thermolysis of $Na_2[MeSb(OH)_3O]_2 \cdot 6\,H_2O$[91]

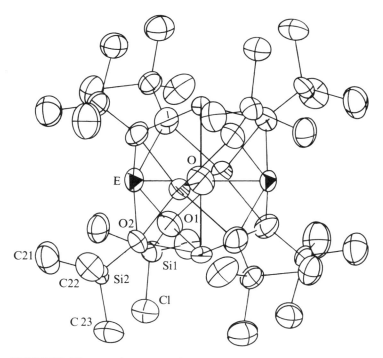

FIGURE 9. The crystal structure of $Na_4[Sb_2O(OSiMe_3)_8]$[95] (ORTEP; 50% probability ellipsoids) with atomic numbering. Atoms E designate disordered Na and Sb atoms (ratio 4:2) at the vertices of a pseudooctahedron; the oxygen atoms of the eight trimethylsilyloxy groups are capping the octahedral faces and thus forming a cube. The hydrogen atoms have been omitted for clarity. See also Figure 10

548

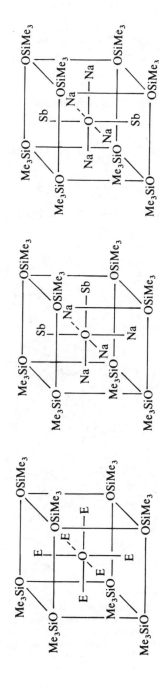

FIGURE 10. Schematic representations of the disordered structure, and of the *cis* and *trans* structures of the heterosiloxane $Na_4[Sb_2O(OSiMe_3)_8]$

Antimony(III) alkoxides and related compounds have been proposed as antimony precursors for MOCVD and sol-gel processes. The sol-gel technique is a method of obtaining a film of known composition on a substrate. The technique consists of dipping the substrate in a solution of the precursor(s) to afford a liquid film. Removal of most or all of the solvent results in a gel or a xerogel of the precursor as a uniform film on the substrate. Controlled thermolysis affords the desired film.

However, it is noteworthy that the thermolysis of $(Me_3SiO)_3Sb$ under a non-oxidizing atmosphere (argon) affords $SbMe_3$ and not the desired Sb_2O_3 in MOCVD processes[92].

$$(Me_3SiO)_3Sb \xrightarrow{\text{220-235 °C/Ar}} SbMe_3 + [Si(Me_2)O]_3 \tag{53}$$

The thermolysis is strongly affected by structural features; e.g. Westin and Nygren[93] investigated the thermolysis of antimony-containing binary alkoxides with transition metals. The compounds approximate the composition $MSb(OEt)_5$, M = Mn, Fe, Co and Ni. These materials are suitable for sol-gel processes. The gels were heated in the temperature range 50–950 °C and quenched at various temperatures to be examined spectroscopically (I.R. and X-ray powder diffraction). The thermolysis was carried out in both N_2 and air. The final products are shown in the equations below:

$$MSb(OEt)_5 \xrightarrow{\text{950 °C/Air}} MSb_2O_6 \tag{54}$$

$$MSb(OEt_5) \xrightarrow{\text{400 °C/N}_2} MSb_2O_4 \xrightarrow{\text{950 °C/N}_2} MO_x + Sb_2O_3 \text{ (sublimes)} \tag{55}$$

Tris(trialkylsilyl)antimonites(III), $(R_3SiO)_3Sb$, have been proposed as modifies for poly(siloxanes)[94] and, more recently, as single-source precursors for the deposition of antimony-containing silicate glasses used in the surface technology of semiconductors[92]. These compounds afford a product with an antimony-to-silicon ratio of 1:3. Schmidbaur and coworkers[95] utilized the available lone pair on these antimonites to prepare $Na_4Sb_2O(OSiMe_3)_8$ with an antimony-to-silicon ratio of 1:4. The material which has a high apparent symmetry (Figures 9 and 10) decomposes at 120 °C. For sol-gel processes intolerant of sodium, the cation would have to be replaced by a moiety such as R_4N, which is volatilized on pyrolysis.

At First sight, it may be thought that the decomposition of mixed alkoxides should be a series of independent reactions. Trunova and coworkers[96] have shown that the decomposition of 'antimony glycerate' $Sb(C_3H_5(OH)_2O)_2Cl$ is not independent in the co-pyrolysis with $Si(OEt)_4$ (see Figure 11). Further, when lanthanum isopropoxide $La(OPr^i)_3$ was added to the antimony and silicon precursors the thermal behaviour was again changed (see Figure 12).

Polymeric materials are ubiquitous and vital to our modern world. However, they often represent a fire hazard and important components in their formulations are various flame and smoke retardants. One of the more important additives in this respect are various inorganic antimony compounds, especially Sb_2O_3. Although these compounds fall outside the scope of this book, this important branch of antimony chemical technology deserves at least a mention and the interested reader is referred to some recent papers.[97-99]

Bismuth-containing superconductors, such as $Bi_2Sr_2CaCu_2O_x$ and $Bi_2Sr_2Ca_2Cu_3O_x$ with transition temperatures 80 and 110 K, respectively, were originally synthesized by grinding and firing the mixed oxides. In order to improve the quality control in their preparation, various routes using metal–organics have been suggested, e.g. via mixed oxalates[100] or poly(acrylic acid)[101] derivatives for bulk quantities of the superconductors.

The mixed oxalates show a gradual weight loss between 100–220 °C and two exotherms at 200 °C and 350 °C associated with the loss of organics. A further exotherm at 800 °C

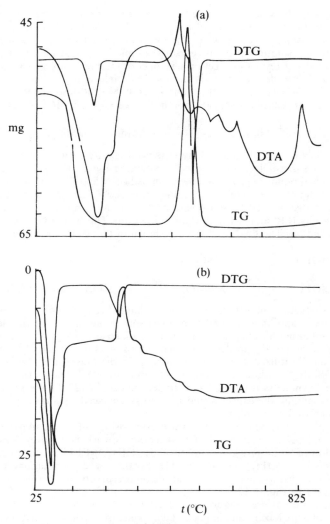

FIGURE 11. Derivatograms of the thermal decomposition of antimony glycerate (a) and of the binary compound $Sb(O_3C_3H_7)_2Cl \cdot Si(OC_2H_5)_4$(b)

due to the decomposition of carbonates $SrCO_3$, etc. is observed, the T_c phase forming at $> 800\,°C$[102].

Rojo and coworkers[103] have explored the use of ethylenediaminetetraacetic acid (EDTA) and diethylenetriaminepentaacetic acid (DTPA) salts as precursors for the preparation of bismuth-containing superconductors. The pyrolysis of the mixed lead–bismuth complex $PbBi(DTPA) \cdot 3\,H_2O$ was shown to proceed in three stages: firstly the dehydration of the complex (32–150 °C) followed by the loss of organics starting at 300 °C, and finally the decomposition of the residual ill-defined carbonates up to 500 °C affording a mixture of PbO and an unidentified new Bi—Pb—O phase.

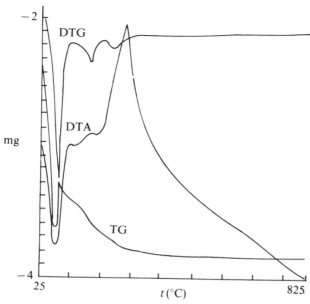

FIGURE 12. Derivatogram of thermal decomposition of the ternary compound $Sb(O_3C_3H_7)_2Cl \cdot Si(OC_2H_5)_4 \cdot La(OC_3H_7)_3$

Interest in bismuth alkoxides has increased owing to their potential as Bi precursors in the MOCVD preparation of bismuth-containing superconductors[104]. The ethoxide $Bi(OEt)_3$ has been used in a sol-gel process with plasma oxidation[105]. $Bi(OC(Me_2)CH_2\,CH_3)_3$ has been proposed as particularly suited to MOCVD processes[106], subliming in the range 60–90 °C at 10^{-4} torr. Thermal decomposition under N_2 at 200 °C followed by annealing at 600 °C gave mainly $\alpha\text{-}Bi_2O_3$ with some $\gamma\text{-}Bi_2O_{3+x}$.

B. Sulfur Ligands

The pyrolysis of arsine sulphides ($R_3As{=}S$) is similar to that of the oxide analogues inasmuch as only low yields of arsinous thioesters are obtained, i.e. it is not a synthetically significant reaction.

$$Et_2PhAs{=}S \xrightarrow{\Delta} PhEtAs{-}SEt \quad \text{yield 6–9\%} \tag{56}$$

The halobis(piperidyldithiocarbamates) of As(III), Sb(III) and Bi(III) have been studied by TG, DTG and DTA[107] both in air and in N_2. The halogens were Cl, Br and I, and the temperature range up to 560 °C. In the case of the arsenic–iodine analogue in air, the first ligand loss was of iodine and the final product was As_2S_3. In a nitrogen atmosphere the final product was As_4. The antimony–bromine analogue in contrast lost, in air, CS_2 and Br initially and gave a residue of Sb_2O_4. The bismuth–iodine analogue, also in air, lost CS_2 and iodine initially and the residue was $(BiO)_2SO_4$.

Antimony(III) diethyldithiocarbamate melts sharply at 136–137 °C; however, after cooling a redetermination of the melting points gives variable results[108]. This behaviour is diagnostic of thermal decomposition. Thermogravimetry revealed a 71% ligand loss in the range 253–298 °C, followed by two ill-defined decomposition processes probably involving antimony sulphides and finally a residue of 13% formed above 525 °C corresponding to the formula stoichiometry.

$SbI_3 \cdot tmtu$ [tmtu = tetramethylthiourea, $(Me_2N)_2C{=}S$] decomposes above its melting point and TG study showed a complete weight loss in the range 155–435 °C[109] (cf the arsenic analogue)[108].

The dithiastibepin **9** eliminates Ph_3Sb at its melting point or in refluxing benzene[110].

$$ \text{(9)} \qquad\qquad\qquad\qquad\qquad\qquad\qquad\qquad + SbPh_3 \qquad (57) $$

Similar reductive thermolyses were observed by Schmidbaur and Mitschke[111].

$$ Me_4SbSR \xrightarrow{\;RT\;} Me_3Sb + RSMe \qquad (58) $$
$$ R = Me, Et, Ph, Bz $$

$$ Me_3Sb(SR)_2 \xrightarrow{\;RT\;} Me_3Sb + RS{-}SR \qquad (59) $$

The thermolytic elimination of stibines was also noted by Wardell and Grant[112].

$$ Ph_4SbSC_6H_4Y \xrightarrow{\;\Delta\;} Ph_3Sb + PhSC_6H_4Y \qquad (60) $$
$$ Y = H, 4\text{-Me}, 4\text{-Br}, 2\text{-MeO}, 4\text{-MeO} $$

Other product were also found, among them PhH, $YC_6H_4SSC_6H_4Y$ and $Ph{-}Ph$.

In the series As, Sb and Bi the mean enthalpy of the pnictogen–sulphur bond, $\bar{D}(E{-}S) = \Delta_{ho}H/6$ (assuming a resonance structure equilizing the $E{-}S$ bonds and neglecting the reorganizing energy of the ligand on complex formation), decreases with increase in the mean $(E{-}S)$ bond length[108].

The thermal behavour of bismuth(III) tri(diethyldithiocarbamate), $Bi(S_2C \cdot NEt_2)_3$, is similar to that of its antimony analogue[108], i.e. although it has a sharp melting range (193–194 °C) the latter is not reproducible when cooled and redetermined. TG analysis showed a 64% ligand loss in the range 283–318 °C. The residue was a mixture of bismuth metal and the sulphide Bi_2S_2[108].

The pyrolysis of bismuth dithiocarbomates has commercial potential for the production of Bi_2S_3 films, which are useful for photodiode arrays or photovoltaics. Normura and coworkers[113,114] developed techniques for solution pyrolysis and its application to the formation of thin films. They investigated four bismuth dithiocarbomates (Table 5) for the production of films of Bi_2S_3[115]. To obtain useful films a solution of the dithiocarbamate in p-xylene or chloroform is heated on a glass substrate at 100 °C for 1 h, cooled to room temperature and finally heated at 250–350 °C

TABLE 5. Thermal analysis data of bismuth dithiocarbamates

Compound	Conventional melting point (°C)	DTA melting endotherm (°C)	DTA decomposition endotherm (°C)
$Bi(S_2CNEt_2)_3$	194–196	192	328
$Bi(S_2CNBu_2{}^n)_3$	98–99	99	317
$Bi(S_2CNHex_2{}^n)_3$	80–81	79	313
$Bi(S_2CN(H)Bu^t)_3$	140 decomp.	146	ca 150

for 1 h. The thin films obtained are smooth and homogeneous with a surface morphology resembling those from the vacuum evaporation of Bi_2S_3.

TG analysis of the adduct $BiI_3 \cdot tmtu$ [tmtu is $(Me_2N)_2C=S$] did not show any steps, just a continuous loss of weight with no residue over the temperature range 203 °C to 697 °C[116].

Wieber and Sauer observed that organo-bismuth sulphur compounds of the type R_2BiSAr undergo ligand redistribution on thermolysis[117]:

$$Me_2BiSPh \xrightarrow{90\,°C/15\,h} Me_3Bi + MeBi(SPh)_2 \tag{61}$$

C. Selenium and Tellurium Ligands

Breunig and Güleç[118,119] used the thermal redistribution reaction of R_2SbER' (E = S, Se, Te) to produce $RSb(ER')_2$. This thermolysis is preparatively useful since the stibines can be removed by distillation, driving the equilibrium to the right.

$$2\,R_2SbER' \xrightarrow{80-145\,°C} R_3Sb + RSb(ER')_2 \tag{62}$$

$R, R' = Me; \; E = Se$
$R = Me, \; R' = Ph; \; E = S, Se$
$R = Et, \; R' = Me, Ph; \; E = S, Se$

Prolonged heating in the case of $Me_2SbSeMe$ afforded $(MeSe)_3Sb$[118]. With the tellurides, the method is not synthetically useful owing to the number of side reactions yielding, among others, SbR_3, R_2^1Te, RR^1Te, $R^1Te-TeR^1$ and $RSb(TeR')_2$[118].

The analogous bismuth–selenium compounds, $R_2BiSeAr$, disproportionate slowly at room temperature and afford preparatively useful yields under thermolytic conditions[117].

$$2\,Me_2BiSePh \xrightarrow{80\,°C/10\,h} Me_3Bi + MeBi(SePh)_2 \quad (100\%) \tag{63}$$

However, although the bismuth–tellurium analogues have been prepared and are stable at $-25\,°C$, their decomposition is complex and not preparatively useful for bismuth derivatives, e.g. $Pr_2BiTeTol$-p affords Pr^nTeTol-p as the main tellurium-containing product.

VI. GROUP 17 LIGANDS

A. Arsenic Compounds

Haloorganopnictogen(III) compounds on heating will, in common with most unsymmetrically substituted pnictogen species, disproportionate to symmetrical molecules. However, with some structures other reaction pathways open up. In the case of compounds of the type $1,2$-$(R(X)AsC_6H_4As(X)R)$ thermolysis affords the aesthetically symmetrical triptycene $As_2(C_6H_4)_3$[120].

$$\tag{64}$$

TABLE 6. The pyrolysis of dihaloarsoranes

Dihaloarsirane	Pyrolysis temperature ($^\circ$C)	Pyrolysis products	Reference
$(CF_3)_3AsCl_2$	125	$(CF_3)_2AsCl$, CF_3AsCl_2	121
Et_3AsBr_2	100	Et_2AsBr	122
$Me_2(1\text{-naphthyl})AsCl_2$	Dry distillation	$Me(1\text{-naphthyl})AsCl$	123
$Me_2(1\text{-naphthyl})AsBr_2$		$Me(1\text{-naphthyl})AsBr$	124
Ph_3AsCl_2	230/vac	Ph_2AsCl	122, 125
Cy_3AsCl_2	230	Cy_2AsCl	122

The dihaloarsoranes are readily available and consequently much investigated. It has been observed that in unsymmetrical species it is the smaller of the alkyl groups that is lost on pyrolysis.

$$CF_3Ph_2AsBr_2 \xrightarrow{120\,^\circ C} Ph_2AsBr + BrCF_3 \qquad (65)$$

$$MePh(m\text{-Tol})AsCl_2 \xrightarrow{140-170\,^\circ C} Ph(m\text{-Tol})AsCl \qquad (66)$$

Further examples are given in Table 6 and may be summarized in the general equation 67.

$$R_3EX_2 \xrightarrow{\Delta} R_2EX + RX \qquad (67)$$

Equation 67 showing the formation of R_2EX from R_3EX_2 invites the intuitive supposition that the mechanism is an intramolecular elimination of RX. However, R_3EX_2 molecules are usually trigonal bipyramidal with the halogens occupying the axial sites, leaving aside those with ionic, $[R_3EX]^+X^-$, and spoke, R_3E—X—X^{126}, structures. Thus, an intramolecular process requires an axial–equatorial elimination which is symmetry-forbidden for a thermal reaction[127,128]. Elbel and coworkers[129], investigating the gas-phase pyrolysis of Me_3EX_2 (E = As, Sb; X = Cl, F), found that HX and not MeX was eliminated under conditions favouring unimolecular processes. They suggested a mechanism involving a loss of HX followed by a rearrangement (equation 68).

$$Me_3ECl_2 \xrightarrow[-HCl]{\Delta} Me_2E(Cl)CH_2 \longrightarrow Me_2ECH_2Cl \qquad (68)$$

It is interesting to note in this context that the reductive pyrolysis of Me_3SbX_2 is synthetically useful only for X = I (see below). for X = Cl, yields of Me_2SbCl are low, and Me_3SbCl_2 has even been sublimed under mild conditions without decomposition.

B. Antimony Compounds

The mobility of halide ligands on antimony has also been used preparatively. The thermolysis of R_2SbX was used by Breunig and Kanig[130] to prepare $MeSbI_2$.

$$Me_2SbI \xrightarrow{\Delta} Me_3Sb + MeSbI_2 \qquad (69)$$

The thermolysis of dihalostiboranes affords broadly similar products to their arsenic analogues; see Table 7.

The quaternary stibonium halides usually afford Sb(III) species[131], e.g.

$$[(CH_2{=}CH)_3SbMe]I \xrightarrow{\Delta} (CH_2{=}CH)_3Sb + MeI \qquad (70)$$

TABLE 7. The pyrolysis of dihalostiboranes

Dihalostiborane	Pyrolysis temperature (°C)	Pyrolysis products	Reference
$(CF_3)_3SbBr_2$	room temp.	$(CF_3)_2SbBr$	136
Me_3SbBr_2	180 °C/90 torr	Me_2SbBr	137, 138
Et_3SbBr_2	220–240 °C/400 torr	Et_2SbBr	138
$Pr^i_3SbBr_2$	170 °C/90 torr	Pr^i_2SbBr	138
Cy_3SbCl_2	220 °C/350 torr	Cy_2SbCl	138

However, the reaction pathway is determined by the halide groups, e.g.[132]

$$[Ph_4As]Br_3 \xrightarrow{180°C} Ph_3SbBr_2 + PhBr \quad (71)$$

The higher halides of antimony are increasingly thermally unstable as the number of halogen ligands increases.

$$Me_2SbCl_3 \xrightarrow{\Delta} MeSbCl_2 + MeCl \quad (72)$$

$$MeSbCl_4 \xrightarrow{RT} SbCl_3 + MeCl \quad (73)$$

$$2\,PhSbCl_4 \xrightarrow{slowly\ at\ RT} Ph_2SbCl_3 + SbCl_3 + Cl_2 \quad (74)$$

The instability of R_2SbX_3 compounds has been used as a synthetic route to $RSbX_2$[133,134].

$$R_2SbX + X_2 \longrightarrow R_2SbX_3 \xrightarrow{\Delta} RSbX_2 + RX \quad (75)$$

Ph_3SbBr_2 has been evaluated as a fire retardant in poly(propene)[135]. The radicals formed on its thermolysis are the quenching agents for the radicals formed in the pyrolysis of the polypropylene.

$$Ph_3SbBr_2 \xrightarrow{\Delta} Ph_3SbBr^{\bullet} + Br^{\bullet} \quad (76)$$

C. Bismuth Compounds

The stability of compounds of the type R_3BiX_2 is strongly structure-dependent, e.g. $(C_6F_5)_3BiF_2$ is stable at -10 °C but Bu_3BiF_2 is only a transient species affording Bu_2BiF_3[139] on decomposition. In contrast, Ph_3BiF_2 is pyrolysed at 190 °C to 200 °C to give *inter alia* Ph_3Bi, Ph_2 and PhF.

Quarternary bismuth compounds $[R_4Bi]^+X^-$ are unstable at ambient temperatures although stable in the low $(-30 °C)$ ranges[140].

$$Ph_4BiCl \xrightarrow{RT} Ph_3Bi + PhCl \quad (77)$$

The thermolysis products are influenced by the thermal history, i.e. the decomposition pathway is thermally determined; e.g.[140] slow warming to $+5$ °C affords a different product to that obtained when warmed rapidly to room temperature.

$$[Ph_4Bi]Br_3 \xrightarrow{-40\ to\ +5\,°C} Ph_3BiBr_2 + PhBr \quad (78)$$

$$[Ph_4Bi]Br_3 \xrightarrow{RT} Ph_2BiBr + 2\,PhBr \quad (79)$$

VII. TRANSITION METAL COMPLEXES WITH GROUP 15 DONORS

Acyclic arsines are superior to phosphines in their ability to retain chirality, with the R_3As inversion barrier, being ca 40 kcal mol^{-1}, compared to that of R_3P, ca 30 kcal mol^{-1}. Where the thermal degradation of P-chiral transition metal catalyst is significant, the substitution of arsenic for phosphorus may lead to an acceptable stability.

The thermal decomposition of transition metal complexes is often reported as a decomposition point in the form of 'm.p. n °C (decomp.)', e.g.[141]

$$(O)PPh_2CH_2CH_2AsPh_2Re(O)(OEt)Cl_2 \text{ m.p. } 230 °C \text{ (decomp.)}$$

although the nature of the products is rarely discussed.

The complex $[Pd(AsMe_3)_2Cl_2]$ loses trimethylarsine on heating to afford the dinuclear, chloride bridged, complex $[Pd(AsMe_3)Cl_2]_2$[142].

The cobalt complex $[\{Cp*Co(CO)_2\}\mu\text{-}\eta^{1:1:1:1}\text{-}As_4]$ on heating in xylene at 140 °C loses carbon monoxide in two stages, 1 mol after 1 h, the other after 6 h. On losing CO, the cobalt adopts four arsenic ligands as shown in Scheme 11[143].

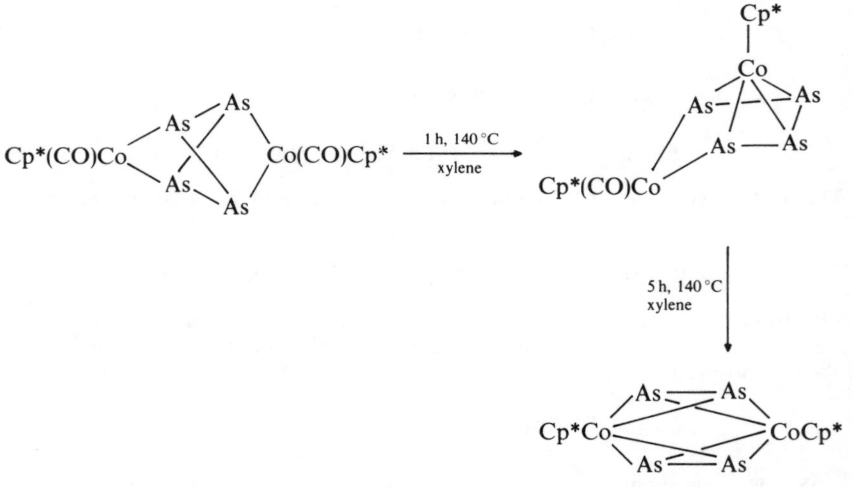

SCHEME 11. The thermolysis of $[Cp*(CO)Co)_2As_4]$ in xylene[143]

Heating the σ-bonded Bi—Mn complex 10 at its metling point for 5 h changed the coordination mode, involving loss of CO from the maganese[144].

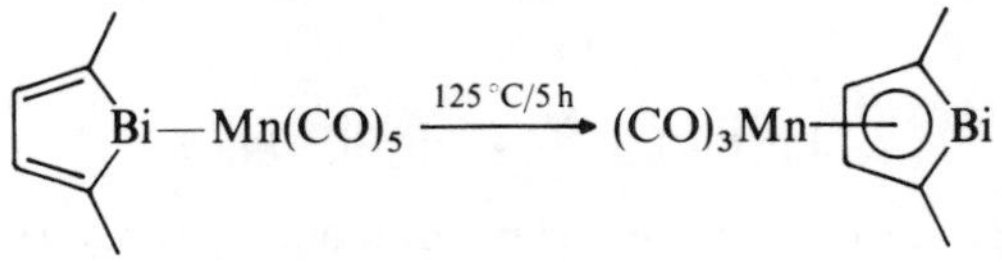

(10)   (80)

The complex $[Ph_2BiMn(CO)_5]$ was synthesized by Whitmire and Cassidy[145] as a candidate single-source precursor for the alloy MnBi, which has interesting magnetic properties. Unfortunately no thermolysis data were presented. Some bismuth–transition metal complexes are surprisingly stable, e.g.[146].

$$\left((CO)_3Mn\!-\!\boxed{}\!\Big\rangle \right)_3 Bi \xrightarrow{300\,°C} \boxed{}\!-\!\boxed{} + Bi$$

$$\underset{(CO)_3}{Mn} \qquad \underset{(CO)_3}{Mn} \qquad\qquad (81)$$

In the case of bismuth the reaction is quantitative. The antimony analogue affords Sb/Mn and volatile products.

VIII. APPENDIX: PYROLYSIS NOMENCLATURE

after I. Ericsson and R. P. Lattimer, *J. Anal. Appl. Pyrolysis*, **14**, 219 (1989)

Analytical pyrolysis: The characterization of a material or a chemical process by the intrumental analysis of its pyrolysate.

Applied pyrolysis: The production of commercially useful materials by means of pyrolysis.

Catalytic pyrolysis: A pyrolysis that is influenced by the addition of a catalyst.

Char: A solid carbonaceous residue.

Coil pyrolyser: A pyrolyser in which the sample (sometimes placed in a tubular vessel) is placed in a metal coil that is heated to induce pyrolysis.

Continuous mode (furnace) pyrolyser: A pyrolyser in which the sample is introduced into a pyrolyser preheated to the final temperature.

Curie-point pyrolyser: A pyrolyser in which a ferromagnetic sample carrier is inductively heated to its Curie point.

Filament (ribbon) pyrolyser: A pyrolyser in which the sample is placed on a metal filament (ribbon) that is resistively or inductively heated to induce pyrolysis.

Final temperature (T_f): The final (steady-state) temperature which is attained by a pyrolyser. (The terms 'equilibrium temperature' and 'pyrolysis temperature' may be used when referring to an isothermal pyrolysis; they are not recommended for use with a non-isothermal pyrolysis, however.)

Flash pyrolysis: A pyrolysis that is carried out with a fast temperature rise time (to a constant final temperature).

Fractionated pyrolysis: A pyrolysis in which the sample is pyrolysed at different temperatures for different times in order to study special fractions of the sample.

Isothermal pyrolysis: A pyrolysis during which the temperature is essentially constant.

Maximum temperature (T_{max}): The highest temperature in a temperature/time profile.

Nonisothermal pyrolysis: A pyrolysis during which the temperature varies significantly.

Oxidative pyrolysis: A pyrolysis that occurs (only) in the presence of an oxidative atmosphere.

Pyrogram: A chromatogram or spectrum of a pyrolysate.

Pyrolysis: A chemical degradation reaction that is induced by thermal energy (alone). (The term 'pyrolysis' generally refers to an inert environment.)

Pyrolysis-gas chromatography (Py-GC): A pyrolysis technique in which the volatile pyrolysates are separated and detected by gas chromatography.

Pyrolysis-gas chromatography/infrared spectroscopy (Py-GCIR): A pyrolysis technique in which the volatile pyrolysates are analysed by on-line gas chromatography/infrared spectroscopy.

Pyrolysis-gas chromatography/mass spectrometry (Py-GC/MS): A pyrolysis technique in which the volatile pyrolysates are analysed by on-line gas chromatography/mass spectrometry.

Pyrolysis-infrared spectroscopy (Py-IR): A pyrolysis technique in which the pyrolysates are analysed by *infrared* spectroscopy.

Pyrolysis-mas spectrometry (Py-MS): A pyrolysis technique in which the volatile pyrolysates are analysed by mass spectrometry.

Pyrolysis reactor: That portion of the pyrolyser in which the pyrolysis takes place.

Pyrolysis residue: That portion of the pyrolysate that does not leave the reactor.

Pyrolysate: The products of pyrolysis.

Pyrolyser: A device for performing pyrolysis.

Pulse mode pyrolyser: A pyrolyser in which the sample is introduced into a cold pyrolyser which is then heated rapidly.

Sequential pyrolysis: A pyrolysis in which the same sample is repetitively pyrolysed under identical conditions (T_f, TRT and THT) until no further products are detected.

Stepwise pyrolysis: A pyrolysis in which the sample temperature is raised stepwise, i.e. pyrolysis is carried out at a selected temperature for a selected period of time after which pyrolyser is stepped rapidly to a higer temperature.

Tar: A liquid pyrolysis residue.

Temperature-programmed pyrolysis: A pyrolysis in which the sample is heated at a controlled rate over a range of temperatures during which pyrolysis occurs.

Temperature rise time (TRT): The time required for a pyrolyser to go from its initial temperature to the final temperature.

Temperature/time profile (TTP): A graphical representation of temperature versus time for a particular pyrolysis experiment or pyrolyser.

Thermal volatilization analysis (TVA): A pyrolysis technique in which the pressure of the volatile pyrolysates is recorded as the sample is heated.

Total heating time (THT): The time between the onset and conclusion of sample heating in a pyrolysis experiment.

Volatile pyrolysate: That portion of the pyrolysate that has appreciable vapour pressure under the conditions of the pyrolysis.

IX. BIBLIOGRAPHY

1. B. J. Aylett, *Organometallic Compounds*, 4th edn., Volume 1, Part 2, Chapman Hall, London, 1979.
2. G. O. Doak and L. D. Freedman, *Organometallic Compounds of Arsenic, Antimony and Bismuth*, Wiley, New York, 1970.
3. M. Dub (Ed.), *Organometallic Compounds*, Vol. III. *Compounds of Arsenic, Antimony and Bismuth*, Springer-Verlag, New York, 1968.
4. H. J. Emeléus, R. S. Nyholm, J. C. Bailor and A. F. T. Dickenson (Eds.), *Comprehensive Inorganic Chemistry*, 1974.
5. *Gmelin Handbook of Inorganic Chemistry, Organoantimony Compounds*, Springer-Verlag, Heidelberg, 1981.
6. F. G. Mann, *The Heterocyclic Derivatives of Phosphorus, Arsenic, Antimony and Bismuth*, 2nd edn., Wiely-Interscience, New York, 1970.
7. C. A. McAuliffe and W. Levason, *Phosphine, Arsine and Stibine Complexes of the Transition Elements*, Amsterdam Elsevier, Amsterdam, 1979.
8. G. Wilkinson, F. G. A. Stone and E. W. Abel (Eds.), *Comprehensive Organometallic Chemistry*, Pergamon, Oxford, 1982.
9. G. Wilkinson, R. D. Gillard and J. A. McCleverty (Eds.), *Comprehensive Coordination Chemistry*, Pergamon, Oxford, 1987.

X. REFERENCES

1. L. K. Krannich, C. L. Watkins, D. K. Srivastava and R. K. Kanjolia, *Coord. Chem. Rev.*, **112**, 117 (1992).
2. B. C. Harrison and E. H. Tompkins, *Inorg. Chem.*, **1**, 951 (1962).
3. A. H. Cowley, R. A. Jones, D. E. Heaton, K. B. Kidd and C. M. Nunn, *Polyhedron*, **7**, 1901 (1988).
4. K. H. Theopold, S. S. Sendlinger, B. S. Haggerty and A. L. Rhiengold, *Chem. Ber.*, **124**, 2453 (1991).
5. A. H. Cowley and R. A. Jones, *Angew. Chem., Int. Ed. Engl.*, **28**, 1209 (1989).
6. D. C. Bradley, M. M. Faktor, M. Scott and E. A. D. White, *J. Cryst. Growth*, **75**, 101 (1986).
7. A. Zaouk, E. Salvetat, J. Sakaya, F. Maury and G. Constant, *J. Cryst. Growth*, **55**, 135 (1981).
8. A. H. Cowley, B. L. Benac, J. G. Ekerdt, R. A. Jones, K. B. Kidd, J. Y. Lee and J. E. Miller, *J. Am. Chem. Soc.*, **110**, 6248 (1988).
9. A. H. Cowley, A. M. Arif, B. L. Benac, R. L. Geerts, R. A. Jones, K. B. Kidd, J. M. Power and S. T. Schwarb, *J. Chem. Soc., Chem. Commun.*, 1543 (1986).
10. A. H. Cowley, R. A. Jones, M. A. Mardones and C. M. Nunn, *Organometallics*, **10**, 1635 (1991).
11. A. H. Cowley, P. R. Harris, R. A. Jones and C. M. Nunn, *Organometallics*, **10**, 652 (1991).
12. R. H. Wells, A. P. Purdy, K. T. Higa, A. T. McPhail and C. G. Pitts, *J. Organometal. Chem.*, **325**, C7 (1987).
13. A. M. Arif, B. L. Benac, A. H. Cowley, R. A. Jones, K. B. Kidd and C. M. Nunn, *New J. Chem.*, **12**, 553 (1988).
14. A. H. Cowley, A. R. Barron, R. A. Jones, C. M. Nunn and D. L. Westmoreland, *Polyhedron*, **7**, 77 (1988).
15. R. H. Fish, F. E. Brinkmann and K. L. Jewett, *Environ. Sci. Technol.*, **16**, 174 (1982).
16. S. W. Benson, J. T. Francis and T. T. Tsotsis, *J. Phys. Chem.*, **92**, 4515 (1988).
17. H. M. Manasevit, *J. Electrochem. Soc.*, **116**, 1725 (1969).

18. G. B. Stringfellow, *Organometallic Vapour Phase Epitaxy: Theory and Practice*, Academic Press, Boston, (1989).
19. D. M. Speckman and J. P. Wendt, *J. Appl. Phys.*, **69**, 3316 (1991).
20. J. Nishizawa and T. Kurabayashi, *J. Electrochem. Soc.*, **130**, 413 (1983).
21. D. H. Reep and S. K. Gandhi, *J. Electrochem. Soc.*, **130**, 675 (1983).
22. D. M. Speckman and J. P. Wendt, *J. Electrochem. Soc.*, **137**, 2271 (1990).
23. K. F. Jensen, P. W. Lee, T. R. Omstead and D. R. McKenna, *J. Cryst. Growth*, **93**, 134 (1988).
24. P. D. Dapkus, A. Melas, S. P. Denbaars and B. Y. Maa, *J. Electrochem. Soc.*, **136**, 2067 (1989).
25. N. Kobayashi, Y. Yamauchi and Y. Horikoshi, *J. Cryst. Growth*, **115**, 353 (1991).
26. G. B. Stringfellow, N. I. Buchan, C. A. Larsen and S. H. Li, *J. Cryst. Growth*, **98**, 309 (1989).
27. A. D. Zorin, I. A. Frolov, S. A. Nosyrev and V. S. Zaburdyaev, *J. Appl. Chem., USSR (Engl. Transl.)*, **47**, 1233 (1974).
28. V. A. Yablokov, I. A. Zelyaev, E. I. Makarov and N. S. Lokhov, *J. Gen. Chem. USSR (Engl. Transl.)*, **57**, 2034 (1987).
29. T. Kauffmann, R. Hamsen, R. Kriegesmann and A. Vahrenhorst, *Tetrahedron Lett.*, 4395, 4399 (1978).
30. T. F. Bertlitz, H. Sinning, L. Lorberth and U. Müller, *Z. Naturforsch.*, **43B**, 744 (1988).
31. C. A. Larsen, S. H. Li and G. B. Stringfellow, *Chem. Mater.*, **3**, 39 (1991).
32. T. F. Keuth and E. Veuhoff, *J. Cryst. Growth*, **68**, 148 (1984).
33. D. S. Cao, G. B. Stringfellow and Z. M. Fang, *Electronic Materials Conf.*, paper F-2, Boulder, (1991).
34. G. B. Stringfellow, K. Y. Ma, Z. M. Fang and R. M. Cohen, *J. Appl. Phys.*, **68**, 4586 (1990).
35. G. B. Stringfellow, K. Y. Ma, Z. M. Fang and R. M. Cohen, *Electronic Materials Conf.*, paper F-7, Boulder, 1991.
36. C. A. Larsen, R. W. Gedridge and G. B. Stringfellow, *Chem. Mater.*, **3**, 96 (1991).
37. C. A. Larsen, R. W. Gedridge, S. H. Li and G. B. Stringfellow, *J. Electron. Mater.*, **20**, 457 (1991).
38. C. H. Chen, K. T. Huang, D. L. Drobeck and G. B. Stringfellow, *J. Cryst. Growth*, **124**, 142 (1992).
39. J. Shin, C. H. Chen, C. T. Chiu, L. C. Lu, K. T. Huang and G. B. Stringfellow, *J. Electron. Mater.*, **22**, 87 (1993).
40. D. G. Hendershot, J. C. Pazik and A. D. Berry, *Chem. Mater.*, **4**, 833 (1992).
41. S. Elbel, H. Egsgaard and L. Carlsen, *J. Chem. Soc., Dalton Trans.*, 195 (1988).
42. E. Amberger, *Chem. Ber.*, **94**, 1447 (1961).
43. S. J. W. Price and A. F. Trotman-Dickenson, *Trans. Faraday Soc.*, **44**, 1630 (1958).
44. N. R. Parikh, T. P. Humphreys, P. K. Chiang and S. M. Bedair, *Appl. Phys. Lett.*, **53**, 142 (1988).
45. G. B. Stringfellow, W. P. Kosar and D. W. Brown, *Appl. Phys. Lett.*, **55**, 2420 (1989).
46. K. Natori, S. I. Yoshizawa, J. Yoshino and H. Kukimoto, *Jpn. J. Appl. Phys.*, **28**, L1578 (1989).
47. N. Hamaguchi, J. Vigil, R. Gardiner and P. S. Kirlin, *Jpn. J. Appl. Phys.*, **29**, L596 (1990).
48. K. Kobayashi, S. Ichikawa and G. Okada, *Jpn. J. Appl. Phys.*, **28**, L2165 (1989).
49. M. Fatemi, D. K. Gaskill, A. D. Berry, R. T. Holm, E. J. Cukauskas and R. Kaplan, *J. Cryst. Growth*, **92**, 344 (1988).
50. B. W. Wessels, T. J. Marks, J. Zhang, J. Zhao, H. O. Marcy, L. M. Tonge. and C. R. Kannewurf, *Appl. Phys. Lett.*, **54**, 1166 (1989).
51. G. Wittig and D. Hellwinkel, *Chem. Ber.*, **97**, 769 (1964).
52. V. A. Dodonov, O. P. Bolotova and A. V. Gushchin, *J. Gen. Chem. USSR (Engl. Transl.)*, **58**, 711 (1988).
53. H. Schmidbaur and W. Tronich, *Inorg. Chem.*, **7**, 168 (1968).
54. N. A. Nesmeyanov, V. V. Pravdina and A. O. Reutov, *J. Org. Chem. USSR (Engl. Transl.)*, **3**, 574 (1967).
55. A. W. Johnson and H. Schubert, *J. Org. Chem.*, **35**, 2678 (1970).
56. D. Lloyd, I. Gosney and R. A. Ormiston, *Chem. Soc., Rev.*, **16**, 45 (1987).
57. F. G. Mann and E. R. H. Jones, *J. Chem. Soc.*, 401, 405, 411 (1955).
58. T. Kikkawa, T. Ohori, H. Tanaka, K. Kasai and J. Komeno, *J. Cryst. Growth*, **115**, 448 (1991).
59. N. S. Vyazankin, G. A. Razuvaev, O. A. Kruglaya and G. S. Semchikova, *J. Organomet. Chem.*, **6**, 474 (1966).

60. G. Zimmermann, H. Protzmann, T. Marschner, O. Zsebok, W. Stolz, E. O. Gobel, P. Gimmnich, J. Lorberth, W. Richter, T. Filz and P. Kurpas, *J. Cryst. Growth*, **129**, 37 (1993).
61. N. C. Norman, W. Clegg, N. A. Compton, R. J. Errington, G. A. Fisher, M. E. Green and D. C. R. Hockless, *Inorg. Chem.*, **30**, 4680 (1991).
62. D. B. Sowerby and D. M. Revitt, *J. Chem. Soc., Dalton Trans.*, 847 (1972).
63. W. T. Reichle, Tetrahedron, *Lett.*, 51 (1962).
64. P. O. Dunstan and C. Airoldi, *J. Chem. Eng. Data*, **33**, 93 (1983).
65. P. O. Dunstan, *Thermochim. Acta*, **197**, 201 (1992).
66. M. Baudler and S. Klautke, *Z. Naturforsch.*, **36B**, 527 (1981).
67. M. Baudler and S. Klautke, *Z. Naturforsch.*, **38B**, 121 (1983).
68. L. R. Smith and J. L. Mills, *J. Organomet. Chem.*, **84**, 1 (1975).
69. B. O. West, P. S. Elmes and S. Middleton, *Aust. J. Chem.*, **23**, 1559 (1970).
70. M. Bundler and P. Bachmann, *Z. Anorg. Allg. Chem.*, **485**, 129 (1982).
71. A. J. Ashe III, *Adv. Organomet. Chem.*, **30**, 77 (1990).
72. S. L. Buchwald, R. E. v. H. Spence and D. P. Hsu, *Organometallics*, **11**, 3492 (1992).
73. W. Axmacher, B. Ross and W. Marzi, *Chem. Ber.*, **113**, 2928 (1980).
74. R. G. Goel and H. S. Prasad, *J. Organomet. Chem.*, **50**, 129 (1973).
75. V. S. Gamayurova, *Russ. Chem. Rev. (Engl. Transl.)*, **50**, 836 (1981).
76. B. E. Abalonin, *Russ. Chem. Rev. (Engl. Transl.)*, **60**, 1346 (1991).
77. M. R. Smith, K. J. Irgolic and R. A. Zingaro, *Thermochim. Acta.*, **4**, 1 (1972).
78. M. Durand and J. P. Laurent, *J. Organomet. Chem.*, **77**, 225 (1974).
79. B. A. Arbuzov and M. K. Saikina, *Zh. Fiz. Khim.*, **34**, 2344 (1960).
80. L. M. Werbel, T. P. Dawson, J. R. Hooton and T. E. Dalbey, *J. Org. Chem.*, **22**, 452 (1957).
81. G. Kamai and B. D. Chernokalskii, *Zh. Obshch. Khim.*, **30**, 1176 (1960).
82. W. R. Cullen and L. G. Walker, *Can. J. Chem.*, **38**, 472 (1960).
83. G. Kamai and N. A. Chadaeva, *Chem. Abstr.*, **54**, 6521 (1960).
84. G. Kamai and B. D. Chernokalskii, *Dokl. Akad. Nauk, SSSR*, **128**, 299 (1959).
85. G. Kamai and B. D. Chernokalskii, *Chem. Abstr.*, **54**, 24345 (1960).
86. F. G. Mann and R. C. Cookson, *J. Chem. Soc.*, 67 (1949).
87. W. E. McEwen, J. L. Lubinkowski and C. G. Arrieche, *J. Org. Chem.*, **45**, 2075 (1980).
88. D. H. R. Barton, J. P. Finet, C. Giannotti and F. Halley, *J. Chem. Soc., Perkin Trans. 1*, 241 (1987).
89. J. Dahlmann and K. Winsel, *Z. Chem.*, **14**, 232 (1974).
90. N. K. Jha, A. C. Misra, P. L. Maurya and P. Bajaj, *Synth. React. Inorg. Met.-Org. Chem.*, **16**, 623 (1986).
91. M. Wieber, U. Simonis and D. Kraft, *Z. Naturforsch*, **46B**, 139 (1991).
92. M. G. Voronkov, S. V. Basenko, V. Y. Vitkovskii, S. M. Nozdrya and R. G. Mirskov, *Organomet. Chem. USSR (Engl. Transl.)*, **2**, 145 (1989).
93. G. Westin and M. Nygren, *J. Mater. Sci.*, **27**, 1617 (1992).
94. M. G. Voronkov and N. F. Orlov, *Zh. Obshch. Khim.*, **36**, 347 (1966).
95. H. Schmidbaur, M. Baeir, P. Bessinger and J. Blumel, *Chem. Ber.*, **126**, 947 (1993).
96. T. A. Makotrik, E. K. Trunova and E. A. Mazurenko, *Ukr. Khim. Zh.*, **56**, 920 (1990).
97. N. Chand and S. Verma, *Fire Safety Journal*, **15**, 325 (1989).
98. D. Sallet, V. Mailhoslefievre and B. Martel, *Polym. Degrad. Stab.*, **30**, 29 (1990).
99. P. J. Gale, *Int. J. Mass Spectrom. Ion Proc.*, **100**, 313 (1990).
100. Y. Zhang, Z. Fang, M. Muhammed, K. V. Rao, V. Skumryev, H. Medelius and J. L. Costa, *Physica C.* **157**, 108 (1989).
101. J. D. Tweed, J. C. McDowell and N. M. D. Brown, *J. Mater. Sci. Lett.*, **12**, 461 (1993).
102. C. Y. Shei, R. S. Liu, C. T. Chang and P. T. Wu, *Inorg. Chem.*, **29**, 3117 (1990).
103. T. Rojo, M. Inisansti, M. I. Arrortua, E. Hernandez and J. Zubillaga, *Thermochim. Acta*, **195**, 95 (1992).
104. H. Kurosawa, H. Iwasaki, Y. Muto, H. Yamane, T. Hirai and N. Kobayashi, *Jpn. J. Appl. Phys.*, **28**, L827 (1989).
105. T. Kobayashi, F. Uchikawa, K. Nomura and T. Masumi, *Jpn. J. Appl. Phys.*, **28**, L2168 (1989).
106. M. A. Matchett, M. Y. Chiang and W. E. Buhro, *Inorg. Chem.*, **29**, 358 (1990).
107. M. Lalia-Kantouri, A. Christofides and G. E. Manousskis, *J. Thermal. Anal.*, **30**, 399 (1985).
108. A. G. de Souza and C. Airoldi, *Thermochim. Acta.*, **130**, 95 (1988).
109. P. O. Dunstan and L. C. R. Santos, *Thermochim. Acta*, **156**, 163 (1989).

110. B. A. Arbuzov, N. R. Fedotova and V. S. Vinogradova, *Izv. Akad. Nauk SSSR, Ser. Khim.*, 1380 (1987).
111. H. Schmidbaur and K. H. Mitschke, *Chem. Ber.*, **104**, 1837, 1842 (1971).
112. J. L. Wardell and D. W. Grant, *J. Organomet. Chem.*, **188**, 345 (1980).
113. R. Nomura, S. J. Inazawa, H. Matsuda and S. Saeki, *Polyhedron*, **6**, 607 (1987).
114. R. Nomura, K. Kanaya and H. Matsuda, *Chem. Lett.*, 1849 (1988).
115. R. Nomura, K. Kanaya and H. Matsuda, *Bull. Chem. Soc. Jpn.*, **62**, 939 (1989).
116. L. C. R. dos Santos, S. F. de Oliveira, J. G. P. Epinola and C. Airoldi, *Thermochim. Acta*, **206**, 13 (1992).
117. M. Wieber and I. Sauer, *Z. Naturforsch.*, **39B**, 1668 (1984).
118. H. J. Breunig and S. Gülec, *Z. Naturforsch.*, **43B**, 998 (1988).
119. H. J. Breunig and S. Gülec, *Phosphorus Sulfur*, **38**, 97 (1988).
120. A. G. Massey, *Adv. Inorg. Chem.*, **33**, 1 (1989).
121. H. J. Emeleus, R. N. Hazeldine and E. G. Walaschewski, *J. Chem. Soc.*, 1552 (1953).
122. H. Hartmann and G. Nowak, *Z. Anorg. Allg. Chem.*, **290**, 348 (1957).
123. A. Spada, *Atti Soc. Nat. Mat. Modena*, **71**, 155 (1940).
124. A. Spada, *Atti Soc. Nat. Mat. Modena*, **72**, 34 (1941).
125. L. Horner, H. Oedinger and H. Hoffmann, *Ann. Chem.*, **626**, 26 (1959).
126. C. A. McAuliffe, A. G. Mackie, S. M. Godfrey, D. G. Kelly, R. G. Pritchard and S. M. Watson, *J. Chem. Soc., Chem. Commun.*, 1163 (1991).
127. R. Hoffman, J. M. Howell and E. L. Muetterties, *J. Am. Chem. Soc.*, **94**, 3047 (1972).
128. W. J. Kutzelnigg and J. Wasilewski, *J. Am. Chem. Soc.*, **104**, 953 (1982).
129. L. Carlsen, S. Elbel and H. Egsgaard, *J. Chem. Soc., Dalton Trans.*, 481 (1987).
130. H. J. Breunig and W. Kanig, *Phosphorus Sulfur*, **12**, 149 (1982).
131. L. Maier, F. G. A. Stone, E. G. Rochow and D. Seyferth, *J. Am. Chem. Soc.*, **79**, 5884 (1957).
132. G. Wittig and D. Hellwinkel, *Chem. Ber.*, **97**, 789 (1964).
133. K. Brodersen, R. Palmer and D. Breitinger, *Chem. Ber.*, **104**, 360 (1971).
134. M. Ates, J. H. Breunig and S. Gulec, *J. Organomet. Chem.*, **364**, 67 (1989).
135. N. K. Jha, A. C. Misra and P. Bajaj, *Polym. Eng. Sci.*, **26**, 332 (1986).
136. H. J. Emeléus, J. W. Dale, R. N. Haszeldine and J. H. Moss, *J. Chem. Soc.*, 3708 (1957).
137. A. B. Burg and L. R. Grant, *J. Am. Chem. Soc.*, **81**, 1 (1959).
138. H. Hartmann and G. Kühl, *Z. Anorg. Allg. Chem.*, **312**, 186 (1961).
139. D. Naumann and W. Tyrra, *Can. J. Chem.*, **67**, 1949 (1989).
140. G. Wittig and K. Clauss, *Ann. Chem.*, **578**, 136 (1952).
141. K. V. Katti and C. L. Barnes, *Inorg. Chem.*, **31**, 4231 (1992).
142. F. G. Mann and A. F. Wells, *J. Chem. Soc.*, 1938, 702 (1938).
143. O. J. Scherer, K. Pfeiffer and G. Wolmershäuser, *Chem. Ber.*, **125**, 2367 (1992).
144. A. J. Ashe III, J. W. Kampf and D. B. Puranik, *J. Organomet. Chem.*, **447**, 197 (1993).
145. K. H. Whitmire and J. M. Cassidy, *Inorg. Chem.*, **30**, 2788 (1991).
146. G. A. Razuvaev, G. A. Domarachev, V. V. Sharutin and O. N. Suvorova, *Dokl. Chem. (Engl. Transl.)*, **237**, 711 (1977).

CHAPTER **14**

Organoarsenic and organoantimony homocycles

HANS JOACHIM BREUNIG

Universität Bremen, Fachbereich 2 (Chemie), Postfach 330440, D-28334 Bremen, Germany

I. INTRODUCTION

Monocyclic and polycyclic compounds with rings built exclusively of arsenic or antimony atoms are the scope of this chapter. The monocycles are known as cyclopolyarsanes $(RAs)_n$ or cyclopolystibanes $(RSb)_n$ with R = alkyl or aryl and $n = 3-6$. The IUPAC nomenclature suggests the following specific names for each ring size:

	E = As	E = Sb
$(RE)_3$	triarsirane cyclotriarsane	tristibirane cyclotristibane
$(RE)_4$	tetraarsetane cyclotetraarsane	tetrastibetane cyclotetrastibane
$(RE)_5$	pentaarsolane cyclopentaarsane	pentastibolane cyclopentastibane
$(RE)_6$	hexaarsane cyclohexaarsane	hexastibane cyclohexastibane

The chemistry of organic arsenic, antimony and bismuth compounds
Edited by S. Patai © 1994 John Wiley & Sons Ltd

General names for systems of rings or chains are composed of the name of the substituent and the element. For example, the name methylarsenic refers to all the systems of the type $(MeAs)_n$. The progress in the field of organoarsenic[1-10] or organoantimony[5, 11-15] mono- and polycycles has been summarized several times in excellent reviews. In this survey stress is laid on recent developments mainly in the field of cyclostibanes, and an attempt is made to compare analogous homocycles of arsenic and antimony.

Differences in the chemical properties of these rings reflect the atomic parameters of the elements. The increase of size but decrease of covalent bond strengths going from arsenic to antimony is responsible for some general trends. The sterical shielding effect of a given organic group is larger on arsenic than on antimony and the As—As bonds are less reactive than the Sb—Sb bonds. Both trends make antimony rings more labile than arsenic rings. This explains why ring–ring equilibria (equation 1) and ring–chain equilibria (equation 2) occur more frequently in the chemistry of antimony homocycles compared with arsenic rings. As a consequence, for a given substituent only one ring size is known for many arsenic rings whereas analogous antimony systems consist of equilibria of several rings of different size. With bulky alkyl substituents, however, the chemical differences of arsenic and antimony become less apparent and four-membered rings are mainly formed with both elements. Similar chemical properties are also observed in the case of phenylarsenic and phenylantimony where, at least in the solid state, the stable hexamers dominate.

$$5(RE)_4 \rightleftharpoons 4(RE)_5 \tag{1}$$

$$\tfrac{1}{n}(RE)_n + Y_2 \rightleftharpoons \tfrac{1}{x}Y(RE)_xY \tag{2}$$

$$R = alkyl, \; E = As, Sb; \; n = 4\text{--}6; \; x \gg 6; \; Y = end \; group$$

II. PREPARATION AND PROPERTIES OF MONOCYCLES

A. Methyl Derivatives

The methyl arsenic system compromises two well-defined species, the pentamer $(MeAs)_5$, a yellow pyrophoric oil (m.p. 12 °C) and the purple black double-chain polymer, $[(MeAs)_2]_x$ (m.p. 204 °C). A red solid that is possibly the linear chain compound $(MeAs)_x$ has also been described. The polymers are formed when samples of the pentamer are exposed to impurities like arsenic halides that are able to react with the arsenic–arsenic bond and may act as end groups. The most widely used method for the preparation of larger amounts of $(MeAs)_5$ is the reduction of methylarsonic acid, $MeAsO(OH)_2$, or its sodium salt with hypophosphoric acid[16-20]. Samples prepared by this method, however, may contain between 5 and 15% of impurities. A method that minimizes impurities is the reaction of $MeAsH_2$ with dibenzylmercury (equation 3)[20].

$$5\,MeAsH_2 + 5\,(PhCH_2)_2Hg \longrightarrow (MeAs)_5 + 10\,PhCH_3 + 5\,Hg \tag{3}$$

The five-membered methylarsenic ring has been characterized by various methods. A single-crystal X-ray study has shown that the solid consists of discrete cyclic pentamers[21] with equal As–As distances. The ring is puckered with mean As—As—As bond angles of 102°. Methyl groups are arranged in such a way that a staggered conformation at the As—As bonds results. Electron diffraction measurements of the vapour of the yellow form are consistent with the cyclic pentameric structure[22].

The proton NMR spectra and the ^{13}C NMR spectra in solution and in the pure liquid indicate that in these phases cyclic pentamers predominate as well. Only three major resonances appear over a wide temperature range with relative intensities in a 2:2:1 ratio, instead of five singlet signals that should appear if the solid state structure would persist in solution[23]. This result suggests that the rings have a single effective conformation with a time-averaged plane of symmetry. Rapid torsional motion of the rings rather than inversion at arsenic is responsible for the formation of the symmetry plane. Additional signals are observed in the spectra of samples of $(MeAs)_5$ obtained by reduction of methylarsonic acid. These signals increase reversibly in intensity with the temperature but not with dilution. They were therefore assigned to other conformations rather than to other cyclooligomers[24]. Indications for rings of other sizes come, however, from the mass spectra of methylarsenic in the gas phase. In addition to peaks of the pentamers, the signals of the tetramers and trimers were observed. It is difficult to decide, however, whether these signals are formed by fragmentation of the pentamers or whether they are molecular ions of tetraarsetanes or triarsiranes in the gas phase. An analytical tool in this context is the variation of electron energy. As the peaks of the smaller oligomers of methylarsenic increase in intensity on lowering the ionization voltage, they are regarded as parent peaks. This result suggests ring–ring equilibria (equation 4) under MS conditions.

$$\tfrac{1}{5}(MeAs)_5 \;\rightleftharpoons\; \tfrac{1}{4}(MeAs)_4 \;\rightleftharpoons\; \tfrac{1}{3}(MeAs)_3 \qquad (4)$$

In sharp contrast to the findings in the methylarsenic system the existence of stable methylantimony cyclooligomers could not be proven. However, solid polymeric forms of methylantimony that are insoluble in hydrocarbons have been described. A purple black polymer has been obtained from $MeSbH_2$ and $(PhCH_2)_2Hg^{25}$. Reaction of $MeSbCl_2$ with S_2Cl_2 in benzene gave a green solid[26]. A black material was obtained by dehalogenation of $MeSbBr_2$[27,28] with magnesium in THF. The oligomers $(MeSb)_3$[27] and $(MeSb)_5$[26] have so far only been observed in the mass spectra of the polymers.

B. Ethyl, *n*-Propyl and *n*-Butyl Derivatives

For ethyl, *n*-propyl and *n*-butyl derivatives a direct comparison of analogous ring systems of arsenic and antimony is possible because not only cycloarsanes but also cyclostibanes bearing these alkyl substituents have been described. A common feature of the ring systems of both elements is the formation of five-membered rings in solution. The chemistry of the antimony rings, however, appears to be more diverse than the chemistry of the arsenic analogues.

The ethyl and *n*-propyl arsenic rings are obtained by reduction of the corresponding arsonic acids with phosphinic acid[19]. Dehalogenation of *n*-$BuAsCl_2$ with sodium in diethyl ether[29] gives $(n\text{-}BuAs)_5$. Recently $(EtAs)_5$ and $(n\text{-}PrAs)_5$ have been obtained by reaction of $(MeAs)_5$ with $EtAsH_2$ or *n*-$PrAsH_2$ (equation 5)[30,31]. This method avoids unwanted side reactions, and gives pure products in high yield.

$$(MeAs)_5 + 5\,RAsH_2 \longrightarrow (RAs)_5 + 5\,MeAsH_2 \qquad (5)$$
$$R = Et, n\text{-}Pr$$

The *n*-alkylarsenic pentamers are yellow oils that may be distilled under reduced pressure. The presence of trimers and tetramers in the gas phase is possible. Mass spectral studies of ethylarsenic gave signals of trimers or tetramers. Pentakis(2,2,2-trideuterioethyl) cyclopentarsane was synthesized by the reaction of 2,2,2-trideuterioethylarsine with dibenzylmercury[32].

The *n*-alkylantimony ring systems were prepared by reduction of the corresponding *n*-alkylantimony dibromides with magnesium in THF (equation 6)[27] or electrochemically[65].

$$RSbBr_2 + Mg \longrightarrow \tfrac{1}{n}(RSb)_n + MgBr_2 \qquad (6)$$
$$R = Et, n\text{-}Pr, n\text{-}Bu; \; n = 4, 5.$$

In contrast to methylantimony the ethyl, n-propyl and n-butyl derivatives are to some extent soluble in hydrocarbons as yellow antimony rings. These rings, however, polymerize to form black or metallically shining solids on evaporation of the solvents. This process is reversible and may be used for the purification of the compounds. Both the black solids and the yellow solution are air-sensitive but stable in sealed tubes for a long time. Slow thermal decomposition gives tetralkyldistibanes and trialkylstibanes. The size of the n-alkylantimony rings in solution has been determined mainly by ^{1}H and ^{13}C NMR spectroscopy. The spin systems as well as the 2:2:1 ratio of the intensities of the different groups were interpreted in favour of pentamers with a maximum of *trans* positions and a time-averaged plane of symmetry of the pentastibolane rings. A probable structure is analogous to the corresponding arsenic rings and contains a maximum of *trans* positions of the substituents with a minimum of sterical interactions of the alkyl groups:

$$E = As, Sb$$

These results show that structures and dynamics of the penta n-alkylpentastibolanes are closely related to those of $(MeAs)_5$ and other pentaarsolanes.

In contrast to the cycloarsane chemistry, however, the NMR spectra of the n-alkylantimony ring systems reveal the presence of a second type of ring with equivalent substituents. As the intensities of these signals increase on dilution, these signals were attributed to tetramers instead of hexamers that should also have equivalent substituents. The chemical situation for n-alkyl antimony systems is described by equation 7.

$$\tfrac{1}{x}(RSb)_x \rightleftharpoons \tfrac{1}{5}(RSb)_5 \rightleftharpoons \tfrac{1}{4}(RSb)_4 \tag{7}$$

$$R = Et, n\text{-Pr}, n\text{-Bu}; x \gg 5$$

Trimers of alkylantimony were not detected by NMR techniques. Trimers are, however, the most abundant species appearing in the EI mass spectra. Like in the mass spectra of the pentaarsolanes, it is difficult to decide whether the trimeric ions are fragments or parent ions, but experiments with variation of electron energy are in favour of the presence of trimers under MS conditions.

A qualitative comparison of analogous arsenic and antimony rings shows, as expected, similarities with respect to ring size distribution and structural aspects. However, the antimony systems seem to be more labile with respect to ring–ring and ring–chain equilibration. Possible mechanisms for these reactions between different rings include ring opening with reagents that may be present as impurities or concerted cyclic reactions that have been proposed in cyclopolyphosphane chemistry[33]. In equation 8 the equilibrium reaction of a five-membered and a four-membered ring giving a nine-membered ring is described as an example for ring enlargement and ring splitting by transannular interaction of element–element bonds.

$$E = As, Sb; R = n\text{-alkyl} \tag{8}$$

Fast reactions of this type make rings of every size accessible and lead to thermo-dynamic control of ring size. Thermodynamic factors may include the different forms of ring strain, stabilizing transannular or intermolecular interactions and the influence of concentration. According to the principle of le Chatelier, lowering of concentration should lead to a general decrease of ring size until the ring strain becomes too high. High concentrations should favour large rings that become insoluble and are stabilized by interactions in the solid state. The chemical differences between analogous alkylarsenic and alkylantimony ring systems indicate lower kinetic stabilization for antimony rings that may result from lower element–element bond energy and stronger van der Waals interactions.

C. Phenyl and p-Tolyl Derivatives

In the chemistry of phenyl substituted homocycles of arsenic or antimony the hexamers are the best characterized species. Synthetic procedures for the preparation of phenyl-arsenic include the reduction of phenyldihaloarsines or phenylarsonic acid with agents like sodium dithionite or hypophosphorous acid[34,35]. Phenylarsenic is also obtained by gentle oxidation of $PhAsH_2$. A recent synthesis is the reaction of $PhAsH_2$ with $(MeAs)_5$[30,31]. Although earlier workers obtained varying values for the molecular weight in solution, it is now established that at least the main form of phenylarsenic is hexameric. The solid is obtained as yellow crystals (m.p. 212 °C) containing cyclic hexameric molecules.

Attempts to extend the reductive preparative procedures for phenylarsenic to phenyl-antimony have met with little success. Reduction of phenylstibonic acid or phenylanti-mony dihalides only led to polymeric materials. The phenylantimony hexamer was, however, obtained as orange or yellow crystalline solvates $(PhSb)_6 \cdot solv$ (solv = benzene, toluene, THF, 1,4-dioxane) by reaction of $PhSbH_2$ with styrene or phenylacetylene[36] (equation 9) or by slow reaction of $PhSb(SiMe_3)_2$ in air[37,38] (equation 10).

$$6\,PhSbH_2 + 6\,PhCH{=}CH_2 \longrightarrow (PhSb)_6 + 6\,PhCH_2CH_3 \qquad (9)$$

$$6\,PhSb(SiMe_3)_2 + 3\,O_2 \longrightarrow (PhSb)_6 + 6\,(Me_3Si)_2O \qquad (10)$$

In the solid state the molecular structures of $(PhAs)_6$ and $(PhSb)_6$ are similar. Both homocycles adopt the structure of highly puckered six-membered rings analogous to the chair form of cyclohexane in which phenyl groups occupy equatorial positions:

E = As, Sb

The packing of the arsenic or antimony rings in the crystals leads to stacks of rings. Structural data are given in Table 1. It is remarkable that the shortest intermolecular contacts between antimony atoms are considerably shorter than between arsenic atoms. With respect to intermolecular association in the solid state, however, $(PhSb)_6$ is a border-line case. The contact distances are close to the van der Waals distances between antimony atoms (440 pm). For the crystallization of the phenylantimony hexamer, the presence of cyclic organic molecules that are incorporated in the solid plays an important role.

TABLE 1. Structural parameters of $(PhAs)_6$ and $(PhSb)_6$

$(PhE)_6$	$E = As^{35}$	$E = Sb^{38}$
Angles E_3	91, 1°	90°
Bond distance E–E	245, 9 pm	283, 6 pm
Torsion angles EE–EE	88, 9°	90, 1°
Angles EEC	97, 7°	95, 6°
E...E[a]	429, 7 pm	419 pm

[a] Shortest intermolecular contact.

Attempts to obtain crystals in non-cyclic solvents like petroleum ether failed. In the $(PhSb)_6$·solvates the gaps between the stacks of antimony rings are filled with the cyclic organic molecules. The structure of phenylarsenic and phenylantimony in solution and in the gas phase is not as clear as in the solid.

The ^1H NMR spectra of samples of $(PhAs)_6$ and $(PhSb)_6$ dissolved in C_6D_6 or $CDCl_3$ are complex. A detailed ring size analysis from NMR data has not been carried out. Like in the case of the alkylarsenic or antimony homocycles, the mass spectra obtained from samples of the phenyl compounds cannot be interpreted straightforwardly. The EI mass spectrum of $(PhAs)_6$[19] contains the hexameric molecular ions but also pentameric, tetrameric and trimeric species appear. It is not clear if the smaller oligomers are fragments or individual rings. The EI mass spectra of crystals of $(PhSb)_6$[37] contain tetramers at highest mass. Trimers and rearrangement products are also observed but no hexamers. It seems possible, at least for $(PhSb)_6$, that the hexameric structure of the crystalline species is not preserved in other phases.

Ring systems that are easier to study by NMR spectroscopy, but should have a similar chemistry, are the p-tolylarsenic or antimony rings. The methods that have been used for the syntheses of $(PhAs)_6$ are also useful for the preparation of $(p\text{-TolAs})_6$. A recent method is the reaction of $p\text{-TolAsH}_2$ with $(MeAs)_5$[30]. $(p\text{-TolAs})_6$, a white crystalline compound, is the only arsenic homocycle with the p-tolyl substituent. The p-tolylantimony system is more complex. Slow oxidation of $p\text{-TolSb}$ $(SiMe_3)_2$ in the air gives yellow crystals[39]. The ring size of p-tolylantimony in the crystalline state is not known. By analogy with $(PhSb)_6$ hexamers are probable. In solution, however, according to ^1H NMR spectra the exclusive presence of the hexamer cannot be confirmed. For the methyl protons a 2:2:1 ratio of intensities is observed as the main signal. This indicates the presence of p-tolylantimony pentamers in solution as the most abundant species. By analogy, pentamers may also be present in solutions of the phenylantimony system.

This observation sheds some light on the question as to why pentamers are formed with n-alkyl substituents but hexamers with slim aryl substituents. A possible explanation, at least for the organoantimony homocycles, is that in solution pentamers dominate both with alkyl and aryl substituents. In the solid state, however, the six-membered form allows a closer packing of the phenylantimony rings than the pentameric form.

D. *t*-Butyl Derivatives

The chemistry of the t-butylarsenic and t-butylantimony rings is similar, thus showing the predominant influence of the bulky organic substituents. In the case of t-butylarsenic the trimer, the tetramer and the pentamer have been described. In this series the tetramer is the most abundant and the most stable species. Pentamers and tetramers are also known for t-butylantimony rings.

The synthesis of the remarkable three-membered ring $(t\text{-BuAs})_3$ is achieved by a selective $2 + 1$ cyclocondensation reaction of t-BuAs(K)—As(K)Bu-t with t-BuAsCl$_2$ at $-78\,°C$ (equation 11)[40].

$$t\text{-BuAs(K)} - \text{As(K)Bu-}t + t\text{-BuAsCl}_2 \longrightarrow (t\text{-BuAs})_3 + 2\text{KCl} \qquad (11)$$

The trimer is rather unstable. Even in absence of light and air it decomposes at temperatures above $-30\,°C$. At room temperature slow formation of the tetramer takes place. This ring–ring conversion (equation 12) is irreversible.

$$4\,(t\text{-BuAs})_3 \longrightarrow 3\,(t\text{-BuAs})_4 \qquad (12)$$

Syntheses of t-butylarsenic homocycles that are unspecific to ring size, like the dehalogenation of t-BuAsCl$_2$ with Na in Et$_2$O[29] or the reaction of t-BuAsLi$_2$ with t-BuAsCl$_2$[41a], give mainly $(t\text{-BuAs})_4$. The pentamer $(t\text{-BuAs})_5$ is formed as a by-product of the tetramer when t-BuAsCl$_2$ is dehalogenated with Mg in THF[42]. Some information on the solution structures of the rings is available from the ^{1}H NMR spectral data. The spectra of the trimer show two singlet signals in the 1:2 intensity ratio that is characteristic for the $cis/trans$ isomer. A singlet is observed for the tetramer and a set of singlets in the 2:2:1 ratio of intensities for the pentamer. Again the assumption that the rings adopt conformations with a maximum of $trans$ positions of the substituents is most probable.

A comparison of the thermodynamic stability of t-butylarsenic rings based on estimations from preparative results gives the sequence: $(t\text{-BuAs})_4 > (t\text{-BuAs})_5 > (t\text{-BuAs})_3$[42]. The combined effects of ring strain and steric cis interactions of the substituents may be responsible for the low stability of the trimer. In the case of the pentamer the ring is not strained, but cis interactions of the t-butyl groups destabilize the ring. In the tetramer all the bulky substituents are in $trans$ positions, avoiding destabilizing contacts.

The preparative chemistry of the t-butylantimony rings is related to the arsenic systems. Dehalogenation of t-BuSbCl$_2$ with Mg in THF gives the tetramer $(t\text{-BuSb})_4$ as the main product[41a] along with small amounts of $(t\text{-BuSb})_5$ as side product [41b]. A t-butylantimony trimer has not yet been described. The ^{1}H NMR spectra of solutions of the t-butylantimony tetramer and pentamer show the same pattern as the arsenic analogues and similar structures may be assumed. The presence of tetramers and pentamers was also observed in solutions of the n-alkylantimony rings. In contrast to the n-alkyl derivatives, the t-butylantimony rings do not easily participate in ring–ring equilibria. At least in solution at ambient temperature no reactions between the rings were observed.

Both $(t\text{-BuAs})_4$ (colourless crystals) and $(t\text{-BuSb})_4$ (yellow crystals) have been analysed by X-ray crystal diffractometry[41a]. Both rings crystallize in the monoclinic space group $P2_1/c$. In the crystals there are folded four-membered rings E_4 (E = As, Sb) and the substituents occupy the pseudo-equatorial positions of the all-$trans$ configurations:

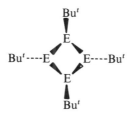

E = As, Sb

It is of interest that the fold angle for the antimony four ring is smaller than for the arsenic ring. Stronger transannular attraction of antimony atoms compared to arsenic may be responsible for these differences. Selected structural data are given in Table 2.

TABLE 2. Distances and angles of $(t\text{-BuE})_4$ $(E = As, Sb)^{41a}$

	E = As	E = Sb
E—E	244 pm	282 pm
E—C	202 pm	221 pm
E—E—E	86°	85°
E—E—C	101°	99°
fold angle	139°	133°

E. Arsenic Monocycles That Have No Counterparts in Antimony Chemistry

As the chemistry of arsenic homocycles is by far more developed than the antimony chemistry, many rings are known that have no analogues in Sb chemistry. In earlier reviews the chemistry of tetraarsetanes $(RAs)_4$ with $R = s\text{-Bu}$, $c\text{-Hex}$, CF_3, C_6F_5, as well as of pentaarsolanes $(RAs)_5$ with $R = i\text{-Bu}$, CF_3, CH_2Cl, SiH_3, GeH_3, and of hexaarsanes $(RAs)_6$ with $R = p\text{-MeOC}_6H_4$, $o\text{-MeC}_6H_4$ has been summarized several times[3,6,8] and will not be discussed here. Recently, a novel pentamer $(Me_3SiCH_2As)_5$ has been synthesized by reduction of the corresponding alkylarsenic dichloride with magnesium in presence of Me_3SiCl^{43}. The molecular structure of the pale yellow crystals of $(Me_3SiCH_2As)_5$ was determined by single-crystal X-ray diffractometry. The ring possesses an envelope-like conformation, although it is somewhat twisted. The substituents adopt a maximum of *trans* positions. The As—As bond lengths vary from 242.4 to 244.6 pm, As_3 bond angles lie between 96.5° and 105.2°. The conformation of $(Me_3SiCH_2As)_5$ is similar to the conformation of $(MeAs)_5$. Again the conformation is not preserved in solution. Like in other pentaarsolanes there is torsional motion[44] generating a time-averaged plane of symmetry and the ^1H NMR spectrum shows the expected three singlet signals in the intensity ratio 2:2:1 for the methyl groups. In the region of the methylene protons two distinct AB patterns were observed.

An unusual cyclotriarsane[45] was recently prepared by reaction of As_4 with tetramesityl-disilene in toluene at 25 °C (equation 13). The ring system was characterized by X-ray diffractometry and other methods. In the triarsirane ring the As—As distances vary from 245.0 pm to 246.3 pm; the As_3 angles are very close to 60°.

$$As_4 + 2\,Mes_2Si{=}SiMes_2 \longrightarrow \quad \begin{array}{c} Mes_2Si\text{——}SiMes_2 \\ \diagdown\;\diagup \\ As \\ | \\ As \\ \diagup\;\diagdown \\ As\text{——}As \\ |\qquad| \\ Mes_2Si\text{——}SiMes_2 \end{array} \tag{13}$$

A cyclotetraarsane bearing two metal and two aryl substituents is the compound $[Fe]_2R_2As_4$ with $[Fe] = (Me_5C_5)(CO)_2Fe$ and $R = 2,4,6\text{-Bu}^t_3C_6H_2$. This ring compound has been obtained by dimerization of the diarsenyl complex $[Fe]$ As=AsR. The As_4 ring has been characterized by spectroscopic techniques including FD (field desorption) mass spectroscopy. Based on NMR data the structure is suggested[46].

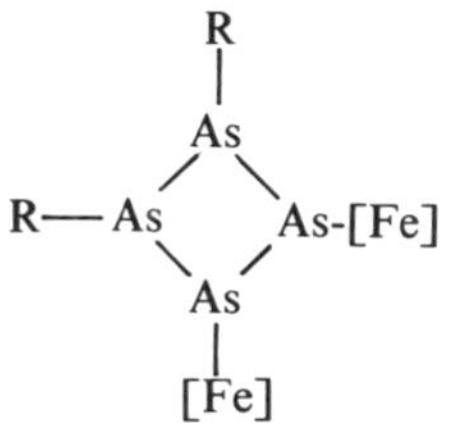

$$R-As \underset{\underset{As}{\overset{As}{\diagdown}}{\nearrow}}{\overset{\overset{R}{|}}{\overset{As}{\diagup}}} \underset{\diagup}{\diagdown} As\text{-}[Fe]$$

$$[\overset{|}{Fe}]$$

F. Antimony Rings That Have No Counterparts in Arsenic Ring Chemistry

The search for antimony monocycles that are protected by bulky substituents has led to the syntheses of tetrastibetanes with bulky substituents like $(MesSb)_4$[27,47], $((Me_3Si)_2CHSb)_4$[28,48] and $(Me_5C_5Sb)_4$[49]. The isopropylantimony[28] system, however, is more closely related to the antimony rings with slim substituents.

Tetramesityltetrastibetane is obtained by reaction of mesitylantimony dibromide with magnesium in THF. Crystallization from toluene or benzene gives the solvates $(MesSb)_4 \cdot$ toluene or $(MesSb)_4 \cdot$ benzene as yellow crystals. In the crystals there are folded Sb_4 rings that are aligned to chains via short intermolecular contacts of opposite antimony atoms in the ring (Figure 1). These contacts lead to distortions of the four-membered ring, and two different fold angles and two different Sb_3 bond angles result. As a consequence of the formation of antimony chains via short intermolecular contacts in the crystal, a bathochromic shift from the solution to the solid form should be expected. There is, however, no such effect and solid $(MesSb)_4$ is yellow like isolated antimony rings. Selected structural data of tetrastibetanes are given in Table 3. In solution the ^{1}H-NMR spectroscopic data show that all the substituents are equivalent and that $(MesSb)_4$ does not take part in ring–ring equilibria.

The reduction of $(Me_3Si)_2CHSbCl_2$ is also achieved with magnesium in THF. After work-up an oily product mixture is obtained. Crystallization of the initial product gives yellow crystals of the tetramers $((Me_3Si)_2CHSb)_4$ that are remarkably stable on heating or exposure to air. In the crystals the tetramers adopt the normal structure of a folded antimony four-ring with the substituents in the all-*trans* configuration. It is remarkable that the Sb—Sb bond distances are not equal. In the ^{1}H NMR spectrum in solution, only

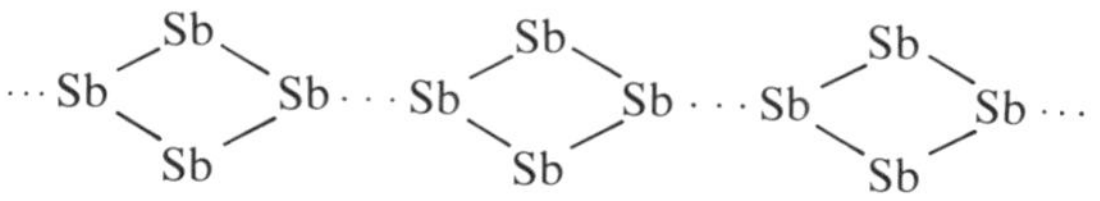

$$\cdots Sb \underset{\overset{Sb}{\diagdown}}{\overset{Sb}{\diagup}} Sb \cdots Sb \underset{\overset{Sb}{\diagdown}}{\overset{Sb}{\diagup}} Sb \cdots Sb \underset{\overset{Sb}{\diagdown}}{\overset{Sb}{\diagup}} Sb \cdots$$

$$\begin{aligned} Sb \cdots Sb \quad & 390 \text{ pm} \\ Sb\text{—}Sb \quad & 285 \text{ pm} \end{aligned}$$

FIGURE 1. Association of Sb atoms in $(MesSb)_4$

TABLE 3. Selected structural data for tetrastibetanes $(RSb)_4$

R	Sb–Sb (pm)	Sb_3 (deg)	Fold angles (deg)
t-Bu[41a]	281.7	85.0	132.7
Me_5C_5[49]	285.6	87.1	144.2
$(Me_3Si)_2CH$[28]	282.2–287.8	80.4	115.4
Mes[27]	285.4	76.9 88.2	119.8 125.2

one singlet signal appears for the methyl protons as expected for a tetrameric all-*trans* structure. Mass spectroscopy shows molecular ions and fragments of the tetramers, but no trimers. Trimeric species are, however, present in the mass spectrum of the initial product mixture before crystallization of the tetramer and also the ^{1}H NMR data of the initial product indicate the formation of $((Me_3Si)_2CHSb)_3$[48]. The isolation and full characterization of this trimer has, however, not yet been accomplished.

A rather unusual tetrastibetane is $(\sigma\text{-}Me_5C_5Sb)_4$. It is formed as a product of the reaction of Me_5C_5K with $SbCl_3$ in diethyl ether at $-78\,^\circ C$ (equation 14)[49].

$$3\,RK + SbCl_3 \longrightarrow \tfrac{1}{4}(RSb)_4 + R_2 + 3\,KCl \qquad (14)$$
$$R = Me_5C_5$$

In the first steps of this reaction $(Me_5C_5)_3Sb$ is probably formed as an intermediate that decomposes at low temperatures to give the tetrastibetane. The structure of the orange yellow crystals contains folded Sb_4 rings and one monohapto bonded pentamethylcyclopentadienyl ring attached to each antimony atom. Relevant distances and angles are given in Table 3. In solution, the structure of an antimony homocycle is not preserved. Instead, monomeric stibinidene molecules RSb are formed. Evidence for this dissociation comes from cryoscopic molecular weight determinations in benzene and cyclohexane solutions.

The isopropylantimony rings, $(i\text{-}PrSb)_4$ and $(i\text{-}PrSb)_5$ have been obtained as yellow air-sensitive and thermolabile solutions in THF or benzene by the usual dehalogenation of the corresponding dibromide with Mg in THF[28]. The rings are unstable with respect to polymerization in absence of a solvent like the n-alkylantimony rings. The ring sizes have been determined by NMR and mass spectroscopy. The relative amount of the tetramer is larger than in the case of the alkylantimony rings with unbranched alkyl substituents, but much smaller than in the case of the t-butylantimony rings.

III. PREPARATION AND PROPERTIES OF CAGE COMPOUNDS

Arsenic homocycles have been incorporated in various polycylic structures that are represented by the following structures:

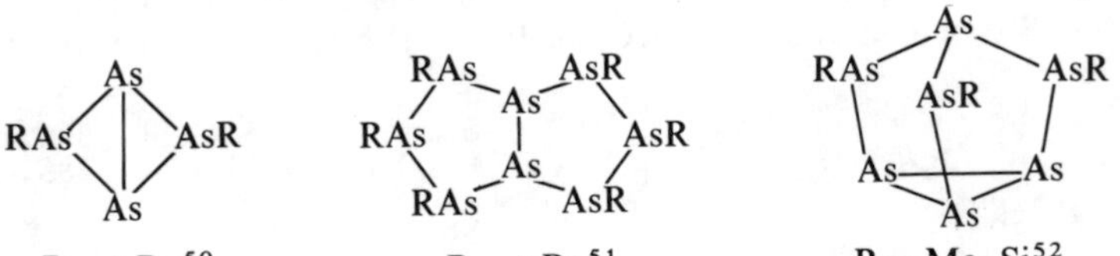

$$R = t\text{-}Bu^{50} \qquad\qquad R = t\text{-}Bu^{51} \qquad\qquad R = Me_3\,Si^{52}$$

The chemistry of these cages has been thoroughly reviewed[8] but it has not been extended to antimony chemistry. For these reasons it will not be discussed here in more detail.

Good examples for a comparison of analogous arsenic and antimony cages are, however, the nortricyclane systems that contain three-membered arsenic or antimony homocycles. The cage $MeC(CH_2As)_3$ is formed as colourless crystals by reduction of the corresponding iodide $MeC(CH_2AsI_2)_3$ with sodium[53,54], or $NaAsPh_2$[55] and by desulphurization of $MeC(CH_2AsS)_3$[55]. The analogous antimony derivative $MeC(CH_2Sb)_3$ is a deep orange solid that crystallizes from a yellow solution in THF after the dehalogenation of the corresponding dichloride with sodium in THF[56,57]. Both cages have been characterized by X-ray diffractometry. The structure of $MeC(CH_2As)_3$ contains molecules of approximate C_{3v} symmetry[53]. The mean As–As distances are 242 pm and the AsAsC bond angles are close to $90°$. The shortest intermolecular distances between arsenic atoms

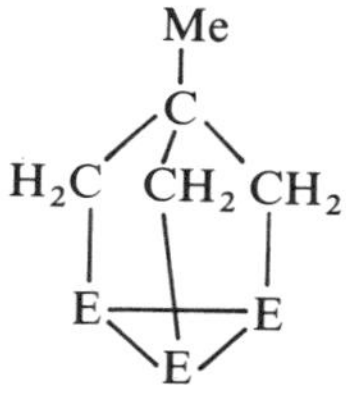

E = As, Sb

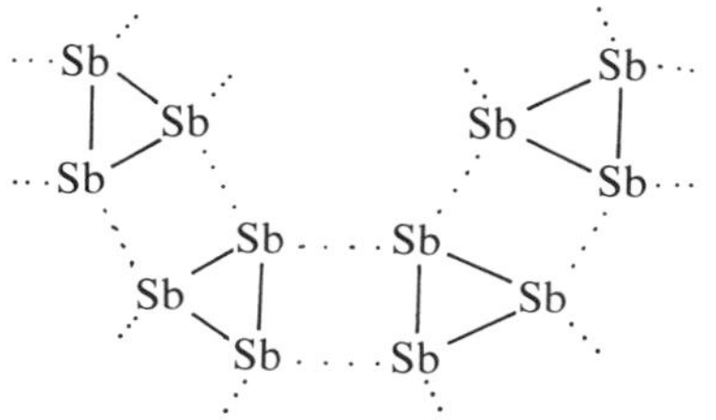

Sb—Sb 280–282 pm, Sb···Sb 397–401 pm

FIGURE 2. Intermolecular association of the Sb skeleton in $MeC(CH_2Sb)_3$

(388 pm) are only slightly shorter than the sum of the van der Waals radii (approximately 400 pm). The structure of the $MeC(CH_2Sb)_3$ molecules in the crystal[58] is similar to the arsenic analogue with a basic Sb_3 triangle (mean Sb–Sb distance 280 pm) and nearly perpendicular Sb—C bonds (mean SbSbC angle 86°). In the crystal the Sb triangles are associated to layers with intermolecular Sb...Sb contacts of 396.6(2) pm and 401.2(3) pm (Figure 2). These contact distances are considerably shorter than the sum of van der Waals radii of antimony atoms (440 pm). Delocalization of electrons along the layers of antimony atoms may be responsible for the bathochromic shift of the solid compared to the solution.

IV. REACTIONS OF ARSENIC AND ANTIMONY HOMOCYCLES

Relatively little is known of the chemistry of antimony rings compared to arsenic rings. The reactions of arsenic homocycles have been reviewed several times[1,3,4,7,8]. They include cleavage of the As—As bonds with halogens and insertion of chalcogens or unsaturated hydrocarbons and reductive cleavage with potassium metal. Representative examples are given in equations 15–18.

$$\tfrac{3}{5}(RAs)_5 + \tfrac{3}{8}Se_8 \longrightarrow (RAsSe)_3 \tag{15}$$
$$R = Me[59]$$

$$\tfrac{1}{4}(RAs)_4 + I_2 \longrightarrow RAsI_2 \tag{16}$$
$$R = t\text{-Bu}[29]$$

$$(RAs)_6 + 2\,PhC{\equiv}CPh \longrightarrow 2 \underset{\displaystyle AsR}{\overset{\displaystyle PhC{=}CPh}{\underset{RAs\qquad AsR}{\diagup\ \diagdown}}} \tag{17}$$

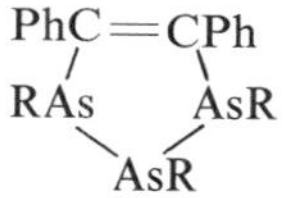

$$R = Ph[60]$$

$$(RAs)_4 + 2K \longrightarrow K(RAs)_4K \tag{18}$$
$$R = t\text{-}Bu^{29}$$

Extensive studies have also been carried out concerning the reactivity of organoarsenic monocycles towards transition metal carbonyls and a rich coordination chemistry of organoarsenic rings or chains as ligands in transition metal complexes has been developed. This field has been covered in excellent reviews[3,61].

Recently, the oxidation of a cycloarsane has been combined with coordination. The methylarsenic pentamer reacts with oxygen in the presence of $Mo(CO)_6$ at $150\,°C$ to form $(MeAsO)_6$ $[Mo(CO)_3]_2$, an alkylarsoxane stabilized in the coordination sphere of $Mo(CO)_3$ units. The complex has been characterized by X-ray diffractometry[62]. The structure contains a 12-membered ring of alternating CH_3As groups and oxygen atoms. The ring atoms form a flattened cuboctahedron which is *trans* bicapped by $Mo(CO)_3$ groups:

$$[Mo] = Mo(CO)_3$$

The high reactivity of the organoantimony homocycles has become apparent during the synthesis of these compounds. Like other Sb(III) derivatives with antimony–antimony bonds, the cycles are air-sensitive in solution and white solids form. In the case of ethylantimony rings these oxidation products[28] analyse as $(EtSbO)_n$ and the oligomers with $n = 3, 4$ have been observed by mass spectroscopy. Air oxidation of $(t\text{-}BuSb)_4$, however, yields $Sb_4O_6^{63}$. The reaction of ethylantimony rings with bromine gives $EtSbBr_2$[28]. A more systematic study has been carried out for the reactivity of selected organoantimony monocycles with diselenides and ditellurides. Only the reactions of the ethylantimony rings in benzene at room temperature were of preparative use, giving products of the type $EtSb(ER)_2$ in good yields. Examples are given in equation 19.

$$\tfrac{1}{x}(EtSb)_x + REER \longrightarrow EtSb(ER)_2 \tag{19}$$
$$ER = SeMe, SePh, TeMe, TePh, Te\text{—}p\text{-}Tol^{64}$$

Only $EtSb(TePh)_2$ was prepared on a larger scale and isolated as crystals in 78% yield. This type of reaction could not be applied to $(MesSb)_4$. Reactions of this stibetane with dichalcogenides resulted in scrambling reactions giving product mixtures[64].

The reactivity of antimony rings towards distibanes has also been studied by NMR spectroscopy in benzene and by mass spectroscopy of neat liquids. With excess of distibanes the formation of short antimony chains (tristibanes) was observed (equation 20) after a few minutes reaction time.

$$R_2SbSbR_2 + \tfrac{1}{n}(R'Sb)_n = R_2Sb(R'Sb)SbR_2 \tag{20}$$
$$R = Me, Et, Ph; \ R' = Et, n\text{-}Pr, t\text{-}Bu, (Me_3Si)_2CH, Ph^{28}$$

An estimation of the equilibrium constants shows that the tristibane formation is favoured when the antimony rings bear sterically less demanding groups like the ethyl or n-propyl substituent. Bulky groups favour the formation of rings in the equilibria. As a product of the reaction of Ph_4Sb_2 with ethylantimony homocycles also a tetrastibane, $Ph_2SbSb(Et)Sb(Et)SbPh_2$ was identified by NMR spectroscopy.

Similar reactions have also been carried out in cycloarsane chemistry by addition of tetramethyldiarsane to $(MeAs)_5$ as neat liquids at room temperature. The complex NMR pattern indicated the formation of catena arsanes[24] $Me_2As(AsMe)_nAsMe_2$ with $n = 1-6$ and $n > 6$ in equilibria with $Me_2AsAsMe_2$ and $(MeAs)_5$. It is difficult to compare quantitatively the corresponding equilibria for arsenic and antimony because of the differences of reaction conditions and organic substituents.

Reactions of antimony monocycles with alkali metals have been studied in the case of the reactions of $(PhSb)_6$ with sodium in liquid ammonia. According to the molar ratio of the reagents two different antimonides, $PhSbNa_2$ and $NaSb(Ph)Sb(Ph)Na$, were obtained. These antimonides proved to be synthetically useful and, with organic halides, the expected stibanes or distibanes were formed. For example, the reaction of the antimonides with EtBr gave $PhSbEt_2$ and $Et(Ph)SbSb(Ph)Et$, respectively[36].

The first examples of cyclostibane complexes[57] were obtained by reaction of the cage compounds $MeC(CH_2Sb)_3$ with $M(CO)_5THF$ (M = Cr, Mo, W) (equation 21).

$$MeC(CH_2Sb)_3 + M(CO)_5THF \longrightarrow Me-C \begin{matrix} CH_2-Sb \\ | \\ CH_2-Sb \\ | \\ CH_2-Sb \end{matrix} M(CO)_5 + THF \quad (21)$$

M = Cr, Mo, W

The cyclostibane complexes were obtained in 50–76% yield as red crystals. The Mo complex is less stable than the Cr and W derivatives. The latter can be stored in the dark for months at room temperature. With these results a perspective is opened for the coordination chemistry of cyclostibane ligands.

V. REFERENCES

1. I. Haiduc, *The Chemistry of Inorganic Rings*, Part 1, Wiley, New York, 1970, p. 102.
2. G. O. Doak and L. D. Freedman, *Organometallic Compounds of Arsenic, Antimony, and Bismuth*, Wiley, New York, 1970, p. 249.
3. L. R. Smith and J. L. Mills, *J. Organometal. Chem.*, **84**, 1 (1975).
4. S. Samaan, *Methoden der Organischen Chemie (Houben-Weyl)*, Band XIII/8, *Metallorganische Verbindungen, As, Sb, Bi*, Thieme, Stuttgart, 1978, p. 154.
5. A. L. Rheingold, in *Homoatomic Rings, Chains and Macromolecules of Main Group Elements* (Ed. A. L. Rheingold), Elsevier, Amsterdam, 1977, p. 385.
6. A. Tzschach and J. Heinicke, *Arsenheterocyclen*, VEB Deutscher Verlag für Grundstoffindustrie, Leipzig, 1978, p. 34.
7. F. Kober, *Chem. Ztg.*, **105**, 199 (1981).
8. I. Haiduc and D. B. Sowerby, *The Chemistry of Inorganic Homo- and Heterocycles*, Vol. 2, Academic Press, London, 1987, p. 701.
9. H. H. Sisler, in *Inorganic Reactions and Methods*, Vol. 7 (Ed. J. J. Zuckerman), VCH Publishers, New York, 1988.
10. B. J. Aylett, *Organometallic Compounds*, Fourth Edition, Volume One, Part Two, *Groups IV and V*, Chapman and Hall, London, 1979.
11. I. Haiduc, *The Chemistry of Inorganic Rings*, Part 1, Wiley, New York, 1970, p. 107.
12. G. O. Doak and L. D. Freedman, *Organometallic Compounds of Arsenic, Antimony, and Bismuth*, Wiley, New York, 1970, p. 402.
13. S. Samaan, *Methoden der Organischen Chemie (Houben-Weyl)*, Band XIII/8, *Metallorganische Verbindungen, As, Sb, Bi*, Thieme, Stuttgart, 1978, p. 479.

14. M. Wieber, *Gmelin Handbook of Inorganic Chemistry, Sb Organoantimony Compounds Part 2*, Springer-Verlag, Berlin, 1981, p. 150.
15. I. Haiduc and D. B. Sowerby, *The Chemistry of Inorganic Homo- and Heterocycles*, Vol. 2, Academic Press, London, 1987, p. 729.
16. V. Auger, *Compt. Rend. acad. Sci. Paris*, **138**, 1705 (1904).
17. C. S. Palmer and A. B. Scott, *J. Am. Chem. Soc.*, **50**, 536 (1928).
18. E. J. Wells, *Can. J. Chem.*, **46**, 2733 (1968).
19. P. S. Elmes, S. Middleton and B. O. West, *Aust. J. Chem.*, **23**, 1559 (1970).
20. A. L. Rheingold, *Organomet. Synth.*, **3**, 613 (1986).
21. J. H. Burns and J. Wazer, *J. Am. Chem. Soc.*, **79**, 859 (1957).
22. J. Wazer and V. Schomaker, *J. Am. Chem. Soc.*, **67**, 2014 (1945).
23. A. L. Rheingold and J. M. Bellama, *J. Organomet. Chem.*, **102**, 445 (1975).
24. F. Knoll, H. C. Marsmann and J. R. Van Wazer, *J. Am. Chem. Soc.*, **91**, 4986 (1969).
25. A. L. Rheingold and P. Choudhury, *J. Organomet. Chem.*, **128**, 155 (1977).
26. P. Choudhury and A. L. Rheingold, *Inorg. Chim. Acta*, **28**, L 127 (1978).
27. M. Ates, H. J. Breunig, S. Gülec, W. Offermann, K. Häberle and M. Dräger, *Chem. Ber.*, **122**, 473 (1989).
28. M. Ates, H. J. Breunig, K. Ebert, S. Gülec, R. Kaller and M. Dräger, *Organometallics*, **11**, 145 (1992).
29. A. Tzschach and V. Kiesel, *J. Prakt. Chem.*, **313**, 259 (1971).
30. V. K. Gupta, L. K. Krannich and C. L. Watkins, *Inorg. Chem.*, **26**, 1638 (1987).
31. V. K. Gupta, L. K. Krannich and C. L. Watkins, *Synth. React. Inorg. Met.-Org. Chem.*, **17**, 501 (1987).
32. A. L. Rheingold and S. Natarajan, *J. Organomet. Chem.*, **231**, 119 (1982).
33. M. Baudler, *Angew. Chem.*, **94**, 520 (1982); *Angew. Chem., Int. Ed. Engl.*, **21**, 492 (1982).
34. J. W. B. Reesor and G. F. Wright, *J. Org. Chem.*, **22**, 382 (1957).
35. A. L. Rheingold and P. J. Sullivan, *Organometallics*, **2**, 327 (1983).
36. K. Issleib and A. Balszuweit, *Z. Anorg. Allg. Chem.*, **419**, 87 (1976).
37. H. J. Breunig, K. Häberle, M. Dräger and T. Severengiz, *Angew. Chem.*, **97**, 62 (1965).
38. H. J. Breunig, A. Soltani-Neshan, K. Häberle and M. Dräger, *Z. Naturforsch.*, **41b**, 327 (1986).
39. H. J. Breunig, S. Gülec and R. Kaller, *Phosphorus, Sulfur, and Silicon*, **64**, 107 (1992).
40. M. Baudler and P. Bachmann, *Angew Chem.*, **93**, 112 (1981); *Angew. Chem., Int. Ed. Engl.*, **20**, 123 (1981).
41. a) O. Mundt, G. Becker, H.-J. Wessely, H. J. Breunig and K. Kischkel, *Z. Anorg. Allg. Chem.*, **4486**, 70 (1982).
    b) H. J. Breunig and H. Kischkel, *Z. Anorg. Allg. Chem.*, **502**, 175 (1983).
42. M. Baudler and P. Bachmann, *Z. Anorg. Allg. Chem.*, **485**, 129 (1982).
43. R. L. Wells, C.-Y. Kwag, A. P. Purdy, A. T. Mc Phail and C. G. Pitt, *Polyhedron*, **9**, 319 (1990).
44. A. L. Rheingold, *J. Organomet. Chem.*, **102**, 445 (1975).
45. R. P. Tan, N. M. Comerlato, D. R. Powell and R. West, *Angew. Chem.*, **104**, 1251 (1992); *Angew. Chem., Int. Ed. Engl.*, **31**, 1217 (1992).
46. L. Weber and D. Bungardt, *J. Organomet. Chem.*, **354**, C 1 (1988).
47. M. Ates, H. J. Breunig, A. Soltani-Neshan and M. Tegeler, *Z. Naturforsch.*, **41b**, 321 (1986).
48. H. J. Breunig and A. Soltani-Neshan, *J. Organomet. Chem.*, **262**, C 27 (1984).
49. T. F. Berlitz, H. Sinning, L. Lorberth and U. Müller, *Z. Naturforsch.*, **43b**, 744 (1988).
50. M. Baudler and S. Weitfeldt-Haltenhoff, *Angew. Chem.*, **96**, 361 (1984); *Angew. Chem., Int. Ed. Engl.*, **23**, 379 (1984).
51. M. Baudler, J. Hellmann, P. Bachmann, K. F. Tebbe, R. Fröhlich and M. Feher, *Angew. Chem.*, **93**, 415 (1981); *Angew. Chem., Int. Ed. Engl.*, **20**, 406 (1986).
52. H. G. von Schnering, D. Fenske, W. Hönle and M. Binnewies, *Angew. Chem.*, **91**, 755 (1979); *Angew. Chem., Int. Ed. Engl.*, **18**, 679 (1979).
53. J. Ellermann and H. Schössner, *Angew. Chem.*, **86**, 646, (1974); *Angew. Chem., Int. Ed. Engl.*, **13** (1974) 601.
54. G. Thiele, G. Zoubek, H. A. Lindner and J. Ellermann, *Angew. Chem.*, **90**, 133 (1978); *Angew. Chem., Int. Ed. Engl.*, **17**, 135 (1978).
55. J. Ellermann and M. Lietz, *Z. Naturforsch.*, **35b**, 1514 (1980).
56. J. Ellermann and A. Veit, *Angew. Chem.*, **94**, 377 (1982); *Angew. Chem., Int. Ed. Engl.*, **21**, 375 (1982).

57. J. Ellermann and A. Veit, *J. Organomet. Chem.*, **290**, 307 (1985).
58. J. Ellermann, E. Köck and H. Burzlaff, *Acta Crystallogr.*, **C 41**, 1437 (1985).
59. D. Herrmann, *Z. Anorg. Allg. Chem.*, **416**, 50 (1975).
60. G. Sennyey, F. Mathey, J. Fischer and A. Mischler, *Organometallics*, **2**, 298 (1983).
61. B. O. West, in *Homoatomic Rings, Chains and Macromolecules of Main Group Elements* (Ed. A. L. Rheingold), Elsevier, Amsterdam, 1977, p. 409.
62. A. L. Rheingold and A.-J. DiMaio, *Organometallics*, **5**, 393 (1986).
63. H. J. Breunig, unpublished results.
64. H. J. Breunig, S. Gülec and R. Kaller, *Phosphorus, Sulfur, and Silicon*, **67**, 33 (1992).
65. Y. Mourad, Y. Mugnier, H. J. Breunig and M. Ates, *J. Organomet. Chem.*, **398**, 85 (1990).

CHAPTER **15**

Syntheses and uses of isotopically labelled compounds of bismuth, antimony and arsenic

MIECZYSLAW ZIELIŃSKI

Isotope Laboratory, Faculty of Chemistry, Jagiellonian University, Cracow, Poland
and

MARIANNA KAŃSKA

Department of Chemistry, The University of Warsaw, Warsaw, Poland

The chemistry of organic arsenic, antimony and bismuth compounds
Edited by S. Patai © 1994 John Wiley & Sons Ltd

I. INTRODUCTION

Organometallics are a very active domain of contemporary chemistry with an interdisciplinary character[A–C]. They are reactive intermediates or catalysts frequently used in organic syntheses, in industrial processes including protection of metal surfaces against corrosion and among them there are products of an applied character with wide prospects for the future. Inorganic compounds of radioactive bismuth are known in radiochemistry and nuclear chemistry as members in the natural uranium and thorium decay series since the basic work of M. Curie-Skłodovska. Organometallic compounds of radioactive Bi, Po, Pb, Sb and As are the objects of studies by contemporary radiochemists for several reasons: (1) Hot atom chemistry provides a direct one-step synthesis of different organometallic compounds of Bi, Po and Pb, which are difficult or impossible to obtain by traditional schemes used in preparative organic chemistry; (2) the relatively short-lived α-emitters found in natural decay series are very toxic for proliferating cancer cells and find various uses, in nuclear medicine; (3) both Bi and As have been known in traditional pharmacy for many years as the constituents of many drugs. The elucidation of their not fully understood behaviour in living organisms on the molecular level is an urgent task for many laboratory and clinical research teams; (4) finally, the impact of arsenic compounds emitted by power plants into the environment imposed on environmental radiochemists the task of detecting arsenic in its diverse forms at very low concentration levels in water reservoirs and in the biosphere in general. Cooperation of radiochemists, inorganic chemists, biochemists and scientists engaged in nuclear medicine resulted in considerable progress in the practical applications of the isotopes described in this chapter. Physicists are also involved in these studies by determining the structure of compounds of As, Sb and Bi of practical importance.

II. SYNTHESIS AND USES OF ISOTOPICALLY LABELLED BISMUTH COMPOUNDS

A. Synthesis of Bismuth Compounds Labelled with Stable Isotopes

1. Synthesis of tris-(o-deuteriophenyl)bismuth and bis(o-deuteriophenyl)bismuth bromide

Tris(o-deuteriophenyl)bismuth **1** has been prepared[1–3], as shown in equation 1, by reacting $BiCl_3$ with Grignard reagent **2** by analogy to the Ph_3Bi synthesis[3,4]. The line at 2247 cm^{-1} observed in I. R. and Raman spectrum of **1** corresponds to the v(C—D) bond formed.

$$o\text{-}C_6H_4BrCl \xrightarrow{Mg, Et_2O} o\text{-}C_6H_4(MgBr)Cl \xrightarrow{CH_3COOD, Et_2O} o\text{-}C_6H_4DCl \xrightarrow{Mg, Et_2O}$$
$$48.3\%$$

$$o\text{-}C_6H_4(MgCl)D \xrightarrow{BiCl_3, Et_2O} \qquad (o\text{-}C_6H_4D)_3Bi \qquad (1)$$
$$\textbf{(2)} \qquad\qquad\qquad \textbf{(1)} \ 69.5\% \ \text{Yield, m.p. } 73.5\text{–}74.0\,^{\circ}C$$

Bis(o-deuteriophenyl)bismuth bromide **3** has been obtained similarly[5] to Ph_2BiBr, in the redistribution reaction of 2 moles of **1** and 1 mole of $BiBr_3$ (equation 2).

$$2(o\text{-}C_6H_4D)_3Bi + BiBr_3 \xrightarrow{Et_2O} \qquad (o\text{-}C_6H_4D)_2BiBr \qquad (2)$$
$$\textbf{(3)} \ 85.3\% \ \text{yield, m.p. } 152\text{–}154\,^{\circ}C$$

Analysis of the ^1H NMR spectrum of Ph_3Bi and of 1 showed that the paramagnetically shifted ^1H NMR signal observed in the phenyl derivatives of bismuth has to be assigned to the *ortho* protons as in the phenyl derivatives of mercury[5,6]. The relation of the integral signal was found to be $[o\text{-}H/(m\text{-}H + p\text{-}H)] = 2/3$ in Ph_3B but $1/3$ in the case of **1**. In the ^1H NMR spectrum of $(o\text{-}C_6H_4D)_2BiBr^1$ the relation for protons $[o\text{-}H/(m\text{-}H + p\text{-}H)]$ was again $1/3$.

The ground-state infrared spectrum of BiH and BiD has been reassessed by Urban, Polomsky and Jones[7].

2. Synthesis of methyl[^{13}C]bis(thiomethanolato)bismuth(III)

The methyl[^{13}C]labelled dithiolate complex **4** has been synthesized[8] (75.7% yield) from methyl[^{13}C]dibromobismuth(III)[8,9] and lithium thiomethanolate (equation 3).

$$^{13}CH_3BiBr_2 + 2CH_3SLi \xrightarrow{RT, AT} {}^{13}CH_3Bi(SCH_3)_2 + 2LiBr \qquad (3)$$
$$\textbf{(5)} \qquad\qquad\qquad\qquad \textbf{(4)}$$

The ^{13}C-labelled **4** has been used to study the ion genesis in the M.S. of unlabelled **4** and for characterization of its bactericidal activity. **4** is more toxic than the aminothiophenol derivatives of bismuth, for instance $C_6H_5Bi(SC_6H_4NH_2\text{-}p)_2$, **6**[10,11].

B. Synthesis and Applications of Bismuth Compounds Labelled with Tritium and Carbon-14

1. Nuclear synthesis of tritium-labelled organometallic compounds of Bi, Sb and As

Tritium multilabelled organic compounds of V–VII Group elements including Ph_3N, Ph_3P, Ph_3As, Ph_3Sb and Ph_3Bi have been synthesized[12,13] employing the highly reactive tritium-labelled phenyl cations[14] generated in 75% yield in the beta decay of the tritium of multilabelled benzene (equations 4–8).

$$C_6{}^3H_6 \xrightarrow{\beta^-} [C_6{}^3H_5He]^+ \longrightarrow C_6{}^3H_5{}^+ + He \qquad (4)$$

In equations 5–7, n denotes the number of the group of the element M (i.e., $n = 5$ for the elements N, P, As, Sb and Bi).

$$C_6{}^3H_5{}^+ + R_{8-n}M \longrightarrow [C_6{}^3H_5R_{8-n}M]^+ \qquad (5)$$

$$[C_6{}^3H_5R_{8-n}M]^+ + X^- \longrightarrow [C_6{}^3H_5R_{8-n}M]^+X^- \qquad (6)$$
$$\textbf{(2)}$$

where $X^- = Cl^-, I^-, BF_4^-, ClO_4^-$.

$$[C_6{}^3H_5R_{8-n}M]^{*+} \xrightarrow{-R^+} C_6{}^3H_5R_{(8-n)-1}M \qquad (7)$$

(7)

$$C_6{}^3H_5{}^+ + 2C_6H_6 \longrightarrow C_6{}^3H_5-Ph + C_6H_6 \cdot H^+ \qquad (8)$$

(8)

The ion–molecule reactions 4–7, yielding the tritium-labelled organometallic compounds 7 and a certain amount of diphenyl compounds 8, were carried out in sealed ampoules containing benzene multilabelled with tritium, the organometallic compound to be labelled and salts of inorganic acids such as KI or KBF_4. The yields of salts 9, $[C_6{}^3H_5(C_6H_5)_3M^+ \cdot BF_4^-$, and of organometallic compounds 7 $C_6{}^3H_5Ph_2M$ (given in parentheses), synthesized in the solid phase in the presence of BF_4^-, were found to be 15.6% (and 20.9%) for M = As; 30.8% (and 17.7%) for M = Sb and 10.7% (and 22.2%) for M = Bi. The yields are the ratios of the activity of the labelled isolated organometallic compound to the total activity of all tritium-labelled compounds obtained.

2. Carbon-14 ligand exchange studies between triphenylbismuth or triphenylantimony and phenyllithium

Ligand exchanges ('redistribution') between two different organometallic compounds are widely used in preparative organometallic chemistry[15-20]. Lithium organic compounds are frequently applied in such syntheses, while organometallic compounds of Sb and Bi are employed as the second reagent.

The rate of organic ligand exchange depends on the nature of the ligand and on the nature of the solution[16,17]. Benzene and hexane inhibit the ligand exchange, but ether accelerates it. This problem has been investigated[20] by undertaking systematic studies of the ligand exchanges in the systems Ph_3Sb—PhLi (I) and Ph_3Bi—PhLi (II) in pentane/ether solutions at 248–307 K with the use of ^{14}C-labelled phenyl ligands (equation 9).

$$(^{14}C_6H_5)_3M + 3C_6H_5Li \rightleftharpoons (C_6H_5)_3M + 3\,^{14}C_6H_5Li \qquad (9)$$

$$M = Sb \text{ (system I), Bi (system II)}$$

Non-labelled and ^{14}C-labelled MPh_3 have been obtained as described earlier[15,16]. The dielectric constants ε of binary solutions employed have been calculated for different temperatures[18]. The rate of the exchange reaction 9 has been followed by determining the radioactivity of benzoic acid separated from the exchange mixture at preset times, calculating the exchange degrees $F_t = A_t/A_\sim$; where A_t and $A_\sim$ are specific activities of benzoic acid at time t and at equilibrium, respectively, and evaluating the exchange rate constants k from the slopes ($\tan \alpha$) of straight lines expressing the dependence of $-\ln(1-F)$ on time t^{20}, $k = 3 \tan \alpha/(a+3b)$, where a and b are the initial concentrations (in $mol\,dm^{-3}$) of PhLi and Ph_3M, respectively.

No ligand exchange has been found between $(^{14}C_6H_5)_3Bi$, $(^{14}C_6H_5)_3Sb$ and C_6H_5Li in pure pentane, but the rate increases with N, the molarity of ether in the solution. The C—Li bond is more polar in the ether complexes than in the pure non-solvated PhLi. A further increase in the rate of ligand exchange at N larger than 0.5 and ε larger than 2.8 is caused by the increase in the dielectric constant of the solution. Thus the specific solvation of PhLi is not the unique factor responsible for the increase in the exchange rates at N larger than 0.5. The ligand exchange reaction of equation 9 is first order with respect to each of the reagents, Ph_3M and PhLi. No departure from linearity of semilogarithmic plots, $-\ln(1-F)$ vs t, has been observed. This means that the reactions are clean ones without side reactions. The exchange rate constants for the Ph_3Sb—PhLi system are by two orders of magnitude higher than for the Ph_3Bi—PhLi system [for instance, k (for I) $= 13.5 \times 10^{-3} \, mol^{-1}$ at 278 °C and at $\varepsilon = 2.988$; k (for II) $= 7.6 \times 10^{-5} \, mol^{-1} \, s^{-1}$ at 293 °C and at $\varepsilon = 2.988$). The experimental dependence, $k = f(\varepsilon)$, of the rate constants k of the exchange reaction 9 on the dielectric constant ε is given by equation 10, describing the interaction between two dipoles:

$$\lg(k) = \lg(k_0) - (1/2.3 \, RT)[(\varepsilon - 1)/(2\varepsilon + 1)]\sum(\mu^2/r^3) \tag{10}$$

where k_0 is the rate constant of the hypothetical reaction 9 in the gase phase,

$$\sum(\mu^2/r^3) = (\mu_1^2/r_1^3) + (\mu_2^2/r_2^3) - [(\mu^{\#})^2/(r^{\#})^3]$$

and $\mu_1, \mu_2, \mu^{\#}$ and $r_2, r_2, r^{\#}$ are dipole moments and effective radiii of the substrates and of the transition state, respectively.

The observed increase in the rate constants k with increase in ε of the solution indicates that the transition state (T.S.) of the reaction 9 is more polar and more solvated than the substrates of the reaction[19] (a preferential non-specific solvation of T.S. does not lead to chemical changes of the solvent molecules). The exchange rate constants are well described by the Arrhenius temperature dependence for both systems; E_{Arrh} and $\lg(A_{Arrh})$ decrease with the increase in the dielectric constant ε for both exchange systems. In the case of the $Sb(^{14}C_6H_5)_3$—LiC_6H_5 system E_{Arrh} and $\lg(A)$ change from $75.0 \, kJ \, mol^{-1}$ and 9.17 at $\varepsilon = 1.804$ to $51.1 \, kJ \, mol^{-1}$ and 5.27 at $\varepsilon = 3.912$ (both at $T = 278 \, K$); $\Delta G^{\#}$ changes negligibly in this interval of ε values from $95.3 \, kJ \, mol^{-1}$ to $93.2 \, kJ \, mol^{-1}$.

In the case of the $Bi(^{14}C_6H_5)_3$—LiC_6H_5 exchange system E_{Arrh}, $\lg(A)$ and $\Delta G^{\#}$ (calculated at 293 K) changed from $50.8 \, kJ \, mol^{-1}$, 7.68 and 77.9 $kJ \, mol^{-1}$ respectively (taken at $\varepsilon = 2.988$) to $43.7 \, kJ \, mol^{-1}$, 6.82 and $75.4 \, kJ \, mol^{-1}$ at $\varepsilon = 4.339$.

Linear correlations 11 and 12 between enthalpy and entropy changes for I- and II-exchange systems have been found:

$$\Delta H_I^{\#} = 94.2 + 0.432 \, \Delta S_I^{\#} \tag{11}$$

$$\Delta H_{II}^{\#} = 97.5 + 0.321 \, \Delta S_{II}^{\#} \tag{12}$$

Increase in the polarity of the T.S. with increase in ε leads to stronger solvation and to decrease in $\Delta S^{\#}$. Additional solvation of the T.S. increases the possibility of charge delocalization in the T.S. and to the decrease in $\Delta H^{\#}$ ($\Delta G^{\#}$ does not change in such a case). The relatively low values of $\Delta S^{\#}$ indicate the high degree of ordering of the T.S.[21]. The dependence of rate constants and activation parameters on ε indicate that, at ε higher than 2.8, the exchange reaction 9 is determined by electrostatic interactions.

C. Radiochemical Syntheses and Uses of Organometallic Compounds of Radioactive Bismuth Isotopes

1. Introduction

As early as in 1934 Martenson and Leighton[22-24] noticed that in the $^{238}U \longrightarrow {}^{206}Pb$ decay series (equation 13), in the final stages of the series initiated by the volatile $^{210}PbMe_4$, after a decay time sufficient for accumulating appreciable amounts of ^{210}Bi(RaE) and ^{210}Po(RaF) (about thirty days) the volatile fraction contained β- and α-radioactivity of RaE and RaF, respectively.

This means that in the decays shown in equation 14 no complete rupture of all chemical bonds with ^{210}Pb and with ^{210}Bi occurred and the production of volatile organometallic compounds of RaE, $^{210}Bi \, Me_3$, and of $^{210}PoMe_2$ took place according to equation 15.

$$^{238}U \xrightarrow{\alpha(E_\alpha = 4.20\,\text{MeV},\ T_{1/2} = 4.51 \times 10^9\,\text{year})} {}^{234}Th \xrightarrow{\beta^- (E_{max} = 0.19\,\text{MeV},\ T_{1/2} = 24.1\,\text{day})}$$

$$^{234}Pa\,[\text{UX}_2,\ 99.4\%,\ E_{max} = 2.29\,\text{MeV},$$

$$T_{(1/2)} = 1.175\,\text{min},\ \text{UZ},\ 0.6\%,\ E_{max} = 0.53\,\text{MeV},\ T_{1/2} = 6.65\,\text{h}]\ \xrightarrow{\beta^-}$$

$$^{234}U \xrightarrow{\alpha(4.77\,\text{MeV},\ 2.48 \times 10^5\,\text{year})} {}^{230}Th \xrightarrow{\alpha(4.68\,\text{MeV},\ 7.52\,10^4\,\text{year})}$$

$$^{226}Ra \xrightarrow{\alpha(4.78\,\text{MeV},\ 1602\,\text{year})} {}^{222}Em\,(Rn) \xrightarrow{\alpha(5.49\,\text{MeV},\ 3.825\,\text{day})}$$

$$^{218}Po^r\,(RaA) \xrightarrow{\alpha(6.00\,\text{MeV},\ 3.05\,\text{min})} {}^{214}Pb\,(RaB) \xrightarrow{\beta^- (1.03\,\text{MeV},\ 26.8\,\text{min})}$$

$$^{214}Bi^r\,(RaC) \xrightarrow{\beta^- (3.26\,\text{MeV},\ 19.7\,\text{min})} {}^{214}Po\,(RaC') \xrightarrow{\alpha(7.69\,\text{MeV},\ 1.64 \times 10^{-4}\,\text{s})}$$

$$^{210}Pb^r\,(RaD) \xrightarrow{\beta^- (0.061,\ \text{MeV},\ 22.0\,\text{year})} {}^{210}Bi^r\,(RaE) \xrightarrow{\beta^- (1.16\,\text{MeV},\ 5.0\,\text{day})}$$

$$^{210}Po\,(RaF) \xrightarrow{\alpha(5.3\,\text{MeV},\ 138.4\,\text{day})} {}^{206}Pb\,(RaG,\ \text{Stable}) \qquad (13)$$

Superscript ^{r} denotes branching decay.

$$^{210}Pb \xrightarrow{\beta^-} {}^{210}Bi \xrightarrow{\beta^-} {}^{210}Po \xrightarrow{\alpha} \qquad (14)$$

$$^{210}PbMe_4 \xrightarrow{\beta^-} {}^{210}BiMe_3^+ \xrightarrow{+e^-} {}^{210}BiMe_3$$

$$\xrightarrow{\beta^-} {}^{210}PoMe_2^+ \xrightarrow{+e^-} {}^{210}PoMe_2 \qquad (15)$$

It has been suggested[22] that the formation of the volatile $^{210}BiMe_3$ probably proceeds according to the thermodynamically more favourable equation16 and that the expulsion of an electron from the nucleus apparently causes the rupture of one bond only.

$$^{210}PbMe_4 \xrightarrow{\beta^-} [^{210}BiMe_4]^+ \longrightarrow {}^{210}BiMe_3 + Me^+,\quad \text{etc.} \qquad (16)$$

The radiochemistry of organometallic compounds of bismuth containing aromatic radicals has been investigated by Nefedov and coworkers[25].

2. Radiochemical studies and radiochemical synthesis with $^{210}Bi(RaE)$

$^{210}PbPh_4$, $^{210}PbPh_3Cl$, $^{210}PbPh_2Cl_2$, $BiPh_3$ and $BiPh_3Cl_2$ have been synthesized[26] as described in Reference 27 and have been used for identification of stable chemical forms assumed by the daughter $^{210}Bi(RaE)$ in benzene solutions. $^{210}BiPh_3$, $^{210}BiPh_2Cl$, $^{210}BiPh_3Cl_2$ and the unidentified forms of $^{210}Bi^{3+}$ have been isolated by paper chromatography and their yields determined. Similar studies have been carried out also[28] with crystalline $^{210}PbPh_4$ and $^{210}PbPh_3Cl$, as well as with $^{210}PbPh_2Cl_2$, $(^{210}PbPh_3)_2$,

$^{210}PbPh_3NO_3$ and $^{210}PbPh_2(NO_3)_2$[29,30]. The yields of $^{210}BiPh_3$ and $^{210}BiPh_3Cl_2$ are very small in the decays of $^{210}PbPh_2Cl_2$. The higher yields of $^{210}BiPh_3Cl_2$ found in crystals compared to those in solutions did not depend on the methods of isolation. The mechanism of formation of small amounts of $^{210}BiPh_3Cl_2$ in the course of β-decay of $^{210}PbPh_4$ or in the course of separation procedures has not been discussed. The phenyl derivatives of Pb, Bi and Sb are stable crystalline compounds which exchange their central M atom very slowly at RT. The half-exchange time of ^{210}Bi between[31] $^{210}BiPh_3$ and $BiPh_3Cl_2$ (equation 17) in 96% EtOH at 100 °C equals 82.5 min.

$$^{210}BiPh_3 + BiPh_3Cl_2 \rightleftharpoons BiPh_3 + {}^{210}BiPh_3Cl_2 \tag{17}$$

The maximum recoil energy of $RaE(^{210}Bi)$ produced in the β-disintegration of $RaD(^{210}Pb)$ equals 0.06 eV at the β- particle emission, 0.006 eV at the emission of the most hard γ-rays and 0.12 eV at the emission of 'electrons of conversion' from the N-shell[32–34]. In all cases the recoil energy is too low to break the chemical bond (about 2.5 eV for one bond). The probability of interaction of the emitted beta particles directly with the electrons of the atom is also small[35]. The non-adiabatic change (increase or decrease) of the nuclear charge in β^-- and β^+-emissions results in ionization of the peripheral shell electrons, when the probability of ionization of K, L and M shells equals 10^{-3}, 10^{-2} and 5×10^{-2} only[36,37].

According to Wu and coworkers[38] the degrees of internal conversion of γ-rays of $RaD(^{208}Pb)$ on different shell electrons are $(N_{eL}/N_\beta) = 64\%$, $(N_{eM}/N_\beta) = 21\%$ and $(N_{eL+M+N}/N_\beta) = 85 \pm 5\%$. Internal conversion of γ-rays results[39,40] in substantial ionization of the atom and in the decomposition of the molecule if the atom emitting the radiation was chemically bound[41,42]. β^+- and β^--emissions resulting in changes of the charge of the nucleus by -1, $+1$ unit enhance also changes of the valency state of the daughter atom[43]. A special study[44] has been devoted to the investigation of the dependence of the distribution of radioactive bismuth isotopes between different stable chemical forms on the decay schemes of natural isotopes of lead, $RaD(^{210}Pb) \xrightarrow{\beta^-} RaE(^{210}Bi)$, $ThB(^{212}Pb) \xrightarrow{\beta^-} ThC(^{212}Bi)$ and $RaB(^{214}Pb) \xrightarrow{\beta^-} RaC(^{214}Bi)$. It has been found that in the β-decay of $^{210}BiPh_3Cl$ $19 \pm 1\%$ of ^{210}Bi stabilizes as $^{210}BiPh_3Cl_2$ and $43 \pm 3\%$ as $^{210}Bi^{3+}$ while in the β-decay of $^{212}PbPh_3Cl$ about $50 \pm 2\%$ of ^{212}Bi is stabilized as $^{212}BiPh_3Cl_2$ and $17 \pm 1\%$ as $^{212}Bi^{3+}$. It is obvious that the difference in chemical behaviour of isotopes of bismuth is related in this case to different schemes of decay of the parent radioisotopes, $RaD(^{210}Pb)$ and $ThB(^{212}Bi)$. In the β-decay of $*PbPh_4$, where $*Pb = {}^{210}Pb$, ^{212}Pb, ^{214}Pb, the yields of $^{210}Bi^{3+}$, $^{212}Bi^{3+}$ and $^{214}Bi^{3+}$ were equal to $45 \pm 1\%$, $25 \pm 2\%$ and $10 \pm 1\%$, respectively. These 'radiochemical isotope effects' are caused by different values of internal conversions[44,45] equal to 60–80% for RaD, to 30% for ThB and to 15% for RaB.

Organometallic compounds of ^{210}Po *(RaF)* are produced as the result of β-decay taking place in crystals of $^{210}BiPh_3$ and $^{210}BiPh_3Cl_2$, which in turn were obtained in the β-decay of ^{210}Pb or $^{210}PbPh_4$ of high activity. $^{210}PoPh_2Cl_2$ $(15 \pm 6\%)$, $^{210}PoPh_2$ $(24 \pm 6\%)$ and $61 \pm 6\%$ of unidentified compounds of ^{210}Po have been isolated[46] from $^{210}BiPh_3$ by chromatography.

Final chemical forms assumed by ^{210}Po *in the beta decays of* $^{210}Bi(p\text{-}Tol)_3$ *and* $^{210}Bi(P\text{-}Tol)_3Cl_2$. Inorganic ^{210}Po $(8 \pm 3\%)$, $^{210}Po(p\text{-}Tol)_2$ $(11 \pm 4\%)$, $^{210}Po(p\text{-}Tol)_2Cl_2$ $(32\text{-} \pm 7\%)$ and ^{210}Po $(p\text{-}Tol)_3Cl$ $(49\% \pm 5\%)$ have been isolated by paper chromatography as the daughter products of RaE undergoing beta disintegration in the mother ^{210}Bi (RaE) Tol_3 crystals. In the β-decay of RaE in $^{210}BiTol_3Cl_2$ crystals, ^{210}Po $(p\text{-}Tol)_2$ $(13 \pm 3\%)$, $^{210}Po(p\text{-}Tol)_2Cl_2$ $(80 \pm 3\%)$ and $^{210}Po(p\text{-}Tol)_3Cl$ $(7 \pm 2\%$ yield) have been separated[47].

TABLE 1 Yields of organometallic compounds of Po-210 produced in the β-decay of ^{210}BiAr$_3$ during 6–7 half-life-times of RaE at room temperature in the crystalline state

Mother compound	% Yield of Po–210 compounds			
	inorganic^{210}Po	^{210}PoAr$_3$X	^{210}PoAr$_2$X$_2$	^{210}PoAr$_2$
Bi(p-ClC$_6$H$_4$)$_3$	—	62 ± 3	—	38 ± 4
Bi(p-BrC$_6$H$_4$)$_3$	11 ± 3	65 ± 3	—	22 ± 1
Bi(o-Tol)$_3$	18 ± 3	47 ± 4	—	35 ± 4

α-*Naphthyl organometallic compounds of bismuth-210.* New organometallic α-naphthyl derivatives of two- and four-valent Po, namely ^{210}Po (1-C$_{10}$H$_7$)$_2$, ^{210}Po(1-C$_{10}$H$_7$)$_2$X$_2$ and ^{210}Po(1-C$_{10}$H$_7$)$_3$X, have been isolated[48] from ^{210}Bi(1-C$_{10}$H$_7$)$_3$ and ^{210}Bi(1-C$_{10}$H$_7$)$_3$Cl$_2$ crystals synthesized from bismuth irradiated in neutron flux of 10^{14} neutron s^{-1}cm^{-2}, by the method described in the literature[49,50]. In the presence of TeR$_2$Cl$_2$ used as a carrier, the redistribution reaction of equation 18 is

$$R_2Po + [R_2Te]^{2+} \rightleftharpoons [R_2Po]^{2+} + R_2Te \qquad (18)$$

responsible for the transformation of large amounts of R$_2$Po into PoR$_2$X$_2$. The excited ions [PoR$_3$X$_2$]$^+$ formed in the β-decay of RaE in ^{210}BiR$_3$Cl$_2$ might also give a small quantity of PoR$_3$Cl + Cl$^+$ found in the products of ^{210}BiR$_3$Cl$_2$ decomposition.

The chemical effects of nuclear transformations have been investigated also[51] in the case of β-decay of ^{210}Bi bound in crystalline compounds: ^{210}Bi(p-ClC$_6$H$_4$)$_3$, ^{210}Bi(p-BrC$_6$H$_4$)$_3$ and ^{210}Bi(o-Tol)$_3$. The respective yields of ^{210}Po-organometallic compounds are presented in Table 1.

A linear correlation (equation 19) has been found[51] between the logarithm of the yield Q (%) of the stabilization product and the value of the Hammett *para*-substituent constants σ_p by taking

$$\lg(Q) = 1.80 - 0.22(\sigma_p) \qquad (19)$$

$\sigma_p = 0$ for p-Cl and p-Br substituents. Assuming that there is a linear relation between the frequency of the symmetric vibration v_s of the Bi—C$_{Ar}$ bond and the dissociation energy D of this bond and that all triaryl derivatives of Bi have flat structure and the anharmonicity for them is the same, the dissociation energies D (in eV) have been calculated for the following Bi—C$_{Ar}$ bonds[52]:

^{210}BiPh$_3$($v_s = 450$cm^{-1}, $D = 2.05$ eV, 63% yield of stabilization products PoAr$_3$X),
^{210}Bi(p-Tol)$_3$(485cm^{-1}, 2.20 eV, 68%),
^{210}Bi(o-Tol)$_3$(437cm^{-1}, 1.98 eV, 47%),
^{210}Bi(m-Tol)$_3$(422cm^{-1}, 2.17 eV, 62%),
^{210}Bi(p-ClC$_6$H$_4$)$_3$480cm^{-1}, 1.91eV, 39%),
^{210}Bi(p-BrC$_6$H$_4$)$_3$(482cm^{-1}, 2.18eV, 65%),
^{210}Bi(p-NH$_2$SO$_2$C$_6$H$_4$)$_3$(425cm^{-1}, 1.93eV, 48%).

Crystalline o-, m- and p-Tolyl derivatives of ^{210}Bi were produced as described by Supniewski and Adams[52b], using radiochemically pure ^{210}BiCl$_3$. These decomposed in the course of storage giving (o-Tol)$_3$ ^{210}PoX(67 ± 4% yield), (m-Tol)$_3$^{210}PoX (39 ± 3%), (p-Tol)$_3$ ^{210}PoX (68 ± 3%) and (o-Tol)$_2$ ^{210}Po (33 ± 2%), (m-Tol)$_2$^{210}Po(61 ± 4%), (p-Tol)$_2$ ^{210}Po(32 ± 2%). The yields of the compounds of the type Ar$_3$^{210}PoX are practically the same for o- and p-tolyl derivatives and distinctly higher than the yield of the analogous m-derivatives. This indicates that the o- and p-methyl groups increasing the electron density at the C-^{210}Po bond increase the strength of this bond. The primary cations $[Ar_3{}^{210}Po]^+$, produced in the crystalline lattice according to Ar_3

$^{210}Bi \xrightarrow{\beta^-} [Ar_3{}^{210}Po]^+$, have a better chance to escape decomposition in the case of o- and p-tolyl derivatives and to produce $Ar_3{}^{210}PoX$ by associating with the counterion X^- in the solution in solvents of low polarity, than the m-tolyl derivative[53]. In the case of $(m\text{-Tol})_3{}^{210}Bi$, the excited molecular cations have higher probability to decompose according to the scheme $[Ar_3{}^{210}Po]^+ \longrightarrow Ar_2{}^{210}Po + Ar^+$ and to produce $(m\text{-Tol})_2{}^{210}Po$ in higher $(61 \pm 4\%)$ yields than the o- and $[(p\text{-Tol})_3{}^{210}Po]^+$ cations. The lower rate of energy dissipation in the case of m-tolyl derivative is also reflected in the lower m.p. $(66\,°C)$ of the m-tolyl derivative compared to the m.p. of the o- and p-derivatives $(131\,°C$ and $120\,°C)$.

p-Xylyl derivatives of ^{210}Bi. $Ar_2{}^{210}Po$ $(33 \pm 2\%$ *yield) and* Ar_3PoX $(67 \pm 4\%$ *yield) have* been isolated[54] from the β-decay of $(p\text{-Xyl})_3{}^{210}Bi$ and Ar_3PoX $(44 \pm 2\%)$, Ar_2PoX_2 $(56 \pm 2\%)$ in the β-decay of $(p\text{-Xyl})_3{}^{210}BiCl_2$ synthesized as described[52a].

p-Anisyl organometallic derivatives of ^{210}Bi. $(p\text{-}An)_3{}^{210}BiCl_2$ *and* $(p\text{-}An)_3{}^{210}Bi(RaE)$ synthesized by the method of Supniewski and Adams[52b,56] using bismuth irradiated in neutron flux of $10^{14}\,ns^{-1}cm^{-2}$, stored in the crystalline state or as 0.05 M solutions in benzene or in acetone for about one month, yielded $(p\text{-}An)_2{}^{210}Po\,(10)$, $(p\text{-}An)_2{}^{210}PoX_2$ (**11**) and $(p\text{-}An)_3{}^{210}PoX$ (**12**) besides inorganic ^{210}Po of high specific activity. The presence of the p-MeO group in the ligand increased the electron density at the metal–carbon bond and consequently increased the yield of **12** to 76% in the crystalline state, and to 95% in benzene in comparison to 63% and 50% yields of the primary molecular ions observed in the case of triphenyl and tri-p-tolyl-bismuth-210 in the crystalline state. The observed lower yield of **12** (61%) in acetone is explained by more efficient transfer of excitation energy of the primary cation to benzene molecules than to acetone molecules and by the interaction of the excited cations with the lone electron pair of oxygen in Me_2CO. Beta disintegration of $R_3{}^{210}BiCl_2$ leads to formation of the primary molecular ion $[R_3{}^{210}PoCl_2]^+$, which decomposes according to equation 20:

$$[R_3{}^{210}PoCl_2]^+ \longrightarrow R_2{}^{210}PoCl_2 + R^+ \tag{20}$$

Only the most excited ions undergo decomposition also according to the energetically less favourable scheme (equation 21):

$$[R_3{}^{210}PoCl_2]^+ \longrightarrow R_3PoCl + Cl^+ \tag{21}$$

The β-*decay of p-phenethyl derivative of trivalent* ^{210}Bi, $(p\text{-}EtOC_6H_4)_3{}^{210}Bi(RaE)$, yielded two new polonium compounds[57], $(p\text{-}EtOC_6H_4)_3{}^{210}PoX$ in 72% yield and $(p\text{-}EtOC_6H_4)_2{}^{210}Po$ (28% yield). The first is the stabilization form of the primary molecular ion $[(p\text{-}EtOC_6H_4)_3{}^{210}Po]^+$, the second is the result of its decomposition according to equation 22:

$$[(p\text{-}EtOC_6H_4)_3Po]^+ \longrightarrow (p\text{-}EtOC_6H_4)_2Po + (p\text{-}EtOC_6H_4)^+ \tag{22}$$

No inorganic forms of ^{210}Po have been noticed using EtOAc/HCOOEt (3/1) and benzene/ethanol (21/4) chromatographic systems.

The crystalline $(1,2,5\text{-}Me_3C_6H_2)_3{}^{210}Bi$, synthesized[58] from $[1,2,5\text{-}Me_3C_6H_2]MgBr$ and $^{210}BiCl_3$, produced in self-β-decay inorganic ^{210}Po (15%), $R_3{}^{210}PoX$ (35%), R_2PoX_2 (33%) and $R_2{}^{210}Po(17\%)$, where R = Mesityl. The yields of the primary products $R_3{}^{210}PoX$ decreased compared to the yields observed in β-decay of tolyl-, p-anisyl- and p-phenethyl derivatives of ^{210}Bi. This is taken as due to the decrease in the degree of conjugation in the primary molecular ion caused by the loss of coplanarity in the system.

In the majority of the cases the β-decay results in the loss of one aromatic radical and formation of R_2Po. The presence of the R_2PoX_2 compounds is the result of ox–redox reactions with one of the carriers, R_2TeX_2 or with air oxygen. The fraction of R_2Po decreased after keeping an aqueous solution of R_3 ^{210}Bi crystals in the presence of air during one hour. The lower degree of conjugation of Ms_3Bi ($Ms = $ Mesityl) and of $[Ms_3Po]^+$ ions is supported by the increase of their oxidation into Ms_3BiX_2 and Ms_2PoX_2. The higher conjugation in the p-anisyl and p-phenethyl derivatives results in a decrease in the rate of formation of oxides of pentavalent bismuth. Ms_2 ^{210}Po, Ms_2 $^{210}PoCl_2$, Ms $^{210}PoBr_2$ and Ms_2PoO have been synthesized earlier[59,60] (equation 23.)

$$Me_3C_6H_2MgBr \xrightarrow[\text{Treat. with water at RT, under argon}]{\text{PoCl}_4,\ \text{ether/benzene, 2h boiling}} \underset{\textbf{(13)}}{(Me_3C_6H_2)_2Po} \qquad (23)$$

$(M_3C_6H_2)_2$ $^{210}PoCl_2$ has been obtained by treating **13** with Cl_2, in benzene. Dimesithyl-polonium oxide has been obtained by air oxidation.

In the case of β-disintegrations[61,62] occurring in *tris(p-phenylsulphonamide) bismuth-210*, $(p\text{-}NH_2SO_2C_6H_4)_3$ ^{210}Bi, possessing the electron-withdrawing sulphonamido group, the electron density in the o- and p-positions of the benzene ring is depleted, the strength of the ^{210}Po—C bond in the excited $[R_3Po]^+$ cation is weakened and the probability of the reaction $[R_3Po]^+ \longrightarrow R_2Po + R^+$ is increased. This results in lower yields of compounds of the type R_3PoX in the β-decay of RaE (49%) compared to p-anisole and p-phenethyl derivatives (73 and 70%, respectively) having electron-donating characters. The compounds $(p\text{-}NH_2SO_2C_6H_4)_2$ ^{210}Po, $(p\text{-}NH_2SO_2C_6H_4)_2$ $^{210}PoX_2$ and $(p\text{-}NH_2SO_2C_6H_4)_3PoX$ have also been isolated in this work[61].

3. Radiochemical studies with ^{212}Bi and related studies with ^{208}Tl

a. ^{212}Bi *(ThC, half-life 60.5 min)* is formed during the β-decay of ThB (half-life 10.6 h). Thus, $PbPh_4$ labelled with ^{212}Pb(ThB), **14**, has been synthesized[29] in ether/toluene solution, as shown in equation 24 by treating $PhMgBr$ with dry lead(II) chloride labelled with ThB. The $^{212}PbPh_4$ (m.p. 224–226 °C) has been used subsequently to synthesize $^{212}PbPh_3Cl$ (m.p. 204–208 °C) as described before[62].

$$4PhMgX + 2\,^{212}PbX_2 \longrightarrow Ph_4\,^{212}Pb + 4MgX_2 + Pb \qquad (24)$$

The course of the reaction has been followed by measuring the hard β-radiation of ThC and ThC'' (see equation 25) by covering the samples with an Al foil (240 mg cm^{-2} thick[63]) which absorbs the soft β-radiation of ^{212}Pb. In the β-decay of $^{212}PbPh_4$, about 70% of the ^{212}Bi is stabilized as $^{212}BiPh_3$ while in the case of decomposition of $^{212}PbPh_3Cl$ only about 40% of ThC is stabilized as $^{212}BiPh_3$ accompanied by about 40% of the pentavalent compound of ThC, $^{212}BiPh_3Cl_2$. This has been explained by the increase in the ionization of the daughter molecule.

In the branched decay of ^{212}Bi, 64.06% occurs in the β^--mode, 35.94% in the α-mode.

In the case of the decay of the parent $^{212}PbPh_4$ compound, the chemical environment does not favour the stabilization of the pentavalent ThC and the $^{212}BiPh_3^{2+}$ ions, if formed, are reduced to $^{212}BiPh_3$ (equation 26).

The recoil energy of ^{212}Bi has been calculated to equal about 2.6 eV[63]. This value is comparable with the energy of a single chemical bond and is insufficient for breaking all chemical bonds in $^{212}PbPh_4$ and in $PbPh_3Cl$. The complete decomposition of these molecules results from internal conversion of 238-keV γ-photons emitted in 96% by ThB[64,65]. The value of the coefficient of this conversion[64–66] about 30%, which coincides with the about 30% fraction of inorganic ThC(^{212}Bi) atoms found in solution. In the α-decay of ^{212}Bi(ThC) the recoil energy is high enough to break chemical bonds and,

$$^{232}\text{Th} \xrightarrow{\alpha(E=4.01\,\text{MeV},\ T_{1/2}=1.41\times10^{10}\,\text{yr})} {}^{228}\text{Ra}\,(\text{MsTh1}) \xrightarrow{\beta^-(E_{\text{max}}=0.05\,\text{MeV},\ T_{1/2}=5.75\,\text{yr})}$$

$$^{228}\text{Ac}(\text{MsTh2}) \xrightarrow{\beta^-(E_{\text{max}}=1.11,1.7,2.11\,\text{MeV},\ T_{1/2}=6.14\,\text{h})} {}^{228}\text{Th}\,(\text{RdTh}) \xrightarrow{\alpha(E_\alpha=5.43\,\text{MeV},\ T_{1/2}=1.913\,\text{yr})}$$

$$^{224}\text{Ra}(\text{ThX}) \xrightarrow{\alpha(E_\alpha=5.68,\ 5.45\,\text{MeV},\ T_{1/2}=3.66\,\text{day})}$$

$$^{220}\text{Rn}(\text{Tn}) \xrightarrow{\alpha(E=6.29\,\text{MeV},\ 99.93\%,\ 5.75\,\text{MeV},\ 0.07\%,\ T_{1/2}=55.6\,\text{s})} \tag{25}$$

$$^{216}\text{Po}(\text{ThA}) \xrightarrow{\alpha(E_\alpha=6.778\,\text{MeV},\ 99.998\%,\ 5.985\,\text{MeV},\ 0.002\%,\ T_{1/2}=0.150\,\text{s})}$$

$$^{212}\text{Pb}(\text{ThB}) \xrightarrow{\beta^-(\langle E_\beta\rangle=0.101\,\text{MeV},\ E_{\text{max},\beta}=0.355\,\text{MeV},\ T_{1/2}=10.64\,\text{h})}$$

$$^{212}\text{Bi}(\text{ThC})^{\text{r}} \overset{\beta^-}{\underset{\alpha}{\rightleftarrows}} \begin{matrix} {}^{212}\text{Po}(\text{ThC}') \overset{\alpha}{\searrow} \\ {}^{208}\text{Tl}(\text{ThC}'') \underset{\beta^-}{\nearrow} \end{matrix} {}^{208}\text{Pb}(\text{Stable})$$

$$^{212}\text{BiPh}_3{}^{2+} + 2e \longrightarrow {}^{212}\text{BiPh}_3 \tag{26}$$

accordingly, about 50% of the ThC''(^{208}Tl) has been identified in inorganic forms and isolated from benzene or ether solutions of ^{212}BiPh$_3$. Thus lead-212 and bismuth-212 phenyl derivatives can be used for the production of radiochemically pure inorganic preparations of ^{212}Bi(ThC) and ^{208}Tl(ThC''). The half-life of ^{208}Tl(3.0 min), measured by analysing the decay curve of β-activity of aqueous solutions, served as an indication of the high radiochemical purity of the ^{208}Tl preparations ($T_{1/2} = 3.053$ min[67]).

b. *The valency states of* ^{208}Tl, produced by the decay of ^{212}Bi (equation 25) taking place in aqueous solution or when implanted in NaCl, KCl or in thalium salts, have been reported[68,69]. The hot atom chemistry of thalium-208 recoil atoms produced by the α-decay of ^{212}Bi in the diethyldithiocarbamate chelate, Bi(DDC)$_3$, has been investigated also[70]. The retention, $R\%$, of ^{208}Tl, after solid-phase ^{212}Bi(DDC)$_3$ storage, was found to equal 48.6% after dissolution of ^{212}Bi(DDC)$_3$ in benzene and 34.6% in CHCl$_3$ solution. These values were higher after storage of ^{212}Bi(DDC)$_3$ in benzene ($R = 64.7\%$) and in CHCl$_3$ solution ($R = 58.4\%$), respectively. Almost no inorganic ^{208}Tl^{3+} has been found at room temperature and or at liquid nitrogen temperature[70]. This suggests that the inorganic thalium consisted exclusively of ^{208}Tl$^+$. The solid-phase experiments with ^{212}Bi(DDC)$_3$ did not yield any evidence for the formation of ^{208}Tl^{3+}. The ^{208}Tl retentions as a function of ^{212}Bi(DDC)$_3$ storage temperature were 60.0% (at $-196\,^\circ$C), 51.8% (at $-78\,^\circ$C) and 48.6% (at 25 °C). The above yields spectrum has been used as evidence that the chemical environment in which ^{212}Bi undergoes radioactive decay governs its chemical state during reaction. The reduced species ^{208}Tl$^+$ is more stable than ^{208}Tl^{3+} in the absence of oxidative surroundings in the above experiments. The isoelectronic structure of ^{208}Tl$^+$ with its mother atom ^{212}Bi^{3+} also facilitates the formation of ^{208}Tl$^+$ instead of Tl^{3+}. α-Decay of ^{212}Bi produces ^{208}Tl hot atoms with 116-keV recoil energy. Besides that, 69.9% of the α-decay of ^{212}Bi results in the formation of an internally converted 39.85-keV excited state of ^{208}Tl[67,71] leading to the creation of inner-shell vacancy and the production of charged ^{208}Tl^{n+} ions. This high positive charge and recoil energy should lead to complete disruption of the parent molecule. The observed high retention of ^{208}Tl stresses the dominant role of the chemical environment played in the final chemical stabilization of ^{208}Tl.

Other chelate parent compounds of ^{212}Bi (e.g., with 8-hydroxyquinoline) have also been investigated and the results compared with the Bi(DDC)$_3$ system[70]. Granulated Th(OH)$_4$ has been used as the source of ^{212}Bi. 'Active deposits' of lead-212 and bismuth-212 have been collected and the ^{212}Bi has been separated from ^{212}Pb with Bi—Hg amalgam[72]. The bismuth has been oxidized to water-soluble Bi^{3+} using Cu(NO$_3$)$_2$.

^{207}BiCl$_3$ and ^{113}SnCl$_4$ are able to form alkyl and alkyl-substituted compounds and have been used to study the effect of metal impurities (Sn, Bi, Cd) on the purification process of Me$_3$Sb by rectification[73]. Bismuthine compounds (BiR$_3$, R = p-Tol) are used for the production of toners for electrostratographic developers[74].

D. Biochemical and Radiotherapeutical Applications of Bismuth Isotopes

1. Synthesis of Bi-205, Bi-206 and Bi-212

α-Emitting radionuclides astatine-211 [produced from ^{209}Bi(α, 2n)^{211}At] and bismuth-212 (descendant of ^{228}Th) have been found[75,76,77] to serve as highly localized energy sources, providing controllable cell-lethal radiation doses potentially useful in the radiotheraphy of cancer. α-Particles deposit high energy per unit path length (100 keV μm^{-1}) and are able to kill cells under hypoxic conditions. DNA damage caused by α-particles is not easily repaired by the cell and cytotoxicity of α-particles is not affected by oxygen. The parent ^{212}Pb is the source of ^{212}Bi[67,78–80] (equation 25) which, together with its short-lived daughter ^{212}Po, deposits 6 and 8.8 MeV, respectively, in the 60–100 μm range in the tissue (about 6–10 cell diameters)[81] and may therefore be considered as a potential radiotherapeutic agent. Non-carrier–added (NCA) bismuth isotopes ^{206}Bi (half-life 6.24 days) and ^{205}Bi (15.3 days), which decay according to equation 27, have been produced[81] by high-energy proton irradiation of natural Pb targets (isotopic composition: ^{204}Pb 1.42%, ^{206}Pb 24.1%, ^{207}Pb 22.1%, ^{208}Pb 52.3%) in nuclear reactions shown in equations 28a and 28b.

$$^{206}\text{Bi} \xrightarrow{\text{K}(\beta^+),\ E_\gamma = 0.803\ \text{MeV},(98.90\%),\ 0.881\ \text{MeV}\,(66.16\%)} {}^{206}\text{Pb} \qquad (27)$$

$$^{205}Bi \xrightarrow{\text{K}(\beta^+),\,E_\gamma = 0.703\ \text{MeV},(\text{probability per decay}\ 31\%)} {}^{206}\text{Pb} \xrightarrow{\text{K},\,T_{1/2} = 1.4\times10^7\ \text{year}} {}^{205}\text{Tl (stable)} \qquad (27a)$$

$$^{206}\text{Pb}(p,n)^{206}\text{Bi},\ {}^{207}\text{Pb}(p,2n)^{206}\text{Bi},\ {}^{208}\text{Pb}(p,3n)^{206}\text{Bi} \qquad (28a)$$

$$^{206}\text{Pb}(p,2n)^{205}\text{Bi},\ {}^{207}\text{Pb}(p,3n)^{205}\text{Bi},\ {}^{208}\text{Pb}(p,4n)^{205}\text{Bi} \qquad (28b)$$

The total cross-sections and yields for production of ^{205}Bi and ^{206}Bi in reactions 28a–28b have been measured[81]. The radiochemical method used to separate the NCA Bi radioactivities from multi-gram Pb targets provided aqueous solutions containing chemically and biologically useful amounts of pure ^{205,206}Bi sufficient for the preparation of labelled protein (antibody). The latter was used to study the tissue and subcellular biological ^{212}Bi/^{212}Pb distribution for better understanding the nature and *in vivo* behaviour of the ^{212}Pb/^{212}Bi parent/daughter system needed to manage properly the therapeutic procedure[82]. Radiogenic or enriched ^{207}Pb or ^{208}Pb target materials have also been used for the production of ^{206}Bi.

2. Application of bismuth-212 labelled antiTac monoclonal antibody[83,84]

This is a vehicle to deliver cytotoxic reagents to leukemia cells. ^{212}Bi, obtained as shown in equation 25, has been successfully conjugated[83] by means of a bifunctional ligand (isobutylcarboxycarbonic anhydride of diethylenetriaminepentaacetic acid[85,86], **15**) to *anti-Tac*, a murine IgG$_{2a}$ monoclonal antibody directed to the human interleukin-2 (IL-2-) receptor[87a] and used for the elimination of IL-2 receptor positive cells. In that way, the specificity of ^{212}Bi-anti-Tac (i.e. ^{212}Bi-labelled DTPA–anti-Tac complex) for the IL-2-receptor-positive (R$^+$) adult T-cell leukemia line HUT-102B2 (but not for IL-2 receptor-negative R$^-$ cell lines) has been demonstrated. 0.5 μCi, equivalent to 12 rad ml^{-1}, of α-radiation, targeted by *anti-Tac*, eliminated more than 98% of the proliferative (protein synthesis) capabilities of HUT-102B2 cells.

$$\text{HOOCCH}_2 \diagdown \quad\quad \overset{\displaystyle \text{CH}_2\text{COOH}}{|} \quad \diagup \text{CH}_2\text{COOH}$$
$$\diagup \text{NCH}_2\text{CH}_2\text{NCH}_2\text{CH}_2\text{N} \diagdown$$
$$\text{HOOCCH}_2 \diagup \quad\quad\quad\quad\quad\quad\quad \text{CH}_2\text{COOH}$$

(15) DTPA (pentetic acid) = N, N-bis[2-[bis(carboxymethyl)-amino]ethyl]glycine[87b].

The specific cytotoxicity of ^{212}Bi-anti-Tac has been blocked by excess of unlabelled anti-Tac but not by human IgG. Anti-Tac alone is unable to inhibit the proliferation or protein synthesis of most leukemic T-cell lines[88]. Toxins conjugated to anti-Tac bound to the surface of the leukemic cells are poorly and slowly internalized into endosomes, and toxin conjugate of anti-Tac does not pass easily from endosome vehicles to the cytosol in order to kill the cells. α-Particles emitted from bismuth-212 bound to the surface of the leukemic cells are much more effective cytotoxic reagents. Hits of one or two α-particles are sufficient for cell inactivation. The one-hour half-life of ^{212}Bi makes it appropriate for rapid targetting of leukemic cells without prolonged exposure of the normal tissue.

The ^{212}Bi-anti-Tac[83,86] was superior in immunotherapy, as has been demonstrated by supplementary bio-radiation experiments which showed that to reduce the protein synthesis by HUT-102B2 cells to 50% of the control value, only a 2-rad dose of ^{212}Bi-anti-Tac bound to the cell surface is necessary, but 45 rad is needed when the ^{212}Bi-anti-Tac is inhibited from binding to HUT-102B2 cells by the presence of an excess of unlabelled anti-Tac, and 400 rad is needed for an equivalent effect by γ-radiation delivered by acesium-137 source. Thus ^{212}Bi is well suited for killing circulating leukemic cells in patients with adult T-cell leukemia, involving large numbers of IL-2 receptors. The normal resting cells are receptor negative. Astatine-211 has been used to label monoclonal antibodies, but its conjugate was unstable and free astatine was taken up by normal tissue[89].

^{211}Pb (descendant of ^{227}Ac), decaying as shown in equation 29, has also been proposed as a candidate for biomedical applications.

$$^{211}\text{Pb} \xrightarrow{\beta^-(E_{max}=1.36\,MeV,\ T_{1/2}=36.1\,min)} {}^{211}\text{Bi} \xrightarrow{\alpha(E_\alpha=6.62\,MeV,\ T_{1/2}=2.15\,min)}$$
$$^{207}\text{Tl} \xrightarrow{\beta^-(E_{max}=1.44\,MeV,\ T_{1/2}=4.79\,min)} {}^{207}\text{Pb (stable)} \tag{29}$$

^{212}Bi-anti-Tac has been found also to be effective[90,91] in eliminating selectively unwanted T cells thought to play a major role in the rejection of transplanted tissues without removal of the majority of T-cells which are not involved in graft rejection.

3. Application of ^{212}Bi-labelled antibody B72.3 for radioimmunotherapy of peritoneal colon cancer xenografts

The antibody B72.3, a murine IgG1 which recognizes the mucin antigen TAG-72 found in many human adenocarcinomas[92], has been labelled[92-95] with ^{212}Bi by conjugation to the chelator linker glycyltyrosyl-lysyl-N-ε-diethylenetriaminepentaacetic acid, GYK-DTPA', by reductive amination to the carbohydrate (oligosaccharide) residue of the antibody. The preparations of antibody-therapeutic agent conjugates by attaching a therapeutic agent to an antibody or antibody fragment directed against a target antigen are described by Goers and coworkers[96]. ^{212}Bi eluted from the ^{224}Ra generator was combined with 100–200 μg of antibody-GYK-DTTA conjugate and the sample incubated at 37 °C for 15 min. More than 75% of the ^{212}Bi added was bound to the antibody conjugates purified by HPLC. The specific activity of the ^{212}Bi-labelled antibody was in the range 5–10 μCi μg^{-1}.

In bioexperiments, Athymic nude mice have been injected with LS174T cells (a human colon carcinoma). Seven to 13 days later the mice were treated with the ^{212}Bi-labelled antibody using single doses of 150–450 μCi or multiple doses of 80–180 μCi on consecutive days. Dissections performed 9–16 days after the end of treatment showed tumor reductions of greater than 90% (when compared to mice which were receiving unlabelled antibody).

4. Study of the bismuth-212-NRLU-10 toxicity in LS174T multicell spheroids

Cell killing produced by densely ionizing α-particles is influenced very little by dose rate or hypoxic conditions. Radioimmunotherapy using α-emitter-labelled antibodies (Abs) in single-cell suspensions in vitro[97,98] has shown excellent cell killing as compared with β-emitters or with X-ray emitters. α-Emitter-labelled Abs directed in vivo against tumor antigens or lymphocyte subsets have also shown selective toxicity to antigen positive cells[97,98]. Bismuth-212-NRLU-10 (an IgG2b Ab to a pancarcinoma antigen), which is very effective in killing single cells, has been found[100] to be ineffective in spheroids of 450-micrometer diameter for 4-hour incubations and 1000 μm for 6-hour incubations due to inadequate penetration of bismuth-212 into the spheroids before the ^{212}Bi decayed. It takes 6–24 h for the Ab to penetrate past the first few cell layers at the Ab concentrations used. After 6 h the ^{212}Bi activity decreases to 0.0162 of that at time zero. In addition, the LS174T human cell line has a very large number of binding sites per cell (1.2×10^6) so that it takes longer to saturate the binding sites on the outer cells and the movement of Ab into the depths of the spheroid is slowed down. Fab fragments[100] have lower affinity than the intact Ab from which they are derived and penetrate more rapidly, but renal toxicity prevents the therapeutic used of radiolabelled Fab. The conclusion has been reached that, although the toxicity of ^{212}Bi-NRLU-10 on multicell spheroids is low, the efficacy of α-emitter-labelled Ab therapy in solid tumors can be improved by using higher Ab concentrations, lower affinity Abs, α-emitters with longer half-lives (^{211}At with $T_{1/2} = 7.2$ h, ^{212}Pb with $T_{1/2} = 10.6$ h) and pretargeting with bifunctional Ab, binding the antigen with one arm and the chelate with the other[101,102].

The bismuth-212 antibody used in the above experiments in vivo has been obtained[100] by conjugation of the Abs with p-isothiocyanatobenzyl DTPA, $\mathbf{16}^{103}$ so (approximately one chelate was attached per Ab molecule), followed by metallation with bismuth-212 eluted. The specific activity of the bismuth-212-NRLU-10 ranged from 4.1–23.6 μCi μg^{-1}.

$$\underset{\text{HOOCCH}_2}{\overset{\text{HOOCCH}_2}{\diagdown}}\text{N}\underset{\underset{\text{CH}_2\text{C}_6\text{H}_4\text{SCN-}p}{|}}{\text{CHCH}_2}\text{NCH}_2\text{CH}_2\text{N}\overset{\overset{\text{CH}_2\text{COOH}}{|}}{\underset{\text{CH}_2\text{COOH}}{\diagup}}\overset{\text{CH}_2\text{COOH}}{\diagup}$$

(16)

for intact Ab and 18.9–24.2 μCi μg^{-1} for its Fab fragment. Both unlabelled NRLU-10 and its Fab fragment have been obtained commercially[104].

Labelling with the isothiocyanate ligands has been effected by reaction with protein amine groups to form thiourea bonds. The coupled antibodies could be stored for months. The use of the cyclic dianhydride (CA-DTPA), **17**[105], has some disadvantage since one ligand carboxylate metal binding site is occupied in an acid amide bond to form a protein-linked diethylenetriaminetetraacetic acid (DTTA).

(17) CA-DTPA

$$17, \text{CA-DTPA} \xrightarrow{\text{PCH}_2\text{CH}_2\text{CH}_2\text{NH}_2}$$

$$\underset{\text{O}}{(\text{PCH}_2\text{CH}_2\text{CH}_2)_2\text{N}\overset{\parallel}{\text{C}}\text{CH}_2\overset{\mid}{\underset{}{\text{N}}}\text{CH}_2\text{CH}_2} \quad \underset{}{\text{NCH}_2\text{CH}_2\text{N(CH}_2\text{COOH)}_2}$$

(with CH$_2$COOH, CH$_2$COOH, CH$_2$COOH substituents)

III. SYNTHESES AND USES OF ISOTOPICALLY LABELLED COMPOUNDS OF ANTIMONY

A. Radiochemical Syntheses and Studies with Antimony Compounds

1. Radiochemical studies with antimony isotopes

a. Methylation of nuclear reaction products ^{116}Sb and ^{63}Zn in acetone gas. These isotopes have been produced in nuclear reactions as described in the literature[106]. ^{116m}SbMe$_3$ and ^{63}ZnMe$_2$ were prepared and their transportation yields with carrier acetone in the pressure interval 0–15 Torr and their thermal decompositions in a TDP (Thermal Decomposition Port) in the temperature interval 300–800 °C were investigated[106]. Gas-phase reactions of isotopes have been used in the past for a rapid and continuous 'on-line' chemical separation of short-lived radionuclides[107–109]. Low-boiling-point methyl compounds[109] have been synthesized by reacting nuclear reaction products with methyl radicals. It has been found that acetone gas itself gives the same results as methyl radicals produced by thermal decomposition of acetone. The radiolysis of acetone by the proton beam from the 110-cm cyclotron at Osaka University's Laboratory of Nuclear Studies produces effectively the methyl radicals, which in turn react with ^{116m}Sb yielding ^{116m}SbMe$_3$. The recoil energy (E_R in keV) of ^{116m}Sb produced in ^{116}Sn(p, n) nuclear reaction is 76.3 keV; the recoil energies of ^{63}Zn produced in ^{60}Ni(α, n) and in ^{63}Cu(p, n) nuclear reactions are 1230 and 138 keV, respectively[106].

b. α-Decay of ^{125}SbR$_3$. The ^{125m}Te obtained in the β-decay of ^{125}SbR$_3$ (R = Ph or PhCH$_2$) in the crystalline state stabilized ^{110}mainly as ^{125m}TeR$_2$ (60–63%) and ^{125m}TeR$_3$Cl (27–29%). In the decay of ^{125}SbR$_3$Cl$_2$ the ^{125m}Te has been indentified as

$^{125m}TeR_2Cl_2$ (83–90%). These results have been interpreted[110] as supporting the mechanism shown in equations 30 and 31:

$$^{125}SbR_3 \xrightarrow{\beta^-} {}^{125m}TeR_3^+ \begin{array}{c} \nearrow {}^{125m}TeR_3Cl \\ \searrow {}^{125m}TeR_2 + R^+ \end{array} \tag{30}$$

$$^{125}SbR_3Cl_2 \xrightarrow{\beta^-} {}^{125m}TeR_3Cl_2{}^+ \longrightarrow {}^{125m}TeR_2Cl_2 + R^+ \tag{31}$$

2. Synthesis of tritium-labelled SbPh₃

Molecular–ion reactions shown in equations 4–7 have been applied also for the synthesis of tritium-labelled organometallic compounds of Sb, Te and I (equations 32 and 33)[111].

$$C_6{}^3H_5{}^+ + SbPh_3 \longrightarrow [(C_6{}^3H_5)Ph_3Sb]^{*+} \longrightarrow (C_6{}^3H_5)Ph_2Sb + C_6H_5{}^+ \tag{32}$$

$$C_6{}^3H_5{}^+ + C_6H_6 \longrightarrow C_6{}^3H_5C_6H_5 + H^+ \tag{33}$$

The yields of radiochemically pure tritium-labelled onium compounds of Ph_3Sb, $TePh_2$ and PhI in the reactions of tritiated phenyl cations with unlabelled Ph_3Sb, Ph_2Te and PhI were equal to 73%, 63% and 51%, respectively.

3. Synthesis and application of ¹⁴C-labelled triethyl oxonium salts of hexachloroantimonate

Carbon-14-labelled triethyloxonium hexachloroantimonate, $[(^{14}C_2H_5)_2OC_2H_5]^+$ $(SbCl_6)^-$, **18** (specific activity 1.15×10^4 Bq g^{-1}), has been obtained[112,113] by treating the ^{14}C-labelled $(^{14}C_2H_5)_2O \cdot SbCl_5$ complex with ethyl chloride. The salt was used to clarify the mechanism of initiation of polymerization of 2, 3, 4-tri-O-methyl-L-glucosane (**19**)[114], since it has been noted[112,115,116] that the incorporation of the radioactive label into the polymer increases with the increase in the degree of conversion of the monomer (at 10.5% of the equilibrium yield of the polymer, 2.5% of the initial radioactivity of **18** has been

(34)

incorporated into the polymer, at 34% conversion 9.3%, at 40% conversion 15% radio-activity and at 100% conversion 24% radioactivity has been incorporated). The degradation of the polymer and chromatographic analyses of the products of fragmentation indicated[112] that the initial interaction of the ethyl cation of the oxonium salt with the monomer proceeds, as shown in equation 34, by attack at the methoxy groups, which precede the formation of the active centre **22** of the polymerization process (equations 35 and 35a). The ^{14}C-labelled ethyl groups are distributed uniformly between the molecules of the monomer. The radioactivity of the polymer increases as the result of subsequent incorporation of the labelled monomers into the chain of the growing polymer.

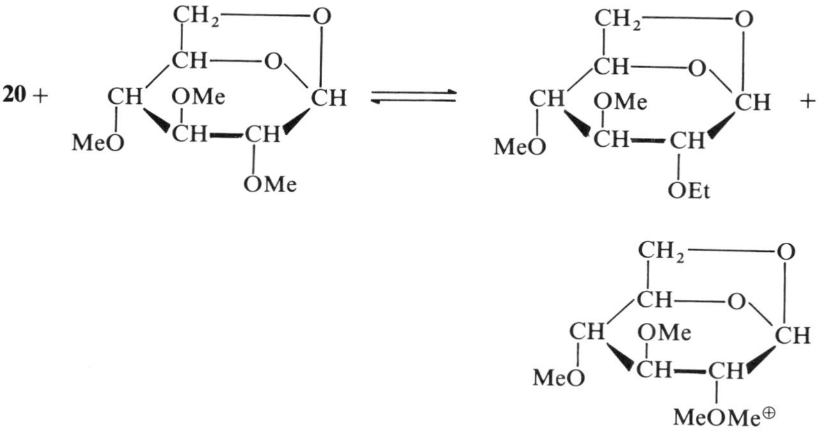

(35)

(21)

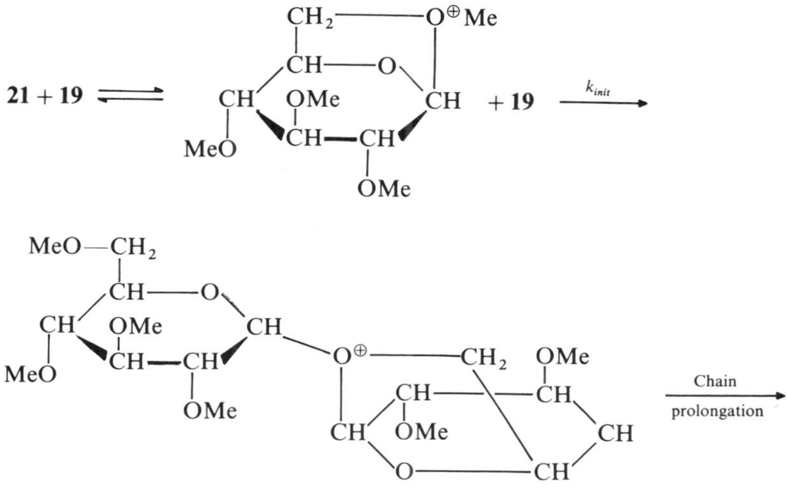

(22)

B. Stable Isotope Syntheses and Applications of Antimony Derivatives

1. Syntheses of deuteriated oxonium hexafluoroantimonates and hexafluoroarsenates

The deuteriated oxonium salt, $^+OD_3SbF_6{}^-$, has been prepared[117] from SbF_5, DF and D_2O. The white solid obtained was mainly $OD_3{}^+SbF_6{}^-$ contaminated with less than 1% of $OD_2H^+SbF_6{}^-$. At room temperature $OD_3{}^+SbF_6{}^-$ has an ordered hydrogen-bonded CsCl-type structure, whose phase transition to a disordered phase occurs above room temperature. The hydrogen-bridge bond strength, deduced from the OH . . . FM stretching mode, was found to be 1.77 kcal mol^{-1}.

$OD_3{}^+AsF_6{}^-$ has been prepared similarly from DF, D_2O and AsF_5. The white solid contained less than 1% of $OD_2H^+AsF_6{}^-$ as impurity.

2. Synthesis, characterization and crystal structure of deuterium-labelled ternary adduct [Ni(CD₃CN)₆] (SbF₆)₂

The $[Ni(CD_3CN)_6](SbF_6)_2$ complex **23** has been prepared[118] from CD_3CN and $Ni(SbF_6)_2$ at $-196\,°C$[119]. Purple crystals of **23** have been grown by slow evaporation of the solvent from the bright blue solution. $Ni(BiF_6)_2$, prepared by treating NiF_2 with BiF_5 in anhydrous HF, also reacted with acetonitrile to produce the ternary adduct $NiF_2 \cdot 2BiF_5 \cdot 6CD_3CN$ (or $6CH_3CN$), in which the Ni^{2+} ion is octahedrally coordinated by six CD_3CN molecules via the nitrogen. These isolate the Ni^{2+} cations from the $SbF_6{}^-$ counteranions diminishing the polarizing strength and hardness of the 'Ni^{2+} acid' found in $Ni(MF_6)_2$ salts, where $M = Bi$ or Sb. The compound has been characterized by elemental analysis, X-ray powder data and vibrational spectroscopy. The I.R. spectra[120] of both $[Ni(CH_3CN)_6]^{2+}(SbF_6{}^-)_2$ and $[Ni(CD_3CN)_6]^{2+}(SbF_6{}^-)_2$ have been determined and assigned.

Combination modes of coordinated CH_3CN and CD_3CN have also been recorded and assigned. The Sb atoms are octahedrally coordinated by six fluorine atoms with two non-equivalent Sb–F distances of 1.80 and 1.83 Å. The distances between the octahedrally coordinated Ni and N atoms are 2.07 Å. The complex **23** crystallizes in the trigonal space group R3 with $a = 11.346$ Å, $c = 17.366$ Å, $V = 1936$ Å^3 and $Z = 3$.

3. Synthesis of deuterium-labelled antimony containing heterocycles. MX8R¹R² (X8 = eight-membered ring)

The pentadeuteriophenylated antimony derivatives $(C_6{}^2H_5)Sb(SCH_2CH_2)_2X$ of the general structure **24**, where $X = O$ or $X = S$, have been synthesized[121–124] in five steps, as shown in equation 36, starting with perdeuterated nitrobenzene. The I.R. and Raman spectra of cyclic dithiolates **24** have been determined in the $3200–100\,cm^{-1}$ region. The

M = Ge, Sn, Pb, As, Sb, Bi

X = O, S

$R^1, R^2 = $ Cl, Br, I, Me, Ph,

(**24**)MX8R¹R²

$$C_6{}^2H_5NO_2 \longrightarrow C_6{}^2H_5NH_2 \longrightarrow C_6{}^2H_5N_2{}^+Cl^- \xrightarrow{SbCl_3} C_6{}^2H_5SbO_3H_2$$

$$\xrightarrow{HCl, SO_2} C_6{}^2H_5SbCl_2 \xrightarrow{(HSCH_2CH_2)_2X} SbX8(C_6{}^2H_5) \tag{36}$$

$$\textbf{(24)}$$

$$X = O \text{ or } S$$

spectra are very similar for both $X = O$ and $X = S$. The deuteriation permitted one to separate signals originating from vibrations of phenyl groups, of the aliphatic ring moieties and of the coordination polyhedra around M respectively and to determine all isotopic shifts ranging from C—H/C—D (I.R. 3049/I.R. 2273 cm^{-1}) to SbPh modes of vibrations (390–100 cm^{-1}, $\Delta = 0$–10 cm^{-1} isotopic shifts).

4. Exchange and decomposition reactions in the system $Sb(CH_3)_3$–$Sb(CD_3)_3Cl_2$–ethanol

It has been suggested[125,126] that the exchange in the systems MR_3–MR_3X_2–solvent (where M is an element of the first subgroup of the Vth group, R is an organic radical and X is halogen) proceeds according to an electron transfer mechanism. This view has been tested using perdeuteriated $Sb(CD_3)_3Cl_2$ as one of the components[127]. If the electron transfer mechanism is operating as shown in equation 37, then the perdeuteriated $Sb(CD_3)_3$ species should appear in the volatile $SbMe_3$ component.

$$Sb(CD_3)_3{}^{2+} + Sb\overleftarrow{(CH_3)_3} \xrightleftharpoons[]{2e \qquad 2e} Sb\overrightarrow{(CD_3)_3} + Sb(CH_3)_3{}^{2+} \tag{37}$$

Perdeuteriated and ^{124}Sb-labelled trimethyl antimony has been synthesized from $SbCl_3$ and a Grignard reagent. Pure $SbMe_3$ has been obtained by reducing $^{121,123}Sb(Me)_3Br_2$ with metallic zinc[128]. The exchange has been studied at 30–80 °C by analysing the isotopic composition of $SbMe_3$ by M.S. At 30 °C the mass spectrum showed mainly the products of electron exchange [i.e. $Sb(CD_3)_3$ and $SbMe_3$], but negligible peaks corresponding to partial exchange of methyl groups[121,123]$Sb(CH_3)_2CD_3$ or $^{121,123}Sb(CD_3)_2CH_3$ appeared also. Their intensities increased by keeping the exchange mixture at 80 °C for longer times. Thus it has been concluded that the antimony exchange proceeds mainly according to an electron transfer mechanism. The exchange of methyl groups proceeds at a much slower rate. The concentrations of $Sb(CH_3)_2CD_3$ and $Sb(CH_3)(CD_3)_2$ species were approximately the same. This indicates that the disproportionation of methyl groups takes place in binary complexes $[(CH_3)_3Sb:Sb(CD_3)_3]^{++}$ or $[(CH_3)_3Sb:Sb(CD_3)_3Cl]^+$. The electron transfer and methyl group transfer might proceed within these complexes. In the second complex the exchange occurs by transfer of chlorine.

The decomposition of R_3Sb compounds (R = Me, vinyl, isopropyl) in a flow tube reactor has been studied using D_2 and He as carrier gases to obtain the deuterium-labelled byproducts and to elucidate the mechanism of the pyrolysis[129].

5. Isotopic studies of the anionotropic rearrangements of cis-t-butyl-trans-t-butyl-D_9-thiiranium hexachloroantimonate 25 and its trans-t-butyl-cis-t-butyl-D_9 isotopomer 26.

The hemideuteriated isotopomeric cations 25 and 26, obtained in equimolar amounts by addition of methylbis(methylthio)sulphonium hexachloroantimonate 27[130–132] to deuteriated olefin, E-t-butyl-t-butyl-D_9-ethylene 28 (equation 38), have been converted to the non-separable hemideuteriated thietanium ions 29 and 30, characterized spectroscopically. These rearrangements have been followed at 25 °C (over periods of 1–16 weeks) by

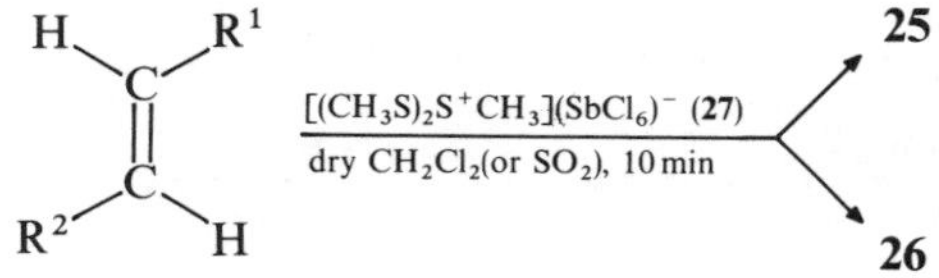

$$\begin{array}{c} H \\ \diagdown \\ R^2 \end{array} C = C \begin{array}{c} R^1 \\ \diagup \\ \diagdown \\ H \end{array} \xrightarrow[\text{dry } CH_2Cl_2(\text{or } SO_2),\ 10\,\text{min}]{[(CH_3S)_2S^+CH_3](SbCl_6)^-\ (27)} \begin{array}{c} \nearrow\ 25 \\ \\ \searrow\ 26 \end{array}$$

(28)

(38)

$R^1 = t\text{-Bu},\ R^2 = t\text{-Bu-D}_9$

$k_H^c = 7.42 \times 10^{-6}\,\text{s}^{-1}$ ⟶ (29)

$k_H^t = 3.83 \times 10^{-7}\,\text{s}^{-1}$

$(k_D^t = 3.02 \times 10^{-7}\,\text{s}^{-1})$

$k_D^c = 5.84 \times 10^{-6}\,\text{s}^{-1}$ ⟶ (30)

(25)

(26)

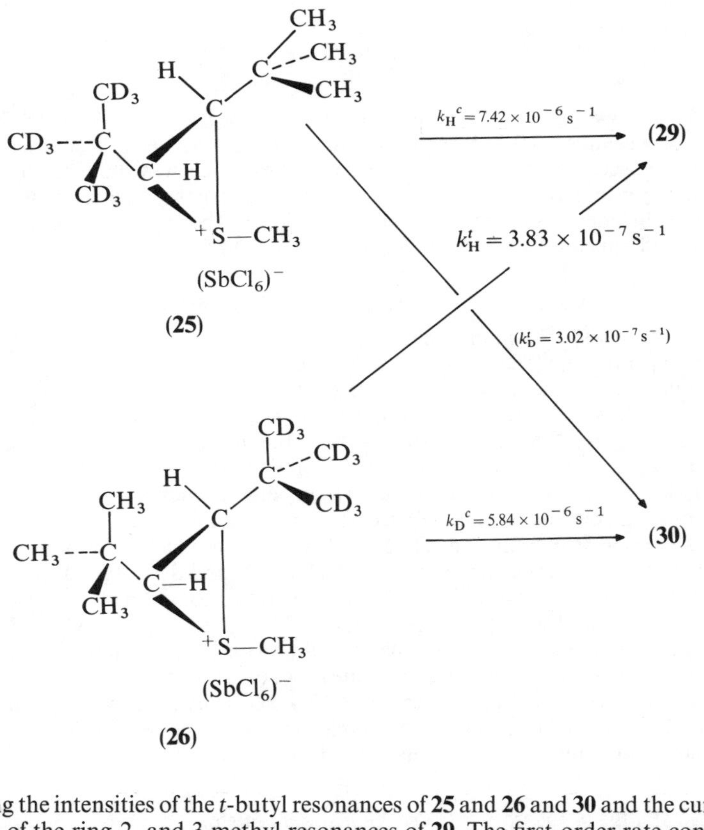

monitoring the intensities of the t-butyl resonances of **25** and **26** and **30** and the cumulative intensities of the ring-2- and 3-methyl resonances of **29**. The first-order rate constants k^c and k^t were optimized by means of the Simplex procedure[133] and show 95.1% preference for rearrangement of the group *cis* to S-methyl. The optimized K.I.E. equals 1.27 ($k_H^c/k_D^c = 7.42 \times 10^{-6}\,\text{s}^{-1}/5.84 \times 10^{-6}\,\text{s}^{-1} = 1.2705$). This deuterium kinetic isotope effect has been classified as an α-effect resulting from concerted breaking of the C—S bond assisted by the migration of the methide group from the t-butyl moiety and formation of the tertiary carbonium ion **31**. The ring closure proceeds by the interaction of the sulphur atom with the electron-deficient carbon as indicated in the T.S.-like structure **31**. The value

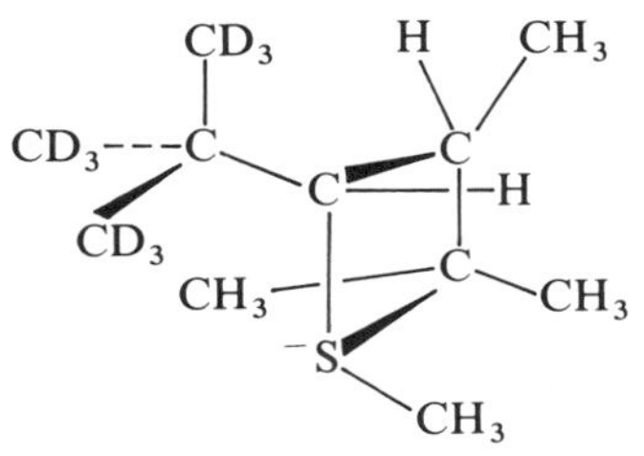

(29) 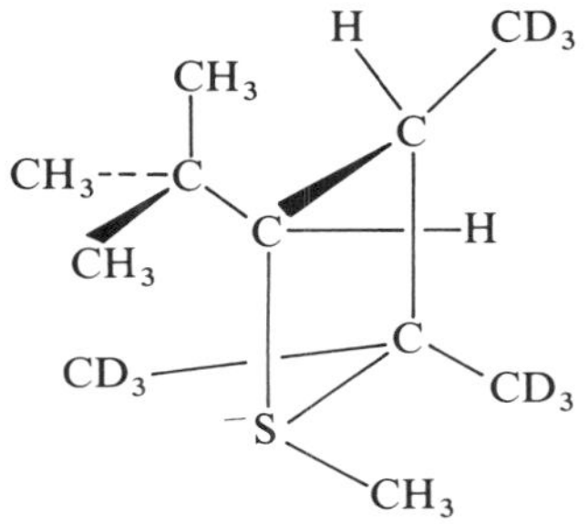 (30)

$$(k_H^t/k_D^t) = (3.83 \times 10^{-7}\,s^{-1}/3.02 \times 10^{-7}\,s^{-1}) = 1.2682$$
$$(k_H^c/k_H^t = (7.42 \times 10^{-6}\,s^{-1}/3.83 \times 10^{-7}\,s^{-1}) = 19.37$$
$$(k_D^c/k_D^t = (5.84 \times 10^{-6}\,s^{-1}/3.02 \times 10^{-7}\,s^{-1}) = 19.34$$

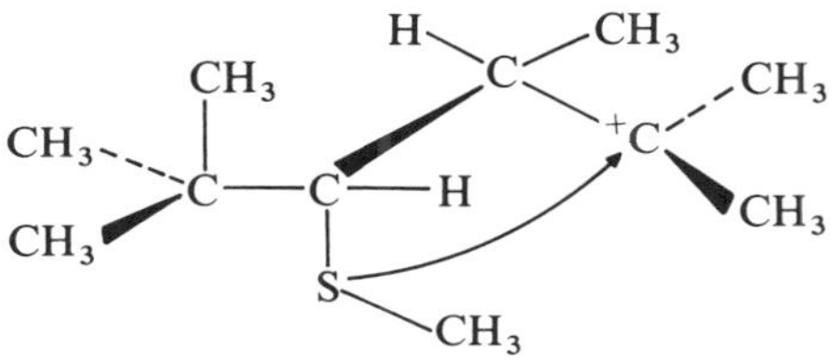

(31)

1.27, larger than the K.I.E. observed in methide migration from a perdeuterated *t*-butyl group[134], may be indicative of a stronger and decisive participation of the methide migration in the C—S bond breaking and of the steric relief on going from **25** and **26** to **31**. Simple mass effect in the methyl group (18 versus 15) alone cannot explain the deuterium K.I.E. determined[130,131].

6. Synthesis of isotopically labelled cationic hexafluoroantimonate complexes

a. Synthesis of trans-bis[1,2-bis(diphenylphosphino)ethane]di[¹³C]carbonyl-ruthenium(II)hexauoroantimonate···CH₂Cl₂. trans $[Ru(^{13}CO)_2\,(dppe)_2][SbF_6]_2$, (**32**) has been obtained[135] by a route involving $[RuCl_2(dppe)_2]$, $Ag[SbF_6]$ and ^{13}CO (96% enriched). The product was purified by recrystallization from CH_2Cl_2–Et_2O at − 30 °C. The yield of **32** (white crystals) was 44% $[\nu(^{13}C{\equiv}O) = 1994\,cm^{-1}$, $\nu(^{12}C{\equiv}O) = 2040\,cm^{-1}$ for *trans*-**32**].

b. Synthesis of trans-bis[1,2-bis(diphenylphosphino)ethane-PP-]carbonyl (deuterio-formyl)ruthenium(II)hexafluoroantimonate dichloromethane (1/1). The deuterium-labelled cationic formyl complex *trans*-$[Ru(CDO)(CO)(dppe)_2][SbF_6]$, **33**, has been synthesized by reduction of **32** of natural isotopic composition with $[Li(DBEt_3)]$ in CH_2Cl_2 solution for 16h at − 35 °C under nitrogen[135]. The ^{13}C analogue of **33**, *trans*-$[Ru(^{13}CHO)(^{13}CO)(dppe)_2][BEt_4]$, has been prepared from *trans*-$[Ru(^{13}CO)_2$ $(dppe)_2][SbF_6]_2$. The complexes are sufficiently stable in the solid state and in solution below $-30\,°C$ for full characterization by spectroscopic means (I.R., N.M.R.) and X-ray structure. At $35\,°C$ trans-$[Ru(CHO)(CO)(dppe)_2]^+$ decomposes with a half-life of 9 min in CH_2Cl_2 and trans- $[Ru(CDO)(CO)(dppe)_2]^+$ with a half-life of 16 min (I.R. evidence).

The following infrared data (in cm^{-1}) have been recorded for ruthenium formyl complexes. trans-$[Ru(CHO)(CO)(dppe)_2][SbF_6]$:$2550(v_{C-H})$, $1978(v_{C\equiv O})$, 1596 $(v_{C=O})$, $1362(\delta_{C-H})$; trans-$[Ru(CDO)(CO)(dppe)_2][SbF_6]$: $1918(v_{C-D})$, $1984(v_{C\equiv O})$, $1595(v_{C=O})$, $1040(\delta_{C-D})$; cis-$[Ru(CHO)(CO)(dppm)_2][SbF_6]$: $2720(v_{C-H})$, 2590(absorption from Fermi resonance between v_{C-H} and $2\delta_{C-H})$, $1970(v_{C\equiv O})$, $1600(v_{C=O})$, $1350(\delta_{C-H})$; cis-$[Ru(CDO)(CO)(dppm)_2][SbF_6]$: $1692(v_{C-D})$, $1977(v_{C\equiv O})$, $1604(v_{C=O})$, $1030(\delta_{C-D})$; trans-$[Ru(CHO)(CO)(dppe)_2][BEt_4]$: $2552(v_{C-H})$, $1979(v_{C\equiv O})$, $1600(v_{C=O})$; trans-$[Ru(^{13}CHO)(^{13}CO)(dppe)_2][BEt_4]$: $2502(v_{C-H})$, $1928(v_{C\equiv O})$, $1558(v_{C=O})$.

cis-$[Ru(CDO)(CO)(dppm)_2][SbF_6]$ has been obtained by reduction of cis-$[Ru(CO)_2(dppm)_2][SbF_6]_2$ with $[Li(DBEt_3)]$–THF; dppm = $PPh_2CH_2PPh_2$, bis(diphenylphosphino)methane.

Structure **34**, has been proposed for cis-$[R(CHO)(CO)(dppm)_2]^+$; PP = dppm, Ru—C bond 2.09Å, Ru—C—O angle $133°$ and resonance forms **35** for ruthenium-bound formyl.

(34)   (35)

The low-temperature hydride reductions of $[Ru(CO)_2(P—P)_2][SbF_6]_2$ (PP = $PPh_2(CH_2)_nPPh_2$; $n = 1$, dppm, $n = 2$, dppe) have been undertaken[136,137] to elucidate the mechanism of the homogeneous hydrogenation of carbon monoxide to produce organic products for the petrochemical industry catalysed by ruthenium formyl complexes[138].

7. Deuterium isotope effect and tracer study of the mechanism of decomposition of the cationic ruthenium complexes

The decomposition reactions of metal formyl complexes have been reviewed by Gladysz and coworkers[139–142].

a. Deuterium K.I.E. studies of the decomposition of **33** *and of the decomposition of* trans-[Ru(CDO)(CO)(dppb)_2][SbF_6], **36**, where dppb = 1,2-bis(diphenylphosphino)benzene, have been carried out[143,144]. Two different pathways have been proposed: Trans-$[Ru(CXO)(CO)(dppe)_2]$ decomposes in CH_2Cl_2 at $+30\,°C$ by first order kinetics with $T_{1/2}$ of 9.24 min ($k = 12.5 \times 10^{-4}\,s^{-1}$) for X = H and with $T_{1/2} = 16.4 \pm 1$ min for X = D to give cis-$[RuX(CO)(dppe)_2][Y](Y = SbF_6$ or $BEt_4)$[145] which isomerizes at room temperature to give the *trans* isomer ($T_{1/2}$ about 72 ± 16 h for X = H and 73 ± 8 h for X = D). The observed deuterium isotope effect of 1.77 on the decomposition of **33** has been interpreted as being primary and supporting the mechanism presented in equation 39, which involves a bent triangular transition state[146] **39** in the rate-determining hydride migration from CO to Ru in a five-coordinate **37**.

The kinetics of decomposition of **33** have been followed by monitoring the height of the $v(C=O)$ vibration while the kinetics of isomerization of cis-$[RuX(CO)(dppe)_2]$ $[SbF_6]$

$$33 \; \rightleftharpoons \; \left[\begin{array}{c} \text{CDO} \\ \text{Ru} \\ \text{CO} \end{array} \; P, P, P, P \right]^{-} \quad \xrightarrow{\text{slow}}$$

(37)

$$\left[\begin{array}{c} \text{CO} \\ \text{D} \; \text{Ru} \\ \text{CO} \end{array} \; P, P, P, P \right]^{+} \quad \xrightarrow[\text{rapid}]{-\text{CO}} \quad \left[\begin{array}{c} \text{CO} \\ \text{D} \; \text{Ru} \\ P \end{array} \; P, P, P \right]^{+}$$

(38)   (39)

$$\begin{array}{c} \text{O} \\ \| \\ \text{C} \\ \diagup \; \diagdown \\ \text{D} \cdots\cdots \text{Ru} \end{array} \qquad \text{T.S. (39)}$$

(X = H or D) in CH_2Cl_2 were followed by recording its ^{31}P N.M.R. spectra (integrations) over 140 h.

An initial pre-equilibrium involving carbon monoxide loss to give transient $[Ru(CDO)(dppe)_2]^+$ has been excluded by carrying out the decomposition of 33 under ^{13}CO, showing no incorporation of ^{13}CO into the product or into the recovered complex 33. No ^{13}C kinetic isotope effect test of mechanism (39) has been attempted.

The intramolecular deuterium exchange observed during low-temperature isomerization of cis-$[RuD(CO)(dppe)_2]^+$ with hydrogen atoms of the dppe ligand occurs at both phenyl ($\nu_{C-D} = 2360$ cm^{-1}) and methylene ($\nu_{C-D} = 2318$ cm^{-1}) groups in the five-coordinate intermediate 37. No exchange of the hydride hydrogen atom with the solvent has been found in the decomposition of trans-$[Ru(CHO)(CO)(dppe)_2]^+$ in CD_2Cl_2 or of trans-33 in CH_2Cl_2.

It has been suggested that the isomerization of cis-$[RuH(CO)(dppe)_2]^+$ in CH_2Cl_2 and its H/D exchange proceed according to equation 40[143]. 33 and trans-$[Ru(^{13}CHO)(^{13}CO)(dppe)][SbF_6]$ have been used also[147] to identify more fully the radicals formed in the decomposition of cationic formyl complexes of ruthenium. The conclusion has been reached that a radical mechanism does not constitute major pathway for these decompositions[147].

b. Synthesis and decomposition of trans-$[Ru(^{13}CHO)(^{13}CO)(dppb)_2][SbF_6]$ **40** *and* *trans-$[Ru(CDO)(CO)(dppb)_2]$* **36**. These compounds were prepared[144] in high yields as shown in equations 41 and 42, and are indefinitely stable in the solid state at room temperature or in CH_2Cl_2 and $MeNO_2$ at $-30\,°C$. They decompose in solution to trans-$[RuX(CO)(dppb)_2][SbF_6]$ exclusively, following first-order kinetics for at least three half-lives. $T_{1/2}$ of the formyl complex **40** in CH_2Cl_2 solution at $30\,°C$, determined by monitoring the intensity of the formyl $\nu(C=O)$ absorption, was 28 min. The K.I.E. (k_H/k_D) of 2.3 has been found in the decomposition of **36** at this temperature. **40** decomposing in a solution of CH_2Cl_2 and methyl methacrylate (under N_2), which is known to undergo

$$cis\text{-}[RuH(CO)(dppe)_2]^+ \longrightarrow \qquad\qquad (40)$$

$$trans\text{-}[RuH(CO)(dppe)_2]^+ \longrightarrow$$

free-radical induced polymerization, produced poly(methyl methacrylate) (pmma, M_N about 18,000) but it has been found that only ca 2% of decomposing molecules of **40** initiate polymerization. No polymer has been found under identical experimental conditions in the presence of unlabelled **33**.

$$[RuCl_2(PPh_3)_3] \xrightarrow[\text{acetone, 2h reflux}]{o\text{-}C_6H_4(PPh_2)_2,\,\text{excess}} trans\text{-}[RuCl_2(dppb)_2]$$
$$100\%$$

$$\xrightarrow[\text{80 °C, 36 h, work up}]{^{13}CO(3\,\text{atm})/CH_2Cl_2,\,AgSbF_6} trans\text{-}[Ru(^{13}CO)_2(dppb)_2][SbF_6]_2$$
$$\textbf{41},\,55\text{–}65\%$$

$$\xrightarrow[\text{5 h stir. at } -35\,°C,\,\text{work up}]{K[BH(OPr\text{-}i)_3]\,\text{in }Ch_2Cl_2/THF} \textbf{40},\,50\% \text{ yield white crystals} \qquad (41)$$

40   (I.R. $v(^{13}C{\equiv}O) = 1940\,s\,cm^{-1}$; $v(^{12}C{\equiv}O) = 1990\,s\,cm^{-1}$)

(I.R. $v(^{13}C{=}O) = 1560\,cm^{-1}$; $v(^{12}C{=}O) = 1602\,s\,cm^{-1}$)

41 (0.4 mmol) $\xrightarrow[\text{8 h at } -35\,°C,\,\text{work up at } -35\,°C]{Li(BDEt_3),\,0.8\,\text{mmol in THF}}$ **36** $\qquad (42)$

unlabelled $\qquad\qquad\qquad\qquad\qquad\qquad\qquad$ 55%, white crystals

36 (I.R.: $v(C{\equiv}O) = 1990\,cm^{-1}$; $v(C{-}D) = 1925\,cm^{-1}$;

$v(C{=}O) = 1598\,cm^{-1}$; $\delta(C{-}D) = 1039\,cm^{-1}$)

The decomposition of **36** in the presence of **40** in CD_2Cl_2 at RT during 1 day provided only trans-[RuH(^{13}CO)(dppb)$_2$]. The cross-over product trans-[RuH(CO)(dppb)$_2$]$^+$ has not been detected.

These results led to the conclusion that the decomposition of **40** and **36** proceeds according to the major pathway presented in equation 43 involving concerted H(D) migration and CO loss in a six-coordinate complex **42** and partly via a less favourable homolytic cleavage of the Ru—CHO bond involving a biradical like T.S. **43**.

$$(42)\ P—P = dppb$$

$$\text{(43)}$$

$$(43)$$

It is concluded also that the use of dppb in place of dppe stabilizes trans-[Ru(CHO)(CO)(P—P)$_2$]$^+$ complexes by suppressing Ru—P bond rupture.

Comparative studies of ligand exchanges between isotopically labelled free dppe and dppe bound in trans-[Ru(CHO)(CO)(dppe)$_2$]$^+$ cation on one side and between isotopically labelled dppb and dppb bound in trans-[Ru(CHO)(CO)(dppb)$_2$]$^+$ cation as well as ^{13}C (and ^{14}C) K.I.E. determinations (equations 30 and 43) for the decompositions of two formyl complexes possessing two different ligands, dppe and dppb, should provide decisive arguments in favour of the new decomposition pathway presented in equation 43, differing largely from the T.S.$^{\#}$ and mechanism of equation 39. The magnitude of ^{13}C K.I.E. will indicate[148-151] which bond is broken or weakened in the T.S.$^{\#}$ of the decarbonylation of the formyl complexes studied, whether it is the C—C (structure **43**) or the C—H bond (structure **38** in equation 39) or whether both the C—C and H—C bonds are partly weakened in the four-centre T.S.$^{\#}$ shown explicitly in the transition structure **42**.

8. Deuterium and carbon-13 labelled cationic hexafluoroantimonate complexes in a study of the catalytic carbonylation of methanol to acetic acid

Deuterium and ^{13}C labelled cationic hexafluoroantimonate complexes have also been used[152] in studies of alkyl migration in rhodium complexes, employed in homogeneous catalytic carbonylation of methanol (equations 44 and 45) to acetic acid[153,154]. COD is displaced in **44** by CO to form **45**. Oxidative addition of ^{13}CH$_3$I to **46** afforded the acyl complexes **47** and **48** containing one or two Rh—O bonds. In the presence of CO and H$_2$O **47** and **48** gave ^{13}CH$_3$COOH with cleavage of one or two Rh—O bonds and back formation of **45**. The 'ether-phosphane' ligands promote the formation of the complexes within the reaction cycle through an 'opening and closing mechanism'. On employing PPrR$_2$ phosphane ligands, oxidative addition of ^{13}CH$_3$I also takes place but no acyl

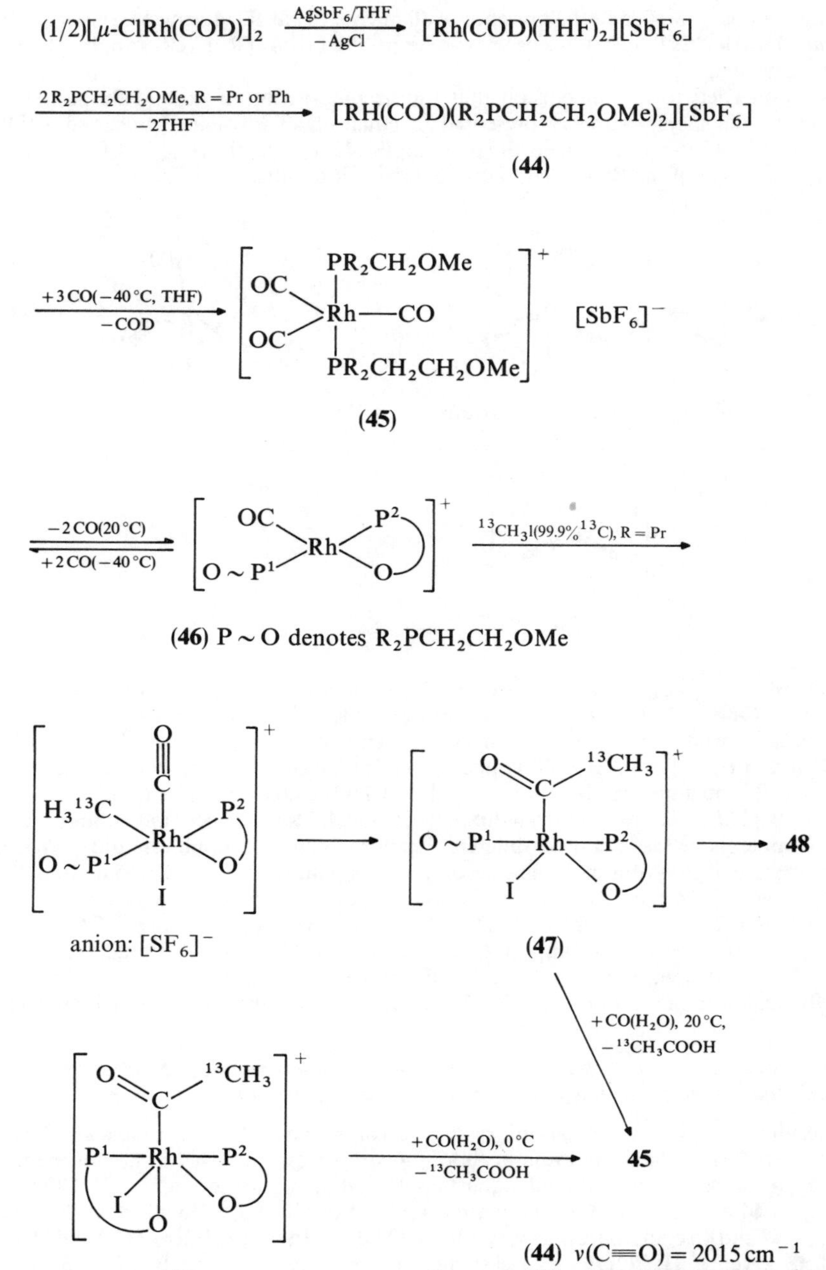

$$(1/2)[\mu\text{-ClRh(COD)}]_2 \xrightarrow[-\text{AgCl}]{\text{AgSbF}_6/\text{THF}} [\text{Rh(COD)(THF)}_2][\text{SbF}_6]$$

$$\xrightarrow[-2\text{THF}]{2\,\text{R}_2\text{PCH}_2\text{CH}_2\text{OMe, R = Pr or Ph}} [\text{RH(COD)(R}_2\text{PCH}_2\text{CH}_2\text{OMe)}_2][\text{SbF}_6]$$

(44)

$$\xrightarrow[-\text{COD}]{+3\,\text{CO}(-40\,^\circ\text{C, THF})} \begin{bmatrix} \text{OC} & \text{PR}_2\text{CH}_2\text{OMe} \\ & \text{Rh—CO} \\ \text{OC} & \text{PR}_2\text{CH}_2\text{CH}_2\text{OMe} \end{bmatrix}^+ \quad [\text{SbF}_6]^-$$

(45)

$$\xrightarrow[+2\,\text{CO}(-40\,^\circ\text{C})]{-2\,\text{CO}(20\,^\circ\text{C})} \begin{bmatrix} \text{OC} & \text{P}^2 \\ & \text{Rh} \\ \text{O} \sim \text{P}^1 & \text{O} \end{bmatrix}^+ \xrightarrow{^{13}\text{CH}_3\text{I}(99.9\%\,^{13}\text{C}),\,\text{R = Pr}}$$

(46) P $\sim$ O denotes R$_2$PCH$_2$CH$_2$OMe

anion: [SF$_6$]$^-$ **(47)**

$$\xrightarrow[-^{13}\text{CH}_3\text{COOH}]{+\text{CO(H}_2\text{O), }20\,^\circ\text{C,}} \mathbf{45}$$

$$\xrightarrow[-^{13}\text{CH}_3\text{COOH}]{+\text{CO(H}_2\text{O), }0\,^\circ\text{C}} \mathbf{45} \qquad (44)$$

(48)

(44) $\nu(\text{C}\equiv\text{O}) = 2015\,\text{cm}^{-1}$
(45) $\nu(\text{C}\equiv\text{O}) = 1957\,\text{cm}^{-1}\,(\text{R = Pr})$
        $\nu(\text{C}\equiv\text{O}) = 1982\,\text{cm}^{-1}\,(\text{R = Ph})$
(46) $\nu(\text{>C}=\text{O}) = 1694\,\text{cm}^{-1}$
(47) $\nu(\text{>C}=\text{O}) = 1689\,\text{cm}^{-1}$.

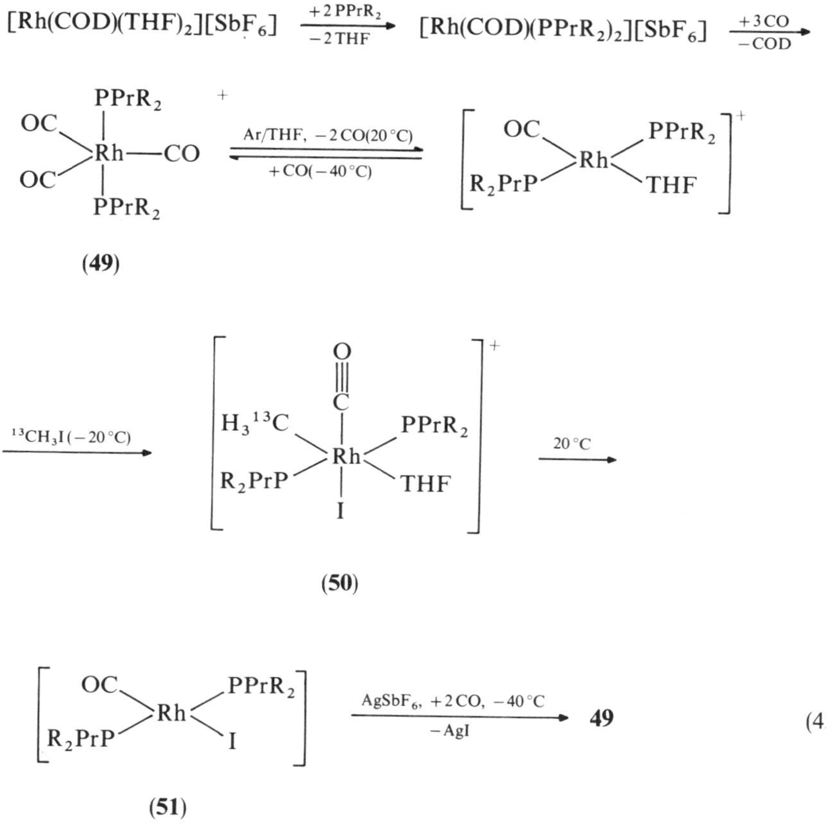

$$[Rh(COD)(THF)_2][SbF_6] \xrightarrow[-2\,THF]{+2\,PPrR_2} [Rh(COD)(PPrR_2)_2][SbF_6] \xrightarrow[-COD]{+3\,CO}$$

OC—|—PPrR₂
Rh—CO
OC—|—PPrR₂

$$\xrightleftharpoons[+CO(-40\,°C)]{Ar/THF,\,-2\,CO(20\,°C)}$$

$\left[\begin{array}{c} OC\diagdown \diagup PPrR_2 \\ Rh \\ R_2PrP \diagup \diagdown THF \end{array} \right]^{+}$

(49)

$$\xrightarrow{^{13}CH_3I\,(-20\,°C)}$$

$\left[\begin{array}{c} O \\ \| \\ C \\ H_3{}^{13}C \diagdown \diagup PPrR_2 \\ Rh \\ R_2PrP \diagup \diagdown THF \\ | \\ I \end{array} \right]^{+}$

$$\xrightarrow{20\,°C}$$

(50)

$\left[\begin{array}{c} OC\diagdown \diagup PPrR_2 \\ Rh \\ R_2PrP \diagup \diagdown I \end{array} \right]$

$$\xrightarrow[-AgI]{AgSbF_6,\,+2\,CO,\,-40\,°C} \quad \textbf{49} \qquad (45)$$

(51)

complexes are formed (equation 45). Oxidative addition of EtI, PrI and i-PrI has not been observed. At 20 °C **50** forms **51** which, on reaction with AgSbF₆ and carbon monoxide, reforms **49**.

9. Synthesis of nitrogen-15 labelled hexafluoroantimonate salts of sulphuranes and their demethylation

Nitrogen-15 has been incorporated[155–157] into the sulphurane compounds **52–55** as shown in equation 46. The rates of demethylation of this series by pyridine-D₅ (equation 47) have been examined by observing the ^1H N.M.R. spectra of the N-methyl ($\delta = 2.77$) and TMS (equation 47)[155]. The apical methoxy ligand coordinated to the sulphurane sulphur (compound **53**) is demethylated faster in perdeuterated pyridine ($k_{25°C} = 4.54 \times 10^{-5}\,s^{-1}$; in the 324.9–298.2 K interval $\Delta H = 17.8 \pm 0.3$ kcal mol^{-1} and $\Delta S = -18.5$ eu) than is the methyl ligand of **52** ($k_{25°C} = 1.48 \times 10^{-6}\,s^{-1}$, $\Delta S = -7.82$ eu, $\Delta H = 23.1$ kcal mol^{-1}). On the other hand, the demethylation of **52** by pyridine-D₅ proceeds 1300 times slower than the demethylation of diphenylmethylsulphonium tetrafluoroborate **57**[156,157]. This has been explained by assuming that in the course of transmethylation of pyridine-D₅ with **52** two bonds (N—S and S—Me) must be cleaved to form **56a** and the transition state barrier is higher in reaction 47 than in the demethylation of **57**. The additional steric hindrances caused by the presence of the

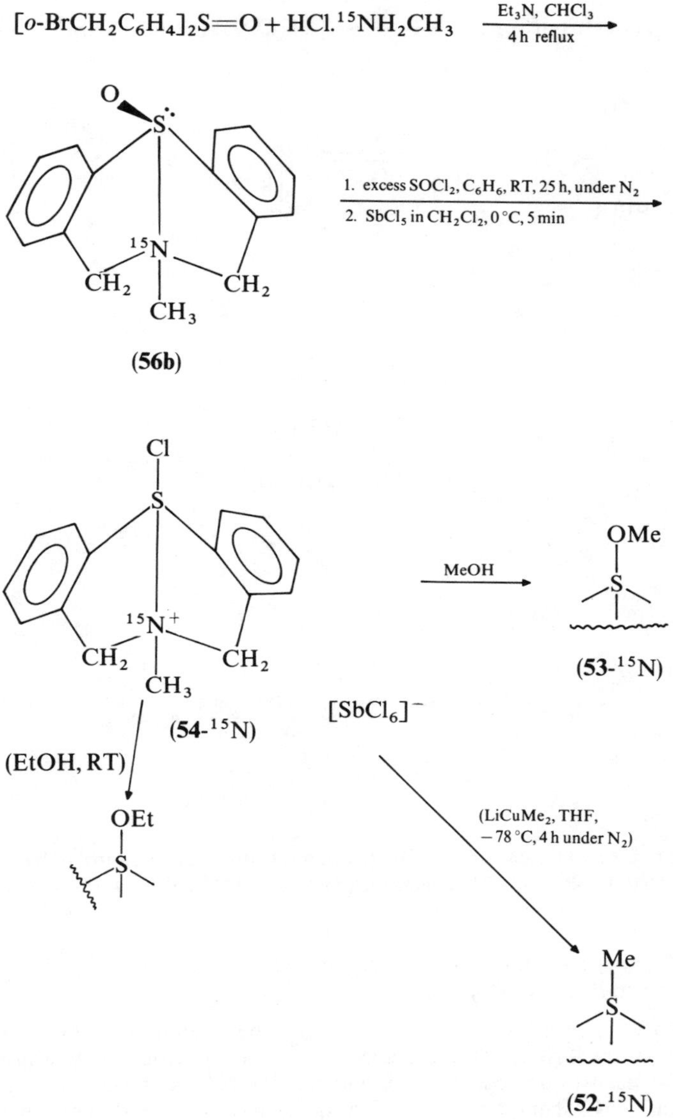

$$[o\text{-}BrCH_2C_6H_4]_2S{=}O + HCl.^{15}NH_2CH_3 \xrightarrow[\text{4 h reflux}]{Et_3N,\ CHCl_3}$$

(56b)

1. excess $SOCl_2$, C_6H_6, RT, 25 h, under N_2
2. $SbCl_5$ in CH_2Cl_2, 0 °C, 5 min

$(54\text{-}^{15}N)$

$[SbCl_6]^-$

MeOH

$(53\text{-}^{15}N)$

(EtOH, RT)

$(LiCuMe_2, THF, -78 °C, 4 h under N_2)$

$(52\text{-}^{15}N)$ (46)

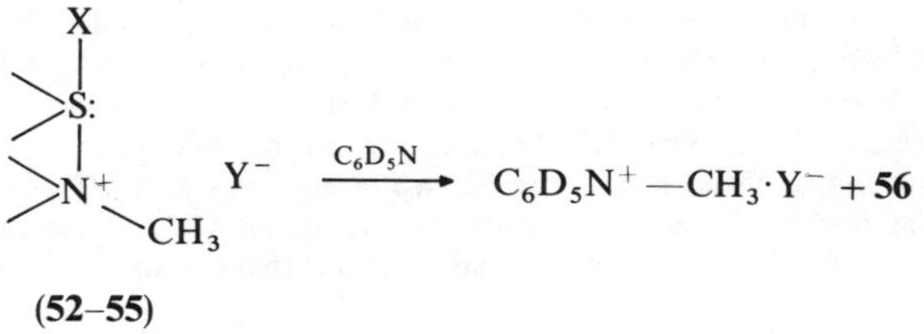

$(52\text{--}55)$ $Y^- \xrightarrow{C_6D_5N} C_6D_5N^+\!\!-CH_3\cdot Y^- + 56$ (47)

neighbouring group in **52** and **53** close to the sulphur atom are not taken into consideration. The methoxysulphurane **53** is demethylated faster in pyridine-D$_5$ than SMe sulphuranes, because the oxygen of **53** remaining on the sulphur after demethylation stabilizes the final sulfurane **56**.

A linear relationship $\delta_N(^{15}N) = 154.7\sigma_m + 47.9$ (48) (equation 48) of the ^{15}N N.M.R. singlet signals, $\delta_N(^{15}N)$ displayed at 34, 66, 69 and 103 ppm from the ^{15}NH$_3$ external standard, against Hammett substituent constants of the groups (Me, Et, MeO and Cl) located at the apical position of the cationic sulphur having the values, $\sigma_m = -0.07$, 0.1, 0.12 and 0.37 has been revealed. The chemical shifts have been measured by using aniline-^{15}N as an external reference ($\delta = 56.5$ ppm) and have been evaluated from a ^{15}NH$_5$ external standard. The N—S distances in the solid state in **52**–**55** equal 2.46 Å. The existence of an N—S bond in the solution is supported by the N—S—CH$_3$ coupling $^2J_{CN}$ and by considerable attractive interaction between the sulphur and nitrogen atoms in solution as shown by the deceleration of the transmethylation of **52** to pyridine-D$_5$. Due to electron transfer from the nitrogen to the sulphonio moiety the rate constant of the methyl transfer from the sulphur of **52** to pyridine at 25 °C was 1300 times slower than the analogous reaction of the methyldiphenylsulphonium salt **57**.

The nucleophilic methyl displacement reactions of compounds **57a**–**57e** in pyridine-d_5 have been examined and discussed[155].

(52) X = CH$_3$, Y = PF$_6$
(53) X = MeO, Y = SbCl$_6$
(54) X = EtO, Y = SbCl$_6$
(55) X = Cl, Y = SbCl$_6$

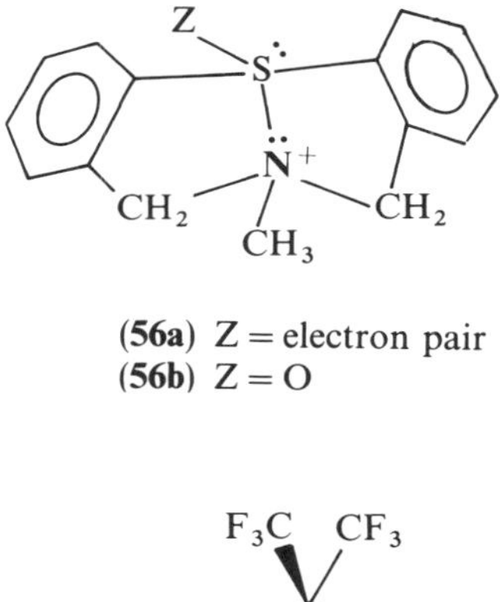

(56a) Z = electron pair
(56b) Z = O

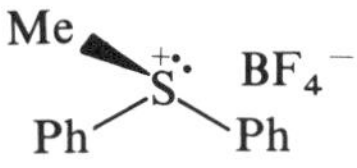

(57)

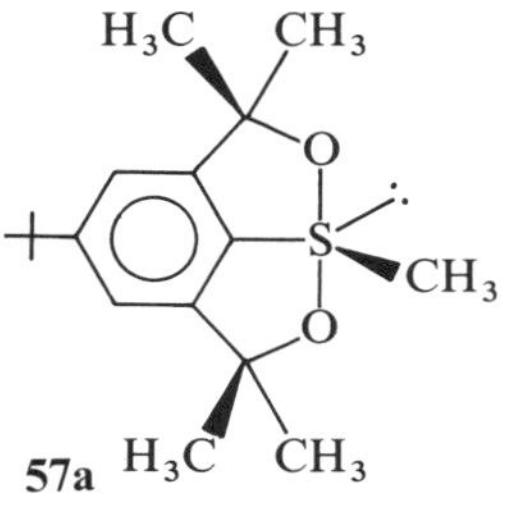

57a

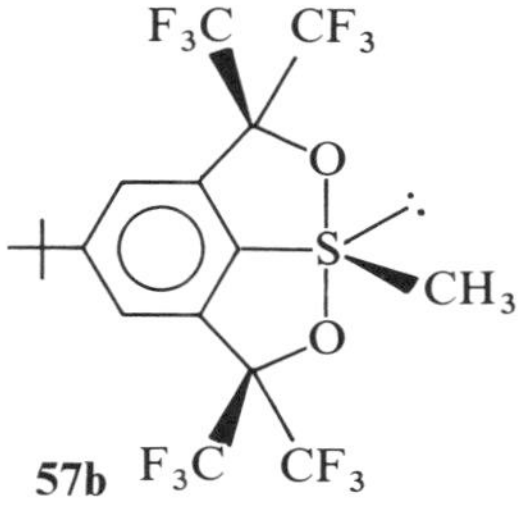

57b

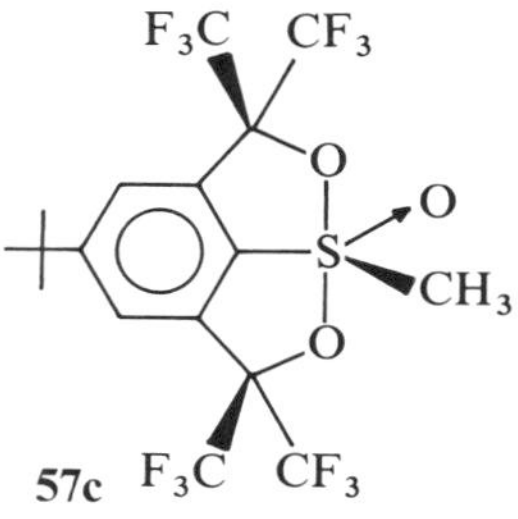

57c

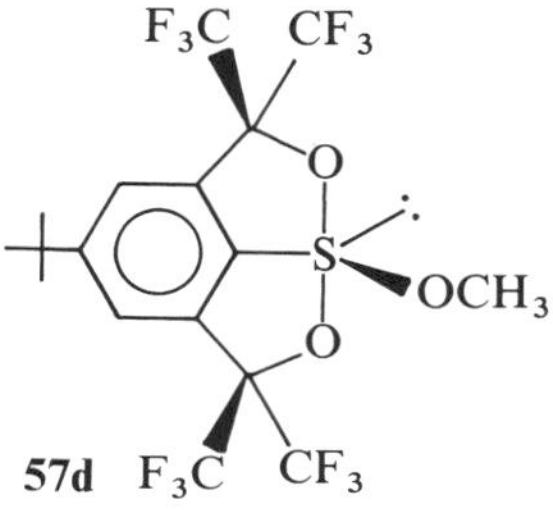

57d

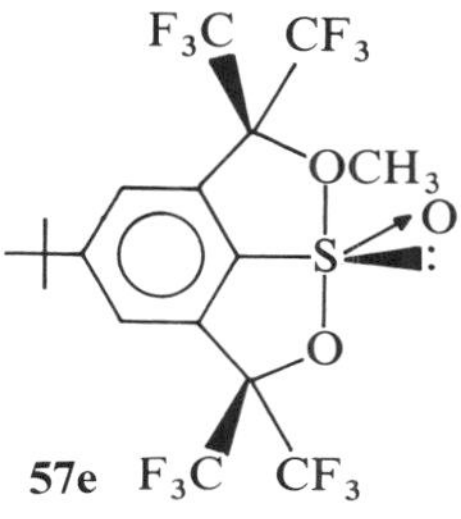

57e

The methoxy exchange reaction of **53** in methanol-D_4 has also been studied (equation 48)[155] by measuring the area of ^1H N.M.R. peaks of MeO of **53** and N—Me of **53** and **58**. The value of $k_{exch.}$ at 50 °C was $3.12 \times 10^{-5}\,s^{-1}$ (half-life about 61.7 h).

$$\text{(53)} \quad + CD_3OD \rightleftharpoons \quad \text{(58)} \quad + CH_3OD \tag{48}$$

(53)   (58)

The transition state structure **59** and an associative displacement at the sulphur has been proposed[155,158] for the mechanism of exchange of methoxy groups in **53**.

(59)

10. Complexes of azoxybenzenes with $SbCl_5$

^{15}N and ^{18}O labelled azoxybenzene, $C_6H_5*N(*O)=*NC_6H_5$ (**60**), and several of its fluoro derivatives have been synthesized[159] and the isotopic shifts in their I.R. and Raman spectra determined[159]. The I.R. spectra of the complexes of compound **60** and of 2,3,4,5,6-pentafluoroazoxybenzene (**61**) with $SbCl_5$ labelled with ^{18}O and the I.R. and N.M.R. spectra of these complexes labelled with ^{15}N have been recorded also and interpreted. The $N=N$ stretching vibration at $1444\,cm^{-1}$ is shifted by $\Delta\nu(^{15}N) = 30\,cm^{-1}$ (the values calculated with the valence force field method are 1413 and $28\,cm^{-1}$). The isotopic shift $\Delta\nu(^{15}N) = 8\,cm^{-1}$ of the I.R. absorption band at $830\,cm^{-1}$ has been ascribed to vibration involving the $N \rightarrow O$ bond in the $4\text{-}HC_6F_4N(O)=NC_6F_4H\text{-}4'$ compound[160]. The presence of fluorine atoms in the phenyl fragments of azoxybenzenes has practically no effect on the stretching frequencies of the azoxy group but decreases the band intensity of the vibrations of the azoxy and phenyl groups in the Raman spectra. Introduction of fluorine into the aromatic rings of azoxybenzene diminishes the complex formation with $SbCl_5$ at the oxygen of the azoxy group because of the decreased basicity of azoxy oxygen in the fluorinated azoxybenzenes as compared with the non-fluorinated ones. The I.R. and ^{15}N N.M.R. data are in agreement with the suggested coordination of $SbCl_5$ at azoxy oxygen.

11. Syntheses and photochemical applications of deuterium-labelled $[SbF_6]^-$
salts of methyl(hydrido)diplatinum(II) complexes

[Pt(CD$_3$)$_2$(dppm)] **(62)** has been synthesized[161] from [PtCl$_2$(dppm)] and CD$_3$Li. [Pt$_2$(μ-D)(CD$_3$)$_2$(μ-dppm)$_2$]$^+$ **(63)** and [Pt$_2$(CD$_3$)$_3$(μ-dppm)$_2$]$^+$ **(64)** have been obtained[161] as [PF$_6$]$^-$ and [SbF$_6$]$^-$ salts in the course of an isotopic study of the mechanism of reductive elimination of H$_2$, CH$_4$ and C$_2$H$_6$ from the complex ions [Pt$_2$H$_n$Me$_{3-n}$(μ-dppm)$_2$]$^+$, where $n = 0$–3 and dppm = Ph$_2$PCH$_2$ PPh$_2$, upon irradiations with a medium pressure Hg lamp. **63** has been prepared in 70% yield by reduction of [Pt(CD$_3$)$_2$(dppm)$_2$]$^+$ with Na[BD$_4$] in MeOD. [Pt$_2$Me(CD$_3$CN)(μ-dppm)$_2$][SbF$_6$] **(65)** has been prepared by photolysis of [Pt$_2$H(μ-H)Me(μ-dppm)$_2$][SbF$_6$] in CD$_3$CN. The complex was stable in CD$_3$CN solution in the absence of oxygen. The same product **65** has been formed by photolysis of [Pt$_2$(μ-H)Me$_2$(μ-dppm)$_2$][SbF$_6$], or as the [PF$_6$]$^-$ salt by photolysis of [Pt$_2$Me$_3$(μ-dppm)$_2$][PF$_6$] **(66)** (equations 49 and 50).

$$\left[\begin{array}{c} \text{CH}_2 \\ \text{Ph}_2\text{P} \diagdown \diagup \text{PPh}_2 \\ | \quad \text{H} \quad | \\ \text{Me-Pt} \diagdown \text{Pt-H} \\ | \quad | \\ \text{Ph}_2\text{P} \diagup \diagdown \text{PPh}_2 \\ \text{CH}_2 \end{array}\right]^+ \xrightarrow[\text{S}]{hv} \left[\begin{array}{c} \text{CH}_2 \\ \text{Ph}_2\text{P} \diagdown \diagup \text{PPh}_2 \\ | \quad | \\ \text{Me-Pt} \text{—} \text{Pt-S} \\ | \quad | \\ \text{Ph}_2\text{P} \diagup \diagdown \text{PPh}_2 \\ \text{CH}_2 \end{array}\right]^+ + \text{H}_2 \qquad (49)$$

(65) S = CD$_3$CN, (CD$_3$)$_2$CO

$$\left[\begin{array}{c} \text{CH}_2 \\ \text{Ph}_2\text{P} \diagdown \diagup \text{PPh}_2 \\ | \quad \text{H} \quad | \\ \text{Me-Pt} \diagdown \text{Pt-Me} \\ | \quad | \\ \text{Ph}_2\text{P} \diagup \diagdown \text{PPh}_2 \\ \text{CH}_2 \end{array}\right]^+ \xrightarrow[\text{S}]{hv} \mathbf{65} + \text{CH}_4 \qquad (50)$$

In the poor donor solvent, acetone, the primary photochemical cleavage of the Pt$_2$(μ-H) linkage by excitation of an electron to a σ^*(Pt–Pt), orbital[162,163] is followed by predominantly intramolecular coupling of methyl groups and ethane formation (bond formation is favoured in the sequence H—H > CH$_3$H > CH$_3$CH$_3$). The mechanism of C—C bond formation could not be rigorously established because it was impossible to synthesize the asymmetrically labelled complex [(CD$_3$)$_2$Pt(μ-dppm)$_2$PtCH$_3$]$^+$ **(66a)** by the reaction of [Pt(CD$_3$)$_2$(dppm)] with [PtClMe(dppm)]. Instead, the compound **66b** was produced with complete scrambling of CH$_3$ and CD$_3$ groups (determined by ^{1}N N.M.R.) as has been observed in other diplatinum complexes[164].

12. Microwave spectroscopic studies of isotopic species of stibabenzene

The syntheses, as well as physical and theoretical studies regarding the aromatic character of heterobenzenes, such as phosphabenzene[165–168], arsabenzene[165,168–176] and bismabenzene[177,178], have been extended also for stibabenzene[179,180]. The benzene-like structure has been attributed to stibabenzene on the basis of experimental study and theoretical interpretation of the microwave spectra of ^{121}SbC$_5$H$_5$, ^{123}SbC$_5$H$_5$, β-dideuterio ^{121}SbC$_5$H$_3$D$_2$ and β-dideuterio ^{123}SbC$_5$H$_3$D$_2$ in the region 26.5–40.0 GHz.

The structure is found to be planar and of C_{2v} symmetry along the a-axis. The constancy of the moments of inertia for any two species differing only in the antimony isotope indicates that Sb lies on the a-axis and $(I_a/u\,\text{Å}^2 = 111.99$ is exactly the same for both $^{121}\text{SbC}_5\text{H}_5$ and $^{123}\text{SbC}_5\text{H}_5$, and similarly $(I_a/u\,\text{Å}^2) = 121.01$ for $^{121}\text{SbC}_5\text{H}_3\text{D}_2$ and 121.00 for $^{123}\text{SbC}_5\text{H}_3\text{D}_2$). A shortening of the normal Sb—C single bond length $[d(\text{Sb—C}) = 2.050 \pm 0.005\,\text{Å}]$ and values for $d(\text{C—C})$ close to the carbon–carbon bond length found in benzene $(1.40 \pm 0.03\,\text{Å})$ are observed. The carbon ring angles are greater than $120°$ in value. The CSbC angle is near $93°$, the smallest one observed for the Group V heterobenzene series. The derived quadrupole coupling constants for the 121 and 123 antimony isotopes support a σ-donating property for the antimony heteroatom. The lone pair orbital is formed primarily from the largest $5p_a$ orbital of Sb (the smallest $5p_c$ orbital donates electrons to the ring)[180]. Further theoretical quantum-mechanical studies are needed to understand the electronic structure of stibabenzene.

3,5-dideuterio-stibabenzene has been prepared[180] by the reaction of 1,4-dihydro-1,1-dibutylstannabenzene-3,5-D_2 with antimony trichloride. The obtained 1,4-dihydro-1-chloro-stibabenzene-3,5-D_2 was treated with DBN (1,5-diazabicyclo[4,3,0]non-5-ene) to yield the desired stibabenzene-3,5-D_2. Both the deuteriated and normal species can been stored in tetraglyme at low temperatures and can be distilled at RT under high vacuum.

IV. SYNTHESES AND USES OF ISOTOPICALLY LABELLED COMPOUNDS OF ARSENIC

A. Radiochemical and Radioanalytical Aspects of Arsenic

1. Synthesis of arsenic radioisotopes for environmental toxicology and biomedical purposes

Tons of As are emitted into the atmosphere from coal-fired power plants in Europe and tons are present in the solid fly ash residue. Understanding of the environmental impact of arsenic on the ground water and the possibility of the accumulation of the metal by flora and fauna exposed to effluent waters requires one to study the metabolism of actual environmental levels of As in humans and in laboratory animals[181] and to establish health criteria for As. Radioarsenic with very high specific activity must be used as a tracer in biochemical studies. Carrier-free radioarsenic is produced at GBq levels by proton bombardment on germanium targets in cyclotrons for use in metallobiochemical programmes. ^{74}As and ^{72}As are of interest in nuclear medicine[182]. $^{70-76}\text{As}$ isotopes are produced by proton bombardment of $^{70-76}\text{Ge}$ isotopes present in germanium targets of natural isotopic composition.

In the (n, γ) reaction with natural arsenic (100% ^{75}As) in a nuclear reactor with a flux of $10^{12}\,\text{n cm}^{-2}\,\text{s}^{-1}$, samples of ^{76}As with specific activity of 920 mCi g^{-1} are obtained only. Cyclotron arsenic isotopes $^{71,72,76}\text{As}$ are utilized in short-term biological experiments, ^{74}As for medium-term experiments.

The excitation function for production of As radioisotopes via $\text{Ge}^{\text{nat}}(p, xn)$ reactions has been determined[183]. Only ^{74}As is obtainable in sufficiently pure form after the appropriate 'cooling time'. Employing ^{74}Ge or ^{76}Ge enriched targets, the ^{74}As could be produced in higher yields via (p, n) or $(p, 3n)$ nuclear reactions. An isotope generator for carrier-free ^{72}As, the daughter of neutron-deficient ^{72}Se, has been proposed[181,184].

2. Synthesis of $^{76}\text{AsCl}_3$

Arsenic trihalides, especially arsenic trichloride, are the most versatile starting radiolabelled arsenicals employed for the synthesis of a variety of organoarsenic compounds applied in chemotherapy[185], environmental studies[186,187] and for 'biochemical mapping'[188]. The ^{76}As used has been produced in the ^{75}As(n,γ)^{76}As nuclear reaction by irradiation of As$_2$O$_3$ sealed in quartz tubes in 6×10^{13}n cm^{-2}s^{-1} neutron flux for about 2 h. The specific activity of the sample was 300 mCi mmol^{-1}. ^{76}AsCl$_3$ has been prepared in 80% yield from ^{76}As$_2$O$_3$ with sulphur monochloride[189] (equation 51). ^{76}AsCl$_3$ and diphenylamine have been used for the synthesis of 10-chloro-5,10-dihydrophenarsazine (10-CPA)[190], 67 (equation 52). 10-CPA is the intermediate for synthesis of arsenic analogues of biologically active phenothiazines.

$$2\,^{76}\text{As}_2\text{O}_3 + 6\,\text{S}_2\text{Cl}_2 \xrightarrow{\text{heat}} 4\,^{76}\text{AsCl}_3 + 3\,\text{SO}_2 + 9\,\text{S} \tag{51}$$

$$^{76}\text{AsCl}_3 + \text{Ph}_2\text{NH} \longrightarrow \qquad \text{(yellow–green crystals,} \atop \text{m.p. 192–194 °C} \tag{52}$$

(67)

3. Synthesis of dimethylchloroarsine (DCA) labelled with ^{76}As

DCA labelled with ^{76}As has been obtained[191] from [^{76}As]arsenic trioxide and potassium acetate at 355 °C for 4 h under inert gas and subsequent trapping of the dimethylarsine oxide produced in a ferric chloride solution (50% hydrated FeCl$_3$ in conc. HCl) (equation 53).

$$*\text{As}_2\text{O}_3 + 4\,\text{CH}_3\text{COOK} \longrightarrow (\text{Me}_2*\text{As})_2\text{O} + \text{K}_2\text{O} + 4\,\text{CO}_2$$

$$(\text{Me}_2*\text{As})_2\text{O} + 2\,\text{HCl} \xrightarrow{\text{FeCl}_3} 2\,\text{Me}_2*\text{AsCl} + \text{H}_2\text{O} \tag{53}$$

The ^{76}As-labelled DCA has been used subsequently for the preparation of labelled dimethylarsinopenicillamine (DAP, 68) and dimethylarsinomercaptoethanol (DAM, 69) (equation 54). DAP and DAM represent a new group of carcinostatic radiopharmaceuticals[192,193] which have been used for tracer studies in experimental animals[194] (CD-1 male mice). This study demonstrated the possibility of developing new radiopharmaceuticals from HS-containing biomolecules.

$$\text{Me}_2*\text{AsCl} + \text{HSR} \xrightarrow{\text{Et}_3\text{N, reflux}} \text{Me}_2*\text{AsSR} \tag{54}$$

R = —CMe$_2$CH(NH$_2$)COOH for DAP, 68
R = —CH$_2$CH$_2$OH for DAM, 69 (colourless liquid, b.p. 87–88 °C at 8 mm Hg).

4. Synthesis of ^{73}As-radiolabelled arsenobetaine and arsenocholine

Arsenobetaine, arsenosugars and arsenocholine are organoarsenicals less toxic for animals and humans than inorganic arsenic compounds, and have been found in certain marine organisms and in seafoods. They are excreted rapidly in urine (about 70% of the dose in 24 h)[195-198]. Natural arsenic is a pure one-isotope element (^{75}As) and had to be labelled with radioactive arsenic for metabolic studies. ^{73}As has been chosen as the most suitable isotope for tracer investigations.

a. [73]As-labelled trimethylarsine. $^{73}AsMe_3$, a key intermediate in preparing radiolabelled arsenobetaine, has been synthesized from $^{73}AsCl_3$ with an excess of MeLi in di-*n*-butyl ether to prevent dimethylchloroarsine or monomethyldichloroarsine formation.

b. [73]As-labelled arsenobetaine, **70**, has been prepared[199] as shown in equation 55. The specific activity of the product **70** was 46 μCi (1.70 MBq) $^{73}As\,mg^{-1}$ (As).

$$Me_3^{73}As + BrCH_2COOEt \xrightarrow{\text{benzene, RT, 24h}} Me_3^{73}As^+CH_2COOEt + Br^-$$

$$\xrightarrow{OH^-} Me_3^{73}As^+CH_2COO^- + EtOH + Br^- \qquad (55)$$
$$\mathbf{70}$$

c. [73]As-labelled arsenocholine, **71**, has been synthesized as shown in equation 56, in 85.6% yield, specific activity 15 μCi (0.56 MBq) mg^{-1}. Preliminary studies showed that more than 95% of the ^{73}As, injected i.v. as [^{73}As]betaine into male albino rats at doses of 4 mg As kg^{-1} body wt., was eliminated during 3 days following administration. This agrees with the high recovery of arsenic in excreta of rats fed with flounders containing high levels of arsenic[200].

$$Me_3^{73}As + BrCH_2CH_2OH \xrightarrow{\text{MeCN, 80°C, 48–72 h reflux}} Me_3^{73}As^+CH_2CH_2OH + Br^- \quad (56)$$
$$\mathbf{71}$$

5. Synthesis of methylarsonic acid, methylarsonous acid, dimethylarsinic acid, dimethylarsinous acid and trimethylarsine labelled with ^{74}As

Methylarsonic acid, $CH_3^{74}AsO(OH)_2$, **72**, has been obtained[201] as shown in equation 57. The intermediate dichloromethylarsine **73** gave, after treatment with water, methylarsonous acid (**74**) or methylarsine oxide $CH_3^{74}AsO$ (**75**). Methylarsonic acid has been prepared by shaking the dichloromethylarsine in benzene with 1 M hydrochloric acid with some drops of bromine water added.

$$Na_2^{74}AsO_4 \xrightarrow{H_2SO_4,\,KI,\,Na_2SO_3} {}^{74}AsI_3 \xrightarrow[\text{2. MeI, 40–50°C, 2–4h}]{\text{1.10M NaOH, RT,}} CH_3^{74}AsO(OH)_2 \xrightarrow{4M\,HCl,\,Na_2SO_3,\,KI}$$
$$\mathbf{(72)}$$

$$CH_3^{74}AsCl_2 \xrightarrow{H_2O} CH_3^{74}As(OH)_2 (or\ CH_3^{74}AsO) \qquad (57)$$
$$\mathbf{(73)} \qquad\qquad \mathbf{(74)} \qquad\qquad \mathbf{(75)}$$

Dimethylarsinic acid, $(CH_3)_2^{74}AsO(OH)$ (**76**), has been obtained by synthesizing first chlorodimethylarsine, $(CH_3)_2^{74}AsCl$, transforming it into dimethylarsinous acid, $(CH_3)_2^{74}AsOH$ (**77**), followed by oxidation with air oxygen to give **76**.

Trimethylarsine, $(CH_3)_3^{74}As$ (**78**), has been synthesized from CH_3MgI and $^{74}AsI_3$. The product **78** is stable when kept in benzene solution. **78** forms with mercury(II) chloride a complex[202] $Me_3^{74}As\cdot HgCl_2$ (**79**) soluble in water. **78** can be liberated from **79** by reduction of the Hg(II) with zinc powder or by addition of an excess of potassium iodide.

The different labelled arsenic compounds have been used to investigate their liquid–liquid distribution in order to develop procedures for the separation of the various arsenic species found in natural waters[203–205] in human urine and plasma[205] and in plants[206a]. Only trimethylarsine, dimethylarsine and methylarsine are completely extracted (more than 98%) into benzene from neutral solutions. All trivalent arsenic species (arsenic trioxide, As_2O_3, methylarsonous acid **74** and dimethylarsinous acid **77**) are quantitatively extracted into 0.01 M diethylammonium salt of diethyldithiocarbamic acid in benzene, etc.[206b].

6. Synthesis and application of ^{76}As and ^{99m}Tc labelled arsonomethylphosphate (AMPA) and arsonoacetate (AAA)

Arsenic-76 labelled AMPA, **80**, an analogue of methylenediphosphonate (MDP), and arsonoacetate (AAA), **81**, an analogue of phosphonoacetate, have been synthesized[207] by reacting [^{76}As]arsenic trioxide (dissolved in sodium hydroxide) with chloromethyl phosphinic acid[208a] or with monochloroacetic acid[209], respectively.

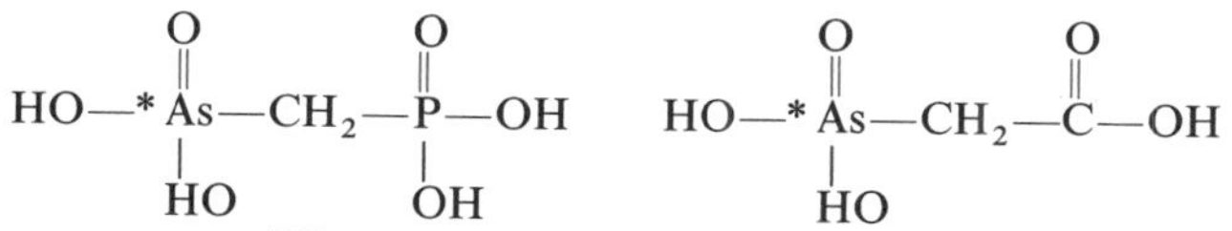

(80) (80) *As = ^{76}As

80 and **81** have been labelled with ^{99m}Tc in the presence of stannous chloride[210a]. The doubly ^{99m}Tc–^{76}As labelled AMPA and AAA have been injected into tail veins of mice and the biodistribution of the two radionuclides has been studied. The (As-76/Tc-99m) ratio was found to be close to 1 for all organs in the case of AMPA, reflecting the similar behaviour of the two labels associated with **80**. The biodistributions of ^{76}As and ^{99m}Tc associated with AAA have been found not to be the same. This might be due to the change of the AAA owing to metabolic activity[207].

7. Synthesis of [2-^{3}H]methyl-2-phenyl[3-^{3}H]oxirane

Methyltriphenylarsonium iodide, acetophenone and [^{3}H]water gave α-methyl styrene oxide **82** (80%) labelled with tritium both in the oxirane ring and on the methyl group[208b], through the formation of a ylide intermediate **83** as shown in equations 58a,b,c. 85% yield colourless oil, b.p. 86 °C/18 torr, radiochemical purity 98%, specific activity 5.44 mCi mmol^{-1}

$$[Ph_3As^+Me]I^- \xrightarrow{NaOH} Ph_3As^+C^-H_2 \underset{H_2O}{\overset{^3H_2O}{\rightleftharpoons}} \underset{(\mathbf{83})}{Ph_3As^+C^-{*H_2}} \qquad (58a)$$

$$PhC(=O)CH_3 \underset{}{\overset{NaO*H(*H_2O)}{\rightleftharpoons}} PhC(=O)C{*H_3} \qquad (58b)$$

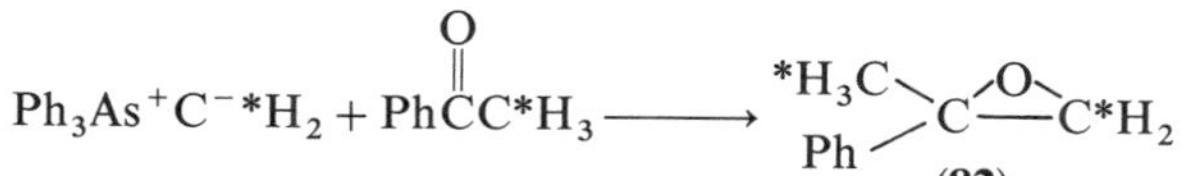

(58c)

(82)

cis-**84** and *trans*-β-methyl styrene **85** oxides have been tritium labelled also[208b] and used for biochemical characterization of the various forms of epoxide hydrolase.

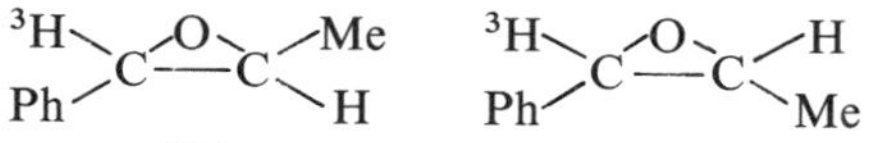

(85) (84)

8. Reaction of recoil tritium with arsine

The gas-phase reactions of recoil tritium atoms (generated in the ^{3}He(n,p)^{3}H nuclear reaction) with gaseous arsine, AsH_3, at high pressure has been studied[210b] in mixtures either with or without the scavenger ethylene. In the absence of scavenger about 67.3% of the tritium has been found as H^3H(HT) and about 31.8% as AsH_2T.

In the presence of the scavenger 63.2% of the tritium stabilized as HT, 30.2% as AsH_2T and 5.8% as ethane-^3H and ethylene-T. A similar investigation of the $(PH_3/^3He)$, $(H_2S/^3He)$ and $(HCl/^3He)$ systems [210b] was compared with earlier studies[211a–215a] of the $(*T + CH_4)$, $(*T + SiH_4)$ and $(*T + GeH_4)$ hydrides. Some regularities have been noticed in the reaction of hot recoil T with lone pair electrons, namely that the substitution yields increase from the IVth group of hydrides (24–27%) to the VIIth group hydrides (63%) while within the same group the substitution decreases as the radius of the central atom increases. The increase in the substitution reactivity for compounds of the same period is ascribed to the increase in the number of lone pairs in the molecule. A relationship between the H—M—H bond angle and the hot substitution yields has been observed. The decrease in the substitution yield within the same group has been related to the properties of the central atom. With the decrease in the H—M—H bond angle (B.A.) the yield of the substitution is lowered: 55% in ammonia (B.A. = 1.87 rad), 34% in phosphine (B.A. = 1.61 rad) and 30% in arsine (B.A. = 1.60 rad). The same trend is observed for H_2O (B.A. = 1.81, yield = 68%) and for H_2S (B.A. = 1.60, yield = 46%). A suggestion has been made that the substitution is a variety of the Walden inversion[216a] (equation 59).

$$*T + :\overset{\displaystyle H}{\underset{\displaystyle H}{M}}{-}H \rightarrow T{-}\overset{\displaystyle H}{\underset{\displaystyle H}{M}}: + H^\cdot \qquad (59)$$

9. Spectroscopic studies with tritiated and deuteriated arsines

A normal coordinate vibrational analysis of isotopically substituted arsines of the pyramidal type, $AsHD_2$, $AsHT_2$, $AsDH_2$, $AsDT_2$, $AsTH_2$ and $AsTD_2$ having six fundamental frequencies, has been carried out[211b] by the method of Decius[212b] and Jones[213b], using the structural parameters and harmonic frequencies taken from the literature[214b,215b]. Besides harmonic potential constants and compliance constants[212b], the mean amplitudes of vibrations for both bonded and non-bonded distances (at 298.16 K) have been calculated and found to be as follows:

For $AsHD_2$ and $AsDH_2$ molecules:

	(As—H)	(As—D)	(H...D)	
$AsHD_2$:	8.8533;	7.2927;	15.1453;	12.7990 (D...D) 10^{-2} Å
$AsDH_2$:	8.8430;	7.9362;	14.7773;	14.7781 (H...H);

For tritium substituted arsines:

	(As—H)	(As—T)	(H...T)	
$AsHT_2$:	8.7714;	6.5792;	14.2793;	11.9297 (T...T) 10^{-2} Å
$AsTH_2$:	8.3665;	6.6569;	13.8342;	15.5342 (H...H);

For doubly labelled arsines:

	(As—D)	(As—T)	(D...T)	
$AsDT_2$:	7.8348;	6.8434;	12.7785;	12.6163 (T...T) 10^{-2} Å
$AsTD_2$:	7.7660;	6.6171;	12.7546;	12.9721 (D...D)

The distance $l_{(X-Z)}$ in these XYZ_2 pyramidal molecules decreases as the mass of the Z atom increases. The distance As–H is slightly longer in $AsHD_2$ than in $AsDH_2$. This trend is observed also in $AsHT_2$ and $AsTH_2$. No data for AsH_3, AsD_3 and AsT_3 have been presented.

10. Study of the ligand exchange in complexes of the uranyl cation with tertiary arsenic oxides

Neutral organoarsenic compounds of the type R_3AsO are widely used for extraction of uranium and other valuable elements from aqueous solutions[216b]. The photolumine-scence method has been applied to study the kinetics of formation of complexes of uranyl cations with organic and inorganic ligands[217].

Introduction of oxides of tertiary arsines, R_3AsO (where R = Me, Et, n-Pr, n-Bu, n-C_5H_{11}—, $Me_2CH(CH_2)_2$—, Ph and $PhCH_2$ into aqueous sulphuric acid solutions of the uranyl cation results in the quenching of the fluorescence intensity I and in the shortening of the life-time τ of $*UO_2^{2+}$ in the excited state[218] (both proportional to the concentration of R_3AsO). The life-time τ of $*UO_2^{2+}$ in the absence of R_3AsO equals $10.5\,\mu s$ at $T = 298$ K in 1 M H_2SO_4 solutions. An increase in the length of the aliphatic chain from R = Me to R = C_5H_{11} caused an increase in the rate constants of the bimolecular quenching process (i.e. of the complex formation) by nearly two orders of magnitude. The protonated R_3AsO molecules enter into the coordination sphere of UO_2^{2+} in acidic medium as $[R_3AsOH]^+$ entities and $[R_3AsOH]_2$ $[UO_2(SO_4)_2]$ complexes are formed. A linear correlation of the type $Y = aX + b$ for R_3AsO has been found, where $X = -\Sigma\sigma^*$, $Y = \lg k_{quench.}$, $a = 9.0$, $b = 5.75$ and σ = Taft constants[218]. The rate constants of ligand exchange in complexes of UO_2^{2+} with R_3AsO depend on the structure and electronic properties of the ligand.

B. Synthesis and Applications of Arsenic Compounds Labelled with Stable Isotopes

1. Deuterium and oxygen-18 infrared study of isotopic arsine–ozone complexes in argon matrix and their reactions

AsD_3 has been obtained from Zn_3As_2 and 30% D_2SO_4 in D_2O and purified by vacuum distillation[219,220]. 98% and 55% ^{18}O-enriched commercial ozone was used. Mixed iso-topic arsines, AsH_xD_{3-x} ($x = 0, 1, 2, 3$), have been prepared using a mixture of isotopic acids. The symmetric bending modes of AsH_3, AsH_2D, $AsHD_2$ and AsD_3 at 906, 813, 690 and 658 cm^{-1}, respectively, provided a measure of the degree of deuterium enrichment.

The $[AsD_3-^{18}O_3]$ complex, accompanied by some cis- and trans-D_2AsOD have been obtained in the condensation of $Ar/^{18}O_3/AsD_3$ samples (during codeposition of AsH_3 and O_3 at high dilution in solid argon during subsequent irradiation). Complex formation increased the probability of the photodissociation for the $^{18}O_3$ moiety. D_3AsO, D_2AsOD and an intermediate formulated as $DAsO$ have been observed[219a] in the red photolysis (equation 60) of arsine–ozone complexes. Subsequent blue and near U.V. irradiations destroyed the DAsO species and produced $DOAsO_2$. A comparison of the results of the $AsD_3/^{18}OH_2/Ar$ system with the one involving PH_3 showed that AsH_3 is slightly more reactive than PH_3 in the PH_3–O_3 complex[221]. The arsine–ozone interaction in the complex markedly increased the oscillator strength for ozone photodissociation by red light. The infrared adsorptions of the compounds **86**, **87** and **88** and their various complexes with ozone, including different isotopic species have been studied in detail and assigned[219–222].

$$[D_3As...{}^{18}O_3] \xrightarrow{h\nu(630-1000\,\text{nm})} [D_3AsO]^* + O_2$$

precursor complex

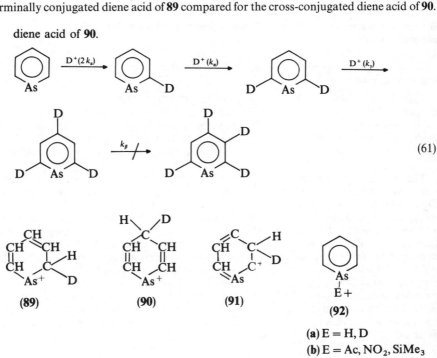

DOAsO$_2$ $\xleftarrow{O_2(220<\lambda<515\,\text{nm})}$ DAsO   D$_2$AsOD $\leftarrow$ D$_3$AsO   **(86)**   (60)

metaarsenic acid   (*cis*- and *trans*-   (arsine oxide)

(87)   **(88)** arsinous acid)[219b,c]

2. Acid catalysed deuterium exchange reactions of arsabenzene and 1,1'-diarsaferrocene

a. The aromatic character of arsabenzene has been explored by studying the deuterium exchange between trifluoroacetic acid (TFAA) and arsabenzene in methylene chloride solvent (TFAA/CH$_2$Cl$_2$ = 1:1 V/V) in sealed ampoules at 100–130 °C. Mass spectral analysis has shown that at 100 °C both mono- and dideuteriation of arsabenzene are taking place. After 20 h at 100 °C the concentrations of deuterioarsabenzenes were 39.0% (d$_0$), 46.2% (d$_1$), 14.7% (d$_2$), 0.2% (d$_3$). After 3 days (also at 100 °C) the deuteriation increased to: 41.5% (d$_1$), 46.3% (d$_2$) and 1.0% (d$_3$), while the amount of the d$_0$ species decreased to 11.0%. At 130 °C, after four days, 59.3% of d$_2$ and 40.6% of d$_3$ and 0.1% of d$_4$ species has been found[223]. The above data have been interpreted[223] as corresponding to fast deuteriation in the two α-positions (equation 61)[223,224] and to slower deuteriation at the unique γ-position (d$_3$-species). Deuteriation at the β-positions is negligible even at 130 °C. The faster α- than γ-deuterium exchange is due to the greater stability of the terminally conjugated diene acid of **89** compared for the cross-conjugated diene acid of **90**.

diene acid of **90**.

(61)

(a) E = H, D
(b) E = Ac, NO$_2$, SiMe$_3$

The intermediate structure **91** has higher energy than **90** and no evidence has been found for protonation of the arsenic atom, **92a**. Acetylation and nitration of arsabenzene gave mixtures of 2- and 4-substituted products. The ^{1}H NMR spectrum revealed that 2-acetylarsabenzene is the minor isomer and 4-acetylarsabenzene the major one (4:1 ratio). It has been suggested that this product ratio is the result of electrostatic repulsion between the electropositive electrophile and the positively charged arsenic atom.

Protodesilylation takes place faster in the α- than in the γ-position. Arsabenzene behaves like a moderately activated *ortho*-, *para*-directing substituted benzene toward electrophilic attack.

b. 1,1-Diarsaferrocene, **93**, (m.p. 195 °C, monoclinic crystals), was obtained in 50% yield as shown in equation 62^{225} and, as arsabenzene, undergoes acid-catalyzed deuterium exchange. At -20 °C in CF_3COOD/CH_2Cl_2 **93** exchanged four protons in one minute, while protons in the β-positions underwent only partial deuterium exchange during seven hours heating at 70 °C. The relative $(\alpha:\beta)$ reactivity of hydrogens in the α- and β-positions has been assessed to be 10^6. Ferrocene undergoes at -20 °C in CH_2Cl_2/CF_3COOD exchange 2.4 times faster than pentamethylbenzene[225,226].

$$(62)$$

(93)

3. Synthesis and vibrational analyses of deuterated methylchlorometallates

The preparations, vibrational spectra and X-ray crystallography of solid salts of the type $[(Me)_4M^V]^+ [(Me)_{4-n}M^{III}Cl_n]^-$, where $M^V = As$ or Sb, $M^{III} = Ga$ or In, and $n = 0, 1, 2, 3$ or $4^{227,228}$, has been supplemented[229] by taking I.R. and Raman spectra of fully deuteriated methylchlorometallates prepared as shown in equations 63–66. The starting deuteriated materials $(CD_3)_3GaOEt_2$ and $(CD_3)_3In$ have been prepared by Grignard reactions from CD_3I and Ga/Mg or In/Mg alloys in diethyl ether[229]. Reactions of these deuteriated derivatives with stoichiometric amounts of $GaCl_3$ or $InCl_3$ provided $(CD_3)_2GaCl$ and $(CD_3)_2InCl$ of high purity after recrystallization from dichloromethylene.

$$(CH_3)_3AsCl_2 + (CD_3)_2M^{III}Cl \longrightarrow [(CH_3)_3As(CD_3)]^+[(CD_3)M^{III}Cl_3]^- \quad (63)$$
$$92\% \; (M^{III} = Ga); \; 88\% \; (M^{III} = In)$$

$$(CH_3)_4AsCl + (CD_3)_3M^{III}Cl \longrightarrow [(CH_3)_4As]^+[(CD_3)_2M^{III}Cl_2]^- \quad (64)$$
$$80\% \; (M^{III} = Ga); \; 84\% \; (M^{III} = In)$$

$$(CH_3)_4AsCl + (CD_3)_3In \longrightarrow [(CH_3)_4As]^+[(CD_3)_3InCl]^- \quad (65)$$
$$74\%$$

$$(CH_3)_4AsCl + (CD_3)_3Ga.OEt_2 \longrightarrow [(CH_3)_4As]^+[(CD_3)_3GaCl_1]^- \qquad (66)$$
$$68\%$$

The normal coordinate analysis carried out using a program written by Hilder-brandt[230] allowed one to deduce the best values for the force constants (within the simple valence molecular force field) for the species under consideration based on the observed vibrational frequencies. The observed and calculated frequencies are compared also. For instance, the observed $\nu_{as}(CH_3)$ and $\nu_s(CH_3)$ for $[(CH_3)_4As]^+[(CH_3)_3GaCl]^-$ are equal to 2950 and 2846 cm^{-1}. The corresponding calculated values are 2976 and 2862 cm^{-1}. In the case of the deuterio-methyl chlorometallate salt $[(CH_3)_3AsCD_3]^+[CD_3GaCl_3]^-$ the observed values for $\nu_{as}(CD_3)$ and $\nu_s(CD_3)$ are 2238 and 2118 cm^{-1}, while the calculated ones are 2234 and 2076 cm^{-1}.

Increase of the M—Cl and M—C stretching force constants (accompanied by a corresponding decrease of the M—Cl and M—C bond distances) with increasing chlorine substitution across the series $[(CH_3)_{4-n}M^{III}Cl_n]^-$, where $n = 0$ to 4, has been observed[229] and rationalized as a combination of inductive and resonance effects. For instance, in the case of M = Ga the bond distances $R(Ga—C)$ in Å and stretching force constants $f(Ga—C)$ in mdyn Å^{-1} for the $[Me_{4-n}GaCl_n]^-$ series are:

2.22 Å and 1.87 mdyn Å^{-1} for $[Me_4Ga]^-$
2.053   2.24   for $[Me_3GaCl]^-$
1.980   2.56   for $[Me_2GaCl_2]^-$
1.934   2.80   for $[MeGaCl_3]^-$

Each M—C bond in the ions $[Me_4M]^-$ represented as $(CH_3)_3M \leftarrow CH_3^-$ should be 75% covalent and 25% dative. Also, the anionic complexes $[(CH_3)_3MCl]^-$ may be described as donor–acceptor complexes[229], $(CH_3)_3M \leftarrow Cl^-$, i.e. $[GaCl_4]^-$ as $Cl_3Ga \leftarrow Cl^-$, etc. The donor–acceptor bond is expected to be shorter and contribute to the average bond distance found, for instance, in the octahedral $[GaCl_4]^-$ ion.

4. Hydrolysis and oxidative hydrolysis of tetrakis[(diiodarsino)methyl]methane

Hydrolysis of $C(CH_2AsI_2)_4$ (95) with aqueous NaOH carried out in THF in sealed glass ampoules under vacuum gave the arsonous acid–organoarsenic oxide compound 94-H (equation 67) in 55.5% yield (m.p. > 300 °C, decomp.). 94 heated with a large excess of D_2O at 150 °C in a sealed ampoule produced $(DO)_2AsCH_2C(CH_2AsO)_3$ (94-D) in 75.4% yield (equation 68)[231–233]. The crystal structure 94-D has been determined by X-ray diffraction (monoclinic, space group P2$_1$/c, with $a = 661.5$, $b = 852.8$, $c = 1922.4$ pm (p = pico = 10^{-12}), $\beta = 97.95°$, and $Z = 4$). Heating of 94-H(D) gives an insoluble polymer 96, $[C(CH_2AsO)_4]_n$. Oxidation of 94 or 96 with H_2O_2 (or D_2O_2) yields

$$C(CH_2AsI_2)_4 + 8\,NaOH \longrightarrow 8\,NaI\cdot 3H_2O \; + \qquad (67)$$

(94-H)

$$\mathbf{94} + D_2O \text{ (excess)} \quad \rightleftharpoons \quad \begin{array}{c} As(OD)_2 \\ | \\ CH_2 \\ \sim\!\!\sim \end{array}$$

$$(68)$$

(94-D)

$C[CH_2AsO(OH)_2]_4$, **97-H** or **97-D**. Reduction of **97** with SO_2 in MeOH regenerates **94**. The I.R. and Raman spectra of **94-H** and **94-D** as well as bond lengths and bond angles of **94** have been presented.

5. Reaction of deuterium-labelled triallyl arsenite with Pd(PPh₃)₄

No thermal Arbuzov rearrangement shown in equation 69 has been found[234–236] in the reaction of allylic arsenite with $Pd(PPh_3)_4$:

$$CH_2=CH-CH_2-OZ \xrightarrow{\;M^\circ\;} (C \overset{C}{\underset{M}{\cdots}} C)(O \cdots\cdots Z) \longrightarrow CH_2=CHCH_2Z \overset{\displaystyle O}{\overset{\|}{}}$$

$$(Z = S, P; M = Pd, Ni) \qquad (69)$$

This lack of rearrangement 69 has been explained by assuming that the arsenite ion, which is produced by oxidative addition, recombines with the π-allyl group by its oxygen atom again instead of by the arsenic atom (equation 70):

$$CH_2=CH-CH_2-OAs \xrightarrow{\;Pd^\circ\;} CH_2 \overset{CH}{\underset{L-Pd^+-L}{\diagup \diagdown}} CH_2 \; ^-OAs \qquad (70)$$

The mechanism of this reversible coupling has been studied by stirring 3-deuteriated triallyl arsenite with $Pd(PPh_3)_4$ and PPh_3 in THF at room temperature for 11 h. After removal of THF the product has been recovered by distillation at $80\,°C/0.3$ torr. The ^1H N.M.R. data showed that half of the deuterium atoms changed their position from $C_{(1)}$ to $C_{(3)}$. This indicated that, after forming the π-allyl palladium complex, the arsenite ion ^-OAs attacks at both positions $C_{(1)}$ and $C_{(3)}$ with the same probability and gives the deuterium-scrambled recoupling product according to equation 71.

$$(HDC=CHCH_2O)_3As + Pd(PPh_3)_4 \longrightarrow$$

$$HDC=CHCH_2O)_n \, As(OCDHCH=CH_2)_{3-n} \qquad (71)$$

6. Synthesis of ^{13}C-labelled tetraphenylarsonium salts

Several ^{13}C-labelled tetraphenyl arsonium salts of Rh(III) containing complex anions have been synthesized[237] to investigate, by the ^{13}C N.M.R. method, the mechanism of the rhodium/iodine-catalysed industrial carbonylation of methanol used for acetic acid manufacture. A revised catalytic cycle for the reaction has been proposed (equation 72)[237,238].

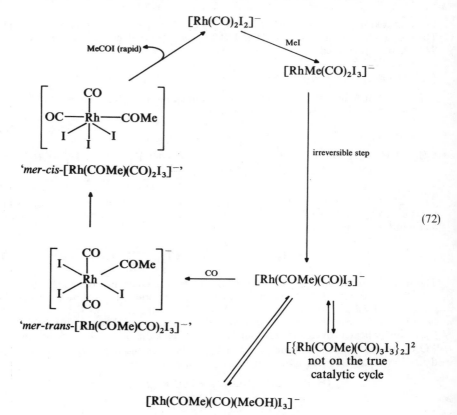

$$(72)$$

a. Di[1-^{13}C]acetyl[^{13}C]dicarbonyldi-μ-iodo-tetraiododirhodate(III), [AsPh$_4$]$_2$[Rh$_2$-(^{13}COMe)$_2$(^{13}CO)$_2$I$_6$], **98A**, has been obtained (75%) from [Rh$_2$(^{13}CO)$_4$Cl$_2$], AsPh$_4$Cl, LiI and methyl iodide. The product showed I.R. absorptions at v(CO) = 2065m, 2020s, 1735vw and 1700cm^{-1} (in CH$_2$Cl$_2$).

Orange crystals [Rh$_2$(^{13}CO)$_4$Cl$_2$] have been prepared from [Rh$_2$(CO)$_4$Cl$_2$] and ^{13}CO (v$_{CO}$ = 2085m, 2048s, 2004s and 991s cm^{-1}).

b. [AsPh$_4$]$_2$[Rh$_2$(^{13}CO^{13}CH$_3$)(^{13}CO)$_2$I$_6$], **98B**, has been obtained from [Rh$_2$-(^{13}CO)$_4$I$_2$], AsPh$_4$I and ^{13}CH$_3$I in liquid N$_2$. The solid product has been recrystallized from dichloromethane and light petroleum (b.p. 40–60 °C) to produce **98B** as a maroon powder (68%).

c. Tetraphenylarsonium acetylcarbonyltri-iodo(pyridine)rhodate(III), [AsPh$_4$]-[Rh(COMe)(CO)(NC$_5$H$_5$)I$_3$] (**99**), has been synthesized by treating **98A** with pyridine in '

CH_2Cl_2 (41%, m.p. 116–119 °C dec., ν_{CO} at 2060s and 1709m cm^{-1}). **99** has been characterized by ^{13}C N.M.R. and by X-ray crystallography (triclinic, $a = 11.17$, $b = 11.58$, $c = 15.87$ Å, $\alpha = 103.72$, $\beta = 98.34$, $\gamma = 95.11°$, $V = 1956$ Å^3, $Z = 2$, $D_c = 1.883$ g cm^{-3}, space group P1 (C_i, no. 2). Some bond lengths and angles for **99** have been determined.

7. Synthesis and ^{31}P N.M.R. spectroscopic study of $[^{107}Ag(\mu\text{-}R_2PCH_2PR_2)_3{}^{107}Ag]^{2+}[AsF_6]_2$ (R = Me or Ph)

These complexes of silver(I), $[^{107}Ag_2(dppm)_3][AsF_6]_2$, **100** (dppm = $Ph_2PCH_2PPh_2$), and $[^{107}Ag_2(dmpm)_3][AsF_6]_2$, **101** (dmpm = $Me_2PCH_2PMe_2$), have been synthesized in 85% and 78%, respectively, by adding $^{107}AgAsF_6$ to a solution of dppm or dmpm in Me_2NO_2 or in Me_2CO[239]. These new complexes have a 2:3 stoichiometry, while previous syntheses yielded $[Ag(\mu\text{-}dppm)Ag]^{2+}$ (with Ag:dppm = 2:1)[240] and $[Ag(\mu\text{-}dppm)_2Ag]$, the 2:2 complex[241]. Addition of more dppm to **100** did not result in production of further silver complexes.

The detailed study of the variable-temperature ^{31}P N.M.R. spectra using the ^{107}Ag-enriched compounds provided evidence for an intramolecular exchange process (faster for R = Ph than for R = Me) involving 'end-over-end' exchange in the $R_2PCH_2PR_2$ moiety (equation 73) with approximate enthalpies of activation equal to 59 kJ mol^{-1} for **100** and 50 kJ mol^{-1} for **101** in Me_2CO. The exchange presented in equation 73 probably involves the short-lived intermediate complex $[Ag(\mu\text{-}dppm)_2Ag(\eta^2\text{-}dppm)]^{2+}$ as shown in equation 74[242].

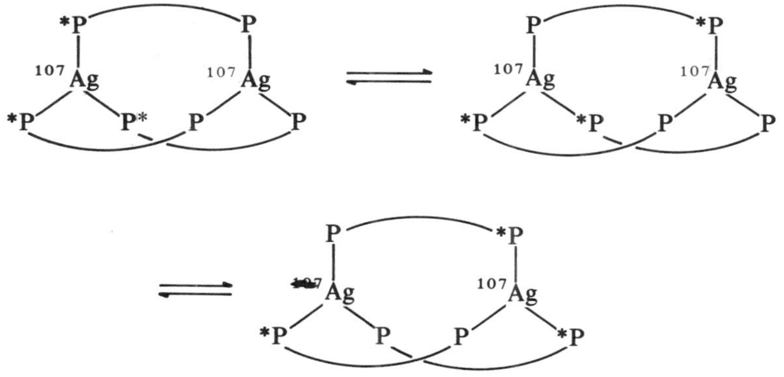

(73)

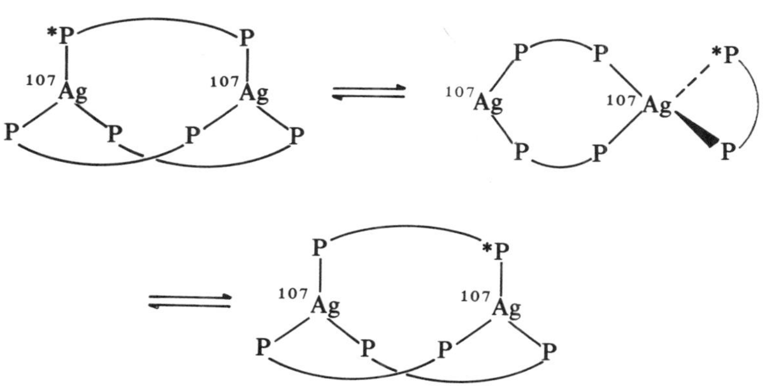

(74)

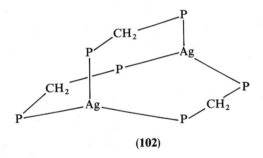

(102)

The proposed binuclear structure **102** for **100** and **101** is based on the analysis of the structure and of the intensities of the ^{31}P N.M.R. spectra of the above two complexes (at 213–327 K) and on the analysis of the N.M.R. spectrum of the two binuclear mixed ligand complexes formed by simple ligand redistribution as in equation 75. At 213 K the ^{31}P N.M.R. spectra show the expected coupling to ^{109}Ag (I = 1/2, 48.161% natural) and ^{107}Ag (I = 1/2, 51.839% natural). The observed (temperature-independent) 321–1 Hz doublet splitting (coupling) from (to) ^{107}Ag ($J(^{107}$Ag–^{31}P) = ~ 309 Hz, etc.) and 52 Hz ^{31}P–^{31}P coupling within the triplets are consistent with the complexes having the 'manxane' structure **102**. The multiplicities of the low-temperature broad N.M.R. signals disappear with rising temperature as a result of intramolecular exchange process shown in equations 74 and 75[239].

$$[^{107}Ag_2(dppm)_3]^{2+} + [^{107}Ag_2(dmpm)_3]^{2+} \rightleftharpoons [^{107}Ag_2(dppm)_{3-x}(dmpm)_x]^{2+} \quad (75)$$

C. Syntheses and Structures of Tetraphenylarsonium Complexes with Counter Anions Containing Technetium Radioisotopes

1. Synthesis and crystal structures of tetraphenylarsonium tetrachlorooxotechnetate(V)

Bright yellow crystals of $[AsPh_4]_2[TcCl_6]$ (**103**) have been obtained[243] (91%) by heating under reflux a mixture of $[AsPh_4][TcOCl_4]$[244] and $AsPh_4Cl$ in $SOCl_2$ for 5 min. The I.R. spectrum of **103** showed peaks at 1578m, 1480s, 1079vs, 996vs, 755vs, 741vs and 689vs cm^{-1}. No peak corresponding to Tc=O was noticed. **103** had the same I.R. and m.p. as $[AsPh_4]_2[TcCl_6]$ prepared by the reduction of NH_4TcO_4 with HCl and precipitation with $AsPh_4Cl$.

$[AsPh_4][TcO_4]$ dissolved in $SOCl_2$ gave, after standing 2.5 h at room temperature and subsequent work-up, grey-green crystals of $[AsPh_4][TcOCl_4]$, **104** (52% yield, contaminated slightly with **103**), showing I.R. peaks at 1482s, 1437vs, 1083vs, 1025vs (Tc = O), 996vs, 745vs, 688vs and 680vs cm^{-1}. Crystals of **104** are tetragonal, space group P4/n, $a = 12.644$, $c = 7.822$ Å, $Z = 2$. **104** consists of discrete $[AsPh_4]^+$ cations and $[TcOCl_4]^-$ anions with the technetium atom in a square-pyramidal environment. The $[TcOCl_4]^-$ anion possesses ideal C_{4v} symmetry (Tc=O bond length of 1.593 Å, Tc—Cl of 2.309 Å) with the chloro ligands occupying the basal position and the oxo ligand the apical position, **105**. The approximate C_{2v} symmetry for the same anion found in $[N(PPh_3)_2]$-$[TcOCl_4]$ is caused by the presence of the large $[N(PPh_3)_2]^+$ cation in the crystal lattice[245]. $[AsPh_4][TcOCl_4]$ is isostructural with the series $[AsPh_4][MYCl_4]$ (Y = N, M = Mo, Re, Ru, or Os; Y = O, M = Mo or Re)[246].

O
‖
Cl—Tc—Cl
  / \
 Cl   Cl

[TcOCl₄]⁻ anion in **104**

(**105**)

2. Synthesis of Tc(V)–dithiol complexes and other new organ-specific radiopharmaceuticals

The use of technetium-99m (half-life = 6.006 h) in diagnostic medicine[247], particularly for brain and heart imaging[248a], has became widespread since the availability of macroscopic quantities of the long-lived technetium-99 isotope from fission products (half-life = 2.13×10^5 yr, weak β^--emiter, $E_{max} = 0.292$ MeV, $\langle \beta - \rangle = 85$ keV. Normal glassware gives adequate protection against this weak β-emission, but all operations have to be carried out in well ventilated hoods and with gloves. Bremsstrahlung has to be taken into consideration when working with ^{99}Tc on a large scale, > 20 mg). Numerous compounds containing technetium have been synthesized and characterized by X-ray diffraction.

a. Electrochemical reduction of TcO_4^-. The Tc(V)–dithiolate complexes produced by the electrochemical reduction of [^{99}Tc]pertechnetate ion in the presence of 1,3-dimercaptopropane (DMP), 2,3-dimercaptopropanol (BAL) and 2,3-dimercaptosuccinic acid (DMSA) in alkaline aqueous media[248b] have been precipitated as tetraphenylarsonium salts and characterized by U.V., visible and I.R. spectrophotometry[248b]. The strong band in the 930–950 cm⁻¹ region, characteristic for the Tc=O stretch of Tc(V) compounds containing the TcO^{3+} core, has been observed in these complexes, and is comparable with the absorptions found in the oxobis(dithiolate) technetate(V) complexes[249-251] and in complexes obtained by titration of TcO_4^- with Sn(II) in the presence of the corresponding dithiols, DMP, BAL, DMSA[248b,252]. The technetium(V)–dithiol complexes are stable to oxygen and do not exhibit a disproportionation reaction[248b].

b. Tetraphenylarsonium bis(benzene-1,2-dithiolato)oxotechnetate(V), [AsPh₄][TcO-(bdt)₂] (**106**) has been synthesized[253,254] from [AsPh₄][TcOCl₄] with benzene-1,2-dithiol. Red-brown monoclinic crystals of **106** were obtained in quantitative yield (m.p. 208–210 °C, I.R. peaks at 1439vs, 995m, 938vs, 748s, 732s, 689m and 679 cm⁻¹). The structure of a distorted square-pyramid about the technetium(V) atom (with four sulphur atom in the basal positions and the oxo ligand in the apical position) has been compared with the structures of aliphatic dithiolate complexes[255,256], particularly[255] with [AsPh₄][TcO(SCH₂CH₂S)₂], possessing also square-pyramidal coordination about the technetium atom with the oxo group in the apical position. In the [AsPh₄]⁺ cation the geometry about the arsenic atom is tetrahedral[257,258], only slightly distorted. The As—C bond lengths are 1.911 Å and the C—As—C angle is 109.5°. The arsenic atom lies at distances ranging from 0.154 and 0.310 Å from the phenyl ring planes.

c. Synthesis of [AsPh₄][TcO(tmbt)₄] (**107**) *and* [AsPh₄][TcO(tibt)₄] (**108**). **107**, where Htmbt = 2,3,5,6-tetramethylbenzenethiol, has been prepared[259,260] from [AsPh₄][TcOCl₄], Htmbt and MeONa in CH₂Cl₂/MeOH (1:1). Recrystallized from CH₂Cl₂/hexane analytically pure, red crystals of **107** have been obtained in 72% yield. **108**, a red-black microcrystalline complex, has been prepared in the same manner in 77% yield[260] (Htibt = 2,4,6-triisopropylbenzenethiol). All four thiolate ligands in **107** and **108** are magnetically equivalent. Both in **107** and **108** the I.R. stretching frequency of the Tc=O group equals 934 cm⁻¹, typical for Tc(V) oxo complexes with thiolate ligands. The compounds **107** and **108** are slowly oxidized by air in alkaline methanolic solutions to form [TcO₄]⁻. **107** has been reduced to Tc(tmbt)₃(PEt₃)₂ **109** (bright blue solid, 40% yield), by oxygen atom abstraction with PEt₃. The diamagnetic compound **109** has trigonal bipyramidal geometry as do other compounds of the general formula Tc^{III}(SAr)₃L₂. The non-thiolate ligands are inequivalent[261]. The oxidative and reductive oxo-transfer reactions have been coupled to complete a catalytic cycle.[260a,b,c]. Pink-purple trans-Tc(III)(tmbt)₃(i-PrNC)₂, **110** (I.R. of v_{CN} = 2108 cm⁻¹), has been obtained from **107** with i-PrNC[259]. Bulky isonitrile ligands (t-Bu, cyclohexyl, i-Pr) are difficult to replace, therefore a synthetically more useful complex. Tc(III)(tmbt)₃(MeCN)₂, **III** (I.R.: v_{NC} = 2255 cm⁻¹), has been prepared[259]. The complex **111** is a trigonal bipyramid with two inequivalent axial MeCN ligands. Two of the aryl groups are equivalent and on the same side of the equatorial plane containing the sulphur atoms. Addition of pyridine N-oxide to a refluxing methanolic solution of **111** in the presence of an excess of Htmbt gives [AsPh₄][TcO(tmbt)₄] in 60% yield.

   *d. Tetraphenylarsonium bis[1,2-dicyanoethenedithiolato]oxotechnetate(V), [AsPh₄] [TcO(mnt)₂], **112** (mnt = 1,2-dicyanoethenedithiolate).* has been obtained[262] from sodium 1,2-dicyanoethene dithiolate with [AsPh₄] [TcOCl₄]. The red-brown monoclinic crystals melting at 156–159 °C showed I.R. peaks at 2203s, 1525m, 1482m, 1440vs, 1150m, 1079s, 996s, 950vs (Tc=O), 738vs and 685 cm⁻¹. The structure of **112** consists of discrete [AsPh₄]⁺ and [TcO(mnt)₂]⁻ ions. As in the case of **106**, the technetium(V) atom in **112** is five-coordinate with the four sulphur atoms in the basal positions and the oxo ligand in the apical position, to give a distorted square-pyramidal environment about the technetium. The Tc^V=O bond distance is 1.655 Å. There appears to be a reciprocal relationship between the Tc=O bond length and the displacement of the technetium atom from the plane of the four basal ligand atoms. The [AsPh₄]⁺ cation has its usual conformation[257].

   e. Tetraphenylarsonium tris(1,2-dicyanoethenedithiolato)technetate(IV), [AsPh₄]₂-[Tc(mnt)₃] **(113)**, has been obtained[263] by treating 1,2-dicyanoethenedithiolate and (NH₄)₂[TcBr₆] with AsPh₄Cl (45%). The red-brown orthorhombic crystals of **113**, melting at 295–298 °C, showed I.R. peaks at 2200vs, 1435s, 1080m, 997m, 760vs and 715vs cm⁻¹. **113** is isomorphous and isostructural with [AsPh₄]₂[M(mnt)₃] (where M = Mo, W)[264] and consists of discrete [AsPh₄]⁺ cations and [Tc(mnt)₃]²⁻ anions. The technetium(IV) atom in the anion is coordinated to six sulphur atoms to give a distorted octahedron.

   f. Tetraphenylarsonium tris(benzene-1,2-dithiolato)technetate(V), [AsPh₄][Tc(bdt)₃] **(114)**, dark brown crystals, have been obtained from benzene-1,2-dithiol (H₂bdt) with [AsPh₄][Tc^{VI}NCl₄][265,266]. It melted at about 280 °C with decomposition (I.R. peaks at 1439s, 1298s, 1082m, 995m, 750s, 738s, and 688m cm⁻¹). **114** consists of discrete [AsPh₄]⁺ cations and [Tc(bdt)₃]⁻ anions. The technetium(V) atom in [Tc(bdt)₃] is coordinated by a nearly perfect trigonal-prismatic array of six sulphur atoms[265].

3. Synthesis of [⁹⁹Tc]chlorobis(N-methylsalicylideneiminato)oxotechnetium(V)[267]

   [AsPh₄][TcOCl₄] with an excess of Hmsal at RT gave a red oil, which dissolved in CH₂Cl₂ produced after slow evaporation deep red crystals of Tc^{V}O(Cl)L₂ in 90% yield. The oxotechnetium(V) complexes [TcO(Cl)L₂] are reduced by PMe₂Ph to give the corresponding technetium(III) complexes, [TcCl₃L(PPh₃)], [TcCl₄(HL)₂] and [TcClL₂(PMe₂Ph)], where HL = Hmsal or N-phenylsalicylideneimine (Hpsal). The structure of [TcCl(psal)₂(PMe₂Ph)] has been determined: a = 9.500, b = 10.596,

$c = 31.000$ Å, $\beta = 95.59°$, monoclinic, space group $P2_1/c$ and $Z = 4$. The coordination around Tc is approximately octahedral[267].

4. Synthesis of bis(tetraphenylarsonium)tris(oxalato)technetate(IV), [AsPh₄]₂ [Tc(ox)₃] (115).

This has been obtained[265] from $[NH_4]_2[TcBr_6]$, oxalic acid and $AsPh_4Cl$. The pale yellow crystals (57%, m.p. *ca* 260 °C with decomposition) showed I.R. peaks at 1712vs, 1670m, 1439s, 1335s, 1081m, 998m, 794m, 741s and 690s cm^{-1}. 115 crystallized in the monoclinic space group C2/c with cell parameters $a = 23.164$, $b = 13.507$, $c = 16.047$ Å and $\beta = 104.90°$, with $Z = 4$. It consists of $[Tc(ox)_3]^{2-}$ anions and $[AsPh_4]^+$ cations. The technetium atom in $[Tc(ox)^3]^{2-}$ is coordinated to six oxygen atoms in a distorted octahedral array. The structure of the anion in the cyclic tetramer $[AsPh_4]_4$-$[Tc_4N_4(O)_2(ox)_6]$ (116) consists of two $[(ox)Tc^{VI}N\!-\!O\!-\!Tc^{VI}N(ox)]$ units joined by two quadridendate oxalato ligands[268a]. The oxalato ligand comprising atoms $C_{(1)}$, $C_{(2)}$, $O_{(1)}$, $O_{(2)}$, $O_{(4)}$ and $O_{(5)}$ is almost planar. The ligand comprising atoms $C_{(3)}$, $C_{(3')}$, $O_{(3)}$, $O_{(3')}$, $O_{(6)}$ and $O_{(6')}$ is considerably distorted from planarity.

(117) The $[Tc(ox)_3]^{2-}$ anion in $[AsPh_4]_2[Tc(ox)_3]$ (115)

5. Synthesis of tetraphenylarsonium hexakis(isothiocyanato)technetate(IV)–dichloromethane (1/1) solvate, 118

$[AsPh_4]_2[Tc(NCS)_6]\cdot CH_2Cl_2$ (118) has been obtained[268b] from $(NH_4)_2[TcCl_6]$, KNCS and $[AsPh_4]Cl$ in MeCN. The I.R. spectrum of 118 in KBr disc showed peaks at 2128vw, 2088w, 2020vs (cyanide stretching region), 1480w, 1435w, 1077w, 993w, 734m and 683w cm^{-1} (due to the tetraphenylarsonium cation). The $[Tc(NCS)]^{2-}$ anion has an octahedral structure with the technetium(IV) atom situated on a site of exact 4/m symmetry. The Tc—N bond lengths are 2.00 and 2.01 Å. All N—Tc—N angles are constrained by symmetry to exactly 90 or 180°. Likewise, two of the Tc—NCS groups are constrained by symmetry to perfect linearity, whereas the remaining four Tc—NCS groups are almost linear with Tc—N(1)—C(1) 175.9° and N(1)—C(1)—S(1) 175.3°. Disorder in the crystal lattices of the salts of $[Tc(NCS)_6]^{2-}$ and $[Tc(NCS)_6]^{3-}$, responsible for the relatively low accuracy of the structure determination, is also manifested by the presence of a solvent CH_2Cl_2 molecules partially occupying four symmetry-related positions throughout the crystal lattice.

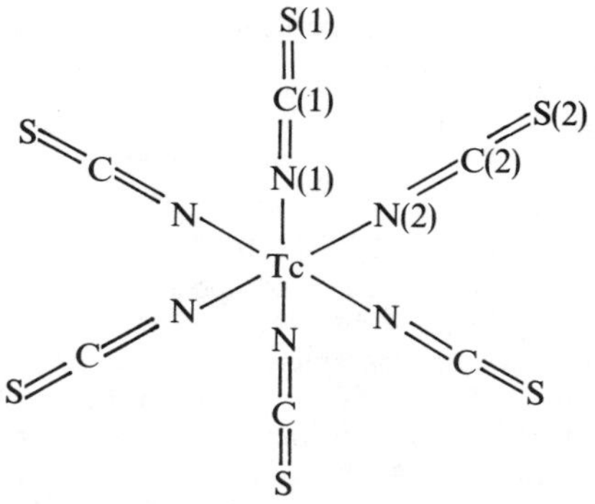

S(1)
‖
C(1)
‖
N(1)

(119) The $[Tc(NCS)_6]^{2-}$ anion in **118**

6. Synthesis of technetium(V) and rhenium(V) complexes of 2,3-bis(mercaptoacetamido)propanoate (map)

Complexes of the general formula $[Ph_4X][MO(map)]$, where X = As, P; M = Tc, Re, were applied for the radiolabelling of monoclonal antibodies[269]. They have been synthesized[270] by reduction of $NH_4{}^{99}TcO_4$ by sodium dithionite at 80 °C (15 min) in the presence of map ligand **120**. The A and B epimeric mixture of $TcO(map)^-$ **121** was separated by HPLC. The separated A epimer was precipitated with tetraphenylarsonium chloride, and the B epimer with tetraphenylphosphonium chloride, since in the latter case the crystals obtained with the tetraphenylarsonium countercation were not suitable for X-ray studies. The structures of the two *syn-* and *anti-*epimers of technetium have been determined by single-crystal X-ray analyses. The complexes crystallize in the space group $P2_1/c$ and have four formula units in the unit cell. The crystal data for $[Ph_4As]$-$[syn\text{-}TcO(map)]$ are as follows: $a = 12.729$ Å, $b = 13.225$ Å, $c = 18.318$ Å, $\beta = 91.67°$, $V = 3082.4$ Å^3. In the case of $[Ph_4P][anti\text{-}TcO(map)]$: $a = 11.260$ Å, $b = 23.622$ Å, $c = 11.834$ Å, $\beta = 94.52°$, $V = 3138$ Å. The crystal data for $[Ph_4As][syn\text{-}ReO(map)]$ have also been determined. In all three complexes the metal (Tc, Re) is coordinated to one oxygen (yl) atom, two sulphur (thiolate) atoms and two nitrogen (amide) atoms in a distorted square-pyramidal geometry. The metal atom is 0.745 ± 0.005 Å above the basal plane (N_2S_2). The mean M—O, M—N and M—S bond lengths are 1.654, 1.979 and 2.283 Å, respectively. There is no bonding between the carboxylic acid group of the ligand and the metal or yl oxygen atom. The epimer A is excreted faster than B in humans. The epimer A is more easily recognized by kidney receptors (the COOH and Tc=O groups are on the same side above the chelate plane and interact more effectively with the transport

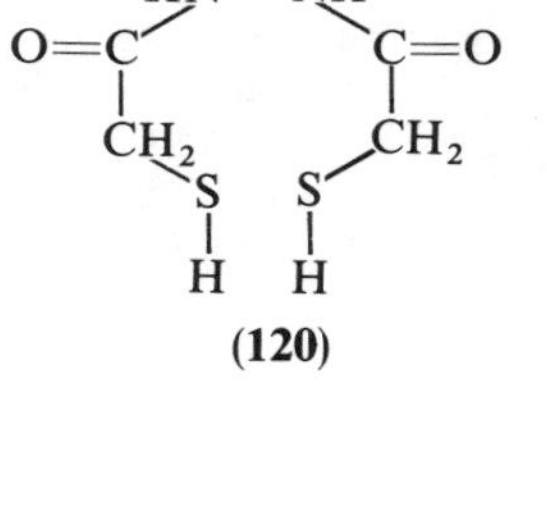

(120) 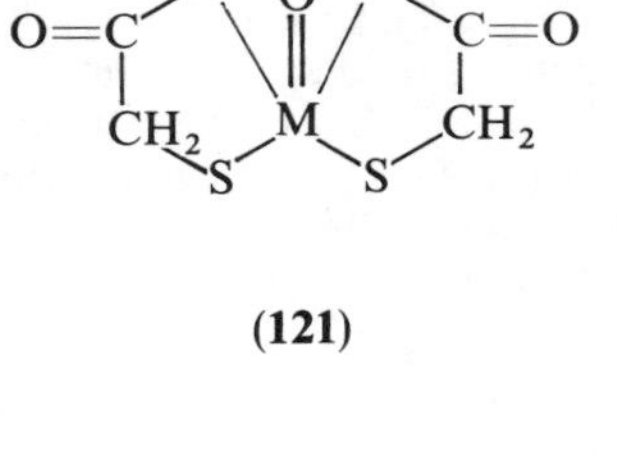   **(121)**

system). The structural characterization of the Tc and Re complexes is useful for the understanding of the active transport of metal complexes by the kidney and other *in vivo* interactions[271]. Monooxo complexes of technetium(V) (M = ^{99}Tc) and of rhenium(V) (M = $^{185/187}$Re, ^{187}Re: 62.60%) with 4,5-bis(mercaptoacetamido)pentanoic acid (mapt, **122**) have been synthesized also[272,273] and characterized by N.M.R., I.R., U.V., visible spectroscopy and by FAB mass spectroscopy[273]. Two chelate ring epimers have been formed upon complexation due to *syn* and *anti* orientation of the COOH group of the ligand relative to the metal oxo group. The epimers have been separated by HPLC and isolated as the salts of Ph_4X (where X = As or P).

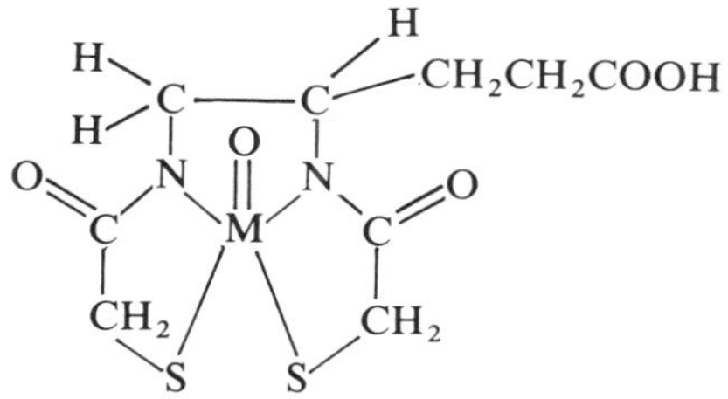

(122) mapt

7. Synthesis and molecular structure of a 'Lantern' dimer
[AsPh$_4$]$_2$[Tc$_2$O$_2$(SCH$_2$CONHCH$_2$CH$_2$NHCOCH$_2$S)$_4$], 123[274]

Reaction of $(Bu_4N)[TcOCl_4]$ (0.26 mmol) with 5-fold excess of N,N'-ethylenebis(2-mercaptoacetamide)(H_4ema) in methanolic sodium methoxide yielded a blue precipitate, which has been redissolved in water and treated with $AsPh_4Cl$ to obtain $[AsPh_4]_2[Tc_2O_2(H_2ema)_4]$ (**123**) as a blue crystalline precipitate (49%). After recrystallization from 50% aqueous MeOH, the pure **123** melted at 200 °C with decomposition. The methanol solution of **123**, when treated with NaOH or when heated in solution, was converted to $[AsPh_4][TcO(ema)]$ of bright yellow colour, **124** (equation 76)[274].

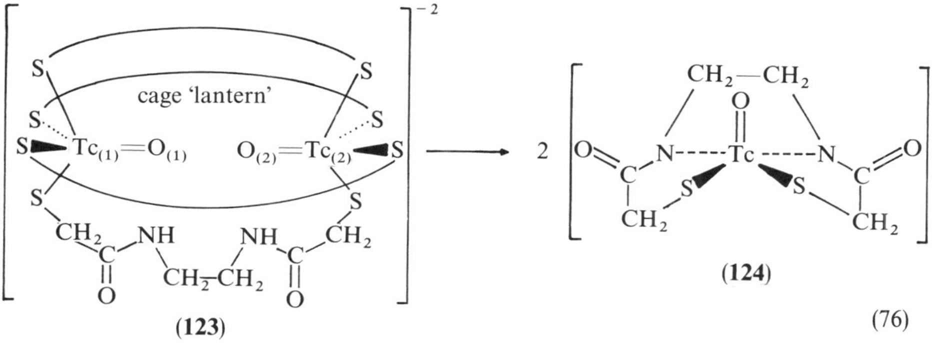

(76)

The dimeric formulation of **123** is based on the negative-mode fast atom bombardment spectrum (FABMS) of **123**. The strong I.R. absorption of **123** at $v_{max} = 962\,cm^{-1}$ is characteristic of a square-pyramidal technetium(V) complex with a multiply bonded oxo group (Tc=O). I.R. absorptions at 1659 and 3312 cm^{-1} have been assigned to protonated and uncoordinated amide groups. The absence of the SH stretch at 2600–2550 cm^{-1} suggested thiolate coordination. Single-crystal X-ray crystallography of **123** showed a centrosymmetric dianion with two square-pyramidal $OTcS_4$ cores bridged 4-fold by the $[H_2em^{2-}]$ dithiolate ligands. The two metal-oxo groups are oriented into the centre of the cage (called 'lantern') indicated in equation 76. The intramolecular $Tc_{(1)}-Tc_{(2)}$ and $O_{(1)}-O_{(2)}$ distances equal 7.175 and 3.96 Å, respectively. The six water molecules per dimer were hydrogen bonded to each other, to the amide NH and to the C=O groups on the ligands. The crystallographic data for monoclinic **123** were: $a = 15.791$ Å, $b = 14.126$ Å, $c = 19.069$ Å, $\beta = 109.07°$, $V = 4020$ Å^3, space group $P2_1/n$ (No. 14), $Z = 2^{274}$.

Synthetic methods employed for the preparation of ligands of the type $R^2SCH_2CONHCH_2CH_2NHCOCH_2SR^2$, $H_2ema(R^1)(R^2)$, where R^1 = alkyl, amino-alkyl and carboxyalkyl and R^2 = H, benzoyl, acetamidomethyl and benzamidomethyl, as well as for the synthesis of the $[TcO(emaR)]$ complexes, and for the quantitative synthesis of $[AsPh_4][TcO(ema)]$ have been described also[275]. The m.p., I.R., ^{1}H N.M.R. and analytical data were identical with earlier published data[276a]. The crystal data for structure determination of $[TcO(ema(morph))]$, and for the characterization of the neutral complex $[TcO(emaR)]$ (R = Me, undec, morph and benzyl) have also been presented[275].

8. Syntheses and crystal structures of tetraphenylarsonium oxotrichloro[2-(2-hydroxyphenyl)benzothiazolate]technetate(V) and rhenate

These compounds, $[AsPh_4][MoX_3(hbt)]$ (**125**), where M = Tc or Re; X = Cl, Br and hbt = 2-(2-hydroxyphenyl)benzothiazole (**126**), have been prepared[276b] from an excess of hbt and $[AsPh_4][MOX_4]$ in isopropyl alcohol at room temperature. The precipitate (red-orange in the case of Tc and green for Re) has been recrystallized from CH_2Cl_2 and EtOH (90% yield). It is soluble in CH_2Cl_2, $CHCl_3$ and acetone, and insoluble in MeOH, EtOH, Et_2O, CCl_4 and pentane. The salt **125** (M = Tc) crystallizes in the orthorhombic space group $P2_1, 2_1, 2_1$ (No. 19) with $a = 12.104$ Å, $b = 13.772$ Å, $c = 20.539$ Å, $V = 3424$ Å^3 and $Z = 4$. The Tc(V) centre is in a distorted-octahedral configuration with the three chlorine atoms and the neutral nitrogen atom of the hbt ligand in the plane normal to the Tc=O linkage and the anionic phenolate oxygen atom of the hbt ligand situated *trans* to the Tc=O linkage.

(126) hbt

The I.R. frequencies $v(M=O) = 945$ cm^{-1}, $v(C=N) = 1585$ cm^{-1}, $v(M-X) = 310$ cm^{-1} and $m/e = 830$ (theoretical value 830.97) are for $[AsPh_4]$ $[TcOCl_3(hbt)]$ and $v(M=O) = 940$ cm^{-1}, $v(C=N) = 1587$ cm^{-1} and $m/e = 963$ (theoretical value 964.33) are for $[AsPh_4][TcOBr_3(hbt)]$.

The reactivity of $[AsPh_4][MOX_4]$ toward hbtH (**127**) has been studied[276b]. When hbtH is mixed with a solution of $[AsPh_4]$ $[MOX_4]$, the former is rapidly converted into hbt (**126**) with the formal loss of two hydrogen atoms. The reaction is carried out in the dark to avoid interference with light-induced (equation 77) dehydrogenation[277,278].

$$\xrightarrow{hv,\ RT} \text{hbt (126)} \tag{77}$$

(127) hbth

The symmetry of the TcO_4^- anions as a function of the nature of the cation M^+ in pertechnetates $MTcO_4$ for $M = Li$, NH_4, Rb, Me_4N, Cs, Et_4N, n-Bu_4N, Ph_4P and Ph_4As^{279} has been investigated[280] by the ^{99}Tc N.M.R. method[281]. Only in the case of the ^{99}Tc N.M.R. spectrum of $LiTcO_4$ has no multiplet structure caused by quadrupole interactions been found at R.T. A review of the quadrupole nuclear interactions of ^{99}Tc in polycrystalline pertechnetates has been presented by Tarasov and coworkers[282,283]. The quadrupole coupling constants K and asymmetric electric field gradient tensors at ^{99}Tc in $MTcO_4$ (where $M = NH_4$, Cs, Et_4N, Me_4N, Ph_4P, Rb, Li, Na, K, ND_4, Bu_4N, Ph_4As and $\frac{1}{2}Mg$) and in $Tc_2O_7 \cdot H_2O$ have been determined by N.M.R. The lattice contribution to K and the temperature dependence of K have been determined also and compared with theoretically calculated values[282].

D. Syntheses and Characterization of Tetraphenylarsonium Salts of Technetium Nitrido Complexes

1. Synthesis of tetraphenylarsonium tetrachloronitridotechnetate

a. The $[AsPh_4][Tc^{VI}NCl_4]$ complex (128), contains the nitrido ligand (N^{3-}), a powerful π-electron donor which stabilizes metals in high oxidation states. 128 can be employed for the preparation of Tc—N complexes by ligand exchange carried out in organic solvents. It is prepared[284] from ammonium pertechnetate concentrated hydrochloric acid and sodium azide in water. Subsequent addition of the tetraphenylarsonium chloride produced an orange precipitate (94%) which recrystallized from benzene–acetonitrile (1:1, V:V), gave orange-red needles melting at 272–274 °C and showing I.R. peaks at 1482s, 1441vs, 1437vs and 1084vs.

Similarly, Tc(VI) nitrido complexes of the general formula $R[TcNX_4]$ (where $R = NBu_4$, $AsPh_4$, $X = Cl$, Br) and $M_2[TcNX_5]$ ($M = Cs$, Rb; $X = Cl$, Br) have been prepared[284,285] and their structures have been assigned on the basis of the microanalytical and E.S.R. data, and of the presence of sharp absorption bands at 1074–1080 cm^{-1} in the solid-state I.R. spectra, characteristic for terminal M≡N groups[286]. The lack of absorption bands in the 1900–2500 cm^{-1} region confirms the absence of coordinated azido groups. The E.S.R. spectra of $[TcNCl_4]^-$ and $[TcNBr_4]^-$ are typical of ions with the $4d^1$ electronic configuration and with an unpaired electron in an orbital with substantial $4d_{xy}$ character. The crystals are tetragonal, having the space group P4/n, with $a = 12.707$ Å, $c = 7.793$ Å and $Z = 2$. The structure consists of discrete $[AsPh_4]^+$ and $[TcNCl_4]^-$ ions. The $[TcNCl_4]^-$ ions possess C_{4v} symmetry with interatomic bond distances equal to (Å) 1.581 (Tc≡N), 2.3220 (Tc—Cl$_{(1)}$), and bond angles equal to (deg) 103.34 (N≡Tc—Cl$_{(1)}$), 86.95 (Cl$_{(1)}$—Tc—Cl$_{(1)}$) and 153.33 (Cl$_{(1)}$—Tc—Cl$_{(1'')}$). 128 is isostructural with the series $[AsPh_4][MNCl_4]$ (where $M = Mo$, W, Re, Ru, or Os)[287] and with $[AsPh_4][MOCl_4]$ ($M = Mo$ or Re)[288].

$$\left[\begin{array}{c} N \\ ||| \\ Tc \\ Cl \diagup \diagdown Cl \\ Cl_{(1)} \quad Cl \end{array} \right]^- \quad [Ph_4As]^+$$

The $[TcNCl_4]$ anion in 128

b. The crystal structure of tetraphenylarsonium tetrabromonitridotechnetate, $[TcNBr_4]$ (129), has been determined by single-crystal X-ray diffraction[289,290]. Crystals of 129 are tetragonal, space group P4/n with $a = 12.857$ Å, $c = 7.992$ Å and $Z = 2$. The structure consists of discrete $[AsPh_4]^+$ and $[TcNBr_4]^-$ ions, with the $[TcNBr_4]^-$ ions possessing ideal C_{4v} symmetry. **129** is isostructural with the series R $[MNBr_4]$, where M = Mo, Ru, Re or Os; R = $[AsPh_4]$ or $[PPh_4]$. The $Tc^{VI}\equiv N$ bond distance is 1.596 Å and the Tc–Br distances are 2.481 Å. The difference in colour between the orange-red $[AsPh_4][TcNCl_4]^{284}$ and the intense blue-black **129** is consistent with the position of chlorine and bromine in the spectrochemical series and is not due to any gross changes in structural geometry[289].

    c. The radioanalytical applications and the E.P.R. spectroscopy of the tetraphenyl-arsonium $[AsPh_4]^+$ salts[291], particularly a single-crystal E.P.R. study of ^{15}N-enriched **128** diamagnetically diluted by the isoelectronic oxo-complex **104**[292] as well as ENDOR (electron nuclear double resonance) and ESEEM (electron spin-echo envelope modulation) studies on ^{35,37}Cl and ^{14,15}N hyperfine and quadrupole interactions in the complex **128**, have been presented and discussed by Abram and coworkers[291–293]. **128** crystallizes in the space group P4/n, Z = 2, with $a = b = 12.670$ Å and $c = 7.824$ Å. The very complicated low-temperature ^{99}Tc, ^{35,37}Cl and ^{15}N ligand hyperfine patterns have been solved by ENDOR and ESEEM techniques and the conclusion has been reached that 20% of the spin density is localized in the 3p orbitals of the Cl atoms.

*2. Synthesis and crystal structures of $[AsPh_4]_4[Tc_4N_4O_2(ox)_6]$ (**130**) and $[AsPh_4]_2[TcO(ox)_2(Hox)]\cdot 3H_2O$ (**131**)*

130 has been obtained[294] by treating acetone solution of $[AsPh_4][TcNCl_4]$ with oxalic acid in water. Red-brown crystals have been formed [38%, m.p. *ca* 250–255 °C (dec.)]. I.R. peaks at 1731s, 1712vs, 1660vs (CO stretches), 1438s, 1326vs, 1080m, 1050s (Tc≡N), 995m, [809m, 795s, δ(O—C=O)], 740s and 688 cm^{-1}. In the $[AsPh_4]_2[Tc(ox)_3]$ salt,

(**130**) structure of $[Tc^{VI}_4N_4O_2(ox)_6]^-$ anion in $[AsPh_4][Tc_4N_4O_2(ox)_6]$
with numbering scheme

where all the oxalato ligands are equivalent, an intense CO stretch at 1711 cm^{-1}, together with a weaker peak at 1670 cm^{-1} and only one sharp peak at 793 cm^{-1}, have been found[295].

The anion in **130** is a cyclic tetranuclear complex, $[Tc_4N_4O_2(ox)_6]$ with Ci point symmetry. Each Tc(VI) atom is coordinated by five oxygen atoms and one nitrogen atom to give a distorted octahedron. In each half of the anion, a quadridendate oxalato ligand forms a bridge between the two octahedra. The technetium atoms are displaced above the plane of the four oxygen atoms by 0.36 Å for $Tc_{(1)}$ $[O_{(1)}, O_{(2)}, O_{(4)}, O_{(9)}]$ and 0.37 Å for $Tc_{(2)}[O_{(5)}, O_{(7)}, O_{(8)}, O_{(9')}]$. Asymmetry of the μ_4-oxalato ligands is pronounced in different Tc—O bond distances (the *trans* $Tc_{(1)}$—$O_{(3)}$ and $Tc_{(2)}$—$O_{(6)}$ distances are longer than the $Tc_{(1)}$—$O_{(4)}$ and $Tc_{(2)}$—$O_{(5)}$ distances.

Tetraphenylarsonium hydrogenoxalatobis(oxalato)oxotechnetate(V) trihydrate (**131**) has been produced from oxalic acid dihydrate $[NBu_4][TcOCl_4]$ and tetraphenylarsonium chloride. The pale green microcrystalline product had m.p. (decomp.) 245–255 °C and peaks in the I.R. spectrum at 1732vs, 1702vs, 1673vs, 1480m, 1438m, 1346s, 1307m, 1079s, 996s, 963vs (Tc=O), 803m, 781m, 740vs and 688vs cm^{-1}.

The Tc(V) in anion **131** is coordinated by six oxygen atoms to give a distorted octahedron and contains a protonated unidentate oxalato ligand, Hox. The hydrogen atom resides most likely at $O_{(6)}$ since the $C_{(6)}$—$O_{(6)}$ and $C_{(6)}$—$O_{(7)}$ bond lengths are similar to those found in other hydrogen oxalates ($[M][HC_2O_4]$, M = Li, Na, K, NH_4)[296].

Structure of $[TcO(ox)_2(Hox)]^{2-}$ anion in **131**, $[AsPh_4]_2[TcO(ox)_2(Hox)] \cdot 3H_2O$

3. Synthesis and crystal structure of bis(tetraphenylarsonium)-bis(1,2-dicyanoethenedithiolato)nitridotechnetate(V), 132

This compound, $[AsPh_4]_2[Tc^VN(mnt)_2]$ (mnt = 1,2-dicyanoethenedithiolate), **132**, has been obtained[297] from $[AsPh_4][Tc^{VI}NCl_4]$ in MeCN with an ethanolic solution of Na_2mnt and addition of $AsPh_4Cl$ in water. The yield of the bright yellow cubic crystals of **132** was 32%, melting at 267–268 °C and showing I.R. peaks at 2205m, 2194s, 1485vs, 1442vs, 1083s, 1060s (Tc≡N), 997s, 740s and 688s cm^{-1}.

The Tc^{VI} oxidation state is retained in the reaction of thionyl chloride with $[AsPh_4]_4$-$[Tc_4N_4(O)_2(ox)_6]$ and $[AsPh_4][TcNCl_4]$ is produced. The compound **132** (TcV), heated under reflux in $SOCl_2$ for 20 min, yielded 61% of yellow crystals of $[AsPh_4]_2[Tc^{IV}Cl_6]$ which had I.R. spectrum and m.p. identical with the sample of $[AsPh_4]_2[TcCl_6]$ prepared by reduction of ammonium pertechnetate with HCl and precipitation with $AsPh_4Cl$. Crystals of **132** are monoclinic, space group P_n with $a = 11.369$, $b = 15.530$, $c = 14.421$ Å. $\beta = 97.58°$ and $Z = 2$, and consist of discrete $[AsPh_4]^+$ and $[TcN(mnt)_2]^{2-}$ ions. The technetium atom in the $[TcN(mnt)_2]^{2-}$ anion has square-pyramidal coordination geometry with the nitrido ligand in the apical position and four sulphur atoms in the basal plane. Some selected bond distances are 1.59 Å ($Tc\equiv N$) and four Tc—S bonds with distances in the interval between 2.367 and 2.419 Å. A comparison of structural data[297] for square-pyramidal complexes of technetium containing oxo and nitrido ligands has been presented. The $Tc\equiv N$ bond distances are significantly shorter than the analogues $Tc=O$ distances.

4. Synthesis and crystal structure of bis(8-quinolinethiolato)nitridotechnetium(V), 134

Tetraphenylarsonium tetrachloronitridotechnetate(VI)·$[AsPh_4][TcNCl_4]$, **133**, is a useful intermediate for the preparation of different $Tc\equiv N$ complexes[298].

a. **133** dissolved in the minimum amount of acetone with anhydrous LiBr gives $[AsPh_4][TcNBr_4]$ in 81% yield (m.p. 291–293 °C, I.R. maxima located at 1482s, 1439s, 1436vs, 1085vs, 1074s ($Tc\equiv N$), 977s, 742vs, 688vs and 680vs cm^{-1}).

b. The $[TcNCl_4]^-$ anion with CsCl in concentrated HCl gives red-brown $Cs_2[TcNCl_5]$ in 81% yield [IR(KBr) shows only one peak at 1027 cm^{-1} due to $Tc\equiv N$].

c. The $[Tc^{VI}NCl_4]^-$ anion reduced in acetonitrile with PPh_3, KNCS, $Na[S_2CNEt_2]$ and with 8-quinolinethiol (C_9H_6NSH) hydrochloride gives $Tc^V\equiv N$ complexes: $[TcNCl_2$-$(PPh_3)_2]$ (90% yield); $[NEt_4]_2[TcN(NCS)_4(MeCN)]$ (55% yield, softening, with decomposition at 160 °C, I.R. ν_{max} at 2300vw, 2272vw, 2110vs, 2085vs, 1482s, 1442m, 1170m, 1081m ($Tc\equiv N$), 998m, 828w and 785m cm^{-1}), $[TcN(S_2CNEt_2)_2]$ (74% yield, m.p. 254–256 °C, I.R. ν_{max} at 1512vs, 1438s, 1284s, 1205s and 1070vs cm^{-1}) and $[TcN(C_9H_6NS)_2]$ (41% yield of orange crystalline precipitate, m.p. 286–289 °C, I.R. ν_{max} at 1494vs, 1454m, 1367m, 1299s, 1212m, 1064s, 1000s, 821vs, 774vs and 690s cm^{-1}.

The single crystals of **134**, grown by slow evaporation at RT of a CH_2Cl_2 solution[298], are monoclinic, space group C2/c with $a = 15.92$ Å, $b = 7.347$ Å, $c = 15.33$ Å, $\beta = 110.89°$, $V = 1675.2$ Å^3, $D_c = 1.72$ Mg m^{-3}. The coordination environment of technetium is a distorted square pyramid with the nitrido nitrogen atom in the apical position. The 8-quinolinethiolato ligands are arranged with the like donor atoms diametrically opposed. The bond distances of Tc—N, Tc—S and Tc—N (quinoline) are 1.623 Å, 2.3559 Å and 2.135 Å, respectively.

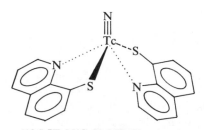

(**134**) $[TcN(C_9H_6NS)_2]$ molecule

5. Synthesis of $[AsPh_4]_2[TcN(dto)_2]$, 135 (dto = dithiooxalate)

Calmanet and Mackay obtained[299a] crystals of $[AsPh_4]_2[TcN(SCOCOS)_2]$ (**135**) from **133**, potassium dithiooxalate and $AsPh_4Cl$. The pale purple microcrystalline product **135** (83% yield) melted with decomposition at *ca* 215 °C and showed I.R. peaks at 1612vs, 1438s, 1079m, 1071m ($Tc\equiv N$), 1039s, 742s, 735s and 685s cm^{-1}. Single crystals of **135** for X-ray diffraction studies have been grown by slow evaporation of an acetonitrile–ethanol solution. Single crystals of $[AsPh_4][TcO(dto)_2]$ (**136**) have been prepared similarly. The crystals of **135** and **136** with the TcN^{2+} core and TcO^{3+} core belong to the triclinic space group. P1 with $Z = 2$ and cell parameters are $a = 14.225\,\text{Å}$, $b = 17.778\,\text{Å}$. $c = 10.993\,\text{Å}$, $\alpha = 101.52°$, $\beta = 111.74°$, $\gamma = 100.68°$ in the case of **135** and $a = 12.294\,\text{Å}$, $b = 12.531\,\text{Å}$, $c = 13.071\,\text{Å}$, $\alpha = 115.10°$, $\beta = 114.22°$, $\gamma = 101.93°$ in the case of **136**. The technetium atoms in the complex anions are five-coordinate, with the four sulphur atoms in the basal position and either the nitrogen or the oxygen in the apical position. A distorted square-pyramidal environment around each technetium is formed. The Tc≡N and Tc=O bond lengths are 1.613 and 1.646 Å in **135** and **136**, respectively.

The first preparation of a nitrido complex of technetium, bis(diethyldithiocarbamato)-nitridotechnetium(V), was described in 1981[299b,300]. The Tc(V) complexes are oxidized by excess chlorine or bromine, yielding finally $[Tc^{VI}NCl_4]^-$ and $[Tc^{VI}NBr_4]^-$ complexes. These Tc(VI) products can be detected easily by E.P.R. spectroscopy due to $4d^1$-configuration (with $S = 1/2$) of Tc^{6+}. *The monoclinic modification of bis(tetraphenyl-arsonium)bis(dithiooxalato)nitridotechnetate(V)* (deep purple crystals, 81%) have been obtained[301] from $TcNCl_2(Ph_3P)_2$, K_2dto and $AsPh_4Cl$. The product melted at 142–144 °C and showed the characteristic I.R. maxima at $1045\,cm^{-1}$ (Tc≡N) and at $1620\,cm^{-1}$ (C=O). The compound crystallized in at least two modifications: one was pale purple recrystallized from MeCN/EtOH melted at 215 °C[299]; the others form[301] is a monoclinic one (space group C2/c, Z = 4, a = 19.423, b = 11.254, c = 24.958 Å, β = 107.68°). The following bonding parameters have been obtained for the roughly square-pyramidal complex anion **135**.

(**135**) complex anion of $[AsPh_4]_2$
$[TcN(dto)_2]$ with numbering scheme

Bond length (Å): Tc≡N (1.606 Å), Tc—$S_{(1)}$(2.390), Tc—$S_{(2)}$(2.398), $S_{(1)}$—$C_{(1)}$(1.735), $S_{(2)}$—$C_{(2)}$ (1.745), $C_{(1)}$—$C_{(2)}$ (1.558), $C_{(1)}$—$O_{(1)}$ (1.211), $C_{(2)}$—$O_{(2)}$ (1.210 Å).

Bond angles (deg): $S_{(1)}$—Tc≡N (104.94°), $S_{(2)}$—Tc—N (104.58), $S_{(1)}$—Tc—$S_{(2)}$ (86.22), $S_{(1)}$—Tc—$S_{(1')}$ (150.13), $S_{(2)}$—Tc—$S_{(2')}$ (150.85), $S_{(1)}$—Tc—$S_{(2')}$ (86.34), $S_{(2)}$—Tc—$S_{(1')}$ (86.22), Tc—$S_{(1)}$—$C_{(1)}$ (106.4), Tc—$S_{(2)}$—$C_{(2)}$ (106.7°).

Pairs of distances and angles in the complex anions are almost identical within experimental error. Samples of **135** dissolved in $CHCl_3$ have been treated with different amounts of Br_2 and of Cl_2, and the reaction mixtures were studied by E.P.R., after 3 and 30 min[301].

The E.P.R. spectrum of an oxidation mixture of **135** with Cl_2 consists of 10-line multiplets due to the hyperfine interaction of the unpaired electron spin with the nuclear spin of technetium $^{99}Tc: I = 9/2$. The E.P.R. spectrum of a frozen solution (at $T = 130\,K$, $CHCl_3$) shows a typical axially symmetric pattern with ^{99}Tc multiplets in parallel and perpendicular parts, described by the spin Hamiltonian

$$H_{sp} = \beta_e[g_\| B_z \hat{S}_z + g_\perp (B_x \hat{S}_x + B_y \hat{S}_y)] + A_\|^{Tc} \hat{S}_z \hat{I}_z + A_\perp^{Tc}(\hat{S}_x I_x + \hat{S}_y \hat{I}_y) \tag{78}$$

where $g_\|, g_\perp, A_\|^{Tc}$ and $A_\perp^{Tc}$ are the principal values of the $\tilde{g}$ and ^{99}Tc-hyperfine interaction tensor $\tilde{A}^{Tc}$.

No mixed-ligand intermediates have been detected by E.P.R. in the experiments with Cl_2 independently of the Cl_2/complex ratio and $TcNCl_4^-$ has been the exclusive oxidation product. On oxidation of the complex with the weaker oxidant, Br_2, signals of five individual species have been recorded in the high-field region of the E.P.R. spectrum at 3 min reaction time corresponding to mixed Br/dto coordination spheres, especially at low $Br_2/[TcN(dto)_2]^{2-}$ ratios. Using an excess of Br_2 and longer reaction times (30–60 min), $TcNBr_4^-$ is the main product.

6. Synthesis of tetrakis(isothiocyanato)nitridotechnetate(V)

$[AsPh_4]_2[Tc^VN(NCS)_4]$ (137) has been prepared[302] (80%) from $[AsPh_4][Tc^{VI}NBr_4]$ with KSCN followed by the addition of Ph_4AsCl. The recrystallized sample melted at 81–85 °C and showed maxima in the I.R. spectrum (in KBr, v in cm^{-1}) at 3068w, 2931w, 2111s ($v_{C\equiv N}$), 2087 (vs, C≡N), 1487m, 1445s, 1390w, 1340w, 1315w, 1188m, 1166w, 1094s (Tc≡N)[303,304], 1084s, 1038w, 1000s (Tc=O), 839m, 766s, 741s, 694s, 675m, 480m and 469s.

The orange-red needles of 137 are well soluble in organic solvents like chloroform or acetone.

7. Nitrido complexes of technetium with tertiary arsines and phosphines.

Diamagnetic technetium(V) complexes of the general formulae $[TcNX_2(Ph_3Y)_2]$ (138) and $[TcNX_2(Me_2PhP)_3]$ (X = Cl, Br; Y = P, As) (139) have been prepared[305] in 85–95% yields by refluxing $(Bu_4N)TcNCl_4$ or $(Bu_4N)TcNBr_4$[306] with arsine or phosphine ligands. The brick-red to red-brown coloured complexes 138 are poorly soluble in organic solvents, whereas the complexes 139 are easily soluble in dichloromethane, benzene and acetonitrile. The yields (in%), m.p. (°C) and I.R. (cm^{-1} for Tc≡N) are 85%, 232–234 °C,

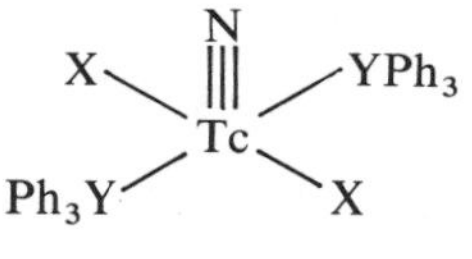

| (138) | (139a) | (139b) |

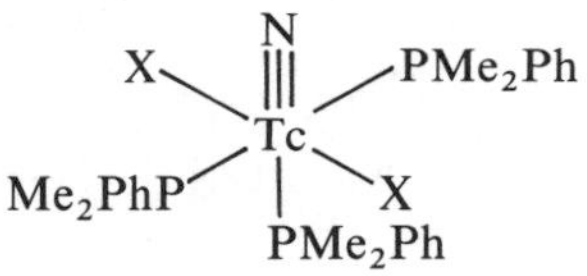

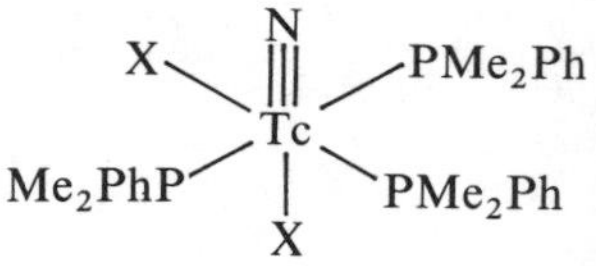

1091 cm^{-1} for the yellow-brown complex $[TcNCl_2(Ph_3As)_2]$ and 90%, 236–237 °C, 1091 cm^{-1} for the red-brown complex $[TcNBr_2(Ph_3As)_2]$. The title complexes 138 or 139 treated with chelating ligands, such as sodium diethyldithiocarbamate, $Na(S_2CNEt_2)$, 140, or with N-(N″-morpholinylthiocarbonyl)benzamidine, Hmorphtcb, 141, in acetone in the presence of NEt_3 gave the well-known Tc(V) complexes $[TcN(et_2dtc)_2]$[306,307] and $[TcN(morphtcb)_2]$[308] in 80–95% yields. 138 and 139 have been found to be good starting materials for the synthesis of further Tc(V) nitrido compounds[305].

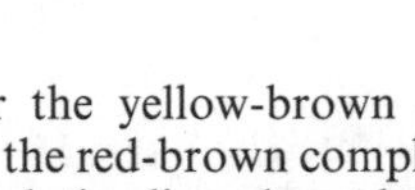

(140) Naet₂dtc

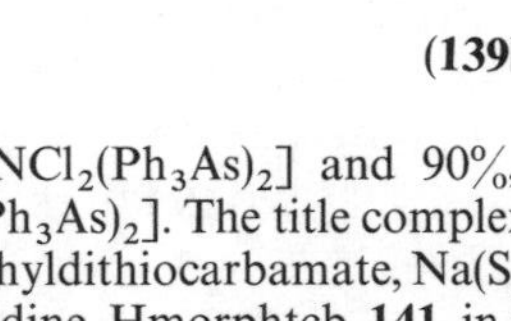

(141)

8. E.S.R. studies of the oxidation of [Tc^VNCl₂(EPh₃)₂] (E = As or P) to

8. E.S.R. studies of the oxidation of $[Tc^VNCl_2(EPh_3)_2]$ (E = As or P) to $[Tc^VNCl_4]^-$ by thionyl chloride

The structure of single crystals of $[Tc^VNCl_2(AsPh_3)_2]$ (**142**) consisting of discrete molecules $(C_{36}H_{30}As_2Cl_2NTc$, $M = 796.30$, monoclinic space group $12/a$, $a = 15.824$, $b = 9.638$, $c = 22.490$ Å, $\beta = 101.91°$ and $Z = 4$) has been determined by Baldas and coworkers[309]. They possess C_2 symmetry with $N{\equiv}Tc{-}Cl$ and $N{\equiv}Tc{-}As$ angles of 110.51 and 99.22°, respectively. Some interatomic distances (Å) are 1.601 ($Tc{\equiv}N$), 2.544 ($Tc{-}As$), 2.373 ($Tc{-}Cl$), 1.9374 ($As{-}C_{(1)}$), 1.944 ($As{-}C_{(13)}$), 1.941 ($As{-}C_{(7)}$).

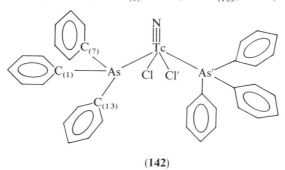

(**142**)

The complex **142** refluxed in $SOCl_2$ for 15 s turned into an orange-red solution which, treated with $AsPh_4Cl$, gave orange-red crystals of $[AsPh_4][Tc^{VI}NCl_4]$ (66% yield, m.p. 270–272°C). E.S.R. spectroscopy has established[309] that this oxidation proceeds via the technetium(VI) intermediate, $[TcNCl_3(AsPh_3)]$, with coordination to Tc of only one As-containing ligand (E.S.R. spectral parameters at 130 K: $g_\| = 2.0246$, $g_\perp = 1.9986$, $A_\| = 0.0271$ cm^{-1}, $A_\perp = 0.0125$ cm^{-1}, $Q = 0.0003$ cm^{-1}, $a_x = a_y = 0.00195$ cm^{-1} and $a_z = 0.0021$ cm^{-1}). A mixture of $[AsPh_4][Tc^{VI}NCl_4]$ and of PPh_3 refluxed in $SOCl_2$ for 30 min, on addition of $AsPh_4Cl$, produced $[AsPh_4]_2[Tc^{IV}Cl_6]$ (77%). In the absence of the reducing ligand the $[AsPh_4][TcNCl_4]$ is unaffected by reflux in $SOCl_2$.

9. E.S.R. study of the equilibrium between tetrahalogeno- and pentahalogeno-nitridotechnetate(VI) ions in solutions

A detailed E.S.R. study[310] (with the use of the spin Hamiltonian equation 79 where $S = 1/2$, $I = 9/2$, the other symbols have their usual meaning and H_{shf} represents the superhyperfine interactions due to the ligand nuclei) has been carried out. The equation describes spectra showing typical axially symmetric patterns with ^{99}Tc-multiplets in parallel and perpendicular parts of the solutions of $[AsPh_4][TcNCl_4]$, $Cs_2[TcNCl_5]$, $[AsPh_4][TcNBr_4]$ and $Cs_2[TcNBr_5]$ in acetonitrile, in dichloromethane (or dibromomethane), in DMSO and in concentrated hydrochloric acid. No E.S.R. spectral evidence has been found for the existence of the equilibrium shown in equation 80 in solution, not even in the presence of a large excess of halide ions, which is expected to favour its establishment.

$$H = g_\| \beta B_z S_z + g_\perp \beta (B_x S_x + B_y S_y) + A_\| S_z I_z + A_\perp (S_x I_x + S_y I_y) + Q[I_z^2 - \tfrac{1}{3} I(I + 1)] + H_{shf} \tag{79}$$

$$[TcNX_4]^{-1} + X^- \rightleftharpoons [TcNX_5]^{2-} \tag{80}$$

The radioligand *X exchange in systems containing tetra- and penta-halogenonitridotechnetium(VI) complexes would provide useful information concerning the nature of the

Tc—X bonds and the lability of X-ligands, but it has not been investigated, unlike tungsten cyanide complexes[311,312] and unlike the ^{36}Cl exchange between gaseous HCl and organic HCl solvates[313]. The equilibrium in dichloromethane has been studied by Thomas and coworkers[314].

$$[TcOCl_4]^- + Cl^- \rightleftharpoons [TcOCl_5]^{2-}$$

E. Synthesis of Technetium(VII) Nitridoperoxo Complexes

1. Syntheses of new technetium(VII) nitridoperoxo complexes

Transition metal peroxo complexes are stoichiometric or catalytic oxidants of organic compounds[315]. Many of them have been structurally characterized[316,317]. Synthesis of several new technetium(VII) nitridoperoxo complexes has been reported recently[318,319]. Thus, $[AsPh_4][TcN(O_2)_2Cl]$ (143) has been prepared[319] from $[AsPh_4][TcNCl_4]^{318}$ in MeCN with 10% hydrogen peroxide. The pale-yellow complex 143 melted with decomposition at 205°C. $[AsPh_4][TcN(O_2)_2Br]$ (144) has been obtained from $[AsPh_4][TcNBr_4]$ in MeCN/10% H_2O_2 (m.p. 110 °C, dec.). Both 143 and 144 dissolved in $SOCl_2$ gave an orange solution which, after removal of $SOCl_2$ and recrystallization from MeCN/C_6H_6 (1:1), gave back $[AsPh_4][TcNCl_4]$ identified by I.R. and m.p.

The pale-yellow $[TcN(O_2)_2(bipy)]$ (145) has been obtained by adding 2,2'-bipyridyl to nitridotechnetic(VI) acid 146 dissolved in 10% H_2O_2. 146 has been prepared by hydrolysis of $Cs_2[TcNCl_5]$ synthesized in turn in the $TcO_4^-/NaN_3/HCl$ reaction[320].

$[TcN(O_2)_2(phen)]$ (147) has been prepared similarly by reaction of $[TcN(O_2)_2(OH_2)_2]$ with 1,10-phenanthroline·HCl. The pale yellow $[AsPh_4]_2[\{TcN(O_2)_2\}_2ox]$ complex 148 has been obtained from $AsPh_4Cl$, $H_2C_2O_4·2H_2O$ and nitridotechnetic(VI) acid in 10% H_2O_2. The complex 148 exploded at about 140 °C on heating.

The dimeric structure 149 has been proposed[318] for nitridotechnetic(VI) acid. Solutions of 149 in 10% H_2O_2 contain the species 150 (equation 81). Solutions of 150 slowly hydrolysed to TcO_4^- at room temperature. A polymeric oxo-bridged hydrate structure $[TcN(OH)_3]_n$ (151) has been suggested also for the nitridotechnetic(VI) acid [showing I.R. peaks at 1054(Tc≡N) and 708(Tc—O—Tc) cm^{-1}], produced by loss of all five chloride ions in hydrolysis of $Cs_2[TcNCl_5]$ by water[321].

$$[Tc_2N_2O_2(OH)_2(H_2O)_2] + 5H_2O_2 \longrightarrow 2[TcN(O_2)_2(H_2O)_2] + 4H_2O \quad (81)$$
$$\qquad\qquad (149) \qquad\qquad\qquad\qquad\qquad\qquad (150)$$

The nitridoperoxo formulations are based on the microanalytical data and on the I.R. spectra. Strong absorptions at 1069–1035 cm^{-1} correspond to the Tc≡N group, absorptions at 912–894 cm^{-1} to v(O—O), weak but characteristic peaks at 665–647 cm^{-1} belong to v_s (TcO$_2$). The synthesis of bromoperoxo complex $[AsCH_3Ph_3][CrO(O_2)_2Br]$ (152) has been reported[322].

2. Synthesis and structure of $Cs[Tc^{VII}N(O_2)_2Cl]$ (153)

This transition metal nitridoperoxo complex is the first reported stable peroxocomplex of a metal in 7+ oxidation state. It has been obtained[317] in high yield by dissolving[320] $Cs_2[TcNCl_5]$ in 10% hydrogen peroxide and evaporating the yellow solution at room temperature. The coordination about technetium atom is a distorted pentagonal pyramid with the nitrogen atom in the apical position (Tc≡N = 1.63 Å) and sideways bound peroxo ligands with the $O_{(1)}-O_{(2)}$ distance = 1.41 Å, $O_{(3)}-O_{(4)}$ = 1.46 Å, and mean Tc–O = 1.95 Å. The $O_{(1)}$—Tc—$O_{(2)}$ angle is 42.6° and the $O_{(3)}$—Tc—$O_{(4)}$ angle is 43.9°.

The Tc–Cl distance equals 2.355 A. The peroxo and chloro ligands lie essentially in a plane with the Tc atom displaced by 0.45 Å above the plane.

Structure of $[TcN(O_2)_2Cl]^-$ in 153

An attempt has been made to determine also the structure of the 143 which showed I.R. peaks at 1069, 897 and 653 cm^{-1} besides those arising from the cation. The complex 153 is very soluble in water and converts to $[TcNCl_4]^-$ when dissolved in conc. HCl. It precipitates by addition of $AsPh_4Cl$ as the insoluble $[AsPh_4][TcNCl_4]$. 153 decomposes at about 110 °C on heating and explodes at 160 °C. The salt 143 is more stable and decomposes at about 205 °C. $Cs_2[TcOCl_5]^{314}$ treated with 10% hydrogen peroxide gave $CsTcO_4$, not $Cs[Tc^{VI}O(O_2)_2Cl]$ as expected.

3. Synthesis and structure of $[AsPh_4]_2[\{TcN(O_2)_2\}_2(ox)] \cdot 2Me_2CO$

This acetone solvate of the bis(tetraphenylarsonium) oxalate-bridged technetium(VII) nitridoperoxo dimer 154 has been obtained[323] by slow evaporation of an acetone solution of $[AsPh_4]_2[\{TcN(O_2)_2\}_2(ox)][ox = oxalate(2-)]$. It crystallizes in the monoclinic space group C2/c with cell parameters $a = 34.49$, $b = 14.684$, $c = 22.776$ Å, $\beta = 107.18°$ and $Z = 8$. The structure is dimeric, consisting of two $TcN(O_2)_2$ units bridged by a quadridendate sideways-bound oxalate. The geometry about each technetium atom is distorted pentagonal–bipyramidal with the nitrido ligand in an apical position and peroxo ligands in equatorial positions.

Structure of the $[\{TcN(O_2)_2\}_2(ox)]^{2-}$ anion in 154

The Tc≡N bond distances are 1.693 and 1.61 Å^{324}, the O–O distances are 1.43 and 1.50 Å [characteristic for sideways-bound (η^2)peroxide]^{325}, the Tc–O distances are in the range 1.91–1.98 Å but $Tc_{(2)}-O_{(12)}$ is only 1.86 Å.

The solvent acetone molecules have been poorly defined in the general lattice positions, but they gave a strong $v(C=O)$ absorption at 1713 cm^{-1} in the I.R. spectrum. The other characteristic I.R. absorptions in the solvate 154 were $v(Tc≡N)$ 1062, $v(O—O)$ 900 and $v(TcO_2)$ 658 cm^{-1} and absorptions due to oxalate at 1665 $v(C=O)$, 1277 and 801 cm^{-1} ^{326}.

Complexes containing the $Tc^{VII}N(O_2)$ core have been prepared also ^{320,327} by treating precursors such as $(NEt_4)_2[TcN(NCS)_4(MeCN)]$ or $[TcN(tu)_4Cl]Cl (tu = thiourea)$ with 10% H_2O_2 and subsequent addition of $AsPh_4Cl$ to give $[AsPh_4][TcN(O_2)_2Cl]$ in high yields^{320,327}.

F. Synthesis and Structure of Perchlorate and Chloride Salts of trans-Dichlorobis[o-phenylenebis(dimethylarsine)]technetium(III)

The structures of the dark-red crystals of the Cl^- salt trans- $[TcCl_2(diars)_2]Cl$ (155) where diars = o-$C_6H_4[AsMe_2]_2$, as well as the structure of the yellow-orange ClO_4^- salt, trans-$[TcCl_2(diars)_2]ClO_4$ (C2, $a = 13.001$, $b = 10.409$, $c = 11.796$ Å, $\beta = 114.50°$, $V = 1452.6$ Å^3, $Z = 2$) have been determined [328a]. The six-coordinate Tc(III) complex 155 has trans-chloride ligands and four As atoms from two nearly coplanar o-phenylenebis-(dimethylarsine) ligands, giving ideal D_{2h} both in Cl^- and ClO_4^- salts. The structure of the complex cation and the Tc–As, As–C and C–C distances in the Cl^- salt and in the ClO_4^- salt are similar. The mean Tc–As distance is 2.512 Å. In the Cl^- salt the two Tc–Cl distances are equal at 2.329 Å. The average Tc–Cl distance in the ClO_4^- salt equals 2.318 Å but Tc–Cl$_{(1)}$ is 2.348 Å (elongation due to intermolecular forces) and Tc–Cl$_{(2)}$ is 2.288 Å (shortening due to intermolecular forces). It has been suggested that the colour difference between the two complexes is likely due to a charge-transfer transition, specifically involving the Cl^- anion which is absent from the ClO_4^- salt.

(156) the complex cation $[TcCl_2(diars)_2]^+$ in 155

G. Quantitative Study of the Structure–Stability Relationship of Tc Complexes

By analysing structural parameters of more than 100 Tc compounds and using the cone packing model, a stability indicator of the Tc compounds has been derived [329]. This is based on the solid angle factor sum (SAS), calculated from the van der Waals radii and bond lengths between all the coordinating atoms and Tc. The average SAS value is 0.97 ± 0.13, which is very close to the theoretically optimal value 1.00.

The quantitative calculation of the SAS is sueful for predicting stability and designing Tc-radiopharmaceuticals used in diagnostic nuclear medicine. The SAS indicators (given in parentheses) for the following As-containing ligands have been calculated: $[AsPh_4]_2[TcCl_6]$ (SAS = 0.99), $[AsPh_4][TcNBr_4]$ (0.887), $[AsPh_4][TcNCl_4]$ (0.901), Tc_2As_3 (0.992), trans-$[Tc(diars)_2Cl_2]^+$ (1.007), trans-$[Tc(diars)_2Cl_2]ClO_4$ (1.007).

H. The Determination of ⁹⁹Tc in Complex Compounds by Short-time Instrumental Neutron Activation Analysis

The importance of technetium-99m radiopharmaceuticals stimulated numerous studies related to the basic chemistry of long-lived technetium ^{99}Tc. A great number of technetium complexes reviewed in the preceding sections has been synthesized and characterized. Exact technetium analyses in these compounds have been needed to confirm their composition. Besides the traditional determinations of Tc by conductometric titrations of pertechnetate or by liquid scintillation counting, neutron activation of ^{99}Tc has also been applied[330-332] (equation 82) for the analysis of pure complexes.

$$^{99}_{43}Tc(n, \gamma)^{100}Tc \xrightarrow{\beta^-, \gamma, T_{1/2} = 15.8\,s} {}^{100}Ru(stable, 12.53\%) \qquad (82)$$

$$E_{max}(\beta^-) = 2.2;\ 2.88;\ 3.38\ MeV;\ \langle\beta\text{-}\rangle = 1315\ keV$$

$$E(\gamma) = 539.53\ keV\ (7\%\ yield);\ 590.76\ keV\ (5.7\%\ yield)$$

A series of compounds $M[TcOL_2]$, where $M = AsPh_4$ and $L = -SCH_2CH_2S-$, $-NEt_4$ $-SCH(COOMe)CH(COOMe)S-$, $-SC(CN)=C(CN)S-$, or $(-S)C(Se^-)=C(CN)_2$ has been prepared[333] by reactions of technetium(V) gluconate with various dithiol ligands and their Tc content was determined by fast activation (the total time for a determination was about 6 min) with good analytical accuracy and reproducibility. New Tc coordination compounds, TcL_3 (L = monothiodibenzoylmethane), $TcOOHL_2$ (L = 6-mercaptopurine), $TcOHL_2$ (L = N-salicylidene-2-hydroxyaniline), $TcOHL_2$ (L = N-salicylidene-2-mercaptoaniline) and $TcOHL_2$ (L = N-salicylidenecysteamine), have been synthesized[330] also and their composition formulated on the basis of measured Tc values.

I. Biomedical Applications of Radiolabelled Arsenicals

1. Cationic Tc-99m complexes as potential myocardial imaging agents

Thallium-201 is widely accepted[334] for myocardial imaging but it is rather expensive, has low-energy photopeaks (68.89-keV, 70.82-keV and 80.12-keV Hg X-rays), a low count rate owing to be low does permissible and a relatively long half-life (3.046 d). These drawbacks are avoided by using Tc-99m regarded as the optimal emitter for diagnostic imaging applications[335].

A continuous search for the Tc-99m myocardial perfusion agents is carried out, based on the assumption that various monovalent cations K^+, Rb^+, Cs^+ and Ti^+ as well as singly charged organic cations accumulate preferentially in the myocardium. Deutsch and coworkers[336] synthesized nineteen cationic Tc-99m complexes (equation 83) and evaluated them as myocardial imaging agents in dogs. All phosphine, arsine and thiol

$$^{99m}TcO_4^- + diars + HCl \longrightarrow {}^{99m}Tc(diars)_2Cl_2^+ + diars(O) \qquad (83)$$

$$diars = o\text{-}C_2H_4(AsMe_2)_2$$

ligands used in their work are sufficiently powerful reductants to reduce pertechnetate to Tc(III) or Tc(II).

The biological distribution of the $[^{99m}Tc(diars)_2X_2]^+$ series in rats has also been determined, as well as the biological distribution of $[^{99m}Tc(diars)_2Br_2]^+$ and of ^{201}Tl in dogs.

The four halogen derivatives of the diars series concentrate in the myocardium. A definite difference in the biodistributions of $[^{99m}Tc(diars)_2Cl_2]^+$ and $[^{99m}Tc(diars)_2Br_2]^+$ in rats has been found despite the similar size, shape, charge and lipophilicity of these complexes. The following compounds have been synthesized by Deutsch and coworkers[336,337]: $Tc(diars)_2F_2^{1+}$, $Tc(diars)_2Cl_2^{1+}$, $Tc(diars)_2Br_2^{1+}$, $Tc(diars)_2I_2^{1+}$, Tc-diars-om (om = n-octyl mercaptan), Tc-diars-mtg (mtg = monothioglycerol), Tc-diars-cys (cys = 2-mercaptoethylamine), Tc-dae-Cl[dae = bis(1,2-diphenylarsino)ethane], Tc-dae-Br, Tc-dae-I, Tc-ppn-Cl[ppn = bis(2-diphenylphosphinoethyl)amine], Tc-ppn-Br, Tc-ppn-I, Tc-ppn-mtg, Tc-tetraphos-Cl [tetraphos = tris(2-diphenylphosphinoethyl)phosphine], Tc-tetraphos-Br, Tc-tetraphos-I, Tc-tetraphos-nor (nor = DL-norepinephrine) and Tc-tetraphos-bcat (bcat = 4-t-butylcatechol). The total preparation time for each complex was about 1.5 h.

When molecular chlorine is introduced into an alcoholic solution of $[Tc(diars)_2Cl_2]ClO_4^{337}$ at room temperature, the solution turns rapidly deep red and a dark brown solid is subsequently deposited. This reaction is defined as an oxidative addition process in which the six-coordinate, d^4, Tc(III) is converted into eight-coordinate, d^2, Tc(V) (equation 84). The product yields, by metathesis with $NaPF_6$, the dark brown $[Tc(diars)_2Cl_4]PF_6$.

$$[Tc(diars)_2Cl_2]ClO_4 + Cl_2 \longrightarrow [Tc(diars)_2Cl_4]ClO_4 \qquad (84)$$

The complexes containing diars, dae, tetraphos, ppn, mtg, cys, om, bcat and nor have been chosen to investigate the effects of varying both the chelating ligands forming the basal plane of the octahedral complex and the monodendate ligands situated *trans* to each other in the octahedral complex. The conclusion has been reached that relatively non-specific cation acceptor sites exist in the myocardial muscle (since inorganic, as well as organic cations accumulate in the myocardium). However, the cationic character by itself is not sufficient to ensure myocardial uptake since the majority of the cationic Tc-99m complexes do not collect in the myocardium. Thus the lipophilicity of the cation and its molecular configuration are also important. The optimal combination of ligands for myocardium imaging has not yet been found.

The ^{99m}Tc-labelled diars and dmpe chloro compounds [dmpe = bis(1,2-dimethyl-phosphino)ethane] have been synthesized also by Strauss and coworkers[338], using the method described elsewhere[336,339,340] and applied for evaluating the myocardial kinetics of these agents in normal and ischemic myocardium. The study demonstrated the superiority of dmpe over diars for clinical imaging. The clearance rate from normal and ischemia myocardium has been found to be similar for both diars and dmpe.

Attenuation of the 78-keV photons recorded when thalium is employed in studies on human subjects makes the diagnosis of diseases in women difficult because of the overlying breast. The technetium-labelled agents do not show these limitations.

2. Myocardial specificity studies of radioiodinated phosphonium, arsonium and ammonium cations

Inefficient detection of low-energy X-rays and redistribution during the imaging period are the main disadvantages of the thalium-201 cationic perfusion agent most widely used for differentiation of ischemia from irreversible myocardial damage. ^{123}I-labelled organic cations have been suggested for evaluation of heart disease by γ-camera imaging[341]. The effect of alkyl and aryl substitution on the myocardial specificity of radioiodinated phosphonium, arsonium and ammonium cations has been studied[342] by synthesizing several (E)-(1-[^{123}I]iodo-1-penten-5-yl)-trisubstituted phosphonium, arsonium and ammonium iodides (157) as shown in equation 85a or by quaternization of 158 with $R^1R^2R^3M$ followed by one-step iodination of the borono analogue 159[342,343]. 157d has been synthesized by method a and by iodination of 159, with the use of sodium [^{125}I] iodide and chloramine T. Method b has been followed in the synthesis of 157 radiolabelled with iodine-123 in view of its shorter half-life ($T_{1/2} = 13.2$ h). The crude products have been treated with excess NaI to insure that iodide was the anion, and have been purified by chromatography.

The distribution of radioactivity in tissues of female Fisher rats demonstrated that the replacement of phosphorus with arsenic or the replacement of a phenyl with a cyclohexyl ring has little apparent effect on heart uptake. However, the replacement of cyclic systems with alkyl groups significantly decreased the heart uptake with both the phosphorus and nitrogen compounds studied. The γ-camera imaging studies with [^{123}I]-157a' and with [^{123}I]-157c' confirmed the decreased heart uptake with alkylation and the apparent hepatobiliary clearance of 157c'.

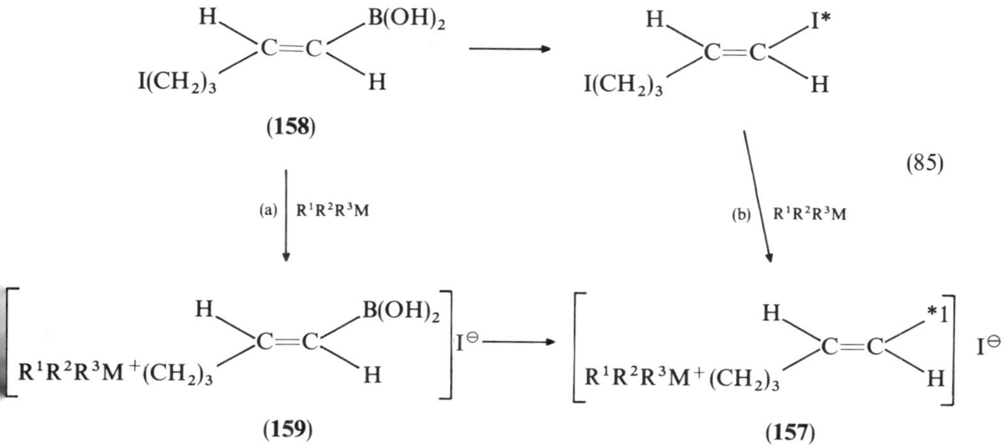

(158)

(85)

(a) $R^1R^2R^3M$

(b) $R^1R^2R^3M$

(159)   (157)

$R^1R^2R^3M = (C_6H_{11})_2C_6H_5P$ (157a); $(C_6H_{11})_3P$ (157b); $Me_2(n\text{-}C_8H_{17})P$
(157c); Ph_3As (157d); Et_3N (157e) *I = ^{123}I or ^{125}I

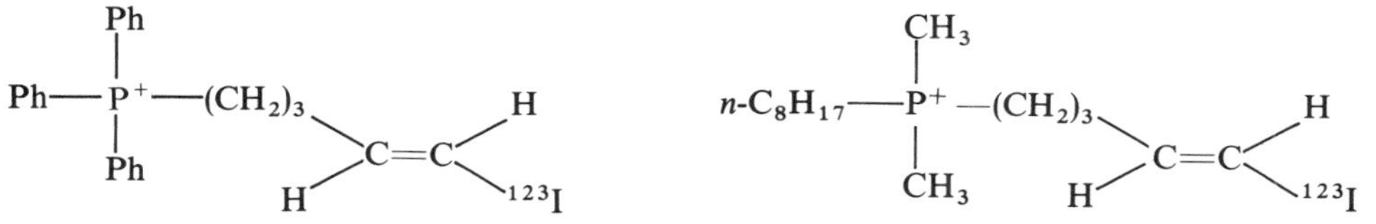

(157a')   (157c')

3. Neutral technetium(II)-99m complexes as potential brain perfusion imaging agents

Molecules which are neutral and lipophilic diffuse passively across the BBB[344]. The ^{99m}Tc brain perfusion imaging agents are lipophilic complexes of technetium(V) which contain Tc=O linkages. Since the biological mechanism of action of the ^{99m}Tc agents is not clear, the question whether either the Tc(V) oxidation state or the Tc=O linkage is necessary in order for a ^{99m}Tc complex to cross the BBB has been probed[345] by synthesizing a series of eight neutral lipophilic technetium(II)-99m complexes of the general formula $tr\text{-}[^{99m}Tc^{II}D_2X_2]^0$, where D represents a chelating ditertiary phosphine or arsine ligand and X represents a halide or pseudohalide ligand, and by evaluating their ability to cross the BBB in rats. Some of the analogous Tc(III) cations, $tr\text{-}[^{99m}Tc^{III}D_2X_2]^+$, have been synthesized also in order to obtain some information about the role of *in vivo* redox reactions in determining the biodistribution of these complexes[345,346]. Several compounds of this series exhibit significant brain uptake in rats, the maximum uptake being exhibited by $tr\text{-}[^{99m}Tc^{II}(diars)_2Cl_2]^0$. Several cationic Tc(III)analogues, $tr\text{-}[^{99m}TcD_2X_2]^+$, also exhibit significant brain uptake via a mechanism involving *in vivo* reduction to the neutral Tc(II) compound. The properties of the $tr\text{-}[^{99m}Tc^{III/II}D_2X_2]^{+/0}$ couples, relevant to their potential application in brain imaging (such as lipophilicity, redox potential, blood binding) depend on the nature of D and X in a complicated interrelated fashion which can be utilized to yield an effective ^{99m}Tc brain perfusion imaging agent. $^{99m}TcO_4^-$ does not cross the BBB, and its brain uptake is very low. In the series $[^{99m}Tc^{II}(dmpe)_2X_2]^0$ (X = Cl, Br, I) increasing the lipophilicity increases brain uptake; in the series X = Cl, Br, SCN the brain uptake is controlled by blood binding (the least tightly bonded Br complex exhibits the largest brain uptake).

The following ^{99m}Tc-labelled complex compounds have been synthesized and used for studying the brain and blood uptakes of ^{99m}Tc agents in rats as a function of time [345]:tr-$[^{99m}Tc^{III}(diars)_2Cl_2]^+$, tr-$[^{99m}Tc^{III}(diars)_2Br_2]^+$, tr-$[^{99m}Tc^{III}(dmpe)_2Cl_2]^+$, and tr-$[^{99m}Tc^{III}(dmpe)_2Br_2]^+$. The neutral $[^{99m}Tc^{II}D_2X_2]^0$ complexes with D = diars or dmpe have been prepared by reduction of the analogous $^{99m}Tc(III)$ compounds. Thus tr-$[^{99m}Tc^{II}(diars)_2Cl_2]^0$ has been obtained by reduction of $[^{99m}Tc^{III}(diars)_2Cl_2]^+$, with $NaBH_4$ at RT at pH 5, or by reduction with the relatively mild reductant, 2-mercapto-ethanol. For a given D ligand, the Tc(III) complexes with X = Br(I) are easier to reduce than corresponding complexes with X = Cl.

tr-$[^{99m}Tc^{II}(diars)_2Br_2]^0$ has been synthesized from $[^{99m}Tc^{III}(diars)_2Br_2]^+$ with 2-mercaptoethanol or with ascorbic acid. Borohydride has been found to be too strong a reductant for the synthesis of the diars and dmpe complexes with X = Br (or I), producing large amounts of unidentified reaction products. Cation exchange chromatography has been used to separate neutral $^{99m}Tc(II)$ complexes exhibiting retention times of 4–5 min, from the cationic $^{99m}Tc(III)$ complexes (exhibiting retention times of 10–13 min), even at the very low 'nca' concentrations of technetium.

4. X-ray photoelectron spectroscopy of potential technetium-based imaging agents

The *in vivo* behaviour of metal complexes depends on the overall charge on the complex and on the charge distribution on the central metal and ligand atoms. Charged technetium compounds[347,348] like NH_4TcO_4, $[Me_4N]TcCl_6$ and $[Tc(diars)_2Br_2]Br$ do not cross the intact blood–brain barrier(BBB) to any appreciable extent whereas neutral technetium species can cross[349]. X-ray photoelectron spectroscopy (XPS) permits one to measure binding energy shifts compared to values from model compounds and thus provides, information concerning the overall charge distribution within the complex. Thompson and coworkers[348] carried out XPS examination of ^{99}Tc chelates including several compounds showing considerable promise as organ imaging agents. They also determined the following technetium $3d_{5/2}$ binding energies and Pauling partial charges[350], q_p 253.9 eV $(q_p = 0)$ for $Tc^{(0)}$, 258.8 eV $(q_p = 1.6)$ for $NH_4Tc^{(IV)}O_4$ 256.9 eV $(q_p = 0.9)$ for $[Me_4N]_2Tc^{IV}Cl_6$, 254.6 eV $(q_p = 0.2)$ for $[Tc^{(III)}(dmpe)_2Br_2]Br$, 254.9 eV $(q_p = 0.3)$ for $[Tc^{III}(diars)_2Br_2]Br$[348].

Ligand and anion binding energies (in eV) were found to be: in NH_4TcO_4, 401 eV (N 1s), 531.1 (O 1s); $[Tc(dmpe)_2Br_2]Br$, 66.2 (Br $3d_{5/2}$), 130.3 (P $2p_{3/2}$); $[Tc(diars)_2Br_2]Br$, 66.5 (Br $3d_{5/2}$), 42.1 (As 3d). Cl 3p spectra of $[Tc(dmpe)_2Cl_2]Cl$ and of $[Tc(dppe)_2Cl_2]Cl$, where dppe = 1, 2-bis(diphenylphosphino)ethane, exhibit two signals with an average split of 2.1 eV in the intensity ratio of 2:1 ascribed to covalent and anionic chlorine, respectively.

The complex $^{99m}TcCl[(cdo)_3mb]$, where cdo = cyclohexyldioxime and mb = methyl-boronic acid, is under clinical trials[348] as a myocardial imaging agent and has allowed the visualization of exercise-induced ischemia. Its Tc oxidation state is 3, the Tc($3 d_{5/2}$) binding energy is 255.0 and $q_p = 0.3$. Other members of the compounds investigated by Thompson[348] have shown brain uptake and retention in animals commensurate with their physiochemical properties.

5. Studies on thioredoxin with organoarsenical reagents

Thioredoxin from *Escherichia coli* is a small ubiquitous protein with M_r of 11,700[351] which contains a redox-active cystine moiety on an exposed β-reverse-turn[352]. It can

occur in the oxidized state as thioredoxin-S_2 (with a 14-membered polypeptide ring, comparising—Cys^{32}—Gly^{33}—Pro^{34}—Cys^{35}—) or in the reduced state as thioredoxin– —S—S—
(SH)$_2$ (with opening and relaxation of the 14-membered polypeptide ring.) As a reductant, thioredoxin functions as an electron transport protein between NADPH and the reduction of ribonucleotides to deoxyribonucleotides, as a general catalyst for the reduction of protein disulphides[353], in other reductive processes[354] and in the regulation of the hormonal response[355].

Thioredoxin, reduced with mercaptoethylamine, has been subjected to covalent chemical modifications employing[356] monofunctional organoarsenical reagents, $H_2NPhAsO$ and $HO(CH_2)_4AsCl_2$, specific for 'spatially close' thiols. The modifications resulted in the formation of stable 15-membered cyclic dithioarsenite ring structures, thioredoxin$<^S_S>$AsR, where R = H_2NPh—, $HO(CH_2)_4$— or $C^3H_3CONHPh$, that readily extrude the arsenic moiety upon treatment with 2, 3-dimercaptopropanol. Bifunctional reagents, such as $BrCH_2CONHPhAsO$, have been used subsequently to obtain the structure **160**, which could function as an active-site-directed inactivator for thioredoxin reductase and ribonucleotide reductase. The modifications have been monitored by the loss in the free thiol content, by the percent incorporation of radiolabelled organoarsenical reagents and by observing the changes in the amounts of various thioredoxins by size exclusion chromatography(SEC)[357] and by fluorescence anisotropy decay measurements. Upon reduction, an increase in the hydrated volume of the protein has been observed, and the volume of the protein swells further upon modification. The stable 15-membered cyclodithioarsenite ring structures with no apparent changes in the secondary structure of the protein are 'relaxed' compared to the 14-membered ring of the oxidized state, but are not as relaxed as the reduced state of the thioredoxin. It has been proposed[357] that the reduction of thioredoxin—S_2 leads to the opening of the reverse turn of the active cystine region with possible increased freedom of motion of the α_2-helix, but the parallel one- and two-dimensional proton N.M.R. studies of oxidized, reduced and organoarsenically modified thioredoxin[358] led to the conclusion that the structural changes between oxidized and reduced thioredoxin are localized in the area of the redox cystine.

The arsenicals caused strong quenching of the tryptophan fluorescence. It has been suggested[356] that the mechanism of the fluorescence quenching in thioredoxin–(SH)$_2$ by covalently bound arsenic involves the formation of an —S—As—S— bridge that acts as a quencher or plays a role in 'locking' the α_2—helix and the organic group affects the efficiency of this —S—As—S— quenching group.

$$\text{thioredoxin}\underset{\diagdown S\diagup}{\overset{\diagup S\diagdown}{\bigcirc}}\text{As-PhNHCOCH}_2\text{Br}$$

(160)

The p-[(bromo[1-^{14}C]acetyl)amino]phenylarsenoxide used in the above study has been synthesized by the method of Adamson and coworkers[359] modified by Holmes and Stevenson[360]. The p-([2-^{3}H$_3$]acetyl)aminophenyl dichloroarsine, p-C^3H_3CONH $C_6H_4AsCl_2$, has been prepared similarly. N-Phenyl-α-bromo[1-^{14}C]acetamide has been synthesized as described by Stevenson and coworkers[361,362], as was also 4-hydroxybutyl-dichloroarsine[356]. The dichloroarsine derivatives have been prepared from arsenoxides and concentrated HCl.

6. *Inhibition of pyruvate dehydrogenase multienzyme complex (PD) from E. coli with radiolabelled bifunctional arsenoxide*

PD catalyses the reaction (86): incubation[359] of PD from *E. coli* with thiamin pyrophos-

$$\text{pyruvate} + \text{NAD}^+ + \text{CoASH} \longrightarrow \text{acetyl-CoA} + \text{NADH} + \text{H}^+ + \text{CO}_2 \qquad (86)$$

phate (vitamin B_1), pyruvate, coenzyme A, Mg^{2+} and the radiolabelled bifunctional p-[(bromo[^{14}C]acetyl)-amino]phenyl arsenoxide, p-BrCH$_2^{14}$CONHC$_6$H$_4$AsO (**161**) at $0\,°C$ resulted in the irreversible loss of lipoamide dehydrogenase E3 activity. The PD complex consists of three different enzymes: E1 (pyruvate dehydrogenase), E2 (lipoate acetyltransferase) and E3 (lipoamide dehydrogenase). E2 forms the core of the complex to which enzymes E1 and E3 bind in a non-covalent manner. The inactivation proceeds by initial 'anchoring' of the reagent via the —AsO group to reduced lipoyl resides on E2 followed by the delivery of the BrCH$_2$^{14}CO-moiety into or near the active site of E3 where the irreversible alkylation takes place[361]. Incubation of the radiolabelled bifunctional reagent with PD complex in the absence of substrates, isolation of the E3 subunits from inhibited and control PD complexes by chromatography, acid hydrolysis of the alkylated E3 and control E3 samples and subsequent isolation, identification and quantitative determinations of the produced amino acids by electrophoresis and radiochemical analysis showed that the inhibited sample contained N^3-(carboxymethyl)histidine and a small amount of S-(carboxymethyl)cysteine. These residues have not been found in significant amounts in the controls. Thus most of the loss of E3 biological activity is caused by alkylation of histidine at or near the active site of E3 and of some cysteine[363].

161 used in this investigation has been prepared[357] (68%) from aminophenyl arsenoxide in acetone with bromo[1-^{14}C acetic anhydride in CH$_2$Cl$_2$, yielding a white solid, decomposing at 196 °C and having a specific activity of 8165 dpm mmol^{-1} (136.1 Bq mmol^{-1}).

The above results were confirmed subsequently by Holmes and Stevenson[360] by determining the amino acid sequence and establishing the position of the ^{14}C-labelled histidine in the peptide B obtained from E3. A reliable amino acid sequence **162** has been obtained for peptide B up to residue 30 which was in agreement with the amino acid sequence determined from the gene sequence of E3. The authors[360] demonstrated that a histidine residue is selectively alkylated within the active site of *E. coli* E3 when the enzyme is a component of the PD complex. The reagent **161** functions as a unique form of an active-site-directed irreversible inhibitor.

GCDAEDIALTIHA*HPTL–EIVGLAAEVFEG
(162)

The studies with radiolabelled **161** were preceded[361] by mechanistic investigations of the inhibition of PD complex from *E. coli* with unlabelled arsenoxide reagents, p-H$_2$NPhAsO, and unlabelled **161**. The bifunctional reagent caused 100% loss of the PD complex activity within 15 min in the presence of pyruvate and coenzyme A. The loss of E3 activity has been found to lag a few minutes behind the loss of the PD complex activity and reached a value of 90% within 20 min. The initial reaction of the bifunctional reagent occurred on E2 via the R—AsO moiety. The inactivation of E3 has been attributed to the subsequent transfer of the bromoacetyl moiety into the active site of E3. At this stage E2 and E3 were crosslinked. Addition of 2,3-dithiopropanol failed to regenerate PD complex activity and E3 activity, indicating that although the reduced lipoic acid group on E2 was regenerated, the alkylation in E3 had not been inversed. The accepted mechanism for the PD complex is shown in this study[361]. A brief review of the literature covering the studies of toxic effects of trivalent arsenicals on proteins possessing thiol groups, for the period since the paper of Ehrlich[364], has been presented[361].

7. Applications of phenylarsine oxide (PAO) and 4-[^{125}I]iodo-PAO (IPAO) in studies of the mechanism of insulin signaling involving phosphotyrosyl turnover

IPAO and PAO inhibit insulin-activated hexose uptake[365] and, in combination with insulin, cause accumulation of phospho-422 (aP2) protein (pp15) in 3T3-L1 adipocytes[366] without affecting basal glucose uptake or insulin receptor autophosphorylation[367]. The concentration of IPAO needed to completely inhibit insulin-stimulated-hexose uptake (64 μM) was about twice that for PAO (35 μM). Protein pp15 accumulates in cells treated with insulin and PAO because the arsenical blocks turnover of the phosphoryl group on pp 15. Trivalent arsenic present in PAO blocks dephosphorylation of pp15 by bridging covalently the vicinal or neighbouring —SH groups on the proteins[368].

PAO of natural isotopic composition has been employed in the biosynthesis of [^{32}P]pp15, involving ^{32}P$_i$. Subsequent use of [^{32}P]pp15 in bioexperiments facilitated the isolation of two (membrane) protein tyrosine phosphatases (PTPases), and permitted one to establish that more than 90% of the pp15 dephosphorylating activity is membrane associated. The ^{32}P lost from [^{32}P]pp15 during the bioreaction has been quantitatively recovered as ^{32}P$_i$.

Two different PTPases, enzymes which catalyse hydrolysis of phospho-Tyr^{19} of pp15, have been isolated from 3T3-L1 and characterized. These enzymes, designated as PTPases HA1 and HA2, inhibited by PAO, have been covalently labelled subsequently by 4-[^{125}I]-iodo–PAO. Labelling of HA1 and HA2 with [^{125}I]IPAO has been reversed by subsequent treatment with a stoichiometric amount of the vicinal thiol, 2,3-DMP. Thus it has been shown that HA1 and HA2 possess adjacent, closely spaced thiol groups that can be bridged by an atom of arsenic. Crosslinking of two —SH groups by oxidation to disulphide or by reaction with PAO inactivates the PTPase, emphasizing the important role of —SH groups in catalysis by HA1 and HA2. A conclusion has been reached that phosphotyrosine is a transient intermediate in insulin signaling, and other factors which determine the intensity and duration of the insulin signal have been studied.

8. Synthesis and bioapplications of [^{125}I]diiodoarsanilic acid and other studies

a. Diazo[^{125}I]diiodoarsanilic acid is a highly polar non-penetrating reagent binding selectively to externally exposed membrane proteins containing tyrosine. It has been used[369] to label selectively proteins of MW = 85,000, which have been isolated subsequently in an affinity column with anti-arsanilic acid antibodies. The general method is based on covalent labelling of membrane proteins in the intact cell, separation of the crude membrane and isolation of the labelled proteins by affinity chromatography, using antibodies against haptenic groups on the labelling reagents.

Diiodoarsanilic acid has been labelled with ^{125}I by the exchange method[370]. This acid gave diazo[^{125}I]diiodoarsanilic acid on reaction with amyl nitrite, CH$_3$(CH$_2$)$_4$ONO. The diazonium salt was added to a suspension of 1 × 10^8 blood platelets, incubated for 10 min at 4 °C and separated by centrifugation. The labelled blood platelets were lysed with dilute buffer (pH 7.2) and the suspension obtained centrifuged at 40,000 × g for 30 min. The resulting particulate protein fraction has been chromatographed on antibody affinity columns[371]. The labelled proteins have been recovered by elution, dialyzed, lyophilized and the pattern of proteins determined by SDS-gel electrophoresis[372]. The main protein labelled with ^{125}I had a molecular weight of 85,000–90,000. Thus the method of separation of the labelled antibodies has been established.

b. The blister agents, bis(2-chloroethyl)sulphide (**163**), and dichloro (2-chloro-vinyl)arsine have been synthesized with ^{14}C at all carbon atoms[373]. Chloro(2-chlorovinyl)mercury prepared from [^{14}C]acetylene has been converted to **163** along with the bis- and tris(chlorovinyl) compounds. The metabolism of methylarsine oxide and sulphide has been studied[374]. They have been found to be more toxic than arsenite. A review of earlier studies on radioactive As as a substitute for P in radiopharmaceuticals has been presented by Hosain and coworkers[375].

J. Brief Review of Recent Physical Studies with Isotopically Labelled Arsenicals

I.R. and Raman spectra of Me_2AsOOD, Me_2AsOOH, $(CD_3)_2$ AsOOH and $(CD_3)_2AsOOD$ have been recorded at different temperatures and their assignments proposed in terms of group vibrations[376]. The dimers dissociate in the symmetrical excited state into two neutral monomers, $Me_2AsOOH(D)$, or into two ionic entities, Me_2AsOO^- and $Me_2AsOOH_2^+$, in the asymmetrical excited state. Force constants, compliance constants and mean amplitudes of vibrations of four sets of 12 isotopic pyramidal XY_3 molecules, including As^3H_3, have been evaluated[377].

The quadrupole interactions of Tc-99 nuclei in polycrystalline pertechnetates $MTcO_4$, where $M = Li$, K, NH_4, ND_4, Me_4N, Et_4N, Bu_4N, Ph_4N or Ph_4As and $Mg(TcO_4)_2$ have been studied[378]; the quadrupole coupling constants(QCC) of ^{99}Tc nuclei in these compounds were determined from N.M.R. spectra and the relations between QCC and their structures discussed.

The dielectric relaxation in KD_2AsO_4 has been studied[379] and the problem of the replacement of P by As in the family of KH_2PO_4 and $NH_4H_2PO_4$ crystals has been discussed. The influence of isotope effects on the propagation of bulk and surface acoustic waves in antiferroelectric crystals with the tetragonal structure [deuterated $(NH_4)H_2AsO_4$ and $(NH_4)H_2PO_4$ salts] has been studied[380] and the measured and theoretically calculated propagations of waves in these crystals compared.

Ferroelectricity in partially deuteriated betaine arsenate has been investigated[381]. Ferroelectric betaine arsenate becomes antiferroelectric by deuteriation and the Curie point shifts monotonically with the D-content. Dielectric and caloric data of some crystals with different deuteriation degrees have been presented.

A thermodynamic investigation of the solubility and extraction of tetraphenylarsonium pertechnetate, Ph_4AsTcO_4, has been carried out[382] in a large number of solvents. The standard free energies of transfer have been used to characterize the difference between the solubility in H_2O and in organic solvents. The concentration dependence of the distribution coefficients is a function of the dissociation and association of Ph_4AsTcO_4 in the two phases.

V. ACKNOWLEDGEMENTS

The work on this chapter has been supported by The Faculty of Chemistry of The Jagiellonian University in Cracow and by The Department of Chemistry of The University of Warsaw. M. Z. thanks his mother again, as several times before, for providing room in her house in Witeradów and for her services facilitating the writing of the chapter during the summer. The help given by Gregory Czarnota in typing and in further preparation of the manuscript in Cracow is also acknowledged. Dr. Kański provided us with photocopies of several valuable radiochemical papers and references utilized in the text. Mgr. Halina Papiernik-Zielińska rendered her advice on drugs containing bismuth and arsenic.

VI. REFERENCES

General references to The Introduction:

A. Yu. A. Alexandrov, *Liquid Phase Autooxidation of Organometallic Compounds*, Science, Moscow, 1978.

B. M. I. Kabachnik, *Metalloorganic Compounds and Radicals,* Science, Moscow, 1985.
C. Ch. Elschenbroich and A. Salzer, *'Organometallics', A Concise Introduction,* Second edition, WCH, Weinheim, 1992.

1. U. Praeckel and F. Huber, *J. Organomet. Chem.,* **240**, 45 (1982).
2. J. L. Hartwell, *Org. Synth.,* Coll. Vol. III, Wiley, New York, 1955, p. 185.
3. F. F. Blicke, U. O.Oakdale and F. D. Smith, *J. Am. Chem. Soc.,* **53**, 1025 (1931).
4. D. M. Hawley and G. Ferguson, *J. Chem. Soc.(A),* 2059 (1968).
5. V. S. Petrosyan and O. A. Reutov, *J. Organomet. Chem.,* **76**, 123 (1974).
6. B. C. Smith and C. B. Waller, *J. Organomet. Chem.,* **32** 11(1971).
7. R. D. Urban, P. Polomsky and H. Jones, *Chem. Phys. Lett.,* **181**, 485 (1991).
8. T. Klapötke, *Monatsh. Chem.,* **119**, 1317 (1988).
9. A. Marquard, *Ber. Dtsch. Chem. Ges.,* **20**, 1516 (1987).
10. T. Klapötke, *J. Organomet. Chem.,* **331**, 299 (1987).
11. T. Klapötke and P. Gowlik, *Z. Naturforsch.,* **42b**, 940 (1987).
12. V. D. Nefedov and M. A. Toropova, N. I. Shchepina, V. V. Avropin, V. E. Zhuravleva and N. I. Trofimova, *Radiochemistry,* **31**, 69 (1989) (Russian).
13. N. I. Shchepina, V. D. Nefedov and M. A. Toropova, *Chem. Abstr.,* **95**, 7413g (1981); *Sintezy na Osnove Magnii-i i Tsinkorganich. Soedin,* Perm, 1980, pp. 116–119 (Russian).
14. T. Carlson, *J. Phys. Chem.,* **32**, 1234 (1960).
15. K. A. Kochetkov, A. P. Skoldinov and N. N. Zemlanskii, *Methods in Metalloorganic Chemistry, Antimony, Bismuth,* Science, Moscow, 1976, p. 483.
16. T. V. Talalaeva and K. A. Kochetkov, *Izv. Akad. Nauk SSSR, Otd. Khim. Nauk,* 126 (1953).
17. G. Wittig and A. Maercker, *J. Organomet. Chem.,* **8**, 491 (1967).
18. A. P. Batalov and G. A. Rostokin, *J. Gen. Chem.,* **43**, 963, 1569 (1973); **44**, 2019 (1974) (Russian).
19. A. P. Batalov, *J. Gen. Chem.,* **44**, 2534 (1974); **46**, 1514 (1976).
20. A. V. Severin and A. P. Batalov, *Radiochemistry,* **30**, 537 (1988) (Russian).
21. B. C. Gowenlock, *Quart. Rev.,* **14**, 133 (1960).
22. R. Martenson and P. Leighton, *J. Am. Chem. Soc.,* **56**, 2397 (1934).
23. C. Wu, F. Bochm and E. Nagel, *Phys. Rev.,* **91**, 319 (1953).
24. F. Rowland, *J. Am. Chem. Soc.,* **80**, 3165 (1958).
25. V. D. Nefedov, V. M. Zaitsev and M. A. Toropova, *Uspekhi Khim.,* **32**, 1367 (1963).
26. V. D. Nefedov and S. A. Grachev, *Radiochemistry (Radiokhimia),* **2**, 464 (1965) (*Russian*).
27. A. N. Nesmeyanov and K. A. Kochetkov, *Synthetic Methods in the Field of Metalloorganic Compounds,* USSR Academy, Moscow, 1945.
28. V. D. Nefedov, V. P. Bykhovcev, U. Czi-Lan and S. A. Grachev, *Radiochemistry,* **3**, 225 (1961) (*Russian*).
29. V. D. Nefedov and M. P. Beldy, *J. Phys. Chem.,* **31**, 986 (1957) (Russian).
30. V. D. Nefedov, L. Yuan-Fan, L. Van-chan and L. Yun-khua, *Acta Chimica Sinica,* **25**, 165 (1959).
31. V. D. Nefedov and V. I. Andreev, *J. Phys. Chem.,* **31**, 563 (1957) (Russian).
32. R. R. Edwards, J. M. Day and R. F. Overman, *J. Chem. Phys.,* **21**, 1555 (1953).
33. K. H. Lieser, *Einfuhrung in die Kernchemie,* Verlag Chemie GmbH, Weinheim, 1969.
34. E. Browne and R. B. Firestone, in *Table of Radioactive Isotopes* (Ed. V. S. Shirley), Wiley, Chichester, 1986.
35. E. L. Feinberg, *J. Phys.,* **4**, 423 (1941) (*Russian*).
36. R. A. Murdal, *J. Phys.,* **4**, 449 (1941) (Russian).
37. H. M. Schwartz, *J. Chem. Phys.,* **21**, 319 (1953).
38. C. S. Wu, F. Boehm and E. Nagel, *Phys. Rev.,* **91**, 319 (1953).
39. L. Cranberg, *Phys. Rev.,* **77**, 155 (1950).
40. P. E. Damona and R. R. Edwards, *Phys. Rev.,* **90**, 280 (1953).
41. D. Devault and W. Libby, *J. Am. Chem. Soc.,* **63**, 3216 (1941).
42. E. Cooper, *Phys. Rev.,* **61**, 1 (1942).
43. W. H. Burgus and J. W. Kennedy, *J. Chem. Phys.,* **18**, 97 (1950).
44. V. D. Nefedov, Yu. A. Riukhin and M. A. Toropova, *Radiochemistry,* **2**, 458 (1960).
45. See ref. 29, and V. D. Nefedov and M. A. Toropova. *Collection of Works on Radiochemistry,* Leningrad, 1955, p. 139.
46. A. N. Murin, V. D. Nefedov, V. M. Zaicev and S. A. Grachev, *Dokl. Akad. Sci. USSR,* **133**, 123 (1960).
47. V. D. Nefedov, M. A. Toropova, S. A. Grachev and Z. A. Grant, *J. Gen. Chem.,* **33**, 15 (1963).
48. V. D. Nefedov and S. A. Gluvka, *J. Gen. Chem.,* **33**, 333 (1963).

49. H. Gilman, H. Yablunky and A. Svigoon, *J. Am. Chem. Soc.*, **61**, 1170 (1939).
50. F. Challenger and A. Goddard, *J. Am. Chem. Soc.*, **42**, 762 (1920).
51. V. D. Nefedov, L. N. Petrov and V. V. Avrorin, *Radiochemistry*, **14**, 3, (1972).
52. (a) V. D. Nefedov, L. N. Petrov and V. V. Avrorin, *Vestn. Leningr. Univ.*, **4**, 396 (1969).
    (b) J. V. Supniewski and R. Adams, *J. Am. Chem. Soc.*, **48**, 4, 507 (1926).
53. V. D. Nefedov, M. Vobecki, E. N. Sinotova and I. Borak, *Radiochemistry*, **7**, 627 (1965).
54. V. D. Nefedov, M. Vobecki and I. Borak, *Radiochemistry*, **7**, 628 (1965).
55. V. D. Nefedov, I. S. Kirin, L. M. Gracheva and S. A. Grachev, *Radiochemistry*, **8**, 1 (1966).
56. J. Supniewski, *Roczniki Chemii.*, **6**, 97 (1926).
57. V. D. Nefedov, L. M. Gracheva, S. A. Grachev and L. N. Petrov, *Radiochemistry*, **7**, 741 (1965).
58. V. D. Nefedov, S. A. Grachev, L. M. Gracheva and L. M. Petrov, *Radiochemistry*, **8**, 376 (1966).
59. V. I. Zhuravlev and N. F. Antipina, *Radiochemistry*, **9**, 6 (1967).
60. K. Lederer, *Chem. Ber.*, **49**, 345 (1916).
61. L. M. Gracheva, V. D. Nefedov and S. A. Grachev, *Radiochemistry*, **9**, 738 (1967).
62. A. N. Nesmeyanov and K. A. Kochetkov, *Synthetic Methods in the Domain of Bismuth and Antimony Metalloorganic Compounds*, 8th ed., Akad. Nauk SSSR, Moscow–Leningrad (St. Petersburg), 1947.
63. L. E. Clendemn, *Nukleonics*, **2**, 1, 12 (1948).
64. D. G. Martin and H. O. W. Richardson, *Proc. R. Soc., London*, **195**, 1042 (1948).
65. C. D. Ellis, *Proc. R. Soc., London*, **138**, A835, 318 (1932).
66. A. Flammersfald, *Z. Phys.*, **114**, 227 (1939).
67. Reference 34, pp. 207–5, 211–5.
68. A. H. W. Aten, Jr., I. Heertje and P. Polak, *J. Inorg. Nucl. Chem.*, **14**, 132 (1960).
69. R. Ackerhalt, P. Ellerbe and G. Harbottle, *Radiochim. Acta.*, **18**, 73 (1972).
70. P. Zhang, H. Qian, Y. Wu, X. Wany, Y. Huang, Z. Xue and Y. Liu, *Radiochim. Acta.*, **34**, 169 (1983).
71. A. G. Maddock, in *Chemical Effects of Nuclear Transformation in Inorganic Systems* (Eds. G. Harbottle and A. G. Maddock), Chap. 21, North-Holland, Amsterdam, 1979, p. 386.
72. W. Li and H. Ma, *Science and Technology of Atomic Energy* (P. R. China), **6**, 401 (1966).
73. V. P. Krasavin, E. E. Grinberg and A. A. Efimov, *Tr. IREA*, **46**, 38 (1984); *Chem. Abstr.*, **103**, 1961762 (1985).
74. K. Tanaka and H. Fukumoto, *Chem. Abstr.*, **105**, 235, 8286 (1986); **106**, 129290f (1987); two Japanese patents: Jpn. Kokai Tokkyo Koho Jp61, 120, 168 [86, 120, 168], 6 pp, and Jpn. Kokai Tokkyo Koho Jp 61, 160, 759 [86, 160, 759], 10 pp.
75. W. F. Bale, I. L. Spar and K. L. Goodland, *Cancer. Res.*, **20**, 1488 (1960).
76. W. D. Bloomer, W. H. McLaughlin, R. D. Neirinckx, S. J. Adelstein, P. R. Gordon, T. J. Ruth and A. P. Wolf, *Science*, **212**, 340 (1981).
77. M. K. Rosenow, G. L. Zucchini, P. M. Bridwell, F. P. Stuart and A. M. Friedman, *Int. J. Nucl. Med. Biol.*, **10**, 189 (1983).
78. Reference 33, pp. 107 and 141.
79. G. L. Zucchini and A. M. Friedman, *Int. J. Nucl. Med. Biol.*, **9**, 83 (1982).
80. R. W. Atcher, A. M. Friedman and J. J. Hines, *Appl. Radiat. Isot.*, **39**, 283 (1988).
81. M. C. Lagunas-Solar, O. F. Carvacho, L. Nagahara, A. Mishra and N. J. Parks, *Appl. Radiat. Isot.*, **38**, 129 (1987).
82. C. Birattari, M. Bonardi and M. C. Gilardi, *Radiochem. Radioanal. Lett.*, **49**, 25 (1981).
83. R. W. Kozak, R. W. Atcher, O. A. Gansow, A. M. Friedman, J. J. Hines and T. A. Waldman, *Proc. Natl. Acad. Sci. USA*, **83**, 474 (1986).
84. G. Kohler and C. Milstein, *Nature (London)*, **256**, 495 (1975).
85. K. Nakamoto, Y. Morimoto and A. E. Martell, *J. Am. Chem. Soc.*, **85**, 309 (1963).
86. G. E. K. Krejcarek and K. L. Tucker, *Biochem. Biophys., Res. Commun.*, **77**, 581 (1977).
87. (a) W. J. Leonard, J. M Depper, T. Uchiyama, K. A. Smith, T. A. Waldman and W. C. Green, *Nature (London)*, **300**, 267 (1982).
    (b) Reference 85 and in *Aldrich-Chemie*, GmbH and Co., W-7924 Steinheim, 1992, p. 457.
88. M. Krönhe, I. M. Depper, W. J. Leonard, E. S. Witteta, T. A. Waldman and W. C. Greene, *Blood*, **85**, 1416 (1985).
89. A. T. M. Vaughan, W. J. Bateman and D. R. Fisher, *J. Radiat. Oncol. Biol. Phys.*, **8**, 1943 (1982).
90. R. W. Kozak, D. P. Fitzgerald, R. W. Atcher, C. K. Goldman, D. L. Nelson, O. A. Gansow, I. Pastan and T. A. Waldman, *J. Immunol.*, **144**, 3417 (1990).

91. J. M. Depper, W. J. Leonard, R. J. Robb, T. A. Waldmann and W. C. Greene, *J. Immunol.*, **131**, 690 (1983).

92. C. A. Szpak, W. W. Johnston, S. C. Lottich, D. Kufe, A. Thor and J. Schlom, *Acta Cytol.*, **29**, 356 (1984).

93. R. B. Simonson, M. E. Ultee, Jo. A. Hauler and V. L. Alvarez, *Cancer Res.*, **50**, 985 (1990).

94. J. D. Rodwell, V. L. Alvarez, C. Lee, A. D. Lopes, J. W. F. Goers, H. D. King, H. J. Powsner and T. J. McKearn, *Proc. Natl. Acad. Sci. USA*, **83** 2632 (1986); *Chem. Abstr.*, **105**, 21060w (1986).

95. V. L. Alvarez, M. L. Wen, C. Lee, A. D. Lopes, J. D. Rodwell and T. J. McKearn, *Nucl. Med. Biol.*, **13**, 347 (1986).

96. J. W. Goers, C. Lee, R. C. Siegel, T. J. McKearn, H. D. King, D. J. Coughlin and J. D. Rodwell, Eur. Pat. Appl. EP. 175, 617; *Chem. Abstr.*, **105**, 102583w (1986).

97. R. M. Macklis, B. M. Kinsey and A. I. Kassis, *Science*, **240**, 1024 (1988).

98. A. T. M. Vaughan, W. J. Bateman, G. Brown and J. Cowan, *Int. J. Nucl. Med. Biol.*, **9**, 167 (1982).

99. C. D. V. Black, R. W. Atcher and J. Barbet, *Antibody, Immunoconj. and Radiopharm.*, **1**, 43 (1988).

100. V. K. Langmuir, R. W. Atcher, J. J. Hines and M. W. Brechbiel, *J. Nucl. Med.*, **31**, 1527 (1990).

101. D. A. Goodwin, C. F. Meares, M. J. McCall, M. McTgue and W. Chaovapong, *J. Nucl. Med.*, **29**, 226 (1988).

102. J. M. Le Doussal, M. Martin, E. Gautherot, M. Delaage and J. Barbet, *J. Nucl. Med.*, **30**, 1358 (1989).

103. M. W. Brechbiel, O. A. Gansow, R. W. Atcher, J. Schlom, J. Esteban, D. E. Simpson and D. Colcher, *Inorg. Chem.*, **25**, 2772 (1986).

104. T. Okabe, T. Kaizu and M. Fujisawa, *Cancer Res.*, **44**, 5273 (1984).

105. D. J. Hnatowich, M. W. Layne and R. L. Childs, *Int. J. Appl. Radiat. Isot.*, **33**, 327 (1982).

106. T. Tsuneyoshi, M. Yamamoto and K. Otozai, *Radiochim. Acta*, **33**, 135 (1983).

107. I. Zvara, Y. T. Chuburkov, R. Calerka, T. S. Zvarova, M. R. Shalaevsky and B. V. Shilov, *At. Energy (USSR)*, **21**, 83 (1966).

108. P. Hoffmann, K. Bachmann, H. Klenk and K. H. Lieser, *Inorg. Nucl. Chem. Lett.*, **7**, 577 (1971).

109. K. Otozai, T. Tsuneyoshi and M. Yamamoto, *OULNS Ann. Rep.*, **81–3**, 128 (1980).

110. V. D. Nefedov, I. S. Kirin and V. M. Zaicev, *Radiochemistry*, **4**, 351 (1962).

111. V. D. Nefedov, M. A. Toropova, V. V. Avrorin, N. I. Shchepina and V. K. Vasil'ev. *Radiochemistry*, **18**, 305 (1976).

112. E. L. Berman, Z. N. Nysenko, A. M. Sakharov, A. B. Rabovskii and V. A. Ponomarenko, *Vysokomol. Soedin, Ser. B*, **26**, 302 (1984).

113. H. Meerwein, E. Battenberg, H. Cold, E. Pfeil and G. Willand, *J. Prakt. Chem.*, **B**, **154**, 83 (1939).

114. J. C. Irvine and J. W. H. Oldham, *J. Chem. Soc.*, **127**, 2903 (1925).

115. D. Vofsi and A. V. Tobolsky, *J. Polymer Sci. A*, **3**, 3261 (1965).

116. Z. N. Nysenko, E. L. Berman, E. B. Ludvig, A. P. Klimov, V. A. Ponomarenko and G. V. Isagulyanc, *Vysokomol. Soedin., Ser. A*, **18**, 1696 (1976).

117. (a) K. O. Christe, P. Charpin, E. Soulier, R. Bougon, J. Fawcett and D. R. Russell, *Inorg. Chem.*, **23**, 3756 (1984).
    (b) K. O. Christe, C. J. Schack and R. D. Wilson, *Inorg. Chem.*, **14**, 2224 (1975).
    (c) J. P. Masson, J. P. Desmoulin, P. Charpin and R. Bougon, *Inorg. Chem.*, **15**, 2529 (1976).
    (d) K. O. Christe, W. W. Wilson and C. J. Schack, *J. Fluorine Chemistry*, **11**, 71 (1978).
    (e) S. Cohen, H. Selig and R. Gut, *J. Fluorine Chem.*, **20**, 349 (1982).

118. R. Bougon, P. Charpin, K. O. Christe, J. Isabey, M. Lance, M. Nierlich, J. Vigner and W. W. Wilson, *Inorg. Chem.*, **27**, 1389 (1988).

119. K. O. Christe, W. W. Wilson, R. Bougon and P. Charpin, *J. Fluorine Chem.*, **34**, 287 (1987).

120. E. L. Pace and L. W. Noe, *J. Chem. Phys.*, **49**, 5317 (1968).

121. H. M. Hoffman, M. Dràger, B. M. Schmidt and N. Kleiner, *Spectrochim. Acta*, **42A**, 1255 (1986).

122. H. H. Jaffe and G. O. Doak, *J. Am. Chem. Soc.*, **72**, 3025 (1950).

123. D. Issleib and A. Balszuweit, *Z. Anorg. Allg. Chem.*, **418**, 158 (1975).

124. H. M. Hoffman and M. Dràger, *J. Organomet. Chem.*, **295**, 33 (1985).

125. V. D. Nefedov, E. N. Sinotova and V. D. Trenin, *Radiochemistry*, **2**, 739 (1960) (Russian).

126. V. D. Nefedov and G. A. Skorobogatov, *Radiochemistry*, **3**, 229 (1961).

127. V. D. Nefedov, I. S. Kirin, V. M. Zaicev, G. A. Semenov and B. E. Dzevicki, *J. Gen. Chem.*, **33**, 2407 (1963) (Russian).

128. G. T. Morgan and V. E. Yarsley, *J. Chem. Soc.*, **127**, 184 (1925).

129. C. A. Larsen, R. W. Gedridge Jr., H. S. Li and H. Gerald, Mat. Res. Soc. Symp. Proc., 1991, pp.

129–134, Utah, Salt Lake City, UT 84112, USA; *Chem. Abstr.*, **116**, 6660a (1992).
130. V. Lucchini, G. Modena and L. Pasquato, *J. Am. Chem. Soc.*, **110**, 6900 (1988).
131. V. Lucchini, G. Modena and L. Pasquato, *J. Am. Chem. Soc.*, **113**, 6600 (1991).
132. R. Weiss and C. Schlierf. *Synthesis*, 323 (1976).
133. J. C. Nash, *Compact Numerical Methods for Computers*, Adam Hilger, Bristol, 1979, p. 141.
134. T. Ando, H. Yamataka, H. Morisaki, J. Yamawaki, J. Kuramochi and Y. Yukawa, *J. Am. Chem. Soc.*, **103**, 430 (1981).
135. G. Smith, D. J. Cole-Hamilton, M. Thornton-Pett and M. B. Hursthouse, *J. Chem. Soc.*, Dalton Trans., 2501 (1983).
136. G. Smith and D. J. Cole-Hamilton, *J. Chem. Soc., Chem. Commun.*, 490 (1982).
137. G. Smith, D. J. Cole-Hamilton, A. C. Gregory and N. G. Gooden, *Polyhedron*, **1**, 97 (1982).
138. B. D. Dombek, *J. Am. Chem. Soc.*, **103**, 3959 (1981).
139. J. A. Gladysz, *Adv. Organomet. Chem.*, **20**, 1 (1982).
140. W. Tam, G. Y. Lin and J. A. Gladysz, *Organometallics*, **1**, 525 (1982).
141. W. Tam, G. Y. Lin, W. K. Wong, W. A. Kiel, V. K. Wong and J. A. Gladysz, *J. Am. Chem. Soc.*, **104**, 114 (1982).
142. W. Tam, M. Marsi and J. A. Gladysz, *Inorg. Chem.*, **22**, 1413 (1983).
143. G. Smith and D. J. Cole-Hamilton, *J. Chem. Soc., Dalton Trans.*, 1203 (1984).
144. D. S. Barratt and D. J. Cole-Hamilton, *J. Chem. Soc., Dalton Trans.*, 2683 (1987).
145. W. R. Roper and L. J. Wright, *J. Organomet. Chem.*, **234**, 5 (1982).
146. M. Zieliński, *Isotope Effects in Chemistry*, Polish Sci. Publ., Warsaw, 1979, pp. 102–111.
147. G. Smith, L. H. Sutcliffe and D. J. Cole-Hamilton, *J. Chem. Soc., Dalton Trans.*, 1209 (1984).
148. M. Zieliński and H. Papiernik–Zielińska, *J. Radioanal. Nucl. Chem.*, **162**, 25 (1992).
149. M. Zieliński, G. Czarnota and H. Papiernik–Zielińska, *J. Radioanal. Nucl. Chem.*, **159**, 305 (1992); lecture presented during the IXth National Meeting of the Polish Society for Radiation and Nuclear Research, Cracow, April 1992.
150. M. Zieliński, H. Papiernik-Zielińska and G. Czarnota, Annual Meeting of The Polish Chem. Soc., Bialystok, September 9–12, 1992, poster S–3 P–21, p. 110; full paper submitted for publication in *Nukleonika*, **37**, 71 (1992).
151. M. Zieliński and M. Kańska in *The Chemistry of Acid Derivatives, Volume 2* (Eds. S. Patai and Z. Rappoport), Chap. 10, Wiley, Chichester, 1992.
152. E. Lindner and H. Norz, *Chem. Ber.*, **123**, 459 (1990).
153. I. Wender, *Catal. Rev. Sci. Eng.*, **26**, 303 (1984).
154. D. Forster, *J. Am. Chem. Soc.*, **98**, 846 (1976).
155. K. Ohkata, M. Ohnishi, K. Yoshinaga, K. Akiba, J. C. Rongione and J. C. Martin, *J. Am. Chem. Soc.*, **113**, 9270 (1991).
156. K. Akiba, K. Takee, Y. Shimizu and K. Ohkata, *J. Am. Chem. Soc.*, **108**, 6320 (1986).
157. D. Gudat, L. M. Damels and J. G. Verkade, *J. Am. Chem. Soc.*, **111**, 8520 (1989).
158. C. W. Perkins, S. R. Wilson and J. C. Martin, *J. Am. Chem. Soc.*, **107**, 3209 (1985).
159. I. K. Korobeinicheva, O. M. Fugaeva and G. G. Furin, *J. Fluorine Chem.*, **39**, 373 (1988).
160. Y. Yamamoto, Y. Nisigaki, M. Umezu and T. Matsura, *Tetrahedron*, **36**, 3177 (1980).
161. K. A. Azam, R. H. Hill and R. J. Puddephatt, *Can. J. Chem.*, **62**, 2029 (1984).
162. R. H. Hill, P. De Mayo and R. J. Puddephatt, *Inorg. Chem.*, **21** 3642 (1982).
163. D. M. Hoffman and R. Hoffmann, *Inorg. Chem.*, **20**, 3543 (1981).
164. K. A. Ayam, M. P. Brown, S. J. Cooper and R. J. Puddephatt, *Organomettallics*, **1**, 1183 (1982).
165. A. J. Ashe, III, *J. Am. Chem. Soc.*, **93**, 3293 (1971).
166. R. L. Kuczkowski and A. J. Ashe, III, *J. Mol. Struct.*, **42**, 457 (1972).
167. T. C. Wong and L. S. Bartell, *J. Chem. Phys.*, **61**, 2840 (1974).
168. A. J. Ashe, III and W. T. Chan, *Tetrahedron Lett.*, 2749 (1975).
169. T. C. Wong, A. J. Ashe, III and L. S. Bartell, *J. Mol. Struct.*, **25**, 65 (1975).
170. R. Lattimer, R. L. Kuczkowski, A. J. Ashe. III and A. L. Meinzer, *J. Mol. Spectrosl.*, **57**, 428 (1975).
171. A. Schweig and H. Oehling, *Tetrahedron Lett.*, 4941 (1970).
172. A. Schweig, H. L. Hase, J. Radloff and H. Hahn, *Tetrahedron*, **29**, 475 (1973).
173. D. T. Clark and I. W. Scanlan, *J. Chem. Soc., Faraday Trans.*, **70**, 1222 (1974).
174. A. J. Ashe, III, *Acc. Chem. Res.*, **11**, 153 (1987).
175. T. C. Wong and I. S. Bartell, *J. Mol. Struct.*, **44**, 169 (1978).
176. A. J. Ashe, III, R. R. Sharp and J. W. Tolan, *J. Am. Chem. Soc.*, **98**, 5451 (1976).

177. A. J. Ashe, III and M. D. Gordon, *J. Am. Chem. Soc.*, **94**, 7596 (1972).
178 A. J. Ashe, III, *Tetrahedron Lett.*, 415 (1976).
179. A. J. Ashe, III, *J. Am. Chem. Soc.*, **93**, 6690 (1971).
180. G. D. Fong, R. L. Kuczkowski and A. J. Ashe, *J. Mol Spectrosc.*, **70**, 197 (1978).
181. E. Sabbioni and F. Girardi, *Total Environ.*, **7**, 145, (1977).
182. M. L. Thakur, *Int. J. Appl. Radiat. Isot.*, **28**, 183 (1977).
183. D. Basile, C. Birattari, M. Bonardi, L. Goetz, E. Sabbioni and A. Salomone, *Int. J. Appl. Radiat. Isot.*, **32**, 403 (1981).
184. S. H. Al-Kouraishi and G. G. J. Boswell, *Int. J. Applied Radiat. Isot.*, **29**, 607 (1978).
185. C. H. Banks, J. R. Daniel and R. A. Zingaro, *J. Med. Chem.*, **22**, 572 (1979).
186. E. A. Crecelius, *Environ. Health Perspect.*, **19**, 147 (1977).
187. J. R. Cannon, J. S. Edmonds, K. A. Francesconi, C. L. Raston, J. B. Saunders, B. W. Skelton and A. H. White., *Aust. J. Chem.*, **34**, 787 (1981).
188. R. P. Spencer, L. A. Spitznagle, P. Hosain and F. Hosain, *Biophys. J.*, **25**, 279a (1979).
189. A. M. Emran, N. M. Shanbaky and R. P. Spencer, *Appl. Radiat. Isot.*, **17**, 545 (1986).
190. H. Weiland and W. Rheinheimer, *Ann. Chem.*, **423**, 1 (1920).
191. F. Hosain, A. Emran, R. P. Spencer and K. S. Clampitt, *Int. J. Appl. Radiat. Isot.*, **33**, 1477 (1982).
192. J. R. Daniel and R. A. Zingaro, *Phosphorus and Sulphur*, **4**, 179 (1978).
193. C. H. Banks, J. R. Daniel and R. A. Zingaro, *J. Med. Chem.*, **22**, 572 (1979).
194. A. Emran, F. Hosain, R. P. Spencer and K. S. Kolstad, *Int. J. Nucl. Med. Biol.*, **11**, 259 (1984).
195. J. S. Edmonds, K. A. Francesconi, J. R. Cannon, C. L. Raston, B. W. Skelton and A. H. White, *Tetrahedron Lett.*, 1543 (1977).
196. I. C. Munro, S. M. Charbonneau, E. Sandi, K. Spencer, F. Bryce and H. C. Grice, *Toxicol. Appl. Pharmacol.*, **24**, 111 (1974).
197. S. M. Charbonneau, K. Spencer, F. Bryce and E. Sandi, *Bull. Environ. Contam. Toxicol.*, **20**, 470 (1978).
198. H. C. Freeman, J. R. Uthe, R. B. Fleming, P. H. Odense, R. G. Ackman, G. Landry and C. Musial, *Bull. Environ. Contam. Toxicol.*, **22**, 224 (1979).
199. L. Goetz and H. Norin, *Int. J. Radiat. Isot.*, **34**, 1509 (1983).
200. T. C. Siewicki and J. S. Sydlowski, *Nutr. Rep. Int.*, **24**, 121 (1981).
201. J. Stray, A. Zeman and K. Kratzer, *Radiochem. Radioanal. Lett.*, **52**, 263 (1982).
202. *Beilsteins Handbuch der organischen Chemie.*, EIII 4, Springer-Verlag, Berlin, 1963, pp. 1974–1975.
203. M. O. Andreae, *Limnol. Oceanogr.*, **24**, 440 (1979).
204. M. O. Andreae and D. Klumpp. *Environ. Sci. Technol.*, **13**, 783 (1979).
205. J. U. Làkso, L. L. Rose, S. A. Peoples and D. Y. Shirachi, *J. Agric. Food Chem.*, **27**, 1229 (1979).
206. (a) J. G. Sanders and H. L. Windom, *Estuarine Coastal Mar. Sci.*, **10**, 555 (1980).
     (b) S. Tagawa, *Bunseki Kagaku*, **29**, 563 (1980).
207. P. Hosain, P. K. Sripada, R. P. Spencer and F. Hosain, *Int. J. Nucl. Med. Biol.*, **8**, 209 (1981).
208. (a) D. Webster, M. J. Sparkes and H. B. F. Dixon, *Biochem. J.*, **169**, 239 (1978).
     (b) F. Oesek, A. J. Sparrow and K. L. Platt, *J. Labelled Compd. Radiopharm.*, **20**, 1297 (1983).
209. D. Hamer and R. G. Leckey, *J. Chem. Soc.*, **I**, 1398 (1961).
210. (a) P. Hosain, R. P. Spencer, F. Hosan and P. K. Sripada, *Int. J. Nucl. Med. Biol.*, **7**, 51 (1980).
     (b) M. Castiglioni and P. Volpe, *Radiochim. Acta*, **34**, 165 (1983).
211. (a) M. A. El-Sayed, P. J. Estrup and R. Wolfgang, *J. Phys. Chem.*, **62**, 1356 (1958).
     (b) S. Mohan, K. G. Ravikumar and S. Gunasekaran, *Indian J. Pure Appl. Phys.*, **21**, 311 (1983).
212. (a) G. Cetini, O. Gambino, M. Castiglioni and P. Volpe, *J. Chem. Phys.*, **46**, 89(1967).
     (b) J. C. Decius, *J. Chem. Phys.*, **38**, 241 (1963).
213. (a) M. Castiglioni and P. Volpe, *Polyhedron*, **2**, 225 (1983).
     (b) L. H. Jones, *Inorganic Vibrational Spectroscopy*, Vol. 1, Marcel Dekker, New York, 1971.
214. (a) E. Tachikawa and Y. Aratono, *J. Inorg. Nucl. Chem.*, **38**, 193 (1975).
     (b) F. P. Reading and G. F. Horning, *J. Chem. Phys.*, **23**, 1053 (1955).
215. (a) J. L. Beauchamp, *Ann. Rev. Phys. Chem.*, **22**, 527 (1971).
     (b) F. Cleveland, S. Sundaram and C. Thiagarajan, *J. Mol. Spectrosc.*, **5**, 307 (1960).
216. (a) M. Castiglioni and P. Volpe, *Gazz. Chim. Ital.*, **105**, 247 (1975).
     (b) B. N. Laskorin, V. V. Yakshin and N. A. Lyubosvetova, *Dokl. Akad. Nauk USSR*, **249**, 651 (1979).
217. B. N. Laskorin, V. V. Yakshin and N. A. Lyubosvetova, *J. Gen. Chem. (USSR)*, **47**, 1118 (1978).

218. V. V. Yakshin, N. A. Lyubosvetova, N. L. Khokholova, V. P. Kazakov and D. D. Afonishev, *Radiochemistry* (*USSR*), **25**, 628 (1983); *Chem. Abstr.*, **100**, 28876x (1984).
219. (a) L. Andrews, R., Withnall and B. W. Moores, *J. Phys. Chem.*, **93**, 1279 (1989).
     (b) M. W. Schmidt, S. Yabushita and M. S. Gordon, *J. Phys. Chem.*, **88**, 382 (1984).
     (c) W. B. Person, J. S. Kwiatkowski and R. T. Bartlett, *J. Mol. Struct.*, **157**, 237 (1987).
220. R. T. Arlinghaus and L. Andrews, *J. Chem. Phys.*, **81**, 4341 (1984).
221. R. Withnall and L. Andrews, *J. Phys. Chem.*, **91**, 784 (1987).
222. J. O' Keefe, B. Domenges and G. V. Gibbs, *J. Phys. Chem.*, **89**, 2304 (1985).
223. A. J. Ashe, III, W. T. Chang, T. W. Smith and K. M. Taba, *J. Org. Chem.*, **46**, 881 (1981).
224. A. J. Ashe, III, R. R. Sharp and J. W. Tolan, *J. Am. Chem. Soc.*, **98**, 5451 (1976).
225. A. J. Ashe, III, S. Mahmoud, C. Elschenbroich and M, Wünsch, *Angew. Chem.*, **99**, 249 (1987).
226. K. E. Richards, A. L. Wilkinson and G. J. Wright, *Aust. J. Chem.*, **25**, 2369 (1972).
227. J. Z. Weidlein, *Z. Anorg. Allg. Chem.*, **435**, 179 (1977).
228. H. D. Hausen, H. J. Guder and W. J. Schwarz, *Organomet. Chem.*, **132**, 37 (1977).
229. A. Haaland and J. Weidlein, *Acta Chem. Scand.*, **A36**, 805 (1982).
230. R. L. Hilderbrandt and J. D. Wieser, *J. Chem. Phys.*, **55**, 4648 (1971).
231. J. Ellerman, L. Brehm and E. Köck, *J. Organomet. Chem.*, **336**, 323 (1987).
232. J. Ellerman, H. Schössner, H. Haagand and H. Schödel, *J. Organomet. chem.*, **65**, 33 (1974).
233. K. Sommer and A. Rothe, *Z. Anorg. Allg. Chem.*, **378**, 303 (1970).
234. X. Lu and L. Lu, *J. Organomet. Chem.*, **307**, 285 (1986).
235. Yu. F. Gatilov, L. B. Ionov, S. S. Molodsov and V. P. Kovyrzina, *Zh. Obshch. Khim.*, **42**, 1959 (1972).
236. X. Lu, J. Huang and J. Zhu, *Acta Chim. Sinica,* **43**, 702 (1985).
237. H. Adams, N. A. Bailey, B. E. Mann, C. P. Manuel, C. M. Spencer and A. G. Kent., *J. Chem. Soc.*, Dalton T rans., 489 (1988).
238. D. Forster, *Adv. Organomet. Chem.*, **17**, 255 (1979).
239. P. A. W. Dean, J. J. Vittal and R. S. Srivastava, *Can. J. Chem.*, **65**, 2628 (1987).
240. A. F. M. J. Van der Ploeg and G. Van Koten, *Inorg. Chem. Acta,* **51**, 225 (1981).
241. D. M. Ho and R. Bau, *Inorg. Chem.*, **22**, 4073 (1983).
242. W. Bensch, M. Prelati and W. Ludwig, *J. Chem. Soc., Chem. Commun.*, 1762 (1986).
243. J. Baldas and S. F. Colmanet, *Aust. J. Chem.*, **42**, 1155 (1989).
244. A. Davison, H. S. Trop, B. V. DePamphilis and A. G. Jones, *Inorg. Synth.*, **21**, 160 (1982).
245. F. A. Cotton, A. Davison, V. W. Day, L. D. Gage and H. S. Trop, *Inorg. Chem.*, **18**, 3024 (1979).
246. F. A. Phillips and A. C. Skapski, *J. Cryst. Mol. Struct.*, **5**, 83 (1975).
247. *Radiopharmaceuticals II, Proc. 2nd Int. Symp. Radiopharmaceuticals* (Eds. V. J. Sodd, D. R. Allen, D. R. Hoogland and R. D. Ice), The Society for Nuclear Medicine, Inc., New York, 1979.
248. (a) M. J. Clarke and L. Podbielski. *Coord. Chem. Rev.*, **78**, 253 (1987).
     (b) I. De Gregori and S. Lobos, *Int. J. Appl. Radiat. Isot.*, **40**, 385 (1989).
249. B. V. De Pamphilis, A. G. Jones and M. A. Davis, *J. Am. Chem. Soc.*, **100**, 5570 (1978).
250. A. Davison and G. J. Lund, *Int. J. Appl. Radiat. Isot.*, **33**, 875 (1982).
251. E. F. Byrne and J. E. Smith, *Inorg. Chem.*, **18**, 1832 (1979).
252. G. Mulder and S. J. Oldenburg, *Int. J. Appl. Radiat. Isot.*, **32**, 675 (1981).
253. S. F. Colmanet and M. F. Mackay, *Aust. J. Chem.*, **40**, 1301 (1987).
254. A. Davison, C. Orvig, H. S. Trop. M. Sohn, B. V. De Pamphilis and A. G. Jones., *Inorg. Chem.*, **19**, 1988 (1980).
255. J. E. Smith, E. F. Bryne, F. A. Cotton and J. C. Sekutowski, *J. Am. Chem. Soc.*, **100**, 5571 (1978).
256. B. V. De Pamphilis, A. G. Jones, M. A. Davis and A. Davison, *J. Am. Chem. Soc.*, **100**, 5570 (1978).
257. U. Müller, *Acta. Crystallogr., Sect. B.*, **36**, 1075 (1980).
258. J. L. Martin and J. Takats, *Inorg. Chem.*, **14**, 1358 (1975).
259. A. Davison. N. De Vries, J. Devan and A. Jones, *Inorg. Chim. Acta*, **120**, 115 (1986).
260. (a) N. de Vries, A. G. Jones and A. Davison, *Inorg. Chem.*, **28**, 3728 (1989).
     (b) J. Berg and R. H. Holm, *J. Am. Chem. Soc.*, **107**, 917, 925 (1985).
     (c) R. H. Holm, *Chem. Rev.*, **87**, 1401 (1987).
261. N. de Vries, J. C. Dewan, A. G. Jones and A. Davison, *Inorg. Chem.*, **27**, 1574 (1988).
262. S. F. Colmanet and M. F. Mackay, *Aust. J. Chem.*, **41**, 151 (1988).
263. S. F. Colmanet and M. F. Mackay, *Aust. J. Chem.*, **41**, 1127 (1988).
264. G. F. Brown and E. I. Stiefel, *Inorg. Chem.*, **12**, 2140 (1973).
265. S. F. Colmanet, G. A. Williams and M. F. Mackay, *J. Chem. Soc., Dalton Trans.*, 2305 (1987).

266. J. Baldas, J. Bonnyman and G. A. Williams, *Inorg. Chem.*, **25**, 150 (1986).
267. A. Duatti, A. Marchi, S. A. Luna, G. Bandoli, U. Mazzi and F. Tisato, *Chem. Soc., Dalton Trans.*, 867 (1987).
268. (a) J. Baldas, S. F. Colmanet and M. F. Mackay, *J. Chem. Soc., Dalton Trans.*, 1725 (1988).
     (b) G. A. Williams, J. Bonnyman and J. Baldas, *Aust. J. Chem.*, **40**, 27 (1987).
269. J. F. Eary, R. W. Schroff, P. G. Abrams, A. R. Fritzberg, A. C. Morgan, S. Kasina, J. M. Reno, A. Srinivasan, C. S. Woodhouse, S. D. Wilbur, R. B. Natale, C. Collins, J. S. Stehlin, M. Mitchell and W. B. Nelp, *J. Nucl. Med.*, **30**, 25 (1989).
270. T. N. Rao, D. Adhikesavalu, A. Camerman and A. R. Fritzberg, *J. Am. Chem. Soc.*, **112**, 5798 (1990).
271. J. F. Vanderheyden, A. R. Fritzberg, T. N. Rao, S. Kasina, A. Srinivasan, J. M., Reno and A. C. Morgan, *J. Nucl. Med.*, **28**, 656 (1987).
272. A. R. Fritzberg, J. L. Vanderheyden, T. N. Rao, S. Kasina, D. Eshima and A. T. Taylor, *J. Nucl. Med.*, **30**, 60 (1989).
273. T. N. Rao, D. I. Brixner, A. Srinivasan, S. Kasina, J. L. Vanderheyden, D. W. Webster and A. R. Fritzberg, *Appl. Radiat. Isot.*, **42**, 525 (1991).
274. N. J. Bryson, D. Brenner, J. Lister-James, A. G. Jones, J. C. Dewan and A. Davison, *Inorg. Chem.*, **28**, 3825 (1989).
275. N. Bryson, J. C. Dewan, J. Lister-James, A. G. Jones and A. Davison, *Inorg. Chem.*, **27**, 2154 (1988).
276. (a) A. Davison, A. G. Jones, C. Orvig and M. Sohn, *Inorg. Chem.*, **20**, 1629 (1981).
     (b) A. Duatti, A. Marchi, R. Rossi, L. Magon, E. Deutsch, V. Bertolasi and F. Belluci, *Inorg. Chem.*, **27**, 4208 (1988).
277. K. H. Grellmann and E. Tauer, *J. Am. Chem. Soc.*, **95**, 3104 (1973).
278. C. A. Rice, C. G. Benson, C. A. McAuliffe and W. E. Hill, *Inorg. Chim. Acta*, **59**, 33 (1982).
279. A. Müller and W. Rittner. *Spectrochim. Acta*, **23A**, 1831 (1967).
280. V. I. Spicyn, V. P. Tarasov, K. E. German, S. A. Petrushin, A. F. Kuzina and S. V. Kriuchkov, *Dokl. Akad. Nauk SSSR*, **290**, 1411 (1986).
281. V. P. Tarasov, V. I. Privalov and S. A. Petrushin, *Dokl. Akad. Nauk SSSR*, **272**, 919 (1983).
282. V. P. Tarasov, S. A. Petrushin, V. I. Privalov, K. E. German, S. V. Kryuchkov and Yu. A. Buslaev, *Koord. Khim. (USSR)*, **12**, 1227 (1986); *Chem. Abstr.*, **106**, 42706x (1987).
283. V. P. Tarasov, V. I. Privalov and Yu. A. Buslaev, *Dokl. Akad. Nauk (USSR)* **262**, 1433 (1982).
284. J. Baldas, J. F. Boas, J. Bonnyman and G. A. Williams, *J. Chem. Soc., Dalton Trans.*, 2395 (1984).
285. J. Baldas, J. Bonnyman and G. A. Williams, *Inorg. Chem.*, **25**, 150 (1986).
286. K. Dehnicke and J. Strähle, *Angew. Chem., Int. Ed. Engl.*, **20**, 413 (1981).
287. W. Liese, K. Dehnicke, R. D. Rogers, R. Shakir and J. L. Atwood, *J. Chem. Soc., Dalton Trans.*, 1061 (1981).
288. T. Lis and B. Jesowska-Trzebiatowska, *Acta Crystallogr., Sect. B*, **33**, 1248 (1977).
289. J. Baldas, J. Bonnyman and G. A. Williams, *Aust. J. Chem.*, **38**, 215 (1985).
290. J. Baldas, J. F. Boas, J. Bonnyman and G. A. Williams, *J. Chem. Soc., Dalton Treans.*, 2395 (1984).
291. U. Abram and R. Kirmse, *J. Radioanal. Nucl. Chem., Articles*, **122**, 311 (1988).
292. R. Kirmse, K. Köhler, U. Abram, R. Böttcher, L. Golic and E. De Boer, *Chem. Phys.*, **143**, 83 (1990).
293. K. Kohler, R. Kirmse, R. Bottcher, U. Abram, M. C. M. Gribnau, C. P. Keijzers and E. De Boer, *Chem. Phys.*, **143**, 83 (1990).
294. J. Baldas and S. F. Colmanet, *J. Chem. Soc., Dalton Trans.*, 1725 (1988).
295. S. F. Colmanet, G. A. Williams and M. F. Mackay, *J. Chem. Soc., Dalton Trans.*, 2305 (1987).
296. H. Küppers, *Acta Crystallogr., Sect. B*, **29**, 318 (1973).
297. G. A. Williams and J. Baldas, *Aust. J. Chem.*, **42**, 875 (1989).
298. J. Baldas, J. Bonnyman and G. A. Williams, *Inorg. Chem.*, **25**, 150 (1986).
299. (a) S. L. Colmanet and M. F. Mackay, *Inorg. Chim. Acta*, **147**, 173 (1988).
     (b) L. Kaden, B. Lorenz, K. Schmidt, H. Sprinz and M. Wahren, *Isotopenpraxis*, **7**, 174 (1981).
300. J. Baldas, J. Bonnyman, P. M. Pojer, G. A. Williams and M. F. Mackay, *J. Chem. Soc., Dalton Trans.*, 1798 (1981).
301. U. Abram and R. Münze, *Inorg. Chim. Acta*, **169**, 49 (1990).
302. U. Abram, H. Spies, S. Abram, R. Kirmse and J. Stach, *Z. Chem.*, **26**, 140 (1986); *Chem. Abstr.*, **105**, 107294n (1986).
303. Reference 299a.

304. Reference 300.
305. U. Abram, B. Lorenz and L. Kaden, *Polyhedron*, **7**, 285 (1988).
306. Reference 290.
307. Reference 299a.
308. U. Abram, J. Hartung, L. Beyer, R. Kirmse and K. Köhler, *Z. Chem.*, **27**, 101 (1987).
309. J. Baldas, J. F. Boas, S. F. Colmanet and G. A. Williams, *J. Chem. Soc., Dalton Trans.*, 2441 (1991).
310. J. Baldas, J. F. Boas and J. Bonnyman, *J. Chem. Soc., Dalton Trans.*, 1721 (1987).
311. M. Zieliński, *Polish J. Chem.*, **52**, 1507 (1978).
312. M. Zieliński, *Polish J. Chem.*, **53**, 1453 (1979).
313. M. Zieliński, H. Torbicki and S. Szpilowski, *Roczniki Chemii*, **48**, 1107 (1974).
314. R. W. Thomas, M. J. Heeg, R. C. Elder and E. Deutsch, *Inorg. Chem.*, **24**, 1472 (1985).
315. K. A. Jorgensen, *Chem. Rev.*, **89**, 431 (1989).
316. E. I. Stiefel, in *Comprehensive Coordination Chemistry*, Vol. 3 (Eds. G. Wilkinson, R. D. Gillard and J. A. McCleverty), Chap. 36.5, Pergamon, Oxford, 1987, p. 1375.
317. J. Baldas, S. F. Colmanet and M. F. Mackay, *J. Chem. Soc., Chem., Commun.*, 1890 (1989).
318. J. Baldas, J. F. Boas and J. Bonnyman, *Aust. J. Chem.*, **42**, 639 (1989).
319. J. Baldas and S. F. Colmanet, *Inorg. Chim. Acta*, **176**, 1 (1990).
320. J. Baldas, J. Bonnyman and G. A. Williams, *Inorg. Chem.*, **25**, 150 (1986).
321. J. Baldas, *Pure Appl. Chem.*, **62**, 1079 (1990).
322. R. Armstrong and N. A. Gibson, *Aust. J. Chem.*, **21**, 897 (1968).
323. J. Baldas, S. F. Colmanet and G. A. Williams, *J. Chem. Soc., Dalton Trans.*, 1631 (1991).
324. S. F. Colmanet and G. A. Williams, in *Technetium and Rhenium in Chemistry and Nuclear Medicine 3* (Eds. M. Nicolini, G. Bandoli and U., Mazzi), Cortina International, Verona, 1990, p. 55.
325. H. A. O. Hill and D. G. Tew, in *Comprehensive Coordination Chemistry* (Eds. G. Wilkinson, R. D. Gillard and J. A. McCleverty), Vol. 2, Chap. 15.2, Pergamon, Oxford, 1987, p. 315.
326. K. Nakamato, *Infrared Spectra of Inorganic and Coordination Compounds*, 2nd ed., Wiley-Interscience, New York, 1970, p. 244.
327. J. Baldas and J. Bonnyman, *Inorg. Chim. Acta*, **141**, 153 (1988).
328. (a) R. C. Elder, R. Whittle, K. A. Glavan, J. F. Johnson and E. Deutsch, *Acta Crystallogr.*, **B36**, 1662 (1980).
   (b) K. Glavan, R., Whittle, J. F. Johnson, R. C. Elder and E. Deutsch, *J. Am. Chem. Soc.*, **102**, 2102 (1980).
329. Y. Wei, B. L. Liu and H. F. Kung, *Int. J. Radiat. Appl. Isot.*, **41**, 763 (1990).
330. W. Görner and H. Spies, *Anal. Chem.*, **58**, 1261 (1986).
331. C. M. Biagini, D. A. Clemente, L. Magon and U. Mazzi, *Inorg. Chim. Acta*, **13**, 47 (1975).
332. F. Houdek, I. Obrusnik and K. Svoboda, *Radiochem. Radioanal. Lett.*, **39**, 343 (1979).
333. H. Spies and B. Johannsen, *Inorg. Chim. Acta*, **48**, 255 (1981).
334. B. Pitt and H. W. Strauss, in *Clinical Application of Myocardial Imaging with Thallium in Cardiovascular Nuclear Medicine*, 2nd ed. (Eds. H. W. Strauss and B. Pitt), C. V. Mosby, St. Louis, 1979, pp. 243–252.
335. W. C. Eckelman and S. M. Levenson, *Int. J. Appl. Radiat. Isot.*, **28**, 67 (1977).
336. E. Deutsch, K. A. Glavan, V. J. Sodd, H. Nishiyama, D. L. Ferguson and S. J. Lukes, *J. Nucl. Med.*, **22**, 897 (1981).
337. J. E. Fergusson and R. S. Nyholm, *Chem. Ind. (London)*, 347 (1960).
338. P. J. Sullivan, J. Werre, D. R. Elmaleh, R. D. Okada, S. Y. Kopiwoda, F. P. Castronovo Jr, K. A. McKusick and H. W. Strauss, *Int. J. Nucl. Med. Biol.*, **11**, 3 (1984).
339. H. Nishiyama, E. Deutsch, R. J. Adolph, V. J. Sodd, K. Libson, E. L. Saenger, M. C. Gerson, M. Gabel, S. J. Lukes, J. L. Vanderheyden, D. L. Fortman, K. L. Scholz, L. W. Grossman and C. C. William, *J. Nucl. Med.*, **23**, 1093 (1982).
340. H. Nishiyama, R. J. Adolph, E. Deutsch, V. J. Sodd, K. Libson, M. C. Gerson, E. L. Saenger, S. J. Lukes, M. Gabel, J. L. Vanderheyden and D. L. Fortman, *J. Nucl. Med.*, **23**, 1102 (1982).
341. P. C. Srivastava, A. P. Callahan and F. F. Knapp, Jr., *J. Nucl. Med.*, **24**, 43 (1983).
342. P. C. Srivastava, H. G. Hay and F. F. Knapp, Jr., *J. Med. Chem.*, **28**, 901 (1985).
343. P. C. Srivastava and F. F. Knapp, Jr., *J. Med. Chem.*, **27**, 978 (1984).
344. W. H. Oldendorf, in *Functional Radionuclide Imaging of the Brain*, Raven Press, New York,

1983, p. 1; in *Handbook of Neurochemistry*, 2nd ed. (Ed. A. Lattha), Plenum Press, New York, 1982, p. 485.

345. M. Neves, K. Libson and E. Deutsch, *Nucl. Med. Biol.*, **14**, 503 (1987).

346. J. L. Vanderheyden, M. J. Heeg and E. Deutsch, *Inorg. Chem.*, **24**, 1666 (1985).

347. E. Deutsch, M. Nicolini and H. N. Wagner, Jr., *Technetium in Chemistry and Nuclear Medicine*, Cortina International, Verona, Italy (distributed by Raven Press, New York), 1983.

348. M. Thompson, A. D. Nunn and E. N. Treher, *Anal. Chem.*, **58**, 3100 (1986).

349. J. Lister-James, in *Radionuclide Imaging of the Brain* (Ed. B. L. Holman), Churchill-Livingstone, New York, 1985, p. 75.

350. V. N. Gerasimov, S. V. Kryuchkov, A. F. Kuzina, V. M. Kulakov, S. V. Pirozhkov and V. I. Spitsyn, *Dokl. Akad. Nauk USSR (Engl. Transl.)*, **266**, 148 (1984).

351. C. J. Lim, D. Geraghty and J. A. Fuchs, *J. Bacteriol.*, **163**, 311 (1985).

352. A. Holmgren, B. O. Soderberg, H. Eklund and C. I. Branden, *Proc. Natl. Acad. Sci. USA*, **72**, 2305 (1975).

353. V. P. Pigiet and B. J. Schuster, *Proc. Natl. Acad. Sci. USA*, **83**, 7643 (1986).

354. A. Holmgren, *Annu. Rev. Biochem.*, **54**, 237 (1985).

355. J. F. Grippo, W. Tienrungroj, M. K. Dahmer, P. R. Housley and W.B. Pratt, *J. Biol. Chem.*, **258**, 658 (1983).

356. S. B. Brown, R. J. Turner, R. S. Roche and K. J. Stevenson, *Biochemistry*, **26**, 863 (1987).

357. S. B. Brown, R. J. Turner, R. S. Roche and K. J. Stevenson, *Biochem. Cell. Biol.*, **67**, 25 (1989).

358. T. Hiraoki, S. B. Brown, K. J. Stevenson and H. J. Vogel, *Biochemistry*, **27**, 500 (1988).

359. S. R. Adamson, J. A. Robinson and K. J. Stevenson, *Biochemistry*, **23**, 1269 (1984).

360. C. M. B. Holmes, and K. J. Stevenson, *Biochem. Cell. Biol.*, **64**, 509 (1986).

361. K. J. Stevenson, G. Hale and R. N. Perham, *Biochemistry*, **17**, 2189 (1978).

362. J. A. Robinson, *MSc. Thesis*, Univ of Calgary, 1980, Calgary, Candada.

363. R. G. Matthews, D. P. Ballou, C. Thorpe and C. H. Williams Jr., *J. Biol. Chem.*, **252**, 3199 (1977).

364. P. Ehrlich, *Ber.*, **42**, 17 (1909).

365. K. Liao, R. D. Hoffman and M. D. Lane, *J. Biol. Chem.*, **266**, 6544 (1991).

366. S. C. Frost and M. D. Lane, *J. Biol. Chem.*, **260**, 2646 (1985).

367. S. C. Frost, R. A. Kohanski and M. D. lane, *J. Biol. Chem.*, **262**, 9872 (1987).

368. J. L. Webb, in *Enzymes and Metabolic Inhibitors*, vol. 3, Academic Press, 1966, New York, pp. 595–793.

369. A. Rotman and S. Linder, *Biochim. Biophys. Acta*, **641**, 114 (1981).

370. R. W. Helmkamp and D. S. Sears, *Int. J. Appl. Radiat. Isot.*, **21**, 430 (1970).

371. E. A. Kabat, *Structural Concepts in Immunology and Immunochemistry*, Holt, Rinehart and Winston, New York, 1968, p. 29.

372. U. K. Laemmli, *Nature*, **227**, 680 (1970).

373. D. G. OH, M. J. Reisfeld, J. Martin and T. W. Whaley, *Synth. Appl. Isot. Labelled Compd. Proc. Int. Symp., 2nd 1985*, (Ed. R. R. Muccino), Elsevier, Amsterdam, 1986, pp. 409–414; *Chem. Abstr.*, **105**, 172599e (1986).

374. W. R. Cullen, B. C. McBridge, H. Manji, A. P. Pickett and J. Reglinski, *Appl. Organomet. Chem.*, **3**, 71 (1989); *Chem. Abstr.*, **112**, 2172n (1990).

375. P. Hosain, A. Erman and R. P. Spencer, in *Radiopharm Struct. Act. Relat.*, Proc. Symp. 1980 (Ed. R. P. Spencer), Grune and Stratton, New York, 1981, pp. 267–274.

376. M. Joaniridou and F. Fillaux, *J. Chim. Phys. Phys.-Chim. Biol.*, **81**, 397 (1984).

377. A. Natarajan and V. A. Chinnappan, *Proc. Indian Acad. Sci. (Ser. Chem. Sci.)*, **92**, 211 (1983).

378. V. P. Tarasov, V. I. Privalov, S. A. Petrushin and G. A. Kirakosyan, *Sovrem. Metody Ya. M. R. EPR Khim. Tverd. Tela [Mat. Vses. Koor. Soveshch.]*, 4th, pp. 112–114 (1985); *Chem. Abstr.*, **104**, 27681a (1986).

379. R. Jakutas, E. Narewski, L. Sobczak and H. Konwent, *Solid State Commun.*, **55**, 831 (1985); *Chem. Abstr.*, **103**, 151963n (1985).

380. J. Berdowski and M. Hajduk, *Acustica*, **60**, 283 (1986); *Chem. Abstr.*, **105**, 181986f (1986).

381. H. J. Rother, J. Albers, A. Kloepperpieper and H. E. Mueser, *Jpn. J. Appl. Phys. Part I*, **24** (Suppl. 24–2, *Proc. Int. Meet. Ferroelectr.* 6th), pp. 384–386 (1985); *Chem. Abstr.*, **105**, 16192m (1986).

382. N. Neck, B. Kanellakopoulos and J. I. Kim., *Ber. KfK.*, 1985, 3998, 65 pp.; *Chem. Abstr.*, **104**, 76086n (1986).

CHAPTER **16**

Arsonium, stibonium and bismuthonium ylides and imines

DOUGLAS LLOYD

Department of Chemistry, Purdie Building, University of St. Andrews, St. Andrews, Fife, KY16 9ST, Scotland

and

IAN GOSNEY

Department of Chemistry, The University of Edinburgh, Edinburgh, EH9 3JJ, Scotland

The chemistry of organic arsenic, antimony and bismuth compounds
Edited by S. Patai © 1994 John Wiley & Sons Ltd

I. INTRODUCTION

A great deal has been published about phosphonium ylides but much less attention has been paid to arsonium ylides. There is probably to some extent a psychological barrier, for the word arsenic has for centuries been associated with poisoning. In the case of triarylarsonium ylides, which are those most commonly used, there are no problems of this sort in their normal handling, but it must be emphasized that great care must be taken in handling alkylarsines and in these cases efficient fume-chambers are essential.

The first account of the preparation of an arsonium ylide (1) was published more than ninety years ago[1] (equation 1), although the correct ylide structure for the product was not provided for nearly half a century[2].

$$Ph_3As + PhCOCH_2Br \longrightarrow Ph_3\overset{+}{As}CH_2COPh \xrightarrow{\text{NaOH}} Ph_3As{=}CHCOPh \quad (1)$$
$$Br^- \qquad\qquad\qquad\qquad\qquad (1)$$

For convenience, throughout this account ylides will normally be represented by covalent structures with double bonds linking the hetero-atom to the ylidic carbon atom, as in 1, but the dipolar contribution to such structures must always be borne in mind (see Section II).

Following the discussion of arsonium ylides there is a short section dealing with arsinimines, the nitrogen analogues of the ylides; the first example of these compounds was prepared in 1937[3].

Until fairly recently relatively little was known about stibonium and bismuthonium ylides, but in the last decade there have been a number of studies of these compounds.

There are earlier reviews of the chemistry of arsonium[4,5], stibonium and bismuthonium ylides[5].

II. STRUCTURE OF ARSONIUM YLIDES

Arsonium ylides may be represented as hybrids of pentacovalent arsenic (2) and dipolar (3) structures.

$$R_3As\!=\!CH_2 \longleftrightarrow R_3\overset{+}{A}s\!-\!\overset{-}{C}H_2 \qquad Ph_3\overset{+}{A}s\!-\!CH\!=\!\overset{\overset{O^-}{|}}{C}CH_3 \qquad Ph_3\overset{+}{A}s\!-\!\underset{\ominus}{\bigcirc}$$

$$(2) \qquad\qquad (3) \qquad\qquad\qquad (4) \qquad\qquad\qquad (5)$$

If there are electron-withdrawing substituents conjugated with the ylidic carbon atom, further dipolar structures such as 4 and 5 may make major contributions to the overall structure. Delocalization of the charge in this way frequently leads to the ylides being isolable. Such ylides are commonly described as *stable* ylides; in this context stable is, in effect, a synonym for isolable. Many ylides are not, however, isolable, because of their high reactivity, in particular their very ready hydrolysis. In this article such ylides are called *reactive* ylides. Some other ylides, notably benzylylides, have reactivity intermediate between those ylides which are obviously *stable* or obviously *reactive*. They are described as *semi-stabilized* ylides.

Experimental evidence suggests that arsonium ylides are more dipolar than their phosphonium and sulphonium analogues.

This is shown, for example, by dipole moment measurements on a series of tetraphenyl-cyclopentadienylides. The moments for the diphenylsulphonium, triphenylphosphonium and triphenylarsonium derivatives are, respectively, 6.69, 7.75 and 8.32 D[6]. Similarly triphenylarsonium fluorenylide was shown to be more polar than its triphenylphosphonium analogue[7].

A. Spectra

Infra-red (IR) spectra of a number of stable arsonium ylides indicate their polarity and show that the negative charge is sited appreciably in the stabilizing substituent groups[8-11]. The stretching frequencies associated with β-carbonyl, β-sulphonyl or β-cyano substituents are all significantly lower than those observed for analogous phosphonium ylides, in keeping with the greater polarity of the arsonium ylides.

Analysis of the ^{1}H-NMR spectra of a series of cyclopentadienylides similarly indicated that the triphenylarsonium ylide is more polar and has greater delocalization of charge in the five-membered ring than its phosphonium or sulphonium analogues[12].

Comparative studies of ^{13}C-NMR spectra of phosphonium and arsonium ylides also suggest more covalent bonding in the former[13,14]. When methyltriphenyl- or tetramethyl-arsonium halides were converted into, respectively, triphenyl- or trimethyl-arsonium methylides there was only a small change in the CH coupling constant at the carbon atom undergoing deprotonation, whereas in the corresponding phosphorus systems. J_{CH} was greatly increased on deprotonation. This was taken to show that in the case of arsenic the bonding of the relevant carbon atom remained virtually unchanged, indicating a pseudo-tetrahedral geometry (6), in contrast to the case of phosphorus, where deprotonation causes an effective sp^3 to sp^2 rehybridization of the ylidic carbon atom (7). Introduction of silyl groups at the ylidic carbon atom causes a flattening of that atom[14] to a more planar structure. In arsonium acylylides the structure of the ylidic carbon atom appears to be nearly planar[15,16].

(6)   (7)

The ^{13}C-NMR spectra of a series of triphenylarsonium p-substituted aroylylides indicated a high electron density on the ylidic carbon atom[17]. The chemical shifts for the ylidic carbon atoms, but not those for the carbonyl carbon atoms, showed a correlation with the electronic character of the p-substituent[17].

The difference between arsonium and phosphonium ylides is commonly ascribed to less efficient $p\pi$–$d\pi$ overlap between the C-sp^2 orbitals and the larger and more diffuse 4d orbitals of arsenic, and to decreased electrostatic interaction across the ylide bond, but it is likely that these are not the only factors involved.

The photoelectron (P.E.) spectra of trimethylarsonium methylide and its phosphorus analogue have been recorded and CNDO/2 calculations were carried out[18]. Results are in accord with a raising of the HOMO levels for the arsenic ylide compared to the phosphorus analogue, thereby lowering the d-population and the ylide bond order, and increasing the charge on the ylidic carbon atom.

B. Conformation of Arsonium Ylides in Solution

Some study has been devoted to the conformational mobility of arsonium β-carbonyl ylides in solution; such compounds could take up either Z or E configurations (8).

(8)

In the ylide (8, R^1 = H or Ph; R^2 = MeO), separate NMR signals for Z and E isomers (with respect to the AsC = CO bond) are observable; coalescence occurs at higher temperatures[19]. The Z configuration is strongly favoured, presumably because of coulombic interaction between As$^+$ and O$^-$, but less so than in the case of phosphorus ylides. It was suggested that phosphorus orbitals overlap with the oxygen 2p-orbitals more effectively than do the more diffuse arsenic orbitals, giving rise to a P–O interaction that is more bonding than the As–O interaction[19]. Similar variation of spectra with temperature was observed when R^2 = SEt[20].

When R^2 is an alkyl or aryl group no such temperature dependence is observed[9]. This can be rationalized in terms of there being a higher negative charge on a keto oxygen atom than on the corresponding atom of an ester group, leading to stronger coulombic interaction in the keto case and a freezing of the ylide into a Z-structure. Some broadening of the methine signal was observed in the spectrum of the benzoylylide (**9**, $R^1 = Ph$, $R^2 = H$), but it was attributed to a protolytic exchange reaction[21].

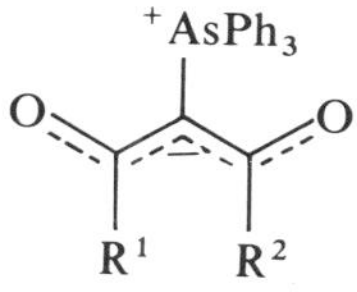

+AsPh3
O
O
R1
R2

(9)

In the case of β,β^1-dicarbonyl substituted ylides (**8**, $R^1 = COR^3$) the situation is more complex, but in nearly all cases NMR spectra suggest that the molecules exist as *Z*, *Z*-isomers[10]. Exceptions may arise if the acyl group is stabilized in an alternative configuration by intramolecular hydrogen bonding[10].

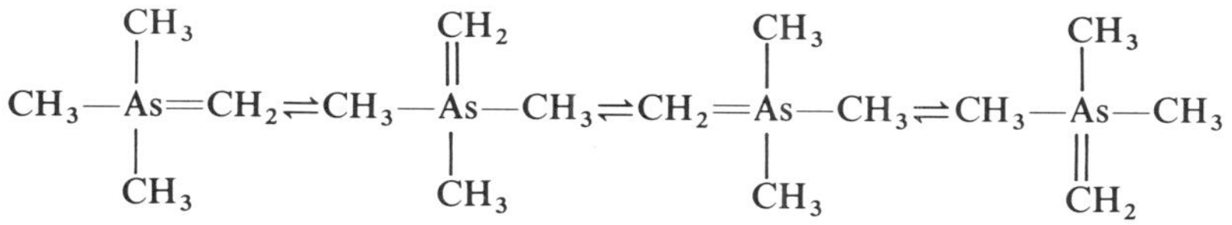

CH3
CH2
CH3
CH3
CH3—As=CH2⇌CH3—As—CH3⇌CH2=As—CH3⇌CH3—As—CH3
CH3
CH3
CH3
CH2

(10)

The ^{1}H-NMR spectrum of trimethylarsonium methylide (**10**) is of interest[22]. At room temperature the expected two singlets are observed, at chemical shifts which indicate a considerable dipolar contribution to the structure of the ylide. At higher temperatures there is considerable line broadening of both signals; a coalescence temperature could not be attained because of the onset of thermal decomposition. This broadening was attributed to fast proton exchange among the groups attached to arsenic, as shown in **10**. Significantly, addition of trace amounts of protic acid caused coalescence induced by catalysis of the proton exchange at room temperature.

C. Crystal Structure of Arsonium Ylides

X-ray crystallographic structure determinations have been carried out on a monobenzoyl[23] and two pentafluoropropionyl arsonium ylides[24,25], on two arsonium cyclopentadienylides (e.g. **11**)[26–28] and on three arsonium ylides each bearing two electron-withdrawing substituents **12**[29].

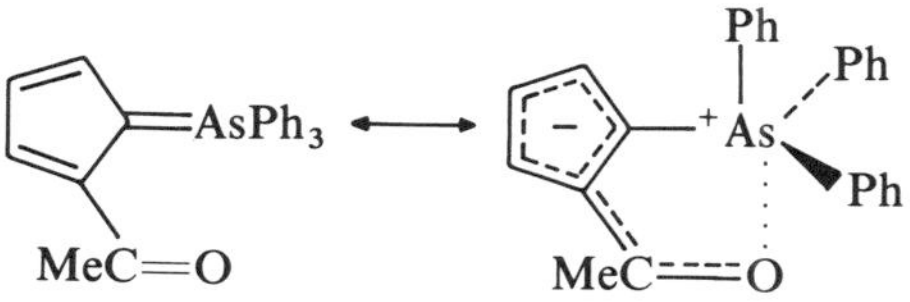

=AsPh3
MeC=O
Ph
Ph
+As
Ph
MeC=O

$Ph_3As{=}CRAc$, R = COOEt, CONHPh, NO_2

(11) **(12)**

The ylidic bonds lengths (~ 1.86–1.88 Å) fall between the sums of the covalent radii of singly and doubly bonded carbon and arsenic (1.98, 1.775 Å), signifying an appreciable amount of single-bond character and consequently of dipolar character. In keeping with delocalization of the negative charge into the electron-withdrawing substituents, the bonds linking these substituents to the ylidic carbon atom all show lengths corresponding to a measure of double-bond character. In all the acylylides the oxygen atoms are directed towards the arsenic atoms, as happens in solution. In the case of ylide 11 there is strong interaction between the oxygen and arsenic atoms resulting in the latter being distorted from a tetrahedral-type geometry towards a trigonal bipyramidal configuration with the oxygen at one of the apical positions[26,27]. That this interaction is also important in solution is evident from its lower dipole moment[6]; dipole moment measurements also confirm that other acylylides take up similar conformations in solution[30].

The structures of the solid ylides 12 are similar to those in solution. Ylides (12, R = COOEt, NO$_2$) take up Z, Z-conformations but ylide (12, R = CONHPh) has a Z, E-conformation because of intramolecular hydrogen bonding between the amide proton and the oxygen of the acetyl group[29].

III. STABILITY OF ARSONIUM YLIDES

Many of the stable arsonium ylides are crystalline solids which may be kept in air without significant decomposition. More reactive arsonium ylides need to be made as required and used *in situ*. Their decomposition usually arises from hydrolytic attack; most arsonium ylides appear to be thermodynamically stable at room temperature.

A. Thermal Decomposition

Some arsonium ylides have been reported to decompose when heated in solution. For example, when triphenylarsonium benzylide was heated in a boiling benzene–ether mixture it decomposed to give triphenylarsine and a mixture of stilbenes. Trimethylarsonium methylide likewise undergoes thermal decomposition to give trimethylarsine and ethylene[22]. A likely mechanism for these reactions involves carbenic decomposition of the ylide followed by attack of the carbene on unchanged ylide with expulsion of an arsine fragment (equation 2).

$$R_3^1As{=}CHR^2 \longrightarrow R_3^1As + R^2HC: \xrightarrow{R_3^1As{=}CHR^2} R_3^1\overset{+}{As}{-}CHR^2 \atop \overset{-}{C}HR^2$$

$$\longrightarrow R_3^1As + \underset{CHR^2}{\overset{CHR^2}{\|}} \qquad\qquad (13) \qquad\qquad (2)$$

An alternative mechanism has, however, been proposed[21], which implicates the presence of some protonated ylide, either residual salt from the mode of preparation, or arising from protonation of the ylide by traces of moisture. This undergoes nucleophilic displacement of its arsine group by a molecule of ylide to give a salt of 13 which then provides the isolated alkene by means of an elimination reaction. Protonated ylide is regenerated so that only catalytic amounts of it need be present. Some support for this latter mechanism as at least a contributing mechanism derives from the observation that thermal decomposition is greatly retarded in the presence of a large amount of base[31].

The more stable triphenylarsonium benzoylylide could be recovered unchanged from prolonged heating in boiling benzene but decomposed in boiling toluene to give triphenylarsine and *trans*-1, 2, 3-tribenzoylcyclopropane[21]. The latter product could arise from conjugate addition of unchanged ylide to alkene formed by thermal decomposition of part of the ylide (equation 3). In support of this mechanism it is known that arsonium ylides can react with conjugated unsaturated ketones[32,33] or esters[33,34] to give cyclopropane derivatives. The formation of cyclopropane derivatives from acylylides but not from a benzylide or methylide presumably reflects the fact that in the latter cases the alkenes which are formed first are not susceptible to nucleophilic attack.

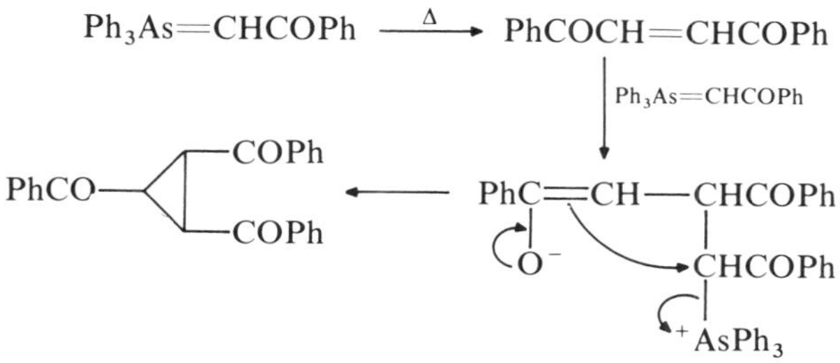

$$Ph_3As=CHCOPh \xrightarrow{\Delta} PhCOCH=CHCOPh$$

$$\downarrow {\scriptstyle Ph_3As=CHCOPh}$$

(3)

Keto-stabilized phosphonium ylides undergo thermal decomposition with extrusion of phosphine oxides[35], but no such reaction takes place with keto-stabilized arsonium ylides, presumably since the driving force to form an arsenic–oxygen bond is much less than that to form a phosphorus–oxygen bond. However, an interesting rearrangement seems to be involved in the mass-spectrometric decomposition of triphenylarsonium nitromethylide (**14**), arising from arsenic–oxygen bond formation, most plausibly explained by a four-centre oxygen transfer reaction (equation 4)[10]. The phosphonium ylide corresponding to **14** decomposes spontaneously at room temperature to triphenylphosphine oxide and fulminic acid[36]. This is a rare case of an arsonium ylide being more stable than its phosphonium analogue and is also attributed to the much greater energetic inducement to produce a P—O bond as compared to an As—O bond.

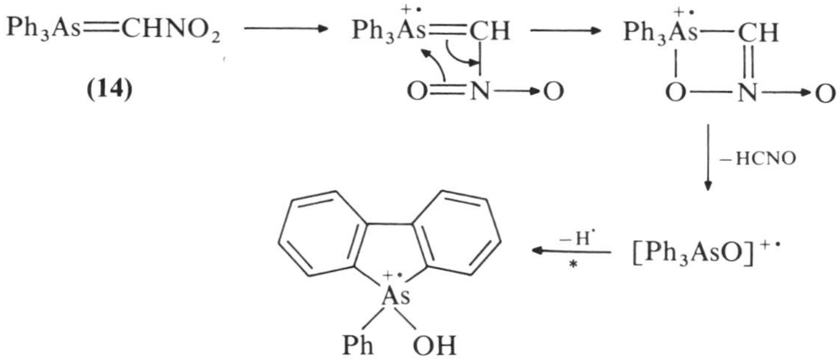

(4)

B. Hydrolysis

Many arsonium ylides are hydrolysed in the presence of moisture to give an arsine oxide and an organic residue. The first reported example described the rapid conversion of the semi-stabilized trimethylarsonium fluorenylide into trimethylarsine oxide and fluorene[37]. More stable arsonium ylides may require heating under reflux with solutions of sodium hydroxide to bring about hydrolysis; some diketo-ylides can be recovered unchanged even under these drastic conditions[10]. Ease of hydrolysis may depend strongly upon the solubility of the ylide in the reaction solvent. Thus, whereas triphenylarsonium 2, 3, 4, 5-tetraphenylcyclopentadienylide was recovered essentially unchanged from ethanolic potassium hydroxide solution in which it is barely soluble[38], it and other triarylarsonium analogues decomposed to provide tetraphenylcyclopentadiene when heated in methanol, in which they are soluble[39].

It has been speculated[40] that, by analogy with the mechanism proposed for the extensively studied hydrolysis of phosphonium ylides, the steps involved in hydrolysis are protonation, followed by formation of a pentacovalent arsenic species and finally loss of a carbanion (equation 5). Presumably for arsonium ylides, as for phosphonium ylides, the

$$Ph_3As{=\!\!=}CXY \xrightarrow{H_2O} Ph_3\overset{+}{As}CHXY \xrightarrow{HO^-} Ph_3As(OH)CHXY$$

$$\xrightarrow{HO^-} Ph_3AsO + {}^-CHXY$$

$$\Big\downarrow {}^{H_2O} \qquad (5)$$

$$CH_2XY$$

group which leaves the arsenic atom in the last step will be the group which provides the most stable carbanion. Thus ethanolysis of triphenylarsonium benzylide provides toluene via the intermediacy of the benzyl anion[32]. If there is no group present able to provide a stable anion, reaction may not proceed to give an arsine oxide. For example, triphenylarsonium methylide reacts with water to give the arsonium hydroxide (15)[41] and trimethylarsonium methylide reacts with methanol to provide a pentacoordinate arsorane (16)[42].

$$Ph_3\overset{+}{As}Me\ HO^- \qquad MeOAsMe_4$$
$$\quad\ (15) \qquad\qquad\ (16)$$

IV. BASICITY OF YLIDES AND ACIDITY OF THEIR CONJUGATE ACIDS

Many arsonium ylides dissolve in acids to form salts, from which they can be re-obtained by treatment with a base, as in the salt method for their preparation. The basicity of the ylides indicates the relative stabilities of the ylides and their salts and in so doing gives some guide to the stability of the ylides. Thus stable ylides are less readily protonated than are reactive ylides and require weaker bases for their generation from salts.

Stabilizing substituent groups are commonly those which can delocalize the negative charge on the ylidic carbon atom. Measurements on a series of triphenylarsonium p-substituted benzoylylides (17) showed that their basicity is lower the more electron-withdrawing the p-substituent[8,43,44]. Similarly for the ylides (18, X = H), the basicity decreases as the group COZ becomes more capable of delocalizing a negative charge; the basicity when Z = alkyl or acyl is much less than when Z = OR or NR_2[44]. Increase in the electron-donating character of X in 18, e.g. from H to Me to OMe, causes an increase in basicity[44], as does replacement of a triphenylarsonium group by a trimethylarsonium group[43]. But, while tri-p-tolylarsonium tri- and tetra-phenylcyclopentadienylides are more basic than their triphenylarsonium analogues, the corresponding o-tolylarsonium ylides are less basic[45]; presumably steric factors are relevant in the latter case.

$$Ph_3As{=\!\!=}CHCOC_6H_4X(p) \qquad (p\text{-}XC_6H_4)_3As{=\!\!=}CHCOZ$$
$$\qquad\quad (17) \qquad\qquad\qquad\qquad\qquad (18)$$

In studies of ylides having different heteronium atoms, e.g. on methoxycarbonylylides[46], benzoylylides[8,43,47], fluorenylides[48], tetraphenylcyclopentadienylides[21,49] and triphenylcyclopentadienylides[50], the arsonium ylides were uniformly more basic than their phosphonium or sulphonium analogues.

These results imply that arsenic plays a smaller part than do phosphorus and sulphur in the distribution of negative charge from the adjacent carbanion. This difference has commonly been attributed (*inter alia*[8,18,19,43,46,47,50,51]) to the lower electronegativity of the arsenic atom, which leads to a lower electrostatic interaction between the arsenic and ylidic carbon atoms, and to a lower effectiveness of $p\pi$–$d\pi$ orbital overlap between these atoms because of the greater size and diffuseness of the arsenic 4d-orbitals compared with the 3d-orbitals of phosphorus or sulphur. But it has been pointed out that other factors in addition to electronegativity and $p\pi - d\pi$ orbital overlap must play a part in determining the relative acidity of the heteronium salts and the stability of the related ylides[46]. The involvement of steric factors has been noted[45,50].

V. REACTIONS OF ARSONIUM YLIDES

By far the most important reactions of ylides are those of the Wittig type, especially those with carbonyl compounds, and they will be dealt with first; other reactions of this sort are those with other $C = X$ functions and with nitroso-compounds. Other carbanionic reactions are then considered and a final section deals with the formation of cyclic compounds from arsonium ylides. Hydrolysis has already been discussion in Section III.B.

A. Reactions with Carbonyl Compounds

The first example of a reaction between an arsonium ylide and a carbonyl compound was recorded in a thesis in 1937[52]. In this thesis it was reported that triphenylarsonium benzoylylide reacted with benzaldehyde to give benzylideneacetophenone. Two publications dealing with the reactions of arsonium ylides appeared in 1960. One described the formation of alkenes in high yield, starting from a fluorenylide[53], and the other reported that from triphenylarsonium methylide and benzophenone both 1, 1-diphenylethylene and phenylacetaldehyde were obtained, with the latter predominating $(1:3.5)$[54]; it was suggested that the aldehyde arose from an initially formed epoxide during an acid work-up of the reaction mixture. Similarly, a reaction between triphenylarsonium ethylide and p-tolualdehyde gave a mixture of a small amount of an alkene and, as principal product, p-tolylacetone, again formed by acid-induced rearrangement of an initially formed epoxide[55].

Thus at an early stage in the study of arsonium ylides it was shown that either alkenes or epoxides might be formed, in contrast to the behaviour of phosphonium ylides, which gave only alkenes, and sulphonium ylides, which gave only epoxides[51]. It was also apparent that arsonium ylides were more reactive than their phosphonium analogues, for while triphenylarsonium fluorenylide reacted with p-dimethylaminobenzaldehyde to give an alkene in high yield, triphenylphosphonium fluorenylide did not react with this aldehyde[53]. This arsonium ylide reacted in high yield with a number of substituted benzaldehydes and with acetaldehyde; it did not react with acetone or acetophenone but did with the more reactive ketone p-nitroacetophenone[59].

The reaction of triphenylarsonium benzylide (19) with p-nitrobenzaldehyde (equation 6) provided an alkene and an epoxide in about equal amounts, together with equimolar amounts of triphenylarsine and triphenylarsine oxide[7].

$$\text{Ph}_3\text{As} = \text{CHPh} + (1,4)\text{-OHCC}_6\text{H}_4\text{NO}_2 \begin{array}{c} \nearrow \text{Ph}_3\text{AsO} + \text{PhCH} = \text{CHC}_6\text{H}_4\text{NO}_2 \\ \\ \searrow \text{Ph}_3\text{As} + \text{PhCH} \overset{\text{O}}{\overset{\diagup \diagdown}{}} \text{CHC}_6\text{H}_4\text{NO}_2 \end{array} \qquad (6)$$

(19)

By contrast, in another investigation of the reactions between arsonium ylides and aldehydes, it was found that *either* an alkene *or* an epoxide was formed, depending upon the identity of the ylide, but not both of them together[32]. Both alkenes and epoxides were always *trans*[32]. This appears usually to be in the case.

1. Comparison of stable, reactive and semi-stabilized ylides

The general pattern which emerged was that stable arsonium ylides provided alkenes whilst reactive arsonium ylides gave epoxides[7,21,32,56]. This was attributed to stabilization of the transition state leading to alkene formation being provided by those same electron-withdrawing groups which stabilized the ylides[7,32].

Thus arsonium ylides stabilized by acyl groups[10,21,56-65], alkoxycarbonyl groups[10,60-63,66], cyano groups[64,67] and cyclopentadiene rings[38,49,50,68] all provide alkenes, predominantly *trans*, as products from reactions with carbonyl compounds. Most ylides with two electron-withdrawing substituents did not, however, take part in Wittig reactions[10]; steric factors may also sometimes inhibit reaction[45].

The arsonium salt $Ph_2(PhCH_2)\overset{+}{As}CH_2CH_2OH\ Br^-$ requires two equivalents of butyllithium for its conversion into an ylide $^-OCH_2CH_2(Ph_2)As=CHPh$; the presence of the alkoxide group does not prevent the ylide reacting with aldehydes to form a *trans*-alkene together with 2-hydroxyethyldiphenylarsine oxide[69]. In contrast, 2-hydroxyethyltriphenylarsonium bromide reacts with aryl aldehydes in the presence of solid potassium hydroxide to give 2-aryl-3-hydroxymethyloxiranes[70].

Reactive ylides give good yields of *trans*-epoxides[71], although stereospecificity may vary with conditions. For example, use of an arsonium tetrafluoroborate as precursor of the ylide, and potassium bis(trimethylsilyl)amide as base gave 100% *trans*-epoxide, whereas with iodide as counterion and butyllithium as base, there was less stereospecificity[71].

Reactive ylides generated from optically active arsonium salts reacted with aryl aldehydes to give *trans*-epoxides which were optically active, with modest optical induction[72,73]. Reaction of (+)-benzylmenthyldiphenylarsonium bromide with sodium ethoxide and a prochiral aryl aldehyde provided 2, 3-diaryloxiranes with optical purities up to 41%; the degree of asymmetric induction appears to depend upon the nature of the substituents on the ylide, the aldehydes and the reaction conditions[74]. Similar results were observed using other optically active arsonium salts with asymmetric arsenic atoms.

It was pointed out that the ylide **20** can be regarded as a β-formylvinyl anion equivalent **21**; it reacts with aldehydes to give oxiranes, which can readily be converted under mild conditions into 4-hydroxy-2-(E)-enals (equation 7)[75]. A vinylogue of **20** has been converted in the same way into a 6-hydroxy-2, 4-(E, E)-dienal; this reaction enables a five carbon atom homologation of aldehydes to be achieved[76].

$$Ph_3As=CHCH_2CH(OPr^i)_2 \longleftrightarrow Ph_3\overset{+}{As}\!\!-\!\!\overset{-}{C}HCH_2CH(OPr^i)_2 \quad \overset{-}{C}H=CH\diagdown CHO$$

$$\text{(20)} \qquad\qquad\qquad\qquad\qquad \text{(21)}$$

$$\Big\downarrow \text{RCHO}$$

$$\underset{O}{\overset{\diagup\ \diagdown}{RCH\!-\!CHCH_2CH(OPr^i)_2}} \quad \xrightarrow{CF_3COOH} \quad \underset{O}{\overset{\diagup\ \diagdown}{RCH\!-\!CHCH_2CHO}} \qquad\qquad (7)$$

$$\xrightarrow[Et_2O]{Et_3N} \quad \underset{H}{\overset{RCHOH}{}}\!C\!=\!C\!\underset{CHO}{\overset{H}{}}$$

Allylic arsonium ylides show a similar pattern of reactivity. Ethoxycarbonylallyl ylides, wherein the ester group is conjugated with the ylidic carbon atom, gave dienes in reactions with aldehydes or ketones[77,78], whereas other allylic ylides lacking such an electron-withdrawing substituent gave vinylic epoxides in high yield[78,79,80a,b], e.g. equation 8. In the latter case it was found that the presence of hexamethylphosphoramide resulted in the formation of a diene instead of the epoxide[80a]. A trifluoromethylallyl ylide, $Ph_3As=CHC(CF_3)=CMe_2$, reacted with aldehydes to give epoxides[81]. The epoxides produced from the reaction of aldehydes with 3-(tri-isopropylsiloxy)allyl ylides were converted by hydrolysis into 2, 3-disubstituted furans[80b].

$$Ph_3As=CHCH=CHCOOEt \xrightarrow[Et_2O]{R^1R^2CO} R^1R^2C=CHCH=CHCOOEt$$

(8)

$$Ph_3As=CHCH=CHPh \xrightarrow[Et_2O]{R^1R^2CO} R^1R^2C-CHCH=CHPh$$

The ethynyl ylide $Ph_3As=CHC\equiv CSiMe_3$ reacts with ketones to give conjugated enynes, giving higher yields than does its phosphonium analogue[82].

The reaction of the hydroxyallyl ylide $Ph_3As=CHCH=CHCH_2OH$ with benzaldehyde takes place under phase-transfer conditions and gives a mixture of E and Z 3-hydroxyallyloxiranes; it gives a higher yield in acetonitrile in the presence of potassium fluoride and aluminium oxide than in the presence of solid potassium hydroxide[83].

Semi-stabilized arsonium ylides are intermediate in behaviour between stable and reactive ylides, and may provide alkenes and/or epoxides[8,32,84-90]. In these cases factors such as the substituent groups on arsenic, and the nature of the solvent and of the base, may become important in determining the nature of the product; this will be considered in more detail later (Section V.A.4). Small changes in the structure of the ylidic moiety may also have a marked effect; for example, whereas triphenylarsonium β-naphthylmethylide reacts to give epoxide, the presence of a bromine atom at the adjacent α-position of the naphthalene ring results instead in the formation of alkenes[86].

These results may be summarized as shown in Figure 1.

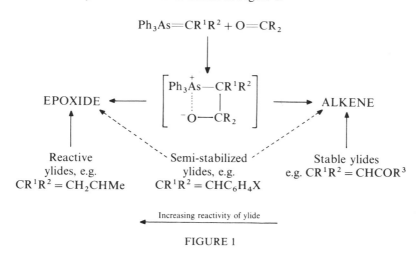

FIGURE 1

2. Reactivities of arsonium ylides compared to those of phosphonium and sulphonium ylides

Comparative studies involving acylylides[10,59], fluorenylides[53] and cyclopentadienylides[38,49,50,90], show that arsonium ylides are markedly more reactive than their phosphonium and sulphonium analogues. In many cases reactions proceed only in the case of the arsonium ylides; this is especially true the more electron-withdrawing are the substituents on the ylidic carbon atom, although some arsonium ylides with two electron-withdrawing substituents will not react even with aldehydes as reactive as 2,4-dinitrobenzaldehyde[10].

The reactions of a series of arsonium ylides with p-nitrobenzaldehyde have been shown to be first order for each reagent and there is a general tendency for the more basic ylides to be the more reactive[44]. The correlation is not, however, complete, since factors other than basicity, e.g. steric, and interactions between ylidic substituents and the arsenic atom, also affect the reactivity[10,11,45,50], but as a generalization it is largely valid and also must be a significant factor in the greater reactivity of arsonium compared to phosphonium and sulphonium ylides. A fair correlation has also been noted between the chemical shift of the signal from the methine proton in a series of ylides and their rates of reaction with p-nitrobenzaldehyde[44].

3. Effects of different substituents on the arsenic atom

The first report of such effects was in a study of the reactions of a series of tris(p-substituted phenyl)arsonium ylides with benzaldehyde; all these ylides gave epoxides in high yield save for the tris(p-dimethylamino) compound, which gave instead the trans-alkene[91]. In further experiments replacement of a triphenylarsonium group by a tris(p-methoxyphenyl)arsonium group was found to have little effect on the ratio of products[32], and inclusion of the arsenic atom in a strained ring also had no effect[92].

A comprehensive investigation of a series of ylides (**22**) with benzaldehyde showed that as X, Y became more electron-donating, so the proportion of alkene to epoxide increased[93]. For example, the ratios epoxide:alkene changed from $\sim 11:1$ (X = Y = H) to $\sim 6.5:1$ (X = Y = MeO), $\sim 6:1$ (X = H, Y = NMe$_2$), $\sim 4:1$ (X = NMe$_2$, Y = H), $\sim 1:1$ (X = Y = NMe$_2$). Even more striking is the effect of replacing these aryl groups by alkyl groups, namely ratios epoxide:alkene were Ph$_3$As, $\sim 11:1$; Et$_2$PhAs, $\sim 1:5$; Et$_3$As, $\sim 1:87$[93]. These results strongly suggest that remote control of the major pathway followed in reactions of semi-stabilized arsonium ylides with carbonyl compounds might be achieved by choice of the appropriate arsonium group. The reactivity of arsonium ylides is also affected by both electronic and steric effects on the arsenic atom[39].

$$(p\text{-}XC_6H_4)_2 (p\text{-}YC_6H_4)As = CHPh$$

$$(\mathbf{22})$$

4. Solvent effects and effects of base

Initial studies of solvent effects, on the reactions of triarylarsonium benzoylylides with p-nitrobenzaldehyde in N,N-dimethylformamide, dimethyl sulphoxide or methanol, indicated little solvent effect in these cases[44], but later studies of the more finely balanced reactions of semi-stabilized ylides have provided examples of strong influences due to the effect of different base and solvent when the ylide is generated in the presence of a carbonyl compound[22,80,88,89,94]. Thus, when benzyltriphenylarsonium bromide or p-chlorobenzyltriphenylarsonium bromide were treated with sodium hydride in benzene in the presence of a variety of p-substituted benzaldehydes the products were alkenes, but if sodium ethoxide in ethanol was used the isolated products were epoxides[94]. Likewise, when triphenylarsonium benzylylide was generated by phenyllithium in the presence of either benzaldehyde or acetaldehyde, the preponderant product was the epoxide whereas use of sodium amide as base provided mostly the alkene[88]. Similar results were obtained when an allyltriphenylarsonium salt was deprotonated using different hexamethyldisilazide salts at $-65°C$ in tetrahydrofuran in the presence of aldehydes[89]. When the potassium salt was used as base the product was *trans*-alkene, but when the lithium salt was used a 2:1 mixture of *trans*- and *cis*-epoxides resulted. It was suggested that when the potassium salt is used, the reactants can equilibrate between the *cis* and *trans* forms of the conjugated ylide, but that the lithium ion enhances production of a betaine (see Section V.A.7), and that this cyclizes to give the two epoxides. Again scope for the control of the reaction, by suitable choice of solvent and/or base, at least of semi-stabilized ylides, is indicated.

5. Reactions with conjugated enones

Arsonium ylides may react with conjugated enones either to give dienes or by conjugate addition to provide cyclopropane derivatives[33,34,59,80,95–97], e.g. equation 9[34]. Formation of a cyclopropane derivative rather than participation in a Wittig reaction appears to be encouraged by the presence of a bulky group adjacent to the carbonyl group of the ketone[34]. Enals and conjugated enals undergo Wittig reactions with arsonium ylides to give alkenes and dienes, respectively[98,99]. Formation of cyclopropane derivatives is discussed further in Section V.D.

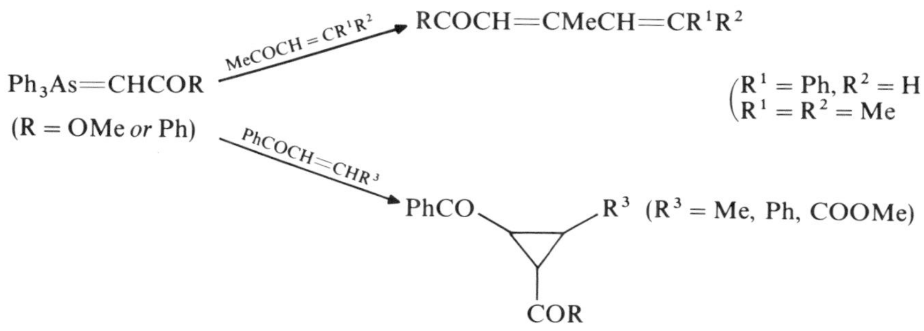

$$Ph_3As=CHCOR \quad (R = OMe \text{ or } Ph)$$

$$\xrightarrow{MeCOCH=CR^1R^2} RCOCH=CMeCH=CR^1R^2$$

$$\left(\begin{matrix} R^1 = Ph, R^2 = H \text{ or} \\ R^1 = R^2 = Me \end{matrix}\right)$$

$$\xrightarrow{PhCOCH=CHR^3} PhCO\triangleleft R^3 \text{ (}R^3 = Me, Ph, COOMe)\text{, } COR$$

(9)

6. Reaction with other C=X functions

Semi-stabilized arsonium ylides react with thioketones; with cyclic thioketones triphenylarsonium benzylides gave only exocyclic alkenes and no thiirans[100]. Benzothiopyrones (and, better, benzopyrones) also react to give exocyclic alkenes, in these cases arylidenebenzopyrans[101].

Reaction of triphenylarsonium benzylide with benzylideneaniline gave 1,2,3-triphenylaziridine[32]. This reaction is analogues to that of reactive arsonium ylides with carbonyl compounds.

7. Mechanism

Since the behaviour of arsonium ylides in Wittig reactions appeared to be intermediate between that of sulphonium and phosphonium ylides, it was inferred that mechanisms similar to those accepted for the respective reactions of the latter ylides were involved[7,51], namely equation 10. (The equilibrium in the top line may sometimes be better represented by a hybrid of the two structures shown.) The energetic driving force to generate an arsenic–oxygen bond is not as strong as that to form a phosphorus–oxygen bond, so that there is not the same compulsion to alkene formation in the case of arsonium ylides, allowing the alternative pathway (b) to compete.

$$
\begin{array}{l}
R_3\overset{+}{As}-CR_2^1 \quad\quad R_3As-CR_2^1 \quad\quad R_3As\cdots CR_2^1 \quad\quad R_3AsO \\
\quad\quad | \quad\quad\quad\quad\;\; |\quad\;\; | \quad\;\; \xrightarrow{(a)} \quad |\quad\;\; | \quad\;\; \longrightarrow \quad\;\; + \\
\quad\;\; {}^-O-CR_2^2 \quad\quad\; O-CR_2^2 \quad\quad\quad O-CR_2^2 \quad\quad R_2^1C\!=\!CR_2^2 \\[2mm]
R_3As\!=\!CR_2^1 \;\rightleftharpoons \\
\quad\; + \quad\quad\quad \rightleftharpoons \quad \Updownarrow \\
O-CR_2^2 \quad\quad R_3\overset{+}{As}-CR_2^1 \quad\quad\quad R_3\overset{+}{As}\cdots CR_2^1 \quad\quad\quad\quad\quad\quad\quad CR_2^1 \\
\quad\quad\quad\quad\quad\quad | \quad\quad \xrightarrow{(b)} \quad\quad / \;\; \backslash \quad\; \longrightarrow \; R_3As + R_2^2C \overset{\diagdown}{\underset{\diagup}{\;}} O \\
\quad\quad\quad\quad\quad\; R_2^2C-O^- \quad\quad\quad R_2^2C-O^-
\end{array}
$$

$$(10)$$

It was suggested that in the case of stable ylides, wherein R^1 is an electron-withdrawing group, the presence of the latter group, which becomes conjugated with the carbon–carbon double bond in the final alkene, also stabilizes the transition state leading to the formation of the alkene, thus promoting pathway (a) with respect to pathway (b)[7,32].

The rates of reactions of acylylides, $Ph_3As\!=\!COX$, with ketones decrease as X is more electron-withdrawing, making the ylide less nucleophilic[102]. This observation, together with the relation between the reactivity and basicity of an ylide[44] and the second-order character of the reaction[44,102], all suggest that the first step, which is slow and reversible, is the rate-determining step. The lack of solvent effect on the rates of reactions between benzoylylides and p-nitrobenzaldehyde led to the suggestion[44] that, in alkene formation, reaction goes directly to a four-membered ring transition state without intermediate formation of a betaine. This was seen to be consistent with the large negative entropy of reaction and the very low activation energies observed[44].

Formation of an epoxide must, however, involve an intermediate betaine which reacts further by intramolecular displacement of an arsine.

When pathways (a) and (b) are followed, the electrons in the arsenic–carbon bond are displaced in an opposite direction in the two mechanisms. In alkene formation displacement of electrons is away from the arsenic atom, and in epoxide formation displacement of electrons is towards the arsenic atom. The change in pathway, depending upon the nature of the substituents at arsenic, could be associated with this, for electron-donating substituents on arsenic should assist displacement of the electrons away from the arsenic and favour alkene formation as observed[93]. For similar reasons electron-withdrawing substituents on the ylidic carbon atom should favour alkene formation.

The observed solvent effects on the type of product formed[90,94] could also be associated with the structure of the intermediate. Formation of alkenes or epoxides necessitates, respectively, *cisoid* and *transoid* arrangements of the arsenic and oxygen atoms, and a *transoid* structure is likely to be much more favoured in a polar protic solvent such as ethanol than in benzene. Similarly, as observed[88,89], the presence of lithium ions should favour epoxide formation compared to the effect of sodium or potassium ions, since the *transoid* intermediate is likely to be stabilized by association of the oxygen with lithium but less so with sodium or potassium.

8. Coda

Reactions of arsonium ylides with carbonyl compounds take place much more readily than with phosphonium or sulphonium ylides. The nature of the products depends upon the character of the substituents on the ylide carbon atom, where electron-withdrawing substituents favour alkene formation, and of substituents on the arsenic atom, where

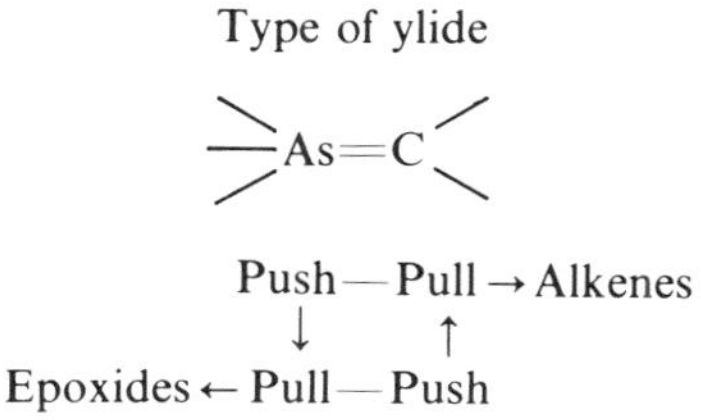

Type of ylide

(Push, pull refer, respectively, to electron-donating or electron-withdrawing effects of substituents on the arsenic or carbon atoms).

$$\nwarrow \diagup As=C \diagdown \diagup$$

Push — Pull → Alkenes
   ↓   ↑
Epoxides ← Pull — Push

FIGURE 2

electron-donating substituents favour alkene formation. This may be summarized as in Figure 2.

However, choice of appropriate base and solvent can, in the case of less stabilized ylides, have an effect on both the product distribution and on the stereospecificity of the product. More detailed analysis is desirable, but it seems likely that, at least in the case of semi-stabilized ylides and possibly for others also, control over the product can be achieved by suitable choice of the substituents on arsenic, and of the solvent and base, thus making a valuable addition to the organic chemist's synthetic armoury.

Reactions which may at first glance appear to involve ylides may not in fact do so. For example, a bromomalonic ester reacted with a variety of aldehydes in the presence of tributylarsine to form alkylidene- or arylidene-malonic esters, but no base was required and it was suggested that reaction proceeded not via an ylide but via an arsonium salt formed by extraction of a bromonium ion from the malonic ester[103]. The malonate anion then makes a nucleophilic attack on the aldehyde (equation 11).

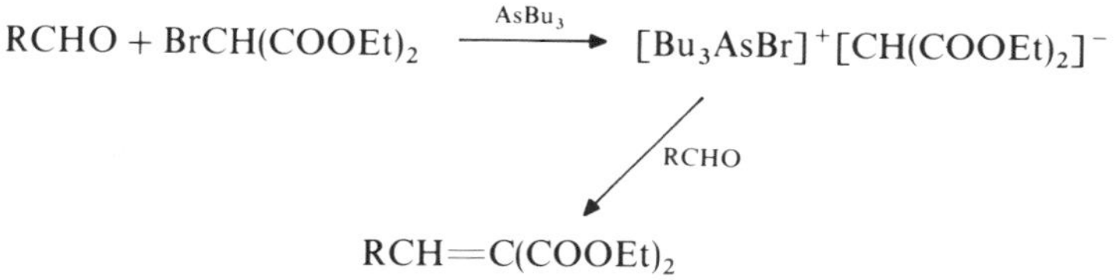

$$RCHO + BrCH(COOEt)_2 \xrightarrow{AsBu_3} [Bu_3AsBr]^+[CH(COOEt)_2]^-$$

$$\diagdown RCHO \qquad (11)$$

$$RCH=C(COOEt)_2$$

B. Reactions with Nitrosobenzene

Ylides may react with nitrosobenzene in a similar fashion to their reactions with carbonyl compounds (equation 12). Sulphonium ylides give nitrones[51], phosphonium

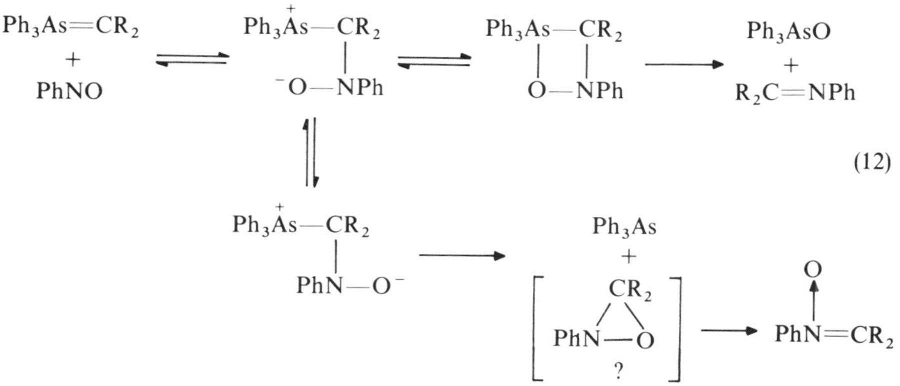

$$Ph_3As=CR_2 \quad\rightleftharpoons\quad \overset{+}{Ph_3As}-CR_2 \quad\rightleftharpoons\quad \overset{+}{Ph_3As}-CR_2 \quad\longrightarrow\quad Ph_3AsO$$
$$+ \qquad\qquad\qquad |\qquad\qquad\qquad |\;\; | \qquad\qquad\qquad +$$
$$PhNO \qquad\qquad ^-O-NPh \qquad\qquad O-NPh \qquad\qquad R_2C=NPh$$

$$(12)$$

$$\overset{+}{Ph_3As}-CR_2 \quad\longrightarrow\quad Ph_3As$$
$$| \qquad\qquad\qquad\qquad\qquad +$$
$$PhN-O^- \qquad\qquad\qquad \left[\begin{array}{c} CR_2 \\ \diagup \;\; \diagdown \\ PhN-O \\ ? \end{array} \right] \quad\longrightarrow\quad PhN=CR_2$$

ylides give anils[51] and, true to their intermediate character, arsonium ylides may give either anils or nitrones or both[7,38,45,49,50]. Similar considerations should apply to these reactions as to those with carbonyl compounds, and again the reactivity of arsonium ylides is much greater than that of corresponding phosphonium (or sulphonium) ylides[38,49,50]. p-Electron-donating substituents in aryl groups attached to arsenic increase the proportion of anil to nitrone in keeping with their effect on reactions with carbonyl compounds[45].

C. Reactions with Other Electrophiles

Because of the partial negative charge on the ylidic carbon atom, arsonium ylides are also attacked by other electrophiles. The first reported examples of such reactions are shown in equation 13[41,57]. Chlorination was achieved using iodobenzene trichloride[104].

$$
\begin{array}{c}
\underset{\text{BF}_3\cdot\text{Et}_2\text{O}}{\nearrow} \quad \overset{+}{\text{Ph}_3\text{As}}\!-\!\overset{-}{\text{CH}_2}\!\cdot\text{BF}_3 \\[2mm]
\text{Ph}_3\text{As}\!=\!\text{CH}_2 \\[2mm]
\underset{\text{Me}_3\text{SiBr}}{\searrow} \quad \overset{+}{\text{Ph}_3\text{AsCH}_2\text{SiMe}_3} \quad \text{Br}^- \\[4mm]
\underset{\text{Br}_2}{\nearrow} \quad \overset{+}{\text{Ph}_3\text{AsCHBrCOPh}} \quad \text{Br}^- \\[2mm]
\text{Ph}_3\text{As}\!=\!\text{CHCOPh} \\[2mm]
\underset{\text{SO}_3}{\searrow} \quad \overset{+}{\text{Ph}_3\text{AsCH}}(\text{SO}_3^-)\text{COPh}
\end{array}
\tag{13}
$$

Alkylation of trimethylarsonium methylide with methyl iodide gave ethyltrimethylarsonium iodide[54], but reactions of triphenylarsonium benzoylylide with ethyl iodide gave an O-ethylated product rather than a C-ethylated product[21]. Phosphonium acyl ylides undergo O-alkylation[105] while sulphonium ylides undergo C-alkylation[106]. It has been suggested that O-alkylation and C-alkylation of arsonium ylides result, respectively, from kinetically and thermodynamically controlled reactions[21]. Alkylation of triphenylarsonium benzylide with methyl chloroformate gave a substituted ylide, formed by C-alkylation (equation 14)[19]. A similar reaction involving 1-iodo-2, 2-di(isopropoxy)ethane was used to make ylide 20[75] (Section V.A.1).

$$
\text{Ph}_3\text{As}\!=\!\text{CHPh} \xrightarrow{\text{ClCOOMe}} [\text{Ph}_3\text{AsCH(Ph)COOMe}]^+\,\text{Cl}^-
$$

$$
\xrightarrow{-\text{HCl}} \text{Ph}_3\text{As}\!=\!\text{CPhCOOMe}
\tag{14}
$$

Arsonium ylides react with fluoroalkenes; the isolated products, after hydrolysis, are disubstituted arsonium ylides[107,108], e.g. equation 15. In some, but not all cases, the

$$
\text{Ph}_3\text{As}\!=\!\text{CHCN} + \text{CF}_2\!=\!\text{CFCF}_3 \longrightarrow [\text{Ph}_3\text{As}\!=\!\text{C(CN)CF}\!=\!\text{CFCF}_3]
$$

$$
\underset{\text{H}_2\text{O}}{\swarrow}
$$

$$
\text{Ph}_3\text{As}\!=\!\text{C(CN)COCHFCF}_3
\tag{15}
$$

prehydrolysis product has been isolated. Triphenylarsonium methylide reacted with hexafluorobenzene at room temperature to give a pentafluorobenzylide, which reacted with aryl aldehydes to give pentafluorostilbene derivatives in high yield (equation 16)[109].

$$2Ph_3As\!=\!CH_2 + C_6F_6 \xrightarrow[0-20\,°C]{Et_2O} Ph_3As\!=\!CHC_6F_5 \xrightarrow{ArCHO} ArCH\!=\!CHC_6F_5 \quad (16)$$

Acylation has frequently been used to convert arsonium ylides into other more stabilized ylides, reagents used being acid chlorides[21,110-113], acid anhydrides[10,21,114,115] or esters[21]. These examples all involve C-acylation but complications can arise, as shown in equation 17[21]. It was suggested that reaction with benzoyl bromide gives a kinetically controlled product which, in the presence of acetate, is converted into the thermodynamically controlled product[21]. In this case delocalization of the negative charge leads to acylation at a site other than the ylidic carbon atom.

$$(17)$$

Similarly with cyclopentadienylides, wherein the charge is delocalized around the ring, acylation takes place at other positions in the ring than the ylidic carbon atoms, but, since a carbon–carbon bond is formed, there is no tendency for the acyl group to migrate (equation 18)[50,89].

$$(18)$$

Aryldiazonium salts also react with cyclopentadienylides to give 2- or 3-phenylazo-derivatives[89,116].

Other reagents which react with arsonium ylides are phenyl sulphine and phenyl sulphene (equation 19)[32]. Ethyl cinnamate is also a product of the latter reaction,

$$Ph_3As\!=\!CHCOC_6H_4Br(p) + PhCH\!=\!SO \longrightarrow Ph_3As\!=\!C(SOCH_2Ph)COC_6H_4Br(p)$$
$$(19)$$

$$Ph_3As\!=\!CHCOOEt + PhCH\!=\!SO_2 \longrightarrow Ph_3As\!=\!C(SO_2CH_2Ph)COOEt$$

presumably formed by the route shown in equation 20.

$$(20)$$

The ylidic carbon atom of a number of acylylides reacted with tropylium bromide to give tropylarsonium salts (23), which decomposed by elimination of triphenylarsine to give β-acylstyrenes and β-acyl-β-tropylstyrenes[117]. Triphenylarsonium benzoylylide was also alkylated with 1-p-nitrobenzoylaziridine in boiling toluene to give 24[118].

$$\text{CH(COR)}\overset{+}{\text{A}}\text{sPh}_3 \quad \text{BF}_4^- \qquad \text{Ph}_3\text{As}=\text{C(COPh)CH}_2\text{CH}_2\text{NHCOC}_6\text{H}_4\text{NO}_2\text{-}p$$

(23)   (24)

A more complex reaction ensues between acylylides and dimethyl acetylenedicarboxylate, the final product presumably arising via a four-membered ring intermediate (equation 21)[32].

$$\begin{array}{c}
\text{Ph}_3\text{As}=\text{CHCOR} \\
+ \\
\text{MeOOCC}\equiv\text{CCOOMe}
\end{array}
\longrightarrow
\begin{array}{c}
\text{Ph}_3\text{As}-\text{CHCOR} \\
|\quad\quad| \\
\text{MeOOCC}=\text{CCOMe}
\end{array}$$

$$\longrightarrow \quad \text{Ph}_3\text{As}=\text{C(COOMe)C(COOMe)}=\text{CHCOR} \quad (21)$$

D. Formation of Cyclic Compounds from Arsonium Ylides

Cyclopropane derivatives have been prepared from reactions of arsonium ylides with conjugated enones[21,32,59,95-97,119a] and α, β-unsaturated esters[33,43,59]. Initial Michael-type reaction is followed by intramolecular elimination of triphenylarsine, e.g. equation 22[32]. These reactions often give high yields and show high stereoselectivity.

$$\text{ArCOCH}=\text{CHPh} + \text{Ph}_3\text{As}=\text{CH}_2 \rightarrow \text{ArCO}\overset{-}{\text{C}}\text{H}-\text{CHPh} \rightarrow \text{ArCOCH}-\text{CHPh}$$

(22)

A variety of heterocycles has been made from arsonium ylides. For example, as mentioned in Section V.A.6, 1,2,3-triphenylaziridine was obtained by reaction of triphenylarsonium benzylide with benzylideneaniline[32].

α-Pyrones result from reactions of acylylides with diphenylcyclopropenone, presumably via attack by the acyl oxygen atom of the ylide on the cyclopropene ring[10].

Reaction of acylylides with a 2-(cyclohexadienylidene)dimedone derivative resulted in the formation of a dihydrofuran ring[119b]. Furans have been made from reactions of 3-alkylsiloxyallylylides with aldehydes followed by acid hydrolysis of the resultant epoxides[80b].

Indoles have been prepared from reactions of o-aminophenylketones with reactive[120] or stable[121-123] arsonium ylides. Oxo-stabilized ylides reacted with 2-chloro-oximes to give trans-5-acyl-Δ^2-isoxazolines[124], and isoxazoles have been obtained from reactive arsonium ylides and α-isonitrosoketones[125], and from triphenylarsonium methylide and nitrile oxides[126]. The latter ylide reacts similarly with nitrile imines to give pyrazoles[126]. With triphenylarsonium benzylides and benzoylylides, benzene diazonium salts give 1,3,4,6-substituted 1,4-dihydro-1,2,4,5-tetrazines in a reaction in which initial coupling of the reagents is followed by a dimerisation[127].

E. Complexes Involving Arsonium Ylides

Either triphenylarsonium benzoylylide or dibenzoylylide form complexes with bis(1,5-cyclo-octadiene)nickel; these oligomerise ethylene under mild conditions[128]. The dibenzoyl derivative is the more effective catalyst of the two, this being ascribed to the greater stabilization provided by the second benzoyl group; the monobenzoyl derivative provides largely linear alkenes, but the dibenzoyl derivative leads to decreased linearity in the products.

VI. PREPARATION OF ARSONIUM YLIDES

A. Salt Method

The first account of the preparation of an arsonium ylide (1) (see Section I) involved what is known as the 'salt method', namely the reaction of a halogeno compound with triphenylarsine to form an arsonium salt, which is then treated with a suitable base to provide the ylide (see Scheme 1). Stable ylides are frequently made and isolated by using aqueous alkali. Reactive ylides need anhydrous conditions and the use of a suitable strong base, and are used *in situ*. Thus triphenylarsonium methylide (25) has been prepared in solution[32,41,54,129]; it was isolated by using as base sodium amide in tetrahydrofuran under an atmosphere of nitrogen[13,14]. Trimethylarsonium methylide (26) has been made indirectly, by desilylation of the trimethylsilyl methylide (27) with trimethylsilanol[22]. Ylide (27) was itself made by the salt method from chloromethyltrimethylsilane[130].

$$\text{Ph}_3\text{As}=\text{CH}_2 \quad \text{Me}_3\text{As}=\text{CH}_2 \quad \text{Me}_3\text{As}=\text{CHSiMe}_3$$
$$\textbf{(25)} \qquad\qquad \textbf{(26)} \qquad\qquad \textbf{(27)}$$

A number of stable arsine ylides (28), wherein COX represents a ketone, ester or amide function, have been prepared by the salt method[8,9,33,43,44,46,47,56,57,66,114,131] as have arsonium cyclopentadienylides[132] (29) and fluorenylides[37,53,58].

$$\text{R}_3\text{As}=\text{CR}^1\text{COX}$$
$$\textbf{(28)}$$

$$\textbf{(29)}$$

A phenyl group attached to the ylidic carbon atom has a smaller stabilizing effect and a number of examples of such semi-stablized ylides have been prepared by the salt method[84-86,94].

A publication[71] discussing the uses of reactive arsonium ylides for the stereospecific preparation of epoxides draws attention to the fact that arsonium salts are less readily prepared than phosphonium salts because of the poorer nucleophilicity of arsenic compared to phosphorus, and suggests methods for obtaining them. Primary salts were made from alkyl triflates, while α-branched salts were prepared from alkyldiphenylarsines, obtained from iodo compounds as, for example, in equation 23. Reaction of alkyl halides with arsines to form arsonium salts is also promoted by the presence of silver tetrafluoroborate[90].

$$\text{ICH}_2\text{CHMe}_2 \xrightarrow{\text{Ph}_2\text{AsLi}} \text{Ph}_2\text{AsCH}_2\text{CHMe}_2 \xrightarrow[\text{2,NaBF}_4]{\text{1, Bu}^t\text{Cl,AlCl}_3} \text{Bu}^t\text{Ph}_2\overset{+}{\text{As}}\text{CH}_2\text{CHMe}_2\,\text{Br}^-$$

$$(23)$$

Variants of this salt method include the use of 1,3-dihalogeno compounds, which undergo both substitution and elimination reactions to provide ylides. This method was used to prepare the cyclopentadienylide (30) (equation 24)[116].

B. Preparation from Arsine Dihalides

In the presence of triethylamine, triphenylarsine dichloride reacts with a variety of compounds having reactive methylene groups to give arsonium ylides (equation 25)[133]. This method is limited to compounds in which X, Y are electron-withdrawing groups, i.e. to the preparation of stable ylides.

$$Ph_3AsCl_2 + CH_2XY \xrightarrow{Et_3N} Ph_3As{=}CXY \qquad (25)$$

C. Preparation from Arsine Oxides

Compounds having reactive methylene groups also react with triphenylarsine oxide, either in acetic anhydride, or in triethylamine with phosphorus pentoxide also present, to give arsonium ylides. First applied to cyclopentadienes bearing either phenyl or acyl substituents[68,89,132,134], its use was extended to prepare a range of stable arsonium ylides[10,45]. When the reaction is carried out in acetic anhydride, acetylation may accompany the condensation reaction, e.g. equation 26[10,132,134].

$$Ph_3As{=}CHNO_2 \xleftarrow[P_2O_5]{Et_3N} CH_3NO_2 + Ph_3AsO \xrightarrow{Ac_2O} Ph_3As{=}CAcNO_2$$

The mechanism[10] involves initial formation of an acetylated or phosphorylated cation, which reacts with a carbanion to form a salt that is strongly acidic because of its substituent electron-withdrawing groups. This salt is hence readily converted into an ylide by loss of a proton, whose removal is assisted by the acetic anhydride or triethylamine (see equation 27). As with method B the arsine oxide method is limited to the preparation of stable ylides, since its success depends upon the acidity of the methylene compound. Almost all examples of this method have utilized triphenylarsine oxide; tri-n-butylarsine oxide has been used in triethylamine but gave only intractable products in acetic anhydride[10]. In a modification of this method, an ylide has been prepared by reaction of acetoacetanilide with diacetoxytriphenylarsorane, the latter compound having been prepared from triphenylarsine and lead tetra-acetate[135].

$$Ph_3AsO + Ac_2O \rightarrow [Ph_3AsOAc]^+$$

$$CH_2XY + Ac_2O \rightarrow [CHXY]^-$$

$$[Ph_3AsCHXY]^+[AcO]^-$$

$$\downarrow Ac_2O \qquad (27)$$

$$Ph_3As{=}CXY$$

X, Y = electron-withdrawing groups

D. Preparation from Diazo Compounds

Stable arsonium ylides can be prepared by heating a diazo compound in the presence of an arsine. Thermogravimetric analysis indicated that the diazo compound first decomposes to give a carbene which adds to the arsine, e.g. equation 28[136]. As first introduced for the preparation of arsonium cyclopentadienylides, the two reagents were simply mixed and heated together[137]. Subsequent improvements included plunging the mixture of the reagents into a preheated bath[68] and, above all, the use of copper or copper salts as catalysts[10,68]. Copper derivatives may not only catalyse the reaction, they may also promote ylide formation which does not take place in the absence of catalyst[10,68]. In particular, the use of catalysts such as copper acetylacetonate enables reactions to be carried out in solution[138], for example in a variety of solvents such as benzene, cyclohexane or ethanol, at temperatures well below the normal decomposition temperature of the diazo compound involved[68,138]. A particularly effective catalyst is copper hexafluoroacetylacetonate[110,139]; arsonium ylides have been prepared even at room temperature in its presence[110,140]. The function of the catalyst appears to be to bring the reactants into close proximity to each other through their co-ordination at the copper; a number of other metals have proved to be ineffective as catalysts[110,138]. This catalytic method greatly increases the scope of the reactants which may be employed. A range of stable arsonium ylides has been prepared in this way but attempts to obtain ylides from monoacyldiazo compounds such as ethyl diazoacetate were unsuccessful[10]

$$(28)$$

Occasionally unexpected products arise. Thus diazo-2,5-diphenylcyclopentadiene gave a 2,4-diphenylcyclopentadienylide, possibly due to rearrangement of the intermediate carbene[68], and 9-diazofluorene gave fluorenone ketazine, resulting from rapid reaction of the first-formed ylide with unchanged diazo compound[10,68].

E. Preparation from Other Ylides

A method which is related to the conversion of diazo compounds into arsonium ylides is the thermal decomposition of iodonium ylides in the presence of triphenylarsine, either by melting them together[141] or heating them in solution[138,142] in the presence of a copper derivative, e.g. equation 29.

$$(RCO)_2C{=}IPh \xrightarrow[Cu\ salt]{Ph_3As} (RCO)_2C{=}AsPh_3 \qquad (29)$$

Preparations of arsonium ylides by reactions of other arsonium ylides with suitable substrates such as acid chlorides[115], acid anhydrides[10,115], chlorosilanes[95], acetylenes[32], sulphines[32] or N-acylaziridines[118] have already been mentioned in Section V.C.

An interesting example of a conversion of a simple ylide into a more complex one is the reaction of triphenylarsonium methylide with a phosphonio derivative of a ketone (equation 30)[81].

$$Ph_3\overset{+}{P}CMe_2COCF_3 + Ph_3As=CH_2 \longrightarrow Ph_3As=CHC(CF)_3=CMe_2 + Ph_3PO \tag{30}$$

The process involves Wittig reaction type attack by the ylide at the carbonyl group followed by preferential elimination of triphenylphosphine oxide instead of triphenylarsine oxide.

F. Preparations Involving a Reverse-Wittig Reaction

Triphenylarsine oxide reacts with a number of electrophilic acetylenes having electron-withdrawing substituents in what are, in effect, reverse-Wittig reactions, thereby providing stable arsonium ylides (equation 31)[111]. Reaction is presumably initiated by Michael-type reaction of the oxide with the alkyne, as exemplified in equation 32. As would be expected from such a mechanism, use of an unsymmetric alkyne, as in the foregoing example, results in virtually regiospecific attack by the oxide to give the product shown.

$$Ph_3AsO + R^1C\equiv CR^2 \longrightarrow Ph_3As=CR^1COCR^2 \tag{31}$$

$$Ph_3AsO + HC\equiv CCOOMe \longrightarrow MeOO\overset{-}{C}\overset{\overset{\overset{+}{Ph_3As}}{|}}{-}C=CH \longrightarrow MeOOCC\overset{\overset{Ph_3As-O}{| \quad |}}{=}CH$$

$$Ph_3As=C(CHO)COOMe \tag{32}$$

G. Some Unusual Arsonium Ylides

A squaric acid derivative which is also an arsonium ylide has been made (equation 33)[143].

$$\text{(squaric acid structure)} + Ph_3As + H_2O \longrightarrow Ph_3As=\text{(structure)}=O \tag{33}$$

A cumulated arsonium ylide has been prepared by reaction of a methoxycarbonyl arsonium ylide with sodium bis(trimethylsilyl)amide (equation 34)[20]. This ylide undergoes

$$Ph_3As=CHCOOMe + NaN(SiMe_3)_2 \longrightarrow Ph_3As=C=C=O \tag{34}$$

reactions similar to its phosphonium analogue, as shown in equation 35, including some providing other arsonium ylides.

The bisylide **31** has been prepared by the salt method (equation 36); it is unstable as a solid and especially in solution[144]. Its central carbon atom could not be detected in its ^{13}C-NMR spectrum. The mixed phosphorus–arsenic ylide (**32**) was also prepared; it formed yellow crystals which decomposed easily[145].

$$Ph_2Me\overset{+}{As}CH_2\overset{+}{As}MePh_2 \ 2X^- \xrightarrow{\ NaNH_2\ } Ph_2MeAs{=}C{=}AsMePh_2$$

$$(31) \qquad\qquad (36)$$

$$Ph_2MeP{=}C{=}AsMePh_2$$

$$(32)$$

Ylide **33** has been prepared by deprotonation of dibenzyldiphenylarsonium bromide with butyllithium[146].

$$[PhCH{=}AsPh_2{-}\bar{C}HPh \longleftrightarrow Ph\bar{C}H{-}\overset{+}{As}Ph_2{-}\bar{C}HPh$$

$$\longleftrightarrow Ph\bar{C}H{-}AsPh_2{=}CHPh]Li^+$$

$$(33)$$

This reacted with hexanal in a Wittig-type reaction to give phenyhept-1-ene.

The arsa-alkyne **34**, although not strictly an ylide, is related to them. It forms stable pale yellow crystals[147].

(**34**)

VII. ARSINIMINES

Arsinimines **35** are the nitrogen analogues of arsonium ylides. They appear to be more resistant to hydrolysis than are ylides, for even the simple non-stabilized example (**35**,

$R^1 = Ph$, $R^2 = H$) can be handled in air, although it is less stable than its phosphonium analogue.

$$R^1_3As{=}NR^2 \longleftrightarrow R^1_3\overset{+}{As}{-}\overset{-}{N}R^2$$

$$(35)$$

A. Structure

An X-ray analysis of the arsinimine **36** has been carried out and confirms that it may be regarded as a hybrid of the three structures **a**, **b** and **c**[148]. The bond angles are of interest; hybridization about the ylidic nitrogen is near-trigonal, which favours interaction of the negative charge in an orthogonal p-orbital with the nitrile function, the N—C—N bonds are almost collinear and the arrangement of the bonds about the arsenic atom is almost tetrahedral. The As—N bond distance is intermediate between that expected for single or double bonding, in contrast to the corresponding bond distance in Ph_3PNCN[149] wherein the bond distance is almost identical with the sum of the covalent radii for doubly bonded $P{=}N$.

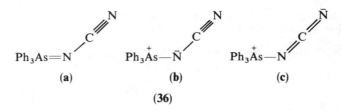

(a)   (b)   (c)

(36)

Infrared spectra of N-acylarsinimines show that the negative charge is delocalized onto the oxygen atom[150].

B. Stability and Reactions

Many arsinimines are stable solids, handleable in air; as in the case of arsonium ylides, the presence of electron-withdrawing substituents on the nitrogen atom increases their stability. When heated, N-benzoyltriphenylarsinimine decomposes to give triphenylarsine oxide and benzonitrile[150].

Arsinimines are hydrolysed by aqueous base[150,151] or acid[150], to triphenylarsine oxide and an amide. With hydrogen chloride they form arsonium halides[152].

Arsinimines have been alkylated, using methyl iodide[150,153], and acylated by acyl halides[152]. When an N-acylarsinimine was methylated, reaction took place at either the oxygen or the nitrogen atoms (equation 37)[150].

$$Ph_3As{=}NCOPh \xrightarrow{\text{MeI}} Ph_3\overset{+}{As}N{=}C(OMe)Ph\ I^- + Ph_3\overset{+}{As}NMeCOPh\ I^- \qquad (37)$$

N-Phenyltriphenylarsinimine reacted rapidly with benzophenone and with p-nitrobenzaldehyde or quinones, in reactions analogous to the Wittig reaction, to form imines (equation 38)[150]. A similar reaction with nitrosobenzene provided azobenzenes[154].

$$Ph_3As{=}NPh + R^1R^2CO \longrightarrow R^1R^2C{=}NPh + Ph_3AsO \qquad (38)$$

Other reactions include those with phenyl isothiocyanate or carbon disulphide to give a carbodiimide, and with sulphur dioxide to give an N-sulphinylaniline[153]. Reactions of N-phenyltriphenylarsinimine with triphenylacetonitrile-N-oxide provided an arsaheterocycle, and with dimethyl acetylenedicarboxylate gave a product which was probably an arsonium ylide (equation 39)[153]. A final reaction of some interest was that of N-tosyl-triphenylarsinimine with phenyllithium to give pentaphenylarsenane, Ph_5As[155].

$$Ph_3As=NPh \tag{39}$$

(probably)
$$Ph_3As=C(COOMe)C(COOMe)=NPh$$

C. Preparation

The type of methods used for the preparation of arsonium ylides, namely the salt method, use of arsine dihalides, condensation reactions with arsine oxides and trapping of carbenes by arsines, have all been applied to the preparation of arsinimines.

In contrast to the common use of the salt method for making ylides, there is only one example of its use for preparing an arsinimine, to provide what is as yet the simplest reported example (equation 40)[152]. Arsinimines have been prepared from arsine dihalides, using reactions between dibromotriphenylarsine and amides[150,156].

$$Ph_3As + ClNH_2 \longrightarrow Ph_3\overset{+}{As}NH_2Cl^- \xrightarrow{NaNH_2, NH_3} Ph_3As=NH \tag{40}$$

The first reported example of the preparation of an arsinimine, in 1937, utilized the reaction of the sodium salt of chloramine T with triphenylarsine (equation 41)[3]. This reaction, which has been repeated by later workers[155], is not straightforward. Earlier work had shown that chloramine T reacts with the arsine to convert it into an arsine oxide[151], which then condenses with the tosylamide to provide the final product. Other arsinimines have been made by the same method, but in the majority of cases were isolated as their water adducts[157]. In a modification of this reaction, chloramine T itself, rather than a salt, underwent an exothermic reaction with triphenylarsine in dry benzene, and the resultant intermediate, which was not isolated, gave, on treatment with copper powder, an arsinimine (equation 42)[158]. Other arsinimes have also been made by reactions of chloramines with arsines[159].

$$Ph_3As + [p\text{-}MeC_6H_4SO_2NCl]^- Na^+ \longrightarrow Ph_3As=NSO_2C_6H_4Me\text{-}p \tag{41}$$

$$p\text{-}MeC_6H_4SO_2NCl_2 \xrightarrow[PhH]{Ph_3As} \xrightarrow{Cu} p\text{-}MeC_6H_4SO_2N=AsPh_3 \tag{42}$$

Triphenylarsine oxide reacts with a variety of nitrogen compounds, namely aryl isocyanates[153,160], acyl isocyanates[150], N-sulphinylamines[160] and N-sulphinylamides[160,161] to give arsinimines. Nitriles that are fully substituted on the α-carbon atom add triphenylarsine oxide to give N-acylarsinimines[162]; other nitriles did not react, or instead condensed with the arsine oxide at the α-carbon atom to give ylides.

Closely related to the arsine oxide method is a ready way for preparing arsinimines by heating together triphenylarsine, an amide and lead tetraacetate in a one-pot reaction[135,163]. The reaction sequence, which is as shown in equation 43, is analogous to that suggested for the formation of arsonium ylides from triphenylarsine oxide in acetic anhydride[10]. Crystalline bisacetoxytriphenylarsine (37) has been isolated and shown to react with amides to give arsinimines[135,163]. Tosyl and mesyl amides and benzamide react with 37 at room temperature, but less nucleophilic amides required heating in boiling 1,2-dichloroethane for reaction to take place[135].

$$Ph_3As + Pb(OAc)_4 \longrightarrow Ph_3As(OAc)_2 \quad \substack{RSO_2NH_2 \\ \nearrow \\ \searrow \\ RCONH_2} \quad \substack{Ph_3As{=}NSO_2R \\ \\ Ph_3As{=}NCOR} \qquad (43)$$

(37)

As carbenes react with arsine to give ylides, so nitrenes react to give arsinimines[163,164]. The nitrenes were generated *in situ* by copper-catalysed decomposition either of azides[163,164] or 3-aryl-1,4,2-dioxazolidin-5-ones[164]. N-Ethoxycarbonyl- and N-p-tolyl-sulphonyl-triphenylarsinimines have been prepared by nitrene capture reactions in which the nitrenes were generated by the action of base on, respectively, a sulphonyloxyurethane and a sulphonamide[164], e.g. equation 44.

$$EtOOCNHOSO_2C_6H_4NO_2\text{-}p \xrightarrow{Et_3N} EtOOCN: \xrightarrow{Ph_3As} Ph_3As{=}NCOOEt \quad (44)$$

VIII. STIBONIUM YLIDES AND STIBINIMINES

A. Preparation of Stibonium Ylides

The first preparation of a stibonium ylide was probably achieved in 1953[37]. When dibenzyldimethylstibonium bromide was treated with phenyllithium, the product obtained was dimethyl(1,2-diphenylethyl)stibine which could arise from a Stephens rearrangement of the expected ylide (see equation 45); addition of the phenyllithium first of all afforded a yellow colour which rapidly disappeared.

$$(PhCH_2)_2\overset{+}{S}bMe_2 \ Br^- \xrightarrow{PhLi} \left[\substack{Ph\bar{C}H \\ \diagdown \\ \diagup \\ PhCH_2} \overset{+}{S}bMe_2 \right] \longrightarrow PhCH_2CHPhSbMe_2$$

(45)

In 1960 triphenylstibonium methylide, $Ph_3Sb{=}CH_2$, was obtained in solution by treating methyltriphenylstibonium tetrafluoroborate with phenyllithium in ether[54]. When the resultant solution was treated with benzophenone, acid work-up provided high yields of triphenylstibine and diphenylacetaldehyde[54], and it has been suggested[165] that the latter product arose by acid-induced rearrangement of initially formed 1,1-diphenylethylene oxide, the most likely product from reaction of this reactive ylide with a carbonyl compound.

The first stibonium ylide to be isolated was triphenylstibonium tetraphenylcyclopentadienylide, obtained by thermal decomposition of diazotetraphenylcyclopentadiene in the presence of triphenylstibine. A problem which arose initially in the preparation of stibonium ylides from thermal decomposition reactions of diazo compounds was that the temperatures required to achieve the latter often led to partial or complete decomposition of the required ylide. However, use of copper hexafluoroacetylacetonate as a catalyst[110,139], with the associated milder conditions that this permits, has enabled the preparation and isolation of a number of stibonium ylides, $Ph_3Sb{=}CX_2$, where $X{=}RCO$ or RSO_2, from diazo compounds in solution in benzene[11,110,140].

The preparation of a stibonium ylide in good yield from indanetrione has been reported (equation 46)[167].

$$\tag{46}$$

A number of tributylstibonium ylides have been prepared in solution by the salt method (equation 47); they were sensitive and could not be isolated in the presence of air[168].

$$Bu_3\overset{+}{S}bCH_2E\ Br^- \xrightarrow[\text{THF}]{\text{KOBu}^t} Bu_3Sb{=}CHE$$

$$\tag{47}$$

$$(E = COOMe, COOEt, CN, CONEt_2, CON\langle\ \rangle\,)$$

Treatment of benzyltributylstibonium bromide with lithium diisopropylamide and an aryl aldehyde led to the formation of a stilbene derivative and the related epoxide, indicating that an ylide had been formed and reacted with the aldehyde, but when the same salt or the corresponding benzyltriethyl salt were treated with an alkyl- or phenyl-lithium in the presence of an aryl aldehyde or cyclohexanone, the final product was a substituted alcohol (38); it was suggested that reaction proceeded via a pentaco-ordinate stiborane that to some extent dissociated into ionic fragments which reacted with the carbonyl compound, as shown in equation 48[169]. It had been reported previously that the reaction of phenyllithium with methyltriphenylstibonium iodide or tetrafluoroborate give a mixture of methyltetraphenylantimony and pentaphenylantimony[170]. Quaternary stibonium salts, $[Bu_3SbCH_2E]^+X^-$, in which E can act as an electron-withdrawing group ($E = CH{=}CH_2$, $CH{=}CHCOOEt$, COOEt, Ph, CN) also react with butyl- or phenyllithium to give penta-alkyl or tetra-alkylphenyl stiboranes, which react with aryl aldehydes to give, after hydrolysis, alcohols e.g. $PhCH(CH_2E)OH$[171].

$$\tag{48}$$

$$(38)$$

B. Stability and Reactions

As in the case of arsonium ylides, the stability of stibonium ylides depends very much upon the substituents present on the ylidic carbon atom. Ylides having two electron-withdrawing-substituents are stable in a dry atmosphere, but other stibonium ylides are less stable than their arsonium or phosphonium analogues, for example triphenyl-stibonium tetraphenylcyclopentadienylide decomposes when heated in ethanol or nitromethane[166]. Even the more stabilized ylides decompose when heated in higher boiling solvents, including aprotic solvents[11]. All stibonium ylides are readily hydrolysed in protic solvents if any moisture is present, giving a stibine oxide[11,172]. The difference between stibonium ylides and arsonium and phosphonium ylides has been attributed to greater polarity of the C=Sb bond compared to either C=P or C=As bonds. Triphenylstibonium tetraphenylcyclopentadienylide is more basic that its phosphonium or arsonium analogues[38].

An interesting property of the stibonium ylides with two electron-withdrawing substituents is that they very readily gain static and adhere to glass surfaces and to spatulae[11].

It was mentioned above in Section VIII.A that triphenylstibonium methylide reacted readily with benzophenone and, as might be expected for a reactive ylide, formed an epoxide which, however, rearranged to diphenylacetaldehyde[54]. Triphenylstibonium tetraphenylcyclopentadienylide was shown to be more reactive than its arsonium or phosphonium analogues towards benzaldehydes and nitrosobenzene in Wittig reactions, in keeping with the greater polarity of its ylidic bonding[38]. As a stable ylide it gave alkenes on reaction with aldehydes but, in contrast, the product from nitrosobenzene was a nitrone.

As mentioned in Section VIII.A, tributylstibonium benzylide reacted with aryl aldehydes and, like arsonium benzylides, gave mixtures of alkene and epoxide[169].

Tributylstibonium ylides having one electron-withdrawing substituent react readily with aldehydes or ketones to give E-alkenes[168]. Since the ylides are sensitive to air they were prepared *in situ* and not isolated. Evidence for their formation was provided by trapping the initial adduct formed from the ylide and aldehyde with trimethylsilyl chloride, as shown in equation 49[168].

$$Bu_3Sb{=}CHCOOMe + ArCHO \xrightarrow{-78\,°C} \begin{array}{c} ArCH{-}CHCOOMe \\ | \qquad\quad | \\ O^- \quad\ ^+SbBu_3 \end{array}$$

$$\xrightarrow[{-78\,°C}]{MeSiCl} \left[\begin{array}{c} Ar{-}CH{-}CHCOOMe \\ | \qquad\qquad | \\ Me_3SiO \quad\ ^+SbBu_3 \end{array} \right] Cl^-$$

$$\Big\downarrow Bu^tOK$$

$$Me_3SiOCHAr{-}C(COOMe){=}CHAr \xleftarrow[{-78\,°C\ to\ r.t.}]{ArCHO} Me_3SiOCHAr{-}C(COOMe){=}SiBu_3 \tag{49}$$

Stibonium ylides bearing two electron-withdrawing substituents did not undergo Wittig reactions[11].

As happens with comparable arsenic compounds, some α-halogeno esters and amides react with oxo compounds in the presence of stibines, but without the need of base, to form arylidene or alkylidene derivatives[169b,173,174]. The mechanism is thought to be the same as for the arsenic analogues, namely via initial formation of a stibonium salt (equation 50).

$$BrCH(COOMe)_2 + R_3^1Sb \longrightarrow [R_3^1SbBr]^+[CH(COOMe)_2]^-$$

$$R^2CHO$$

$$R^2CH{=}C(COOMe)_2$$

(50)

Cyclopropane derivatives have been prepared similarly from dibromo compounds, $Br_2CXCOOR$, where X is another electron-withdrawing group, and conjugated alkenes, $R^1CH{=}CHY$ (Y = electron-withdrawing group), in the presence of tributylstibine and again formation of a stibonium salt as an intermediate is postulated (equation 51)[169b,175].

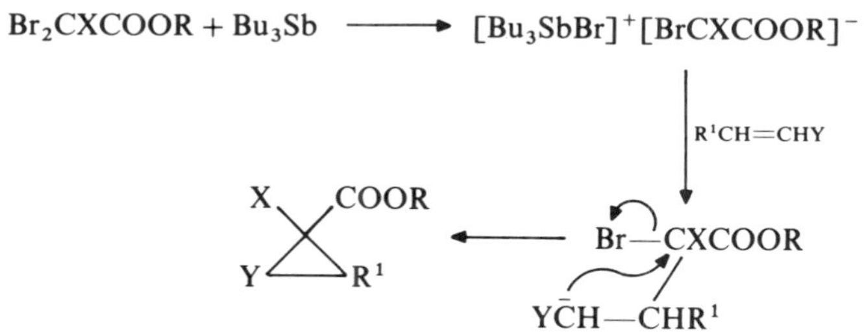

$$Br_2CXCOOR + Bu_3Sb \longrightarrow [Bu_3SbBr]^+[BrCXCOOR]^-$$

$$R^1CH{=}CHY$$

(51)

It has also proved possible to make alkenes in a one-pot process from diazo compounds N_2CR_2, where R is an electron-withdrawing group, and aldehydes or ketones by heating them in the presence of tributylstibine; no base was required[176]. Reaction was assumed to proceed through a stibonium ylide as intermediate, although it proved impossible to isolate this ylide, which probably reacted further too rapidly to allow this.

C. Spectra and Structure

The electronic spectra of the triphenylphosphonium, triphenylarsonium and triphenyl-stibonium tetraphenylcyclopentadienylides confirmed the impression derived from their stabilities, basicities and reactivities in Wittig-type reactions, that the polarity of the ylidic bonding increased in the order P, As, Sb and concomitantly, the double-bond character decreased. Thus the longest-wavelength absorption peaks were at 288 nm (P), 291 nm (As), 349 nm (Sb); this was attributed to the less efficient overlap between the 2p-orbitals of the ylidic carbon atom and the d-orbitals of antimony, because of the greater size and diffuseness of the d-orbitals on going down the Periodic Table[38,49]. None of these compounds was solvatochromic[38].

When the stibonium ylides (39) became available, they provided a more complicated picture[11]. In their IR spectra the frequencies of the CO and SO_2 groups become steadily lower on going from phosphonium to arsonium to stibonium ylides, consistent with an expected increase in polarity of ylides on going down the Periodic Table.

$$Ph_3Sb{=}CX_2 \quad (X = RCO \text{ or } RSO_2)$$

(39)

However, the electronic spectra of the stibonium ylides were very similar to those of their arsonium analogues, both in solution and in the solid state, and their dipole moments were almost the same, indicating an overall similarity in the electronic structure of the two sets of ylides. Furthermore the stibonium ylides appeared to be no more reactive than the arsenic ylides. ^{13}C-NMR spectra even suggested that there is more negative charge on the ylidic carbon atom in the arsonium ylides than in the stibonium ylides. Thus the lone piece of evidence, which at first sight appears to indicate greater polarity in the stibonium ylides, comes from their IR spectra[11].

The difference between the ylides (39) and the tetraphenylcyclopentadienylide is that the former, but not the latter, have substituent groups, carbonyl or sulphonyl, which can interact intramolecularly with the antimony or arsenic atom. This interaction has been clearly shown by X-ray crystal structures[11].

These X-rays studies show that the interaction is greater in the case of antimony than in the case of arsenic. Despite the greater size of antimony than arsenic, the $Sb\cdots O$ distance is shorter than the $As\cdots O$ distance. In addition the Sb—C(ylidic) and As—C(ylidic) bond distances indicate that the Sb—C bond has *more* double-bond character than the As—C bond. In all cases these ylides take up Z,Z-conformations, both in solution (as observed by NMR spectra) and in the solid state, but in the solid state they are not symmetric, only one of the substituent groups being involved in the intramolecular interaction. This can be represented as in 40 and 41. In accord with such structures one of the bonds linking substituents to the ylidic carbon atom is shorter than the other.

(40)   (41)   M = As, Sb

In the case of the tetraphenylcylopentadienylides the comparative properties of the different ylides depend almost entirely on the nature of the ylidic bond alone; this is no longer so in the case of the acyl- and sulphonyl-substituted ylides. Here the interaction, and the extent of the interaction, between the hetero atom and the substituent oxygen atoms plays a crucial role. Three canonical structures (42, 43, 44) must be considered. The X-ray studies indicate that structure 44 is more important for the stibonium ylides than for arsonium ylides. This is presumably because of the greater intrinsic polarity of stibonium ylides. Structures 42 contribute little; the increased charge is, however, dissipated by the contribution of 44. The greater polarity of the ylidic bond in triphenylstibonium tetra-phenylcyclopentadienylide compared to its phosphonium and arsonium analogue no longer obtains. Thus there is little difference in the apparent dipolarity, as indicated by physical and chemical properties, between the stibonium and arsonium ylides in the cases of ylides 40 and 41. For the arsonium ylides the intrinsic polarity is less, but the contribution of structure 43 as compared with 44 is greater.

(42)   (43)   (44)

The one physical feature which shows greater polarity of the stibonium ylides, namely their IR spectra, does so just because it is concerned with the CO or SO groups.

This chelation between substituent carbonyl or sulphonyl groups and the hetero atom also contributes to the unreactivity of such ylides towards electrophiles, for example in Wittig reactions, because it leads to a diminished negative charge on the ylidic carbon atom. Furthermore, these X-ray studies show that these ylides already possess structures closely resembling the transition states for Wittig-type reactions, which will also inhibit their participation in intermolecular Wittig reactions. It may be noted that some arsonium ylides having powerful electron-withdrawing groups which, for electronic or steric reasons, will interact less with the arsenic atom [e.g. $Ph_3As=C(COOEt)_2$, $Ph_3As=C(CN)_2$] do undergo Wittig reactions[10].

D. Stibinimines

A stibinimine has been prepared by reaction of triphenylstibine with chloramine T under conditions analogous to those used to prepare its arsenic counterpart[177]. This stibinimine is stable in air, but is hydrolysed by water. Like its arsenic analogue it reacted with phenyllithium to give pentaphenylantimony. Other related stibinimines have been prepared similarly[13].

IX. BISMUTHONIUM YLIDES AND IMINES

A. Preparation of Bismuthonium Ylides

The first bismuthonium ylide to be prepared and isolated was triphenylbismuthonium tetraphenylcyclopentadienylide, made by thermal decomposition of diazotetraphenylcyclopentadiene in the presence of triphenylbismuth[178]. More recently some other bismuthonium ylides ($Ph_3Bi=CX_2$, X = RCO or RSO_2) have been prepared and isolated by the diazo method, using copper hexafluoroacetylacetonate as catalyst[140].

Stabilized bismuthonium ylides have also been made from compounds having reactive methylene groups, CH_2CXY, X, Y = RCO or RSO_2, either by making their sodium salts and letting the latter react with dichlorotriphenylbismuth, or by reaction with triphenylbismuth oxide[179–181].

Triphenylbismuthonium 4,4-dimethyl-2,6-dioxocyclohexylide has also been made by heating dimedone with triphenylbismuth carbonate in refluxing dichloromethane[182].

The latter ylide has also been made in a transylidation reaction from the corresponding bismuthonium diacetylylide and dimedone in the presence of triethylamine; no reaction took place in the absence of the amine[181].

B. Stability and Reactions

Triphenylbismuthonium tetraphenylcyclopentadienylide is stable for some time as a solid but decomposes rapidly in solution; it decomposed rapidly when treated with acid or with base[177]. Acyl and sulphonyl bismuthonium ylides are moderately stable, but less so than their arsonium and stibonium analogues; some, while stable in solution in inert solvents, decompose when isolation is attempted[11,181].

The bismuthonium ylides (**45**, X = CH_2, O) react with aryl or alkyl aldehydes to give a variety of products, depending on the ylide, the aldehyde (RCHO) and the conditions. The products include cyclopropane derivatives (**46**), furan derivatives analogous to **46** but with a furan ring replacing the cyclopropane ring, and in some cases (from **45**, X = O and p-methoxybenzaldehyde or cinnamaldehyde) the straightforward product from a Wittig reaction, **47**[183].

(45)   (46)   (47)

Ylide (**45**, X = CH$_2$) reacts with a number of alkyl and aryl alkynes, in the presence of copper(I) chloride as catalyst, to form bicyclic furan derivatives (equation 52)[184]. This reaction probably involves a carbene as intermediate.

$$\text{(52)}$$

Reactions of some bismuthonium diacylylides with methanesulphonyl chloride and triethylamine give rise to 1, 3-oxathiolane derivatives, e.g. equation 53[180].

$$(MeCO)_2 C = BiPh_3 \xrightarrow{MeSO_2Cl, Et_3N} \quad \text{(53)}$$

A number of triphenylbismuthonium diacylylides and also the diphenylsulphonylylide undergo transylidation when treated with dimethyl sulphide in the presence of copper(I) chloride in benzene at room temperature, giving the corresponding dimethylsulphonium ylides; no such reaction took place with diphenyl sulphide[181]. Some of these bismuthonium ylides reacted with triphenylarsine or triphenylphosphine and were converted thereby into arsonium or phosphonium ylides; they did not react with triphenylstibine[181]. The conversion of the diacetylylide into the 4,4-dimethyl-2,6-dioxocyclohexylide as mentioned in Section IX.A should also be noted.

C. Structure of Bismuthonium Ylides

Triphenylbismuthonium tetraphenylcyclopentadienylide is an intensely blue solid, unlike its phosphonium, arsonium or stibonium analogues which are yellow. Its electronic spectrum closely resembles that of pyridinium tetraphenylcyclopentadienylide[38], and, like the latter compound but unlike its P, As and Sb analogues, it is solvatochromic, giving, for example, blue solutions in ether or benzene but red-purple solutions in alcohols[38,178]. In nitrogen ylides pπ–dπ orbital overlap is not possible; the spectroscopic similarity between the triphenylbismuthonium and pyridinium tetraphenylcyclopentadienylides and the huge difference between their spectra and those of the related stibonium, arsonium, phosphonium and sulphonium analogues can be attributed to the fact that the 6d-orbitals or bismuth are likely to be much too large and too diffuse to provide any effective overlap, and lends substance to the concept of the influence of $p\pi-d\pi$ orbital overlap on the properties of ylides. If such $p\pi-d\pi$ overlap between the ylidic carbon atom and the heteronium atom contributes to the stability of an ylide, then it is understandable that this bismuthonium ylide is much less stable than its analogues with other elements.

The bismuthonium acyl- and sulphonyl-ylides have a completely different appearance and electronic spectra, all of them being yellowish in colour. This is because in their cases, as in those of similar arsonium and stibonium ylides, interaction between an oxygen atom and the bismuth atom is possible, leading to structures such as **48**; in accord with this

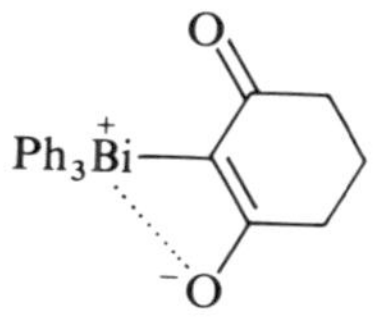

(48)

formulation the carbonyl group provides a broad peak at 1440 cm^{-1} in the IR spectrum, indicative of a very polarized carbonyl group[11]. An X-ray crystal structure of the 4,4-dimethyl derivative of **48** confirms this picture; the ylidic Bi—C bond is only slightly shorter than a Bi—C(phenyl) bond, indicating a largely ionic character and little double-bond character; the molecule is not symmetric, but one oxygen atom is nearer than the other to the bismuth atom[185].

D. Bismuthimines

Bismuthimines have been prepared by the reaction of chloramine T with triphenylbismuth[163,186] and other triarylbismuths[186]. When it was first prepared, it was not isolated but treated at once with phenyllithium to give pentaphenylbismuth[163]. These imines are moderately stable under dry nitrogen but decompose readily in air or in hydroxylic solvents. The following reactions of the triphenylbismuthimine **49** were reported (equations 54)[186].

X. REFERENCES

1. A. Michaelis, *Ann. Chem.*, **321**, 174 (1902).
2. F. Krohnke, *Chem. Ber.*, **83**, 291 (1950).
3. F. G. Mann and E. J. Chaplin, *J. Chem. Soc.*, 527 (1937).
4. Huang Yaozeng and Shen Yanchang, *Adv. Organomet. Chem.*, **20**, 115 (1982).
5. D. Lloyd, I Gosney and R. A. Ormiston, *Chem. Soc. Rev.*, **16**, 45 (1987).
6. H. Lumbroso, D. Lloyd and G. S. Harris, *C. R. Seances Acad. Sci.*, **C278**, 219 (1974).
7. A. W. Johnson and J. O. Martin, *Chem. Ind. (London)*, 1726 (1965).

8. N. A. Nesmeyanov, V. V. Mikulshin and O. A. Reutov, *J. Organomet. Chem.*, **13**, 263 (1968).
9. A. J. Dale and P. Frøyen, *Acta Chem. Scand.*, **24**, 3772 (1970).
10. I. Gosney and D. Lloyd, *Tetrahedron*, **29**, 1697 (1973).
11. G. Ferguson, C. Glidewell, I. Gosney, D. Lloyd, S. Metcalfe and H. Lumbroso, *J. Chem. Soc. Perkin Trans.*, **2**, 1829 (1988).
12. E. E. Ernstbrunner and D. Lloyd, *Chem. Ind. (London)*, 1332 (1971).
13. Y. Yamamoto and H. Schmidbaur, *J. Chem. Soc., Chem. Commun.*, 668 (1975).
14. H. Schmidbaur, W. Richter, W. Wolf and F. H. Köhler, *Chem. Ber.*, **108**, 2649 (1975).
15. G. Fronza, P. Bravo and C. Ticozzi, *J. Organomet. Chem.*, **157**, 299 (1978).
16. G. Ferguson, I. Gosney, D. Lloyd and B. Ruhl, *J. Chem. Res. (S)* 260, *(M)* 2140 (1987).
17. P. Frøyen and D. G. Morris, *Acta Chem. Scand., Ser. B*, **30**, 435 (1976).
18. K-H. A. O. Starzewski, W. Richter and H. Schmidbaur, *Chem. Ber.*, **109**, 473 (1976).
19. A. J. Dale and P. Frøyen, *Acta Chem. Scand.*, **25**, 1452 (1971).
20. H. J. Bestmann and R. K. Bansal, *Tetrahedron Lett.*, 3839 (1981).
21. A. W. Johnson and H. Schubert, *J. Org. Chem.*, **35**, 2678 (1970).
22. H. Schmidbaur and W. Tronich, *Inorg. Chem.*, **7**, 168 (1968).
23. Meiching Yaozeng, Xianglin Jin, Yougi Tang, Quichen Huang and Yaozeng Huang, *Tetrahedron Lett.*, 5343 (1982).
24. Xia Zong-Xiang and Zhang Zhi-ming, *Acta Chim. Sinica, Engl. Ed.*, **41**, 148 (1983).
25. Fan Zhao-Chang and Shen Yan-Chang, *Acta Chim. Sinica, Engl. Ed.*, **42**, 759 (1984).
26. G. Ferguson, D. F. Rendle, D. Lloyd and M. I. C. Singer, *J. Chem. Soc., Chem. Commun.*, 1647 (1971).
27. G. Ferguson and D. F. Rendle, *J. Chem. Soc., Dalton Trans.*, 1284 (1975).
28. G. Ferguson and D. F. Rendle, *J. Chem. Soc., Dalton Trans.*, 171 (1976).
29. G. Ferguson, I. Gosney, D. Lloyd and B. L. Ruhl, *J. Chem. Res. (S)* 260, *(M)* 2140 (1987).
30. H. Lumbroso, D. M. Bertin and P. Frøyen, *Bull. Soc. Chim. Fr.*, 819 (1974).
31. N. A. Nesmeyanov, V. V. Pravdina and O. A. Reutov, *Zh. Org. Khim.*, **3**, 598 (1967); *J. Org. Chem. USSR (Engl. Transl.)*, **3**, 574 (1967).
32. S. Trippett and M. A. Walker *J. Chem. Soc., C*, 1114 (1971).
33. Huang Yao-tseng, Shen Yan-Chang, Ma Jing-ji and Xin Yuan-kong, *Acta Chim. Sinica*, **38**, 185 (1980).
34. Huang Yao-tseng, Shen Yan-Chang, Xin Yuan-kong and Ma Jing-ji, *Acta Chim. Sinica, Engl. Ed.*, **23**, 1396 (1980).
35. A. W. Johnson, *Ylid Chemistry*, Academic Press, New York, 1966, p. 105.
36. S. Trippett and M. A. Walker, *J. Chem. Soc.*, 3874 (1959).
37. G. Wittig and H. Laib, *Ann. Chem.*, **580**, 57 (1953).
38. B. H. Freeman, D. Lloyd and M. I. C. Singer, *Tetrahedron*, **28**, 343 (1972).
39. I. Gosney, D. Lloyd and W. A. MacDonald, unpublished work.
40. Reference 35, p. 292.
41. D. Seyferth and H. M. Cohen, *J. Inorg. Nucl. Chem.*, **20**, 73 (1961).
42. H. Schmidbaur and W. Richter, *Angew. Chem.*, **87**, 204 (1975); *Angew. Chem., Int. Ed. Engl.*, **14**, 183 (1975).
43. G. Asknes and J. Songstad, *Acta Chem. Scand.*, **18**, 655 (1964).
44. P. Frøyen, *Acta Chem. Scand.*, **25**, 2541 (1971).
45. G. S. Harris, D. Lloyd, W. A. MacDonald and I. Gosney, *Tetrahedron*, **39**, 297 (1983).
46. K. Isslieb and R. Lindner, *Ann. Chem.*, **707**, 120 (1967).
47. A. W. Johnson and R. T. Amel, *Can. J. Chem.*, **46**, 461 (1968).
48. A. W. Johnson and R. B. LaCount, *Tetrahedron*, **9**, 130 (1960).
49. D. Lloyd and M. I. C. Singer, *Chem. Ind. (London)*, 1277 (1968).
50. D. Lloyd and M. I. C. Singer, *Tetrahedron*, **28**, 353 (1972).
51. Reference 35, pp. 284–299.
52. W. Heffe, Dissertation, University of Berlin, 1937; quoted by G. Wittig, *Pure Appl. Chem.*, **9**, 249 (1964) (W. Heffe was a student with F. Krohnke).
53. A. W. Johnson, *J. Org. Chem.*, **25**, 183 (1960).
54. M. C. Henry and G. Wittig, *J. Am. Chem. Soc.*, **82**, 563 (1960).
55. A. Maccioni and M. Secci, *Rend. Seminario Fac. Sci. Univ. Cagliari*, **34**, 328 (1964).
56. N. A. Nesmeyanov, V. V. Pravdina and O. A. Reutov, *Izv. Akad. Nauk SSSR, Ser. Khim.*, 1474 (1965); *Bull. Acad. Sci. USSR, Div. Chem. Sci.*, 1434 (1965).

57. N. A. Nesmeyanov, V. V. Pravdina and O. A. Reutov, *Dokl. Akad. Nauk SSSR*, **155**, 1364 (1964); *Proc. Acad. Sci. USSR*, **155**, 424 (1964).
58. R. S. Tewari and K. C. Gupta, *Indian J. Chem., Sect. B.*, **16**, 623 (1978).
59. Yaozeng Huang, Yuanyao Xu and Shong Li, *Org. Prep. Proced. Int.*, **14**, 373 (1982).
60. P. Bravo, C. Ticozzi and A. Cezza, *Gazz. Chim. Ital.*, **105**, 109 (1975).
61. R. S. Tewari and K. C. Gupta, *Indian J. Chem., Sect. B*, **14**, 419 (1976); **17**, 637 (1979).
62. R. S. Tewari and S. C. Chaturvedi, *Synthesis*, 616 (1978).
63. Yaozeng Huang, Lilan Shi and Jianhua Yang, *Tetrahedron Lett.*, **26**, 6447 (1985).
64. Y. Z. Huang, N. T. Hsing, L. L. Shi, F. L. Ling and Y. Y. Xu, *Acta Chim. Sinica*, **39**, 348 (1981).
65. W. Y. Ting, H. S. Sheng, W. Y. Shen and Y. T. Huang, *Bull. Nat. Sci. Univ. Chem. Eng. Sect.*, 540 (1965).
66. Y. T. Huang, W. Y. Ting and H. Sheng, *Acta. Chim. Sinica*, **31**, 38 (1965).
67. Huang Yao-zeng, Shi Li-lan, Li Bin-quan and Ling Fang-le, *Acta Chim. Sinica*, **41**, 269 (1983).
68. B. H. Freeman and D. Lloyd, *Tetrahedron*, **39**, 297 (1983).
69. B. M. Boubia, A. Mann, F. D. Bellamy and C. Mioskowski, *Angew. Chem.*, **102**, 1522 (1990); *Angew. Chem., Int. Ed. Engl.*, **29**, 1454 (1990).
70. Lilan Shi, Weibo Wang and Yaozeng Huang, *Tetrahedron Lett.*, **29**, 5295 (1988).
71. W. C. Still and V. J. Novack, *J. Am. Chem. Soc.*, **103**, 1283 (1981).
72. D. G. Allen, N. K. Roberts and S. B. Wild, *J. Chem. Soc., Chem. Commun.*, 346 (1978).
73. D. G. Allen, C. L. Raston, B. W. Skelton, A. H. White and S. B. Wild, *Aust. J. Chem.*, **37**, 1141 (1984).
74. D. G. Allen and S. B. Wild, *Organometallics*, **2**, 394 (1983).
75. P. Chabert, J. B. Ousset and C. Mioskowski, *Tetrahedron Lett.*, **30**, 179 (1989).
76. P. Chabert and C. Mioskowski, *Tetrahedron Lett.*, **30**, 6031 (1989).
77. Yaozeng Huang, Yanchang Shen, Jianhua Zheng and Shixiang Zhang, *Synthesis*, 57 (1985).
78. Lilan Shi, Wengjuan Xiao, Xueging Wen and Yaozeng Huang, *Synthesis*, 370 (1987).
79. J. B. Ousset, C. Mioskowski and G. Solladié, *Tetrahedron Lett.*, 4419 (1983).
80. (a) J. B. Ousset, C. Mioskowski and G. Solladie, *Synth. Commun.*, **13**, 1193 (1983).
    (b) Sunggak Kim and Yong Gil Kim, *Tetrahedron Lett.*, **32**, 2913 (1991).
81. Yanchang Shen, Qimu Liao and Weiming Qia, *J. Chem. Soc., Chem. Commun.*, 1309 (1988).
82. Yanchang Shen and Qimu Liao, *J. Organomet. Chem.*, **346**, 181 (1988).
83. Wei-bo Wang, Li-lan Shi, Zhi-qun Li and Yaozeng Huang, *Tetrahedron Lett.*, **32**, 3999 (1991).
84. P. S. Kendurkar and R. S. Tewari, *J. Organomet. Chem.*, **60**, 247 (1973); **85**, 173 (1975).
85. N. Kumari, P. S. Kendurkar and R. S. Tewari, *J. Organomet. Chem.*, **96**, 237 (1975).
86. P. S. Kendurkar and R. S. Tewari, *J. Organomet. Chem.*, **108**, 175 (1976).
87. R. S. Tewari and S. Gupta, *J. Organomet. Chem.*, **112**, 279 (1976).
88. R. Broos and M. J. Anteunis, *Bull. Soc. Chim. Belg.*, **97**, 271 (1988).
89. J. D. Hsi and M. Koreeda, *J. Org. Chem.*, **54**, 3229 (1989).
90. D. Lloyd and N. W. Preston, *J. Chem. Soc. (C)*, 2464 (1969).
91. S. G. Dwyer, Ph.D. Thesis, University of North Dakota, 1970.
92. D. W. Allen and G. Jackson, *J. Organomet. Chem.*, **110**, 315 (1976).
93. I. Gosney, T. J. Lillie and D. Lloyd, *Angew. Chem.*, **89**, 502 (1977); *Angew. Chem., Int. Ed. Engl.*, **16**, 487 (1977).
94. R. S. Tewari and S. C. Chaturvedi, *Tetrahedron Lett.*, 3843 (1977); *Indian J. Chem., Sect. B*, **18**, 859 (1979).
95. Yanchang Shen, Zhengziang Cu, Weiyn Ding and Yaozeng Huang, *Tetrahedron Lett.*, **25**, 4425 (1984).
96. Yanchang Shen and Qimu Liao, *Synthesis*, 321 (1988).
97. Yanchang Shen and Qimu Liao, *J. Organomet. Chem.*, **371**, 31 (1989).
98. Lilan Shi, Wenjuan Xiao, Jianhua Yang, Xueging Wen and Yaozeng Huang, *Tetrahedron Lett.*, **28**, 2155 (1987).
99. Lilan Shi, Jianhua Yang, Xueging Wen and Yaozeng Huang, *Tetrahedron Lett.*, **29**, 3949 (1988).

100. R. S. Tewari, S. K. Suri and K. C. Gupta, *Z. Naturforsch., B*, **35**, 95 (1980).
101. R. S. Tewari, S. K. Suri and K. C. Gupta, *Synth. Commun.*, **10**, 457 (1980).
102. N. A. Nesmeyanov, E. V. Binshtok, O. A. Rebrova and O. A. Reutov, *Izv. Akad. Nauk SSSR, Ser. Khim.*, 2113 (1972); *Bull. Acad. Sci. USSR, Div. Chem. Sci.*, 2056 (1972).
103. Yanchang Shen and Baozhen Yang, *J. Organomet. Chem.*, **375**, 45 (1989).

104. R. M. Moriarty, I. Prakash and W. A. Freedman, *J. Am. Chem. Soc.*, **106**, 6082 (1984).
105. F. Ramirez and S. Dershowitz, *J. Org. Chem.*, **22**, 41 (1957).
106. A. W. Johnson and R. T. Amel, *J. Org. Chem.*, **34**, 1240 (1969).
107. Y. T. Huang, W. Y. Ding, W. Cai, J. J. Ma and Q. W. Wang, *Sci. Sinica*, **24**, 189 (1981); *Chem. Abstr.*, **95**, 97923a (1981).
108. Ding Wei-yu, Cai Wen, Dai Jin-Shan, Huang Yaozeng and Zhen Jian-hua, *Acta Chim. Sinica*, **41**, 67 (1983).
109. Y. Shen and M. Qiu, *Synthesis*, 65 (1987).
110. C. Glidewell, D. Lloyd and S. Metcalfe, *Tetrahedron*, **42**, 3887 (1986).
111. E. Ciganek, *J. Org. Chem.*, **35**, 1725 (1970).
112. K. C. Gupta and R. S. Tewari, *Indian J. Chem.*, **13**, 834 (1975).
113. P. S. Kendurkar and R. S. Tewari, *J. Organomet. Chem.*, **102**, 141 (1975).
114. R. K. Bansal and G. Bhagchandani, *J. Prakt. Chem.*, **323**, 49 (1981).
115. R. S. Tewari and D. K. Nagpal, *Z. Naturforsch., B*, **35**, 99 (1980).
116. B. H. Freeman and D. Lloyd, *J. Chem. Soc. (C)*, 3164 (1971).
117. G. Covicchio, M. D'Antonio, G. Gaudiano, V. Marchetti and P. P. Ponti, *Tetrahedron Lett.*, 3493 (1977).
118. H. W. Heine and G. D. Wachob, *J. Org. Chem.*, **37**, 1049 (1972).
119. (a) N. A. Nesmeyanov and V. V. Mikul'shina, *Zh. Org. Khim.*, **7**, 696 (1971).
     (b) V. G. Kharitonov, V. I. Nikanorov, S. V. Sergeev, M. Galakhov, S. O. Yakushin, V. V. Mikul'shina, V. I. Rozenberg and O. A. Reutov, *Dokl. Akad. Nauk SSSR*, **319**, 177 (1991).
120. P. Bravo, G. Gaudiano, P. P. Ponti and M. G. Zubiani, *Tetrahedron Lett.*, 4535 (1970).
121. R. K. Bansal and S. K. Sharma, *Tetrahedron Lett.*, 1923 (1977).
122. R. K. Bansal and S. K. Sharma, *J. Organomet. Chem.*, **149**, 309 (1978).
123. R. K. Bansal and G. Bhagchandari, *Bull. Chem. Soc. Jpn.*, **53**, 2423 (1980).
124. P. Bravo, G. Gaudiano, P. P. Ponti and C. Ticozzi, *Tetrahedron*, **28**, 3845 (1972).
125. P. Bravo, G. Gaudiano and C. Ticozzi, *Gazz. Chem. Ital.*, **102**, 395 (1972).
126. G. Gaudiano, C. Ticozzi, A. Umani-Ronchi and A. Selva, *Chem. Ind. (Milan)*, **49**, 1343 (1967).
127. R. K. Bansal and S. K. Sharma, *J. Organomet. Chem.*, **155**, 293 (1978).
128. W. Keim, A. Behr, B. Limbäcker and C. Krüger, *Angew. Chem.*, **95**, 505 (1983); *Angew. Chem. Int. Ed. Engl.*, **22**, 503 Suppl. 655 (1983).
129. S. O. Grim and D. Seyferth, *Chem. Ind. (London)*, 849 (1959).
130. N. E. Miller, *Inorg. Chem.*, **4**, 1458 (1965).
131. R. S. Tewari and K. C. Gupta, *Indian J. Chem., Sect. B*, **17**, 637 (1979).
132. D. Lloyd and M. I. C. Singer, *J. Chem. Soc., C*, 2941 (1971).
133. L. Horner and H. Oediger, *Chem. Ber.*, **91**, 437 (1958); *Ann. Chem.*, **627**, 142 (1959).
134. G. S. Harris, D. Lloyd, N. W. Preston and M. I. C. Singer, *Chem. Ind. (London)*, 1483 (1968).
135. J. I. G. Cadogan and I. Gosney, *J. Chem. Soc., Perkin Trans., 1*, 466 (1974).
136. B. H. Freeman, G. S. Harris, B. W. Kennedy and D. Lloyd, *Chem. Commun.*, 912 (1972).
137. D. Lloyd and M. I. C. Singer, *Chem. Ind. (London)*, 510 (1967).
138. J. N. C. Hood, D. Lloyd, W. A. MacDonald and T. M. Shepherd, *Tetrahedron*, **38**, 3355 (1982).
139. D. Lloyd and S. Metcalfe, *J. Chem. Res. (S)*, 292 (1983).
140. C. Glidewell, D. Lloyd and S. Metcalfe, *Synthesis*, 319 (1988).
141. K. Friedrich, W. Amann and H. Fritz, *Chem. Ber.*, **112**, 1267 (1979).
142. L. Hadjiarapoglou and A. Varvoglis, *Synthesis*, 913 (1988).
143. A. H. Schmidt, R. Aimène and M. Hoch, *Synthesis*, 754 (1984).
144. H. Schmidbaur and P. Nusstein, *Organometallics*, **4**, 344 (1987).
145. H. Schmidbaur and P. Nusstein, *Chem. Ber.*, **120**, 1281 (1987).
146. B. Boubia, C. Mioskowski and F. Bellamy, *Tetrahedron Lett.*, **30**, 5263 (1989).
147. G. Märkl and H. Sejpka, *Angew. Chem.*, **98**, 286 (1986); *Angew. Chem. Int. Ed. Engl.*, **25**, 264 (1986).
148. K. Bailey, I. Gosney, R. O. Gould, D. Lloyd and P. Taylor, *J. Chem. Res. (S)*, 386, (*M*) 2950 (1988).
149. H. Hartung and R. Richter, *Z. Anorg. Allg. Chem.*, **469**, 188 (1980).
150. P. Frøyen, *Acta Chem. Scand.*, **23**, 2935 (1969).
151. F. G. Mann, *J. Chem. Soc.*, 958 (1932).
152. R. Appel and D. Wagner, *Angew. Chem.*, **72**, 209 (1960).
153. P. Frøyen, *Acta Chem. Scand.*, **27**, 141 (1973).

154. P. Frøyen, *Acta Chem. Scand.*, **25**, 2781 (1971).
155. G. Wittig and D. Hellwinkel, *Chem. Ber.*, **97**, 769 (1964).
156. R. D. Chernokal'skii, S. S. Nasybullina, R. R. Shagidullin, I. A. Lamonova and G. Kamai, *Izv. Vyssh. Uchebn. Zaved., Khim. Khim. Tekhnol.*, **9**, 768 (1966); *Chem. Abstr.*, **66**, 76112 (1967).
157. D. S. Tarbell and J. R. Vaughan, *J. Am. Chem. Soc.*, **67**, 41 (1945).
158. A. Schönberg and E. Singer, *Chem. Ber.*, **102**, 2557 (1969).
159. J. I. Shah, *Can. J. Chem.*, **83**, 2381 (1975).
160. P. Frøyen, *Acta Chem. Scand.*, **25**, 983 (1971).
161. A. Senning, *Acta Chem. Scand.*, **19**, 1755 (1965).
162. G. Gadeau, A. Fouchaud and P. Merot, *Synthesis*, 73 (1981).
163. J. I. G. Cadogan and I. Gosney, *J. Chem. Soc., Chem. Commun.*, 586 (1973).
164. J. I. G. Cadogan and I. Gosney, *J. Chem. Soc., Perkin Trans., 1*, 460 (1974).
165. A. W. Johnson, *Ylid Chemistry*, Academic Press, New York, 1966, p. 301.
166. D. Lloyd and M. I. C. Singer, *Chem. Ind. (London)*, 787 (1967).
167. M. R. Mahran, W. M. Abdou, M. M. Abd-El-Rahman and M. M. Sidky, *Egypt. J. Chem.*, **30**, 401 (1987).
168. Yi Liao, Yaozeng Huang, Li-Jun Zhang and Chen Chen, *J. Chem. Res. (S)*, 388 (1990).
169. (a) Yaozeng Huang, Yi Liao and Chen Chen, *J. Chem. Soc., Chem. Commun.*, **85** (1990).
    (b) Yaozeng Huang, *Acc. Chem. Res.*, **25**, 182 (1992).
170. G. Doleshall, N. A. Nesmeyanov and O. A. Reutov, *J. Organomet. Chem.*, **30**, 369 (1971).
171. Yaozeng Huang and Yi Liao, *J. Org. Chem.*, **56**, 1381 (1991).
172. G. Ferguson, C. Glidewell, B. Kaitner, D. Lloyd and S. Metcalfe, *Acta Crystallogr.* **C43**, 824 (1987).
173. Yaozeng Huang, Chen Chen and Yanchang Shen, *J. Organomet. Chem.*, **366**, 87 (1989).
174. Chen Chen, Yaozeng Husang, Yanchang Shen and Yi Liao, *Heteroat. Chem.*, **1**, 49 (1990).
175. Chen Chen, Yi Liao and Yaozeng Huang, *Tetrahedron*, **45**, 3011 (1989).
176. Yi Liao and Yaozeng Huang, *Tetrahedron Lett.*, **31**, 5897 (1990).
177. G. Wittig and D. Hellwinkel, *Chem. Ber.*, **97**, 789 (1964).
178. D. Lloyd and M. I. C. Singer, *Chem. Commun.*, 1042 (1967).
179. H. Suzuki, T. Murafuji and T. Ogawa, *Chem. Lett.*, 247 (1988).
180. T. Ogawa, T. Murafuji and H. Suzuki, *J. Chem. Soc., Chem. Commun.*, 1749 (1989).
181. H. Suzuki and T. Murafuji, *Bull. Chem. Soc. Jpn.*, **63**, 950 (1990).
182. D. H. R. Barton, J. C. Blazejewski, B. Charpiot, J.-P. Finet, W. B. Motherwell, M. T. P. Papoula and S. P. Stanforth, *J. Chem. Soc. Perkin Trans. 1.* 2667 (1985).
183. T. Ogawa, T. Murafuji and H. Suzuki, *Chem. Lett.*, 849 (1988).
184. T. Ogawa, T. Murafuji, K. Iwata and H. Suzuki, *Chem. Lett.*, 325 (1989).
185. M. Yasui, T. Kikuchi, F. Iwasaki, H. Suzuki, T. Murafuji and T. Ogawa, *J. Chem. Soc., Perkin Trans. 1*, 3367 (1990).
186. H. Suzuki, C. Nakaya, Y. Matano and T. Ogawa, *Chem. Lett.*, 108 (1991).

CHAPTER **17**

The biochemistry of arsenic, bismuth and antimony

KILIAN DILL*

The Department of Chemistry, Clemson University, Clemson, South Carolina 29634-1905, USA

and

EVELYN L. McGOWN

The Chemistry Branch, Blood Research Division, Letterman Army Institute of Research, Presidio of San Francisco, California 94129-6800, USA

I. GENERAL CONSIDERATIONS

The biological effects of arsenic have been studied for years because of its (1) long history as a poison, (2) uses in industry, (3) clinical applications in protozoal and neoplastic diseases and (4) use in chemical warfare. In contrast, the biochemical effects of bismuth and antimony are not as well documented. Although most of the known compounds derived from the three metals are man-made, some adducts also occur naturally[1-6]. It is clear that these compounds in their ionic states are potent metabolic inhibitors and are lethal at sufficiently high doses. In the elemental state, arsenic, at least, is relatively nontoxic and is presumably the form consumed in years past by the 'arsenic eaters' of the Alps[3].

* *Present address*: Molecular Devices Corporation, 4700 Bohannon Drive, Menlo Park, California 94025, USA.

The chemistry of organic arsenic, antimony and bismuth compounds
Edited by S. Patai   © 1994 John Wiley & Sons Ltd

The metals under consideration all belong in group V with nitrogen and phosphorous. Their outer electronic configurations ($s^2 p^3$) are identical in that they both consist of the outer s orbital and the outer p orbital. The orbitals are not hybridized and therefore the two s-electrons can be considered a true lone pair. As expected, the atomic size increases in going from As (2.2 Å) to Sb (2.4 Å) to Bi (2.6 Å). The ionic radii vary according to the oxidation state, but for the $+3$ state, the corresponding radii are 0.55, 0.75 and 0.94. The oxidation state and ionic radius in turn affect the coordination sphere—an important factor which influences whether a metal ion or its derivative can fit into a particular molecular pocket (e.g. in a protein) and thus interact with the biological system.

The valencies of these metal atoms are limited to the tri- and pentavalent states. As can be seen in the earlier chapters in this volume, their true oxidation states can vary, depending on the ligands attached to the central atom. However, in nonorganic solvents and in the presence of air, the pentavalent state dominates. The biological effects of the trivalent and pentavalent species are entirely different and especially well-documented when it comes to arsenic. The trivalent compounds are almost always more toxic than the corresponding pentavalent derivatives[7].

In contrast to synthetic organic methodology where any number and type of ligand may be attached to the metal-ion center, there are only a few potential ligands under physiological conditions: oxygen (from carboxyl groups, hydroxyl groups), nitrogen (from amides, amines) and thiols (e.g. cysteine and lipoic acid). As we will see, thiols are highly preferred by these compounds and also provide the basis for antidote work.

II. ARSENIC

Arsenic has a long history as a poison. Its applications include (or have included) antimicrobials, herbicides, insecticides and rodenticides (and homicides). Arsenicals (as well as antimony and bismuth compounds) have been used clinically in protozoal, venereal and neoplastic diseases. In World War I, arsenicals were used as chemical warfare agents (see References 1–21 and references cited therein).

One area that is of interest and will not be covered in this review is the carcinogenesis, mutagenesis, teratology of arsenic. There have been a considerable number of studies and the exact mechanism is not understood[1,4,22]. However, it is known that exposure over a long period of time will cause mutagenesis and carcinogenesis and it has been thought to result from the inhibition of DNA repair. A number of the articles cited are review articles.

To review the older literature dealing with the biochemistry of arsenic, the reader is referred to two books[23,24] and the excellent reviews by Webb[25], Fowler[1] and Zingaro and Bottino (an article in Reference 4).

Arsenic is ubiquitous in the environment and ranks 14th in overall abundance among trace metals[1,3,4]. Arsenic can be detected in all living matter. Seafoods have the highest concentrations (2.7 to 8.9 ppm) and the human body contains an average of ~ 20 mg arsenic[3].

In nature, the pentavalent form is most abundant. Organisms at the bottom of the ecosystems accumulate arsenic from the surroundings. For example, algae, seaweed and marine animals have considerably higher concentrations of arsenic than does the surrounding water (see References 1–5, especially the article by Zingaro and Bottino in Reference 4). Arsenate enters these cells via the phosphate transport system located on the cell membranes. Uptake of arsenate is greatest in low-phosphate environments. Once inside the cell, arsenate is detoxified by alkylation, primarily methylation. Methylated arsenic derivatives are less toxic and more readily excreted than arsenate. Other metabolic fates of arsenate (depending on the species) include conversion to a number of organoarsenic derivatives of natural metabolites in which it replaces nitrogen (see Zingaro and Bottino in Reference 4, and Reference 24). For example, organic arsenicals identified in algae include arsenobetaine, dimethylarsinate, methyl arsonate and phospholipid derivatives. If organisms are exposed to high enough concentrations of arsenate, the detoxification systems are overwhelmed. Arsenate then competes with phosphate in metabolic reactions and can also replace phosphate in sugar metabolites.

Trivalent arsenic is less abundant in nature, but considerably more toxic. Detoxification of arsenite results in the same methylated species as those arising from arsenate. However, because trivalent arsenic binds tightly to thiol-containing molecules in tissues, it is much less easily detoxified and excreted.

Biomethylation of inorganic arsenic has been documented in a variety of organisms ranging from microorganisms to humans (see Zingaro and Bottino in Reference 4, and Reference 24). In mammals, the major metabolites of arsenate and arsenite are dimethylarsinic acid (sometimes called cacodylic acid) and variable amounts of methylarsonic acid[26-31]. Humans appear to excrete more monomethylarsenic than other animals. The monomethylated species is presumably a precursor to the dimethyl form[32,33]. Aquatic plants and animals (especially marine organisms) combine methylation with additional alkylation to form relatively nontoxic organoarsenic derivatives of natural metabolites in which arsenic replaces quaternary nitrogen (see Zingaro and Bottino in Reference 4, and Reference 24). Numerous organoarsenicals have been detected[34-37]; those that have been identified include arsenobetaine, trimethylarsonium lactate and arsenolipids, especially O-phosphatidyltrimethylarsonium lactate. Biomethylation of arsenic proceeds to a greater extent in lower organisms than in higher life forms. Microorganisms can combine methylation with reduction to generate trimethylarsine, which readily diffuses through cell walls and escapes.

Methylated arsenic is also found in compounds in which arsenic replaces phosphorus of natural biomolecules. Several arsenic-containing ribofuranosides have been identified in brown kelp[38,39]. The dimethylated arsenic is linked to the C-4 of ribose by a methylene bridge.

The mechanisms of biomethylation of arsenic have not been documented. However, Cullen and coworkers[40,41] have proposed a plausible mechanistic model based on oxidative methylation of arsenic(III) by S-adenosylmethionine and reduction by a thiol such as lipoic acid.

Thiol-containing molecules will reduce pentavalent arsenic to the trivalent species (see Zingaro and Bottino in Reference 4, and Reference 40). Because of the abundance of intracellular thiols in biological systems, one might expect such reductions to occur in vivo. In fact, arsenite has been detected in the blood or urine of rats, mice and rabbits after administration of arsenate[7,42].

A. Arsenic Consumed

Ore smelters, the electronic industries and the use of herbicides, fungicides, antibacterial agents, antimicrobial agents, growth promoters, fertilizers and drugs containing arsenic have added arsenic to the environment. Many of these compounds have been discontinued (with other alternatives found) or their use has been limited. For these compounds there is a critical limiting quantity in their use: a little helps but a larger dose has extreme adverse effects. Some of the compounds that have been (and may be) in use are given below.

$$CH_3-\overset{\overset{\displaystyle O}{\|}}{\underset{\underset{\displaystyle CH_3}{|}}{As}}-OH \qquad NaO-\overset{\overset{\displaystyle O}{\|}}{\underset{\underset{\displaystyle CH_3}{|}}{As}}-ONa \qquad 10,10'\text{-oxybisphenoxarsine}$$

'Cacodylic Acid'   Herbicides for grass (23 million lb used per year)   Fungi control of cotton. Fabric, vinyl films

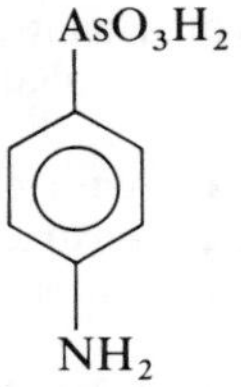

AsO$_3$H$_2$

NH$_2$

'*Arsanilic acid*'
'Control of swine dysentery.
Growth promoter for
chicken and swine.
Improve pigmentation
of chickens.

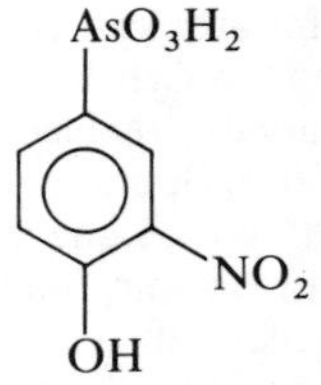

AsO$_3$H$_2$

NO$_2$

OH

'*Roxarsone*'
Control of coccidiosis
in chickens.

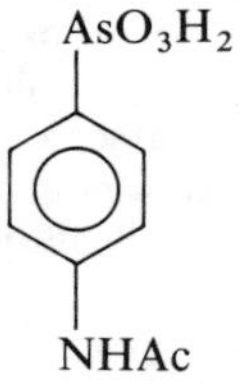

AsO$_3$H$_2$

NHAc

'*Carbarsone*'
Antiamebic.
Antihistomonad in turkeys.
Antisyphlitic.

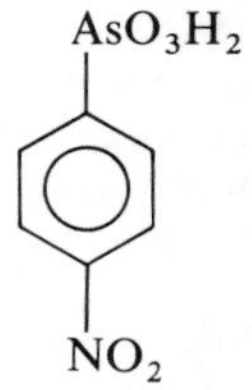

AsO$_3$H$_2$

NO$_2$

'*Nitrasone*'
Antihistomonad in turkeys.

AsO$_3$H$_2$

NHCOOCH$_2$OH

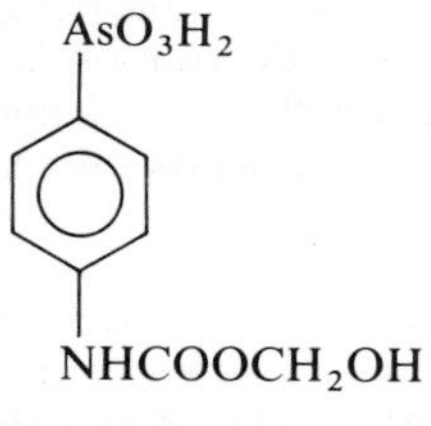

Glycarsamide-entozon crystals'
Antimicrobial

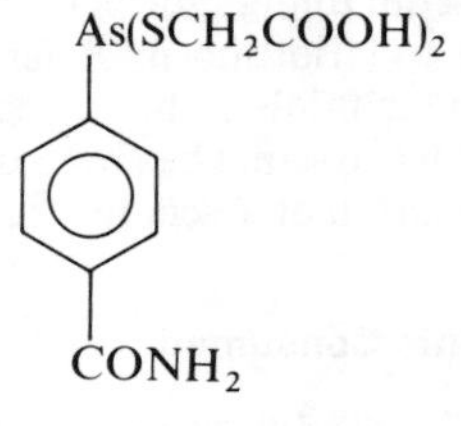

As(SCH$_2$COOH)$_2$

CONH$_2$

'*Arsenamide*'
Anthelmic
Heartworm infection in dogs

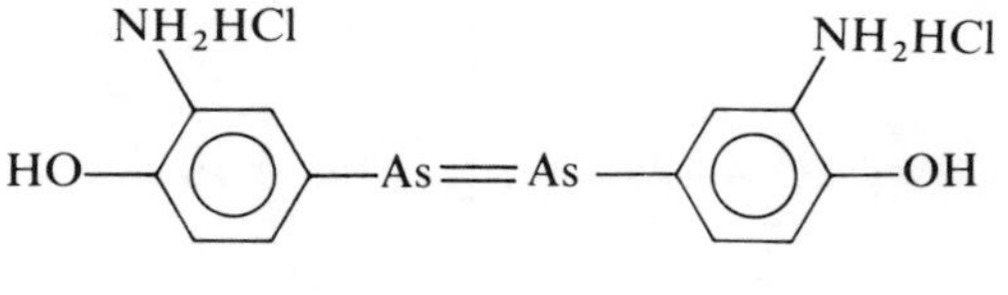

NH$_2$HCl   NH$_2$HCl

HO—⬡—As＝As—⬡—OH

Arsphenamine
Antisyphlitic

As$_2$O$_5$ CaHAsO$_4$

Wood preservative   Herbicide.
Fungicide.   Control of measles on grapes.
Weed control.

'*Amomonical copper arsenate*'
'*Chromated copper arsenate*'
Wood preservatives.

Methylarsonic Acid   Arsenobetaine

B. Toxicity and Biochemical Mechanism

The toxicity of arsenic is well-documented. Trivalent arsenic (arsenite) is almost always more toxic than the pentavalent species (arsenate)[1–4]. The differential toxicity is due to their different modes of action. The primary biological targets of trivalent arsenic are thiol-containing molecules, e.g. in active sites of enzymes (see below). Arsenate, on the other hand, is better tolerated (detoxified) at low concentrations and, at higher levels, competes with phosphate, an ubiquitous biological anion. Indeed the toxicity of pentavalent arsenic in vivo may be partly due to its intracellular reduction to the trivalent form[7,40,43].

Organic arsenicals are more toxic than inorganic arsenic. Because of their lipophilicity, they readily penetrate cells and have greater access to hydrophobic regions of macromolecules than does arsenite[25,44]. Monosubstituted trivalent organic arsenic forms exceedingly stable cyclic adducts with certain dithiol-containing molecules (e.g. lipoic acid and some proteins)[25]. Disubstituted trivalent arsenicals cannot form cyclic structures and thus their thiol adducts are much less stable. Trivalent organic arsenicals have been used in many mechanistic studies of biological effects of arsenic.

In aqueous media the arsenical must chelate with some portion of a biomolecule in order to inhibit biological function. The limited number of possibilities include oxygen and sulfur atoms. Although there are some reports indicating that arsenic may bind to oxygen as a ligand[45], sulfur has always been envisioned as the ligand of choice for the arsenical[46–61]. The effects of arsenicals have been measured on many complex enzyme systems with thiol groups at the active sites[25]. These include pyruvate dehydrogenase[58], cardiac adenylate cyclase[59], lipoamide dehydrogenase[56,57], glutathione reductase[61], urease[49], pyridine nucleotide dehydrogenase[54] and thioredoxin[55]. It has been shown that phenylarsine oxide can inhibit insulin-stimulated glucose transport[53,60]. Furthermore, phenylarsine oxide seems to regulate arachidonic acid metabolism and amplifies prostaglandin I$_2$ in rat liver cells[50]. All the data from these publications suggests that a thiol (or vicinal thiols) at the active site plays a role in binding the arsenical.

Other studies indicate that inorganic arsenic as well as organic arsenicals stimulate the Factor-B activity[51]. These results indicate that uncoupling of oxidative phosphorylation occurs and can be reversed by addition of dithiol groups. Moreover, arsenicals can also inhibit ATPase because of their oxidative abilities[52]. Two recent reports by Simons and coworkers indicate that arsenite (but not arsenate) blocks the binding of steroids to the glucocorticoid receptors[62,63]. Again, detailed studies by other authors all indicate that thiol groups (vicinal dithiols being preferred) are direct targets for these arsenicals.

As seen in the previous work, arsenic will bind to biological molecules and hinder enzymic activity. The main mode of action seems to be to bind to systems that contain thiol as a functional group. For proteins, this can be either cysteine or cofactors that contain thiol groups. Our recent research efforts have elucidated the interaction with some of the target molecules. The arsenical that we have used is phenyldichloroarsine, an organic trivalent species that can penetrate cells easily.

The simplest studies deal with the interaction of phenyldichloroarsine (PDA, 1) with L-cysteine[64]. Figure 1 shows the effects of the addition of L-cysteine on the ^{13}C NMR

PDA

(1)

spectrum of PDA in methanol. As we have previously shown, the reaction results in the formation of a 1:1 intermediate adduct (as a marker we monitored the nonprotonated aromatic carbon of PDA which exhibits a resonance at 154 ppm whereas the adduct shows a resonance at 145 ppm) and then goes on to form a 2:1 final product when additional L-cysteine is added (resonance formed at 139 ppm). We also observed that when the carboxyl or amino functional groups of L-Cys are blocked, identical results are obtained. In aqueous media, the results are somewhat different. A dynamic equilibrium results and it favors the formation of the 1:2 species (PDA/Cys) with little of the 1:1 species to be found. These data show that the sulfhydryl group was indeed the preferred point of chelation by arsenic (and not the amino or carboxyl group) and the reactions in aqueous media would differ from reactions and equilibria in organic media.

One of the simplest biological systems that can be studied is the red blood cell, which readily binds trivalent arsenic[25]. In comparison to other species, the rat has a high binding capacity for arsenic[25,31] and dimethylarsinate[26]. Sulfhydryl groups in the red cell arise primarily from hemoglobin and glutathione (GSH). GSH is a tripeptide whose sequence is Glu(γ)CysGly and is one of the major intracellular reducing agents. In most species, hemoglobin contains two exposed sulfhydryl groups per hemoglobin tetramer. (Others are present, but buried within the molecule and thus not readily accessible without dissociation of the hemoglobin subunits.) Rat hemoglobin differs from most other species in that it has four exposed sulfhydryl groups per tetramer.

To test the hypothesis that hemoglobin may in some cases bind organic arsenicals, we studied the uptake of various red blood cells with radiolabeled PDA[65]. The results are shown in Figures 2 and 3. Clearly, rat red blood cells did absorb the largest amount of the arsenical. In most species, the capacity of PDA binding correlated to the GSH content

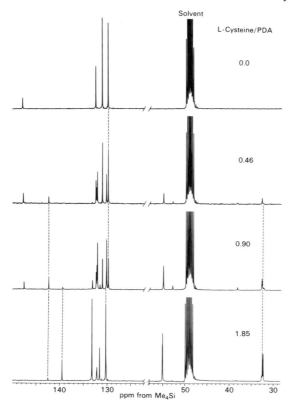

FIGURE 1. The effects of added L-cysteine on the proton decoupled, natural abundance ^{13}C NMR spectrum of PDA. The concentration of PDA was 125 mM in methanol-d_4. Solid cysteine was added to give the molar ratios of L-Cys/PDA as indicated in the right-hand portion of each trace

corresponding to Scheme 1, but PDA binding by the rat red cell greatly exceeded the GSH content. A Scatchard plot analysis of PDA binding by rat red cells indicated 2.3 strong binding sites/mol and that the K_d for this association was 5.2×10^{-6}. In order to further our studies and to fully characterize the reacted species, we chromatographed the radiolabeled hemolysates. Figure 4 shows that in rat preparations, the radiolabeled material eluted with both glutathione and hemoglobin. For the other species, the radiolabeled material eluted only with glutathione.

$$PhAsCl_2 + GSH \xrightleftharpoons{-HCl} PhAsCl(GS) + GSH \xrightleftharpoons{-HCl} PhAsCl(GS)$$

$$PhAs(OH)_2 + GSH \xrightleftharpoons{-H_2O} PhAsOH(GS) + GSH \xrightleftharpoons{-H_2O} PhAs(GS)_2$$

SCHEME 1

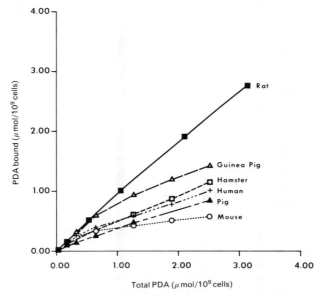

FIGURE 2. Uptake of PDA by erythrocytes of several species. Erythrocyte suspensions (10^9 cells/ml) were exposed for 10 min to graded concentrations of [^{14}C] PDA and centrifuged, and the radioactivity of the supernatant was measured

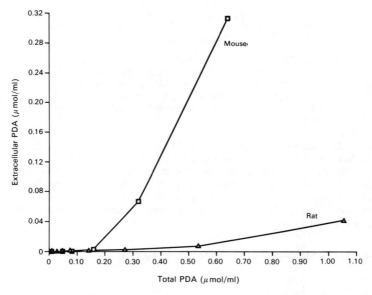

FIGURE 3. Low level PDA binding by rat and mouse erythrocytes (10^9 cells/ml). The number of primary binding sites was taken to be the threshold where the plot rose above the x-axis

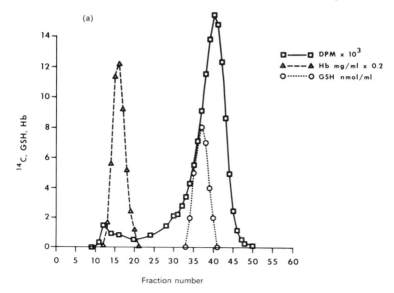

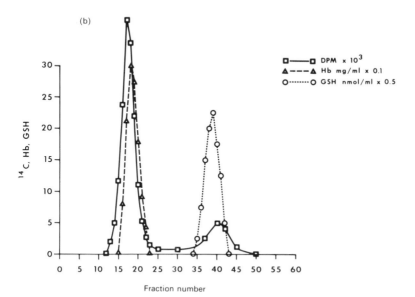

FIGURE 4. (A) Gel filtration chromatogram of hemolysate from mouse erythrocytes that had been exposed to $0.24\,\mu\mathrm{mol}\;[^{14}\mathrm{C}]\;\mathrm{PDA}/10^9$ cells. (B) Gel filtration of hemolysate from rat erythrocytes that had been exposed to $0.24\,\mu\mathrm{mol}\;[^{14}\mathrm{C}]\;\mathrm{PDA}/10^9$ cells

Although the additional PDA binding by rat red cells is presumably due to the additional exposed sulfhydryls (on residue 13 of each alpha chain), questions still remain. PDA preferentially forms 1:2 adducts with thiol molecules[66]. One might presume that PDA does not bind to most species' hemoglobins because of the lack of a second sulfhydryl ligand. The puzzling question then is: why does rat hemoglobin bind PDA when the additional cysteines (α-13) are not located near other sulfhydryls? Possibly a carboxyl amino acid (e.g. glutamic acid -116) is situated near the α-cysteine-13 such that PDA can chelate to a sulfur and an oxygen atom.

Another method to prove the structure of our proposed adducts was to synthesize and characterize the species[66,67]. We synthesized the PDA–glutathione complex (2) and

(2)

characterized the species spectroscopically. We can clearly show that a 2:1 (glutathione/PDA) adduct is formed (favored) in aqueous media (Figures 5–7). As can be seen from the data, due to slow inversion and movement about the As atom, diastereomers have been formed and thus the resonances of the carbon signals near the arsenic atom are doublets. Spectral assignments were confirmed by 2D NMR[66,67]. In order to establish that the arsenical is indeed penetrating the red blood cell, proton spin-echo NMR experiments were undertaken[68]. This was necessary because the normal 1D proton NMR spectrum of red blood cells is broad and shows resonances pertaining to the lipid bilayer and the larger, more abundant, proteins (Figure 8). The results of the spin-echo experiments are shown in Figure 9. Clearly, as PDA is added, the signals pertaining to GSH disappear when a GSH/PDA ratio of 2:1 is reached, indicating complexation and internal viscosity changes.

Complexes of glutathione with diphenylchloroarsine were also attempted, but they were found to be unstable. We found them to be unstable to the atmosphere, and the samples decomposed in the purification process[69].

Other potentially important biological species that may react with arsenicals are thio-sugars. We found that these reacted with PDA and readily formed 2:1 complexes;

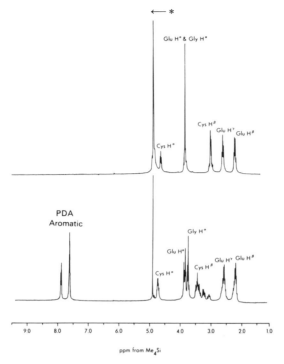

FIGURE 5. ^{1}H NMR spectrum of glutathione (upper trace) and PDA (GSH)$_2$ * residual partially deuterated water

these compounds were stable over long periods of time as long as oxygen was excluded from the samples[70]. The NMR data for these systems also exhibited similar diastereotopic chemical shift differences. The published relaxation data gave an indication about the relative mobility of the various groups about the arsenic atom (diffusion coefficients); the phenyl moiety was 6.4 times more mobile than the sugar rings.

Another biological complex that was studied was the arsenical lipoic acid complex[71]. This has a great degree of relevance, because lipoic acid is a cofactor in many enzyme

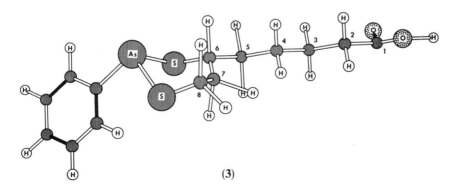

(3)

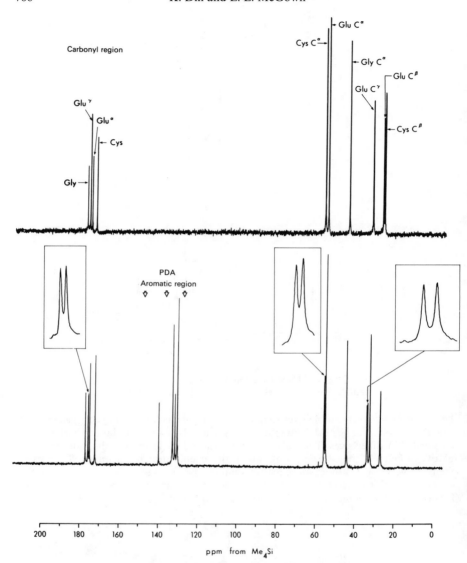

FIGURE 6. Proton decoupled, ^{13}C-NMR spectra of GSH and PDA(GSH)$_2$[66]

systems arsenic sensitive. This can only occur if the arsenical reacts with the two thiol groups in order to form a six-membered ring (**3**). Figures 10 and 11 show the spectral data associated with the complex and prove that the dominant structure present is that depicted in **3**. Again multidimensional NMR was required in the resonance assignments because of the complexity of the 1D spectra.

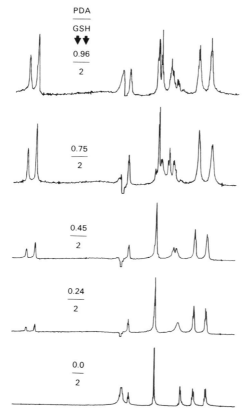

FIGURE 7. The efects of graded additions of PDA on the ^{1}H-NMR spectrum of GSH[66]

Thus, all the biological systems that we have studied to date confirm that the arsenical reacts with the thiol groups of proteins, predominantly, and most likely does damage by shutting down enzymes that have thiol groups at the active sites.

III. ANTIMONY

In general in the literature, the biological aspects of antimony are somewhat limited and closely follow the information gathered for bismuth and arsenic[5]. Most of the information given below is based on articles during the period 1922–1991 for the development and use of drug-based compounds[1,15−20,72,73].

A. Chemistry

The chemistry of this metal has been studied the least. Like all the other metals studied it exists in the trivalent or pentavalent state. Numerous compounds are known containing ligands such as aryl and alkyl groups, amides, oximes, carboxylates and amino acids. Most of the work has been in the area of drug development. The biological targets for antimony

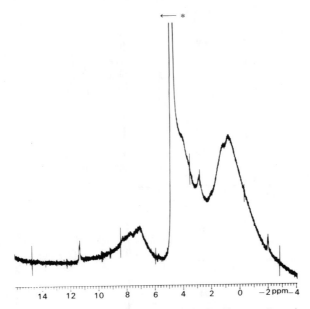

FIGURE 8. ^{1}H-NMR spectrum of deuterated intact guinea pig erythrocytes * residual partially deuterated water

are not well characterized, although they most likely contain thiol. In some recent work involving cell 'rescue' work, BAL (British Anti-Lewisite) was used as a therapeutic agent in order to prevent cellular damage due to subsequent antimony exposure[72,73]. In order to make the drugs more water-soluble, counterions present (for salts) or the functional groups aid in this endeavor. They are given as urea complexes of stibenic acid or as stibamine. Furthermore, the more lipophilic the compound, the more dangerous it is with respect to tissue and cell damage, rather than being therapeutically valuable. Details will be given below.

B. Biological aspects

The drug-based use of antimony dates back to as early as the 1920s[20,21]. Early work in India using antimonial compounds for the treatment of Kala-Azar disease was based on p-aminophenyl stibenic acid with tartrate derivatives and urea stibamines. Initial animal studies were undertaken and, later, patients tested. The later trials with patients indicated that quite a few could be cured without a relapse. The basic animal studies investigated toxicity and the limits of their use as drugs. It was found that there was severe systemic damage in pathological lesions consisting of hemorrhages in the internal organs and necrosis of several organs. Organs that were the primary targets were lungs, kidneys, liver, spleen and various glands. Some had severe damage, others only showed slight damage. The general trend was that lipophilic compounds exerted a greater toxicity.

Since the early study, a number of papers have appeared in the literature dealing with the use of antimony-based drugs [Glucantime, Triostib] that have been used in several tropical diseases, such as schistosomiasis and leishmaniasis, but not without side effects such as arthralgia or possible antimony intoxication. Many of these articles are based on

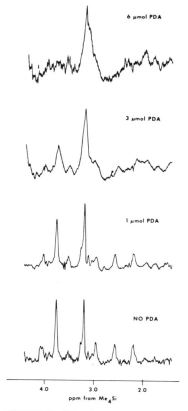

FIGURE 9. ^{1}H spin-echo NMR spectra of intact guinea pig erythrocytes upon graded additions of PDA

drug toxicity and drug efficacy and not the molecular basis of the action of antimony.

A few articles in the early forties do give some insight into the molecular targets of these compounds. Furthermore, the BAL 'cellular' rescue work also provides details about the targets and confirms the functional groups and enzymes involved. Work by Campello and coworkers[72] with isolated rat liver mitochondria investigated the effects of sodium antimony gluconate (Triostib) on the oxidative phosphorylation of succinate, gluconate and α-ketoglutarate. They showed that oxidative phosphorylation and oxidation of NAD^+–linked substrates by liver mitochondria were suppressed by Triostib. It was hypothesized that Triostib acts at the NAD^+–oxidase segment of the chain, its site being localized between NAD^+ and the flavoprotein. Other reports present data on the prevention of these molecular events by BAL.

IV. BISMUTH

In general, the biological concepts of bismuth chemistry are rather limited and are once again based upon earlier work and also dealing with the latest data involving arsenic and the related chemistries[5]. Furthermore, much of the published work is once again based

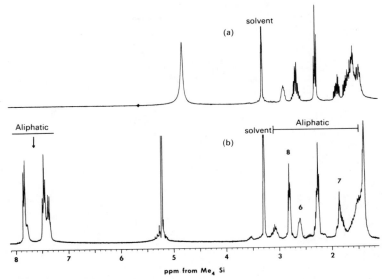

FIGURE 10. ^{1}H-NMR spectrum of (a) lipoic acid and (b) the lipoic acid–PDA adduct[71]

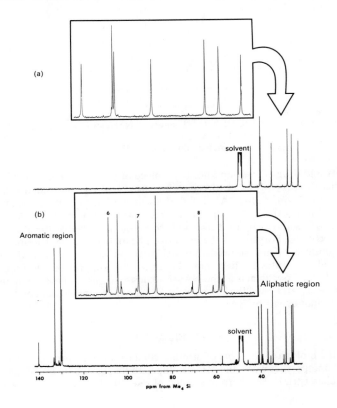

upon microbial, antitumor and antifungal work and related rescue work and not much is based upon the actual biochemistry of biological molecules[5,11-14,74,75].

A. Chemistry

There is considerable published work on some of the chemistry of bismuth versus the related atomic species of arsenic and antimony. Much of the recent work can be obtained from an excellent review published by Klapötke[5]. Bismuth is not a biological essential element. Like arsenic, it also has two valency states, 3 and 5. The carbon–bismuth bonds have a very low degree of polarity. Their thermodynamics indicate that for the formation of $E(Me)_3$ (where E is either As, Sb or Bi) the enthalpy of formation is largely endothermic; this is due to the very weak Bi—C bond, the weakest of the main group metals. Bismuth–carbon bonds can decompose upon heating to produce intermediate radicals. This affects their stability and toxicity. The weak polar bonds also produce compounds that have a very small dipole moment. In effect, the will not be very soluble in most media, especially water, unless a functional group is added. Many ligands, like those found for arsenic, contain sulfur as ligands.

B. Biochemistry

A host of bismuth compounds has been used for studies involving antibacterial and antitumor activities and warfare agents. It should be noted that many of the compounds are no longer in use, because with the advent of other natural compounds these metallic species are used very little. The most common compounds used to date are bismuth subcitrate[11] or bismuth subsalicylate[15] as bacterial intestinal agents and for action against ulcers (Pepto-bismol; Proctor and Gamble). In both cases these were provided as colloidal agents which had limited water solubility.

V. REFERENCES

1. B. F. Fowler, *Toxicology of Environmental Arsenic*, in *Advances of Modern Toxicology*, Vol. 2: *Toxicology of Trace Metals* (Eds. R. A. Goyer and M. A. Mehlman), Hemisphere, Washington, 1977, pp. 79–122.
2. F. W. Sunderman, Jr., *Biol. Trace Elem. Res.*, **1**, 63 (1979).
3. H. A. Schroeder and J. J. Balassa, *J. Chronic Dis.*, **19**, 85 (1966).
4. W. H. Lederer and R. J. Fensterheim (Eds.), *Arsenic: Industrial, Biomedical, Environmental Perspectives*, Van Nostrand Reinhold, New York, 1983.
5. T. Klapötke, *Biol. Metals*, **1**, 69 (1988).
6. S. Budavari (Ed.), *The Merck Index*, Merck & Co, Inc., Rahway, NJ, 1989.
7. M. Vahter and J. Envall, *Environ. Res.*, **32**, 14 (1983).
8. A. Cohen, H. King and W. I. Strangeways, *J. Chem. Soc.*, 3043 (1931).
9. B. A. Fowler, J. S. Woods and C. M. Schiller, *Environ. Health Perspect.*, **19**, 197 (1977).
10. P. H. Schlesinger, D. J. Krogstad and B. L. Herwaldt, *Antimicrob. Agents Chemother.*, **32**, 793 (1988).
11. S. J. Konturek, A. Dembinski, Z. Warzecha, W. Bielanski, T. Brzozowski and D. Drozdowicz, *Gut*, **29**, 894 (1988).
12. T. Klapötke, *J. Organomet. Chem.*, **331**, 299 (1987).
13. T. Klapötke, *Monats. Chem.*, **119**, 1317 (1988).
14. T. Klapötke and P. Gowik, *Z. Naturforsch.*, **42b**, 940 (1987).
15. T. E. Sox and C. A. Olson, *Antimicrob. Agents Chemother.*, **33**, 2075 (1989).

FIGURE 11. Proton decoupled, ^{13}C-NMR spectrum of (a) lipoic acid and (b) the lipoic acid–PDA adduct[71]

16. T. R. Navin, B. A. Arana, F. E. Arana, A. M. DeMerida, A. L. Castillo and J. L. Pozuelo, *Am. J. Trop. Med. Hyg.*, **42**, 43 (1990).
17. E. M. Netto, P. D. Marsden, E. A. Llanos-Cuentas, J. M. L. Costa, C. C. Cuba, A. C. Barreto, R. Badaro, W. D. Johnson and T. L. Jones, *Trans. R. Soc. Trop. Med. Hyg.*, **84**, 367 (1990).
18. C. Castro, R. N. Sampaio and P. D. Marsden, *Trans. R. Soc. Trop. Med. Hyg.*, **84**, 362 (1990).
19. K. Singhal, R. Rastogi and P. Raj, *Ind. J. Chem.*, **26A**, 146 (1987).
20. U. N. Brahmachari, *Ind. J. Med. Res.*, **10**, 492 (1922).
21. U. N. Brahmachari, *Ind. J. Med. Res.*, **11**, 393 (1923).
22. A. Leonard, *Tox. Environ. Chem.*, **7**, 241 (1984).
23. *Organometallics and Organometalloids*, ACS Symp. Ser. No. 82 (1978) Washington, DC.
24. J. S. Thayer, *Organometallic Compounds and Living Organisms* (a volume in the monograph series *Organometallic Chemistry*, Eds. P. M. Maitlis, F. G. A. Stone and R. West), Academic Press, New York, 1984.
25. J. L. Webb, in *Enzyme and Metabolic Inhibitors*, Vol. 3, Academic Press, New York, 1966, pp. 595–793.
26. S. Lerman and T. W. Clarkson, *Fundam. Appl. Toxicol.*, **3**, 309 (1983).
27. M. Vahter and E. Marafante, *Chem. -Biol. Interact.*, **47**, 29 (1983).
28. H. Yamauchi and Y. Yamamura, *Bull. Environ. Contam. Toxicol.*, **31**, 267 (1983).
29. H. Yamauchi and Y. Yamamura, *Toxicol. Appl. Pharmacol.*, **74**, 134 (1984).
30. J. U. Lakso and S. A. Peoples, *J. Agric. Food Chem.*, **4**, 674 (1975).
31. Y. Odanaka, O. Matano, and S. Goto, *Bull. Environ. Contam. Toxicol.*, **24**, 452 (1980).
32. J. P. Buchet, R. Lauwerys, P. Mahieu and A. Geubel, *Arch. Toxicol.*, Suppl. **5**, 326 (1982).
33. G. K. H. Tam, S. M. Charbonneau, F. Bryce, C. Pomroy and E. Sandi, *Toxicol. Appl. Pharmacol.*, **50**, 319 (1979).
34. J. B. Luten, G. Riekwel-Booy and A. Rauchbaar, *Environ. Health Perspect.*, **45**, 165 (1982).
35. J. S. Edmonds and K. A. Francesconi, *Mar. Pollut. Bull.*, **12**, 92 (1981).
36. J. R. Cannon, J. B. Saunders and R. F. Toia, *Sci. Total Environ.*, **31**, 181 (1983).
37. J. S. Edmonds and K. A. Francesconi. *Tetrahedron Lett.*, 1,543 (1977).
38. J. S. Edmonds and K. A. Francesconi, *J. Chem. Soc., Perkin Trans. 1*, 2375 (1983).
39. J. S. Edmonds, K. A. Francesconi, P. C. Healy and A. H. White, *J. Chem. Soc., Perkin Trans. 1*, 2989 (1982).
40. W. R. Cullen, B. C. McBride and J. Reglinski, *J. Inorg. Biochem.*, **21**, 45 (1984).
41. W. R. Cullen, B. C. McBride and J. Reglinski, *J. Inorg. Biochem.*, **21**, 179 (1984).
42. I. R. Rowland and M. J. Davies, *J. Appl. Toxicol.*, **2**, 294 (1982).
43. T. B. B. Crawford and G. A. Levvy, *Biochem. J.*, **41**, 333 (1947).
44. E. L. McGown, T. van Ravenswaay and C. R. Dumlao, *Toxicol. Pathol.*, **15**, 149 (1987).
45. A. G. Douen and M. N. Jones, *Bio Factors*, **2**, 153 (1990).
46. F. C. Knowles and A. A. Benson, 111–114, 4th Spurenelement Symposium (1983), Leipzig-Jena.
47. F. C. Knowles and A. A. Benson, 115–118, 4th Spurenelement Symposium (1983), Leipzig-Jena.
48. S. C. Frost and M. S. Schwalbe, *Biochem. J.*, **269**, 589 (1990).
49. J. J. Gordon and J. H. Quastel, *Biochem. J.*, **42**, 337 (1948).
50. L. Levine, *Biochem. Biophys. Res. Commun.*, **178**, 641 (1991).
51. S. Joshi and J. B. Hughes, *J. Biol. Chem.*, **256**, 11112 (1981).
52. T. Beeler, *Biochim. Biophys. Acta*, **1027**, 264 (1990).
53. K. Pettengell and S. C. Frost, *Biochem. Biophys. Res. Commun.*, **161**, 633 (1989).
54. G. Voordouw, S. M. Van Der Vies, C. Veeger and K. J. Stevenson, *Eur. J. Biochem.*, **118**, 541 (1981).
55. S. B. Brown, R. J. Turner, R. S. Roche and K. J. Stevenson, *Biochemistry*, **26**, 863 (1987).
56. M. J. Danson, A. McQuattie and K. J. Stevenson, *Biochemistry*, **25**, 3880 (1986).
57. S. R. Adamson, J. A. Robinson and K. J. Stevenson, *Biochemistry*, **23**, 1269 (1984).
58. K. J. Stevenson, G. Hale and R. N. Pernman, *Biochemistry*, **17**, 2189 (1978).
59. G. I. Drummond, *Arch. Biochem. Biophys.*, **211**, 30 (1981).
60. S. C. Frost and M. D. Lane, *J. Biol. Chem.*, **260**, 2646 (1985).
61. F. C. Knowles, *Arch. Biochem. Biophys.*, **242**, 1 (1985).
62. S. S. Simons, Jr., P. K. Chakraborti and A. H. Cavanaugh, *J. Biol. Chem.*, **265**, 1398 (1990).
63. S. Lopez, Y. Miyashita and S. S. Simons, Jr., *J. Biol. Chem.*, **265**, 16039 (1990).
64. K. Dill, R. J. O'Connor and E. L. McGown, *Inorg. Chim. Acta*, **168**, 11 (1990).
65. S. Chong, K. Dill and E. L. McGown, *J. Biochem. Toxicol.*, **4**, 39 (1989).

66. K. Dill, E. R. Adams, R. J. O'Connor, S. Chong and E. L. McGown, *Arch. Biochem. Biophys.*, **257**, 293 (1987).
67. R. J. O'Connor, E. L. McGown and K. Dill, USAMRDC-LAIR Report # 222 (1986).
68. K. Dill, R. J. O'Connor and E. L. McGown, *Inorg. Chim. Acta.*, **138**, 95 (1987).
69. E. R. Adams, J. W. Kolis and K. Dill *Inorg. Chim. Acta*, **152**, 1 (1988).
70. K. Dill, S. Hu, R. J. O'Connor and E. L. McGown, *Carbohydr. Res.*, **196**, 141 (1990).
71. K. Dill, E. R. Adams, R. J. O'Connor and E. L. McGown, *Chem. Res. Toxicol.*, **2**, 181 (1989).
72. A. P. Campello, D. Brandao, M. Baranski and D. O. Voss, *Biochem. Pharmacol.*, **19**, 1615 (1970).
73. H. Braun, L. M. Lusky and H. O. Calvery, *J. Pharmacol. Exp. Ther., Suppl.*, **87**, 119 (1946).
74. H. Eagle, F. G. Germuth, Jr., H. J. Magnuson and R. Fleishman, *J. Pharmacol. Exp. Ther.*, **89**, 196 (1947).
75. R. W. Lewis and G. E. Oliver, *Res. Commun. Chem. Pathol. Pharmacol.*, **18**, 377 (1977).

CHAPTER 18

Pharmacology and toxicology of organic bismuth, arsenic and antimony compounds

URI WORMSER and ISAAC NIR

Department of Pharmacology, School of Pharmacy, Faculty of Medicine, The Hebrew University of Jerusalem, Jerusalem, Israel

I. PHARMACOLOGY AND TOXICOLOGY OF ORGANIC BISMUTH COMPOUNDS (OBC)

Organic bismuth salts have long been employed for treatment of duodenal and gastric ulcers. The most widely used of these compounds are bismuth subcitrate and bismuth subsalicylate. Both are effective in the treatment of ulcerous gastritis, travellers' diarrhoea and relief of pain. They are insoluble compounds, which, together with bismuth subgallate, commonly used in post colostomy or post ileostomy patients to control the consistency and odour of the stools, comprise the most accepted group of organic bismuth drugs in use.

The chemistry of organic arsenic, antimony and bismuth compounds
Edited by S. Patai   © 1994 John Wiley & Sons Ltd

A. Pharmacology of OBC

The antiulcerative activity of OBC is related to several pharmacological activities demonstrated in experimental studies. The first effect is associated with the strong affinity of OBC to mucosal glycoproteins, particularly in the ulcerative areas of the gastro-intestinal wall[1]. If we take as an example the subcitrate salt of bismuth, the formula of which is approximately $[Bi(OH)_3C_6H_5O_7]_n$, this compound forms a layer of polymer–glycoprotein complex coating the ulcer, thus protecting the tissue from adverse activity of hydrochloric acid and proteolytic enzymes[1,2]. Another mechanism of action of OBC is connected with changes in prostaglandin activity in the gastro-intestinal wall, bringing about an enhanced production of mucus protecting the stomach wall.

1. The antibacterial activity of OBC

An additional beneficial effect of OBC is their antibacterial activity[3,4] leading to prevention of microbial growth. Bismuth subcitrate was shown to have an efficient bactericidal activity against slow-growing helicobacter pylori[5]. Examination of the microbial activity in human volunteers, as assessed by intragastric urea activity[6], revealed that bismuth compounds suppress the bacterial action in the upper gastro-intestinal tract. This chemotherapeutic activity might be attributed to inhibition of the microbial capsular polysaccharide production, slowing down the bacterial growth, and in the case of Klebsiella pneumoniae potentiation of aminoglycoside inhibition[7]. Interestingly, in vitro studies showed that exposure of helicobacter pylori to colloidal bismuth subcitrate impaired bacterial adherence to Hep-2 cells as demonstrated by electron microscopy[8]. Moreover, bismuth salts were shown to enhance opsonophagocytosis of Klebsiella pneumoniae by human peripheral white blood cells or rat alveolar macrophages[9]. Although the mechanism of action of bactericidal activity of OBC has not been fully elucidated, the beneficial effects in both travellers' diarrhoea and gastritis led to the wide clinical use of these compounds.

2. Cytoprotection and eicosanoid production by OBC

Apart from the above-discussed pharmaceutical features, the antiulcerative effect of bismuth compounds is attributed to cytoprotective/morphological changes produced in the stomach wall. Rats treated with colloidal bismuth subcitrate demonstrated reduced age-related degeneration of the stomach[10]. There was a decline in the number of deep erosions in the gastric mucosa and an increase in the number and mucin content of the mucosal epithelial cells. It was also shown that OBC administration caused a significant increase in the proliferation of the stem cell population[10]. A similar phenomenon was observed in humans treated with OBC[12]. Gastroscopic and histological assessment of antral and duodenal mucosal biopsy specimens taken from 18 patients with duodenal ulcers showed complete ulcer healing in 15 out of 18 patients and improved recovery in 3 of them after 4 weeks of treatment with colloidal bismuth subcitrate. The beneficial effect of this compound might be related to alterations in mucosal levels of arachidonic acid metabolites, i.e. prostaglandins E_2[11] and leukotriene C_4[12]. However, this assumption must be further clarified. Additional intracellular alteration induced by bismuth is enhanced metabolism of mucosal phosphoinositides. In vitro and in vivo studies[13] have demonstrated that elevated mucin synthesis caused by colloidal bismuth subcitrate is accompanied by a decrease in PI and PI_2 and increase in IP1 and IP3. It was also shown that the detrimental effects evoked by ethanol on mucin content were inhibited by bismuth. Interestingly, the effect of OBC was not prevented by indomethacin, indicating that arachidonic acid metabolites are not directly involved in the protective effect of OBC.

Nevertheless, the involvement of various cellular functions in the mechanism of action of bismuth compounds needs further clarification in both experimental animals and humans.

The interesting phenomenon of protection induced by bismuth salts against toxicity of antineoplastic and other hazardous agents will not be discussed in this chapter because all the relevant reports have concerned the inorganic form of bismuth, mainly bismuth subnitrate[14-21].

In general, organic bismuth compounds can be considered safe oral drugs when used appropriately in low therapeutic doses[3]. However, some of the bismuth salts are partially absorbed via the gastro-intestinal tract and enter the blood circulation[4]. Higher doses producing enhanced plasma concentrations, especially of the more soluble OBC, may bring about toxic manifestations[3] to be discussed below.

B. Toxicology of OBC

One of the most serious hazardous effects of bismuth compounds is encephalopathy. An epidemic of OBC-related neurotoxicity in France in the previous decade raised the question of safety assessment of the drug[22]. The neurologic effects of bismuth intoxication in humans included problems in walking, standing or writing, deterioration of memory, behavioural alterations, insomnia, muscle cramps and psychiatric symptoms. During manifest phase changes in awareness, myoclonia astasia and/or abasia and dysarthria become evident[23]. Recovery was observed following discontinuation of the drug[23]. Experimental animals showed myoclonus, ataxia, tremors and convulsions[24]. Characterization of the neuropathologic alterations included hydrocephalus and axonal swellings in the spinal cord[24]. It should be stressed that most of the neurotoxic effects appeared in patients after ingestion of very large quantities of bismuth for prolonged periods[22], and in therapeutic dose levels for certain periods of time OBC appear to be relatively safe for clinical use[25]. Additional toxic manifestations which should be considered in safety evaluation of OBC are nephropathy, osteoarthropathy, gingivitis, stomatitis and colitis[23]. In spite of these possible side effects, the OBC can be safely used for their acute indications and for up to 3–4 weeks of extended dosing[25]. Regarding intoxication, in experimental animals some chelating agents, such as dimercaptosuccinic acid and dimercaptopropanesulphonic acid, have been shown to be beneficial in detoxification of bismuth[26]. However, more animal studies and human experimentation are needed before recommending their clinical applications in cases of human bismuth poisoning.

In summing up the possible clinical applications of OBC in gastro-intestinal diseases, there is no doubt a place for these compounds as useful remedies for certain gastrointestinal conditions. They cannot, however, be considered the drugs of choice in treatment of acute duodenal and gastric ulcers. Nevertheless, a better understanding of the mechanisms governing their biological actions may lead to the development of more potent and safer OBC with better and perhaps new medical applications.

II. PHARMACOLOGY AND TOXICOLOGY OF ORGANIC ARSENICALS

Arsenicals have been used throughout history starting as a poison in Greece and Rome. The discovery by Ehrlich, that syphilis could be successfully treated by arsenical compounds (salvarsan and neosalvarsan), stimulated worldwide interest in arsenic-containing substances. In the modern clinic, arsenicals are still employed in chemotherapy of several tropical parasitical diseases[27]. Organic arsenicals are also used as pesticides[28], and thus they can contaminate drinking water, fish and related products, soil, fruits and vegetables. Arsenicals are sometimes added to food of livestock to promote growth[29]. Thus, due to the environmental dispersion of organic arsenicals, this group of compounds has to be considered as a serious risk to human public health in certain areas of the world. Our interest, therefore, in organic arsenicals stems from rather toxicological than pharmaco-logical aspects. The toxicological problems associated with arsenical exposure in the Western world derive mainly from environmental and industrial contamination.

Although many of the arsenic-containing compounds used are inorganic salts, trivalent and pentavalent, both are biotransformed in the body to their methylated forms[30,31]. Therefore, concerning human exposure to some of the inorganic arsenicals, these should be considered as organic compounds. Nevertheless, the present discussion will be focussed on the pharmacological and toxicological effects, mechanisms of actions and basic molecular characteristics of organic arsenicals and the implications of exposure to these agents.

A. Mechanism of Action

Organic arsenicals share similar mechanisms of action with the inorganic arsenic-containing compounds. Arsenicals are considered as sulphydryl reagents, which attack cellular compartments possessing available SH groups[28]. This kind of functional chemical group is ubiquitous throughout all kinds of cells of a wide variety of biological systems, including those of the mammalian body and of low organisms like parasites. Therefore, the mechanism of action responsible for the pharmacological effects of organic arsenicals, i.e. chemotherapy of protozoal infections, share similarities with that responsible for the toxicity to the host tissues. The main organs adversely affected by arsenicals are the nervous system, kidney, liver and gastro-intestinal tract[27]. Nevertheless, the beneficial therapeutic effects of some organic arsenicals in chemotherapy of protozoal and helmin-thic infections stem from their preferential toxic influences on the parasites rather than on the mammalian cells. The pharamacological and toxicological implications of this phenomenon will be discussed in the following paragraphs.

B. Pharmacology

The trypanocidal properties of arsenical melamine-containing compounds were dis-covered fifty years ago. One of the well-known drugs of this family is melarsoprol (Mel B, Arsobal).

Melarsoprol

During recent years other newly synthesized organic arsenicals have been developed for treatment of several types of trypanosomiasis[32-34] and helminthic diseases[35]. Similarly to microbial and cancer cells, drug-resistant parasites are one of the most serious problems facing modern chemotherapy. A combination of arsenicals with other agents such as difluoromethylornithine was beneficial in both experimental animals[36] and humans[37]. The mechanism of action underlying this activity is related to the attack of arsenicals on the sulphydryl moieties. The apparent parasite target molecule of arsenicals is trypanothione[38]. In addition, the compounds were shown to inhibit pyruvate kinase of both trypanosomal and mammalian origin[27]. Mammalian tissues biotransform the drug to the oxidized, non-toxic form, while this process occurs at a slower rate in protozoa. The effect or organic arsenicals may also be explained by the relative rapid penetration of the drug into the parasite as compared to the mammalian cells[27].

Apart from their pharmaceutical applications, the organic arsenicals are also used as herbicides. The main compounds are monosodium methanarsonate (MSMA) and hydroxydimethylarsine oxide (cacodylic acid). The relatively wide use of these agents is one of the main reasons for concern about their potential hazard to public health. Additionally, bioaccumulation of organic arsenicals in aquatic organisms such as seaweeds, freshwater algae and crustacea[39] plays a major role in the evaluation of the toxicity of these materials. Consequently, human exposure to organic arsenicals due to the combination of environmental and therapeutic applications is the main cause of the variety of toxic manifestations of these compounds.

C. Toxicology

Recent reports have indicated that treatment of patients suffering from sleeping sickness with melarsoprol may result in a Guillain–Barre-like syndrome[40]. The neurological symptoms include myalgias, distal paresthesias and progressive weakness in all four limbs. Neuropathological examinations have shown massive distal Wallerian degeneration in peripheral nerves and structural alterations in dorsal ganglia and spinal cord. Vacuolation of anterior horn cells and axonal neurofilamentous masses were observed. Interestingly, high concentrations of arsenic were found in the spinal cord whereas its level in peripheral nerves were undetectable[40]. In a case report[41], the possible role of organic arsenicals in causing encephalopathy in humans is described. A patient receiving melarsoprol for the treatment of trypanosomiasis developed convulsions, coma and hemiplegia after three days.

Several organic arsenicals are vesicants which result in skin blisters and necrosis. The military agent chlorovinylarsine dichloride (Lewisite) is a typical arsenic compound for eliciting blisterogenic and ulcerative effects by topical application as well as by other routes of exposure[42]. Although the specific mechanism of action of the vesicating activity of arsenicals is unclear, the involvement of heat shock proteins in this process has to be considered. The levels of these unique stress proteins are elevated in cultured human epidermal keratinocytes following exposure to sodium arsenite and, to a lesser extent, to the organic arsenical phenyldichloroarsine[43]. In contrast, another blisterogen, mechlorethamine (nitrogen mustard), has an inhibitory effect on synthesis of heat shock proteins in human keratinocytes[43]. These findings indicate that the blisterogenic activity of arsenicals may be associated with induction of heat shock proteins and that different types of vesicating agents may exert their activities by different molecular mechanisms.

Some organic arsenicals are particularly harmful to the liver. The antitrypanosomal, tryparsamide, can be hepatotoxic at therapeutic doses[44]. In general, the parenchymal cells are affected but damage may also appear as bile duct occlusion. The hepatic injury caused by organic arsenicals may be attributed to their blockage of insulin-stimulated activity. Frost and coworkers have shown that phenylarsine oxide, a trivalent arsenical, which binds neighbouring dithiols, blocked not only insulin-stimulated fluid phase endocytosis, but basal endocytotic and synthetic activities were also inhibited by the toxin[45,46]. Preferential inhibitory effect of phenylarsine oxide on protein internalization without significant deleterious effect on energy balance and oxygen consumption was observed in cultured hepatocytes[47]. High concentrations of the toxic agent adversely affected vital cellular functions[47]. In isolated muscle cells from adult rat, heart insulin internalization and insulin action on glucose transport were blocked by phenylarsine oxide[48].

In spite of the limited amount of information available on the mechanism underlying the pharmacological and toxicological effects of organic arsenicals, the given data throw some light on the mode of action of these toxic compounds.

The pharmacological tools to detoxify arsenicals are sulphydryl reagents. The classic drug for that purpose is 2,3-dimercapto-1-propanol, dimercaprol (British antilewisite, BAL). Since this compound has side effects and a low safety ratio, new analogs with less adverse effects have been developed, e.g. dimercaptopropanesulphonate, dimercaptosuccinic acid[49] and dithioerythritol[50]. The efficiency and toxicity of these agents need to be further evaluated.

An interesting endogenous detoxification mechanism for arsenical compounds has been described by Rosen and coworkers[51]. Bacterial cells resistant to these toxic compounds were able to form an active efflux of oxyanions, preventing their concentration from reaching toxic levels. This pump exhibited ATPase activity dependent on the presence of arsenite and antimonate[51,52]. This type of mechanism was also found in drug-resistant tumour cells in which P-glycoprotein was considered as an active translocator of drugs whereby cancerous cells as well as parasites become refractory to chemotherapeutic agents[53].

Long-term exposure to arsenicals might increase their oncogenic risks[54]. Agricultural workers suffer relatively higher mortality from several sources of tumours where the suspected carcinogens are arsenical compounds[54]. The carcinogenic potential of arsenicals was observed in cell cultures. In vitro cell transformation and cytogenetic effects, including endoreduplication, chromosomal aberrations and sister chromatid exchanges, were induced by arsenicals[55]. Furthermore, dimethylarsinic acid, one of the main metabolites of arsenics in mammals, was shown to be mutagenic in bacterial tests[56]. Although most of these effects were induced by inorganic forms of arsenic (arsenate and arsenite sodium salts) the mutagenic and carcinogenic potential of organic arsenicals in prolonged use should not be neglected when assessing the risks of these compounds.

The relationship between the intake of arsenicals and malignant neoplasms have been discussed by Reyman and coworkers[57]. They have indicated that medical treatment with arsenicals may lead to the development of intestinal as well as skin malignancies. Although in all instances described in the literature the causative agents seem to be inorganic arsenic salts, one should keep in mind the occurrence of pharmacokinetic transformation of the various organic arsenicals in the human organism. Moreover, some of these compounds may reach the human body after their transformation in animals or plants into carcinogenic arsenicals[57].

Using the same reasoning it should also be recognized that drug arsenicals could have potential toxic effects on the cardiovascular system which may result in ECG abnormalities and arrhythmias[58]. Organic arsenicals have been listed among the drugs that may cause polyarteritis[58].

III. PHARMACOLOGY AND TOXICOLOGY OF ORGANIC ANTIMONIALS

Similarly to arsenicals, the therapeutic and toxicologic characteristics of antimonials apparently stem from their high affinity to sulphydryl groups. In general, there is limited information on the molecular aspects of the biological activity of organic antimonials and their mechanism of action needs to be thoroughly elucidated.

The pentavalent antimony compound, sodium stibogluconate, is used as an antiparasitic drug. Treatment of Leishmania braziliensis has been reported to be successful with sodium stibogluconate[59] whereas in Leishmania mexicana, the response was lower and the preferred drug for this disease was ketoconazole[59].

A clinical study demonstrated the beneficial effect of a combination of stibogluconate and aminosidine[60]. The antimonial drugs were also shown to be effective against experimental trypanosomiasis in combination with difluoromethylornithine[61]. One of the proposed modes of action of this group of drugs is blockade of type 1 DNA topoisomerase, resulting in inhibition of the relaxation of supercoiled plasmid pBR322[62]. Additional

$$
\begin{array}{ccc}
\text{CH}_2\text{OH} & & \text{HOH}_2\text{C} \\
| & & | \\
\text{CHOH} & & \text{HOHC} \\
| & & | \\
\text{CHO} \quad \text{OH} & \text{O}^- \quad \text{OHC} \\
| \diagdown \; | & | \; \diagup | \\
\text{CHO} {\rightarrow} \text{Sb} - \text{O} - \text{Sb} {\leftarrow} \text{OHC} & 3\text{Na}^+ \\
| \diagup & \diagdown | \\
\text{CHO} & \text{OHC} \\
| & | \\
\text{COO}^- & {}^-\text{OOC}
\end{array}
$$

Sodium stibogluconate

mechanistic studies suggest that antimonials selectively inhibit metabolic pathways of glucose and fatty acids of the parasite[63]. The toxic manifestations of organic antimonial compounds are mainly gastro-intestinal disturbances including vomiting and diarrhoea[27,64]. Abnormalities in ECG, renal and hepatic functions may also occur[27].

Here too the lack of data on the mechanism of action of these compounds is evident. Further studies at the molecular levels to elaborate their activities and mode of action should contribute to our knowledge on the effectiveness, adverse effects and drug resistance of the organic antimonials.

IV. REFERENCES

1. R. S. D'Souza and V. G. Dhume, *Indian J. Physiol. Pharmacol.*, **35**, 88 (1991).
2. S. P. Lee, *Scand. J. Gastroenterol.*, **185**, 1 (1991).
3. F. Caron and B. Rouveix, *Therapie*, **46**, 393 (1991).
4. G. Treiber, U. Gladziwa, T. H. Ittel, S. Walker, F. Schweinsberg and U. Klotz, *Aliment. Pharmacol. Ther.*, **5**, 491 (1991).
5. M. R. Millar and J. Pike, *Antimicrob. Agents Chemother.*, **36**, 185 (1992).
6. E. J. Prewett, Y. W. Luk, A. G. Fraser, W. M. Lam and R. E. Pounder, *Aliment. Pharmacol. Ther.*, **6**, 97 (1992).
7. P. Domenico, D. R. Landolphi and B. A. Cunha, *J. Antimicrob. Chemother.*, **28**, 801 (1991).
8. J. A. Armstrong, M. Cooper, C. S. Goodwin, J. Robinson, S. H. Wee, M. Burton and V. Burke, *J. Med. Microbiol.*, **34**, 181 (1991).
9. P. Domenico, R. J. Salo, D. C. Straus, J. C. Hutson and B. A. Cunha, *Infection*, **20**, 66 (1992).
10. S. M. Hinsull, *Gut*, **32**, 355 (1991).
11. P. A. Aho and I. B. Linden, *Scand. J. Gastroenterol.*, **27**, 134 (1992).
12. A. Ahmed, P. R. Salmon, C. R. Cairns, M. Hobsley and J. R. Hould, *Gut*, **33**, 159 (1992).
13. B. L. Slomiany, X. Y. Wang, D. Palecz, K. Okazaki and A. Slomiany, *Alcohol Clin. Exp. Res.*, **14**, 580 (1990).
14. Y. Kondo, M. Satoh, N. Imura N and M. Akimoto, *Cancer Chemother. Pharmacol.*, **29**, 19 (1991).
15. P. J. Boogaard, A. Slikkerveer, J. F. Nagelker and G. J. Mulder, *Biochem. Pharmacol.*, **41**, 369 (1991).
16. T. Morikawa, E. Kawamura, T. Komiyama and N. Imura, *Nippon Gan. Chiryo Gakkai Shi.*, **25**, 1138 (1990).
17. T. Hamada, Y. Nishiwaki, T. Kodama, A. Hayashibe, N. Nukariya, H. Sasaki, T. Morikawa, T. Hirosawa and T. Natsuyama, *Gan. To. Kagaku. Ryoho.*, **16**, 3587 (1989).
18. A. Naganuma, M. Satoh and N. Imura, *Jpn. J. Cancer Res.*, **79**, 406 (1988).
19. I. Naruse and Y. Hayashi, *Teratology*, **40**, 459 (1989).
20. N. Satomi, A. Sakurai, R. Haranaka and K. Haranaka, *J. Biol. Response Mod.*, **7**, 54 (1988).
21. A. Naganuma, M. Satoh and N. Imura, *Cancer Res.*, **47**, 983 (1987).
22. J. P. Bader, *Digestion*, **37**, suppl. 2, 53 (1987).

722   U. Wormser and I. Nir

23. A. Slikkerveer and F. A. de-Wolff, *Med. Toxicol. Adverse Drug Exp.*, **4**, 303 (1989).
24. J. F. Ross, Z. Sahenk, C. Hyser, J. R. Mendell and C. L. Aldern, *Neurotoxicology*, **9**, 581 (1988).
25. D. W. Bierer, *Rev. Infect. Dis.*, **12**, Suppl. 1, 93 (1990).
26. A. Slikkerveer, H. B. Jong, R. B. Helmich and F. A. de Wolff, *J. Lab. Clin. Med.*, **119**, 529 (1992).
27. L. T. Webster Jr., in *Goodman and Gilman's The Pharmacological Basis of Therapeutics*, 7th ed. (Eds. A. G. Gilman, L. S. Goodman, T. W. Rall and F. Murad), Macmillan, New York, 1985, pp. 1058–1065.
28. L. El-Bahri and S. Ben-Romdane, *Vet. Hum. Toxicol.*, **33**, 259 (1991).
29. P. A. VanderKop and J. D. MacNeil, *Vet. Hum. Toxicol.*, **31**, 209 (1989).
30. C. D. Klaassen, in *Goodman and Gilman's The Pharmacological Basis of Therapeutics*, 7th ed. (Eds. A. G. Gilman, L. S. Goodman, T. W. Rall and F. Murad), Macmillan, New York, 1985, pp. 1605–1627.
31. J. P. Buchet and R. Lauwerys, *Toxicol. Appl. Pharmacol.*, **91**, 65 (1987).
32. G. Dreyfuss, P. Loiseau, J. G. Wolf, C. Bories, P. Gayral and J. A. Nicolas, *Bull. Soc. Pathol. Exot. Filiales*, **81**, 561 (1988).
33. E. Zweygarth and R. Kaminsky, *Trop. Med. Parasitol.*, **41**, 208 (1990).
34. B. T. Wellde, M. J. Reardon, D. A. Chumo, R. M. Muriithi, S. Towett and J. Mwangi, *Ann. Trop. Med. Parasitol.*, **83**, 161 (1989).
35. D. A. Denham, I. T. Midwinter and E. A. Friedheim, *J. Helmintol.*, **64**, 100 (1990).
36. P. W. Jennings, *Bull. Soc. Pathol. Exot. Filiales*, **81**, 595 (1988).
37. G. L. Kazyumba, J. F. Ruppol, A. K. Tshefu and N. Nkanga, *Bull. Soc. Pathol. Exot. Filiales*, **81**, 591 (1988).
38. N. Yarlett, B. Goldberg, H. C. Nathan, J. Garofalo and C. J. Bacchi, *Exp. Parasitol.*, **72**, 205 (1991).
39. R. E. Menzer and J. O. Nelson, in *Casarett and Doull's Toxicology, The Basic Science of Poisons*, 3rd ed. (Eds. C. D. Klaassen, M. O. Amdur and J. Doull), Macmillan, New York, 1986, pp. 825–853.
40. K. Gherardi, P. Chariot, M. Vanderstigel, D. Malapert, J. Verroust and A. Astier, *Muscle Nerve*, **13**, 637 (1990).
41. G. Pialoux, S. Kernbaum and F. Vachon, *Bull. Soc. Pathol. Exot. Filiales*, **81**, 555 (1988).
42. N. I. Sax and R. J. Lewis, Sr., in *Dangerous Properties of Industrial Substances*, 7th ed. (Eds. N. I. Sax and R. J. Lewis, Sr.), Van Nostrand Reinhold, New York, 1989, pp. 900–901.
43. M. A. Deaton, P. D. Bowman, G. P. Jones and M. C. Powanda, *Fundam. Appl. Toxicol.*, **14**, 471 (1990).
44. C. D. Klaassen, in *Goodman and Gilman's The Pharmacological Basis of Therapeutics*, 7th ed. (Eds. A. G. Gilman, L. S. Goodman, T. W. Rall and F. Murad), Macmillan, New York, 1985, pp. 1592–1604.
45. K. Pettengell and S. C. Frost, *Biochem. Biophys. Res Commun.*, **161**, 633 (1989).
46. S. C. Frost, M. D. Lane and E. M. Gibbs, *J. Cell Physiol.*, **141**, 467 (1989).
47. A. E. Gibson, R. J. Noel, J. T. Herlihy and W. F. Ward, *Am. J. Physiol.*, **257**, C182 (1989).
48. J. Eckel and H. Reinauer, *Biochem. J.*, **249**, 111 (1988).
49. R. H. Inns, P. Rice, J. E. Bright and T. C. Marrs, *Hum. Exp. Toxicol.*, **9**, 215 (1990).
50. V. L. Boyd, J. W. Harbell, R. J. O'Connor and E. L. McGown, *Chem. Res. Toxicol.*, **2**, 301 (1989).
51. B. P. Rosen, U. Weigel, C. Karkaria and P. Gangola, *J. Biol. Chem.*, **263**, 2067 (1988).
52. M. H. Ching, P. Kaur, C. E. Karkaria, R. F. Steiner and B. P. Rosen, *J. Biol. Chem.*, **266**, 2327 (1991).
53. M. Grogl, R. K. Martin, A. M. Oduola, W. K. Milhous and D. E. Kyle, *Am. J. Trop. Med. Hyg.*, **45**, 98 (1991).
54. P. Vineis, L. Settimi and A. Seniori-Costantini, *Med. Lav.*, **81**, 363 (1990).
55. J. C. Barrett, P. W. Lamb, T. C. Wang and T. C. Lee, *Biol. Trace Elem. Res.*, **21**, 421 (1989).
56. K. Yamanaka, H. Ohba, A. Hasegawa, R. Sawamura and S. Okada, *Chem. Pharm. Bull. Tokyo*, **37**, 2753 (1989).
57. E. Reyman, R. Moller and A. Nidsen, *Arch. Dermatol.*, **114**, 378 (1978).
58. Z. Fastner, in *Meyler's Side Effects of Drugs*, 9th ed. (Ed. M. N. G. Dukes), Excerpta Medica, Amsterdam, 1980, pp. 368–369.
59. T. R. Navin, B. A. Arana, F. E. Arana, J. D. Berman and J. F. Chajon, *J. Infect. Dis.*, **165**, 528 (1992).

60. C. N. Chunge, J. Owate, H. O. Pamba and L. Donno, Trans. R. Soc. Trop. Med. Hyg., **84**, 221 (1990).
61. F. W. Jennings, *Trop. Med. Parasitol.*, **42**, 135 (1991).
62. A. K. Chakraborty and H. K. Majumder, *Biochem. Biophys. Res. Commun.*, **152**, 605 (1988).
63. J. D. Berman, J. V. Gallalee and J. M. Best, *Biochem. Pharmacol.*, **36**, 197 (1987).
64. T. Naitoh, M. Imamura and R. J. Wassersug, *Comp. Biochem. Physiol.*, C, **100**, 353 (1991).

CHAPTER **19**

Safety and environmental effects

SHIGERU MAEDA

Department of Applied Chemistry and Chemical Engineering, Faculty of Engineering, Kagoshima University, 1-21-40 Korimoto, Kagoshima 890, Japan

The chemistry of organic arsenic, antimony and bismuth compounds
Edited by S. Patai © 1994 John Wiley & Sons Ltd

I. INTRODUCTION

A number of different factors can influence the toxicity of heavy metal (metalloid) elements, including chemical structure, physical properties, mode of administration and the nature of the species affected. The toxic effects of the elements on organisms are very complicated. Some organisms are very sensitive to one chemical compound of an element, but have high resistance to another compound of the same element or to other toxic metals.

The mechanisms of resistance to toxic metals for organisms are varied. In order to simplify the consideration of the mechanism, Wood and Wang presented several modes of algal resistance to toxic elements and summarized them as follows[1].

a. The development of energy-driven efflux pumps that keep toxic element levels low in the interior of the cell.

b. Oxidation state change by which a more toxic form of a metal can enzymatically and intracellularly be converted to a less toxic form, e.g. mercury [Hg(II) to Hg(0)] by *Chlorella*[2]; molybdenum [Mo(VI) to Mo(III)] by *Chlorella*[3]; gold [Au(III) to Au(I) or Au(0)] by *Chlorella*[4].

c. Binding of metal ions on the cell surfaces, e.g. *Eucheuma striatum* and *E. spinosum* with lead[5], or *Nitzschia* with iron[6].

d. Precipitation of insoluble metal complexes on the cell wall surface, e.g. *Cyanidium caldarium* converting chromium and aluminum into their sulfides[7].

e. Complexing of metal ions with excreted metabolites (extracellular products), which can extracellularly mask a toxic metal, e.g. *Skeltonema* etc. in the case of zinc[8].

f. Vaporization and elimination by means of converting a toxic metal to a volatile chemical species, e.g. *Chlorella* converting mercury(II) to volatile methylmercury or mercury(0)[2].

g. Binding of metal ions with protein or polysaccharides in the interior of the cell, which may deactivate the metal ion's toxicity, e.g. *Chlorella* with cadmium protein[9,10]; *Fucus* with technetium polysaccharides and protein[11].

h. Methylation of the element, which can enzymatically and intracellularly prevent a toxic element from reacting with SH groups, e.g. many algae convert arsenic to dimethyl- and trimethylarsenic compounds[12–15].

When the impact of the toxic metal exceeds the capacity of the resistance mechanism of an organism and/or of an environment, the organism and/or the environment may suffer toxic effects.

A considerable number of organometallic species of arsenic, antimony and bismuth have been detected in the natural environment in different manners. A number of these are nonmethyl compounds which have entered the environment after manufacture and use [e.g. butyltin and phenyltin compounds for antifouling paints on boats, and arsanilic acid (Figure 2, **5**) and phenylarsonic acids (Figure 2, **6–8**) for animal husbandry]. Only a few methyl compounds are now manufactured and used (e.g. methyltin compounds for oxide film precursors on glass and methylarsenic compounds for desiccants or defoliants).

It is now well established that certain organometallic compounds are formed in the environment including derivatives of mercury, arsenic, selenium, tellurium and tin, as well as (deduced on the basis of analytical evidence) including those of lead, germanium, antimony and thallium[16]. A list of elements which may undergo methylation is given in Table 1. Biological methylation of arsenic has been demonstrated directly by use of experimental organisms. Methylantimony compounds were found in the environment but the biomethylation has not been confirmed experimentally. Methylbismuth compounds have not been observed in the environment.

The main interest in methylation is the change in properties resulting from the attachment of methyl groups to the inorganic elements or compounds. Lipid solubility, volatility and persistence of metals in biological systems may be increased in the methyl

TABLE 1. Elements forming methyl derivatives in the environment (modified from
Reference 16)

Metals	Metalloids
Mercury	Arsenic
Lead	Antimony
Tin	Selenium
Thallium	Tellurium
Cobalt	Germanium

derivatives. Most organometallic compounds are more toxic than those in inorganic forms, but sometimes the reverse in the case (particularly for arsenic).

This review describes factors concerning the safety and environmental effects of organic arsenic, antimony and bismuth compounds. The factors involve the production and use of the elements, toxicity, pollution, metabolism (alkylation), health effect assessment, fate and so on.

Regarding antimony and bismuth, only a few studies exist involving organometallic compounds with metal–carbon bonds. Hence in these cases inorganic and organic compounds with metal—O—C bonds are also described.

II. ARSENIC

A. Introduction

Arsenic is ubiquitous in the Earth's crust: arsenic ranks 20th among the elements in abundance. In nature, arsenic is widely but sparsely distributed. It is associated with igneous and sedimentary rocks, particularly with sulfidic ores. Arsenic enters the aquatic environment indirectly from industrial and other air emission, and directly from localized effluent discharges. There is general agreement that most man-caused atmospheric input is due to smelting operations and fossil-fuel combustion. Arsenic emission to the atmosphere was calculated with the factors listed in Table 2[17].

The majority of natural and man-produced arsenic in seawater is present as inorganic forms. Arsenic is susceptible to methylation by microorganisms forming volatile methyl-ated arsines and nonvolatile organoarsenic acids, with subsequent release to the water. Marine organisms can accumulate arsenic by direct water contact and ingestion of food.

TABLE 2. Arsenic emission factors[a]

Arsenic source	Arsenic concentration
Mining and milling	0.45 ton per million tons of copper, lead, zinc, silver, gold or uranium ore
Smelting and refining	955 tons per million tons of copper produced 591 tons per million tons of zinc produced 364 tons per million tons of lead produced
Coal	1.4 tons per million tons of coal burned
Petroleum	5.2 kg per million barrels of petroleum burned

[a] Reference 17, p. 71.

B. Production and Use

Arsenic was recovered as arsenic trioxide in about 20 countries from the smelting or roasting of nonferrous metal ores or concentrates. The most common separation methods are volatilization from the ores, and sublimation of arsenic trioxide occurs during the roasting stage of these ores.

World production of arsenic trioxide is about 50,000–53,000 tons per year. Table 3 shows world production by country[18]. The three leading arsenic-trioxide-producing capacities were in Sweden, France and the USSR each with about 10,000 tons per year. Metallic arsenic accounts for 3% of the world demand for arsenic and was produced by the reduction of arsenic trioxide.

The major use of arsenic is for agricultural pesticides, wood preservatives and animal feed additives.

Estimated end-use distribution of arsenic in the United States[18] was 70% (16,000 tons as elemental arsenic) in industrial chemicals (mainly as wood preservatives), 22% (5200 tons) in agricultural chemicals (mainly as herbicides and desiccants), 4% (900 tons) in glass manufacture, 3% (700 tons) in nonferrous alloys and 1% (300 tons) for other purposes (animal feed additives, pharmaceuticals etc.) in 1989. Estimated end-use of arsenic in Japan[19,20] was 35% (230 tons) for refining of zinc, 34% (220 tons) for glass manufacture, 15% (100 tons) for electronics as ultrapure arsenic metal and 15% (100 tons) for wood preservatives and agricultural chemicals in 1988.

The major use of arsenical chemicals in agriculture in the United States is in cotton production, where they are used as herbicides, and plant desiccants and defoliants. The arsenical herbicides, monosodium methylarsonate (MSMA, **1**) and disodium methylarsonate (DSMA, **2**), are produced from arsenic trioxide. The use of arsenic in agricultural chemicals has been under close scrutiny by the US Environmental Protection Agency (EPA) for many years. A large number of pesticide uses for arsenical chemicals have been

TABLE 3. World production of arsenic trioxide[a] by country (metric tons)[18]

Country	1985	1986	1987	1988	1989
Belgium[c]	3 000	3 000	3 500	3 500	3 500
Bolivia[c]	361	2 412	132	191	350
Canada[c]	3 000	3 000	2 000	2 000	2 000
Chile	[c]4 000	[c]4 000	3 616	3 207	3 400
France[c]	8 000	10 000	10 000	10 000	10 000
Germany, Federal Republic of[c]	360	360	360	360	360
Japan[c]	500	500	500	500	500
Mexico	4 782	5 315	5 304	5 164	5 100
Namibia	2 471	2 208	1 864	2 983	2 900
Peru	1 254	1 273	1 757	828	1 000
Philippines[c]	5 000	5 000	5 000	5 000	5 000
Portugal	204	176	218	214	180
Sweden[b]	10 000	10 000	10 000	10 000	10 000
USSR[c]	8 100	8 100	8 100	8 100	8 100
United States	2 200	—	—	—	—
Total	53 235	53 173	52 351	52 047	52 390

[a] Including calculated arsenic trioxide equivalent of metallic arsenic and of arsenic compounds other than arsenic trioxide, where inclusion of such materials would not duplicate reported arsenic trioxide production. The table includes data available through May 25, 1990.
[b] Based on arsenic trioxide exported plus the arsenic trioxide equivalent of the metallic arsenic exported.
[c] Estimated.

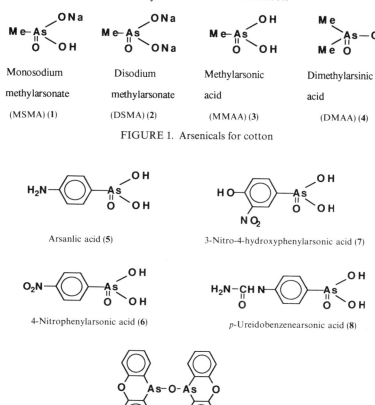

FIGURE 1. Arsenicals for cotton

FIGURE 2. Organoarsenicals for feed additives (5–8) and fungicide (9)

banned by the EPA. The use of the arsenicals disodium methylarsonate (DSMA), monosodium methylarsonate (MSMA), methylarsonic acid (MMAA), dimethylarsinic acid (DMAA) and arsenic acid on cotton is expected to remain stable, or rather to decline, over the next five years[18]. The chemical structures are shown in Figure 1.

A small amount of organoarsenicals is used for animal feed additives. Four arsenic compounds are now used in animal husbandry (5–8) in the USA. The heterocycle 10,10′-oxybisphenoxarsine (9) is manufactured and marketed as an antimicrobial agent which is particularly useful in conjunction with plastics. These organoarsenicals are shown in Figure 2.

C. Concentration and Speciation of Organoarsenic Compounds in the Natural Environment

In nature, arsenic is widely but sparsely distributed. Exceptionally higher levels of total arsenic are commonly found in commercially available marine seafood products[21]. These levels are usually in the range of 1–45 ppm, and sometimes reach levels higher than 100 ppm.

Arsenic enters the aquatic environment indirectly from industrial and utility air emissions, and directly from localized effluent discharge. Open ocean water normally contains a low range of arsenic concentration at a few ppb. The majority of natural arsenic in seawater is present as inorganic compounds. Lunde found that a large part of arsenic in marine organisms was in organic form[21]. Many investigators reported high concentrations of organoarsenic compounds in marine organisms. It has been known for many years that the concentration of arsenic species in marine and freshwater animals is considerable above the background concentrations in the surrounding water and that 'fish arsenic' is chemically and physiologically different from arsenate and arsenite[21–25].

Cullen and Reimer[26] summarized data on organoarsenicals isolated from seventy-two marine animals and their organs. According to their data, the arsenic concentrations range

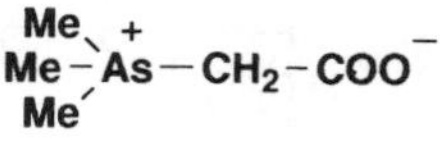

$$\overset{Me}{\underset{Me}{\overset{|}{Me-As}}}{}^{+}-CH_2-COO^{-}$$

Arsenobetaine (**10**)

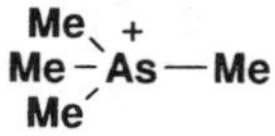

$$\overset{Me}{\underset{Me}{\overset{|}{Me-As}}}{}^{+}-Me$$

Tetramethylarsine (**11**)

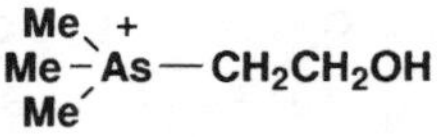

$$\overset{Me}{\underset{Me}{\overset{|}{Me-As}}}{}^{+}-CH_2CH_2OH$$

Arsenocholine (**12**)

$$\overset{Me}{\underset{Me}{As}}=O$$

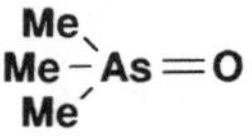

Trimethylarsine oxide (**13**)

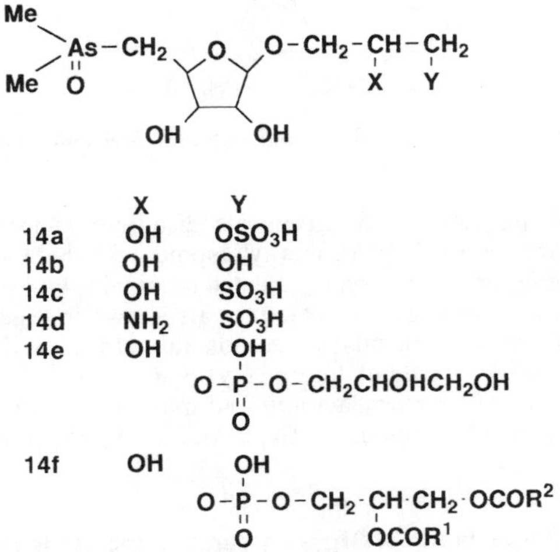

Me–As(=O)–CH₂–[tetrahydrofuran ring with OH, OH]–O–CH₂–CH(X)–CH₂(Y)

	X	Y
14a	OH	OSO₃H
14b	OH	OH
14c	OH	SO₃H
14d	NH₂	SO₃H
14e	OH	OH
14f	OH	OH

14e Y: $O-\overset{\overset{O}{\|}}{P}-O-CH_2CHOHCH_2OH$

14f Y: $O-\overset{\overset{O}{\|}}{\underset{}{P}}-O-CH_2\cdot CH\cdot CH_2\cdot OCOR^2$, with $OCOR^1$

Arsenosugars (**14**)

FIGURE 3. Organoarsenicals isolated from marine animals and plants

from 0.31 ppm in salmon to a high of 340 ppm in the midgut gland of a carnivorous gastropod. Although substantial contributions about the distribution and nature of arsenicals in the marine environment had been made by Lunde[21-25], it was not until 1977 that arsenobetaine (10) was isolated from the rock lobster[27,28]. Since this discovery, arsenobetaine has been shown to be present, and to be the most abundant arsenical, in most marine animals so far investigated. Other organoarsenicals were also found as shown in Figure 3.

Arsenosugars (14) involving a dimethylarsenic moiety were mainly observed in marine plants.

Concentration and chemical forms of arsenic in freshwater environment have been reviewed in detail by the present author[29]. When comparing freshwater and marine organisms in the natural environment, there seems to be a clear difference in the total arsenic concentration, which in freshwater organisms is lower than in marine organisms. Trimethyl-, dimethyl- and monomethylarsenic compounds were detected in freshwater organisms whose chemical structure in vivo have not been confirmed[29].

D. Biological Methylation and Transformation

It is over 40 years since Challenger reviewed his work on the identification of 'Gosio gas' as trimethylarsine (Me_3As)[30]. Gosio gas is the volatile, toxic, arsenic species produced by molds growing on wallpaper colored with arsenic-containing pigments such as copper arsenite. Experimental results on methylation of arsenic by fungi, bacteria, algae and other organisms have been reported by a number of researchers and reviewed by Cullen and Reimer[26].

Laboratory studies have shown that the microorganisms present in natural marine sediments from British Columbia (Canada) and sediments contaminated with mine-tailings were capable of methylating arsenic under either aerobic or anaerobic conditions. Incubation of sediments with culture media produced volatile arsines (including AsH_3, $MeAsH_2$ and Me_3As) as well as the methylarsenic(V) compounds $Me_nAs(O)(OH)_{3-n}$ ($n = 1,2,3$)[31].

The presence of organoarsenicals in marine organisms is commonly assumed to be due to the accumulation of compounds that have been synthesized from arsenate at low trophic levels.

Many papers on biotransformation of arsenic via the marine food chain have been published. For example, three trophic levels of marine organisms (phytoplankton— *Dunaliella marina*, zooplankton— *Artemia salina* and shrimp— *Lysmata seticaudata*) were tested regarding their arsenic metabolism by Wrench and coworkers[32]. The experimental results led to the conclusion that organic forms of arsenic in marine foods were derived from an in vivo synthesis by primary producers and were efficiently transferred along the marine food chain. The shrimp, the highest trophic level organism in this food chain, could not form organic arsenic by itself. In this case, arsenate taken up from water was converted largely to arsenite. Similar conclusions were obtained from the experimental results on a phytoplankton–mussel–crab[33], a phytoplankton–grazer snail–carnivore snail system[34,35] and a phytoplankton–lobster system[36]. More information about biotransformations of arsenic in marine ecosystems is available in a review by Andreae[37].

Only a few papers on biotransformation of arsenic in freshwater ecosystems were published. Isensee and coworkers[38] examined the distribution of dimethylarsenic compounds among freshwater organisms in model ecosystems (water–algae–snail or water–diatoms–daphnia–fish). The results showed that lower food chain organisms (algae and daphnia) bioaccumulated more dimethylarsenic compounds than did higher food chain organisms (snails and fish). Amounts accumulated indicate that dimethylarsenic compounds did not show a high potential to biomagnify in the environment.

TABLE 4. Arsenic accumulation and metabolism in three-step freshwater food chain[12]

Organisms	Accumulation route (As concn. in $\mu g \, As \, g^{-1}$)	Arsenic in organisms[a] $\mu g \, As \, g^{-1}$ (%)				
		Total	MA	MMA	DMA	TMA
Chlorella sp.	1. water (30)	745 (100)	12.7 (1.7)	tr —	12.7 —	tr —
	2. water (100)	2850 (100)	23.1 (0.8)	tr —	14.7 (0.5)	8.4 (0.3)
Moina sp.	1. water (0.1)	9.5 (100)	4.7 (49.5)	1.0 (10.5)	3.7 (39.0)	tr —
	2. water (2)	17.9 (100)	4.4 (24.0)	1.1 (6.1)	3.3 (18.4)	tr tr
	3. food [*Chlorella* (2)]	225 (100)	37.1 (16.5)	21.5 (9.6)	15.6 (6.9)	tr tr
Carassius sp.	1. water (0.5)	33.2 (100)	1.1 (3.3)	tr tr	0.2 (0.6)	0.9 (2.7)
	2. water (1.0)	51.3 (100)	14.4 (28.1)	2.9 (5.7)	1.0 (1.9)	10.5 (20.5)
	3. food [*Chlorella* (2) ↑ *Moina* (3)]	37.0 (100)	13.5 (36.5)	6.9 (18.6)	1.3 (3.5)	5.3 (14.3)

[a] MA denotes methylated arsenic (MMA + DMA + TMA), MMA denotes monomethyl arsenic, DMA denotes dimethylarsenic and TMA denotes trimethylarsenic.

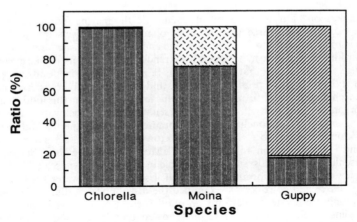

FIGURE 4. Relative concentration of arsenic derivatives in organisms via food chain (Chlorella– Moina–Guppy). TMA (▨): trimethylarsenic; DMA (▨): dimethylarsenic; MMA (▨): momethylarsenic

Maeda and coworkers examined transformations of arsenic via some freshwater food chains[12-14]. For example, the experimental result[12] in a three-step freshwater food chain (phytoplankton *Chlorella vulgaris*–grazer *Moina macrocopa*–carnivore *Carassius carassius auratus* juvenile) is shown in Table 4. The latter shows that the grazer and carnivore also accumulated arsenic from water and methylated a part of it. The arsenic accumulation from food decreased one order of magnitude and the biomethylation ratio of the arsenic increased, successively with an elevation in the trophic level. This result means that arsenic concentrations are not magnified in the aquatic food web. Del Riego[39], Penrose and coworkers[40] and Seydel[41] have also obtained similar results. Arsenic levels in pelagic fish (in the order of 0.3–3 ppm) are significantly lower than those of bottom feeders and shellfish (in the order of 1–55 ppm).

A similar result was obtained from an experiment in which another food chain (phytoplankton *C. vulgaris*–grazer *Moina* sp.–carnivorous guppy *Poecilia* sp. adult) was examined[14]. Figure 4 shows the relative concentration of arsenic species accumulated and demonstrates that the relative concentration of dimethyl arsenic and trimethyl arsenic compounds dramatically increased successively with an elevation in the trophic level.

E. Toxicity and Environmental Effects

1. Inorganic arsenicals

A number of different factors can influence the toxicity of arsenicals, including chemical structure, physical properties, mode of administration and species of organism.

The trivalent forms of arsenic apparently exerts their toxic effects chiefly by reacting with the sulfhydryl groups of vital cellular enzymes. Pyruvate dehydrogenase seems to be a particularly vulnerable site in the metabolism, because it contains the thiol lipoic acid that is especially reactive with trivalent arsenicals.

The biochemical basis of the toxic action of pentavalent inorganic compounds is known with less certainty. It is estimated that such arsenicals may well compete with phosphate in phosphorylation to form unstable arsenyl esters that spontaneously hydrolyze and thereby short-circuit energy-yielding bioenergetic processes. The pentavalent arsenicals may exert their effects in many cases after being reduced in vivo to trivalent forms.

2. Organic arsenicals

Arsenic can form chemically and physiologically stable bonds with carbon in a large number of organic derivatives. There are significant toxicological differences among various organic arsenicals and different species. From the perspective of likely exposure and potential risk to human health, the following four groups of derivatives are worthy of special attention[42].

a. Methyl derivatives of arsenic acid. The most important members of this group are methylarsonic acid (MMAA, **3**), dimethylarsinic acid (DMAA, **4**) and their salts (**1, 2** and sodium salt of **4**). These compounds have been widely used as pesticides, herbicides and defoliants.

Both MMAA and DMAA are well absorbed in animals and humans. By the oral route, these compounds produce symptoms of gastrointestinal irritation and renal and hepatic injury similar to some of the effects produced by inorganic arsenic, but the potency of these methyl derivatives is much lower.

Both MMAA and DMAA are excreted primarily in urine. Buchet and coworkers administered an oral dose of 500 μg arsenic in the form of arsenite, MMAA and DMAA to human volunteers. After four days, the excretion of urinary arsenic was 46, 78 and 75% of the dose of arsenite, MMAA and DMAA, respectively[43].

The methyl derivatives are also formed in the body by metabolism of inorganic arsenic[44], and this is generally viewed as a detoxification pathway.

Environmental and health effects of MSMA applied along the Louisiana Highway, USA were examined[45,46]. The abstract of the reports is as follows. The use of MSMA as a grass control herbicide along the highway is desirable because of its economic advantage in chemical mowing and Johnson grass control. The objective of this project was to obtain baseline arsenic residue data along the highway rights-of-way of five selected highways in Louisiana. In summary, 432 or 77.3% of all surface-soil samples collected contained 0 to 50 ppm of arsenic residues while 127 or 22.7% contained over 50 ppm. Most (14 out of 18) water samples contained arsenic residues below 50 ppb, which is the EPA Primary Drinking Water Standard. The background (baseline) arsenic-residue concentrations in surface-soil samples ranged from 3.72 ppm to 13.9 ppm of arsenic. The report concluded that the environmental and health risks due to the accumulation of arsenic residues in soil and water appeared to be minimal.

Abdelghani and coworkers examined the uptake and excretion of MSMA in workers applying the herbicide to highway rights-of-way[47]. Urine, blood and hair samples were collected from workers as well as air samples taken from the workers' breathing zone. Arsenic concentrations in air samples ranged from $0.001-1.086\,\mu g\,m^{-3}$ with no significant difference in air arsenic concentrations over time. Blood arsenic values ranged from $0.0-0.2\,mg\,l^{-1}$, well within the limits considered normal. Arsenic urine values ranged from $0.002-1.725\,mg\,l^{-1}$. This is above the normal range but is consistent with arsenic levels in urine following a seafood meal. The mean arsenic concentration in urine and hair increased during the week but returned to low levels on weekends. There was no accumulation of arsenic in urine over the spraying season.

b. Phenyl derivatives of arsenic acid. Several phenyl arsenates (**5–8**) and their salts have been widely used as feed additives to improve weight gain and prevent enteric disease in poultry and swine. The mechanism of these beneficial effects is not known but it may have an effect on the intestinal microogranisms rather than a direct effect on the animal. Excessive arsenical residues in poultry and pork tissues occur only when the arsenicals are fed at excessive dosages for long periods leading to losses of animal weight. Exposure of animals to high levels of phenylarsonate compounds results primarily in sensory and peripheral nerve injury.

These compounds are for the most part absorbed and excreted without any metabolic change[17]. The toxicity of these compounds in humans has not been investigated extensively.

c. 'Fish arsenic'. Fish and shellfish often accumulate rather high tissue levels of arsenic. Depending on the species, most of this accumulation exists in the form of arsenobetaine (**10**) or arsenocholine (**12**).

Cannon and coworkers administered **10** by intraperitoneal injection to mice and found that doses up to $500\,mg\,kg^{-1}$ produced no symptoms of toxicity[48]. Similarly, Kaise and coworkers administered oral doses of **10** as high as $10,000\,mg\,kg^{-1}$ to mice[49] and hamsters[50] and observed no toxic symptoms. Kaise and coworkers[51] and Marafante and coworkers[52] examined the metabolism of **12** in experimental animals and reported that it was converted to **10** and rapidly excreted in urine.

Maeda and coworkers[12] fed *Moina* sp. with arsenic-accumulated *Chlorella vulgaris*. Figure 5 shows the growth and arsenic concentration in *Moina* sp. and demonstrates that *Moina* sp. might prefer *C. vulgaris* accumulating arsenic at higher concentrations up to $1270\,\mu g\,As\,g^{-1}$ and arsenic concentration in *Moina* sp. increased with an increase of

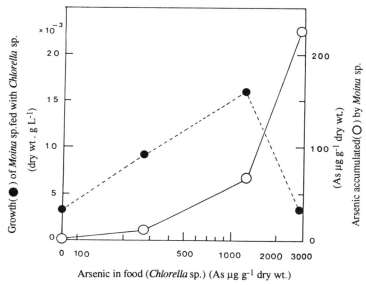

FIGURE 5. Growth of *Moina* sp. fed with *Chlorella* sp. accumulating different levels of arsenic and accumulation of arsenic by *Moina* sp. after 7 days culture[12]

arsenic concentration in the food, *C. vulgaris*. Although the chemical forms of arsenic in *C. vulgaris* have not been identified, the former result suggests that the arsenic in *C. vulgaris* may have a stimulating effect on the growth for *Moina* sp.[12].

Available data indicate that these organic derivatives have low toxicity, and ingestion of arsenic in this form is not generally considered to be a health hazard.

d. Methyl derivatives of arsine. Methylarsines are not widely used in industry but might be encountered at some waste sites. In addition, they may be formed in the environment from other arsenic compounds by the action of microbial organisms or by the inadvertent chemical reactions that generate strong reducing conditions.

Methylarsines (mono-, di- and trimethylarsine) are strong irritants, but these compounds are less powerful than arsine as hemolytic agents. No quantitative data on the toxicokinetics and dose–response relationships for the methylarsines were located[42].

F. Biodegradation of Organoarsenic Compounds

Both MMAA and DMAA are stable in the aqueous environment. Andreae has reported successful long-term storage under sterile conditions[53]. The observed loss of these compounds must therefore be due to bacterial demethylation.

The first example of biological arsenic–carbon bond cleavage was described by Challenger[54]. Trimethylarsine is produced when *S. brevicaulis* and *P. notatum* act on $ClCH_2CH_2AsO(OH)_2$. Because *P. notatum* does not methylate arsenate, it seems likely that loss of the $ClCH_2CH_2$ group takes place after at least one methylation step.

Demethylation of sodium methylarsonate to arsenate was reported by Shariatpanahi and coworkers[55]. One organism isolated from soil, *Alcaligenes*, produces only arsenate;

others isolated from the same environment produce arsine as well. Wine yeast is reported to demethylate dimethylarsinate to methylarsonate[56].

Methyl cleavage reactions have also been observed in broken-cell homogenates of *C. humicola*[57]. Thus [^{14}C]dimethylarsinate is metabolized to [^{14}C]methylarsonate in the presence of *S*-adenosylmethionine and NADPH. *C. humicola* seems to be able to cleave the aryl group from 2-OH-4-NH$_2$C$_6$H$_3$AsO(OH)$_2$ because trimethylarsine is a product[58].

Klebsiella oxytoca and *Xanthomonas* sp., which grew well in the medium containing arsenate at concentrations higher than 1000 mg As l^{-1}, were able to methylate arsenate and methylarsonate to di- and trimethylarsenic compounds, and also demethylate dimethylarsinate and arsenobetaine to mono- and dimethylarsenic compounds, respectively[15].

Hanaoka and coworkers studied biological degradation of arsenobetaine by microorganisms in the bottom sediments of coastal waters and by microorganisms associated with marine macro-algae. Microorganisms living in the sediments converted arsenobetaine to trimethylarsine oxide and trimethylarsine oxide to less methylated metabolites including monomethylarsenic compounds[59-62]. Microorganisms associated with macro-algae also converted arsenobetaine to trimethylarsine oxide or dimethylarsinic acid and further into inorganic arsenic[63].

G. Health Effect Assessment and Safety

The most serious hazard associated with the handling of arsenicals is the generation of arsine gas from contact of inorganic arsenical solutions with active metals, such as zinc and aluminum[64].

A potential hazard in the handling of methylated arsenical herbicides is generation of methylarsines when exposed to active metals. Contact of aluminum with a water solution of disodium methylarsonate is particularly hazardous because of the relatively high alkalinity of this arsenical, which makes it reactive with aluminum[64].

In general, organic derivatives of arsenic are less toxic than inorganic forms. The apparent order of toxicity is phenylarsonates > methylarsonates > 'fish arsenic' (arsenobetaine). The most characteristic effect of the phenylarsonates is neurotoxicity, with effects occurring at oral exposure levels about 4 mg kg^{-1} day^{-1} or higher.

Methylarsonates are primarily associated with irritation of the gastrointestinal tract or the skin, usually at exposure levels in excess of 10 mg kg^{-1} day^{-1}.

Arsenobetaine has not been found to be toxic in animals even at very high doses (up to 10,000 mg kg^{-1}).

Arsine, that breaks down to inorganic arsenic in the body, is one of the powerful hemolytic agents whereas its methylated derivatives are of less toxicity[42].

H. Summary

Arsenic is susceptible to methylation by organisms. Methylated arsenic compounds are less toxic against most organisms than inorganic arsenic compounds. On the other hand, organoderivatives of many other metals (mercury, lead, tin etc.) are more highly toxic against algae[65] and most organisms than the inorganic species. The heavy metals apparently exert their toxic effect chiefly by reacting with the sulfhydryl groups of vital cellular enzymes.

It is reasonable to consider the increase in toxicity of heavy metal compounds due to their methylation as attributable to the increasing membrane permeability of the methylated metal compounds and also to their remaining reactivity with sulfhydryl groups.

However, highly methylated arsenic compounds [e.g. arsenobetaine (10)] have no sites to react with sulfhydryl groups because all four covalent sites of arsenic are fully occupied by four carbon atoms, and the methyl–arsenic bond is quite stable both chemically and physiologically.

III. ANTIMONY

A. Introduction

Trace elements like antimony and arsenic enter the atmosphere as part of natural biological and geochemical cycling processes. There are various natural sources of these trace elements, including soil, seawater and volcanic eruptions. Human activities, such as power generation and other industrial procedures, also lead to substantial emission of some elements.

Global emission values for antimony and arsenic are listed in Table 5, for global emissions of the two elements from both natural and human sources. These data were taken from detailed studies of inventories carried out in mid- to the late 1970s, but they have subsequently been adapted to incorporate more recent information[66,67]. Table 5 shows that the human sources of both antimony and arsenic exceed the natural sources.

B. Production and Use

1. Properties and production of antimony oxides

Antimony occurs in nature mainly as Sb_2S_3 (stibnite, antimonite); Sb_2O_3 (valentinite) occurs as a decomposition product of stibnite. Antimony is commonly found in ores of copper, silver and lead. The metal antimonides NiSb (breithauptite), NiSbS (ullmannite) and Ag_2Sb (dicrasite) also are found naturally; there are numerous thioantimonates such as Ag_3SbS_3 (pyrargyrite). Antimony is produced either by roasting the sulfide with iron or by roasting the sulfide and reducing the sublimate of Sb_4O_6; high-purity antimony is produced by electrolytic refining[69].

World production of antimony is about 68,000 tons per year. Table 6 shows world mine production by main countries[70]. China, Bolivia, Mexico, the Republic of South Africa and the USSR accounted for 84% of the total world estimated mine production during 1989.

Antimony trioxide (Sb_2O_3), a white, odorless, crystalline solid at room temperature, is produced commercially by oxidation of molten antimony trisulfide ore and/or antimony metal in air at 600–800 °C, and alkaline hydrolysis of antimony trichloride followed by dehydration[71].

Antimony tetroxide (Sb_2O_4) is a white, crystalline solid at room temperature. Antimony

TABLE 5. Global emissions ($kt\ y^{-1}$) of antimony and arsenic to the atmosphere[68]

Human source	Antimony	Arsenic	Natural source	Antimony	Arsenic
Energy production[a]	1.3	2.2	Wind-borne dust[b]	0.8	2.6
Mining	0.1	0.1	Sea salt spray	0.6	1.7
Smelting and refining	1.4	12.3	Volcanic activity	0.7	3.8
Manufacturing	—	2.0	Forest fires	0.2	0.2
Commercial	—	2.0	Biogenic sources	0.3	3,9
Waste incineration	0.7	0.3			
Total (A)	3.5	18.9	Total (B)	2.6	12.2
Overall total (A + B)	6.1	31.1			

[a] Energy production includes coal, oil and gas.
[b] Includes industrial sources of dust.

TABLE 6. World mine production of antimony by country (metric tons[a] (revised from Reference 70)

Country	1985	1986	1987	1988	1989
Australia[b]	1 458	1 131	1 231	1 200	1 200
Bolivia	8 925	10 243	10 635	9 943	9 100
Canada[c]	1 075	3 805	3 706	3 171	2 882
China[d]	15 000	15 000	27 000	30 000	30 000
Czechoslovakia[d]	900	1 000	1 000	1 000	1 000
Guatemala	1 239	1 839	1 881	921	1 000
Mexico[e]	4 266	3 337	2 839	2 185	2 500
South Africa	7 390	6 816	6 673	6 264	6 100
Thailand	1 240	1 019	409	445	500
Turkey	982	2 752	2 344	1 945	2 000
USSR[d]	9 400	9 500	9 600	9 600	9 600
United States	W	W	—	W	W
Yugoslavia	1 088	859	834	725	750
Other	3 761	3 714	2 725	3 258	2 750
Total	55 824	60 015	69 857	67 992	68 362

[a] Antimony content of ore unless otherwise indicated
[b] Antimony content of antimony ore and concentrate, lead concentrate and lead–zinc concentrate.
[c] Partly estimated on basis of reported value of total production.
[d] Estimated
[e] Antimony content of ores for export plus antimony content of antimonial lead and other smelter products produced.
W: Withheld to avoid disclosing company proprietary data; not included in 'Total'.

TABLE 7. Selected physical properties of antimony and some of its compounds[72]

Compound	Empirical formula	Molecular weight	From	mp (°C)	bp (°C)	Solubility in water
Antimony	Sb	121.75	silver-white	630.5	1750	insoluble
Antimony trioxide	Sb_2O_3	291.50	white	656	1550	$5\ mg\,l^{-1}$
Antimony tetroxide	Sb_2O_4	307.50	white	930	NA	slightly soluble
Antimony pentoxide	Sb_2O_5	323.50	yellow	390^a, 930^b	NA	slightly soluble
Antimony trisulfide	Sb_2S_3	339.69	black, yellow	550	1150	$1.75\ mg\,l^{-1}$

[a] At 390 °C, antimony pentoxide loses one oxygen atom to become antimony tetroxide.
[b] At 390 °C, antimony pentoxide loses two oxygen atoms to become antimony trioxide.
NA: not available.

tetroxide is a mixture of antimony(III) oxide and antimony(V) oxide (i.e. $Sb_2O_3 \cdot Sb_2O_5$). It is formed by heating antimony trioxide in air at 460–540 °C[71].

Antimony pentoxide (As_2O_5) is a yellowish, crystalline solid that is always somewhat hydrated. There is no satisfactory evidence for the existence of an anhydrous form of antimony pentoxide. Antimony pentoxide is produced commercially by oxidation of antimony trioxide with nitrates or peroxides[71].

Some of the physical properties of antimony, antimony oxides and antimony trisulfide are listed in Table 7.

In Japan, total 1985 and 1988 imports and production of antimony trioxide was 8000 and 9400 tons, respectively[73].

2. Usage

Table 8 shows industrial consumption of antimony in the United States[70] and Japan[74] in 1985 and 1989.

a. Antimony metal. Commercial-grade antimony is used in many alloys (1–20%), especially lead alloys, which become much harder and mechanically stronger than pure lead; the use for batteries, cable sheathing, antifriction bearing and type metal consume almost half of all the antimony metal produced.

TABLE 8. Industrial consumption (in metric tons) of antimony in the United States and Japan (modified from References 70 and 74)

Product	United States		Japan	
	1985	1989	1985	1989
Metal products:				
Ammunition	372	521	—	—
Antimonial lead	515	1 872	516	363
Bearing metal and bearing	161	129	—	—
Cable covering	W	W	—	—
Casting	10	8	60	50
Collapsible tubes and foil	W	W	82	63
Sheet and pipe	W	157	3	3
Solder	305	245	—	—
Type metal	28	4	1	4
Other	95	80	319	187
Total	1 486	3 016	981	670
Nonmetal products:				
Ammunition primers	24	20	—	—
Ceramics and primers	1 077	1 050	310	253
Fireworks	4	5	—	—
Pigments	133	196	—	—
Plastics	905	1 141	—	—
Rubber products	23	97	—	—
Other	128	159	267	278
Total[a]	2 294	2 668	—	—
Flame-retardant:				
Adhesives	281	219	—	—
Paper	101	W	—	—
Pigments	7	926	—	—
Plastics	5 016	5 842	—	—
Rubber	286	174	—	—
Textiles	1 140	558	—	—
Other	—	21		
Total[a]	6 831	7 740	7 986[b]	9 263[b]
Grand total[a]	10 611	13 424	8 967	9 933

W: Withheld to avoid disclosing company proprietary data; included with 'Other'.

[a] Data may not add to totals shown because of independent rounding.

[b] Sum of Nonmetal and Flame-retardant.

b. Antimony trioxide. Antimony trioxide (Sb_2O_3) has many uses including flameproofing (flame retardants) of textiles, paper and plastics; as a paint pigment, ceramic opacifier, catalyst, mordant and glass decolorizer.

Of the antimony trioxide consumed annually both in the United States[71] and Japan[73], *ca* 90% is used as a flame retardant.

Antimony trioxide is used also to produce antimony tetroxide (Sb_2O_4) and pentoxide (Sb_2O_5) as well as tartar emetic (**15**), stibophen (**16**) and astiban (**17**) which are used as drugs for schistosomiasis (see Figure 6).

c. Antimony pentoxide. Antimony tetroxide is used as an oxidation catalyst, particularly for oxidative dehydrogenation of olefins. Antimony pentoxide is used to produce antimony tetroxide, antimonates and other antimony compounds, and as an ion-exchange material.

d. Antimony potassium tartrate. Antimony potassium tatrate (**15**), popularly known as tartar emetic, is widely utilized in the treatment of various parasitic infections as leishmaniasis, schistosomiasis, ascariasis, trypanosomasis and bilharziasis in various animal species[75]. It is also used as an expectorant for cattle, horses, sheep, goats, swine and dogs. As an insecticide, it has been found to be effective against thrips and has been recommended as a spray on onions, gladioli and citrus trees.

Bai and Majumder[75] investigated the optimum concentration of **15** that could be incorporated in rodenticidal baits (zinc phosphide), which will not affect the toxicity or palatability of the baits, but which will induce vomiting when it is ingested accidently by the nontarget species. The experimental results are as follows: Zinc phosphide alone at 1.0% resulted in 100% mortality of test rats within 18 hours of feeding. Although the

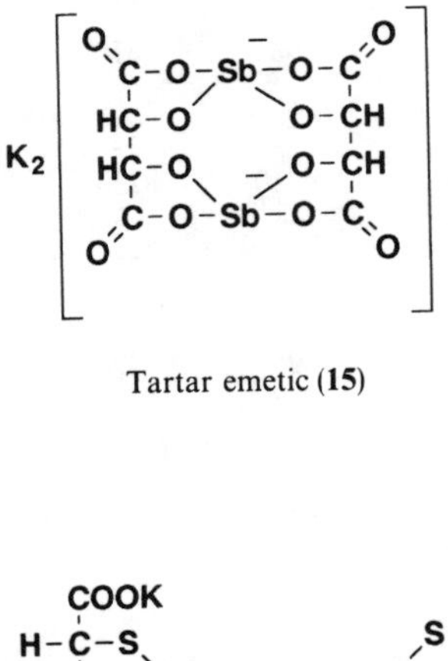

Tartar emetic (**15**)

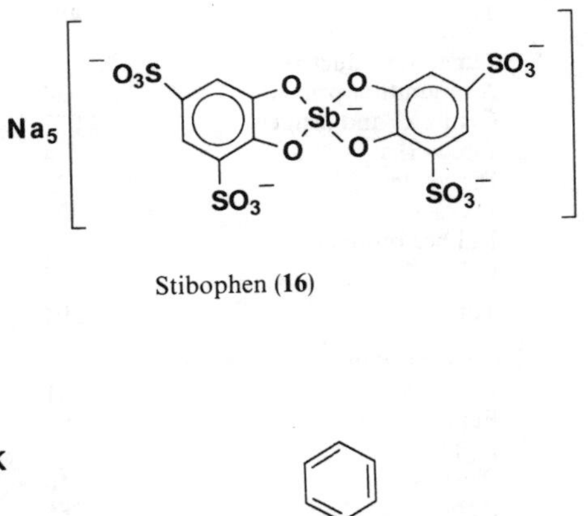

Stibophen (**16**)

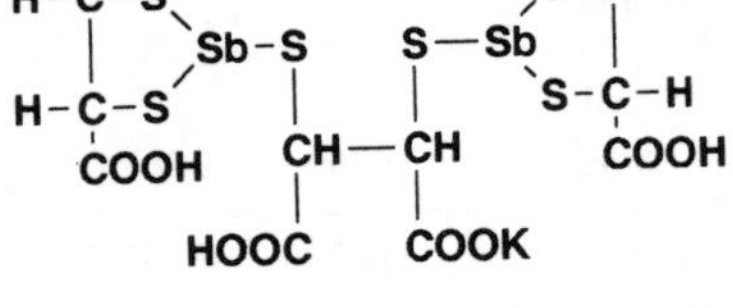

Astiban (**17**)

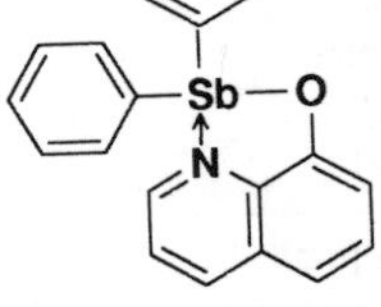

Diphenylantimony (III) (**18**)

FIGURE 6. Organic antimony compounds as drugs

addition of less than 2% of **13** to zinc phosphide (1%) reduced the bait intake slightly, addition of **15** at lower concentrations (0.5 or 0.25%) did not affect the mortality or acceptability of the bait when compared to zinc phosphide alone. The addition of **15** at 0.25% level to zinc phosphide baits seems to be sufficient to cause vomiting it pets and humans in cases of accidental poisoning[76,77]. Hence, it is suggested that the optimum concentration of **15** to be added to zinc phosphide baits should be restricted to 0.25–0.5%.

C. Concentration and Speciation in the Natural Environment

Antimony occurs widely in the environment in concentrations which are generally low but nevertheless significant owing to the very high potential toxicity of antimony[78]. Typical environmental levels of antimony are: precipitation 0.03–0.31 ppb[79], river water (Thames, UK) 0.09–0.86 ppb[80], river water (five rivers, Japan) 0.07–0.29 ppb[81], river suspended particulate (Thames, UK) 1.3–127 ppm[80], lake water (Biwa and Hamanako, Japan) 0.09–0.46 ppb[81], seawater (China coasts) 0.8–0.9 ppb[82], seawater (Japan coasts) 0.18 ppb[81], rain water (Japan) 0.10 ppm[81], house dust 1.8–30.6 ppm[83], street dust 2.6–6.8 ppm[83], soil 4.3–7.9 ppm[83], human hair 0.03–1.63 ppm[84], human milk 0.05–12.9 ppb[85], human lung (wet tissue) 3.5–48.2 ppb[86], and ambient particulates (rural, urban area and petroleum industrial area, Japan) 37–103 ppm[87], ambient particulates (iron, steel and nonferrous industrial area, Japan) 58–1170 ppm[87].

Abbasi[78] conducted a submicro determination (down to 10^{-2} ppb levels) of Sb(III) and Sb(V) in natural and polluted waters and biological materials. The Sb(III) and Sb(V) concentrations obtained were as follows: surface sample of reservoir water 0 and 0, near-bottom sample of reservoir water 0.17 and 0.16 ppb, sea water (India) 0 and 0.28 ppb, and polluted water (rubber industry) 0.85 and 1.91 ppb. Total antimony concentrations of goat liver and frog muscle were 0.094 and 0.027 ppb, respectively[78].

Concentration of antimony in mollusc shells found in the coastal waters of British Columbia, Canada, reported by Cullen and coworkers[88], are: soft-shell clam from Hastin Arm 0.51 ppm, from Anyox 0.14 ppm, gastropods 0.2–1.4 ppm, butter clam Sb(III) 4.5, Sb(V) 114 ppb, bent-nose clam Sb(III) not detected, Sb(V) 35 ppb.

TABLE 9. Antimony species concentration (ng Sb 1^{-1}) in fresh, estuarine and marine waters[89]

Water samples		Sb(III)	Sb(V)	MSA[a]	DMSA[a]
Ochlockonee River		3.2	22.9	0.5	ND
Trinity River		0.9	145.0	0.8	ND
Mississippi River		0.3	148.0	2.3	ND
Escambia River		0.3	12.8	0.8	ND
Apalachicola River		0.4	55.0	0.6	ND
Ochlockonee Bay Estuary					
salinity (‰):	4.3	2.8	42.0	0.8	ND
	12.3	6.6	81.5	1.4	ND
	18.9	8.3	113.0	2.2	ND
	23.8	5.4	122.0	5.1	0.6
	30.2	8.8	126.0	10.9	1.1
	33.4	11.1	136.0	12.6	1.5
Gulf of Mexico		4.4	149.0	5.3	3.2

[a] MSA denotes methylstibonic acid [MeSbO(OH)$_2$]; DMSA denotes dimethylstibinic acid [Me$_2$SbO(OH)].

Methylantimony compounds together with inorganic antimony compounds in river waters and seawater were determined by Andreae and coworkers[89]. Results of the analyses of water samples from five rivers draining into the Gulf of Mexico, from the estuary of the Ochlockonee River, Fl, USA, and from the Gulf of Mexico are presented in Table 9. As shown, antimony(V) was the predominant species in all cases, but Sb(III) was found in most samples, and the methylantimonials were always present in the marine and estuarine environment. In the estuary, concentrations of methylantimonials increased with increasing salinity. This result suggests that the methylantimony acids are probably due to marine biological activity.

D. Antimony Pollution

1. Waters and sediments

Antimony concentrations in surface soils were found to decrease with increasing distance from an antimony smelter situated on the north bank of the River Tyne in northeast England. This pattern was also found in Sphagnum moss exposed in the same area. At three sites close to the smelter, antimony concentrations in soil and vegetation were much higher than at a rural control site. Maximum soil and plant concentrations on a dry weight basis of 1489 ppm and 336 ppm, respectively, were found. Field exposure of grass in pots of uncontaminated soil and a laboratory experiment using soils from near the smelter suggested that the antimony in vegetation was largely due to continued aerial deposition and not to uptake from soil[90].

Water samples from the Main Stem and the South Fork of the Coeur d'Alene River, Idaho, USA, which were contaminated with Sb and As, and the other heavy metals from the local mining operation, showed high levels of Sb (0.23–8.25 ppb) and As (0.11–1.64 ppb). The major inorganic Sb species was Sb(V) in all three branches of the river. Leaching of Sb and As species from the contaminated Main Stem sediments depended on the pH values of the water as well as on the free iron oxides and manganese oxides present in the sediments[91].

Andreae and coworkers analyzed two polluted rivers in Germany for inorganic and methylated antimony compounds[89]. Sb(III), Sb(V) and methylstibonic acid in the Rhine River at Oppenheim and the Main River at Frankfurt were 0.4, 231.0 and 1.2, and 0.3, 311.0 and 1.8 ppb, respectively. Dimethylstibinic acid was not detected.

2. Organisms

Concentrations of antimony in invertebrates and small mammals from grasslands in the vicinity of an antimony smelter[90] were significantly elevated compared to control sites as shown in Table 10[92]. Higher concentrations of antimony were recorded in liver, lung and kidney tissue of herbivorous and insectivorous mammals from the contaminated sites. However, there was little evidence of biomagnification of antimony in food chains represented by the soil–vegetation–invertebrate–insectivore pathway of grasslands, and little indication of significant accumulation by herbivorous mammals despite marked contamination of their diet[92].

E. Toxicity

Antimony, which is considered a nonessential element, is comparable in its toxicological behavior to arsenic and bismuth. In analogy to arsenic, trivalent antimony compounds generally are more toxic than the pentavalent compounds. Poisoning with antimony and its compounds can result from acute and chronic exposure, especially from exposure to

TABLE 10 Antimony (mg kg^{-1} dry weight) in invertebrates and small mammals from contaminated and control sites (mean ± standard error; $n = 6$) (modified from Ainsworth and coworkers[92])

Organisms	Control sites	100 m site	250 m site	450 m site
Invertebrates				
Isopoda	0.16 ± 0.03	118 ± 35	11.1 ± 1.9	NA
Diplopoda (liliformia)	0.06 ± 0.01	127 ± 28	21.5 ± 2.6	9.5 ± 1.6
(Polydesmidae)	0.13 ± 0.03	NA	28.3 ± 5.3	11.0 ± 1.7
Lepidoptera	NA	17.8 ± 6.0	11.5 ± 2.2	2.7 ± 1.7
Diptera	NA	NA	55.6 ± 5.3	24.1 ± 2.8
Coleoptera (Carabidae)	< 0.03	17.4 ± 1.6	NA	NA
(adult Staphylinidae)	NA	33.8 ± 7.9	24.1 ± 10.5	3.6 ± 0.4
(larval Staphylinidae)	NA	50.6 ± 7.5	NA	NA
(Elateridae)	NA	38.9 ± 4.9	NA	2.4 ± 0.6
(Coccinellidae)	NA	70.9 ± 14.3	NA	NA
(Circulionidae)	NA	49.3 ± 16.2	NA	NA
Lycosidae	NA	31.1 ± 6.8	30 ± 12.8	4.6 ± 0.4
Oligochaeta	NA	398 ± 94	213 ± 65	109 ± 28
Small mammals				
Microtus agrestis				
liver	NA	0.3 ± 0.04	NA	NA
lung	NA	0.31 ± 0.03	NA	NA
kidney	NA	0.18 ± 0.02	NA	NA
Oryctolagus				
liver	NA	NA	0.68 ± 0.17	NA
lung	NA	NA	0.31 ± 0.03	NA
kidney	NA	NA	0.28 ± 0.05	NA
Sorex araneus				
liver	0.04 ± 0.01	NA	NA	0.49 ± 0.19
lung	0.15 ± 0.04	NA	NA	0.11 ± 0.19
kidney	0.15 ± 0.05	NA	NA	0.33 ± 0.07

NA: data not available.

airborne particles in the workplace; to a minor extent, exposure occurs through treatment of tropical diseases with antimony compounds[93].

Most antimony compounds, mainly those with poor water solubility, are absorbed only slowly from the gastrointestinal tracts. Trivalent compounds especially tend to accumulate in the human body, because they are excreted very slowly via urine and feces. Antimony and its compounds react with —SH groups in various cellular constituents, especially in enzymes, blocking their activity. After acute and chronic exposure the highest concentrations are found in liver, kidney, adrenals and thyroid.

1. Volatile antimony compound

In Vienna a case of chronic antimony poisoning occurred in a house with silk curtains containing a mordant with a compound of antimony. A series of experiments was performed to detect the possible formation of volatile compounds of antimony, but the speciation of the volatile antimony compound has been unsuccessful[94].

Stibine, SbH$_3$, a highly toxic and relatively unstable gas with an unpleasant odor, causes symptoms of the central nervous and circulatory systems, such as nausea, vomiting, headache, slow breathing, weak pulse, hemolysis, hematuria, abdominal pain and death.

Stibine is generated if nascent hydrogen can react with antimony in an acid environment, e.g. in lead-acid batteries, where antimony may be a component of the battery plates[93].

2. Antimony chlorides

Acute respiratory exposure of seven workers to $70–80 \, \text{mg m}^{-3}$ of antimony trichloride, $SbCl_3$, resulted in irritation of the upper respiratory tract[93]. Antimony pentachloride, $SbCl_5$, caused severe pulmonary edema in three cases, two of them being lethal[95].

Completed inhalation studies have demonstrated that inhalation of antimony trioxide induces lung tumors in female rats[96].

3. Antimony trioxide

Most cases of chronic respiratory intoxication result from exposure to airborne particles containing antimony trioxide, Sb_2O_3. The symptoms were reported to be soreness, nosebleeds, rhinitis, pharyngitis, pneumonitis and tracheitis.

4. Fungitoxicity of organoantimony compounds

Organic derivatives of tin, germanium and lead were shown to be active against many fungal species, and derivatives of tetravalent metal proved to be the most toxic. Studies on

TABLE 11. Fungitoxicity of organoantimony compounds for four fungi grown on potato dextrose agar incubated at $30\,°C$ for 72 hours[97]

Compounds	EC_{50} values (mg l^{-1}) for fungi			
	F. oxysporum[a]	Pestalotia sp.	P. hirsutus	S. homeocarpa[b]
Ph_2SbCl	20 ± 1.8	>30	>30	>30
$Ph_2Sb(MeCOO)$	>30	>30	>30	>30
$Ph_2Sb(acac)^c$	24 ± 2.0	>30	>30	>30
$Ph_2Sb(oxine)$	12 ± 0.8	1.8 ± 0.2	15 ± 0.6	1.8 ± 0.2
Ph_3Sb	>30	>30	>30	>30
Ph_3SbCl_2	>30	>30	>30	>30
$Ph_3Sb(MeCOO)_2$	>30	>30	>30	>30
$Ph_3Sb(CH_2ClCOO)_2$	>30	>30	>30	>30
$Ph_3Sb(CHCl_2COO)_2$	>30	>30	>30	>30
$Ph_3Sb(CCl_3COO)_2$	>30	>30	>30	>30
Me_3SbCl_2	>30	>30	>30	>30
$Me_3Sb(MeCOO)_2$	>30	>30	>30	>30
$Me_3Sb(CH_2ClCOO)_2$	>30	>30	>30	>30
$Me_3Sb(CHCl_2COO)_2$	>30	>30	>30	>30
$Me_3Sb(CCl_3COO)_2$	>30	>30	>30	>30
NaN_3	0.57	0.57	0.57	0.57
Benomyl[d]	1.05 ± 0.02	0.09 ± 0.003	3.0 ± 0.14	0.22 ± 0.02
Captan[e]	>30	7.6 ± 0.4	10.6 ± 0.8	17.8 ± 1

[a] Fusarium sp.
[b] Sclerotina sp., 8-h incubation period.
[c] Diphenylantimony(III) acetylacetonate.
[d] Benomyl: commercial fungicide, methyl 1-(butylcarbamoyl)-benzimidazol-2-ylcarbamate.
[e] Captan: commercial fungicide, N-[(trichloromethyl)thio]cyclohex-4-ene-1,2-dicarboximide.

the antimicrobial activity of organic bismuth and antimony compounds are limited in number.

Burrell and coworkers determined fungitoxicity of organoantimony and organo-bismuth compounds[97]. The effective concentration (EC_{50}) values obtained with the organoantimony compounds for the tested fungi are summarized in Table 11[97]. Only 3 of 15 compounds listed had EC_{50} values less than $30 \, mg \, l^{-1}$ for *Fusarium oxysporum*. The organoantimony compounds with the exception of diphenylantimony(III) oxinate (**18**, Figure 6) were found to be not fungitoxic for other fungi. The antifungal activity of the most active compound (i.e., **18**) was compared with those of the reference fungicides benomyl and captan, and found to be inferior to the former but superior to the latter[97].

F. Biological Methylation, Transformation and Fate

1. Biological methylation

A considerable number of organometallic species have been detected in the natural environment in recent years. A number of these are nonmethyl compounds which have entered the environment after manufacture and use (e.g. butyltin and phenyltin compounds in antifouling paints for boats). Only a few methyl compounds are now manufactured and used (e.g. methyltin compounds for oxide film precursors on glass and methylarsenic for compounds used as desiccants or defoliants). It is now well established that certain organometallic compounds are formed in the environment, unequivocally so for those of mercury, arsenic, selenium, tellurium and tin, and deduced on the basis of analytical evidence for lead, germanium, antimony and thallium[16].

Since the case of chronic antimony poisoning which occurred in Vienna, as mentioned before, many attempts to obtain evidence of biological methylation of antimony were performed[94]. The mould *Scopulariopsis brevicaulis*, which was found to biologically methylate inorganic arsenic compounds, was grown on media containing one percent of tartar emetic (**15**), and the volatile products were aspirated through concentrated nitric acid which was tested for antimony with negative results. Similar studies by many researchers were also unsuccessful.

Parris and Brinckman[98] stated: 'At this time, it has not been demonstrated that methylstibines are metabolites of microorganisms acting on inorganic antimony compounds, but the extensive similarity of the chemistry of arsenic and antimony gives reason to believe that antimony can be biologically methylated'.

The experimental results in Table 9 imply that inorganic antimony compounds are methylated by marine organisms[89].

2. Quaternarization

Methylarsines and methylstibines are subject to a number of reactions such as oxidation, quaternization and complex formation, which could facilitate or inhibit their dispersal in the environment[98]. It has been reported that environmentally important concentrations of halocarbons (MeI, MeBr and MeCl) are produced naturally and accumulate in the oceans and the atmosphere. Parris and Brinckman[98] reported quantitative measurements of the rate of quaternization of trimethylstibine and trimethylarsine by alkyl halides in polar solvents.

The orders of reactivity of the alkyl halides (MeI $\gg$ EtI > PrI and EtI > EtBr) are typical of S_N2-type reactions. Parris and Brinckman estimated the rate constant for reaction of Me_3Sb and Me_3As with MeI in water to be 7×10^{-4} and $3 \times 10^{-3} \, M^{-1} \, s^{-1}$, respectively[98].

3. Oxidation of trimethylstibine

Parris and Brinckman [99] further examined the oxidation of trimethylarsine and trimethylstibine by atmospheric oxygen, and semiquantitative rate constants were calculated. In the gas phase the rate constants are estimated as 10^3 and 10^{-6} M^{-1} s^{-1} for the reactions of trimethylstibine and trimethylarsine with oxygen, respectively. A scheme based on PMR evidence for reactive intermediates is suggested to account for the products of oxygen, Me_3EO and Me_2EO_2H, of the compounds Me_3E ($E = Sb$, As). Parris and Brinckman concluded from these results that even if biological methylation of antimony occurs in nature, analogous to that of arsenic, the rapidity with which Me_3Sb is oxidized would probably prevent hazardous concentrations from building up in aerated surroundings[99].

4. Fixation and mobilization of antimony in sediments

Brannon and Patrick studied the mobility of sediment antimony during sediment–water interactions[100]. Ten sediment samples from the east, west, south and north coasts in USA were tested. Sequential selective extraction procedures revealed that most native and added Sb in the sediment was associated with relatively immobile iron and aluminum compounds. In sediments amended with Sb, the Sb concentrations of interstitial water and exchangeable phase were also high. Initial and long-term (6 months) release of Sb were much higher from Sb-amended sediments than from sediments containing no added Sb, apparently as a result of Sb concentrations in more mobile sediment phases. In most sediments, the largest amount of Sb release occurred early in the leaching experiment. These results suggest that Sb release from contaminated sediments is more likely to occur during the first few months of sediment–water interaction. During aerobic leaching, Sb moved into a less available sediment phase, decreasing the potential for further releases. The equivalent Fe and $CaCO_3$ concentrations of the sediments were found to affect releases of added Sb. Evolution of volatile Sb compounds was also noted from sediments under anaerobic conditions[100].

G. Health Effect Assessment and Safety

The Environmental Protection Agency (Cincinnati, OH, USA) reported investigations regarding the health effects of antimony and its compounds[72]. The conclusion of the report is as follows: Oral reference dose values (RfDo) were derived for antimony and selected compounds based on the LOAEL (lowest-observed-adverse-effect-level) for antimony of $350\,\mu g kg^{-1}$ day^{-1} associated with potassium antimony tartrate (15) in the drinking water of rats for lifetime exposure. Reduced lifespan was observed in both sexes and altered blood biochemical characters in males. The only concentration tested was of 5 ppm antimony. RfDo values for antimony of $24.5\,\mu g\, day^{-1}$, for antimony potassium tartrate of $65.5\,\mu g\, day^{-1}$ and for antimony tri-, tetra- and pentoxides of 29.3, 30.9 and $32.5\,\mu g\, day^{-1}$, respectively, were calculated. It should be noted that orally administered antimony has been inadequately tested for carcinogenicity.

The report[72] also showed that a statistically significant increase in the incidence of lung tumors was observed in rats exposed to antimony trioxide by inhalation. It was also noted that there is an increased incidence of lung cancer in rats exposed by inhalation to antimony trisulfide. This observation, coupled with indications that occupational exposure to antimony processing is associated with lung cancers in humans, is qualitative evidence for the carcinogenicity of antimony by inhalation. However, an earlier U.S. EPA[101] analysis concluded that the animal data were insufficient for quantitative estimation of the carcinogenic potency of antimony.

The health and environmental effects profile for antimony oxides was prepared by the Office of Health and Environmental Assessment, Environmental Criteria and Assessment Office, Cincinnati, OH, USA[71]. Acceptable daily intakes, defined as the amount of a chemical to which humans can be exposed on a daily basis over an extended period of time (usually a lifetime) without suffering deleterious effects, were 24.5, 29.3, 30.9 and 32.5 μg day^{-1}, for oral exposure, for antimony, antimony trioxide, antimony tetroxide and antimony pentoxide, respectively.

H. Summary

Biological methylation of antimony has not been demonstrated directly by use of experimental organisms. However, methylstibonic acid and dimethylstibinic acid were found in fact in marine and estuarine environment[89]. There is no obvious thermodynamic barrier to biomethylation. The chemical similarity between Sb and Sn, Pb, As, Se and Te, which surround Sb in the periodic table, and all of which have been shown to be subject to biomethylation, would suggest the existence of a biomethylation pathway also for antimony. This possibility assumes importance when the current extensive use (ca 2×10^7 kg y^{-1} in United State and ca 10^7 kg y^{-1} in Japan) of inorganic and organic antimony compounds in conjunction with halocarbons in fire retardant systems is considered. Products containing antimony-based fire retardant systems include textiles, plastics, elastomers, paper, wood, paints and coatings[2].

If methylation of antimony should occur during the biodegradation of discarded or poorly maintained consumer items protected with antimony compounds, the antimony could be transformed into a much more water-soluble form (Me_3SbO) which would allow leaching of the antimony, reducing the flame retardancy of the material and contributing to pollution of waterways[99].

When the waste-containing fire retardant is incinerated, the antimony is emitted to the atmosphere as shown in Table 5. The emission of antimony to the atmosphere and to waterways should continuously increase in the future with the increased use of the product. This necessitates the careful watching of the antimony concentration in the environment.

Methylated arsenic compounds are less toxic against most organisms than inorganic arsenic compounds. However, organic derivatives of many other heavy metals (mercury, lead, tin and so on) are much more toxic against algae[65] and other organisms than the inorganic species. According to Table 11, some organoantimony compounds are not considered fungitoxic. There is a possibility that inorganic antimony compounds may be detoxified by transformation to organic forms.

Further studies on biological methylation or alkylation and on the toxicity of these organoantimony compounds are expected.

IV. BISMUTH

A. Introduction

Bismuth ranks 64th in abundance among the elements in the earth's crust; like antimony and cadmium, its abundance is estimated to be 0.17 to 0.2 ppm. Most bismuth sulfides occur associated with lead, copper and silver. A few deposits in which bismuth is a major mineral are also known[102].

Inorganic bismuth compounds usually have a $+3$ or $+5$ oxidation state. The $+5$ oxidation state is a strong oxidation agent, e.g. $NaBiO_3$ or BiF_5. In gases, such as BiCl or BiO, in $Bi(AlCl_4)$ and in alloy-like compounds, such as BiS or BiSe, bismuth can rarely have a $+1$ or $+2$ oxidation state, and $+1$, 0 and -1 are present in polynuclear ionic species. On dissolution, bismuth compounds hydrolyze easily, yielding nearly insoluble basic salts of the type BiOX. Only a few inorganic bismuth compounds are produced commercially.

Organobismuth compounds have stable trivalent and pentavalent states. The compounds are more subject to thermal decomposition than the corresponding phosphorus, arsenic and antimony compounds, and they are more toxic. Both statements are especially true for halogeno- or pseudohalogenoorganobismuth compounds. Triarylbismuthines are stable, but all trialkylbismuthines except trimethylbismuthine are self-igniting in air. The triarylbismuth dichlorides and dibromides are the most important organobismuth compounds[102].

B. Production and Use

Bismuth is recovered mainly as a by-product of processing lead and copper ores and only in Bolivia was it mined as the principal product. Table 12 shows world refinery production of bismuth[74,103].

Bismuth is one of the few metals which expand on solidification, the expansion being 3.35%. The thermal conductivity of bismuth is the lowest among all metals, with the exception of mercury. The main use of bismuth is in the manufacture of low-melting alloys, many of them melting below 100 °C. Another important use of bismuth is in the manufacture of pharmaceutical compounds. Various bismuth preparations have been employed in the treatment of skin injuries, alimentary diseases such as diarrhea and ulcers, and syphilis. The oxide and basic nitrate are perhaps the most widely used compounds of bismuth. The trioxide is used in the manufacture of glass and ceramic products, while the basic nitrate is used in porcelain painting to fire on gilt decoration. Bismuth telluride is used extensively for thermoelectric cooling and for low-temperature thermoelectric power production[103].

The United States and Japan were the largest consumers of bismuth and were also the only countries for which published data on end uses were available. Table 13 shows bismuth metal consumed in the United States and Japan in 1985 and 1989[104,105].

Chemical use in the United States rose and accounted for almost one-half of the total

TABLE 12. World refinery production of bismuth (modified from References 74 and 103)

Country	1985	1986	1987	1988	1989
Belgium[b]	610	1000	865	795	750
Bolivia[b]	—	—	—	30	130
Canada	180	212	218	225	272
China[b]	260	260	260	275	275
Germany[b], Federal Republic of	400	400	400	—	—
Japan	642	640	546	524	500
Korea, Republic of	135	136	145	140	140
Mexico	925	749	1012	958	1000
Peru	738	569	387	310	290
United Kingdom[b]	150	150	180	135	150
United States	W	W	W	W	W
Others[a]	285	251	277	205	218
Total	4325	4367	4290	3597	3725

W: Withheld to avoid disclosing company proprietary data; excluded from 'Total'.
[a] Including Italy, Romania, USSR and Yugoslavia where productions were below 100 tons per year.
[b] Estimated.

TABLE 13. Bismuth metal consumed in the United State and Japan in 1985 and 1989 (modified from References 104 and 105)

Use	United States		Japan[a]	
	1985	1989	1985	1989
Chemicals[b]	601	659	—	—
Ferrites[c]	—	—	108	242
Metallurgical additivies	303	395	120	82
Fugible alloys	277	272	35	35
Other alloys	10	11	—	—
Pharmaceuticals	—	—	28	22
Others	9	14	27	75
Total[d]	1199	1352	318	464

[a] Reported shipments from smelters. Does not include imports.
[b] Includes industrial and laboratory chemicals, cosmetics and pharmaceuticals.
[c] Includes varistors and capacitors.
[d] Data may not add to totals shown because of independent rounding.
(—) dash means not recorded.

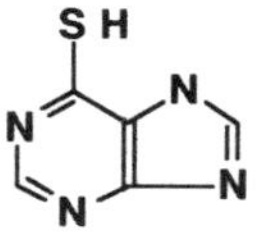

FIGURE 7. 6-mercaptopurine (19)

domestic consumption. Major chemical uses included compounds for treating digestive tract ailments, *pearlescent* pigments for cosmetics and plastics, and varistors and capacitors. The sum of pharmaceutical and ferrite uses in Japan accounted for one-half or more of the total domestic consumption.

6-Mercaptopurine (19) (Figure 7) complexed with Bi metal was used at various dosage levels to treat leukemia in mice. Anticarcinogenic activity of the 19–Bi complex was observed[106,107].

C. Concentration and Speciation in the Natural Environment

Lee[108] analyzed natural waters, shells, marine algae and sediments for total bismuth concentration. The results for natural waters are shown in Table 14.

Generally, coastal waters contained more bismuth by about an order of magnitude than did open ocean waters. A significant fraction of the total bismuth, *ca* 70%, was in particulate forms and was rather constant, although bismuth concentrations changed by an order of magnitude from one location to another. A vertical profile of dissolved bismuth in North Pacific Ocean waters (water depth, 3550 m) is shown in Figure 8. The maximum bismuth concentration, 0.26 pM (0.053 ng Bi l^{-1}), was found in the surface water. The high bismuth concentration in the surface water may originate either from aeolian inputs or from fluvial sources. The second bismuth maximum occurred at the oxygen minimum depth, and might result from Bi regeneration from organic matter.

Bismuth concentrations in marine sediment cores from Narragansett Bay (water depth, 8 m) and Pacific Ocean (I) (35°31′N, 123° 19′W; water depths, 4300 m) and Pacific Ocean (II) (55°35′N. 169°23′W; water depth, 1598 m) are shown in Table 15, which shows that the Narragansett Bay sediments contained higher bismuth concentrations than their North

TABLE 14. Bismuth concentration in environmental waters (ng Bi l^{-1}) (modified from reference 108)

Sample	Dissolved[a]	Particulate	Total
[Seawater]			
Pacific Ocean (water depth 3550 m)			
surface	0.053	—	—
2500 m below surface	<0.003	—	—
Scripps Pier, CA	0.052	0.078 (60%)	0.13
	0.085	0.205 (71%)	0.29
San Diego Bay, CA	0.63	1.37 (69%)	2.0
Mission Bay, CA	0.46	1.14 (71%)	1.6
[Lake water]			
Lake Miramar, CA	<0.15	—	<0.15
[Rain water]			
La Jolla, CA	0.62	2.58 (81%)	3.2

[a] Passed through 0.45 μm Millipore filter.

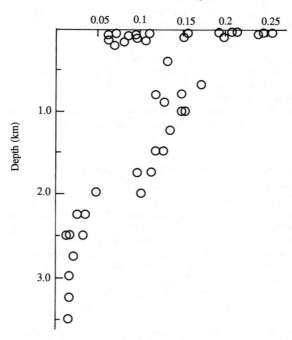

FIGURE 8. Vertical profile of dissolved bismuth at North Pacific Ocean (17° 30′N, 109° 00′W; water depth, 3550 m) (modified from Reference **108**)

TABLE 15. Bismuth in sediments from Narragansett Bay and from Pacific Ocean (modified from Reference 108)

Depth in core (cm) Narragansett Bay	Depositional period[a]	Bi (μg g^{-1}) (dry wt)
0–1	1972–1974	0.44
3–4	1965–1967	0.42
8–9	1952–1954	0.57
12–14	1939–1944	0.54
20–22	1919–1924	0.64
26–29	1902–1909	0.59
35–38	pre-1900	0.49
49–54	pre-1900	0.27
Pacific Ocean (I) surface		0.10
Pacific Ocean (II) surface		0.12

[a]Determined by measurement of ^{210}Pb

Pacific Ocean counterparts. The vertical distribution of bismuth had a peak near the 20-cm depth.

Similar results were obtained on the sediments of Tokyo Bay, Japan by Ohyama and coworkers[109]. Bismuth concentrations of sediment cores were as follows: 0–5 cm, 1.9 μg g^{-1}; 5–10 cm, 2.3 μg g^{-1}; 10–15 cm, 3.6 μg g^{-1}; 15–20 cm, 4.6 μg g^{-1}; 20–25 cm, 1.8 μg g^{-1}; 25–30 cm, 0.8 μg g^{-1}; 30–35 cm, 0.6 μg g^{-1}; 35–40 cm, 0.4 μg g^{-1}; 40–45 cm, 0.5 μg g^{-1}; 45–50 cm, 0.4 μg g^{-1}.

Other elements of anthropogenic origin, such as mrcury, lead and tin, were found to have peaks around similar core depths of the sediments. These results showed that the concentration of bismuth in the sediment core may serve as one of the promising indices of anthropogenic activities.

Soil samples were collected from pollution-free sites (6 paddy fields, 7 upland fields and 12 forests) in Hokkaido, Fukushima, Shizuoka, Saga Prefectures in Japan and analyzed for bismuth[110]. Average, maximum and minimum values of bismuth and antimony concentrations obtained in dry base were 0.34, 0.12 and 0.91 μg Bi g^{-1}, and 0.37, 0.13 and 0.91 μg Sb g^{-1}, respectively.

Analytical results for macroalgae and shell are as follows: kelp (unknown) 5.3; macrocrysis (unknown) 8.9; mussel shell (M. californianus) 0.74; mussel shell (M. edulis) 2.3; oyster shell (C. virginica) 4.2 ng g^{-1} (dry weight)[108].

D. Bismuth Pollution

Komura[111] measured. ^{207}Bi in water filters used at the Scott Base in Antarctica and in the surface soils in Japan. ^{207}Bi is though to be a fallout from atomic bomb tests. The levels of ^{207}Bi in these samples were found to be nearly the same or a little higher than those of fallout ^{60}Co and the ^{207}Bi ^{137}Cs^{-1} activity ratios were in the range of 0.001–0.018. Contamination of bismuth by ^{207}Bi was found in high-purity bismuth on sale and its level was measured to be 1.9 mBq g^{-1} Bi.

Kubota and coworkers[112] examined metal pollutants in rice paddy soils near a metal processing factory in Annaka city (Japan). In the vicinity of the factory, average and maximum concentrations of metals in the soils were Bi 0.47 and 1.56; Sb 0.59 and 1.86; Cd 1.84 and 8.7; Zn 173 and 692; Pb 36 and 143, and Cu 48 and 78 μg g^{-1}.

Kubota and coworkers[113] determined metal concentration in soil cores at sites once

TABLE 16. Bismuth and some other metals in soil ores at sites polluted by copper or cadmium (modified from References 113)

Site	Depth in core (cm)	Metal concentration ($\mu g\, g^{-1}$)			
		Bi	Cu	Cd	Sb
A.	0–14	3.52	265	4.7	0.88
polluted with	14–32	7.73	793	14.7	1.35
copper	32–50	0.20	86	1.5	0.32
	50–70	0.20	84	1.5	0.31
	70–	0.14	43	2.1	0.27
B.	0–12	1.41	112	3.2	1.10
polluted with	12–55	0.97	95	4.6	0.78
cadmium	55–	0.48	41	3.1	0.31
C.	0–16	1.21	161	3.0	1.10
polluted with	16–25	1.10	177	3.2	0.85
cadmium	25–36	1.09	187	3.1	0.73
	36–	1.27	182	2.0	0.60
Natural concn.		0.34	19	0.30	0.37

polluted by copper or cadmium in Akita Prefecture, Japan. Experimental results are summarized in Table 16.

At site A in Table 16, concentrations of all metals in the second soil-core depth were higher than those in the first soil-core depth. Bismuth and antimony concentrations in the soil core under the second core depth were less than those of natural concentrations, 0.34 and 0.37 $\mu g\, g^{-1}$, respectively. At sites B and C, the profiles of metal concentrations were somewhat different from those in site A[113]. The difference is probably due to dissimilarities in pollution history, surrounding circumstances and properties of soil.

Kubota and coworkers[114] also analyzed soils and sediments from bismuth-polluted sites at which a bismuth smelter had been operated since 1970. Twenty-four soils at sites within 2 km from the smelter were analyzed. Geometric mean, maximum and minimum values of bismuth and antimony concentrations in the soils were 4.2, 122 and 0.45, and 3.2, 37.3 and 0.61 $\mu g\, g^{-1}$ in the dry base, respectively. The bismuth and antimony levels from sediments of the exhaust port and downstream of the river were 200–700 and 100–200 times higher than natural concentrations, respectively[114]. Effects of the pollution on the ecosystem were not considered.

E. Toxicity

Poisoning by bismuth and bismuth compounds has occurred more frequently during medical therapy than by exposure at workplaces. It resembles poisoning caused by lead and mercury and their compounds. After oral administration, water-insoluble bismuth compounds, such as bismuth nitrate oxide, $BiO(NO_3)$, are hardly absorbed, and acute poisoning is seldom induced. Water-soluble bismuth compounds are absorbed quickly, and acute poisoning is likely to occur.

The main hazard of bismuth is the chronic exposure that took place during long-term therapy. Serfontein and Mekel[115] surveyed a number of documented cases of bismuth toxicity. Sixty patients, including ten cases of encephalopathy, were associated with the ingestion of inorganic bismuth compounds. Neuropathological symptoms were also described caused by bismuth subgallate, and some were case histories of patients with

TABLE 17. EC_{50} values of organobismuth compounds for four fungi grown on potato dextrose agar incubated at 30 °C for 72 hours (modified from Burrel and coworkers[97])

	EC_{50} values $(mg\,l^{-1})$ for fungi			
Compounds	F. oxysporum[a]	Pestalotia sp.	P. hirsutus	S. homeocarpa[a]
Ph_3Bi	>30	>30	>30	20 ± 1.0
Ph_3BiCl	>30	>30	>30	22 ± 0.9
$Ph_3Bi(MeCOO)_2$	>30	>30	>30	28 ± 1.2
$Ph_3Bi(CH_2ClCOO)_2$	>30	>30	>30	>30
$Ph_3Bi(CHCl_2COO)_2$	>30	>30	>30	>30
$Ph_2BiCl_2Pyr_2$	7.5 ± 0.4	7.0 ± 0.3	28 ± 1.6	12.0 ± 0.9
Ph_3BiCN	15 ± 0.8	5.5 ± 0.07	17 ± 0.9	15 ± 0.7
$NEt_4BiPh_2Cl_2$	17.0 ± 1.2	10.0 ± 0.5	22.0 ± 2.3	16.0 ± 1.8
NEt_4Cl	>30	>30	>30	>30
Ph_2BiN_3	>30	>30	>30	>30
$NEt_4BiPh(N_3)_2$	1.8 ± 0.04	7.5 ± 0.05	10 ± 0.04	2.6 ± 0.02
NaN_3	<0.57	<0.57	<0.57	<0.57

[a] See footnote of Table 11; NEt_4 denotes tetraethylammonium.

renal failure after administration of water-soluble bismuth compounds[115]. The toxicity of bismuth compounds greatly depends upon their chemical forms. 'Organobismuth carboxylate' (i.e., Coscat 83 catalyst)[116] has low toxicity.

Burrell and coworkers determined fungitoxicity of organobismuth compounds[97]. The EC_{50} values obtained for the test fungi are summarized in Table 17.

Triphenylbismuth(III) and the four triphenylbismuth(V) derivatives showed rather limited toxicity. The compound $NEt_4BiPh_2Cl_2$ [tetraethylammonium diphenyldichlorobismuthate(III)] gave EC_{50} values less than $30\,\mu g\,ml^{-1}$ for all fungi tested, while the cationic group $NEt_4{}^+$, supplied as the chloride salt (NEt_4Cl), was not inhibitory. Diphenylbismuth(III) cyanide (Ph_2BiCN) and the phenylbismuth dichloride–dipyridine complex ($PhBiCl_2Pyr_2$) were also inhibitory to all fungal growth.

Sodium azide as a reference compound caused complete inhibition at less than $0.57\,\mu g\,ml^{-1}$. Diphenylbismuth(III) azide (Ph_2BiN_3) exhibited low antifungal activity with EC_{50} values of $30\,\mu g\,ml^{-1}$. When this molecule was modified by the addition of a tetraethylammonium moiety and a second N_3-ion to increase lipid solubility, antifungal activity was greatly increased. These data were further evidence of the stability of the organometallic compounds in the test system.

Low toxicity of triphenylbismuth was also shown by Schafer and Bowles[117].

F. Metabolism

Gregus and Klaassen[118] carried out a comparative study of fecal and urinary excretion and tissue distribution of eighteen metals in rats after intravenous injection. Total (fecal + urinary) excretion was relatively rapid (over 50% of the dose in 4 days) for cobalt, silver and manganese; between 50 and 20% for copper, thallium, bismuth, lead, cesium, gold, zinc, mercury, selenium and chromium; and below 20% for arsenic, cadmium, iron, methylmercury and tin. Feces was the predominant route of excretion for silver, manganese, copper, thallium, lead, zinc, cadmium, iron and methylmercury whereas urine was the predominant route of excretion of cobalt, cesium, gold, selenium, arsenic and tin. Most of the metals reached the highest concentration in liver and kidney. However, there was no direct relationship between the distribution of metal to these excretory organs and their primary route of excretion[118].

cis-Diamminedichloroplatinum (cisplatin) is a potent antitumor agent containing the heavy metal platinum. However, its application in large doses is limited by several toxic side effects, of which its nephrotoxicity is the most common dose-limiting factor. The toxic side effects of cisplatin have been found to be prevented by bismuth subnitrate. Prevention of toxic side effects of cisplatin by bismuth subnitrate without compromising its antitumor activity was found in mice by Naganuma and coworkers[119,120], in golden hamster by Kawata and coworkers[121] and in man by Morikawa and Kawamura[122], Hamada and coworkers[123] and Shibuya and coworkers[124].

The ability of bismuth subnitrate to reduce specifically the toxicity of cisplatin and adriamycin was ascribed to the fact that bismuth induces metallothionein by Naganuma and coworkers[120,125,126] and Boogaard and coworkers[127]. Naganuma and coworkers[126] also found that preinduction of metallothionein by oral administration of bismuth subnitrate significantly decreased the lethal toxicity, cardiotoxicity and bone-marrow toxicity observed with a single subcutaneous injection of adriamycin.

Both Szymańska and coworkers[128,129] and Naganuma and coworkers[125] found that bismuth is a potent inducer for accelerating metallothionein synthesis especially in the kidney.

Kinetics and distribution of ^{206}Bi in rats and rabbits were measured[130]. Urinary excretion of ^{206}Bi was 75 and 70% of the dose in rats and rabbits, respectively. Fecal excretion, only partially of biliary origin, was ca 16% in both species. The major sites of accumulation were the kidneys, the liver, the muscles and the intestines. Only a small amount of ^{206}Bi was found in the brain[130].

Chaleil and coworkers[131] measured bismuth level in blood and tissues of rats after a daily oral dose of 50 mg of bismuth subnitrate during 30 days. Rats which had been colonized with methanogenic bacteria of human origin were compared to germ-free controls. A slight increase of the blood level and a small decrease of the muscular bismuth concentration were observed in colonized rats. The concentration in the brain was very low in both groups. These data show that the colonization by methanogenic bacteria does not increase the tissue deposition of bismuth in rats. Therefore, the presence of similar methanogenic bacteria in the human digestive tract does not seem to be a risk for the development of bismuth encephalopathy[131].

In rats, the effects of various thiol compounds (mercaptopropionic acid, D-penicillamine, L-cysteine, DL-homocysteine, 2-mercaptoethylamine and mercaptoethane) on the absorption and elimination of bismuth subnitrate given orally were determined[132]. All of the thiol substances, and particularly cysteine, homocysteine and mercaptopropionic acid, enhanced Bi absorption and elimination. Moreover, the acute toxicity of Bi was enhanced when Bi was given as a complex with cysteine ($LD_{50} = 156$ mgkg^{-1}). Studies of ^{13}C and ^{15}N NMR confirmed complexation of Bi and cysteine. The enhancement of the absorption, elimination and acute toxicity of Bi may be caused by the increases of solubility and permeability of Bi arising from complexation with thiol compounds.

G. Health Effects and Safety

Serfontein and Mekel[115] found that a survey of the literature on bismuth toxicity in man in relation to blood-level data shows the necessity of distinguishing between lipid-soluble and water-soluble organic complexes of bismuth on the one hand, and the simple inorganic salts of bismuth on the other. A characteristic feature of the former, illustrated by the water-soluble bismuth complex triglycollamate, is the high bismuth levels and the nephrotoxic properties of the compound in man. Bismuth absorption after administration of the simple inorganic salts of bismuth is postulated to occur in the form of ionic bismuth, with low bismuth levels being a characteristic feature of such compounds. A new anti-ulcer drug (bismuth–protein complex: Bicitropeptide) behaves pharmocologically in a manner similar to the inorganic bismuth salts in man, showing low bismuth blood levels and the absence of toxic side effects. It was suggested that bismuth blood level values below 50 μg ml^{-1} were unlikely to be toxic in man[115].

The following toxicological data are available for bismuth and bismuth compounds: lowest published lethal dose for bismuth, 221 mg kg^{-1} (humans, presumably oral)[133], bismuth chloride oxide, BiOCl, LD$_{50}$ 22 g kg^{-1} (rat, oral); trimethylbismuth, BiMe$_3$, LD$_{50}$ 484 mg kg^{-1} (rabbit, oral), LD$_{50}$ 182 mg kg^{-1} (rabbit, subcutaneous), LD$_{50}$ 11 mg kg^{-1} (rabbit, intravenous)[134].

H. Summary

Poisoning by bismuth and bismuth compounds has occurred more frequently during medical therapy than by exposure at the workplace as mentioned before. However, today the use of bismuth compounds in medicine is decreasing and the use of bismuth in pharmaceutical products is being viewed more and more as a questionable practice[102]. France, which in 1972 consumed more than 1000 ton of bismuth, making it the largest bismuth consumer, has restricted the use of bismuth in pharmaceuticals, which resulted in a sharp drop in bismuth consumption. Bismuth compounds have been used because of their astringent, antiphlogistic, bacteriostatic and disinfecting action. Medicines for depigmentation can no longer contain bismuth. Thus the hazard of bismuth during long-term therapy will be decreasing.

On the other hand, the uses of nonorganic compounds of bismuth in alloys, metallurgical additives and chemicals are still increasing, especially in pearlescent pigments for cosmetics and plastics, and in varistors and ceramic capacitors.

Organobismuth compounds have stable trivalent and pentavalent states. They are more subject to thermal decomposition than the corresponding phosphorus, arsenic and antimony compounds, and they are more toxic. Triarylbismuthines are stable, but all trialkylbismuthines except trimethylbismuthine are self-igniting in air. Triarylbismuth halides are known as stable pentavalent organobismuth compounds. Only the chloride and bromide can be prepared; the other triarylbismuth halides cannot be produced because of instant decomposition. Few investigations on the fate of organobismuth compounds and their environmental effects were carried out. Neither biodegradation of organobismuth compounds nor biomethylation of inorganic bismuth compounds were observed. Further investigations concerning these subject are expected.

V. ACKNOWLEDGMENT

The author is sincerely grateful to Professor A. Inoue (Research Center for the South Pacific, Kagoshima University) for many helpful discussions and useful suggestions.

VI. REFERENCES

1. J. M. Wood and H. Wang, *Environ. Sci. Technol.*, **17**, 582A (1983).
2. D. Ben-Bassat and A. M. Mayer, *Physiol. Plant.*, **33**, 128 (1975).
3. T. Sakaguchi, A. Nakajima and T. Horikoshi, *Eur. J. Appl. Microbiol. Biotechnol.*, **12**, 84 (1981).
4. B. Greene, M. Hosea, R. McPherson, M. Henzl, M. D. Alexandar and D. W. Darnall, *Environ. Sci. Technol.*, **20**, 627 (1986).
5. R. L. Veroy, N. Montano, M. L. B. de Guzman, E. C. Laserna and G. J. B. Cajipe, *Bot. Mar.*, **23**, 59 (1980).
6. J. L. Stauber and T. M. Florence, *Aquat. Toxicol.*, **6**, 297 (1985).
7. J. M. Wood and F. E. Engle, *Prepr. Soc. Min. Eng. AIME* (SME-86-48), **i**, 1 (1986).

8. N. S. Fisher and J. G. Fabris, *Mar. Chem.*, **11**, 245 (1982).
9. T. Nagano, Y. Watanabe, K. Hida, Y. Suketa and S. Okada, *Eisei Kagaku*, **28**, 83 (1982).
10. T. Nagano, Y. Watanabe, K. Hida, Y. Suketa and S. Okada, *Eisei Kagaku*, **28**, 114 (1982).
11. S. Bonotto, G. B. Gerber, Jr., C. T. Garten, C. M. Vandecasteele, C. Myttenaere, J. V. Baelen, M. Cogneau and D. van der Ben, *Eur. Rep. Comm. Eur. Commun.* **EUR-9214**, 381 (1984).
12. S. Maeda, R. Inoue, T. Kozono, T. Tokuda, A. Ohki and T. Takeshita, *Chemosphere*, **20**, 101 (1990).
13. S. Maeda, A. Ohki, T. Tokuda and M. Ohmine, *Appl. Organomet. Chem.*, **4**, 251 (1990).
14. S. Maeda, A. Ohki, K. Kusadome, T. Kuroiwa, I. Yoshifuku and K. Naka, *Appl. Organomet. Chem.*, **6**, 213 (1992).
15. S. Maeda, A. Ohki, K. Miyahara, K. Naka and S. Higashi, *Appl. Organomet. Chem.*, **6**, 415 (1992).
16. P. J. Craig, in *The Chemistry of the Metal–Carbon Bond*, Vol. 5 (Ed. F. R. Hartley), Chap. 10, Wiley, Chichester, 1989, pp. 437–463.
17. National Research Council, Ed., *Arsenic: Medical and Biological Effects of Environmental Pollutants*, National Academy of Sciences, Washington, D.C., 1977, pp. 1–332.
18. J. R. Loebenstein, in *Mineral Yearbook–1989*, Vol. 1, US Interior Bureau of Mines, Washington, DC, 1991, pp. 123–126.
19. S. Ishiguro, in *Shin-kinzoku-detabukku–1991* (*New Metals Data Book—1991*), Homatto-Ado, Ltd., Tokyo, 1991, pp. 511–521.
20. S. Ishiguro, *Appl. Organomet. Chem.*, **6**, 323 (1992).
21. G. Lunde, *Environ. Health Perspect.*, **19**, 47 (1977).
22. G. Lunde, *Nature (London)*, **224**, 186 (1969).
23. G. Lunde, *J. Sci. Food Agric.*, **26**, 1247 (1975).
24. G. Lunde, *Acta Chem. Scand.*, **27**, 1586 (1973).
25. G. Lunde, *J. Sci. Food Agric.*, **24**, 1021 (1973).
26. W. R. Cullen and K. J. Reimer, *Chem. Rev.*, **89**, 713 (1989).
27. J. S. Edmonds, K. A. Francesconi, J. R. Cannon, C. L. Raston, B. W. Skelton and A. H. White, *Tetrahedron Lett.*, **18**, 1543 (1977).
28. J. R. Cannon, J. S. Edmonds, K. A. Francesconi, C. L. Raston, J. B. Saunders, B. W. Skelton and A. H. White, *Aust. J. Chem.*, **34**, 787 (1981).
29. S. Maeda, chapter titled *Biotransformation of Arsenic in the Freshwater Environment*, in *Arsenic in the Environment* (Ed. J. O. Nriagu), Wiley, New York, 1993 (in press).
30. F. Challenger, *Chem. Rev.*, **36**, 315 (1945).
31. K. J. Reimer, *Appl. Organomet. Chem.*, **3**, 475 (1989).
32. J. Wrench, S. W. Fowler and M. Y. Ünlü, *Mar. Pollut. Bull.*, **10**, 18 (1979).
33. M. Y. Ünlü, *Chemosphere*, **5**, 269 (1979).
34. D. W. Klumpp, *Mar. Biol.*, **58**, 265 (1980).
35. D. W. Klumpp and P. J. Peterson, *Mar. Biol.* **62**, 297 (1981).
36. R. V. Coony and A. A. Benson, *Chemosphere*, **9**, 335 (1980).
37. M. O. Andreae, in *Arsenic: Industrial, Biomedical, Environmental Perspectives* (Eds. W. H. Lederer and R. J. Fensterheim), Van Nostrand Reinhold, New York, pp. 378–392 (1983).
38. A. R. Isensee, P. C. Kearney, E. A. Woolson, G. E. Jones and V. P. Williams, *Environ. Sci. Technol.*, **7**, 841 (1973).
39. A. F. del Riego, *Bol. Inst. Esp. Oceanogr.*, **134**, 3 (1968).
40. W. R. Penrose, H. B. S. Conacher, R. Black, J. C. Meranger, W. Miles, H. M. Cunninham and W. R. Squires, *Environ. Health Perspect.*, **19**, 53 (1977).
41. I. S. Seydel, *Arch. Hydrobiol.*, **71**, 17 (1972).
42. Life Systems, Inc., *PB Rep. (USA)*, **PB-89-185706**, Environmental Protection Agency, Washington, DC, 1989, pp. 1–131.
43. J. P. Buchet, R. Lauwery and H. Roels, *Int. Arch. Occup. Environ. Health*, **48**, 71 (1981).
44. For man see: (a) E. A. Crecelius, *Environ. Health Perspect.*, **19**, 147 (1977); (b) E. A. Crecelius, *NBS Special Publication*, **464**, 495 (1977); (c) E. A. Crecelius, *ERDA Symp. Ser.*, **42**, 63 (1979); (d) J. P. Buchet, R. Lauwerys and H. Roels, *Int. Arch. Occup. Environ. Health*, **48**, 111 (1981). For algae, moina and fishes see: References 12–14.
45. D. L. Perry, R. D. Germany and R. W. Flournoy, Environ-Med Labs., Inc., *PB Rep.*, **PB-85-219145**, Federal Highway Administration, Baton Rouge, LA, 1984, pp. 1–97.
46. D. L. Perry, R. D. Germany and R. W. Flournoy, Enviro-Med Labs., Inc., *PB Rep.*, **PB-85-219152**, Federal Highway Administration, Baton Rouge, LA, 1984, pp. 1–21.

47. A. A. Abdelghani, A. C. Anderson, M. Jaghabir, F. Mather, M. Palmgren, L. White and A. J. Engkande, *PB Rep.*, **PB-86-115763**, Federal Highway Administration, Baton Rouge, LA, 1984, pp. 1–28.
48. J. R. Cannon, J. B. Saunders and R. F. Toia, *Sci. Total Environ.*, **31**, 181 (1983).
49. T. Kaise, S. Watanabe and K. Itoh, *Chemosphere*, **14**, 1327 (1985).
50. H. Yamauchi, T. Kaise and Y. Yamamura, *Bull. Environ. Contam. Toxicol.*, **36**, 350 (1986).
51. T. Kaise, Y. Horiguchi, S. Fukui, K. Shiomi, M. Chino and T. Kikuchi, *Appl. Organomet. Chem.*, **6**, 369 (1992).
52. E. Marafante, M. Vahter and L. Dencker, *Sci. Total Toxicol.*, **34**, 223 (1984).
53. M. O. Andreae, in *Organoarsenic Compounds in the Environment* (Ed. P. J. Craig), Wiley, New York, 1986, pp. 198–228.
54. F. Challenger, *Adv. Enzymol.*, **12**, 429 (1951).
55. M. Shariatpanahi, A. C. Anderson and A. A. Abdelghani, *Trace Subst. Environ. Health*, **16**, 170 (1982).
56. E. A. Crecelius, *Bull. Environ. Contam. Toxicol.*, **18**, 227 (1977).
57. W. R. Cullen, B. C. McBride and A. W. Pickett, *Can. J. Microbiol.*, **25**, 1201 (1979).
58. W. R. Cullen, A. E. Erdman, B. C. McBride and A. W. Pickett, *J. Microbiol. Methods*, **1**, 297 (1983).
59. K. Hanaoka, T. Matsumoto, S. Tagawa and T. Kaise, *Chemosphere*, **16**, 2545 (1987).
60. K. Hanaoka, H. Yamamoto, K. Kawashima, S. Tagawa and T. Kaise, *Appl. Organomet. Chem.*, **2**, 371 (1988).
61. K. Hanaoka, K. Ueno, S. Tagawa and T. Kaise, *Comp. Biochem. Physiol.*, **94B**, 379 (1989).
62. K. Hanaoka, S. Hasegawa, N. Kawabe, S. Tagawa and T. Kaise, *Appl. Organomet. Chem.*, **4**, 239 (1990).
63. K. Hanaoka, S. Tagawa and T. Kaise, *Appl. Organomet. Chem.*, **6**, 139 (1992).
64. L. O. Moore and P. J. Ehman, in *Encyclopedia of Chemical Processing and Design*, Vol. 3 (Ed. J. J. McKetta), Marcel Dekker, Inc., New York, 1977, pp. 396–415.
65. S. Maeda and T. Sakaguchi, in *Introduction to Applied Phycology* (Ed. I. Akatsuka), SPB Academic Publishing BV, The Hague, The Netherlands, 1990, pp. 109–136.
66. J. O. Nriagu and J. M. Pacyna, *Nature*, **333**, 134 (1988).
67. J. O. Nriagu, *Environment*, **32**, 7 (1990).
68. L. B. Clarke and L. L. Sloss, *IEA Coal Research*, **49** 'Trace Elements', 15 (1992).
69. J. L. T. Waugh, in *Encyclopedia of Science & Technology*, Vol. 1, McGraw-Hill, Inc., New York, 1987, pp. 646–648.
70. T. O. Llewellyn, in *Minerals Yearbook–1989*, Vol. 1, US Interior Bureau of Mines, Washington, DC, 1991, pp. 113–121.
71. Environmental Protection Agency, OH, USA, *PB Rep. (USA)*, **PB-88-175039**, 1985, pp. 1–119.
72. Environmental Protection Agency, OH, USA *PB Rep. (USA)*, **PB-88-179445**, 1987, pp. 1–34.
73. T. Murata, *Gekkan-Haikibutsu*, 222 (1991).
74. M. Takano, in *Shin-kinzoku-detabukku–1991 (New Metals Data Book–1991)*, Homatto-Ado, Ltd., Tokyo, 1991, pp. 67–78.
75. K. M. Bai and S. K. Majumder, *Bull. Environ. Contam. Toxicol.*, **29**, 107 (1982).
76. R. B. D. Gradwohl, in *Clinical Laboratory Methods and Diagnosis*, Vol. 2, 5th edn., Henry Kramton, London, 1956, p. 2236.
77. T. Murata, **Gekkan-Haikibutsu**, 209 (1991).
78. S. A. Abbasi, *Anal. Lett.*, **22**, 237 (1989).
79. T. N. Mahadevan, S. Sadashivan and U. C. Mishra, *Sci. Total Environ.*, **24**, 275 (1982).
80. S. Habib, *Sci. Total Environ.*, **22**, 253 (1982).
81. S. Tanaka, M. Nakamura, H. Yokoi, M. Yumura and Y. Hashimoto, *Bunseki Kagaku*, **35**, 116 (1986).
82. L. Xiankun, L. Jing, C. Shuzhu and D. Guosheng, *Acta Oceanologica Sinica*, **9**, 255 (1990).
83. J. E. Fergusson, E. A. Forbes and R. J. Schroeder, *Sci. Total Environ.*, **50**, 217 (1986).
84. G. Chittleborough, *Sci. Total Environ.*, **14**, 53 (1980).
85. G. F. Clemente, G. Ingrao and G. P. Santaroni, *Sci. Total Environ.*, **24**, 255 (1982).
86. C. Vanoeteren and R. Cornelis, *Sci. Total Environ.*, **54**, 217 (1986).
87. T. Yamashige, M. Yamamoto, Y. Shigetomi and Y. Yamamoto, *Bunseki Kagaku*, **34**, 646 (1985).
88. W. R. Cullen, M. Dodd, B. U. Nwata, D. A. Reimer and K. J. Reimer, *Appl. Organomet. Chem.*, **3**, 351 (1989).

89. M. O. Andreae, J. F. Asmodé, P. Foster and L. Van't dack, *Anal. Chem.*, **53**, 1766 (1981).
90. N. Ainsworth, J. A. Cooke and M. S. Johnson, *Environ. Pollut.*, **65**, 65 (1990).
91. W. M. Mok and C. M. Wai, *Environ. Sci. Technol.*, **24**, 102 (1990).
92. N. Ainsworth, J. A. Cooke and M. S. Johnson, *Environ. Pollut.*, **65**, 79 (1990).
93. K. A. Herbst, G. Rose, K. Hanusch, H. Schumann and H. U. Wolf, in *Ullman's Encyclopedia of Industrial Chemistry*, Vol. A3, VCH Verlagsgesellschaft, Weinheim, Germany, 1985, pp. 55–76.
94. F. Challenger, in *ACS Symposium Series*, **82** (Eds. F. E. Brinckman and J. M. Bellama), Am. Chem. Soc., Washington, D.C., 1978, pp. 1–22.
95. E. M. Cordasco and F. D. Stone, *Chest*, **64**, 182 (1973).
96. R. M. James and B. D. Dinman, *Trans Am. Foundrymen's Soc.* (AFS Transactions), **95**, 883 (1987).
97. R. E. Burrell, C. T. Corke and R. G. Goel, *J. Agric. Food Chem.*, **31**, 85 (1983).
98. G. E. Parris and F. E. Brinckman, *J. Org. Chem.*, **40**, 3801 (1975).
99. G. E. Parris and F. E. Brinckman, *Environ. Sci. Technol.*, **10**, 1128 (1976).
100. J. M. Brannon and W. H. Patrick, Jr., *Environ. Pollut. (Ser. B)*, **9**, 107 (1985).
101. U. S. EPA, *Federal Register*, **48**, 717 (1983).
102. J. Krüger, P. Winkler, E. Lüderitz, M. Lück and H. U. Wolf, in *Ullman's Encyclopedia of Industrial Chemistry*, Vol. A4, VCH Verlagsgesellschaft, Weinheim, Germany, 1985, pp. 171–189.
103. S. J. Yosim, in *Encyclopedia of Science & Technology*, Vol 2, McGraw-Hill, Inc., New York, 1987, pp. 596–598.
104. S. M. Jasinski, in *Minerals Yearbook–1989*, Vol. 1, US Interior Bureau of Mines, Washington, DC, 1991, pp. 185–188.
105. M. Horikoshi, in *Shin-kinzoku-detabukku–1991 (New Metals Data Book–1991)*, Homatto-Ado, Ltd., Tokyo, 1991, pp. 523–528.
106. S. M. Skinner and R. W. Lewis, *Res. Commun. Chem. Pathol. Pharmacol.*, **16**, 183 (1977).
107. S. M. Skinner, J. M. Swatzell and R. W. Lewis, *Res. Commun. Chem. Pathol. Pharmacol.*, **19**, 165 (1978).
108. D. S. Lee, *Anal. Chem.*, **54**, 1682 (1982).
109. J. Ohyama, F. Maruyama and Y. Dokiya, *Anal. Sci.*, **3**, 413 (1987).
110. T. Asami, M. Kubota and K. Minamisawa, *Nippon Dojo Hiryougaku Zasshi*, **59**, 197 (1988).
111. K. Komura, *Radioisotopes*, **34**, 555 (1985).
112. M. Kubota, T. Asami, K. Minamisawa, M. Matsuki and M. Yasuda, *Ningen to Kankyou (Man and Environment)*, **14**, 2 (1988).
113. M. Kubota, T. Asami and M. Matsuki, *Ningen to Kankyou (Man and Environment)*, **15**, 28 (1989).
114. M. Kubota, T. Asami, M. Matsuki and A. Kashimura, *Nippon Dojo Hiryougaku Zasshi (Jpn. J. Soil Sci., Plant Nutr.)*, **61**, 190 (1990).
115. W. J. Serfontein and R. Mekel, *Res. Commun. Chem. Pathol. Pharmacol.*, **26**, 391 (1979).
116. A. R. Leckart and L. S. Slovin, *J. Elastomers Plast.*, **19**, 313 (1987).
117. E. W. Schafer, Jr. and W. A. Bowles, Jr., *Arch. Environ. Contam. Toxicol.*, **14**, 111 (1985).
118. Z. Gregus and C. D. Klaassen, *Toxicol. Appl. Pharmacol.*, **85**, 24 (1986).
119. A. Naganuma, M. Satoh and N. Imura, *Taisha (Metabolism)*, **24**, 191 (1987).
120. A. Naganuma, M. Satoh and N. Imura, *Cancer Res.*, **47**, 983 (1987).
121. M. Kawata, Y. Orita and Y. Sato, *Ear Res. Jpn.*, **21**, 331 (1990).
122. T. Morikawa and E. Kawamura, *Gan to Kagakuryouhou (Jpn. J. Cancer Chemother.)*, **16**, 1094 (1989).
123. T. Hamada, Y. Nishiwaki, T. Kodama, A. Hayashibe, N. Nukariya, H. Sasaki, T. Morikawa, T. Hirosawa and T. Matsuyama, *Gan to Kagakuryouhou (Jpn. J. Cancer Chemother.)*, **16**, 3587 (1989).
124. M. Shibuya, A. Hirosawa and H. Niitani, *Gan to Kagakuryouhou (Jpn. J. Cancer Chemother.)*, **17**, 950 (1990).
125. A. Naganuma and N. Imura, *Tanpakushitsu, Kakusan, Kouso (Protein, Nucleic Acid, and Enzyme)*, **32**, 1031 (1987).
126. A. Naganuma, M. Satoh and N. Imura, *Jpn. J. Cancer Res. (Gann)*, **79**, 406 (1988).
127. P. J. Boogaard, A. Slikkerveer, J. F. Nagelkerke and G. J. Mulder, *Biochem. Pharmacol.*, **41** 369 (1991).

128. J. A. Szymanska, M. Zychowicz, A. J. Zelazowski and J. K. Piotrowski, *Arch. Toxicol.*, **40**, 131 (1978).

129. J. A. Szymanska, A. J. Zelazowski and S. Kawiorski, *Clin. Toxicol.*, **18**, 1291 (1981).

130. R. Vienet, P. Bouvet and M. Istin, *Int. J. Appl. Radiat. Isot.*, **34**, 747 (1983).

131. D. Chaleil, J. P. Regnault, P. Allain, R. Motta, G. Raynaud and P. Bouvet, *Ann. Pharm. Fr.*, **46**, 133 (1988).

132. D. Chaleil, F. Lefevre, P. Allain and G. J. Martin, J. Inorg. Biochem., **15**, 213 (1981).

133. J. M. Arena, in *Poisoning; Toxicology, Symptoms, Treatments*, 2nd ed., C. C. Thomas, Springfield, Ill., 1970, p. 73.

134. T. Sollmann and J. Seifter, *J. Pharmacol. Exp. Ther.*, **67**, 17 (1939).

CHAPTER **20**

Syntheses of organoantimony and organobismuth compounds

KIN-YA AKIBA and YOHSUKE YAMAMOTO

Department of Chemistry, University of Hiroshima, 1-3-1 Kagamiyama, Higashi-Hiroshima, 724, Japan

The chemistry of organic arsenic, antimony and bismuth compounds
Edited by S. Patai © 1994 John Wiley & Sons Ltd

I. INTRODUCTION

Progress in organoantimony and organobismuth compounds has been reviewed annually by Freedman and Doak in *Journal of Organometallic Chemistry*[1] and by Wardell in *Organometallic Chemistry*[2]. In addition, *Annual Report*[3] and *Coordination Chemistry Reviews*[4] cover Group V elements annually. Compilation of the methods of preparation and properties of organoantimony and -bismuth was given in *Gmelin Handbuch* in 1977 (Bi)[5] and 1981–1990 (Sb)[6]. Another compilation of organoarsenic, -antimony and -bismuth compounds was given by Samaan in the Houben-Weyl Series in 1978[7]. In addition, some general reviews have appeared in *Comprehensive Organic Chemistry* on organoantimony and -bismuth derivatives (1979)[8], *Comprehensive Organometallic Chemistry* on Group(V) (1982)[9] and *Comprehensive Coordination Chemistry* (1987)[10]. Aylett reviewed the organic chemistry of Group V elements in 1979[11] and Freedman and Doak reviewed the organic chemistry of bismuth in 1982[12]. More specialized reviews were written by Freedman and Doak on the synthetic application of organoantimony and -bismuth compounds in this series in 1989[13], by Huang on the synthetic application of organoantimony compounds in 1992[14], by Barton in 1988[15] and in 1990[16] and by Finet in 1989[17] on arylation by use of Bi(V), as well as by Hellwinkel on pentaorganyl derivatives of the Group V elements in 1983[18], by Seppelt on the preparation and colors of pentaaryl-bismuth in 1990[19] and in 1992[20], by Ashe on thermochromic distibines and dibismuthines in 1990[21], by Arduengo and Dixon on their tridentate system in 1990[22], by Norman in 1988[23] and in 1990[24] and by Huttner and Evertz in 1986[25] on transition metal complexes incorporating atoms of the heavier Main-group elements.

In the present review, the methods of synthesis of organoantimony and organobismuth compounds will be surveyed. Typical and recent methods have been selected from a large number of reported methods.

II. PREPARATION OF ORGANOANTIMONY(III) AND -BISMUTH(III) COMPOUNDS

A. Preparation of Triorganoantimony(III) and -bismuth(III) Compounds

1 Transmetallation by Organometallic reagents

Reactions of Grignard reagents with metal halides, usually chlorides or bromides, have been extensively used for the preparation of symmetrical triorganoantimony and

-bismuth derivatives (equation 1). A wide variety of triorganoantimony and -bismuth(III) compounds have been prepared in good to excellent yields in ether or in THF. The introduced group can be aliphatic or aromatic[5–7]. Experimental procedures for the preparation of Ph_3Sb^{26}, $(2,4,6\text{-}Me_3C_6H_2)_3Sb^{27}$ and Et_3Bi^{28} have been described in detail.

$$MX_3 \xrightarrow{\;3\,RMgX\;} R_3M \tag{1}$$

$$(M = Sb, Bi) \quad (R = alkyl, alkenyl, alkynyl, aryl)$$

Fairly bulky alkyl substituents such as neopentyl and (trimethylsilyl)methyl are also introduced by this method (equation 2)[29]. Unsymmetrical compounds have also been prepared from the appropriate halide and Grignard reagents. For example the synthesis of $\mathbf{1}^{30}$ and $\mathbf{2}^{31}$ has been reported (equations 3 and 4).

$$SbCl_3 \xrightarrow{\;3\,RMgCl\;} R_3Sb \tag{2}$$

$$\left(\begin{array}{l} R = Me_3CCH_2, 80\% \text{ yield} \\ R = Me_3SiCH_2, 80\% \text{ yield} \end{array}\right)$$

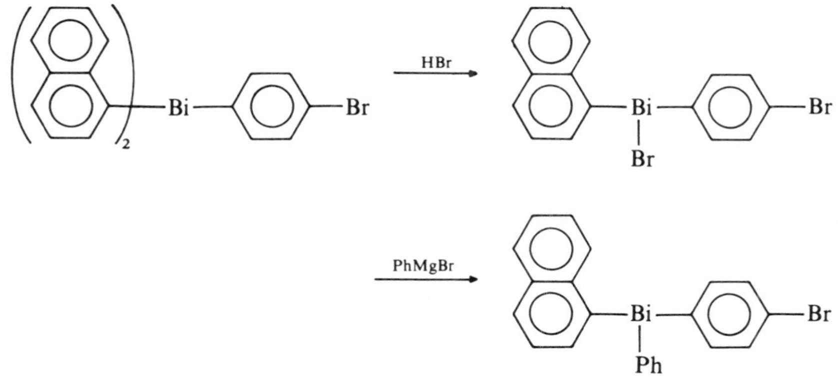

$$(3)$$

(1) 44%

$$(4)$$

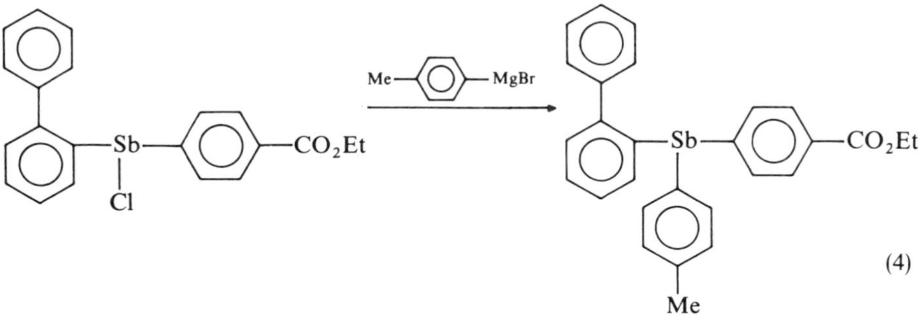

(2) 50–55%

Organolithium reagents have also been used frequently in recent studies, especially for the preparation of cyclic compounds. The preparation of biphenylene compounds (**3,** equation 5^{32}) and stibole (**4**, equation 6^{33}) has been reported.

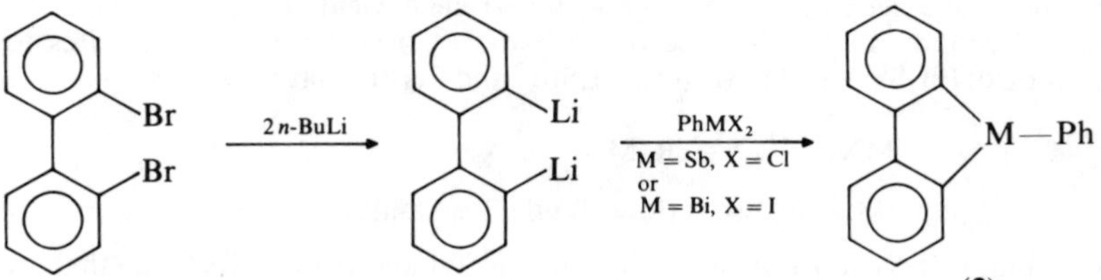

(5)

(**3**)

M = Sb, 64%
M = Bi, 72%

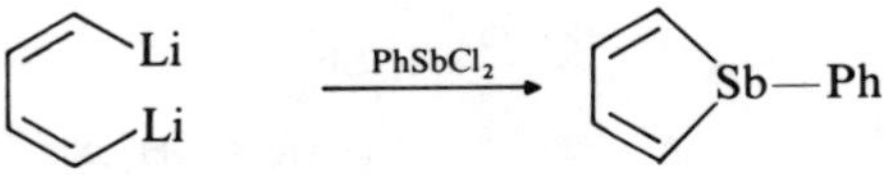

(6)

(**4**) 10%

Three units of the sterically crowded substituent, $2,4,6\text{-}(CF_3)_3C_6H_2$, have been introduced by the lithium method on the bismuth atom to form **5** (equation $7)^{34}$.

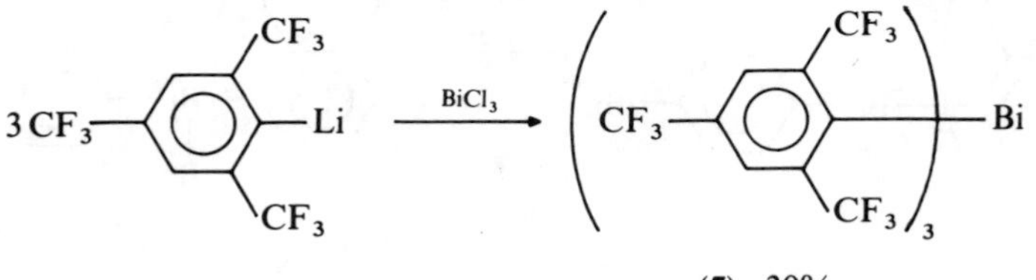

(7)

(**5**) 30%

Although organolithium reagents have been reported to undergo ligand exchange at the bismuth atom (equations 8^{35} and 9^{36}), synthesis of the unsymmetrical bismuthine **6** has been performed by use of (2-*t*-butylsulfonyl)phenyl group as a leaving group in the reaction with lithium reagents (equation $10)^{37}$.

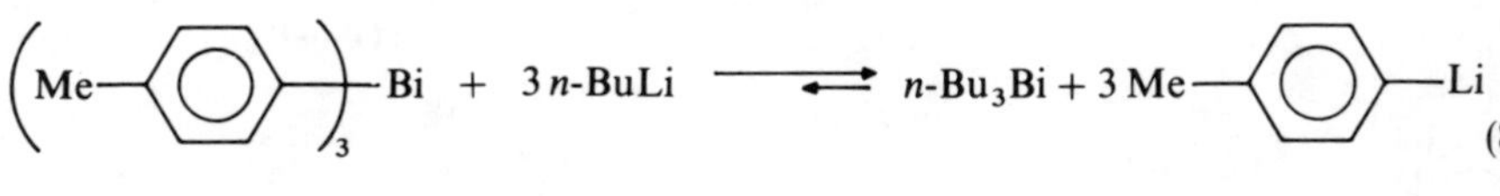

(8)

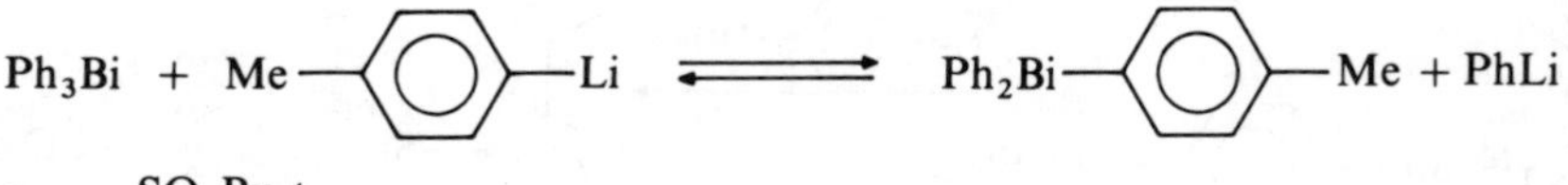

(9)

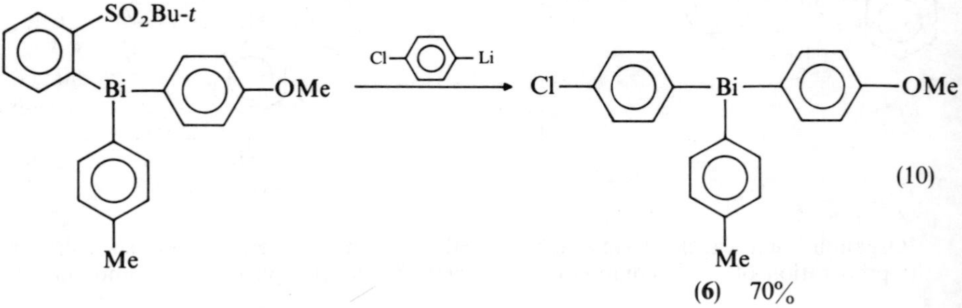

(10)

(**6**) 70%

Sodium reagents are sometimes useful for the preparation of alkynyl compounds (**7**, equation 11)[38] or cyclopentadienyl derivatives (**8**, equation 12)[39].

$$R_2MCl \xrightarrow{HC\equiv C^-Na^+} R_2MC\equiv CH \tag{11}$$
$$(\mathbf{7})$$

$$\begin{pmatrix} M = Sb, R = alkyl \\ or \\ M = Bi, R = aryl \end{pmatrix} \begin{pmatrix} M = Sb, R = Et, 36\% \\ M = Bi, R = p\text{-}MeC_6H_4, 43\% \end{pmatrix}$$

$$MCl_3 \xrightarrow{C_5H_5Na(CpNa)} Cp_3M \tag{12}$$
$$(\mathbf{8})$$

$$(M = Sb, Bi) \qquad \begin{pmatrix} M = Sb, 89\% \\ M = Bi, 80\% \end{pmatrix}$$

A cadmium trifluoromethylation reagent $(CF_3)_2Cd \cdot MeCN$ (**9**) has been used for the preparation of $(CF_3)_3M$ (**10**, $M = Sb$, Bi) (equation 13)[40] and $Ph_nBi(CF_3)_{3-n}$ ($n = 1,2$) (equation 14)[40,41].

$$MX_3 \xrightarrow{(CF_3)_2Cd \cdot MeCN\ (9)} (CF_3)_3M \tag{13}$$
$$(\mathbf{10})$$

$$\begin{pmatrix} M = Sb, Bi \\ X = Cl, Br, I \end{pmatrix} \qquad \begin{pmatrix} M = Sb, 82\% \\ M = Bi, 92\% \end{pmatrix}$$

$$Ph_nBiCl_{3-n} \xrightarrow{9} Ph_nBi(CF_3)_{3-n} \tag{14}$$

$$(n = 1, 2) \qquad (n = 2, 58\%)$$

Zirconocenes (**11**) have been reported to be good precursors for the preparation of some heterocyclic compounds (**12**) containing Sb and Bi (equations 15)[42]. Tin heterocycles (**13**) also have been used for the synthesis of **14** (equation 16)[43].

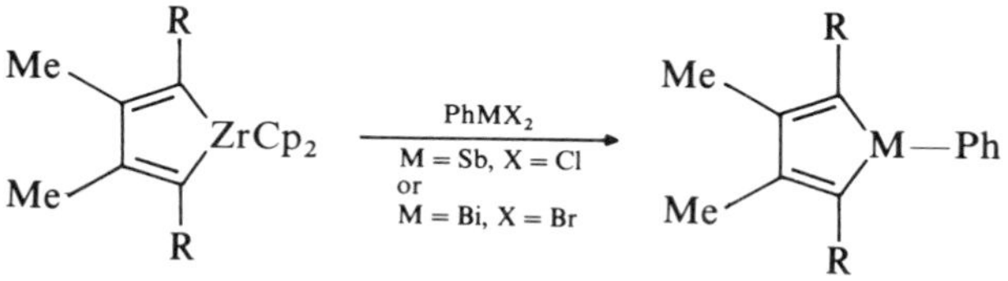

$$\begin{array}{c} \text{(structures for equation 15)} \end{array} \tag{15}$$

(**11**)   (**12**)

(R = Me, Me_3Si)

$$\begin{pmatrix} M = Sb, R = Me_3Si, 94\% \\ M = Bi, R = Me, 95\% \end{pmatrix}$$

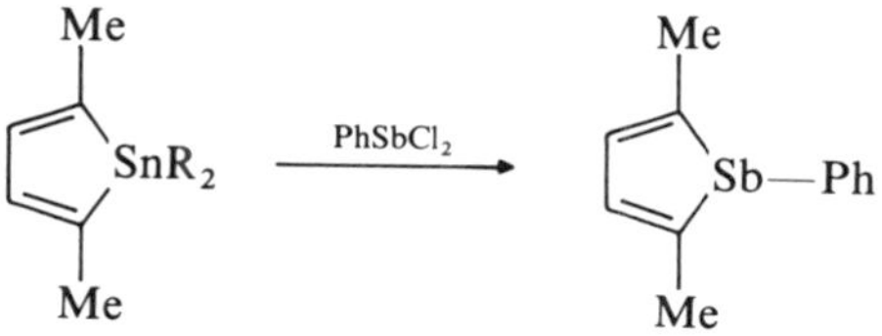

$$\begin{array}{c} \text{(structures for equation 16)} \end{array} \tag{16}$$

(**13**)   (**14**) 65%

Other organometallic reagents such as organoaluminum reagents (equation 17)[44] or mercury reagents (equation 18)[45] yield Sb(III) and Bi(III) compounds, but these reagents have not been used frequently in recent years.

$$Et_4Al^-Li^+ \xrightarrow{MX_3} Et_3M \qquad (17)$$

$$\begin{pmatrix} M = Sb, X = F \\ or \\ M = Bi, X = Cl \end{pmatrix} \quad \begin{pmatrix} M = Sb, 65\% \\ M = Bi, 85\% \end{pmatrix}$$

$$Ph_2Hg \xrightarrow{BiBr_3} Ph_3Bi \qquad (18)$$
$$\sim 100\%$$

2. Reactions of free radicals

Functionalized bismuthines possessing functional groups such as carboxyl or nitro groups are generally difficult to prepare. However, $(p\text{-}O_2NC_6H_4)_3Bi$ (15) has been prepared by the radical exchange reaction between $(p\text{-}CH_3C_6H_4)_3Bi$ and $p\text{-}O_2NC_6H_4$ radical prepared from $p\text{-}O_2NC_6H_4N_2{}^+BF_4^-$ with copper metal in DMF although in low (16%) yield (equation 19)[46].

$$(15) \quad 16\%$$

The diphenylantimony radical (16), which is prepared from the homolysis of tetraphenyldistibine by irradiation, reacts with alkyl iodide to form alkyl(diphenyl)stibine (17) by a radical chain reaction in high yield (equation 20)[47].

$$Ph_2SbSb\,Ph_2 \xrightarrow{h\nu} [Ph_2Sb\cdot] \xrightarrow{RI} Ph_2SbR \qquad (20)$$
$$\qquad\qquad\qquad (16) \qquad\qquad (17)$$

$$80\text{-}88\%$$

3. Reactions of Group V alkali metal salts with organic halides

This method has also been a general method for the synthesis of triorganoantimony and -bismuth compounds. The antimony or bismuth anion $(R_2M^-, M = Sb, Bi)$ reacts with organic halides, which can be either aryl halides such as 18 in equation 21[48], or alkyl halides (equations 22[49] and 23[50]). Dimetallated triorganometallic compounds such as 19 which are useful as ligands for transition metals are available from this method (equation 24)[51].

$$\text{(18)} \xrightarrow[\text{lig NH}_3]{\text{PhMeSb}^-\text{Na}^+} \text{(MeSbPh, 45\%)} \qquad (21)$$

$$\text{Ph}_2\text{Bi}^-\text{K}^+ \xrightarrow{\text{EtBr}} \text{Ph}_2\text{BiEt} \qquad (22)$$
$$39\%$$

$$\text{PhMeSb}^-\text{Na}^+ \xrightarrow{\text{Cl(CH}_2)_n\text{Cl}} [\text{Ph(Me)Sb}]_2(\text{CH}_2)_n \qquad (23)$$
$$\sim 60\%$$

$$\xrightarrow{\text{Me}_2\text{Sb}^-\text{Na}^+} \qquad (24)$$
$$\text{(19)} \quad 22\%$$

Reaction of elemental antimony with sodium in liquid ammonia forms Sb^{3-} that reacts with aryl halides under irradiation. The reaction gives symmetrical triarylstibines[3] by the $S_{RN}1$ mechanism (equation 25)[52].

$$\text{PhCl} \xrightarrow[\text{liq NH}_3, h\nu]{\text{Sb/Na}} \text{Ph}_3\text{Sb} \qquad (25)$$
$$40\%$$

4. Miscellaneous methods

Fairly acidic protons can be substituted by amides of Sb(III) or Bi(III). For example, Me_2MNMeR (20, M = Sb, R = Me; 21, M = Bi, R = $SiMe_3$) react with cyclopentadiene to form thermally labile Me_2MCp (22, M = Sb, Bi) (equation 26)[53]. Diazoalkanes also react with 20 or 21 to form $(Me_2M)_2CN_2$ (23, M = Sb, Bi) (equation 27)[54].

$$\text{Me}_2\text{MN(Me)R} \begin{cases} \xrightarrow{\text{C}_5\text{H}_6 \text{ (CpH)}} \text{Me}_2\text{MCp} \begin{pmatrix} \text{M} = \text{Sb}, 76\% \\ \text{M} = \text{Bi}, 50\% \end{pmatrix} & (26) \\ \text{(22)} \\ \xrightarrow{\text{CH}_2\text{N}_2} (\text{Me}_2\text{M})_2\text{CN}_2 \begin{pmatrix} \text{M} = \text{Sb}, \sim 100\% \\ \text{M} = \text{Bi}, \sim 100\% \end{pmatrix} & (27) \\ \text{(23)} \end{cases}$$

20, M = Sb, R = Me
21, M = Bi, R = Me_3Si

Reductive elimination from tetra- and penta-organometal(V) compounds affords trioganometal(III) compounds. Since pentavalent compounds are usually prepared from

$$\left(R-\bigcirc\right)_3\text{Sb(C}\equiv\text{CPh)}_2 \xrightarrow{110\,°C} \left(R-\bigcirc\right)_{3-n}\text{Sb(C}\equiv\text{CPh)}_n + (\text{PhC}\equiv\text{C})_2$$

$$(n = 0, 1)$$
$$+ \text{PhC}\equiv\text{C}-\bigcirc-\text{R} \qquad (28)$$

trivalent compounds, the procedure probably has little synthetic potential although the mechanism has attracted much attention (equation 28)[55].

Diazomethane adds across the Bi—Bi bond of $Ph_2BiBiPh_2$ and the Bi—E bond of Ph_2BiEPh (E = Se, Te) (equations 29[56] and 30[57]).

$$Ph_2BiBiPh_2 \xrightarrow{CH_2N_2} Ph_2BiCH_2BiPh_2 \tag{29}$$
$$49\%$$

$$Ph_2BiEPh \xrightarrow{CH_2N_2} Ph_2BiCH_2EPh \tag{30}$$
$$(E = Se, Te)$$

B. Preparation of Diorganoantimony(III) and -bismuth(III) Halides

1. Redistribution reactions

Diorganoantimony and -bismuth(III) chlorides or bromides have been conveniently prepared by the redistribution reaction of a 2:1 mixture of the corresponding tri-organo compounds and trihalides. This method has been generally used for alkyl and aryl substituted bismuth halides, and aryl substituted antimony halides (equation 31). Yields are generally good to excellent. Usually, organic solvents such as ether[58], dichloromethane and benzene are used; but the redistribution of a 2/1 molar mixture of Ph_3Sb and SbX_3 (X = Cl or Br) without solvent has been reported to be complete in 3 h at $25\,°C$[59]. For the preparation of alkylantimony halides this method has not been so useful since redox reactions occur between R_3Sb and SbX_3 (equation 32)[60]. Thermolysis of trioganoantimony dihalides has been frequently used for the preparation of alkylantimony halides (*vide infra*).

$$2\,R_3M + MX_3 \longrightarrow 3\,R_2MX \tag{31}$$
$$\begin{pmatrix} M = Sb, R = aryl \\ M = Bi, R = aryl, alkyl \end{pmatrix} (X = Cl, Br)$$

$$3\,R_3Sb + 2SbX_3 \longrightarrow 3\,R_3SbX_2 + 2\,Sb \tag{32}$$
$$(R = alkyl) \qquad\qquad (X = Cl, Br)$$

2. Transmetallation

It is difficult to obtain acyclic diorganoantimony and -bismuth(III) halides in good yields by selective transmetallation using Grignard or organolithium reagents. However, organomagnesium or -lithium reagents bearing sterically bulky substituents can be introduced to yield diorganoantimony and -bismuth(III) halides under controlled conditions. Several examples are shown in equations 33[61], 34[62] and 35[63].

$$t\text{-BuMgCl} \xrightarrow[-25\,°C]{SbCl_3} t\text{-Bu}_2SbCl \tag{33}$$
$$71\%$$

$$\tag{34}$$

$$\begin{pmatrix} M = Sb, 58\% \\ M = Bi, 50\% \end{pmatrix}$$

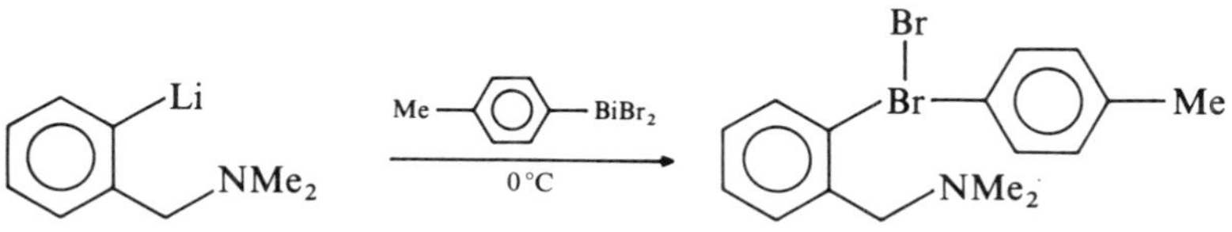

$$(35)$$

(25) 70%

Usually, less nucleophilic organometallic reagents such as tin (**26**[64] or **27**[65]), zirconium (**28**[66]) and cadmium[67] have been used for selective transmetallation (equations 36[64], 37[65], 38[66] and 39[67]). $C_6H_5SiF_3$ has been used to prepare $(C_6H_5)_2SbF$ from SbF_3 (equation 40)[68].

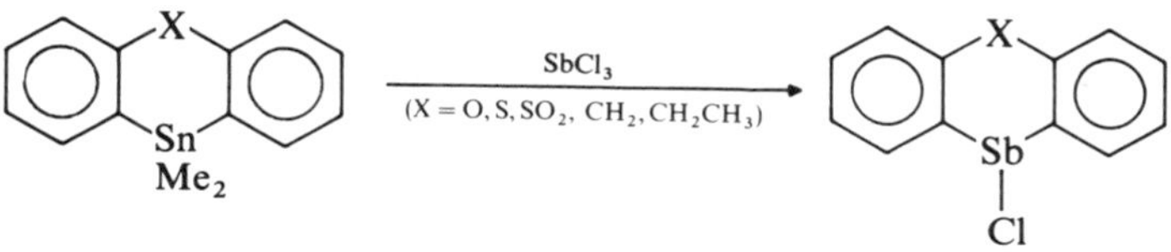

$$(36)$$

(X = O, S, SO$_2$, CH$_2$, CH$_2$CH$_3$)

(26)   50–70%

$$(37)$$

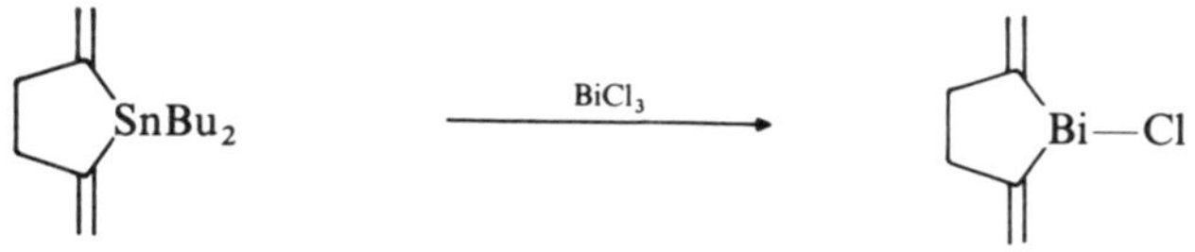

(27)   99%

$$(38)$$

$$\left(\begin{array}{l} Me = Sb, 78\% \\ M = Bi, 70\% \end{array}\right)$$

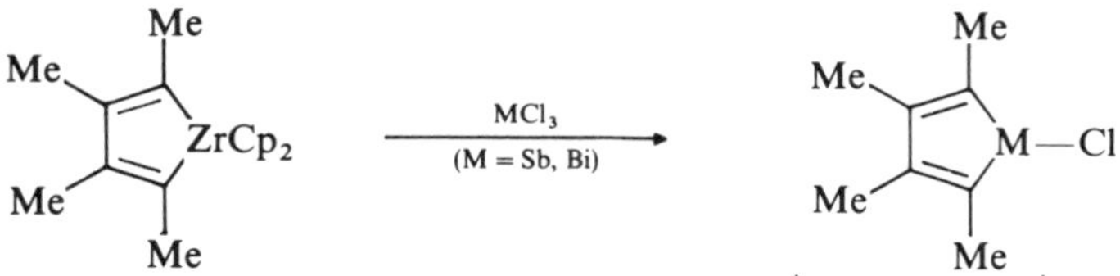

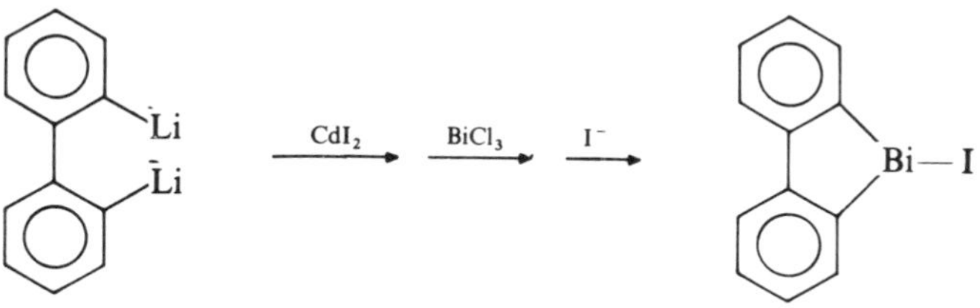

(28)

$$(39)$$

63%

$$PhSiF_3 \xrightarrow{\ SbF_3\ } Ph_2SbF \qquad (40)$$

68%

3. Thermolysis of triorganometal(V) dihalides

The method has been useful for the preparation of alkyl derivatives of antimony which are difficult to prepare by the redistribution reaction as described above. This method has also been useful for the preparation of bismuth derivatives. The ease of thermolysis of R_3MX_2 decreases in the order $X = I > Br > Cl > F$, $R = alkyl > aryl$ and $M = Bi > Sb$. Me_2SbI^{69}, Et_2SbBr^{38d} and 29^{37} are shown as examples (equations 41^{69}, 42^{38d} and 43^{37}).

$$Me_3SbI_2 \xrightarrow[\text{70 mm Hg}]{140\,°C} Me_2SbI \tag{41}$$

$$75\%$$

$$Et_3SbBr_2 \xrightarrow[\text{400 mm Hg}]{220\text{-}240\,°C} Et_2SbBr \tag{42}$$

$$79\%$$

$$(29)\quad 100\%$$

4. Halogen exchange from chlorides or bromides to iodides

Diorganoantimony and -bismuth(III) iodides have been prepared by the reaction of the corresponding chlorides or bromides with sodium or potassium iodide (equation $44)^{70}$.

$$R_2MX \xrightarrow[\text{acetone}]{\text{NaI}} R_2MI \tag{44}$$

$$\begin{pmatrix} M = Sb, Bi \\ X = Cl, Br \end{pmatrix}$$

5. Reduction of diorganometal(V) trihalides and stibinic acids

The method has been used for the preparation of antimony derivatives (equation $45)^{71}$, However, the method has not been used frequently in recent years.

$$95\%$$

6. Protonolysis of metal–carbon bonds.

Protonolysis of trioganoantimony and -bismuth compounds by mineral acids gives diorganoantimony and -bismuth(III) halides under controlled conditions (equations 46[30] and 47[72]).

$$(46)$$

$$Ph_3Sb \xrightarrow[\text{MeOH/reflux}]{HCl} Ph_2Sb\,Cl \qquad (47)$$
$$50\%$$

7. Halogenolysis of the Sb—Sb bond

Distibines react readily with SO_2Cl_2, bromine or iodine to form R_2SbX ($X = Cl$, Br, I) (equation 48)[73].

$$R_2SbSbR_2 \xrightarrow[\text{(X = Br, I)}]{SO_2Cl_2 \text{ or } X_2} 2\,R_2SbX \qquad (48)$$

$$\left(\begin{array}{l} R = Me, X = Br, 90\% \\ R = Me, X = I, 97\% \end{array} \right)$$

C. Preparation of Monoorganoantimony(III) and -bismuth(III) Dihalides

1. Redistribution reactions

This has been the most frequently used method for the preparation of mono-organoantimony and -bismuth(III) chlorides and bromides. They have been conveniently prepared by the reaction of a 1:2 mixture of the corresponding triorgano compounds and trihalides (equation 49) (cf Section II.B)[58,59]. The experimental procedure for the preparation of $PhSbBr_2$[74] has been described in detail. Sterically bulky $(2,4,6\text{-}Me_3C_6H_2)_{3-n}SbCl_n$ ($n = 1,2$) has been prepared by this method although higher temperature (80–100 °C) is necessary[75]. Alkylantimony dihalides are difficult to prepare from the redistribution reaction between R_3Sb and SbX_3 ($X = Cl$, Br) as described above (equation 32). However, $RSbX_2$ compounds have been obtained from the redistribution reaction between R_2SbX and SbX_3 ($X = Cl$, Br) (equation 50)[60].

$$R_3M + 2\,MX_3 \longrightarrow 3\,RMX_2 \qquad (49)$$

$$\left(\begin{array}{l} M = Sb, R = aryl \\ M = Bi, alkyl, aryl \end{array} \right) \quad (X = Cl, Br)$$

$$R_2SbX + SbX_3 \longrightarrow 2\,RSbX_2 \qquad (50)$$

$$\left(\begin{array}{l} R = Me, Et, Pr, Bu \\ X = Cl, Br \end{array} \right) \left(\begin{array}{l} R = Me, X = Cl, 85\% \\ R = Me, X = Br, 90\% \\ R = Et, X = Br, 90\% \end{array} \right)$$

2. Transmetallation

Although a successful preparation of $MeSbI_2$ from the reaction of MeMgI with $SbCl_3^{74}$ has been reported, generally only weak nucleophiles or sterically bulky nucleophiles can produce RMX_2 (M = Sb, Bi; X = Cl, Br) selectively under controlled conditions. For example, $(C_5H_5)Na^{77}$, $t\text{-}BuMgBr^{78}$, Me_4Sn^{79}, R_4Pb (R = Me[76] and Et[80]) have been used to prepare the corresponding halides (equations 51[77], 52[78], 53[79] and 54[76,80]).

$$C_5H_5Na\,(CpNa) \xrightarrow[-30\,°C]{BiCl_3} \underset{77\%}{CpBiCl_2} \tag{51}$$

$$t\text{-}BuMgCl \xrightarrow[-50\,°C]{SbCl_3} \underset{64\%}{t\text{-}BuSbCl_2} \tag{52}$$

$$Me_4Sn \xrightarrow[(X\,=\,Cl,\,Br)]{SbX_3} MeSbX_2 \quad \begin{pmatrix} X = Cl, 70\% \\ X = Br, 55\% \end{pmatrix} \tag{53}$$

$$R_4Pb \xrightarrow[PhH]{SbCl_3} \underset{71\%}{RSbCl_2} \tag{54}$$

3. Thermolysis of diorganometal(V) trihalides

Since trihalides are thermolized under milder conditions than dihalides, several alkylstibines such as $MeSbCl_2$[81] and $EtSbBr_2$[60] have been obtained in good yields (equation 55). Bismuth(V) trihalides are generally unstable even at room temperature, thus $PhBiBr_2$ has been obtained from the reaction of Ph_2BiBr and Br_2 in good yields at room temperature[82].

$$R_2SbX_3 \longrightarrow RSbX_2 + RX \tag{55}$$
$$\begin{pmatrix} R = Me, X = Cl \\ 110–115\,°C \end{pmatrix} \begin{pmatrix} R = Me, X = Cl, 83\% \\ R = Et, X = Br, 90\% \end{pmatrix}$$

4. Reduction of stibonic acids

Stibonic acids, which can be prepared from diazonium salts and $SbCl_3$, have been reduced with SO_2/HCl or $SnCl_2/HCl$ to give monoorganoantimony(III) dichlorides. This method has been especially useful for the preparation of antimony derivatives bearing polar substituents which are incompatible with organometallic reagents such as Grignard reagents. For example, $p\text{-}EtO_2CC_6H_4SbCl_2$[83] and 30[84] (equations 56[83] and 57[84]) have been prepared by this method.

$$EtO_2C\!-\!\!\!\left\langle\bigcirc\right\rangle\!\!-\!\!\overset{\overset{\displaystyle OH}{|}}{\underset{\overset{\|}{O}}{Sb}}\!-\!OH \xrightarrow[EtOH]{SnCl_2/HCl} EtO_2C\!-\!\!\!\left\langle\bigcirc\right\rangle\!\!-\!SbCl_2 \tag{56}$$

$$\tag{57}$$

(30) 78%

5. Miscellaneous methods

Monoorganoantimony dialkoxides react with acetyl chloride to form **31** in good yield (equation 58[85]).

$$(EtO)_2Sb\!-\!\bigcirc\!-\!Sb(OEt)_2 \quad \xrightarrow{\text{MeC(O)Cl}} \quad Cl_2Sb\!-\!\bigcirc\!-\!SbCl_2 \quad (58)$$

$$\textbf{(31)} \quad 83\%$$

The mixed triorganoantimony derivative **32** reacts with HCl to afford **33** with selective cleavage of the Sb—Ph bonds (equation 59[86]).

$$\underset{\textbf{(32)}}{MeC(CH_2SbPh_2)_3} \quad \xrightarrow[CH_2Cl_2]{HCl} \quad \underset{\textbf{(33)}}{MeC(CH_2SbCl_2)_3} \tag{59}$$

The cleavage of a silacyclobutane by SbCl$_3$ affords **34** (equation 60[87]).

$$\boxed{}\!-\!SiMe_2 \quad \xrightarrow{SbCl_3} \quad \underset{\underset{Cl}{|}}{Me_2Si(CH_2)_3SbCl_2} \tag{60}$$

$$\textbf{(34)} \qquad 60\%$$

D. Preparation of Organoantimony(III) and -bismuth(III) Hydrides

Since these hydrides, especially bismuth hydrides, are thermally unstable, the preparation should be carried out at a low temperature. Among bismuth hydrides only MeBiH$_2$[88] and Me$_2$BiH[88] are actually known.

1. Reduction of halides

Organoantimony and -bismuth(III) halides are reduced with hydride reagents at low temperatures (equations 61[88], 62[88], 63[89] and 64[90]).

$$MeBiCl_2 \quad \xrightarrow[-110\,°C]{LiAlH_4} \quad MeBiH_2 \tag{61}$$

$$Me_2BiBr \quad \xrightarrow[-60\,°C]{LiAlH_4} \quad Me_2BiH \tag{62}$$

$$Me_2SbBr \quad \xrightarrow[-40\,°C]{LiBH(OMe)_3} \quad \underset{35\%}{Me_2SbH} \tag{63}$$

$$PhSbCl_2 \quad \xrightarrow{LiBH_4} \quad \underset{96\%}{PhSbH_2} \tag{64}$$

2. Hydrolysis of Sb—Si compounds

Recently, PhSbH$_2$[91] and Ph$_2$SbH[91] have been prepared from the corresponding silylated compounds by hydrolysis or methanolysis (equations 65 and 66).

$$PhSb(SiMe_3)_2 \quad \xrightarrow[-50\,°C]{MeOH} \quad \underset{70\%}{PhSbH_2} \tag{65}$$

$$Ph_2SbSiMe_3 \xrightarrow[\text{rt}]{\text{MeOH}} Ph_2SbH \qquad (66)$$
$$\phantom{Ph_2SbSiMe_3 \xrightarrow[\text{rt}]{\text{MeOH}}} 70\%$$

3. Protonation of the Sb anion

A stable hydride Mes_2SbH (Mes = mesityl) **35** has been obtained by protonation of **36** with $(Me_3NH)^+Cl^-$ (equation 67)[92].

$$(67)$$

E. Preparation of Organoantimony(III) and -bismuth(III) Nitrogen Compounds

1. Nucleophilic displacement of halides with nitrogen nucleophiles

Amide anions replace the halides of the M—halogen bond to form the corresponding M—NR_2 bond. For example, Me_2SbNMe_2[93], $Me_2BiNMeSiMe_3$[81], $p\text{-}BrC_6H_4Sb(NR_2)_2$[85] are obtained as shown in equations 68[93], 69[81] and 70[85].

$$Me_2SbCl \xrightarrow{Me_2NLi} Me_2SbNMe_2 \qquad (68)$$
$$\phantom{Me_2SbCl \xrightarrow{Me_2NLi}} 46\%$$

$$Me_2BiBr \xrightarrow{Me_3Si(Me)NLi} Me_2BiN(Me)SiMe_3 \qquad (69)$$
$$\phantom{Me_2BiBr \xrightarrow{Me_3Si(Me)NLi}} 38\%$$

$$Br\text{—}\langle\bigcirc\rangle\text{—}SbCl_2 \xrightarrow{Pr^i_2NLi} Br\text{—}\langle\bigcirc\rangle\text{—}Sb(NPr^i_2)_2 \qquad (70)$$
$$42\%$$

Azido and amide anions also replace halides to form the corresponding derivatives (equations 71[94] and 72[95]).

$$Ph_2MCl \xrightarrow[\text{pyridine}]{NaN_3} Ph_2MN_3 \qquad (71)$$

$$(M = Sb, 94\%)$$

$$(72)$$

$$(85\%)$$

2. Transmetallation

Some amide chlorides such as **37**[85], **38**[96] can be transformed into amino derivatives by organolithium reagents (equations 73[85] and 74[96]).

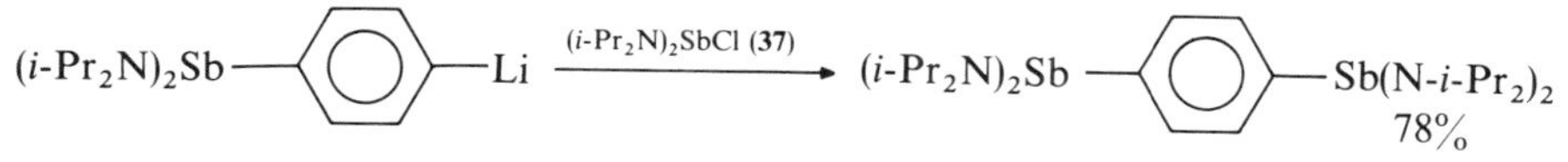

$$(i\text{-}Pr_2N)_2Sb\text{—}\langle\bigcirc\rangle\text{—}Li \xrightarrow{(i\text{-}Pr_2N)_2SbCl\ (37)} (i\text{-}Pr_2N)_2Sb\text{—}\langle\bigcirc\rangle\text{—}Sb(N\text{-}i\text{-}Pr_2)_2$$
$$78\%$$

(73)

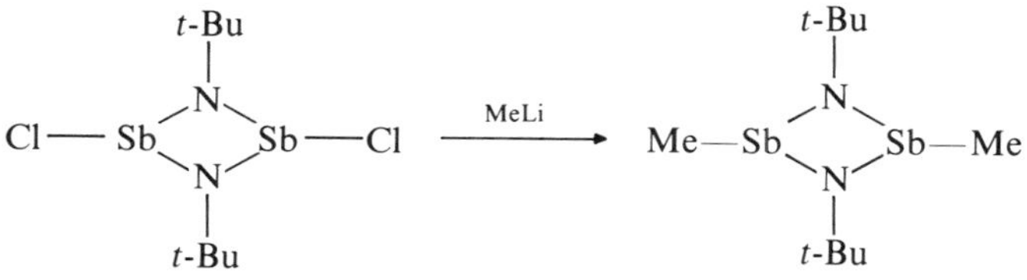

(74)

(38)   28%

3. Miscellaneous methods

Triorganobismuth compounds give azido derivatives such as **39**[97] by reaction with HN_3 (equations 75[97]).

$$Me_3Bi \xrightarrow{HN_3} Me_2BiN_3$$
$$\textbf{(39)}\quad 75\%$$

(75)

Reaction of $(Ph_2Sb)_2O$ wih HCN gives Ph_2SbCN (equation 76[98]).

$$(Ph_2Sb)_2O \xrightarrow{HCN} Ph_2SbCN$$
$$40\%$$

(76)

Me_2MNMe_2 (M = Sb, Bi) gives Me_2MN_3 after treatment with HN_3, and the same paper describes the reaction of Me_2SbBr with AgN_3 (equations 77[99] and 78[99]).

$$Me_2MNR_2 \xrightarrow{HN_3} Me_2MN_3$$

(77)

$$\left(\begin{array}{l} M = Sb, R = Me \\ M = Bi, R = Me_3Si \end{array}\right) \quad (M = Bi, 86\%)$$

$$Me_2SbBr \xrightarrow{AgN_3} Me_2SbN_3$$

(78)

Arduengo and coworkers described formation of the diorganoantimony amide **41**, by the reaction of **40** with an activated acetylene (equation 79[100]).

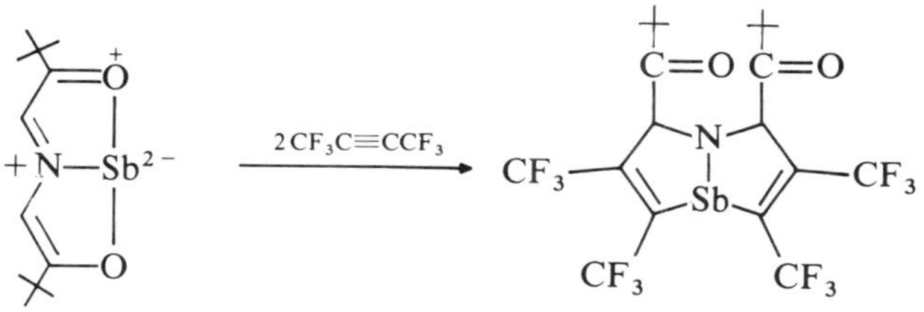

(79)

(40)   **(41)**   71%

F. Preparation of Organoantimony(III) and -bismuth(III) Oxygen Compounds

1. Nucleophilic displacement of halides with oxygen nucleophiles

Alkoxide anions replace the halides of the M—halogen bond to form the corresponding M—OR bond. For example, compounds $Me_2MOSiMe_3$[101] (M = Sb, Bi) are obtained as shown in equation 80[101].

$$Me_2MBr \xrightarrow{\text{Me}_3\text{SiONa}} Me_2MOSiMe_3 \tag{80}$$

$$(M = Sb, Bi) \qquad \begin{pmatrix} M = Sb, 79\% \\ M = Bi, 10\% \end{pmatrix}$$

$ArMCl_2$ (M = Sb, Bi) react with **42** to afford **43** (equation 81[102]).

$$(81)$$

$$\begin{pmatrix} M = Sb, Ar = p\text{-Tol}, 40\% \\ M = Bi, Ar = p\text{-Tol}, 35\% \end{pmatrix}$$

2. Protonation of the Bi—C bond

Phenols and carboxylic acids cleave Bi—C bonds under controlled conditions to afford the corresponding compounds **44** and **45** (equations 82[103] and 83[104]).

$$Et_3Bi \xrightarrow{\text{ArOH}} Et_2BiOAr \tag{82}$$
$$\textbf{(44)}$$

$$Ph_3Bi \xrightarrow{\text{PhCO}_2\text{H}} PhBi(OCOPh)_2 \tag{83}$$
$$\textbf{(45)} \quad 44\%$$

3. Miscellaneous methods

Reaction of $(p\text{-Tol}_2Sb)_2O$ with AcOH gives $p\text{-Tol}_2SbOAc$ (equation 84[105]).

$$(84)$$

$t\text{-Bu}_2SbNMe_2$ gives the antimony enolate **46** on treatment with cyclohexanone (equation 85[106]).

Diols can react with acyclic dialkoxy compounds to effect an exchange reaction (equation 86[107]).

Reaction of R_2SbCl with aq. Na_2CO_3 or NaOH gives $(R_2Sb)_2O$. $(R_2M)_2O$ can also be obtained by controlled oxidation of $(R_2M)_2$ with O_2 (equations 87[108] and 88[109]).

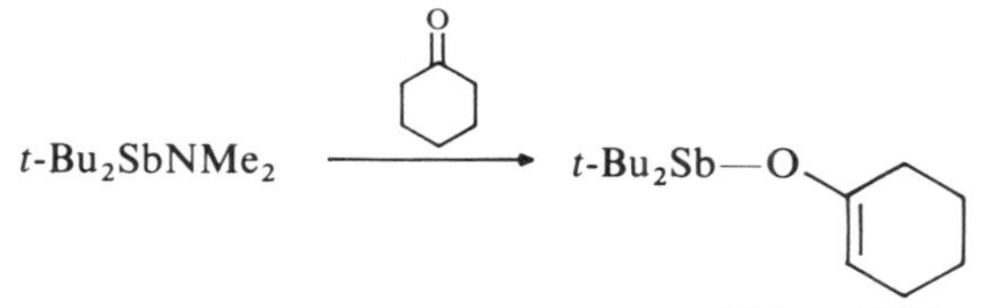

$$t\text{-Bu}_2\text{SbNMe}_2 \quad \xrightarrow{\quad} \quad t\text{-Bu}_2\text{Sb}-\text{O} \qquad (85)$$

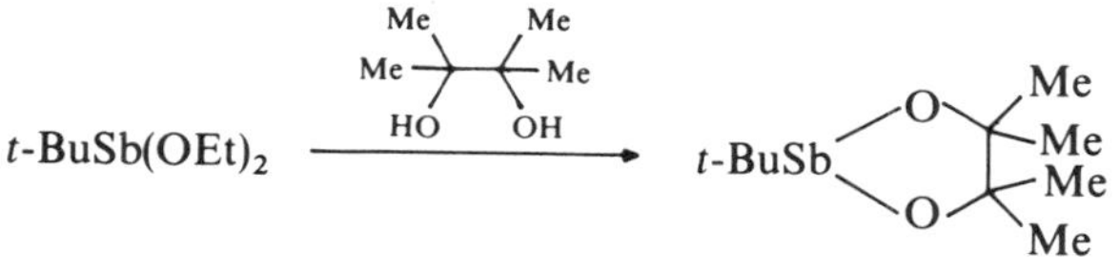

(**46**)   79%

$$t\text{-BuSb(OEt)}_2 \quad \xrightarrow{\quad \text{HO} \quad \text{OH} \quad} \quad t\text{-BuSb} \qquad (86)$$

71%

$$t\text{-Bu}_2\text{SbCl} \quad \xrightarrow{\text{aq NaOH}} \quad (t\text{-Bu}_2\text{Sb})_2\text{O} \qquad (87)$$

$$(\text{R}_2\text{M})_2 \quad \xrightarrow{1/2\,\text{O}_2} \quad (\text{R}_2\text{M})_2\text{O} \qquad (88)$$

$$\left(\begin{array}{l} \text{R} = \text{Me, Et, } i\text{-Pr} \\ \text{M} = \text{Sb, Bi} \end{array} \right) \quad (\text{M} = \text{Bi, R} = \text{Me, 95\%})$$

Reaction of PhSbCl_2 with aq. NaOH gives PhSbO, which is a polymeric oxygen-bridged substance (equation 89[110]).

$$\text{PhSbCl}_2 \quad \xrightarrow{\text{aq NaOH}} \quad \text{PhSbO} \qquad (89)$$

$$\sim 90\%$$

Reaction of $\text{Ph}_2\text{SbSiMe}_3$ with O_2 gives $\text{Ph}_2\text{SbOSiMe}_3$ under controlled conditions (equations 90[111]).

$$\text{Ph}_2\text{SbSiMe}_3 \quad \xrightarrow{1/2\,\text{O}_2} \quad \text{Ph}_2\text{SbOSiMe}_3 \qquad (90)$$

$$65\%$$

G. Preparation of Organoantimony(III) and -bismuth(III) Compounds Containing Phosphorus or Arsenic

1. Reaction of organoantimony anionic species with phosphorus and arsenic halides

Diphenylantimony lithium has been reported to react with Ph_2ECl (E = P, As) to afford $\text{Ph}_2\text{SbEPh}_2$, which undergoes redistribution above $80\,^\circ\text{C}$ (equation 91[112]).

$$\text{Ph}_2\text{Sb}^-\text{Li}^+ \quad \xrightarrow[\text{(E = P, As)}]{\text{Ph}_2\text{ECl}} \quad \text{Ph}_2\text{Sb}-\text{EPh}_2 \qquad (91)$$

$$\left(\begin{array}{l} \text{E} = \text{P, 69\%} \\ \text{E} = \text{As, 50\%} \end{array} \right)$$

2. Reaction of organoantimony halides with phosphorus anions

The three-membered P_2Sb heterocycle **47** has been prepared by this method as shown in equation 92[113].

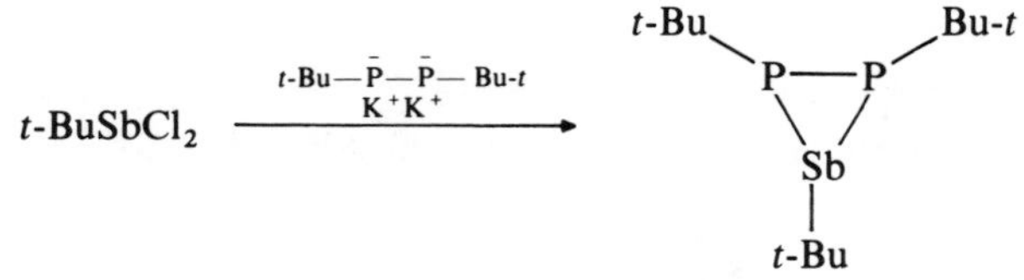

$$t\text{-BuSbCl}_2 \xrightarrow[\text{K}^+\text{K}^+]{t\text{-Bu}-\overset{-}{\text{P}}-\overset{-}{\text{P}}-\text{Bu-}t}$$

(92)

(47)   53%

3. Redistribution reaction

Distibines and dibismuthines (R_2MMR_2: M = Sb, Bi) react with diarsines ($R'_2AsAsR'_2$) to give equilibrium mixtures of $R_2MAsR'_2$ (equation 93[114]).

$$R_2MMR_2 + R'_2AsAsR'_2 \rightleftharpoons 2R_2MAsR'_2 \tag{93}$$

(M = Sb, Bi)
$$\left(\begin{array}{l} R = R' = Me, M = Sb, K = 0.9 \\ R = R' = Me, M = Bi, K = 0.09 \end{array} \right)$$

H. Preparation of Organoantimony(III) and -bismuth(III) Compounds Containing Sulfur, Selenium or Tellurium

1. Nucleophilic displacement with sulfur and selenium nucleophiles

Thiols or selenols react with alkoxy compounds, halides and acetates to give $R_nM(ER')_{3-n}$ (E = S, Se). Several examples are shown in equations 94[115a], 95[104], 96[116] and 97[117]

$$MeM(OEt)_2 \xrightarrow{\text{EtSH}} MeM(SEt)_2 \tag{94}$$

(M = Sb, Bi)
$$\left(\begin{array}{l} M = Sb, 71\% \\ M = Bi, 67\% \end{array} \right)$$

$$Ph_2BiCl \xrightarrow{\text{PhSH}} \underset{80\%}{Ph_2BiSPh} \tag{95}$$

$$Ph_2SbOAc \xrightarrow{\text{ArSNa}} \underset{\sim 80\%}{Ph_2SbSAr} \tag{96}$$

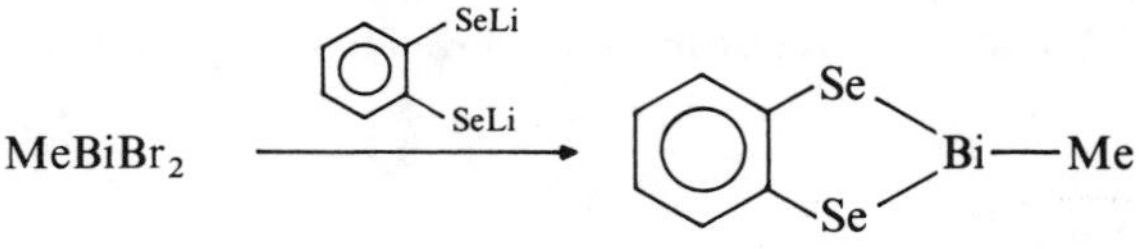

62%

(97)

The thiocynate anion also replaces halides to form the corresponding derivatives (equation 98[118]).

$$Ph_2SbCl \xrightarrow{\text{KSCN}} Ph_2SbSCN \tag{98}$$

2. Protonolysis with RSH

Ph$_3$Sb reacts with PhSH to form Ph$_2$SbSPh (equation 99[119]).

$$Ph_3Sb \xrightarrow{\text{PhSH}} \underset{100\%}{Ph_2SbSPh} \tag{99}$$

Triorganobismuth compounds give thiocyanate derivatives **48** by reaction with HSCN (equation 100[120]).

$$Ph_3Bi \xrightarrow{\text{HSCN}} \underset{(\textbf{48})\quad 33\%}{Ph_2BiSCN} \tag{100}$$

3. Transmetallation

The zirconium heterocycle **49** reacts with PhSbCl$_2$ to form **50** (equation 101[121]).

$$\tag{101}$$

(R = H, 86%)

4. Miscellaneous methods

Recently, alkyl radical substitution of the SPh group in (PhS)$_3$M (M = Sb, Bi) to afford RM(SPh)$_2$ has been reported (equation 102[122]).

$$(R = (PhCH_2)_2CH, \ M = Bi, 80\%)$$

$$\tag{102}$$

Distibines and dibismuthines (R$_2$MMR$_2$: M = Sb, Bi) react with disulfides, diselenides and ditellurides (R'EER': E = S, Se, Te) to form R$_2$MER' (equation 103[123]). The experimental procedure for the preparation of Me$_2$SbTe(Tol-p)[124] has been described in detail.

$$R_2MMR_2 + R'EER' \longrightarrow R_2MER' \tag{103}$$

(M = Sb, Bi) (E = S, Se, Te)
$$\begin{pmatrix} M = Bi, R = Me, E = S, R' = Ph, 100\%^{123a} \\ M = Bi, R = p\text{-Tol}, E = Se, R' = Ph, 92\%^{123b} \\ M = Bi, R = n\text{-Pr}, E = Te, R' = p\text{-Tol}, 100\%^{123c} \\ M = Sb, R = Et, E = S, R' = Ph, 100\%^{123d} \\ M = Sb, R = Me, E = Se, R' = Me, 74\%^{123e} \end{pmatrix}$$

Sulfur, selenium and tellurium atoms insert into the M—M (M = Sb, Bi) bond to form R$_2$MEMR$_2$ (E = S, Se, Te) (equation 104 [123b,123d,123e]).

$$R_2MMR_2 \xrightarrow[\text{(E = S, Se, Te)}]{E} R_2M\text{---}E\text{---}MR_2 \tag{104}$$

$$\text{(M = Sb, Bi)} \qquad \begin{pmatrix} M = Bi, R = p\text{-Tol}, E = S, 82\%^{123b} \\ M = Bi, R = p\text{-Tol}, E = Se, 74\%^{123b} \end{pmatrix}$$

Sulfur atom inserts into the Sb—Si bond in $Ph_2SbSiMe_3$ to form $Ph_2SbSSiMe_3$ (equation 105[91]).

$$Ph_2SbSiMe_3 \xrightarrow{1/8\,S_8} Ph_2SbSSiMe_3 \tag{105}$$
$$48\%$$

$RSb(ER')_2$ (E = S, Se) have been reported to form by the redistribution reaction of R_2SbER' (equation 106[125]).

$$2\,R_2SbER' \longrightarrow RSb(ER')_2 + R_3Sb \tag{106}$$

$$\text{(E = S, Se)} \qquad \begin{pmatrix} E = S, R = Me, R' = Ph, 84\% \\ E = Se, R = R' = Me, 74\% \end{pmatrix}$$

I. Preparation of Organoantimony(III) and -bismuth(III) Compounds containing Lithium, Sodium, Potassium or Magnesium

1. Reduction with metal

Triorganoantimony and -bismuth compounds, distibines, diorganoantimony and -bismuth halides are reduced to afford diorganoantimony alkali metal compounds as shown in equations 107[126], 108[127], 109[50], 110[124], 111[128], 112[129], 113[130] and 114[49]. The selectivity in the Sb—C cleavage of unsymmetrical triarylstibines is complicated[130b]. The reactions shown in equations 109, 110 and 111 suggest that the cleavage of the isopropyl or benzyl group is preferred to those of the phenyl and methyl groups, and the phenyl group in $PhMe_2Sb$ is reluctant to be cleaved.

Antimony halides R_nSbX_{3-n} (n = 1, 2) react with magnesium to form Grignard-type reagents $R_nSb(MgX)_{3-n}$ (n = 1, 2) (equations 115[131] and 116[131]), but the yield is usually low because the method also gives Sb—Sb compounds (*vide infra*).

$$Ph_3Sb \xrightarrow[\text{THF}]{Li} Ph_2Sb^-Li^+ \tag{107}$$

$$n\text{-Bu}_2SbBr \xrightarrow[\text{THF}]{Li} n\text{-Bu}_2Sb^-Li^+ \tag{108}$$

$$Ph_2MeSb \xrightarrow[\text{liq NH}_3]{Na} Ph(Me)Sb^-Na^+ \tag{109}$$

$$PhMe_2Sb \xrightarrow[\text{liq NH}_3]{Na} Ph(Me)Sb^-Na^+ \tag{110}$$

$$Ph_2(R)Sb \xrightarrow[\text{liq NH}_3]{Na} Ph_2Sb^-Na^+ \tag{111}$$

$$\text{(R = } i\text{-Pr, PhCH}_2\text{)}$$

$$\text{(112)}$$

structures with M—Ph → M⁻Li⁺ (M = Sb, Bi), reacting with Li in THF

$$Ph_2BiI \xrightarrow[\text{liq NH}_3]{\text{Na}} Ph_2Bi^-Na^+ \tag{113}$$

$$Ph_3Bi \xrightarrow[\text{THF}]{\text{K}} Ph_2Bi^-K^+ \tag{114}$$

$$t\text{-}Bu_2SbBr \xrightarrow[\text{THF}]{\text{Mg}} t\text{-}Bu_2SbMgBr \tag{115}$$

$$t\text{-}BuSbBr_2 \xrightarrow[\text{THF}]{\text{Mg}} t\text{-}BuSb(MgBr)_2 \tag{116}$$

2. From Ph₂SbH

Organolithium reagents react with Ph_2SbH to give Ph_2SbLi (equation 117)[132]. The lithium salt has been structurally characterized by X-ray crystallography as the ion-pair $[Li(12\text{-crown-}4)_2][SbPh_2]\cdot1/3\,THF$[132b].

$$Ph_2SbH \xrightarrow{\text{RLi}} Ph_2Sb^-Li^+ \tag{117}$$

3. Electroreduction

Electroreduction of $(Ph_2Sb)_2O$[133], $(Ph_2Sb)_2$[133,134], Ph_2SbR (R = Ph, Bu)[133] in THF gives Ph_2Sb^- (equation 118).

$$Ph_2SbR \xrightarrow{\text{electroreduction}} Ph_2Sb^- \tag{118}$$
$$(R = OSbPh_2, SbPh_2, Ph, n\text{-}Bu)$$

J. Preparation of Compounds Containing Antimony(III)–Antimony(III) and Bismuth(III)–Bismuth(III) Bond

1. Reaction of diorganoantimony and -bismuth anionic species with halides

Reaction of diorganoantimony and -bismuth halides with Li, Na, Mg or Zn gives tetraorganoantimony–antimony or -bismuth–bismuth compounds as shown in equations 119[135,136], 120[137], 121[123b] and 122[65]. These can be considered as reactions between diorganoantimony and -bismuth species formed upon reduction of the halides.

$$Me_2SbBr \xrightarrow{\text{M}} Me_2SbSbMe_2 \tag{119}$$
$$\begin{pmatrix} M = Li, 55\%^{135} \\ M = Na, 75\%^{135} \\ M = Mg, 85\%^{136a} \end{pmatrix}$$

$$(CF_3)_2SbI \xrightarrow{\text{Zn}} (CF_3)_2SbSb(CF_3)_2 \tag{120}$$
$$\sim 100\%$$

$$\left(\text{Me}-\text{C}_6\text{H}_4-\right)_2\text{BiBr} \xrightarrow{\text{Na}} \left(\text{Me}-\text{C}_6\text{H}_4-\right)_2\text{Bi}-\text{Bi}\left(-\text{C}_6\text{H}_4-\text{Me}\right)_2$$

$$45\% \qquad\qquad (121)$$

$$\text{(Bi-Cl)} \xrightarrow{\text{Na}} \text{(Bi-Bi)} \qquad (122)$$

$$83\%$$

Reaction of diorganoantimony and -bismuth anionic species with 1,2-dibromoethane or 1,2-dichloroethane is a variation of this method as shown in equations 123[132], 124[138] and 125[73a].

$$\text{Ph}_2\text{Sb}^-\text{Li}^+ \xrightarrow{\text{XCH}_2\text{CH}_2\text{X}} \text{Ph}_2\text{SbSbPh}_2 \qquad (123)$$

$$\begin{pmatrix} X = \text{Cl}, 79\% \\ X = \text{Br}, 82\% \end{pmatrix}$$

$$\text{Et}_2\text{Bi}^-\text{Na}^+ \xrightarrow{\text{BrCH}_2\text{CH}_2\text{Br}} \text{Et}_2\text{BiBiEt}_2 \qquad (124)$$

$$63\%$$

$$\text{(Sb}^-\text{)} \quad \text{Na}^+ \xrightarrow{\text{ClCH}_2\text{CH}_2\text{Cl}} \text{(Sb}-\text{Sb)} \qquad (125)$$

$$80\%$$

Reaction of monoorganoantimony halides with metals gives compounds containing antimony–antimony bonds as shown in equation 126[139] and 127[86].

$$t\text{-BuSbCl}_2 \xrightarrow{\text{Mg}} \tfrac{1}{4}(t\text{-BuSb})_4 \qquad (126)$$

$$20\%$$

$$\text{MeC(CH}_2\text{SbCl}_2)_3 \xrightarrow[\text{THF}]{\text{Na}} \text{Me}-\text{C}(\text{CH}_2\text{Sb})_3 \qquad (127)$$

$$61\%$$

2. Miscellaneous methods

A hexamer $(\text{PhSb})_6 \cdot \text{PhH}$ has been prepared by heating PhSbH_2 with styrene in benzene under an argon atmosphere (equation 128[140]).

$$\text{PhSbH}_2 \xrightarrow{\Delta} (\text{PhSb})_6 \cdot \text{PhH} \qquad (128)$$

$$97\%$$

Recently, another hexamer $(\text{PhSb})_6 \cdot (1,4\text{-dioxane})$ has been prepared by slow oxidation of $\text{PhSb(SiMe}_3)_2$ in 1,4-dioxane and the crystal structure has been reported (equation 129[141].

$$\text{PhSb(SiMe}_3)_2 \xrightarrow[\text{1,4-dioxane}]{O_2} \quad \underset{\text{16\%}}{\text{[structure]}} \cdot \text{1,4-dioxane} \quad (129)$$

Reduction of $(\text{PhSb})_6 \cdot \text{PhH}$ by sodium in liquid ammonia gives $(\text{PhSbNa})_2$, which produces **51** upon addition of 1,3-dichloropropane (equation 130[140]).

$$(\text{PhSb})_6 \cdot \text{PhH} \xrightarrow[\text{liz NH}_3]{\text{Na}} \left(\begin{array}{c} \text{Ph}-\bar{\text{S}}\text{b}-\bar{\text{S}}\text{b}-\text{Ph} \\ \text{Na}^+ \ \text{Na}^+ \end{array} \right) \xrightarrow{\text{Cl(CH}_2)_3\text{Cl}} \quad \underset{\textbf{(51)} \ 81\%}{\text{[structure]}}$$

$$(130)$$

Reduction of Ph_2BiI with Cp_2Co gives $\text{Ph}_2\text{BiBiPh}_2$ (equation 131[57]).

$$\text{Ph}_2\text{BiI} \xrightarrow{\text{Cp}_2\text{Co}} \underset{80\%}{\text{Ph}_2\text{BiBiPh}_2} + \text{Cp}_2\text{CoI} \quad (131)$$

The redistribution reaction between $\text{Me}_2\text{SbSbMe}_2$ and $\text{Me}_2\text{BiBiMe}_2$ has been reported to give $\text{Me}_2\text{SbBiMe}_2$ (equation 132[142]).

$$\text{Me}_2\text{SbSbMe}_2 + \text{Me}_2\text{BiBiMe}_2 \underset{C_6D_6}{\overset{K=1,2}{\rightleftharpoons}} 2\,\text{Me}_2\text{SbBiMe}_2 \quad (132)$$

The structure of $[\text{Ph}_4\text{Sb}_3][\text{Li}(12\text{-crown-4})_2]\text{THF}$, which has been formed as a byproduct in the reduction of Ph_3Sb with Li, has been characterized by X-ray structural analysis (equation 133[132b]).

$$\text{Ph}_3\text{Sb} \xrightarrow[\text{THF}]{\text{Li}} \xrightarrow{12\text{-crown-4}} \underset{9\%}{[\text{Ph}_2\text{Sb}-\text{Sb}-\text{SbPh}_2]^- \cdot \text{Li}(12\text{-crown-4})_2^+ \cdot \text{THF}} \quad (133)$$

K. Preparation of Compounds Containing Antimony(III)–Main Group Metal or Bismuth(III)–Main Group Metal Bonds

1. Reaction of organoantimony and -bismuth anionic species with metal halides

This method has been the most useful for the preparation of metal complexes. Several compounds containing main-group elements such as boron, gallium, indium, silicon, germanium, tin and lead as the metal are shown in equations 134[143], 135[144], 136[145], 137[91,146], 138[131], 139[147], 140[126] and 141[148].

$$\text{Et}_2\text{Sb}^- \text{Li}^+ \xrightarrow{(\text{Me}_2\text{N})_2\text{BCl}} \underset{60\%}{\text{Et}_2\text{SbB(NMe}_2)_2} \quad (134)$$

$$\text{R}_2\text{Sb}^- \text{Li}^+ \xrightarrow[(E = \text{Ga, In})]{\text{Me}_2\text{ECl}} \text{R}_2\text{SbEMe}_2 \quad (135)$$

$$t\text{-Bu}_2\text{SbSiMe}_3 \xrightarrow[(E = \text{Ga, In})]{\text{ECl}_3} (t\text{-Bu}_2\text{Sb})_n\text{ECl}_{3-n} \quad (136)$$

$$\left(\begin{array}{c} E = \text{Ga}, n = 1 \\ E = \text{In}, n = 2 \end{array} \right)$$

$$\text{Ph}_n\text{SbCl}_{3-n} \xrightarrow{\text{Me}_3\text{SiCl/Mg}} \text{Ph}_n\text{Sb(SiMe}_3)_{3-n} \tag{137}$$

$$\begin{pmatrix} n = 1, 58\% \\ n = 2, 63\% \end{pmatrix}$$

$$t\text{-Bu}_2\text{SbCl} \xrightarrow{\text{Me}_3\text{SiCl/Mg}} \underset{54\%}{t\text{-Bu}_2\text{SbSiMe}_3} \tag{138}$$

$$\text{PhCH}_2\text{Sb(SiMe}_3)_2 \xrightarrow{\text{Me}_3\text{SnCl}} \underset{72\%}{\text{PhCH}_2\text{Sb(SnMe}_3)_2} \tag{139}$$

$$\text{Ph}_2\text{Sb}^-\text{Li}^+ \xrightarrow{\text{Me}_2\text{ECl}_2} (\text{Ph}_2\text{Sb})_2\text{EMe}_2 \tag{140}$$

$$\begin{pmatrix} \text{E} = \text{Si}, 38\% \\ \text{E} = \text{Sn}, 33\% \\ \text{E} = \text{Pb}, 32\% \end{pmatrix}$$

$$\text{Ph}_2\text{Sb}^-\text{Na}^+ \xrightarrow{\text{Ph}_3\text{PbCl}} \text{Ph}_2\text{SbPbPh}_3 \tag{141}$$

2. Miscellaneous methods

Alkylation of $(\text{R}_3\text{E})_3\text{Sb}$ (R = Me, E = Sn[149] or R = Et, E = Ge[150]) with alkyl halides gives substituted metal complexes **52**[149] and **53**[150] (equations 142 and 143).

$$(\text{Me}_3\text{Sn})_3\text{Sb} \xrightarrow{t\text{-BuI}} \underset{\textbf{(52)} \quad 62\%}{t\text{-Bu}_2\text{SbSnMe}_3} \tag{142}$$

$$(\text{Et}_3\text{Ge})_3\text{Sb} \xrightarrow[100\,°\text{C}]{\text{PhCH}_2\text{Br}} \underset{\textbf{(53)} \quad 58\%}{\text{PhCH}_2\text{Sb(GeEt}_3)_2} \tag{143}$$

Et_3Bi reacts with $(\text{C}_6\text{F}_5)_3\text{GeH}$ to give $\text{Et}_2\text{BiGe}(\text{C}_6\text{F}_5)_3$ (equation 144[151]).

$$\text{Et}_3\text{Bi} \xrightarrow{(\text{C}_6\text{F}_5)_3\text{GeH}} \underset{75\%}{\text{Et}_2\text{BiGe}(\text{C}_6\text{F}_5)_3} \tag{144}$$

L. Preparation of Compounds with Antimony(III)–Transition Metal and Bismuth(III)-Transition Metal Bonds

1. Reaction of organoantimony and -bismuth anionic species with metal halides

This has been one of the useful methods for the preparation of transition metal complexes. Organoantimony and -bismuth anions have also been utilized for the preparation. Several transition metal bonded compounds with vanadium and copper as the transition metal are shown in equations 145[152] and 146[92].

$$\text{Ph}_2\text{Sb}^-\text{Li}^+ \xrightarrow{\text{Cp}_2\text{VCl}} \text{Ph}_2\text{SbVCp}_2 \tag{145}$$

$$\tag{146}$$

2. Reaction of organoantimony and -bismuth halides with metal anions

This has been the most frequently used method for the preparation of transition metal complexes with chromium, molybdenum, tungsten, maganese, iron, osmium, cobalt and nickel as the transition metal (equations 147[153], 148[154], 149[155], 150[156], 151[157], 152[156], 153[158], 154[159] and 155[160]).

$$Me_2SbBr \xrightarrow[\text{(M = Cr, Mo, W)}]{CpM(CO)_3Na} Me_2SbM(CO)_3Cp \qquad (147)$$
$$70\%$$

$$RMBr_2 \xrightarrow{CpM'(CO)_3Na} RM\{M'(CO)_3Cp\}_2 \qquad (148)$$

$$\left(\begin{array}{l} M = Sb, R = Me, M' = Cr, 49\% \\ M = Sb, R = Me, M' = Mo, W, 60\% \end{array} \right)$$

$$RSbCl_2 \xrightarrow[\text{(M = Mo, W)}]{Na_2M_2(CO)_{10}} RSb[M(CO)_5]_2 \qquad (149)$$

$$\left(\begin{array}{l} R = t\text{-Bu}, M = Cr, 30\% \\ R = t\text{-Bu}, M = W, 6\% \end{array} \right)$$

$$Ph_2BiCl \xrightarrow{Mn(CO)_5Na} Ph_2BiMn(CO)_5 \qquad (150)$$
$$33\%$$

$$Me_2SbBr \xrightarrow{CpFe(CO)_2Na} Me_2SbFeCp(CO)_2 \qquad (151)$$
$$52\%$$

$$Ph_2BiCl \xrightarrow{Fe(CO)_4Na_2} [Ph_2BiFe(CO)_4]^- \qquad (152)$$
$$50\%$$

$$Ph_2SbCl \xrightarrow{[Os_3(CO)_{11}H]^-} Ph_2SbOs_3(CO)_{10}H \qquad (153)$$

$$PhBiBr_2 \xrightarrow{Co(CO)_3PPh_3^-K^+} PhBi[Co(CO)_3PPh_3]_2 \qquad (154)$$
$$80\%$$

$$PhSbCl_2 \xrightarrow{[Ni_6(CO)_{12}]^{2-}} [Ni_{10}(SbPh)_2(CO)_{18}]^{2-} \qquad (155)$$
$$50 \sim 60\%$$

Monoorganoantimony dihalides are reductively coupled by interaction with $Na_2Cr_2(CO)_{10}$ and $Cp(CO)_2Mn$. THF to give **54** and **55** (equations 156[161] and 157[161]).

$$t\text{-BuSbCl}_2 \xrightarrow[\substack{Zn \\ \text{(M = Cr, W)}}]{M_2(CO)_{10}Na_2}$$

$$(54) \quad 11\%$$

$$RSbCl_2 \xrightarrow[\substack{Zn}]{Cp(CO)_2Mn\cdot THF}$$

$$(55) \quad (R = Me, 28\%)$$

3. Nucleophilic displacement of halides of transition metal bonded compounds by organic nucleophiles

Mixed halides such as **56** and **57** give organobismuth compounds bearing Bi–metal bonds upon treatment with nucleophiles (equations 158[159] and 159[162]).

$$[Cp(CO)_3Mo]_2BiCl \xrightarrow{Et_3BH^-K^+} EtBi[MoCp(CO)_3]_2 \qquad (158)$$
$$\textbf{(56)} \qquad\qquad\qquad\qquad 80\%$$

$$[Cp(CO)_2Fe]_2BiBr \xrightarrow{Et_3Al} EtBi[FeCp(CO)_2]_2 \qquad (159)$$
$$\textbf{(57)} \qquad\qquad\qquad\qquad 32\%$$

4. Miscellaneous methods

$Ph_2BiCo(CO)_3(PPh_3)$ has been prepared by the reaction of $Ph_2BiBiPh_2$ with $Co_2(CO)_8$ in the presence of PPh_3 (equation 160[163]).

$$Ph_2BiBiPh_2 \xrightarrow[2.\ PPh_3]{1.\ Co_2(CO)_8} Ph_2BiCo(CO)_3(PPh_3) \qquad (160)$$
$$65\%$$

Diphenylantimony lithium reacts with Cp_2ZrCl_2 to give **58** (equation 161[164]).

$$Ph_2Sb^-Li^+ \xrightarrow{Cp_2ZrCl_2} (Ph_2Sb)_2ZrCp_2 \qquad (161)$$
$$\textbf{(58)} \quad 62\%$$

M. Preparation of Tetracoordinate Organoantimony(III) and -bismuth(III) Anions

Organoantimony(III) and -bismuth(III) compounds containing electronegative atoms such as halogen, oxygen and so on show Lewis acidic character. Thus, tetracoordinate anions and even pentacoordinate dianions have been prepared from the corresponding halides or pseudohalides, and several of the compounds have recently been fully characterized by X-ray crystallography. Several examples are shown in equations 162[165], 163[166], 164[167], 165[58,168], 166[169], 167[170], 168[171] and 169[172].

$$PhBiBr_2 \longrightarrow PhBiBr_2 \cdot \qquad (162)$$

$$Ph_2BiX \xrightarrow{MX} (Ph_2BiX_2)^-M^+ \qquad (163)$$
$$\begin{pmatrix} X = Cl, Br, I, SCN \\ M = Me_4N, Ph_4As \end{pmatrix}$$

$$(Ph_2BiCl_2)^-Et_4N^+ \xrightarrow{2\ NaX} (Ph_2BiX_2)^-Et_4N^+ \qquad (164)$$
$$(X = CN, SCN, N_3)$$

$$PhBiX_2 \xrightarrow{R_4N^+X^-} (PhBiX_3)^-R_4N^+ \qquad (165)$$
$$(X = Br, I) \qquad \begin{pmatrix} X = Br, R = Bu \\ X = I, R = Et \end{pmatrix}$$

$$(PhSbCl_5)^- \overset{+}{N}H_4 \xrightarrow{HX} \xrightarrow{Na_2SO_3} \xrightarrow{HX} (PhSbX_3)^- M^+ \qquad (166)$$
$$\begin{pmatrix} X = Cl, Br, I \\ M = Me_4N, Ph_4As \end{pmatrix}$$

$$PhSbCl_2 \xrightarrow{MCl_2} (PhSbCl_3)^- M^+ + (PhSbCl_4)^{2-} 2M^+ \qquad (167)$$
$$(M^+ = Me_4N^+, \text{[ring]})$$

$$PhBi(OCOCF_3)_2 \xrightarrow{Ph_5Bi} [Ph_2Bi(OCOCF_3)_2]^- Ph_4Bi^+ \qquad (168)$$
$$86\%$$

$$(169)$$

$$\begin{pmatrix} M = Sb, 36\% \\ M = Bi, 64\% \end{pmatrix}$$

N. Preparation of Stibabenzenes, Bismabenzenes and Compounds with Double Bonds between Sb or Bi and Group 14 or 15 Elements

1. Stibabenzenes and bismabenzenes

The parent stibabenzene (**60**) and bismabenzene (**61**) have been prepared by dehydrohalogenation of **59a** and **59b**, respectively (equation 170[173]). While stibabenzene (**60**) has been isolated, it is a labile compound and rapidly polymerizes at room temperature. Bismabenzene (**61**) exists mainly as a dimer, but has been detected spectroscopically and via chemical trapping by hexafluorobutyne (equation 171[174]).

$$(170)$$

$$\begin{array}{cc} (\textbf{59a}) \ M = Sb & (\textbf{60}) \ M = Sb \\ (\textbf{59b}) \ M = Bi & (\textbf{61}) \ M = Bi \end{array}$$

$$(171)$$

$$(\textbf{60}) \ M = Sb$$
$$(\textbf{61}) \ M = Bi$$

4-Alkyl-substituted derivatives **62** and **63** have also been synthesized and have shown marked stability toward polymerization (equation 172[175]).

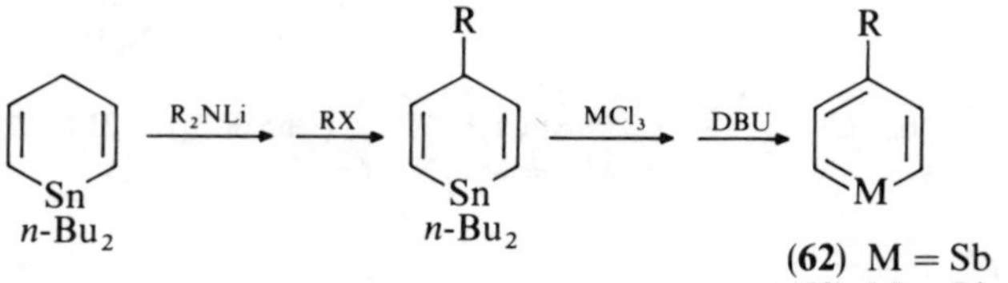

$$(172)$$

(62) M = Sb
(63) M = Bi

2. Compounds featuring double bonding between Sb or Bi and group 14, 15 elements

Since the first isolation and characterization of a diphosphene compound (**64**, equation 173[176]), compounds featuring double bonding between group 15 elements have attracted very much interest[177]. However, due to the instability of double bonding between the heavier group 15 elements only **65** containing a P=Sb double bond has been synthesized (equation 174[178]).

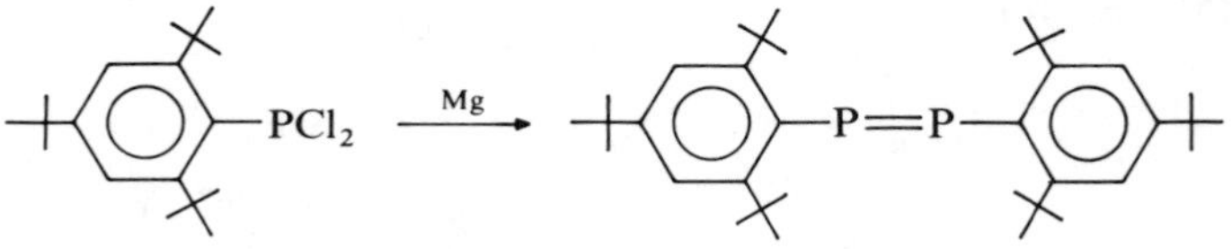

$$(173)$$

(64)   54%

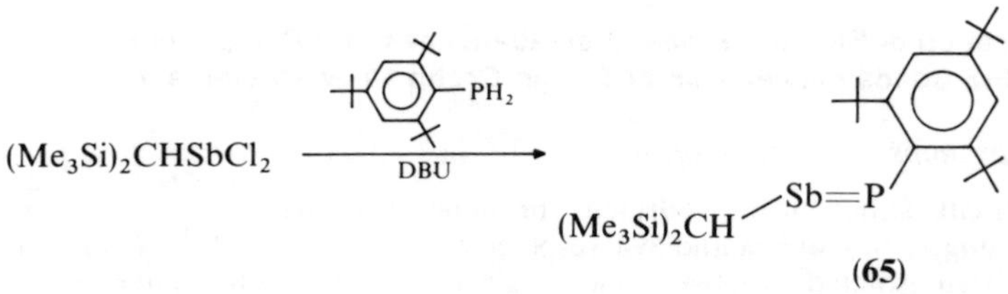

$$(174)$$

(65)

Attempts to prepare distibenes (RSb=SbR) and dibismuthenes (RBi=BiR) have been unsuccessful, although calculations have suggested the stability of these compounds[179]. However, several transition-metal-coordinated distibenes have been characterized (equations 175[180] and 176[181]).

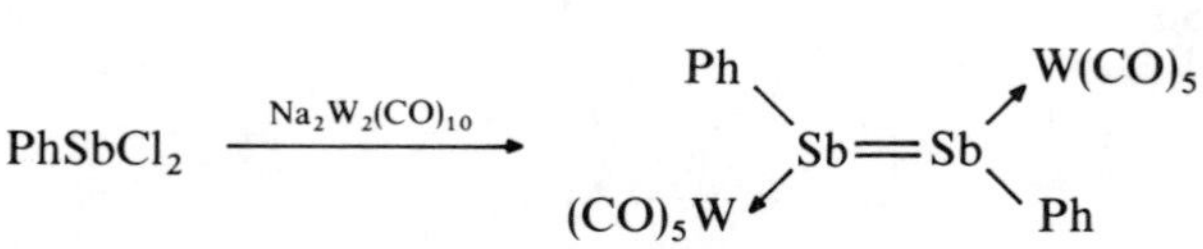

$$(175)$$

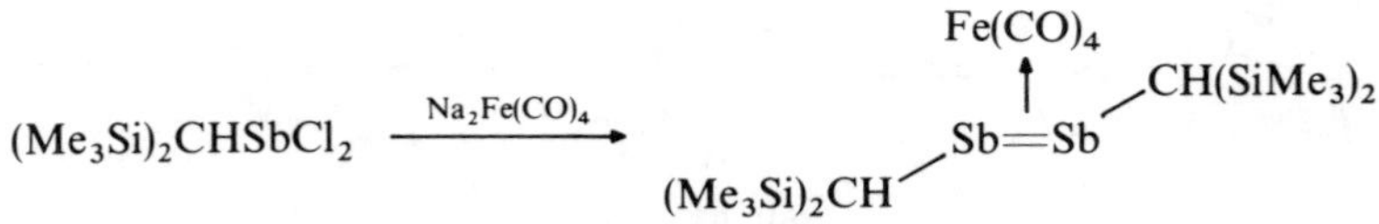

$$(176)$$

20%

III. PREPARATION OF ORGANOANTIMONY(V) AND -BISMUTH(V) COMPOUNDS

A. Preparation of Hexaorganoantimony(V) and -bismuth(V) Compounds

Only a few compounds containing six carbon–antimony or –bismuth bonds are known. The first compound of this type was obtained as colorless crystals by Wittig and Clauss from Ph_5Sb and $PhLi$ (equation 177[182]). The ate complex 66 reacts with water forming Ph_5Sb.

$$Ph_5Sb + PhLi \longrightarrow Ph_6Sb^- Li^+ \qquad (177)$$
$$(66) \quad 68\%$$

The corresponding bismuth ate complex 67, however, is only observable at low temperatures. Equilibration between 67 and $Ph_5Bi + PhLi$ has been considered to take place (equation 178[183]).

$$Ph_5Bi + PhLi \longrightarrow Ph_6Bi^- Li^+ \qquad (178)$$
$$(67)$$

A cyclic bismuth ate complex 68 has been prepared, but it is only observable at low temperatures (equation 179[184]).

$$(68)$$

B. Preparation of Pentaorganoantimony(V) and -bismuth(V) Compounds

Various alkyl and aryl substituted pentaorganoantimony compounds are known, however pentaorganobismuth compounds have been limited to aryl substituted compounds. Most of this type of compound has been synthesized by transmetallation from Grignard reagents or organolithium reagents to antimony and bismuth centers.

1. Transmetallation

Symmetrical pentaorganoantimony and -bismuth compounds have been synthesized from triorganoantimony and -bismuth compounds R_3MX_2 (M = Sb, Bi; X = F, Cl, Br, I, OAc, OR) and tetraorganoantimony and -bismuth compounds R_4MX (M = Sb, Bi; X = F, Cl, Br, I, OAc, OR) in good to excellent yields. Several examples are shown in equations 180[185], 181[182], 182[186], 183[186], 184[187], 185[188], 186[189] and 187[190]. Tosylimine derivatives such as 69 and $SbCl_5$ have also been used for the preparation (equations 188[189] and 189[182]).

$$Me_3SbBr_2 \xrightarrow{MeLi} Me_5Sb \qquad (180)$$
$$60\%$$

$$Ph_3SbCl_2 \xrightarrow{PhLi} Ph_5Sb \qquad (181)$$
$$90\%$$

$$Ph_3BiCl_2 \xrightarrow{\text{PhLi}} \underset{81\%}{Ph_5Bi} \tag{182}$$

$$Ph_4BiCl \xrightarrow{\text{PhLi}} \underset{67\%}{Ph_5Bi} \tag{183}$$

$$Et_4SbCl \xrightarrow{\text{EtLi}} \underset{85\%}{Et_5Sb} \tag{184}$$

$$Ph_3SbF_2 \xrightarrow{\text{PhMgBr}} \underset{75\%}{Ph_5Sb} \tag{185}$$

$$Ph_3Sb(OAc)_2 \xrightarrow{\text{PhLi}} \underset{80\%}{Ph_5Sb} \tag{186}$$

$$Ph_4SbOMe \xrightarrow{\text{PhLi}} \underset{50\%}{Ph_5Sb} \tag{187}$$

$$\underset{(69)}{Ph_3Sb{=}NTs} \xrightarrow{\text{PhLi}} \underset{85\%}{Ph_5Sb} \tag{188}$$

$$SbCl_5 \xrightarrow{\text{PhLi}} \underset{50\%}{Ph_5Sb} \tag{189}$$

Preparation of unsymmetrically substituted compounds in pure form is rather difficult, since the formation of hexaorganoantimony and -bismuth anions takes place during the reactions described above, leading to mixtures of compounds by protonolysis. For example, a mixture of Me_nSbEt_{5-n} has been obtained in the reaction of Me_4SbI with EtMgBr (equation 190[191]). However, by use of the appropriate combination of reagents, preparation of unsymmetrically substituted compounds in pure form can be achieved under controlled conditions. For example, several mixed organobismuth compounds have been synthesized and have been structurally characterized by X-ray crystallography (equation 191[192]).

$$Me_4SbI + EtMgBr \longrightarrow Me_nSbEt_{(5-n)} \quad \text{(mixture)} \tag{190}$$

$$Ar_3BiCl_2 \xrightarrow[-78\,°C,\,Et_2O]{\text{Ar'Li}} Ar_3BiAr'_2 \tag{191}$$

$$\left(\begin{array}{l} Ar = p\text{-Tol},\ Ar' = 2,6\text{-}F_2C_6H_3,\ 29\%^{[192a]} \\ Ar = p\text{-Tol},\ Ar' = C_6F_5,\ 51\%^{[192b]} \\ Ar = Ph,\ Ar' = o\text{-}FC_6H_4,\ 45\%^{[192c]} \end{array} \right)$$

For the preparation of mixed organoantimony compounds the combination of antimony fluorides and Grignard reagents is the method of choice. By use of this method Ph_4SbMe and all combinations of $(p\text{-}MeC_6H_4)_nSb(p\text{-}CF_3C_6H_4)_{5-n}$ have been prepared (equations 192[188a], 193[193] and 194[193]).

$$Ph_4SbF \xrightarrow{\text{MeMgBr}} \underset{75\%}{Ph_4SbMe} \tag{192}$$

$$p\text{-Tol}_n\!-\!\text{Sb}\!\left(\!\!\text{—}\!\!\left\langle\!\!\bigcirc\!\!\right\rangle\!\!\text{—CF}_3\right)_{3-n} \xrightarrow[0\,^\circ\text{C, 20 min}]{p\text{-Tol MgBr}} p\text{-Tol}_{n+2}\!-\!\text{Sb}\!\left(\!\!\text{—}\!\!\left\langle\!\!\bigcirc\!\!\right\rangle\!\!\text{—CF}_3\right)_{3-n}$$

$$\left(\begin{array}{c} n = 0,\ 58\% \\ n = 1,\ 55\% \\ n = 2,\ 49\% \end{array}\right)$$

$$(193)$$

$$\left(\text{CF}_3\text{—}\!\!\left\langle\!\!\bigcirc\!\!\right\rangle\!\!\text{—}\right)_4\!\!\text{SbF} \xrightarrow{p\text{-Tol Mg Br}} p\text{-Tol—Sb}\!\left(\!\!\text{—}\!\!\left\langle\!\!\bigcirc\!\!\right\rangle\!\!\text{—CF}_3\right)_4 \quad (194)$$

$$67\%$$

Weak nucleophiles such as the acetylene anion have been introduced cleanly to form $R_3Sb(C\!\equiv\!CR')_2$ (equation 195[55,194]).

$$R_3SbCl_2 \xrightarrow{R'C\equiv CM} R_3Sb(C\!\equiv\!CR')_2 \qquad (195)$$

$$\left(\begin{array}{c} R = Me,\ R' = Me^{[194]} \\ R = p\text{-Tol},\ R' = Ph^{[55]} \end{array}\right)$$

2. SO₂ elimination from Ar₄SbOS(O)R

Thermolysis of sulfinate derivatives $Ph_4SbOS(O)R$ has been reported to give Ph_4SbR in refluxing benzene (equation 196[195]). By this method, however, unsymmetrically substituted compounds in pure form may not be obtained because a redistribution reaction between the mixed pentaaryl compounds takes place even at ca 60 °C (equation 197[196]).

$$Ph_4SbO\!-\!\underset{\underset{O}{\|}}{S}\!-\!R \xrightarrow[\text{reflux}]{\text{benzene}} Ph_4SbR \qquad (196)$$

$$p\text{-Tol}_5Sb \rightleftharpoons p\text{-Tol}_4Sb\!-\!\!\left\langle\!\!\bigcirc\!\!\right\rangle\!\!\text{—CF}_3 \rightleftharpoons p\text{-Tol}_3Sb\!\left(\!\!\left\langle\!\!\bigcirc\!\!\right\rangle\!\!\text{—CF}_3\right)_2$$

$$\rightleftharpoons p\text{-Tol}_2Sb\!\left(\!\!\left\langle\!\!\bigcirc\!\!\right\rangle\!\!\text{—CF}_3\right)_3 \rightleftharpoons p\text{-TolSb}\!\left(\!\!\left\langle\!\!\bigcirc\!\!\right\rangle\!\!\text{—CF}_3\right)_4$$

$$\rightleftharpoons Sb\!\left(\!\!\left\langle\!\!\bigcirc\!\!\right\rangle\!\!\text{—CF}_3\right)_5 \qquad (197)$$

C. Preparation of Tetraorganoantimony(V) and -bismuth(V) Halides

Since tetraorganobismuth halides are thermally unstable, only Ph_4BiCl is actually known (Ph_4BiBr is stable only at $-70\,^\circ\text{C}$[186]). In contrast, a variety of alkyl and aryl substituted tetraorganoantimony halides (F, Cl, Br and I) have been isolated.

1. Cleavage of M—C bond by electrophiles

Pentaorganoantimony and -bismuth compounds react with electrophiles such as HF,

HCl, HBr, Cl_2, Br_2, I_2, Me_3SnF and Et_2NSF_3. Several examples are shown in equations 198[197], 199[186], 200[198], 201[186], 202[185], 203[182], 204[199] and 205[200]).

$$Me_5Sb \xrightarrow{\text{HF}} Me_4SbF \qquad (198)$$
$$75\%$$

$$Ph_5Bi \xrightarrow{\text{HCl}} Ph_4BiCl \qquad (199)$$
$$54\%$$

$$Ph_5Sb \xrightarrow{\text{HBr}} Ph_4SbBr \qquad (200)$$
$$53\%$$

$$Ph_5Sb \xrightarrow{\text{Cl}_2} Ph_4SbCl \qquad (201)$$
$$76\%$$

$$Me_5Sb \xrightarrow{\text{Br}_2} Me_4SbBr \qquad (202)$$
$$72\%$$

$$Ph_5Sb \xrightarrow{\text{I}_2} Ph_4SbI \qquad (203)$$
$$86\%$$

$$Me_5Sb \xrightarrow{\text{Me}_3\text{SnF}} Me_4SbF \qquad (204)$$
$$85\%$$

$$\left(CF_3-\!\!\left\langle\bigcirc\right\rangle\!\!-\right)_5 Sb \xrightarrow{\text{Et}_2\text{NSF}_3} \left(CF_3-\!\!\left\langle\bigcirc\right\rangle\!\!-\right)_4 SbF \qquad (205)$$
$$66\%$$

2. Protonolysis of tetraorganoantimony oxygen compounds

Tetraorganoantimony oxygen compounds react with HCl or HBr to form the corresponding chlorides or bromides, respectively (equation 206[201]).

$$Ph_4SbOR \xrightarrow{\text{HX}} Ph_4SbX \qquad (206)$$
$$\left(\begin{array}{l} R = H, X = Cl, \sim 100\% \\ R = H, X = Br, \sim 100\% \end{array}\right)$$

3. Halogen exchange

Tetraorganoantimony fluorides and iodides have been prepared by the reaction of the corresponding bromides with potassium fluoride or iodide (equation 207[188a,202]).

$$Ph_4SbBr \xrightarrow[\text{(X = F, I)}]{\text{KX}} Ph_4SbX \qquad (207)$$
$$\left(\begin{array}{l} X = F, 73\% \\ X = I, 99\% \end{array}\right)$$

By use of silver chloride and fluoride tetraorganoantimony iodides (or bromides) have been converted to the corresponding chlorides and fluorides, respectively (equations 208[203] and 209[204]).

$$Me_4SbI \xrightarrow{\text{AgCl}} Me_4SbCl \qquad (208)$$

$$Ph_4SbBr \xrightarrow{\text{AgF}} Ph_4SbF \qquad (209)$$

4. Reaction of triorganoantimony compounds with electrophiles

Trialkylantimony compounds react with alkyl halides to form the corresponding tetraalkylantimony salts (equations 210[185,205,206] and 211[207]).

$$R_3Sb \xrightarrow{R'X} R_3(R')SbX \tag{210}$$

$$(R = alkyl) \begin{pmatrix} R = R' = Me, X = I^{185} \\ R = n\text{-}Bu, R' = PhCH_2, X = Br, 95\%^{205c} \\ R = Me, R' = Me_3SiCH_2, X = I, 100\%^{206} \\ R = Et, R' = EtO_2CCH_2, X = Br, > 71\%^{205d} \end{pmatrix}$$

$$n\text{-}Bu_3Sb \xrightarrow{HC\equiv CCH_2Br} n\text{-}Bu_3Sb\diagdown_{Br} \tag{211}$$

With trialkyloxonium salts aryl substituted stibines react to give tetraorganoantimony compounds (equations 212[188a,208] and 213[209]).

$$Ph_3Sb \xrightarrow{Me_3O^+ BF_4^-} Ph_3\overset{+}{Sb}Me\ \overset{-}{BF_4} \xrightarrow{MX} Ph_3(Me)SbX \tag{212}$$

$$\underset{\underset{Me}{|}}{Ph-Sb}-i\text{-}Pr \xrightarrow{Et_3O^+ BF_4^-} \xrightarrow{KI} \underset{\underset{Me}{|}}{Ph-\overset{\overset{Et}{|}}{Sb^+}}-i\text{-}Pr\ I^- \tag{213}$$

Triarylantimony compounds do not react with PhCOCl or ClCH₂COCl, but react with 3,5-(O₂N)₂C₆H₃COCl to give the corresponding benzoyl derivatives (equations 214[210]).

(Ar = Ph, p-Tol)   (>90%)   (214)

D. Preparation of Triorganoantimony(V) and -bismuth(V) Dihalides

Since trialkylbismuth dihalides are thermally unstable, only triarylbismuth dihalides have actually been prepared.

1. Oxidation of triorganoantimony and -bismuth(V) compounds with elemental halogen or the equivalent

Triorganoantimony and -bismuth compounds react readily with X₂ to form the corresponding dihalides in high yields (equations 215[186,211] and 216[212]).

Other oxidants such as SO₂Cl₂, PhICl₂, XeF₂, Et₂NSF₃, IF₅ and PhIF₂ have also been quite useful for the preparation (equations 217[211d], 218[213], 219[214], 220[215], 221[212a,216] and 222[217]).

$$R_3M \xrightarrow[\substack{(X = F, Cl, Br, I)}]{X_2} R_3MX_2 \atop \sim 100\% \tag{215}$$

$$(C_6F_5)_3Sb \xrightarrow[\substack{(X = F, Cl, Br)}]{X_2} (C_6F_5)_3SbX_2 \tag{216}$$

$$Ph_3Sb \xrightarrow{SO_2Cl_2} Ph_3SbCl_2 \atop 90\% \tag{217}$$

$$\left(Me_2N-\!\!\left\langle \bigcirc \right\rangle\!\!- \right)_3 M \xrightarrow[\substack{(M = Sb, Bi)}]{PhICl_2} \left(Me_2N-\!\!\left\langle \bigcirc \right\rangle\!\!- \right)_3 MCl_2 \tag{218}$$

$$\left(\begin{matrix} M = Sb, 88\% \\ M = Bi, 83\% \end{matrix} \right)$$

$$R_3M \xrightarrow{XeF_2} R_3MF_2 \tag{219}$$

$$\left(\begin{matrix} R = Me, M = Sb, 96\%^{214b} \\ R = C_6F_5, M = Sb, 100\%^{241a} \\ R = C_6F_5, M = Bi, 100\%^{192b,214a} \end{matrix} \right)$$

$$p\text{-Tol}-Sb\!\!\left(\!\!\left\langle \bigcirc \right\rangle\!\!-CF_3 \right)_2 \xrightarrow{Et_2NSF_3} p\text{-TolSb}\!\!\left(\!\!\left\langle \bigcirc \right\rangle\!\!-CF_3 \right)_2 \atop F_2$$

$$65\% \tag{220}$$

$$Ph_3Bi \xrightarrow{IF_5} Ph_3BiF_2 \atop 47\% \tag{221}$$

$$Ph_3Sb \xrightarrow{PhIF_2} Ph_3SbF_2 \atop 70\% \tag{222}$$

Due to the facile exchange reactions between R_3SbX_2, $R_3SbX(Y)$ and R_3SbY_2 (equation 223[218]), mixed halides are generally difficult to obtain. However, several triorgano-antimony mixed halides $R_3SbX(Y)$ has been prepared by the reaction of triorgano-antimony with interhalogens at low temperatures as shown in equation 224[219].

$$R_3SbX_2 \rightleftharpoons R_3SbX(Y) \rightleftharpoons R_3SbY_2 \tag{223}$$

$$Ph_3Sb \xrightarrow[\substack{(X = Cl, Br)}]{IX} Ph_3Sb(I)X \atop (X = Br, 72\%) \tag{224}$$

Counter anion exchange reactions give stable $R_3MX^+Y^-$ ($M = Sb$, Bi) type compounds as shown in equations 225[220] and 226[221].

$$Me_3SbCl_2 \xrightarrow{SbCl_5} Me_3\overset{+}{Sb}Cl \overset{-}{SbCl_6} \tag{225}$$

$$Ph_3MI_2 \xrightarrow{AgAsF_6} Ph_3\overset{+}{M}I AsF_6^- \tag{226}$$

$$(M = Sb, Bi) \qquad \left(\begin{matrix} M = Sb, 88\% \\ M = Bi, 85\% \end{matrix} \right)$$

2. Halogenolysis of triorganoantimony oxygen compounds

Triorganoantimony oxygen compounds react with electrophiles to give the corresponding halides. Several examples are shown in equations 227[188c] and 228[222].

$$Ph_3SbO \xrightarrow{SF_4} \underset{92\%}{Ph_3SbF_2} \tag{227}$$

$$Me_3Sb(OSiMe_3)_2 \xrightarrow{SO_2Cl_2 \text{ or } HCl} Me_3SbCl_2 \tag{228}$$

3. Halogen exchange

Triorganoantimony and -bismuth fluorides have been prepared by the reaction of the corresponding bromides (or chlorides) with potassium fluoride (equation 229[188a,213,223]). By use of silver fluoride, triorganoantimony bromides have been converted to the corresponding fluorides (equation 230[211d]).

$$Ph_3MX_2 \xrightarrow{KF} Ph_3MF_2 \tag{229}$$

$$\left(\begin{array}{l} M = Sb, X = Br \\ M = Bi, X = Cl \end{array} \right)$$

$$Me_3SbBr_2 \xrightarrow{AgF} Me_3SbF_2 \tag{230}$$

E. Preparation of Diorganoantimony(V) Trihalides

Diorganobismuth trihalides have not been isolated because of their thermal instability. Although diorganoantimony trihalides can be prepared, they are also generally unstable and decompose at elevated temperatures. In the case of diorganoantimony triiodides, only **70** has been isolated.

1. Oxidation of diorganoantimony compounds with elemental halogen or the equivalent

Diorganoantimony compounds react readily with X_2 to form the corresponding trihalides in high yields (equations 231[224] and 232[225]). Other oxidants such as SO_2Cl_2, XeF_2 and $PhIF_2$ have also been quite useful for the preparation (equations 233[226], 234[227] and 235[217]).

$$R_2SbX \xrightarrow[(X = Cl, Br)]{X_2} R_2SbX_3 \tag{231}$$

$$\left(\begin{array}{l} R = Me, X = Cl, 100\%[224a] \\ R = Ph, X = Br, 83\%[224b] \end{array} \right)$$

$$\tag{232}$$

(**70**) 73%

$$Ph_2SbF \xrightarrow{XeF_2} Ph_2SbF_3 \qquad (234)$$
$$98\%$$

$$Ph_2SbF \xrightarrow{PhIF_2} Ph_2SbF_3 \qquad (235)$$
$$70\%$$

Several diorganoantimony mixed halides have been prepared by the reaction of diorganoantimony halides with elemental halogens as shown in equations 236[224b,228] and 237[224b,228].

$$Ph_2SbBr \xrightarrow[-90\,°C\,to\,rt]{Cl_2} Ph_2Sb(Br)Cl_2 \qquad (236)$$
$$70\%$$

$$Ph_2SbCl \xrightarrow{Br_2} Ph_2Sb(Cl)Br_2 \qquad (237)$$
$$85\%$$

Reaction of bromine with Ph_2SbF does not give pure $Ph_2Sb(F)Br_2$; instead a small amount of a diantimony complex $Ph_2Sb(F)Br_2 \cdot Ph_2SbBr_3$ has been obtained (equation 238[229]).

$$Ph_2SbF \xrightarrow{Br_2} Ph_2Sb(F)Br_2 \cdot Ph_2SbBr_3 \qquad (238)$$

2. Halogenolysis of diphenylstibinic acid

Diphenylstibinic acid reacts with SF_4 or HCl to give the corresponding fluoride or chloride as shown in equations 239[188c] and 240[230].

$$\underset{\underset{O}{\overset{\|}{}}}{Ph_2Sb}{-}OH \xrightarrow{SF_4} Ph_2SbF_3 \qquad (239)$$
$$98\%$$

$$\underset{\underset{O}{\overset{\|}{}}}{Ph_2Sb}{-}OH \xrightarrow{HCl} Ph_2SbF_3 \qquad (240)$$

3. Halogen exchange

Dimethylantimony trifluoride has been prepared by the reaction of the corresponding chloride with silver fluoride (equation 241[218b]).

$$Me_2SbCl_3 \xrightarrow{AgF} Me_2SbF_3 \qquad (241)$$
$$70\%$$

4. Transmetallation

Tetraorganotin compounds react with antimony pentachloride to give the corresponding trichlorides (equation 242[321]).

$$R_4Sn \xrightarrow{SbCl_5} R_2SbCl_3 \tag{242}$$
$$(R = Ph, 79–88\%)$$

F. Preparation of Monoorganoantimony(V) Tetrahalides

Since monoorganoantimony tetrahalides show strong Lewis acid character, they readily form hexacoordinate complexes (see Section IIIM). Only $PhSbF_4$ and $PhSbCl_4$ have been isolated. The methods of preparation are shown in equations 243[188c] and 244[232].

$$PhSb(OH)_2 \xrightarrow{SF_4} \underset{92\%}{PhSbF_4 \cdot SF_4} \xrightarrow[1 \text{ mm Hg}]{65-70\,°C} \underset{94\%}{PhSbF_4} \tag{243}$$
$$\overset{\|}{O}$$

$$PhSb(OH)_2 \xrightarrow{HCl} PhSbCl_4 \tag{244}$$
$$\overset{\|}{O}$$

G. Preparation of Organoantimony(V) and -bismuth(V) Nitrogen Compounds

1. Nucleophilic displacement of halides with nitrogen nucleophiles

Reactions of organoantimony and -bismuth dihalides with various nitrogen nucleophiles give the corresponding nitrogen compounds. Several examples are shown in equations 245[233], 246[233b,234], 247[235] and 248[236].

$$R_3MCl_2 \xrightarrow{NaN_3} R_3M(N_3)_2 \tag{245}$$
$$\begin{pmatrix} M = Sb, R = Me, Ph \\ M = Bi, R = Ph \end{pmatrix}(R = Me, M = Sb, 80\%)$$

$$R_3SbCl_2 \xrightarrow{AgNCO} R_3Sb(NCO)_2 \tag{246}$$
$$(R = Me, Ph)$$

$$R_3MX_2 \xrightarrow{AgNO_3} R_3M(NO_3)_2 \tag{247}$$
$$\begin{pmatrix} M = Sb, R = Me, X = Br \\ M = Sb, R = C_6F_5, X = Cl \\ M = Bi, R = Ph, X = Br \end{pmatrix} \begin{pmatrix} M = Sb, R = Me,^{235a} \\ M = Sb, R = C_6F_5, 40\%^{235b} \\ M = Bi, R = Ph^{235c} \end{pmatrix}$$

$$Ph_{3+n}SbBr_{2-n} \xrightarrow[Et_3N]{} Ph_{3+n} Sb \begin{pmatrix} N \end{pmatrix}_{2-n} \tag{248}$$
$$(n = 0, 1) \qquad\qquad (n = 1, 85\%^{236b})$$

2. Protonolysis of pentaorganoantimony compounds with HN_3

Tetraorganoantimony azides have been prepared from the reaction of pentaorgano-antimony compounds with HN_3 (equation 249)[237].

$$R_5Sb \xrightarrow{HN_3} R_4SbN_3 \qquad (249)$$
$$\begin{pmatrix} R = Me, 92\% \\ R = Ph, 89\% \end{pmatrix}$$

3. Oxidation by chloramine-T and sulfonyl azides

Triorganoantimony and -bismuth react with chloramine-T or with sulfonyl azides to give the corresponding imines (equations 250[189,238] and 251[239]).

$$R_3M \xrightarrow[\text{(chloramine-T)}]{p\text{-Tol}\overset{\overset{O}{\|}}{\underset{\|}{S}}-N\overset{Cl}{\underset{Na}{\diagup}}} R_3M{=}NSO_2Tol\text{-}p \qquad (250)$$

$$\begin{pmatrix} R = Ph, M = Sb, 81\% \\ R = Ph, M = Bi, \sim 100\% \end{pmatrix}$$

$$R_3M \xrightarrow{R'SO_2N_3} R_3Sb{=}NSO_2R' \qquad (251)$$

4. Oxidation of organoantimony(III) compounds containing Sb—N bonds with SO_2Cl_2 (equation 252[240])

$$MeSb\{N(SiMe_3)_2\}_2 \xrightarrow{SO_2Cl_2} \overset{\overset{Cl}{|}}{\underset{\underset{Cl}{|}}{MeSb}}\{N(SiMe_3)_2\}_2 \qquad (252)$$
$$75\%$$

H. Preparation of Organoantimony(V) and -bismuth(V) Oxygen Compounds

1. Nucleophilic displacement of halides with oxygen nucleophiles

Reaction of organoantimony and -bismuth halides with various oxygen nucleophiles gives the corresponding oxygen compounds. Several examples are shown in equations 253[198a], 254[211d], 255[241], 256[242], 257[243], 258[244], 259[245], 260[246], 261[247], 262[248] and 263[211d,249].

$$Ph_4SbBr \xrightarrow{MeONa} Ph_4SbOMe \qquad (253)$$
$$70\%$$

$$Me_3SbCl_2 \xrightarrow{H_2O} Me_3Sb(OH)_2 \qquad (254)$$
$$85\%$$

$$Ph_3BiCl_2 \xrightarrow{Ag_2O} Ph_3BiO \qquad (255)$$
$$40\%$$

$$Ph_3BiCl_2 \xrightarrow{K_2CO_3} Ph_3BiCO_3 \tag{256}$$
$$100\%$$

$$Ph_3SbX_2 \xrightarrow{t\text{-BuOONa}} Ph_3Sb(OOBu\text{-}t)_2 \tag{257}$$
$$77\%$$

$$PhSbCl_2 \xrightarrow[\text{CuBr}]{p\text{-FC}_6\text{H}_4\text{N}_2^+ \text{ BF}_4^+} [Ph(p\text{-FC}_6H_4)SbX_3]$$

$$\xrightarrow{\text{NaOH}} Ph(p\text{-FC}_6H_4)\overset{\overset{\displaystyle O}{\|}}{Sb}OH \tag{258}$$
$$38\%$$

$$PhSbCl_2 \xrightarrow{SO_2Cl_2} PhSbCl_4$$
$$\xrightarrow{AgO_2CCH_3} PhSb(O_2CCH_3)_4 \tag{259}$$

$$[EtO_2C\text{—}\langle\bigcirc\rangle\text{—}SbCl_5]^- \quad \underset{\overset{|}{\underset{H}{N^+}}}{\langle\bigcirc\rangle} \xrightarrow{Na_2CO_3} EtO_2C\text{—}\langle\bigcirc\rangle\text{—}\overset{\overset{\displaystyle}{Sb}}{\underset{\underset{\displaystyle O}{\|}}{}}(OH)_2$$
$$60\% \tag{260}$$

(261)

$$\begin{pmatrix} M = Sb,\ R = Me,\ 33\% \\ M = Sb,\ R = p\text{-Tol},\ 71\% \\ M = Bi,\ R = p\text{-Tol},\ 55\% \end{pmatrix}$$

$$R_3SbX_2 \xrightarrow{1\,eq\ \ R'ONa} R_3SbX \tag{262}$$
$$\overset{|}{OR'}$$

$$\begin{pmatrix} R = Ph,\ X = Cl,\ R' = Me,\ 86\%^{248a} \\[2ex] R = Me,\ X = Br,\ R' = \text{(structure)},\ 42\%^{248} \end{pmatrix}$$

$$R_3SbX_2 \xrightarrow{H_2O} (R_3Sb\text{—})_2O \tag{263}$$
$$\overset{|}{X}$$
$$(R = Ph,\ X = Br,\ 34\%^{249a})$$

2. Reaction of oxygen nucleophiles with $R_3Sb(X)Y$ compounds

Several oxygenated organoantimony compounds have been prepared by reaction of oxygen nucleophiles with triorganoantimony alkoxides, imines and oxides. Several examples are shown in equations 264[250], 265[251], 266[252] and 267[253].

$$Ph_3Sb{=}NSO_2Ph \xrightarrow{\text{PhOH}} \underset{90\%}{Ph_3Sb(OPh)_2} \tag{264}$$

$$Et_2Sb(OMe)_3 \xrightarrow{\text{AcOH}} \underset{\sim 100\%}{Et_2Sb(OAc)_3} \tag{265}$$

$$\tag{266}$$

$$Me_3SbO \xrightarrow{\text{HCO}_2\text{H}} \underset{\sim 100\%}{Me_3Sb(OC(O)H)_2} \tag{267}$$

3. Protonolysis of pentaorganoantimony and -bismuth compounds

Reaction of pentaorganoantimony and -bismuth compounds with acids gives the corresponding oxygen compounds. Several examples are shown in equations 268[199,242,254] and 269[194].

$$R_5M \xrightarrow{\text{R'OH}} R_4MOR' \tag{268}$$

$$\begin{pmatrix} M = Sb, R = R' = Me, 85\%^{199} \\ M = Bi, R = Ph, R' = SO_2CF_3, 90\%^{242} \\ M = Bi, R = Ph, R' = Ph_3Si^{254a} \\ M = Sb, R = Ph, R' = C(O)CH_3^{254d,e} \end{pmatrix}$$

$$Et_3SbMe_2 \xrightarrow{\text{MeOH}} Et_3Sb(Me)OMe \tag{269}$$

4. Oxidation of organoantimony(III) compounds

Trivalent organoantimony compounds react with oxidants to give the corresponding organoantimony(V) compounds. Several examples are shown in equations 270[255], 271[256], 272[257], 273[258] and 274[259].

$$R_3Sb \xrightarrow[\text{(R = Me, Ph)}]{\text{H}_2\text{O}_2} R_3SbO \tag{270}$$

$$Ph_2SbY \xrightarrow[\text{(Y = Cl, Br)}]{\text{\textit{t}-BuOOH}} Ph_2SbY \quad (271)$$
$$\underset{O}{\overset{\|}{}}$$

$$Me_3Sb \xrightarrow{\text{\textit{t}-BuOOH}} \underset{86\%}{Me_3Sb(OOBu\text{-}t)_2} \quad (272)$$

$$Ph_3Sb \xrightarrow[\text{(X = Cl, Br)}]{\text{\textit{t}-BuOX}} Ph_3Sb(OBu\text{-}t)X \quad (273)$$

$$(272) \quad (274)$$

(**72**) 57%

I. Preparation of Organoantimony(V) and -bismuth(V) Compounds Containing Sulfur and Selenium

1. Nucleophilic displacement with sulfur and selenium nucleophiles

Sulfur and selenium nucleophiles react with organoantimony and -bismuth halides or alkoxides. Several examples are shown in equations 275[260], 276[261], 277[262] and 278[263].

$$Ph_4SbBr \xrightarrow[\text{18-crown-6}]{\text{KSeCN}} \underset{84\%}{Ph_4SbSeCN} \quad (275)$$

$$R_3MCl_2 \xrightarrow{R'SH} R_3M(SR')_2 \quad (276)$$
$$\left(\begin{array}{l} M = Sb, R = R' = Me, 74\% \\ M = Bi, R = R' = Ph, 35\% \end{array} \right)$$

$$Me_3Sb(OEt)_2 \xrightarrow{MeC(O)SH} \underset{71\%}{Me_3Sb(SC(O)Me)_2} \quad (277)$$

$$Ph_3SbBr_2 \xrightarrow{H_2S} Ph_3SbS \quad (278)$$

2. Oxidation of triorganoantimony compounds with elemental sulfur and selenium

Trialkylantimony compounds react with elemental sulfur and selenium to give trialkylantimony sulfides and selenides, respectively (equation 279[264]).

$$R_3Sb \xrightarrow{E} R_3SbE \quad (279)$$
$$(R = Alkyl), (E = S, Se) \quad \begin{array}{l} E = S, R = i\text{-}Pr, 70\% \\ E = Se, R = Me \end{array}$$

J. Preparation of Organoantimony and -bismuth Ylides

Recently, efficient methods to prepare organoantimony and -bismuth ylides have been developed and X-ray structural determination has been carried out for the antimony ylide **71**[266a] and the bismuth ylide **72**[265d]. Reaction of active methylene compounds with triphenylantimony and -bismuth dihalides (or carbonates) in the presence of base gives the corresponding ylides (equation 280[265]). An alternative way is to react triphenylantimony with diazo compounds in the presence of bis(hexafluoroacetylacetonato)copper(II) as a catalyst (equation 281[266]).

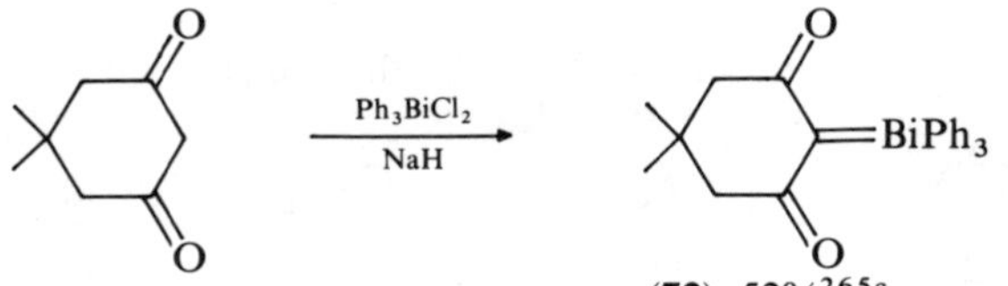

(280)

(**72**) 52%[265c]

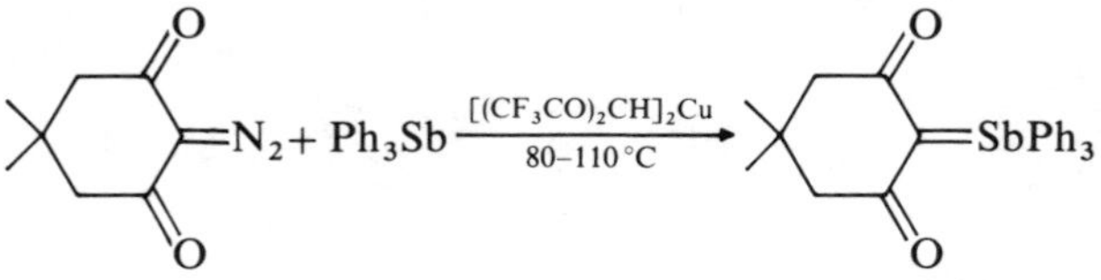

(281)

(**71**) 40%[266a]

K. Preparation of Compounds with Antimony(V)–Antimony(III) Bonds

Recently a compound (**73**) containing Sb(V)—Sb(III) bonds has been prepared as shown in equation 282[267].

$$p\text{-}Tol_3SbBr_2 \xrightarrow{Ph_2SbLi} \underset{(\mathbf{73})}{p\text{-}Tol_3Sb(SbPh_2)_2} \qquad (282)$$

L. Preparation of Compounds with Antimony(V)–Transition Metal Bonds

1. Oxidation of organoantimony(III) compounds containing antimony(III)–transition metal bonds

Several organoantimony(III) compounds containing antimony(III)–transition metal bonds have been oxidized by elemental bromine and Ph_2PCl to give the corresponding dihalides (equations 283[268] and 284[269]).

$$Me_2SbM(CO)_3Cp \xrightarrow{Br_2} Me_2\overset{\overset{\displaystyle Br}{|}}{\underset{\underset{\displaystyle Br}{|}}{Sb}}\text{—}M(CO)_3Cp \qquad (283)$$

(M = Cr, Mo, W)

$$Me_2SbFe(CO)(PPh_3)Cp \xrightarrow{Ph_2PC} Me_2\overset{\overset{\displaystyle Cl}{|}}{\underset{\underset{\displaystyle Cl}{|}}{Sb}}\text{—}Fe(CO)(PPh_3)Cp \quad 51\% \qquad (284)$$

2. Miscellaneous methods

Reaction of **74** with alkyl halides gives **75** (equation 285[270]). Triorganoantimony compounds containing allyl groups react with the iron complex **76** with elimination of the allyl moiety to give **77** (equation 286[271]). Reaction of the tetracoordinate antimony ate complex **78** with the iron complex **79** gives **80**, which has been characterized by X-ray analysis (equation 287[272]).

$$MeSb\{M(CO)_3Cp\}_2 \xrightarrow{MeX} [Me_2Sb\{M(CO)_3Cp\}_2]X \qquad (285)$$
$$\textbf{(74)} \qquad\qquad \textbf{(75)}$$
$$\left(\begin{array}{l} M = Mo,\ X = Br,\ 30\% \\ M = W,\ X = I,\ 83\% \end{array}\right)$$

$$Me_2SbCH_2-CH\!=\!CH_2 \xrightarrow{CpFe(CO)_2Cl\,(76)}$$

$$\xrightarrow{Ph_4B^-Na^+} Me_2Sb\{Fe(CO)_2Cp\}_2^+ Ph_4B^- \qquad (286)$$
$$\textbf{(77)} \quad 70\%$$

$$(287)$$

M. Preparation of Hexacoordinate Organoantimony(V) and -bismuth(V) Anions

Hexaorganoantimony and -bismuth anions have been described in Section III.A. Monoorganoantimony tetrahalides and diorganoantimony trihalides react readily with halide anions to form the corresponding hexacoordinate anions. Several examples are shown in equations 288[273], 289[274], 290[273b] and 291[275].

$$PhSb(OH)_2 \underset{\underset{\displaystyle O}{\|}}{\overset{}{}} \xrightarrow[(X = F, Cl, Br)]{HX} PhSbX_4 \xrightarrow{Me_4N^+X^-} PhSbX_5^- Me_4N^+ \qquad (288)$$

$$MeSbCl_2 \xrightarrow{SO_2Cl_2} MeSbCl_4 \xrightarrow{Me_4N^+N_3^-} MeSb(N_3)Cl_4^- Me_4N^+ \qquad (289)$$
$$89\%$$

$$Me_2SbCl_3 \xrightarrow{Me_4N^+Cl^-} Me_2SbCl_4^- Me_4N^+ \xrightarrow{NaF} Me_2SbF_4^- Me_4N^+ \qquad (290)$$

$$Ph_2SbCl_3 \xrightarrow{HBr} \xrightarrow{Ph_4As^+Br^-} Ph_2SbBr_4{}^- \ Ph_4As^+ \qquad (291)$$

By use of the bidentate ligand **81**, hexacoordinate antimony anions **82** containing five Sb—C bonds and bismuth anions **83** containing four Bi—C bonds have been characterized spectroscopically (equations 292[247b] and 293[276]). Neutral compounds **84** have been prepared and their X-ray analysis has shown that they have a distorted octahedral structure (equation 294[276]).

$$(292)$$

$$(293)$$

$$(294)$$

$$\left(\begin{array}{l} X = OMe, \ R = R' = CF_3 \\ X = NMe_2, \ R = H, \ R' = CF_3 \\ X = NMe_2, \ R = H, \ R' = CF_3 \end{array} \right)$$

IV. REFERENCES

1. Most recent review covering the year 1990: L. D. Freedman and G. O. Doak, *J. Organometal. Chem.*, **442**, 1 (1992) (Sb); **442**, 61 (1992) (Bi).
2. Most recent review covering the year 1990: J. L. Wardell, *Organomet. Chem.*, **21**, 130 (1992).
3. Most recent review covering the years 1989–91: D. A. Armitage, *Annu. Rep. Prog. Chem., Sect. A*, **89**, 49 (1993).
4. Most recent review covering the years 1986–87: D. B. Sowerby, *Coord. Chem. Rev.*, **103**, 1 (1990).

5. M. Wieber (Ed.), in *Gmelin Handbuch der Anorganischen Chemie. Bismut-Organische Verbindungen*, Springer-Verlag, Berlin, 1977.
6. (a) M. Wieber (Ed.), in *Gmelin Handbook of Inorganic Chemistry. Organoantimony Compounds*, Springer-Verlag, Berlin, Part 1 (1981). (b) M. Wieber (Ed.), *ibid.*, part 2 (1981). (c) M. Wieber (Ed.), *ibid.*, part 3 (1982). (d) M. Wieber (Ed.), *ibid.*, part 4 (1986). (e) U. Krüerke and M. Mirbach (Eds), *ibid.*, part 5 (1990).
7. Samaan (Ed.), *Methoden der Organischen Chemie. Metallorganische Verbindungen As, Sb, Bi*, Band XIII, Teil 8, Georg Thieme Verlag, Stuttgart, 1978.
8. P. C. Poller, in *Comprehensive Organic Chemistry* (Ed. D. N. Jones), Vol. 3, Chapter 15.5, Pergamon Press, Oxford, 1979, p. 1111.
9. J. L. Wardell, in *Comprehensive Organometallic Chemistry* (Eds G. Wilkinson, F. G. A. Stone and E. W. Abel), Vol. 2, Chapter 13, Pergamon Press, Oxford, 1982, p. 681.
10. (a) C. F. McAuliffe, in *Comprehensive Coordination Chemistry* (Eds. G. Wilkinson, R. D. Guillard, J. A. McCleverty and E. W. Abel), Vol. 2, Chapter 14, Pergamon Press, Oxford, 1987, p. 989. (b) C. F. McAuliffe, *ibid.*, Vol. 3, Chapter 28, Pergmon Press, Oxford, 1987, p. 237.
11. B. J. Aylett, *Organometallic Compounds*, Vol. I, Part 2, Methuen, New York, 1979, p. 387.
12. L. D. Freedman and G. O. Doak, *Chem. Rev.*, **82**, 15 (1982).
13. L. D. Freedman and G. O. Doak, in *The Chemistry of the Metal–Carbon Bond* (Ed. F. R. Hartley), Vol. 5, Wiley-Interscience, Chichester, 1989, p. 397.
14. Y.-Z. Huang, *Acc. Chem. Res.*, **25**, 182 (1992).
15. R. A. Abramovitch, D. H. R. Barton and J.-P. Finet, *Tetrahedron*, **44**, 3039 (1988).
16. D. H. R. Barton, in *Heteroatom Chemistry* (Ed. E. Block), Chapter 5, VCH Publishers, New York, 1990, p. 95.
17. J.-P. Finet, *Chem. Rev.*, **89**, 1487 (1989).
18. D. Hellwinkel, *Top. Curr. Chem.*, **109**, 1 (1983).
19. K. Seppelt, in *Heteroatom Chemistry* (Ed. E. Block), Chapter 19, VCH Publishers, New York, 1990, p. 335.
20. K. Seppelt, *Adv. Organometal. Chem.*, **34**, 207 (1992).
21. A. J. Ashe, III, *Adv. Organometal. Chem.*, **30**, 77 (1990).
22. A. J. Arduengo III and D. A. Dixon, in *Heteroatom Chemistry* (Ed. E. Block), Chapter 3, VCH Publishers, New York, 1990, p. 47.
23. N. C. Norman, *Chem. Soc. Rev.*, **17**, 269 (1988).
24. N. A. Compton, R. J. Errington and N. C. Norman, *Adv. Organometal. Chem.*, **31**, 91 (1990).
25. G. Huttner and K. Evertz, *Acc. Chem. Res.*, **19**, 406 (1986).
26. G. S. Hiers, *Org. Synth., Coll. Vol.*, **1**, 550 (1941).
27. H. J. Breunig, M. Ates and A. Soltani-Neshan, *Organometal. Synth.*, **4**, 593 (1988).
28. H. J. Breunig and D. Moller, *Organometal. Synth.*, **3**, 638 (1986).
29. (a) D. G. Hendershot, J. C. Pazik, C. George and A. D. Berry, *Organometallics*, **11**, 2163 (1992). (b) D. Seyferth, *J. Am. Chem. Soc.*, **80**, 1336 (1958).
30. P. Bras, A. Van der Gen and J. Wolters, *J. Organometal. Chem.*, **256**, C1 (1983).
31. I. G. M. Campbell, *J. Chem. Soc.*, 3116 (1955).
32. (a) R. R. Holmes, R. O. Day, V. Chandrasekhar and J. M. Holmes, *Inorg. Chem.*, **26**, 157 (1987). (b) G. Wittig and D. Hellwinkel, *Chem. Ber.*, **97**, 789 (1964).
33. A. J. Ashe III and F. J. Drone, *Organometallics*, **4**, 1478 (1985).
34. K. H. Whitmire, D. Labahn, H. W. Roesky, M. Noltemeyer and G. M. Sheldrick, *J. Organometal. Chem.*, **402**, 55 (1991).
35. H. Gilman, H. L. Yablunky and A. C. Svigoon, *J. Am. Chem. Soc.*, **61**, 1170 (1939).
36. G. Wittig and A. Maercker, *J. Organometal. Chem.*, **8**, 491 (1967).
37. H. Suzuki and T. Murafuji, *J. Chem. Soc., Chem. Commun.*, 1143 (1992).
38. (a) Bi: H. Hartmann, G. Habenicht and W. Reiss, *Z. Anorg. Allg. Chem.*, **317**, 54 (1964). (b) K. Rudolph and M. Wieber, *Z. Naturforcsch., Teil B*, **46**, 1319 (1991). (c) Sb: H. Hartmann and G. Kühl, *Angew. Chem.*, **68**, 619 (1956). (d) H. Hartmann and G. Kühl, *Z. Anorg. Allg. Chem.*, **312**, 186 (1961).
39. (a) Bi: E. O. Fischer and S. Schreiner, *Chem. Ber.*, **93**, 1417 (1960). (b) Sb: B. Deubzer, M. Elian, E. O. Fischer and H. P. Fritz, *Chem. Ber.*, **103**, 799 (1970).
40. (a) Bi: D. Naumann and W. Tyrra, *J. Organometal. Chem.*, **334**, 323 (1987). (b) Sb: D. Naumann, W. Tyrra and F. Leifeld, *J. Organometal. Chem.*, **333**, 193 (1987).

41. S. Pasenok, D. Naumann and W. Tyrra, *J. Organometal. Chem.*, **417**, C47 (1991).
42. R. E. v. H. Spence, D. P. Hsu and S. L. Buchwald, *Organometallics*, **11**, 3492 (1992).
43. A. J. Ashe III and T. R. Diephouse, *J. Organometal. Chem.*, **202**, C95 (1980).
44. R. S. Dickson and B. O. West, *Aust. J. Chem.*, **15**, 710 (1962).
45. F. Challenger and C. F. Allpress, *J. Chem. Soc.*, **119**, 913 (921).
46. D. H. R. Barton, N. Y. Bhatnagar, J.-P. Finet and W. B. Motherwell, *Tetrahedron*, **42**, 3111, (1986).
47. A. G. M. Barrett and L. M. Melcher, *J. Am. Chem. Soc.*, **113**, 8177 (1991).
48. E. Shewchuk and S. B. Wild, *J. Organometal. Chem.*, **210**, 181 (1981).
49. H. J. Breunig and D. Müller, *J. Organometal. Chem.*, **253**, C21 (1983).
50. S. Sato, Y. Matsumura and R. Okawara, *J. Organometal. Chem.*, **43**, 333 (1972).
51. W. Levason, K. G. Smith, C. A. McAuliffe, F. P. McCullough, R. D. Sedgwick and S. G. Murray, *J. Chem. Soc., Dalton Trans.*, 1718 (1979).
52. E. R. Bornancini, R. A. Alonso and R. A. Rossi, *J. Organometal. Chem.*, **270**, 177 (1984).
53. (a) P. Krommes and J. Lorberth, *J. Organometal. Chem.*, **88**, 329 (1975).
    (b) M. Birkhahn, P. Krommes, W. Massa and J. Lorberth, *J. Organometal. Chem.*, **208**, 161 (1981).
54. P. Krommes and J. Lorberth, *J. Organometal. Chem.*, **93**, 339 (1975).
55. K.-y. Akiba, T. Okinaka, M. Nakatani and Y. Yamamoto, *Tetrahedron Lett.*, **28**, 3367 (1987).
56. F. Calderazzo, R. Poli and G. Pelizzi, *J. Chem. Soc., Dalton Trans.*, 2365 (1984).
57. F. Calderazzo, A. Morvillo, G. Pelizzi, R. Poli and F. Ungari, *Inorg. Chem.*, **27**, 3730 (1988).
58. W. Clegg, R. J. Errington, G. A. Fisher, D. C. R. Hockless, N. C. Norman, A. G. Orpen and S. E. Stratford, *J. Chem. Soc., Dalton Trans.*, 1967 (1992).
59. M. Nunn, D. B. Sowerby and D. M. Wesolek, *J. Organometal. Chem.*, **251**, C45 (1983).
60. M. Ates, H. J. Breunig and S. Gülec, *J. Organometal. Chem.*, **364**, 67 (1989).
61. K. Issleib, B. Hamann and L. Schmidt, *Z. Anorg. Allg. Chem.*, **339**, 298 (1965).
62. K. Ohkata, S. Takemoto, M. Ohnishi and K.-y. Akiba, *Tetrahedron Lett.*, **30**, 4841 (1989).
63. P. Bras, H. Herwijer and J. Wolters, *J. Organometal. Chem.*, **212**, C7 (1981).
64. H. A. Meinema, C. J. R. C. Romão and J. G. Noltes, *J. Organometal. Chem.*, **55**, 139 (1973).
65. A. J. Ashe III and C. M. Kausch, *Organometallics*, **6**, 1185 (1987).
66. P. J. Fagan and W. A. Nugent, *J. Am. Chem. Soc.*, **110**, 2310 (1988).
67. D. Hellwinkel and M. Bach, *J. Organometal. Chem.*, **17**, 389 (1969).
68. R. Müller and C. Dathe, *Chem. Ber.*, **99**, 1609 (1966).
69. H. J. Breunig and W. Kanig, *Phosphorus Sulfur*, **12**, 149 (1982).
70. F. F. Blicke, U. O. Oakdale and F. D. Smith, *J. Am. Chem. Soc.*, **53**, 1025 (1931).
71. I. G. M. Campbell, *J. Chem. Soc.*, 3109 (1950).
72. K. Issleib and B. Hamann, *Z. Anorg. Allg. Chem.*, **343**, 196 (1966).
73. (a) H. A. Meinema, H. F. Martens, J. G. Noltes, N. Bartazzi and R. Berbieri, *J. Organometal. Chem.*, **136**, 173 (1977).
    (b) H. J. Breunig and H. Jawad, *J. Organometal. Chem.*, **243**, 417 (1983).
74. D. B. Sowerby and H. J. Breunig, *Organomet. Synth.*, **4**, 585 (1988).
75. M. Ates, H. J. Breunig, A. Soltani-Neshan and M. Tegeler, *Z. Naturforsch., Teil B*, **41**, 321 (1986).
76. A. L. Rheingold, *Organomet. Synth.*, **3**, 633 (1986).
77. Bi: W. Frank, *J. Organometal. Chem.*, **386**, 177 (1990).
78. Sb: R. W. Gedridge, Jr., *Organometallics*, **11**, 967 (1992).
79. M. Wieber, D. Wirth and I. Fetzer, *Z. Anorg. Allg. Chem.*, **505**, 134 (1983).
80. M. S. Kharasch, E. V. Jensen and S. Weinhouse, *J. Org. Chem.*, **14**, 429 (1949)
81. O. J. Scherer, P. Hornig and M. Schmidt, *J. Organometal. Chem.*, **6**, 259 (1966).
82. F. Challenger, *J. Chem. Soc.*, **105**, 2210 (1914).
83. I. G. M. Campbell, *J. Chem. Soc.*, 4 (1947).
84. B. R. Cook, C. A. McAuliffe and D. W. Meek, *Inorg. Chem.*, **10**, 2676 (1971).
85. R. Stricker-Lennartz and H. P. Latscha, *Z. Naturforsch., Teil B*, **40**, 1045 (1985).
86. (a) J. Ellermann and A. Veit, *J. Organometal. Chem.*, **290**, 307 (1985).
    (b) J. Ellermann and A. Veit, *Angew. Chem., Int. Ed. Engl.*, **21**, 375 (1982).
87. N. Auner and J. Grobe, *Z. Anorg. Allg. Chem.*, **500**, 132 (1983).
88. E. Amberger, *Chem. Ber.*, **94**, 1447 (1961).
89. R. Müller and C. Dathe, *Chem. Ber.*, **99**, 1609 (1966).

90. K. Issleib and A. Balszuweit, Z. Anorg. Allg. Chem., **418**, 158 (1975).
91. M. Ates, H. J. Breunig and S. Gülec, Phosphorus, Sulfur, and Silicon, **44**, 129 (1989).
92. A. H. Cowley, R. A. Jones, C. M. Nunn and D. L. Westmoreland, Angew. Chem., Int. Ed. Engl., **28**, 1018 (1989).
93. J. Koketsu, M. Okamura and Y. Ishii, Bull. Chem. Soc. Jpn., **44**, 1155 (1971).
94. W. T. Reichle, J. Organometal. Chem., **13**, 529 (1968).
95. P. Raj, A. K. Aggarwal and N. Misra, Polyhedron, **8**, 581 (1989).
96. B. Ross, J. Belz and M. Nieger, Chem. Ber., **123**, 975 (1990).
97. J. Müller, Z. Anorg. Allg. Chem., **381**, 103 (1971).
98. W. Steinkopf, I. Schubart and J. Roch, Chem. Ber., **65**, 409 (1932).
99. J. Müller, U. Müller, A. Loss, J. Lorberth, H. Donath and W. Massa, Z. Naturforsch., Teil B., **40**, 1320 (1985).
100. A. J. Arduengo III, C. A. Stewart, F. Davidson, D. A. Dixon, J. Y. Becker, S. A. Culley and M. B. Mizen, J. Am. Chem. Soc., **109**, 627 (1987).
101. (a) Bi: H. Schmidbaur and M. Bergfeld, Z. Anorg. Allg. Chem., **363**, 84 (1968).
    (b) Sb: H. Schmidbaur, H.-S. Arnold and E. Beinhofer, Chem. Ber., **97**, 449 (1964).
102. Y. Yamamoto, X. Chen. and K.-y. Akiba, J. Am. Chem. Soc., **114**, 7906 (1992).
103. K. H. Whitmire, J. C. Hutchison, A. L. McKnight and C. M. Jones, J. Chem. Soc., Chem. Commun., 1021 (1992).
104. H. Gilman and H. L. Yale, J. Am. Chem. Soc., **73**, 2880 (1951).
105. F. F. Blicke and U. O. Oakdale, J. Am. Chem. Soc., **55**, 1198 (1933).
106. V. L. Foss, N. M. Semenenko, N. M. Sorokin and I. F. Lutsenko, J. Organometal Chem., **78**, 107 (1974).
107. M. Wieber and N. Baumann, Z. Anorg. Allg. Chem., **418**, 167 (1975).
108. (a) H. J. Breunig and H. Kischkel, Z. Naturforsch., Teil B, **36**, 1105 (1981).
    (b) A. Michaelis and A. Günther, Chem. Ber., **44**, 2316 (1911).
109. (a) Bi: M. Wieber and I. Sauer, Z. Naturforsch., Teil B, **39**, 887 (1984).
    (b) Bi: H. J. Breunig and D. Müller, Z. Naturforsch., Teil B, **41**, 1129 (1986).
    (c) Sb: H. J. Breunig and H. Jawad, Z. Naturforsch., Teil B, **37**, 1104 (1982).
110. G. O. Doak and H. H. Jaffé, J. Am. Chem. Soc., **72**, 3025 (1950).
111. H. J. Breunig, A. Soltani-Neshan, K. Häberle and M. Dräger, Z. Naturforsch., Teil B, **41**, 327 (1986).
112. N. Kuhn and M. Winter, Chem. -Ztg., **107**, 342 (1983).
113. M. Baudler and S. Klautke, Z. Naturforsch., Teil B, **38**, 121 (1983).
114. (a) A. J. Ashe III and E. G. Ludwig Jr, J. Organometal. Chem., **303**, 197 (1986).
    (b) A. J. Ashe III and E. G. Ludwig Jr, J. Organometal. Chem., **308**, 289 (1986).
115. (a) Bi: M. Wieber and U. Baudis, Z. Anorg. Allg. Chem., **423**, 40 (1976).
    (b) Sb: N. Baumann and M. Wieber, Z. Anorg. Allg. Chem., **408**, 261 (1974).
116. D. N. Kravtsov, B. A. Kvasov, S. I. Pombrik and E. I. Fedin, J. Organometal. Chem., **86**, 383 (1975).
117. T. Klapötke, Polyhedron, **6**, 1593 (1987).
118. G. E. Forster, I. G. Southerington, M. J. Begley and D. B. Sowerby, J. Chem. Soc. Chem. Commun., 54 (1991).
119. A. G. Davies, S. C. W. Hook, J. Chem. Soc. (B), 735 (1970).
120. F. Challenger, A. L. Smith and F. J. Paton, J. Chem. Soc., **123**, 1046 (1923).
121. R. A. Fisher, R. B. Nielsen, W. M. Davis and S. L. Buchwald, J. Am. Chem. Soc., **113**, 165 (1991).
122. D. H. R. Barton, D. Bridon and S. Z. Zard, Tetrahedron, **45**, 2615 (1989).
123. (a) Bi: M. Wieber and I. Sauèr, Z. Naturforsch., Teil B, **39**, 1668 (1984).
    (b) Bi: M. Wieber and I. Sauèr, Z. Naturforsch., Teil B, **42**, 695 (1987).
    (c) Bi: W.-W. du Mont, T. Severengiz, H. J. Breunig and D. Müller, Z. Naturforsch., Teil B, **40**, 848 (1984).
    (d) Sb: H. J. Breunig and S. Gülec, Z. Naturforsch., Teil B, **41**, 1387 (1986).
    (e) H. J. Breunig, Phosphorus and Sulfur, **38**, 97 (1988).
124. H. J. Breunig, W.-W. du Mont and T. Severengiz, Organomet. Synth., **4**, 587 (1988).
125. H. J. Breunig and S. Gülec, Z. Naturforsch., Teil B, **43**, 998 (1988).
126. H. J. Breunig and T. P. Knobloch, Z. Anorg. Allg. Chem., **446**, 119 (1978).
127. S. Herbstman, J. Org. Chem., **29**, 986 (1964).

128. S. Sato, Y. Matsumura and R. Okawara, *Inorg. Nucl. Chem. Lett.*, **8**, 837 (1972).
129. A. J. Ashe, III, J. W. Kampf, S. M. Al-Taweel, *J. Am. Chem. Soc.*, **114**, 372 (1992).
130. H. Gilman and H. L. Yablunky, *J. Am. Chem. Soc.*, **63**, 212 (1941).
131. (a) H. J. Breunig and T. Severengiz, *Z. Naturforsch., Teil B*, **37**, 395 (1982).
    (b) G. O. Doak and L. D. Freedman, *Synthesis*, 328 (1974).
132. (a) K. Issleib and B. Hamann, *Z. Anorg. Allg. Chem.*, **343**, 196 (1966).
    (b) R. A. Bartlett, H. V. R. Dias, H. Hope, B. D. Murray, M. M. Olmstead and P. P. Power, *J. Am. Chem. Soc.*, **108**, 6921 (1986).
133. Y. Mourad, Y. Mugnier, H. J. Breunig and M. Ates, *J. Organometal. Chem.*, **398**, 85 (1990).
134. Y. Mourad, Y. Mugnier, H. J. Breunig and M. Ates, *J. Organometal. Chem.*, **388**, C9 (1990).
135. A. B. Burg and L. R. Grant, *J. Am. Chem. Soc.*, **81**, 1 (1959).
136. (a) H. J. Breunig, *Organomet. Synth.*, **3**, 625 (1986).
    (b) O. Mundt, H. Riffel, G. Becker and A. Simon, *Z. Naturforsch., Teil B*, **39**, 317 (1984).
    (c) A. J. Ashe III, E. G. Ludwig Jr., J. Oleksyszyn and J. C. Huffman, *Organometallics*, **3**, 337 (1984).
137. J. W. Dale, H. J. Eméléus, R. N. Haszeldine and J. H. Moss, *J. Chem. Soc.*, 3708 (1957).
138. H. J. Breunig and D. Müller, *Organomet. Synth.*, **3**, 636 (1986).
139. (a) H. J. Breunig, W. Kanig and H. Kischkel, *Organomet. Synth.*, **3**, 621 (1986).
    (b) O. Mundt, G. Becker, H.-J. Wessely, H. J. Breunig and H. Kischkel, *Z. Anorg. Allg. Chem.*, **486**, 70 (1982).
140. K. Issleib and A. Balszuweit, *Z. Anorg. Allg. Chem.*, **419**, 87 (1976).
141. H. J. Breunig, K. Häberle, M. Dräger and T. Severengiz, *Angew. Chem., Int. Ed. Engl.*, **24**, 72 (1985).
142. A. J. Ashe III and E. G. Ludwig, Jr., *J. Organometal. Chem.*, **303**, 197 (1986).
143. W. Becker and H. Nöth, *Chem. Ber.*, **105**, 1962 (1972).
144. A. H. Cowley, R. A. Jones, C. M. Nunn and D. L. Westmoreland, *Chem. Mater.*, **2**, 221 (1990).
145. (a) In: A. H. Cowley, R. A. Jones, K. B. Kidd, C. M. Nunn and D. L. Westmoreland, *J. Organometal. Chem.*, **341**, C1 (1988).
    (b) In: A. R. Barton, A. H. Cowley, R. A. Jones, C. M. Nunn and D. L. Westmoreland, *Polyhedron*, **7**, 77 (1988).
146. H. J. Breunig and T. Severengiz, *Organomet. Synth.*, **4**, 589 (1988).
147. G. Becker, M. Meiser, O. Mundt and J. Weidlein, *Z. Anorg. Allg. Chem.*, **569**, 62 (1989).
148. H. Shumann and M. Schmidt, *Inorg. Nucl. Chem. Lett.*, **1**, 1 (1965).
149. H. J. Breunig, *Z. Naturforsch., Teil B*, **33**, 990 (1978).
150. N. S. Vyazankin, G. A. Razuvaev, O. A. Kruglaya and G. S. Semchikova, *J. Organometal. Chem.*, **6**, 474 (1966).
151. (a) M. N Bochkarev, N. I. Gurév and G. A. Razuvaev, *J. Organometal. Chem.*, **162**, 289 (1978).
    (b) G. A. Razuvaev, *J. Organometal. Chem.*, **200**, 243 (1980).
152. L. I. Vyshinskaya, S. P. Korneva, G. P. Kulikova and T. A. Vishnyakova, *Metalloorg. Khim.*, **2**, 1163 (1989); *Chem. Abstr.*, **123**, 975 (1990).
153. G. N. Schrauzer and G. Kratel, *Chem. Ber.*, **102**, 2392 (1969).
154. (a) W. Malisch and P. Panster, *Chem. Ber.*, **108**, 700 (1975).
    (b) R. Shemm and W. Malisch, *J. Organometal. Chem.*, **288**, C9 (1985).
155. (a) W: U. Weber, G. Huttner, O. Scheidsteger and L. Zsolnai, *J. Organometal. Chem.*, **289**, 357 (1985).
    (b) Cr: U. Weber, L. Zsolnai and G. Huttner, *J. Organometal. Chem.*, **260**, 281 (1984).
156. J. M. Cassidy and K. H. Whitmire, *Inorg. Chem.*, **30**, 2788 (1991).
157. (a) P. Panster and W. Malisch, *Chem. Ber.*, **109**, 692 (1976).
    (b) H.-A. Kaul, D. Greissinger, M. Luksza and W. Malisch, *J. Organometal. Chem.*, **228**, C29 (1982).
    (c) D. Greissinger, W. Malisch and H.-A. Kaul, *J. Organometal. Chem.*, **252**, C23 (1983).
158. B. F. G. Johnson, J. Lewis, A. J. Whitton and S. G. Bott, *J. Organometal. Chem.*, **389**, 129 (1990).
159. (a) W. Clegg, N. A. Compton, R. J. Errington, G. A. Fisher, N. C. Norman and N. Wishart, *J. Organometal. Chem.*, **399**, C21 (1990).
    (b) W. Clegg, N. A. Compton, R. J. Errington, N. C. Norman, A.-J. Tucker and M. J. Winter, *J. Chem. Soc., Dalton Trans.*, 2941 (1988).
160. R. E. DesEnfants II, J. A. Gavney Jr., R. K. Hayashi, A. D. Rae, L. F. Dahl and A. Bjarnason, *J. Organometal. Chem.*, **383**, 543 (1990).

161. U. Weber, L. Zsolnai and G. Huttner, *Z. Naturforsch., Teil B*, **40**, 1430 (1985).
162. J. M. Wallis, G. Müller and H. Schmidbaur, *J. Organometal. Chem.*, **325**, 159 (1987).
163. F. Calderazzo, R. Poli and G. Pelizzi, *J. Chem. Soc., Dalton Trans.*, 2535 (1984).
164. S. R. Wade, M. G. H. Wallbridge and G. R. Willey, *J. Organometal. Chem.*, **267**, 271 (1984).
165. R. Okawara, K. Yasuda and M. Inoue, *Bull. Chem. Soc. Jpn.*, **39**, 1823 (1966).
166. G. Faraglia, *J. Organometal. Chem.*, **20**, 99 (1969).
167. T. Allman, R. G. Goel and H. S. Prasad, *J. Organometal. Chem.*, **166**, 365 (1979).
168. W. Clegg, R. J. Errington, G. A. Fisher, R. J. Flynn and N. C. Norman, *J. Chem. Soc., Dalton Trans.*, 637 (1993).
169. G. Alonzo, N. Bertazzi, F. Di Bianca and T. C. Gibb, *J. Organometal. Chem.*, **210**, 63 (1981).
170. M. Hall and D. B. Sowerby, *J. Organometal. Chem.*, **347**, 59 (1988).
171. D. H. R. Barton, B. Charpiot, E. T. H. Dau, W. B. Motherwell. C. Pascard and C. Pichon, *Helv. Chim. Acta*, **67**, 586 (1984).
172. (a) Sb: K.-y. Akiba, H. Nakata, Y. Yamamoto and S. Kojima, *Chem. Lett.*, 1563 (1992).
    (b) Bi: X. Chen, Y. Yamamoto, K.-y. Akiba, S. Yoshida, M. Yasui and F. Iwasaki, *Tetrahedron Lett.*, **33**, 6653 (1992).
173. A. J. Ashe III, *Acc. Chem. Res.*, **11**, 153 (1978).
174. (a) A. J. Ashe III, *J. Am. Chem. Soc.*, **93**, 6690 (1971).
    (b) A. J. Ashe III and M. D. Gordon, *J. Am. Chem. Soc.*, **94**, 7596 (1972).
175. A. J. Ashe III, T. R. Diephouse and M. Y. El-Sheikh, *J. Am. Chem. Soc.*, **104**, 5693 (1982).
176. M. Yoshifuji, I. Shima, N. Inamoto, K. Hirotsu and T. Higuchi, *J. Am. Chem. Soc.*, **103**, 4587 (1981).
177. A. H. Cowley and N. C. Norman, *Prog. Inorg. Chem.*, **34**, 1 (1986).
178. (a) A. H. Cowley, J. E. Kilduff, J. G. Lasch, S. K. Mehrotra, N. C. Norman, M. Pakulski, B. R. Whittlesey, J. L. Atwood and W. E. Hunter, *Inorg. Chem.*, **23**, 2582 (1984).
    (b) A. H. Cowley, J. G. Lasch, N. C. Norman, M. Pakulski and B. R. Whittlesey, *J. Chem. Soc., Chem. Commun.*, 881 (1983).
179. S. Nagase, S. Suzuki and T. Kurakake, *J. Chem. Soc., Chem. Commun.*, 1724 (1990).
180. G. Huttner, U. Weber, B. Sigwarth and O. Scheidsteger, *Angew. Chem., Int. Ed. Engl.*, **21**, 215 (1982).
181. A. H. Cowley, N. C. Norman and M. Pakulski, *J. Am. Chem. Soc.*, **106**, 6844 (1984).
182. G. Wittig and K. Clauss, *Justus Liebigs Ann. Chem.*, **577**, 26 (1952).
183. D. Hellwinkel and G. Kilthau, *Justus Liebigs Ann. Chem.*, **705**, 66 (1967).
184. D. Hellwinkel and M. Buch, *Justus, Liebigs Ann. Chem.*, **720**, 198 (1968).
185. G. Wittig and K. Torssell, *Acta Chem. Scand.*, **7**, 1293 (1953).
186. G. Wittig and K. Clauss, *Justus Liebigs Ann. Chem.*, **578**, 136 (1952).
187. Y. Takashi, *J. Organometal. Chem.*, **8**, 225 (1967).
188. (a) G. Doleshall, N. A. Nesmeyanov and O. A. Reutov, *J. Organometal. Chem.*, **30**, 369 (1971).
    (b) B. Raynier, B. Waegell, R. Commandeur and H. Mathais, *Nouv. J. Chim.*, **3**, 393 (1979).
    (c) L. M. Yagupol'skii, N. V. Kondratenko and V. I. Popov, *J. Gen. Chem. USSR*, **46**, 618 (1976).
189. G. Witing and D. Hellwinkel, *Chem. Ber.*, **97**, 789 (1964).
190. K.-w. Shen, W. E. McEwen and A. P. Wolf, *J. Am. Chem. Soc.*, **91**, 1283 (1969).
191. H. A. Meinema and J. G. Noltes, *J. Organometal. Chem.*, **22**, 653 (1970).
192. (a) A. Schmuck, D. Leopold, S. Wallenhauer and K. Seppelt, *Chem. Ber.*, **123**, 761 (1990).
    (b) A. Schmuck and K. Seppelt, *Chem. Ber.*, **122**, 803 (1989).
    (c) A. Schmuck, P. Pyykkö and K. Seppelt, *Angew. Chem. Int. Ed. Engl.*, **29**, 213 (1990).
193. G. Schröder, Y. Yamamoto and K.-y. Akiba, unpublished results.
194. N. Tempel, W. Schwarz and J. Weidlein, *J. Organometal. Chem.*, **154**, 21 (1978).
195. I.-P. Lorenz and J. K. Thekumparampil, *Z. Naturforsch. Teil B*, **33**, 47 (1978).
196. Y. Mimura, G. Schröder, A. Hasuoka, F. Kondoh, T. Matsuzaki, T. Okinaka, Y. Yamamoto and K.-y. Akiba, unpublished results.
197. H. Schmidbaur, K.-H. Mitschke, W. Buchner, H. Stühler and J. Weidlein, *Chem. Ber.*, **106**, 1226 (1973).
198. (a) W. E. McEwen, G. H. Briles and B. E. Giddings, *J. Am. Chem. Soc.*, **91**, 7079 (1969).
    (b) For the crystal structure of Ph_4SbBr: O. Knop, B. R. Vincent and T. S. Cameron, *Can. J. Chem.*, **67**, 63 (1989).
199. H. Schmidbaur, J. Weidlein and K.-H. Mitschke, *Chem. Ber.*, **102**, 4136 (1969).

200. G. Schröder, Y. Yamamoto and K.-y. Akiba, unpublished results.
201. (a) G. Kalfoglou and L. H. Bowen, *J. Phys. Chem.*, **73**, 2728 (1969).
    (b) G. O. Doak, G. G. Long and L. D. Freedman, *J. Organometal. Chem.*, **12**, 443 (1968).
202. J. Chatt and F. G. Mann, *J. Chem. Soc.*, 1192 (1940).
203. H. Siebert, *Z. Anorg. Allg. Chem.*, **273**, 161 (1953).
204. R. G. Goel, *Can. J. Chem.*, **47**, 4607 (1969).
205. (a) Y.-Z. Huang and Y. Liao, *J. Org. Chem.*, **56**, 1381 (1991).
    (b) Y.-Z. Huang, Y. Liao and C. Chen, *J. Chem. Soc., Chem. Commun.*, 85 (1990).
    (c) L.-J. Zhang, Y.-Z. Huang, H.-X. Jiang, J. Duan-Mu and Y. Liao, *J. Org. Chem.*, **57**, 774 (1992).
    (d) Y.-Z. Huang, C. Chen and Y. Shen, *J. Organometal. Chem.*, **366**, 87 (1989).
206. H. Schmidbaur and G. Haßlberger, *Chem. Ber.*, **111**, 2702 (1978).
207. L.-J. Zhang, Y.-Z. Huang and Z.-H. Huang, *Tetrahedron Lett.*, **32**, 6579 (1991).
208. (a) F: J. Bordner, B. C. Andrews and G. G. Long, *Cryst. Struct. Commun.*, **5**, 801 (1976).
    (b) Br: G. E. Parris, G. G. Long, B. C. Andrews and R. M. Parris, *J. Org. Chem.*, **41**, 1276 (1976).
209. S. Sato, Y. Matsumura and R. Okawara, *J. Organometal. Chem.*, **60**, C9 (1973).
210. A. Asthana and R. C. Srivastava, *J. Organometal. Chem.*, **366**, 281 (1989).
211. (a) Bi: X = Cl; K. A. Jensen, *Z. Anorg. Allg. Chem.*, **250**, 257 (1943).
    (b) Bi: X = I; J. F. Wilkinson and F. Challenger, *J. Chem. Soc.*, **125**, 854 (1924).
    (c) Sb: X = F; I. Rippert and V. Bastian, *Angew. Chem.*, **90**, 226 (1978). (d) Sb: X = Cl; A. J. Banister and L. F. Moore, *J. Chem. Soc. (A)*, 1137 (1968); G. G. Long, G. O. Doak and L. D. Freedman, *J. Am. Chem. Soc.*, **86**, 209 (1964).
    (e) Sb: X = Br; A. G. Davies and S. C. W. Hook, *J. Chem. Soc. (C)*, 1660 (1971).
    (f) Sb: X = I; A. D. Beveridge, G. S. Harris and F. Inglis, *J. Chem. Soc.(A)*, 520 (1966); G. O. Doak, G. G Long and M. E. Key, *Inorg. Synth.*, **9**, 92 (1967).
212. (a) X = F; R. Kasemann and D. Naumann, *J. Fluorine Chem.*, **41**, 321 (1988).
    (b) X = Cl, Br, G. S. Harris, A. Khan and I. Lennon, *J. Fluorine Chem.*, **37**, 247 (1987).
213. J.-M. Keck and G. Klar, *Z. Naturforsch., Teil B*, **27**, 591 (1972).
214. (a) W. Tyrra and D. Naumann, *Can. J. Chem.*, **67**, 1949 (1989).
    (b) A. M. Forster and A. J. Downs, *Polyhedron*, **4**, 1625 (1985).
215. G. Schröder, Y. Yamamoto and K.-y. Akiba, unpublished results.
216. H. J. Frohn and M. Maurer, *J. Fluorine Chem.*, **34**, 129 (1986).
217. V. I. Popov and N. V. Kondratenko, *J. Gen. Chem. USSR*, **46**, 2477 (1976).
218. (a) R = Me: C. G. Moreland, M. H. O'Brien, C. E. Douthit and G. G. Long, *Inorg. Chem.*, **7**, 834 (1968).
    (b) R = Ph; E. L. Muetterties, W. Mahler, K. J. Packer and R. Schmutzler, *Inorg. Chem.*, **3**, 1298 (1964).
219. (a) S. N. Bhattacharya and M. Singh, *Indian J. Chem., A.*, **16**, 778 (1970).
    (b) G. O. Doak and G. G. Long, *Trans New York Acad. Sci.*, **28**, 402 (1966).
    (c) A. D. Beveridge, G. S. Harris and F. Inglis, *J. Chem. Soc. (A)*, 520 (1966).
220. (a) A. Schmidt, *Chem. Ber.*, **102**, 380 (1969).
    (b) W. Dorsch and A. Schmidt, *Z. Anorg. Allg. Chem.*, **508**, 149 (1984).
221. I. Tornieporth-Oetting and T. Klapötke, *J. Organometal. Chem.*, **379**, 251 (1959).
222. H. Schmidbaur, H.-S. Arnold and E. Beinhofer, *Chem. Ber.*, **97**, 449 (1964).
223. F. Challenger and J. F. Wilkinson, *J. Chem. Soc.*, **121**, 91 (1922).
224. (a) R = Me, X = Cl: H. G. Nadler and K. Dehnicke, *J. Organometal. Chem.*, **90**, 291 (1975).
    (b) R = Ph, X = Br: S. P. Bone and D. B. Sowerby, *J. Chem. Soc., Dalton Trans.*, 715 (1979).
225. D. Hellwinkel and M. Bach, *J. Organometal. Chem.*, **17**, 389 (1969).
226. H. A. Meinema, H. F. Martens, J. G. Noltes, N. Bertazzi and R. Barbieri, *J. Organometal. Chem.*, **136**, 173 (1977).
227. L. M. Yagupol'skii, V. I. Popov, N. V. Kondratenko, B. L. Korsunskii and N. N. Aleinikov, *J. Org. Chem. USSR*. **11**, 454 (1975).
228. S. P. Bone and D. B. Sowerby, *J. Chem. Soc. Dalton Trans.*, 718 (1979).
229. S. P. Bone, M. J. Begley and D. B. Sowerby, *J. Chem. Soc., Dalton Trans.*, 2085 (1992).
230. H. Schmidt, *Justus Liebigs Ann. Chem.*, **429**, 123 (1922).
231. (a) I. Haiduc and C. Silvestru, *Inorg. Synth.*, **23**, 194 (1985).
    (b) K. B. Dillon and G. F. Hewitson, *Polyhedron*, **3**, 957 (1984).

232. H. Schmidt, *Justus Liebigs Ann. Chem.*, **421**, 174 (1920).
233. (a) Sb: A. Schmidt, *Chem. Ber.*, **101**, 3976 (1968).
    (b) Sb: R. G. Goel and D. R. Ridely, *Inorg. Nucl. Chem. Lett.*, **7**, 21 (1971).
    (c) Bi: R. G. Goel and H. S. Prasad, *J. Organometal. Chem.*, **50**, 129 (1973).
234. G. Ferguson, R. G. Goel and D. R. Ridley, *J. Chem. Soc., Dalton Trans.*, 1288 (1975).
235. (a) G. G. Long, G. O. Doak and L. D. Freedman, *J. Am. Chem. Soc.*, **86**, 209 (1964).
    (b) A. Otero and P. Royo, *J. Organometal. Chem.*, **154**, 13 (1978).
    (c) J. V. Supniewski and R. Adams, *J. Am. Chem. Soc.*, **48**, 507 (1926).
236. (a) K. Bajpai and R. C. Srivastava, *Synth. React. Inorg. Met.-Org. Chem.*, **11**, 7 (1981).
    (b) P. Raj, A. Ranjan, K. Singhal and R. Rastogi, *Synth. React. Inorg. Met.- Org. Chem.*, **14**, 269 (1984).
237. H. Schmidbaur, K.-H. Mitschke, J. Weidlein and S. Cradock, *Z. Anorg. Allg. Chem.*, **386**, 139 (1971).
238. M = Bi: H. Suzuki, C. Nakaya, Y. Matano and T. Ogawa, *Chem. Lett.*, 105 (1991).
239. Z. I. Kuplennik, Z. N. Belaya and A. M. Pinchuk, *J. Gen. Chem. USSR*, **51**, 2339 (1981).
240. W. Kolondra, W. Schwarz and J. Weidlein, *Z. Naturforsch., Teil B*, **40**, 872 (1985).
241. R. G. Goel and H. S. Prasad, *J. Organometal. Chem.*, **36**, 323 (1972).
242. D. H. R. Barton, N. Y. Bhatnagar, J.-C. Blazejewski, B. Charpiot, J.-P. Finet, D. J. Lester, W. B. Motherwell, M. T. B. Popoula and S. P. Stanforth, *J. Chem. Soc., Perkin Trans., I*, 2657 (1985).
243. (a) A. Rieche, J. Dahlmann and D. List, *Justus Liebigs Ann. Chem.*, **678**, 167 (1964).
    (b) Z. A. Starikova, T. M. Shchegoleva, V. K. Trunov, I. E. Pokrovskaya and E. N. Kannikova, *Soviet Phys. Cryst.*, **24**, 694 (1979).
244. (a) G. O. Doak and J. M. Summy, *J. Organometal. Chem.*, **55**, 143 (1978).
    (b). L. H. Bowen and G. G. Long, *Inorg. Chem.*, **15**, 1039 (1976).
245. M. Wieber, I. Fetzer-Kremling, H. Reith and C. Burschka, *Z. Naturforsch., Teil B*, **27**, 591 (1972).
246. (a) G. O. Doak and H. G. Steinman, *J. Am. Chem. Soc.*, **68**, 1989 (1946).
    (b) G. O. Doak and H. H. Jaffé, *J. Am. Chem. Soc.*, **72**, 3025 (1950).
247. (a) M = Sb: K.-y. Akiba, H. Fujikawa, Y. Sunaguchi and Y. Yamamoto, *J. Am. Chem. Soc.*, **109**, 1245 (1987).
    (b) M = Sb: Y. Yamamoto, H. Fujikawa, H. Fujishima and K.-y. Akiba, *J. Am. Chem. Soc.*, **111**, 2276 (1989).
    (c) M = Bi: K.-y. Akiba, K. Ohdoi and Y. Yamamoto, *Tetrahedron Lett.*, **29**, 3817 (1988).
    (d) M = Bi: X. Chen, K. Ohdoi, Y. Yamamoto and K.-y. Akiba, *Organometallics*, **12**, 1857 (1993).
248. (a) L. Kolditz, M. Gitter and E. Rösel, *Z. Anorg. Allg. Chem.*, **316**, 270 (1962).
    (b) D. M. Joshi and N. K. Jha, *Synth. React. Inorg. Met. -Org. Chem.*, **17**, 961 (1987).
249. (a) A. Ouchi and S. Sato, *Bull. Chem. Soc. Jpn.*, **61**, 1806 (1988).
    (b) G. Ferguson, F. C. March and D. R. Ridley, *Acta Cryst. Sect. B*, **31**, 1260 (1975).
    (c) E. R. T. Tiekink, *J. Organometal Chem.*, **333**, 199 (1987).
250. A. M. Pinchuk, Z. I. Kuplennik and Z. N. Belaya, *J. Gen. Chem. USSR*, **46**, 2155 (1976).
251. H. A. Meinema and J. G. Notles, *J. Organometal. Chem.*, **36**, 313 (1972).
252. (a) M. Wieber and N. Baumann, *Z. Anorg. Allg. Chem.*, **418**, 279 (1975).
    (b) M. Wieber, N. Baumann, H. Wunderlich and H. Rippstein, *J. Organometal. Chem.*, **133**, 183 (1977).
253. (a) M. Shindo and R. Okawara, *J. Organometal. Chem.*, **5**, 537 (1966).
    (b) T. Westhoff, F. Huber and H. Preut, *J. Organometal. Chem.*, **348**, 185 (1988).
254. (a) G. A. Razuvaev, N. A. Osanova and V. V. Sharutin, *Dokl. Akad. Nauk SSSR*, **225**, 581 (1975).
    (b) H. Schmidbaur, K.-H. Mitschke and J. Weidlein, *Z. Anorg. Allg. Chem.*, **386**, 147 (1971).
    (c) S. P. Bone and D. B. Sowerby, *J. Chem. Res. (S)*, 82 (1979).
    (d) H. Schmidbaur and K.-H. Mitschke, *Angew. Chem., Int. Ed. Engl.*, **10**, 136 (1971).
    (e) S. P. Bone and D. B. Sowerby, *Phosphorus, Sulfur, and Silicon*, **45**, 23 (1989).
255. (a) R = Ph: J. Bordner, G. O. Doak and T. S. Everett, *J. Am. Chem. Soc.*, **108**, 4206 (1986).
    (b) R = Ph: D. L. Venezky, C. W. Sink, B. A. Nevett and W. F. Fortescue, *J. Organometal. Chem.*, **35**, 131 (1972).
    (c) R = Ph: W. E. McEwen, G. H. Briles and D. N. Schulz, *Phosphorous*, **2**, 147 (1972).
    (d) R = Me: W. Morris, R. A. Zingaro and J. Laane, *J. Organometal. Chem.*, **91**, 295 (1975).

256. D. M. Wesolek, D. B. Sowerby and M. J. Begley, *J. Organometal. Chem.*, **293**, C5 (1985).
257. A. G. Davies and S. C. W. Hook, *J. Chem. Soc. (C)*, 1660 (1971).
258. J. Dahlmann and L. Austenat, *Justus Liebigs Ann. Chem.*, **729**, 1 (1969).
259. R. R. Holmes, R. O. Day, V. Chandrasekhar and J. M. Holmes, *Inorg. Chem.*, **26**, 157 (1987).
260. P. Raj, R. Rastogi, K. Singhal and A. K. Saxena, *Polyhedron*, **5**, 1581 (1986).
261. (a) M = Sb: H. Schmidbaur and K.-H. Mitschke, *Chem. Ber.*, **104**, 1842 (1971).
    (b) M = Bi: H. Gilman and H. L. Yale, *J. Am. Chem. Soc.*, **73**, 4470 (1951).
262. Y. Matsumura, M. Shindo and R. Okawara, *Inorg. Nucl. Chem. Lett.*, **3**, 219 (1967).
263. W. J. Lile and TR. J. Menzies, *J. Chem. Soc.*, 617 (1950).
264. (a) E = S: G. N. Chremos and R. A. Zingaro, *J. Organometal. Chem.*, **22**, 637 (1970).
    (b) E = Se: R. A. Zingaro and A. Merijanian, *J. Organometal. Chem.*, **1**, 369 (1964).
    (c) N. Kuhn and H. Schumann, *J. Organometal. Chem.*, **288**, C51 (1985).
265. (a) H. Suzuki, T. Murafuji and T. Ogawa, *Chem. Lett.*, 847 (1988).
    (b) T. Ogawa, T. Murafuji and H. Suzuki, *Chem. Lett.*, 849 (1988).
    (c) T. Ogawa, T. Murafuji and H. Suzuki, *J. Chem. Soc., Chem. Commun.*, 1749 (1989).
    (d) M. Yausi, T. Kikuchi, F. Iwasaki, H. Suzuki, T. Murafuji and T. Ogawa, *J. Chem. Soc., Perkin Trans. 1*, 3367 (1990).
    (e) H. Suzuki and T. Murafuji, *Bull. Chem. Soc. Jpn.*, **63**, 950 (1990).
    (f) D. H. R. Barton, J. Blazejewski, B. Charpiot, J.-P. Finet, W. B. Motherwell, M. T. B. Papoula and S. P. Stanforth, *J. Chem. Soc., Perkin Trans. 1*, 2667 (1985).
266. (a) C. Glidewell, D. Lloyd and S. Metcalfe, *Tetrahedron*, **42**, 3887 (1986).
    (b) D. Lloyd and M. I. C. Singer, *Chem. Ind. (London)*, 787 (1967).
    (c) C. Glidewell, D. Lloyd and S. Metcalfe, *Synthesis*, 319 (1988).
    (d) G. Ferguson, C. Glidewell, I. Gosney, D. Lloyd, S. Metcalfe and H. Lumbroso, *J. Chem. Soc., Perkin Trans. 2*, 1829 (1988).
    (e) G. Ferguson, C. Glidewell, D. Lloyd and S. Metcalfe, *J. Chem. Soc., Perkin. Trans. 2*, 731 (1988).
267. N. K. Jha and P. Sharma, *J. Chem. Soc., Chem. Commun.*, 1447 (1992).
268. W. Malish and P. Panster, *Angew. Chem., Int. Ed. Engl.*, **13**, 670 (1974).
269. W. Malish, H.-A. Kaul, E. Groß and U. Thewalt, *Angew. Chem., Int. Ed. Engl.*, **21**, 549 (1982).
270. W. Malish and P. Panster, *Chem. Ber.*, **108**, 700 (1975).
271. Y. Matsumura, M. Harakawa and R. Okawara, *J. Organometal. Chem.*, **71**, 403 (1974).
272. M. Okazaki, Y. Yamamoto and K.-y. Akiba, *Organometallics*, to be published.
273. (a) N. Bertazzi, M. Airoldi and L. Pellerito, *J. Organometal. Chem.*, **97**, 399 (1975).
    (b) N. Bertazzi, T. C. Gibb and N. N. Greenwood, *J. Chem. Soc., Dalton Trans.*, 1153 (1976).
274. K. Dehnicke and H. G. Nadler, *Z. Anorg. Allg. Chem.*, **426**, 253 (1976).
275. N. Bertazzi, *J. Organometal. Chem.*, **110**, 175 (1976).
276. (a) K.-y. Akiba, K. Ohdoi and Y. Yamamoto, *Tetrahedron. Lett.*, **30**, 953 (1989).
    (b) Y. Yamamoto, K. Ohdoi, X. Chen, M. Kitano and K.-y. Akiba, *Organometallics*, **12**, in press (1993).

CHAPTER **21**

Syntheses of organoarsenic compounds

YOHSUKE YAMAMOTO and KIN-YA AKIBA

Department of Chemistry, Hiroshima University, 1-3-1 Kagamiyama, Higashi–Hiroshima 724, Japan

The chemistry of organic arsenic, antimony and bismuth compounds
Edited by S. Patai © 1994 John Wiley & Sons Ltd

I. INTRODUCTION

Progress in organoarsenic compounds has been reviewed annually in *Organometallic Chemistry*[1], *Annual Reports*[2] and *Coordination Chemistry Reviews*[3]. The methods of preparation and properties of organoarsenic compounds were compiled in the series of Houben-Weyl in 1978[4]. In addition, some general reviews have appeared in *Comprehensive Organometallic Chemistry* (1982)[5] and *Comprehensive Coordination Chemistry* (1987)[6]. Aylett reviewed the organic chemistry of Group V elements in 1979[7]. More specialized reviews have been written by Bohra and Roesky on pentacoordinated arsenic(V) compounds in 1984[8], Hellwinkel on pentaorganyl derivatives of the Group V elements in 1983[9], Huang and coworkers on the synthetic application of arsonium ylides in 1982[10] and 1990[11], Lloyd and coworkers on arsonium ylides in 1987[12], Kauffmann on arsenic compounds as reagents for organic synthesis in 1980[13], Kober on aminoarsines in 1982[14], Ashe on arsabenzenes in 1982[15], Weber on diphosphenes and diarsenes in 1992[16], Cowley in 1984[17] and Cowley and Norman in 1986[18] on double bonded compounds between Group 14 and Group 15 elements, Arduengo and Dixon on tridentate system in 1990[19], Doak and Freedman on Group V anions in 1974[20], Huttner and Evertz on arsinidene–transition metal complexes in 1986[21], DiMaio and Rheingold on transition metal complexes containing arsenic–arsenic bonds in 1990[22], Omae on intramolecularly-coordinated arsenic compounds in 1986[23], Krannich and coworkers on aminoarsine–borane complexes in 1992[24], Cowley and Jones in 1989[25] and Wells in 1992[26] on gallium–arsenic compounds and Cullen and Reimer on arsenic speciation in the environment in 1989[27].

In the present review, the methods of synthesis of organoarsenic compounds will be surveyed. Typical and recent methods have been selected from a large number of reported methods.

II. PREPARATION OF ORGANOARSENIC(III) COMPOUNDS

A. Preparation of Triorganoarsenic(III) Compounds

1. Transmetallation by use of organometallic reagents

Reactions of Grignard reagents or organolithium compounds with arsenic halides, usually chlorides, have been one of the most frequently used method for the preparation of triorganoarsenic compounds. A wide variety of triorganoarsenic(III) compounds have been prepared as shown in equations 1[28–31], 2[32], 3[33], 4[34], 5[35], 6[36], 7[37] and 8[38].

$$RMgBr \xrightarrow{\text{AsCl}_3} R_3As \tag{1}$$

R = Me, 60–63%[28]
R = cyclohexyl, 87%[29]
R = vinyl, 62%[30]
R = Ph, 90%[31]

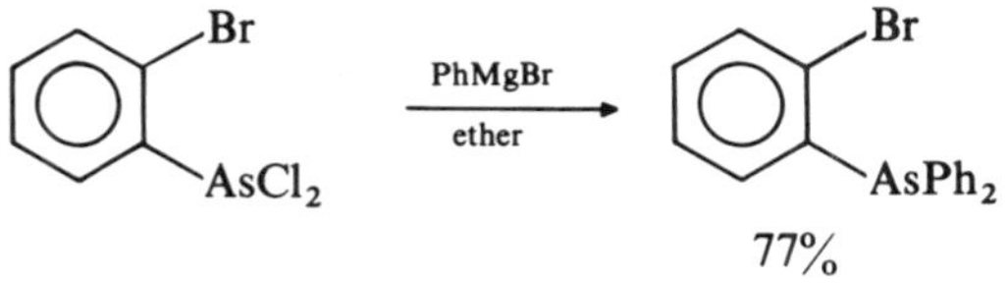

$$\text{(2)}$$

77%

$$t\text{-BuAsCl}_2 \xrightarrow[\text{ether}]{\text{BrMg(CH}_2)_n\text{MgBr}} \quad \text{(CH}_2)_n \quad \text{AsBu-}t \qquad \text{(3)}$$

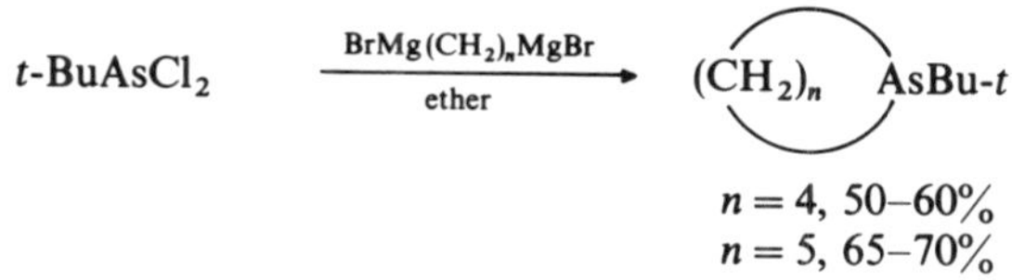

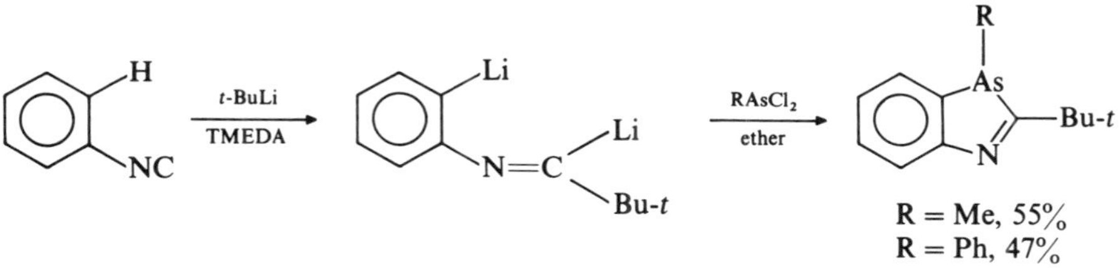

$n = 4,\ 50\text{–}60\%$
$n = 5,\ 65\text{–}70\%$

$$\text{(4)}$$

R = Me, 55%
R = Ph, 47%

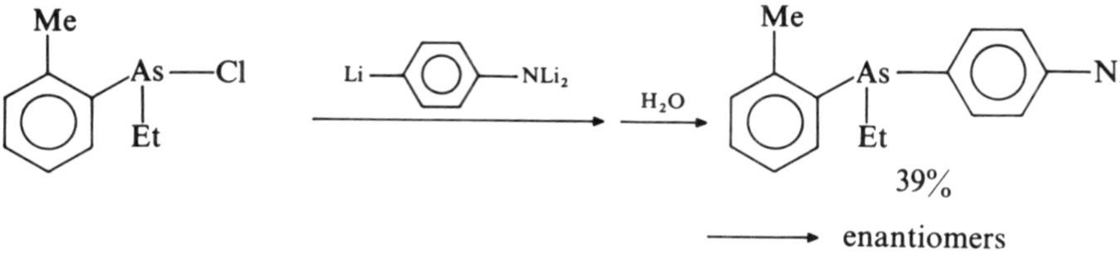

39%

$$\xrightarrow{\qquad} \text{enantiomers} \qquad \text{(5)}$$

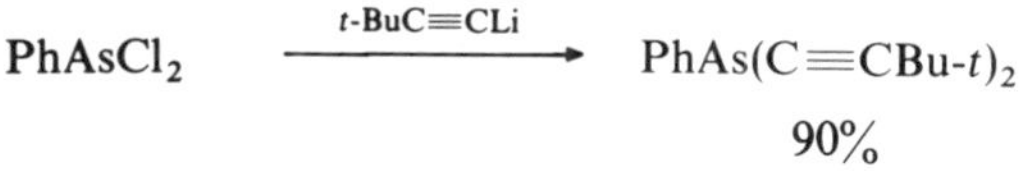

$$\text{PhAsCl}_2 \xrightarrow{t\text{-BuC}\equiv\text{CLi}} \text{PhAs(C}\equiv\text{CBu-}t)_2 \qquad \text{(6)}$$

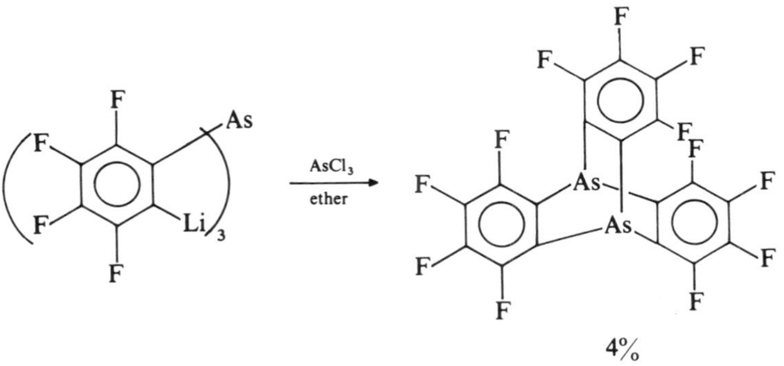

90%

$$\text{(7)}$$

4%

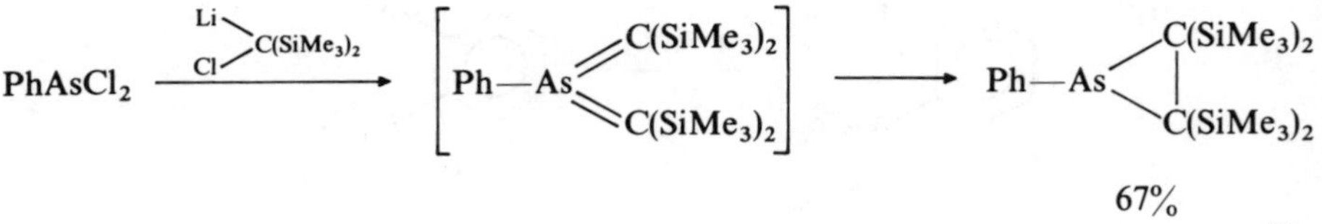

$$PhAsCl_2 \xrightarrow[\text{Cl}]{\overset{\text{Li}}{\diagdown}\text{C(SiMe}_3)_2} \left[Ph-As\overset{C(SiMe_3)_2}{\underset{C(SiMe_3)_2}{\diagup}} \right] \longrightarrow Ph-As\overset{C(SiMe_3)_2}{\underset{C(SiMe_3)_2}{\diagup\diagdown}}$$

$$67\%$$

$$(8)$$

Organosodium, zirconium, titanium, aluminum, tin and mercury compounds have also been used for the preparation of triorganoarsenic compounds in recent papers as shown in equations 9[39], 10[40], 11[41], 12[42], 13[43], 14[44] and 15[45].

$$C_5H_5Na \xrightarrow{AsCl_3} Cp_3As \qquad (9)$$

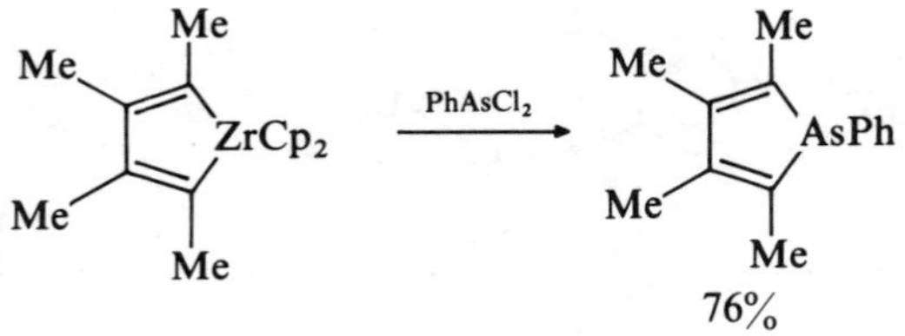

$$(CpNa) \qquad\qquad 71\%$$

$$(10)$$

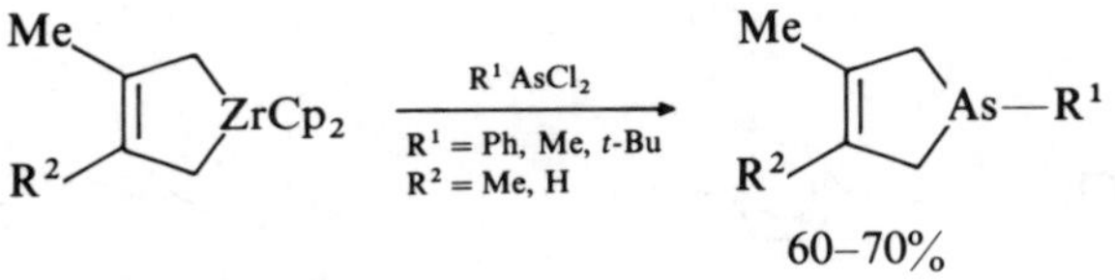

76%

60–70%

$$(11)$$

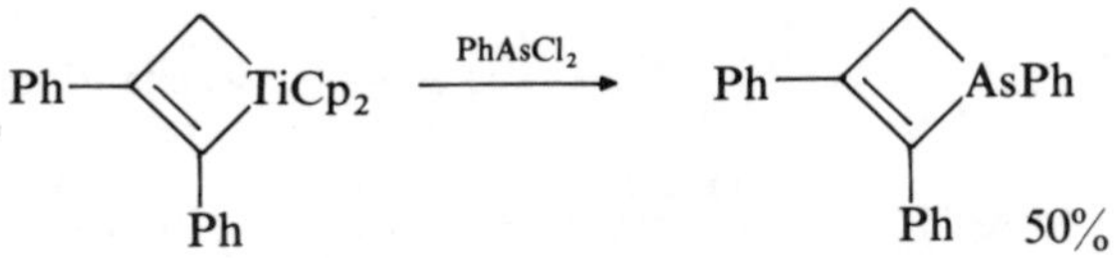

$$(12)$$

Ph   Ph   50%

$$Et_4Al^-Li^+ \xrightarrow{AsCl_3} Et_3As \qquad (13)$$

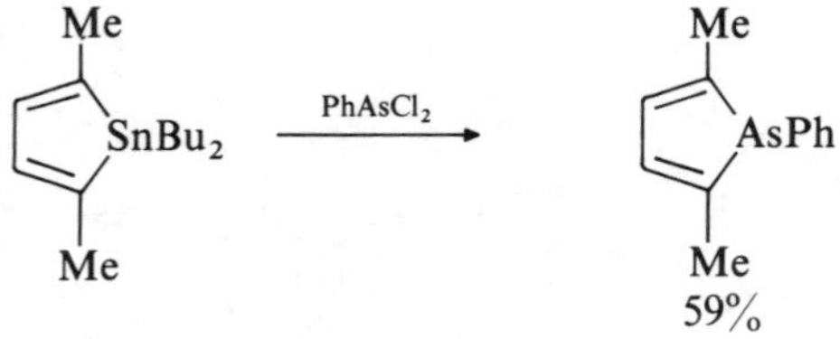

$$31\%$$

$$(14)$$

59%

$$(CF_3)_2Hg \xrightarrow{AsI_3} (CF_3)_3As \qquad (15)$$

$$75\%$$

The use of chlorodioxarsolane (**1**) has been reported to be superior in the preparation of highly pure tertiary arsines (equation 16)[46].

$$\text{AsCl}_3 \xrightarrow[\text{Et}_3\text{N}]{\text{HO} \quad \text{OH}} \underset{\textbf{(1)} \ 62\%}{\left[\text{As—Cl}\right]} \xrightarrow[]{\text{Organometallic reagents}} \text{R}_3\text{As} \qquad (16)$$

$$R = Et$$
$$\begin{pmatrix} \text{Grignard, 82\%} \\ \text{zinc, 84\%} \\ \text{aluminium, 71\%} \end{pmatrix}$$

Aminoarsines have also been useful for the preparation of sterically unhindered tertiary arsines. This reaction is especially useful for preparation of trimethylarsine (equation 17)[47].

$$\text{As(NMe}_2)_3 \xrightarrow{\text{Me}_3\text{Al}} \underset{84\%}{\text{Me}_3\text{As}} \qquad (17)$$

2. Reaction of arsenide anions with organic electrophiles

This method has also been quite useful for the synthesis of triorganoarsenic compounds. Arsenide anions, which can be prepared by several methods (*vide infra*), react with alkyl and aryl halides. Several examples are shown in equations 18[48], 19[49], 20[50], 21[51], 22[52], 23[53], 24[54] and 25[55]. In the reaction with alkyl halides inversion of configuration at the sp[3] carbon atom takes place and in the reaction of vinyl halides retention of configuration has been observed (equations 18 and 21). Photostimulated reaction with aryl halides has been reported to proceed via the $S_{RN}1$ mechanism; thus, the ketone group is compatible with the reaction conditions as shown in equation 23[53].

$$(18)$$
$$41\%$$

$$\text{C(CH}_2\text{Cl)}_4 \xrightarrow[\text{liq NH}_3]{\text{Ph}_2\text{As}^-\text{Na}^+} \underset{96\%}{(\text{Ph}_2\text{AsCH}_2)_3\text{CCH}_2\text{Cl}} \qquad (19)$$

$$\text{CH}_2\text{Cl}_2 \xrightarrow[\text{THF}]{\text{Ph}_2\text{As}^-\text{K}^+} \underset{55\%}{(\text{Ph}_2\text{As})_2\text{CH}_2} \qquad (20)$$

$$(21)$$
$$61\%$$

$$\text{(22)}$$

$$\text{(23)}$$

$$\text{(24)}$$

$$\text{(25)}$$

In addition to alkyl halides alkyl tosylates, acyl chlorides, isocyanates and the three-membered ring compounds including some cyclopropanes, oxiranes, thiiranes and aziridines can be used as electrophiles to afford a variety of triorganoarsenic compounds (equations 26[56], 27[57], 28[58], 29[59], 30[60], 31[61] and 32[62]).

$$\text{(26)}$$

$$\text{(27)}$$

$$c\text{-}C_6H_{11}\text{---}N{=}C{=}O \xrightarrow[\text{ether}]{(c\text{-}C_6H_{11})_2As^-Li^+} \xrightarrow{\text{EtOH}} (c\text{-}C_6H_{11})_2As\text{---}\underset{\underset{O}{\|}}{C}NHc\text{-}C_6H_{11} \quad 81\%$$

$$(28)$$

$$(29)$$

$$(30)$$

$$(31)$$

$$(32)$$

Trimethylsilyl- and trimethylstannyl-arsines react with reactive electrophiles to give triorganoarsenic compounds (equations 33[63a] and 34[63b]).

$$RAs(SiMe_3)_2 \xrightarrow[\text{DME}]{\overset{\overset{O}{\|}}{t\text{-BuCCl}}} RAs(\underset{\underset{O}{\|}}{C}Bu\text{-}t)_2$$

$$(33)$$

$$R = t\text{-Bu}, 63\%$$
$$R = Ph, 94\%$$

$$Me_2AsMMe_3 \xrightarrow{CF_3I} Me_2AsCF_3$$

$$(34)$$

$$M = Si, 40\%$$
$$M = Sn, 45\%$$

Reaction of elemental arsenic with sodium in liquid ammonia forms As^{3-} that reacts with aryl halides under irradiation. The reaction gives symmetrical triarylarsine by the $S_{RN}1$ mechanism (equation 35)[64].

$$PhBr \xrightarrow[\text{liq NH}_3\ h\nu]{As/Na} \underset{75\%}{Ph_3As}$$

$$(35)$$

3. Reaction of primary and secondary arsines with electrophiles

Primary and secondary arsines react with reactive carbon-electrophiles to give some triorganoarsenic compounds (equations 36[65] and 37[66]).

$$\text{PhAsH}_2 \xrightarrow{\overset{\overset{\text{O}}{\|}}{\text{MeCH}}} \qquad 47\% \qquad (36)$$

$$\text{Me}_2\text{AsH} \xrightarrow{\text{CF}_2=\text{CClCF}_2\text{Cl}} \text{Me}_2\text{AsCF}_2-\text{CCl}=\text{CF}_2 \qquad (37)$$
$$91\%$$

4. Addition of primary and secondary arsines to alkenes or alkynes

Compounds bearing As—H bonds add across carbon–carbon double bonds or triple bonds (equations 38[67] and 39[68]).

$$\xrightarrow{\text{R}^2\text{AsH}_2} \qquad (38)$$

$$\text{R}^1 = t\text{-Bu, R}^2 = \text{Ph, }55\%$$
with AIBN, 62%
with KOH-18-Crown-6, 79%

$$\text{Ph}_2\text{PCH}=\text{CH}_2 \xrightarrow[t\text{-BuOK}]{\text{Ph}_2\text{AsH}} \text{Ph}_2\text{PCH}_2\text{CH}_2\text{AsPh}_2 \qquad (39)$$
$$77\%$$

5. Reaction of arsenic halides with alkynes or aromatic compounds

Friedel-Crafts reaction of arsenic halides has been reported to give aromatic compounds as shown in equation 40[69]. Reactions with alkynes usually give mixtures of

$$\xrightarrow{250\,°\text{C}} \qquad 30\% \qquad (40)$$

compounds (equation 41[70]), but nucleophilic alkynes react cleanly with arsenic halides (equation 42[71]).

$$HC\equiv CH \xrightarrow[AlCl_3]{PhAsCl_2} \underset{\underset{Cl}{|}}{PhAsCH}=CHCl + PhAs(CH=CHCl)_2 + Ph_2AsCH=CHCl \tag{41}$$

$$Ph_2AsC\equiv COR \xrightarrow{Ph_2AsCl} \underset{Ph_2As}{\overset{Ph_2As}{>}}C=C=O \tag{42}$$

R = Me, Et   78%

6. Addition of arsinidenes across dienes

Some compounds having As—As bonds have been reported to react with dienes (equation 43[72]).

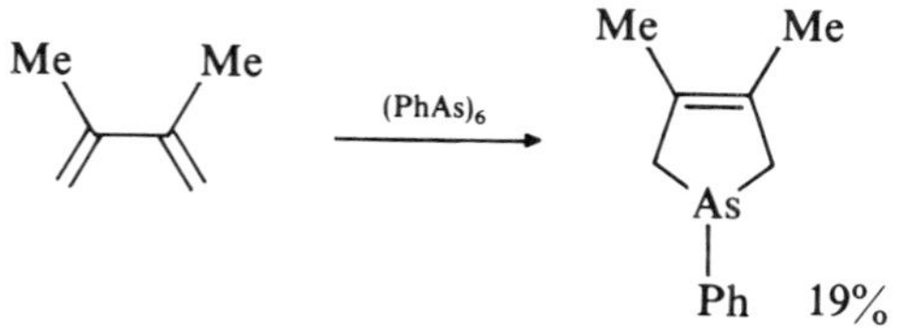

$$\tag{43}$$

Ph   19%

7. Dimerization of As=C bonded compounds and addition of As=C bonded compounds to dienes and acetylenes

Photoirradiation of As=C bonded compounds gives dimerized products (equation 44[73]). In addition Diels-Alder type reactions with dienes and acetylenes take place to give cyclic compounds (equations 45[74] and 46[75]).

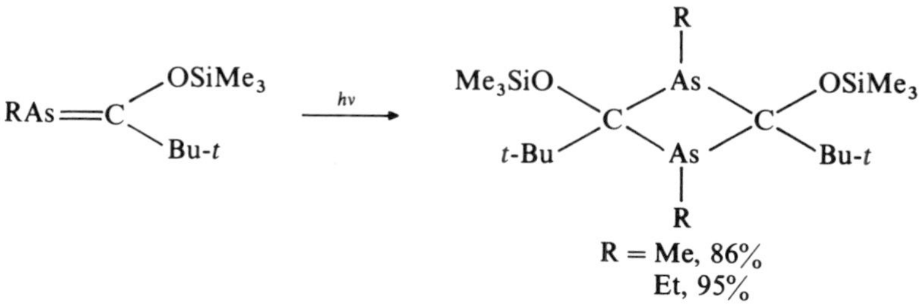

$$\tag{44}$$

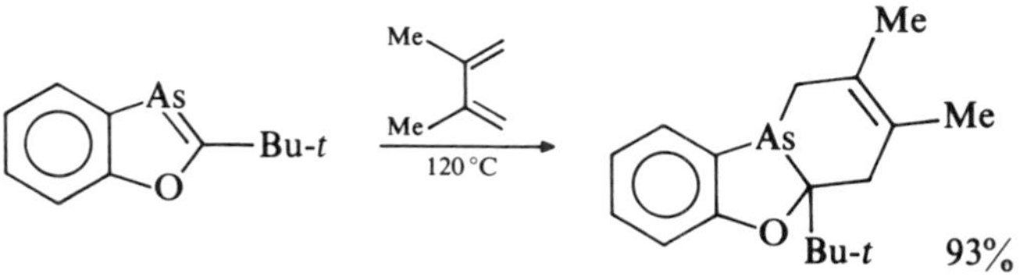

R = Me, 86%
Et, 95%

$$\tag{45}$$

Bu-t   93%

$$\tag{46}$$

8. Reduction of triorganoarsenic oxides

Reduction of triorganoarsenic oxides by use of several reductants has been reported (equations 47[76], 48[77] and 49[78]).

$$Ph_3As{=}O \xrightarrow{\text{TiCl/LiAlH}_4} Ph_3As \quad 79\% \tag{47}$$

$$\tag{48}$$

$$PhAs(CH_2CO_2H)_2 \xrightarrow[\text{HCl/KI}]{SO_2} PhAs(CH_2CO_2H)_2 \tag{49}$$

9. Reactions of arsenic with various electrophiles

Elemental arsenic reacts with bis(trifluoromethyl)tellurium, 1,2-diiodotetrafluorobenzene and trifluoromethyl iodide at high temperatures to give symmetrical arsines (equations 50[79], 51[80] and 52[81]).

$$(CF_3)_2Te \xrightarrow[\text{220 °C}]{\text{As}} (CF_3)_3As \tag{50}$$
$$46\%$$

$$\tag{51}$$

$$CF_3I \xrightarrow[220-240\,^\circ C]{As} (CF_3)_3As \qquad (52)$$
$$70\%$$

10. Cleavage of As—C bond in tetraorganoarsonium salts

Some tetraorganoarsonium salts bearing nucleophilic counter anions afford triorganoarsines at high temperatures (equation 53)[82]. Tetraorganoarsonium salts bearing benzyl or allyl groups give triorganoarsines after treatment with $LiAlH_4$, KCN or upon electrolysis (equations 54[83a-c] and 55[83d]).

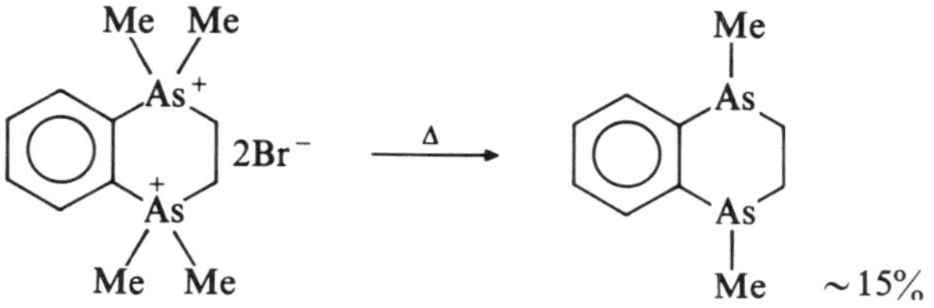

$$(53)$$

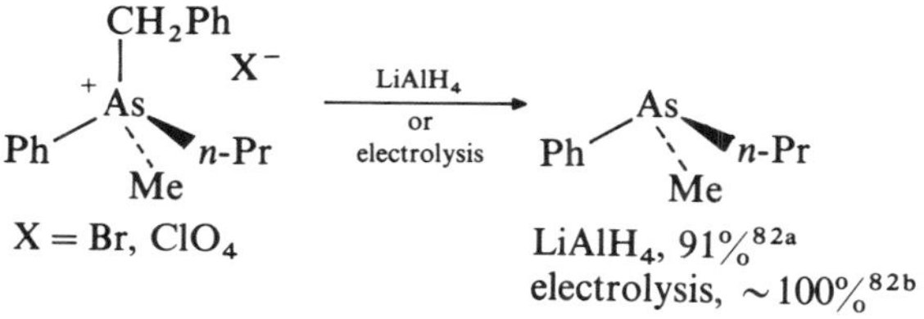

$\sim 15\%$

$$(54)$$

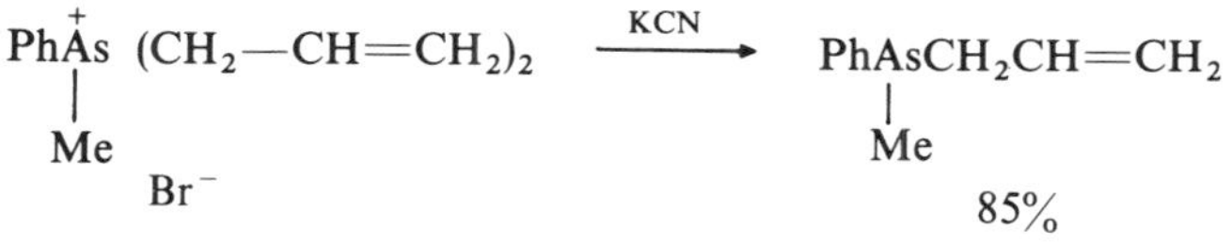

$X = Br, ClO_4$

$LiAlH_4$, 91%[82a]
electrolysis, $\sim 100\%$[82b]

$$\overset{+}{Ph\underset{\underset{Br^-}{\overset{|}{Me}}}{As}} (CH_2-CH=CH_2)_2 \xrightarrow{KCN} \underset{\overset{|}{Me}}{PhAsCH_2CH=CH_2} \qquad (55)$$

85%

11. Miscellaneous methods

Reduction of triorganoarsenic sulfides with n-PrLi has been reported (equation 56)[84]. In the reduction step retention of configuration at the arsenic atom has been observed.

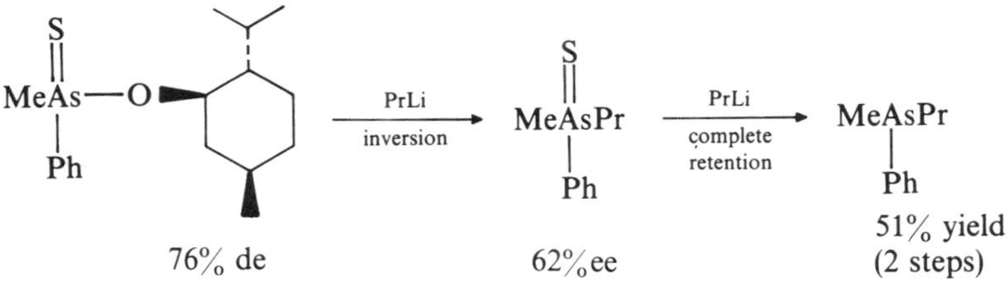

$$(56)$$

76% de   62% ee   51% yield (2 steps)

Diazomethanes react with Me_2AsNMe_2 to give diazo compounds, and elimination of nitrogen has been reported in the reaction of diazomethanes with Me_2AsH (equations 57[85] and 58[86]).

$$Me_2AsNMe_2 \xrightarrow[Me_3SnCl]{CH_2N_2} (Me_2As)_2CN_2 + Me_3SnCl \cdot HNMe_2 \qquad (57)$$
$$78\%$$

$$Me_2AsH \xrightarrow{(CF_3)_2CN_2} Me_2AsCH(CF_3)_2 \qquad (58)$$
$$67\%$$

Reductive elimination from pentaorganoarsenic(V) compounds affords triorgano-arsenic(III) compounds (equation 59)[87].

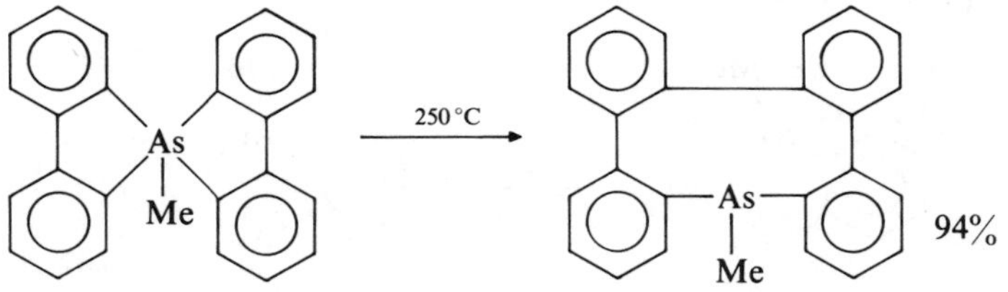

$$(59)$$
$$94\%$$

Ethyl-*n*-propylarsinic acid has been converted to ethylmethyl-*n*-propylarsine by the mould *Scopulariopsis brevicaulis* (equation 60)[88].

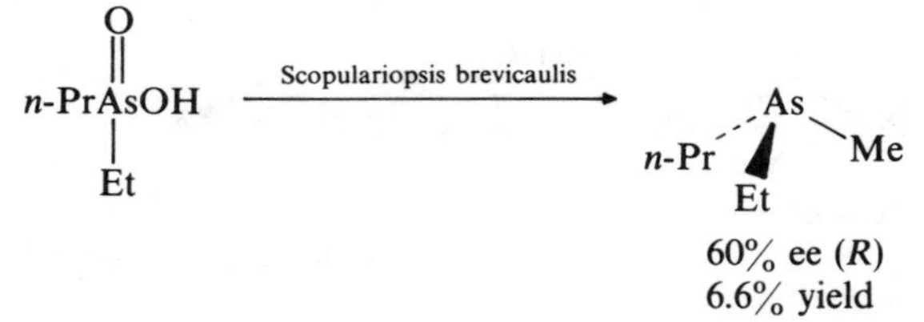

$$(60)$$
$$60\% \text{ ee } (R)$$
$$6.6\% \text{ yield}$$

The enantiomerically pure secondary arsine–iron complex **2** has been used for assymmetric synthesis of a tertiary arsine (equation 61)[89].

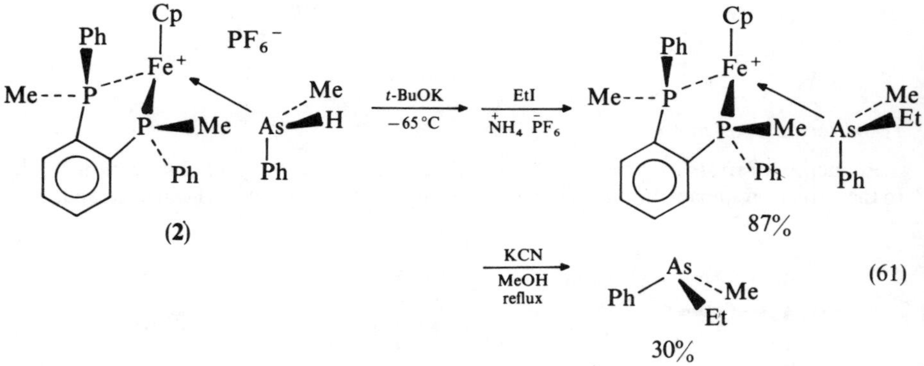

$$(61)$$
$$87\%$$
$$30\%$$

B. Preparation of Diorganoarsenic(III) Halides

1. Transmetallation

It is usually difficult to obtain acyclic diorganoarsenic(III) halides in good yields by selective transmetallation using Grignard or organolithium reagents. However, organomagnesium or -lithium reagents bearing sterically bulky substituents can be introduced to yield diorganoarsenic(III) halides under controlled conditions. Several examples are shown in equations 62^{90}, 63^{91} and 64^{92}.

$$t\text{-BuMgBr} \xrightarrow[-10\,°C]{\text{AsCl}_3} t\text{-Bu}_2\text{AsCl} \qquad (62)$$
$$56\%$$

$$(Me_3Si)_2CHLi \xrightarrow[0\,°C]{\text{AsCl}_3} \{(Me_3Si)_2CH\}_2AsCl \qquad (63)$$
$$61\%$$

$$(64)$$

$$67\%$$

Less nucleophilic organometallic reagents, such as derivatives of tin[93], zirconium[94], lead[95] and cadmium[87], have usually been used for selective transmetallation (equations 65^{93}, 66^{94}, 67^{95} and 68^{87}). Enamines have also been used (equation 69)[96].

$$(CH_2{=}CH)_2SnBu_2 \xrightarrow[100\,°C]{\text{AsBr}_3} (CH_2{=}CH)_2AsBr \qquad (65)$$
$$81\%$$

$$(66)$$

$$80\%$$

$$Et_4Pb \xrightarrow[120\,°C]{\text{EtAsCl}_2} Et_2AsCl \qquad (67)$$
$$78\%$$

$$(68)$$

$$82\%$$

$$(69)$$

$$81\%$$

2. Reduction of diorganoarsenic acids

Reduction has been achieved by use of SO_2/HX or PX_3 (equations 70[97] and 71[98]).

$$MeAsI_2 \xrightarrow{\text{NaOH}} \xrightarrow{\text{MeI}} \underset{\underset{O}{\overset{\|}{}}{Me_2As-ONa} \xrightarrow[\text{HCl/NaI}]{SO_2} \underset{50\%}{Me_2AsI} \qquad (70)$$

$$(71)$$

3. Thermolysis of triorganoarsenic(V) dihalides

The method has been useful for the preparation of alkyl and aryl derivatives, and some unsymmetrical compounds as shown in equations 72[99], 73[100], 74[101] and 75[103]. In the reaction of unsymmetrically substituted compounds such as 3, cleavage of the alkyl–arsenic bond is preferred to that of the aryl–arsenic bond (equation 75[102]). The ease of thermolysis of R_3AsX_2 decreases in the order $X = I > Br > Cl > F$, $R = $ alkyl $>$ aryl.

$$(72)$$

$$(Me_3CCH_2)_3As \xrightarrow{Br_2} (Me_3CCH_2)_3AsBr_2 \xrightarrow[4-5h]{140-150\,°C} (Me_3CCH_2)_2AsBr \qquad (73)$$

$$\underset{99\%}{}$$

$$Ph_3AsCl_2 \xrightarrow{230\,°C} Ph_2AsCl \qquad (74)$$

$$36\%$$

$$(75)$$

4. Redistribution reactions

Diorganoarsenic(III) chlorides or bromides have been prepared by the redistribution reaction of a 2:1 mixture of the corresponding triorganoarsenic compounds and trihalides.

The method is applicable for triorganoarsenic compounds bearing aryl or vinyl substituents (equations 76[103], 77[93] and 78[104]), while for the preparation of alkylarsenic halides this method has not been useful since only adduct formation and redox reactions have been reported between R_3As and AsX_3 (equations 79a and b)[105]. Reaction of tris(trifluoromethyl) arsine with arsenic iodide, however, has been reported to afford a mixture of halides (equation 80)[106]. In the reaction of unsymmetrically substituted compounds such as **4**, cleavage of the aryl–arsenic bond is preferred to that of the alkyl–arsenic bond (equation 78[104]). The cleavage of **4** by this method is in contrast to that of **3** in thermolysis (equation 75).

$$2\,Ph_3As + AsCl_3 \xrightarrow{350\,°C} 3\,Ph_2AsCl \qquad (76)$$
$$63\%$$

$$2\,(CH_2{=}CH)_3As + AsBr_3 \xrightarrow{130\,°C} 3\,(CH_2{=}CH)_2AsBr \qquad (77)$$
$$34\%$$

$$\underset{Ph}{\overset{Me}{>}}As{+}CH_2{)}_nAs\underset{Ph}{\overset{Me}{<}} \xrightarrow[260\,°C]{AsCl_3} \underset{Cl}{\overset{Me}{>}}As{+}CH_2{)}_nAs\underset{Cl}{\overset{Me}{<}} \qquad (78)$$
$$\textbf{(4)} \qquad\qquad\qquad\qquad n = 1,66\%$$
$$n = 2,70\%$$

$$Me_3As + AsCl_3 \longrightarrow Me_3As{\cdot}AsCl_3 \quad 60\% \qquad (79a)$$

$$(Me_3SiCH_2)_3As + AsCl_3 \longrightarrow (Me_3SiCH_2)_3AsCl_2 + (As) \qquad (79b)$$
$$17\%$$

$$(CF_3)_3As + AsI_3 \longrightarrow (CF_3)_2AsI + CF_3AsI_2 \qquad (80)$$

5. Protonolysis of arsenic–carbon bonds in triorganoarsenic compounds

Protonolysis of triorganoarsenic compounds by mineral acids gives diorganoarsenic(III) halides (equation 81)[107].

$$(81)$$

6. Halogenolysis of As—N, As—O, As—S, As—H, As—As bonds

Compounds bearing As—N, As—O, As—S, As—H, As—As bonds react with various halogen-based electrophiles to give the corresponding halogenated compounds (equations 82[108], 83[109], 84[110], 85[111], 86[112] and 87[113]).

$$(C_6F_5)_2AsNMe_2 \xrightarrow{HCl} (C_6F_5)_2AsCl \qquad (82)$$
$$77\%$$

$$Me_2AsOAsMe_2 \xrightarrow{SOCl_2} Me_2AsCl \qquad (83)$$
$$85\%$$

$$(CF_3)_2AsOMe \xrightarrow{BF_3} (CF_3)_2AsF \atop 80\%} \tag{84}$$

$$(85)$$

$$\underset{\substack{| \\ H}}{PhAs} \overbrace{(CH_2)_2}^{} \underset{\substack{| \\ H}}{AsPh} \xrightarrow{Br_2} \underset{\substack{| \\ Br}}{PhAs} \overbrace{(CH_2)_2}^{} \underset{\substack{| \\ Br}}{AsPh} \tag{86}$$

$$Me_2As\!-\!AsMe_2 \xrightarrow{I_2} Me_2AsI \tag{87}$$

7. Halogen exchange

Diorganoarsenic(III) iodides and fluorides have been prepared by the reaction of the corresponding chlorides or bromides with sodium or silver iodide, and ammonium fluoride (equations 88[114] and 89[115]).

$$Ph_2AsCl \xrightarrow{NaI} Ph_2AsI \atop 85\%} \tag{88}$$

$$Me_2AsCl \xrightarrow[SbF_3]{\overset{+}{NH_4}F^-} Me_2AsF \atop 40\%} \tag{89}$$

8. Friedel–Crafts reaction of aromatics with AsCl₃

Electron-rich aromatics react with $AsCl_3$ in the presence or absence of Lewis acids (equation 90)[116].

$$(90)$$

C. Preparation of Monoorganoarsenic(III) Dihalides

1. Transmetallation

Only weak nucleophiles or sterically bulky nucleophiles can produce $RAsX_2$ selectively under controlled conditions (equations 91[90], 92[91], 93[117], 94[118], 95[119], 96[95] and 97[120]).

$$t\text{-BuMgCl} \xrightarrow[-30\,to\,-35\,°C]{AsCl_3} t\text{-BuAsCl}_2 \atop 53\%} \tag{91}$$

$$(Me_3Si)_2CHLi \xrightarrow[0\,°C]{AsCl_3} (Me_3Si)_2CHAsCl_2 \qquad (92)$$
$$70\%$$

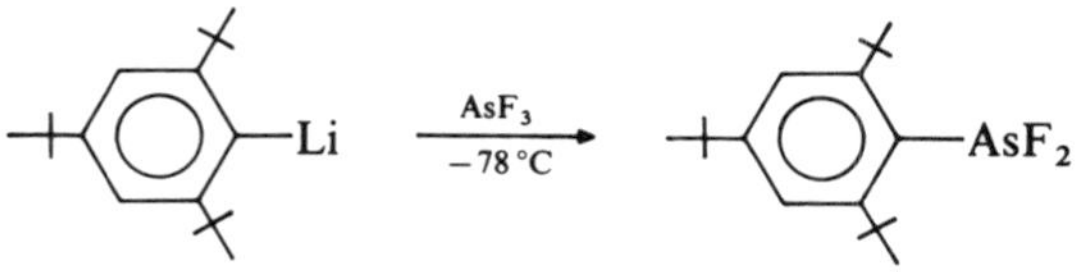

(93)

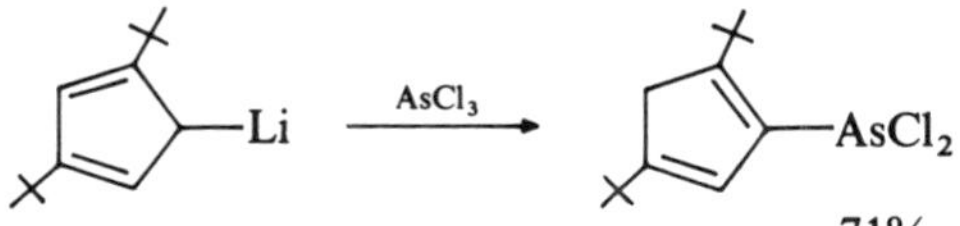

(94)

$$71\%$$

$$(ClCH_2)Hg \xrightarrow{AsCl_3} ClCH_2AsCl_2 \qquad (95)$$
$$60–70\%$$

$$Et_4Pb \xrightarrow{AsCl_3} EtAsCl_2 \qquad (96)$$
$$95\%$$

$$Me_3SiCH=C\begin{smallmatrix}OSiMe_3\\OEt\end{smallmatrix} \xrightarrow[20\,°C]{AsCl_3} \begin{smallmatrix}Me_3Si\\EtO_2C\end{smallmatrix}CHAsCl_2 \qquad (97)$$
$$\sim 100\%$$

2. Reduction of organoarsonic acids

Organoarsenic acids have been reduced by SO_2/HX or PX_3 (equations 98[121], 99[122] and 100[123]).

$$As_2O_3 \xrightarrow{NaOH} As(ONa)_3 \xrightarrow{MeI} MeAs(ONa)_2 \xrightarrow[HCl/NaI]{SO_2} MeAsI_2 \qquad (98)$$
$$78\%$$

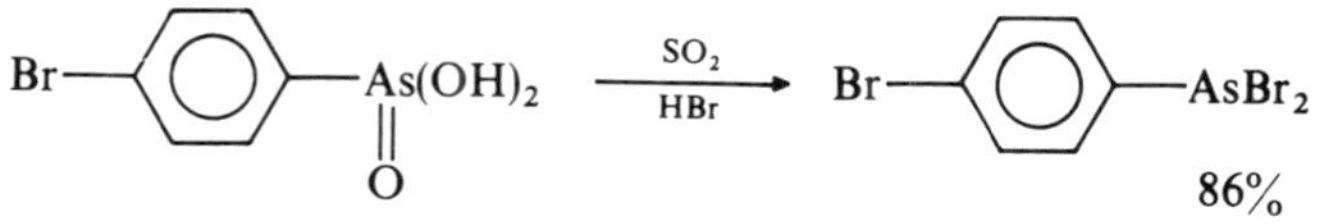

(99)

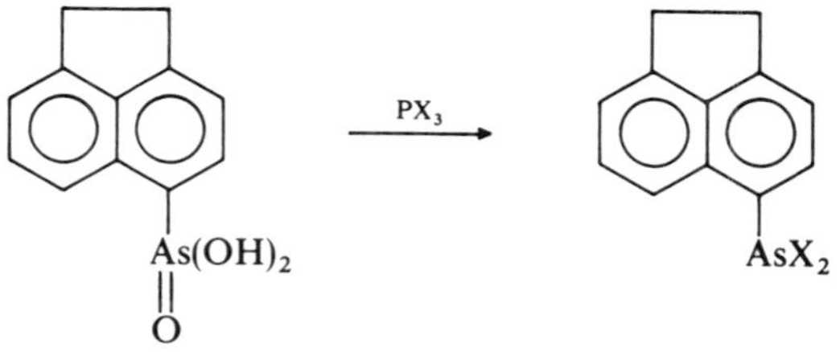

$$86\%$$

(100)

X = Cl, 55%
X = Br, 56%

3. Thermolysis of diorganoarsenic(V) trihalides

Since trihalides are thermolized under milder conditions than dihalides, several alkylarsenic compounds have been obtained in good yields (equations 101[99] and 102[100]) by this reaction.

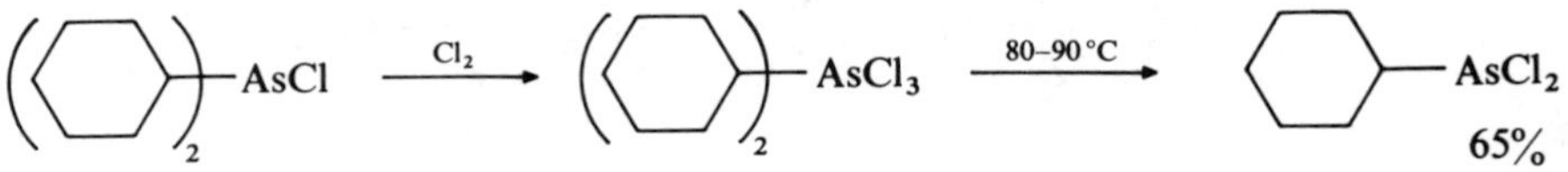

$$\left(\!\left\langle\!\!\left\langle\bigcirc\right\rangle\!\!\right\rangle\!\!-\!\!AsCl\right)_2 \xrightarrow{Cl_2} \left(\!\left\langle\!\!\left\langle\bigcirc\right\rangle\!\!\right\rangle\!\!-\!\!AsCl_3\right)_2 \xrightarrow{80-90\,°C} \left\langle\bigcirc\right\rangle\!\!-\!\!AsCl_2$$

65%

(101)

$$(Me_3CCH_2)_2AsBr \xrightarrow{Br_2} (Me_3CCH_2)_2AsBr_3 \xrightarrow{45-50\,°C} Me_3CCH_2AsBr_2 \quad (102)$$

83%

4. Redistribution reactions

The method has sometimes been useful for the preparation of monoorganoarsenic(III) chlorides and bromides (equations 103[93] and 104[24]).

$$(CH_2{=}CH)_3As + 2\,AsBr_3 \longrightarrow (CH_2{=}CH)AsBr_2 \tag{103}$$

38%

$$Ph_2AsCH_2AsPh_2 + 4\,AsCl_3 \longrightarrow Cl_2AsCH_2AsCl_2 + 4\,PhAsCl_2 \tag{104}$$

5. Protonolysis of arsenic–carbon bonds in triorganoarsenic compounds

Protonolysis of triorganoarsenic compounds by mineral acids gives organoarsenic(III) dihalides (equation 105)[125].

$$Ph_2As(CH_2)_3AsPh_2 \xrightarrow{HI} I_2As(CH_2)_3AsI_2 \tag{105}$$

76%

6. Halogenolysis of As—N, As—O, As—H, As—As bonds

Compounds bearing As—N, As—O, As—H, As—As bonds react with several electrophiles (equations 106[126], 107[127], 108[128] and 109[129]).

$$MeAs(NMe_2)_2 \xrightarrow{HX} MeAsX_2 \quad {\sim}100\% \tag{106}$$

$$X = F, Cl, Br, I$$

$$O{=}As{+}(CH_2)_2As{=}O \xrightarrow{SOCl_2} Cl_2As(CH_2)_2AsCl_2 \tag{107}$$

$$PhAsH_2 \xrightarrow{I_2} PhAsI_2 \quad {\sim}100\% \tag{108}$$

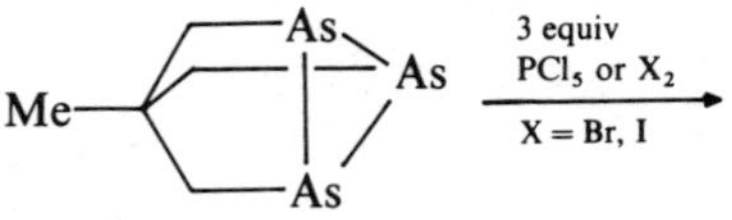

$$Me\!-\!\!\left\langle\begin{matrix}As\\As\\As\end{matrix}\right\rangle \xrightarrow[X = Br, I]{\substack{3\ equiv \\ PCl_5\ or\ X_2}} MeC(CH_2AsX_2)_3 \tag{109}$$

X = Cl, 90%
X = Br, 88%
X = I, 82%

7. Halogen exchange

Halogen exchange reaction takes place by use of metal halides or ammonium fluoride (equations 110[30] and 111[131]).

$$MeAsCl_2 \xrightarrow[80\,°C]{NH_4^+F^-} MeAsF_2 \qquad (110)$$
$$\sim 100\%$$

$$CF_3AsI_2 \xrightarrow{AgCl} CF_3AsCl_2 \qquad (111)$$
$$57\%$$

D. Preparation of Organoarsenic(III) Hydrides

1. Reduction of arsenic halides

Organoarsenic(III) halides are reduced with hydride reagents to give primary and secondary arsines in good yields (equations 112[132], 113[105b,133] and 114[134]).

$$Me_3SiCH_2AsCl_2 \xrightarrow{LiAlH_4} Me_3SiCH_2AsH_2 \qquad (112)$$
$$81\%$$

$$(Me_3SiCH_2)_2AsCl \xrightarrow[HCl]{Zn/Cu/Hg} (Me_3SiCH_2)_2AsH \qquad (113)$$
$$86\%$$

$$CF_3AsI_2 \xrightarrow{HI/Hg} CF_3AsH_2 \qquad (114)$$
$$92\%$$

2. Reduction of As—As compounds

Arsenic–arsenic bonds can be reduced by several reductants to give arsenic hydrides (equations 115[134] and 116[135]).

$$(CF_3)_2As—As(CF_3)_2 \xrightarrow{HI/Hg} (CF_3)_2AsH \qquad (115)$$
$$97\%$$

$$Ph_2As—AsPh_2 \xrightarrow{LiAlH_4} Ph_2AsH \qquad (116)$$
$$86\%$$

3. Protonolysis of arsenide anions

Reaction of arsenide anions with protic acids such as water gives the corresponding hydrides (equations 117[136] and 118[137]).

$$Ph_3As \xrightarrow[liq\,NH_3]{Na} Ph_2As^-Na^+ \xrightarrow{H_2O} Ph_2AsH \qquad (117)$$
$$79\%$$

93% (118)

4. Alkylation of arsenide anions derived from primary arsines

Primary arsines give arsenide anions which react with alkyl halides to afford secondary arsines (equation 119)[138].

$$PhAsH_2 \xrightarrow[\text{liq NH}_3/\text{ether}]{\text{Na}} \xrightarrow{\text{RX}} \begin{array}{c} PhAsH \\ | \\ R \end{array} \qquad (119)$$

$$R = Et, 31\%$$
$$R = Bu, 68\%$$

5. Protonolysis of As—Sn compounds

Phenylarsine has been obtained from reaction of phenyl(trimethylstannyl)arsine and hydrogen bromide (equation 120)[139].

$$\begin{array}{c} PhAsSnMe_3 \\ | \\ H \end{array} \xrightarrow{\text{HBr}} \begin{array}{c} PhAsH_2 \\ \sim 100\% \end{array} \qquad (120)$$

6. Reduction of arsonic and arsinic acids

Primary arsines and secondary arsines have been reduced from arsonic and arsinic acids, respectively (equations 121[58] and 122[140]).

$$\begin{array}{c} BuAs(OH)_2 \\ \| \\ O \end{array} \xrightarrow[\text{HCl}]{\text{Zn/Hg}} \begin{array}{c} BuAsH_2 \\ 41\% \end{array} \qquad (121)$$

$$\begin{array}{c} Ph_2AsOH \\ \| \\ O \end{array} \xrightarrow[\text{HCl}]{\text{Zn/Hg}} \begin{array}{c} Ph_2AsH \\ 82\% \end{array} \qquad (122)$$

E. Preparation of Organoarsenic(III) Nitrogen Compounds

1. Nucleophilic displacement of halides with nitrogen nucleophiles

Amines replace the halides of the As–halogen bond to form the corresponding As—NR_2 bond as shown in equations 123[141], 124[142] and 125[143].

$$Me_2AsI \xrightarrow[-60\,^\circ\text{C}]{\text{MeNH}_2} \begin{array}{c} (Me_2As)_2NMe \\ 60\% \end{array} \qquad (123)$$

$$PhAsCl_2 \xrightarrow{\text{Me}_2\text{NH}} \begin{array}{c} PhAs(NMe_2)_2 \\ 80\% \end{array} \qquad (124)$$

$$PhAsCl_2 \xrightarrow{\text{Et}_2\text{NH}} \begin{array}{c} PhAsNEt_2 \\ | \\ Cl \ 58\% \end{array} \qquad (125)$$

The azido anion also replaces halides to form the corresponding derivatives (equations

126^{144} and 127^{145}).

$$Me_2AsI \xrightarrow{AgN_3} \underset{90\%}{Me_2AsN_3} \tag{126}$$

$$PhAsBr_2 \xrightarrow[toluene]{LiN_3} \underset{\underset{72\%}{Br}}{\overset{|}{Ph-AsN_3}} \tag{127}$$

2. Transmetallation

Amide chlorides have been transformed into amino derivatives by Grignard reagents (equation 128)^{146}.

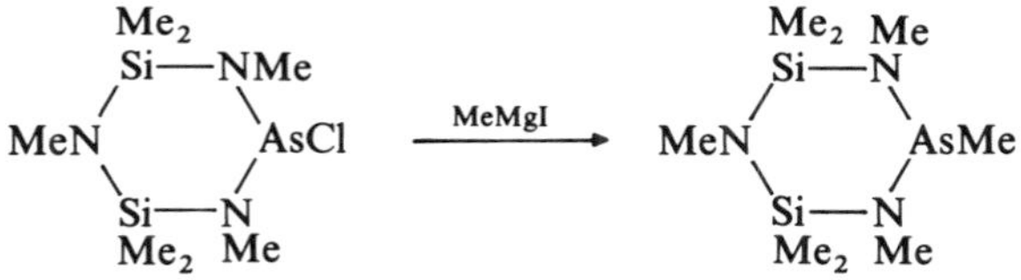

$$\tag{128}$$

60%

3. Miscellaneous methods

Arduengo and coworkers described formation of diorganoarsenic amide **6**, by the reaction of **5** with an activated acetylene (equation 129)^{147}. Reaction of As(NMe$_2$)$_3$ with CF$_3$I gives CF$_3$As(NMe$_2$)$_2$ (equation 130)^{148}.

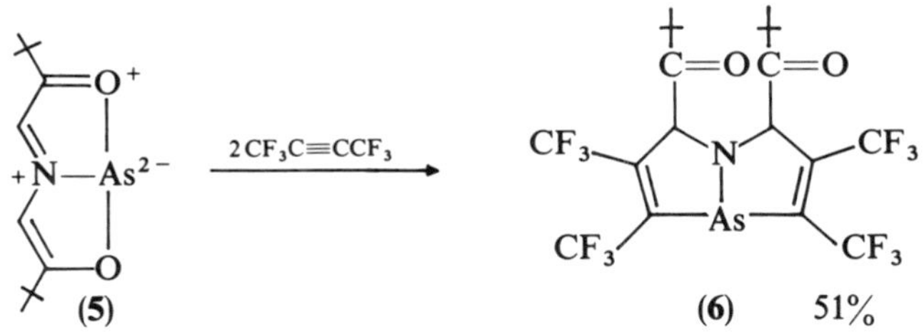

$$\tag{129}$$

(5)   (6)   51%

$$As(NMe_2)_3 \xrightarrow[145\,°C]{CF_3I} \underset{15\%}{CF_3As(NMe_2)_2} \tag{130}$$

Reaction of (CF$_3$)$_2$AsN$_3$ with PPh$_3$ gives (CF$_3$)$_2$AsN=PPh$_3$ (equation 131)^{149}.

$$(CF_3)_2AsN_3 \xrightarrow[rt]{PPh_3} \underset{98\%}{(CF_3)_2AsN=PPh_3} \tag{131}$$

F. Preparation of Organoarsenic(III) Oxygen Compounds

1. Nucleophilic displacement of halides with oxygen nucleophiles

Oxygen nucleophiles replace the halides of the As–halogen bond to form the corresponding As—OR bond as shown in equations 132[150], 133[151], 134[152] and 135[153]. It is also possible to replace only one X^- in $RAsX_2$ with RO^- as shown in equation 136[154].

$$Me_2AsCl \xrightarrow{Me_3SiONa} \underset{84\%}{Me_2AsOSiMe_3} \tag{132}$$

$$MeAsCl_2 \xrightarrow[Et_3N]{c\text{-}C_6H_{11}OH} \underset{84\%}{MeAs(O\text{—}c\text{-}C_6H_{11})_2} \tag{133}$$

$$\tag{134}$$

$$\left(R = Me\text{—}\underset{Me}{\overset{Me}{\bigcirc}}\text{—} \right)$$

$$PrAsCl_2 \xrightarrow{MeCOOAg} \underset{51\%}{PrAs(OCOMe)_2} \tag{135}$$

$$PhAsCl_2 \xrightarrow{RONa} PhAs\overset{OR}{\underset{Cl}{\diagdown}} \tag{136}$$

$$R = Me, 47\%$$
$$R = Et, 80\%$$

2. Alcoholysis of As—N bonds

Compounds bearing As—N bonds react with alcohol to give compounds bearing As—OR bonds as shown in 137[155], 138[156] and 139[157].

$$Me_2AsNMe_2 \xrightarrow{HOCH_2CH_2OH} Me_2AsOCH_2CH_2OAsMe_2 \tag{137}$$

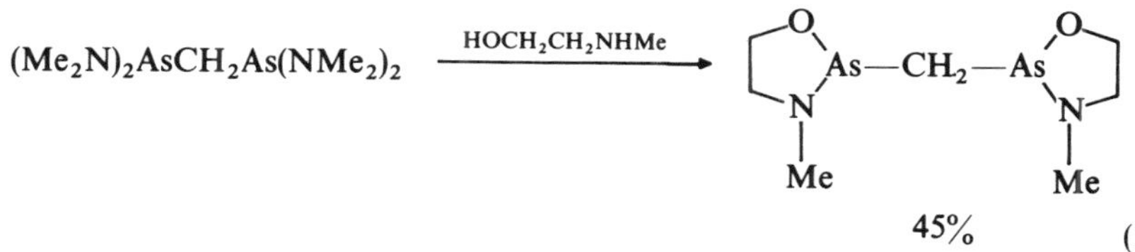

$$(Me_2N)_2AsCH_2As(NMe_2)_2 \xrightarrow{HOCH_2CH_2NHMe}$$

As—CH$_2$—As   45%   (138)

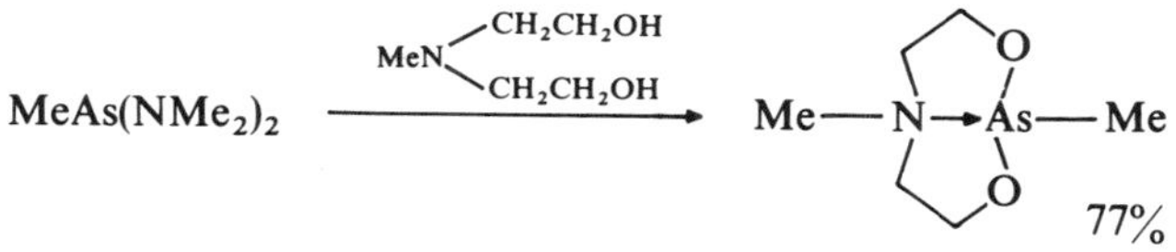

Me   Me

$$MeAs(NMe_2)_2 \xrightarrow[\substack{CH_2CH_2OH \\ MeN \\ CH_2CH_2OH}]{}$$

Me—N→As—Me   (139)

77%

3. Miscellaneous methods

Dialkoxy compounds react with several halogen-based electrophiles to give the corresponding halogenated compounds (equations 140[158] and 141[159]).

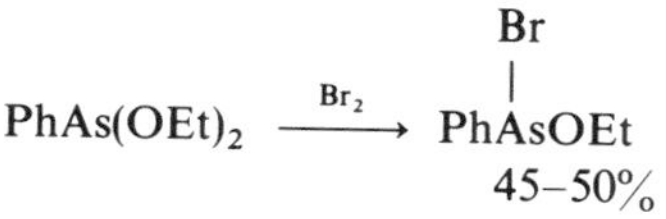

$$PhAs(OEt)_2 \xrightarrow{Br_2} Ph\underset{Br}{\overset{Br}{As}}OEt$$

45–50%   (140)

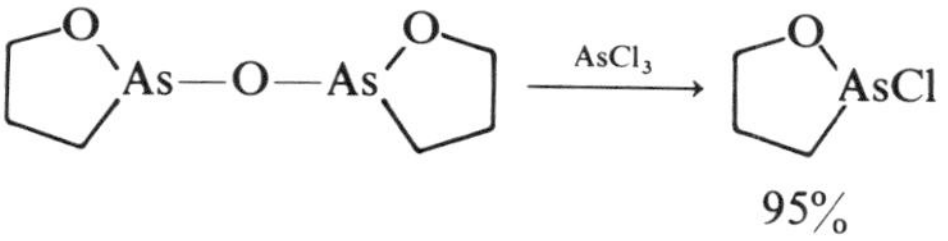

As—O—As $\xrightarrow{AsCl_3}$ AsCl   (141)

95%

Monoalkoxy halogenated compounds react with organometallic reagents to give the corresponding diorgano monoalkoxycompounds (equation 142[159]).

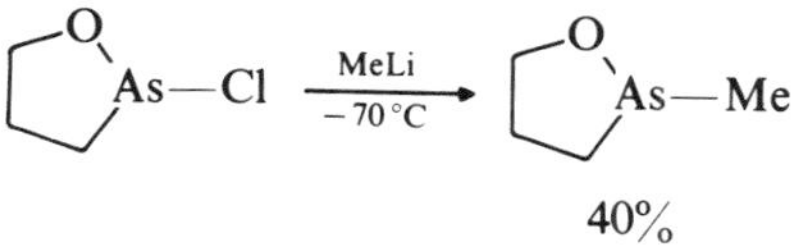

As—Cl $\xrightarrow[-70\,°C]{MeLi}$ As—Me   (142)

40%

Reduction of arsonic acids by use of SO_2 in the presence of protic acids/water gives the corresponding arsinic acids (equations 143[160] and 144[161]).

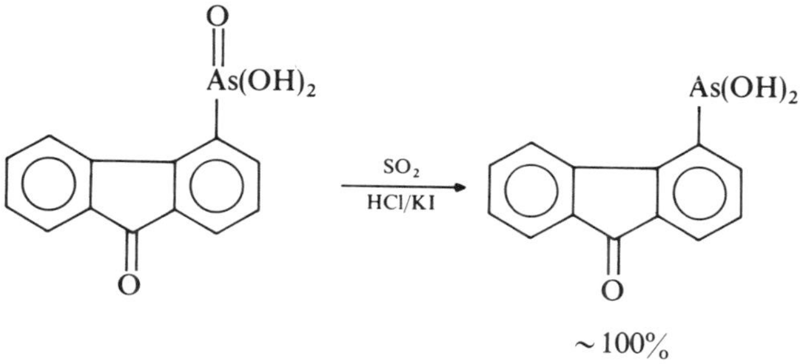

$$\xrightarrow[HCl/KI]{SO_2}$$

(143)

~100%

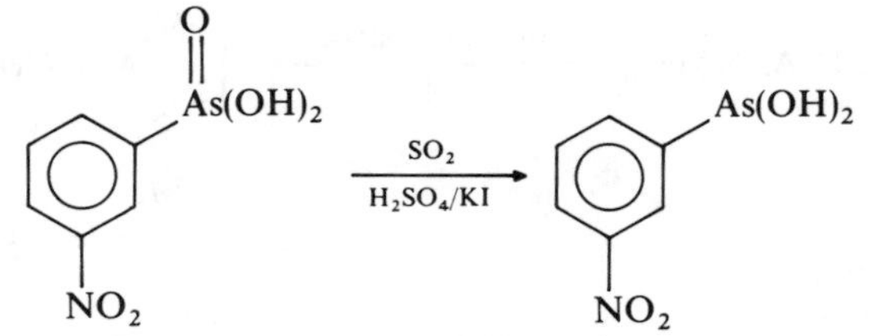

(144)

A tetrameric arsenic–oxygen heterocycle cyclo-$(MeAsO)_4$ (7) has been obtained from aerobic oxidation of cyclo-$(MeAs)_5$ via a metal–carbonyl template reaction using $Mn_2(CO)_{10}$ (equation 145[162]).

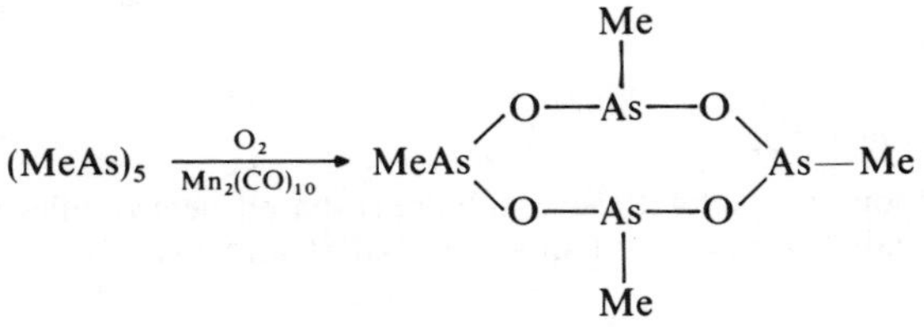

(145)

(7)

G. Preparation of Organoarsenic(III) Compounds Containing Arsenic–Phosphorus, –Antimony or –Bismuth Bonds

1. Exchange reaction of tetraorganodipnictogens

Diarsines (R_2AsAsR_2) react easily with diphosphines, distibines and dibismuthines $(R'_2MMR'_2: M = P, Sb, Bi)$ to give equilibrium mixtures containing $R_2AsMR'_2$ (equation 146[163]).

$$R_2AsAsR_2 + R'_2MMR'_2 \underset{25\,°C}{\overset{K}{\rightleftharpoons}} 2\,R_2AsMR'_2$$

(146)

$M = P, Sb, Bi$

$R = R' = Me, M = P, K = 0.26\ (C_6D_6)^{163a}$
$M = Sb, K = 0.9\ (C_6D_6)^{163a}$
$M = Bi, K = 0.009\ (C_6D_6)^{163a}$
$R = R' = Ph, M = P, K = 0.27\ (CHCl_3)^{163b}$

Red crystals of 10 have been reported to be separable manually from a mixture of biarsole 8 and bistibole 9 (equation 147[163a]).

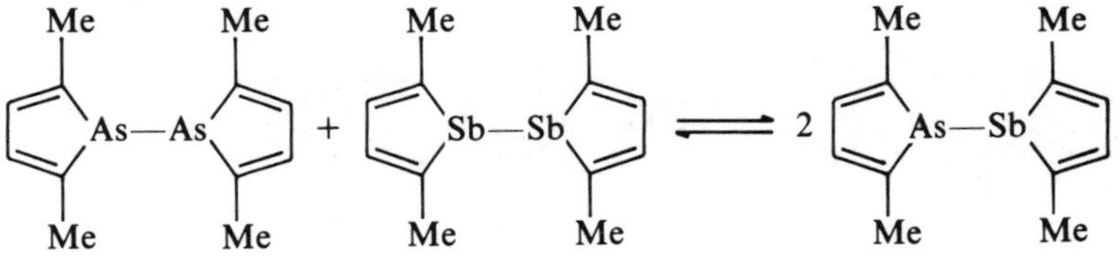

(147)

(8)   (9)   (10)

Usually, mixed tetraorganyls of R_2AsMR_2' can only be obtained in a pure state when the R groups attached to arsenic and Group V elements are very different in their electronic nature as shown in equations 148[164] and 149[165].

$$(CF_3)_2PP(CF_3)_2 \xrightarrow{\text{Me}_2\text{AsH}} \underset{96\%}{(CF_3)_2PAsMe_2} \tag{148}$$

$$(CF_3P)_4 \xrightarrow{\text{Me}_2\text{AsAsMe}_2} \underset{100\%}{(Me_2As)_2PCF_3} \tag{149}$$

2. Miscellaneous methods

Reaction of diazomethane with As═P double bond gives a three-membered arsaphosphacyclopropane (equation 150[166]).

$$(150)$$

The arylarsenic dichloride **11** reacts with bis(trimethylsilyl)phosphidoiron complex **12** to give diphosphaarsirane **13** (equation 151[167]). Cyclic 1,3,2-benzodiphospharsoles have been obtained as shown in equation 152[168].

$$(151)$$

$$(152)$$

H. Preparation of Organoarsenic(III) Compounds Containing Arsenic–Sulfur, –Selenium or –Tellurium Bonds

1. Exchange reaction of tetraorganodipnictogens with diorganodichalcogenides

Diarsines (R_2AsAsR_2) react easily with diorganodichalcogenides (R'EER': M = S, Se, Te) to give equilibrium mixtures with R_2AsER' being the predominant product (equation 153[169,170a]).

$$Me_2AsAsMe_2 + MeEEMe \xrightleftharpoons[25\,°C]{K} 2\,Me_2AsEMe \qquad (153)$$

$$E = S,\ K = 77\ (C_6D_6)$$
$$E = Se,\ K > 10^3\ (C_6D_6)$$
$$E = Te,\ K = 63\ (C_6D_6)$$

2. Nucleophilic displacement with sulfur and selenium nucleophiles

Sulfur and selenium nucleophiles react with arsenic halides, amides and alkoxides to give the corresponding substituted products as shown in equations 154[170], 155[171], 156[172], 157[173], 158[14,174], 159[175], 160[176], 161[177] and 162[178].

$$Me_2AsCl \xrightarrow[Me_3N]{MeEH} Me_2AsEMe \qquad (154)$$

$$E = S,\ 80\%\ ^{170a}$$
$$E = Se,\ 72\%^{169b}$$

$$ArAsCl_2 \xrightarrow{} \qquad (155)$$

$$Ar = Ph,\ 60\%$$
$$Ar = p\text{-}CH_3C_6H_4,\ 32\%$$

$$PhAsCl_2 \xrightarrow[R = Me,\ Et,\ n\text{-}Pr,\ i\text{-}Pr]{\substack{1\ equiv \\ ROCS^-\ K^+ \\ \| \\ S}} \quad \underset{\underset{Cl}{|}}{PhAs}\overset{\overset{S}{\|}}{-}SCOR \qquad (156)$$

$$(CF_3)_2AsI \xrightarrow[rt]{(CF_3E)_2Hg} (CF_3)_2AsECF_3 \qquad (157)$$

$$E = S,\ 90\%$$
$$E = Se,\ 84\%$$

$$MeAs(NMe_2)_2 \xrightarrow{EtEH} MeAs(EEt)_2 \qquad (158)$$

$$E = S,\ 92\%$$
$$E = Se,\ 40\%$$

$$RAs(OMe)_2 \xrightarrow{HSCH_2CH_2OH} RAs(SCH_2CH_2OH)_2 \qquad (159)$$

R = Et, 99%
R = Ph, 100%

$$PhAs(OR)_2 \xrightarrow{CS_2} PhAs(SCOR)_2 \qquad (160)$$

$$\underset{S}{\overset{\|}{}}$$

R = Et, 82%
R = i-Pr, 84%

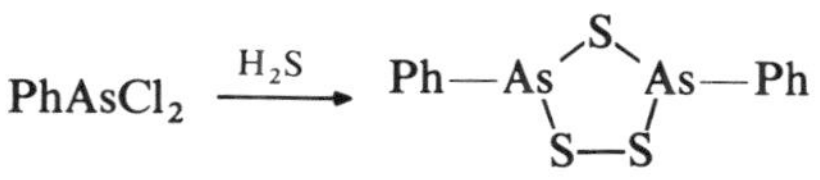

$$PhAsCl_2 \xrightarrow{H_2S} Ph-As\overset{S}{\underset{S-S}{\diagdown}}As-Ph \qquad (161)$$

$$\xrightarrow{H_2Se} O\diagup As-Se-As\diagdown O \qquad (162)$$

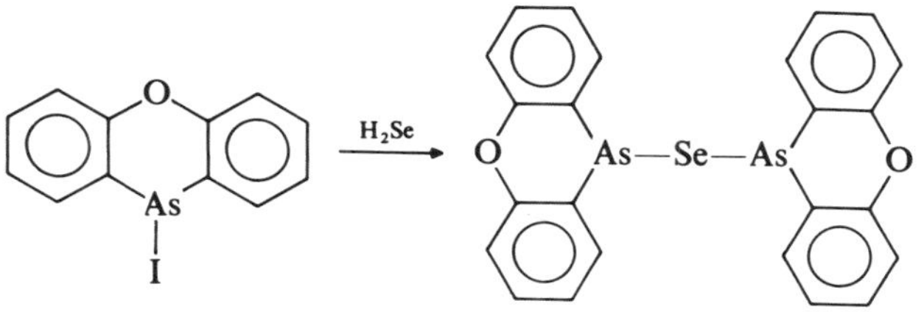

3. Miscellaneous methods

Elemental sulfur and selenium insert into the As—As bond to form $R_2As-E-AsR_2$ (E = S, Se) (equations 163[179] and 164[180]).

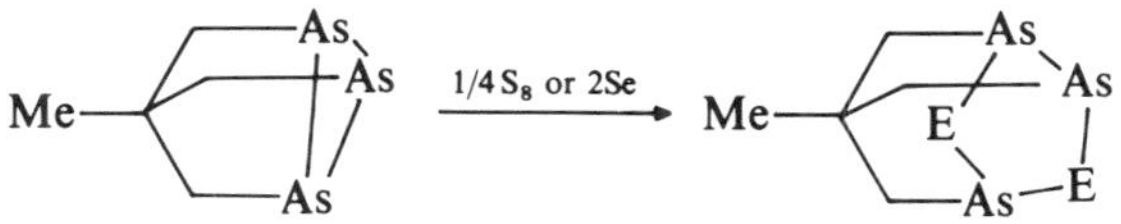

$$\qquad (163)$$

E = S, 88%
E = Se, 69%

$$\text{cyclo-}(MeAs)_5 \xrightarrow[Mo(CO)_6]{S_8} \text{cyclo-}(MeAsS)_n \qquad (164)$$

n = 3, 35%
n = 4, 9%

Elemental sulfur reacts with secondary arsines to give As—S—As compounds (equation 165[181]).

$$(CF_3)_2AsH \xrightarrow{1/4\,S_8} (CF_3)_2AsSAs(CF_3)_2 \qquad (165)$$

Elemental sulfur adds to As=P double bonds to give three-membered phospha-arsathiirane (equation 166[182]).

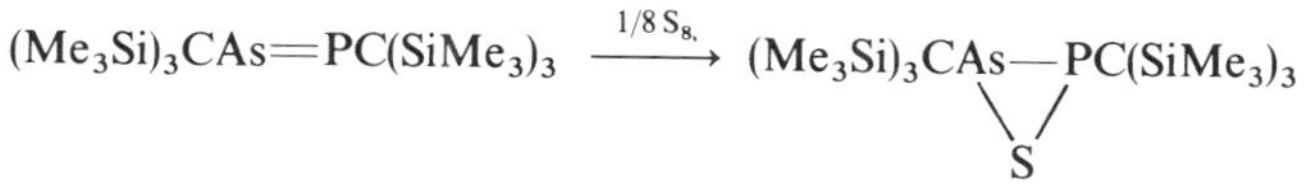

$$(Me_3Si)_3CAs=PC(SiMe_3)_3 \xrightarrow{1/8\,S_8} (Me_3Si)_3CAs\overset{\diagdown\,\diagup}{\underset{S}{}}PC(SiMe_3)_3 \qquad (166)$$

100%

Dialkylarsine sulfides rearrange to alkyl(alkylthio)arylarsine upon heating (equation 167[183]).

$$\underset{\substack{\| \\ S}}{PhAsEt_2} \xrightarrow[\text{or EtI}]{\Delta} \underset{\substack{| \\ SEt}}{PhAsEt} \tag{167}$$

I. Preparation of Organoarsenic(III) Lithium, Sodium, Potassium or Magnesium Compounds

1. Reduction with metal

Triorganoarsenic compounds, diarsines, diorganoarsenic halides are reduced by alkali metals to afford the corresponding diorganoarsenic alkali metal compounds as shown in equations 168[51], 169[138b], 170[20,184], 171[163b], 172[55a], 173[61b,185] and 174[59a]. The selectivity in the As—C cleavage of unsymmetrical tertiary arsines of the type $(R^1)_2R^2As$ has been reported to be thermodynamically controlled[20,184]. The relative stability of possible pairs of anions, i.e. the sum of the pK_a values of their conjugate acids, determines the products. The smaller the sum, the more stable the products as shown in equation 170.

$$Ph_3As \xrightarrow[\text{THF}]{Li} Ph_2As^- \ Li^+ \tag{168}$$

$$\tag{169}$$

$$Ph_2AsEt \xrightarrow[\text{dioxane}]{K} Ph\bar{A}sEt + Ph^- \tag{170}$$
$$(pK_a = 23.5) \ (pK_a = 37)$$
$$\left(\begin{array}{c} \text{no} \quad Ph_2As^- + Et^- \\ (pK_a = 20.3) \quad (pK_a = 42) \end{array} \right)$$

$$Ph_2AsAsPh_2 \xrightarrow[\text{THF}]{Na} Ph_2As^- \ Na^+ \tag{171}$$

$$\tag{172}$$

$$\tag{173}$$

$$Ph_2AsCl \xrightarrow{Li} Ph_2As^- \ Li^+ \tag{174}$$

2. Proton abstraction from arsenic hydrides

Organolithium, Grignard reagents and alkali metals react with arsenic hydrides to give the corresponding arsenide anions as shown in equations 175[60,62,186], 176[187], 177[54,61a], 178[57a,188], 179[58] and 180[138b-e].

$$Ph_2AsH \xrightarrow[\text{THF}]{\text{BuLi}} Ph_2As^- \ Li^+ \tag{175}$$

$$(CF_3)_2AsH \xrightarrow[\text{liq NH}_3]{\text{M}} (CF_3)_2As^- \ M^+ \tag{176}$$

$$M = Na, Cs$$

$$PhMeAsH \xrightarrow[\text{THF}]{\text{Na}} Ph\bar{A}sMe \ Na^+ \tag{177}$$

$$Ph_2AsH \xrightarrow[\text{ether}]{\text{PhMgBr}} Ph_2As^- \ MgBr^+ \tag{178}$$

$$BuAsH_2 \xrightarrow[\text{petroleum ether}]{\substack{2\,\text{equiv} \\ n\text{-BuLi}}} BuAs^{2-} \ 2\,Li^+ \tag{179}$$

$$PhAsH_2 \xrightarrow[\text{liq NH}_3]{1.1\,\text{equiv Na}} PhAsH^- \ Na^+ \tag{180}$$

3. Cleavage of As—As bonds by organolithium reagents

PhLi reacts with As—As bonds to give monolithium arsenides (equation 181[59c,189]).

$$\tag{181}$$

J. Preparation of Compounds Containing Arsenic(III)–Arsenic(III) Bonds

Although an unsymmetrical tetraorganodiarsine, $(CF_3)_2AsAsMe_2$, has been reported to be isolated[164], symmetrization reactions of compounds formulated $R_2AsAsR'_2$ usually occur to give a mixture of $R_2AsAsR'_2$, R_2AsAsR_2 and $R'_2AsAsR'_2$ (equation 182[190]).

$$Me_2AsNMe_2 \xrightarrow{Ph_2AsH} Me_2AsAsPh_2 \underset{-25\,°C}{\rightleftharpoons} \tfrac{1}{2}Me_2AsAsMe_2 + \tfrac{1}{2}Ph_2AsAsPh_2$$

$$\tag{182}$$

1. Reaction of arsenide anions or arsines with As–halogen bonds

Reaction of organoarsenic halides with metals or with metal hydrides gives diarsines as shown in equations 183[191], 184[192], 185[193], 186[194], 187[195], 188[196], 189[197], 190[198], 191[199] and 192[200]. These can be considered as reactions between the organoarsenic anionic species formed upon the reduction and the halides.

$$Me_2AsI \cdot \xrightarrow[\text{Et}_2O]{\text{Li}} Me_2AsAsMe_2 \tag{183}$$

$$80\%$$

$$81\%   84\%$$

$$(184)$$

$$Ph_2As^- K^+ \xrightarrow[\text{dioxane}]{BrCH_2CH_2Br} Ph_2AsAsPh_2 \qquad (185)$$
$$71\%$$

$$AsH_2^- Li^+ \xrightarrow[\text{THF}]{t\text{-BuCCl}} (t\text{-BuC})_2As^- Li^+$$
$$68\%$$

$$\xrightarrow{HBF_4} (t\text{-BuC})_2AsAs(Ct\text{-Bu})_2 \qquad (186)$$
$$64\%$$

$$(187)$$

$$(188)$$

$$54\%$$

$$CF_3AsI_2 \xrightarrow{Hg} (CF_3As)_4 \qquad (189)$$

$$(190)$$

$$46\%$$

$$t\text{-BuAsCl}_2 \xrightarrow{\text{Mg}} \underset{1\%}{(t\text{-BuAs})_5} \tag{191}$$

$$2\,t\text{-BuAsCl}_2 + 2\,\text{AsCl}_3 + 10\,\text{LiH} \longrightarrow t\text{-BuAs}\overset{\text{As}}{\underset{\text{As}}{\diagup\!\!\diagdown}}\text{AsBu-}t \tag{192}$$

$$51\%$$

Primary and secondary arsines have also been used for reactions with arsenic halides (equations 193[201] and 194[201]).

$$\text{PhAsH}_2 + 1.3\,\text{Cl}_2 \xrightarrow{\text{Et}_3\text{N}} \underset{73\%}{(\text{PhAs})_6} \tag{193}$$

$$14.5\,\text{Me}_2\text{AsH} + \text{NH}_2\text{Cl} \longrightarrow \underset{96\%}{\text{Me}_2\text{AsAsMe}_2} \tag{194}$$

2. Reduction of arsonic acids by hypophosphorous acid

This reaction has long been known for the preparation of linear and cyclic polyarsines as shown in equations 195[202] and 196[203].

$$\underset{\underset{\text{O}}{\overset{\|}{}}}{\text{Me}_2\text{AsOH}} \xrightarrow{\text{H}_3\text{PO}_2} \text{Me}_2\text{AsAsMe}_2 \tag{195}$$

$$\underset{\underset{\text{O}}{\overset{\|}{}}}{\text{PhAs(OH)}_2} \xrightarrow{\text{H}_3\text{PO}_2} \begin{array}{c}\text{Ph}\quad\text{Ph}\\ \text{As}\!\!-\!\!\text{As}\\ \text{PhAs}\diagup\qquad\diagdown\text{AsPh}\\ \diagdown\text{As}\!\!-\!\!\text{As}\diagup\\ \text{Ph}\quad\text{Ph}\end{array} \tag{196}$$

3. Reaction of polyarsines with dienes or acetylenes

Cyclic polyarsines add across dienes or acetylenes to give the corresponding heterocycles having As—As bonds (equations 197[204] and 198[205]).

$$(\text{MeAs})_5 \xrightarrow[h\nu]{\begin{array}{c}\text{Me}\quad\text{Me}\\ \diagup\!\!=\!\!\diagdown\end{array}} \begin{array}{c}\text{Me}\quad\text{Me}\\ \\ \text{As}\!\!-\!\!\text{As}\\ \text{Me}\quad\text{Me}\end{array} \tag{197}$$

$$65\%$$

$$(\text{PhAs})_6 \xrightarrow[160\,°\text{C}]{\text{PhC}\!\equiv\!\text{CPh excess}} \begin{array}{c}\text{Ph}\qquad\text{Ph}\\ \\ \text{As}\!\!-\!\!\text{As}\\ \text{Ph}\qquad\text{Ph}\end{array} + \begin{array}{c}\text{Ph}\qquad\text{Ph}\\ \text{PhAs}\diagup\quad\diagdown\text{AsPh}\\ \diagdown\text{As}\diagup\\ \text{Ph}\end{array} \tag{198}$$

$$13.5\% \qquad\qquad 43.5\%$$

4. Miscellaneous methods

Reaction of $(MeAs)_5$ with $RAsH_2$ has been reported to be very efficient for the preparation of highly pure cyclic polyarsines (equation 199[206]).

$$n(MeAs)_5 + 5nRAsH_2 \longrightarrow \underset{80-85\%}{5(RAs)_n} + 5n\,MeAsH_2 \tag{199}$$

$$R = Et, Pr, n = 5$$
$$R = Ph, p\text{-}Tol, n = 6$$

Reaction of Me_2AsI with Ga_2I_4 gives $Me_2AsAsMe_2I^+$ GaI_4^- (equation 200[207]).

$$Me_2AsI \xrightarrow{\quad Ga_2I_4 \quad} \underset{\underset{I}{|}}{Me_2As\overset{+}{As}Me_2} \quad GaI_4^- \tag{200}$$

Reaction of $t\text{-}Bu_2AsLi$ with $MgBr_2$ gives compound **14** having an As—As bond (equation 201[208]).

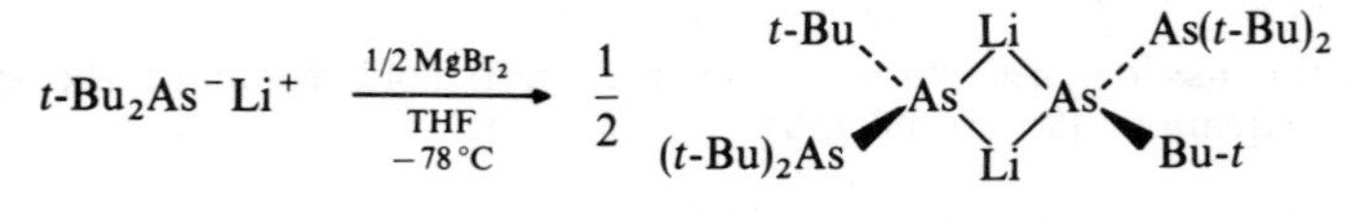

$$t\text{-}Bu_2As^-Li^+ \xrightarrow[\substack{THF \\ -78\,°C}]{1/2\,MgBr_2} \frac{1}{2} \tag{201}$$

(14) 65%

K. Preparation of Compounds Containing Arsenic(III)–Main Group Metal Bonds

In recent years, the area of gallium arsenide has received much attention in relation to Group III/V semiconductors. Two reviews have appeared on the topic[25,26].

1. Reaction of organoarsenic anions with metal halides

This method has been quite useful for the preparation of metal complexes. Several main-group element bonded compounds with boron, gallium, indium, silicon, germanium, tin, lead as the element are shown in equations 202[209], 203[210], 204[211], 205[212], 206[213], 207[214], 208[215], 209[216], 210[217], 211[218], 212[219], 213[220], 214[221] and 215[222].

$$Ph_2BCl \xrightarrow[THF]{Ph_2As^-Na^+} \underset{57\%}{Ph_2AsBPh_2} \tag{202}$$

$$Et_2NBBr_2 \xrightarrow{Et_2As^-Li^+} \underset{\underset{\substack{Br \\ 42\%}}{|}}{Et_2AsBNEt_2} + \underset{21\%}{(Et_2As)_2BNEt_2} \tag{203}$$

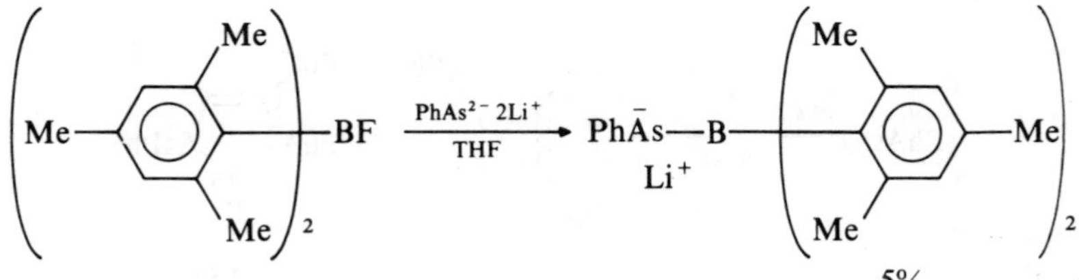

$$\tag{204}$$

5%

$$t\text{-Bu}_2\text{GaCl} \xrightarrow[\text{benzene}]{t\text{-Bu}_2\text{As}^-\text{Li}^+} t\text{-Bu}_2\text{As}\!-\!\text{Ga}(t\text{-Bu})_2 \qquad (205)$$
$$\text{(monomer)} \quad 97\%$$

$$t\text{-Bu}_2\text{GaCl} \xrightarrow{[(\text{Me}_3\text{Si})_2\text{CH}]\bar{\text{A}}\text{s}(\text{SiPh}_3)\text{Li}^+} \underset{\underset{\text{SiPh}_3}{\mid}}{[(\text{Me}_3\text{Si})_2\text{CH}]\,\text{As}\!-\!\text{Ga}(t\text{-Bu})_2} \qquad (206)$$
$$\text{(monomer)} \quad 55\%$$

$$2\,\text{GaCl}_3 + 2(t\text{-Bu})_2\text{As}^-\text{Li}^+ + 4\,\text{MeLi} \xrightarrow[\text{or THF}]{\text{toluene}} \qquad (207)$$

$$\xrightarrow[\text{THF}]{\text{PhAs}^{2-}2\text{Li}^+} \qquad 21\% \qquad (208)$$

$$\text{GaCl}_3 \xrightarrow[n\text{-hexane}]{(\text{Me}_3\text{SiCH}_2)_2\bar{\text{A}}\text{sLi}^+} \{[(\text{Me}_3\text{SiCH}_2)_2\text{As}]_3\text{Ga}\}_2 \qquad (209)$$
$$\text{(dimer)} \quad 22\%$$

$$\text{GaCl}_3 \xrightarrow[\text{THF}]{4\,\text{equiv Ph}_2\text{As}^-\text{Li}^+} (\text{Ph}_2\text{As})_4\text{Ga}^- \quad \text{Li(THF)}_4^+ \qquad (210)$$
$$70\%$$

$$\underset{\underset{\text{Me}}{\mid}}{\overset{\overset{\text{Me}}{\mid}}{\text{Me}_2\text{AsCH}_2\text{CH}_2\text{Si}\!-\!\text{Cl}}} \xrightarrow[n\text{-hexane}]{\text{Me}_2\text{As}^-\text{Li}^+} \underset{\underset{\text{Me}}{\mid}}{\overset{\overset{\text{Me}}{\mid}}{\text{Me}_2\text{As}\!-\!\text{Si}\!-\!\text{CH}_2\text{CH}_2\text{AsMe}_2}} \qquad (211)$$
$$85\%$$

$$\text{Me}_3\text{SiCl} \xrightarrow[\text{ether}]{t\text{-BuAs}^{2-}2\,\text{Li}^+} t\text{-BuAs(SiMe}_3)_2 \qquad (212)$$
$$82\%$$

$$\text{Me}_3\text{GeCl} \xrightarrow[n\text{-hexane/ether}]{\text{Me}_2\text{As}^-\text{Li}^+} \text{Me}_2\text{AsGeMe}_3 \qquad (213)$$
$$91\%$$

$$\text{R}_3\text{SnBr} \xrightarrow[\text{liq NH}_3]{\text{Ph}_2\text{As}^-\text{Na}^+} \text{Ph}_2\text{AsSnR}_3 \qquad (214)$$
$$\text{R} = \text{Et, Pr, Bu, } 60\text{--}62\%$$

$$Ph_3MCl \xrightarrow[\text{liq NH}_3]{Ph_2As^-Na^+} Ph_2AsMPh_3 \tag{215}$$

$$M = Ge, 64\%$$
$$M = Pb, 42\%$$

Primary and secondary arsines have also been used as nucleophiles for reactions with tin halides (equation 216[223]).

$$2\,R^1_3SnCl \xrightarrow[\text{Et}_3N]{R^2AsH_2} R^2As(SnR^1_3)_2 \tag{216}$$

$$R^1 = R^2 = Ph, 79\%$$
$$R^1 = Me, R^2 = Ph, 72\%$$
$$R^1 = R^2 = Me, 63\%$$

Silylarsines have been very useful for the preparation of As—B, As—Ga and As—Sn compounds (equations 217[224], 218[225], 219[226] and 220[227]).

$$Me_2AsSiMe_3 \xrightarrow{BF_3} (Me_2As)_2BF \tag{217}$$

$$(Me_3SiCH_2)_2AsSiMe_3 \xrightarrow{GaBr_3} [(Me_3SiCH_2)_2AsGaBr_2]_3 \tag{218}$$
$$\text{(trimer)}$$

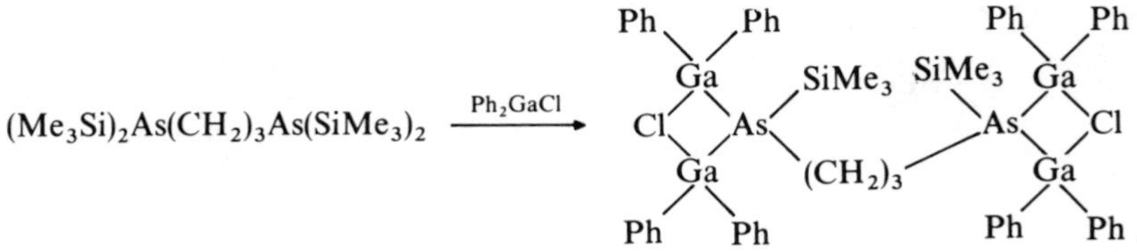

$$(219)$$

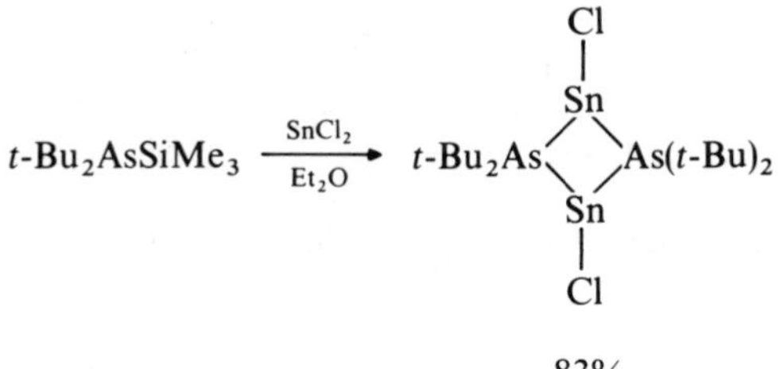

$$(220)$$

$$83\%$$

2. Reaction of arsenic hydrides with trialkylaluminum, -gallium and -indium derivatives

Reaction of arsenic hydrides with trialkylaluminum, -gallium and -indium has been useful for preparation of As—Al, As—Ga and As—In compounds with elimination of an alkane as shown in equations 221[228], 222[228a], 223[229] and 224[230].

$$Me_2AsH \xrightarrow[\substack{M = Al, 200-215\,^\circ C \\ M = Ga, 30-50\,^\circ C \\ M = In, 15-25\,^\circ C}]{Me_3M} (Me_2AsMMe_2)_n \tag{221}$$

$$M = In, n = 3^{228b}$$

$$MeAsH_2 \xrightarrow[\substack{M = Al, 200-210\,°C \\ M = Ga, 95-100\,°C \\ M = In, 0-130\,°C}]{Me_3M} (MeAsMMe)_m \quad \text{(polymer)} \tag{222}$$

$$i\text{-}Pr_2AsH \xrightarrow[130\,°C]{Me_3Ga} (i\text{-}Pr_2AsGaMe_2)_3 \tag{223}$$

$$PhAsH_2 \xrightarrow[50\,°C]{(Me_3SiCH_2)_3Ga} [(PhAsH)(R_2Ga)(PhAs)_6(RGa)_4] \tag{224}$$
$$R = Me_3SiCH_2$$

3. Reaction of arsenic halides with metal anions

Silyl and stannyl anions react with arsenic halides to give the corresponding As—Si and As—Sn compounds, respectively (equations 225[231] and 226[222b,232]).

$$t\text{-}BuAsCl_2 \xrightarrow[\text{ether}]{(Me_3Si)_2Mg(DME)} t\text{-}BuAs(SiMe_3)_2 \tag{225}$$
$$61\%$$

$$Ph_2AsCl \xrightarrow[\text{THF}]{Ph_3SnLi} Ph_2AsSnPh_3 \tag{226}$$
$$81\% \text{ and}$$

4. Miscellaneous methods

Primary and secondary arsines react with boron hydrides to give cyclic trimers or polymers (equation 227[233]).

$$6(CF_3)_2AsH \xrightarrow{3\,B_2H_6} 2[(CF_3)_2AsBH_2]_3 \tag{227}$$

Arsasilene **15** reacts with benzophenone to give the corresponding four-membered heterocycle **16** (equation 228[234]).

$$\tag{228}$$

The lithium salt **17** reacts with $AsCl_3$ to yield the arsacarbollyl derivative **18**, which has been transformed to **19** with $i\text{-}PrMgBr$ (equation 229[235]).

$$Li_2(Me_2C_2B_9H_9) \xrightarrow{AsCl_3} (Me_2C_2B_9H_9)AsCl \xrightarrow{i\text{-}PrMgBr} (Me_2C_2B_9H_9)AsPr\text{-}i$$
$$\textbf{(17)} \qquad\qquad \textbf{(18)} \quad 21\% \qquad\qquad \textbf{(19)} \quad 40\%$$
$$\tag{229}$$

L. Preparation of Compounds with Arsenic(III)–Transition Metal Bonds

1. Reaction of organoarsenic halides with metal anions

The utilization of organoarsenic halides has been a useful method for the preparation of transition metal complexes. Several compounds with Cr, Mo, W, Re, Fe, Co and Ni as the transition metal are shown in equations 230[236], 231[237], 232[238], 233[239a], 234[239a], 235[240], 236[236a,236c,241], 237[240] and 238[242].

$$[CpM(CO)_3]\,Na \xrightarrow[\text{cyclohexane}]{R_2AsCl} R_2AsM(CO)_3Cp \qquad (230)$$

$$M = Cr, Mo, W \qquad\qquad
\begin{aligned}
&R = C_6F_5,\ M = Mo,\ 65\%^{236a}\\
&R = Me,\ M = Cr,\ 41\%^{236b}\\
&R = Me,\ M = Mo,\ 72\%^{236b}\\
&R = Me,\ M = W,\ 85\%^{236b}
\end{aligned}$$

$$[CpW(CO)_3]Na \xrightarrow{t\text{-BuAsCl}_2} t\text{-BuAsW(CO)}_3Cp \xrightarrow{MeLi} t\text{-BuAsW(CO)}_3Cp$$
$$\qquad\qquad\qquad\qquad\qquad\quad |\qquad\qquad\qquad\qquad\qquad |$$
$$\qquad\qquad\qquad\qquad\qquad\quad Cl\qquad\qquad\qquad\qquad\qquad Me$$

$$(231)$$

$$Cr_2(CO)_{10}Na_2 \xrightarrow{t\text{-Bu}_2AsCl} t\text{-BuAs}\{Cr(CO)_5\}_2 \qquad (232)$$
$$60\%$$

$$W_2(CO)_{10}Na_2 \xrightarrow[\text{THF}]{PhAsCl_2}
\begin{array}{c}
Ph \diagdown \quad \diagup W(CO)_5 \\
As \\
\diagup \quad \diagdown \\
(OC)_5W\!-\!\!-W(CO)_5
\end{array} \qquad (233)$$
$$5\%$$

$$W_2(CO)_{10}Na_2 \xrightarrow[\text{CH}_2\text{Cl}_2]{PhAsCl_2}
\begin{array}{c}
Ph \diagdown \qquad \diagup W(CO)_5 \\
\diagup As\!=\!As\diagdown \\
(OC)_5W \qquad\qquad Ph
\end{array} \qquad (234)$$
$$24\%$$

$$Re(CO)_5K \xrightarrow[\text{pentane}]{Me_2AsCl} Me_2AsRe(CO)_5 \qquad (235)$$
$$49\%$$

$$[Fe(CO)_2Cp][Bu_3PMe] \xrightarrow[\text{benzene}]{Me_2AsCl} Me_2AsFe(CO)_2Cp \qquad (236)$$
$$55\%$$

$$Co(CO)_3PMe_3K \xrightarrow[\text{Pentane}]{Me_2AsCl} Me_2AsCo(CO)_3PMe_3 \qquad (237)$$
$$55\%$$

$$[Ni_6(CO)_{12}][Me_4N]_2 \xrightarrow{MeAsBr_2} [Ni_{10}(AsMe)_2(CO)_{18}][Me_4N]_2 \qquad (238)$$
$$40\%$$

2. Reaction of organoarsenic anions with metal halides

This has been useful for the preparation of transition metal complexes with Ti, Zr and Rh as the transition metal (equations 239[243], 240[244], 241[245] and 242[246]).

$$Cp_2TiCl_2 \xrightarrow{(EtAs)_5K_2} Cp_2Ti(AsEt)_3 \qquad (239)$$

$$Cp_2ZrCl_2 \xrightarrow[THF]{Ph_2AsLi} \underset{92\%}{Cp_2Zr(AsPh_2)_2} \qquad (240)$$

$$[Rh(CO)_2Cl]_2 \xrightarrow{t\text{-}Bu_2AsLi} \underset{35\%}{Rh_6(CO)_9(\mu\text{-}CO)_2(\mu\text{-}t\text{-}Bu_2As)_2(\mu_4\text{-}t\text{-}BuAs)}$$

$$+ \underset{25\%}{[(OC)_2RhAs(t\text{-}Bu)_2]_2} \qquad (241)$$

$$CpRhCH_2I(CO) \underset{I}{\big|} \xrightarrow{PhAsHLi} \underset{\underset{Ph}{\diagup}}{\overset{CpRh(CO)}{\overset{\diagup\diagdown}{As\!-\!CH_2}}} \qquad (242)$$

$$37\%$$

3. Reaction of arsenic hydrides with Co and Ir compounds

The arsenic–transition metal bond has been formed by reaction of arsenic hydrides with Co and Ir compounds (equations 243[247], 244[248] and 245[249]).

$$HCo(CO)_4 \xrightarrow{Me_2AsH} Me_2AsCo(CO)_3 \qquad (243)$$

$$Co(BF_4)_2\cdot 6H_2O \xrightarrow[tripod]{PhAsH_2} \underset{50\%}{[(tripod)Co\text{-}\mu_2(\eta^3\text{-}PhAs_3)Co(tripod)]^{2+}2\,BF_4^-} \qquad (244)$$

$$tripod = CH_3C(CH_2PPh_2)_3$$

$$Ir_4(CO)_{12} \xrightarrow[toluene\ reflux]{4t\text{-}Bu_2AsH} \underset{43\%}{[Ir(\mu\text{-}t\text{-}Bu_2As)]_4(CO)_2(\mu\text{-}CO)_2}$$

$$+ \underset{12\%}{[(OC)_2IrAs(t\text{-}Bu)_2]_2} \qquad (245)$$

4. From As—As bonded compounds

Reaction of compounds with an As—As bond with M—M bonds has been useful for preparation of As—M bonded compounds as shown in equations 246[250], 247[251], 248[252], 249[250], 250[253], 251[254] and 252[255].

$$[CpMo(CO)_3]_2 \xrightarrow[toluene]{(CF_3)_2AsAs(CF_3)_2} \underset{64\%}{(CF_3)_2AsMo(CO)_3Cp} \qquad (246)$$

$$Mo(CO)_6 \xrightarrow{cyclo\text{-}(MeAs)_5} \underset{60\%}{cyclo\text{-}(MeAs)_{10}Mo_2(CO)_6} \qquad (247)$$

$$Re_2(CO)_{10} \xrightarrow{cyclo\text{-}(MeAs)_5} cyclo\text{-}[(MeAs)_7(As)Re(CO)_4]Re_2(CO)_6 \qquad (248)$$
$$26\%$$

$$[CpFe(CO)_2]_2 \xrightarrow{(CF_3)_2AsAs(CF_3)_2} (CF_3)_2AsFe(CO)_2Cp \qquad (249)$$
$$65\%$$

$$[CpRu(CO)_2]_2 \xrightarrow[\substack{R=Me,\,n=5 \\ R=Ph,\,n=6}]{cyclo\text{-}(RAs)_n} [CpRu(CO)]_2(\mu\text{-}CO)(\mu\text{-}AsR) \qquad (250)$$
$$R = Me, 70\%$$
$$R = Ph, 80\%$$

$$Co_2(CO)_8 \xrightarrow[benzene]{Ph_2AsAsPh_2} Ph_2AsCo(CO)_3PPh_3 \qquad (251)$$
$$83\%$$

$$Co_2(CO)_8 \xrightarrow{cyclo\text{-}(PhAs)_6} [Co_8(\mu_6\text{-}As)(\mu_4\text{-}As)(\mu_4\text{-}AsPh)_2(CO)_{16}]_2$$
$$21\%$$
$$(252)$$

Cp_2TiCl_2 reacts with cyclo-$(AsPh)_6$ in the presence of Mg to give $Cp_2Ti(AsPh)_3$ (equation 253[256]).

$$Cp_2TiCl_2 + Mg \xrightarrow{1/2\,cyclo\text{-}(PhAs)_6} Cp_2Ti(AsPh)_3 \qquad (253)$$
$$60\%$$

5. Miscellaneous methods

$HAs[CpMn(CO)_2]_2$ reacts with diazomethane to give the insertion product $MeAs[CpMn(CO)_2]_2$ (equation 254[257]).

$$CpMn(CO)_2 \cdot THF \xrightarrow{AsH_3} HAs[CpMn(CO)_2]_2 \xrightarrow{CH_2N_2} MeAs[CpMn(CO)_2]_2$$
$$80\% \qquad\qquad\qquad 70\%$$
$$(254)$$

$PhAsH_2Cr(CO)_5$ has been metallated with butyllithium to give $PhAsLi_2Cr(CO)_5$, which can be converted to $PhAs[Cr(CO)_5]_2$ by treatment with N,N-dichlorocyclo-hexylamine (equation 255[258]).

$$PhAsH_2 \cdot Cr(CO)_5 \xrightarrow[-78\,°C]{BuLi} PhAsLi_2 \cdot Cr(CO)_5 \xrightarrow[\substack{-50\,°C \\ pentane}]{} PhAs[Cr(CO)_5]_2$$
$$11\%$$
$$(255)$$

$PhAsCl_2[CpMn(CO)_2]$ reacts with t-BuLi[259] or $Fe_2(CO)_9$[258b] to give $PhAs[CpMn(CO)_2]_2$ (equation 256[259]).

$$PhAsCl_2 \cdot [CpMn(CO)_2] \xrightarrow[TMEDA]{t\text{-}BuLi} PhAs[CpMn(CO)_2]_2 \qquad (256)$$
$$90\%$$

$ClAs[Cr(CO)_5]_2$ reacts with Ph_3Bi to give the phenylated product $PhAs[Cr(CO)_5]_2$

(equation 257[260]).

$$ClAs[Cr(CO)_5]_2 \xrightarrow[CH_2Cl_2]{Ph_3Bi} PhAs[Cr(CO)_5]_2 \qquad (257)$$
$$90\%$$

M. Preparation of Arsabenzenes and Compounds with Double Bonding Between As and Group 14 or 15 Elements

1. Arsabenzenes and compounds having As=C bonds

Arsabenzenes (**21**) have been prepared by dehydrohalogenation of **20** (equation 258[261]).

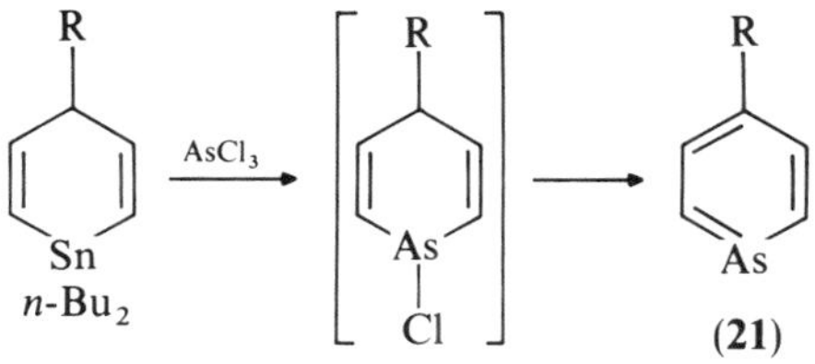

$$(258)$$

(**20**) $R = CH_2CO_2Et$, 68%[261b]

Rearrangement of **22** with protic acids gives 2-aryl substituted arsabenzenes (equation 259[262]).

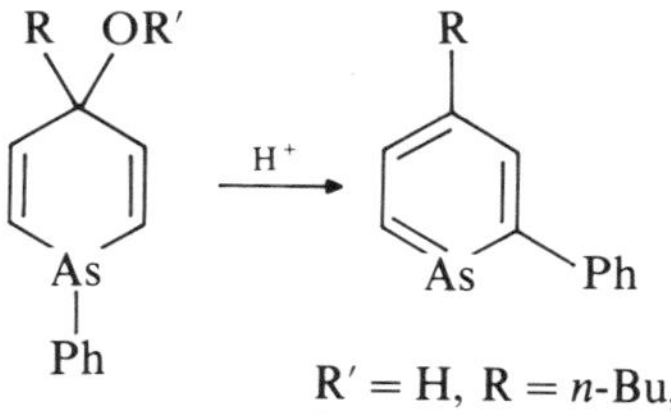

$$(259)$$

$R' = H$, $R = n$-Bu, 46%
$R' = H$, $R = Ph$, 61%
(**22**)

Elimination of methanol followed by a t-Bu group from **23** by use of protic acids gives 4-substituted arsabenzenes (equation 260[263]).

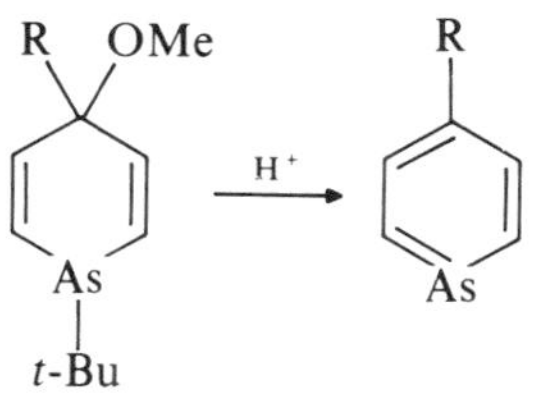

$$(260)$$

(**23**)

Diels-Alder reaction of **24** with Danishefsky's diene gives the substituted arsabenzene **25** (equation 261[120a]).

Diels-Alder reaction of **26** with acetylenes gives substituted arsabenzenes (equation 262[264]).

Reaction of **27** with tetrazine **28** gives arsanaphthalene (equation 263[265]).

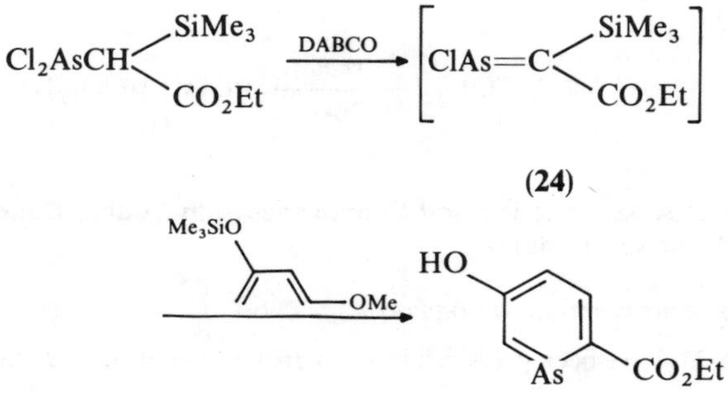

(24)

(261)

(25) 20%

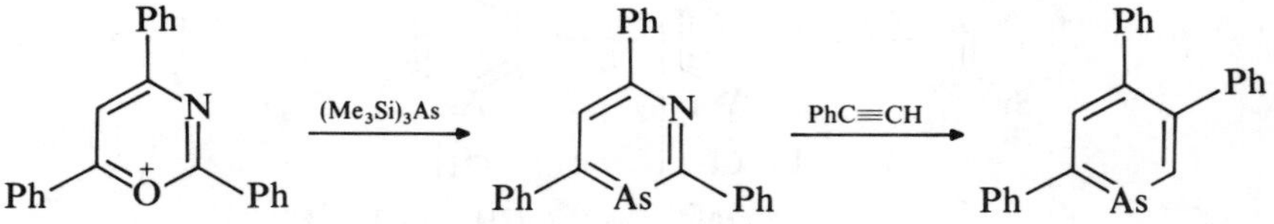

(26) 25%   85%

(262)

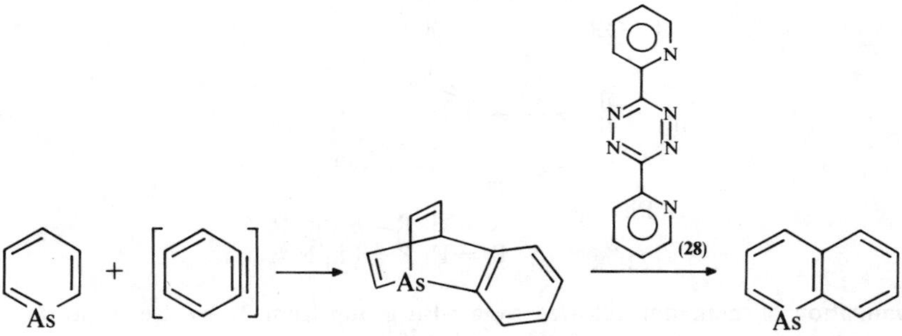

(27)   (263)

33%

Arsacyclopentadienyl anion has been prepared by reductive cleavage of the As—Ph bond in **29** and the crystal structure has been reported (equation 264[94]).

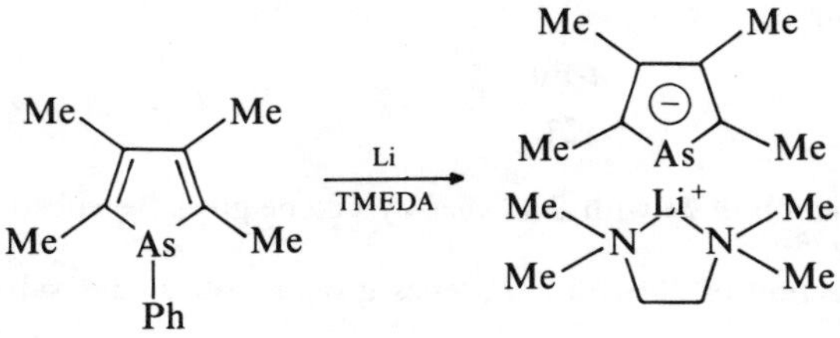

(264)

(29)

Arsadiazoles have been prepared by treating $AsCl_3$ with phenylhydrazone in the presence of triethylamine (equation 265[266]).

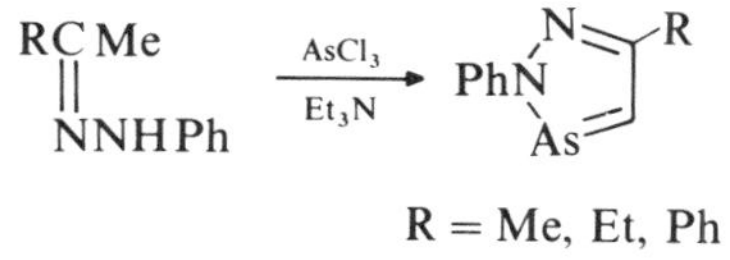

$$
\begin{array}{c}
\underset{\underset{NNHPh}{\|}}{RCMe} \xrightarrow[Et_3N]{AsCl_3} PhN\underset{As}{\overset{N}{\diagup}}\diagdown R
\end{array}
\tag{265}
$$

$$R = Me,\ Et,\ Ph$$

1,3-Dipolar cycloadditions of ethyl diazoacetate with arsaalkenes gives diazaarsoles (equations 266[267] and 267[120b]).

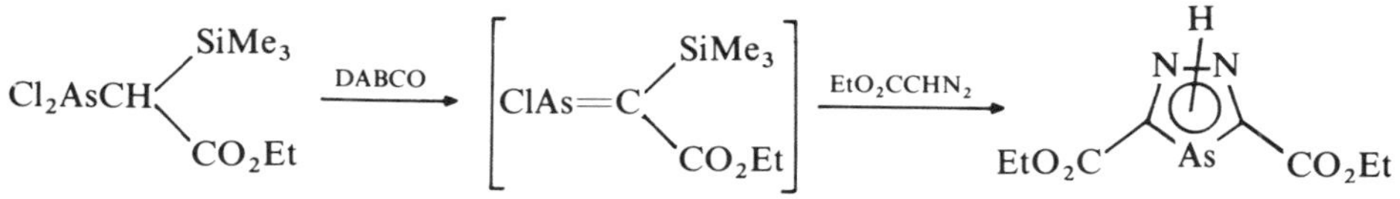

$$
Cl_2AsCH\overset{SiMe_3}{\underset{CO_2Et}{\diagup}} \xrightarrow{DABCO} \left[ClAs\!=\!C\overset{SiMe_3}{\underset{CO_2Et}{\diagup}} \right] \xrightarrow{EtO_2CCHN_2} \underset{EtO_2C}{\overset{H}{\underset{As}{\diagdown}}}\ CO_2Et
\tag{266}
$$

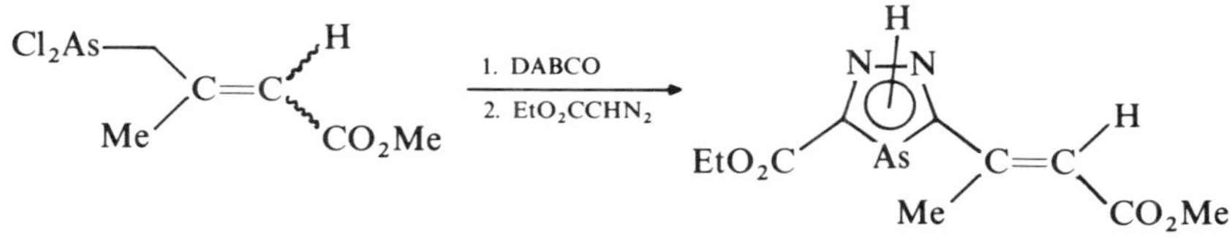

$$
Cl_2As\!-\!\underset{Me}{\overset{}{C}}\!=\!\underset{CO_2Me}{\overset{H}{C}} \xrightarrow[\text{2. EtO}_2\text{CCHN}_2]{\text{1. DABCO}} \underset{EtO_2C}{\overset{H}{\underset{As}{\diagdown}}}\underset{Me}{\overset{}{C}}\!=\!\underset{CO_2Me}{\overset{H}{C}}
\tag{267}
$$

Heterocycles having an onium group react with $(Me_3Si)_3As$ to give corresponding heterocycles having As=C bonds (equations 268[268] and 269[269]).

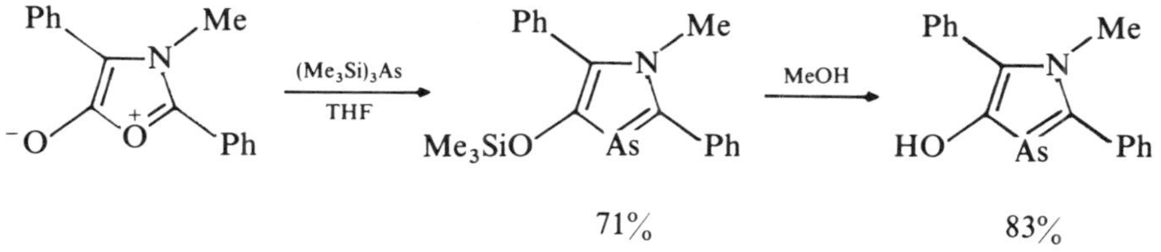

$$
\underset{{}^-O}{\overset{Ph}{\diagup}}\underset{O}{\overset{N}{\diagup}}\overset{Me}{\diagdown}\underset{Ph}{} \xrightarrow[THF]{(Me_3Si)_3As} \underset{Me_3SiO}{\overset{Ph}{\diagup}}\underset{As}{\overset{N}{\diagup}}\overset{Me}{\diagdown}\underset{Ph}{} \xrightarrow{MeOH} \underset{HO}{\overset{Ph}{\diagup}}\underset{As}{\overset{N}{\diagup}}\overset{Me}{\diagdown}\underset{Ph}{}
\tag{268}
$$

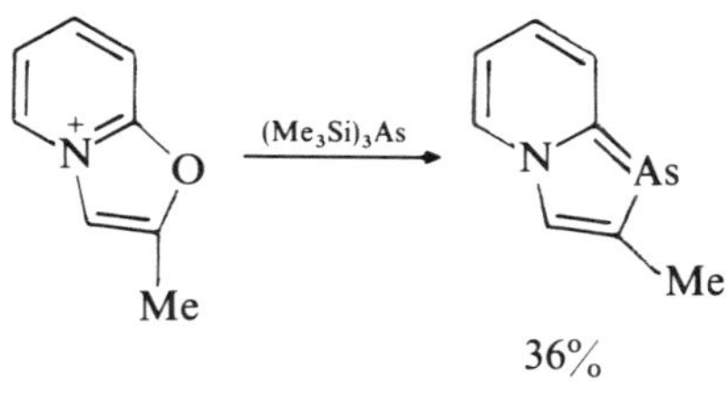

$$71\% \qquad\qquad 83\%$$

$$
\xrightarrow{(Me_3Si)_3As}
\tag{269}
$$

$$36\%$$

Reaction of arsinophenols with imidoyl chlorides gives benzoxarsoles **30** (equation 270[270]).

1,3-Migration of a trimethylsilyl group in $RAs(R'C\!=\!O)SiMe_3$ takes place easily to give $RAs\!=\!C(OSiMe_3)R'$ (equations 271[73a] and 272[271]).

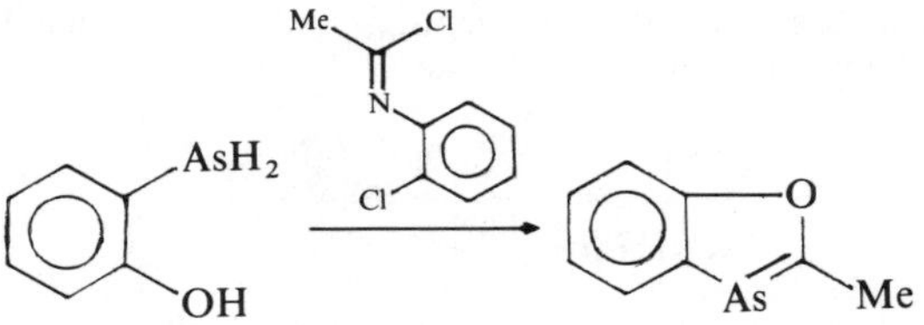 (270)

(30)

$$MeAs(C(O)Bu\text{-}t)(SiMe_3) \xrightarrow{80\,^{\circ}C} MeAs{=}C(Bu\text{-}t)(OSiMe_3) \quad (271)$$

$$(Me_3Si)_2AsFe(CO)_2Cp \xrightarrow[THF]{t\text{-}BuC(O)Cl} CpFe(CO)_2As{=}C(Bu\text{-}t)(OSiMe_3) \quad (272)$$

27%

Bis(trifluoromethyl)stannylarsine (**31**) has been thermolyzed to give $CF_3As{=}CF_2$ (equation 273[272]).

$$\underset{(\mathbf{31})}{(CF_3)_2AsSnMe_3} \xrightarrow[10^{-3}\,Torr]{300\text{–}340\,^{\circ}C} CF_3As{=}CF_2 \quad (273)$$

2. Compounds with double bonds between As and group 14 or 15 elements

Since the first isolation and characterization of a diphosphene compound **32** (equation 274[273]), compounds featuring double bonding between group 15 elements have attracted very much interest[16–18].

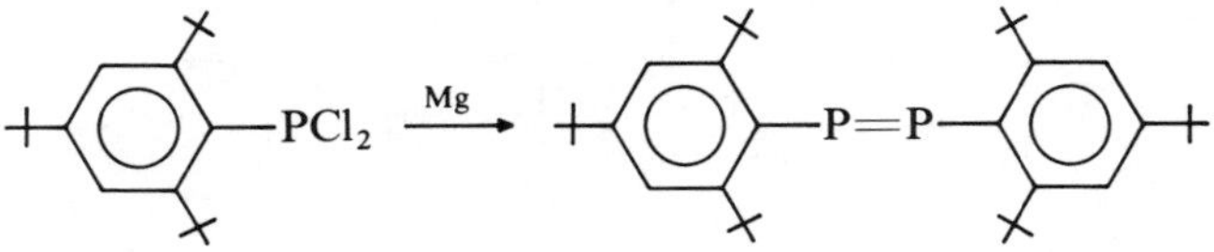 (274)

(32)

In a similar way, reductive coupling of $RAsCl_2$ has been useful for the preparation of symmetrical As=As bonded compounds (equations 275[274] and 276[275]).

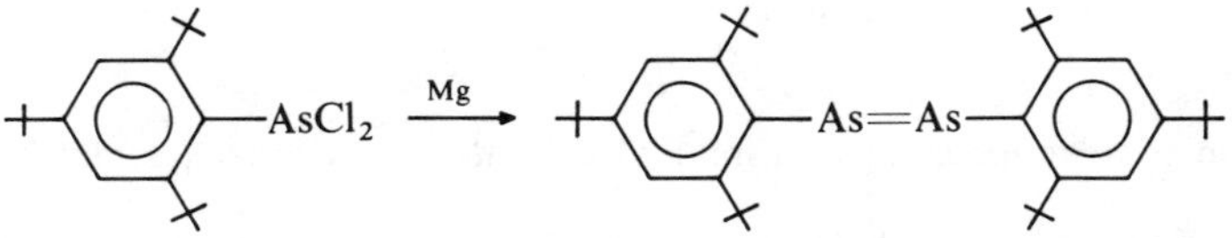 (275)

61%

$$(Me_3Si)_3CAsCl_2 \xrightarrow{t\text{-BuLi}} (Me_3Si)_3CAs=AsC(SiMe_3)_3 \quad (276)$$

For the preparation of unsymmetrical As=P or As=As bonded compounds, reaction of REH_2 [or $REH_n(SiMe_2)_{2-n}$ $(n = 0, 1)$] with $R'ECl_2$ (E = As or P) has generally been used (equations 277[117,276], 278[166,277], 279[278] and 280[167,279]).

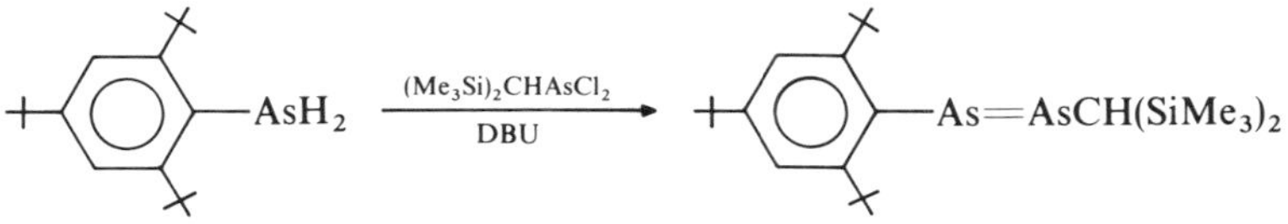

$$(277)$$

72%

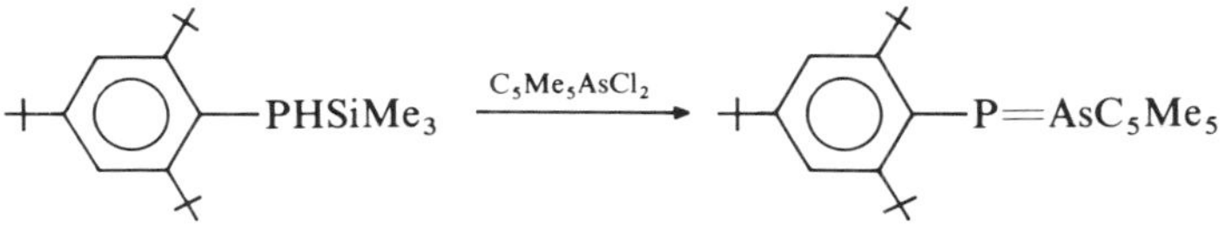

$$(278)$$

54–57%

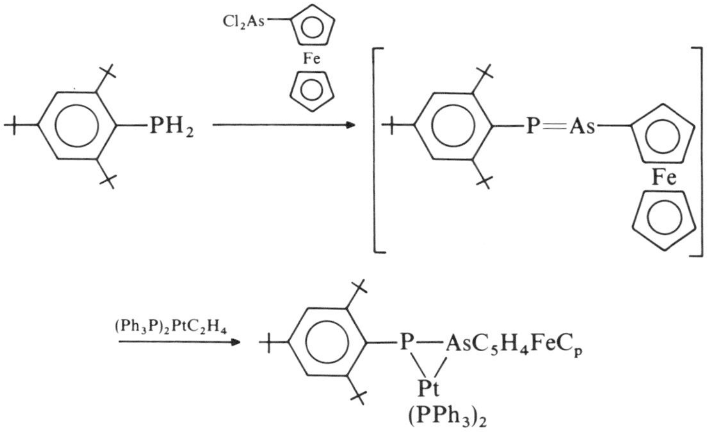

$$(279)$$

60%

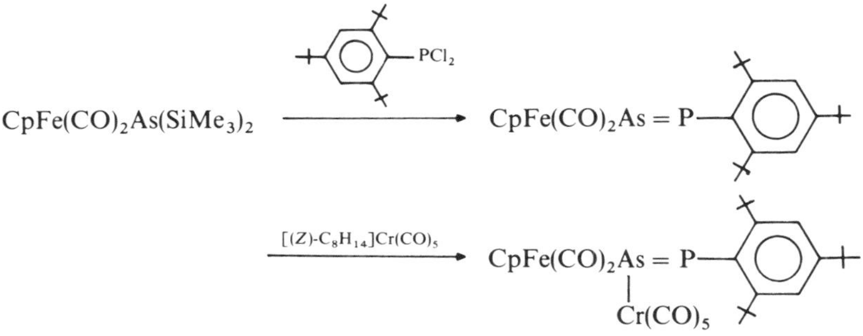

$$(280)$$

Recently the As=Si compound **33** has been prepared by a similar method (equation 281[234]).

$$Is_2SiF_2 \xrightarrow{H_2AsLi} \begin{matrix} Is_2Si-As-H \\ | \quad | \\ F \quad Li \end{matrix} \xrightarrow{i\text{-}Pr_3SiOTf} \xrightarrow{BuLi} \begin{matrix} Is_2Si-As-Li \\ | \quad | \\ F \quad Si\,i\text{-}Pr_3 \end{matrix}$$

$$\xrightarrow{80\,°C} \underset{(33)}{Is_2Si=AsSi(i\text{-}Pr)_3} \tag{281}$$

$$Is = 2,4,6\text{-}i\text{-}Pr_3C_6H_2$$

Reaction of the (pentamethylcyclopentadienyl)-substituted arsaphosphene **34** with $M(CO)_3(CH_3CN)_3$ (M = Cr, Mo) gives $(\eta^5\text{-}C_5H_5)(CO)_3MAs=PAr$ **35** (equation 282[280]).

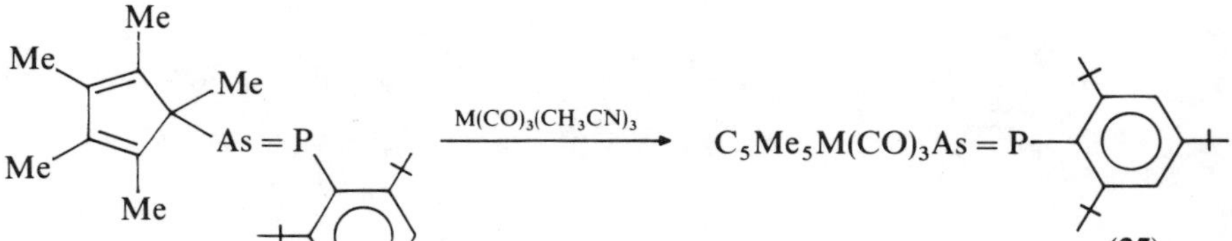

$$M = Cr, 55\%$$
$$M = Mo, 44\%$$

(34)   (35)   (282)

III. PREPARATION OF ORGANOARSENIC(V) COMPOUNDS

A. Preparation of Hexaorganoarsenic(V) Compounds

Among the compounds containing six carbon–arsenic bonds only the spirocyclic compound **36** is known (equation 283[281]).

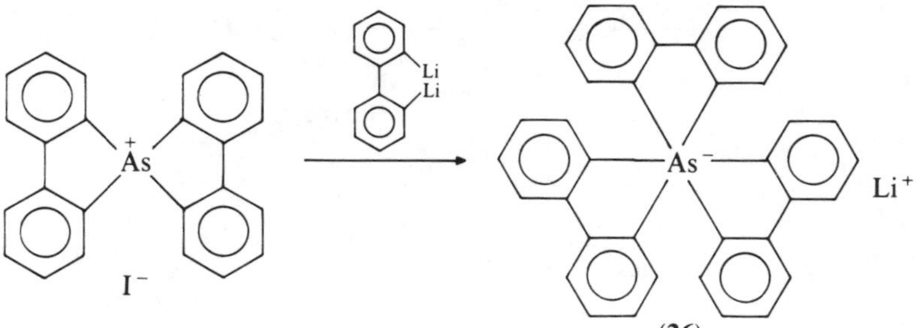

(283)

(36)

B. Preparation of Pentaorganoarsenic(V) Compounds

Various alkyl and aryl substituted pentaorganoarsenic compounds are known[8,9]. Most of these have been synthesized by transmetallation from Grignard reagents or organolithium reagents to arsenic centers.

1. Transmetallation

Symmetrical pentaorganoarsenic compounds have been synthesized from triorgano-arsenic halides R_3AsX_2 and tetraorganoarsenic halides R_4AsX as shown in equations 284^{282} and 285^{283}. Tosylimine derivatives such as **37** have also been used for the preparation of pentaorganoarsenic compounds (equation 286^{284}).

$$Ph_4AsBr \xrightarrow{PhLi} \underset{65\%}{Ph_5As} \qquad (284)$$

$$Me_3AsCl_2 \xrightarrow{MeLi} \underset{80\%}{Me_5As} \qquad (285)$$

$$(286)$$

C. Preparation of Tetraorganoarsenic(V) Halides

1. Reaction of tertiary arsines with organoelectrophiles

Trialkyl- and triaryl-arsenic compounds react with alkyl halides and acyl halides to form the corresponding tetraorganoarsenic salts (equations 287^{285}, 288^{286} and 289^{287}).

$$\longrightarrow \text{enantiomers} \qquad (287)$$

$$(288)$$

$$(289)$$

Aryl halides also react with tertiary arsines in the presence of Lewis acids such as $AlCl_3$ or $AlCl_3/NiBr_2$ (equation 290[288]).

$$Ph_3As \xrightarrow[AlCl_3/NiBr_2]{Me-\text{\textcircled{}}-Br} Ph_3\overset{+}{As}-\text{\textcircled{}}-Me \qquad (290)$$

$$Br^- \quad 31\%$$

2. Reactions of primary arsines, secondary arsines and compounds having As—As bonds with organoelectrophiles

These arsines also react readily with alkyl halides to give the corresponding tetra-organoarsonium salts as shown in equations 291[289] and 292[290].

$$\underset{\overset{|}{H}}{PhAs}(CH_2)_4 \underset{\overset{|}{H}}{AsPh} \xrightarrow{MeI} \underset{\overset{|}{Me}}{Ph\overset{+}{As}}\!\!+\!\!(CH_2)_4\!\!+\!\!\overset{+}{\underset{\overset{|}{Me}}{AsPh}} \quad 2I^- \qquad (291)$$

$$> 50\%$$

$$69\%$$

3. Cleavage of As(V)–carbon bonds by electrophiles

Pentaorganoarsenic compounds react with electrophiles such as HX and RX as shown in equations 293[282] and 294[284,291].

$$Ph_5As \xrightarrow{I_2} Ph_4As^+ \quad I_3^- \qquad (293)$$

$$(294)$$

4. Halogenolysis of tetraorganoarsenic oxygen compounds

Tetraorganoarsenic oxygen compounds react with HCl to form the corresponding chlorides (equation 295[292]).

$$Ph_3As{=}O \xrightarrow{PhMgBr} Ph_4AsOMgBr \xrightarrow{HCl} Ph_4AsCl \qquad (295)$$

D. Preparation of Triorganoarsenic(V) Dihalides

1. Oxidation of triorganoarsenic compounds with elemental halogen or the equivalent

Triorganoarsenic compounds react readily with X_2 to form the corresponding compounds R_3AsX_2 as shown in equations 296^{293}, 297^{294}, $298^{294,295}$ and $299^{295,296}$. The geometry of the compounds has been reported to be trigonal–bipyramid for difluorides and dichlorides, and to be ionic tetrahedral for diiodides[294,296–298]. The geometry for dibromides is dependent on the substituents R, thus Me_3AsBr_2[294] and Et_3AsBr_2[299] have been shown to be ionic with tetrahedral geometries in contrast to the trigonal–bipyramid structure for $(Me_3CCH_2)_3AsBr_2$[300]. The X-ray structure of Ph_3AsI_2, which has been prepared from $Mn[(OAsPh_3)_3I_2(SO_2)_2]$, has recently been shown to be charge transfer type complex (equation 311)[301]. Reaction of an excess amount of iodine with Ph_3As gives $[(AsPh_3I)_2I_3]^+I_3^-$ (equation 300)[302].

Other oxidants such as SO_2Cl_2, $PhICl_2$, XeF_2 and IF_5 have also been useful for the preparation of dihalides (equations 301^{303}, 302^{295}, 303^{304} and 304^{305}).

$$(C_6F_5)_3As \xrightarrow[CFCl_3]{F_2/N_2} (C_6F_5)_3AsF_2 \qquad (296)$$
$$94\text{--}95\%$$

$$Me_3As \xrightarrow[CCl_4]{Cl_2} Me_3AsCl_2 \qquad (297)$$

$$\left(Me_2N\!-\!\!\left\langle\bigcirc\right\rangle\!-\right)_3\!\!As \xrightarrow[CH_2Cl_2 \text{ or } CHCl_3]{Br_2} \left(Me_2N\!-\!\!\left\langle\bigcirc\right\rangle\!-\right)_3\!\!AsBr_2 \qquad (298)$$
$$95\%$$

$$Ph_3As \xrightarrow{I_2} Ph_3As\cdot I_2 \qquad (299)$$

$$Ph_3As \xrightarrow{I_2} Ph_3As\cdot I_2 \xrightarrow{I_2} [(Ph_3AsI)_2I_3]^+I_3^- \qquad (300)$$
$$\sim 100\%$$

$$Ph_3As \xrightarrow{SO_2Cl_2} Ph_3AsCl_2 \qquad (301)$$
$$95\%$$

$$\left(Me_2N\!-\!\!\left\langle\bigcirc\right\rangle\!-\right)_3\!\!As \xrightarrow{PhICl_2} \left(Me_2N\!-\!\!\left\langle\bigcirc\right\rangle\!-\right)_3\!\!AsCl_2 \qquad (302)$$
$$92\%$$

$$Me_3As \xrightarrow{XeF_2} Me_3AsF_2 \qquad (303)$$
$$98\%$$

$$Bu_3As \xrightarrow{IF_5} Bu_3AsF_2 \qquad (304)$$
$$64\%$$

Due to the facile exchange reactions between R_3AsX_2, $R_3AsX(Y)$ and R_3AsY_2 (equation

305[306]), mixed halides are generally difficult to obtain.

$$\text{Me}_3\text{AsF}_2 + \text{Me}_3\text{AsCl}_2 \rightleftharpoons 2\,\text{Me}_3\text{AsFCl} \qquad (305)$$

$$K = 5.4 \pm 0.6 \text{ at } -53\,^\circ\text{C}$$

Counter anion exchange reactions give stable $\text{R}_3\text{AsX}^+\text{Y}^-$ type compounds as shown in equations 306[307], 307[308] and 308[309].

$$\text{Ph}_3\text{As} \xrightarrow{\text{I}_2} \text{Ph}_3\text{AsI}_2 \xrightarrow{\text{AgAsF}_6} [\text{Ph}_3\text{AsI}]^+\text{AsF}_6^- \qquad (306)$$
$$81\%$$

$$\text{Me}_3\text{AsCl}_2 \xrightarrow{\text{SbCl}_5} [\text{Me}_3\overset{+}{\text{As}}\text{Cl}]\,\text{SbCl}_6^- \qquad (307)$$

$$\left(\text{Me}_2\text{N}-\!\!\!\left\langle\bigcirc\right\rangle\!\!\!-\right)_{\!\!3}\!\!\text{AsCl}_2 \xrightarrow{2\,\text{SbCl}_5} \left[\left(\text{Me}_2\text{N}-\!\!\!\left\langle\bigcirc\right\rangle\!\!\!-\right)_{\!\!3}\!\!\text{As}\right]^{2+} 2\,\text{SbCl}_6^- \qquad (308)$$
$$52\%$$

2. Halogenolysis of triorganoarsenic oxygen compounds

Triorganoarsenic oxygen compounds react with electrophiles to give the corresponding halides as shown in equation 309[310].

$$(309)$$

3. Halogen exchange

Triorganoarsenic fluorides have been prepared by the reaction of the corresponding chlorides with silver fluoride (equation 310[306a,311]).

$$[\text{PhCH}_2]_3\text{AsCl}_2 \xrightarrow[\text{MeCN}]{\text{AgF}} (\text{PhCH}_2)_3\text{AsF}_2 \qquad (310)$$

4. Miscellaneous methods

Ph_3AsI_2 has been formed in the thermal decomposition of $\text{Mn}[(\text{OAsPh}_3)_3\text{I}_2(\text{SO}_2)_2]$ (equation 311).[301]

$$\text{Ph}_3\text{As}{=}\text{O} \xrightarrow[\text{SO}_2]{\text{MnI}_2} [\text{Mn}(\text{OAsPh}_3)_3\text{I}_2(\text{SO}_2)] \xrightarrow{100\,^\circ\text{C}} \text{Ph}_3\text{As}\cdot\text{I}_2 \qquad (311)$$

E. Preparation of Diorganoarsenic(V) Trihalides

Although diorganoarsenic trihalides can be prepared, they are generally unstable and decompose at elevated temperatures.

1. Oxidation of diorganoarsenic compounds with elemental halogen or the equivalent

Diorganoarsenic compounds react readily with X_2 or the equivalent to form the corresponding trihalides in high yields (equations 312[300] and 313[305]).

$$(Me_3CCH_2)_2AsBr \xrightarrow[\text{pentane}]{Br_2} (Me_3CCH_2)_2AsBr_3 \xrightarrow{rt} Me_3CCH_2AsBr_2 \quad (312)$$

$$(C_6F_5)_2AsCl \xrightarrow[\substack{\text{MeCN} \\ -30\,^\circ C}]{IF_5} (C_6F_5)_2AsF_3 \atop 74\% \quad (313)$$

2. Halogen exchange

Diorganoarsenic fluorides have been prepared by the reaction of the corresponding chlorides with silver fluoride (equation 314[312]).

$$(CF_3)_2AsCl_3 \xrightarrow{AgF} (CF_3)_2AsF_3 \quad (314)$$

F. Preparation of Monoorganoarsenic(V) Tetrahalides

Since monoorganoarsenic tetrahalides are unstable, only fluorides and some chlorides are known. The methods of preparation are shown in equations 315[313] and 316[314].

$$PhAsCl_2 \xrightarrow{Cl_2 \atop CH_2Cl_2} PhAsCl_4 \quad (315)$$

$$\underset{\underset{O}{\|}}{PhAs(OH)_2} \xrightarrow[70\,^\circ C]{SF_4} PhAsF_4 \atop 45\% \quad (316)$$

G. Preparation of Organoarsenic(V) Nitrogen Compounds

1. Oxidative addition of chloroamines, (SCN)₂ and S₄N₄ to tertiary arsines

Reaction of tertiary arsines with chloroamines, $(SCN)_2$ and S_4N_4 gives the corresponding As(V) compounds as shown in equations 317[315], 318[316] and 319[317].

$$Ph_3As \xrightarrow{(SCN)_2} Ph_3As(NCS)_2 \quad (317)$$

$$Et_3As \xrightarrow{MeNHCl} Et_3\overset{+}{A}sNHMe \quad Cl^- \quad (318)$$
$$68\%$$

$$Ph_3As \xrightarrow{S_4N_4} Ph_3As{=}N{-}S{\underset{N-S}{\overset{N=S}{\diagup\diagdown\,N}}} \quad (319)$$

$$81\%$$

2. Nucleophilic substitution of halogen or oxygen with nitrogen nucleophiles

Reaction of organoarsenic(V) halides or oxides with several nitrogen nucleophiles gives the corresponding nitrogen compounds. Several examples are shown in equations

320[318], 321[319], 322[320] and 323[321].

$$\text{Me}_3\text{AsBr}_2 \xrightarrow{\text{NH}_3} \underset{76\%}{[\text{Me}_3\overset{+}{\text{As}}\text{NH}_2]\ \text{Br}^-} \xrightarrow[-70\,^\circ\text{C}]{\text{KNH}_2} \underset{83\%}{\text{Me}_3\text{As}{=}\text{NH}} \qquad (320)$$

$$\text{Ph}_3\text{As}{=}\text{O} \xrightarrow{\text{PhN}{=}\text{C}{=}\text{O}} \text{Ph}_3\text{As}{=}\text{NPh} \qquad\qquad\qquad (321)$$

$$(322)$$
82%

$$(323)$$

3. Oxidation of organoarsenic(III) compounds containing As—N bonds

Oxidation of organoarsenic(III) compounds containing As—N bonds by chloranil or Cl$_2$ gives the corresponding arsenic(V) nitrogen compounds as shown in equations 324[322] and 325[323].

$$(324)$$
44%

$$(\text{CF}_3)_2\text{AsN}(\text{SiMe}_3)_2 \xrightarrow[\text{CH}_2\text{Cl}_2]{\text{Cl}_2} [(\text{CF}_3)_2\text{As}{=}\text{N}(\text{SiMe}_3)]$$

$$[(\text{CF}_3)_2\text{AsN}]_3 + [(\text{CF}_3)_2\text{AsN}]_4$$

trimer   tetramer
(25%)   (30%)   (325)

H. Preparation of Organoarsenic(V) Oxygen Compounds

1. Preparation of arsonic acids {RAs(=O)(OH)₂}

1. Preparation of arsonic acids $\{RAs(=O)(OH)_2\}$

Arsonic acids have been prepared from readily available compounds such as arsenic(III) oxide and arsenic trichloride. Reaction of sodium arsenite with alkyl halides has been named the Meyer reaction, and reaction with arenediazonium salts the Bart reaction. In addition, some modifications have been made for the preparation of arsonic acids. Some examples are shown in equations 326[121] and 327[324].

$$As_2O_3 \xrightarrow{\text{NaOH}} Na_3AsO_3 \xrightarrow{\text{MeI}} \underset{\underset{O}{\|}}{MeAs(ONa)_2} \qquad (326)$$

$$Na_3AsO_3 \xrightarrow[\text{HCl}]{\text{PhN}_2^+\text{Cl}^-} \underset{\underset{O}{\|}}{PhAs(OH)_2} \qquad (327)$$

$$39\text{--}45\%$$

Reaction of arenediazonium salts with arsenic trichloride gives arylarsonic acids in the presence of cuprous bromide (equation 328)[325].

$$(328)$$

$$76\%$$

The use of arenediazonium tetrafluoroborates has been reported to be useful in the preparation of arylarsonic acids (equation 329)[326].

$$(329)$$

$$54\%$$

Oxidation of monoorganoarsenic(III) compounds has been used for the preparation of some organoarsonic acids (equations 330[327], 331[328] and 332[329]).

$$(330)$$

$$73\text{--}85\%$$

$$(331)$$

$$72\%$$

$$(Et_2N)_2AsCl \xrightarrow{RMgCl} RAs(NEt_2)_2 \xrightarrow[\text{2. } H_2O_2]{\text{1. aq HCl}} RAs(OH)_2$$

with the structure showing As double-bonded to O below the $RAs(OH)_2$ group:

$$\underset{\underset{O}{\|}}{RAs(OH)_2} \qquad (332)$$

R = 2-ethylhexyl, 78%
R = 1-octynyl, 21%

2. Preparation of arsinic acids {$R_2As(=O)(OH)$}

Arsinic acids have been prepared by similar methods as for the preparation of arsonic acids. Thus, reaction of sodium organoarsenites with alkyl halides (Meyer reaction) and reaction of sodium organoarsenite with arenediazonium salts (Bart reaction) are available for the preparation of arsinic acids. Several examples are shown in equations 333[330] and 334[331].

$$RAsCl_2 \xrightarrow[H_2O]{NaOH} RAs(ONa)_2 \xrightarrow{R'X} R-\underset{\underset{R'}{|}}{\overset{\overset{O}{\|}}{As}}-ONa \qquad (333)$$

R = Me, Et, ClCH$_2$CH$_2$, 2-methylbutyl

$$PhAsCl_2 \xrightarrow{NaOH} PhAs(ONa)_2 \xrightarrow[H_2O]{ClCH_2COONa} Ph-\underset{\underset{CH_2CO_2H}{|}}{\overset{\overset{O}{\|}}{As}}-OH \qquad (334)$$

72%

Oxidation of diorganoarsenic(III) compounds has been used for the preparation of some organoarsinic acids (equations 335[284], 336[332] and 337[329]).

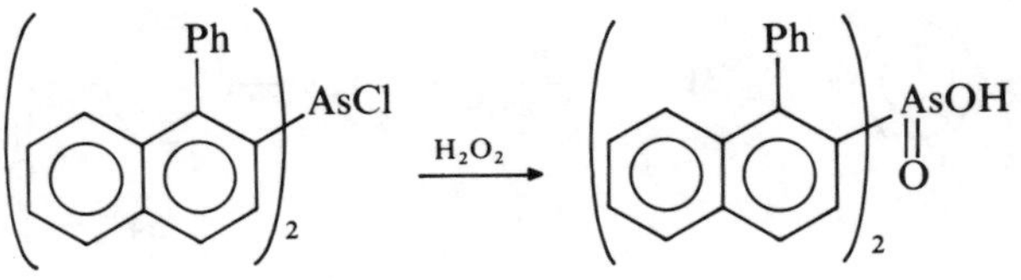

$$\qquad (335)$$

88%

$$\underset{Ph}{\overset{Me}{>}}As-As\underset{Ph}{\overset{Me}{<}} \xrightarrow[rt]{O_2} \underset{Ph}{\overset{Me}{>}}As-OH \qquad (336)$$

$$\underset{\underset{O \quad 60\%}{\|}}{}$$

$$Et_2NAsCl_2 \xrightarrow{RMgBr} R_2AsNEt_2 \xrightarrow[H_2O]{aq\ HCl} \xrightarrow{H_2O_2} R_2AsOH \qquad (337)$$

with structure showing As double-bonded to O:

$$\underset{}{\overset{\overset{O}{\|}}{R_2AsOH}}$$

R = 2-ethylhexyl, 52%
R = 1-octynyl, 34%

Cyclization of diphenyl-2-arsonic acid **38** gives 9-arsafluorenic acid **39** by use of sulfuric acid (equation 338)[333].

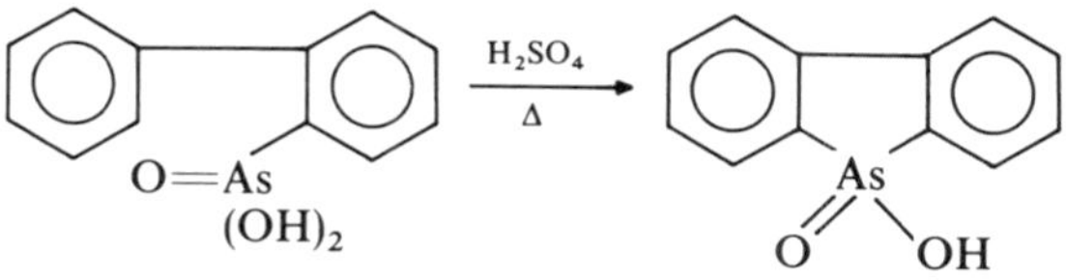

(338)

(**38**)   (**39**)

80%

3. Preparation of triorganoarsine oxides

Various methods of oxidation of triorganoarsines have been used for the preparation of triorganoarsine oxides as shown in equations 339[334], 340[335] and 341[324a].

$$Pr_3As \xrightarrow{\text{HgO}} Pr_3As{=}O \qquad (339)$$

$$Et_2AsCH{=}CH_2 \xrightarrow{\text{MnO}_2} \underset{\underset{O}{\|}}{Et_2AsCH{=}CH_2} \qquad (340)$$

$$Ph_3As \xrightarrow{\text{H}_2\text{O}_2} Ph_3As{=}O \qquad (341)$$

Cyclization of arsinic acid **40** gives arsine oxide **41** by use of phosphoric acid (equation 342)[346].

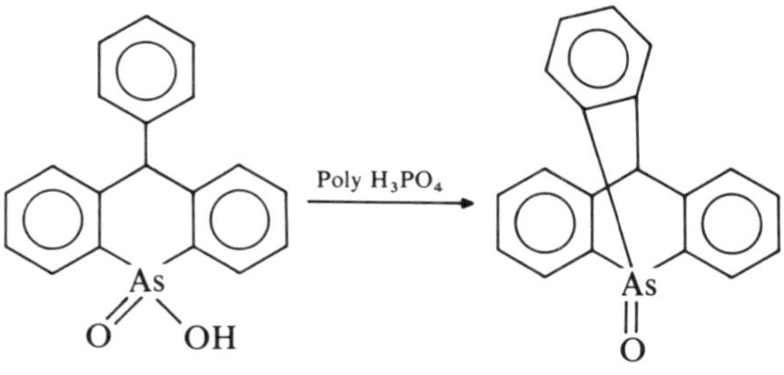

(342)

(**40**)   (**41**)

4. Preparation of other compounds having As(V)—O bonds

Reaction of compounds having As–halogen bonds and As—O bonds such as arsonic acids and arsinic acids with oxygen nucleophiles gives compounds having As(V)—O bonds as shown in equations 343[337], 344[338], 345[339] and 346[340].

Oxidative addition of compounds having arsenic(III)–oxygen bonds by o-benzoquinones gives cyclic As(V)—O bonded compounds as shown in equations 347[341] and 348[342].

Reaction of arsenic ylides **42** with methanol or with ethylene oxide gives alkoxy(tetramethyl)arsoranes as shown in equations 349[343] and 350[344].

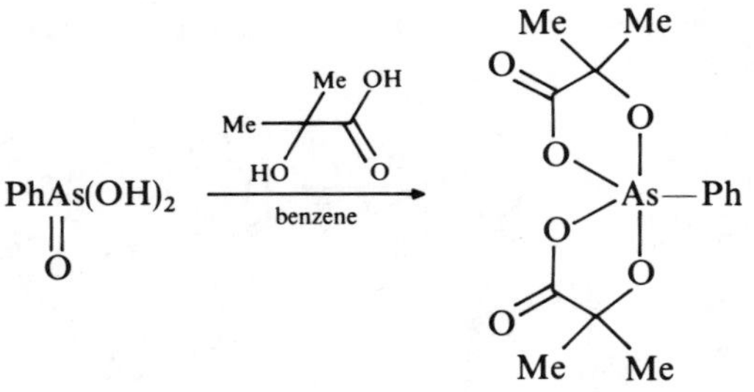

$$(343)$$

94%

$$MeAsI_2 \xrightarrow{Br_2} \xrightarrow{MeOH} MeAs(OMe)_4 \tag{344}$$

$$Ph_3AsCl_2 \xrightarrow[Et_3N]{ArOH} Ph_3AsCl(OAr) \tag{345}$$

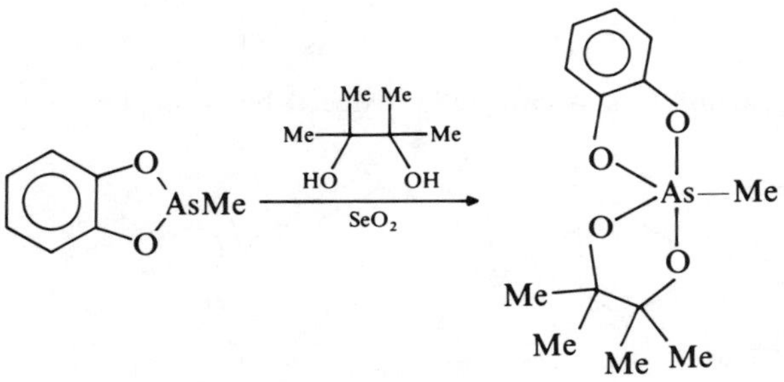

$$(346)$$

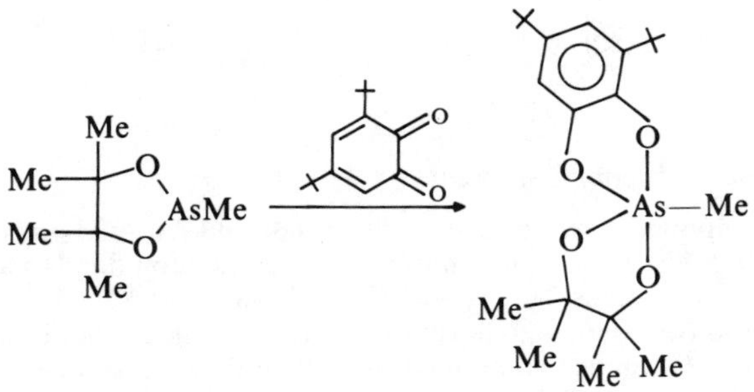

81%

$$(347)$$

95%

$$Me_3As=CH_2 \xrightarrow[-20\,°C]{MeOH} Me_4AsOMe \qquad (349)$$
$$(42)$$

$$(350)$$

$$62\%$$

Protonolysis of pentaorganoarsenic compounds gives tetraorganoarsenic alkoxides as shown in equations 351[284] and 352[345].

$$(351)$$

$$Me_5As \xrightarrow{NH_2OH} Me_4AsONH_2 \qquad (352)$$

I. Preparation of Organoarsenic(V) Compounds Containing Sulfur or Selenium Bonds

1. Oxidation of triorganoarsenic compounds with elemental sulfur or selenium

Triorganoarsenic compounds react with elemental sulfur or with selenium to give the corresponding sulfides and selenides, respectively (equations 353[346] and 354[347]).

$$Me_3As \xrightarrow{S_8} Me_3As=S \qquad (353)$$

$$Pr_3As \xrightarrow[EtOH]{Se} Pr_3As=Se \qquad (354)$$

$$43\%$$

2. Nucleophilic displacement with sulfur nucleophiles

Sulfur nucleophiles react with organoarsenic halides or acids. Some examples are shown in equations 355[348] and 350[349].

$$Me_2AsOH \xrightarrow{H_2S} Me_2AsSAsMe_2 \rightleftharpoons Me_2As—S—S—AsMe_2 \qquad (355)$$

with $=O$ on first arsenic and $=S$ on second arsenic.

$$\begin{array}{cc} 90\% & K = 0.147 \pm 0.009 \text{ at } 350.8\,K \\ \text{solid state} & \text{in } C_2H_2Cl_4 \end{array}$$

$$Ph_2AsCl \xrightarrow{S} [Ph_2AsCl] \xrightarrow{KHS} Ph_2AsSK \qquad (356)$$

with $=S$ groups, 20%

3. Miscellaneous methods

Oxidative addition of compounds having arsenic(III)–sulfur bonds by o-benzoquinones gives cyclic As(V)—S bonded compounds as shown in equation 357[350].

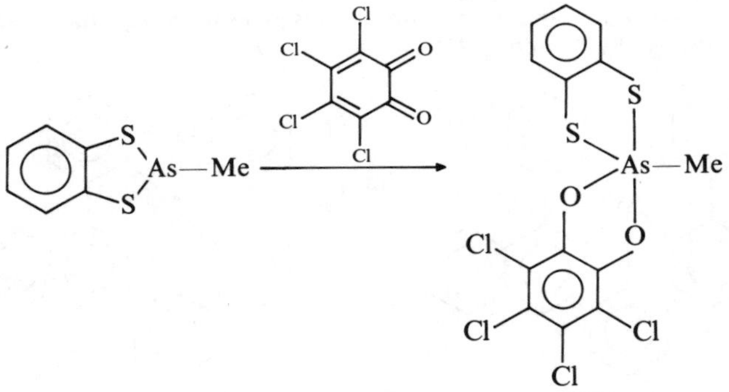

$$\qquad (357)$$

93%

Reaction of menthyl methylphenylthiosulfinate with propyllithium proceeds with inversion of configuration at arsenic to give enantiomerically enriched methyl(phenyl)—(propyl)arsine sulfide as shown in equation 56[84].

J. Preparation of Organoarsonium Ylides

1. Reaction of arsonium salts

With a suitable base such reaction gives arsonium ylides as shown in equations 358[10,11,351], 359[352], 360[353] and 361[354].

$$Ph_3\overset{+}{A}sMe \quad I^- \xrightarrow[-70\,°C]{PhLi} Ph_3As{=}CH_2 \qquad (358)$$

$$Ph_3\overset{+}{A}sCH_2CR \quad Br^- \xrightarrow[MeCN/H_2O]{K_2CO_3} Ph_3As{=}CHCR \qquad (359)$$

with $\overset{\|}{O}$ groups on both.

$$\underset{\underset{Me}{|}}{Ph_2\overset{+}{As}}CH_2AsPh_2 \quad I^- \xrightarrow{\text{NaNH}_2} \underset{\underset{Me}{|}}{Ph_2As}=CHAsPh_2 \tag{360}$$

$$Ph_3As=CH-\overset{\overset{O}{\|}}{C}OMe \xrightarrow{\text{NaN(SiMe}_3)_2} \underset{50\%}{Ph_3As=C=C=O} \tag{361}$$

2. Reaction of diazo compounds with tertiary arsines

Reaction of diazo compounds with tertiary arsines gives stable arsonium ylides as shown in equation 362[12,355].

$$\underset{\text{hfa-hexafluoroacetyl a}}{(MeO_2C)_2C=N_2} \xrightarrow[\underset{100\,°C}{Cu(hfa)_2}]{Ph_3As} \underset{62\%}{Ph_3As=C(CO_2Me)_2} \tag{362}$$

3. Miscellaneous methods

Thermal decomposition of iodonium ylides in the presence of tertiary arsines gives arsonium ylides as shown in equation 363[356].

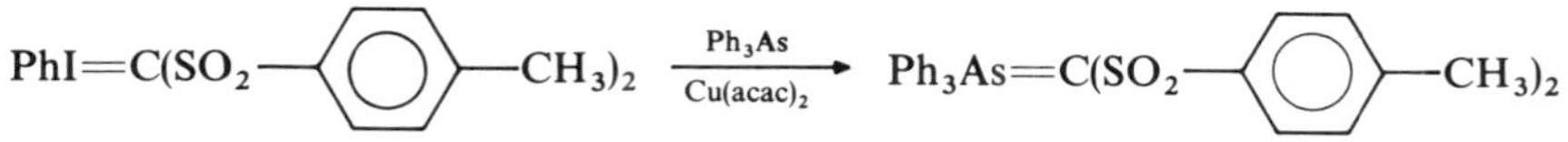

$$PhI=C(SO_2-\!\!\left\langle\bigcirc\right\rangle\!\!-CH_3)_2 \xrightarrow[Cu(acac)_2]{Ph_3As} Ph_3As=C(SO_2-\!\!\left\langle\bigcirc\right\rangle\!\!-CH_3)_2$$

$$59\% \tag{363}$$

Reaction of dichloromaleic anhydride **43** with triphenylarsine gives arsonium ylide **44** as shown in equation 364[357].

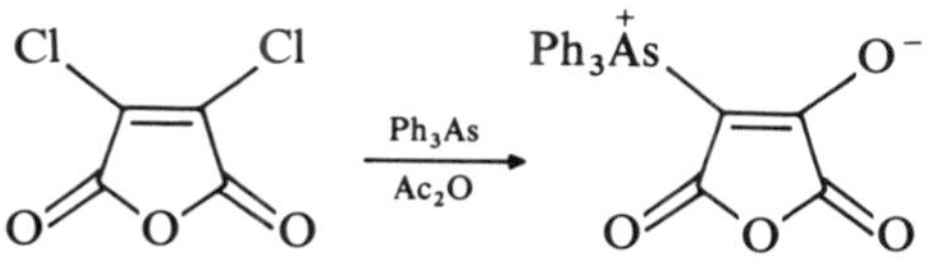

$$\tag{364}$$

(43)   **(44)**
29%

K. Preparation of Compounds with Arsenic(V)–Transition Metal Bonds

Several organoarsenic(III) compounds containing arsenic(III)–transition metal bonds have been oxidized by halogen-based electrophiles or by elemental sulfur to give the corresponding dihalides or sulfides (equations 365[236b], 366[358] and 367[237,359]).

$$Me_2AsMCp(CO)_3 \xrightarrow{X_2} \underset{\underset{X}{|}}{\overset{\overset{X}{|}}{Me_2AsMCp(CO)_3}} \tag{365}$$

$$M = Mo, X = Cl, 66\%$$
$$M = W, \ \ X = Cl, 72\%$$
$$M = W, \ \ X = Br, 68\%$$

$$Me_2AsFeCp(CO)(PMe_3) \xrightarrow{Ph_2PCl} \underset{\underset{Cl}{|}}{\overset{\overset{Cl}{|}}{Me_2AsFeCp(CO)(PMe_3)}} \qquad (366)$$

$$(i\text{-Pr})_2AsWCp(CO)_2 \xrightarrow{1/8\ S_8} \underset{\underset{S}{\parallel}}{(i\text{-Pr})_2AsWCp(CO)_2} \qquad (367)$$

L. Preparation of Hexacoordinate Organoarsenic(V) Anions

Diorganoarsenic trifluorides and triorganoarsenic difluorides react readily with fluoride anions to form the corresponding hexacoordinate anions as shown in equations 368[293b] and 369[312].

$$(C_6F_5)_3AsF_2 \xrightarrow[MeOH]{CsF} \underset{68\%}{(C_6F_5)_3\bar{A}sF_3 \quad Cs^+} \qquad (368)$$

$$(CF_3)_2AsF_3 \xrightarrow[MeCN]{CsF} (CF_3)_2\bar{A}sF_4 \quad Cs^+ \qquad (369)$$

IV. REFERENCES

1. Most recent review covering the year 1990: J. L. Wardell, *Organomet. Chem.*, **21**, 130 (1992).
2. Most recent review covering the years 1989–91: D. A. Armitage, *Annu. Rep. Prog. Chem., Sect. A*, **89**, 49 (1993).
3. Most recent review covering the years 1986–87: D. B. Sowerby, *Coord. Chem. Rev.*, **103**, 1 (1990).
4. S. Samaan (Ed.), *Methoden der Organischen Chemie. Metallorganische Verbindungen As, Sb, Bi*, Band XIII, Teil 8, Georg Thieme Verlag, Stuttgart, 1978.
5. J. L. Wardell, in *Comprehensive Organometallic Chemistry* (Eds. G. Wilkinson, F. G. A. Stone and E. W. Abel), Vol. 2. Chapter 13, Pergamon Press, Oxford, p. 681 (1982).
6. (a) C. F. McAuliffe, in *Comprehensive Coordination Chemistry* (Eds. G. Wilkinson, R. D. Guillard, J. A. McCleverty and E. W. Abel), Vol. 2, Chapter 14, Pergamon Press, Oxford, 1987, p. 989.
   (b) C. F. McAuliffe, in *Comprehensive Coordination Chemistry* (Eds. G. Wilkinson, R. D. Guillard, J. A. McCleverty and E. W. Abel), Vol. 3, Chapter 28, Pergamon Press, Oxford, 1987, p. 237.
7. B. J. Aylett, *Organometallic Compounds*, Vol. I, Part 2, Chapman and Hall, New York, 1979, p. 387.
8. R. Bohra and H. W. Roesky, *Adv. Inorg. Radiochem.*, **28**, 203 (1984).
9. D. Hellwinkel, *Top. Curr. Chem.*, **109**, 1 (1983).
10. H. Yaozeng, Y.-Z. Huang, S. Yanchang and Y. C. Shen, *Adv. Organometal. Chem.*, **20**, 115 (1982).
11. Y.-Z. Huang, L.-L. Shi, J.-H. Yang, W.-J. Xiao, S.-W. Li and W.-B. Wang, in *Heteroatom Chemistry* (Ed. E. Block), Chapter 10, VCH Publishers, New York, 1990, p. 189.
12. D. Lloyd, I. Gosney and R. A. Ormiston, *Chem. Soc. Rev.*, **16**, 45 (1987).
13. T. Kauffmann, *Top. Curr. Chem.*, **92**, 109 (1980).
14. F. Kober, *Synthesis*, 173 (1982).
15. A. J. Ashe, III, *Top. Curr. Chem.*, **105**, 125 (1982).
16. L. Weber, *Chem. Rev.*, **92**, 1839 (1992).
17. A. H. Cowley, *Polyhedron*, **3**, 389 (1984).
18. A. H. Cowley and N. C. Norman, *Prog. Inorg. Chem.*, **34**, 1 (1986).
19. A. J. Arduengo, III and D. A. Dixon, in *Heteroatom Chemistry* (Ed. E. Block), Chapter 3, VCH Publishers, New York, 1990, p. 47.
20. G. O. Doak and L. D. Freedman, *Synthesis*, 328 (1974).

21. G. Huttner and K. Evertz, *Acc. Chem. Res.*, **19**, 406 (1986).
22. A.-J. DiMaio and A. L. Rheingold, *Chem. Rev.*, **90**, 169 (1990).
23. I. Omae, *Organometallic Intramolecular-Coordination Compounds*, Elsevier, Amsterdam, 1986, p. 169.
24. L. K. Krannich, C. L. Watkins, D. K. Srivastava and R. K. Kanjolia, *Coord. Chem. Rev.*, **112**, 117 (1992).
25. A. H. Cowley and R. A. Jones, *Angew. Chem., Int. Ed. Engl.*, **28**, 1208 (1989).
26. R. L. Wells, *Coord. Chem. Rev.*, **112**, 273 (1992).
27. W. R. Cullen and K. J. Reimer, *Chem. Rev.*, **89**, 713 (1989).
28. R. A. Zingaro and A. Merijanian, *Inorg. Chem.*, **3**, 580 (1964).
29. R. Appel and D. Rebhan, *Chem. Ber.*, **102**, 3955 (1969).
30. L. Maier, D. Seyferth, F. G. A. Stone and E. G. Rochow, *J. Am. Chem. Soc.*, **79**, 5884 (1957).
31. (a) R. Armstrong, N. A. Gibson, J. W. Hosking, D. C. Weatherburn, *Aust. J. Chem.*, **20**, 2771 (1967).
    (b) P. Pfeiffer, *Chem. Ber.*, **37**, 4620 (1904).
32. R. Talay and D. Rehder, *Z. Naturforsch., Teil B*, **36**, 451 (1981).
33. J. W. Pasterczyk and A. R. Barron, *Phosphorus, Sulfur, and Silicon*, **48**, 157 (1990).
34. J. Heinicke, *J. Organometal. Chem.*, **364**, C17 (1989).
35. (a) F. D. Yambushev, Y. F. Gatilov, N. K. Tenisheva and V. I. Savin, *J. Gen. Chem. USSR*, **44**, 1698 (1974).
    (b) F. D. Yambushev, Y. F. Gatilov, N. K. Tenisheva and V. I. Savin, *J. Gen. Chem. USSR*, **44**, 1701 (1974).
    (c) Y. F. Gatilov, F. D. Yambushev and N. K. Tenisheva, *J. Gen. Chem. USSR*, **43**, 2659 (1973).
36. X. Li, C. M. Lukehart and L. Han, *Organometallics*, **11**, 3993 (1992).
37. N. A. A. Al-Jabar and A. G. Massey, *J. Organometal. Chem.*, **287**, 57 (1985).
38. R. Appel, T. Gaitzsch and F. Knoch, *Angew. Chem., Int. Ed. Engl.*, **24**, 419 (1985).
39. B. Deubzer, M. Elian, E. O. Fischer and H. P. Fritz, *Chem. Ber.*, **103**, 799 (1990).
40. P. J. Fagan and W. A. Nugent, *J. Am. Chem. Soc.*, **110**, 2310 (1988).
41. W. A. Schenk and E. Voss, *J. Organometal. Chem.*, **396**, C8 (1990).
42. W. Tumas, J. A. Suriano and R. L. Harlow, *Angew. Chem., Int. Ed. Engl.*, **29**, 75 (1990).
43. R. S. Dickson and B. O. West, *Aust. J. Chem.*, **15**, 710 (1962).
44. A. J. Ashe, III, W. M. Butler and T. R. Diephouse, *Organometallics*, **2**, 1005 (1983).
45. E. A. Ganja, C. D. Ontiveros and J. A. Morrison, *Inorg. Chem.*, **27**, 4535 (1988).
46. (a) D. K. Srivastava, L. K. Krannich and C. L. Watkins, *Inorg. Chem.*, **29**, 3502 (1990).
    (b) G. K. Kamai and Z. L. Khiisamova, *J. Gen. Chem. USSR*, **23**, 1387 (1954).
47. (a) L. K. Krannich, C. L. Watkins and D. K. Srivastava, *Polyhedron*, **9**, 289 (1990).
    (b) C. J. Thomas, L. K. Krannich and C. L. Watkins, *Polyhedron*, **12**, 89 (1993).
    (c) C. J. Thomas, L. K. Krannich and C. L. Watkins, *Synth. React. Inorg. Met.-Org. Chem.*, **22**, 461 (1992).
48. D. G. Allen and S. B. Wild, *Organometallics*, **2**, 394 (1983).
49. J. Ellermann and L. Brehm, *Chem. Ber.*, **118**, 4794 (1985).
50. T. Kauffmann, B. Altepeter, N. Klas and R. Kriegesmann, *Chem. Ber.*, **118**, 2353 (1985).
51. A. M. Aguiar and T. G. Archibald, *J. Org. Chem.*, **32**, 2627 (1967).
52. R. D. Feltham and W. Silverthorn, *Inorg. Synth.*, **10**, 159 (1967).
53. (a) R. A. Rossi, R. A. Alonso and S. M. Palacios, *J. Org. Chem.*, **46**, 2498 (1981).
    (b) R. A. Alonso and R. A. Rossi, *J. Org. Chem.*, **47**, 77 (1982).
    (c) R. A. Rossi, S. M. Palacios and A. N. Santiago, *J. Org. Chem.*, **47**, 4654 (1982).
54. D. G. Allen, G. M. McLaughlin, G. B. Robertson, W. L. Steffen, G. Salem and S. B. Wild, *Inorg. Chem.*, **21**, 1007 (1982).
55. (a) T. Kauffmann and J. Ennen, *Chem. Ber.*, **118**, 2692 (1985).
    (b) J. Ennen and T. Kauffmann, *Angew. Chem., Int. Ed. Engl.*, **20**, 118 (1981).
56. B. A. Murrer, J. M. Brown, P. A. Chaloner, P. N. Nicholson and D. Parker, *Synthesis*, 350 (1979).
57. (a) Y. F. Gatilov, L. B. Ionov, L. G. Kokorina and I. P. Mukanov, *J. Gen. Chem. USSR*, **44**, 1695 (1974).
    (b) H. Albers, W. Künzel and W. Schuler, *Chem. Ber.*, **85**, 239 (1952).
58. A. Tzschach and R. Schwarzer, *Justus Liebigs Ann. Chem.*, **709**, 248 (1967).
59. (a) T. Kauffmann, K. Berghus and A. Rensing, *Chem. Ber.*, **118**, 4507 (1985).

(b) T. Kauffmann, K. Berghus, A. Rensing and J. Ennen, *Chem. Ber.*, **118**, 3737 (1985).

(c) T. Kauffmann, J. Ennen and K. Berghus, *Tetrahedron Lett.*, **25**, 1971 (1984).

(d) T. Kauffmann, J. Ennen, H. Lhotak, A. Rensing, F. Steinseifer and A. Woltermann, *Angew. Chem., Int. Ed. Engl.*, **19**, 328 (1980).

(e) K. Berghus, A. Hamsen, A. Rensing, A. Woltermann and T. Kauffmann, *Angew. Chem., Int. Ed. Engl.*, **20**, 117 (1981).

60. A. Tzschach and W. Deylig, *Chem. Ber.*, **98**, 977 (1965).

61. (a) P.-H. Leung, G. M. McLaughlin, J. W. L. Martin and S. B. Wild, *Inorg. Chem.*, **25**, 3392 (1986).

(b) P. G. Kerr, P.-H. Leung and S. B. Wild, *J. Am. Chem. Soc.*, **109**, 4321 (1987).

62. Y. Shigetomi, M. Kojima and J. Fujita, *Bull. Chem. Soc. Jpn.*, **58**, 258 (1985).

63. (a) G. Becker and G. Gutekunst, *Z. Anorg. Allg. Chem.*, **470**, 131 (1980).

(b) J. Apel and J. Grobe, *Z. Anorg. Allg. Chem.*, **453**, 28 (1979).

64. E. R. Bornancini, R. A. Alonso and R. A. Rossi, *J. Organometal. Chem.*, **270**, 177 (1984).

65. P. J. Busse and K. J. Irgolic, *J. Organometal. Chem.*, **93**, 107 (1975).

66. H. Goldwhite, D. G. Rowsell and C. Valdez, *J. Organometal. Chem.*, **12**, 133 (1968).

67. (a) G. Märkl, W. Weber and W. Weiss, *Chem. Ber.*, **118**, 2365 (1985).

(b) G. Märkl and G. Dannhardt, *Tetrahedron Lett.*, 1455 (1973).

(c) G. Märkl and H. Hauptmann, *Tetrahedron Lett.*, 3257 (1968).

68. (a) R. B. King and P. N. Kapoor, *J. Am. Chem. Soc.*, **93**, 4158 (1971).

(b) R. C. Cookson and F. G. Mann, *J. Chem. Soc.*, 618 (1947).

69. (a) I. G. M. Campbell and R. C. Poller, *J. Chem. Soc.*, 1195 (1956).

(b) G. H. Cookson and F. G. Mann, *J. Chem. Soc.*, 2888 (1949).

70. C. K. Banks, F. H. Kahler and C. S. Hamilton, *J. Am. Chem. Soc.*, **69**, 933 (1947).

71. G. Himbert and L. Henn, *Tetrahedron Lett.*, **25**, 1357 (1984).

72. G. Thiollet and F. Mathey, *Tetrahedron Lett.*, 3157 (1979).

73. (a) G. Becker and G. Gutekunst, *Z. Anorg. Allg. Chem.*, **470**, 144 (1980).

(b) G. Becker and G. Gutekunst, *Z. Anorg. Allg. Chem.*, **470**, 157 (1980).

74. J. Heinicke and A. Tzschach, *Tetrahedron Lett.*, **24**, 5481 (1983).

75. (a) G. Märkl and S. Dietl, *Tetrahedron Lett.*, **29**, 539 (1988).

(b) G. Märkl and R. Liebl, *Justus Liebigs Ann. Chem.*, 2095 (1980).

76. Y. D. Xing, X. L. Hou and N. Z. Huang, *Tetrahedron Lett.*, **22**, 4727 (1981).

77. K. Mislow, A. Zimmerman and J. T. Melillo, *J. Am. Chem. Soc.*, **85**, 594 (1963).

78. (a) J. T. Braunholtz and F. G. Mann, *J. Chem. Soc.*, 3285 (1957).

(b) E. R H. Jones and F. G. Mann, *J. Chem. Soc.*, 294 (1958).

79. E. A. Ganja and J. A. Morrison, *Inorg. Chem.*, **29**, 33 (1990).

80. C. M. Woodard, G. Hughes and A. G. Massey, *J. Organometal. Chem.*, **112**, 9 (1976).

81. (a) G. R. A. Brandt, H. J. Emeléus and R. N. Haszeldine, *J. Chem. Soc.*, 2552 (1952).

(b) E. G. Walaschewski, *Chem. Ber.*, **86**, 272 (1953).

82. (a) F. G. Mann and F. C. Baker, *J. Chem. Soc.*, 4142 (1952).

(b) E. R. H. Jones and F. G. Mann, *J. Chem. Soc.*, 405 (1955).

83. (a) L. Horner and M. Ernst, *Chem. Ber.*, **103**, 318 (1970).

(b) L. Horner and H. Fuchs, *Tetrahedron Lett.*, 1573 (1963).

(c) L. Horner and H. Fuchs, *Tetrahedron Lett.*, 203 (1962).

(d) L. Horner and W. Hofer, *Tetrahedron Lett.*, 3321 (1966).

84. J. Stackhouse, R. J. Cook and K. Mislow, *J. Am. Chem. Soc.*, **95**, 953 (1973).

85. (a) P. Krommes and J. Lorberth, *J. Organometal. Chem.*, **93**, 339 (1975).

(b) E. Glozbach and J. Lorberth, *J. Organometal. Chem.*, **191**, 371 (1980).

86. W. R. Cullen and M. C. Waldman, *Can. J. Chem.*, **48**, 1885 (1970).

87. D. Hellwinkel and G. Kilthau, *Chem. Ber.*, **101**, 121 (1968).

88. P. Gugger, A. C. Willis and S. B. Wild, *J. Chem. Soc., Chem. Commun.*, 1169 (1990).

89. G. Salem and S. B. Wild, *J. Organometal. Chem.*, **370**, 33 (1989).

90. A. Tzschach and W. Deylig, *Z. Anorg. Allg. Chem.*, **336**, 36 (1965).

91. M. J. S. Gynane, A. Hudson, M. F. Lappert, P. P. Power and H. Goldwhite, *J. Chem. Soc., Dalton Trans.*, 2428 (1980).

92. M. Scholz, H. W. Roesky. D. Stalke, K. Keller and F. T. Edelmann, *J. Organometal. Chem.*, **366**, 73 (1989).

93. L. Maier, D. Seyferth, F. G. A. Stone and E. G. Rochow, *J. Am. Chem. Soc.*, **79**, 5884 (1957).

94. S. C. Sendlinger, B. S. Haggerty, A. L. Rheingold and K. H. Theopold, *Chem. Ber.*, **124**, 2453 (1991).
95. M. S. Kharasch, E. V. Jensen and S. Weinhouse, *J. Org. Chem.*, **14**, 429 (1949).
96. N. Gamon, C. Reichardt, R. Allmann and A. Waśkowska, *Chem. Ber.*, **114**, 3289 (1981).
97. I. T. Millar, H. Heaney, D. M. Heinekey and W. C. Fernelius, *Inorg. Synth.*, **6**, 116 (1960).
98. (a) R. J. Garascia, A. A. Carr and T. R. Hauser, *J. Org. Chem.*, **21**, 252 (1956).
    (b) D. Hellwinkel, A. Wiel, G. Sattler and B. Nuber, *Angew. Chem., Int. Ed. Engl.*, **29**, 689 (1990).
    (c) D. Hamer and R. G. Leckey, *J. Chem. Soc.*, 1398 (1961).
99. W. Steinkopf, H. Dudek and S. Schmidt, *Chem. Ber.*, **61**, 1911 (1928).
100. J. C. Pazik and C. George, *Organometallics*, **8**, 482 (1989).
101. H. Hartmann and G. Nowak, *Z. Anorg. Allg. Chem.*, **290**, 348 (1957).
102. K. Henrick and S. B. Wild, *J. Chem. Soc., Dalton Trans.*, 1506 (1975).
103. (a) W. J. Pope and E. E. Turner, *J. Chem. Soc.*, **117**, 1447 (1920).
     (b) G. D. Parkes, R. J. Clarke and B. H. Thewlis, *J. Chem. Soc.*, 429 (1947).
     (c) A. G. Evans and E. Warhurst, *Trans. Faraday Soc.*, **44**, 189 (1944).
104. K. Sommer, *Z. Anorg. Allg. Chem.*, **377**, 278 (1970).
105. (a) J. C. Summers and H. H. Sisler, *Inorg. Chem.*, **9**, 862 (1970).
     (b) R. L. Wells, A. P. Purdy and C. G. Pitt, *Phosphorus, Sulfur, and Silicon*, **57**, 1 (1991).
106. H. J. Emeléus, R. N. Haszeldine and E. G. Walaschewski, *J. Chem. Soc.*, 1552 (1953).
107. M. H. Beeby, G. H. Cookson and F. J. Mann, *J. Chem. Soc.*, 1917 (1950).
108. (a) M. Green and D. Kirkpatrick, *J. Chem. Soc. (A)*, 483 (1968).
     (b) M. Green and D. Kirkpatrick, *J. Chem. Soc., Chem. Commun.*, 57 (1967).
     (c) A. Tzschach and W. Lange, *Z. Anorg. Allg. Chem.*, **326**, 280 (1964).
     (d) K. Mödritzer, *Chem. Ber.*, **92**, 2637 (1959).
109. (a) H. Schmidbaur, H.-S. Arnold and E. Beinhofer, *Chem. Ber.*, **97**, 449 (1964).
     (b) R. D. Gigauri, G. N. Chachava, B. D. Chernokal'skii and M. M. Ugulava, *J. Gen. Chem. USSR*, **44**, 1689 (1974).
110. A. B. Burg and J. Singh, *J. Am. Chem. Soc.*, **87**, 1213 (1965).
111. (a) Y. F. Gatilov and L. B. Ionov, *J. Gen. Chem. USSR*, **38**, 2039 (1968).
     (b) N. A. Chadaeva, G. K. Kamai and K. A. Mamakov, *J. Gen. Chem. USSR*, **37**, 1331 (1967).
112. A. Tzschach and G. Pacholke, *Chem. Ber.*, **97**, 419 (1964).
113. C. T. Mortimer and H. A. Skinner, *J. Chem. Soc.*, 4331 (1952).
114. W. Steinkopf and G. Schwen, *Chem. Ber.*, **54**, 1437 (1921).
115. E. G. Claeys, *J. Organometal. Chem.*, **5**, 446 (1966).
116. W. L. Lewis, C. D. Lowry and F. H. Bergeim, *J. Am. Chem. Soc.*, **43**, 891 (1921).
117. A. H. Cowley, J. G. Lasch, N. C. Norman and M. Pakulski, *J. Am. Chem. Soc.*, **105**, 5506 (1983).
118. S. T. Abu-Orabi and P. Jutzi, *J. Organometal. Chem.*, **347**, 307 (1988).
119. (a) K. Sommer, *Z. Anorg. Allg. Chem.*, **377**, 128 (1970).
     (b) C. Spang, F. T. Edelmann, M. Noltemeyer and H. W. Roesky, *Chem. Ber.*, **122**, 1247 (1989).
120. (a) S. Himdi-Kabbab, P. Pellon and J. Hamelin, *Tetrahedron Lett.*, **30**, 349 (1989).
     (b) P. Pellon, S. Himdi-Kabbab, I. Rault, F. Tonnard and J. Hamelin, *Tetrahedron Lett.*, **31**, 1147 (1990).
121. I. T. Millar, H. Heaney, D. M. Heinekey and W. C. Fernelius, *Inorg. Synth.*, **6**, 113 (1960).
122. F. F. Blicke and S. R. Safir, *J. Am. Chem. Soc.*, **63**, 575 (1941).
123. R. J. Garascia, G. W. Batzis and J. O. Kroeger, *J. Org. Chem.*, **25**, 1271 (1960).
124. K. Sommer, *Z. Anorg. Allg. Chem.*, **377**, 120 (1970).
125. J. Ellermann, L. Brehm, E. Lindner, W. Hiller, R. Fawzi, F. L. Dickert and M. Waidhas, *J. Chem. Soc., Dalton Trans.*, 997 (1986).
126. F. Kober, *Z. Anorg. Allg. Chem.*, **397**, 97 (1973).
127. K. Sommer, *Z. Anorg. Allg. Chem.*, **376**, 150 (1970).
128. F. F. Blicke, R. A. Patelski and L. D. Powers, *J. Am. Chem. Soc.*, **55**, 1161 (1993).
129. J. Ellermann, M. Moll and L. Brehm, *Z. Anorg. Allg. Chem.*, **539**, 50 (1986).
130. L. H. Long, H. J. Emeléus and H. V. A. Briscoe, *J. Chem. Soc.*, 1123 (1946).
131. E. G. Walashewski, *Chem. Ber.*, **86**, 272 (1953).
132. R. L. Wells, C.-Y. Kwag, A. P. Purdy, A. T. McPhail and C. G. Pitt, *Polyhedron*, **9**, 319 (1990).
133. R. L. Wells, A. P. Purdy, A. T. McPhail and C. G. Pitt, *J. Organometal. Chem.*, **308**, 281 (1986).
134. R. G. Cavell and R. C. Dobbie, *J. Chem. Soc. (A)*, 1308 (1967).

135. K. Issleib, A. Tzschach and R. Schwarzer, Z. Anorg. Allg, Chem., **338**, 141 (1965).
136. F. G. Mann and M. J. Pragnell, J. Chem. Soc., 4120 (1965).
137. J. W. L. Martin, F. S. Stephens, K. D. V. Weerasuria and S. B. Wild, J. Am. Chem. Soc., **110**, 4346 (1988).
138. (a) F. G. Mann and B. B. Smith, J. Chem. Soc., 4544 (1952).
     (b) T. R. Carlton and C. D. Cook, Inorg. Chem., **10**, 2628 (1971).
     (c) T. Kauffmann and J. Ennen, Chem. Ber., **118**, 2714 (1985).
     (d) T. Kauffmann and J. Ennen, Tetrahedron Lett., **22**, 5035 (1981).
     (e) P. B. Chi and F. Kober, Z. Anorg. Allg. Chem., **501**, 89 (1983).
139. J. W. Anderson and J. E. Drake, Can. J. Chem., **49**, 2524 (1971).
140. R. C. Cookson and F. G. Mann, J. Chem. Soc., 618 (1947).
141. F. Kober, Z. Anorg. Allg. Chem., **401**, 243 (1973).
142. E. Fluck and G. Jakobson, Z. Anorg. Allg. Chem., **369**, 178 (1969).
143. Y. F. Gatilov and L. B. Ionov, J. Gen. Chem. USSR, **40**, 127 (1970).
144. J. Müller, Z. Anorg. Allg. Chem., **381**, 103 (1971).
145. D. M. Revitt and D. B. Sowerby, J. Chem. Soc., Dalton Trans., 847 (1972).
146. (a) U. Wannagat, E. Bogusch and R. Braun, J. Organometal. Chem., **19**, 367 (1969).
     (b) C. F. McBrearty, Jr., K. Irgolic and R. A. Zingaro, J. Organometal. Chem., **12**, 377 (1968).
147. A. J. Arduengo, III, C. A. Stewart, F. Davidson, D. A. Dixon, J. Y. Becker, S. A. Culley and M. B. Mizen, J. Am. Chem. Soc., **109**, 627 (1987).
148. H. G. Ang, G. Manoussakis and Y. O. El-Nigumi, J. Inorg. Nucl. Chem., **30**, 1715 (1968).
149. H. G. Ang, W. L. Kwik, Y. W. Lee and A. L. Rheingold, J. Chem. Soc., Dalton Trans., 663 (1993).
150. H. Schmidbaur, H.-S. Arnold and E. Beinhofer, Chem. Ber., **97**, 449 (1964).
151. (a) L. M. Werbel, T. P. Dawson, J. R. Hooton and T. E. Dalbey, J. Org. Chem., **22**, 452 (1957).
     (b) B. L. Chamberland and A. G. MacDiarmid, J. Am. Chem. Soc., **83**, 549 (1961).
152. A. M. Arif, A. H. Cowley and M. Pakulski, J. Chem. Soc., Chem. Commun., 165 (1987).
153. C. K. Banks, J. F. Morgan, R. L. Clark, E. B. Hatlelid, F. H. Kahler, H. W. Paxton, E. J. Cragoe, R. J. Andres, B. Elpern, R. F. Coles, J. Lawhead and C. S. Hamilton, J. Am. Chem. Soc., **69**, 927 (1947).
154. G. Kamai and R. G. Miftakhova, J. Gen. Chem. USSR, **33**, 2831 (1963).
155. (a) F. Kober and W. J. Rühl, Z. Anorg. Allg. Chem., **406**, 52 (1974).
     (b) K. Sommer, Z. Anorg. Allg. Chem., **383**, 136 (1971).
156. P. Aslanidis and F. Kober, Z. Anorg. Allg. Chem., **605**, 151 (1991).
157. F. Kober and W. J. Rühl, Z. Anorg. Allg. Chem., **420**, 74 (1976).
158. A. Schultze, S. Samaan and L. Horner, Phosphorus, **5**, 265 (1975).
159. K. Sommer, Z. Anorg. Allg. Chem., **375**, 55 (1970).
160. F. E. Ray and R. J. Garascia, J. Org. Chem., **15**, 1233 (1950).
161. E. J. Cragoe, Jr., R. J. Andres, R. F. Coles, B. Elpern, J. F. Morgan and C. S. Hamilton, J. Am. Chem. Soc., **69**, 925 (1947).
162. (a) A.-J. DiMaio and A. L. Rheingold, Organometallics, **10**, 3764 (1991).
     (b) A. L. Rheingold and A.-J. DiMaio, Organometallics, **5**, 393 (1986).
163. (a) A. J. Ashe, III, and E. G. Ludwig, Jr., J. Organometal. Chem., **303**, 197 (1986).
     (b) A. Belforte, F. Calderazzo, A. Morvillo, G. Pelizzi and D. Vitali, Inorg. Chem., **23**, 1504 (1984).
     (c) N. Kuhn and M. Winter, J. Organometal. Chem., **256**, C5 (1983).
164. R. G. Cavell and R. C. Dobbie, J. Chem. Soc. (A), 1406 (1968).
165. A. H. Cowley and D. S. Dierdorf, J. Am. Chem. Soc., **91**, 6609 (1969).
166. P. Jutzi and S. Opiela, Z. Anorg. Allg. Chem., **610**, 75 (1992).
167. L. Weber, D. Bungardt, U. Sonnenberg and R. Boese, Angew. Chem., Int. Ed. Engl., **27**, 1537 (1988).
168. F. G. Mann and A. J. H. Mercer, J. Chem. Soc., Perkin Trans. 1, 2548 (1972).
169. (a) A. J. Ashe, III, and E. G. Ludwig, Jr., J. Organometal. Chem., **308**, 289 (1986).
     (b) W. R. Cullen and P. S. Dhaliwal, Can. J. Chem., **45**, 379 (1967).
170. (a) P. Dehnert, J. Grobe, W. Hildebrandt and D. le Van, Z. Naturforsch., Teil B, **34**, 1646 (1979).
     (b) B. Holz and R. Steudel, J. Organometal. Chem., **406**, 133 (1991).
171. (a) E. Adams, D. Jeter, A. W. Cordes and J. W. Kolis, Inorg. Chem., **29**, 1500 (1990).
     (b) T. Klapötke, J. Organometal. Chem., **331**, 299 (1987).
172. R. K. Gupta, A. K. Rai and R. C. Mehrotra, Indian J. Chem., A, **24**, 752 (1985).

173. H. J. Eméléus, K. J. Packer and N. Welcman, *J. Chem. Soc.*, 2529 (1962).
174. (a) L. S. Sagan, R. A. Zingaro and K. J. Irgolic, *J. Organometal. Chem.*, **39**, 301 (1972).
    (b) F. Kober, *Z. Anorg. Allg. Chem.*, **401**, 243 (1973).
    (c) F. Kober, *Z. Anorg. Allg. Chem.*, **412**, 202 (1975).
175. N. A. Chandaeva, K. A. Mamakov, R. R. Shagidullin and G. K. Kamai, *J. Gen. Chem. USSR*, **43**, 825 (1973).
176. R. K. Gupta, A. K. Rai, R. C. Mehrotra and V. K. Jain, *Polyhedron*, **3**, 721 (1984).
177. (a) A. W. Cordes, P. D. Gwinup and M. C. Malmstrom, *Inorg. Chem.*, **11**, 836 (1972).
    (b) K. Volka, P. Adámek, H. Schulze and H. Barber, *J. Mol. Struct.*, **21**, 457 (1974).
178. E. A. Meyers, C. A. Applegate and R. A. Zingaro, *Phosphorus Sulfur*, **29**, 317 (1987).
179. (a) J. Ellermann and L. Brehm, *Chem. Ber.*, **117**, 2675 (1984).
    (b) G. Thiele, H. W. Rotter, M. Lietz and J. Ellermann, *Z. Naturforsch., Teil B*, **39**, 1344 (1984).
180. A.-J. DiMaio and A. L. Rheingold, *Inorg. Chem.*, **29**, 798 (1990).
181. W. R. Cullen, *Can. J. Chem.*, **41**, 2423 (1963).
182. J. Escudie, C. Couret, H. Ranaivonjatovo and J.-G. Wolf, *Tetrahedron Lett.*, **24**, 3625 (1983).
183. Y. F. Gatilov and V. P. Kovyrzina, *J. Gen. Chem. USSR*, **41**, 562 (1971).
184. (a) A. Tzschach and W. Lange, *Z. Anorg. Allg. Chem.*, **330**, 317 (1964).
    (b) K. Issleib, *Pure Appl. Chem.*, **9**, 205 (1964).
    (c) K. Issleib and R. Kümmel, *J. Organometal. Chem.*, **3**, 84 (1965).
    (d) T. Birchall and W. L. Jolly, *Inorg. Chem.*, **5**, 2177 (1966).
    (e) R. Batchelor and T. Birchall, *J. Am. Chem. Soc.*, **104**, 674 (1982).
185. (a) J. R. Phillips and J. H. Vis, *Can. J. Chem.*, **45**, 675 (1967).
    (b) W. Levason, K. G. Smith, C. A. McAuliffe, F. P. McCullough, R. D. Sedgwick and S. G. Murray, *J. Chem. Soc., Dalton Trans.*, 1718 (1979).
186. (a) H. Hope, M. M. Olmstead, P. P. Power and X. Xu, *J. Am. Chem. Soc.*, **106**, 819 (1984).
    (b) R. A. Bartlett, H. V. R. Dias, H. Hope, B. D. Murray, M. M. Olmstead and P. P. Power, *J. Am. Chem. Soc.*, **108**, 6921 (1986).
    (c) A. Tzschach and G. Pacholke, *Z. Anorg. Allg. Chem.*, **336**, 270 (1965).
187. R. Minkwitz and A. Liedtke, *Inorg. Chem.*, **28**, 1627 (1989).
188. F. F. Blicke and J. F. Oneto, *J. Am. Chem. Soc.*, **57**, 749 (1935).
189. T. Kauffmann, K. Berghus and J. Ennen, *Chem. Ber.*, **118**, 3724 (1985).
190. (a) V. K. Gupta, L. K. Krannich and C. L. Watkins, *Inorg. Chim. Acta.*, **150**, 51 (1988).
    (b) V. K. Gupta, L. K. Krannich and C. L. Watkins, *Polyhedron*, **6**, 1229 (1987).
    (c) V. K. Gupta, L. K. Krannich and C. L. Watkins, *Inorg. Chem.*, **25**, 2553 (1986).
191. (a) J. R. Phillips and J. H. Vis, *Can. J. Chem.*, **45**, 675 (1967).
    (b) A. J. Downs, N. I. Hunt, G. S. McGrady, D. W. H. Rankin and H. E. Robertson, *J. Mol. Struct.*, **248**, 393 (1991).
192. F. Nief, L. Ricard and F. Mathey, *Polyhedron*, **12**, 19 (1993).
193. A. Tzschach and W. Lange, *Chem. Ber.*, **95**, 1360 (1962).
194. G. Becker, M. Schmidt and M. Westerhausen, *Z. Anorg. Allg. Chem.*, **607**, 101 (1992).
195. M. Baudler and P. Bachmann, *Angew. Chem., Int. Ed. Engl.*, **20**, 123 (1981).
196. (a) J. Ellermann and H. Schössner, *Angew. Chem., Int. Ed. Engl.*, **13**, 601 (1974).
    (b) G. Thiele, G. Zoubek, H. A. Lindner and J. Ellermann, *Angew. Chem., Int. Ed. Engl.*, **17**, 135 (1978).
    (c) R. Gleiter, H. Köppel, P. Hofmann, H. R. Schmidt and J. Ellermann, *Inorg. Chem.*, **24**, 4020 (1985).
    (d) J. Ellermann, E. Köck and H. Burzlaff, *Acta Crystallogr., Sect. C*, **42**, 727 (1986).
197. A. H. Cowley, A. B. Burg and W. R. Cullen, *J. Am. Chem. Soc.*, **88**, 3178 (1966).
198. O. Mundt, G. Becker, H.-J. Wessely, H. J. Breunig and H. Kischkel, *Z. Anorg. Allg. Chem.*, **486**, 70 (1982).
199. M. Baudler and P. Bachmann, *Z. Anorg. Allg. Chem.*, **485**, 129 (1982).
200. (a) M. Baudler and S. Wietfeldt-Haltenhoff, *Angew. Chem., Int. Ed. Engl.*, **23**, 379 (1984).
    (b) For the preparation of t-Bu$_6$As$_8$ (20% yield): M. Baudler, J. Hellmann, P. Bachmann, K.-F. Tebbe, R. Fröhlich and M. Fehér, *Angew. Chem., Int. Ed. Engl.*, **20**, 406 (1981).
    (c) For the preparation of t-Bu$_8$As$_{12}$ (7% yield): M. Baudler and S. Wietfeldt-Haltenhoff, *Angew. Chem., Int. Ed. Engl.*, **24**, 991 (1985).
201. L. K. Krannich and H. H. Sisler, *Inorg. Chem.*, **8**, 1032 (1969).
202. (a) J. Waser and V. Schomker, *J. Am. Chem. Soc.*, **67**, 2014 (1945).

(b) F. Knoll, H. C. Marsmann and J. R. Van Wazer, *J. Am. Chem. Soc.*, **91**, 4986 (1969).
(c) For the crystal structure: O. Mundt, H. Riffel, G. Becker and A. Simon, *Z. Naturforsch., Teil B.*, **43**, 952 (1988).
203. (a) J. W. B. Reesor and G. F. Wright, *J. Org. Chem.*, **22**, 382 (1957).
(b) For the crystal structure: A. L. Rheingold and P. J. Sullivan, *Organometallics*, **2**, 327 (1983).
(c) For the preparation of (alkylAs)₅: P. S. Elmes, S. Middleton and B. O. West, *Aust. J. Chem.*, **23**, 1559 (1970).
(c) For the preparation of (MeAs)₅: E. J. Wells, R. C. Ferguson, J. G. Hallett and L. K. Peterson, *Can. J. Chem.*, **46**, 2733 (1968).
204. U. Schmidt, I. Boie, C. Osterroht, R. Schröer and H.-F. Grützmacher, *Chem. Ber.*, **101**, 1381 (1968).
205. G. Sennyey, F. Mathey, J. Fischer and A. Mitschler, *Organometallics*, **2**, 298 (1983).
206. V. K. Gupta, L. K. Krannich and C. L. Watkins, *Inorg. Chem.*, **26**, 1638 (1987).
207. A. Boardman, R. W. H. Small and I. J. Worrall, *Inorg. Chim. Acta.*, **121**, L35 (1986).
208. A. M. Arif, R. A. Jones and K. B. Kidd, *J. Chem. Soc., Chem. Commun.*, 1440 (1986).
209. G. E. Coates and J. G. Livingstone, *J. Chem. Soc.*, 1000 (1961).
210. W. Becker and H. Nöth, *Chem. Ber.*, **105**, 1962 (1972).
211. (a) M. A. Petrie, M. M. Olmstead, H. Hope, R. A. Bartlett and P. P. Power, *J. Am. Chem. Soc.*, **115**, 3221 (1993).
(b) M. A. Petrie, S. C. Shoner, H. V. R. Dias and P. P. Power, *Angew. Chem., Int. Ed. Engl.*, **29**, 1033 (1990).
212. K. T. Higa and C. George, *Organometallics*, **9**, 275 (1990).
213. M. A. Petrie and P. P. Power, *J. Chem. Soc., Dalton Trans.*, 1737 (1993).
214. A. M. Arif, B. L. Benac, A. H. Cowley, R. Geerts, R. A. Jones, K. B. Kidd, J. M. Power and S. T. Schwab, *J. Chem. Soc., Chem. Commun.*, 1543 (1986).
215. M. A. Petrie and P. P. Power, *Inorg. Chem.*, **32**, 1309 (1993).
216. (a) R. L. Wells, A. P. Purdy, K. T. Higa, A. T. McPhail and C. G. Pitt, *J. Organometal. Chem.*, **325**, C7 (1987).
(b) For the preparation of tris(dimesitylarsino)gallane: C. G. Pitt, K. T. Higa, A. T. McPhail and R. L. Wells, *Inorg. Chem.*, **25**, 2483 (1986).
217. C. J. Carrano, A. H. Cowley, D. M. Giolando, R. A. Jones, C. M. Nunn and J. M. Power, *Inorg. Chem.*, **27**, 2709 (1988).
218. (a) J. Grobe, S. Göbelbecker and D. Syndikus, *Z. Anorg. Allg. Chem.*, **608**, 43 (1992).
(b) J. Grobe, S. Göbelbecker, B. Krebs and M. Läge, *Z. Anorg. Allg. Chem.*, **611**, 11 (1992).
219. G. Becker, G. Gutekunst and H. J. Wessely, *Z. Anorg. Allg. Chem.*, **462**, 113 (1980).
220. (a) E. W. Abel, R. Honigschmidt-Grossich and S. M. Illingworth, *J. Chem. Soc. (A)*, 2623 (1968).
(b) J. W. Anderson and J. E. Drake, *J. Chem. Soc. (A)*, 3131 (1970).
221. I. G. M. Campbell, G. W. A. Fowles and L. A. Nixon, *J. Chem. Soc.*, 3026 (1964).
222. (a) H. Schumann and M. Schmidt, *Inorg. Nucl. Chem. Lett.*, **1**, 1 (1965).
(b) H. Schumann, *Angew. Chem., Int. Ed. Engl.*, **8**, 937 (1969).
223. H. Schumann and A. Roth, *Chem. Ber.*, **102**, 3713 (1969).
224. C. R. Russ and A. G. MacDiarmid, *Angew. Chem., Int. Ed. Engl.*, **5**, 418 (1966).
225. (a) R. L. Wells, A. P. Purdy, A. T. McPhail and C. G. Pitt, *J. Organometal. Chem.*, **354**, 287 (1988).
(b) C. G. Pitt, A. P. Purdy, K. T. Higa and R. L. Wells, *Organometallics*, **5**, 1266 (1986).
226. W. K. Holley, J. W. Pasterczyk, C. G. Pitt and R. G. Wells, *Heteroat. Chem.*, **1**, 475 (1990).
227. A. H. Cowley, D. M. Giolando, R. A. Jones. C. M. Nunn, J. M. Power and W.-W. du Mont, *Polyhedron*, **7**, 1317 (1988).
228. (a) O. T. Beachley and G. E. Coates, *J. Chem. Soc.*, 3241 (1965).
(b) For the crystal structure of (Me₂AsInMe₂)₃: A. H. Cowley, R. A. Jones, K. B. Kidd, C. M. Nunn and D. L. Westmoreland, *J. Organometal. Chem.*, **341**, C1 (1988).
(c) For the reaction of Ph₂AsH with Me₃M (M = Al, Ga, In): G. E. Coates and J. Graham, *J. Chem. Soc.*, 233 (1963).
229. (a) A. H. Cowley, R. A. Jones, M. A. Mardones and C. M. Nunn, *Organometallics*, **10**, 1635 (1991).
(b) For the reaction of AsH₃ with *t*-Bu₃Ga: A. H. Cowley, P. R. Harris, R. A. Jones and C. M. Nunn, *Organometallics*, **10**, 652 (1991).

230. R. L. Wells, A. P. Purdy, A. T. McPhail and C. G. Pitt, *J. Chem. Soc., Chem. Commun.*, 487 (1986).
231. H. Schumann and K.-H. Köhricht, *J. Organometal. Chem.*, **373**, 307 (1989).
232. H. Schumann and T. Östermann and M. Schmidt, *Chem. Ber.*, **99**, 2057 (1966).
233. (a) A. P. Lane and A. B. Burg, *J. Am. Chem. Soc.*, **89**, 1040 (1967).
     (b) F. G. A. Stone and A. B. Burg, *J. Am. Chem. Soc.*, **76**, 386 (1954).
234. M. Driess and H. Pritzkow, *Angew. Chem., Int. Ed. Ed. Engl.*, **31**, 316 (1992).
235. P. Jutzi, D. Wegener and M. Hursthouse, *J. Organometal. Chem.*, **418**, 277 (1991).
236. (a) M. Cooke, M. Green and D. Kirkpatrick, *J. Chem. Soc. (A)*., 1507 (1968).
     (b) W. Malisch, M. Kuhn, W. Albert and H. Rössner, *Chem. Ber.*, **113**, 3318 (1980).
     (c) W. Malisch and M. Kuhn, *Angew. Chem., Int. Ed. Engl.*, **13**, 84 (1974).
237. M. Luksza, K. Jörg and W. Malisch, *Inorg. Chim. Acta*, **85**, L49 (1984).
238. R. A. Jones and B. R. Whittlesey, *Organometallics*, **3**, 469 (1984).
239. (a) H. Lang, G. Huttner, B. Sigwarth, U. Weber, L. Zsolnai, I. Jibril and O. Orama, *Z. Naturforsch., Teil B*, **41**, 191 (1986).
     (b) G. Huttner, H.-G. Schmid, A. Frank and O. Orama, *Angew. Chem., Int. Ed. Engl.*, **15**, 234 (1976).
240. R. Müller and H. Vahrenkamp, *Chem. Ber.*, **113**, 3517 (1980).
241. (a) W. Malisch, H. Blau, H. Rößner and G. Jäth, *Chem. Ber.*, **113**, 1180 (1980).
     (b) W. Deck and H. Vahrenkamp, *Z. Anorg. Allg. Chem.*, **598/599**, 83 (1991).
242. D. F. Rieck, R. A. Montag, T. S. McKechnie and L. F. Dahl, *J. Am. Chem. Soc.*, **108**, 1330 (1986).
243. H. Kopf and U. Georges, *Z. Naturforsch., Teil B*, **36**, 1205 (1981).
244. S. R. Wade, M. G. H. Wallbridge and G. R. Willey, *J. Organometal. Chem.*, **267**, 271 (1989).
245. (a) A. M. Arif, R. A. Jones, M. H. Seeberger, B. R. Whittlesey and T. C. Wright, *Inorg. Chem.*, **25**, 3943 (1986).
     (b) R. A. Jones and B. R. Whittlesey, *J. Am. Chem. Soc.*, **107**, 1078 (1985).
246. (a) H. Werner, W. Paul, J. Wolf, M. Steinmetz, R. Zolk, G. Müller, O. Steigelmann and J. Riede, *Chem. Ber.*, **122**, 1061 (1989).
     (b) H. Werner, W. Paul and R. Zolk, *Angew. Chem., Int. Ed. Engl.*, **23**, 626 (1984).
247. Y. L. Baay and A. G. MacDiarmid, *Inorg. Nucl. Chem. Lett.*, **3**, 159 (1967).
248. A. Barth, G. Huttner, M. Fritz and L. Zsolnai, *Angew. Chem., Int. Ed. Engl.*, **29**, 929 (1990).
249. (a) A. M. Arif, D. E. Heaton, R. A. Jones, K. B. Kidd, T. C. Wright, B. R. Whittlesey, J. L. Atwood, W. E. Hunter and H. Zhang, *Inorg. Chem.*, **26**, 4065 (1987).
     (b) A. M. Arif, R. A. Jones, S. T. Schwab and B. R. Whittlesey, *J. Am. Chem. Soc.*, **108**, 1703 (1986).
250. W. R. Cullen and R. G. Hayter, *J. Am. Chem. Soc.*, **86**, 1030 (1964).
251. (a) A. L. Rheingold, M. E. Fountain and A.-J. DiMaio, *J. Am. Chem. Soc.*, **109**, 141 (1987).
     (b) A. L. Rheingold, M. J. Foley and P. J. Sullivan, *J. Am. Chem. Soc.*, **104**, 4727 (1982).
252. A.-J. DiMaio and A. L. Rheingold, *Organometallics*, **6**, 1138 (1987).
253. A.-J. DiMaio, T. E. Bitterwolf and A. L. Rheingold, *Organometallics*, **9**, 551 (1990).
254. F. Calderazzo, R. Poli and G. Pelizzi, *J. Chem. Soc., Dalton Trans.*, 2535 (1984).
255. A. L. Rheingold and P. J. Sullivan, *J. Chem. Soc., Chem. Commun.*, 39 (1983).
256. P. Mercando, A.-J. DiMaio and A. L. Rheingold, *Angew. Chem., Int. Ed. Engl.*, **26**, 244 (1987).
257. (a) W. A. Herrmann, B. Koumbouris, A. Schäfer, T. Zahn and M. L. Ziegler, *Chem. Ber.*, **118**, 2472 (1985).
     (b) W. A. Herrmann, B. Koumbouris, T. Zahn and M. L. Ziegler, *Angew. Chem., Int. Ed. Engl.*, **23**, 812 (1984).
258. (a) G. Huttner and H.-G. Schmid, *Angew. Chem., Int. Ed. Engl.*, **14**, 433 (1975).
     (b) J. von Seyerl, U. Moering, A. Wagner, A. Frank and G. Huttner, *Angew. Chem., Int. Ed. Engl.*, **17**, 844 (1978).
259. H. Lang, G. Mohr, O. Scheidsteger and G. Huttner, *Chem. Ber.*, **118**, 574 (1985).
260. J. von Seyerl, B. Sigwarth, H.-G. Schmid, G. Mohr, A. Frank, M. Marsili and G. Huttner, *Chem. Ber.*, **114**, 1392 (1981).
261. (a) A. J. Ashe, III, *J. Am. Chem. Soc.*, **93**, 3293 (1971).
     (b) A. J. Ashe, III and S. T. Abu-Orabi, *J. Org. Chem.*, **48**, 767 (1983).
     (c) R. V. Hodges, J. L. Beauchamp, A. J. Ashe, III and W.-T. Chan, *Organometallics*, **4**, 457 (1985).

(d) A. J. Ashe, III, W.-T. Chan, T. W. Smith and K. M. Taba, *J. Org. Chem.*, **46**, 881 (1981).

(e) C. Elschenbroich, J. Kroker, W. Massa, M. Wünsch and A. J. Ashe, III, *Angew. Chem., Int. Ed. Engl.*, **25**, 571 (1986).

262. (a) G. Märkl, A. Bergbauer and J. B. Rampal, *Tetrahedron Lett.*, **24**, 4079 (1983).

(b) G. Märkl, R. Liebl and H. Baier *Justus Liebigs Ann. Chem.*, 1610 (1981).

263. (a) G. Märkl and R. Liebl, *Justus Liebigs Ann. Chem.*, 2095 (1980).

(b) G. Märkl and F. Kneidl, *Angew. Chem., Int. Ed. Engl.*, **13**, 667 (1974).

264. G. Märkl and S. Dietl, *Tetrahedron Lett.*, **29**, 535 (1988).

265. A. J. Ashe, III, D. J. Bellville and H. S. Friedman, *J. Chem. Soc., Chem. Commun.*, 880 (1979).

266. (a) G. Märkl and C. Martin, *Tetrahedron Lett.*, 4503 (1973).

(b) R. Carrié and Y. Y. C. Y. L. Ko, *J. Chem. Soc., Chem. Commun.*, 1131 (1981).

(c) J. Högel, A. Schmidpeter and W. S. Sheldrick, *Chem. Ber.*, **116**, 549 (1983).

267. S. Himdi-Kabbab and J. Hamelin, *Tetrahedron Lett.*, **32**, 2755 (1991).

268. G. Märkl and S. Pflaum, *Tetrahedron Lett.*, **29**, 3387 (1988).

269. G. Märkl and S. Pflaum, *Tetrahedron Lett.*, **28**, 1511 (1987).

270. J. Heinicke, B. Raap and A. Tzschach, *J. Organometal. Chem.*, **186**, 39 (1980).

271. L. Weber, G. Meine, R. Boese and D. Bungardt, *Z. Anorg. Allg. Chem.*, **549**, 73 (1987).

272. J. Grobe and D. le Van, *Angew. Chem., Int. Ed. Engl.*, **23**, 710 (1984).

273. M. Yoshifuji, I. Shima, N. Inamoto, K. Hirotsu and T. Higuchi, *J. Am. Chem. Soc.*, **103**, 4587 (1981).

274. L. Weber and U. Sonnenberg, *Chem. Ber.*, **122**, 1809 (1989).

275. (a) C. Couret, J. Escudie, Y. Madaule, H. Ranaivonjatovo and J.-G. Wolf, *Tetrahedron Lett.*, **24**, 2769 (1983).

(b) A. H. Cowley, N. C. Norman and M. Pakulski, *J. Chem. Soc., Dalton Trans.*, 383 (1985).

276. (a) A. H. Cowley, J. E. Kilduff, J. G. Lasch, S. K. Mehrotra, N. C. Norman, M. Pakulski, B. R. Whittlesey, J. L. Atwood, and W. E. Hunter, *Inorg. Chem.*, **23**, 2582 (1984).

(b) A. H. Cowley, J. E. Kilduff, J. G. Lasch, S. K. Mehrotra, N. C. Norman, M. Pakulski and C. A. Stewart, *Phosphorus Sulfur*, **18**, 3 (1983).

277. P. Jutzi and U. Meyer, *J. Organometal. Chem.*, **326**, C6 (1987).

278. F. Edelmann, C. Spang, H. W. Roesky and P. G. Jones, *Z. Naturforsch., Teil B*, **43**, 517 (1988).

279. (a) L. Weber, D. Bungardt and R. Boese, *Chem. Ber.*, **121**, 1535 (1988).

(b) L. Weber, D. Bungardt, A. Müller and H. Bögge, *Organometallics*, **8**, 2800 (1989).

(c) L. Weber, G. Meine, R. Boese and D. Bläster, *Chem. Ber.*, **121**, 853 (1988).

(d) L. Weber and G. Meine, *Chem. Ber.*, **120**, 457 (1987).

(e) L. Weber, K. Reizig, D. Bungardt and R. Boese, *Organometallics*, **6**, 110 (1987).

(f) A. H. Cowley, J. E. Kilduff, J. G. Lasch, N. C. Norman, M. Pakulski, F. Ando and T. C. Wright, *Organometallics*, **3**, 1044 (1984).

280. (a) P. Jutzi and U. Meyer, *Chem. Ber.*, **121**, 559 (1988).

(b) L. Weber, D. Bungardt and R. Boese, *Z. Anorg. Allg. Chem.*, **578**, 205 (1989).

281. D. Hellwinkel and G. Kilthau, *Justus Liebigs Ann. Chem.*, **705**, 66 (1967).

282. G. Wittig and K. Clauss, *Justus Liebigs Ann. Chem.*, **577**, 26 (1952).

283. K. H. Mitschke and H. Schmidbaur, *Chem. Ber.*, **106**, 3645 (1973).

284. G. Wittig and D. Hellwinkel, *Chem. Ber.*, **97**, 769 (1964).

285. D. G. Allen, C. L. Raston, B. W. Skelton, A. H. White and S. B. Wild, *Aust. J. Chem.*, **37**, 1171 (1984).

286. F. G. Mann and F. H. C. Stewart, *J. Chem. Soc.*, 1269 (1955).

287. A. Asthana and R. C. Srivastava, *J. Orgnometal. Chem.*, **366**, 281 (1989).

288. (a) L. Horner and J. Haufe, *Chem. Ber.*, **101**, 2903 (1968).

(b) J. Chatt and F. G. Mann, *J. Chem. Soc.*, 1192 (1940).

289. A. Tzschach and G. Pacholke, *Chem. Ber.*, **97**, 419 (1964).

290. A. Tzschach and V. Kiesel, *J. Prakt. Chem.*, **313**, 259 (1971).

291. D. Hellwinkel and B. Knabe, *Chem. Ber.*, **104**, 1761 (1971).

292. R. L. Shriner and C. N. Wolf, *Org. Synth., Coll. Vol.*, **4**, 910 (1963).

293. (a) H. Preut, R. Kasemann and D. Naumann, *Acta Crystallogr., Sect. C*, **42**, 1875 (1986).

(b) R. Kasemann and D. Naumann, *J. Fluorine Chem.*, **41**, 321 (1988).

294. M. B. Hursthouse and I. A. Steer, *J. Organometal. Chem.*, **27**, C11 (1971).

295. J. -M. Keck and G. Klar, *Z. Naturforch., Teil B*, **27**, 591 (1972).

296. Y. R. Zhang, I. Solomon and S. Aronson, *Can. J. Chem.*, **69**, 606 (1991).

297. A. D. Beveridge, G. S. Harris and F. Inglis, *J. Chem. Soc. (A)*, 520 (1966).
298. T. B. Brill and G. G. Long, *Inorg. Chem.*, **9**, 1980 (1970).
299. L. Vendonck and G. P. VanDer Kelen, *Spectrochem. Acta, Sect. A*, **33**, 601 (1977).
300. J. C. Pazik and C. George, *Organometallics*, **8**, 482 (1989).
301. C. A. McAuliffe, B. Beagley, G. A. Gott, A. G. Mackie, P. P. MacRory and R. G. Pritchard, *Angew. Chem., Int. Ed. Engl.*, **26**, 264 (1987).
302. F. A. Cotton and P. A. Kibala, *J. Am. Chem. Soc.*, **109**, 3308 (1987).
303. A. J. Banister and L. F. Moore, *J. Chem. Soc. (A)*, 1137 (1968).
304. (a) A. M. Forster and A. J. Downs, *Polyhedron*, **4**, 1625 (1985).
    (b) A. J. Downs, M. J. Goode, G. S. McGrady, I. A. Steer, D. W. H. Rankin and H. E. Robertson, *J. Chem. Soc., Dalton Trans.*, 451 (1988).
305. H. J. Frohn and H. Maurer, *J. Fluorine. Chem.*, **34**, 129 (1986).
306. (a) C. G. Moreland, M. H. O'Brien, C. E. Douthit, and G. G. Long, *Inorg. Chem.*, **7**, 834 (1968).
    (b) M. F. Ali and G. S. Harris, *J. Chem. Soc., Dalton Trans.*, 1545 (1980).
    (c) K. B. Dillon and J. Lincoln, *Polyhedron*, **8**, 1445 (1989).
    (d) S. N. Bhattacharya and M. Singh, *Indian J. Chem., A*, **16**, 778 (1978).
307. I. Tornieporth-Oetting and T. Klapötke, *J. Organometal. Chem.*, **379**, 251 (1989).
308. (a) A. Schmidt, *Chem. Ber.*, **102**, 380 (1969).
    (b) W. Dorsch and A. Schmidt, *Z. Anorg. Allg. Chem.*, **508**, 149 (1984).
309. J. -M. Keck and G. Klar, *Z. Naturforsch., Teil B*, **27**, 596 (1972).
310. F. G. Holliman and F. G. Mann, *J. Chem. Soc.*, 547 (1943).
311. M. H. O'Brien, G. O. Doak and G. G. Long, *Inorg. Chim. Acta*, **1**, 34 (1967).
312. S. S. Chan and C. J. Willis, *Can. J. Chem.*, **46**, 1237 (1968).
313. (a) K. B. Dillon, R. J. Lynch and T. C. Waddington, *J. Chem. Soc., Dalton Trans.*, 1478 (1976).
    (b) W. La Coste and A. Michaelis, *Ann. Chem.*, **201**, 222 (1880).
314. (a) W. C. Smith, *J. Am. Chem. Soc.*, **82**, 6176 (1960).
    (b) W. C. Smith, C. W. Tullock, E. L. Muetterties, W. R. Hasek, F. S. Fawcett, V. A. Engelhardt and D. D. Coffman, *J. Am. Chem. Soc.*, **81**, 3165 (1959).
315. T. Weizmann, H. Müler, D. Seybold and K. Dehnicke, *J. Organometal. Chem.*, **20**,˙211 (1969).
316. (a) L. K. Krannich, R. K. Kanjolia and C. L. Watkins, *Inorg. Chim. Acta*, **114**, 159 (1986).
    (b) H. H. Sisler and C. Stratton, *Inorg. Chem.*, **5**, 2003 (1966).
    (c) L. K. Krannich and H. H. Sisler, *Inorg. Chem.*, **11**, 1226 (1972).
317. (a) T. Chivers, A. W. Cordes, R. T. Oakley and P. N. Swepston, *Inorg. Chem.*, **20**, 2376 (1981).
    (b) H. W. Roesky, M. Witt, W. Clegg, W. Isenberg, M. Noltemeyer and G. M. Sheldrick, *Angew. Chem., Int. Ed. Engl.*, **19**, 943 (1980).
318. B. Ross, W. Marzi and W. Axmacher, *Chem. Ber.*, **113**, 2928 (1980).
319. (a) P. Frøyen, *Acta Chem. Scand.*, **25**, 983 (1971).
    (b) P. Frøyen, *Acta Chem. Scand.*, **27**, 141 (1973).
320. J. I. G. Cadogan and I. Gosney, *J. Chem. Soc., Perkin Trans. 1*, 466 (1974).
321. (a) R. O. Day, J. M. Holmes, A. C. Sau, J. R. Devillers, R. R. Holmes and J. A. Deiters, *J. Am. Chem. Soc.*, **104**, 2127 (1982).
    (b) T. Mallon and M. Wieber, *Z. Anorg. Allg. Chem,*, **454**, 31 (1979).
    (c) H. Wunderlich, *Acta Crystallogr., Sect. B*, **36**, 1492 (1980).
322. J. Götz and M. Wieber, *Z. Anorg. Allg. Chem.*, **423**, 239 (1976).
323. R. Bohra, H. W. Roesky, J. Lucas, M. Noltemeyer, and G. M. Sheldrick, *J. Chem. Soc., Dalton Trans.*, 1011 (1983).
324. (a) R. H. Bullard and J. B. Dickey, *Org. Synth., Coll. Vol.*, **2**, 494 (1943).
    (b) R. J. Garascia and R. J. Overberg, *J. Org. Chem.*, **19**, 27 (1954).
325. G. O. Doak, *J. Am. Chem. Soc.*, **62**, 167 (1940).
326. (a) G. O. Doak and L. D. Freedman, *J. Am. Chem. Soc.*, **73**, 5656 (1951).
    (b) A. W. Rudy, E. B. Starkey and W. H. Hartung, *J. Am. Chem. Soc.*, **64**, 828 (1942).
327. (a) A. Binz and O. v. Schickh, *Chem. Ber.*, **69**, 1527 (1936).
    (b) F. Popp, *Chem. Ber.*, **82**, 152 (1949).
328. C. L. Hewett, *J. Chem. Soc.*, 1203 (1948).
329. K. J. Irgolic, L. R. Kallenbach and R. A. Zingaro, *J. Inorg. Nucl. Chem.*, **33**, 3177 (1971).
330. E. J. Cragoe, Jr., R. J. Andres, R. F. Coles, B. Elpern, J. F. Morgan and C. S. Hamilton, *J. Am. Chem. Soc.*, **69**, 925 (1947).
331. (a) A. J. Quick and R. Adams, *J. Am. Chem. Soc.*, **44**, 805 (1922).

(b) L. D. Freedman and G. O. Doak, *J. Am. Chem. Soc.*, **77**, 6374 (1955).

(c) F. G. Mann and J. Watson, *J. Chem. Soc.*, 505 (1947).

(d) H. J. Emeléus, R. N. Haszeldine and R. C. Paul, *J. Chem. Soc.*, 881 (1954).

332. J. W. B. Reesor and G. F. Wright, *J. Org. Chem.*, **22**, 382 (1957).

333. (a) B. N. Feitelson and V. Petrow, *J. Chem. Soc.*, 2279 (1951).

(b) J. A. Aeschlimann, N. D. Lees, N. P. McCleland and G. N. Nicklin, *J. Chem. Soc.*, **127**, 66 (1925).

(c) R. J. Garascia, A. A. Carr and T. R. Hauser, *J. Org. Chem.*, **21**, 252 (1956).

(d) R. J. Garascia and I. V. Mattei, *J. Am. Chem. Soc.*, **75**, 4589 (1953).

334. A. Merijanian and R. A. Zingaro, *Inorg. Chem.*, **5**, 187 (1966).

335. M. A. Weiner and G. Pasternack, *J. Org. Chem.*, **32**, 3707 (1967).

336. (a) H. Vermeer, P. C. J. Kevenaar and F. Bickelhaupt, *Justus Liebigs Ann. Chem.*, **763**, 155 (1972).

(b) E. R. H. Jones and F. G. Mann. *J. Chem. Soc.*, 1719 (1958).

(c) I. G. M. Campbell and R. C. Poller, *J. Chem. Soc.*, 1195 (1956).

337. (a) R. R. Holmes, R. O. Day and A. C. Sau, *Organometallics*, **4**, 714 (1985).

(b) R. H. Fish and R. S. Tannous, *Organometallics*, **1**, 1238 (1982).

338. (a) A. J. Dale and P. Frøyen, *Acta Chem. Scand., Sect. B*, **29**, 362 (1975).

(b) A. J. Dale and P. Froøyen, *Acta Chem. Scand., Sect. B*, **29**, 741 (1975).

339. N. Sharma, B. S. Golen, R. K. Mahajan and S. C. Chaudhry, *Polyhedron*, **10**, 789 (1991).

340. M. Wieber, B. Eichhorn and J. Götz, *Chem. Ber.*, **106**, 2738 (1973).

341. C. A. Poutasse, R. O. Day, J. M. Holmes and R. R. Holmes, *Organometallics*, **4**, 708 (1985).

342. P. Maroni, M. Holeman, J. G. Wolf, L. Ricard and J. Fischer, *Tetrahedron Lett.*, 1193 (1976).

343. H. Schmidbaur and W. Richter, *Angew. Chem., Int. Ed. Engl.*, **14**, 183 (1975).

344. H. Schmidbaur and P. Holl, *Chem. Ber.*, **109**, 3151 (1976).

345. B. Eberwein, R. Ott and J. Weidlein, *Z. Anorg. Allg. Chem.*, **431**, 95 (1977).

346. R. A. Zingaro and E. A. Meyers, *Inorg. Chem.*, **1**, 771 (1962).

347. (a) R. A. Zingaro and A. Merijanian, *Inorg. Chem.*, **3**, 580 (1964).

(b) P. B. Chi and F. Kober, *Z. Anorg. Allg. Chem.*, **466**, 183 (1980).

348. R. A. Zingaro, K. J. Irgolic, D. H. O'Brien and L. J. Edmonson, Jr., *J. Am. Chem. Soc.*, **93**, 5677 (1971).

349. (a) A. Müller and P. Werle, *Chem. Ber.*, **104**, 3782 (1971).

(b) W. Kuchen, M. Förster, H. Hertel and B. Höhn, *Chem. Ber.*, **105**, 3310 (1972).

350. (a) J. Götz and M. Wieber, *Z. Anorg. Allg. Chem.*, **423**, 235 (1976).

(b) M. Wieber and J. Götz, *Z. Anorg. Allg. Chem.*, **424**, 56 (1976).

351. (a) Y. Shen, Z. Gu, W. Ding, and Y. Huang, *Tetrahedron Lett.*, **25**, 4425 (1984).

(b) W. C. Still and V. J. Novack, *J. Am. Chem. Soc.*, **103**, 1283 (1981).

(c) K. C. Gupta, R. K. Nigam, N. Srivastava and S. Malik, *Indian J. Chem., B*, **21**, 241 (1982).

(d) J. B. Ousset, C. Mioskowski and G. Solladie, *Tetrahedron Lett.*, **24**, 4419 (1983).

(e) P. Chabert, J. B. Ousset and C. Mioskowski, *Tetrahedron Lett.*, **30**, 179 (1989).

(f) B. Boubia and C. Mioskowski, *Tetrahedron Lett.*, **30**, 5263 (1989).

(g) B. Boubia, C. Mioskowski, S. Manna and J. R. Falck, *Tetrahedron Lett.*, **30**, 6023 (1989).

(h) Y. Shen and Q. Liao, *J. Organometal. Chem.*, **371**, 31 (1989).

(i) Y. Shen and Y. Xiang, *Tetrahedron Lett.*, **31**, 2305 (1990).

(j) S. Kim and Y. G. Kim, *Tetrahedron Lett.*, **32**, 2913 (1991).

(k) W.-B. Wang, L.-L. Shi, Z.-Q. Li and Y.-Z. Huang, *Tetrahedron Lett.*, **32**, 3999 (1991).

352. (a) Y.-Z. Huang, L.-L. Shi and S.-W. Li, *Synthesis*, 975 (1988).

(b) M. Shao, X. Jin, Y. Tang, Q. Huang and Y. Huang, *Tetrahedron Lett.*, **23**, 5343 (1982).

(c) Y. Huang, L. Shi and J. Yang, *Tetrahedron Lett.*, **26**, 6447 (1985).

(d) L. Shi, W. Xia, J. Yang, X. Wen and Y. Z. Huang, *Tetrahedron Lett.*, **28**, 2155 (1987).

(e) Y. Z. Huang, L. Shi, J. Yang and J. Zhang, *Tetradron Lett.*, **28**, 2159 (1987).

(f) L. Shi, J. Yang, X. Wen and Y.-Z. Huang, *Tetrahedron Lett.*, **29**, 3949 (1988).

(g) L. Shi, W. Wang and Y.-Z. Huang, *Tetrahedron Lett.*, **29**, 5295 (1988).

(h) C. Gravier-Pelletier, J. Dumas, Y. Le Merrer and J. C. Depezay, *Tetrahedron Lett.*, **32**, 1165 (1991).

353. H. Schmidbaur and P. Nusstein, *Organometallics*, **4**, 344 (1985).

354. H. J. Bestmann and R. K. Bansal, *Tetrahedron Lett.*, **22**, 3839 (1981).

355. (a) C. Glidewell, D. Lloyd and S. Metcalfe, *Synthesis*, 319 (1988).
 (b) J. N. C. Hood, D. Lloyd, W. A. MacDonald and T. M. Shepherd, *Tetrahedron*, **38**, 3355 (1982).
 (c) G. S. Harris, D. Lloyd, W. A. MacDonald and I. A. Gosney, *Tetrahedron*, **39**, 297 (1983).
 (d) C. Glidewell, D. Lloyd and S. Metcalfe, *Tetrahedron*, **42**, 3887 (1986).
 (e) G. Ferguson, C. Glidewell, I. Gosney, D. Lloyd, S. Metcalfe and H. Lumbroso, *J. Chem. Soc. Perkin Trans. 2*, 1829 (1988).
356. L. Hadjiarapoglou and A. Varvoglis, *Synthesis*, 913 (1988).
357. A. H. Schmidt, W. Goldberger, M. Dümmler and A. Aimène, *Synthesis*, 782 (1988).
358. (a) W. Malisch, H.- A. Kaul, E. Gross and U. Thewalt, *Angew. Chem., Int. Ed. Engl.*, **21**, 549 (1982).
 (b) W. Malisch, H.-A. Kaul, E. Gross and U. Thewalt, *Angew. Chem. Suppl.*, 1281 (1982).
359. (a) M. Luksza, S. Himmel and W. Malisch, *Angew. Chem., Int. Ed. Engl.*, **22**, 416 (1983).
 (b) R. Janta, M. Luksza, W. Malisch, D. Kempf and G. Kuenzel, *J. Organometal. Chem.*, **266**, C22 (1984).

Author index

This author index is designed to enable the reader to locate an author's name and work with the aid of the reference numbers appearing in the text. The page numbers are printed in normal type in ascending numerical order, followed by the reference numbers in parentheses. The numbers in *italics* refer to the pages on which the references are actually listed.

Index compiled by K. Raven

Subject index

Index compiled by P. Raven